CERAMIC INTEGRATION AND JOINING TECHNOLOGIES

CERAMIC INTEGRATION AND JOINING TECHNOLOGIES

From Macro to Nanoscale

EDITED BY

Mrityunjay Singh
Tatsuki Ohji
Rajiv Asthana
Sanjay Mathur

The American Ceramic Society

WILEY

A JOHN WILEY & SONS, INC., PUBLICATION

Published by John Wiley & Sons, Inc., Hoboken, New Jersey.
Published simultaneously in Canada.

Library of Congress Cataloging-in-Publication Data:

Ceramic integration and joining technologies : from macro to nanoscale / editors: Mrityunjay Singh . . . [et al.].
p. cm.
Includes index.
ISBN 978-0-470-39122-8 (hardback)
1. Ceramic materials. 2. Manufacturing processes. 3. Joints (Engineering)
I. Singh, M. (Mrityunjay)
TA455.C43C4626 2011
620.1'4–dc22

2010053092

oBook ISBN: 978-1-118-05677-6
ePDF ISBN: 978-1-118-05675-2
ePub ISBN: 978-1-118-05676-9

Printed in the United States of America.

10 9 8 7 6 5 4 3 2 1

CONTENTS

Preface ix

Contributors xi

PART I INTRODUCTION 1

1 CERAMIC INTEGRATION ACROSS LENGTH SCALES: TECHNICAL ISSUES, CHALLENGES, AND OPPORTUNITIES 3
Mrityunjay Singh, Tatsuki Ohji, Rajiv Asthana, and Sanjay Mathur

PART II SCIENCE AND TECHNOLOGY FOR MACROSCALE INTEGRATION 15

2 CERAMIC COMPONENT INTEGRATION BY ADVANCED BRAZING TECHNOLOGIES 17
Jolanta Janczak-Rusch

3 JOINING AND INTEGRATION ISSUES OF CERAMIC MATRIX COMPOSITES FOR THE NUCLEAR INDUSTRY 39
Monica Ferraris, Milena Salvo, and Valentina Casalegno

4 AIR BRAZING: A NEW METHOD OF CERAMIC–CERAMIC AND CERAMIC–METAL JOINING 91
K. S. Weil, J. T. Darsell, and J. Y. Kim

5 DIFFUSION BONDING OF SILICON CARBIDE AS AN ENABLING TECHNOLOGY FOR THE FABRICATION OF COMPLEX-SHAPED CERAMIC COMPONENTS 143
Michael C. Halbig and Mrityunjay Singh

6 INTEGRATION OF CARBON–CARBON COMPOSITE TO METALLIC SYSTEMS FOR THERMAL MANAGEMENT APPLICATIONS 163
Mrityunjay Singh and Rajiv Asthana

7 CONTACT INTERACTION IN CARBON–METAL SYSTEMS FOR JOINING AND INTEGRATION 193
V. M. Perevertailo and O. B. Loginova

PART III INTEGRATION ISSUES IN ENERGY GENERATION AND DEVICE FABRICATION 231

8 INTEGRATION TECHNOLOGIES FOR FERRITES AND POWER INDUCTORS IN CERAMIC CIRCUIT BOARDS 233
Richard Matz

9 OXIDE THERMOELECTRIC POWER GENERATION 267
Ryoji Funahashi, Saori Urata, Atsuko Kosuga, and Delphine Flahaut

10 INTEGRATION TECHNOLOGIES FOR SOLID OXIDE FUEL CELLS (SOFCS) AND OTHER ELECTROCHEMICAL REACTORS 297
Yoshinobu Fujishiro, Toshio Suzuki, Toshiro Yamaguchi, Koichi Hamamoto, Masanobu Awano, and Nigel Sammes

11 INTEGRATION TECHNOLOGIES FOR SENSORS 323
Woosuck Shin, Maiko Nishibori, and Ichiro Matsubara

12 ON-CHIP INTEGRATION OF FUNCTIONAL HYBRID MATERIALS AND COMPONENTS IN NANOPHOTONICS AND OPTOELECTRONICS 339
Talha Erdem and Hilmi Volkan Demir

13 INTEGRATION OF MULTIFUNCTIONAL PROPERTIES IN THERMAL BARRIER COATINGS BY CHEMICAL VAPOR DEPOSITION 393
Takashi Goto

14 THE CHANGING PHYSICS IN METAL INTERCONNECT RELIABILITY 415
Cher Ming Tan and Yuejin Hou

15 INTEGRATION ISSUES OF BARIUM STRONTIUM TITANATE THIN FILM FOR TUNABLE MICROWAVE APPLICATIONS 449
Ashok Kumar, Supriya Ketkar, and Venkataraman Gurumurthy

16 AEROSOL DEPOSITION (AD) INTEGRATION TECHNIQUES AND THEIR APPLICATION TO MICRODEVICES 489
Jun Akedo

PART IV NANO- AND BIOINTEGRATION 521

17 ADVANCES IN NANOINTEGRATION METHODOLOGIES: PATTERNING, POSITIONING, AND SELF-ASSEMBLY 523
Yoshitake Masuda and Kunihito Koumoto

18 INTEGRATION OF NANOWIRES IN NEW DEVICES AND CIRCUIT ARCHITECTURES: RECENT DEVELOPMENTS AND CHALLENGES 575
F. Hernández-Ramírez, J. D. Prades, A. Romano-Rodriguez, S. Barth, H. Shen, and S. Mathur

19 INTEGRATING DIAMOND-LIKE CARBON INTO NANOSTRUCTURE DESIGNS (FABRICATING MICROSCALE AND NANOSCALE ARCHITECTURES OF DIAMOND-LIKE CARBON FILMS) 641
Xijun Li and Daniel H. C. Chua

20 SYNTHESIS, PROPERTIES, INTEGRATION, AND APPLICATIONS OF VERTICALLY ALIGNED CERAMIC NANOSTRUCTURES 671
D. Pliszka, S. Sundarrajan, and S. Ramakrishna

21 NANOINTEGRATION BASED ON THIN-FILM TECHNOLOGY 699
C. Jin, W. Wei, R. Aggarwal, and R. J. Narayan

22 MASS-MANUFACTURABLE NANOWIRE INTEGRATION: CHALLENGES AND RECENT DEVELOPMENTS 721
Ataur Sarkar and M. Saif Islam

23 USABILITY OF INK-JET PRINTING TECHNOLOGY AND NANOMATERIALS IN ELECTRICAL INTERCONNECTIONS, ELECTRONIC PACKAGING, AND SYSTEM INTEGRATION FOR MICROELECTRONICS APPLICATIONS 743
Umur Caglar, Ville Pekkanen, Jani Valkama, Pauliina Mansikkamäki, and Jussi Pekkanen

24 BIOINTEGRATION OF PROSTHETIC DEVICES 777
Masakazu Kawashita, Toshiki Miyazaki, and Chikara Ohtsuki

Index 803

PREFACE

There is an ever-increasing interest in research and development on integration technologies across length scales driven by the discovery of new materials and demonstration of their application potential. New and emerging materials require adaptive or innovative approaches to integrate them into components, assemblies, devices, and systems. Integration is, therefore, pervasive and cuts across disciplinary boundaries. It is recognized as a key enabling technology that connects innovation in materials with manufacturing. Yet, at present, there is no book-length treatment of integration of advanced materials, particularly ceramics offering authoritative and comprehensive coverage of innovative research on integration bridging various research domains. Much of the technical know-how on integration is either privy to the practitioners within such domains or those with special interest in it.

It is this pervasive nature of integration and current scarcity of a comprehensive resource dealing with it that motivated us to present the state of the art on ceramic integration in diverse fields in a single volume. The collective knowledge presented in the 24 chapters of this volume captures the diversity and unity of joining and integration methodologies across disciplines and length scales. The book addresses joining and integration issues at the macro-, micro- and nanoscales in diverse areas such as aerospace, nuclear and thermoelectric power, microelectromechanical systems, solid oxide fuel cells, multichip modules, prosthetic devices, and many others. The book envisions integration in a broad sense as a key enabler of advanced technology.

Our contributors, all frontline researchers and practitioners in their respective technical areas, represent universities, industry, and government and private research organizations of 12 different nations. They present contemporary perspectives on how integration is viewed and implemented in key application areas within their respective fields. We hope that this volume will stimulate fresh thinking and exchange of ideas and information on approaches to ceramic integration across fields and offer seminal insights into transformative solutions to challenges of integrating new and emerging materials into advanced technology components and systems.

New or transformative knowledge is not discovered in the same logical and structured manner as organized knowledge that gives identity to a discipline. New pathways often emerge from synergy between innovative and evolving approaches within diverse fields that conceivably lack commonality. It is our belief that such synergies may be facilitated by bringing a truly diverse group of practitioners on a common platform. World-class educational and research institutions across continents have begun to incorporate knowledge about integration into their professional studies curricula to train students in this niche area of materials and manufacturing. We believe that the book will serve the educational and research needs of both practitioners and graduate students involved and interested in joining and integration. Whereas the book does not aim to establish the pedagogical foundation in the science and technology of joining and integration, it provides in-depth cutting-edge treatment of selected areas that are widely recognized to be of critical importance.

We are grateful to all of our revered authors for their valuable contribution. We are indebted to John Wiley & Sons and their publication staff, in particular Rebekah Amos, for her wonderful support during the preparation of the book.

MRITYUNJAY SINGH
Ohio Aerospace Institute, NASA Glenn Research Center, USA

TATSUKI OHJI
National Institute of Advanced Industrial Science and Technology (AIST), Japan

RAJIV ASTHANA
University of Wisconsin-Stout, USA

SANJAY MATHUR
University of Cologne, Germany

CONTRIBUTORS

R. Aggarwal, Department of Materials Science and Engineering, North Carolina State University, Raleigh, NC

Jun Akedo, National Institute of Advanced Industrial Science and Technology (AIST), Tsukuba, Ibaraki, Japan

Rajiv Asthana, University of Wisconsin-Stout, Department of Engineering and Technology, Menomonie, WI

Masanobu Awano, National Institute of Advanced Industrial Science and Technology (AIST), Shimo-Shidami, Moriyama-ku, Nagoya, Japan

S. Barth, Department of Chemistry, University College Cork, Cork, Ireland

Umur Caglar, Department of Electronics, Tampere University of Technology, Tampere, Finland

Valentina Casalegno, Politecnico di Torino, Dipartimento di Scienza dei Materiali e Ingegneria, Chimica-DISMIC, Torino, Italy

Daniel H. C. Chua, Department of Materials Science and Engineering, National University of Singapore, Singapore

J. T. Darsell, Energy Science and Technology Division, Pacific Northwest National Laboratory, Richland, WA

Hilmi Volkan Demir, Department of Electrical and Electronics Engineering, and Department of Physics, Nanotechnology Research Center, and Institute of Materials Science and Nanotechnology, Bilkent University, Ankara, Turkey

Talha Erdem, Department of Electrical and Electronics Engineering, Department of Physics, Nanotechnology Research Center, and Institute of Materials Science and Nanotechnology, Bilkent University, Ankara, Turkey

Monica Ferraris, Politecnico di Torino, Dipartimento di Scienza dei Materiali e Ingegneria Chimica-DISMIC, Torino, Italy

Delphine Flahaut, National Institute of Advanced Industrial Science and Technology (AIST), Ikeda, Osaka, Japan

Yoshinobu Fujishiro, National Institute of Advanced Industrial Science and Technology (AIST), Shimo-Shidami, Moriyama-ku, Nagoya, Japan

Ryoji Funahashi, National Institute of Advanced Industrial Science and Technology (AIST), Ikeda, Osaka, Japan

Takashi Goto, Institute for Materials Research, Tohoku University, Aoba-ku, Sendai, Japan

Venkataraman Gurumurthy, Electrical Engineering Department and Nanotechnology Research and Education Center, University of South Florida, Tampa, FL

Michael C. Halbig, U.S. Army Research Laboratory, Vehicle Technology Directorate-Glenn Site, NASA Glenn Research Center, Cleveland, OH

Koichi Hamamoto, National Institute of Advanced Industrial Science and Technology (AIST), Shimo-Shidami, Moriyama-ku, Nagoya, Japan

F. Hernández-Ramírez, EME/XaRMAE/IN2UB, Department of Electronics, University of Barcelona, Barcelona, Spain

Yuejin Hou, Division of Circuits and Systems, School of EEE, Nanyang Technological University, Singapore

M. Saif Islam, Integrated Nanodevices and Systems Research, Electrical and Computer Engineering, University of California, Davis, CA

Jolanta Janczak-Rusch, EMPA, Swiss Federal Laboratories for Materials Science and Technology, Dübendorf, Switzerland

C. Jin, Joint Department of Biomedical Engineering, University of North Carolina, Chapel Hill, NC

Masakazu Kawashita, Graduate School of Biomedical Engineering, Tohoku University, Aoba-ku, Sendai, Japan

Supriya Ketkar, Electrical Engineering Department and Nanotechnology Research and Education Center, University of South Florida, Tampa, FL

J. Y. Kim, Energy Science & Technology Division, Pacific Northwest National Laboratory, Richland, WA

Atsuko Kosuga, National Institute of Advanced Industrial Science and Technology (AIST), Ikeda, Osaka, Japan

Kunihito Koumoto, Graduate School of Engineering, Nagoya University, Furo-cho, Chikusa-ku, Nagoya, Japan

Ashok Kumar, Mechanical Engineering Department and Nanotechnology Research and Education Center, University of South Florida, Tampa, FL

Xijun Li, Department of Materials Science and Engineering, National University of Singapore, Singapore

O. B. Loginova, V.N. Bakul Institute for Superhard Materials, National Academy of Sciences of Ukraine, Kiev, Ukraine

Pauliina Mansikkamäki, Department of Electronics, Tampere University of Technology, Tampere, Finland

Yoshitake Masuda, National Institute of Advanced Industrial Science and Technology (AIST), Shimo-Shidami, Moriyama-ku, Nagoya, Japan

S. Mathur, Institute of Inorganic Chemistry, University of Cologne, Cologne, Germany

Ichiro Matsubara, National Institute of Advanced Industrial Science and Technology (AIST), Shimo-Shidami, Moriyama-ku, Nagoya, Japan

Richard Matz, Siemens Corporate Technology, Munich, Germany

Toshiki Miyazaki, Graduate School of Life Science and Systems Engineering, Kyushu Institute of Technology, Wakamatsu-ku, Kitakyushu, Japan

R. J. Narayan, Joint Department of Biomedical Engineering, University of North Carolina, Chapel Hill, NC

Maiko Nishibori, National Institute of Advanced Industrial Science and Technology (AIST), Shimo-Shidami, Moriyama-ku, Nagoya, Japan

Tatsuki Ohji, National Institute of Advanced Industrial Science and Technology (AIST), Shimo-Shidami, Moriyama-ku, Nagoya, Japan

Chikara Ohtsuki, Graduate School of Engineering, Nagoya University, Furo-cho, Chikusa-ku, Nagoya, Japan

Jussi Pekkanen, Department of Electronics, Tampere University of Technology, Tampere, Finland

Ville Pekkanen, Department of Electronics, Tampere University of Technology, Tampere, Finland

V. M. Perevertailo, V.N. Bakul Institute for Superhard Materials, National Academy of Sciences of Ukraine, Kiev, Ukraine

D. Pliszka, Nanoscience and Nanotechnology Initiative, National University of Singapore, Singapore, and Institute of Physics, Pomeranian University in Slupsk, Slupsk, Poland

J. D. Prades, EME/XaRMAE/IN2UB, Department of Electronics, University of Barcelona, Barcelona, Spain

S. Ramakrishna, Nanoscience and Nanotechnology Initiative, Department of Mechanical Engineering, and Division of Bioengineering, National University of Singapore, Singapore

A. Romano-Rodriguez, EME/XaRMAE/IN2UB, Department of Electronics, University of Barcelona, Barcelona, Spain

Milena Salvo, Politecnico di Torino, Dipartimento di Scienza dei Materiali e Ingegneria Chimica-DISMIC, Torino, Italy

Nigel M. Sammes, Department of Metallurgical and Materials Engineering, Colorado School of Mines, Golden, CO

Ataur Sarkar, Integrated Nanodevices and Systems Research, Electrical and Computer Engineering, University of California, Davis, CA

H. Shen, Inorganic and Materials Chemistry, University of Cologne, Cologne, Germany

Woosuck Shin, National Institute of Advanced Industrial Science and Technology (AIST) Shimo-Shidami, Moriyama-ku, Nagoya, Japan

Mrityunjay Singh, Ohio Aerospace Institute, NASA Glenn Research Center, Ceramics Branch, Cleveland, OH

S. Sundarrajan, Nanoscience and Nanotechnology Initiative, National University of Singapore, Singapore

Toshio Suzuki, National Institute of Advanced Industrial Science and Technology (AIST), Shimo-Shidami, Moriyama-ku, Nagoya, Japan

Cher Ming Tan, Division of Circuits and Systems, School of EEE, Nanyang Technological University, Singapore

Saori Urata, National Institute of Advanced Industrial Science and Technology (AIST), Ikeda, Osaka, Japan

Jani Valkama, Department of Electronics, Tampere University of Technology, Tampere, Finland

W. Wei, Department of Materials Science and Engineering, North Carolina State University, Raleigh, NC

K. S. Weil, Energy Science & Technology Division, Pacific Northwest National Laboratory, Richland, WA

Toshiro Yamaguchi, National Institute of Advanced Industrial Science and Technology (AIST), Shimo-Shidami, Moriyama-ku, Nagoya, Japan

PART I

INTRODUCTION

1

CERAMIC INTEGRATION ACROSS LENGTH SCALES: TECHNICAL ISSUES, CHALLENGES, AND OPPORTUNITIES

Mrityunjay Singh,[1] Tatsuki Ohji,[2]
Rajiv Asthana,[3] and Sanjay Mathur[4]

[1]*Ohio Aerospace Institute, NASA Glenn Research Center, Cleveland, Ohio*
[2]*National Institute of Advanced Industrial Science and Technology (AIST), Nagoya, Japan*
[3]*University of Wisconsin-Stout, Menomonie, Wisconsin*
[4]*University of Cologne, Cologne, Germany*

INTRODUCTION

The discovery of new and innovative materials has been known to culminate in major turning points in human history. The Bronze Age, the Iron Age, and, in our own times, the age of silicon are all considered as historical benchmarks that have transformed human civilization and have opened theretofore unforeseen possibilities for economic growth and societal impact. These progressive and defining periods in the history of humankind are marked not only by materials innovations (e.g., Damascus steel, used to make swords during 1100–1700AD) but also, more importantly, by the transformation of new materials into goods usable for war, the arts, and commerce. The transformative impact and functional manifestation of new materials have been demonstrated in every historical era by their integration into new products, systems, assemblies, and devices.

Ceramic Integration and Joining Technologies: From Macro to Nanoscale, First Edition.
Edited by Mrityunjay Singh, Tatsuki Ohji, Rajiv Asthana, Sanjay Mathur.

INTEGRATION ISSUES IN ADVANCED TECHNOLOGY SYSTEMS

In modern times, the integration of new materials into usable products has a special relevance for the technological development and economic competitiveness of industrial societies. Current and evolving integration issues span such diverse areas as aeronautics, space, energy, nuclear power, thermoelectric (TE) power, nanoelectromechanical and microelectromechanical systems (MEMS), solid oxide fuel cells (SOFCs), multichip modules (MCMs), prosthetic devices, and many others.

MICROELECTRONICS AND NANOELECTRONICS

Integration is critically important in microelectronics at the wafer, chip, and package levels and is a means to achieving compact designs and cost reduction. Integration technology is used in the manufacture of MEMS, display devices, radio frequency (RF) components, and a number of other microelectronic components. In such applications, integration challenges are manifested in soldering, metallization, service reliability, joint degradation, vacuum seals, and other areas. Highly complex and sophisticated integration technologies are used in the construction of devices with semiconductor chips and in the development of methods to connect MEMS components.

MCMs integrate a number of unique functions into a system and consist of a group of advanced functional electronic devices that permit size reduction. MCMs combine integrated circuits (ICs) based on different materials and provide reliable low-cost integration technology to combine several ICs with a functional substrate. MCMs are designed based on thin-film multilayer structures on ceramic, silicon or metal that can offer the highest integration density per layer. They are usually built using joining and surface modification technology, such as sputtering and plating. The biggest advantage of joining is the ability to join any dissimilar materials if their surface mechanical properties, in terms of flatness, smoothness, and cleanliness, are sufficiently good.

Silicon carbide-based ceramics are increasingly being used in semiconductor switches for their lower losses and improved ability to operate at higher temperatures than silicon. Due to lower losses and higher operating temperature, a smaller heat sink can be used, thus saving on size and system cost. Integration issues in such applications center on joining SiC to other materials, such as metals. Additionally, future microprocessor technologies might increasingly utilize nanoporous organosilicate glass materials as dielectrics, and these can create integration challenges such as controlling the metal–nanoporous glass interface.

Many applications of semiconductor technology rely upon heterostructures in which the integration of dissimilar materials is realized through epitaxial growth. There is interest in creating heterostructures through joining of dissimilar materials, which permits modulation of the functional properties (e.g., electronic work functions) at the interfaces instead of through epitaxial growth. Joining can be used to create heterostructures comprising semiconductors, dielectrics, metals, or ceramics in combinations that would be difficult to achieve via epitaxial growth. In fact, wafer bonding is already being used to integrate dissimilar materials together as an alternative technique to heteroepitaxial growth. Bonding can also be used to create substrates that enable the growth of higher-quality heteroepitaxial films.

Nanotechnology has revolutionized device concepts by offering a range of functionalities available in the nanometer range. To achieve these functionalities, however, nanostructures need to be integrated in electronic, photonic, optoelectronic, and sensing systems and devices. Controlled synthesis and self-assembly of functionally and morphologically distinct nanostructures and their integration into manufacturable devices such as field-effect transistors (FETs) and gas sensors or biosensors demand new design paradigms to overcome such challenges as coupling different forces (e.g., mechanical–electrical), interfaces (e.g., metal–semiconductor), and interactions (e.g., biological–nonbiological components) in integrated systems. In addition, the novel functionalities in hybrid materials such as polymer–ceramic nanocomposites are also based on a molecular-level integration of nanoscopic ceramic particles in polymer matrices, which demands better understanding of interfacial phenomena and integration (synthetic) pathways.

ENERGY

Integration of advanced ceramics plays a critical role in all aspects of energy production, storage, distribution, conservation, and efficiency. Among the alternative energy systems, fuel cell technology is particularly important. Advanced ceramics (e.g., yttria-stabilized zirconia, lanthanum strontium manganese oxide, and Ni–YSZ cermet) play a key role in various components of SOFCs. Ceramic or metal interconnects and sealing components are needed for system integration.

TE devices with high energy-conversion efficiencies and with the capability for prolonged operation are needed for various applications. Electrical and thermal properties of novel electrode materials and interfaces, as well as joint durability, need to be evaluated and optimized. Thermal expansion mismatch coupled with high temperature at the hot shoe-leg joint makes a robust design and the development of integration technologies critical. At each interface in a TE device, the composition, operating temperatures, and thermal mismatch issues are different, and these could become particularly important in segmented legs with multiple interfaces. Integration issues are also important in micro-TE power generators based on ceramic catalyst combustors that employ integration of a thermopile of thin-film metals and thick-film ceramics on dielectric membranes. Another illustration of integration in the energy sector is the development of MEMS-based fuel injectors for use in gas turbine engines.

AERONAUTICS AND GROUND TRANSPORTATION

Many advanced ceramics and ceramic matrix composites (CMCs) have been developed for applications in thermal structures, exhaust nozzles, turbopump blades, combustor liners, radiant burners, heat exchangers, and a number of other applications that involve extremely harsh conditions. For example, carbon–carbon composites containing SiC

(C/C–SiC) show promise for lightweight automotive and aerospace applications. In the automotive industry, C/C–SiC brake disks are already being used in some car models in Europe. Carbon–silicon carbide (C/SiC) composites are being developed for hypersonic thermal structures and advanced propulsion components. Similarly, SiC/SiC composites are being developed for applications in combustor liners, exhaust nozzles, reentry thermal protection systems, hot gas filters, and high-pressure heat exchangers, as well as components for nuclear reactors. Integration issues are important when developing components based on such materials. Mechanical joining and attachment technologies, brazing, and diffusion bonding have emerged as key technologies for integrating CMCs for a number of such applications.

INTEGRATION ACROSS DOMAINS AND LENGTH SCALES

Next, we present a brief overview of key issues in ceramic integration science and technology across length scales and technical fields to set the stage for more focused discussions in subsequent chapters. We summarize the main points of each chapter as presented by their authors to give the reader a bird's-eye view before reading the chapters.

SCIENCE AND TECHNOLOGY FOR MACROSCALE INTEGRATION

In Chapter 2, Janczak-Rusch presents the state of the art of the brazing of ceramics and their composites. The chapter focuses on methods to overcome poor wettability, relieve stresses, and improve joint reliability by designing and developing brazing filler alloys with tailored properties. The author discusses a number of interesting systems: Si_3N_4–TiN, mullite–mullite, and SiC fiber-reinforced glass. An approach for holistic joint investigation is recommended that combines experimental testing, fractography, and microstructural study with numerical simulation to understand and optimize joint behavior and performance.

Carbon–carbon composites have found use in a number of demanding applications, such as in the nose cones of rockets and missiles and in aircraft brakes. One emerging application of C/C is in components used in nuclear reactors. For example, C/C is used as a plasma-facing material because it can mitigate heat flux owing to its plasma tolerance. Even under off-normal plasma events, vapor shielding can protect C/C from erosion at high power fluxes. Owing to the absence of melting and to their excellent resistance to thermal shock and thermal fatigue, C/C targets have demonstrated proven compatibility with plasma conditions, particularly at low densities. These materials perform well under pulsed high heat fluxes and have a low neutron absorption cross-section. Besides, they retain mechanical strength at elevated temperatures and have a low atomic number, which induces low power losses in the plasma. Integration issues for these materials are important both for the next generation of thermonuclear fusion reactors and for fission-reactor components. In both cases, extreme thermomechanical stresses on the joined component must be taken into account, together with material modification related to the presence of neutrons. Joining and integration issues of C/C and CMCs for the nuclear industry, with a focus on brazing technology, are reviewed by Ferraris et al. in Chapter 3. C/C joining is revisited in a later chapter for thermal management applications.

Brazing is a low-cost and industrially proven technology to reliably integrate ceramics in components. Brazing of ceramics demands use of ultrapure atmospheres or cover fluxes in order to suppress the adhesion-limiting effects of atmospheric contaminants. This usually adds to the processing cost and slows production. In Chapter 4, Weil et al. describe an air brazing method that has emerged for joining ceramics. The method was originally developed by the authors to produce oxidation-resistant hermetic joints for use in SOFCs and in oxygen and hydrogen concentrators. The key to developing a successful filler metal composition for air brazing is to identify a metal oxide wetting agent that is mutually soluble in a molten noble metal solvent. For example, near-eutectic Ag–CuO filler metal compositions are promising in joining ceramics such as YSZ, ferrites, alumina, and magnesia. Ternary additions can further improve the filler wettability, raise the use temperature, and increase the joint strength. The authors discuss process mechanisms, braze metallurgy, and performance of air brazed joints in a number of systems.

Among advanced ceramics, silicon carbide is particularly interesting owing to its potential for use in high-temperature, structural applications. It has high strength, creep resistance, corrosion resistance, and high-temperature capability. However, limitations that are in part geometry inherent in hot pressing and chemical vapor deposition (CVD), as well as difficulty in machining have restricted the wider use of SiC. One cost-effective solution for fabricating complex-shaped SiC components is through the joining of simple-shaped ceramics. In Chapter 5, Halbig and Singh discuss joining of SiC for a lean direct ceramic injector proposed for use in jet engines by NASA. Techniques for bonding the SiC laminates of the injector to one another and to Kovar tubes are enabling technologies for developing such injectors. For bonding SiC laminates, diffusion bonding has been proposed; for attaching Kovar tubes, brazing has been proposed. The authors describe the technical challenges to be overcome in diffusion bonding and brazing, such as nonuniformity of bond formation, chemical incompatibility, and residual stresses. In an earlier development, silicate glass was used as the bonding layer between the SiC laminates. However, difficulty in achieving a uniform glass layer, with the resulting lack of hermeticity, prompted diffusion bonding using titanium foils and physical vapor deposited (PVD) Ti coatings. Process optimization was conducted to obtain diffusion bonds that were uniform, chemically stable, and crack free. The authors present the outcomes of studies on microstructure, phase analysis, nondestructive evaluation, and tensile pull tests.

Advanced C/C composites composed of carbon fiber-reinforced carbon matrix are fabricated using either resin infiltration and pyrolysis or chemical vapor infiltration (CVI) approaches. A wide variety of C/C composites have been developed using different types of carbon fibers, fiber weave patterns, fiber coatings, carbon matrices, and fabrication technologies. The fibers make the composite stronger, tougher, and thermal shock resistant and make it highly conductive when high-conductivity C fibers with the basal planes of carbon oriented parallel to the fiber axis are used. These high-conductivity fibers can rapidly spread heat in the direction of the fiber. For integration in components, C/C needs to be joined to other materials. Joining and integration of C/C composite to metals, especially for thermal management applications, is reviewed in Chapter 6 by Singh and Asthana. In particular, they present research in vacuum brazing of C/C composites to titanium and copper-clad molybdenum for thermal management. Technical issues such as wettability and thermomechanical compatibility of joined materials are addressed, and the role of joining atmosphere, filler chemistry, surface roughness, and residual stresses is discussed together with joint microstructure, mechanical properties, and broader joint design issues.

A fundamental requirement for brazing of ceramics using liquid filler is the wettability of solids by liquids. For example, integration of ceramics to metals by brazing requires that molten braze spread on and cover the surfaces to be bonded. Thus, braze composition and joining conditions are designed to facilitate spreading, often with the aid of chemically active additives that favorably modify the surfaces to be bonded. Study of brazing thus represents a confluence of classical surface science, adhesion phenomena, high-temperature chemistry, phase equilibria, and metallurgy, among other disciplines. In Chapter 7, Pervertailo and Loginova examine physicochemical regularities of wetting and contact phenomena in carbon–metal systems under vacuum as well as under high pressure. Such phenomena include, among others, dissolution, adsorption, and reaction, all of which are critical for braze performance.

For example, the authors show that "nonreactive" Ni–C melts contain clusters of weakly deformed tetrahedra and octahedra of Ni atoms, elongated carbon chains, and closed carbon fragments. Carbon chains penetrate the nickel matrix and uniformly distribute throughout the melt. The atomic spacing in chains is similar to that in planar graphite networks, meaning that covalent bonds between carbon atoms in molten Ni are partially retained. As the melt composition approaches the eutectic, the size and number of clusters of weakly deformed tetrahedra and octahedra are increased. These clusters serve as precursors to the adsorbed species and thus influence the wetting behavior in nonreactive high-temperature systems.

INTEGRATION ISSUES IN ENERGY GENERATION AND DEVICE FABRICATION

The next nine chapters focus on integration issues in energy generation and device fabrication.

For several decades, there has been a push toward miniaturization of microelectronic components. However, such miniaturization has been less successful with inductive components. In Chapter 8, Matz reviews the progress made in the design, fabrication, and performance of magnetically coupled inductors using NiZnCu and MnZn ferrite multilayers. In particular, he focuses on low-temperature cofiring technology used to produce mixed dielectric–ferrite multilayered inductors. Magnetic losses are lower in magnetic ceramics than in amorphous magnetic metals, and this has promoted the use of ferrite and dielectric ceramics in multilayer boards. Low-temperature cofiring of dielectric and ferrite ceramic layers in multilayer boards is desirable because magnetic flux leakage is high and electric insulation is low between the turns of a coil when the turns are immediately surrounded by ferrite. Currently, there is strong interest in cofired ceramics, and the push is to reduce inductor line and space widths to low values (<100 μm) by printing methods. However, the challenge to fabricating a ferrite core around a dielectric board lies in both the technology and the thermal expansion mismatch of the materials. Magnetic permeability is sensitive to the mechanical stress that occurs in multilayer structures due to expansion mismatch. This has inhibited the development of a useful cofiring technology for dielectric and ferrite tapes, even though the ferrites are amenable to low-temperature sintering. Matz's results support the conclusion that integrated ceramic transformers with mixed dielectric–ferrite multilayers will be feasible once a few steps in design and technology development have been perfected and the challenges related to cost-efficient sintering and shaping, in combination with low-dielectric multilayer boards, have been overcome.

TE power generated from vast amounts of waste heat emitted by automobiles and factories promises to revolutionize the energy landscape. TE materials such as oxide compounds can convert waste heat into electrical energy without using moving parts such as turbines and without producing carbon dioxide gas, radioactive substances, or other regulated emissions. To achieve realistic TE power generation, a high TE figure of merit and chemical stability are required. In Chapter 9, Funahashi et al. discuss the TE properties at high temperature of a number of oxide compounds mainly based on Co with layered structure, such as Co-349 and BC-222. The temperature dependencies of electrical resistivity, Seebeck coefficient, thermal conductivity, thermal expansion, three-point bend strength, and fracture toughness are presented together with X-ray phase analysis and scanning electron micrographs (SEMs) of microstructures. The authors conclude that even though the layered oxides Co-349 and BC-222 have good TE conversion efficiency, it is currently insufficient for widespread application. New materials possessing higher figures of merit even at low temperatures are necessary, and the authors make suggestions about possible materials and approaches to designing and synthesizing such materials.

SOFC reactors based on ceramics have high efficiency and have the ability to operate in the intermediate temperature range. Power densities in excess of 2 kW/L are possible in auxiliary power units and small generators by improving materials, accumulating small parts, and assembly into high-performance modules. In Chapter 10, Fujishiro et al. discuss integration technologies for SOFCs and other electrochemical reactors, particularly tubular SOFCs with submillimeter diameters when they are accumulated into cubes. The authors show that hundreds of submillimeter SOFC tubes can be precisely mounted in porous electrodes in small volumes of one cubic centimeter. The authors fabricated and tested such microtubular SOFCs and demonstrated their excellent power densities of 1–3 W/cm^3. The authors show how novel ceramic fabrication processes can integrate SOFCs into prototype modules of microhoneycomb cell stacks with a cell integration density of 250 multilayered tubes per cubic centimeter in porous electrode cubes. The authors also discuss a new concept of a nanoscale electrocatalytic reactor for NOx decomposition based on NiO and YSZ. Electrochemically formed nanograins of Ni surrounded by nanopores, in a NiO/YSZ interface of the electrode, can lead to a remarkable improvement in efficiency.

In chemical microsystems including sensors, functional materials such as oxides need to be integrated into the silicon technology. However, functional materials cannot always be obtained by the standard *complementary metal oxide semiconductor* (CMOS) technology; for such cases, transducer materials are obtained separately and are implanted onto microsystems. Recent research has focused on chemical routes,such as precipitation methods, to produce high-quality powders, as opposed to thin films that require the use of vacuum systems. In Chapter 11, Shin et al. present their research on the synthesis of platinum–alumina catalyst pastes from powders and how these pastes are utilized in microsensors and other devices. Sensors were also fabricated by dispensing the pretreated ceramic materials onto the microdevice using microprinting technology. The deposition of functional films, either by screen printing or by the more sophisticated drop-deposition techniques, such as ink-jet systems, was performed after combining the functional material with organic carriers. The authors demonstrate that such a dispensing technique can be successfully employed for the preparation of a ceramic catalyst combustor with nanoparticles for gas-sensing applications.

Optoelectronic devices, such as photodiodes, solar cells, light-emitting diodes (LEDs), and laser diodes, are fundamental to a wide variety of high-technology systems. Generally, optoelectronic devices are composed of discrete elements that perform different functions; these elements interface with one another via fiber connections. Unfortunately, lack of efficient coupling often results in optical losses and high costs. Integration of monolithic components can eliminate problems inherent in device coupling, such as mechanical movement, thus reducing the packaging cost and size. Effective integration demands that each component should function as if it were discrete. Incorporation of nanomaterials into nanophotonic and optoelectronic devices permits this to be achieved, and it increases the range of functionalities for applications such as light generation, displays, modulation, sensing, imaging, and communications. Innovative nanodevices have been developed using combinations of nanostructures that can be embedded in hybrid architectures for "on-chip" integration of components. Such devices require integration of functional materials and components using methods developed for chip-scale integration at the range of micrometers to nanometers, including monolithic integration, hybrid integration, layer-by-layer assembly, and directed assemblies. In Chapter 12, Erdem and Demir describe the state-of-the-art and innovative integration approaches that are being perfected for cutting-edge optoelectronics and nanophotonics.

Extension of lifetime under severe operating conditions is fundamentally important for gas turbines to achieve high efficiencies and low energy consumption. The current operating temperatures of 1500°C have been achieved mainly from development of air cooling and thermal barrier coating (TBC) technologies. TBCs of YSZ together with bond coats of MCrAlY (where M is Ni, Co, etc.) are the most common coating material for turbine blades. However, significant thermal stress during prolonged severe heat cycles causes cracking and fatal delamination. The coatings are deposited using plasma spray, electron-beam PVD, or CVD. Conventional CVD combined with laser heating can accelerate chemical reactions to deposit films. In Chapter 13, Goto describes work on integrating laser technology and conventional CVD to achieve dramatic acceleration of coating deposition on turbine blades.

Metal interconnects are essential for microelectronic device integration. Early ICs frequently failed at interconnections, chiefly by electromigration. The electron wind force and the triple points in the grain structures of interconnections were discovered to lead to such failure. However, with progression toward very large scale integration (VLSI), the scaling laws for interconnect failure subsumed additional factors such as multiple driving forces, multiple diffusion paths, and stress-induced migration. It is noteworthy that, with new failure mechanisms coming into play, the underlying physics changes, and the corresponding stochastic processes, such as the statistical distribution of failure time, are also modulated. In Chapter 14, Tan and Hou present experiments and models to describe how the changing physics of electromigration and stress migration affect the failure probability of interconnections in microelectronic circuits.

Miniaturization of microelectronic components continues to be a major driving force for innovation in industry. Miniaturization of microwave systems relies on thin-film ferroelectrics because of their ability to produce tunable RF and microwave circuits with a broad range of tunability. This enables the designer to meet the stringent frequency and power requirements of wireless communications systems. Tunable circuits can compensate for the effects of aging and temperature excursions in RF circuits. The ferroelectric material barium strontium titanate (BST) exhibits an electric field-dependent dielectric constant. This allows capacitors with BST as the dielectric to have adjustable capacitances. BST is tunable and has high dielectric constant, high power handling capability, and ease of integration with other thin-film devices. In Chapter 15, Kumar et al. present a critical review of BST's material properties and address issues relevant to its integration, such as interdiffusion in the substrate layers and formation of voids and hillocks. Platinum is the preferred electrode material for BST because it is nonreactive to BST and forms an interface possessing favorable electrical properties. Integrating Pt, however, is challenging because of poor adhesion, difficulty in dry etching, and diffusion in the BST–Pt layers. The authors' research suggests that thin films of nanocrystalline diamond can be used as an efficient diffusion barrier layer between BST and Pt.

Ceramic films and coatings for advanced applications can be deposited using a number of techniques, including the newly developed aerosol deposition (AD) method in which submicrometer oxide and nonoxide ceramic particles are accelerated by gas flow up to 100–500 m/s followed by impact on a substrate. A thick, dense, uniform, and hard ceramic coating can form at room temperature without additional energy consumption for melting of ceramic powders as is required in thermal spray processes. The process is simple, energy-effective, and relatively inexpensive, and it can be done under low vacuum. AD can reduce the fabrication steps in the manufacture of electronic devices such as MEMS, RF components, and optoelectronic devices. It is particularly useful for integrating "on-demand" microscale parts. In Chapter 16, Akedo presents the mechanisms and features of AD and its applications to a number of devices.

INTEGRATION ISSUES AT THE NANOSCALE AND IN BIOLOGICAL SYSTEMS

The last group of eight chapters deals with integration issues at nanoscale and in biological systems and devices.

In Chapter 17, Masuda and Koumoto discuss nanointegration and liquid-phase patterning of ceramic thin films and particle assemblies. Micropatterning is attractive for photonic crystals, solar cells, and molecular sensors, among others. The authors describe the procedures used to fabricate nanopatterns and micropatterns of ceramic thin films such as TiO_2, Fe_3O_4, and ZnO, and colloidal crystals using environment-friendly solution chemistry approaches. They show that micropatterns of randomly deposited nanoparticles can be created using capillary, gravitational, and electrostatic forces. In fact, many kinds of patterning techniques have been developed to prepare patterns of thin films, for example, photolithography, microcontact printing, wet etching, and ink-jet printing. However, etching or "liftoff" is required in many such methods, which impairs the performance and increases waste and energy consumption. The deposition of thin films only on desired areas of a substrate is thus required for the pattering of ceramic thin films, and solution synthesis approaches enable this to be readily accomplished.

In the first decade of the twenty-first century, controlled manipulation of nanomaterials progressed to the point where construction and characterization of proof-of-concept nanodevices became feasible. 1-D nanostructures such as nanowires and nanotubes came to be used as building blocks in prototype components such as interconnects, gas sensors, biosensors, photodetectors, lithium-ion batteries, and TE generators. The next key challenge is the scale-up to the production of large-scale components that integrate functional nanostructures in devices. In Chapter 18, Mathur and coworkers review the progress in controlled growth of 1-D nanostructures, structure–property relationships; fabrication of nanowire-based FET, and higher-level integration of nanowires into complex nanodevice architectures. They present both conceptual and prototypical progress in harvesting nanomaterials for use in devices. The authors discuss technical challenges to scale-up, including fabricating better electrical contacts with nanowires, developing and interfacing low-cost nanowire-based electronic components, and meeting the industrial standards for precision and reliability. The authors also make recommendations to solve such challenges and point out the importance of reliability issues that accompany the integration of nanostructures into a conventional device and the necessity of systematic investigations in this regard.

Diamond-like carbon (DLC) has a number of attractive properties, such as high Young's modulus, high hardness, chemical inertness, and hydrophobicity. DLC films are proposed to be used to build microscale and nanoscale architectures possessing resistance to degradation in humid and harsh environments. In Chapter 19, Li and Chua highlight the potential of DLC films versus Si-based materials in nanostructure design. They discuss the synthesis and physical properties of DLC, its potential for use in micromechanical and nanomechanical devices, and fabrication technology such as focused ion beam (FIB) and FIB-assisted CVD to design and build DLC architectures. The authors also highlight the technical challenges in DLC film design and fabrication, such as film etching and the optimization of nanoarchitecture design. Owing to their biocompatibility, nanostructured DLC films can be used in biosensors and nanofluidic systems for single-molecular sensing and detection. The authors suggest that combining bottom-up and top-down approaches will enable sophisticated nanoarchitectures with enhanced performance to be created for DLC-based devices. The theme of nanointegration based on thin-film growth is further discussed in a later chapter by Jin et al.

In special cases, high-performance nanoscale devices demand highly oriented and ordered arrays of nanostructures to be created in order for their anisotropic properties to be profitably harvested or for the areal density of discrete elements (e.g., transistors) of the device to be increased. Highly oriented, 1-D nanowire structures permit enhanced areal density to be achieved, together with electrical and optical properties that can be tailored. Thus, low-dimensional nanowire structures serve as an ideal platform to probe properties that may be inaccessible in large devices. In Chapter 20, Pliszka et al. present synthesis, properties, and applications of vertically aligned ceramic nanowires made from oxides, carbides, and nitrides.

In Chapter 21, Jin et al. discuss advances in nanointegration based on thin-film growth with 2-D and 3-D ordered nanostructures of nanoparticles and nanowires. Ordered nanostructures in thin films can form either spontaneously (e.g., by long-range elastic interactions) or by nucleation and growth on prepatterned templates. The authors discuss both approaches: strain-induced self-organization to develop ordered surface nanostructures, such as ripples and islands, and ordered nanowires on prepatterned surfaces that could yield defect-free nanostructures for use in optoelectronic devices such as nanowire-based FETs. Such applications demand integration of functional nanostructures into progressively larger assemblies that are needed in practical devices. Currently, nanointegration can be done using lithography, nanoimprint lithography, and thin-film growth. The authors focus on nanointegration based on thin-film technology, which is readily adaptable to industrial practices because it is widely used in the semiconductor industry. It is conceivable that thin-film-based nanointegration will play a key role in developing next-generation nanostructured devices for optoelectronic and biomedical applications.

Scaling of electronic device density using bottom-up approaches offers the potential for low-cost, high-density integration of nanoscale devices. Using such approaches, functional nanostructures can be assembled from chemically synthesized nanoscale building blocks in a manner similar to the complex architecture of biological systems such as proteins and biomolecules. Considerable progress has been achieved in synthesizing 1-D nanowires with controlled composition, structure, size, morphology, and electrical and optical properties. A high surface-to-volume ratio, the benefits of quantum confinement, and low-cost synthesis are the major drivers for developing prototypes of functional nanowire devices such as p–n junction diodes, lasers, and photovoltaics. Successful integration of nanowires requires reproducible interfaces, interconnections, and the transfer of nanowires from the mother substrate to the device platform, which remain as major challenges in the large-scale production of nanodevices. In Chapter 22, Sarkar and Islam review the common approaches to controllably aligning and interfacing nanowires with bulk photonic devices and circuits, and they address the challenges and issues in obtaining mass manufacture and reproducible integration of nanowires. The authors propose that formation of "nanobridges" and "nanocolonnades" can create robust contacts to nanowires with low contact resistance on production scales.

Printable electronics extensively utilize nanotechnology to expedite manufacture and to save energy in producing microelectronic interconnects. In particular, piezo printhead-type ink-jet printing technology offers substantial manufacturing advantages in providing promising products for tomorrow, namely, flexibility of produced components, less waste material in manufacturing, environmentally friendly production, and lower-cost products. In Chapter 23, Caglar et al. discuss issues concerned with demonstrating the integration of nanomaterials-assisted ink-jet printing in electronic manufacturing for mass production. Process challenges in designing and patterning printable structures, printing process optimization, and sintering or curing of conductive or dielectric nanomaterials using laser sintering, a process that is faster than conventional furnace sintering, have been discussed. Ink-jet printing technology offers additive on-demand material deposition on substrates. It is possible to print very precise structures without the need for a mask-and-etching process. Line/space widths of $50\,\mu m/50\,\mu m$ are possible and further reductions are feasible. The process minimizes wasted material and, therefore, manufacturing cost and is suitable for integration into a product development value chain.

Perhaps nowhere is the power of integration revealed as remarkably as in the biointegration of prosthetic devices, in which an inorganic substance (ceramic) is integrated with an organic, living tissue. The bonding between living tissue and implanted devices is usually aided by bioceramic coatings. Since the 1970s, biocompatible ceramics have played an increasingly important role in repairing living tissues and organs and in promoting the regeneration of cells. In the last chapter of the volume, Kawashita et al. introduce ceramics for a number of biological applications, including bone repair and artificial joints, and discuss bioactive ceramics, bioceramic–polymer composites, bioactive cements, and bioactive inorganic–organic hybrids. They also discuss the requirements for artificial materials to form apatite—a major bone ingredient—and the role of functional groups that promote apatite nucleation in the joint. In many such applications, long-term stability of biointegrated prosthetic devices and sensors is needed. Conversely, detachability of biomodules for flexibility in repair or to accommodate add-on features could also be a consideration. Thus, integration issues could center on somewhat conflicting requirements. Other developments include bio-MEMS devices such as the lab-on-a-chip (LOC) module that can consolidate all the complicated laboratory procedures onto a single chip for specific biomedical applications, and ink-jet bioprinting technology for manufacturing 2-D and 3-D patterns of immobilized hormones to direct cell behavior. Integration of complex designs at the micrometer and nanometer scales is the basis of these emerging technologies.

The collective state-of-the-art knowledge about integration gathered from diverse fields and presented in these 24 chapters represents the diversity and unity of integration science and technology. Emerging applications of new materials with engineering performance far superior to the current generation of materials will require new developments in integration technology to manufacture devices, components, assemblies, and systems based on materials and structural features at multiple-length scales. The following chapters develop these themes for a variety of advanced and emerging materials.

PART II

SCIENCE AND TECHNOLOGY FOR MACROSCALE INTEGRATION

2

CERAMIC COMPONENT INTEGRATION BY ADVANCED BRAZING TECHNOLOGIES

Jolanta Janczak-Rusch

Empa, Swiss Federal Laboratories for Materials Science and Technology, Dübendorf, Switzerland

INTRODUCTION

Structural ceramics, such as silicon carbide, silicon nitride, zirconia, diamond, ceramic matrix composites (CMC), and ultra high-temperature ceramics, are developing rapidly, stimulated by their potential for many applications. They are being exploited as cutting tools, wear parts, electronic devices, parts of energy-conversion and energy-production systems, sensors, and biomaterials. However, all those elements need to be attached to other components or structural parts to create an assembly. The practical realization of such designs depends on appropriate joining processes.

Ceramics are inherently difficult to join, both to themselves and to metal structures, a consequence of their strong ionic and covalent bonding. The main available, well-established technologies are mechanical attachment techniques, joining by preceramic polymers, soldering/brazing, diffusion bonding, and glass–metal sealing methods (Nicholas 1990). There are also other application-specific processes, such as friction welding, microwave bonding, ultrasonic welding, and adhesive bonding. Ceramic joining techniques are in constant development; new joining methods and modified approaches to conventional methods have been developed over the years, aiming both

Ceramic Integration and Joining Technologies: From Macro to Nanoscale, First Edition.
Edited by Mrityunjay Singh, Tatsuki Ohji, Rajiv Asthana, Sanjay Mathur.

at requirements of new materials and at improved reliability. Transient liquid phase bonding for high-temperature ceramic joints (Locatelli et al. 1995) and laser joining of metals with ceramics (Reinecke and Exner 2001) are some of the most recent advances. All of these processes have their own advantages and shortcuts. The appropriate process is usually chosen according to the base materials' characteristics, requirements of the joint (service temperature, strength, corrosion, etc.), ease of implementation, and functionality. Of the many joining processes available, probably the main and most adaptable technique used to integrate ceramics is brazing. Brazing allows low-cost, large-scale joining of intricate geometries and can be customized to the mass production of components, such as those used in the electronics and automotive industries. The main advanced structural ceramics, that is, SiC, Si_3N_4, and Al_2O_3, have been brazed to a variety of metals and alloys of engineering interest (Park and Eagar 2002; Schwartz 2003). In particular, vacuum brazing (e.g., active brazing) has been recognized as a reliable and cost-effective technology for production of metal–ceramic joints, and is the most widely used joining process for mechanically reliable vacuum tight joints, able to operate at relatively high temperatures. It is also suitable for a wide range of CMCs (Singh 1999). Some application examples of brazing technology for ceramic integration in components are shown in Table 2.1.

WETTABILITY, RESIDUAL STRESSES, AND JOINT RELIABILITY

Ceramic brazing relies on the ability of a filler metal or alloy to wet the ceramic surface (see Fig. 2.1), which is often hindered by the covalent nature and the low surface energy of the ceramics.

There are two basic approaches to encourage wetting: the modification of the ceramic surface and the modification of the braze itself. In either case, the actual brazing operation takes place either in a controlled atmosphere, such as nitrogen or argon, or in a vacuum (pressure lower than 10^{-4} Pa, depending on the alloys used and the temperature range).

Surface treatments include metallization (e.g., the moly-manganese method, which is extensively used in the electronic and electrical industries), metal coating, and metal hydride treatment, while braze modification involves a process known as active metal brazing. Coatings of nickel, silver, copper, and chromium have been applied to the surface of ceramics to enhance wettability (Li 1994), while the sodium process or coatings of Ti layers are used to provide Al_2O_3 with wetting by metallic liquids.

Active brazing (the one-step processing route), which is the focus of this chapter, is defined as the use of activated braze alloy, where an active element alters the surface chemistry of the ceramics by the formation of intermediate reaction layer and lowers the wetting angle of the molten braze on the ceramics (Schwartz 2003). The active elements used for this process include Ti, Zr, Hf, V, and Al. The best known of these active elements is Ti, which is used in many available braze alloys. Typical active braze alloys are based on the Ag, Cu, or Ag-Cu eutectic systems. They are represented by several commercial alloys as, for example, CB6®Brazetec (98.4 Ag, 1.0 In, 0.6 Ti),

TABLE 2.1. Application of Brazing Technology for Ceramic Integration in Devices and Components

Application	Joining Partner 1: Ceramic	Joining Partner 2: Metal or Ceramic	Examples	Reference
Automotive parts	Si_3N_4, Al_2O_3, ZrO_2	Kovar, Mo, steel, Si_3N_4, Al_2O_3, ZrO_2	Turbocharger	(Nicholas 1990)
Hi-tech structure ceramics	B_4C, SiC	Ti, Cu. Ni alloys, metal based composites (Al, Ti-based MMC)	Military armor application	Commercially available
Abrasive	Diamond, cBN, Si_3N_4 + TiN WC TiCN, TiC, TiN	Steel, TiCN, TiC, TiN	Grinding tools, drilling tools, saws, cutting tools	(Blugan et al. 2007) (Lu and Kwon 2002) commercially available
High-temperature resistance joints	SiC, Si_3N_4	High-temperature alloys (Incoloy), C/C composites	Gasturbines parts: guide nozzle Fusion reactor (ITER): plasma facing components	(Ferraris et al. 1994)
Biomedical	Al_2O_3, ZrO_2 YSZ	Ti alloys	Prosthetic devices hermetic seal essential for implantable medical devices	Commercially available
Vacuum hermetic seals	Al_2O_3	Ti alloys, steels, Cu-Au-Ni-Ti,	ToF mass spectrometer neutron generators	(Zigerlig et al., 2002) Sandia National Laboratories
Power electronic applications	AlN	Cu	High-power, high-frequency conductor	(Reicher et al. 2001)
Electronic packaging/ microelectronics	AlN, BeO	Cu, Si (wafer)	Heat sink electronic components: laser diodes, substrates	U.S. Patent 6229208 commercially available
	Al_2O_3	Kovar (Fe-Ni) Ti alloys, Inconels and hastelloys	multichip, MCM package microchannel applications, strain gauges downhole LWD	
MEMS	Si_3N_4	Inconel 718 with thin multilayers of Ti-Cu, Ti-Ni and Ti-Cu-Ni		(Park and Eagar 2002)
Energy	YSZ	Steel	SOFC	(Weil et al. 2004)

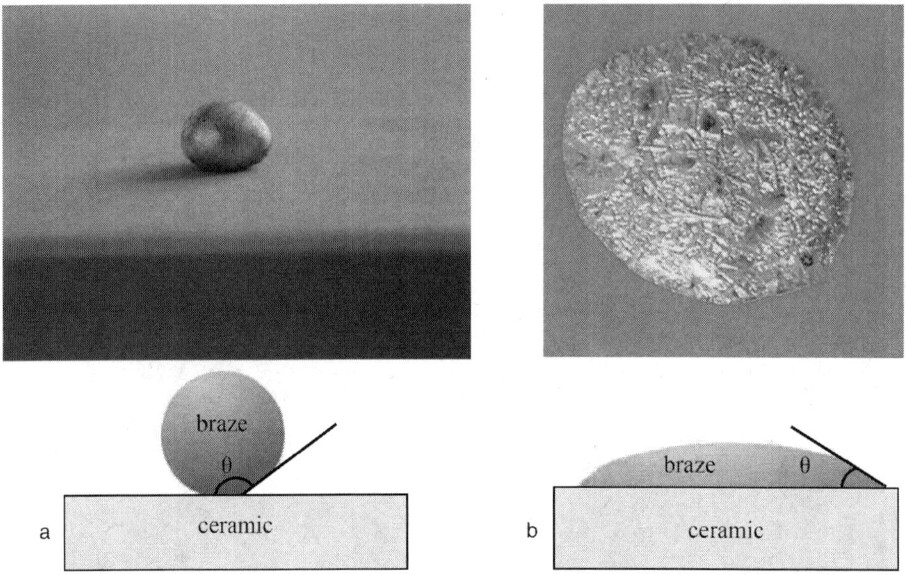

Figure 2.1. Two wetting experiments showing (a) poor and (b) good wetting.

Incusil®-ABA Wesgo (59.0 Ag, 27.3 Cu, 12.5 In, 1.25 Ti), and Ticusil®Wesgo (68.8 Ag, 26.7 Cu, 4.5 Ti).

Besides the wetting characteristics, the other important issue that needs to be overcome when brazing ceramics to metal is the significant differences in the physical properties of the materials to be joined. Their extremely different thermal expansion coefficients ($\Delta\alpha \approx 11 \times 10^{-6}$/K) and Young's moduli lead to high residual stresses when they are cooled down from the brazing temperature. Even if a strong metallurgical bond is achieved, residual stresses induced during processing can become intolerable and cause the deterioration of the joint's mechanical integrity and its failure. Thus, additional steps are needed to relax the residual stresses in the joint and consequently to improve its mechanical performance. The typical approach to deal with this problem is either to use interlayers (ductile metal layers or layers with a thermal expansion coefficient close to that of the ceramic partner) (Kim and Park 2000; Park et al. 2002) or to use more recently developed composite brazing fillers (Zhu and Chung 1997; Klose 1999; Janczak-Rusch et al. 2002; Janczak-Rusch 2005). While the application of multiple interlayers or functionally graded materials involves many processing steps that are usually needed to minimize residual stresses down to an acceptable level, the one-step processing may be achieved by using the composite approach (Galli et al. 2006). In this concept, the properties of the brazing filler material are modified by embedding a second phase, that is, ceramic particles or fibers, in order to reach the desired physical or thermomechanical properties. Figure 2.2 shows how the thermal expansion coefficient and the Young's modulus of active filler metals can be modified by adding SiC particles.

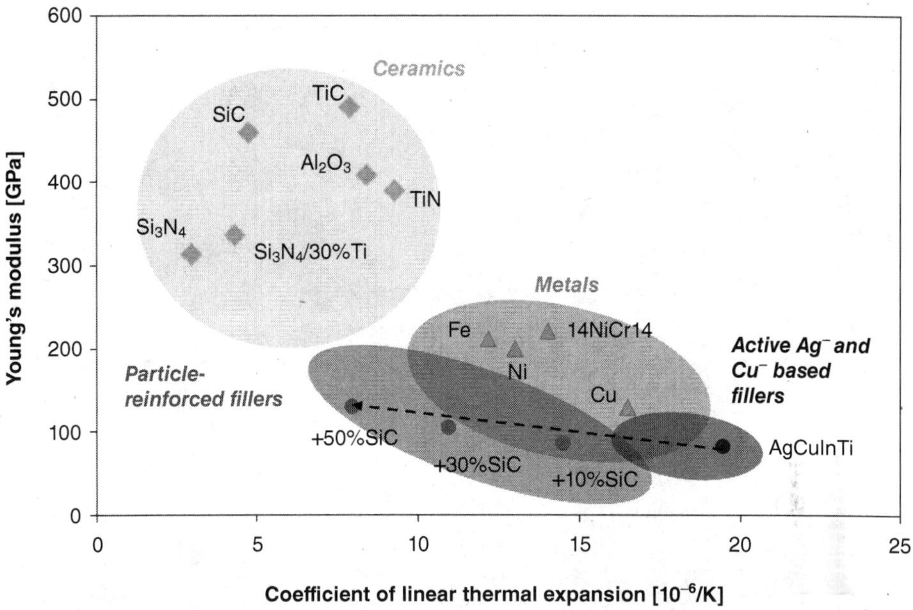

Figure 2.2. Modification of the physical properties of a brazing filler by particle addition.

The composite approach offers the significant advantage of wide gap brazing. For particle-reinforced brazing fillers, the brazing gap width can be adjusted to a value at which the minimum residual stresses are predicted, and therefore the full potential of residual stresses' accommodation by the brazing filler can be utilized. The optimal gap width is needed for the maximal relief of residual stress in the braze zone, and often cannot be reached when brazing with unreinforced fillers. Furthermore, the composite concept allows improving the reliability of the joint, increasing strength at the elevated temperatures, and thus enhancing joint's service temperature. The opportunities of the composite approach will be addressed later in the example of joining of Si_3Ni_4-TiN ceramics with steels (see the section on "Joining of CMCs").

JOINT DESIGN

Joint design is a very complex task. Many relationships have to be considered to produce reliable joints with properties tailored exactly to the requirements. Different concepts may be applied to optimize the interfaces between the materials (especially between the brazing filler and single joining partners) and to relax residual stresses, which are of critical importance. In any case, special attention has to be paid to the selection of the brazing filler with regard to the brazing filler system (including filler/interlayer layout) when considering the properties of the created interfaces.

For active brazing, powder metallurgy may be used to develop new brazing filler compositions with an optimized amount of the active element. It is expected that the

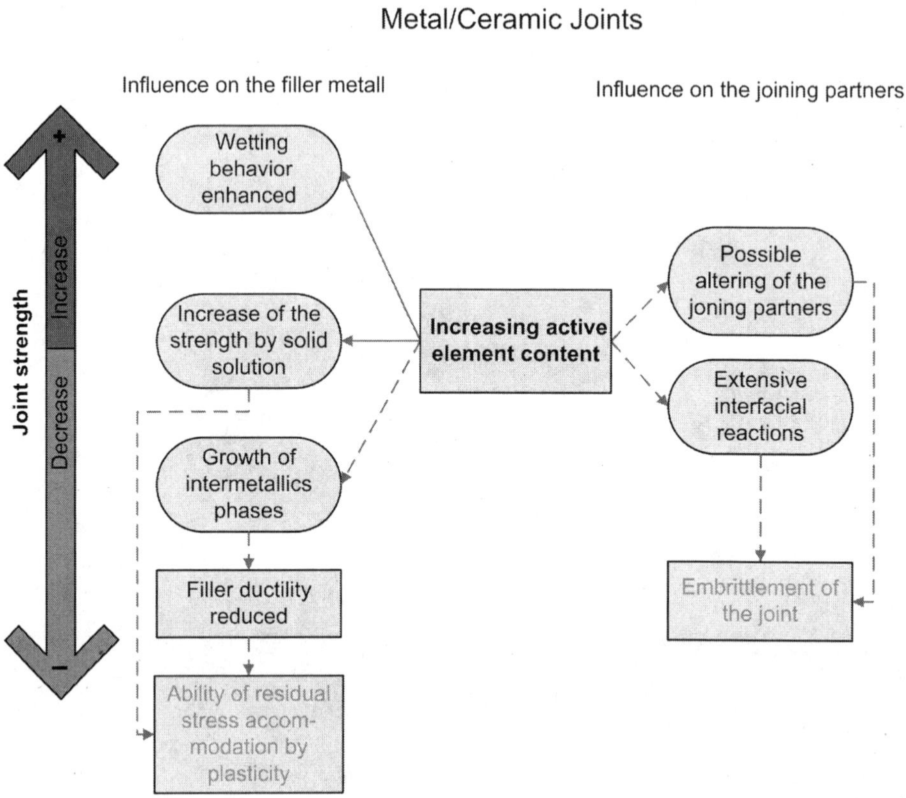

Figure 2.3. The influence of the active element on the joint strength: complex interactions.

increased amount of active element should improve the wetting of the melted filler material on the ceramic, and therefore increase the strength of the bond. Nevertheless, many other accompanying effects may negatively and often unexpectedly affect the joint strength. The interaction of all these effects has to be taken into account when determining the optimum content of the active element for maximum joint strength, as shown in Figure 2.3.

Besides the control of wetting behavior, the active element may significantly change the thermomechanical properties of the filler metal, which will affect its ability to relax the residual stresses, and may also significantly change the properties of base materials. Figure 2.4 presents an example of the change of the stress-strain characteristics of the AgCuIn filler (Incusil 15) with the addition of 1.25% Ti. An increase of brazing filler strength by over 30% was measured due to hardening of the filler alloy with formation of hard intermetallic phases (Bissig et al. 2006). This in turn led to the reduced ability of the filler to accommodate the residual stresses and to the decrease in joint strength in a similar range.

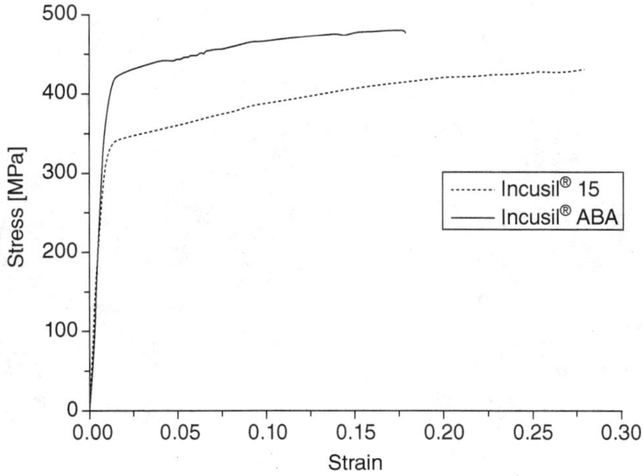

Figure 2.4. The influence of Ti addition on mechanical properties of an AgCuIn brazing filler metal. Stress-strain characteristics of Incusil 15 ($Ag_{61.5}Cu_{23.5}In_{15}$) and Incusil-ABA ($Ag_{59}Cu_{27}In_{12}Ti_{1.25}$) in comparison.

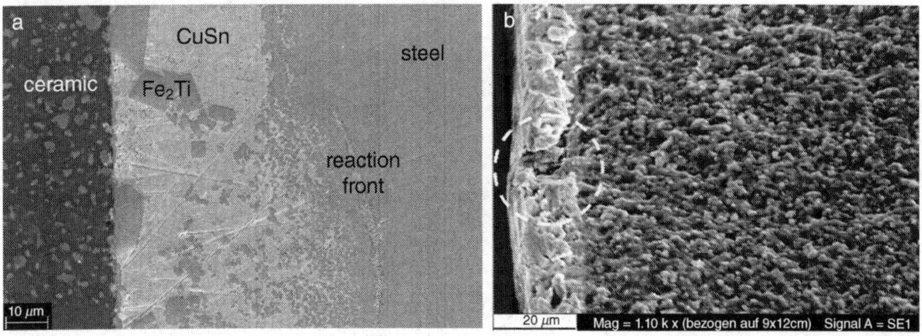

Figure 2.5. The negative effect of the too-high Ti content in the brazing filler metal leading to (a) the formation of the brittle Laves phases (Fe_2Ti) in joints with steel as a base material and (b) diffusion in the Si_3N_4-TiN ceramic.

The negative effect of the too-high Ti content in the filler metal on the base material properties is shown in Figure 2.5.

A combined theoretical–experimental approach is indispensable for a proper joint design, especially when more complex concepts, such as the use of interlayers or composite fillers, are applied. For simple component shapes and preliminary joint design, analytical methods can be used. However, analytical models are usually developed from precise and strong hypotheses, both on the joint geometry and on the constitutive laws of the components (Hsueh and Evans 1985; Iancu et al. 1990), and have limited applicability for the prediction of residual stresses developed in the joint. In the majority of cases, the component or joint is complex and the simulation of its behavior requires

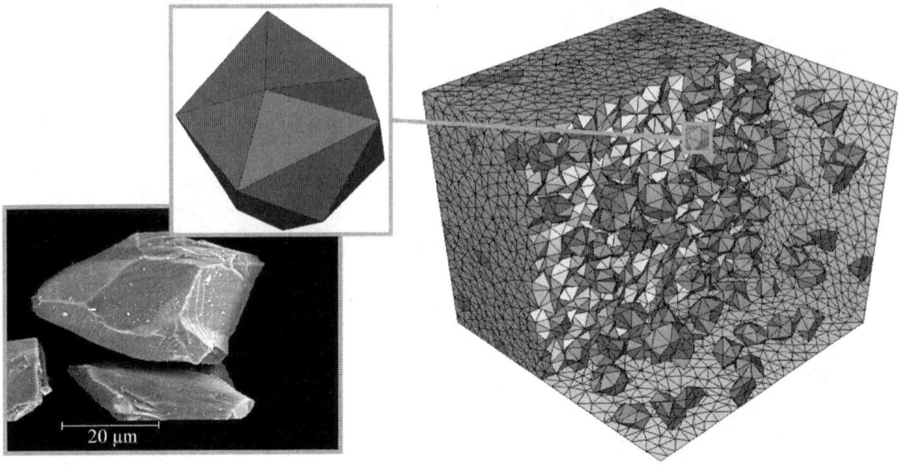

Figure 2.6. FE model of a particle-reinforced brazing filler metal describes well the real microstructure (left: SEM micrograph of SiC-reinforced brazing filler, right: FEM representation of the real structure).

numerical methods such as finite element analysis (FEA). Finite element modeling also provides a powerful way to investigate residual stresses (Mackerle 2001), and has been extensively used to study joined assemblies with the increasingly large calculating powers of computers. Thus, complex material behavior (thermoelastoplasticity, hardening, etc.), complex geometry, and boundary conditions (loading, constraints) can be calculated. Even the microstructure of the filler metal can be taken into account, as shown in Figure 2.6, an example of a particle-reinforced filler metal (Galli et al. 2008). Many different joint designs can be theoretically compared, thus reducing the number of brazing trials required. The summary effect of all parameters even with opposite directions can be evaluated.

To study the mechanical integrity of the ceramic–metal joint, the four-point bending (4PB) test configuration is among the simplest and the most commonly used (Wielage and Ashoff 1990; DVS 3102 2005; Zhang et al. 2002a). In such configuration, the joint is found in the pure bending region of the beam. In the first approximation, the elasticity expressions render the flexural modulus of elasticity and the ultimate strength of beam specimens. The same configuration is used for fatigue studies with a smooth or a precracked specimen (Lee et al. 1995). The tensile tests deliver direct information about the braze behavior under pure tensile load, although it is difficult to carry out such for joints with a ceramic joining partner. To determine the shear strength of a joint test, configurations that simulate simple shear on the plane of the interface have been reported (Park et al. 2002).

Complementary to the mechanical testing with fractographic study, the origin and reason of failure can be determined. In the Figure 2.7, two typical failure patterns observed in metal–ceramic joints are shown. A curved crack (see Fig. 2.7a) in the

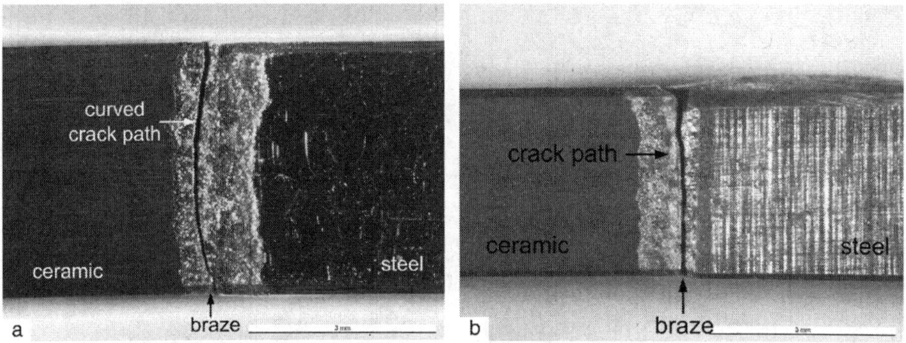

Figure 2.7. Fracture of metal–ceramic joints brazed at different conditions. (a) A curved crack due to high residual stresses;(b) Straight crack path in the ceramic (fracture through the ceramics due to material defects).

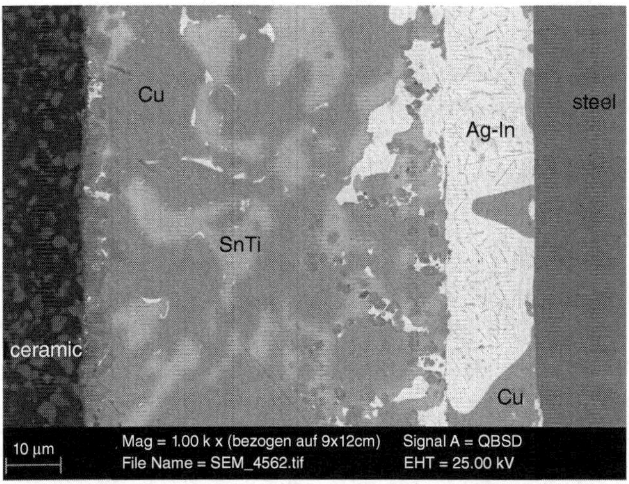

Figure 2.8. SEM image of an optimized Si_3N_4-TiN/steel joint brazed in two steps.

ceramic partner is typical for the systems, where the ceramics experiences high residual stresses (Blugan et al. 2005).

The traditional method of bond quality assessment is the metallographic investigation. It is crucial for understanding the interfacial processes, the brazing metallurgy, and the failure behavior. In the Figure 2.8 the optimized interfaces in a Si_3N_4-TiN/steel joint brazed in two steps (ceramic metallization: CuSnTiZr, brazing of premetalized ceramic with steel: Incusil 15) is shown.

Integral joint investigation combining numerical and experimental methods, such as thermomechanical tests, fractography work, and microstructural investigation, is needed to understand behavior of the joint and to optimize its performance.

JOINING OF CMCS

Only a few of the procedures developed for joining traditional ceramics or glasses are directly transferable to composites. In addition to common problems, which usually appear when joining ceramics with metals, such as poor wetting behavior of ceramics and differences in physical and thermal properties of metals and ceramics, there are also other issues that arise with application of composites. For example, the different wetting behavior and reactivity of the composite constituents (e.g., fiber and matrix) and the compatibility of the joining process with the composite fabrication technique have to be addressed when joining ceramic or glass matrix composites with metals. When joining continuous fiber-reinforced ceramic and glass matrix composites, some additional parameters have to be taken into account when compared with the joining of particle- or short fiber-reinforced composites. The fiber arrangement plays the critical role in regard to the joint design and the components integrity, when joining long fiber-reinforced composites. There are also different boundary conditions (e.g., morphology of the interfaces between the composite material and the metallic counterpart, wetting as well as the surface reaction behavior, load transfer) for the joint design depending on the fiber direction possible. Joints brazed parallel to the fibers' direction are usually well adapted to the relevant loading conditions and thus favorable. In such configuration, the thermomechanical properties of the composite can be fully exploited. On the other hand, when brazing perpendicular to the fiber direction, standardized mechanical tests can be easily applied to assess the joint strength.

The joining of SiC-based ceramic composites to similar SiC/SiC composites or to high-temperature alloys has been reported extensively in the literature (Ferraris et al. 1994; Singh 1999; Zhang et al. 2002). However, there is very little information concerning joining of ceramic composites with nitride or oxide matrices, including silicate matrix composites, to metal parts (Haber and Greenhut 1991; Dixon 1995; Nakamura et al. 1999; Weil et al. 2001). In one of the few studies published on silicate matrix composites, Dixon (Dixon 1995) reported on brazed joints of silicon nitride fiber-reinforced cordierite glass-ceramic with titanium and stainless steel parts. Different interlayer materials were used in order to minimize residual stresses and to prevent the damage of fiber–matrix interfaces in the composite.

In the following, the application of brazing technology to integrate different CMCs will be shown on example of TiN particle-reinforced Si_3N_4, SiC fiber-reinforced borosilicate glass matrix composite (molybdenum counterparts), and fiber-reinforced mullite–mullite composite (Haynes 230 counterparts). The properties of the materials used in the investigation are summarized in Table 2.2. All brazing experiments were performed in a TORVAC vacuum furnace, under processing pressure lower than 10^{-5} mbar. Detailed information on sample preparation, processing, and testing may be found elsewhere (Janczak-Rusch et al. 2005; Bissig et al. 2007; Blugan et al. 2007).

Joining of Si_3N_4-TiN (30 wt%)

Si_3N_4-TiN ceramics exhibit an excellent thermomechanical and tribological behavior even when exposed to the corrosive environments. In particular, they offer a unique set

TABLE 2.2. Characteristics of Brazing Materials Used in Brazing Experiments (Joining Partners, Brazing Filler Metals)

Base Material	E (GPa)	α (RT-600°C) (10^{-6}/K)	Strength (MPa)
14 NiCr 14 Steel	210	14.1	600 (tensile)
Mo	325	5.1	620 (tensile)
Haynes 230 superalloy Ni-22Cr-14W-5Co-3Fe-2Mo	211	16.5	860 (tensile)
Si_3N_4-TiN 70/30	330	3.8	785 (±51) (bending)
Duran borosilicate glass/SiC_f	200	3	2750 (tensile)
Duran® borosilicate glass	63	3.25	60 (tensile)
Mullite–mullite (V_f = 11%)	234	6.8	475 (bending)
Brazing Filler	**Composition (wt%)**	**Liquidus Temperature (°C)**	
CB6	Ag 98.4, In 1, Ti 0.6	948–959	
CuSnTiZr	Cu 74.5, Sn 14, Ti 10, Zr 1.5	795	
Cusil-ABA	Ag 63, Cu 65.25, Ti 1.75	780–815	
Incusil-ABA	Ag 59, Cu 27.25, In 12.5, Ti 1.25	605–715	
Incusil 15	Ag 61.5, Cu 23.5, In 15	605–725	

of properties for different purposes when combined with high-load bearing metallic materials like steels. The Si_3N_4-TiN/steel material combination is believed to have significantly large potential for middle-temperature range applications, especially in production of cutting tools. At the same time, due to pronounced mismatch of physical properties ($\Delta\alpha = 11 \times 10^{-6}$ 1/K), it is a very challenging materials combination for joining purposes.

The brazeability of Si_3N_4-TiN/steel material system, the information being also highly beneficial and transferable to other material couples (e.g., joining of Si_3N_4 with low thermal expansion alloys such as Invar or Kovar alloys), was studied in detail (Bissig et al. 2007, Blugan et al. 2007; Janczak-Rusch 2007). The preliminary investigations have shown that the criteria of joint design used for unreinforced Si_3N_4 ceramics may as well be correct for the Si_3N_4-TiN composites. The uniform particle distribution and the relatively good wetting behavior of TiN particles with the molten brazing filler lead to the quasi-homogenous brazing behavior of the particle-reinforced ceramics, allowing a significant simplification.

The representative Ag- and Cu-based fillers (see Table 2.2) were used to braze Si_3N_4 to itself ("ceramic/ceramic joints") and to 14NiCr steel ("ceramic/metal joints"). Different brazing approaches were realized and compared, like ceramic metallization, use of an active filler metal, ductile interlayer and composite brazing filler system. The experimental results are summarized in Table 2.3. The dimensions of the joining

TABLE 2.3. Four-Point Bending Strength of Si_3N_4-TiN Ceramics Joints Brazed with Different Brazing Fillers or Filler Systems

Brazing filler	Ceramic/Ceramic Joint	Metal/Ceramic Joint		Brazing Temperature (°C)	Processing Steps
	Bending Strength (MPa) at Room Temperature Average (Min–Max)	Test Temp. (°C)	Bending Strength (MPa)		
CB6	299 (241–338)	RT	361 (275–430)	1010	1
		350	457 (425–497)		
		400	375 (290–423)		
		450	423 (347–479)		
		500	326 (295–267)		
		600	219 (166–253)		
CuSnTiZr	333 (286–370)	RT	105 (92–113)	950	1
Cusil-ABA	—	RT	228 (142–336)	800	1
Incusil-ABA	413 (378–442)	RT	329 (259–387)	740	1
		200	431 (423–439)		
		250	485 (456–509)		
		300	476 (404–485)		
		350	454 (408–478)		
		400	256 (182–323)		
CuSnTiZr/ Incusil 15	320 (260–360)	RT	398 (362–438)	740	2
CuSnTiZr/ Incusil-ABA	413 (442–378)	RT	315 (428–227)	740	2
CuSnTiZr/ Cu interlayer	—	RT	354 (304–406)	950	2
Incusil-ABA + 10% SiC_p	425 (394–465)	RT	284 (201–349)	750	1
		250	523 (505–556)		
		300	475 (463–485)		
		350	414 (404–428)		
		400	308 (301–320)		
Incusil-ABA + 20% SiC_p	—	RT	195 (179–212)	750	1
Incusil-ABA + 30% SiC_p	—	RT	39 (28–52)	750	1
Incusil-ABA + 30%SiC_p Sandwich layout	—	RT	403 (342–440)	750	1
		250	459 (489–422)		
		300	505 (572–403)		
		350	387 (485–288)		
		400	291 (324–242)		

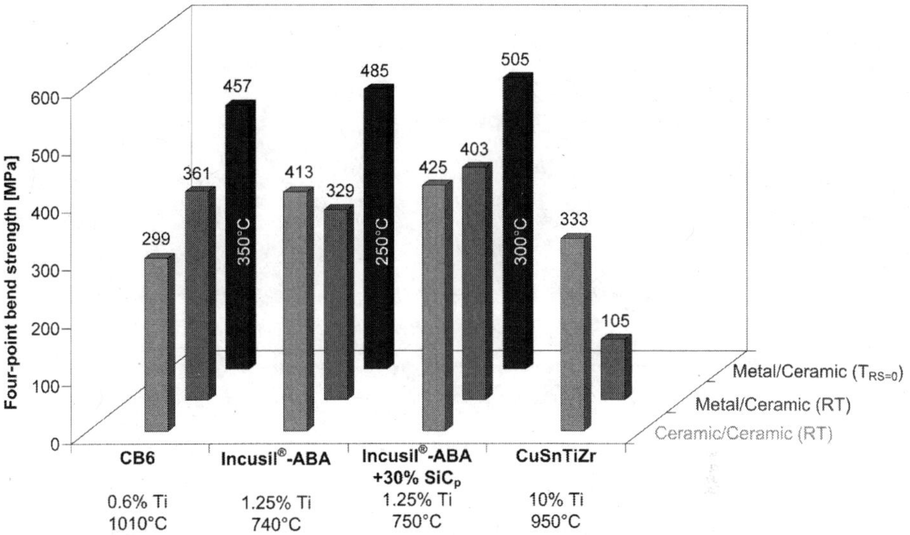

Figure 2.9. Four-point bending strength (average) of Si_3N_4-TiN/steel joints brazed using different commercially available active brazing fillers and a composite brazing filler metal.

partners were 3 × 4 × 25 mm, and at least eight specimens of each kind were tested in four-point bending test.

The study shown that several active brazing fillers, such as CB6, Incusil-ABA, and CuSnTiZr, can be used to produce reliable ceramic/ceramic joints (see Fig. 2.9). For those brazing fillers, the joints strength is in the range between 299 and 413 MPa, being approximately 38–53% of the ceramic strength (785 MPa). The critical issues are found to be the Ti content and the brazing filler gap. Already, 0.6% Ti being the nominal value for CB6 filler was enough to obtain good wetting behavior. However, a significant strength increase of the ceramic/ceramic joints can be achieved when using brazing filler with Ti amount higher than 1.0% (like Incusil-ABA with 1.25% Ti). There is a critical value of Ti content (<10%) at which a negative effect on the ceramic material is observed. In all cases, a very thin brazing gap (<50 μm) is advantageous when brazing Si_3N_4-TiN ceramics to itself.

The brazing fillers that allow producing high-strength ceramic joints may not be suitable for ceramic/metal joints, as happens with CuSnTiZr braze. In the case of ceramic/metal joints, the critical parameter besides the Ti content is the ductility of the brazing filler, since it controls the joint ability to accommodate the residual stress. The highest strength values for Si_3N_4 ceramic/metal joints with commercially available active brazing filler were achieved with CB6 braze (361 MPa). Joints brazed with CB6 showed the lowest state of residual stresses among the investigated materials with nonmodified single-metal fillers, as confirmed by FE simulation (Bissig et al. 2007). Even the relatively high brazing temperature of CB6 (1010°C) is less critical for the joint stress then the limited plasticity of the brazing filler (as in the case of Incusil-ABA with much lower brazing temperature of 740°C). Additionally, the optimal gap width

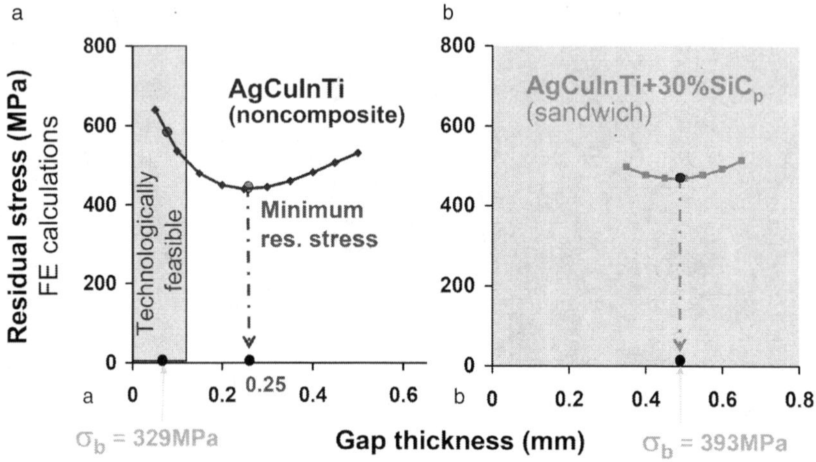

Figure 2.10. The effect of the brazing gap width on the residual stress relief as simulated by FEM for Si_3N_4-TiN/steel joints brazed (a) with Incusil-ABA and (b) with a particle reinforced Incusil-ABA (sandwich layout).

needed for the maximal relief of residual stress in the braze zone when brazing with Incusil-ABA could not be reached due to the technological limits. As shown by numerical calculation (see Fig. 2.10a), the minimum residual stresses and thus the highest strength of ceramic/metal joints would be achieved with a brazing gap of 0.25 mm, which is simply not feasible. For particle-reinforced Incusil-ABA, the optimal width of the brazing gap was adjusted to a predicted value of 0.5 mm, corresponding to the minimum residual stresses (see Fig. 2.10b), and joints with improved strength were achieved (see Fig. 2.11).

When compared with other multi-steps methods of reducing the residual stresses, such as use of the interlayer or suitable brazing filler system, with composite concept the same (or higher) strength increase may be achieved in the one-step process (see Fig. 2.11).

In example of the Si_3N_4-TiN/steel joints brazed with Incusil-ABA 30% SiC (sandwich), the RT-bending strength was increased by 20%, the bending strength at the working temperature of 300°C by 10% when compared with joints brazed with unreinforced Incusil-ABA. The maximum service temperature for the joints was shifted about 50°C, and the joint reliability measured by the scatter in bending strength values (the difference between the maximum and minimum strength value of a testing sample series) was reduced by 35% with the filler reinforcement (see Fig. 2.12).

A prototype of a cutting tool was successfully brazed with the composite brazing (see Fig. 2.13).

Joining of SiC Fiber-Reinforced Borosilicate Glass

SiC fiber (Nicalon® NL 202)-reinforced borosilicate glass (Duran®, Schott Glaswerke) is of high practical relevance, especially when combined with metallic parts, for

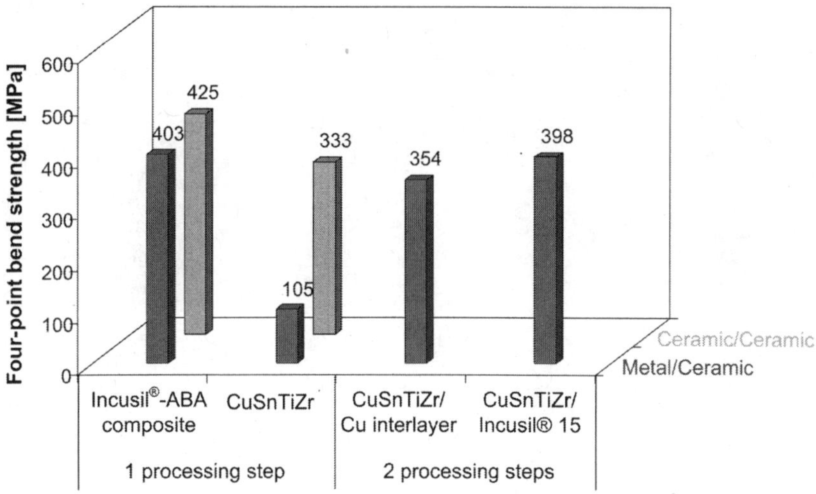

Figure 2.11. Different approaches of residual stress relief and the obtained strength value for the Si_3N_4-TiN/steel joints.

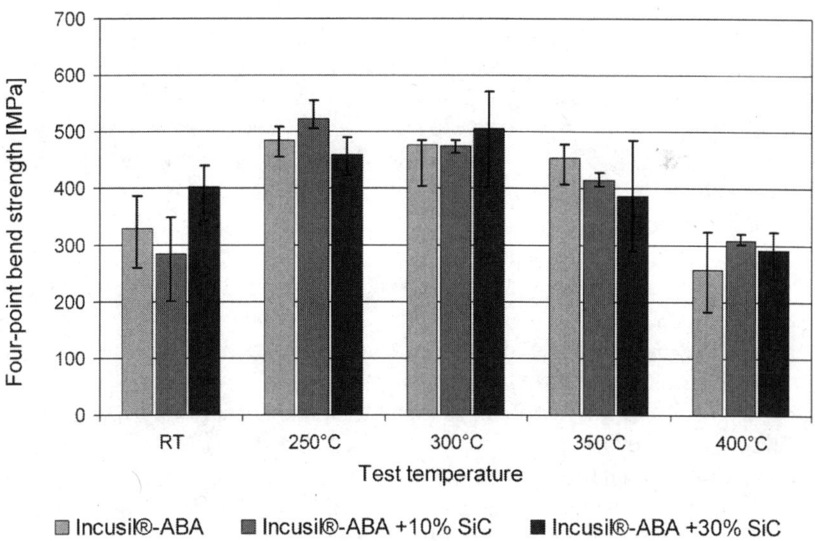

Figure 2.12. The elevated temperature properties of Si_3N_4-TiN/steel joints brazed with composite and noncomposite brazing filler (Incusil-ABA).

Figure 2.13. Ceramic integration in cutting tools via vacuum brazing using composite filler.

example, in components for handling of hot glassware. For such applications molybdenum is a good metallic partner under consideration of system requirements, for example, thermomechanical stability at temperatures of interest (500–750°C), and physical properties, for example, coefficient of thermal expansion close to that of the glass matrix composite.

However, the relatively low-glass transformation temperature, T_g of Duran (550°C), limits the brazing temperature applicable to join the composite. For this temperature range (RT-550°), there are no brazing alloys that would wet borosilicate glass. The only brazing filler materials ensuring wetting of the glass matrix composite at such relatively low temperatures are glass brazes. They are successfully used to join glasses by silicate brazing, that is, for bulk silicate-glass components, such as cathodic ray tubes. However, when brazing Duran/SiC_f composite to Mo with glass solders (e.g., composite glass braze, Schott G018-174, $T_{brazing}$ = 430°C), only joints with relatively low strength are achieved, and the full potential of the glass matrix composites cannot be utilized, even when special steps as surface roughening are applied (Janczak-Rusch et al. 2005). While searching for the other joining solutions, the question was posed, whether it is possible to avoid the degradation of the glass matrix when brazing above the glass transformation temperature (Janczak-Rusch et al. 2005). It was known that SiC fiber/borosilicate glass matrix composites can be for instance exposed to oxidizing environments for short times (up to ~20 hours) up to temperatures of 700°C without major degradation of their mechanical properties (Boccacini et al. 1998). If the degradation of the properties of the glass matrix composite during the brazing process could be avoided, the application temperature of the composite/metal joint component could be increased.

The active brazing of a commercially available SiC fiber-reinforced glass matrix composite with molybdenum counterparts was investigated, assessing different joining configurations (Janczak-Rusch et al. 2005). Joints parallel and perpendicular to the fiber direction have been manufactured (see Fig. 2.14). Incusil-ABA-active filler metal with a relatively low brazing temperature (740°C) was chosen to investigate the possibility of brazing the glass matrix composites far above the glass matrix transformation temperature.

The joint configuration parallel to the fiber direction corresponds to a typical load configuration in which the favorable thermomechanical properties of the glass matrix

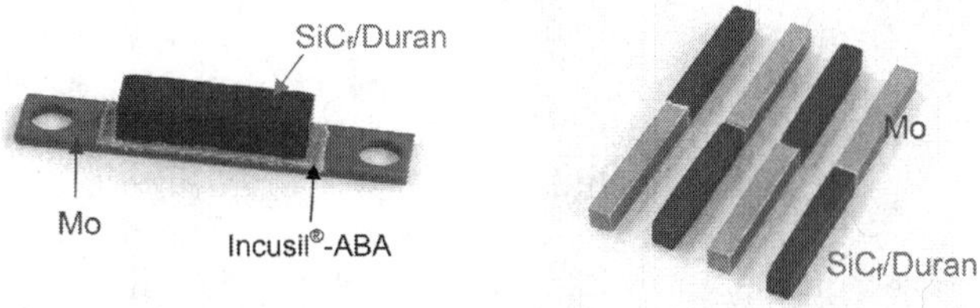

Figure 2.14. Brazed joints of SiC fiber-reinforced borosilicate glass matrix composites with Mo (a) parallel and (b) perpendicular to joint direction.

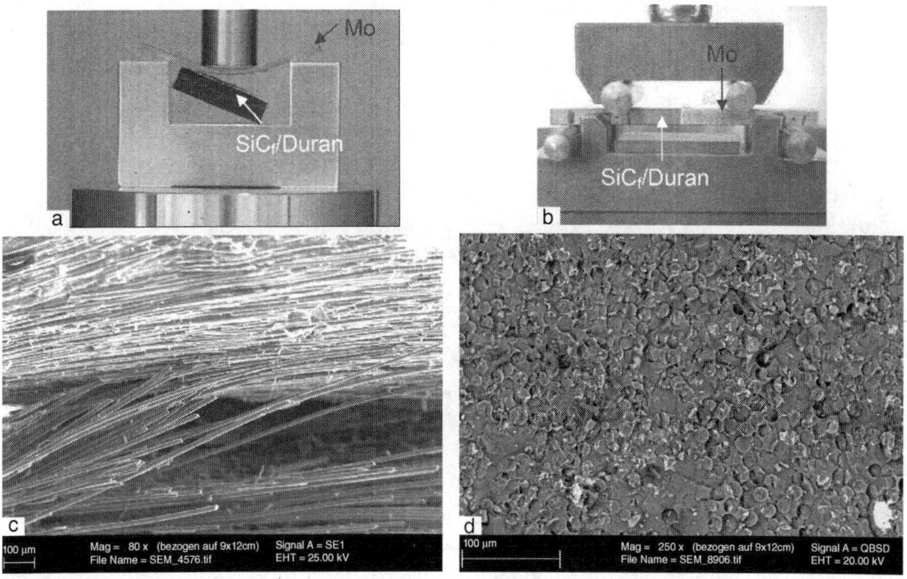

Figure 2.15. Evaluation of the mechanical performance of SiC fiber-reinforced Duran borosilicate glass matrix joints with Mo brazed with Incusil-ABA and SEM micrographs of a typical fracture surface (a) and (c) joints parallel to the fiber direction, (b) and (d) joints perpendicular to the fiber direction.

composite can be exploited. Such type of composite joints is relevant for the actual use of the composites in tools for the handling of hot glassware and nonferrous metals at working temperatures of ~500–750°C. Configuration perpendicular to the fiber direction was mainly used to asses the bending strength of such joints by applying standard mechanical tests.

The use of Incusil-ABA resulted in the glass composite/Mo joints with relatively high loading capacity. In configuration parallel to the fiber direction, failure has always occurred through the glass matrix composite by delamination (see Fig. 2.15a/c). The brazing zone, as well as the Mo substrate remained intact. Moreover, no degradation of the glass matrix composite material due to thermal effects during the brazing process was observed. This indicated that the joint's bond strength is higher than the interlaminar strength of the glass matrix composite. A thin (ca. 1–2 μm) reaction zone (composed of titanium silicides and nitrides) was formed between the Duran-borosilicate glass matrix and the Incusil-ABA brazing filler, which provided a good bonding strength of the joint.

For joints perpendicular to fiber direction (see Fig. 2.15b/d), an average bending strength of 85 MPa was measured. However, the failure behavior was completely different to that observed in case of configuration parallel to the fiber direction (see Fig. 2.15a/c). The joints failed at the interface between brazing filler and the glass matrix composite. Composite delamination was not observed, and fracture occurred by brittle failure at the interface (see Fig. 2.15d). The brazing zone was the weakest link of the joint. The SEM investigation performed on the joint cross-section confirmed that there was an intact bonding between the borosilicate glass matrix and the Incusil-ABA. A brittle reaction layer between the borosilicate glass matrix and the Incusil-ABA brazing filler was found in the brazing zone. However, no reaction between the SiC fiber and the filler metal was observed. This fact indicates a weak bond between the fiber and the filler. Therefore, the composite/filler metal bond strength in this configuration was primarily determined by the strength of the bond between the glass matrix and the filler (which accounts for ~60% of the cross-section area). The load capacity of the glass matrix composite in parallel to the fiber direction could not be fully exploited.

Joining of Mullite–Mullite Ceramic Composites

Oxide ceramics are excellent candidates for high temperature, lightweight applications. In most cases, they need to be combined with metallic parts to create structures. In spite of technical relevance, the materials connections of oxide ceramics with high temperature resistant metals have not been yet fully investigated. However, the preliminary work proves that the oxide ceramics composites (like mullite–mullite composites) can be brazed to high-temperature alloys as Haynes® 230® superalloy. Especially, the two-step process involving premetallization of the ceramic with Ti- and Zr-activated Cu-Zn alloy, followed by brazing (premetallized ceramic to metal) with a Cu-filler, brings promising results (Piazza et al. 2003) when joining mullite–mullite plates with rings of Hayenes 230 superalloy (see Fig. 2.16).

The joints show relatively high strength, and their failure happens via delamination when the structure is overloaded. Certain design steps (see Fig. 2.17) may be undertaken to limit this behavior and improve the joint strength by keeping its thermal isolation performance (Piazza et al. 2003).

CONCLUSIONS

Brazing is a reliable, low-cost, and large-scale technology to integrate ceramics in components with a growing significance. As shown on chosen examples, it can be well

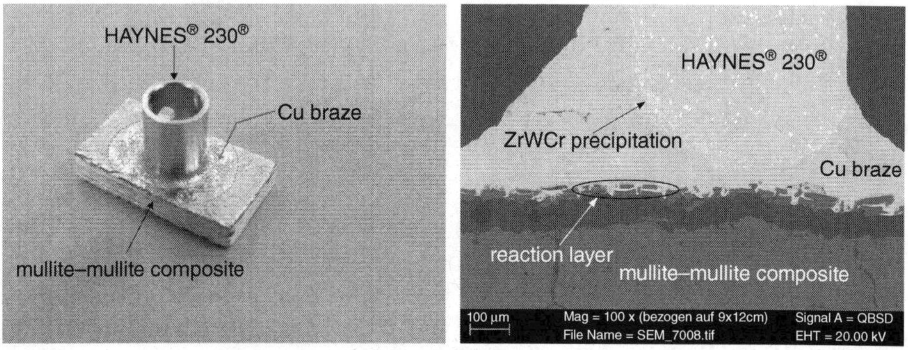

Figure 2.16. Brazed joint of mullite–mullite composite with Hayenes 230 superalloy (two-step brazing using CuSnTiZr as metallization and Cu-filler).

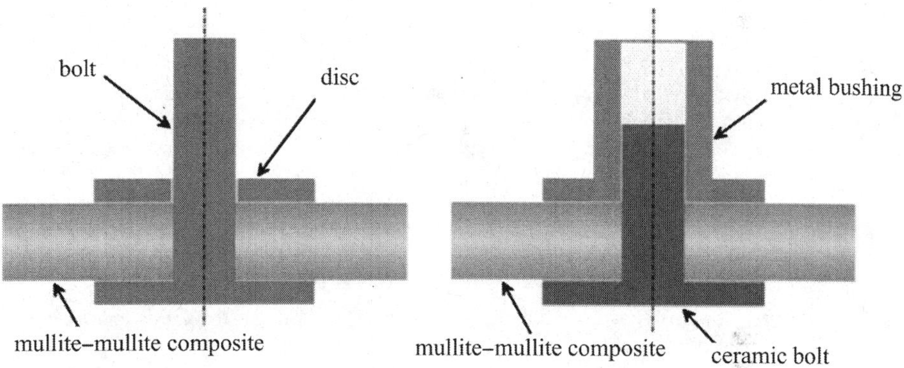

Figure 2.17. Example of joint design optimization for brazing of mullite–mullite composites with a superalloy.

adapted to join ceramic to itself as well as to metals or various composite materials. However, with the increasing requirements the joint design becomes very complex. A combined theoretical–experimental approach is indispensable for a proper joint design, especially when the more complex design concepts, such as the use of interlayers or composite fillers, are adopted. The numerical methods allows a "virtual laboratory" to be set up for a preliminary joint design, simulation and optimization of its behavior, as well as investigation of residual stresses being the critical factor when joining ceramics to other materials. New concepts such as composite approach are needed to increase the joint performance by keeping the number of processing steps low.

ACKNOWLEDGMENTS

Many people have contributed to this work. My special thanks are directed to the doctoral students Vinzenz Bissig and Karolina Sosnowska. I also gratefully acknowledge the contribution of Matteo Galli, John Botsis, Jakob Kübler, Gurdial Blugan, Aldo R. Boccaccini, Hans-Rudolf Elsener and Daniele Piazza. The work was financially supported by Gerbert Rüf Stiftung as well as Empa and EPFL.

REFERENCES

Bissig V, Janczak-Rusch J, Galli M. 2006. Comparison of three different active filler metals used for brazing ceramic-to-ceramic and ceramic-to-metal. Adv Eng Mat 8(3):191–196.

Bissig V, Janczak-Rusch J, Galli M. 2007. Selection and design of brazing fillers for metal-ceramic joints. Mat Sci Forum 539–543:4008–4012.

Blugan G, Hadad M, Janczak-Rusch J, Kübler J, Graule T. 2005. Microstructure, mechanical properties and fractography of commercial silicon nitride-titanium nitride composites for wear applications. J Am Ceram Soc 88:926–935.

Blugan G, Kübler J, Bissig V, Janczak-Rusch J. 2007. Brazing of silicon nitride ceramic composite to steel using SiC-particle-reinforced active brazing alloy. Ceram Int 33: 1033–1039.

Boccacini AR, Strutt A, Vecchio KS, Mendoza D, Chawla KK, Ponton CB, Pearce DH. 1998. Behavior of nicalon-fiber reinforced glass matrix composites under thermal cycling conditions. Composites Part A 29:1343–1352.

Dixon DG. 1995. Ceramic matrix composite-metal brazed joints. J Mater Sci 30:1539–1544.

DVS 3102. 2005. Herstellen von Keramik-Keramik- und Keramik-Metall-Verbindungen durch Aktivlöten.

Ferraris M, Badini C, Montorsi M, Appendino P, Scholz HW. 1994. Joining of SiC_f/SiC composites for thermonuclear fusion-reactors. J Nucl Mater 215:1613–1616.

Galli M, Botsis J, Janczak-Rusch J. 2006. On the relief of the residual stresses in ceramic-metal joints by a layered braze structure. Adv Eng Mater 8(3):197–201.

Galli M, Botsis J, Janczak-Rusch J. 2008. An elastoplastic three-dimensional homogenization model for particle reinforced composites. Comput Mater Sci 41:312–321.

Haber RA, Greenhut VA. 1991. Aluminum/alumina joining. In: *Proc of Metal-Ceramic Joining Symposium TMS Fall Meeting*, Warrendale, PA, 119–131.

Hsueh CH, Evans AG. 1985. Residual stresses in metal/ceramic bonded strips. J Am Ceram Soc 68(5):241–248.

Iancu OT, Munz D, Eigenmann B, Scholtes B, Macherauch E. 1990. Residual stress state of brazed ceramic/metal compounds determined by analytical and X-ray residual stress measurements. J Am Ceram Soc 73(5):1144–1149.

Janczak-Rusch J. 2005. Brazing and characterisation of advanced ceramic joints. Cer Eng Sci 26:391–398.

Janczak-Rusch J. 2007. Design, brazing and characterization of metal-ceramic joints. Inzynieria Materiałowa 157–158:547–551.

Janczak-Rusch J, Zigerlig B, Klotz U, Brem F, Kübler J, Blugan G. 2002. Joining of Si_3N_4/TiN ceramics with steel components using particle reinforced active brazing alloys. In: *Proc of 6th Intern Conf on Joining Ceramics, Glass and Metal*, pp. 41–48.

Janczak-Rusch J, Piazza D, Boccaccini AR. 2005. Joining of SiC fibre reinforced borosilicate glass matrix composites to molybdenum by metal and silicate brazing. J Mater Sci 40: 3693–3701.

Li JG. 1994. Wetting of ceramic materials by liquid silicon, aluminium and metallic melts containing titanium and other reactive elements: A review. Cer Inter 20:391–412.

Kim TW, Park SW. 2000. Effects of interface and residual stress on mechanical properties of ceramic/metal system. Key Eng Mater 183–187:1297–1284.

Klose H. 1999. *PhD Thesis TU Chemnitz, Beitrag zur Berechnung, Herstellung und Charakterisierung von verstärkten Aktivloten.*

Lee SB, Kobayashi H, Huh JW. 1995. Fatigue strength and fracture mechanism of ceramic ± metal joints under cyclic bending. Int J Fatigue 17(6):427–435.

Locatelli MR, Nakashima AP, Tomsia AP, Dalgleish BJ, Glaeser AM. 1995. New strategies for joining ceramics for high-temperature applications. Key Eng Mater 111–112:157–190.

Lu SP, Kwon OY. 2002. Microstructure and bonding strength of WC reinforced Ni-base alloy brazed composite coating. Surf Coat Technol 153:40–48.

Mackerle J. 2001. Finite elements in analysis and design. FEM and BEM analysis and modelling of residural stresses—A bibliography (1998–1999). 37:253–262.

Nakamura M, Mabuchi M, Saito N, Yamada Y, Nakanishi M, Shimojima Y, Shigematsu I. 1999. Joining of Si-Ti-C-O fiber reinforced ceramic composite and Fe-Cr-Ni stainless steel. Key Eng Mater 164-1:435–438.

Nicholas MG. 1990. *Joining of Ceramics*. London: Chapman & Hall.

Park JW, Eagar TW. 2002. Application of the transient liquid phase bonding to microelectronics and MEMS packaging. In: *Proc. of 8th International Symposium on Advanced Packaging Material*:30–38.

Park JW, Mendez PF, Eagar TW. 2002. Strain Energy Distribution in Ceramic/Metal Joints. Acta Mater 50:883–899.

Piazza D, Boccaccini AR, Kaya C, Janczak-Rusch J. 2003. Hochtemperatur-Löten von einem Mullit-Mullit Verbundwerkstoff, in "Verbundwerkstoffe". Hrsg Degischer:702–707.

Reicher R, Smetana W, Schuster JC, Adlaßnig A. 2001. A fritless copper conductor system for power electronic applications. Microelectronics Reliability 41(4):491–498.

Reinecke AM, Exner H. 2001. A new promising joining technology. J Cer Proc Res 2:45–50.

Schwartz MM. 2003. *Ceramic Joining*. Materials Park, OH: ASM International.

Singh M. 1999. Design, fabrication and characterization of high temperature joints in ceramic composites. Key Eng Mater 164-1:415–419.

Weil SK, Coyle CA, Hardy JS, Kim JY, Xia G-G. 2004. Alternative planar SOFC sealing concepts. Fuel Cells Bull 5:11–16.

Weil S, Yang SG, Paxton Dean M. 2001. Joining of mixed-conducting oxides to heat resistant metals. In: *Proc. of Joining of Advanced and Specialty Materials*, ASM, USA, 69–78.

Wielage B, Ashoff D. 1990. Technische Keramik, 3.9.2.1.

Zhang JX, Chandel RS, Chen YZ, Seow HP. 2002a. Effect of residual stress on the strength of an alumina–steel joint by partial transient liquid phase (PTLP) brazing. J Mater Proc Technol 122:220–225.

Zhang JJ, Li SJ, Duan HP, Zhang Y. 2002. Joining of C/SiC to Ni-based superalloy with Zr/Ta composite interlayers by hot-pressing diffusion welding. Rare Metal Mater Eng 31(Suppl 1): 393–396.

Zhu M, Chung DDL. 1997. Improving the strength of brazed joints to alumina by adding carbon fibres. J Mater Sci 32:5321–5333.

Zigerlig B, Elsener HR, Piazza D, Kiser M. 2002. Brazing, welding and design aspects of a multifunctional titanium-alumina ceramic component for a space application. In: *Proc. of 6th International Conference on Joining Ceramics, Glass and Metal*:66–73.

3

JOINING AND INTEGRATION ISSUES OF CERAMIC MATRIX COMPOSITES FOR THE NUCLEAR INDUSTRY

Monica Ferraris, Milena Salvo, and Valentina Casalegno

Politecnico di Torino, Torino, Italy

INTRODUCTION

Ceramic matrix composites (CMCs) are attractive materials for many "extreme" applications, and nuclear energy production has one of the most severe environments, involving thermomechanical loads together with neutron irradiation. Materials with the greatest potential for these applications are fiber-reinforced CMCs—mainly carbon fiber-reinforced–carbon matrix (C/C) composites and silicon carbide (SiC) fiber-reinforced–silicon carbide matrix (SiC/SiC) composites.

Comprehensive books and reviews are available for CMCs (Matthews and Rawlings 1994), C/C composites (Savage 1993; Buckley and Edie 1993; Jortner 2000; Shapiro 2007), and SiC/SiC composites (Di Carlo et al. 2005).

General information about energy production by thermonuclear fusion (Internet resources 1–4; Dietz and the ITER Joint Central Team 1999; Peacock et al. 2004; Ioki et al. 2004) and fission (in particular, Generation IV) reactors can be found in the following: U.S. DOE (2002), Carré (2003), Kotake et al. (2005), and Internet resources 13 and 14.

One of the most critical points for the use of CMCs is joining them to one another or to metals; integration of CMCs is reviewed in Nicholas and Mortimer (1985),

Ceramic Integration and Joining Technologies: From Macro to Nanoscale, First Edition.
Edited by Mrityunjay Singh, Tatsuki Ohji, Rajiv Asthana, Sanjay Mathur.

Akselsen (1992), Shapiro (2007), in Internet resource 5, and in this book by M. Singh et al., M. Halbig, K. Scott Weil, and Jolanta Janczak-Rusch.

This chapter will focus on *joining and integration of CMCs* used in *nuclear applications*, in particular, for thermonuclear fusion (Barabash et al. 2000; Dig et al. 2000; Merola et al. 2002b) and for a new generation of fission reactors (U.S. DOE 2002; Jenkins 2006).

Reliable characterization of joined CMCs has always been an issue, particularly for the strict safety requirements of nuclear applications. Mechanical, thermal, and nondestructive tests (NDTs) are discussed in the following sections.

ITER

Carbon–carbon composites are being considered for some components in the design of the following nuclear reactors: ITER (Merola and Wu 2004), Joint European Torus (JET) (Internet resource 3; Piazza et al. 2007), Tore Supra (Mitteau 2005), and Wendelstein 7-X (Boscary et al. 2007) in Europe, Tokamak Fusion Test Reactor (TFTR) in the United States (Hawryluk et al. 1998; Ying et al. 2006), and JT 60U and JT60SU in Japan (Ando et al. 1992; Masaki et al. 1995; Onozuka et al. 1999; Hosogane et al. 2002). Silicon-doped carbon is being used as first-wall materials in HT-7 (a Chinese reactor) (Guo et al. 2003).

ITER is the most important project in the field of thermonuclear fusion energy. After years of discussions, on June 28, 2005, it was decided to build ITER in France (Cadarache) (EFDA 2005), and the related Agreement on the Establishment of the ITER International Fusion Energy Organization for the Joint Implementation of the ITER Project was signed on November 21, 2006 by the European Atomic Energy Community (EURATOM) and by the governments of the People's Republic of China, the Republic of India, Japan, the Republic of Korea, the Russian Federation, and the United States.

ITER will be the world's largest *tokamak* (Russian for "toroidal magnetic chamber") of the world (Fig. 3.1a) with a plasma volume of 837 m^3 operating at about 100 million degrees; it is expected to produce fusion energy of about 500 MW. The development of ITER components and the research activity about new materials for nuclear environment represents one of the most challenging engineering efforts ever made in the scientific field. A description of nuclear fusion and ITER can be found in Internet resource 1; this section summarizes joining and integration issues of C/C composites in ITER (Fig. 3.1a–d). Comprehensive reviews of this subject are given in Barabash et al. (2000), Daenner et al. (2002), Merola et al. (2002b), Peacock et al. (2004), and Ioki et al. (2004).

As of this writing, the first option for ITER is a compromise—a full beryllium main wall with C/C composites at the strike points and tungsten divertor baffles and dome.

Where Are the Joined CMCs in ITER?

The divertor (Fig. 3.1) is one of the most technically challenging components of ITER (Schlosser et al. 2005): It exhausts the flow of energy from charged particles

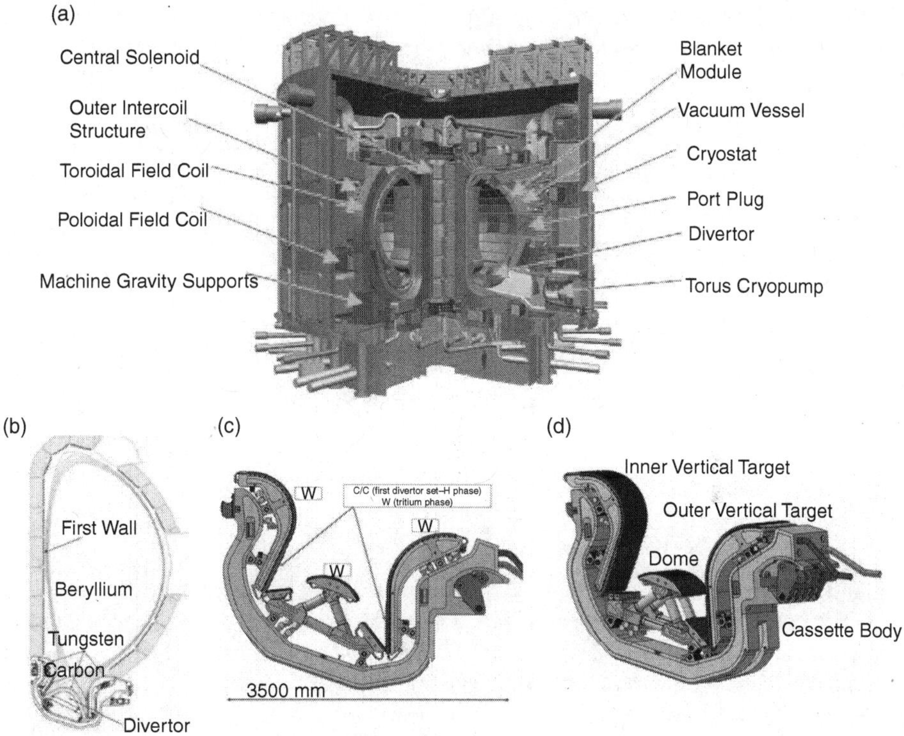

Figure 3.1. (a) Cross-section through ITER (to have an idea of the size, note the man sketched below, right). (b) The divertor in the plasma chamber. (c, d) The materials used in the divertor and the divertor component sizes (courtesy of ITER Organization).

produced in the fusion reactions and removes helium and other impurities from the plasma. The plasma particles that would lead to an unacceptable erosion of the first wall are guided by the magnetic field lines in order to impact the divertor plates located at the bottom of the vacuum vessel, allowing the neutralization of plasma ions.

The divertor consists of 54 modules (cassettes) and it is located at the bottom of the plasma chamber (Fig. 3.1b–d): It has C/C composite (or tungsten) as plasma-facing material joined on a water-cooled precipitation-hardened copper-based alloy (CuCrZr). These plasma-facing units are mounted on a water-cooled stainless steel structure, which is mechanically attached to the so-called cassette body.

Carbon–carbon composites have been selected for the lower part of the vertical targets (Fig. 3.1c, d) for the first phase of the ITER operation. ITER requirements for the lower part of vertical targets (Table 3.1) indicate that they should withstand heat fluxes up to 10 MW/m^2 for at least 3000 cycles of 400 s and up to 20 MW/m^2 (transient events) for about 300 cycles of 10 s (ITER DOCa; ITER DOCc).

TABLE 3.1. Operational Conditions for C/C Joined to Copper Alloy in the Lower Vertical Target, Component Subjected to Glancing Incidence of Heat Flux (ITER, DOCa)

Heat flux (MW/m^2)	Number of cycles	Damage (dpa)	T_{max} of the C/C–CuCrZr joint (°C)	
			Steady state	Transient
10/20[a]	3×10^3	0.2	300	400

[a]20 MW/m^2 transient events, duration 10 s, number 10% of normal shots.

Material damage due to neutron irradiation up to a tenth of displacement per atom (dpa, i.e., the number of times each atom is dislodged from its place in the crystal by radiation) will also occur, if the use of C/C composites is continued during the initial stage of the tritium operation of ITER (see the next subsection).

Why C/C Composites in ITER?

Operation with a C/C composite target is considered to have advantages for the start of ITER operations given its proven range of compatibility with a number of plasma conditions in present devices, particularly at low densities with significant additional heating. C/C also promises to make the development of techniques for heat flux mitigation easier, taking advantage of the larger tolerance of the plasma to C versus W and its absence of melting. Finally, even if the onset of material damage during plasma off-normal events in C/C is similar to that of W, vapor shielding effects are more effective in protecting the C/C from further erosion at high power fluxes.

On the other hand, W suffers from macroscopic cracking, melting, and possible melt-layer loss, thus making the potential damage to the divertor components more serious for W. Conditions of low densities with additional heating and a high frequency of off-normal events will constitute a significant fraction of ITER's initial operation. On the basis of these considerations, C/C composite is the present reference design solution for the lower part of the vertical targets at the start of the ITER operation, when problems related to tritium retention are not relevant.

In fact, C/C composites have several disadvantages, including a relatively high physical sputtering rate and a chemical reactivity with the hydrogenic fuel. The resultant high erosion rate could potentially limit machine operation in the deuterium–tritium phase (due to codeposition with tritium). Excessive codeposition raises safety concerns with respect to tritium inventory limits. While the exact magnitude of the fuel retention associated with C/C composite use under ITER operating conditions remains uncertain, its foreseen magnitude in the deuterium–tritium phase may require frequent intervention to remove tritium with techniques that remain to be developed. These features of carbon have spurred interest in W for the later stages of ITER operation and, in particular, for its "nuclear" phase with a deuterium–tritium plasma.

Carbon–carbon composites have been selected as plasma-facing components for the lower part of vertical targets in the divertor design (Fig. 3.1c, d) due to their absence of melting, their high thermal-shock and thermal-fatigue resistance (low crack propagation), and their high thermal conductivity in comparison with conventional graphite (Merola and Wu 2004). C/C-based components have proven performance under pulsed high heat fluxes and have a low neutron absorption cross-section. Besides, they retain good mechanical strength at high temperature and have a low atomic number that induces low power losses in the plasma, which is valid when the material is eroded or sublimated. There are, of course, some drawbacks related to the use of C/C composites, for example, high affinity for hydrogen, which results in a high chemical sputtering by the formation of hydrocarbon species, strong chemical erosion, and tritium retention due to the high specific surface area.

Neutron irradiation can induce significant changes in all of the properties of carbon-based materials (Wu et al. 1994; Dig et al. 2000; Snead et al. 2005). Many researchers (Bonal and Wu 1996; Wu et al. 1998; Snead et al. 2005) in nuclear applications have studied the superiority of C/C composites as low-activation materials (LAMs) and their behavior under irradiation. Material damage due to neutron irradiation up to a few tenths of displacement per atom will also occur during the tritium operation of ITER (Wu et al. 2000; Linke 2006). C/C composites shrink under neutron irradiation: 1 dpa typically results in 1% volume change. For instance, in carbon materials, 1 dpa is produced by a neutron dose of around $1 \times 10^{25}\,n/m^2$. The neutron flux to the wall of ITER will be of the order of $5 \times 10^{17}\,n/m^2/s$. The extent of the damage is reduced at elevated temperatures due to annealing (Wu et al. 2000). In the current ITER divertor design, the carbon–carbon composites are used relatively far from the burning plasma, and the total neutron damage is expected to be up to 0.2 dpa if C/C is retained for the tritium phase (Table 3.1) (ITER DOCa; Linke 2006).

The use of C/C seems compulsory for the first phase of ITER operation since the tritium issues are not relevant at this stage. Some future considerations arising from the study and observation of C/C during the first phase of ITER operation might influence the decision of whether or not to switch to a full-W divertor (Wu et al. 1994; Bolt et al. 2002; Snead et al. 2005).

Several carbon–carbon composites have been considered as possible candidates for the ITER divertor; at this writing, CFC NB31, produced by Snecma Propulsion Solide (France) (ITER DOCe; ITER DOCf), is the European reference material for the ITER high-heat-flux components, and Dunlop Concept 2 is the alternative one.

As a consequence, most of the literature is on joining C/C, often with the commercial names of CFC SEP NB31, a 3D C/C with pitch precursor fiber and polyacrylonitrile (PAN) fiber.

A few data are available on CFC SEP NS31, a silicon-doped C/C composite, which should have an enhanced resistance to plasma chemical erosion and a lower tritium retention (Grote et al. 1997; ITER DOCa; ITER DOCb). However, Si-doped C/C composites are no longer being considered for ITER due to high cost and manufacturing difficulties.

The properties of C/C composite to be used for ITER (CFC SEP NB31) can be found in ITER DOCa, b, d, e, f, g, h and in Visca et al. (2007). Those of particular interest with respect to this section are given in Table 3.2.

About $50\,m^2$ of C/C composite will be used in the lower vertical target.

TABLE 3.2. Properties of C/C to Be Used for ITER (CFC SEP NB31)

Properties CFC SEP NB31	T (°C)	x (pitch fibers) (high thermal conductivity)	y (PAN fibers)	z (needled PAN fibers)
Coefficient of thermal expansion, CTE (10^{-6} C^{-1})	100	−0.339	−1.376	−0.018
	800	0.777	−0.290	1.013
	1000	0.992	−0.112	1.201
Tensile strength (MPa)	20	130	30	19
Shear strength (MPa)	RT	15 (xz)	9 (yz)	

The z direction is parallel to the tube heat sink, while the x direction is perpendicular to the tube and is parallel to plasma heat flux since it shows the maximum thermal conductivity.

Design Issues for Joined C/C Composite in ITER

As requested by the ITER design, C/C composite must be joined to a CuCrZr-IG (ITER Grade) (Daenner et al. 2002; Merola et al. 2002a; Peacock et al. 2004; Ioki et al. 2004) in order to transfer the heat loads. This joint is one the most critical manufacturing steps upon which the heat removal capabilities and the thermal-fatigue performance of the whole component mainly depend. C/C composite (or tungsten) is the high-heat-flux plasma-facing material, which has to be joined to a precipitation-hardened copper-based alloy (CuCrZr) and a stainless steel structure, which are mechanically attached to the cassettes.

There are two different designs for this component based on C/C composites: flat tile (Fig. 3.2a) and monoblock (Fig. 3.2b).

A pure copper interlayer (1–2 mm thick) of oxygen-free high-conductivity copper (OFHC, 99.95%; coefficient of thermal expansion [CTE], 15.4 at room temperature [RT], up to 20.6 at 700°C and up to 21.6 at 800°C) is used to relieve, by plastic deformation, the stresses due to mismatch in CTE and to the temperature gradient between the C/C composite (CTE C/C in Table 3.2) and the copper alloy (CTE CuCrZr = 16–17 × 10^{-6} K^{-1} and ~19 × 10^{-6} K^{-1} at 700°C) (ITER DOCa; ITER DOCb; ITER DOCd; ITER DOCe; ITER DOCf; ITER DOCg; ITER DOCh; Visca et al. 2007).

Among several possible options, the flat-tile and monoblock configurations (Fig. 3.2a, b) have given the most promising results for the design of the component. In particular, the monoblock gives a more robust solution in comparison with flat tile for the vertical target, and it is now considered the ITER reference geometry (Merola and Vieider 1998; Tivey et al. 2001).

The monoblock design requires drilled blocks of C/C composite into which a CuCrZr tube is inserted and joined; also, a thin layer of pure copper is necessary here, between the C/C and the copper alloy in order to relax high joint interface stress; the monoblock is preferred over the more easily manufactured flat-tile design because of better heat flux performance and because of the observed tendency for flat tiles to suddenly and totally detach in high-heat-flux conditions.

Nevertheless, it must be taken into account that the thermal conductivity of C/C composites is decreased significantly by neutrons, and the monoblock design should be more affected than the flat-tile design.

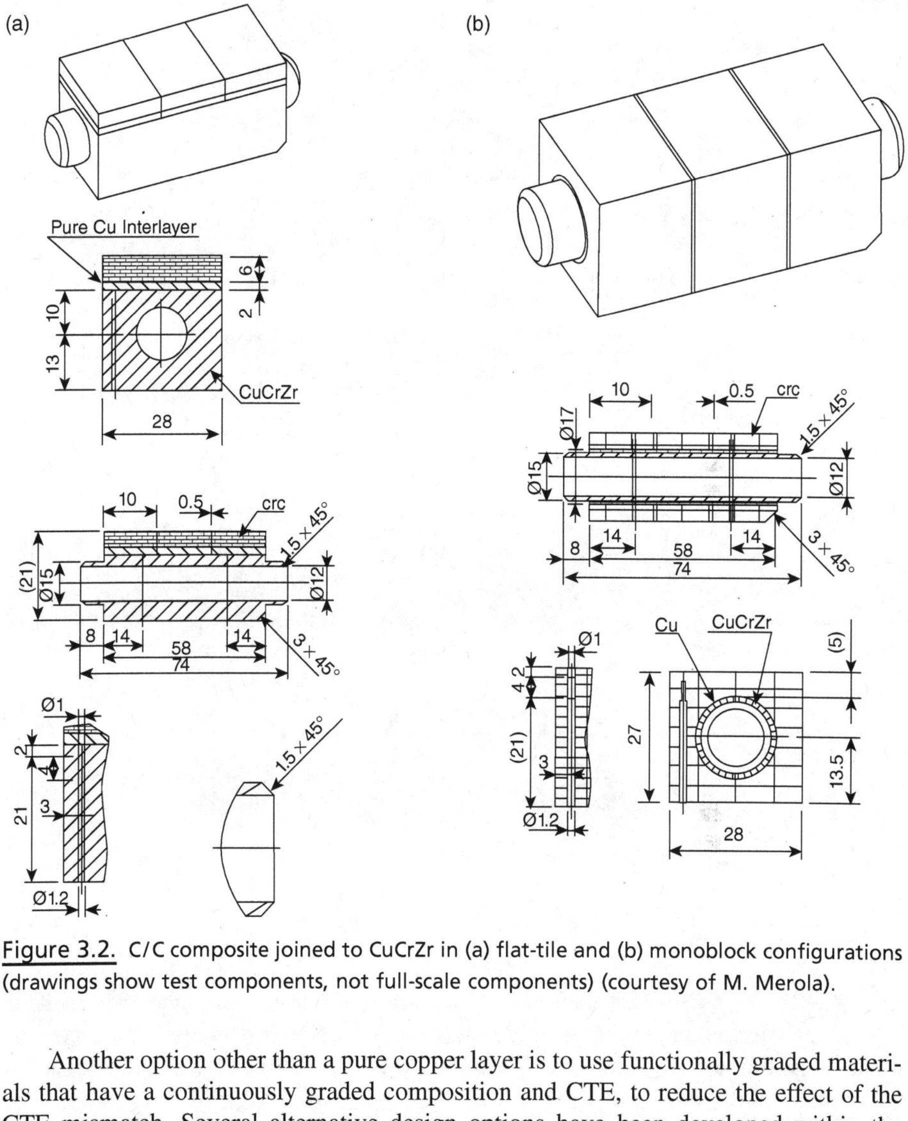

Figure 3.2. C/C composite joined to CuCrZr in (a) flat-tile and (b) monoblock configurations (drawings show test components, not full-scale components) (courtesy of M. Merola).

Another option other than a pure copper layer is to use functionally graded materials that have a continuously graded composition and CTE, to reduce the effect of the CTE mismatch. Several alternative design options have been developed within the European Union (EU) project ExtreMat (Internet resource 12); for example, Mo interlayers are not soft, but strong and with a CTE intermediate between C/C and CuCrZr.

An example of a full-scale component, produced by Plansee, Austria, is given in Figure 3.3.

Joining Techniques for C/C Composites in ITER

Since the design requires the introduction of a thin copper interlayer between the two materials to be joined, there are two joints in this component: C/C to Cu and Cu to

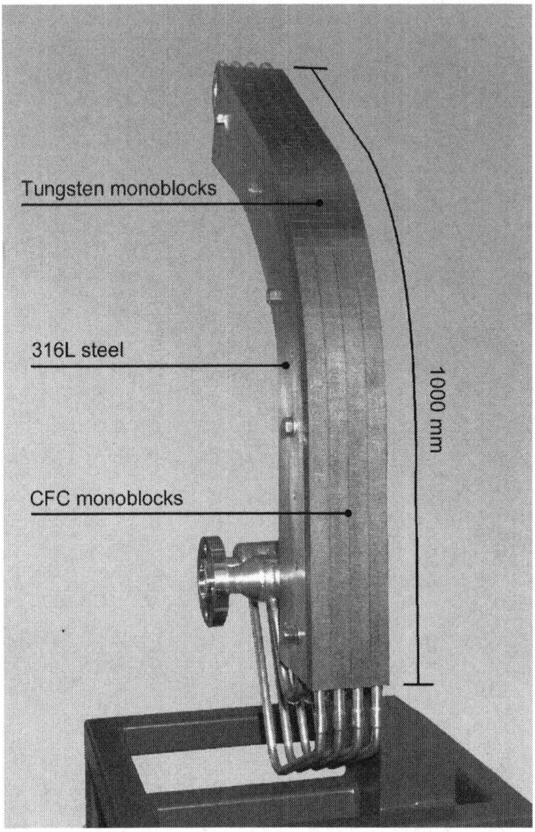

Figure 3.3. Vertical target full-scale prototype manufactured by Plansee GmbH. The top part of the mock-up consists of W brush armoring; the C/C composite is located at the bottom part of the mock-up (courtesy of EFDA). CFC, carbon fiber composite.

CuCrZr (Fig. 3.4). Only the first joint is within the scope of this chapter; the second is a brazing or hot isostatic pressing (HIPing) joint between pure copper and a copper-based alloy. Some references about this second joint are in Merola et al. (2002a, 2003) and Bisio et al. (2005).

Essentially, joining methods used for CMCs originate from techniques developed to join ceramic components to each other or to metals. Brazing is a well-known joining process, recommended for joining dissimilar materials: If properly done, it has good mechanical strength, high fatigue performance, and adequate thermal conductivity, which is required here to remove the expected heat load (Nicholas and Mortimer 1985; Akselsen 1992, Arroyave and Eagar 2003; Shapiro 2007; Internet resource 6)

In the case of joining of CMCs for nuclear applications, some additional restrictions must be taken into account: The joining materials should be preferably LAMs, even if the small volume occupied by these materials is negligible with respect to the total volume of the structural materials in the reactor (Wu et al. 1998). Furthermore,

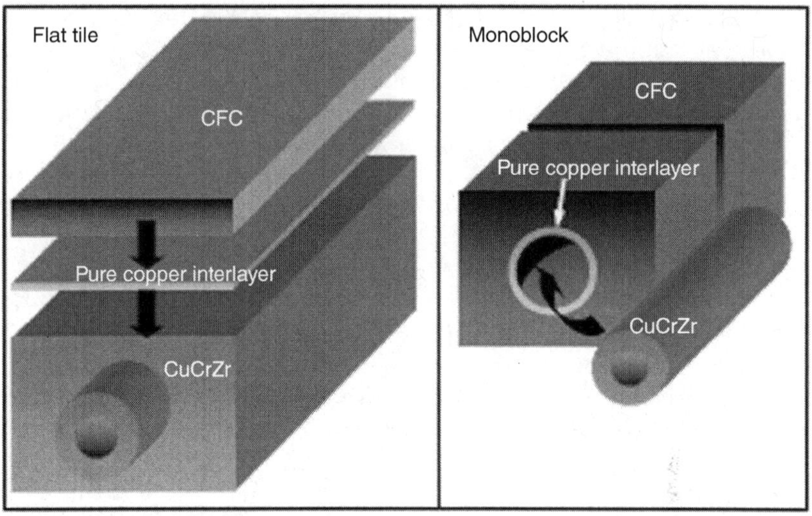

Figure 3.4. A pure Cu interlayer is put between C/C composite and Cu alloy heat sink in both flat-tile (left) and monoblock (right) geometries.

the use of joining techniques that involve high pressure should be minimized since large parts have to be assembled. Some joining materials are not allowed, for example, elements with high vapor pressure (e.g., zinc or cadmium), to avoid poisoning the cryopumps, or those with unfavored nuclear properties, such as silver or gold; in addition to these general requirements, joining materials and techniques to be used for ITER must satisfy the working conditions in ITER DOCa. It requires the joint to have a thermodynamical and mechanical stability up to at least 800°C under vacuum in order to satisfy the requirements of Table 3.1; it must survive the thermal, mechanical, and neutron loads faced by the component, and it is expected to operate in a cyclic mode with an acceptable reliability and lifetime. The joining technology must also be compatible with the overall component manufacture process and, in particular, with the preservation of the thermomechanical properties of the precipitation-hardened CuCrZr alloy (Merola et al. 2002a). If during the joining process the alloy is subjected to additional thermal cycles (e.g., welding, brazing, and HIPing), they should not affect the physical and mechanical properties of CuCrZr by dissolution and/or coarsening of the precipitates (ITER DOCd).

Wettability is a key factor; it is not within the scope of this chapter to review this subject, and an overview can be found in Eustathopoulos et al. (1999), Dezellus and Eustathopoulos (1999), Akselsen (1992), and in this book (V. M. Perevertailo, Chapter 7). We just summarize a few concepts of interest here. The Cu interlayer cannot be obtained by direct casting of copper on the C/C surface because Cu does not wet C/C at all—the contact angle of molten copper on carbon substrate is about 140° (Eustathopoulos et al. 1999; Dezellus and Eustathopoulos 1999). The poor wettability of C/C is related to the nonmetallic character of its bonding, whereas the bonding electrons in copper are delocalized (Nicholas and Mortimer 1985; Akselsen 1992; Eustathopoulos et al. 1999).

Copper *can be cast directly on C/C composite when* the C/C surface is modified to form carbides by a solid-state chemical reaction between the composite surface and elements such as Si, Al, Ti, Zr, Cr, Mo, or W; some metals can form carbides with "metal-like" behavior (Eustathopoulos et al. 1999) and are usually well wetted by molten metals. Several patents refer to joints between carbon-based materials and copper (Friedrich et al. 1992; Grill et al. 1996; Rainer and Reheis 1996; Weeks and Sommer 1998a, b, 1999; Tashiro and Kasahara 1999), for example, pure Cr and Ti react with C to form carbides. Cr and Ti wet modified carbon surfaces very well, with a contact angle of 35°–40° at 1775°C and of 50°–60° at 1740°C, respectively, in argon (Eustathopoulos et al. 1999).

Surface modification of C/C composites and casting of copper have two main advantages when compared with active brazing: a high remelting point (1083°C) due to pure Cu, much higher than that of most active brazing alloys, and no potentially detrimental active elements prone to form brittle intermetallics. The modified surface must be thermodynamically stable, without diffusion or any other detrimental reaction during the subsequent Cu casting.

The Active Metal Casting (AMC®) technique, based on modification of the C/C composite surface, was originally developed in the 1980s by the Austrian company Plansee for different purposes; then it was extended to joining C/C composites for nuclear applications. "Active" indicates the activation of the C/C surface to make it be wetted by Cu: The physical vapor deposition (PVD) or chemical vapor deposition (CVD) of a Ti coating on the C/C surface is followed by a high-temperature treatment to form TiC, which improves the wettability of C/C by molten copper. AMC consists of casting a pure Cu layer onto a laser-textured and TiC-modified C/C surface (Rainer and Reheis 1994; Friedrich et al. 1995; Schlosser et al. 2001). The laser texturing enhances Cu infiltration into the C/C, and the TiC-modified C/C surface improves the wetting. The special laser treatment of the C/C surface produces a large number of closed conical holes (diameter ~50–500 μm, depth 100–750 μm), thus increasing the joined area and providing additional crack-growth resistance. Due to the open porosity of the TiC-modified C/C and laser machining, the cast Cu penetrates into the C/C up to 2 mm (ITER DOCa). Figure 3.5a shows the laser-machined C/C surface; in Figure 5b, a typical view of the AMC joint is given (Rainer and Reheis 1994; Friedrich et al. 1995).

AMC has been successfully applied to both flat-tile and monoblock geometries. On the other hand, AMC technology requires laser machining of the C/C surface, which might be not economically attractive for large-scale production. Laser-induced stresses in the joined area are under evaluation.

Plansee has since improved AMC by using silicon and titanium to modify the C/C surface (TiSi-AMC) (Schedler et al. 2007a).

Ansaldo Ricerche, Genoa, Italy, proposed a joining technique based on Cu-ABA, a *Cu–Ti-based (Cu-ABA) commercial alloy, reinforced by 2-D randomly oriented carbon fibers* uniformly distributed in the brazing alloy. The joining is done at about 1000°C; the Ti reacts with carbon to form a thin TiC layer that promotes wetting (Bisio

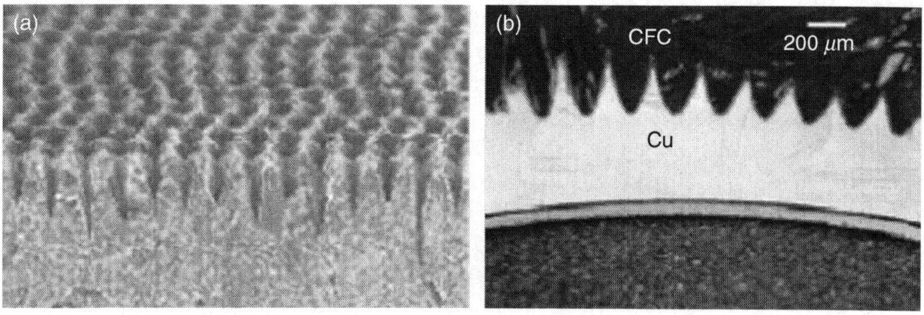

Figure 3.5. (a, b) Laser structuring of C/C in the AMC process (right) and cross-section of the AMC C/C–Cu joint (courtesy of Plansee AG, Austria).

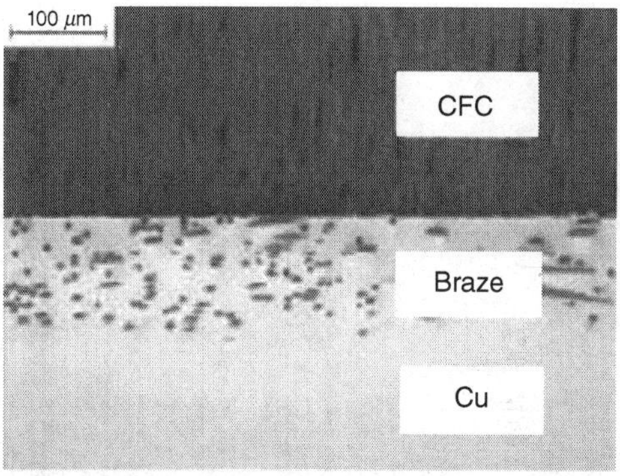

Figure 3.6. C/C–Cu active brazing with Cu–Ti alloy and dispersed carbon fibers in the braze (as in Bisio et al. 2005) (C/C composites are also referred as CFC in some figures).

et al. 2005). The carbon fibers are expected to mitigate thermal expansion mismatch between C/C and the braze (Fig. 3.6) and to react with titanium in the brazing alloy, with beneficial effects on the thermal-fatigue strength of the joint. This technique was successfully tested on C/C NB31-Cu flat-tile joints.

Within ExtreMat, many other solutions have been investigated to modify the C/C composite surface, for example, by TiN or TiC (Internet resource 12).

A method patented by Politecnico di Torino, Italy (Ferraris et al. 2005) is based on the modification of the *C/C surface by the reaction of Cr, Mo, or W.* Such elements, in particular, Cr and Mo, have been extensively used as active elements in brazing alloys for copper active brazing, as well as in patents referring to other applications (Friedrich et al. 1992; Grill et al. 1996; Rainer and Reheis 1996; Weeks and Sommer 1998a, b, 1999; Tashiro and Kasahara 1999). The basis of this method is to use these "active elements" to modify the surface, thus avoiding the use of brazing alloys. It has been shown (Mortimer and Nicholas 1970; Mortimer and Nicholas 1973) that the addition of a small amount of chromium in a copper alloy promotes wetting of carbon, while the same effect is obtained only with massive amounts of titanium, manganese, or vanadium.

Tungsten, molybdenum, and chromium powders were deposited on C/C composite (CFC NB31, Snecma Propulsion Solide, Le Hallian, France) by a slurry technique, a suspension of powders in ethanol, followed by heat treatment under argon atmosphere (<50 mbar) or in vacuum (5×10^{-1} mbar) to form carbides. Both Mo and W formed their respective carbides (Mo_2C, WC) but also residual unreacted metal. Furthermore, WC does not provide good adhesion to the C/C substrate. The copper uniformly wetted the interface, but the adhesion was poor due to the weak carbide interface at the C/C surface. Details on the process can be found in Appendino et al. (2003, 2004, 2006).

Chromium carbide-modified C/C gave the bests results (Fig. 3.7): A 15-μm-thick carbide layer formed and was identified by X-ray diffraction (XRD), $Cr_{23}C_6$, Cr_7C_3 on C/C; the CTE of the carbides lies between that of C/C and that of the copper (given earlier in this section; the CTE of Cr_7C_3 is $10 \times 10^{-6}\,K^{-1}$) (Pedzich et al. 1998).

This is important in order to obtain a mechanically resistant component; the joined region can be considered as a functionally graded material in which there are three components characterized by a thermal expansion gradient.

Figure 3.7. (a) Wettability of pure copper on chromium carbide-modified C/C at 1100°C for 30 min. (b) Optical micrograph of the cross-section of a chromium carbide-modified C/C–Cu joint; the total thickness of the carbide layer is about 20 µm. (c) Picture of chromium carbide-modified C/C–Cu joined sample ($19 \times 22 \times 6\,m^3$). (d) Chromium carbide-modified C/C–Cu-CuCrZr mock-up.

A commercial brazing alloy (Gemco®, 87.75 wt% Cu, 12 wt% Ge, and 0.25 wt% Ni; Wesgo Metals, Hayward, CA) has been used to braze Cr-modified C/C to Cu and to CuCrZr in a single-step process (Salvo et al. 2008), which is an advantage in comparison with other technologies that require two steps to join first C/C to Cu, then C/C–Cu to CuCrZr. Furthermore, this brazing alloy does not contain any active element potentially dangerous for the formation of brittle intermetallics. Flat-tile configurations and mock-ups have been obtained by this technique and have been tested (Fig. 3.8a, b).

ENEA, Italy, has manufactured several actively cooled mock-ups of flat-tile and monoblock types by using different technologies: A new one has been patented, based on *prebrazed casting* and hot radial pressing (PBC + HRP), for the production of monoblocks (Fig. 3.9). The C/C surface modification is obtained by a titanium–copper–nickel commercial brazing alloy, followed by a Cu casting, then a radial diffusion bonding between the cooling tube and the C/C by pressurizing only the internal tube and by keeping the joining zone in vacuum at the required bonding temperature (Libera and Visca 2006; Visca et al. 2003, 2005, 2007).

Complete manufacturing and testing of this vertical target medium-scale mock-up (Fig. 3.9) can be considered as a success for both PBC and HRP processes that, in the survey of the manufacturing technologies for the ITER divertor, has become a valid alternative to current techniques (courtesy of Fusion for Energy and ENEA).

Several joining techniques are based on active brazing, which does not require any C/C surface modification: Active brazing alloys for C/C to Cu should include elements (e.g., Ti, Zr, Cr, and Si) that activate the process of bonding the otherwise not wettable C/C surface and avoid elements (e.g., Cd and Zn) that are not permitted for use in a vacuum environment due to their high evaporation rate; the brazing temperature should be higher than the operational temperature as in Table 3.1 and in (Ise et al. 1998).

There are two main concerns about active brazing: Active elements may form brittle intermetallics or compounds with a low melting point; braze stability after the second joint (Fig. 3.4), when C/C–Cu is to be joined to CuCrZr.

In Appendino et al. (2003), a TiCuNi brazing alloy produced by Wesgo Metals in the form of sheets (*70Ti-15Cu-15Ni*) was used to join C/C and silicon-doped C/C composites to pure copper, but the presence of brittle Ni and Ti intermetallics at the joint interface was detrimental to thermal-fatigue resistance in tests of the joined component. In addition, due to the brazing alloy liquidu's temperature, a second joint higher than 960°C was not possible.

In Japan, a brazing technology was developed in the past (ITER DOCa) to join C/C composite to Cu by a *NiCrP brazing alloy* then to join the C/C–Cu to the CuCrZr by low-temperature HIP (500°C). The HIP pressure was selected to be ~60 MPa, to avoid any damage to the C/C. More recently, a new brazing process was developed, based on a NiCuMn alloy, after metallization of the C/C surface.

Several other brazing alloys have been developed for C/C–Cu joints: Ag-based (63Ag-35Cu-2Ti, 59Ag-27Cu-13In-1Ti) and Cu-based (Cu-3Si-2Al-2Ti, Cu–Mn, Cu–Ti) brazing alloys. Silver was definitely discarded because of transmutation issues (Akiba and Suzuki 1998; Merola et al. 2002b).

Rapidly solidified ribbon-type Ti-based amorphous alloys have been used to join silicon-doped carbon to copper for first-wall components in the Chinese fusion reactor.

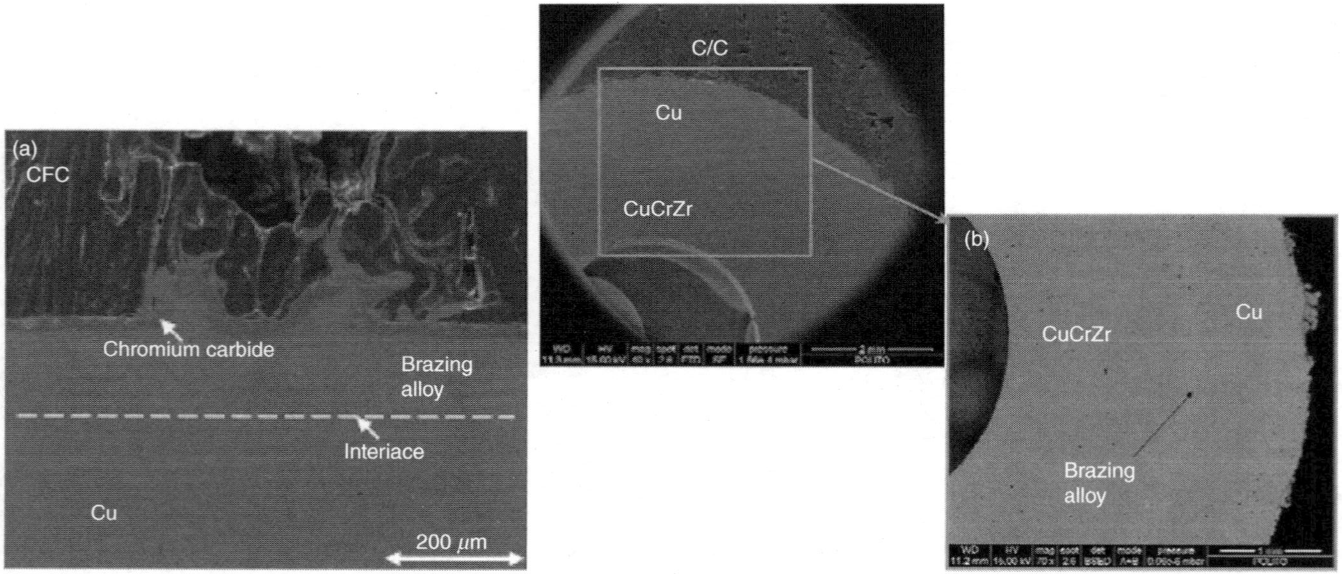

Figure 3.8. Scanning electron microscopy image of the cross-section of C/C–Cu-CuCrZr sample brazed by Gemco alloy. (a) Flat-tile configuration and (b) monoblock configuration.

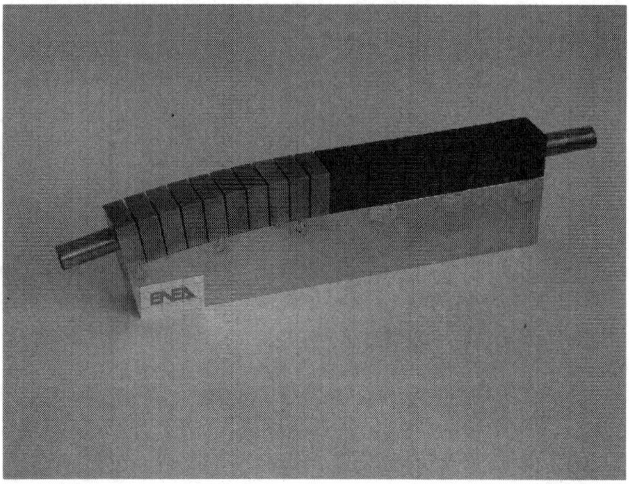

Figure 3.9. Final mock-up assembled on steel support (Visca et al. 2007) (courtesy of Fusion for Energy and ENEA).

TiC and ZrC have been detected near the carbon–metal interface. Using very thin Cu and Mo foil as multiple interlayers, thermal stress between carbon and copper can effectively be relaxed after brazing. (Zhou et al. 2007)

Finally, the galvanic metallization process (GMP) is a process developed by the European Aeronautic Defence and Space Company (EADS) for the joint between C/C and copper. As of this writing, there has been no published work about this process (ITER DOCa).

Mechanical Tests on Joined C/C–Cu

Shear strength measurement is always an issue, in particular when a composite is joined to a metal. There is not an American Society for Testing and Materials (ASTM) test available for pure shear strength of C/C joined to metals. Several laboratories have developed their own tests for C/C-to-metal joints (Morscher et al. 2006), making comparison of results almost impossible.

Joints obtained by AMC have been extensively tested (details on their characterizations can be found in Schedler et al. (2007a) particularly for shear strength, which is ~20–60 MPa. At 600°C, the shear strength dramatically decreases to ~20 MPa, but the fracture is mostly in the cast Cu layer, at the bottom of the C/C cones, rather than in the C/C–Cu interface (ITER DOCa).

The shear strength and tensile strength of the improved AMC (TiSi-AMC) joint are in the range of 54–73 MPa and 39–64 MPa, respectively (Schedler et al. 2007a, b).

Another kind of mechanical characterization performed by Plansee is the fracture toughness of the interface (Schedler et al. 2007a); all mechanical tests are coupled to a fractographic analysis.

The Cr-modified C/C–Cu joined samples (Ferraris et al. 2005; Appendino et al. 2003, 2004, 2006; Casalegno 2006, 2008; Salvo et al. 2008) have been measured by single-lap test (adapted from ASTM C1292, C1425) and offset single-lap test (adapted from ASTM D905) at RT. Independently of C/C surface machining and different casting processes, as described in Appendino et al. (2004) and in Casalegno (2006), results obtained for Cr-modified C/C on more than 50 samples gave average values of apparent shear strength ranging from 26 to 32 MPa. Influence in improving the average shear strength seems to be due to the slight pressure (1 kPa) applied during some Cu casting. The average shear strength is, in any case, higher than the interlaminar shear strength of the C/C (15 MPa). This result can be explained with the structure of the joint at the C/C–Cu interface: Cu penetrates in the C/C bulk, thus forming a sort of metal–ceramic composite. As a consequence, C/C interlaminar shear strength is enhanced near the joint surface. These results are in agreement with those obtained by the AMC process (Schedler et al. 2007a) for the same joined materials.

Molybdenum and tungsten carbide-modified C/C–Cu samples show significantly lower shear strength than chromium carbide-modified C/C–Cu samples. In the case of Mo-modified C/C–Cu joints, mechanically machined C/C joined to copper has higher average shear strength than a nonmachined composite: In case of weak joints, mechanical machining is important for the joint strength, and its contribution is higher than chemical bonding between, for instance, molybdenum carbide and C/C or molybdenum carbide and Cu. For Cr-modified C/C–Cu joints, chemical bonding at the joint interface gives a more important contribution to mechanical strength than the contribution of surface machining to the overall mechanical strength, which is negligible (Casalegno 2006).

Several joints did not fail during single-lap tests (Fig. 3.10) because of the failure of the epoxy adhesive used to glue the joined samples to the fixtures. An alternative test, the offset single-lap test, gives the apparent shear strength, as does the single-lap test, but the samples are not to be glued to the fixtures, and it can be used at high temperature. The C/C and copper have a 2-mm misalignment. The setup of this mechanical test, adapted from ASTM D905, is shown in Figure 3.11 (load is applied on the copper).

Some problems were found with this test: The lateral fixtures affect the apparent shear strength by compression, and there is a significant friction at fixture–sample interface, which cannot be completely eliminated by putting a lubricant foil in between.

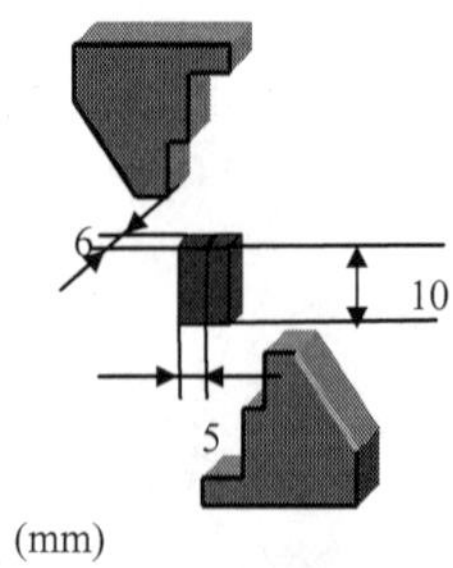

Figure 3.10. Sample and grip configuration for the mechanical shear test. The sample is glued to the grips by an epoxy glue (Ferraris et al. 1998).

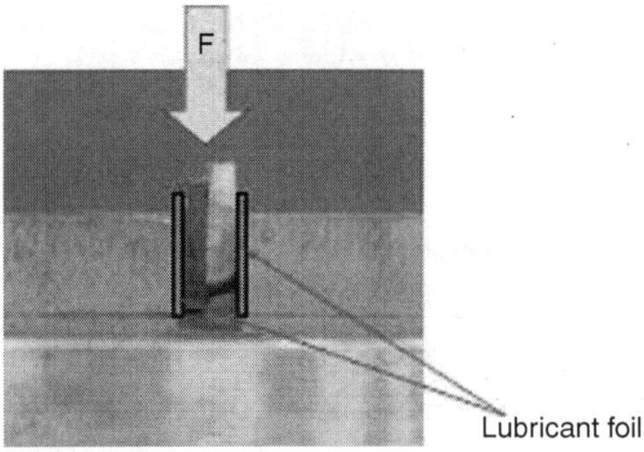

Figure 3.11. Setup of the offset single-lap test, adapted from ASTM D905.

Moreover, this test must be performed in the elastic field, in order to avoid plastic deformation of Cu.

There has not yet been any finite element analysis done on these two tests applied to C/C–Cu joints, in order to compare them and to state which of the two (single-lap and single-lap offset) underestimates or overestimates the joining strength (Casalegno 2006). Work is in progress on comparison with other tests.

The shear strength of the C/C–Cu joints (flat-tile geometry) obtained by using a commercial Gemco brazing alloy to braze C/C to pure copper was 34 ± 4 MPa, measured by single-lap tests at RT. It is comparable to the values obtained by other joining processes and higher than the intrinsic C/C shear strength (Salvo et al. 2008)

Mechanical tests on monoblocks require other designs: Some of them are adapted from ASTM D4562-01, "Standard Test Method for Shear Strength of Adhesives Using Pin-and-Collar Specimen" (Fig. 3.12), as in the compression test used by Plansee AG (Schedler 2003) (Fig. 3.13), but the joint is not stressed in uniform pure shear state.

Torsion tests (Fig. 3.14) on monoblocks are recommended because samples are stressed in pure shear. Further developments are expected from finite element analysis on such complex systems, made of several interfaces between different materials.

The apparent shear strength of the C/C–Cu joints (monoblock geometry, Fig. 3.15) obtained by using a commercial Gemco brazing alloy to braze C/C to pure copper was measured by compression test (see Fig. 3.13).

The average apparent shear strength measured for these samples is 20.5 MPa. The fracture always occurred in the C/C, with the exception of one sample for which the failure occurred within the carbide–braze interface or in the braze itself at 27 MPa.

NDTs

Reliable NDTs are essential for quality assurance of components for nuclear fusion applications, especially for high-heat-flux plasma-facing components. NDTs allow inspection of a component without impairing serviceability; it is important to detect and characterize defects (type, size, and position) as well as the setup of acceptance

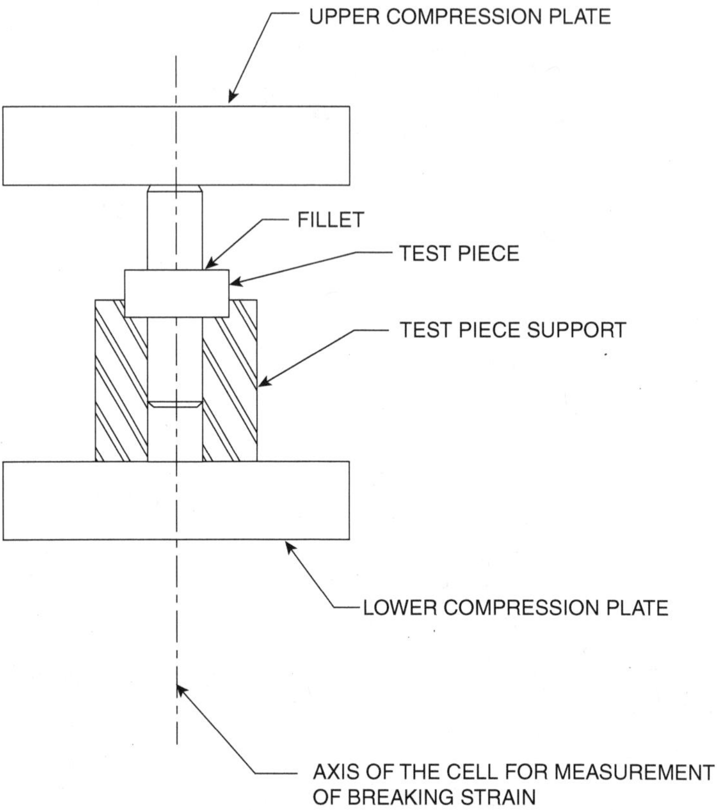

Figure 3.12. ASTM D4562-01, Standard Test Method for Shear Strength of Adhesives Using Pin-and-Collar Specimen (reprinted with permission, copyright ASTM International, Conshohocken, PA).

standards in order to predict their influence on component performance in service conditions. NDTs on joints between carbon fiber-reinforced–carbon matrix composites and copper are complex because of the different behavior of C/C and Cu with regard to physical excitations used to test the component; furthermore, the response to this input must be accurately assessed to identify the detachment of C/C tiles from Cu.

The aim of NDTs is to identify and localize defects in the joined components before submitting them to high-heat-flux tests. It is also important to identify the maximum acceptable defect size, as a function of its position, defined as the largest defect stable under specific loads in ITER (Ezato et al. 2002; Merola et al. 2002c). Several techniques are used as NDTs (Casalegno et al. 2008): X-ray microradiography, X-ray microtomography, ultrasonic inspection (Krautkrämer and Krautkrämer 1990; Kar et al. 2007), lock-in thermography (Datong Wu and Busse, 1998), transient infrared thermography (SATIR, a French acronym for infrared acquisition and data processing device); SATIR is a dedicated facility developed in Cadarache, France, at CEA (Internet resource 2).

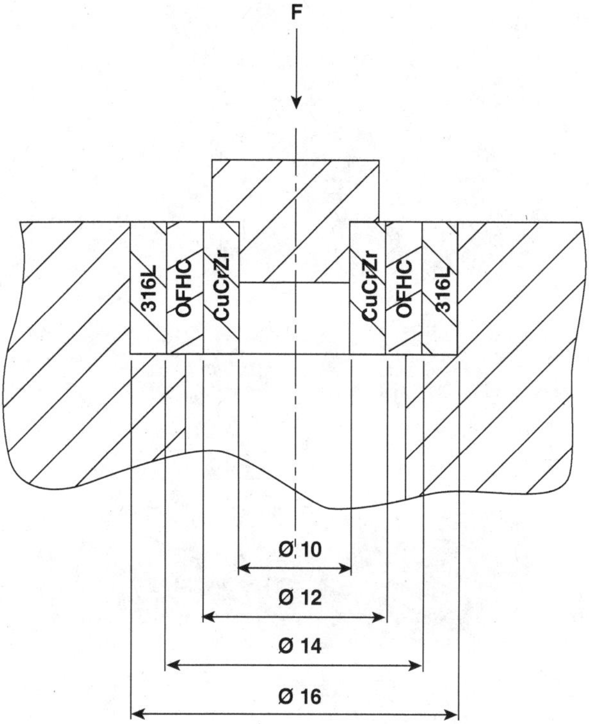

Figure 3.13. Compression test for a tubular joint used by Plansee AG, Austria.

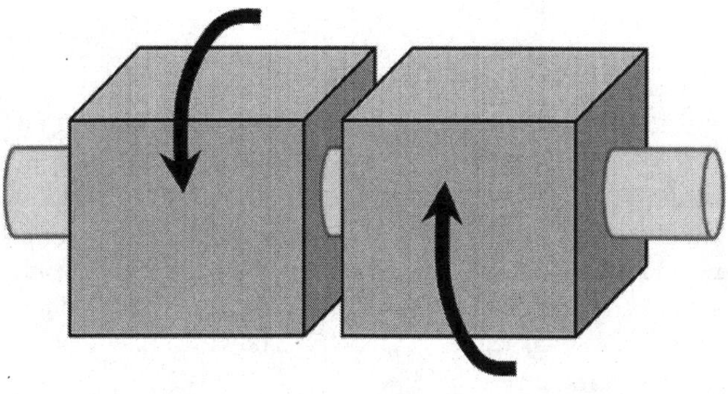

Figure 3.14. Torsion test configuration on monoblock C/C–Cu joined samples.

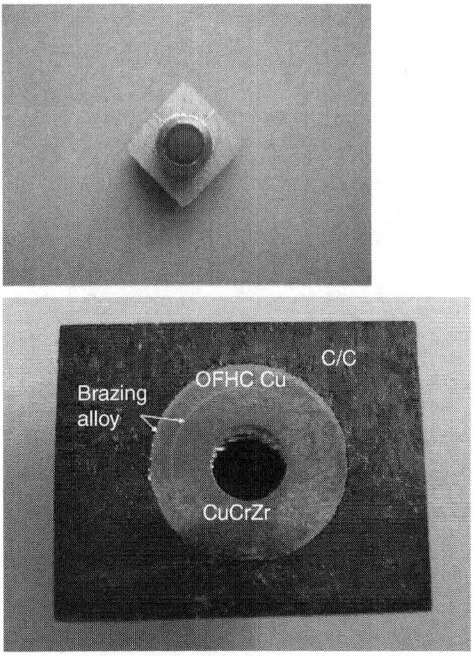

Figure 3.15. C/C–Cu joints (monoblock geometry) obtained by using a commercial Gemco brazing alloy to braze C/C to pure copper.

Ultrasonic inspection has been applied to the flat-tile and monoblock designs (ITER DOCa). Defects of 3 mm between AM copper and the Cu alloy tube are certainly detected, and defects of ~1 mm can be detected with a good probability, but in the case of joints between materials having very different acoustic impedance (e.g., copper and C/C), a high reflected echo is generated that impairs the detectability of defects (Merola and Wu 2004).

Lock-in thermography consists of applying a series of heat flashes on the C/C. This method is giving promising results in the case of flat-tile geometry, but its applicability to a monoblock geometry still needs to be demonstrated. The main advantage of this technique is that there is no need for an active cooling of the component. Therefore, it can also be envisaged as an inspection method all along the manufacturing process (Wu and Busse, 1998; Escourbiac et al. 2007).

Transient infrared thermography–SATIR consists of recording the surface temperature evolution of the component with an infrared device during the circulation of hot (~95°C) and cold (~5°C) water fronts into the cooling channel of the component. Then, the transient thermal response is compared with a "defect-free" component; defects, for instance, debonding of C/C tiles from the heat sink, are detected by a slower surface temperature response.

Several NDTs have been performed on Cr-modified C/C–Cu joined samples (Casalegno et al. 2008) to test joint performance, but also to verify the reliability of these tests on a C/C–Cu interface. It was concluded that the reliability of NDTs of joints should be validated by destructive tests, such as morphological evidence of the detected defect, and mechanical testing:

- Ultrasonic inspection of C/C–Cu joints gave unreliable results. This can be explained considering that C/C and copper have significantly different acoustic impedance; therefore, a high ultrasonic echo exists even if the joint is good. As a consequence, defects in the CuCrZr-Cu brazed joints can be identified, but those at the C/C–Cu interface can hardly be detected. Another disadvantage of ultrasonic inspection is that it is sensitive to irregularly shaped (e.g., mechanically structured) joined surfaces.
- X-ray tomography can only offer qualitative information on C/C–Cu interface, since carbon is significantly less X-ray sensitive than copper.
- Lock-in thermography offers information on thermal continuity at an interface and can predict the component behavior under critical heat-flux events, since it gives global information about the soundness of the heat path, but not necessarily on the chemical continuity at the interface. The advantage of lock-in thermography is that it is not sensitive to irregularly shaped jointed surfaces, and it can also be used for machined C/C–Cu joints.
- Shear tests (e.g., single-lap test) are reliable in measuring the joining strength, which is related to the presence of defects in the joint. Unfortunately, it is a destructive test; however, together with microscopy on the fracture surface, it should be used to validate each proposed NDT.

Finally, to predict the lifetime of a C/C composite, it is necessary to analyze the damage mechanisms and to model damage propagation when the component is exposed to thermal cycling loads. Work has been performed to identify a constitutive law for the C/C composites and parameters to model crack propagation from the edge singularity. The aim is to predict damage rates and to propose geometric or material improvements to increase the strength and the lifetime of the interfacial bond. Research activity is in progress to better understand the manufacturing residual stresses and to model the mechanisms of degradation of the tile bond (Schlosser et al. 2007)

Thermal-Shock and High-Heat-Flux Tests on C/C–Cu for ITER

A preliminary test usually done on joined samples prior to high-heat-flux tests is thermal fatigue. In a typical thermal-fatigue test, the samples are heated in 60 s to a temperature of 450°C with a dwelling time of 5 s and then quenched to RT in water; this heating–quenching sequence is repeated several times. The progressive joint degradation is verified by examination of cracks in the interface. Additional shear tests can be performed on quenched samples.

For instance, Cr-modified C/C–Cu joined samples and Gemco-joined C/C–Cu (Salvo et al. 2008) survived thermal-shock tests performed by heating the samples as previously described, up to 50 times.

Depending on the survival after thermal cycling, mock-ups (reduced-scale final components) are prepared with the most promising joining technology and submitted to heat flux tests.

The suitability of joining techniques to withstand the relevant thermomechanical conditions in nuclear environments is evaluated by testing small mock-ups under relevant heat fluxes in electron or ion beam facilities, because high-heat-flux testing is thus far the only reliable way to validate a component design and its manufacturing technology.

The main facilities used are JUDITH I and II (FZ Juelich, Germany), GLADIS (Garching, Germany), JEBIS (JAERI, Naka, Japan), TSEFEY (Efremov Institute, St. Petersburg, Russia), EBTS (SNLA, Albuquerque, NM), FE 200 (CEA Le Cresout, France), and ENEA at Brasimone (Italy).

The aim of the high-heat-flux test is to check the behavior of the joined components under thermal fatigue, varying the thermal load (MW/m^2). Through observation of the samples after several cycles, it is possible to detect damage or the failure limit of the joints. A typical high-heat-flux test is organized in two phases. The first is a screening to select the most suitable among different joining technologies; this phase is carried out on small, actively cooled mock-ups including one to three cooling channels. The second step is performed on medium-scale mock-ups with cross-section identical to that of the real components, fabricated with the joining technologies that had passed the first step.

The results of high-heat-flux tests on C/C–Cu joints can be summarized as follows.

The C/C mock-up developed by Ansaldo Ricerche (flat tile, C/C NB31 joined to Cu by means of brazing with dispersed C fibers) sustained 1000 cycles at $10\,MW/m^2$ followed by 100 cycles at $12.5\,MW/m^2$ and other 400 cycles at $15\,MW/m^2$ (Merola et al. 2000; Bisio et al. 2005).

For C/C NB31 flat-tile mock-ups produced by Plansee AG by AMC and for the Cu–CuCrZr joint by HIPing, the failure limits were found to be at $10\,MW/m^2$, approximately. The C/C flat-tile mock-ups produced by AMC and electron beam welding showed failure beyond $20\,MW/m^2$. C/C NB31 monoblock produced by AMC sustained $20\,MW/m^2$ for 2000 cycles in case of a vertical target medium-scale prototype; a vertical target full-scale prototype C/C monoblock sustained $10\,MW/m^2$ for 1000 cycles, $20\,MW/m^2$ for 1000 cycles, and $23\,MW/m^2$ for 1000 cycles (Casalegno 2006).

Some C/C monoblocks developed by ENEA (PBC + HRP process) with three NB31 C/C tiles were successfully manufactured and tested for thermal fatigue using electron beam facilities. They all reached at least 1000 cycles at $20\,MW/m^2$ without suffering any damage. The manufactured medium-scale vertical target mock-up is being tested at the FE2000 (France) facility or by e-beam in the JUDITH (FZJ) and TSEFEY-M (Efremov) facilities (Visca et al. 2007).

Chromium-modified C/C flat-tile mock-ups were tested at Juelich FZ, Germany, within the ExtreMat project; results were originally expected by the end of 2008

(Internet resource 12). The full divertor component has been tested in Juelich, but results are unpublished.

Finally, a joint project of the EU, Japan, the Russian Federation, and the United States under the auspices of the International Energy Agency (IEA) has been agreed: the International Fusion Materials Irradiation Facility (IFMIF). The mission of IFMIF is to provide an accelerator-based, D–Li neutron source to test materials and components.

JET

The JET is the largest fusion device in the world (Internet resource 7; Piazza et al. 2007; Ruset et al. 2007). It is a so-called tokamak and is located in the United Kingdom at the Culham Science Centre, near Oxford. The JET machine is a large device of approximately 15 m in diameter and 12 m in height. At the heart of the machine, there is a toroidal vacuum vessel with a major radius of 2.96 m with a D-shaped cross-section (2.5 × 4.2 m). The construction of JET began in 1978, and it started operating in 1983. The project was coordinated by the EURATOM. The JET program was set up to make it possible to carry out fusion tests under conditions that closely resemble those of a commercial fusion power plant. JET was the first fusion facility in the world to achieve a significant production of controlled fusion power (nearly 2 MW) with a deuterium–tritium experiment in 1991.

After 1991, JET was enhanced by the installation of a divertor to handle higher levels of exhaust power. Deuterium experiments in the ITER geometry have made essential contributions to the ITER divertor design and have provided key data on heating, confinement, and fuel purity. In 1997, JET operations reached a record 16.1 MW of fusion power. Further experiments carried out since then have provided useful information about the physical and technical parameters needed for an effective operation of ITER. The basic idea for JET realization was to obtain plasma closest to ITER physics. For instance, ITER divertor design received a significant contribution from the ITER-like geometry studied in JET.

One of the most interesting projects for the purposes of this chapter is the ITER-like wall (ILW); it started at the beginning of 2005 and consists of the replacement of the existing first wall and divertor of JET by materials that will be used for the construction of the ITER reactor. The ILW project reference option is a fully tungsten-coated C/C divertor, whereas an alternative option includes C/C tiles for the strike point areas of the divertor (equivalent to the current ITER reference configuration) (Figure 3.16a, b; Piazza et al. 2007).

Design Issues and Joining Techniques for Joined C/C Composites in JET

The design (Figs. 3.17 and 3.18) consists of at least a 10-mm tungsten layer brazed onto a C/C composite base plate.

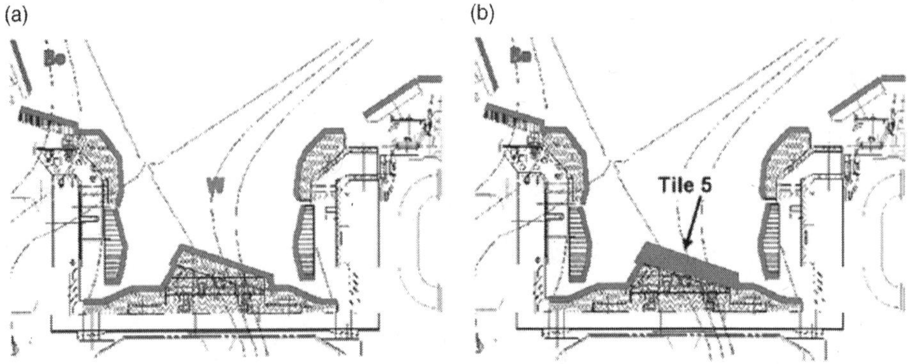

Figure 3.16. JET divertor region with the two options: (a) all-W coatings on CFC; (b) W coating on CFC and bulk W for tile 5 (as in Piazza et al. 2007).

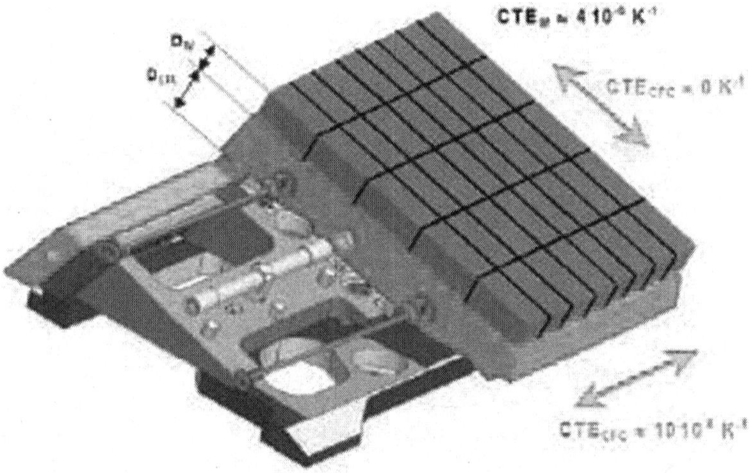

Figure 3.17. Representation of the brazing design; indicated C/C CTEs are those for DMS 780.

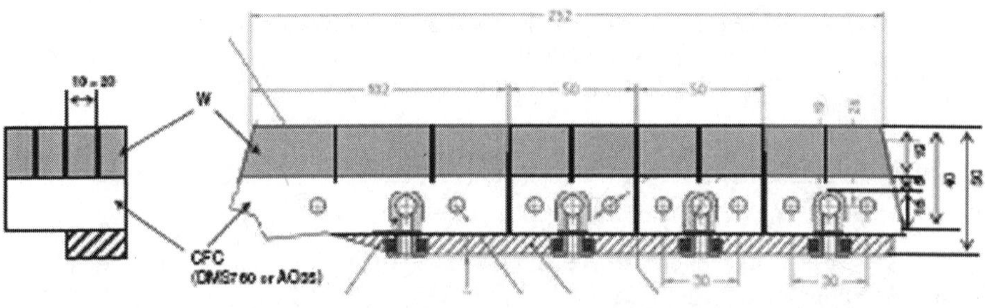

Figure 3.18. Design of the W brazing concept.

In this design, the critical points are the temperature at the brazing interface and the thermomechanical performance of the brazing. Since JET design does not have an active cooling system, the brazed joint between tungsten and C/C has to withstand very high temperatures at the interface when the divertor components are exposed to high heat loads (up to 7 MW/m^2 for 10 s).

Tungsten-coated C/C composites are within the scope of this chapter and are briefly discussed here. Depending on the position, tungsten coatings with different thicknesses might be necessary. The feasibility of this solution has been proven in the tungsten program of the Axially Symmetric Divertor Experiment (ASDEX) Upgrade tokamak in Garching (Germany). The following reference thicknesses have been chosen: thin coatings, 4 and 10 μm; thick coatings, 200 μm. Five European research institutions (CEA, ENEA, IPP, MEdC, and TEKES) have developed and optimized different coating technologies, mostly in cooperation with industry. Thin coatings, with thicknesses of 4 and 10 μm, were produced by CVD using WF6 and ReF6 as reactive gases and by various methods of PVD. Thicker coatings of 200 μm were produced by CVD and vacuum plasma spray (VPS). Characterization and thermal-load testing have been done to select the most promising combinations of thickness and coating method. The high-heat-flux device GLADIS at IPP Garching was used for the testing. A selected number of coating types underwent a thermal cycling process at fixed pulse conditions. This thermal cycling is necessary, since the thermal expansion mismatch between thick coating and the substrate leads to crack formation. Infrared-based nondestructive testing methods (lock-in infrared thermography and flash lamp thermography) and X-ray radiography have been done to examine the uniformity of the manufactured coatings (Piazza et al. 2007; Ruset et al. 2007).

Combined magnetron sputtering and ion implantation (CMSII) is a deposition technique involving simultaneous magnetron sputtering and high-energy ion bombardment: This method was applied to produce nanostructured W coatings on the carbon fiber composite substrate. These coatings have been characterized in terms of adhesion, thickness, structure, and resistance to high thermal loads (up to 23.5 MW/m^2). Based on the results of these tests, CMSII technology was selected for coating about 1100 tiles with a 10-μm tungsten layer (Ruset et al. 2007).

Several works have been published on brazing W on C/C composites, but this option was canceled because of the danger of a W tile falling. A short summary of results on this subject follows.

Brazing was considered a suitable technique to coat C/C by W: Brazes for "high-" (1700°), "medium-" (1500°C), and "low-" (1100–1200°C) temperature brazes have been investigated (Ellett 2005). Low-temperature brazes are Cu-based alloys; medium-temperature brazes contain Pd, while high-temperature brazes are Ti- or Zr-based alloys. Some patents (Internet resource 7; Friedrich et al. 1992; Grill et al. 1996; Rainer and Reheis 1996; Weeks and Sommer 1998a, b, 1999; Tashiro and Kasahara 1999) relating to joining of tungsten and carbon-based materials report reaction brazing of tungsten or molybdenum to carbon-based supports. The C/C composites used for most of this literature are 2-D-C/C: DMS780, from Dunlop Aerospace (Internet resource 15), and AO35 (C/C AEROLOR) produced by Aerospatiale et Le Carbone-Lorraine (France).

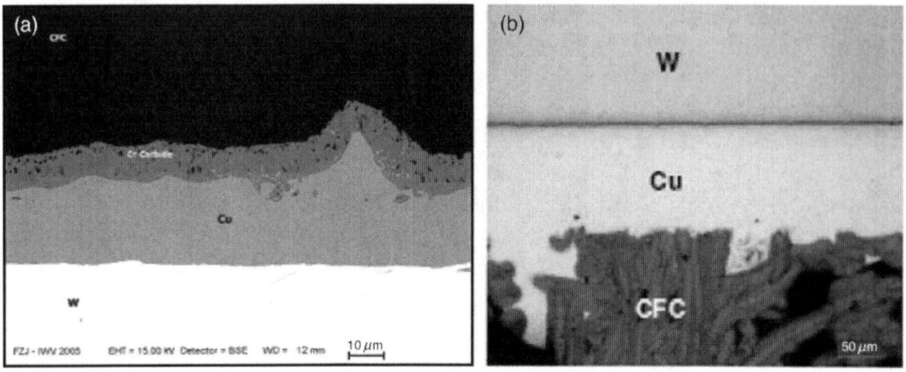

Figure 3.19. Scanning electron microscopy of C/C(DMS780)–Cu–W joint cross-sections. (a) Cr-modified C/C and (b) galvanic deposition of CuCr on C/C surface (Casalegno 2006; Koppitz et al. 2007).

A comprehensive research project is investigating C/C–W joints for the JET ILW application, at the Jülich Research Centre, Germany (Forschungszentrum Jülich, FZJ). Chromium-modified C/C composites (AO35 and DMS780) were successfully joined to $15 \times 16 \times 16\,mm^3$ tungsten tiles by pure copper (Fig. 3.19) (Casalegno 2006; Koppitz et al. 2007), by pure palladium, and by several brazing alloys, produced by IOT Institute in Aachen, Germany, in conventional or in arc-melting furnaces. The Cr modification (Ferraris et al. 2005) and galvanic deposition of CuCr were used to modify the C/C surface (Koppitz et al. 2007)

Palladium brazing seems to be a promising solution, provided that intermetallic formation can be avoided, either by keeping the brazing step in a conventional furnace to a minimum amount of time at a temperature above the melting temperature of Pd or by fast brazing by direct heating with an electron beam.

CONCLUSION

Most of the available literature on joining and integration issues of CMCs for nuclear applications, in particular, for thermonuclear fusion, deals with C/C composites for ITER; some work has been done within JET and with Wendelstein 7-X (Schlosser et al. 2007) in Germany, TFTR in the United States, JT60SU in Japan, and Tore Supra in France (Mertens et al. 2007; Hirai et al. 2007).

At present, AMC seems to be the most reliable technique for C/C–Cu joints for ITER: Several tens of square meters of AMC joints have been produced (Fig. 3.5), as the flat-tile or the monoblock design, not just for ITER but also for Wendelstein 7-X.

For the future, although the copper-compliant layer is sufficient to accommodate residual stresses (shear) by plastic deformation but can be critical for thermal fatigue at high heat flux, a better, smooth transition across the C/C–Cu joint is still needed.

Within the ExtreMat project, some alternative concepts have been developed:

- New interlayer with low CTE to transfer stresses from C/C–metal to a metal–metal interface (e.g., Mo interlayers)
- New graded interlayers with graded CTE from Cu to C/C
- New metal–matrix composite interlayer (W fiber-reinforced Cu) with CTE intermediate between Cu and C/C.

Results were expected by the end of 2009 (Internet resource 12).

FUSION REACTORS BEYOND ITER

Together with efforts to build ITER, researchers are designing new reactors able to produce energy from fusion, not just to demonstrate its feasibility. Composites of silicon carbide fibers in a silicon carbide matrix (SiC/SiC) have been considered not only as parts of components but also as structural material in several conceptual design studies beyond ITER.

Some of these are TAURO in the EU, ARIES-AT in the United States, and DREAM in Japan.

General reviews of issues regarding designs using SiC/SiC composites are given elsewhere (Nishio et al. 1998; Raffray et al. 2001a, b; Raffray et al. 2007; Giancarli et al. 2002; Maisonnier et al. 2005; Katoh et al. 2007; Internet resource 11) and are briefly summarized here where they are of interest for this chapter.

Both TAURO and ARIES-AT blankets are essentially formed by a SiC/SiC box with an indirectly cooled first wall that acts as a container for the Pb-17Li, which has the simultaneous functions of coolant, tritium breeder, neutron multiplier, and, finally, tritium carrier. The TAURO blanket is characterized by 2-m-high single modules, which are reinforced by SiC/SiC stiffeners (see Fig. 3.20). The ARIES-AT concept is characterized by a coaxial Pb-17Li flow, which occurs in two 8-m-high boxes inserted one into the other (see Fig. 3.21). The main design issues (besides the materials issues) still under discussion for this kind of reactor are the development of an appropriate joining technique for SiC/SiC composites and the development of appropriate models, design rules, and defined methods for improving hermeticity, especially against high-pressure He (Ehrlich et al. 2001).

Designs of SiC/SiC components are available (Figs. 3.20–3.24) (Tillack et al. 2000; Ehrlich et al. 2001), but not yet with the detail reported for ITER: the materials scientists are working on new techniques and materials to join SiC/SiC for reactors beyond ITER with approximate working conditions.

The European Fusion Development Agreement (EFDA) launched a call for the development of SiC_f/SiC for structural applications in fusion reactors to assess the consequences of the properties actually achieved on the integrity of a blanket made of SiC/SiC, to reassess the requirements on in-service properties for SiC_f/SiC as a structural material under fusion reactor conditions, and to update the blanket geometry and the type of joints allowing possible application of SiC_f/SiC.

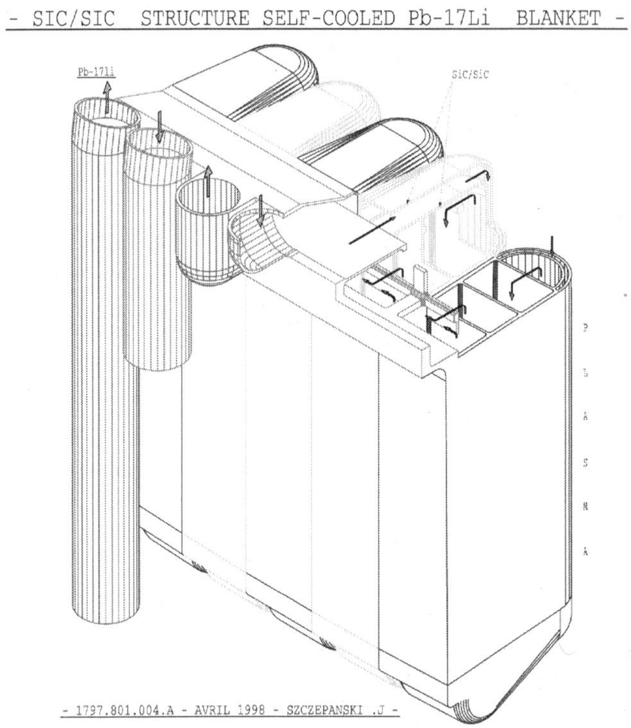

Figure 3.20. TAURO blanket concept (courtesy of G. Aiello, CEA, Saclay, France).

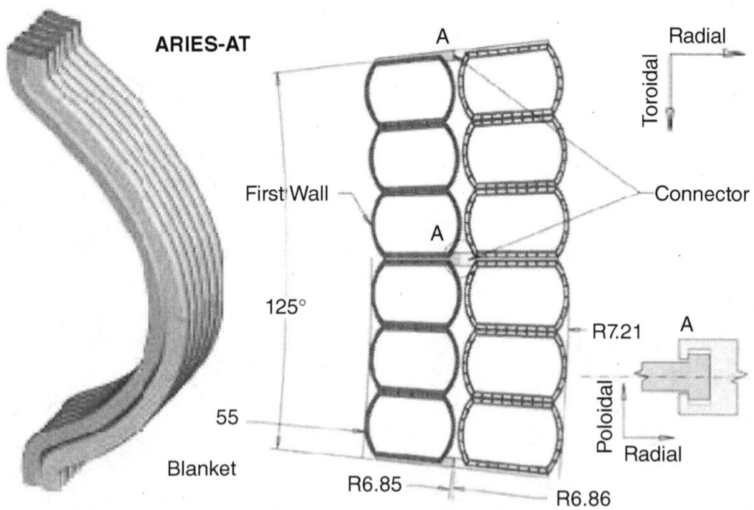

Figure 3.21. The high-temperature ARIES-AT blanket design utilizing SiC_f/SiC as structural material and Pb-17Li as coolant and breeder (from Raffray et al. 2007; courtesy of A.R. Raffray).

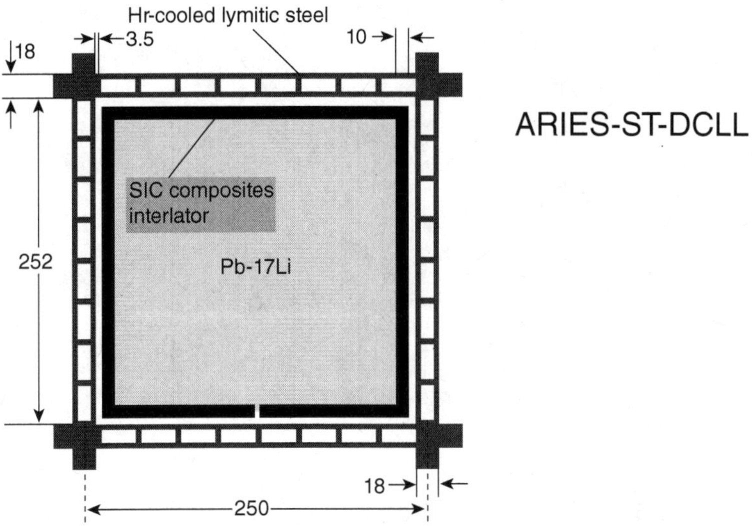

Figure 3.22. ARIES-ST-DCLL concept M. (from Tillack et al. 2000; courtesy of M. Tillack).

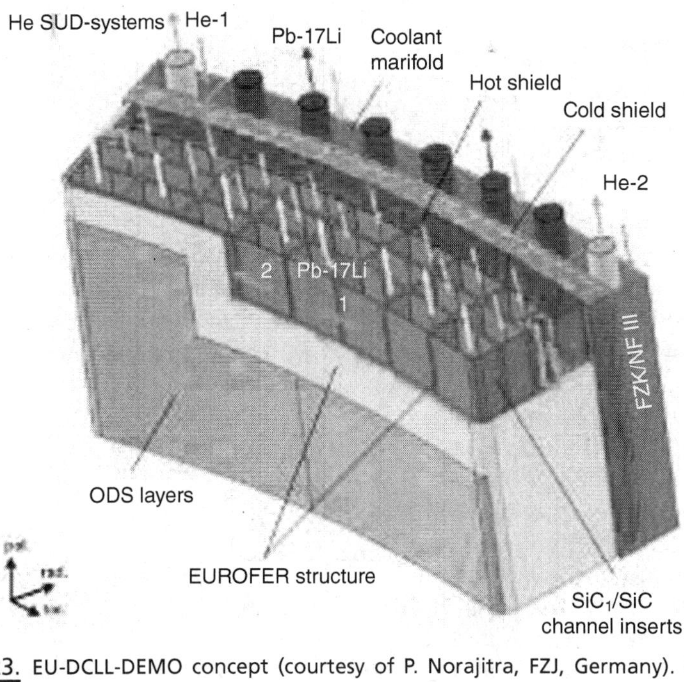

Figure 3.23. EU-DCLL-DEMO concept (courtesy of P. Norajitra, FZJ, Germany). ODS, oxide dispersion strengthened.

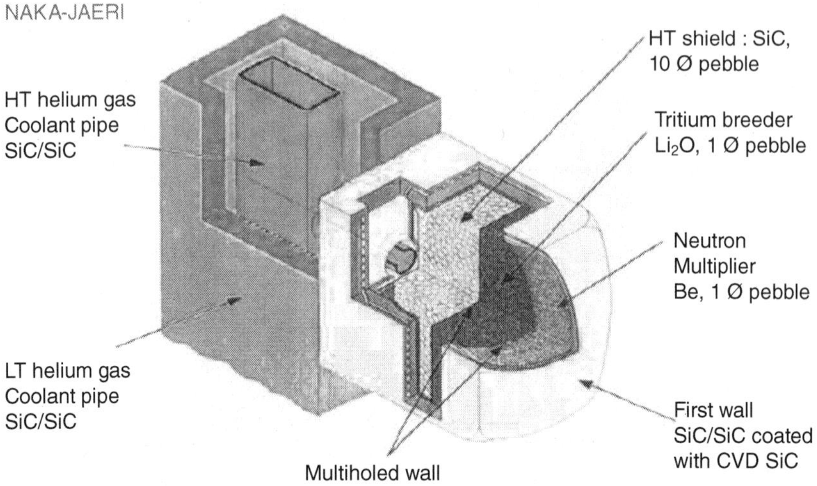

Figure 3.24. DREAM blanket concept (courtesy of N. Satoshi, JAERI, Japan). HT, high temperature; LT, low temperature.

Also, alternative concepts involving SiC in combination with other materials will be addressed, together with their possible improvement in joining and hermeticity (Internet resource 11).

Although it is almost decided that the design of joined components and their working conditions will use C/C in near-term fusion devices, applications of SiC-based composites in fusion reactors beyond ITER and in next-generation nuclear power (Fig. 3.25) (U.S. DOE 2002; Carré 2003; Kotake et al. 2005) are still uncertain.

Why SiC/SiC Composites?

Silicon carbide-based ceramic composites have been studied for fusion applications for more than 15 years. Their potential is mainly due to their unique thermomechanical properties and low-activation characteristic, which reduces the radiological issues of the whole reactor, their well-known behavior against neutron irradiation up to high doses and high temperatures, and their low tritium permeability (Giancarli et al. 1998; Jones et al. 2002; Riccardi et al., 2004a).

A comprehensive paper (Katoh et al. 2007) reviews the status of SiC/SiC composites in application development and the design requirements for use in fusion and advanced fission energy systems (see Section 3.6).

The main drawback of SiC/SiC composites used as structural materials is their thermal conductivity: It decreases with temperature and irradiation. Its value depends on the fabrication procedure and on the fiber characteristics, and SiC/SiC manufacturing is optimized mainly for thermal conductivity (Katoh et al. 2007)

To enable the use of SiC/SiC composites in fusion energy applications, a method of joining components is needed that satisfies the requirements of radiation resistance, mechanical integrity, desirable thermal properties, safety during operation and during

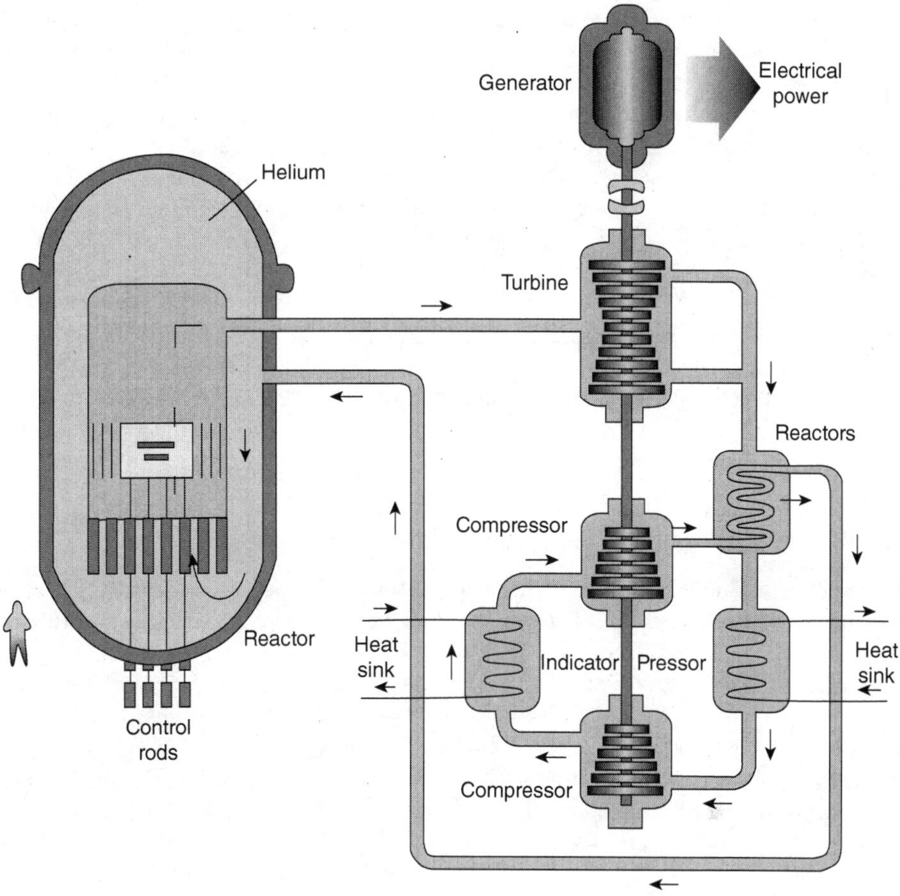

Figure 3.25. CMC tube as control-rod guides in advanced (Generation IV) gas-cooled fast reactors (Jenkins 2006).

maintenance or accident, and acceptable waste management characteristics. Joints made from silicon carbide satisfy these criteria, but practical and reliable methods of producing them must be developed.

Moreover, practical methods for joining simple geometrical shapes are essential for fabricating large or complex-shaped parts made from SiC/SiC composites.

The SiC/SiC composites have no melting phase; therefore, it is impossible to join them by conventional fusion welding, and diffusion bonding does not seem suitable as the interdiffusion of SiC is very low, even at high temperature. A well-known method for joining SiC ceramic parts is the cosintering of ceramic components (Krenkel et al. 1996), but this method needs high pressure and high processing temperature.

Furthermore, several indirect joining techniques suitable for SiC/SiC or SiC ceramics have been investigated: laser-assisted joining of SiC/SiC with oxide-based solders (Lippmann et al. 2004), joining by preceramic polymers (Colombo et al. 2000; Henager et al. 2007), joining by glass and glass–ceramic (Lemoine et al. 1995; Katoh et al. 2000), reaction forming (Singh 1999), metallic brazing (Gasse et al. 1997; Riccardi et al., 2004b), and microwave-assisted joining (Ahmad et al. 1995). Other methods not mentioned here include laser- or electron beam-assisted CVD and diffusion bonding with active metal agent (Harrison and Marcus 1999; Cockeram 2005).

In the field of nuclear energy production (fusion and fission), requirements for SiC/SiC joints are extremely severe: A joining material must be compatible with a neutron environment and the joining technique must comply with the fusion nuclear reactor design, where SiC/SiC components several meters long and 3 mm thick must be reliably and feasibly joined (Giancarli et al. 2002; Riccardi et al. 2000). The very few available solutions for SiC/SiC joining in the field of nuclear applications are reported or reviewed in Gasse et al. (1997), Singh (1999), Colombo et al. (2000), Katoh et al. (2000), Riccardi et al. (2000), Ehrlich et al. (2001), Tavassoli (2002), Lewinsohn et al. (2002), and Hinoki and Kohyama (2005) and are summarized here. None of them has yet been tested in a nuclear environment up to now.

Joining Materials and Techniques for SiC/SiC Composites

For SiC/SiC composites, as for other CMCs, brazing is one of the most promising joining techniques: BraSiC® (Riccardi et al. 2000; Gasse et al. 1997, 2001; Benoit et al. 2007; Gasse 2008) is one of the few commercially available brazing alloys proposed for joining CMCs for nuclear applications, in particular, SiC/SiC; they are M–Si alloys (M=Cr, Rh, Ti, V, etc.) with SiC and/or C as strengthening agents. This brazing alloy contains a sufficient amount of silicon to promote good wetting and to induce some infiltration in the composites. The reactive elements (M) improve the joining strength. By using different alloys and compositions (BraSiC H2 and V2 or V3) and brazing atmosphere (in vacuum or in an inert atmosphere), it is possible to control the infiltration of the alloy in the composites. This joining method allows remarkable mechanical properties, at RT and at high temperatures.

A new brazing alloy (Benoit et al. 2007) based on $PrSi_2$ and Si (eutectic at 1212°C) has been patented for the joining of SiC-based material to metal alloys: The authors claim the absence of detrimental intermetallics.

Eutectic alloys based on Si-16Ti (at%) and the Si-18Cr (at%), with melting points of 1330 and 1305°C, respectively (Colombo et al. 2000; Riccardi et al. 2004c), have been prepared by melting, powdered, and used to join SiC/SiC. The joints are continuous and defect free, with eutectic structures comparable to those of the as-prepared alloys. No diffusion or formation of brittle phases has been detected. Joint specimens tested in apparent shear showed failure in SiC/SiC.

The main concern for these joining materials is the presence of free silicon, which is detrimental to joint strength at temperatures higher than 1000°C and under neutron irradiation (Matthews and Hutchings 1973; Price and Hopkins 1982; Katoh et al. 2007). In particular, the presence of free silicon in SiC-based materials deteriorates strength during neutron irradiation due to extensive development of microcracking induced by differential swelling (Matthews 1974) A similar irradiation response is anticipated to occur in SiC-based joint materials involving a free silicon phase. However, the behavior of such joints under practical working conditions and neutron irradiation has not been experimentally examined.

Several *glasses and glass–ceramics* have been proposed as joining (and coating) materials for CMCs for nuclear applications, mainly because of their hermeticity and their tailorable properties (Lemoine et al. 1995, 1996; Katoh et al. 2000; Zucchetti et al. 2001; Ferraris et al. 1994, 2000, 2002, 2008). They show the unique property of being self-healing joining materials. They are mechanically reliable up to their softening point and can be crystallized to obtain high-temperature-resistant glass–ceramic joints. Since the joining process is performed above the glass softening point, no pressure is required during the joining, which can be significant with large joined structures as for nuclear applications.

However, a glass-based joint for a nuclear application has to resist high temperatures and neutron irradiation while keeping good mechanical strength and remaining leak tight. Research on the different joining technologies involving glass in a severe nuclear environment is in progress (Jacobs 2008). Amorphous phase behavior under irradiation is the main concern for this proposed solution (Coghlan and Clinard 1991); results of a test campaign on glass–ceramic joined SiC/SiC were expected by the end of 2009 as a result of ExtreMat IP (Internet resource 12).

A glass–ceramic based on SiO_2–Al_2O_3–Y_2O_3 has been designed to be used as a low-activation joining material for SiC/SiC (CVD–SiC coated 2-D composites provided by MT Aerospace, Augsburg, Germany) (Ferraris et al. 2008).

According to neutron-induced radioactivity of elements present in the composition, simulated by the European Activation System EASY-2007 code package (Forrest 2007; Internet resource 8), the glass–ceramic should be a LAM.

Two low-activation criteria are used in Ferraris et al. (2008): The first one (contact dose rate lower than 10lSv/h) is more restrictive, with hands-on recycling of the component after an adequate cooling period (less than 100 years as a maximum); the second permits remote-handling recycling operations with the use of simple shielding apparatus: In that case, the contact dose rate must be lower than 2 mSv/h, again after less than 100 years of cooling (Zucchetti et al. 2007).

Results of the simulation are detailed in Ferraris et al. (2008) and are briefly summarized here: According to the first low-activation criterion, C, Mg, O, Si, and Y can be considered fully low-activation elements; Ca and Ti are low-activation elements, too, provided that their concentration is below 30% and 12%, respectively; Al, Ba, Co, Ni, and Pd and their oxides are not LAMs according to the first criterion, and their use should be restricted.

The first low-activation criterion, however, can hardly be applicable to a plasma-facing joining material, since the whole first wall and blanket materials of TAURO, for instance, can fulfill—at the best—the second criterion. Therefore, it does not make sense to apply the first low-activation criterion to the joining material only. If we apply the second low-activation criterion to this glass–ceramic, it fulfills the requirements for a LAM.

It was reported in Brun (1998) that a simple butt joint is not expected to fail in a tough manner, but lapped joints are required to obtain tough behavior. Therefore, SiC/SiC joined samples were thus manufactured and tested not only in the typical sandwichlike

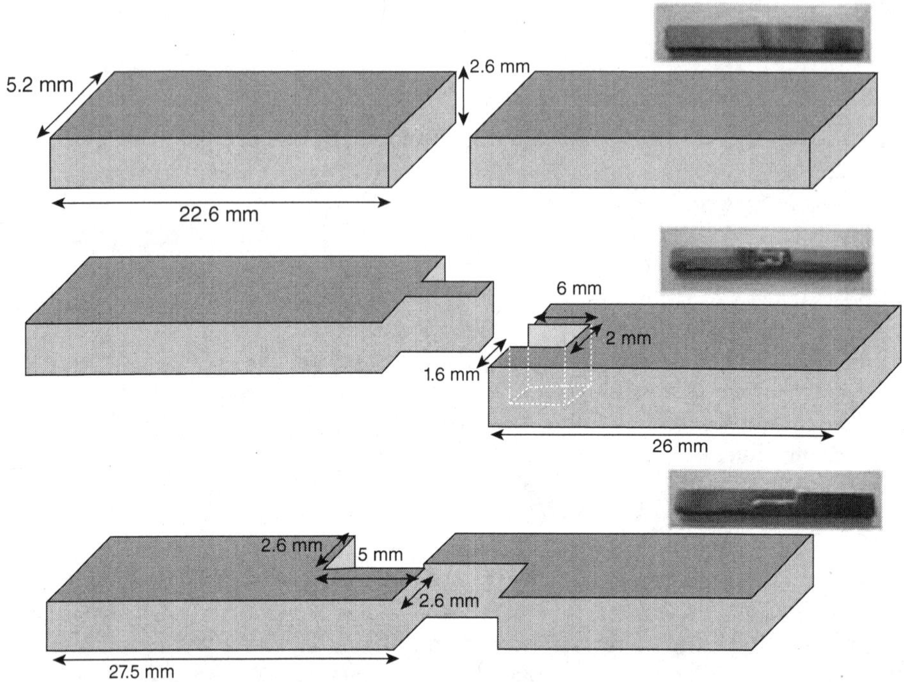

Figure 3.26. Three different joint configurations of joined SiC/SiC composites tested by a four-point bending test (Ferraris et al. 2008).

joined structure but also in several geometries, as shown in Figure 3.26, to test a machined joint with sealant properties, given by the glass–ceramic.

The mechanical strength of joined SiC/SiC was measured at RT by four-point bending (sample size adapted from ASTM C1341-00): The fracture of the joint occurred at 24 and 150 MPa for the first and third geometries, respectively. The composite failed with the second geometry and fracture of the joined area never occurred. Irradiation test results on this joining material and joined samples were expected by the end of 2009, within ExtreMat (Internet resource 12).

Techniques called *transient liquid-phase bonding* (TLPB) can produce a bond at a lower temperature than the one finally used. The technology is currently being adapted for a number of ceramics using interlayers based either on glasses (such as oxynitrides) (Glass et al. 1998) or pure metals or alloys (such as Ge and Ge–Si for joining SiC and SiC/SiC composites). The weakness of the TLPB method is that a favorable reaction between the interlayer and the substrates is required and that their behavior under irradiation can be unsuitable for nuclear applications. Some joining techniques adapted from TLPB to join SiC/SiC composites for nuclear applications are summarized here.

The *nanoinfiltration and transient eutectic phase (NITE)* process was developed by employing a transient liquid-phase sintering method utilizing a small amount of sintering

Figure 3.27. Scanning electron microscopy of SiC/SiC composites joined by the NITE process (Hinoki et al. 2005).

additives such as Al_2O_3, Y_2O_3, and SiO_2. The current status of NITE SiC/SiC composites and their joining process are reviewed in Katoh et al. (2004) and in Hinoki and Kohyama (2005). The same SiC nanopowder and sintering additive as with NITE-processed composites is used to join SiC/SiC (Fig. 3.27), silicon carbide, W, or sialon plates: The process requires hot pressing for 1 h under 15–30 MPa at 1700–1900°C, in Ar; the "NITE joint" for SiC/SiC was evaluated by the tensile test and asymmetric four-point flexural test (ASTM C1469) (Hinoki and Kohyama 2005). Tensile tests gave fractured samples on the side of the SiC/SiC composites, not at the joint interface. The asymmetric four-point flexural test of the NITE-joined SiC/SiC resulted in fractures outside the joined area, thus making it impossible to measure the joint shear strength with this test (Hinoki and Kohyama 2005; Katoh et al. 2004, 2007). Since another ASTM test is not available up to now to measure the pure shear strength of joined CMCs, this is a typical case of having promising joining techniques but being unable to give a value for the obtained joints.

One possible drawback of the NITE joining process for large-scale applications as in nuclear reactors is that it requires high pressure at high temperature in a controlled environment.

The NITE process has also been successfully used to join W to SiC/SiC by applying W, SiC powder, and sintering additives on SiC/SiC to initiate a basically identical microstructure with the matrix of SiC/SiC composites. Tungsten coatings were formed on both SiC and SiC/SiC composites by hot pressing. It was found that the reaction layer including W_5Si_3 was the weakest link, but shear strength can be improved by reducing the reaction layer (Katoh et al. 2004; Hinoki and Kohyama 2005).

A few examples of *mechanical joints made of SiC/SiC* composites are available: Their main advantages are that they allow replacement in case of damage to the component, and they do not need high temperature or pressure to join components; they do not need any surface preparation of the materials to be joined. As drawbacks, the lack of hermeticity and the stress concentration due to the use of mechanical fasteners, such as screws and hooks, are to be considered.

Mechanical fastening of NITE SiC/SiC using screw-threaded tubes and manifolds has been successfully demonstrated: NITE SiC/SiC composites have excellent

Figure 3.28. Screw-type mechanical joint of 10-mm inner diameter (12-mm outer diameter) NITE SiC/SiC tube (courtesy of A. Kohyama and T. Hinoki, Kyoto University).

machinability. The mechanical joint can be easily made when the fiber architecture is appropriately designed. The high density and high Young's modulus of NITE SiC/SiC make it possible to machine even sharp screws. Figure 3.28 is an example of production of a screw-type mechanical joint for a gas piping system. Additional applications of the NITE process to mechanical joints can produce more reliable joints with a sufficient hermeticity.

Another important feature of NITE composites is their excellence in gas leak tightness. The helium permeability test was conducted and the result indicated that monolithic SiC by the NITE process keeps its permeability at the level of 10–12 m^2/s, which is near the level of ordinary metallic materials.

Finally, preliminary neutron irradiation data have been acquired for NITE SiC/SiC composite: The composite was irradiated in the High Flux Isotope Reactor (HFIR) at Oak Ridge National Laboratory (ORNL) to 4.2×10^{25} n/m^2 at 1000°C. The observed mechanical and dimensional stability suggested an irradiation tolerance of the matrix material (Katoh et al. 2007).

The behavior of the joined NITE SiC/SiC composites under operating conditions and neutron irradiation is still under evaluation.

Silicon-based preceramic polymer joining is based on polymers able to infiltrate into and react with the bulk materials to be joined. In the infiltration processes, a mixture of polymer precursor (a source of carbon) and a mixture containing several elements, such as aluminum, boron, and silicon, is applied to the joint surfaces (in a tape, paste, or slurry) and then heated, generally to 1200°C, in an inert atmosphere. The joint forms through pyrolysis of the polymer precursor material, which subsequently reacts with the silicon in the presence of aluminum and boron sintering aids to form in situ, high-density SiC. Initial work demonstrated that polyhydromethylsiloxane (PHMS) systems with appropriate filler materials were able to provide a flexible polymer chemistry and high ceramic yield, particularly as a field repair technology (Lewinsohn et al. 2001, 2002).

An example of an infiltration process used to join SiC/SiC components for fusion applications is given in Colombo et al. (2000) through the application of a layer of a Si-based preceramic polymer. During pyrolysis at high temperature, the polymer transforms into a ceramic, yielding an adhesive bonding with the SiC/SiC. One of the major problems encountered is the volumetric shrinkage of the bonding layer occurring during the polymer-to-ceramic conversion, which reduces the strength of the joints. Several modifications of the basic process have been proposed, including the use of inert or active powder fillers or fibers, the application of pressure during pyrolysis, and the performance of multiple impregnation–pyrolysis cycles. In any case, the mechanical strength of the joints was low (Colombo et al. 2000; Lewinsohn et al. 2001) and their behavior under operating conditions and neutron irradiation is unknown.

Reaction-bonded and reaction-formed SiC-based joints for SiC/SiC were developed by means of a process called Affordable, Robust Ceramic Joining Technology (ARCJoinT™) (Singh 1999), which is based on the reaction-forming approach, and it permits joints with tailorable microstructures to be obtained. The joining process begins with the application of a carbonaceous mixture in the joint area, holding the items to be joined in a fixture, and curing at 110–120°C for 10–20 min. This step fastens the pieces together. Then, silicon or a silicon alloy in tape, paste, or slurry form is applied around the joint region and is heated to 1250–1425°C (depending on the type of infiltrant) for 10–15 min. The molten silicon or silicon-refractory metal alloy reacts with carbon to form silicon carbide with controllable amounts of silicon and other phases as determined by the alloy composition. Joint thickness can be readily controlled through adjustments of the properties of the carbonaceous paste and the applied fixturing force.

The behavior of ARCJoinT joined SiC/SiC under operating conditions and neutron irradiation is unknown.

A relatively new technique for joining ceramic materials is the *laser joining process* (Lippmann et al. 2004), which is based on the local melting of a solder by use of laser radiation. The laser joining technique was used successfully to join SiC ceramic, and the joints obtained were temperature resistant at 1600°C and above; a special solder made from Al_2O_3, Y_2O_3, and SiO_2 was used. This process was developed in order to make possible a kind of "welding" between ceramic pieces, but its application to SiC/SiC joining is not available yet. This process is essentially similar to NITE joining as the low eutectic melting temperature for the alumina–yttria–silica system is utilized. Local heating by laser or electrons is always beneficial to avoid the undesirable effect of heating other locations, but controlling the process condition is typically more difficult than heating (and pressurizing) the whole components.

Another well-known method for joining SiC ceramic parts is the subsequent *cosintering of ceramic components* (Schulte-Fischedick et al. 2002; Internet resource 9). It is known as liquid silicon infiltration (LSI) and consists of a codensification of preforms by means of "green body" infiltration. The steps of the process are as follows: green ceramic bodies glued by a phenolic resin adhesive and graphite powder (or woven C fabric) and cured at 135°, then LSI (capillary forces) and in situ reaction with porous ceramic perform. This method allows a fast, flexible, near-net shape infiltration process without shrinkage, but it needs high pressure, high temperature, and long time intervals. Moreover, the presence of unreacted silicon in the joint makes it, for now, unsuitable for nuclear applications, as previously discussed (Matthews and Hutchings 1973; Matthews 1974; Price and Hopkins 1982).

Finally, some work has been done on *microwave joining of SiC/SiC* composites (Silberglitt et al. 1991; Yiin et al. 1991; Ahmad et al. 1993, 1995, 1997). SiC/SiC composites have been joined by a commercial active brazing alloy and by allyhydridopolycarbosilane (AHPCS) (Starfire Systems, Inc., Watervliet, NY). SiC powders were added to the AHPCS to form a slurry mixture. The joints were obtained by heating with a 1- to 6-kW maximum power continuous wave microwave source operating at 2.45 GHz, in a single-mode cavity, under inert atmosphere (Ahmad et al. 1995).

Characterization of Joined SiC/SiC

As already discussed for joined C/C for applications in fusion reactors, the characterization of joined SiC/SiC is also an issue.

Of particular concern are component geometries that may require specific test methods.

Even without a precise design, defined working conditions, and a characterization standard available, several techniques have been developed to test the properties of SiC/SiC joints. Most of the tests are about hermeticity, temperature stability, thermal shock, atmosphere resistance (oxidation, moisture, etc.), and mechanical strength. Few standards for mechanical testing of SiC/SiC joints are available or widely accepted (Lara-Curzio and Jenkins 1999). The most common tests on joined SiC/SiC composites are tensile tests, bending tests, and shear tests. Shear strength is measured by several tests, including asymmetrical four-point bending (ASTM C 1469-00), single lap in compression (simple and offset, ASTM D905-98), and double-notched samples in compression (ASTM C1292-00 and ASTM C1425-99).

Most of these testing methods were not specifically developed for joined SiC/SiC, but they are modified ASTM tests.

Moreover, only an asymmetrical four-point test can measure a pure shear stress in the joined area: Other tests give "apparent shear" in a nonhomogeneous state of stress. Since the strength of a joint is largely influenced by the geometry of the joint and the test method for evaluating the strength, comparison of the results obtained for joined SiC/SiC is almost impossible. Detailed information on the stress field is provided in some studies, but little information on the criteria of fracture is given. To describe deformation and fracture behavior more precisely, a new and simple simulation method has been developed, and a function representing the density of surface energy to be supplied for the formation of new surface has been introduced into the finite element method using cohesive elements or interface elements (Serizawa et al. 2001, 2004, 2007a, 2007b).

Moreover, all the test methods just cited have been designed to measure the strength resistance of flat joints. If the geometry of the joint is not flat, new tests are required. Stress analysis of a curved joint is completely different from that of a flat joint. In addition, there are considerable differences in the design of the fixtures and in the way of fixing the specimens.

A project involving the United States, Japan (Titan), and the EU is currently running in order to develop shear strength tests for joined SiC/SiC to be tested before and after nuclear irradiation: A torsion test (ASTM F1362-97) has been considered as

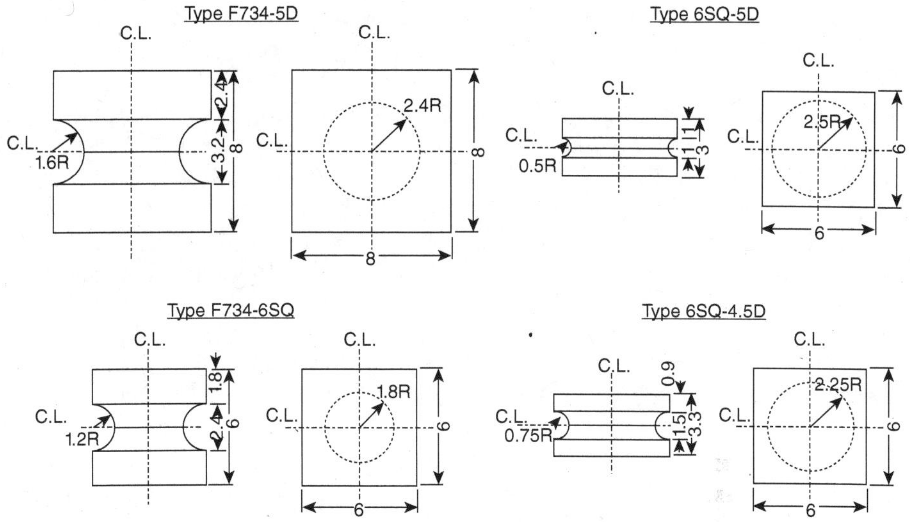

Figure 3.29. Miniature torsional specimen of ceramic joint (from Katoh et al. 2008; courtesy of Y. Katoh, ORNL).

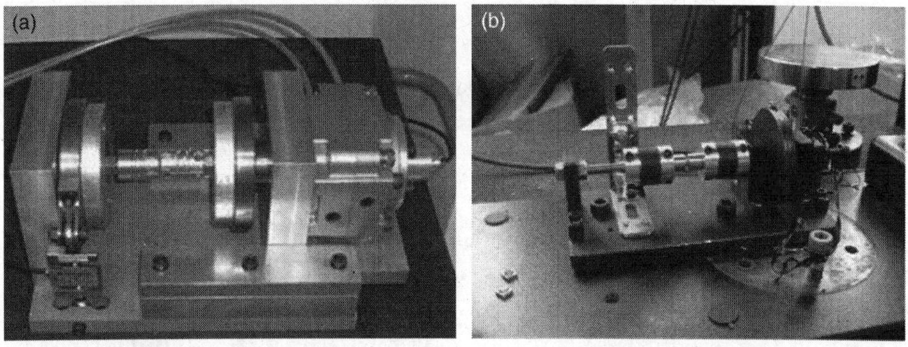

Figure 3.30. Torsion test apparatuses for ceramic joint built at Politecnico di Torino, Italy (a) and at Kyoto University, Japan (b) (courtesy of T. Hinoki, Kyoto University).

an alternative to asymmetrical four-point bending to measure the pure shear strength of the joints (Figs. 3.29 and 3.30). Several low-activation joining materials will be used to join SiC and SiC/SiC; then, joined specimens will be tested by torsion before and after neutron irradiation. There will be the first irradiation campaign, using the HFIR, dedicated to studying SiC joints. Results are expected by the end of 2012.

Finally, the determination of joining material properties is generally difficult because of the small thickness of joints (about 100 μm) and because test specimens of bulk joining material are usually not available. A new technique has been developed for the determination of elastic properties of the joining material: Optical extensometry based on a digital image correlation (DIC) technique (Puyo-Pain and Lamon 2004; Puyo-Pain et al. 2005; Hild et al. 2006) has been proven to measure displacement and strain in thin joints. Four-point bending tests have been done on bulk SiC bars joined by BraSiC. The elastic properties of BraSiC are measured from the joint strain state by using DIC and are calculated by finite element analysis. The BraSiC elastic properties were found to be 164 GPa for Young's modulus and about 0.25 for the Poisson coefficient. These values are in good agreement with those determined by nanoindentation tests.

CMCS IN ADVANCED FISSION REACTORS

The use of CMCs, either C/C or SiC/SiC composites, is being considered for several advanced fission systems, including high-temperature gas-cooled reactor (HTGR)–very high-temperature reactor (VHTR), gas-cooled fast reactor (GFR), and advanced light water reactor (A-LWR). The U.S. Next Generation Nuclear Plant (NGNP) is a type of HTGR in which both C/C and SiC/SiC composites are considered: C/C is needed for components of the graphite reflector block support, whereas SiC/SiC is considered for high-radiation dose applications.

In many fission reactors, several graphite parts are still in use and CMCs may prove to be durable replacements for them; moreover, programs considering potential utilization of SiC/SiC for core and/or in-vessel components in advanced fission energy systems have been initiated during the last few years. These programs are associated with the effort for Generation IV Nuclear Power Plants, which is internationally coordinated by the Generation IV International Forum (GIF).

In the United States, the use of SiC/SiC composites is considered as an option for control rod sleeves in a VHTR–next-generation nuclear power system. Preliminary design and operating conditions are about 800–950°C and a neutron dose of about 1 dpa. This option is based primarily on the properties of SiC/SiC demonstrated by the fusion materials program (U.S. DOE 2002; Carré 2003; Kotake et al. 2005).

A GFR will require very extensive use of SiC/SiC composites because the high-flux neutron irradiation in the GFR core and the surrounding locations will not allow the use of graphite or C/C. GFR designs have not yet reached the level of maturity to identify joining requirements. Neutron irradiation effects will certainly be the primary consideration for R&D of materials and application technologies for GFRs because of anticipated fast neutron irradiation on the order of 100 dpa. There is an A-LWR fuel concept with uranium carbide or nitride fuel encapsulated in silicon carbide cladding. SiC/SiC is preferred over monolithic SiC because of toughness. SiC/SiC cladding is advantageous over present Zircaloy cladding in terms of high-temperature–high-burnup capability, and hence it is anticipated to enhance safety, economic, and environmental features of LWRs. The main functions of fuel cladding are to confine fission products (both corrosive metals and gases) and to retain the exterior dimensions of fuel pellets or pins while transferring heat, generated in the fuel material by fission reactions, to coolant (Feinroth 2005; Barringer et al. 2007).

Development of reliable joining materials and techniques compatible with fuels (in terms of processing conditions), fission products, coolants, radiation environment, and ability to provide hermeticity is a key issue that would enable the use of SiC/SiC cladding of fuel.

All the composite components in NGNP or any other current VHTR designs avoid the use of adhesive joining, but the development of a compatible and reliable joining technique may broaden the component design options.

CONCLUSIONS

In proposed near-term fusion and advanced fission systems involving SiC-based components, present designs assume the unavailability of joining techniques that are reliable in nuclear environments. However, those systems will benefit if such techniques are developed, particularly for hermetic joints.

For the long-term approach, it is likely that a reliable joining technique will be required both for fusion blanket structures and for components of advanced fission reactors: In most cases, a precise joining requirement has not been identified yet.

Several potential techniques that may be viable for joining SiC-based components for nuclear applications have been identified.

Enhancing cross-fertilization of advanced fission and fusion research programs should be emphasized for mutual advantage to avoid duplication and to increase efficiency.

ACKNOWLEDGMENTS

Y. Katoh (ORNL, Oak Ridge, TN), M. Merola (ITER), and G. Pintsuk (FZ Juelich, Germany), as well as colleagues at Politecnico di Torino, Italy, are sincerely acknowledged for their valuable help in writing this chapter.

REFERENCES

Ahmad I, Silberglitt R, Black WM, Sa'adaldin HS, Tian YL, Katz JD. 1993. In: Clark DE, Laia JR, Tinga W, editors. *Microwaves: Theory and Application in Materials Processing II*. Ceram Trans 36. Westerville, OH: American Ceramic Society, pp. 409–415.

Ahmad I, Silberglitt R, Tian YL, Shan HS, Cozzens RC. 1995. In: Clark DE, Folz D, Oda S, Silberglitt R, editors. *Microwaves: Theory and Application in Materials Processing, III*. Ceram. Trans 59. Westerville, OH: American Ceramic Society, pp. 357–365.

Ahmad I, Silberglitt R, Tian YL, Katz JD. 1997. Microwave joining of SiC ceramics and Composites. 1st World Congress on Microwave Processing, Walt Disney Village, FL January 5–9.

Akiba M, Suzuki S. 1998. Overview of the Japanese mock-up tests for ITER high heat flux components. Fusion Eng Des 39–40:219–225.

Akselsen OM. 1992. Review: advances in brazing of ceramics. J Mater Sci 27:1989–2000.

Ando T, Kodama K, Yamamoto M, Arai T, Kaminaga A, Horiike H, Eto M, Fukaya K, Kiuchi T, Teruyama K, Nanai I, Hanai S, Ninomiya S, Tezuka M. 1992. Quality evaluation of graphites and carbon/carbon composites during production of JT-60U plasma facing materials. J Nucl Mater 191–194:1423–1427.

Appendino P, Casalegno C, Ferraris M, Grattarola M, Merola M, Salvo M. 2003. Joining of C/C composites to copper. Fusion Eng Des 66–68:225–229.

Appendino P, Ferraris M, Casalegno V, Salvo M, Merola M, Grattarola M. 2004. Direct joining of CFC to copper. J Nucl Mater 329–333:1563–1566.

Appendino P, Ferraris M, Casalegno V, Salvo M, Merola M, Grattarola M. 2006. Proposal for a new technique to join CFC composites to copper. J Nucl Mater 348:102–107.

Arroyave R, Eagar TW. 2003. Metal substrate effects on the thermochemistry of active brazing interfaces. Acta Mater 51:4871–4880.

Barabash V, Akiba M, Cardella A, Mazul I, Odergard BC, Plochl L, Tivey R, Vieider G. 2000. Armor and heat sink materials joining technologies development for ITER plasma facing components. J Nucl Mater 283–287:1248–1252.

Barringer E, Faiztompkins Z, Feinroth H, Allen T, Lance M, Meyer H, Walker L, Lara-Curzio E. 2007. Corrosion of CVD Silicon Carbide in 5001C Supercritical Water. J Am Ceram Soc 90(1):315–318.

Benoit JM, Fromentin JF, Chaumant V, Gillia O, Eustathopoulos N, Hodaj F, Koltsov A. 2007. Composition de brasure et procede d'assemblage par brasage utilisant cette composition. Patent application FR 2907448-A1.

Bisio M, Branca V, Di Marco M, Federici A, Grattarola M, Gualco G, Guardone P, Luconi U, Merola M, Ozzano C, Pasquale G, Poggi P, Rizzo S, Varon F. 2005. Manufacturing and testing in reactor relevant conditions of brazed plasma facing components of the ITER divertor. Fusion Eng Des 75–79:277–283.

Bolt H, Barabash V, Federici G, Linke J, Loarte A, Roth J, Sato K. 2002. Plasma facing and high heat flux materials- needs for ITER and beyond. J Nucl Mater 307–311:43–52.

Bonal JP, Wu CH. 1996. Neutron irradiation effects on the thermal conductivity and dimensional stability of carbon fiber composites at divertor conditions. J Nucl Mater 228:155–161.

Boscary J, Böswirth B, Greuner H, Missirlian M, Schedler B, Scheiber K, Schlosser J, Streibl B. 2007. Results of the examinations of the W7-X pre-series target elements. Fusion Eng Des 82(15):1634–1638.

Brun MK. 1998. Formation of tough composite joints. J Am Ceram Soc 81(12):3307–3319.

Buckley JD, Edie DD. 1993. *Carbon-Carbon Materials and Composites*. Park Ridge, NJ: William Andrew Publishing Noyes.

Carré F. 2003. *Proceedings of GIF Policy Group Meeting at Cape Town* (CEA/DEN/DDIN, 2003) 5.

Casalegno V. 2006. Joining of composite materials for nuclear fusion applications. PhD thesis, Politecnico di Torino, Torino, Italy.

Casalegno V, Salvo M, Ferraris M, Smeacetto F, Merola M, Bettuzzi M. 2008. Non-destructive characterization of carbon fiber composite/Cu joints for nuclear fusion applications. Fusion Eng Des 83:702–712.

Cockeram BV. 2005. Flexural strength and shear strength of silicon carbide to silicon carbide joints fabricated by a molybdenum diffusion bonding technique. J Am Ceram Soc 88: 1892–1899.

Coghlan WA, Clinard FW. 1991. Damage to Macor glass-ceramic from high-dose 14 MeV neutrons. J Nucl Mater 179–181:391–394.

Colombo P, Riccardi B, Donato A, Scarinci G. 2000. Joining of SiC/SiCf ceramic matrix composites for fusion reactor blanket applications. J Nucl Mater 278:127–135.

Daenner W, Merola M, Lorenzetto P, Peacock A, Bobin-Vastra I, Briottet L, Bucci P, Conchon D, Erskine A, Escourbiac F, Febvre M, Grattarola M, Hjorth CG, Hofmann G, Ilzhoefer A, Lill K, Lind A, Linke J, Richards W, Rigal E, Roedig M, Saint-Antonin F, Schedler B, Schlosser J, Tahtinen S, Visca E. 2002. Status of fabrication development for plasma facing components in the EU. Fusion Eng Des 61–62:61–70.

Datong W, Busse G. 1998. Lock in thermography for nondestructive evaluation of materials. Rev Gen Therm 37:693–703.

Dezellus O, Eustathopoulos N. 1999. The role of van der Waals interactions on wetting and adhesion in metal/carbon systems. Scr Mater 40(11):1283–1288.

DiCarlo JA, Yun HM, Morscher GN, Bhatt RT. 2005. SiC/SiC composites for 1200°C and above. In: Balsam P, editor. *Handbook of Ceramic Composites*. Dordrecht: Kluwer Academic Publishers, pp. 77–98.

Dietz J, the ITER Joint Central Team. 1999. The ITER fusion experiment. Fusion Eng Des 45:377–407.

Dig MR, Conrad R, Derz H, Duwe R, Linke J, Lodato A, Merola M, Pott G, Vieider G, Wiechers B. 2000. Neutron-irradiation effects on high heat flux components -examination of plasma-facing materials and their joints. J Nucl Mater 283-287:1161–1165.

EFDA Newsletter. 2005. August 15, http://www.efda.org.

Ehrlich K, Gasparotto M, Giancarli L, Le Marois G, Malang S, van der Schaaf B. 2001. European Material Assessment Meeting June 2001 EFDA-T-RE-2.0.

Ellett A. 2005. Development and characterization of material joints for current and future fusion devices. University Degree Thesis, Université Pierre et Marie Curie, Paris.

Escourbiac F, Constans S, Courtois X, Durocher A. 2007. Application of lock-in thermography non destructive technique to CFC armoured plasma facing components. J Nucl Mater 367–370:1492–1496.

Eustathopoulos N, Nicholas MG, Drevet B. 1999. *Wettability at High Temperature*. Oxford: Pergamon.

Ezato K, Dairaku M, Taniguchi M, Sato K, Akiba M. 2002. Non-destructive testing of CFC monoblock divertor mock-ups. J Nucl Mater 307–311:144–148.

Feinroth H. 2005. A multi-layered ceramic composite for impermeable fuel cladding for commercial water reactors. Gamma Engineering Report GN-54-03.

Ferraris M, Badini C, Montorsi M, Appendino P, Scholz HW. 1994. Joining Of SiCf/SiC composites for thermonuclear fusion reactors. J Nucl Mater 212–215:1613–1616.

Ferraris M, Salvo M, Isola C, Appendino Montorsi M, Kohyama A. 1998. Glass-ceramic joining and coating of SiC/SiC for fusion applications. J Nucl Mater 258–263:1546–1550.

Ferraris M, Montorsi M, Salvo M. 2000. Glass coating for SiCf/SiC composites for high temperature application. Acta Mater 48(18–19):4721–4724.

Ferraris M, Salvo M, Smeacetto F. 2002. Cordierite-mullite coatings for SiCf/SiC composites. J Eur Ceram Soc 22:2343–2347.

Ferraris M, Casalegno V, Salvo M. 2005. Process to join carbon based materials to metals and its applications. Patent WO2005/037734. Politecnico di Torino.

Ferraris M, Salvo M, Casalegno V, Ciampichetti A, Smeacetto F, Zucchetti M. 2008. Joining of machined SiC/SiC composites for thermonuclear fusion reactors. J Nucl Mater 375: 410–415.

Forrest RA. 2007. The European Activation System: EASY-2007 Overview, UKAEA Report UKAEA FUS 533, January.

Friedrich C, Reheis N, Thalmann W. 1992a. Method of making high-strength brazed joints. U.S. Patent 5160090. Schwarzkopf Technologies Corporation.

Friedrich C, Reheis N, Thalmann W. 1992b. Verfahren zur Herstellung hochfester Lötverbindungen. European Patent EP0476 772.

Gasse A. 2008. Method for assembling parts made of materials based on SiC by non-reactive refractory brazing, brazing composition and joint and assembly obtained by said method. U.S. Patent 7318547. Commissariat a l'Energie Atomique.

Gasse A, Saint Antonin F, Coing Boyat G. 1997. *Specific non reactive BraSiC alloys for SiC–SiC joining*. Report CEA-Grenoble. DEM n.DR 25.

Gasse A, Coing-Boyat G, Bourgeois G. 2001. Method using a thick joint for joining parts in SiC-based materials by refractory brazing and refractory thick joint thus obtained. U.S. Patent 6,221,499. Commissariat a l'Energie Atomique.

Giancarli L, Bonal JP, Caso A, Le Marosi G, Morley NB, Salavy JF. 1998. Design requirements for SiC/SiC composites structural material in fusion power reactor blankets. Fusion Eng Des 41:165–171.

Giancarli L, Golfier H, Nishio S, Raffray AR, Wong CPC, Yamada R. 2002. Progress in blanket designs using SiCf/SiC composites. Fusion Eng Des 61–62:307–318.

Glass J, Mahoney FM, Quillan B, Pollinger JP, Loehman RE. 1998. Refractory oxynitride joints in silicon nitride. Acta Mater 46(6):2393–2399.

Grill R, Reheis N, Witwer M. 1996. Heavily thermally stressable component. U.S. Patent 5580670. Schwarzkopf Technologies Corporatio.

Grote H, Bohmeyer W, Reiner HD, Fuchs T, Kornejew P, Steinbrink J. 1997. Comparison of chemical sputtering yields for different graphites at high ion flux densities. J Nucl Mater 241–243:1152–1156.

Guo QG, Li JG, Noda N, Kubota Y, Chen JL, Liu ZJ, Liu L, Song JR. 2003. Selection of candidate doped graphite materials as plasma facing components for HT-7U device. J Nucl Mater 313–316:144–148.

Harrison S, Marcus HL. 1999. Gas-phase Selective Area Laser Deposition (SALD) joining of SiC. Mater Des 20:147–152.

Hawryluk RJ, et al. 1998. Fusion plasma experiments on TFTR: a 20 year retrospective. Phys Plasmas 5:1577.

Henager CH, Shin Y, Blum Y, Giannuzzi LA, Kempshall BW, Schwarz SM. 2007. Coatings and joining for SiC and SiC-composites for nuclear energy systems. J Nucl Mater 367–370: 1139–1143.

Hild F, Perié JN, Lamon J, Puyo-Pain M. 2006. Use of digital image correlation to analyze the mechanical properties of brittle matrix composites. Ceram Eng Sci Proc 175:14–21.

Hinoki T, Kohyama A. 2005. Current status of SiC/SiC composites for nuclear applications. Ann Chim Sci des matériaux 30(6):659–671.

Hinoki T, Eiza N, Son S, Shimoda K, Lee J, Kohyama A. 2005. Development of joining and coating technique for sic and sic/sic composites utilizing nite processing. Ceram Eng Sci Proc 26(2):399–405.

Hirai T, Bondarchuk E, Borovkov AI, Koppitz T, Linke J, Mertens P, Neubauer O, Panin A, Philipps V, Pintsuk G, Sadakov S, Steinbrech RW, Schweer B, Uytdenhouwen I, Vaßen R, Samm U, Sievering R. 2007. Development and testing of a bulk tungsten tile for the JET divertor. Phys Scr T128:144–149.

Hosogane N, Ninomiya H, Matsukawa M, Ando T, Neyatani Y, Horiike H, Sakurai S, Masaki K, Yamamoto M, Kodama K, Sasajima T, Terakado T, Ohmori S, Ohmori Y, Okano J. 2002. Development and Operational Experiences on the JT-60U Tokamak and Power Supplies. Fusion Sci Technol 42:368–385.

Ioki K, Akiba M, Barabaschi P, Barabash V, Chiocchio S, Daenner W, Elio F, Enoeda M, Ezato K, Federici G, Gervash A, Grebennikov D, Jones L, Kajiura S, Krylov V, Kuroda T, Lorenzetto P, Maruyama S, Merola M, Miki N, Morimoto M, Nakahira M, Ohmori J, Onozuka M, Rozov V, Sato K, Strebkov Y, Suzuki S, Tanchuk V, Tivey R, Utin Y. 2004. ITER nuclear components, preparing for the construction and R&D results. J Nucl Mater 329–333:31–38.

Ise H, Satoh S, Yamazaki S, Akiba M, Nakamura K, Suzuki S, Araki M. 1998. Development of fabrication technologies for the ITER divertor. Fusion Eng Des 39–40:513–519.

ITER Carbon Fiber Composites. 2001. DOCb. ITER doc. G74MA1001-07-11W0.2 (internal project document distributed to the ITER participants).

ITER Carbon Fibre Composites. 2001. Materials Assessment Report. DOCc. ITER doc. G74MA10 W0.3 (internal project document distributed to the ITER participants).

ITER Materials Assessment Report. 2001. Armour/Heat Sink Joining Technologies. DOCa. ITER doc. GAOFDR101-07-13R1.0 (internal project document distributed to the ITER participants).

ITER Materials Properties Handbook. 2001. DOCd. ITER doc. G74MA900-11-10W0.1 (internal project document distributed to the ITER participants).

ITER Materials Properties Handbook. 2001. DOCe. ITER doc. ITER-CB03-3114 (internal project document distributed to the ITER participants).

ITER Materials Properties Handbook. 2001. DOCf. ITER doc. ITER-CB02-3114 (NB31) (internal project document distributed to the ITER participants).

ITER Materials Properties Handbook. 2001. DOCg. ITER doc. ITER-CB02-2104 (internal project document distributed to the ITER participants).

ITER Materials Properties Handbook. 2001. DOCh. ITER doc. ITER-CB03-2104 (internal project document distributed to the ITER participants).

Jacobs M. 2008. Glass-metal joining in nuclear environment: the state of the art. PhD thesis, Ghent University.

Jenkins MG. 2006. Wanted: CMC test standards, design codes, for next-generation nuclear power plants. Am Ceram Soc Bull 85(11):16–20.

Jones RH, Giancarli L, Hasegawa A, Katoh Y, Kohyama A, Riccardi B, Snead L, Weber W. 2002. Promise and challenges of SiCf/SiC composites for fusion energy applications. J Nucl Mater 307–311:1057–1072.

Jortner J. 2000. Applications of carbon-carbon composites. In: Kelly A, Zweben C, editors. *Comprehensive Composite Materials*, Vol. 6, Amsterdam: Elsevier Science, pp. 29–45.

Kar A, Palit Sagar S, KumarRay A. 2007. Characterization of the ceramic–metal brazed interface using ultrasonic technique. Mater Lett 61:4169–4172.

Katoh Y, Totani M, Kohyama A, Montorsi M, Salvo M, Ferraris M. 2000. Microstructure and Mechanical properties of low-activation glass-ceramic joining and coating for SiC/SiC composites. J Nucl Mater 283–287:1262–1266.

Katoh Y, Kohyama A, Nozawa T, Sato M. 2004. SiC/SiC composites through transient eutectic-phase route for fusion applications. J Nucl Mater 329–333:587–591.

Katoh Y, Snead LL, Henager CH, Hasegawa A, Kohyama A, Riccardi B, Hegeman H. 2007. Current status and critical issues for development of SiC composites for fusion applications. J Nucl Mater 367–370:659–671.

Katoh Y, Hinoki Y, Jung HC, Park JS, Konishi S, Ferraris M. 2008. Development and evaluation of silicon carbide joints for applications in radiation environment. Fusion Materials 44, Semiannual Progress Report for Period Ending June 30, 2008, Oak Ridge National Laboratory.

Koppitz T, Pintsuk G, Reisgen U, Remmel J, Hirai T, Sievering R, Rojas Y, Casalegno V. 2007. High-temperature brazing for reliable tungsten–CFC joints. Phys Scr T128:175–181.

Kotake S, Sakamoto Y, Sagayama Y. 2005. Commercialized fast reactor cycle systems and reactor core performance of the promising fast reactors. Prog Nucl Energy 47(1-4):300–313.

Krautkrämer J, Krautkrämer H. 1990. *Ultrasonic Testing of Materials*, 4th edition. New York: Springer-Verlag, pp. 27–55.

Krenkel W, Henke T, Mason N. 1996. In-situ joined CMC component. In: San Sebastian E, editor. *Proceedings of CMMC 96, the First International Conference on CMMC*. Trans Technol Pupl Key Eng Mater B 127–131:S313.

Lamon J. 2005. Chemical vapor infiltrated SiC/SiC composites (CVI SiC/SiC). In: Balsam P, editor. *Handbook of Ceramic Composites*. Dordrecht: Kluwer Academic Publishers, pp. 55–76.

Lara-Curzio E, Jenkins MG. 1999. Development of test standards for continuous fiber ceramic composites in the United States. Composites Part A 30:561–567.

Lemoine P, Ferraris M, Salvo M, Montorsi M, Scholz H. 1995. Reaction of SiCf/SiC composites with a zinc borate glass. J Am Ceram Soc 78(6):1691–1694.

Lemoine P, Ferraris M, Salvo M, Appendino Montorsi M. 1996. Vitreous joining process of SiC/SiC composites. J Eur Ceram Soc 16:1231–1236.

Lewinsohn CA, Colombo P, Reimanis I. 2001. Stresses occurring during joining of ceramics using preceramic polymers. J Am Ceram Soc 84(10):2240–2244.

Lewinsohn CA, Jones RH, Colombo P, Riccardi B. 2002. Silicon carbide-based materials for joining silicon carbide composites for fusion energy applications. J Nucl Mater 307–311:1232–1236.

Libera S, Visca E. 2006. Junction process for a ceramic material and a metallic material with the interposition of a transition material. Patent application WO2006024971. ENEA.

Linke J. 2006. Plasma facing materials and components for future fusion devices—development, characterization and performance under fusion specific loading conditions. J Phys Scr T123:45–53.

Lippmann W, Knorr J, Wolf R, Rasper R, Exner H, Reinecke AM, Nieher M, Schreiber R. 2004. Laser joining of silicon carbide: a new technology for ultra-high temperature resistant joints. Nucl Eng Des 231:151–161.

Maisonnier D, Cook I, Sardain P, Andreani R, Di Pace L, Forrest L, Giancarli L, Hermsmeyer S, Norajitra P, Taylor N, Ward D. 2005. EFDA-RP-RE-5.0, European Fusion Development Agreement.

Masaki K, Ando T, Kodama K, Arai T, Neyatani Y, Yoshino R, Tsuji S, Yagyu J, Kaminaga A, Sasajima T, Ouchi Y, Koike T, Shimizu M. 1995. Investigation of plasma facing components in JT-60U operation. J Nucl Mater 220–222:390–394.

Matthews FL, Rawlings RD. 1994. *Composites Materials: Engineering and Science*. London: Chapman & Hall.

Matthews RB. 1974. Irradiation damage in reaction-bonded silicon carbide. J Nucl Mater 51: 203–208.

Matthews RB, Hutchings WG. 1973. Neutron irradiation reduces the mechanical properties of reaction-bonded silicon carbide. J Nucl Mater 45(4):341–342.

Merola M, Vieider G. 1998. On the use of flat tile armour in high heat flux components. J Nucl Mater 258–263:672–676.

Merola M, Wu CH, the EU ITER Participating Team. 2004. Development of carbon materials and plasma facing components for ITER. Phys Scr T111:152–156.

Merola M, Plochl L, Chappuis P, Escourbiac F, Grattarola M, Smid I, Tivey R, Vieider G. 2000. Manufacturing and testing of a prototypical divertor vertical target for ITER. J Nucl Mater 283–287:1068–1072.

Merola M, Orsini A, Visca E, Libera S, Moreschi LF, Storai S, Panella B, Campagnoli E, Rustica G, Bosco C. 2002a. Influence of the manufacturing heat cycles on the CuCrZr properties. J Nucl Mater 307–311:677–680.

Merola M, Akiba M, Barabash V, Mazul I. 2002b. Overview on fabrication and joining of plasma facing and high heat flux materials for ITER. J Nucl Mater 307–311:1524–1532.

Merola M, Chappuis P, Escourbiac F, Grattarola M, Jeskanen H, Kauppinen P, Plochl L, Schedler B, Schlosser J, Smid I, Tahtinen S, Vesprini R, Visca E, Zabernig A. 2002c. Non-destructive testing of divertor components. Fusion Eng Des 61–62:141–146.

Merola M, Danner W, Palmer J, Vielder G, Wu CH, The EU ITER Participating Team. 2003. European contribution to the development of the ITER divertor. Fusion Eng Des 66–68: 211–217.

Mertens P, Hirai T, Linke J, Neubauer O, Pintsuk G, Philipps V, Sadakov S, Samma U, Schweer B. 2007. Conceptual design for a bulk tungsten divertor tile in JET. Fusion Eng Des 82:1833–1838.

Mitteau R, Tore Supra team. 2005. Steady state heat exhaust in Tore Supra: operational safety and edge parameters. J Nuc Mater 337–339:795–801.

Morscher GN, Singh M, Shpargel T, Asthana R. 2006. A simple test to determine the effectiveness of different braze compositions for joining Ti-tubes to C/C composite plates. Mater Sci Eng A 418:19–24.

Mortimer DA, Nicholas M. 1970. The wetting of carbon by copper and copper Alloys. J Mater Sci 5:149–155.

Mortimer DA, Nicholas M. 1973. The wetting of carbon and carbides by copper alloys. J Mater Sci 8:640–648.

Nicholas MG, Mortimer DA. 1985. Ceramic/metal joining for structural applications. Mater Sci Technol 11:657–665.

Nishio S, Ueda S, Aoki I, Kurihara R, Kuroda T, Miura H, Kunugi T, Seki Y, Nagashima T, Ohta M, Adachi J, Yamazaki S, Kawaguchi I, Hashimoto T, Shinya K, Murakami Y, Takase H, Nakamura T. 1998. Improved tokamak concept focusing on easy maintenance. Fusion Eng Des 41:357–364.

Onozuka M, Morimoto M, Inoue M, Umakoshi T, Shimizu K, Sakurai S, Hosogane N, Masaki K. 1999. Structural evaluation of a compact, semi-closed W-shaped divertor system for JT-60U. Fusion Eng Des 45:41–53.

Peacock AT, Barabash V, Danner W, Rodig M, Lorenzetto P, Marmy P, Merola M, Singh BN, Tahtinen S, van der Laan J, Wu CH. 2004. Overview of recent European materials R&D activities related to ITER. J Nucl Mater 329–333:173–177.

Pedzich Z, Haberko K, Babiarz J, Faryna M. 1998. The TZP-chromium oxide and chromium carbide composites. J Eur Ceram Soc 5:1939–1943.

Piazza G, Matthews GF, Pamela J, Altmann H, Coad JP, Hirai T, Lioure A, Maier H, Mertens P, Philipps V, Riccardo V, Rubel M, Villedieu E, collaborators of the JET ITER-like project. 2007. R&D on tungsten plasma facing components for the JET ITER-like wall project. J Nucl Mater 367–370:1438–1443.

Price RJ, Hopkins GR. 1982. Flexural strength of proof-tested and neutron-irradiated silicon carbide. J Nucl Mater 108–109:732–738.

Puyo-Pain M, Lamon J. 2004. Determination of elastic properties of ceramic-based joint using a digital image correlation method. Ceram Eng Sci Proc 25(4):247–253.

Puyo-Pain M, Lamon J, Hild F. 2005. Determination of elastic moduli and Poisson coefficient of thin silicon-based joint using digital image correlation. Ceram Eng Sci Proc 26(2): 143–148.

Raffray AR, El-Guevaly L, Gordeev S, Malang S, Mogahed E, Najmabadi F, Sviatoslavsky I, Sze DK, Tillack MS, Wang X, Aries Team. 2001a. High performance blanket for ARIES-AT power plant. Fusion Eng Des 58–59:549–553.

Raffray AR, Jones R, Aiello G, Billone M, Giancarli L, Golfier H, Hasegawa A, Katoh Y, Kohyama A, Nishio S, Riccardi B, Tillack MS. 2001b. Design and material issues for high performance SiC_f/SiC-based fusion power cores. Fusion Eng Des 55:55–95.

Raffray AR, El-Guebaly L, Malang S, Sviatoslavsky I, Tillack MS, Wang X, and the ARIES Team. 2007. Advanced power core system for the ARIES-AT power plant. Fusion Eng Design 82:217–236.

Rainer F, Reheis N. 1994. Method for setting-up a cooling system. Patent EP0663670.

Rainer F, Reheis N. 1996. Process for the manufacturing of a cooling unit. U.S. Patent 5533258. Schwarzkopf Technologies Corp.

Riccardi B, Fenici P, Frias Rebelo A, Fiancarli L, Le Marois G, Philippe E. 2000. Status of the European R&D activities on SiCf/SiC composites for fusion reactors. Fusion Eng Des 51–52:11–22.

Riccardi B, Giancarli L, Hasegawa A, Katoh Y, Kohyama A, Jones H, Snead LL. 2004a. Issues and advances in SiC_f/SiC composites development for fusion reactors. J Nucl Mater 329–333:56–65.

Riccardi B, Nannetti CA, Petrisor T, Woltersdorf J, Pippel E, Libera S, Pilloni L. 2004b. Issues of low activation brazing of SiC_f/SiC composites by using alloys without free silicon. J Nucl Mater 329–333:562–566.

Riccardi B, Nannetti CA, Woltersdorf J, Pippel E, Petrisor T. 2004c. Joining of SiC based ceramics and composites with Si-16Ti and Si-18Cr eutectic alloys. Int J Mater Prod Tech 20(5/6):440–451.

Ruset C, Grigore E, Maier H, Neu R, Li X, Dong H, Mitteau R, Courtois X, JET EFDA Contributors. 2007. Tungsten coatings deposited on CFC tiles by the combined magnetron sputtering and ion implantation technique. Phys Scr T128:171–174.

Salvo M, Casalegno V, Rizzo S, Smeacetto F, Ferraris M, Merola M. 2008. One-step brazing process to join CFC composites to copper and copper alloy. J Nucl Mater 374:69–74.

Savage G. 1993. *Carbon-Carbon Composites*. London: Chapman & Hall.

Schedler B, Huber T, Eidenberger E, Scheu C, Pippan R, Clemens H. 2007a. Methods to determine the joint strength of C/C to copper joints. Fusion Eng Des 82(15–24):1786–1792.

Schedler B, Huber T, Friedrich T, Eidenberger E, Kapp M, Scheu C, Pippan R, Clemens H. 2007b. Characteristics of an optimized active metal cast joint between copper and C/C. Phys Scr T128:200–203.

Schedler B. 2003. Manufacture of carbon armoured plasma facing components using hot isostatic pressing. Invited talk at 10th Carbon Workshop, Juelich, September 17–19, 2003.

Schlosser J, Chappuis P, Durocher A, Moncel L, Garin P. 2001. Development of actively cooled componete for tore supra toroidal pump limiter. Phys Scr T91:94–97.

Schlosser J, Escourbiac F, Merola M, Fouqet S, Bayetti P, Cordier JJ, Grosman A, Missirlian M, Tivey R, Rödig M. 2005. Technologies for the ITER divertor vertical target plasma facing components. Nucl Fusion 45:512–518.

Schlosser J, Martin E, Henninger C, Boscary J, Camus G, Escourbiac F, Leguillon F, Missirlian M, Mitteau R. 2007. CFC/Cu bond damage in actively cooled plasma facing components. Phys Scr T128:204–208.

Schulte-Fischedick J, Zern A, Mayer J, Ruhle M, Frieß M, Krenkel W, Kochendorfer R. 2002. The morphology of silicon carbide in C/C–SiC composites. Mater Sci Eng A332: 146–152.

Serizawa H, Lewinsohn C, Murakawa H. 2001. Fem evaluation of asymmetrical four-point bending test of sic/sic composite joints. Trans JWRI 30(1):119–125.

Serizawa H, Lewinsohn CA, Singh M, Murakawa H. 2004. Evaluation of fracture behavior of ceramic composite joints by using a new interface potential. Mater Sci Forum 502: 69–74.

Serizawa H, Fujita D, Lewinsohn CA, Singh M, Murakawa H. 2007a. Numerical analysis of mechanical testing for evaluating shear strength of SiC/SiC composite joints. J Nucl Mater 367–370:1223–1227.

Serizawa H, Lewinsohn CA, Singh M, Murakawa H. 2007b. Numerical analysis of test methods for evaluating shear strength of ceramic composite joints using interface element. Mater Sci Forum 539–543:2143–2148.

Shapiro A. 2007. Graphite and carbon-carbon composites. In: *Aws Brazing Handbook*, 5th edition. American Welding Society, pp. 381–408.

Silberglitt R, Palaith D, Black WM, Sa'adaldb HS, Katz JD, Blake RD. 1991. In: Clark DE, Gac FD, Sutton WH, editors. *Microwaves: Theory and Application in Materials Processing*. Ceram Trans 21. Westerville, OH: American Ceramic Society, pp. 487–492

Singh M. 1999. Design, fabrication and characterization of high temperature joints in ceramic composites. Key Eng Mater 164–165:415–420.

Snead LL, Zinkle SJ, White DP. 2005. Thermal conductivity degradation of ceramic materials due to low temperature, low dose neutron irradiation. J Nucl Mater 340:187–202.

Tashiro M, Kasahara A. 1999. Method of bonding graphite to metal. U.S. Patent 5904287. NEC Corporation.

Tavassoli AAF. 2002. Present limits and improvement of structural materials for fusion reactors—a review. 2002. J Nucl Mater 302:73–88.

Tillack M, et al. 2000. Fusion Eng Des 49–50:689–695.

Tivey R, Akiba M, Driemeyer D, Mazul I, Merola M, Ulrickson M. 2001. ITER R&D: vacuum vessel and in-vessel components: divertor cassette. Fusion Eng Des 55:219–229.

U.S. DOE Nuclear Energy Research Advisory Committee. 2002. *A Technology Roadmap for Generation IV Nuclear Energy Systems*. GIF-002-00, Washington, DC: U.S. DOE.

Visca E, Testani C, Libera S, Sacchetti M. 2003. Manufacturing of small scale W monoblock mockups by hot radial pressing. Fusion Eng Des 66–68:295–299.

Visca E, Libera S, Mancini A, Mazzone G, Pizzuto A, Testani C. 2005. Hot radial pressing: an alternative technique for the manufacturing of plasma-facing components. Fusion Eng Des 75–79:485–489.

Visca E, Libera S, Mancini A, Mazzone G, Pizzuto A, Testani C. 2007. Pre-brazed casting and hot radial pressing: a reliable process for the manufacturing of CFC and W monoblock mock-ups. Fusion Eng Des 82(15–24):1651–1656.

Weeks JK, Sommer JL. 1998a. Heat transfer apparatus and method for tubes incorporated in graphite or carbon/carbon composites. U.S. Patent 5806588. Technical Research Associates, Inc.

Weeks JK, Sommer JL. 1998b. Metal and carbonaceous materials composites. U.S. Patent 5834115. Technical Research Associates, Inc.

Weeks JK, Sommer JL. 1999. Method for coating a carbonaceous material with a molybdenum carbide coating. U.S. Patent 5914156. Technical Research Associates, Inc.

Wu CH, Bonal JP, Thiele B, Tsotridis G, Kwast H, Werle H, Coad JP, Federici G, Vieider G. 1994. Neutron irradiation effects on the properties of carbon materials. J Nucl Mater 212–215:416–420.

Wu CH, Bonal JP, Kwast H, Moons F, Pott G, Werle H, Vieider G. 1998. EU results on neutron effects on PFC materials. Fusion Eng Des 39–40:263–273.

Wu CH, Bonal JP, Werle H. 2000. Neutron effects on properties and annealing of low-Z materials. Fusion Eng Des 49–50:383–388.

Yiin TY, Varadan VV, Varadan VK, Conway JC. 1991. Microwave joining of Si-SiC/Al/Si-SiC. In: Clark DE, Gac FD, Sutton WH, editors. *Microwaves: Theory and Application in Materials Processing*. Ceram Trans 21. Westerville, OH: American Ceramic Society, pp. 507–512.

Ying A, Abdou M, Wong C, Malang S, Morley N, Sawan M, Merrill B, Sze DK, Kurtz K, Willms S, Ulrickson M, Zinkle S. 2006. An overview of US ITER test blanket module program. Fusion Eng Des 81:433–441.

Zhou ZJ, Zhong ZH, Ge CC. 2007. Silicon doped carbon/Cu joints based on amorphous alloy brazing for first wall application. Fusion Eng Des 82:35–40.

Zucchetti M, Ferraris M, Salvo M. 2001. Safety aspects of joints and coatings for plasma facing components with composite structures. Fusion Eng Des 58–59:939–943.

Zucchetti M, El-Guebaly L, Forrest RA, Marshall TD, Taylor TD, Tobita K. 2007. The feasibility of recycling and clearance of active materials from a fusion power plant. J Nucl Mater 367–370:1355–1360.

INTERNET RESOURCES

1. http://www.iter.org
2. http://www.cea.fr

3. http://www.jet.efda.org
4. http://www.jaeri.go.jp
5. http://www.twi.co.uk (World Centre for Materials Joining Technology)
6. http://www.wesgometals.com
7. http://www.jet.efda.org/
8. European Activation System EASY-2007 code package; see Web site: http://www.fusion.org.uk/easy2007/index.html
9. http://www.dlr.de (German Aerospace Agency, DLR)
10. PPCS overall report with annexes (April 2005) in http://www.efda.org/eu_fusion_programme/scientific_and_technical_publications.htm
11. http://fusionforenergy.europa.eu/
12. ExtreMat Web site (http://www.extremat.org)
13. http://www.gen-4.org/
14. http://www.gnep.energy.gov
15. http://www.dunlop-aerospace.com

4

AIR BRAZING: A NEW METHOD OF CERAMIC–CERAMIC AND CERAMIC–METAL JOINING

K. S. Weil, J. T. Darsell, and J. Y. Kim

Pacific Northwest National Laboratory, Richland, WA

INTRODUCTION

Engineering ceramics such as alumina, zirconia, and magnesia are often employed in critical high-temperature components because of their excellent mechanical properties and wear and corrosion resistance at elevated temperatures [1–3]. However, more ubiquitous use of these materials has been limited in part by the current inability to economically manufacture large or complex-shaped components that exhibit reliable performance. One alternative is to fabricate small, simple-shaped ceramic parts that can be assembled and joined to form the larger, more complex structure, similar to what is routinely done with metallic materials. Unfortunately, ceramics are inherently difficult to join either to themselves or to metal structures; a consequence of their strong ionic/covalent bonding. Of the established and emerging technologies for joining ceramics, there are several that can be considered for high-temperature applications, including glass–ceramic joining, diffusion bonding, reaction bonding, metallization, and brazing [4–20]. However, as with any fabrication process, intrinsic to each technique is some form of trade-off in terms functional properties, ease of processing, and/or cost.

Ceramic Integration and Joining Technologies: From Macro to Nanoscale, First Edition.
Edited by Mrityunjay Singh, Tatsuki Ohji, Rajiv Asthana, Sanjay Mathur.

Glass joining, for example, is a cost-effective and relatively simple method of bonding ceramics [6, 14–17]. However, the maximum temperature to which a glass joint can be exposed is limited by the softening point of the glass. Additional complications arise if devitrification occurs during service; the original, carefully engineered thermomechanical properties of the material begin to change, sometimes drastically, as the material crystallizes [14, 17]. An alternative joining technique, diffusion bonding, generally imparts the same material properties (i.e., high-temperature strength, creep resistance, and oxidation resistance) to the joint as found in the base substrate materials [7, 8]. While joining is conducted at temperatures below the melting points of the components, the process is reliant on solid-state diffusion across the bond interface. Thus, the diffusion bonding process is often slow, generally conducted at high temperatures and pressures, and very sensitive to the degree of mating between the joining surfaces. These conditions place severe constraints on component geometry because of concerns with excessive flow, deformation, and grain growth in the material [18, 19], as well as problems with fixturing, for example, accurately positioning and holding in place the individual parts that make up the desired structure.

A third process, known as reaction bonding, relies on an intermediary chemical reaction to form the bond between adjacent components [9]. Depending on the nature of the substrates and the joining reactants, reaction bonding can be conducted at a relatively moderate temperature, which generally facilitates componentry fixturing and reduces processing costs [10]. However, there can be issues with residual porosity, unconverted reactants, and undesired secondary product phases that can degrade the material properties of the joint. For example, when using a polymer-based precursor in reaction bonding, it is important to account for the large volume shrinkages that occur within the joint during the pyrolysis and precursor conversion steps; otherwise, internal cracks can form, which will compromise joint strength [20]. The addition of inert or reactive ceramic filler materials can mitigate this problem, but often in these cases, the joint retains a significant amount of porosity or secondary phases that reduce its strength and reliability [21]. However, there are a number of potential variants of this process yet to be explored and the concept remains promising.

METHODS OF CERAMIC BRAZING

One of the most reliable and adaptable materials joining processes is brazing. A common method of joining metals, brazing is readily scalable to mass production and is routinely employed in producing large volumes of componentry for the electronics and automotive industries. In this technique, a filler metal with a liquidus well below that of the materials to be joined is heated to a point at which it becomes molten and, under capillary action, spreads throughout and fills the gap between the sealing surfaces [22]. The latter phenomenon is commonly referred to as wetting and is essential in forming high-quality joints. As the filler metal cools and solidifies, a permanent bond forms. With respect to the brazing of ceramic materials, there are two key obstacles that must be overcome to form an acceptable joint: (1) Molten metals tend to poorly wet ceramic surfaces, a condition that leads to insufficient bonding, and (2) metals and

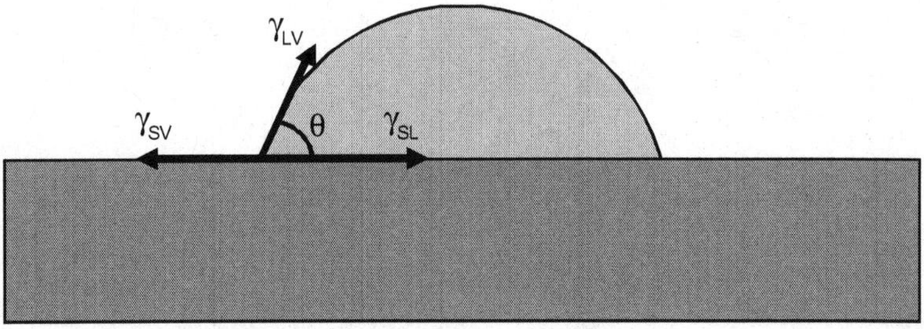

Figure 4.1. Equilibrium shape of a liquid sessile drop formed on a solid surface in the presence of a vapor. γ_{SV}, γ_{LV}, and γ_{SL} are the respective energies along the solid-liquid, solid-vapor, and liquid-vapor interfaces and θ is the contact angle formed by the sessile drop at the position noted in the figure.

ceramics typically exhibit significantly different degrees of thermal expansion when heated and cooled. When the two are bonded, this difference can generate unacceptably high tensile stresses in the ceramic and can lead to direct or impending joint failure.

In the case of metal-to-metal brazing, molten braze alloys tend to wet the substrate surfaces because of the similarity in chemical bonding between the substrate and filler metal. According to the Young model of wetting, a liquid will spread over a given solid surface to the point at which the intervening forces between the liquid, the solid, and the adjacent ambient vapor phases balance [23]. As shown in Figure 4.1, these forces derive from the surface energies between the three phases. Their relative magnitudes define a contact angle, θ, measured from the tangent at the liquid–solid interface to the tangent at the liquid–vapor interface:

$$\cos\theta = \frac{\gamma_{SV} - \gamma_{SL}}{\gamma_{LV}}, \tag{4.1}$$

where γ_{SL}, γ_{SV}, and γ_{LV} are the respective surface energies along the solid–liquid, solid–vapor, and liquid–vapor interfaces [24]. A contact angle of $\theta < 90°$ corresponds to a condition known as wetting, in which $\gamma_{SV} > \gamma_{SL}$ and the liquid spreads over the solid surface to replace a high-energy solid–vapor interface with a lower-energy solid–liquid interface and thereby minimizes the overall energy of the system. A contact angle of $\theta > 90°$ is considered to be a "nonwetting" or poorly wetting condition.

For a molten alloy to wet and spread over the joining or faying surfaces, there must be a degree of chemical interaction between the filler metal and the substrate materials. In the case of metal substrate materials with clean surfaces, this interaction generally means erosion and dissolution of the substrate, which leads to alloying and potential compound formation [25]. The presence of dirt or any layer on the metal surface that prevents good metal-to-metal contact will tend to prohibit wetting. This includes the native oxide scale that naturally forms on the external surfaces of most metals during typical open air handling. Many braze filler metals will not wet these oxide surfaces due to the dissimilarity in chemical bonding, and thus special surface cleaning methods

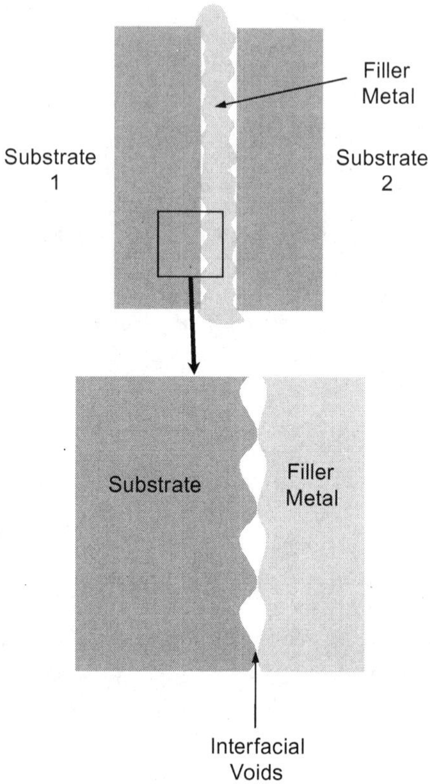

Figure 4.2. An example of filler metal sweating due to poor wetting (adapted from Reference 30).

are often required. These include surface pickling prior to brazing and/or the use of a fluxing agent during brazing, which dissolves or reduces the scale layer [22, 26, 27]. In the case of nickel and copper, the surface will chemically reduce to a metallic composition under moderately low p_{O2} environments, such as found in a vacuum furnace [28] or when using inert gas shielding [29]. While these are proven methods for metal-to-metal brazing, they are not adequate to join ceramic materials.

Without sufficient wetting during brazing, the molten filler metal will flow or "sweat" out of the gap between the components. Any remaining filler metal within the gap is typically found to be nonconformal with the microscopic features of the substrate surfaces and instead forms an array of voids along the filler metal/substrate interface as shown in Figure 4.2. Eustathopoulos et al. [30] have suggested that the wetting behavior of pure liquid metal/solid oxide wetting systems can be divided into three categories based on the degree to which the oxide substrate tends to dissolve into or react with the molten filler metal: (1) nonreactive systems, wherein the molar fraction of oxygen dissolved in the filler metal in less than 10^{-6}, which results in a contact angle on the order of 110–140°; (2) nonreactive systems, wherein the molar fraction of dissolved oxygen is $>10^{-5}$, resulting in contact angles lower than 90° but usually well above 60°; and (3) systems in which the filler metal and oxide substrate react to form an interfacial product that is readily wetted by the molten filler metal, resulting in contact angles typically less than 60°. The latter two types of behavior have been exploited to form the basis for several methods of direct ceramic brazing.

Prior to the development of direct brazing techniques, it was common to circumvent the wetting problem entirely by first metallizing the ceramic, such that brazing would subsequently occur on a metallic, not ceramic, surface. Currently employed in niche applications, metallization can be carried out by a variety of processes including physical and chemical vapor deposition, plating, and thermal spraying [31–33]. A standardized method of metallization developed in the 1940s specifically for joining alumina and metal components employed in vacuum tubes and switches is the Mo–Mn process [34, 35]. The approach is still used to seal spark plug insulators [36]. This is a multiple step process in which the ceramic sealing surface is first coated with a thin layer of a metal-loaded glass frit consisting of powders of molybdenum, manganese, and their respective oxides in a liquid binder. When heat treated in moist hydrogen at 1500°C, the metal–oxide mixture sinters into an interlocking structure consisting of a porous metal network infiltrated by a continuous glassy phase, which bonds to the underlying ceramic substrate. A nickel overcoat is subsequently applied and the component is reheated to 950°C under reducing gas to ensure that the nickel sinters to the Mo–Mn metallization layer. The final joint is prepared by brazing the nickelized ceramic using a conventional low-temperature brazing alloy, for example, Ag-28Cu. Note that brazing must be carried out under vacuum or reducing conditions to avoid oxidizing the nickel or Mo–Mn metallization layers. Similarly, subsequent exposure of the joined components to high-temperature, oxidizing environments, such as air or CO_2, must also be avoided.

Because the Mo–Mn process and other metallization techniques are typically multistep joining approaches, simpler and more cost-effective one-step methods have been sought for ceramic brazing. As discussed earlier, one means of achieving this is by identifying or developing a metal-based system that can dissolve oxygen at concentrations $>10^{-5}$. Copper metal is known to do this at moderate temperatures and in fact displays a eutectic with the Cu_2O scale that forms on its surface, as reported in detail by Baldwin et al. [37]. Burgess and Neugebauer et al. were the first to exploit this metal–oxide system to bond alumina using thin copper interlayers [38–40]. Referred to as direct copper bonding (DCB), the technique has become the standard means of fabricating copper laminated alumina substrates for the power electronics industry. The DCB process relies on the eutectic liquid that forms between Cu and Cu_2O at 1067°C. This liquid wets Al_2O_3 as well as a number of other ceramic materials. The laminates are prepared by fixturing thin copper foils above and below an Al_2O_3 ceramic substrate and heating the assembled stack to 1073 + 2°C under a precisely regulated p_{O2} atmosphere to control the degree of copper oxidation. A thin layer of eutectic liquid forms on the surfaces of the copper foils, wets the ceramic, fills the residual gap between the foils and the ceramic, and solidifies during cooling [41]. The solidified zone contains a small volume percentage of Cu_2O precipitates distributed within a copper matrix and a thin oxide diffusion zone that forms along the original alumina faying surfaces:

$$Al_2O_3 + Cu_2O \rightarrow (Cu_2O \bullet Al_2O_3)_{\text{solid solution}}. \tag{4.2}$$

Baldwin et al. [41] demonstrated that (1) above the melting point of copper (1084°C), a copper aluminate ($CuAlO_2$) reaction product forms along the alumina faying surfaces, and (2) increases in the oxygen content of the melt significantly improve spreading of the copper-based liquid across the alumina surface, resulting in a reduction of the contact angle from 170° to 90° with increasing oxygen concentration [42–45]. Similarly, the addition of oxygen to liquid silver improves its wetting characteristics on the alumina, resulting in a reduction of the contact angle from 159° to 102° with increasing oxygen partial pressure [46, 47], though not quite low enough to make an effective filler metal for ceramic brazing.

A second means of promoting direct wetting of the ceramic substrate is to incorporate into the filler metal a reactive element, such as titanium, zirconium, or molybdenum. These constituents cause chemical transformation of the ceramic at the joining interface via a reduction or displacement reaction, and thereby create an intermediate layer that is in chemical equilibrium with both the underlying ceramic and the molten braze filler metal. Typically more metallic in nature than the underlying parent material, this reacted surface is then readily wetted by the remaining filler material. Referred to as active metal brazing, and employing active braze alloy (ABA) filler metals, the process was originally examined by several groups and was commercialized by GTE Products Corporation [48–50]. Due to the high oxidation potential of the added reactive species, ABA filler metals typically require a stringent firing atmosphere, $p_{O2} < 10^{-5}$. Like the Mo–Mn and DCB processes, active metal brazing is scalable to high-volume production. However, unlike the Mo–Mn process, active metal brazing can be used to join a wide range of ceramics beyond alumina and sapphire, such as carbides, nitrides, silicides, zirconia, and diamond [51–57]. These materials can be joined to themselves or to metal substrates such as stainless steel, nickel-based alloys, tool steels, and Kovar. Active metal brazing has been widely investigated for a number of potential applications and has been recently reviewed in detail by Foley and Andrews [58] and Nicholas [59].

THE AIR BRAZING CONCEPT

There is a growing number of circumstances in which neither the Mo–Mn process, DCB, nor active metal brazing offers a satisfactory ceramic joining solution. Many of these are relevant to high-temperature power generation applications including solid oxide fuel cell (SOFC) stacks, oxygen or hydrogen gas separation for coal gasification, and high-temperature electrochemical sensors, of the type shown in Figure 4.3. Two commonly encountered problems are (1) the joint is not sufficiently oxidation resistant under the target operating conditions of the application, undergoing rapid deterioration when exposed to oxidizing conditions at temperatures beyond 500°C [60, 61], and (2) the need to carry out brazing under vacuum, controlled p_{O2} inert gas, or reducing gas establishes an environment that can cause irreversible chemical reduction and structural/property degradation in a number of electrochemically active and electrically conduc-

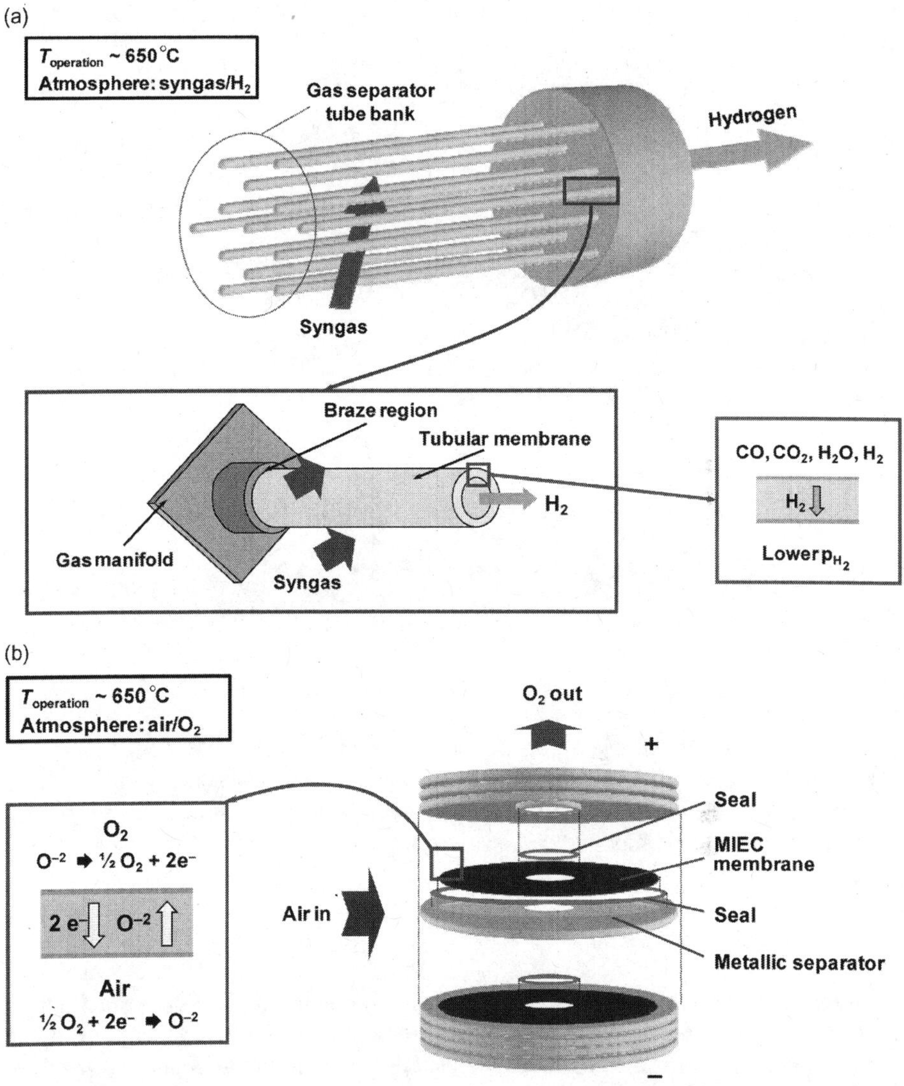

Figure 4.3. Schematics of joined componentry in (a) a hydrogen-from-syngas separator tube bank, (b) an oxygen-from-air concentrator, and (c) a cell-to-frame pSOFC cassette design, exploded and assembled views. MIEC, mixed ionic-electronic conductor.

tive ceramic substrate(s) of interest for these applications, including various perovskites, pyrochlores, and stabilized zirconias (e.g., yttria- and calcium-stabilized variants) [62].

To overcome these issues, a new ceramic brazing technique has been developed. Referred to as air brazing (or, in the case in which the filler metal reacts with the ceramic substrate, reactive air brazing [RAB]), the approach employs a filler metal consisting of (1) a noble metal matrix and (2) an oxide compound that partially or fully dissolves

(c)

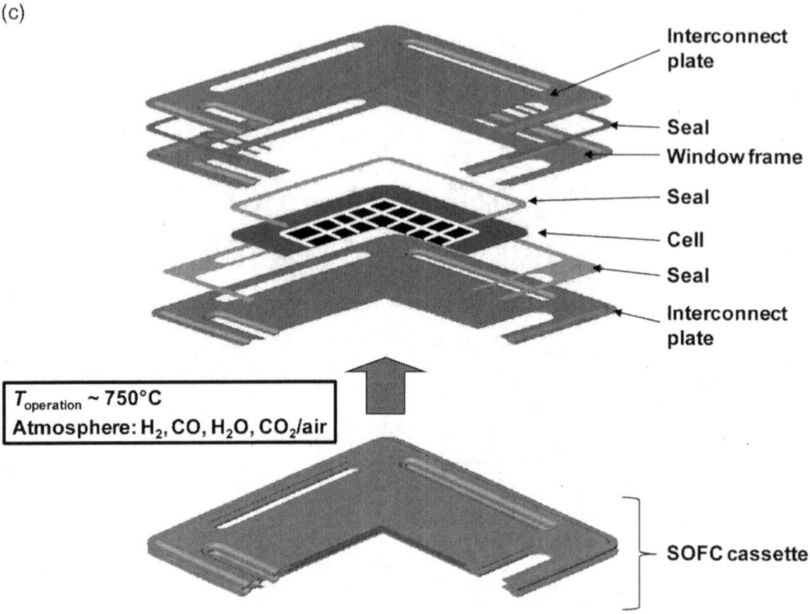

Figure 4.3. *(Continued)*

in the noble metal solvent when it is molten. The oxide compound serves as an in-situ oxygen buffer in the noble metal liquid, raising the chemical activity of dissolved oxygen, which enhances the metal's wetting characteristics on various oxide substrates. In addition, the oxide compound can also often modify the faying surface of the substrate such that the newly formed surface is readily wetted by the molten noble metal filler material. In both respects, air brazing shares similarities with DCB, except that it can be conducted directly in air with no need for fluxing. Because of this, the final joint is resistant to oxidation at moderate-to-high temperatures. In the case of air brazing of oxidation-resistant metal substrates, the bond that forms takes place between the filler metal and an oxide scale that grows on the structural metallic component during the joining operation. Thus, metal alloy substrates that exhibit a high degree of metal-to-scale adhesion afford the highest strength joints.

The concept of brazing alumina directly in air was first reported in 1965 by Naidich and Eremenko [63]. They deposited a thin layer of cupric oxide in slurry form on the faying surfaces of their alumina components. After fusing this layer in air at 1050–1070°C and cooling to room temperature, a silver interlayer was placed in the desired joining region, melted in air, and solidified to form the final joint. They also suggested that it should be possible to use a master alloy of silver and copper or a mixture of cupric oxide in silver as a filler metal for one-step brazing. Schüler et al. [64] independently rediscovered Naidich and Eremenko's approach and adopted it to prepare silver laminated alumina substrates for power semiconductor packaging applications. The Schüler work examined several one-step methods of silver–alumina bonding, including

(1) depositing a CuO slurry layer onto the alumina and bonding the silver foil to this layer in air at 946 ± 2°C, (2) electroless plating Cu onto the silver foil prior to bonding, and (3) depositing a slurry mixture of AgO and CuO onto the alumina prior to bonding with silver foil. Erskine et al. [65] examined the CuO paste-on-silver foil method for preparing metal electrodes on lead magnesium niobate and for brazing these types of perovskite ceramics. They reported joint strengths as high as 40% of the base material when brazing was conducted in air at 1000°C. Weil et al. [66–68] demonstrated Naidich and Eremenko's original suggestion for one-step brazing by applying a paste containing a mixture of silver and copper powders to the faying surfaces of the yttria-stabilized zirconia (YSZ) under study and by allowing the copper to oxidize in situ during brazing. They also examined this approach in preparing joints between YSZ and various oxidation-resistant metal alloys, such as FeCrAlY and 430 and 446 stainless steels [68].

AIR BRAZE FILLER METAL DESIGN: THE Ag–CuO SYSTEM

There are potentially a number of metal oxide-noble metal systems that can be considered for air brazing. To date, the Ag–CuO system has been the one explored in greatest detail. As will be discussed, a key factor in determining whether a given metal/oxide system is potentially viable for air brazing is the degree of mutual solubility between the metal and oxide constituents in the liquid state. However, with the exception of Assal et al.'s computational analysis of phase equilibria in the Ag–CuO system [69], no one has yet developed nor employed a predictive method of identifying mutual solubility within binary metal/metal oxide systems that potentially could form the basis for an air braze filler metal. Experimental evidence suggest that the following characteristics are likely indicators for this behavior: (1) measurable oxygen solubility (i.e., oxygen activity) in the liquid metal constituent, (2) compatible melting points between the metal and oxide to be alloyed, and (3) multiple valence states in the cation species of the oxide constituent, allowing it to serve as an efficient oxygen buffer in the noble metal. In addition to Ag–CuO, likely air braze filler metal systems include combinations of silver and various late transition metal oxides (e.g., Ag–Nb_2O_5, Ag–V_2O_5, Ag–Ta_2O_5, Ag–WO_3, and Ag–MoO_3), silver and several easily reducible early transition metal (such as Ag–NiO and Ag–CoO), and platinum and various late transition metal oxides (e.g., Pt–Nb_2O_5, Pt–V_2O_5, and Pt–WO_3), although the latter may require alloying with Ag to increase the metal's oxygen solubility.

Beyond the experience with the Ag–CuO system, which will be discussed in detail next, recent experimental work has shown the Ag–V_2O_5 system to be viable in joining porous and monolithic alumina [70], as well as lanthanum strontium cobalt ferrite (LSCoF) [71]. The wetting behavior on both substrates varies with the V_2O_5 concentration but, in general, exhibits wetting at small vanadia concentrations (<5 mol%) [70, 71]. Additionally, preliminary bend testing of LSCoF joining specimens brazed with various Ag–V_2O_5 filler metal compositions indicated poor strength, <40% of the base material [71]. Hardy and Weil [72] have found that Co_3O_4, Nb_2O_5, and Ta_2O_5 additions also improve the wetting characteristics of silver.

In designing a new filler metal alloy, the following information is typically desired:

- *Phase Formation and Phase Equilibria.* The filler metal must be appropriate for the application of interest. It must exhibit a liquidus temperature that is below the solidus temperatures of the materials to be joined, in fact, low enough to avoid property degradation (e.g., creep) in these materials during the joining operation. Conversely, the solidus temperature must be high enough to avoid remelting or substantial softening in the filler metal during continued use. That is, the temperature range over which the filler metal alloy melts must be known in order to define the proper processing conditions and subsequent service temperature. In addition, the degree of solidification shrinkage, the potential formation of secondary phases, and the possibility for excessive substrate erosion should be well understood so that the joint can be designed in a way that avoids undue stress concentrations.
- *Wetting.* As discussed previously, the filler metal alloy must properly wet the substrate(s) of interest to ensure gap filling and hermeticity within the joint and to promote a high degree of adhesion between all of the joint components. The wetting phenomenon may depend on interfacial reactions between the filler metal and substrates—understanding these can be critical to optimizing joint quality. There may also be trends in the wetting behavior due to changes in phase equilibrium associated with substrate erosion, dissolution, and/or reaction. Thus, filler metal composition, joining temperature, and time at temperature all play important roles in defining filler metal wettability, as well as joint quality.
- *Joint Strength.* Often the final measure of filler metal functionality is its ability to remain adherent and/or hermetic when placed under service conditions or exposed to operational excursions. Invariably, both properties are dependent in large part on the phase equilibria and wetting factors discussed earlier.

As discussed next, the recent work on the Ag–CuO system illustrates the development of potential filler metals for air brazing.

Phase Equilibria

Generally, it is desired that a braze filler metal transforms to a single-phase liquid when heated to the molten state to take full advantage of the melting point depression that occurs at an invariant composition such as a eutectic. Thus, in designing a new filler metal material, it is important that the phase boundaries in the system of interest are well-defined, particularly as it relates to defining braze temperature and/or identifying possible alloying agents to modify some of the filler metal's characteristics. As Nishiura et al. [73, 74] discuss, the Ag–CuO system is actually a pseudobinary system and is more accurately described as Ag–CuO_x (the nomenclature that will be employed here when referring to the silver–copper oxide phase diagram) because cupric oxide will undergo spontaneous reduction in ambient air at temperatures above 1027°C [75]. Thus, it is informative to consider the Ag–O and Cu–O binaries in addition to the Ag–CuO_x system. For example, metallic Ag is metastable in air at room temperature. The equilibrium phase is Ag_2O (Ag : O = 67:33), which decomposes in air at 147°C to form metallic Ag and oxygen gas [76]. The high solubility of oxygen in both solid and molten silver has been well studied [76–79]. An invariant reaction between silver and oxygen occurs at 939°C, forming a liquid containing 2.1 at% oxygen in silver [79]. Under inert gas, metallic silver melts at 961°C and in air at 951°C [80]. Other silver oxide compounds such as AgO and Ag_3O_4 exist under higher partial pressures of oxygen and low temperatures [81, 82].

In the Cu–O system, two compound oxides exist: Cu_2O (κ; Cu : O = 67:33) and CuO (τ; Cu : O = 50:50) [83]. Cu_4O_3 (π; Cu : O = 57:43) has also been shown to exist [84], although the region of stability of this compound cannot be defined because its thermodynamic properties are not known [83]. In air at ambient temperature and pressure (T ~ 25°C and p_{O2} ~ 0.21 atm), the equilibrium phase is CuO, which decomposes to Cu_2O when heated to 1027°C [75]. Upon further heating in air, Cu_2O melts to form a liquid and excess oxygen at 1124°C [75]. Under a controlled partial pressure of oxygen less than ambient, copper displays a eutectic reaction with Cu_2O at 1067°C and a monotectic reaction at 1223°C, which lead to a liquid miscibility gap at higher temperatures [75].

A ternary Cu–O–Ag isothermal section (1230°C) was originally constructed by Chang and Hsieh [83] based on the prior work of Leroux and Fröhlich [85] and Kohlmeyer and Sprenger [86]. Subsequent thermal analysis performed independently by Shao et al. [87] and Suzuki et al. [74] demonstrated that the pseudobinary CuO_x–Ag system contains both a eutectic and a monotectic reaction joined by a liquid-phase miscibility gap. The two data sets differ slightly, although a more recent investigation conducted by Weil et al. [88] confirms the latter data set for the two invariant points in the system: (1) a eutectic reaction that occurs at 942 ± 1°C between CuO, Ag, and a Ag-rich liquid phase L_1 with a concentration of $x_{Ag}/(x_{Cu} + x_{Ag}) = 0.99 \pm 0.005$ and (2) a monotectic reaction that occurs at 969 ± 1°C between CuO_x, the L_1 liquid phase, and a second CuO_x-rich liquid phase L_2 that has a composition of $x_{Ag}/(x_{Cu} + x_{Ag}) = 0.10 \pm 0.03$. Note that in neither this work nor the theoretical studies of the Cu–O–Ag system [88, 89] were intermediate compounds found nor subsequently considered to form between CuO_x and Ag. The work of Weil et al. also accurately identifies the boundaries of the miscibility gap found in the Ag–CuO_x system up to 1250°C. As discussed in the next section, this knowledge can be important in controlling the wetting properties of these types of air braze filler metals. Shown in Figure 4.4 is a composite phase diagram for the pseudobinary Ag–CuO_x system at p_{O2} based on References 73 and 88.

Substrate Wetting

In their report on brazing using Ag–CuO, Naidich and Eremenko included a table of contact angles for Ag–CuO wetting on various substrates, including aluminum oxide, magnesium oxide, zirconium oxide, quartz, porcelain, and steatite [63]. While the details about how the experiments were conducted (e.g., the composition of the filler metal or whether the contact was measured by in-situ methods) were not reported, contact angles generally ranging from 20° to 40° were observed. Meier et al. [90] were the first to carry out a systematic analysis of the CuO effect on wetting. Contact angle

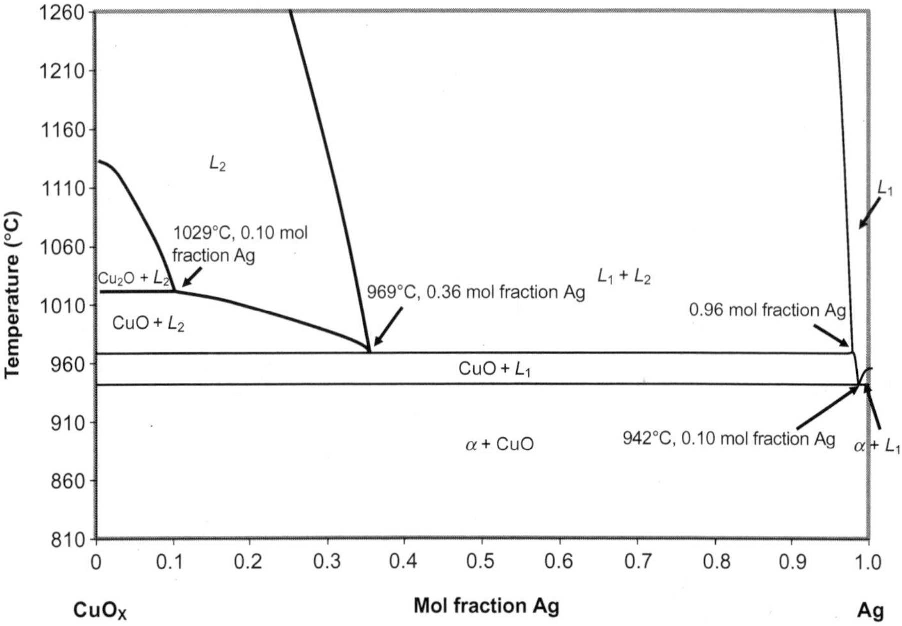

Figure 4.4. A pseudobinary phase diagram for the Ag–CuO_x system at p_{O2} ~ 0.21 atm (adapted from References 73 and 88).

measurements were conducted under a controlled p_{O2} atmosphere on alumina substrates using the sessile drop technique. Each sessile drop was formed by initially placing a known amount of CuO on a known mass of silver foil and heating to temperatures between 970 and 1250°C. The system was overpressured with argon to ~10 kPa and the oxygen partial pressure in the vicinity of the sample was established by the oxygen in the liquid drop, that is, at $p_{O2} < 0.21$ atm typical in ambient air. At a CuO concentration of 1.5 wt%, the sessile drop exhibits a nonwetting 108 ± 8° contact angle. Further addition of CuO produced wetting, which Meier et al. concluded was approximately invariant of composition at an average contact angle of 64 ± 7°.

Several groups have reexamined Ag–CuO wetting on a series of substrate materials, including YSZ [91], Al_2O_3 [92], FeCrAlY, and other stainless steels [65, 68, 93, 94], lead magnesium niobate [65], lead zirconate titanate [95], $La_{0.6}Sr_{0.4}Co_{0.2}Fe_{0.8}O_{3-\delta}$ (LSCoF) [67], and $La_{1-x}Sr_xMnO_3$ (lanthanum strontium manganite [LSM]) [72], and found differences in wetting behavior both between the various sets of substrate materials and the earlier work of Meier et al. [90]. The latter can be attributed largely to two key differences in the way the experiments were conducted: (1) In an effort to simulate the new brazing technique, Weil et al.'s work was conducted in ambient air, not under a controlled p_{O2} environment, and (2) silver and copper powders were generally premixed, comilled, and cold pressed to form a pelletized samples for the sessile drop experiments, as opposed to heating known amounts of CuO on silver foil. Darsell and Weil [96] showed that the copper fully oxidizes during the conditions of the experiment prior to liquid formation. Additionally, they found no significant difference in wetting behavior attributable to the starting form of the copper constituent, that is, Cu versus CuO. Differences in wetting behavior observed for the various substrates are due to differences in the degree of interaction between the filler metal and the substrate; that is, some substrate materials are inert or nonreactive with the filler metal, while others react or undergo a small amount of dissolution, which ultimately affects the wetting characteristics.

Nonreactive Substrates. YSZ is an example of a substrate material that does not appear to react or dissolve in Ag–CuO-based filler metals. Shown in Figure 4.5a is the dependence of the contact angle on temperature for a series of Ag–CuO filler metals on polished polycrystalline 8YSZ (i.e., zirconia stabilized with 8 mol% Y_2O_3) [91, 97]. As reported in several sessile drop studies conducted on YSZ, the measured contact angles appear to achieve stable values within 5 min of reaching the target soak temperature [91, 97, 98]. In general, the contact angle remains nearly constant above 950°C. However, there are two notable discrepancies. The first is a decrease in contact angle that occurs from 950 to 1000°C for filler metals with CuO concentrations greater than 4 mol%. The effect takes place because these compositions are found within the miscibility gap for the Ag–CuO_x system, as seen in Figure 4.4. Since 950°C lies below the monotectic temperature, some solid CuO remains in the filler metal and is likely the source of wetting impediment. When the temperature is raised above the monotectic line (969°C), the CuO melts and the contact angle assumes a value that then remains invariant with temperature, displaying the same behavior observed for the high-CuO content filler metals.

The second discrepancy occurs for the filler metal containing 3 mol% CuO. The contact angle between this liquid and YSZ increases with increasing temperature, a phenomenon that can be explained by replotting the contact angle as a function of composition for each hold temperature, as shown in Figure 4.5b–d [98]. The composition of the silver-rich miscibility gap boundary determined from the data in the Ag–CuO_x phase diagram is noted on each plot as a dotted line. The 3 mol% CuO filler metal composition is also labeled in each figure for reference. Two distinct regions of wetting behavior are readily observed in all three graphs. As measured in air, the contact angle for pure silver is approximately 90°. As copper oxide is added, the contact angle decreases in near-linear fashion until reaching a value of ~40°. As seen in the phase diagram, the filler metal exists as a single-phase, silver-rich liquid over the compositional range of $x_{CuO} = 0$ to the silver-rich miscibility gap boundary. Thus, the reduction in contact angle is directly attributable to a reduction in surface energy with increasing copper oxide addition. This is due to an increase in oxygen activity in the liquid alloy with increasing copper oxide content [99]. It is well-known that the addition of oxygen to liquid copper or silver improves the wettability of these liquids on ceramic oxide surfaces by lowering the liquid–vapor interfacial energy, γ_{LV} [47, 99, 100]. However, beyond the miscibility gap boundary, the exact composition of which depends on temperature, the relationship between filler metal composition and contact angle changes. For filler metal compositions lying within the miscibility gap, two immiscible liquids coexist, one rich in copper oxide and the other in silver. Note from the phase diagram

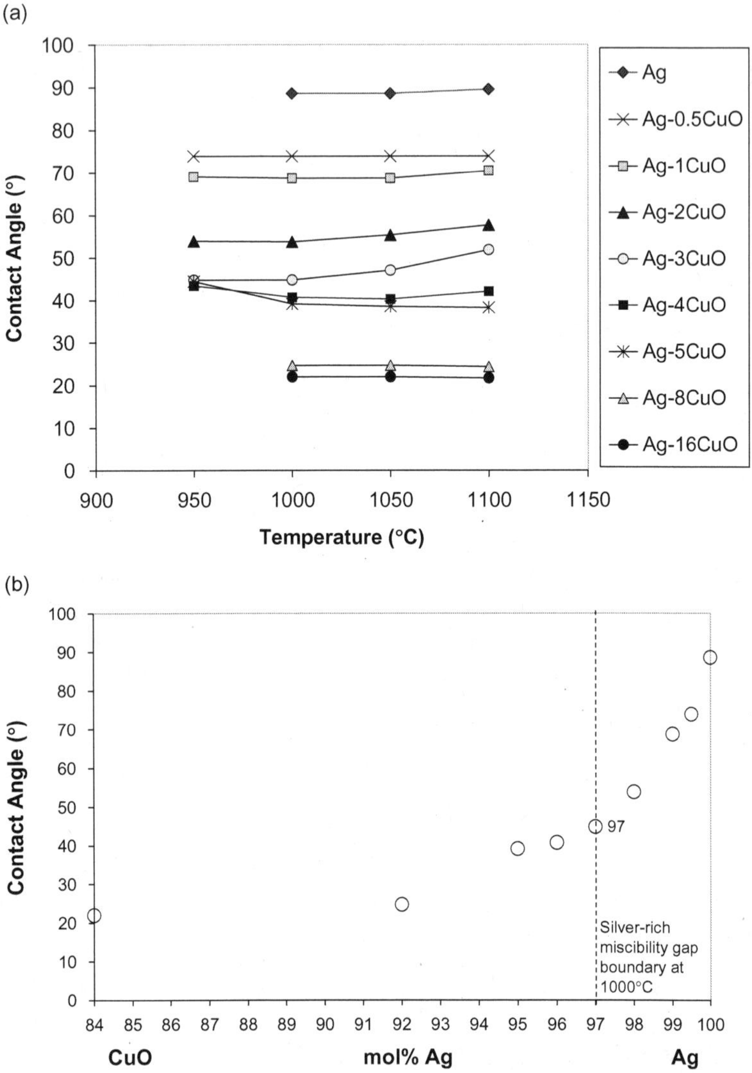

Figure 4.5. (a) Contact angle versus temperature between Ag–CuO and YSZ. The contact between Ag–CuO_x and YSZ at (b) 1000°C, (c) 1050°C, and (d) 1100°C as a function of filler metal composition (from Reference 98).

that as temperature increases, the 3 mol% CuO composition transitions from the silver-rich boundary of the two-liquid phase miscibility gaps to a single liquid (L_1) phase region.

This wetting transition is more clearly seen in the series of plots shown in Figure 4.6a–b, in which the derivative of the contact angle with respect to composition is plotted as a function of composition. The two different regions of wetting behavior are

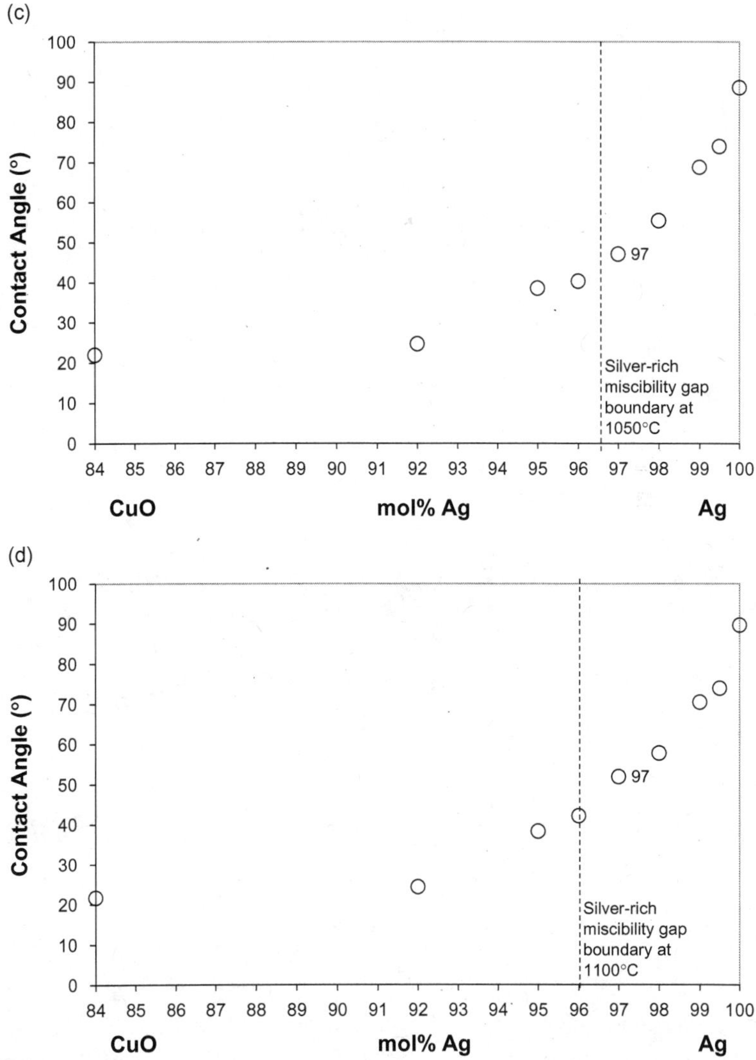

Figure 4.5. *(Continued)*

separated by a discontinuity that occurs directly at the composition corresponding to the silver-rich miscibility gap boundary for each temperature, as seen in Figure 4.6a. A similar first-order wetting transition is observed at 1050 and 1100°C, as shown in Figure 4.6b. In addition, for compositions that fall within the miscibility gap, $d\theta/dx_{Cu}$, begins to approach zero. Cahn's critical point wetting theory predicts that a transition in the wetting behavior from partial to complete wetting will occur for a multicomponent liquid below a critical temperature at which liquid–liquid immiscibility occurs [101]; that is, upon cooling through a single liquid phase region (e.g., L_1) into the

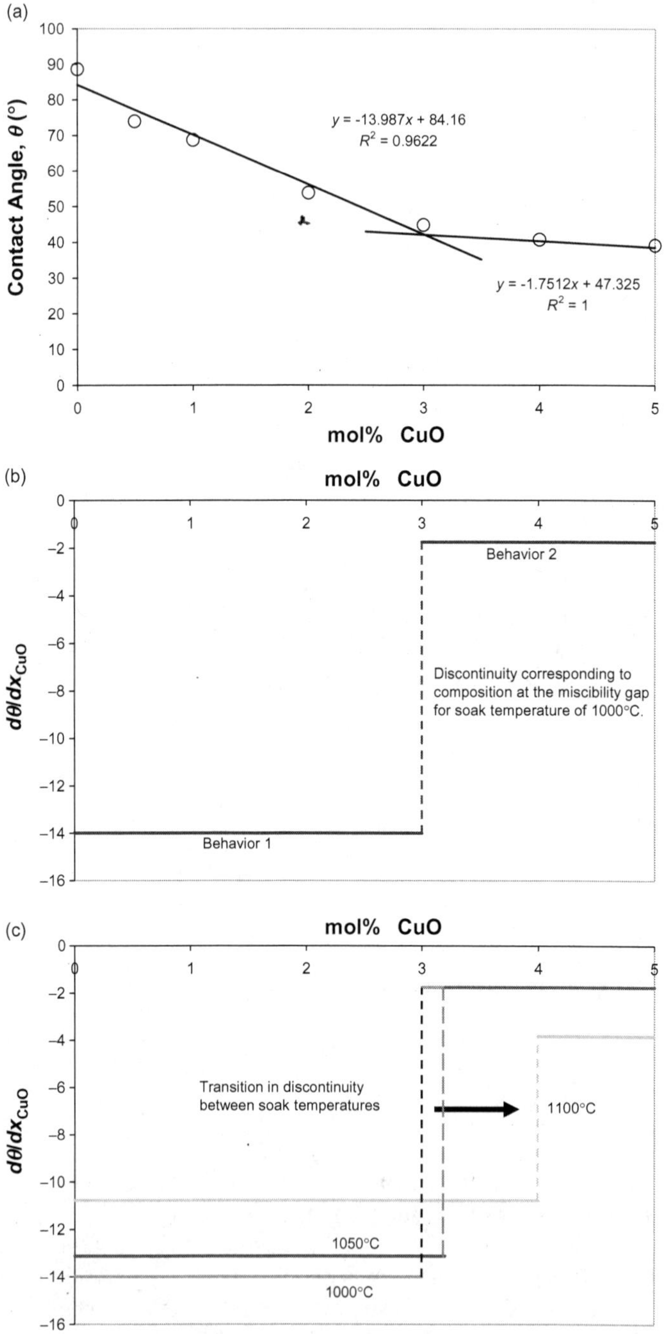
(a)
Contact Angle, θ (°)
y = -13.987x + 84.16
R² = 0.9622
y = -1.7512x + 47.325
R² = 1
mol% CuO
(b)
mol% CuO
dθ/dx_CuO
Behavior 2
Discontinuity corresponding to composition at the miscibility gap for soak temperature of 1000°C.
Behavior 1
(c)
mol% CuO
dθ/dx_CuO
Transition in discontinuity between soak temperatures
1100°C
1050°C
1000°C

Figure 4.6. (a) Contact angle versus copper oxide composition for silver-rich compositions of Ag–CuO on YSZ at 1000°C. Linear functions can be fit to the two distinct regions of contact angle behavior. (b) The derivative of contact angle with respect to copper oxide composition plotted as a function of CuO concentration clearly shows a discontinuity that corresponds to the miscibility gap composition. (c) A comparison of the derivatives of contact angle versus copper oxide content shows that there is a shift in the discontinuity over the temperature range of 1000–1100°C.

miscibility gap, the previously homogeneous liquid separates into two immiscible liquids, only one of which will wet the adjacent solid substrate. In the case of a Ag–CuO filler metal with a composition lying in the miscibility gap, the CuO-rich L_2 liquid preferentially wets the YSZ and thereby excludes the original silver-rich liquid from contact with the substrate. This phenomenon accounts for the temperature-dependent behavior of the 3 mol% CuO filler metal since it transitions from the miscibility gap boundary to a single liquid phase over the 1000–1100°C temperature range.

Evidence of perfect wetting in the Ag–CuO system is seen in the series of micrographs shown in Figure 4.7a–d, taken on sessile drop specimens that were quenched from 1000°C to room temperature in a water bath [102]. In the specimen wetted by 1 mol% CuO in Ag and quenched from an equilibrium temperature of 1000°C, shown in Figure 4.7a, the copper oxide is observed as fine-scale precipitates both in the bulk of the solidified sessile drop and along the interface with the YSZ. The similarity in the bulk and interfacial microstructures in the filler metal indicates that cooling occurred from a single-phase silver-rich liquid, L_1. By comparison, when a sessile drop with a composition at the miscibility gap boundary (3 mol% CuO) is quenched, the sessile drop's internal bulk and interfacial microstructures are distinctly different, as shown in Figure 4.7b. As seen in the high-magnification image shown in Figure 4.7d, the thin interfacial phase exhibits a fine-scale, interwoven structure indicative of monotectic decomposition; that is, it is the CuO-rich L_2 liquid that preferentially wets the YSZ substrate at temperature, even though it is the lighter of the two liquids present in the molten drop. Compositions that lie further within the Ag–CuO_x miscibility gap (5 mol% CuO), such as those shown in Figure 4.7c, exhibit remnants of a thicker L_2 liquid interfacial layer, as expected from the lever rule; that is, for compositions at the miscibility gap or within the gap itself, the silver-rich L_1 liquid is hindered from wetting the YSZ substrate by an intruding L_2 liquid film. Because of this and because the compositions of the L_1 and L_2 liquids remain constant within the miscibility gap, even though the ratio of the two liquids does not, the wetting angles for these compositions remain constant. A similar phenomenon has been observed for the immiscible liquid of the copper–oxygen system, in which L_1 is a Cu-rich liquid with oxygen in solution and L_2 is liquid Cu_2O [40–42].

Reactive Substrates. Alumina and various perovskite compounds such as LSM, LSCoF, and $Ba_{0.5}Sr_{0.5}Co_{0.8}Fe_{0.2}O_{3-\delta}$ (barium strontium cobalt ferrite [BSCoF]) are examples of reactive substrates in air brazing (thus the origin of RAB, based on early work

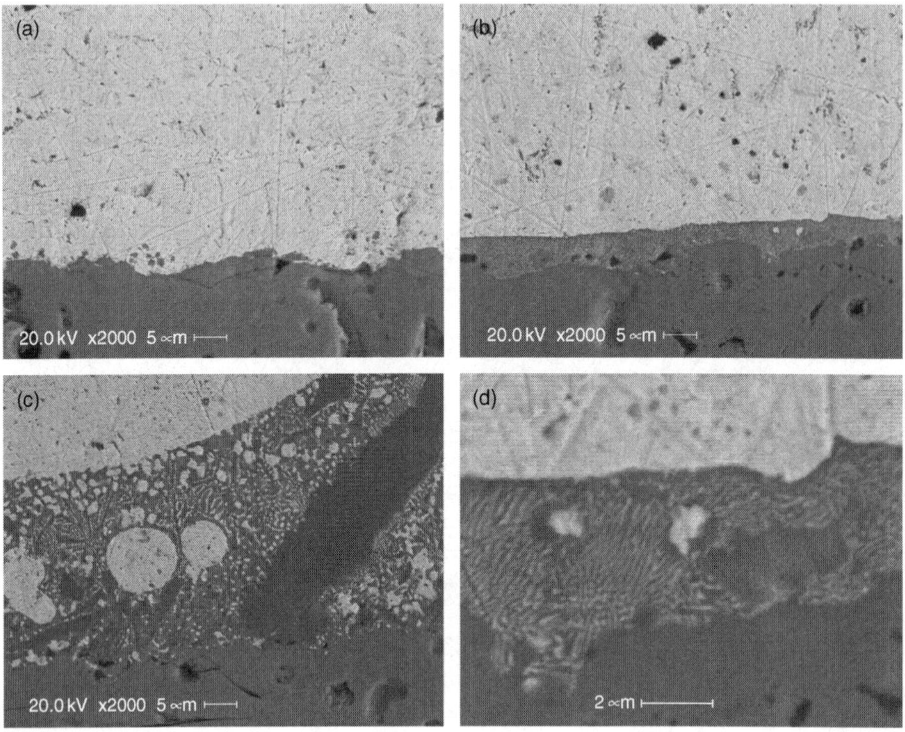

Figure 4.7. Cross-sectional backscattered electron (BSE) micrographs of the Ag–CuO/YSZ interface after quenching the following sessile drop specimens from 1000°C: (a) 1 mol% CuO in Ag, (b) 3 mol% CuO in Ag, and (c) 5 mol% CuO in Ag. A higher-magnification micrograph of the interfacial region in the 3 mol% CuO in the Ag specimen is shown in (d) (from Reference 102). The YSZ appears as the darker-gray, homogeneous material in the lower half of each picture, while silver has a lighter-gray contrast. The copper oxide appears as the darker-gray phase in the solidified material. The black regions are voids.

conducted on Al_2O_3-based substrates [103]). Plotted in Figure 4.8 is the contact angle between a series of Ag–CuO filler metals and polished polycrystalline alumina as a function of copper oxide concentration [104]. As found with YSZ, the steepest decline in the curve occurs for filler metal compositions lying in the L_1 region of the Ag–CuO_x phase diagram when molten. With the exception of pure silver, which displays a non-wetting contact angle of 105° on the alumina, nearly all of the contact angles measured for this composition range are smaller on the alumina than on YSZ by 5°–10° [67, 91, 92, 97, 104]. Additionally, wetting continues to improve, albeit more gradually, with further addition of CuO (forming compositions lying within the Ag–CuO_x miscibility gap) [104]. The reason for this appears to be due to an interfacial CuO–Al_2O_3 reaction product that forms along the original faying surface, as seen in Figure 4.9. At a filler metal composition of 8–10 mol% CuO, remnants of the CuO-rich L_2 liquid are readily apparent along the $CuAlO_2$ reaction layer. Thus, the transition in wetting behavior

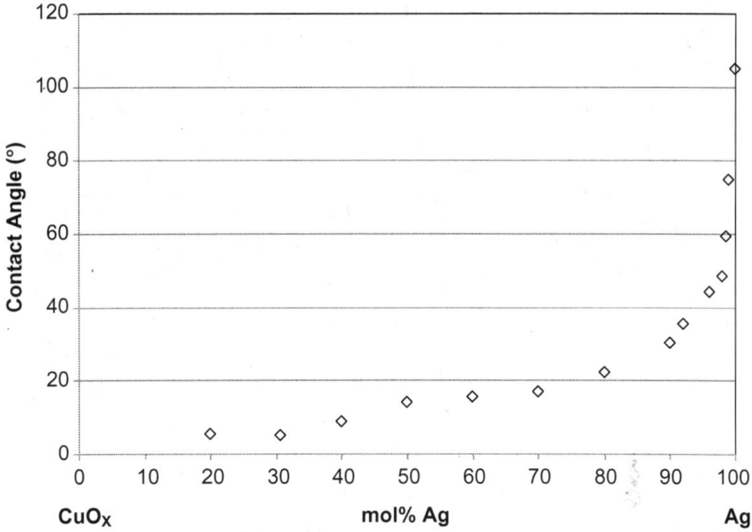

Figure 4.8. Contact angle versus copper oxide composition for Ag–CuO filler metal compositions on Al_2O_3 at 1100°C.

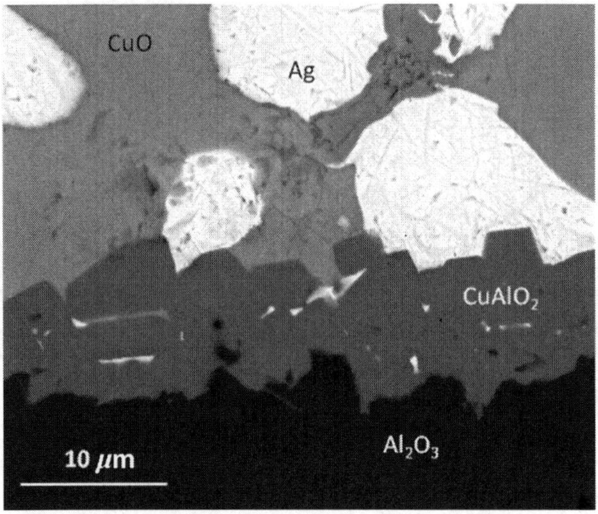

Figure 4.9. Cross-sectional BSE image of the filler metal/substrate interface in a Ag–CuO/Al_2O_3 sessile drop specimen. The filler metal is composed of 40% CuO in Ag, and the experiment was conducted in air at 1100°C.

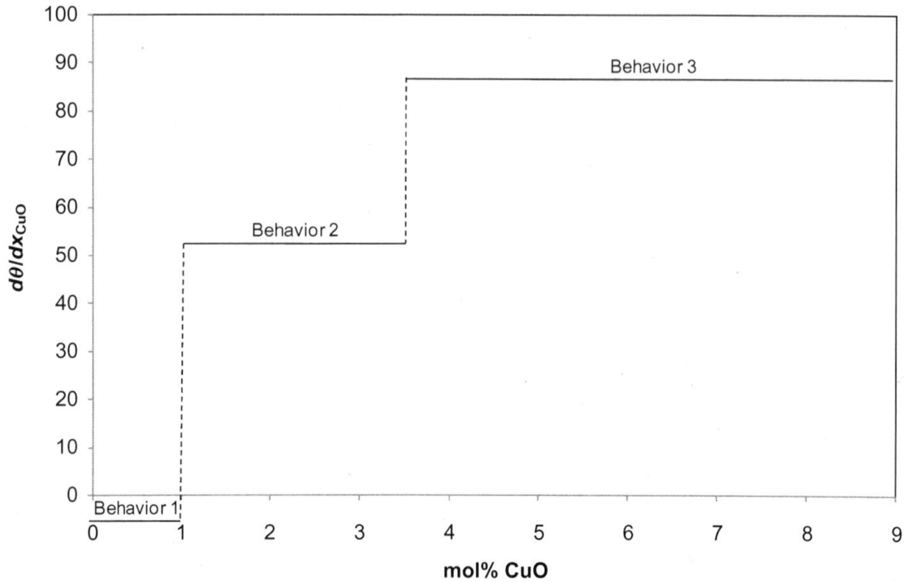

Figure 4.10. The derivative of the contact angle with respect to copper oxide composition as a function of copper oxide concentration for the original wetting plot in Figure 4.9.

should be expected to be "delayed" to higher CuO concentrations, which in fact happens. Plotted in Figure 4.10 is the derivative of the contact angle with respect to CuO concentration as a function of filler metal composition. Three regions of wetting behavior are apparent from this graph:

- Region 1 corresponds to the behavior of the single-phase silver-rich liquid.
- Region 2 is defined by compositions lying within the miscibility gap region that fully react with the substrate (via the newly formed CuO-rich L_2 liquid) but generate little to no "free" L_2 liquid.
- Region 3 corresponds to stable retention of some L_2 liquid that lies in contact with the aluminate reaction layer (in accordance with the critical point wetting theory).

Prior work on Cu–Cu_2O wetting of Al_2O_3 also found $CuAlO_2$ forms at the original faying surface and leads to a reduction in the solid–liquid interfacial energy γ_{SL} [41]. Given the constant decline in wetting angle with increasing CuO concentration in this last wetting regime, it might be expected that the reaction layer somehow changes with filler metal composition. However, neither this nor the kinetics of the aluminate reaction has been explored in detail. Both should be readily tractable using the same approaches employed in investigating reaction layer formation in the active metal brazing of ceramics [14, 48, 51, 52, 105].

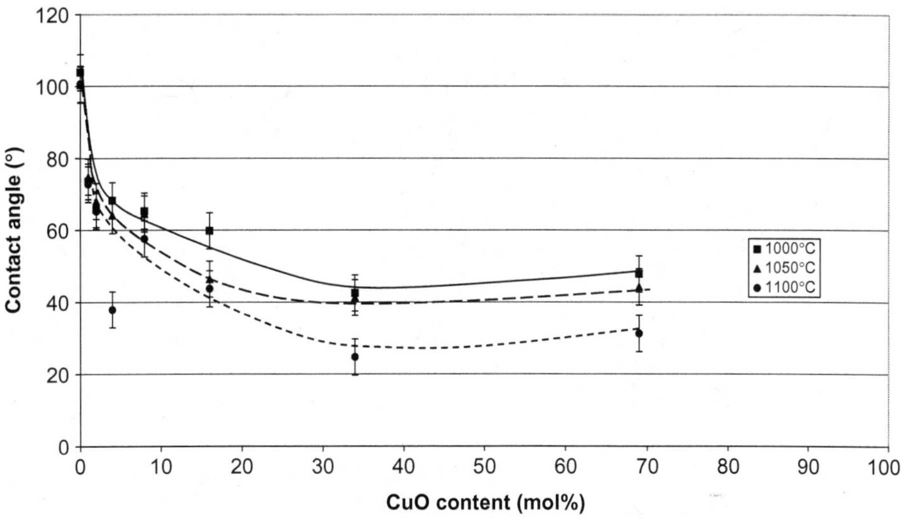

Figure 4.11. Contact angle versus copper oxide composition for Ag–CuO filler metal compositions on LSCoF at three temperatures, 1000, 1050, and 1100°C (from Reference 67).

The wetting of LSCoF substrates by Ag–CuO is also similar to that of YSZ. As shown in Figure 4.11, a steep decrease in contact angle with increasing CuO content is again observed for filler metal compositions lying in the L_1 region of the Ag–CuO_x phase diagram [73, 88]. However, for miscibility gap compositions, wetting on LSCoF is somewhat poorer than that of YSZ or alumina; with contact angles as high as 25°–30° at very high CuO concentrations. As with wetting of the alumina, the behavior is due to the interaction between the LSCoF substrate and the CuO-rich liquid in the two-phase molten filler metal. However, rather than a distinct chemical reaction, the liquid dissolves the substrate, predominantly along surface-accessible grain boundaries. In essence, this forms a series of pores or pits that the silver-rich L_1 phase subsequently infiltrates, as seen in the cross-sectional micrographs of Figure 4.12a. A similar phenomenon occurs with BSCoF substrates, as shown in Figure 4.12b. The surfaces of other "dissolution prone" substrates such as chromia tend to dissolve more uniformly. Note that dissolution occurs when the filler metal is molten (i.e., during brazing), not upon subsequent elevated temperature use. As will be explained in greater detail, the degree of dissolution can be controlled by tailoring the brazing conditions. Whereas with Al_2O_3 a new interfacial phase forms that is more readily wetted by the remaining filler metal, here, partial dissolution of the substrate into the filler metal causes a reduction in wettability.

Analogous reaction and dissolution effects occur between Ag–CuO filler metals and oxidation-resistant metal substrates. As outlined by Yang et al. [106], non-noble versions of these alloys form a stable, adherent oxide scale that protects the underlying metal substrate from further oxidation. These scales are typically alumina, chromia, or silica based. The former two have been examined extensively for alloys used in

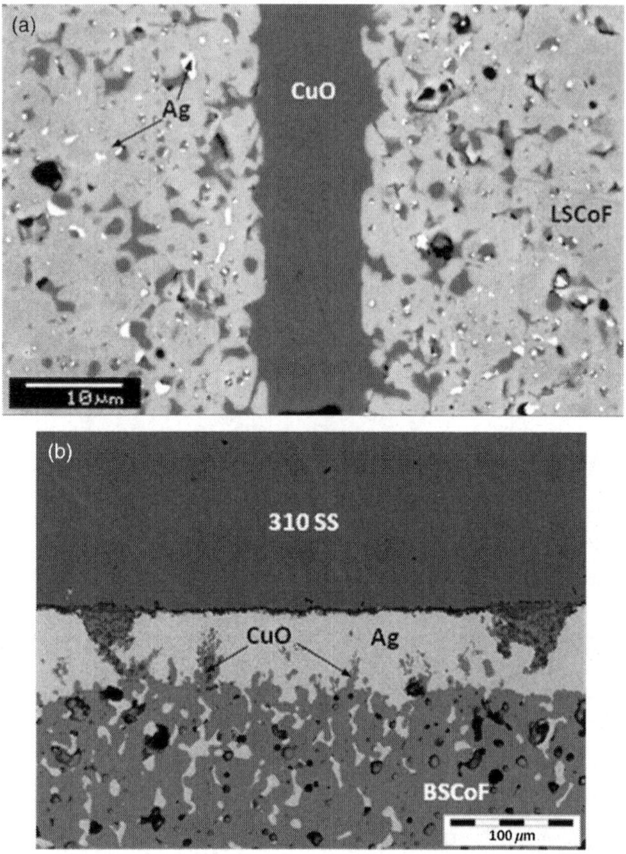

Figure 4.12. Cross-sectional BSE images of joining specimens prepared from (a) LSCoF substrates and (b) BSCoF and 310 stainless steel (SS) substrates. The filler metal employed in (a) was composed of 69 mol% CuO in Ag, while that employed in (b) was 8 mol% CuO in Ag. Both joints were prepared by brazing in air at 1000°C for 10 min under a heating rate of 5°C/min.

high-temperature electrochemical applications and form the experience base in air brazing metal-to-ceramic joints. Alloys that have been joined using air brazing include 430 stainless steel (SS), 441 SS, 446 SS, 310SS, Fecralloy™, FeCrAlY, Kanthal, DuraFoil™, Nicrofer 6025, Al 29-4, Haynes 230, Haynes 214, Inconel 690, E-brite 26-1, and Ducrolloy. The composition-dependent wetting behaviors of Ag–CuO on typical alumina-scale formers such as Fecralloy and Haynes 214 tend to be similar to that of polycrystalline alumina [91, 93]. Analogously, wetting on chromia formers is closer to that of chromia and perovskites substrates; that is, a small amount of Cr_2O_3 dissolution occurs over an average 5- to 15-min braze cycle, leading to wetting curves similar to Figure 4.11. Shown in Figure 4.13 are examples of YSZ/FeCrAlY (alumina-scale former) and YSZ/446 SS (chromia-scale former) joints. It should be noted that longer

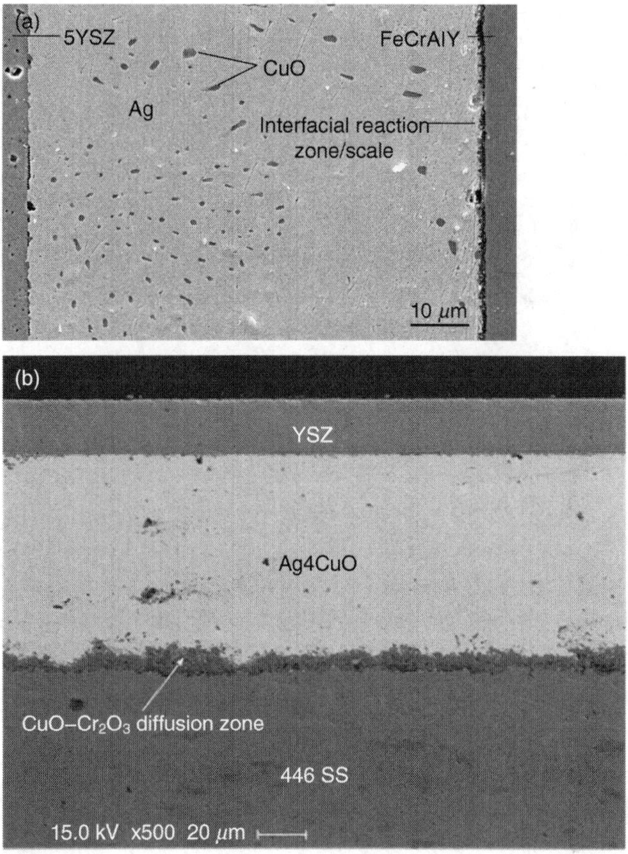

Figure 4.13. Cross-sectional BSE images of air brazed: (a) YSZ/FeCrAlY and (b) YSZ/446 SS joints. The filler metal employed in both cases was 4 mol% CuO in Ag. In the YSZ/FeCrAlY joint, brazing was conducted in air at 1000°C for 15 min, whereas in the YSZ/446 SS joint brazing was conducted at 980°C for 5 min. In both cases, heating and cooling rates of 10°C/min were used.

cycles can lead to a significant consumption of the protective scale in the liquid filler metal, causing oxidation defects of the type shown in Figure 4.14. In this case, dissolution of the chromia scale leads to local depletion of Cr in the underlying ferritic alloy. As shown, when subsequently exposed to a high-temperature oxidation environment, the resulting iron-rich metal forms a rapid-growing, nonprotective porous iron oxide scale that eventually will eventually spall or crack and ruin the integrity of the joint.

Joint Strength

While understanding the phase equilibria and the wetting characteristics for a given filler metal system is important in its design and application, joint strength and, in many

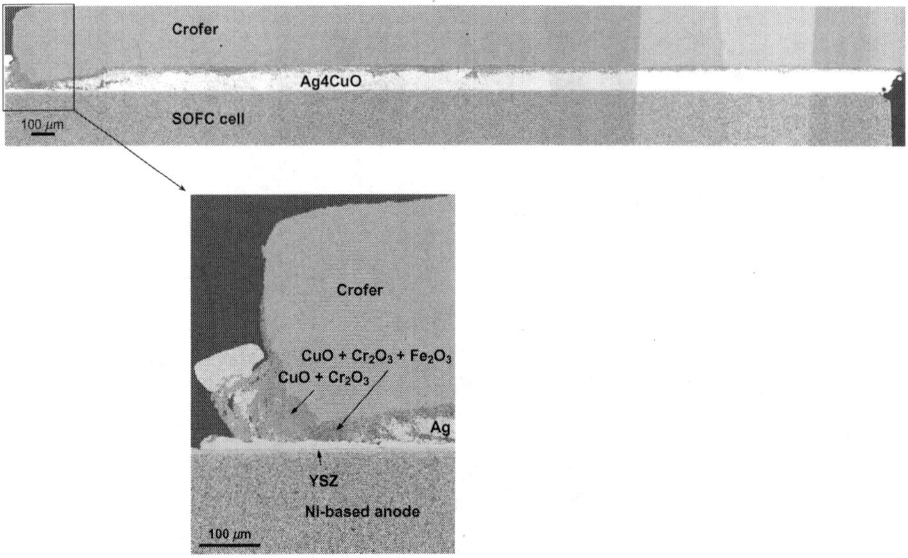

Figure 4.14. Cross-sectional BSE images of an air brazed YSZ/Crofer joint in a SOFC fuel cell assembly. The filler metal employed was 8 mol% CuO in Ag and brazing was conducted in air at 1000°C for 90 min. Heating and cooling rates of 3°C/min were used.

electrochemical applications, hermeticity are the true tests of material functionality. Bonded joints are frequently the weakest link in many assemblies. Therefore, it is critical to understand how material and process parameters affect joint strength and joint quality. This is typically determined by experimental testing and subsequent failure analysis. Modeling analysis is useful in applying this information to subsequent joint design. As discussed for wetting, the success of a ceramic–ceramic joint depends predominantly on the nature of the interactions between the filler metal and substrates. While the generation of an interfacial reaction layer generally promotes wetting with the filler metal [58, 59], the interfacial product may itself be mechanically weak [59, 107]. Alternatively, an interfacial layer may partly relieve residual stresses that can arise due to a thermal expansion mismatch between the filler metal and substrate. This is primarily true if the coefficient of thermal expansion (CTE) of the interfacial product lies between those of the adjacent materials [58, 108]. Thermal mismatch stresses can also be reduced via material deformation, for example, in a thermally softened silver-based filler metal.

Joint integrity is also influenced by the extent of filler metal coverage between the two assembled components [58, 59, 109]. As discussed in the previous section, this phenomenon is related to the ability of the filler metal to wet the components and to fill the interjacent gap via capillarity. Poor wetting leads to poor substrate coverage, poor capillary filling, and a porous joint. Porosity can also arise from the generation of gases during the brazing process and/or from the volumetric shrinkage of the filler metal during cooling and solidification. Gases can evolve during joining if fluxes or binders are employed due to the thermal decomposition of these materials. Trapped gases will expand upon heating according to the ideal gas law and may escape if the path length through the filler metal is sufficiently short. While the quantitative value of "sufficiently short" is material and temperature dependent, an estimate of 1 mm is often used [22]. In applications that require the use of a filler metal paste, a low-temperature bake-out step is generally incorporated into the brazing cycle to eliminate the binder.

Typical conditions used in air brazing a simple joining couple are (1) fix the joint under a loading pressure of ~0.2 psi, (2) heat in air at 5°C/min to a soak temperature of 1000°C and hold for 30 min or less, and (3) cool at 5°C/min. As will be discussed, this is often modified to match a specific application. Shown in Figure 4.15a–c are representative cross-sectional backscattered electron images of YSZ–YSZ joints brazed, respectively, using a silver-rich composition lying in the single liquid phase region of the Ag–CuO_x phase diagram, a composition from the miscibility gap region, and a CuO-rich composition from the single liquid phase region on the monotectic side of the miscibility gap [73, 88]. As seen in Figure 4.15a, specimens joined using the first composition exhibit little to no interfacial porosity. Small interfacial copper oxide precipitates, as well as a small amount of copper oxide within the bulk silver matrix of the joint, are also typically found. Joints prepared from a miscibility gap filler metal

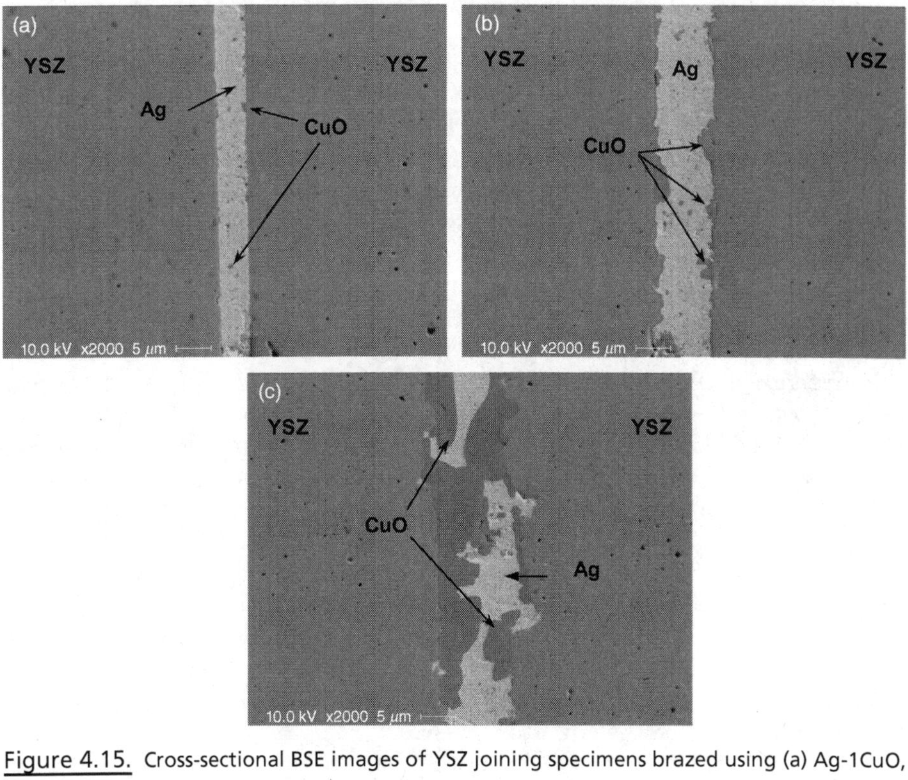

Figure 4.15. Cross-sectional BSE images of YSZ joining specimens brazed using (a) Ag-1CuO, (b) Ag-4CuO, and (c) Ag-32CuO (from Reference 97).

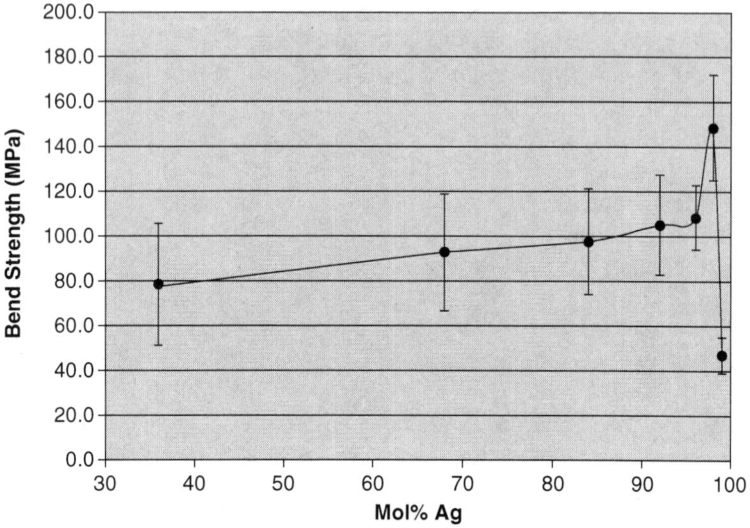

Figure 4.16. Four-point bend strengths for Ag–CuO air brazed 8YSZ–8YSZ joints as a function of CuO content (from Reference 97).

composition (Fig. 4.15b) are characterized by a microstructure that results from liquid-phase separation during brazing. Again, the copper oxide-rich liquid phase preferentially wets the two ceramic substrates in the joint and correspondingly forms a thin copper oxide-rich interfacial layer upon cooling. The joint prepared using the CuO-rich filler metal composition (Fig. 4.15c) exhibits a microstructure composed of both large, continuous CuO- and Ag-rich regions that span across the two substrates.

Shown in Figure 4.16 is a plot of four-point bend strength as a function of filler metal composition for a set of YSZ–YSZ joints tested at ambient temperature [97, 110]. The bend strength is lowest for joints brazed with the 1 mol% CuO filler metal composition and rises to as high as 157 ± 24 MPa, within 81% of that measure in the monolithic YSZ (reported as 193 ± 19 MPa). As the copper oxide content is increased to 4 mol% CuO, the average bend strength decreases to 109 ± 17 MPa and from there declines to ~80 MPa as additional CuO is incorporated into the filler metal. Thus, for nonreactive substrates, there are three regions of composition-dependent joint strength behavior that can be identified: (1) a sharp increase in bend strength as CuO is added to pure silver up to (2) a critical composition at which bend strength is maximized, followed by (3) a sharp decline in bend strength to a value that remains approximately constant over a wide compositional range. A comparison of Figure 4.16 with the Ag–CuO_x phase diagram in Figure 4.4 indicates that the optimal filler metal composition corresponds to the miscibility gap boundary composition at the temperature of brazing, that is, the composition at which wetting is maximized and the amount of residual CuO found along the filler/metal YSZ interface is minimized.

The reason for this is apparent from the general fracture characteristics of joints representing each of the three categories of mechanical behavior. Shown in Figure

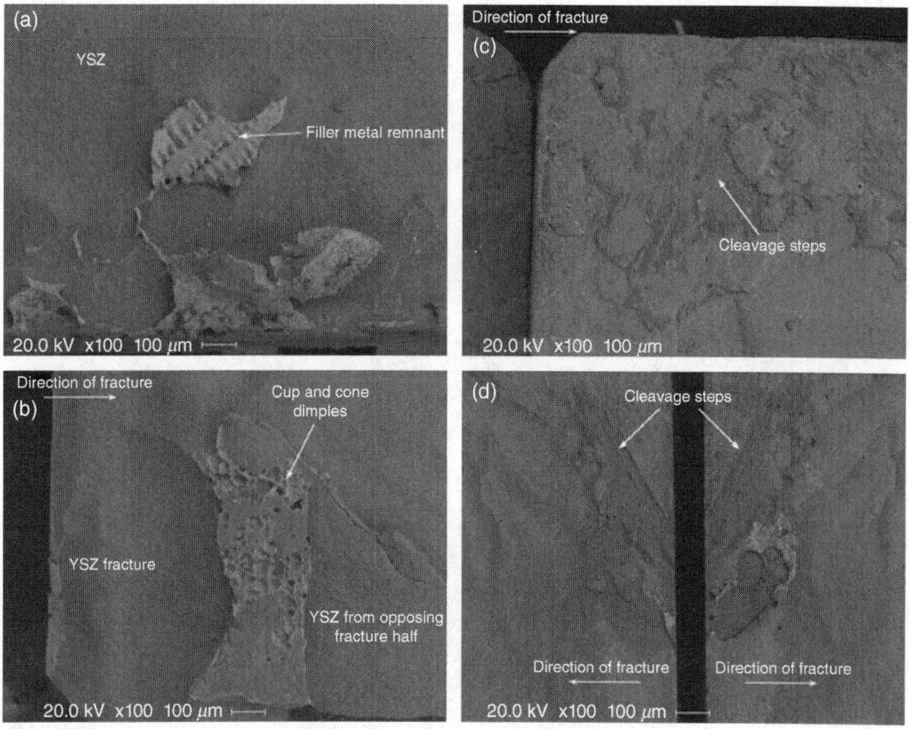

Figure 4.17. BSE images of the fracture surfaces of 8YSZ–8YSZ four-point bend specimens brazed with (a) Ag1CuO, (b) Ag2CuO, (c) Ag8CuO, and (d) Ag32CuO. All of the specimens were joined under the same conditions: heat in static air at 5°C/min to 1000°C, hold at the soak temperature for 60 min, and cool to room temperature at 5°C/min (from Reference 97).

4.17a–d are secondary electron images of the fracture surfaces for a series of YSZ–YSZ four-point bend specimens tested to failure. In specimens brazed with an unoptimized Ag-rich filler metal, there is evidence of ductile peeling between the filler metal and YSZ substrate, of the type seen in Figure 4.17a. This indicates a lack of sufficient adhesion (weak bonding) across the filler metal/YSZ interface. In the case of the specimen brazed using a miscibility gap composition very near the Ag-rich boundary (Fig. 4.17b), failure originates within one of the YSZ substrates and propagates through the thickness of the filler metal and into the second YSZ substrate. The increase in the strength of this joint is due not only to an improved level of wetting but also to a high degree of adhesion across the filler metal/YSZ interface.

The fracture surfaces of specimens brazed with filler metals containing 8 and 32 mol% CuO are shown in Figure 4.17c, d, respectively. The first is a composition that lies well within the miscibility gap when molten and the second a composition from the single-phase, CuO-rich L_2 region at braze temperature. In both cases, signs of cleavage fracture are clearly evident. This, coupled with the general lack of silver present on each of the surfaces, indicates brittle failure of the joints, ostensibly through a nearly

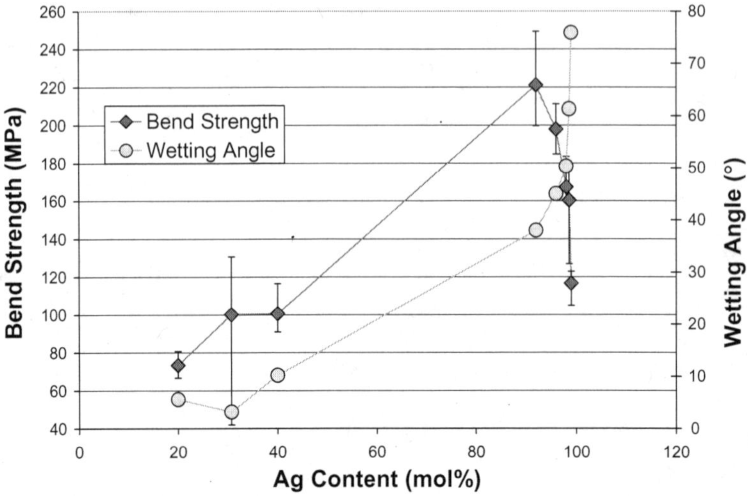

Figure 4.18. Room-temperature four-point bend strength and contact angle for a series of Al_2O_3 specimens as a function of filler metal composition. All of the specimens except Ag80CuO were joined under the same conditions: heat in static air at 5°C/min to 1000°C, hold at the soak temperature for 30 min, and cool to room temperature at 5°C/min. The Ag80CuO filler metal specimens were heated to a soak temperature of 1100°C (from Reference 92).

continuous layer of copper oxide that forms due to the liquid–liquid immiscibility phenomenon previously described. Thus, the strengths of joints prepared using filler metal compositions found well within the Ag–CuO miscibility gap or the L_2 region are dominated by the mechanical behavior of CuO, which accounts for the statistically flat region of the corresponding bend strength curve in Figure 4.16.

The joint strength/composition curves for reactive substrates exhibit trends that are similar to those found for the nonreactive YSZ substrate. As seen in Figure 4.18, the strengths of Ag–CuO brazed Al_2O_3–Al_2O_3 joints reach a maximum value that is nearly 80% that of the monolithic alumina material [92]. This occurs at a filler metal composition that is higher in CuO concentration than that found at the miscibility gap boundary at the temperature of brazing. According to the data in Figure 4.10, this corresponds to the transition between Regions 2 and 3 of the Ag–CuO/Al_2O_3 wetting curve; that is, the composition at which a continuous $CuAlO_2$ reaction layer has formed, the reaction with the CuO-rich liquid is curtailed, and an intruding layer of CuO along the interface with the aluminate reaction layer begins to appear in the joint microstructure. At higher CuO concentrations, joint strength declines eventually to a value that is only ~25% that of the substrate material at [CuO] = 0.80. Fracture analysis indicates that filler metal debonding occurs in specimens brazed with filler metals containing 8 mol% CuO or less (Region 1 in Fig. 4.10), while CuO fracture occurs in joints brazed using filler metals with CuO concentrations greater than ~8 mol% CuO (Region 3 in Fig. 4.10) [93]. Strength is maximized when joint failure occurs via ductile fracture of the silver matrix in the filler metal [93].

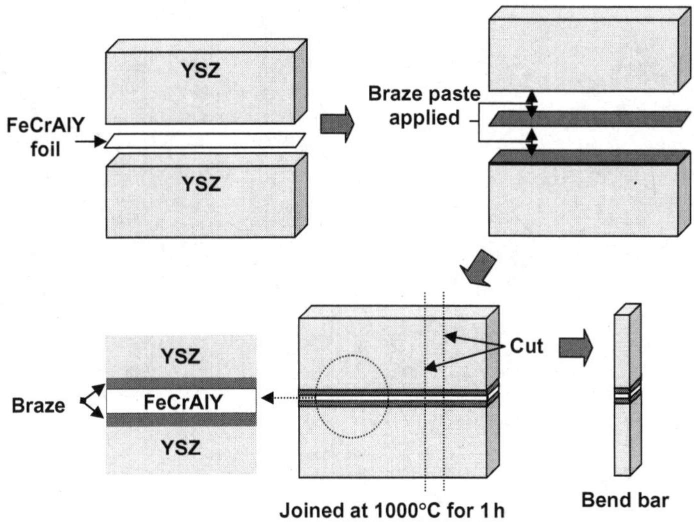

Figure 4.19. Schematic of a YSZ/FeCrAlY/YSZ four-point bend test joint.

Bend test results from joints fabricated with both nonreactive and reactive substrates exhibit trends similar to the Al_2O_3–Al_2O_3 joints. Tests conducted by Kim et al. [93] on YSZ/Fecralloy/YSZ joints of the type shown in Figure 4.19 yielded the strength-composition curve shown in Figure 4.20. As found with Al_2O_3 joining, joint strength is optimized at a filler metal composition just inside of the Ag–CuO_x miscibility gap at the temperature of brazing. However, in comparison with YSZ or Al_2O_3 joining, the YSZ/Fecralloy/YSZ joints exhibit substantially lower bend strengths (~60% of the reported monolithic strength for the YSZ material used in these specimens [93]), particularly at low CuO concentrations, indicating a difference in failure mechanism. A cross-sectional BSE image of a joint brazed with the Ag8CuO composition is shown in Figure 4.21. Note the formation of a continuous $CuAlO_2$ reaction layer (2–5 μm thick) adjacent to the thin Al_2O_3 scale layer on the FeCrAlY. Subsequent fracture analysis of a corresponding set of fracture specimens shows that failure is complex and occurs by competing mechanisms, including fracture and crack propagation along the interface between the alumina scale and underlying Fecralloy substrate, fracture of the YSZ substrate, and, to a smaller extent, fracture occurring between local regions of braze filler material and the copper aluminate reaction product. Whereas the strength of an air brazed ceramic–ceramic joint is dependent on the following:

- the magnitude of filler metal/substrate adhesion,
- the deformation behavior of the silver-based filler metal (i.e., its ductility and yield strength),
- the strength of the ceramic substrate and/or any continuous ceramic layers within the joint, and
- the magnitude and nature of residual stresses in the joining components,

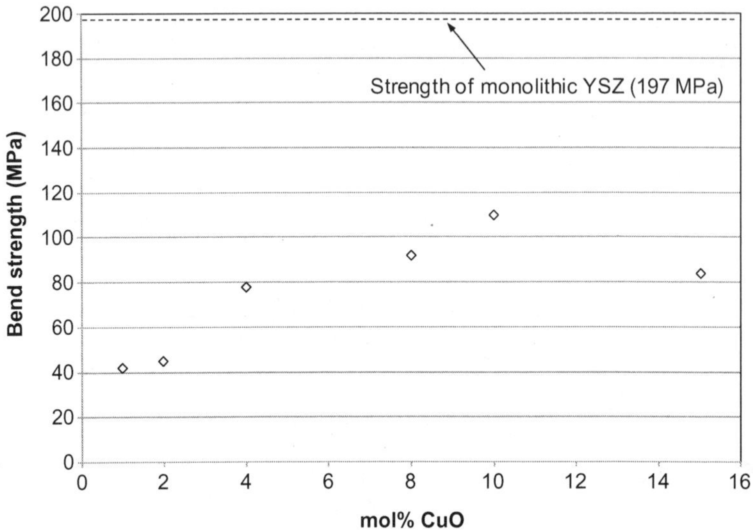

Figure 4.20. Room-temperature four-point bend strength of air brazed YSZ/FeCrAlY/YSZ specimens as a function of filler metal composition. All of the specimens were joined under the same conditions: in air at 1000°C for 60 min using 3°C/min heating and cooling rates (from Reference 93).

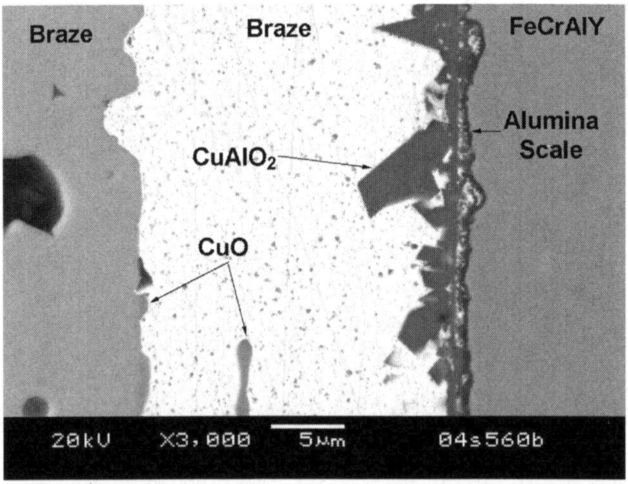

Figure 4.21. Cross-sectional BSE image of an air brazed YSZ/FeCrAlY/YSZ joint brazed using an 8 mol% CuO in Ag filler metal in air at 1000°C for 60 min. Heating and cooling rates of 3°C/min were used (from Reference 93).

the strength of an air brazed ceramic–metal joint additionally depends on the following:

- the strength of bonding at the scale–metal interfaces,
- the CTE contribution of the metal substrate (i.e., the residual stress gradient across the joint is not symmetrical), and
- the deformation behavior of the metal substrate.

In addition to filler metal composition, the conditions of brazing (e.g., the soak temperature, soak time, and rates of heating and cooling used in the braze cycle) can affect joint quality and strength. Shown in Figure 4.22a–d are a series of micrographs taken on a set of BSCoF–Nicrofer joints brazed using Ag-8CuO at several different conditions [72]. The corresponding brazing conditions are listed in Table 4.1. As seen here and in the LSCoF wetting trials, cobalt-containing ceramics can be particularly sensitive to dissolution in the Ag–CuO filler metal. The joint shown in Figure 4.22a was produced using an Ag-8CuO filler metal under air brazing conditions often employed for nonreactive substrates. Note that excessive grain boundary dissolution occurs in the BSCoF substrate and the porosity that subsequently forms wicks away much of the filler metal producing behind an extremely porous, low-quality joint. In an attempt to minimize substrate dissolution, the soak time and temperature were both reduced (the latter to zero), producing the joint found in Figure 4.22b. Unfortunately, this leads to only partial liquation of the filler metal and, as a result, large pores remain inside the joint where portions of the filler metal did not melt and wet properly. It was found that substrate dissolution essentially could be eliminated by increasing the heating and cooling rates to minimize the amount of time that the substrate is exposed to the molten braze. However, as seen in Figure 4.22c, an excessive rate of cooling can lock in high residual stresses at the filler metal/substrate interfaces (by not allowing appreciable annealing or plastic deformation to take place in the filler metal) and/or induce thermal shock conditions causing the ceramic component to crack. This can be mitigated by adjusting the cooling rate. In the case shown in Figure 4.22d, this was done by rapid cooling (100°C/min) to 930°C, a temperature where the filler metal remains

TABLE 4.1. Conditions Employed in Air Brazing the BSCoF/Nicrofer 6025 Joints Shown in Figure 4.23a–d

Joint in Figure 4.23	Soak temperature (°C)	Soak time (min)	Heating rate (°C/min)	Cooling rate
(a)	1000	15	3	3°C/min
(b)	970	0	10	10°C/min
(c)	990	0	10	100°C/min to 900°C and 10°C/min from 900 to 25°C
(d)	990	0	10	100°C/min to 930°C and 10°C/min from 930°C to 25°C

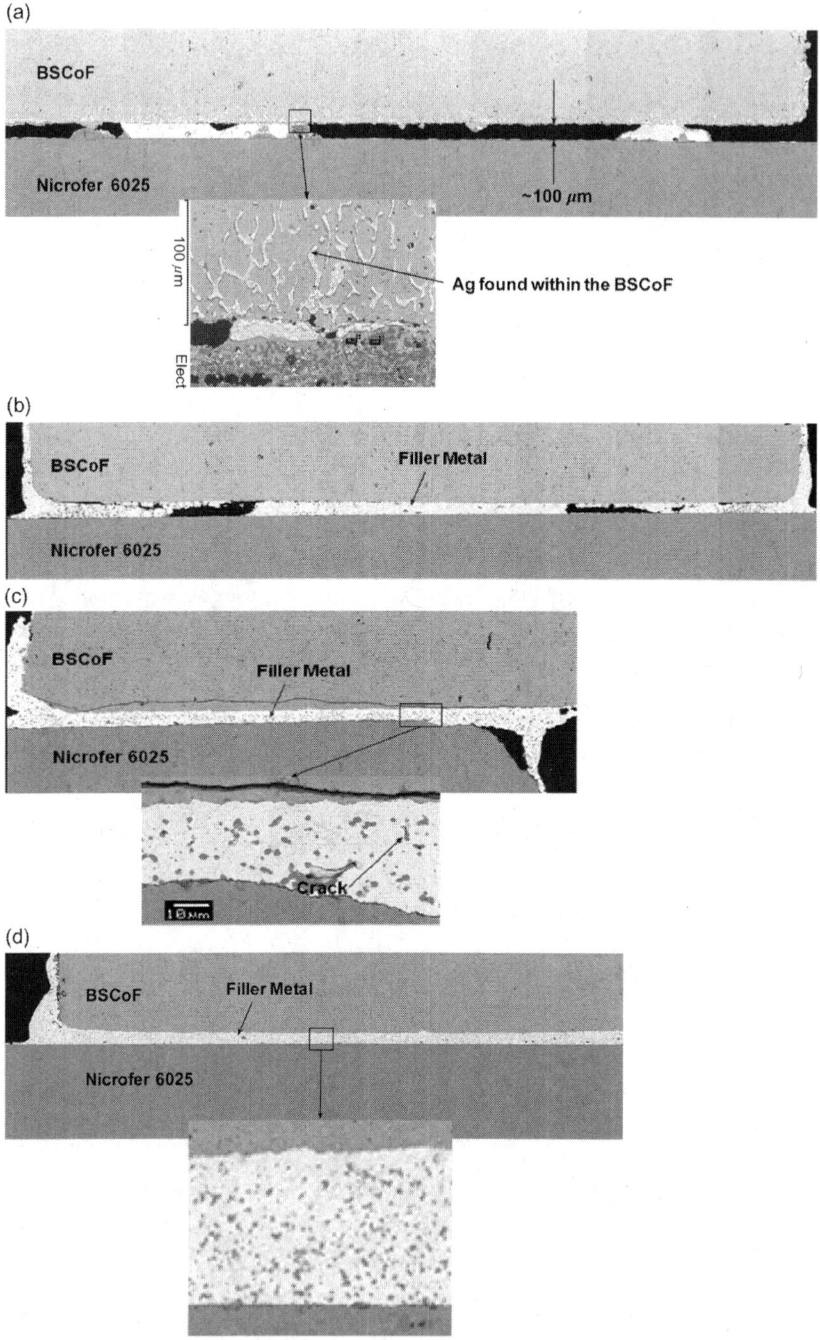

Figure 4.22. Cross-sectional BSE images of BSCoF/Nicrofer 6025 joints air brazed using Ag8CuO. The conditions employed in preparing the specimens in (a)–(d) are listed in Table 4.1.

TABLE 4.2. Recommended Air Braze Filler Metal Compositions and Conditions for Nonreactive, Reactive, and Soluble Substrates

Type of substrate	Filler metal composition (mol% CuO in Ag)	Soak temperature (°C)	Soak time (min)	Heating rate (°C/min)	Cooling rate (°C/min)
Nonreactive	3	1050	<5	5–10	5–10
Reactive	8–10	1100	15	5–10	5
Soluble	6–10	990	<5	10	10[a]

[a]A faster rate of cooling may need to be employed if dissolution is particularly rapid, such as in cobalt-containing oxides. In these cases, a cooling rate of 100°C/min to 930°C (as in Table 4.1) followed by slower cooling has been found to be suitable.

quite soft and ductile (and residual stresses are low), then reducing the cooling rate thereafter.

Recommended Air Brazing Conditions

The previous data suggest that for a given set of components to be joined, the highest joint strength achievable occurs when the wetting properties and resulting microstructure of the filler metal are simultaneously optimized. On this basis, a series of recommended filler metal compositions and brazing conditions can be identified for joining nonreactive and reactive substrates. These are listed in Table 4.2. Joints containing a dissimilar set of materials (e.g., YSZ and FeCrAlY) should employ the conditions listed for the more reactive of the two substrates (i.e., FeCrAlY in the previous example). Once the appropriate materials set have been identified, finite element modeling (FEM) can be used to optimize component design and geometry. Examples of this for air brazed components are the recent work of Weil and Koeppel [111, 112] in comparing potential SOFC sealing options for a specific window frame design and in designing the joint for a brazed YSZ electrochemical probe. As discussed in References [111] and [112], FEM analysis can be used not only to identify the highest stresses, strains, and deflections in each component (an example of which is shown in Figure 4.23) but also for the probability of joint failure via Monte Carlo simulation.

COMPOSITIONAL MODIFICATIONS TO THE Ag–CuO SYSTEM

As is done with active metal brazes, the composition of the base Ag–CuO air braze filler metal can be alloyed to enhance specific properties, such as the melting range, viscosity, wetting characteristics, and joint strength behavior. As shown by the subsequent examples, alloying can be carried out by adding soluble metal and/or nonmetal (typically oxide) constituents. In addition, composite filler metals can be produced by adding nondissolving particulate phases of various morphologies.

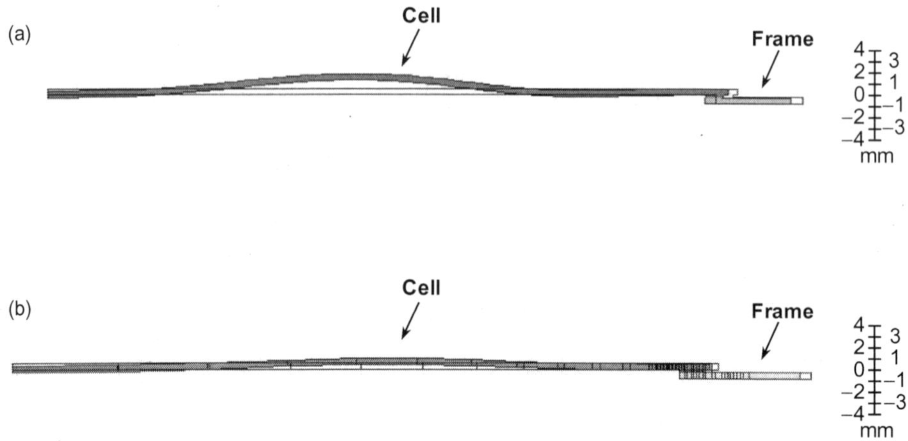

Figure 4.23. Cross-sectional representations of cell deflection (in millimeters) upon cooling from a stress-free state at 1273–273 K for an air brazed SOFC cell in a window frame design of the type shown in Figure 4.4c: (a) along the diagonal of the ceramic cell and (b) along the cell's mid-plane. The frame material considered in this analysis was Haynes 214 (CTE ~ $15.7 \times K^{-1}$). Note that the original size/shape of the components at the stress-free state are denoted by the solid black lines (from Reference 112).

Metal Alloying: Pd and Al

Because the high-temperature application of the Ag–CuO brazes is ultimately limited by the eutectic transformation that occurs in this system, a means of extending the use of temperature has been sought by adding higher-melting point noble metal elements such as palladium. As with the original development of the binary Ag–CuO filler metal system, investigation of potentially useful ternary Pd–Ag–CuO filler metals involves understanding the effects of Pd addition on phase equilibria, wetting characteristics, and joint strength behavior. Darsell and Weil [96] constructed a series of isopleths for the Pd–Ag–CuO_x system at constant concentrations of palladium, an example of which is shown in Figure 4.24, and from this extracted the liquidus and solidus information shown in Figure 4.25 [104]. The addition of palladium can increase both the liquidus and solidus temperatures by as much as 350–390°C relative to the comparable binary compositions. Similar effects would be expected in alloying Ag–CuO with platinum or gold, although the latter would only raise the range for melting very modestly.

Wetting experiments conducted on both YSZ and Al_2O_3, shown in Figure 4.26a, b, respectively, indicate that the addition of palladium to Ag–CuO *does* degrade the wettability of the resulting filler metal relative to comparable binary filler metals at equivalent temperatures [97, 98, 104]. However, as was observed with the Ag–CuO filler metals, contact angle decreases and therefore wettability improves as copper oxide content is increased for each of the filler metal families examined. As a result, nearly all of the resulting compositions are wetting, with the exception of 50PdAg–CuO filler metals containing less than 10 mol% CuO. Additionally, wetting is essentially invariant

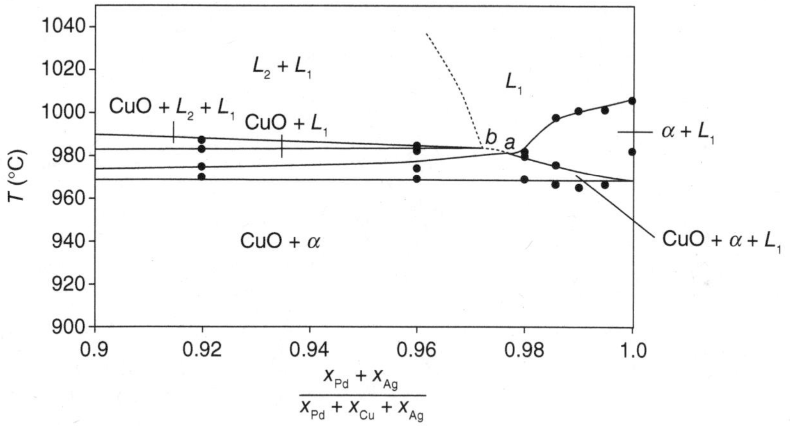

Figure 4.24. An isopleth of the pseudoternary Pd–Ag–CuO_x system at a palladium concentration of 5 mol% (from Reference 96).

with temperature for both filler metal/substrate systems once the liquidus temperature has been exceeded [93, 98, 104]. Transitions in wetting behavior for nonreactive (YSZ) substrates follow the same patterns observed for the binary filler metal, occurring at the miscibility gap boundary for each alloy system: ~2 mol% CuO for the binary system, at ~4 mol% CuO for the 5Pd system, and at ~8 mol% CuO for the 15Pd system [104]. The same does not appear to be true for reactive (Al_2O_3) wetting. For both Pd–Ag–CuO filler metal families, wetting remains approximately constant at low CuO concentrations, then undergoes rapid improvement as CuO is added beyond a critical concentration. The reasons for this phenomenon have yet to be explored.

Palladium alloying also lowers joint strength, as shown in Figure 4.27, particularly when low copper oxide-containing filler metals are employed [110]. Like the binary filler metal series shown on the same plot, the two Pd–Ag–CuO series each display an increase in average bend strength with copper oxide content up to a transition composition, the same one identified from the wetting curves. Beyond this CuO concentration, bend strength either drops slightly or approximately maintains a constant value. Fractographic analysis indicates that distinct changes in failure mechanism occur at each transition, the origins of which depend primarily on joint microstructure and the magnitude of filler metal/substrate adhesion. For example, failure in joints prepared using the 5Pd–AgCuO filler metals can be summarized as follows: (1) Over a compositional range of 0–4 mol% CuO, bend strength increases due to improved filler metal wetting and greater interfacial adherence; (2) at ~4 mol% CuO, maximum bend strength is observed; and (3) beyond ~4 mol% CuO, bend strength gradually declines with copper oxide additions due to brittle failure through a greater fraction of interfacial copper oxide. The 15Pd–AgCuO filler metals are characterized by (1) an increase in bend strength with copper oxide composition over a compositional range of 0–8 mol% CuO, primarily due to improved wetting with copper oxide addition, and (2) a near-constant bend strength over the compositional range of 8–64 mol% CuO, with failure

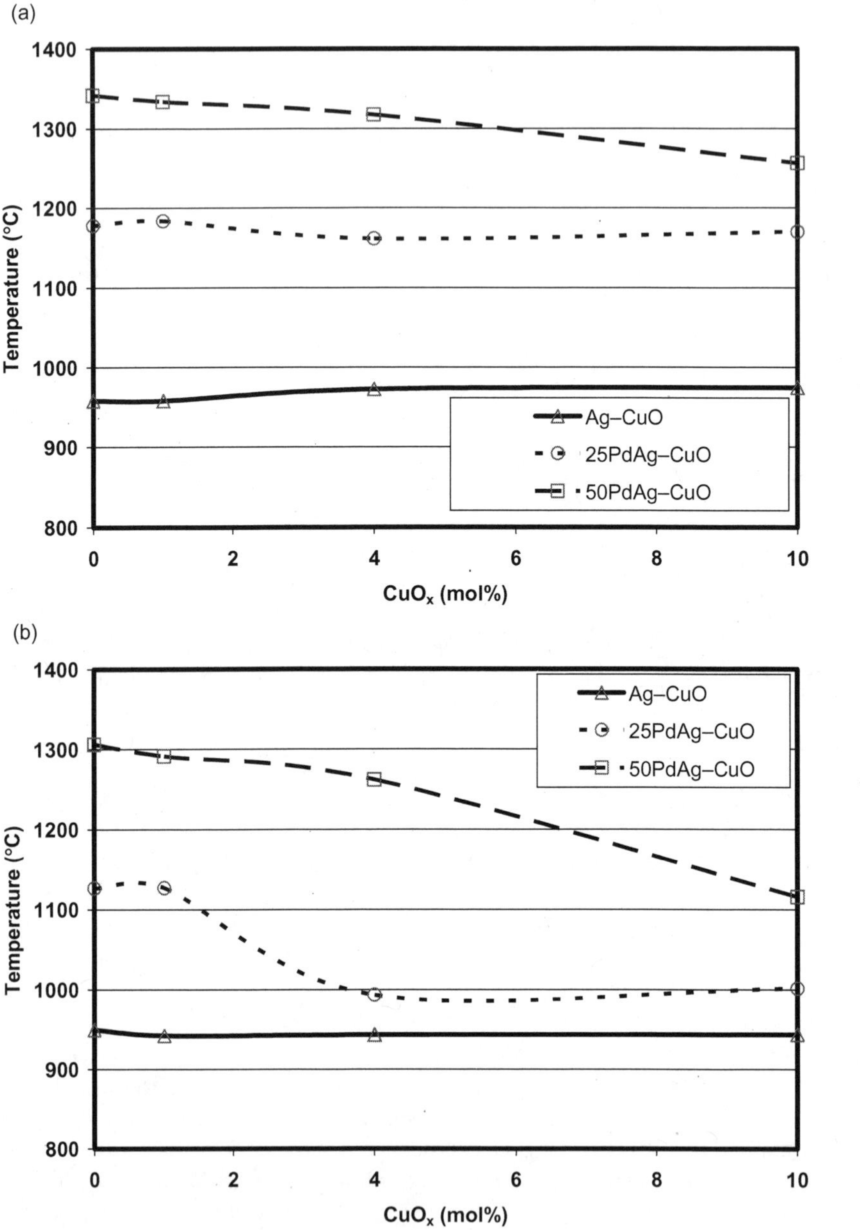

Figure 4.25. Changes in the (a) liquidus and (b) solidus temperatures as a function of CuO_x content for Ag–CuO_x with Pd addition (from Reference 104).

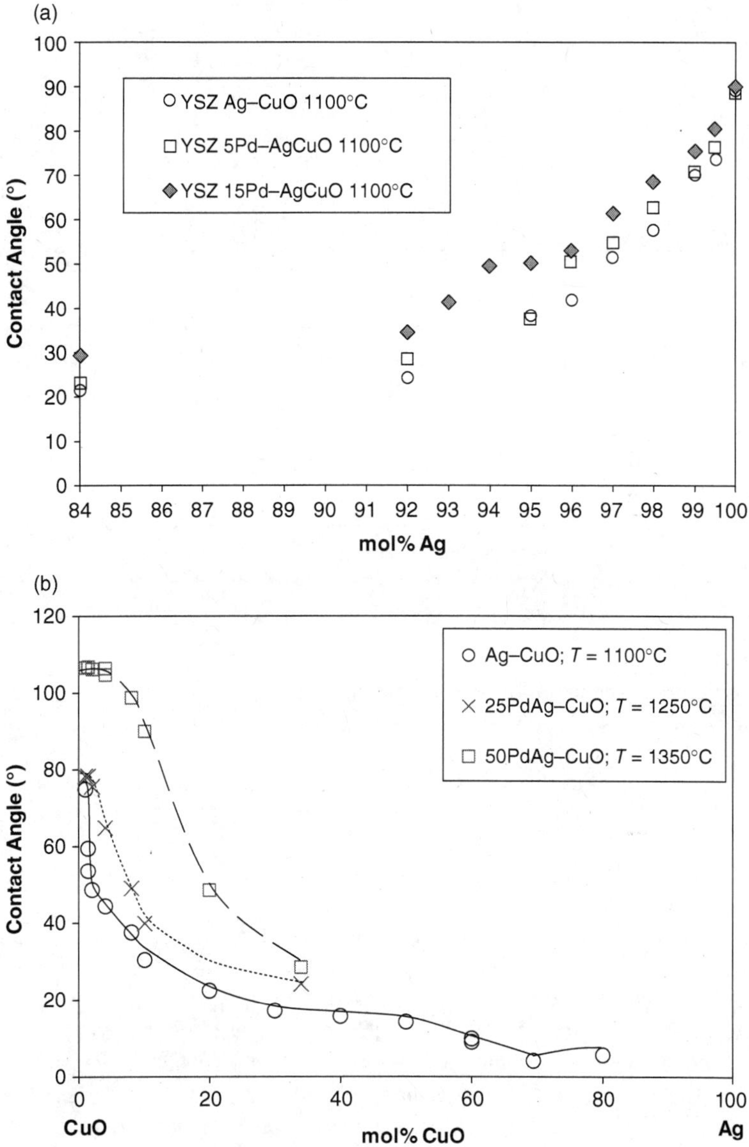

Figure 4.26. Contact angle versus copper oxide content for (a) Ag–CuO, 5Pd–Ag–CuO, and 15Pd–Ag–CuO filler metals on YSZ and for (b) Ag–CuO, 25Pd–Ag–CuO, and 50Pd–Ag–CuO filler metals on Al_2O_3 (from Reference 97).

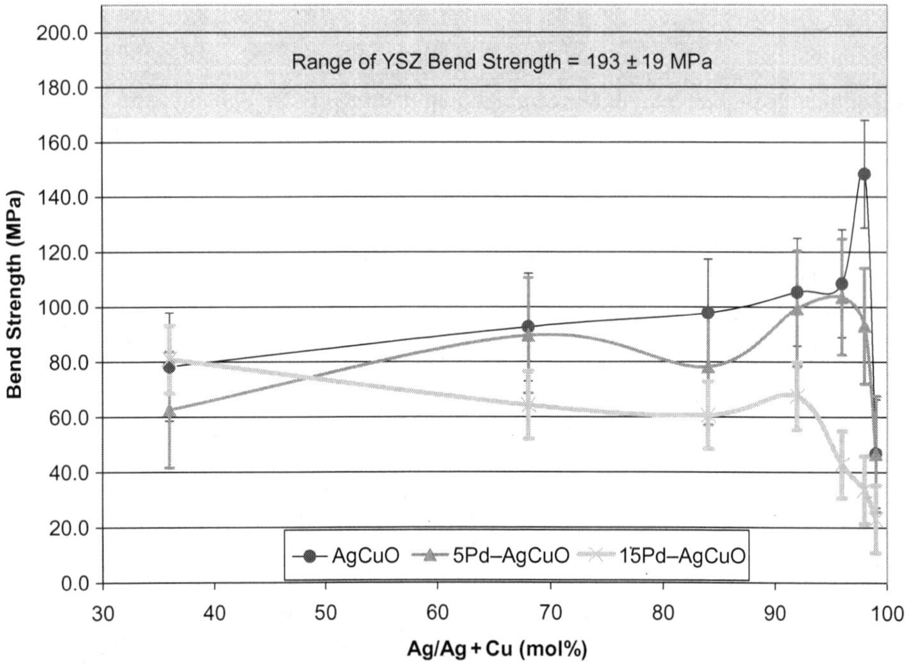

Figure 4.27. Room-temperature four-point bend strength of air brazed YSZ–YSZ as a function of filler metal composition for the Ag–CuO, 5Pd–Ag–CuO, and 15Pd–Ag–CuO series of braze filler metals (from Reference 97).

attributable predominantly to fracture through both the interfacial copper oxide-rich and substrate phases.

An additional alloying agent that has been considered in some detail is aluminum [113, 114]. It has been investigated primarily as a means of improving the dual-atmosphere tolerance of air brazed SOFC seals. In this application, the seal is exposed to fuel (H_2) on one side and air (O_2) on the other. Because both hydrogen and oxygen are relatively soluble in silver and display rapid rates of diffusion, there is concern that reaction of the two diffused species, primarily within grain boundaries, will create internal porosity and will eventually cause a loss of seal hermeticity during stack operation. Excessive pore formation has been reported in pure silver and, to a smaller extent, has also been observed in Ag–CuO-based seals after long-term dual-atmosphere exposure [115–117]. In an attempt to eliminate this phenomenon, Kim et al. [113, 114, 118] examined the effects of Al alloying in Ag and Ag–CuO on mechanical strength and dual-atmosphere tolerance. In the proper concentrations, aluminum can form a protective alumina scale on the air-facing surface of the joint (and potentially on the water saturated H_2 side as well) that can serve as a barrier to oxygen and hydrogen ingress. In the case of alloyed Ag–Al braze foils containing more than 2 at% Al, metallic aluminum was found to persist through the air brazing process and could be detected within the filler metal. Unfortunately, the flexural strengths of specimens joined using the binary Ag–Al filler metal compositions were found to decrease significantly with increasing aluminum content. This was due to the formation of interfacial aluminum oxide, which leads to interfacial debonding between the braze filler and the alumina substrate, and the formation of a majority Al–Ag intermetallic phase (β-Ag_3Al and α-Ag_3Al) for several filler metal compositions [114]. However, the existence of metallic aluminum in the filler metal matrix enhanced the high-temperature dual-atmosphere tolerance of the resulting joint. The size of the dual atmosphere-induced porosity was measurably smaller and the amount of porosity was significantly smaller in the aluminum alloyed filler metals than in those containing no aluminum. The addition of copper to the Ag–Al air braze filler metals dramatically increases joint strength to high as 143 MPa for a Ag-33Al-8CuO filler metal composition but also leads to noticeable degradation in dual-atmosphere tolerance [118,119]. Additional work is needed to optimize the composition for both acceptable strength and dual-atmosphere tolerance.

An interesting concept similar to developing Al-containing air braze filler metals has been forwarded by Kim et al. [117] and by Bengis and Fahrenholtz [120] for Al_2O_3 joining. A high-purity aluminum foil is placed in contact with the alumina work pieces and the entire assembly heated to a temperature well above the melting point of aluminum. The original surfaces of the foil convert to alumina, which then sinter to the adjacent substrates. Depending on the application of the joined material, the joining process can be tailored such that a thin, continuous aluminum layer remains within the joint, thereby providing a measure of ductility. Alternatively joining can be conducted such that the resulting joint microstructure is nearly indistinguishable from that of the substrates, except for a dispersion of residual aluminum that enhances the toughness of the joint. Prior work on the direct metal oxidation (DIMOX) process indicates that additional microstructural modification can be achieved through proper compositional design/selection of the aluminum filler metal [121]. Choi et al. [122] have extended this concept to the use of a paste containing aluminum powder that converts to γ-Al_2O3 under a joining cycle of 1000°C for 2 h.

Nonmetal Alloying: TiO_2

There are a number oxides that are known to form eutectics with CuO and/or Cu_2O [123]. For example, TiO_2–CuO exhibits a eutectic reaction at ~900°C and on this basis was selected as a potential melting point depressant for Ag–CuO_x [124]. While the phase equilibrium effect was ultimately found to be minimal, it was observed that small additions of titania display a dramatic effect on the wetting behavior of filler metal alloys containing less than ~34 mol% copper oxide as indicated by the plot of wetting angle versus composition in Figure 4.28a for a series of Ag–CuO–TiO_2 of filler materials on Al_2O_3. Similar findings have also been reported for both YSZ, Fecralloy, and LSCoF [91, 93, 125]. When the data are replotted as the differential wetting angle ($\Delta\theta$) between each ternary composition and its corresponding binary counterpart as a function of composition (Fig. 4.28b), it is apparent that the effect of TiO_2 on silver–copper oxide wetting behavior strongly depends on copper oxide content. For each curve, three regions of behavior can be identified. For example, in all three filler metal series at

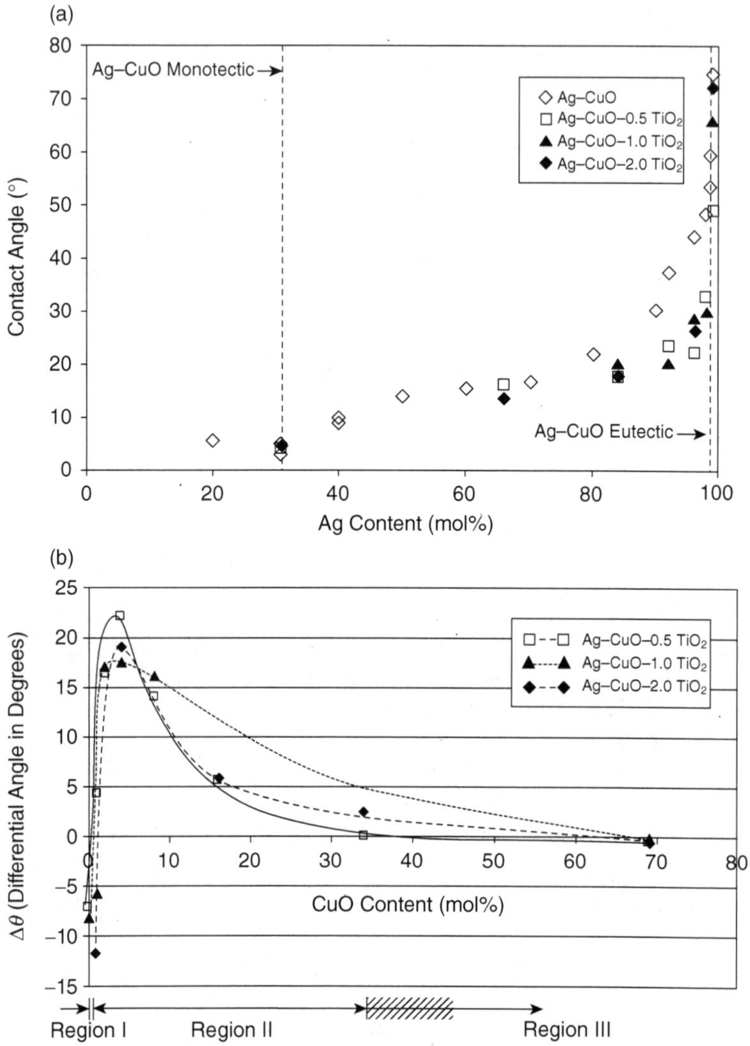

Figure 4.28. (a) Contact angle as a function of silver content for Ag–CuO filler metals with additions of 0.5, 1.0, and 2.0 mol% TiO_2. (b) Differential contact angle between each ternary filler metal composition and its corresponding binary counterpart as a function of copper oxide content. Measurements were recorded after holding at 1100°C for 15 min (from Reference 124).

copper oxide concentrations below 1% (Region I), a detrimental effect is observed. Above this threshold, wetting is synergistically enhanced, which defines the behavior in Region II. The copper oxide concentration at which the phenomenon is maximized occurs at ~4 mol% CuO. This is very close to the silver-rich miscibility gap boundary composition at the temperature considered (1100°C), which suggests that TiO_2 likely increases oxygen activity in the ternary alloy. After reaching this maximum, the degree of wetting enhancement declines and eventually disappears at copper oxide concentrations above ~34 mol%, defining the trend observed in Region III.

In the case of Al_2O_3 wetting, cross-sectional micrographs indicate that the addition of TiO_2 leads to intergranular dissolution of the substrate as shown in Figure 4.29a. If the joint is held at the brazing temperature for a period of time longer than ~5 min, filler metal penetration of the substrate can be quite extensive, up to 200–300 μm into the substrate, as seen in Figure 4.29b. As was found in brazing BSCoF, it is anticipated

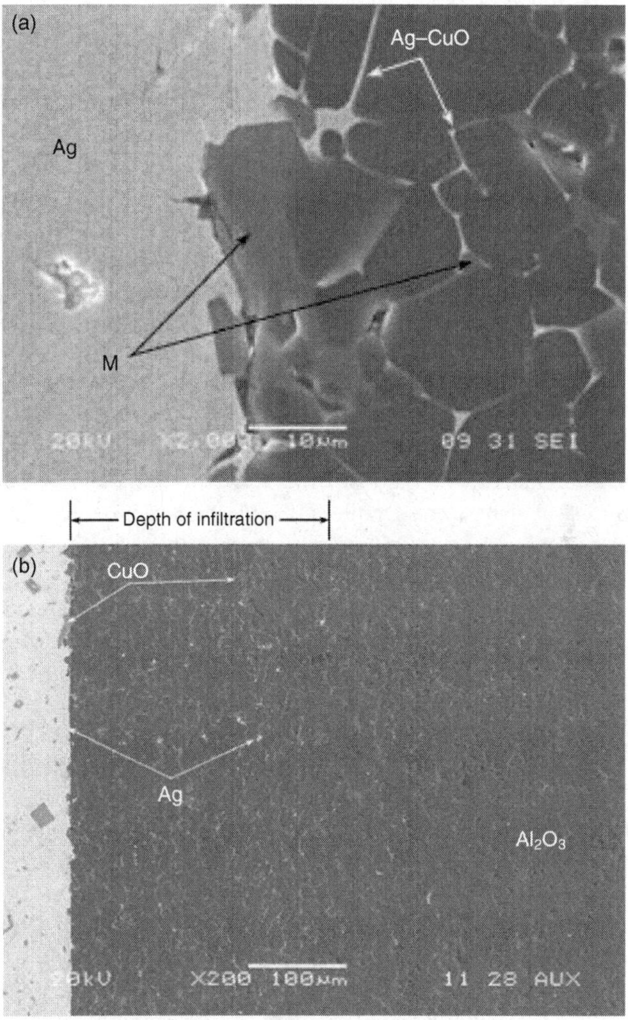

Figure 4.29. BSE images of the filler metal/alumina interface in an Al_2O_3–Al_2O_3 joint brazed using a Ag-4CuO-2TiO_2 filler metal (a) at high magnification and (b) at low magnification. Brazing was conducted in air at 1100°C for 15 min using 5°C/min heating and cooling rates.

that the use of TiO_2 modified filler metal will require short braze times (<15 min) and relatively low brazing temperatures (<1000°C) to prevent excessive substrate dissolution. Preliminary mechanical testing indicates that the strength of Al_2O_3–Al_2O_3 joints can be improved by 20–40% when as little as 0.5 mol% TiO_2 is incorporated in the filler metal. In the case of YSZ and LSCoF, the addition of TiO_2 leads to the formation of an interfacial reaction layer, titanium zirconate for the former substrate, and strontium titanate for the latter, and no substrate dissolution is observed [91, 125]. Again, preliminary joint strength testing on these substrates indicates that not only is wetting improved on each of these substrates but joint strength can also be increased as well [91], although additional work is needed to further understand and refine the effect.

Particulate Additions

A limiting constraint in any dissimilar material joint is thermal expansion mismatch. The use of a soft interlayer material can often mitigate the thermal stresses that are generated when the joint is cooled from the joining temperature, particularly if the CTE of the interlayer closely matches that of the less robust component, that is, the ceramic component(s) in a ceramic–ceramic or ceramic–metal joint. In the case of the silver-based air braze fillers, the silver constituent exhibits a substantially higher CTE than do most oxide ceramics. For example, the CTE of Ag is $22.3 \times 10^{-6}\,K^{-1}$ over 23–700°C, while the CTE of Al_2O_3 and YSZ over the same temperature range are $10.5 \times 10^{-6}\,K^{-1}$ and $8.2 \times 10^{-6}\,K^{-1}$, respectively [126]. This difference can generate high levels of mixed-mode residual stress in the ceramic joint component(s) that in turn weaken the strength of the joint. One means of lowering the thermal mismatch stresses is to reduce the CTE of the filler metal by alloying or by adding a nonsoluble, low CTE constituent such as ceramic particulate. For example, a simple rule of mixture estimate indicates that the CTE of a predominantly silver filler metal can be reduced 20% by adding ~25 vol% Al_2O_3 particulate.

Kim et al. [127] have explored this concept by examining the effects of modest particulate addition, 5 and 10 vol% Al_2O_3 (~1.6-μm average particle size) on the resulting microstructures and strengths of joints air brazed using three baseline filler metal compositions: pure Ag, 2 mol% CuO in Ag, and 8 mol% CuO in Ag. Shown in Figure 4.30a–d are cross-sectional BSE images of four representative joints. The addition of alumina particulate to pure silver simply leads to the formation of pores in the filler metal region (the black regions seen within the silver filler metal region in Fig. 4.30a, b) because of poor wetting between silver and Al_2O_3. However, when sufficient CuO is added, both the alumina particulate and alumina substrates are wetted, resulting in the filler microstructures shown in Figure 4.30c, d. It is interesting to note that the alumina particulate is uniformly dispersed throughout the filler metal, despite the difference in density between Al_2O_3 (3.97 g/cc) and Ag (10.5 g/cc).

The combination of both CuO and Al_2O_3 particulate addition to silver has a significant effect on joint strength, which is plotted as a function of filler metal composition (i.e., CuO and Al_2O_3 content) in Figure 4.31. Joints brazed using non-CuO-containing filler metal compositions all display bend strengths that are extremely low, ranging from 0 to 63 MPa. Again, this is not surprising given the poor wettability displayed between

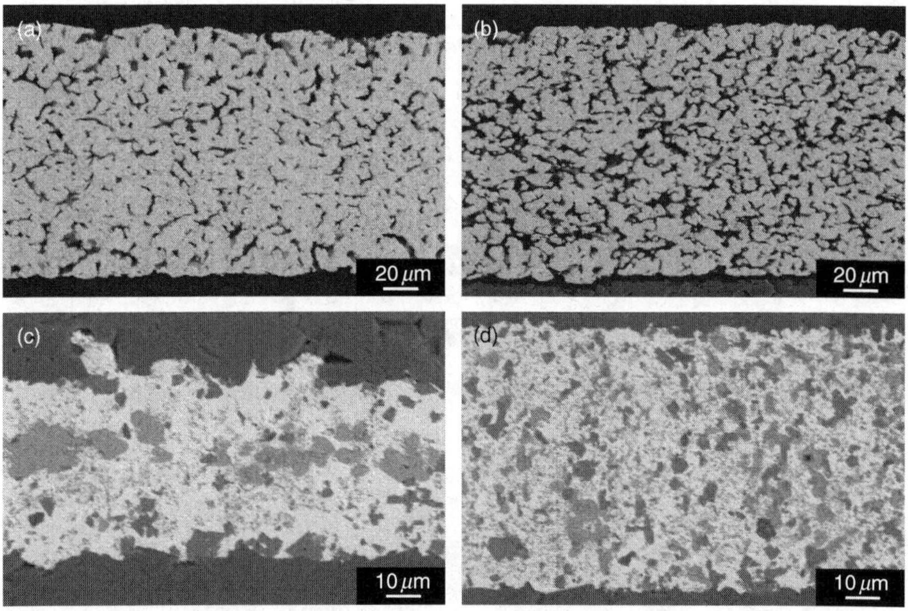

Figure 4.30. Cross-sectional scanning electron micrograph (SEM) images of the braze/alumina interfaces in Al_2O_3–Al_2O_3 joining specimens brazed with (a) 5 vol% Al_2O_3 in Ag, (b) 10 vol% Al_2O_3 in Ag, (c) 5 vol% Al_2O_3 and 8 mol% CuO in Ag, and (d) 10 vol% Al_2O_3 and 8 mol% CuO in Ag. Brazing was conducted in air at 1000°C for 15 min using 5°C/min heating and cooling rates.

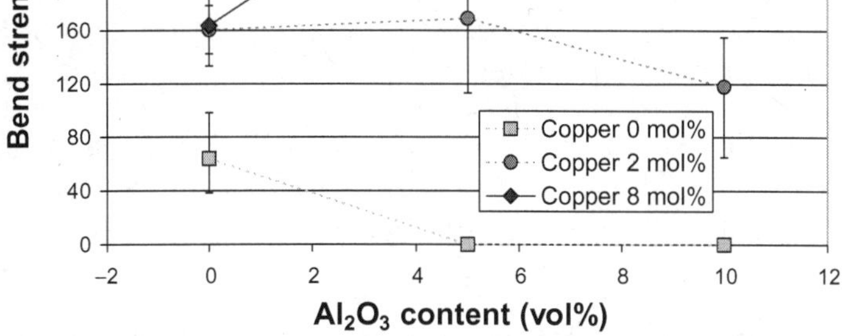

Figure 4.31. Room-temperature four-point bend strength of air brazed Al_2O_3–Al_2O_3 joining specimens as a function of alumina particulate and copper oxide content. Brazing was conducted in air at 1000°C for 15 min using 5°C/min heating and cooling rates.

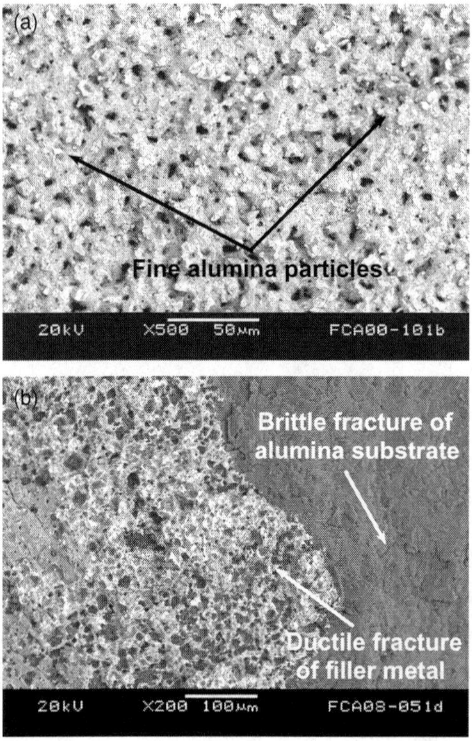

Figure 4.32. Fracture surfaces of Al_2O_3–Al_2O_3 joining specimens brazed with (a) 5 vol% Al_2O_3 in Ag and (b) 5 vol% Al_2O_3 and 8 mol% CuO in Ag. Brazing was conducted in air at 1100°C for 15 min using 5°C/min heating and cooling rates.

Ag and Al_2O_3. The two series of specimens joined using CuO-containing composite filler metals exhibit significantly higher strengths. In fact, specimens brazed with 8 mol% CuO in Ag + 10 vol% Al_2O_3 exhibit joint strengths that are essentially equivalent to the strength of monolithic alumina.

The differences in four-point strength are exemplified by differences in the corresponding fracture surfaces. As seen in Figure 4.32a, when Al_2O_3 particulate is added to a filler metal consisting only of silver, the resulting failed joint shows class signs of peeling due to poor wetting and adhesion between the filler metal and substrate. In the Ag8CuO-5 Al_2O_3 brazed specimens, shown in Figure 4.32b, fracture occurred through both the substrate and the filler metal. In addition, a significant amount of necking around the ceramic particulate can be observed on the filler metal portion of the surface, indicating that this region fails via ductile fracture. While the addition of small amounts of particulate to the binary Ag–CuO filler metals has a demonstrably large effect on joint strength, the mechanisms for this are not yet known. As discussed earlier, better CTE matching plays a role but is likely not the entire reason for improvement. Additional study is needed to understand the phenomenon and to examine possible differences between reactive and inert particulate additions, the effects of particle size and content, and possible combination with alloyed filler metals.

SUMMARY

Air brazing is an emerging ceramic–ceramic/ceramic–metal brazing technology for high-temperature applications and is currently used to bond electrically and electrochemically active ceramics that are otherwise difficult to join. In its present embodiment (i.e., using Ag–CuO-based filler metals), the technique offers several benefits including the ability to join/seal directly in air without the use of fluxes or inert cover gases and the formation of bond that is hermetic, strong, ductile, and inherently resistant to oxidation at high temperature. The development of optimal Ag–CuO filler metals is based on the knowledge of the phase equilibria for this system and of the wetting and strength characteristics for several different types of substrates, including nonreactive, reactive, and "soluble" materials. In general, for a given materials set, the strength of an air brazed joint is maximized for filler metal compositions that lie at a wetting transition point for the most reactive substrate in the joint. For nonreactive substrates such as YSZ, this the Ag–CuO_x miscibility gap boundary composition at the temperature of brazing. For reactive and soluble substrates, the optimal composition lies within the Ag–CuO_x miscibility gap (recommended filler metal compositions are listed in Table 4.2). Brazing of soluble substrates such as BSCoF and LSCoF can be tricky as soak temperature, soak time, and rates of heating/cooling significantly affect the quality of the final joint. Success in air brazing these types of materials is improved when the time for dissolution is minimized (i.e., by using a moderately low soak temperature, holding at temperature for a short period of time, and rapidly cooling to the monotectic or eutectic temperature). More recent work has shown that Ag–CuO filler metals can be readily modified by alloying with a metal or nonmetal constituent and/or by compounding with nonsoluble particulate to improve high-temperature properties, wettability, resistance to dual-atmosphere degradation, and bond strength. To expand the realm of application, new filler metal systems need to be identified and developed, and the mechanisms by which the modifiers for the Ag–CuO system need to be better understood. However, it is clear that air brazing affords a number of advantages in joining advanced multifunctional ceramics.

REFERENCES

1. Belmonte M. 2006. Advanced ceramic materials for high temperature applications. Adv Eng Mater 8:693.
2. Stachowiak GB, Stachowiak GW. 1995. Fretting wear and friction behaviour of engineering ceramics. Wear 190:212.
3. Kim D-J, Choi D-J. 1996. High-temperature corrosion resistance of chemically vapor deposited silicon carbide against hydrogen chloride and hydrogen gaseous environments. J Am Ceram Soc 79:503.

4. Locatelli MR, Tomsia AP, Nakashima K, Dalgleish BJ, Glaeser AM. 1995. Key Eng Mater 111–112:157.
5. Kakegawa K, Glaeser AM. 1997. Transient FGM joining of silicon carbide ceramics: a feasibility study. Composites Part B 28:85.
6. Mackerle J. 1999. Finite element modelling of ceramics and glass: a bibliography (1977–1998). Eng Comput 16:510.
7. Gopal M, De Jonghe LC, Thomas G. 1998. Sintering nitride: from sintering to joining. Acta Mater 46:2401.
8. Nicholas MG, Peteves SD. 1994. Reactive joining; chemical effects on the formation and properties of brazed and diffusion bonded interfaces. Scr Metall Mater 31:1091.
9. Peteves SD, Paulasto M, Ceccone G, Stamos V. 1998. Reactive route to ceramic joining: fabrication, interfacial chemistry and joint properties. Acta Mater 46:2407.
10. Akselsen OM. 1992. Advances in brazing of ceramics. J Mater Sci 27:1989.
11. Arroyave R, Eagar TW. 2003. Metal substrate effects on the thermochemistry of active brazing interfaces. Acta Mater 51:4871.
12. Hanson WB, Ironside KI, Fernie JA. 2000. Active metal brazing of zirconia. Acta Mater 48:4673.
13. Stephens JJ, Hosking FM, Headley TJ, Hlava PF, Yost FG. 2003. Reaction layers and mechanisms for a Ti-activated braze on sapphire. Metall Mater Trans A 34:2963.
14. Lahl N, Singh K, Singheiser L, Hilpert K, Bahadur D. 2000. Crystallisation kinetics in AO-Al_2O_3-SiO_2-B_2O_3 glasses (A = Ba, Ca, Mg). J Mater Sci 35:3089.
15. Tomsia AP, Pask JA, Loehman RE. 1991. Glass/metal and glass-ceramic/metal seals. In: SJ Schneider and RE Loehman, editors. *Engineered Materials Handbook, Ceramics and Glasses*, Vol. 4, Metals Park, OH: ASM International.
16. Haanappel VAC, Shemet V, Gross SM, Koppitz T, Menzler NH, Zahid M, Quadakkers WJ. 2005. Behaviour of various glass-ceramic sealants with ferritic steels under simulated SOFC stack conditions. J Power Sources 150:86.
17. Chou Y-S, Stevenson JW, Singh P. 2008. Effect of pre-oxidation and environmental aging on the seal strength of a novel high-temperature solid oxide fuel cell (SOFC) sealing glass with metallic interconnect. J Power Sources 184:238.
18. Nieh TG, Wadsworth J. 1994. Superplasticity and superplastic forming of ceramics. Mater Sci Forum 170–172:359–368.
19. Yen TF, Chang YH, Yu DL, Yen FS, Tsai DS, Lin I-N. 1991. Diffusion bonding of MgF_2 optical ceramics. Mater Sci Eng A Struct Mater Prop Microstruct Process 147:121.
20. Odeshi AG, Mucha H, Wielage B. 2006. Manufacture and characterisation of a low cost carbon fibre reinforced C/SiC dual matrix composite. Carbon 44:1994.
21. Lewinsohn CA, Colombo P, Reimanis I, Unal O. 2001. Stresses occurring during joining of ceramics using preceramic polymers. J Am Ceram Soc 84:2240.
22. Humpston G, Jacobson DM. 1993. *Principles of Soldering and Brazing*. Metals Park, OH: ASM International.
23. Kumar G, Prabhu KN. 2007. Review of non-reactive and reactive wetting of liquids on surfaces. Adv Colloid Interface Sci 133:61.
24. Clay MA, Miksis MJ. 2004. Effects of surfactant on droplet spreading. Phys Fluids 16:3070.
25. Ikeshoji T, Watanabe Y, Suzumura A, Yamazaki T. 2005. Solubility and dissolution rate of Ni base alloy to molten Ag-Cu-Pd brazing filler. JSME Int J A 48:346.

26. Kempster A, Czech N. 1995. Novel method for refurbishing used hot section gas turbine blades. J Eng Gas Turbines Power Trans ASME 117:166.
27. Liu S, Suzumura A, Ikeshoji T, Yamazaki T. 2005. Brazing of stainless steel to various aluminum alloys in air. JSME Int J A 48:420.
28. Swaney OW, Trace DE, Winterbottom WL. 1986. Brazing aluminum automotive heat exchangers in vacuum: process and materials. Welding J 65:49.
29. Callcut V, Brown L. 1996. Joining of copper and copper alloys. Welding Metal Fabrication 64:3.
30. Eustathopoulos N, Nicholas MG, Drevet B. 1999. *Wettability at High Temperatures.* Oxford: Pergamon.
31. Choy KL. 2003. Chemical vapour deposition of coatings. Prog Mater Sci 48:57.
32. Navinsek B, Panjan P, Milosev I. 1999. PVD coatings as an environmentally clean alternative to electroplating and electroless processes. Surf Coating Tech 116–119:476.
33. Kamota S, Sakai M, Tagashira K. 1989. Bonding of sintered alumina and mild steel using thermal spray coatings. J Mater Sci Lett 8:553.
34. Nolte HJ, Spurck RF. 1950. Metallized-ceramic sealing with manganese. Telev Eng 1:14.
35. Nolte HJ. 1954. Metallized ceramic. U.S. Patent 2,667,432.
36. Twentyman ME. 1975. High-temperature metallizing. III. The use of metallizing paints containing glass or other inorganic bonding agents. J Mater Sci 10:765.
37. Baldwin MD, Chidambaram PR, Edwards GR. 1994. Spreading and interlayer formation at the copper-copper oxide/polycrystalline alumina interface. Metall Mater Trans A 25:2497.
38. Babcock GL, Bryant WM, Neugebauer CA, Burgess JF 1973. Method of direct bonding metals to non-metallic substrates. U.S. Patent 3,766,634.
39. Burgess JF, Neugebauer CA. 1974. Direct bonding of metals with a metal-gas eutectic. U.S. Patent 3,854,892.
40. Burgess JF, Neugebauer CA, Flanagan G, Moore RE. 1975. Hybrid packages by the direct bonded copper process. Solid State Technol 18:42.
41. Beraud C, Courbiere M, Esnouf C, Juve D, Treheux D. 1989. Study of copper-alumina bonding. J Mater Sci 24:4545.
42. Chaklader ACD, Armstrong AM, Misra SK. 1968. Interface reactions between metals and ceramics: IV. Wetting of sapphire by liquid copper-oxygen alloys. J Am Ceram Soc 51:630.
43. O'Brien TE, Chaklader ACD. 1974. Effect of oxygen on the reaction between copper and sapphire. J Am Ceram Soc 57:329.
44. Ownby PD, Liu J. 1988. Surface energy of liquid copper and single-crystal sapphire and the wetting behavior of copper on sapphire. J Adhes Sci Technol 2:255.
45. Mehorta SP, Chaklader ACD. 1985. Interfacial phenomena between molten metals and sapphire substrate. Metall Trans B 16:567.
46. McDonald JE, Eberhart JG. 1965. Adhesion in aluminum oxide-metal systems. Trans AIME 233:512.
47. Chatain D, Chabert F, Ghetta V, Fouletier J. 1994. New experimental setup for wettability characterization under monitored oxygen activity: II, wettability of sapphire by silver-oxygen melts. J Am Ceram Soc 77:197.
48. Ramsey MJ, Lewis MH. 1984. Interfacial reaction mechanism in syalon ceramic bonding. Mater Sci Eng 71:113.

49. Mizuhara H, Mally K. 1985. Ceramic-to-metal joining with active brazing filler metal. Welding J 64:27.
50. Ballard EO, Meyer EA, Brennan GM. 1985. Brazing of large-diameter ceramic rings to niobium using an active metal TiCuSil Process. Welding J 64:37.
51. Carim AH, Loehman RE. 1990. Microstructure at the interface between AlN and a Ag-Cu-Ti braze alloy. J Mater Res 5:1520.
52. Paulasto M, Kivilahti J. 1998. Metallurgical reactions controlling the brazing of Al_2O_3 with Ag-Cu-Ti filler alloys. J Mater Res 13:343.
53. Hammond JP, David SA, Santella ML. 1988. Brazing ceramic oxides to metals at low temperatures. Welding J 67:227.
54. Chattopadhyay AK, Chollet L, Hintermann HE. 1991. Induction brazing of diamond with Ni-Cr hardfacing alloy under argon atmosphere. Surf Coating Tech 45:293.
55. Loehman RE, Tomsia AP. 1988. Joining of ceramics. Am Ceram Soc Bull 67:375.
56. Naidich YV, Zhuravlev VS, Chuprina VG, Strashinskaya LV. 1973. Adhesion, wetting, and formation of intermediate phases in systems composed of a titanium-containing melt and an oxide. Sov Powder Metall Metal Ceram 12:895.
57. Loehman RE, Tomsia AP. 1994. Wetting and joining of mullite ceramics by active-metal braze alloys. J Am Ceram Soc 77:271.
58. Foley AG, Andrews DJ. 1994. Active metal brazing for joining ceramics to metals. GEC ALSTHOM Tech Rev 13:49.
59. Nicholas MG. 1990. *Joining of Ceramics*. London: Chapman and Hall.
60. Weil KS, Rice JP. 2005. Substrate effects on the high-temperature oxidation behavior of a gold-based braze filler metal. Scr Mater 52:1081.
61. Singh M, Shpargel TP, Asthana R. 2007. Brazing of stainless steel to yttria-stabilized zirconia using gold-based brazes for solid oxide fuel cell applications. Int J Appl Ceram Technol 4:119.
62. Weil KS, Hardy JS. 2002. Brazing a mixed ionic/electronic conductor to an oxidation resistant metal. In: CA Lewinsohn, M Singh, RE Loehman, editors. *Advances in Joining of Ceramics*. Ceramic Transactions Vol. 138, The American Ceramic Society, p. 185.
63. Naidich YV, Eremenko VN. 1965. Theory and technology of sintering, heat treatment and chemicothermal treatment processes. Sov Powder Metall Metal Ceram 4:272.
64. Schüler CC, Stuck A, Beck N, Keser H, Tack U. 2000. Direct silver bonding—an alternative for substrates in power semiconductor packaging. J Mater Sci Mater Electron 11:389.
65. Erskine KM, Meier AM, Pilgrim M. 2002. Brazing perovskite ceramics with silver/copper oxide braze alloys. J Mater Sci 37:1705.
66. Weil KS, Paxton DM. 2002. Development of an oxidation resistant ceramic-to-metal braze for use in YSZ-based electrochemical devices. Ceram Eng Sci Proc 23:785.
67. Hardy JS, Kim JY, Weil KS. 2004. Joining mixed conducting oxides using an air-fired electrically conductive braze. J Electrochem Soc 151:J43.
68. Weil KS, Kim JY, Hardy JS. 2005. Reactive air brazing: a new method of sealing solid oxide fuel cells and other solid State electrochemical devices. Electrochem Solid State Lett 8:A133.
69. Assal J, Hallstedt B, Gauckler LJ. 1998. Thermodynamic assessment of the Ag-Cu-O system. J Phase Equilibria 19:315.

70. Lamb MC, Camardello SJ, Meier AM, Weil KS, Hardy JS. 2005. Brazing of porous alumina to monolithic alumina with Ag-CuO and Ag-V_2O_5 alloys. Ceram Eng Sci Proc 26:391.
71. Zink NM, Meier AM, Weil KS, Hardy JS. 2005. Reactive air brazing of LSCoF and alumina with Ag-V_2O_5 alloys for SOFC applications. Ceram Eng Sci Proc 26:341.
72. Hardy JS, Weil KS. Unpublished data, 2008.
73. Nishiura H, Suzuki RO, Ono K, Gauckler LJ. 1998. Experimental phase diagram in the Ag-Cu_2O-CuO system. J Am Ceram Soc 81:2181.
74. Suzuki RO, Nishiura H, Ono K, Gauckler LJ. 1995. Thermodynamics and phase diagram of the Ag-Cu-O system. In: K Yamafuji and T Morishita, editors. *Advances in Superconductivity VII, Proceedings of the 7th International Symposium on Superconductivity.* Tokyo, Japan: Springer-Verlag.
75. Hallstedt B, Risold D, Gauckler LJ. 1994. Thermodynamic assessment of the copper-oxygen system. J Phase Equilibria 15:483.
76. Assal J, Hallstedt B, Gauckler LJ. 1997. Thermodynamic assessment of the silver-oxygen system. J Am Ceram Soc 80:3054.
77. Sieverts A, Haganacker J. 1909. The solubility of hydrogen and oxygen in solid and molten silver. Z Phys Chem 68:115.
78. Eichenauer W, Muller G. 1962. Diffusion and solubility of oxygen in silver. Z Metallkde 53:321.
79. Karayaka I, Thompson WT. 1992. The Ag-O (silver-oxygen) system. J Phase Equilibria 13:137.
80. Allen NP. 1932. The effect of pressure on the liberation of gases from metals (with special reference to silver and oxygen). J Inst Met 49:317.
81. Graff WS, Stadelmaier HH. 1958. Higher oxides of silver. J Electrochem Soc 105:446.
82. Fischer P, Jansen M. 1990. Cyclic voltammetry and X-ray powder techniques applied to anodic crystallization of higher silver oxides. Solid State Ionics 43:61.
83. Chang YA, Hsieh KC. 1989. *Phase Diagrams of Ternary Copper-Oxygen-Metal System.* Metals Park, OH: ASM International.
84. O'Keeffe M, Bovin JO. 1978. The crystal structure of paramelaconite, Cu_4O_3. Am Mineral 63:180.
85. Leroux JA, Fröhlich KW. 1931. The quasi-binary section of silver-cuprous oxide in the system silver-copper-oxygen. Z Metallkde 23:250.
86. Kohlmeyer EJ, Sprenger KV. 1948. The oxidation of liquid silver-copper alloys. Z Anorg Chem 257:199.
87. Shao ZB, Liu KR, Liu LQ. 1993. Equilibrium phase diagrams in the systems PbO-Ag and CuO-Ag. J Am Ceram Soc 76:2663.
88. Weil KS, Kim JY, Hardy JS, Darsell JT. 2007. Experimental determination of phase equilibria in the silver-copper oxide system at high temperature. Scr Mater 56:1111.
89. Hallstedt B, Gauckler LJ. 2003. Revision of the thermodynamic descriptions of the Cu-O, Ag-O, Ag-Cu-O, Bi-Sr-O, Bi-Ca-O, Bi-Cu-O, Sr-Cu-O, Ca-Cu-O and Sr-Ca-Cu-O systems. CALPHAD 27:177.
90. Meier AM, Chidambaram PR, Edwards GR. 1995. A Comparison of the wettability of copper-copper oxide and silver-copper oxide on polycrystalline alumina. J Mater Sci 30:4781.

91. Kim JY, Hardy JS, Weil KS. 2005. Silver-copper oxide based reactive air braze (RAB) for joining yttria-stabilized zirconia. J Mater Res 20:636.
92. Kim JY, Hardy JS, Weil KS. 2005. Effects of CuO content on the wetting behavior and mechanical properties of a Ag-CuO braze for ceramic joining. J Am Ceram Soc 88:2521.
93. Kim JY, Hardy JS, Weil KS. 2005. Novel metal-ceramic joining for planar solid oxide fuel cells. J Electrochem Soc 152:J52–J58.
94. Weil KS, Hardy JS, Kim JY. 2006. Oxidation ceramic to metal braze seals for applications in high temperature electrochemical devices and method of making. U.S. Patent 055733.
95. Pavlina EJ, Meier AM, Monteleone P, Weigner JD. 2007. Development of silver-metal oxide reactive air braze alloys for electroding PZT ceramics. J Mater Sci 42:705.
96. Darsell JT, Weil KS. 2006. The effect of Pd additions on the invariant reactions in the Ag-CuO_x system. J Phase Equilibria Diffusion 27(1):92.
97. Darsell JT. 2007. High temperature Ag-Pd-CuO_x Air braze filler metal. PhD thesis, Washington State University.
98. Darsell JT, Weil KS. Unpublished. High temperature wetting transition in a metal—metal oxide system, in review.
99. Gallois B, Lupis CHP. 1981. Effect of oxygen on the surface tension of liquid copper. Metall Trans B 12:549.
100. Bernard C, Lupis CHP. 1971. Surface tension of liquid silver alloys. I. Silver-gold alloys. Metall Trans 2:555.
101. Cahn J. 1977. Critical point wetting. J Chem Phys 66:3667.
102. Weil KS, Darsell JT, Kim JY. 2006. Evidence of a high-temperature wetting transition between Ag-CuO liquid alloys and yttria-stabilized zirconia substrates. Ceram Trans 191:45.
103. Kim JY, Weil KS. 2002. Development of a copper oxide-silver braze for ceramic joining. Ceram Trans 138:119.
104. Darsell JT, Weil KS. 2007. The effect of palladium additions on the solidus/liquidus temperatures and wetting properties of Ag-CuO based air brazes. J Alloys Compounds 433:184.
105. Meier A, Chidambaram PR, Edwards GR. 1998. Modelling of the spreading kinetics of reactive brazing alloys on ceramic substrates: copper-titanium alloys on polycrystalline alumina. Acta Mater 46:4453.
106. Yang ZG, Weil KS, Paxton DM, Stevenson JW. 2003. Evaluation and selection of heat resistant alloys for SOFC interconnect applications. J Electrochem Soc 150:A1188.
107. He P, Feng JC, Zhou H. 2005. Microstructure and strength of brazed joints of Ti_3Al-base alloy with TiZrNiCu filler metal. Mater Sci Eng A 392:81.
108. Carim AH, Mohr CH. 1997. Brazing of alumina with Ti_4Cu_2O and Ti_3Cu_3O interlayers. Mater Lett 33:195.
109. Liu CF, Zhang J, Zhou Y, Meng QC, Naka M. 2008. Effect of Ti content on microstructure and strength of Si_3N_4/Si_3N_4 joints brazed with Cu-Pd-Ti filler metals. Mater Sci Eng A 491:483.
110. Darsell JT, Weil KS. 2008. The effect of filler metal composition on the strength of YSZ joints brazed with Pd-Ag-CuO_x. Metall Mater Trans A 39:2095.
111. Weil KS, Koeppel BJ. 2008. Thermal stress analysis of the planar SOFC bonded compliant seal design. Int J Hydrogen Energy 33:3976.
112. Weil KS, Koeppel BJ. 2008. Comparative finite element analysis of the stress–strain states in three different bonded solid oxide fuel cell seal designs. J Power Sources 180:343.

113. Kim JY, Hardy JS, Weil KS. 2007. Ag-Al based air braze for high temperature electrochemical devices. Int J Hydrogen Energy 32:3754.

114. Kim JY, Choi J-P, Weil KS. 2008. Mechanical properties and dual atmosphere tolerance of Ag-Al based braze. Int J Hydrogen Energy 33:3952.

115. Klueh RL, Mullins WW. 1968. Some observations on hydrogen embrittlement of silver. Trans Metall Soc AIME 242:237.

116. Singh P, Yang ZG, Viswanathan V, Stevenson JW. 2004. Observations on the structural degradation of silver during simultaneous exposure to oxidizing and reducing environments. J Mater Eng Perform 13:287.

117. Kim JY, Hardy JS, Weil KS. 2006. High-temperature tolerance of the silver-copper oxide braze in reducing and oxidizing atmospheres. J Mater Res 21:1434.

118. Kim JY, Weil KS. 2007. Unpublished data.

119. Kim JY, Hardy JS, Weil KS. 2004. Use of aluminum in air brazing aluminum oxide. J Mater Res 19:1717.

120. Bengis M, Fahrenholtz WG. 2004. Reactive joining of alumina by oxidation of Al interlayers. Key Eng Mater 264–268:655.

121. Newkirk MS, Urquhart AW, Zwicker HR. 1986. Formation of Lanxide™ ceramic composite material. J Mater Res 1:81.

122. Choi J-P, Weil KS, Kim JY, Canfield NL, Ellefson CA, Bradley BR. 2008. A novel method of metal/ceramic transition brazing for ceramic/ceramic joining at high/low temperature electrochemical devices, U.S. Patent, Pending.

123. Lu FH, Fang F-X, Chen Y-S. 2001. Eutectic reaction between copper oxide and titanium dioxide. J Eur Ceram Soc 21:1093.

124. Weil KS, Kim JY, Hardy JS, Darsell JT. 2006. The effect of TiO_2 additions on the wetting behavior of Ag-CuO_x air brazes. Scr Mater 54:1071.

125. Hardy JS, Kim JY, Thomsen EC, Weil KS. 2007. Improved wetting of mixed ionic/electronic conductors used in electrochemical devices with ternary air braze alloys. J Electrochem Soc 154:P32.

126. Shackelford JF, Alexander W, Park JS, editors. 1994. *CRC MSE Handbook*, 2nd edition. Boca Raton, FL: CRC Press.

127. Kim JY, Choi J-P, Weil KS. Unpublished. On the addition of Al_2O_3 particulate to Ag-CuO air braze filler metals, in review.

5

DIFFUSION BONDING OF SILICON CARBIDE AS AN ENABLING TECHNOLOGY FOR THE FABRICATION OF COMPLEX-SHAPED CERAMIC COMPONENTS

Michael C. Halbig[1] and Mrityunjay Singh[2]

[1]*NASA Glenn Research Center, Cleveland, Ohio*
[2]*Ohio Aerospace Institute, NASA Glenn Research Center, Cleveland, Ohio*

INTRODUCTION

Silicon carbide (SiC) is a ceramic material that offers great potential for use in high-temperature, structural applications due to its high strength, creep resistance, corrosion resistance, and high-temperature capability. Silicon carbide is being developed for a wide range of applications. Monolithic silicon carbide is proposed for an injector application [1]. Silicon carbide fiber-reinforced–silicon carbide ceramic matrix composites (SiC/SiC CMCs) are proposed for such applications as combustion liners [2, 3], nuclear and fusion reactors [4–6], and turbine engine applications [6, 7]. In addition to the high-temperature applications, monolithic SiC is also currently finding use in armor [8] and optical applications [9]. However, limitations in the geometries that can be produced in current processing methods (i.e., hot pressing and chemical vapor deposition) hinder the wider utilization of silicon carbide-based ceramic materials. New advances in designs and/or processing methods are needed. One cost-effective solution for fabricating large, complex-shaped components is through the joining of simpler-shaped

Ceramic Integration and Joining Technologies: From Macro to Nanoscale, First Edited.
Edited By Mrityunjay Singh, Tatsuki Ohji, Rajiv Asthana, Sanjay Mathur.

ceramics. One cost-effective solution for fabricating large, complex-shaped components is through the joining of simpler-shaped ceramics. Such a component, which is the focus here, is a lean direct injector [10].

Several methods have been investigated and developed for joining ceramics. The methods have included adhesive bonding, diffusion bonding, brazing, fusion welding, and friction welding [11]. Many of these approaches, as well as other innovative approaches, have been used to join silicon carbide. Reaction-formed joints have been fabricated through the use of a carbonaceous mixture reacted with silicon paste to form silicon carbide [12, 13]. Another method uses preceramic polymers at the interface, which undergo a step to convert the polymer to a ceramic [14] or to a silicon oxycarbide glass [15]. Locatelli et al. [16] have investigated brazing-like processes in which multiple interlayers are used to form joints through the formation of a thin, transient liquid-phase layer at temperatures that are typically several hundred degrees lower than those required for more traditional joining methods. Diffusion bonding of silicon carbide has also been performed with the use of interlayers of Inconel 600 [17], nickel [18], and titanium [18, 19].

The targeted application for the joining work presented here is for a ceramic lean direct injector to be used in jet engines. The ceramic injector, which is to be fabricated from SiC laminates, is illustrated in Figure 5.1a. Through controlled fuel flow, such an injector can provide greater fuel efficiency and reduced emissions during the different stages of flight, that is, take-off, cruise, and landing. Each laminate section contains its own distinct hole pattern for channeling the fuel and combustion air separately. When the laminates are stacked in order, fluid circuits for the flow of the fuel and combustion air are formed. The integration of sensors and actuators into the injector will allow for improved flow monitoring, active combustion control, lean burning, and correction of combustion instabilities. At the exiting surface, the fuel and combustion air mix to provide efficient combustion with low emissions and low NOx. The actual injector that will be used for bonding and leak testing is shown in Figure 5.1b. This is a simplified three-laminate injector compared to the more complex version shown in Figure 5.1a. As pointed out in the figure, physically vapor deposited (PVD) titanium (Ti) coatings have been applied to two of the surfaces. Titanium was applied to all surfaces including those that are recessed. However, diffusion bonding will occur only at the contacting surfaces of mated surfaces. Excess titanium applied to the lower-profile, recessed surfaces will not have an effect on the parts or the bonding.

The bonding of SiC laminates to one another and the attachment of Kovar tubes to SiC are enabling technologies for developing the lean direct injector (Fig. 5.1a). For bonding SiC laminates, diffusion bonding will be used. For attaching the Kovar tubes, brazing with the aid of active braze alloys will be used. Diffusion bonding enables the fabrication of a multilayered component with complex internal passages. The brazing technology allows the injector component to be integrated into the surrounding engine system through the adjoining fuel system. Requirements of the SiC laminate bonding technology include the ability to join relatively large, flat geometries (i.e., 10.16-cm-diameter disks), leak-free operation, and chemical and mechanical stability for long durations at operation conditions. Technical challenges to be overcome in developing the diffusion bonding and brazing approaches include nonuniform bond formation

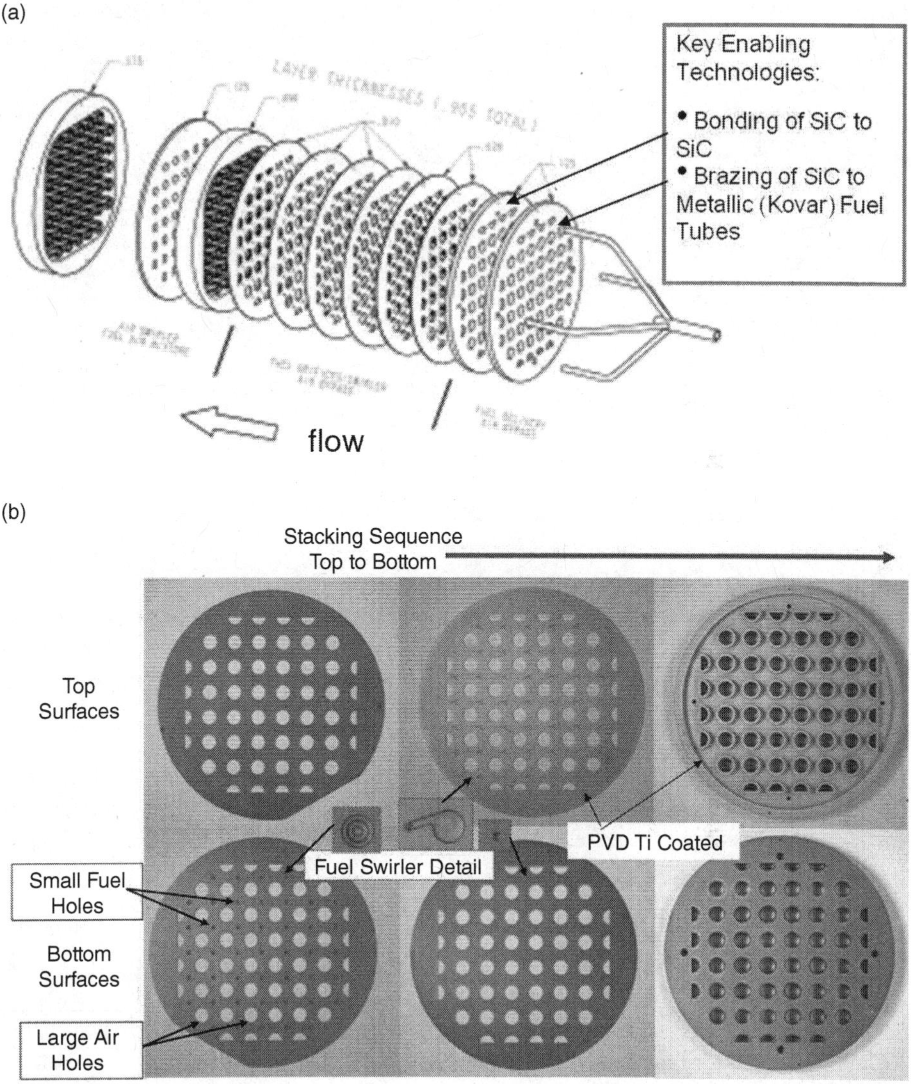

Figure 5.1. (a) Illustration of the lean direct injector design (LDI) [10]. (b) The three-laminate injector parts for demonstration of component fabrication.

across the two substrates, chemical incompatibility, and the formation of residual stresses during the cool-down after processing. The residual stresses can develop during cooling after joint formation due to thermal expansion mismatches between the two joined materials and/or the bonding interlayers. The proper selection of joining interlayers and processing conditions can help alleviate these challenges. Current development of SiC joining has focused on optimization and characterization of the bonding technology that meets the requirements for the injector.

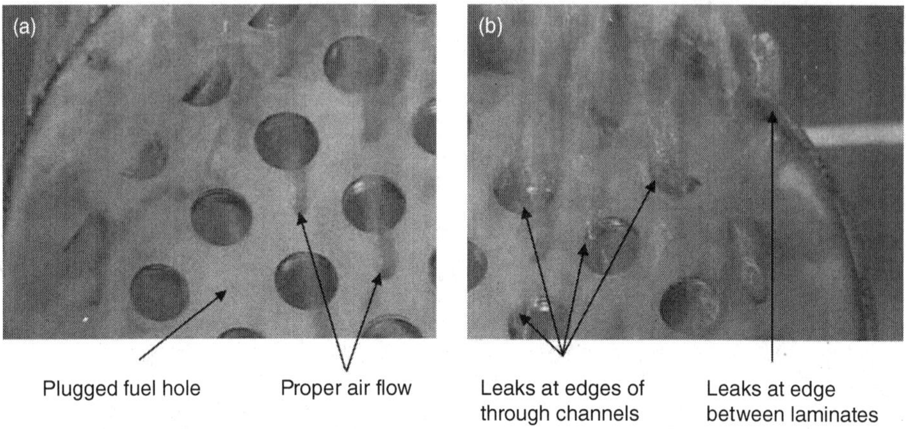

Figure 5.2. Leak test of SiC laminates joined with silicate glass.

In the previous joining method development, silicate glass was used as the bonding layer between the SiC laminates. Silicate glass possessed the temperature capabilities and stability required by the application. However, the use of a manually applied and flowable interlayer is challenging when utilized in an injector design. It was difficult to achieve a uniform glass layer. This was demonstrated during the leak testing of a silicate glass bonded injector as shown in Figure 5.2. During the leak test, air flowed into the fuel tubes and through the fuel circuit. Air was expected to flow out of the small fuel swirler holes on the exit surface. However, some of the fuel holes were blocked, presumably due to the difficulty in preventing the glass from being applied at or near the holes (Fig. 5.2a). Also, interconnected and nonbonded regions allowed for leaks between the laminates. Leaks were observed at the edges of the through holes for the combustion air, which should have been completely sealed off and at the outer edges of the laminates (Fig. 5.2b).

To develop a more suitable joining method, titanium was selected as the interlayer for aiding in the formation of diffusion bonds between SiC substrates. The reaction-formed phases resulting from diffusion bonding are stable and have high-temperature capability. Also, there is no risk of the interlayer flowing during processing as was the case for the silicate glass. Titanium in the form of foils and coatings was used as interlayers for joining development and optimization studies. For joining the actual injector parts, coatings will be used. They offer advantages in that the thickness can be controlled and they can be applied directly where desired through masking. Process optimization was conducted to produce diffusion bonds that were uniform, chemically stable, and free of microcracks. The microstructure of the diffusion bonds was characterized by microstructural and elemental analyses. The bonds were further characterized through nondestructive evaluation (NDE) and tensile pull tests.

EXPERIMENTAL

Two sets of diffusion bonds were fabricated using two different interlayers. The first set of diffusion bonds was formed with the use of an alloyed Ti foil. The second set of diffusion bonds was formed with PVD coatings of pure Ti applied to the bonding surface of the SiC. For the first set of diffusion bonds, with the alloyed Ti foil as the interlayer, three types of silicon carbide substrates were used. The substrates were α-SiC (referred to as CRYSTAR) from Saint-Gobain, chemical vapor composite (CVC) processed SiC from Trex Enterprises, and chemically vapor deposited (CVD) SiC from Rohm and Haas. The Ti interlayer was a 38-μm-thick Ti alloy foil, which also contained aluminum and vanadium. The Ti alloy is referred to as Ti-6Al-4V with 6 wt% aluminum and 4 wt% vanadium. It is an alpha-beta Ti alloy that is commonly used in aeronautics applications. Prior to joining, the materials were ultrasonically cleaned in acetone for 10 min. Baseline processing conditions consisted of a temperature of 1250°C, a clamping pressure of 24 MPa, a 2-h hold in a vacuum environment, and a cooling rate of 5°C/min. Additional bonds were fabricated with slight variations in the processing conditions in an attempt to alleviate microcracking. Only one condition was changed at a time. The variations included a processing temperature of 1300°C, a clamping pressure of 50 MPa, and a slower cooling rate of 2°C/min.

The second set of diffusion bonds were fabricated using a PVD Ti coating as the interlayer. In this series of diffusion bond fabrication, only the CVD SiC substrate from Rohm and Haas was used. In one case, a 10-μm-thick Ti coating was applied to one of the two matching surfaces. In another case, both matching surfaces of the SiC substrates had a 10-μm coating so that together, the interlayer was 20 μm thick. The processing conditions were identical to the baseline conditions stated, that is, 1250°C, and a 2-h hold in a vacuum environment except the cooling rate was 2°C/min and the clamping pressure was 31 MPa. The first and second sets of diffusion bonds were mounted in epoxy and were polished in preparation for the microstructural analysis of the cross-sections using optical and scanning electron microscopes. Elemental analysis and phase identification was conducted with energy dispersive analysis and a scanning electron microprobe.

Two additional diffusion bonds were formed on two sets of disks for NDE testing and for pull testing. The diameter of the disks was 2.54 cm. The two paired sets of SiC disks had different surface finishes. In one case, the matching disks were highly polished and had a black, mirrorlike finish. In the other case, the matching disks were left unpolished and had a dull, nonreflective, gray finish. A 10-μm-thick PVD coating was applied to the inner 1.66-cm diameter of one of the matching disks. The sets of disks are shown in Figure 5.3. Before the disks were bonded, the surfaces and applied coatings were evaluated under an optical microscope. After bonding at the "second set" conditions (stated earlier), these two sets of joined disks were nondestructively evaluated using the methods of pulsed thermography and ultrasonic immersion. Illustrations of the NDE equipment and methods are shown in Figure 5.4. In flash thermography (Fig. 5.4a), flash lamps are used to heat the sample surface. After the light pulse, an infrared camera is used to record the thermal energy that is emitted from the surface.

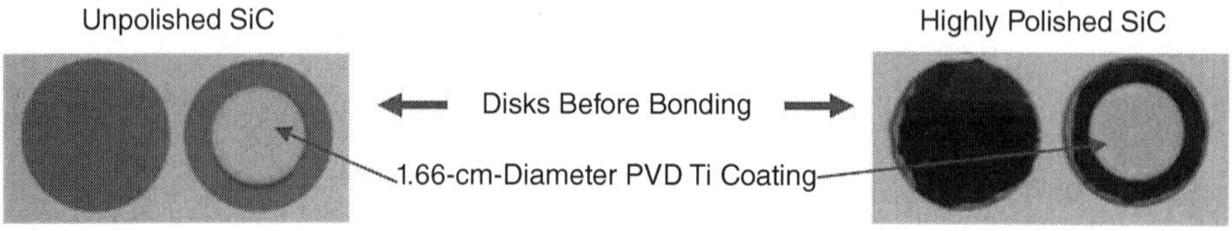

Figure 5.3. Photos of the coated and uncoated sets of SiC disks before bonding. The less polished disks are on the left and the more polished disks are on the right.

At defect locations, there will be an infrared radiation peak. For ultrasonic immersion (Fig. 5.4b), an ultrasonic pulse is emitted and enters the specimen. The difference between the amount of signal that reflects back at the interface and at the backside of the sample determines bond quality. Less deflection at the bond interface correlates with good bonding. After NDE analysis, the disks were fracture tested to determine the strength of the bonds. The fracture surfaces were analyzed under an optical microscope.

RESULTS AND DISCUSSION

Microscopy was conducted on the as-received Ti-6Al-4V foil. A micrograph of the polished cross-section is shown in Figure 5.5. The width of the alloyed foil is 38 μm. Two phases were observed. These phases were a gray-colored primary phase and a secondary phase that appears as small, elongated white islands that are distributed throughout the primary phase. Results of electron microprobe analysis for the two phases are shown in Table 5.1. The alloyed Ti foil was confirmed to have the alpha and beta phases that make up the Ti alloy of Ti-6Al-4V (weight percent). Alpha phases of Ti have a hexagonal close-packed structure and beta phases have a body-centered cubic structure. The gray phase in the micrograph is the alpha Ti phase, which contains about 6.0 wt% aluminum and 3.3 wt% vanadium. The white phase in the micrograph is the beta Ti phase, which contains about 2.7 wt% aluminum, 18.0 wt% vanadium, and 2.2 wt% iron.

In the first set of diffusion bonds that were processed with the alloyed Ti foil and the three different types of SiC substrates, microscopy was conducted to determine the bond quality. Microscopy revealed bonds that were well adhered to the SiC substrates with no delaminations. An example of the adhesion is shown in Figure 5.6, which is a secondary electron image (contrast is provided by topography changes). This is the diffusion bond formed between the Trex SiC with the alloyed Ti foil. The results show good adhesion between the interlayer and SiC substrates and were typical for the all bonds. However, microcracks were observed, which traversed across the width of the diffusion bonds in the direction perpendicular to the substrate surfaces. All additional scanning electron micrograph images to be presented are micrographs taken in the backscattered electron image mode (contrast is due to average atomic number differences). A backscattered image of the same diffusion bond from Figure 5.6 is shown in

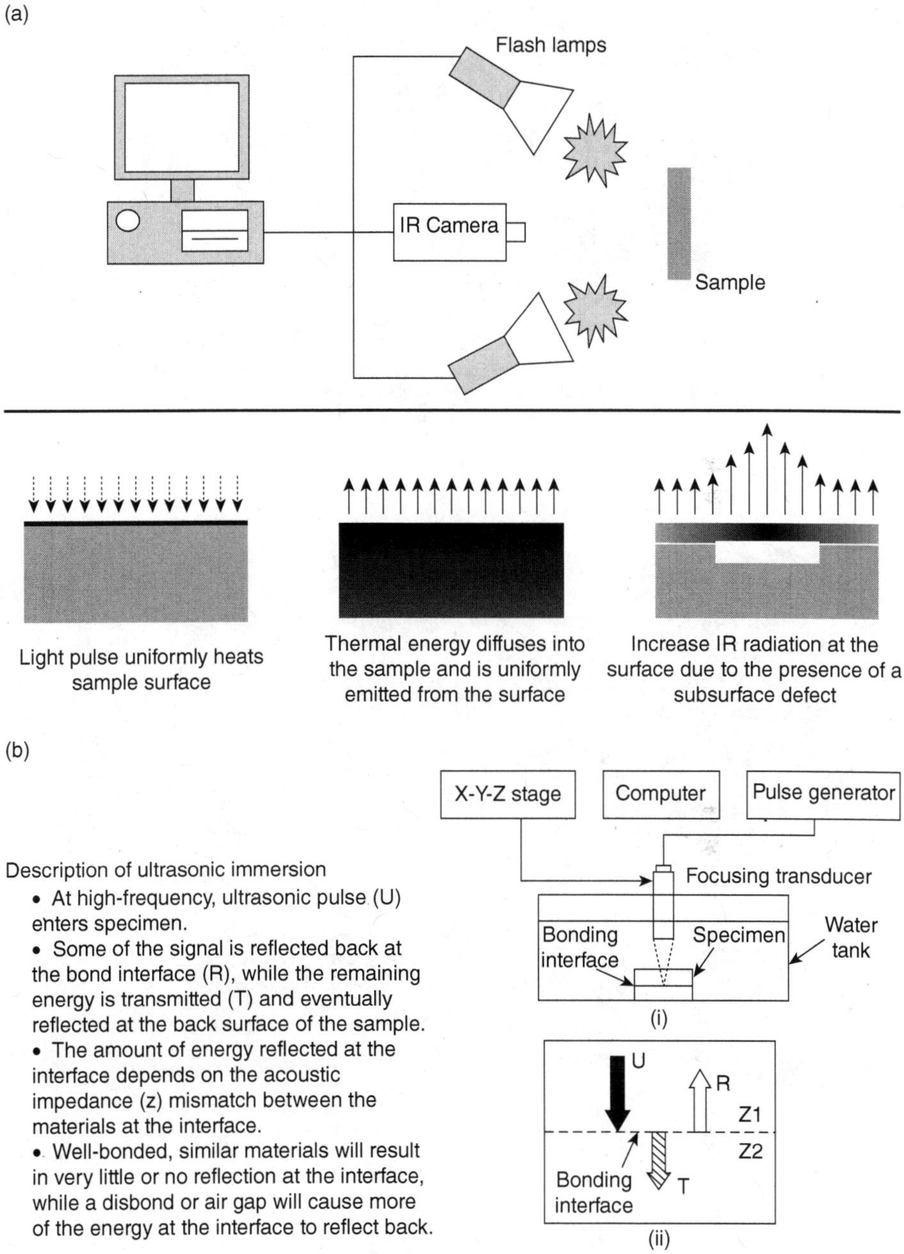

Figure 5.4. (a) The equipment (top) and the method description (bottom) of pulsed thermography. (b) The equipment and the method description of ultrasonic immersion.

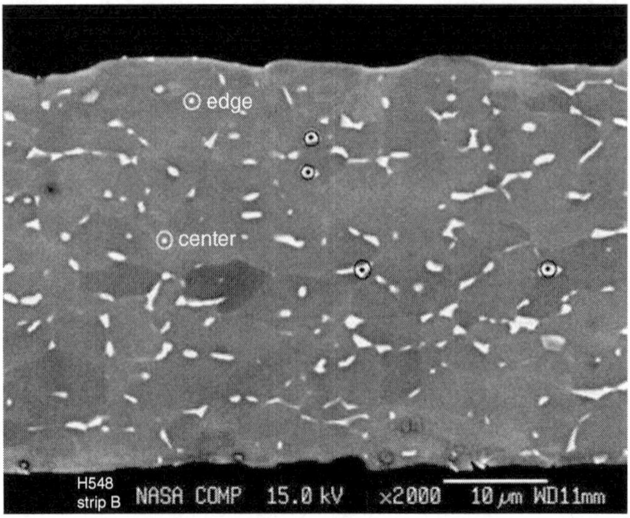

Figure 5.5. Micrograph of the cross-section of the as-received alloyed Ti foil.

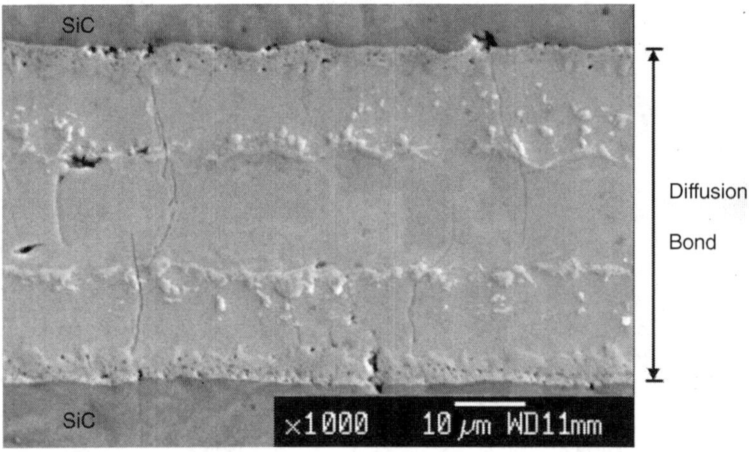

Figure 5.6. Secondary electron image of the diffusion bond formed with the alloyed Ti foil between the Trex CVD SiC substrates.

TABLE 5.1. Microprobe Analysis Results of the Two Phases within the as-Received Alloyed Ti Foil as Shown in Figure 5.5

	Phase	Al	Fe	Ti	V	Total
Atomic ratio	Gray phase	10.196	0.042	86.774	2.988	100.000
Weight (%)	Gray phase	5.999	0.051	90.632	3.318	100.000
Atomic ratio	White phase	4.841	1.850	76.507	16.803	100.000
Weight (%)	White phase	2.748	2.172	77.084	17.997	100.000

Values are an average among two data locations for the dark-gray phase and three for the white phase.

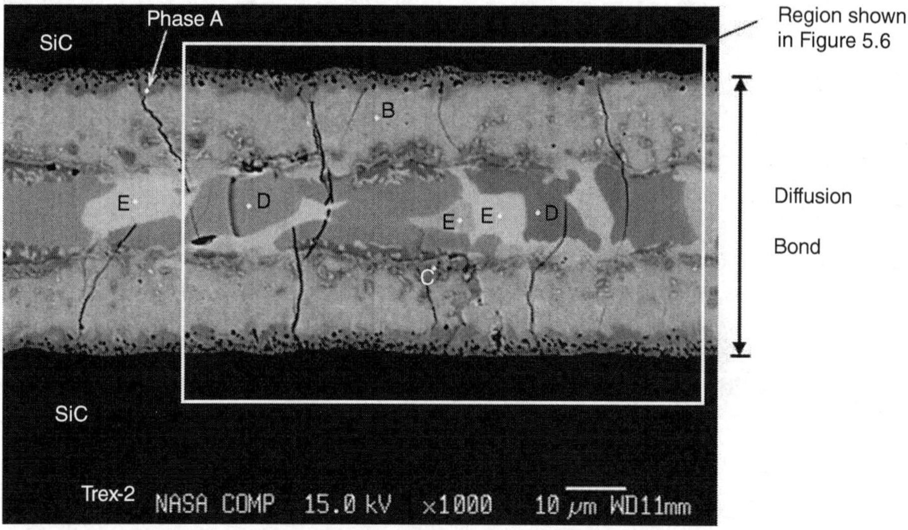

Figure 5.7. Micrograph of the cross-section of the diffusion bond formed between Trex CVD SiC with the alloyed Ti foil as the interlayer.

TABLE 5.2. Microprobe Analysis of the Atomic Ratios for the Reaction-Formed Phases in the Diffusion Bond between Trex SiC as Shown in Figure 5.7

Phase	Al	V	Cr	Si	C	Ti	Fe	Total
Phase A	0.101	0.578	0.026	18.733	24.317	56.243	0.004	100.000
Phase B	0.384	1.220	0.025	34.639	6.904	56.794	0.034	100.000
Phase C	0.604	0.715	0.023	3.148	35.366	60.124	0.020	100.000
Phase D	19.887	1.424	0.026	0.537	13.719	64.381	0.025	100.000
Phase E	26.941	14.935	0.396	0.588	0.747	55.684	0.710	100.000

Atomic ratios are an average from five locations for each phase.

Figure 5.7. Higher contrast can be seen between the many distinct reaction-formed phases in the diffusion bond than in the secondary electron image (Fig. 5.6). The SiC substrates appear black above and below the diffusion bond. Within the diffusion bond, three distinct regions were observed, which were parallel to the substrates. There is a central core region and there are the two regions located adjacent to the SiC substrates. The corresponding microprobe analysis of the phases in the diffusion bond region is given in Table 5.2.

Similar results were obtained for diffusion bonds formed between the other SiC materials with the alloyed Ti foil as an interlayer. For the material combination of the alloyed Ti foil and the CRYSTAR α-SiC, an image of the diffusion bond is shown in Figure 5.8 and the results from the microprobe analysis are given in Table 5.3. An image

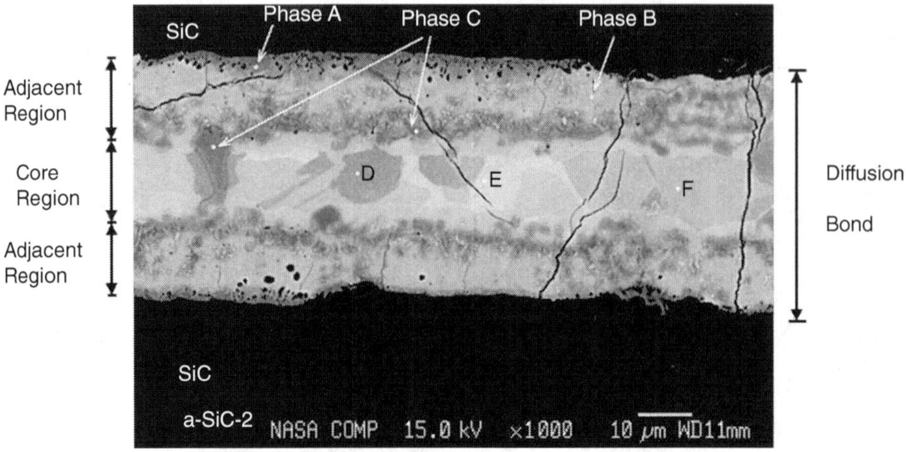

Figure 5.8. Micrograph of the cross-section of the diffusion bond formed between CRYSTAR α-SiC with the alloyed Ti foil as the interlayer.

TABLE 5.3. Microprobe Analysis of the Atomic Ratios for the Reaction-Formed Phases in the Diffusion Bond between α-SiC as Shown in Figure 5.8

Phase	Al	V	Cr	Si	C	Ti	Fe	Total
Phase A	0.020	0.970	0.023	7.662	34.787	56.531	0.007	100.000
Phase B	0.076	0.923	0.020	34.151	6.502	58.220	0.108	100.000
Phase C	0.223	0.760	0.019	6.015	32.269	60.690	0.025	100.000
Phase D	21.862	2.363	0.050	0.432	10.394	64.865	0.036	100.000
Phase E	27.102	21.318	0.611	0.473	1.901	46.456	2.141	100.000
Phase F	25.975	5.962	0.145	1.186	1.898	64.666	0.168	100.000

Atomic ratios are an average from five locations for each phase.

of the material combination of the alloyed Ti foil and the Rohm and Haas SiC is shown in Figure 5.9, and the results from the microprobe analysis are given in Table 5.4.

For all three sets of joints with the different SiC substrates, the central core region is primarily made up of the constituents of the alloyed foil (Ti, Al, and V). Alpha- and beta-type phases have become enriched with Al and V and are more segregated. Some of the Ti has migrated out of this central core region. The two outer regions have phases that contain high concentrations of Si and/or C with high concentrations of Ti. One phase in particular, labeled as phase B in all three sets of micrographs and tables, was identified as $Ti_5Si_3C_x$. Microcracks were present in the diffusion bond whenever the alloyed Ti foil was used as an interlayer. This was despite the use of three different SiC substrates and different processing conditions, such as a slower cooling rate of 2°C/min, a higher processing temperature of 1300°C, and a higher processing stress of 50 MPa.

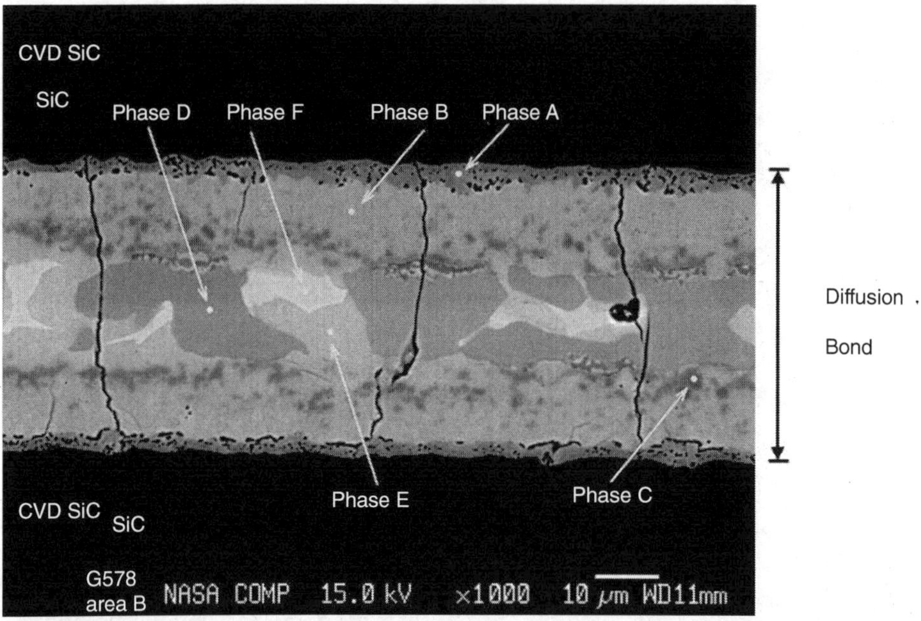

Figure 5.9. Micrograph of the cross-section of the diffusion bond formed between Rohm and Haas SiC with the alloyed Ti foil as the interlayer.

TABLE 5.4. Microprobe Analysis of the Atomic Ratios for the Reaction-Formed Phases in the Diffusion Bond between Rohm and Haas SiC as Shown in Figure 5.9

Phase	Al	V	Cr	Si	C	Ti	Fe	Total
Phase A	0.070	0.848	0.030	17.085	25.802	56.164	0.001	100.000
Phase B	0.303	0.961	0.018	33.766	6.675	58.212	0.064	100.000
Phase C	0.293	0.661	0.010	5.234	35.671	58.122	0.010	100.000
Phase D	19.002	1.392	0.039	0.557	12.707	66.288	0.014	100.000
Phase E	23.244	7.649	0.213	0.758	3.970	64.104	0.062	100.000
Phase F	19.833	20.784	0.634	0.514	4.622	52.302	1.313	100.000

Atomic ratios are an average from five locations for each phase.

The microcracks may be due to a single factor or a combination of factors. However, the contributing factors appear to be the formation of the $Ti_5Si_3C_x$ phase in the regions adjacent to the SiC substrates and the formation of high concentrations of alpha- and beta-type Ti alloy phases in the central core region. The titanium silicide (Ti_5Si_3) phase is anisotropic in its thermal expansion [20, 21]. The coefficient of thermal expansion (CTE) in the *a*-direction is 6.11 ppm/K and 16.62 ppm/K in the *c*-direction, with a ratio of CTE (*c*)/CTE (*a*) equal to 2.72. Other researchers have reported the anisotropy with a ratio as high as 4.39 [22]. The formation of alpha and beta Ti alloy phases is also believed to be a contributor to cracking. The central core of the diffusion bonds had concentrated alpha and beta Ti alloy phases. The alpha phase has an anisotropic thermal expansion, which is 20% greater along the *c*-axis [23]. Also, the beta phase has a thermal expansion that is six times higher in the temperature range of 600–1000°C (5.8×10^{-5}/°C) compared to the thermal expansion below 600°C (9.2×10^{-6}/°C) [24]. These anisotropy and CTE mismatches could cause the microcracking from thermal stresses forming during the cool-down step after hot pressing. The alloyed Ti foil was not considered for further investigation due to the formation of complex diffusion bonds containing several reaction-formed phases and the formation of microcracks.

In the second set of diffusion bonds which were processed with PVD Ti coating interlayer, mixed results were obtained. Diffusion bonds with Ti interlayer thicknesses of 10 and 20 μm were fabricated. Figure 5.10 shows the diffusion bond from the Rohm and Haas CVD SiC and the 20-μm-thick PVD Ti interlayer. Microcracks were observed in the diffusion bond. The corresponding microprobe analysis of the reaction-formed phases is given in Table 5.5. Similar to the bonds formed using the alloyed Ti foil, the presence of $Ti_5Si_3C_x$ and microcracking was observed when a relatively thick layer of PVD Ti (20 μm) was used as the interlayer. Naka et al. [25] suggested that $Ti_5Si_3C_x$ is an intermediate phase that will not be present when phase reactions have gone to completion. This appears to be the case as seen in the bond formed from the thinner 10-μm PVD Ti interlayer. The micrograph in Figure 5.11 shows a well-formed diffusion bond. The bond is much less complex than those from the alloyed Ti foil, which had five to seven phases or the thicker PVC Ti coating (20 μm), which had three phases. Only two phases have formed, as can be seen from the microprobe analysis in Table 5.6. The absence of microcracking when a 10-μm PVD Ti interlayer is used is due to the absence of the detrimental $Ti_5Si_3C_x$ phase. The source of the dark pores in the bond has not yet been determined; however, they may be due to the formation of denser phases during the diffusion bonding process. The presence of the pores may not have a significant effect on the mechanical and leakage properties of the bond since the pores are very small and isolated.

The sets of bonded disks were analyzed by NDE followed by strength tests. The disks had different surface finishes as shown in Figure 5.3. The surface detail of each disk and the applied coatings are shown in Figure 5.12. The highly polished substrate has a denser surface, which allowed for a denser PVD Ti coating to be applied. On the less polished substrate, it was more difficult to distinguish the transition between the coated and uncoated surfaces due to the rough surface. The first NDE method investigated was pulsed thermography (flash thermography). In this NDE technique, light from flash lamps are pulsed uniformly onto the sample surface. The thermal energy diffuses into the sample and is uniformly emitted from the surface. Subsurface defects are detected by an increase in infrared radiation at the surface. Since the coating was only applied to the inner 1.66-cm diameter of the 2.54-cm-diameter disks, it would be expected that NDE would detect the bonded inner diameter and a nonbonded outer ring. However, pulsed thermography did not clearly detect the bonded and nonbonded areas as shown in Figure 5.13. The general region where bonding was presumed to have

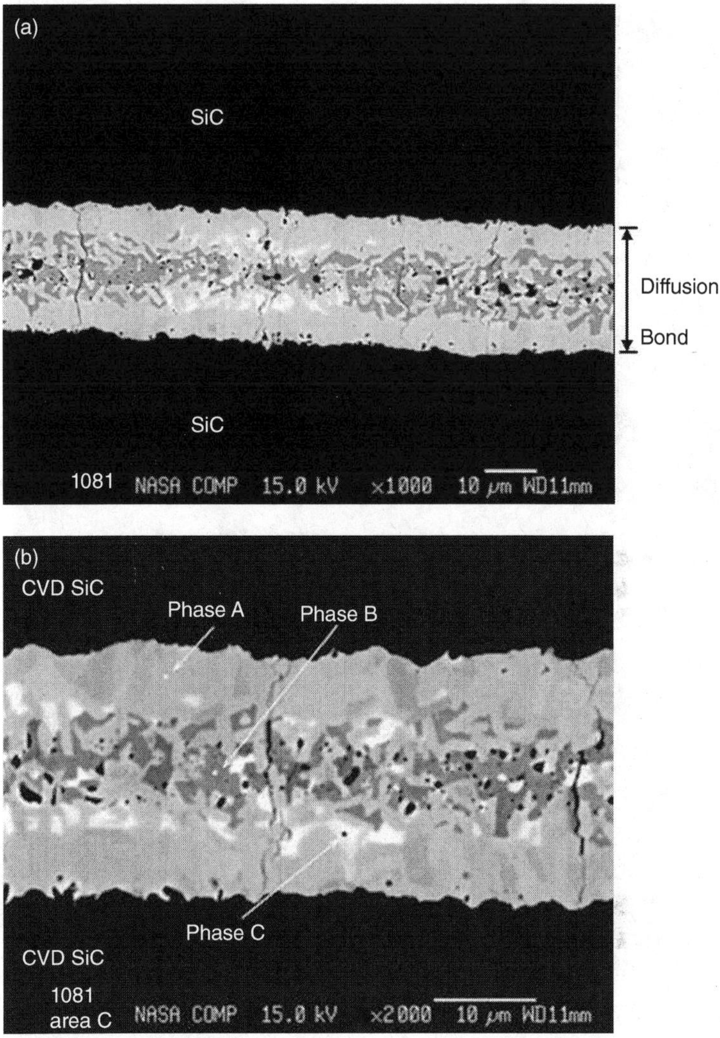

Figure 5.10. Diffusion bond when a 20-μm-thick PVD Ti coating was used as the interlayer between the Rohm and Haas SiC substrates. The top is at a magnification of ×1000, and the bottom micrograph is at a magnification of ×2000. Microcracks and the presence of three phases are observed.

TABLE 5.5. Microprobe Analysis of the Atomic Ratios for the Reaction-Formed Phases in the Diffusion Bond as Shown in Figure 5.10

Phase	Al	Fe	Ti	Si	C	Cr	Total
Phase A	0.011	0.001	56.426	17.792	25.757	0.014	100.000
Phase B	0.007	0.005	35.794	62.621	1.570	0.003	100.000
Phase C	0.027	0.153	58.767	33.891	7.140	0.023	100.000

Atomic ratios are an average from five locations for each phase.

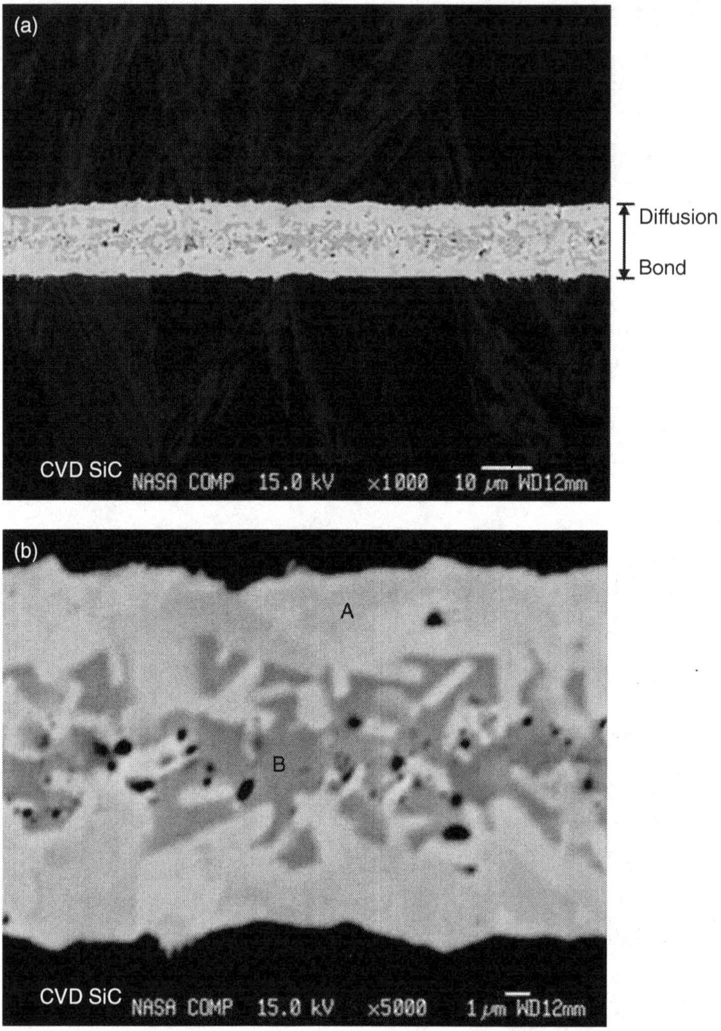

Figure 5.11. Diffusion bond when a 10-μm-thick PVD Ti coating was used as the interlayer between the Rohm and Haas SiC substrates. The top is at a magnification of ×1000, and the bottom micrograph is at a magnification of ×5000. No microcracks and the presence of two phases are observed.

TABLE 5.6. Microprobe Analysis of the Atomic Ratios for the Reaction-Formed Phases in the Diffusion Bond as Shown in Figure 5.11

Phase	C	Si	Ti	Al	Cr	Total
SiC	45.890	54.096	0.011	0.000	0.004	100.000
Phase A	24.686	18.690	56.621	—	0.003	100.000
Phase B	3.028	61.217	35.752	—	0.003	100.000

Atomic ratios are an average from five locations for each phase.

Figure 5.12. Optical views of the SiC substrate and PVD Ti coating for the less polished (left) and the more polished SiC (right). Macroviews of the disks are shown for reference.

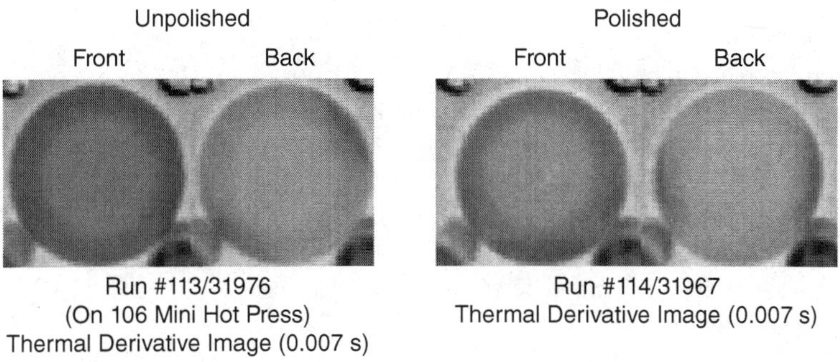

Figure 5.13. Results from pulsed thermography for the less polished pair of joined substrates (left) and for the more polished pair (right).

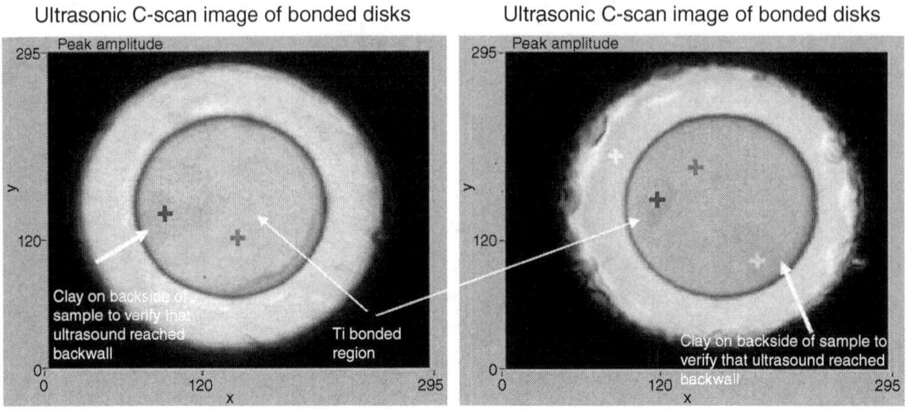

Figure 5.14. Results from ultrasonic immersion evaluation of the less polished pair of joined substrates (left) and for the more polished pair (right).

occurred was detected; however, the image was not very sharp. The NDE analysis of the bond quality was inconclusive due to the high surface reflectivity of the SiC.

Next, ultrasonic immersion was investigated. In this NDE technique, the sample is placed in a water tank and a transducer emits a high-frequency ultrasonic pulse that enters the specimen. Part of the signal is reflected back at the bond interface, while the remaining energy is transmitted through the second substrate and reflected at the back surface of the sample. Nonbonded regions of the joined substrates reflect more energy than from well-bonded regions. As seen in Figure 5.14, the ultrasonic immersion technique provides a clearer image of the inner diameter of the bonded surface and the outer ring of a nonbonded area. For the more polished set of substrates, a sharper image was obtained; the less polished set of substrates appeared as slightly blurred, suggesting that the more polished substrates gave a denser and higher-quality bond.

After completing the NDE analysis, the two sets of joined disks were tensile tested and the fracture surfaces were analyzed under an optical microscope. An illustration of the sample geometry and load condition is shown in Figure 5.15a. The stress versus strain curves are shown in Figure 5.15b. The less polished set of joined disks had a strength of 13.4 MPa and the more polished set had a strength of 15.0 MPa. The strengths are much higher than those required for the application, which is 3.45–6.89 MPa. Although the limited testing may not be able to identify a true trend in strength, the higher strength for the more polished sample could be due to such effects as stronger bonds due to the smoother finish and denser coating or due to the more polished SiC material having fewer surface flaws. Optical micrographs of the fracture surfaces are shown in Figure 5.16. For the less polished set of joined disks, failure was in the SiC substrate rather than in the bond. The bond was pulled out intact from the failing SiC substrate. For the more polished set of joined disks, failure was primarily in the SiC as failure started in one substrate crossed through the bond region and continued in the other SiC substrate.

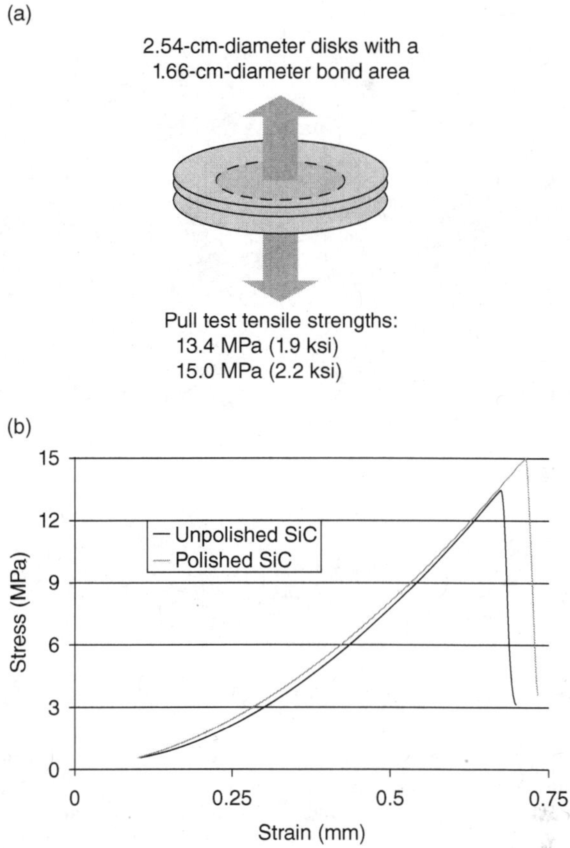

Figure 5.15. (a) Illustration of the tensile tested samples and the resulting strengths (left). (b) Stress versus strain curves for the two sets of joined disks (right).

CONCLUSIONS

A joining approach was developed for bonding silicon carbide using titanium interlayers to form diffusion bonds. The diffusion bonds were well adhered to the SiC substrate. No delaminations were observed between the SiC substrates. For the case of diffusion bonding with an alloyed Ti foil (Ti-6Al-4V), complex diffusion bonds were formed, which consisted of five to seven reaction-formed phases, and microcracking was observed. For the case of the thicker PVD Ti coatings ($20\,\mu m$), microcracking was also observed. A contributor to the microcracking for both cases was the formation of the detrimental phase of $Ti_5Si_3C_x$, which has high anisotropy in its thermal expansion. With the use of the alloyed Ti foil, microcracking also resulted from the formation of concentrated alpha and beta Ti phases, which have anisotropic and high thermal expansions. The best diffusion bonds were obtained when 10-μm-thick PVD Ti coatings were used as interlayers between CVD SiC. The bonds were uniform, did not have the

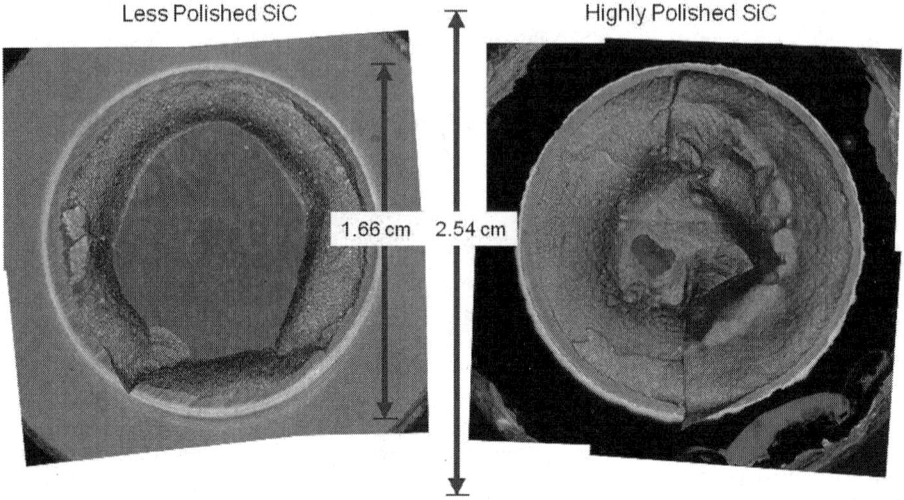

Figure 5.16. Optical micrographs of the failed surfaces of the tensile tested disks. Failure on the less polished set of disks is shown on the left and the failure for the more polished set of disks is on the right.

intermediate phase of $Ti_5Si_3C_x$, and were free of microcracks. The ultrasonic immersion NDE method proved to be a very good method for analyzing the bond quality and for detecting small flaws and debonded areas. The optimization of the processing of the diffusion bond and the resulting properties suggest that diffusion bonding is a good processing approach for fabricating complex-shaped components that can be utilized for various applications.

ACKNOWLEDGMENTS

This effort was supported by the NASA Glenn Research Center (GRC) under the Subsonic Fixed Wing Project. The authors would like to thank James Smith for conducting microprobe analysis and Dr. Robert Okojie of NASA GRC for applying PVD Ti coatings. The authors would also like to thank Richard E. Martin and Laura M. Cosgriff of Cleveland State University for conducting NDE and for providing descriptions of the methods, flash thermography and ultrasonic immersion.

REFERENCES

1. Okojie RS, Tacina R, Wey C, Blaha C. 2007. Micro fabrication of SiC mesoscale lean direct injector array: toward active combustion control. *Solid-State Sensors, Actuators and Microsystems Conference, TRANSDUCERS 2007–2007 International*, Lyon, France, pp. 2433–2436.

2. Kimmel J, Miriyala N, Price J, More K, Tortelli P, Eaton H, Linsey G, Sun E. 2002. Evaluation of CFCC liners with EBC after field Testing in a gas turbine. J Eur Ceram Soc 22:2769–2775.
3. van Roode M, Price J, Kimmel J, Miriyala N, Leroux D, Fahme A, Smith K. 2007. Ceramic matrix composite combustion liners: a summary of field evaluations. J Eng Gas Turbines Power 129:21–30.
4. Hino T, Hayashishita E, Yamauchi Y, Hashiba M, Hirohata Y, Kohyama A. 2005. Helium gas permeability of SiC/SiC composite used for in-vessel components of nuclear fusion reactor. Fusion Eng Des 73:51–56.
5. Hirohata Y, Jinushi T, Yamauchi Y, Hashiba M, Hino T, Katoh Y, Kohyama A. 2002. Gas permeability of SiC/SiC composites as fusion reactor material. Fusion Eng Des 61–62:699–704.
6. Naslain R. 2004. Design, preparation and properties of non-oxide CMCs for application in engines and nuclear reactors: an overview. Compos Sci Technol 64:155–170.
7. Spriet P, Habarou G. 1997. Application of CMCs to turbojet engines: overview of the SEP experience. Key Eng Mater 127–131:1267–1276.
8. Lundberg P, Lundberg B. 2005. Transition between interface defeat and penetration for tungsten projectiles and four silicon carbide materials. Int J Impact Eng 31:781–792.
9. Yao W, Zhang Y, Han J, Zuo H. 2006. Fabrication and test of reaction bonded silicon carbide for optical applications. Trans Nonferrous Met China 16(2):409–413.
10. Tacina R, Wey C, Laing P, Mansour A. April 2002. A Low Lean Direct Injection, Multipoint Integrated Module Combustor Concept for Advanced Aircraft Gas Turbines. NASA/TM-2002-211347.
11. Gottselig B, Gyarmati E, Naoumidis A, Nickel H. 1990. Joining of ceramics demonstrated by the example of SiC/Ti. J Eur Ceram Soc 6:153–160.
12. Singh M. 1997. A reaction forming method for joining of silicon carbide-based ceramics. Scr Mater 37(8):1151–1154.
13. Martínez Fernandez J, Munoz A, Valera-Feria FM, Singh M. 2000. Interfacial and thermal characterization of reaction formed joints in silicon carbide-based materials. J Eur Ceram Soc 20:2641–2648.
14. Colombo P, Donato A, Riccardi B, Woltersdorf J, Pippel E, Silberglitt R, Danko G, Lewinsohn C, Jones R. 2002. Joining SiC-based ceramics and composites with preceramic polymers. Ceram Trans 144:323–334.
15. Pippel E, Woltersdorf J, Colombo P, Donato A. 1997. Structure and composition of interlayers in joints between SiC bodies. J Eur Ceram Soc 17:1259–1265.
16. Locatelli MR, Dalgleish BJ, Nakashima K, Tomsia AP, Gleaser AM. 1997. New approaches to joining ceramics for high-temperature applications. Ceram Int 23:313–322.
17. Li J, Xiao P. 2004. Fabrication and characterization of silicon carbide/superalloy interfaces. J Eur Ceram Soc 24:2149–2156.
18. Bhanumurthy K, Schmid-Fetzer R. 1996. Solid-state bonding of silicon carbide (HIP-SiC) below 1000°C. Mater Sci Eng A220:35–40.
19. Cockeram BV. 1999. The diffusion bonding of silicon carbide and boron carbide using refractory metals. In: *Proceedings from the Material Solutions '99 on Joining of Advanced and Specialty Materials*, Cincinnati, OH.

20. Schneibel JH, Rawn CJ, Payzant EA, Fu CL. 2004. Controlling the thermal expansion anisotropy of Mo_5Si_3 and Ti_5Si_3 silicides. Intermetallics 12:845–850.
21. Schneibel JH, Rawn CJ. 2004. Thermal expansion anisotropy of ternary silicides based on Ti_5Si_3. Acta Mater 52:3843–3848.
22. Zhang L, Wu J. 1998. Thermal expansion and elastic moduli of the silicide based intermetallic alloys Ti_5Si_3(X) and Nb_5Si_3. Scr Mater 38(2):307–313.
23. Boyer R, Welsch G, Colling EW. 1994. *Material Properties Handbook: Titanium Alloys.* Materials Park, OH: ASM International, p. 516.
24. Elmer JW, Palmer TA, Babu SS, Specht ED. 2005. In situ observations of lattice expansion and transformation rates of α and β phases in Ti-6Al-4V. Mater Sci Eng A 391:104–113.
25. Naka M, Feng JC, Schuster JC. 1997. Phase reaction and diffusion path of the SiC/Ti system. Metall Mater Trans A 24A:1385–1390.

6

INTEGRATION OF CARBON–CARBON COMPOSITE TO METALLIC SYSTEMS FOR THERMAL MANAGEMENT APPLICATIONS

Mrityunjay Singh[1] and Rajiv Asthana[2]

[1]*Ohio Aerospace Institute, NASA Glenn Research Center, Cleveland, Ohio*
[2]*University of Wisconsin-Stout, Menomonie, Wisconsin*

INTRODUCTION

Thermal management—efficient transport and removal of heat—is critical for the performance of a number of modern devices and systems both large and small. It is concerned with heat dissipation, waste heat reduction, waste heat utilization and energy harvesting. Thermal management needs are pervasive and are critically required to maintain device temperatures below the maximum operating temperature in personal computers, laptops, cell phones, optoelectronics (lasers and LEDs), automobiles, microrefrigeration, and innumerable other consumer and industrial systems and devices. Most functional devices generate heat because of power consumption during usage, and this heat must be dissipated to maintain functionality, to improve reliability, and to prevent failure.

Thermal management technologies are diverse and utilize a wide variety of approaches. Some key examples include heat pipes, air and liquid cooling, phase change, thermal straps, spray cooling, synthetic jet cooling, thermoelectric cooling and a number of other techniques. The underlying process physics for different thermal management techniques may be different, although their functional requirements are

Ceramic Integration and Joining Technologies: From Macro to Nanoscale, First Edition.
Edited by Mrityunjay Singh, Tatsuki Ohji, Rajiv Asthana, Sanjay Mathur.

quite similar. Thus, heat sinks, widely used in refrigeration and electronic devices, are made of thermally conductive materials with one flat surface for good contact with the component to be cooled, and a number of fins on the other surface to increase the surface area for rapid heat dissipation. They may either use forced flows to enhance the cooling or phase change materials that can store energy from phase transformation. Thermoelectric cooling systems utilize the Peltier effect (or the inverse Seebeck effect), which occurs when a voltage applied to a thermocouple creates a temperature difference. Such devices serve as a low-efficiency heat pipe. Fuel cell operation to generate electrical energy involves exothermic or endothermic reactions that demand innovative thermal management solutions. In some applications, localized heating is done with reactive foils to generate or consume thermal energy. Such diverse applications involve varying levels of complexity of thermal management approaches.

As a general rule, effective thermal management usually demands high thermal conductivity materials with low coefficients of thermal expansion (CTEs) for thermal stability and compatibility with other materials [1]. Additionally, the nature of physical contact at interface/interphase between different materials is important. Customarily, the resistance to thermal transport across contiguous surfaces is characterized in terms of a heat transfer coefficient (h); a large value of h represents good interfacial conductance. With a bonded interlayer (as in brazed, soldered, or adhesively bonded joints), the performance of the thermal interface materials (TIMs) needs to be characterized. It depends upon the quality of the thermal conduction through the TIMs and the quality of contact between the TIMs and contiguous surfaces. The effective performance of the TIM is characterized by the parameter θ_{TIM} given from [2]

$$\theta_{\mathrm{TIM}} = \frac{BLT}{K_{\mathrm{TIM}}} + R_c,$$

where K_{TIM} is the thermal conductivity of the TIM, bond line thickness (BLT) is the thickness of the interface material, and R_c is the contact resistance between the TIM and the mating surfaces. The contact resistance is mainly due to the irregularities or roughness of the surfaces, so the resistance is negligible, if the surfaces are perfectly smooth.

For high thermal management efficiency, θ_{TIM} should be low. This can be accomplished by increasing the conductivity of the TIM (K_{TIM}), by reducing the BLT of the interface material, and by reducing R_c by filling the microcrevices of the surfaces with wettable and conductive materials. Polishing can also reduce R_c and can improve thermal conductance, but overall flatness rather than mirrorlike finish is more critical for heat transfer at an interface.

The complexity of a myriad of thermal systems and devices usually requires development and application of thermal modeling approaches for the analysis, design, prediction, control, and evaluation of thermal performance. Thermal modeling can predict how diverse materials, components based on them, and the arrangement of the components and materials in a device affect the thermal dynamics. Thermal modeling can map heat distribution, simulate convective flows, and develop robust design guidelines for thermal management systems (e.g., heat sinks). Innovative materials play a central enabling role in component design and process development for effective thermal management. Techniques to integrate such materials for thermal management applications constitute the primary focus of this chapter.

MATERIALS FOR THERMAL MANAGEMENT

Among the most common conventional thermal management materials are conductive metals such as copper and aluminum. Copper is a better thermal conductor than aluminum ($K_{Al} = 205\,W/m\text{-}K$ and $K_{Cu} = 400\,W/m\text{-}K$) but also heavier ($\rho_{Cu} = 8900\,kg/m^3$, $\rho_{Al} = 2200\,kg/m^3$). It is more expensive and less amenable to extrusion and creation of complex shapes than aluminum.

Both Al and Cu have relatively high CTEs, and their use as thermal management materials requires design compromises that may impair thermal efficiency. For example, polymeric materials are often used to attach Al and Cu heat spreaders to microprocessor chips; the low thermal conductivity of most polymeric materials adversely affects the interfacial heat transfer in such devices. Direct solder attachments overcome this problem, but issues related to thermal stress come into play. Thermal stress and distortion arise from residual stresses due to CTE differences and may occur even when liquid cooling is employed. Soft solders are generally used to reduce stresses, but issues of thermal fatigue and metallurgical stability under service conditions could become critical. Thus, design compromises imposed by the availability of a limited family of thermal management materials provide an impetus to develop new thermal management materials and innovative technologies to seamlessly integrate these materials in devices and systems.

In a historical and developmental sense, modern thermal management materials may be classified into three generations [3, 4]. The first-generation, high-conductivity, low-CTE materials included Cu–W, Kovar (a Ni–Fe alloy), and Cu–W and Cu–Mo alloys as well as clad metals such as Cu-clad invar and Cu-clad Mo. However, these materials have high densities and only moderate thermal conductivity. For lightweight thermal management applications, the conductivity-to-density ratio (specific conductivity) is of central importance. In this regard, Kovar, Cu–W, and Cu–Mo are poor candidates on account of their low specific conductivities. Figure 6.1 shows the specific thermal conductivities of a variety of thermal management materials as a function of temperature together with the limiting use temperatures for some materials.

Among the second-generation thermal management materials, SiC/Al composites attracted the most attention in the microelectronic and optoelectronic packaging industry around mid-1980s. In addition, E-glass fiber-reinforced polymers and ceramic- and metal-particle-filled polymers have also been used in microelectronic applications [5, 6]. All such polymer-based materials are limited in their conductivity and temperature capability. For higher-use temperatures, besides SiC/Al, composites such as C/Cu, SiC/Cu, C/Al, diamond/Cu, B/Al, and BeO/Be have been considered [7–12]. Carbon fiber-reinforced aluminum composite microwave packages are used in spacecraft antenna and diamond-particle-reinforced SiC is used in IBM-server heat spreaders. Composite

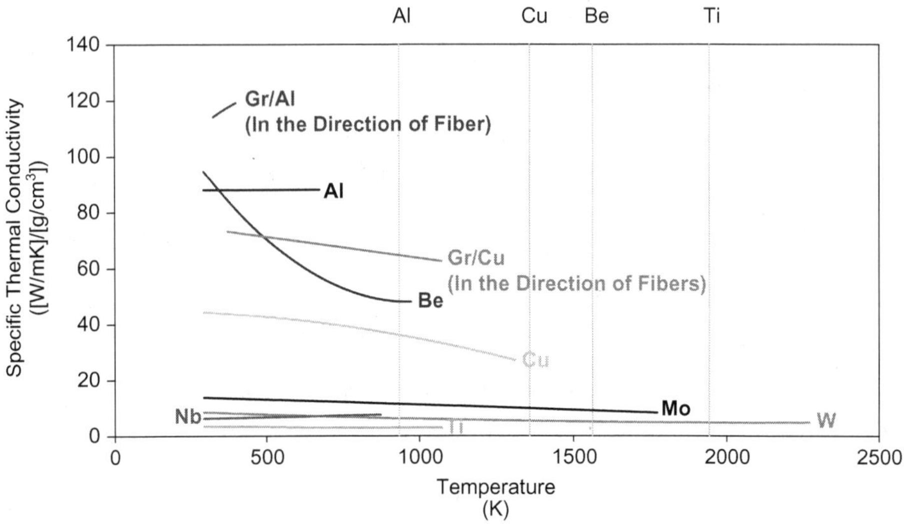

Figure 6.1. Specific thermal conductivities of a variety of thermal management materials as a function of temperature and the limiting use temperatures for some materials. *Source:* SP-100 Program and various standard electronics modules.

solders with low CTEs and high conductivity also have been developed to integrate various material systems.

Composites such as C/C [13] represent the third-generation high-performance thermal management materials. A wide variety of C/C composites have been developed using different types of C fibers, fiber weave patterns, fiber coatings, carbon matrices, and fabrication technologies. The C/C composite material is thermally conductive with a low CTE for better compatibility with a variety of ceramic and nonmetallic materials. It is an outstanding heat spreader and, therefore, excellent to avoid hot spots. Low-CTE, high-conductivity C fibers (K = 1100 W/m-K) are light, stiff, and strong, and are being used in printed circuit boards.

Recently, porous ceramics and porous carbon have also attracted considerable interest in thermal management applications. The base material and pore attributes influence the heat transfer in such materials. Fine pores decrease the permeability (i.e., restrict flow) but increase the interfacial contact for effective heat transfer. Fine pores may be microstructurally less stable during service but provide adequate mechanical properties that may be required to withstand steep thermal gradients and residual stresses during service. Thus, thermal management using porous materials involves complex interplay among several different factors, and an understanding of these factors and their mutual interactions can help in designing and developing new types of porous materials.

Poco foam (HTPoco) is a new ultralight and strong graphite foam for thermal management applications in heat sinks and heat exchangers. It is a graphitic carbon porous foam with an open microcellular structure (75% total porosity, pore diameter ~ 350 μm) that makes it lightweight (density ~ 0.55 g/cm^3) and useful for aerospace thermal management systems. The thermal conductivity of Poco foam is relatively high (out-of-plane: 135–245 W/m-K, in-plane: 45–70 W/m-K [14]). In Poco foam, the combination of open pore structure, dense graphitic matrix, and high conductivity is beneficial to high heat transfer effectiveness.

Still other exotic thermal management materials are beginning to emerge. For example, nanocomposites based on carbon nanotubes (CNTs) with exceptional thermal conductivity (K_{CNT} ~ 3000–6000 W/m-K) are remarkable thermal management materials. CNT dispersions in solders and brazes may provide excellent thermal and mechanical properties. Graphitic nanoplatelets and diamond nanoparticles are other attractive (and in the case of platelets, cheaper) thermal management materials. Besides the benefits of outstanding thermal conductivity and mechanical properties, nanocomposites could potentially offer considerably reduced CTE. Already, there is evidence that microfin structures made of aligned multiwalled CNT arrays can dissipate heat as effectively as the heavier Cu. In addition, CNTs are flexible and resilient, and maintain their strength, low weight, and excellent conductivity at small size scales. Thick CNT films may be soldered to various other substrates for thermal management in microelectronic packages. Finally, graphene, a single atomic-layer-thick plane of carbon atoms arranged in a honeycomb lattice, discovered only in 2004, has exceptionally high thermal conductivity (3500–5300 W/m-K), much higher than that of diamond (1000–2200 W/m-K). All such advanced materials and structures represent the next generation of emerging thermal management materials.

CARBON–CARBON COMPOSITES

Advanced C/C composites (ACCs) fabricated using C fibers and carbon matrices are attractive thermal management materials. High-conductivity carbon fibers can spread heat in the direction of the fiber. Carbon fiber heat sinks are most efficient in systems that do not involve air cooling such as aircraft, missile, and space shuttle. Thermally conductive carbon fibers reduce the CTE mismatch and increase the thermal conductivity of printed circuit boards in packages. These materials improve the system reliability and simplify designs by eliminating a range of complex thermal management techniques such as liquid or air cooling, heat pipes, and fins. Figure 6.2 shows the specific strength and specific modulus of ACCs and other high-temperature materials as a function of test temperature.

Typically, C/C composites have been fabricated using resin, pitch, or carbon deposition via chemical vapor infiltration (CVI). The manufacturing of composites starts with a carbon fiber (derived from polyacrylonitrile [PAN], pitch, or rayon) and the carbon matrix is deposited within the porous perform either by impregnating with a precursor (a thermosetting resin or treated/untreated pitch) or via CVI, in which reactions between gaseous precursors yield the carbon matrix. The resin-impregnated perform is pyrolyzed at 1073–1373 K followed by graphitization, and the cycle is repeated to achieve the desired level of conversion and densification.

The availability of a wide variety of C fibers and the ability to tailor the matrix microstructure and create various types of fiber weave patterns enable tailoring of the

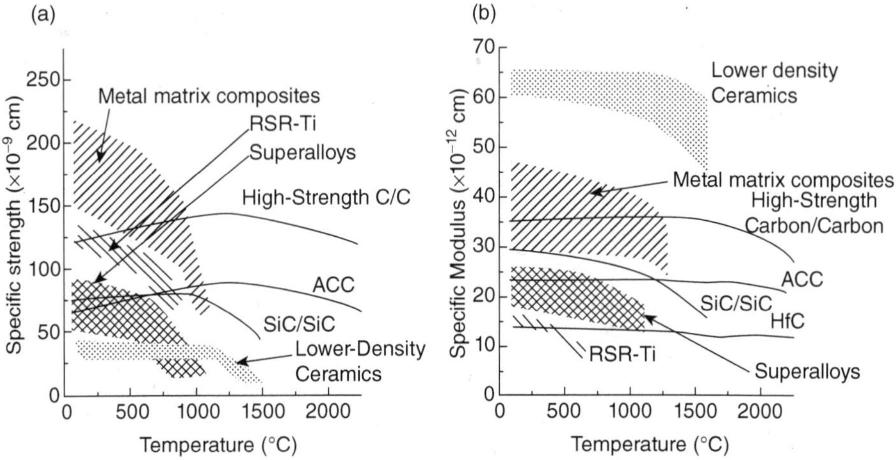

Figure 6.2. (a) Specific strength and (b) specific modulus of C/C composites and selected high-temperature materials as a function of test temperature. *Source:* Composite Materials Handbook, Vol. 5: Ceramic Matrix Composites, MIL-HDBK-17-5, Department of Defense, June 17, 2002, p. 75.

thermal, physical, and mechanical properties of the C/C composite. The composite properties are highly anisotropic and sensitive to porosity in the matrix. Interestingly, monolithic graphite itself is highly anisotropic in many of its properties, such as thermal conductivity. It is an excellent conductor along its basal planes and an insulator along prismatic planes. Pitch base C fibers can approach a thermal conductivity of 1100 W/m-K along their length when the basal planes of carbon are oriented parallel to the fiber axis. A comparison of the specific thermal conductivity of two C/C composites with P100 and K1100 carbon fibers with a number of other materials (Fig. 6.3a) shows the exceptional thermal benefits of C/C composites together with the dramatic influence of anisotropy. The fiber type and fiber architecture also have a significant influence on the thermal conductivity of C/C composites as the data shown in Figure 6.3b clearly reveal.

The thermal expansion of graphite single crystals is also highly anisotropic. The CTE is negative along the basal planes (and increases to 1×10^{-6}/K at higher temperatures) and is ~26×10^{-6}/K along prismatic planes. Carbon fibers grown with the basal planes parallel to the fiber axis have a CTE of nearly zero, but in a perpendicular direction, the CTE is ~10×10^{-6}/K. In C/C composites, the fiber type, fiber architecture, and matrix structure and properties strongly influence the CTE and conductivity.

INTEGRATION OF CARBON AND C/C COMPOSITES WITH METALS

Wettability

Among methods to join C/C composites to metals, brazing (Fig. 6.4a) is the simplest and most promising method [15–19]. Although carbon and graphite have been soldered

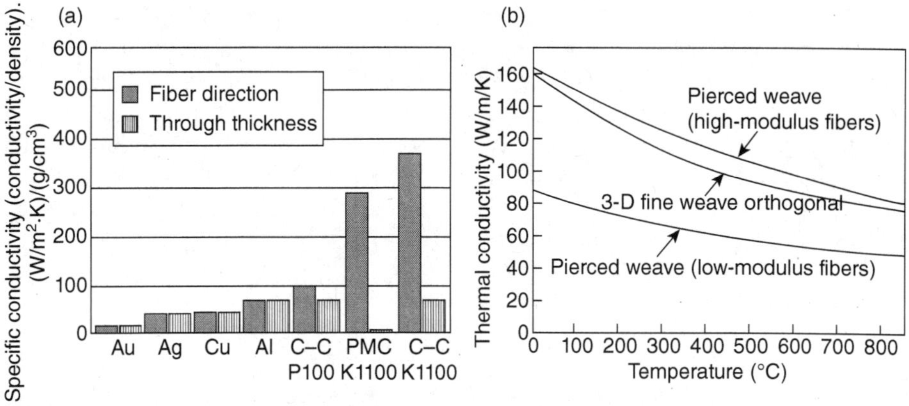

Figure 6.3. Thermal conductivity of C/C composites: (a) comparison of specific conductivity of C/C and other materials, and (b) thermal conductivity of C/C composites with various types of fibers and weave patterns. *Source:* Kearns K. 2002. *Composites, ASM Handbook*, Vol. 21. Materials Park, OH: ASM International, pp. 1067–1070.

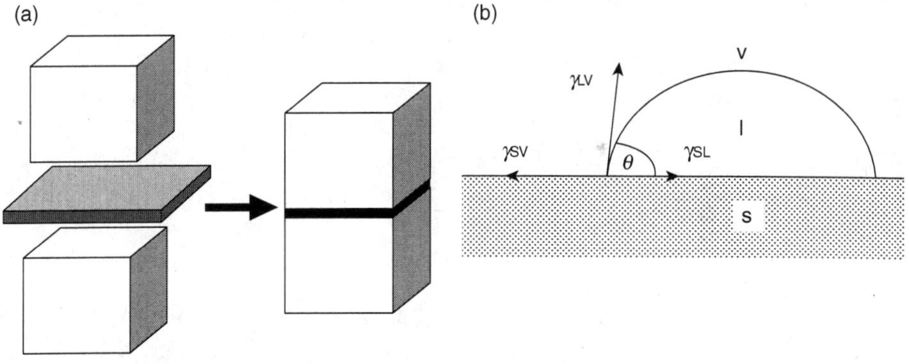

Figure 6.4. (a) Conceptual scheme of brazing and (b) definition of contact angle.

for low-use temperatures for many years, integration of C/C composites to Cu-clad Mo and titanium represents a new thermal management opportunity for relatively higher-use temperatures. Clearly, new challenges and issues become important because of the higher temperatures involved.

Brazing utilizes tapes, powders, pastes, or rods of metallic or glassy filler alloys containing an active metal, such as Ti, that wet and spread in the joint region and solidify to form joint interfaces. Alternatively, premetallized surfaces that promote wettability and bonding can be used. Brazing is generally done under vacuum or high-purity inert atmospheres, and characterization of wettability of solid surfaces is usually done under similar conditions.

Wettability is a fundamental requirement for joining C/C composites to metals via brazing because it governs the spreading and capillary flow processes. High-temperature wettability data can serve as a useful index for brazeability of materials. Although the wettability test standards are currently lacking and the measurements reveal extreme sensitivity to test conditions, wettability data for a wide variety of metals, alloys, and glasses on carbon and carbon-bearing materials are available. Wettability of solid surfaces by liquids is characterized in terms of an angle of contact, θ (Fig. 6.4b), which is defined by the Young–Dupre equation:

$$\sigma_{\mathrm{lv}} \cos\theta + \sigma_{\mathrm{ls}} = \sigma_{\mathrm{sv}},$$

where σ's are the specific interfacial energies of the interfaces between solid (s), vapor (v), and liquid (l). For $\theta < 90°$, the solid is wet by the liquid, and for $\theta > 90°$, the solid is not wet by the liquid. The limits $\theta = 0°$ and $\theta = 180°$ define complete wetting and complete nonwetting, respectively. The interfacial energy σ_{sl} is related to Gibb's free energy of formation (ΔG) of the solid, its stoichiometry, interfacial reactions, temperature, and atmosphere. ΔG is a measure of the chemical stability of the solid, and large negative ΔG values correlate with large contact angles. In other words, the wettability of solids by liquid metals decreases as the stability of the solid increases.

Both graphite and diamond are covalent, high-melting point solids that are characterized by closed stable electron configurations and high-strength interatomic bonds. Metals of the secondary B-subgroups in IV, V, and VI periods of the periodic table (Cu, Ag, Au, Zn, Cd, In, Ge, Sn, Pb, Bi, etc.) exhibit little chemical affinity for carbon. Pure Ag and Cu, base metals for a large number of brazes, do not wet graphite ($\theta \sim 137°$–$140°$). In contrast, transitions metals show considerable interaction with carbon at high temperatures and a large work of adhesion (~20–25 Kcal/mol). Alloying copper with Cr, V, Co, and Fe reduces the contact angle of Cu on graphite [20], with Cr and V showing the greatest improvements in wettability.

The very large values of the surface tension σ_{lv} of the base filler metals Ag, Ni, and Cu at their respective melting points show that these metals in pure state will probably not wet the C–C. For example, the σ_{lv} of Ni is 1796 N/m (at 1728 K), that of Ag is 925 N/m (at 1233 K), and that of Cu is 1330 N/m (at 1359 K), respectively [21]. The temperature-corrected surface tension data also do not show any significant drop in the σ_{lv} of these metals at higher temperatures. The addition of Ti to braze improves the wetting and spreading. The wetting and spreading of Ti-containing braze on C–C is excellent even at relatively small Ti contents [20, 22, 23], which lead to good bonding.

Figure 6.5a, b show the effect of the Ti and Cr contents of some alloys on their contact angle with different types of carbon substrates. These data reveal that Ag and Cu braze alloys containing active metals such as Ti or Cr wet the carbon and, in most cases, equilibrium contact angles approach acute (e.g., near-zero) values from initially large obtuse values [20, 22–26]. Titanium and chromium promote reaction-enhanced wettability and braze flow, spreading, and bonding.

The literature data on contact angles also show that usually, short brazing times suffice for maximum spread; for example, θ approaches 0° in 5 min for Cu-12%Ti melt

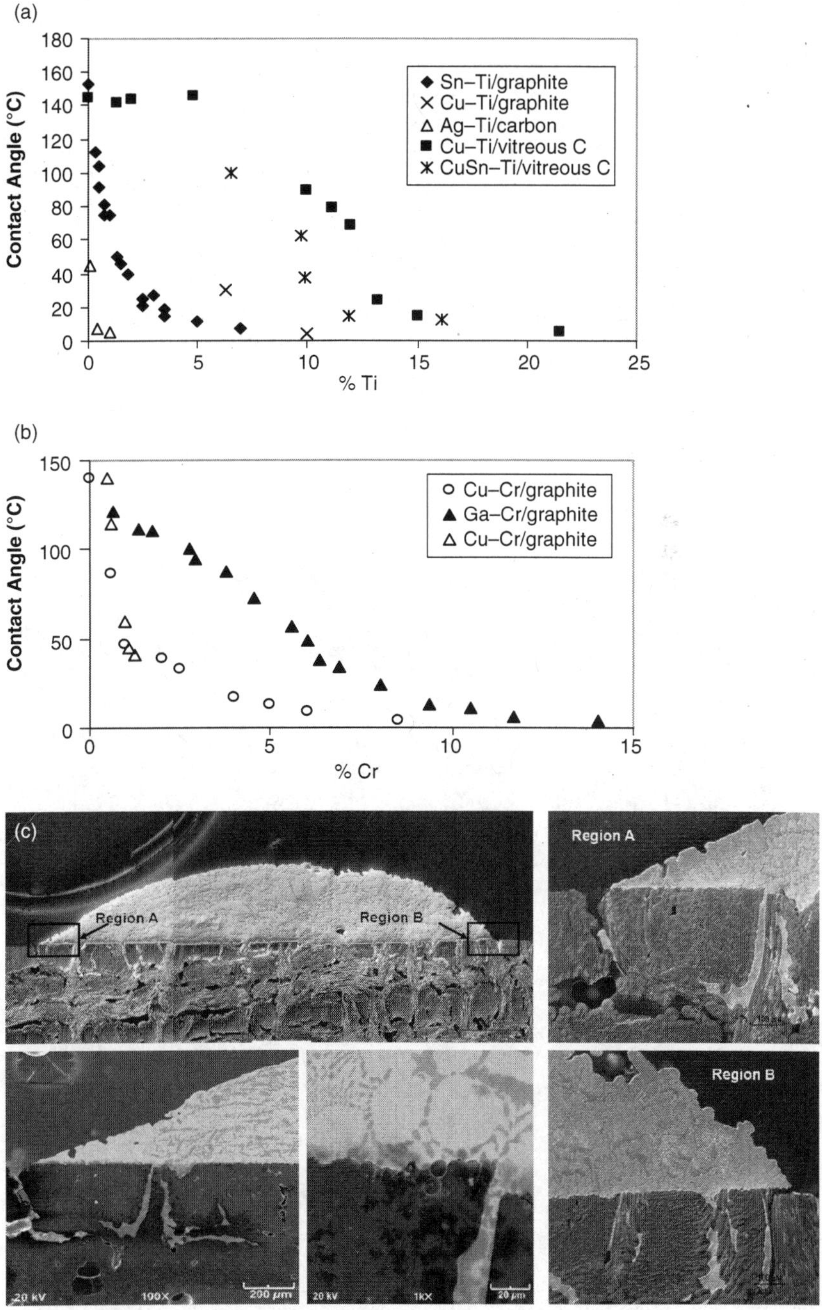

Figure 6.5. Contact angle versus active metal content for various alloys containing (a) Ti and (b) Cr on graphite substrates ([22–24, 28]). (c) A solidified Ticusil braze sessile drop on C/C composite (1073 K, 5 min): clockwise (from top left)—overall view, images of regions "A" and "B" near droplet periphery, interface near droplet center, and interface near edge [31].

TABLE 6.1. Selected Properties of Braze Alloys Used to Join C/C to Metals

Braze	Composition (%)	T_L (K)	T_S (K)	E (GPa)	YS (MPa)	CTE ($\times 10^{-6}$/K)	K (W/m·K)	%El
Cu-ABA[a]	92.8Cu-3Si-2Al -2.25Ti	1297	1231	96	279	19.5	38	42
MBF-30[b]	Ni-4.61Si-2.8B-0.02Fe-0.02Co	1257	1327	—	—	—	—	—
MBF-20[b]	Ni-6.48Cr-3.13Fe -4.38Si-3.13B	1242	1297	—	—	—	—	—
Ticuni[a]	15Cu-15Ni-70Ti	1233	1183	144	—	20.3	—	—
Ticusil[a]	68.8Ag-26.7Cu -4.5Ti	1173	1053	85	292	18.5	219	28
Palcusil-15[a]	65Ag-20Cu-15Pd	1173	1123		379	—	98	23
Silcoro 75[a]	75Au-20Cu-5Ag	1168	1158	—	—	—	—	—
Cusil-ABA[a]	63Ag-35.3Cu-1.75Ti	1088	1053	83	271	18.5	180	42
Cusil[a]	72Ag-28Cu	1053	1053	83	272	19.6	371	19

[a]Morgan Advanced Ceramics.

[b]Metglas, Inc.

T_L, liquidus temperature; T_S, solidus temperature; E, Young's modulus; YS, yield strength; CTE, coefficient of thermal expansion; K, thermal conductivity; %El, percent elongation.

in contact with vitreous C [23]. Besides titanium, silicon also improves wettability via the formation of silicon carbide at the interface [27, 28]. In commercial brazes such as Cu-ABA (Table 6.1) that contain both Si and Ti, the formation of TiC and substoichiometric titanium carbides ($TiC_{0.95}$, $TiC_{0.91}$, $TiC_{0.80}$, $TiC_{0.70}$, $TiC_{0.60}$, and $TiC_{0.48}$) together with SiC may take place. Pure Si reacts with and dramatically lowers the contact angle on carbon ($\theta \sim 0°$ for Si/C [27, 28]). The formation of TiC and SiC is thermodynamically favored at the brazing temperatures for AgCu brazes. Gibb's free energy change (ΔG) for TiC formation via $Ti + C \rightarrow TiC$ in the temperature range 1193–1323 K is −174 to −169 kJ and ΔG for SiC formation via $Si + C \rightarrow SiC$ over the same temperature range is −62.4 to −61.1 kJ. Besides carbides, wettability-enhancing, metal-like interfacial oxides such as TiO may form from the reaction of Ti in the braze with the residual oxygen in furnace atmosphere (TiO is a stable oxide of Ti that forms even at very low oxygen partial pressures, $\sim 10^{-28}$ atm).

Nickel in braze alloys (e.g., Ticuni, Table 6.1) has a higher affinity for carbon than Cu, Au, and Ag, and segregates at the C/C/metal interface. Even though there are no stable carbides in the Ni–C system, the formation of metastable nickel carbide due to interaction of liquid Ni with solid C has been observed; this leads to a reasonably good wettability with relatively low contact angles (~68°–90°) of Ni on C. However, as the principal constituent of Ticuni is Ti (~70%), which strongly reacts with the carbon, the formation of nickel carbide is less probable. Similar considerations apply to the alloying elements in other brazes.

The braze metallurgy and phase transformations influence the reactions between active metals in braze and carbon. In the case of the AgCuTi braze, Ticusil (Table 6.1), the alloy system contains a liquid miscibility gap. The ternary Ag–Cu–Ti phase diagram shows that eutectic Ag–Cu alloys containing more than 5 at% Ti divide the alloy into a Ti-depleted liquid and a Ti-rich liquid, the latter containing greater than 5 at% Ti. This Ti-rich liquid actually reacts with the carbon to form a metallurgical bond. The Ti-depleted liquid forms a Ag(Cu,Ti) solid solution upon solidification and eutectic-type phase mixtures. The Ti-rich liquid solidifies to yield Ag(Cu) precipitates and intermetallics such as AgTi, Ti_2Cu_3, and $TiCu_2$.

The spreading and bonding behaviors depend also upon the reaction layer morphology, which is sensitive to porosity in the substrate. For example, in the case of porous graphite, Cu–Ti alloys both wet and impregnate the graphite. With Ti in braze, the TiC reaction layer is discontinuous with a nonhomogeneous structure [26] that permits continued penetration of porous carbon by the melt until either all the liquid supply is exhausted or the porous carbon becomes saturated with molten braze.

The chemically and structurally inhomogeneous surfaces of C/C composites render the prediction of brazeability of the composite from knowledge of the wettability of individual constituents quite challenging. Surface roughness, fiber arrangement, fiber weave pattern, surface chemical inhomogeneity, and defect populations can significantly influence the contact angle and braze flow characteristics. Scant measurements of contact angle on C/C composites currently exist, and this represents an important research opportunity in the joining area.

Surface roughness can remarkably influence the wettability (via contact angle hysteresis) and stress concentration in joints. As a general rule, roughness may increase the contact (bonded) area and chemical interactions in a wettable system besides promoting frictional bonding, but it may also enhance stress concentration due to notch effects. Generalizations of the roughness effects may be difficult because surface preparation (e.g., machining, grinding, and polishing) not only reduce roughness but also might introduce surface and subsurface damages (grain or fiber pullout and void formation). In addition, residual stresses may be introduced by machining and grinding. In some ceramic materials, damage induced by grinding may be healed by refiring the ground material prior to brazing. However, in the case of carbon, secondary roughening may occur during vacuum heating after polishing. This is because even though some porosity is closed by substrate grains detached during polishing, during subsequent heating under vacuum, these grains are removed along with evacuating gas, resulting in increased surface roughness as is actually observed [29, 30].

The surface chemical and structural inhomogeneity originating in different chemical phases and fiber architecture in composites demands direct practical assessment of braze wettability and spreading behavior for joining applications. Sessile-drop wettability tests with Ag- and Pd-based active brazes on C/C and SiC/SiC composites reveal [31] nonuniform, anisotropic spreading, copious braze infiltration, and Ti enrichment at interfaces together with dissolution and redistribution of elements in and around the joint, and some tendency toward interlaminar shear cracking within the composite substrates having low interlaminar shear strength. Figure 6.5c shows a solidified Ticusil braze droplet on a C/C composite where good wetting was accompanied by braze infiltration and edge cracking due to residual stresses. Contact angle data extracted from such high-temperature sessile-drop tests provide insight into braze spreading on real, heterogeneous composite surfaces.

Thus far, we had considered only the wettability of carbon by brazes. A few comments are in order for the wettability of the metallic substrates joined to C/C composites by brazes. Most active braze alloys may be expected to wet metal substrates (e.g., Ti and Cu-clad Mo) because of metal-to-metal contact. Generally, the braze/metal wettability is characterized in terms of a wettability index (WI), where WI = (area covered by the braze metal) × cos θ, where θ is the contact angle. The higher the value of WI, the better is the wetting and spreading of the braze alloy [32]. WI is used only for a relative assessment of braze spreading behavior and depends upon the volume of the braze metal at the joint. Generally, WI > 0.05 indicates good spreading, and WI > 0.10 indicates excellent spreading characteristics. For Ni-based brazes such as MBF alloys, good wetting is expected on Ti substrates; this is because these brazes show excellent wetting with steels and superalloys and lead to very low contact angles (<10°–15° [33]). High vacuum levels and reducing or inert gas atmosphere (e.g., H_2, Ar, N_2, or their mixtures) yield low contact angles (e.g., less than 6°) between MBF brazes and steels in pure Ar [33]. The WI of a number of commercial brazes on metals and alloys has been characterized [32].

Brazing

Both graphite [34] and C/C composites [34–40] have been integrated with metals and alloys such as Cu, Ti, and Cu-clad Mo. Carbon–carbon composites have been integrated with titanium tubular subcomponents (Fig. 6.6) for use in Brayton and thermoelectric power conversion systems. Titanium is light (density: 4500 kg/m^3) and has a relatively low CTE (~8.6 × 10^{-6}/K) that is moderately different from the CTE of C/C composites (~0–1.0 × 10^{-6}/K over 20–250°C, and ~2.0–4.0 × 10^{-6}/K over 293–2773 K). These properties make Ti suitable for tubular components carrying hot or cold fluid and for integrating Ti with high-conductivity C/C composites for thermal management. Even though the thermal conductivity of Ti is low (K_{Ti} ~ 22 W/m-K), thin-walled Ti tubes offer little resistance to heat transfer from a fluid flowing through Ti tubular subelements joined to C/C composites using thermally conductive Ag and Cu active brazes.

Copper-clad molybdenum allows for an excellent combination of thermal conductivity and controlled CTE to be obtained. The conductivity and CTE of Cu-clad Mo depend upon the clad layer thickness [41]. For example, the thermal conductivity of Cu-clad Mo (27% Cu thickness per side of Mo substrate) is ~224 W/m-K [13], which is better than that of Al (K_{Al} = 205 W/m-K) but slightly inferior to that of pure Cu

Figure 6.6. C/C composites joined to titanium tubes using a saddle material (carbon foam).

(K_{Cu} = 400 W/m-K); however, the CTE of Cu-clad Mo (~8.2 × 10^{-6}/K for 27% clad layer thickness) is about half that of pure Cu (16.5 × 10^{-6}/K) and only moderately high compared to the CTE of C–C (~2.0–4.0 × 10^{-6}/K). Additionally, the high ductility of the copper-clad layer is beneficial from the standpoint of residual stresses.

Recently, the integration of controlled expansion Cu-clad Mo with high-conductivity C/C composites has been explored for increased functionality at reduced weight in thermal management applications [35, 37]. Acting in combination, Cu-clad Mo and C/C can provide excellent heat dissipation capability. By controlling the clad layer thickness on Mo via rolling, the CTE mismatch between C/C and Cu-clad Mo can be designed to mitigate residual stresses during joining and service while maintaining high levels of thermal conductivity for heat conduction.

Braze interlayers play a key role in seamless thermal integration of a system. In general, braze alloys suitable for joining C/C to Ti and Cu-clad Mo must satisfy a number of criteria, which essentially represent a compromise. First, brazes essentially serve as high-performance metallic TIMs and their thermal properties (*K* and CTE) are important. Figure 6.7 shows the *K* and CTE values of a number of off-the-shelf commercial brazes including several that have been used to join C/C to metals. Besides high conductivity, brazes must show good wetting and adherence to the substrates; be ductile and resistant to grain growth, creep, and oxidation; have closely matched CTE with the joined materials; and have melting points greater than the operating temperature of the joint but lower than the joined materials' melting temperature. Many commercially available brazes are based on Ag, Au, Cu, Ni, and Pd and are available in paste or foil form. A number of rapidly solidified Ni-, Co-, Cu-, Ti-, and Zr-based metallic glass braze alloys with high strength, leak tightness, and resistance to shock, vibration, and corrosion are also available. Most commercial brazes contain an active metal, such as Ti, that promotes wetting and bonding. Ti additions to Cu- and Ag-based brazes decrease the latter's melting point, which makes braze spreading easier, leading to the formation of a good joint. However, Ti alloying usually decreases the thermal conductivity; as a result, the Ti content of brazes is kept low (<3 wt%), so there is minimal detriment to the braze conductivity.

The brazing process to join C/C composites to Ti and Cu-clad Mo used by the authors consists of placing the braze foil or paste between the metal and the composite substrates, heating the assembly to 15–20 K above the braze liquidus in vacuum (10^{-6} Torr) under a minor normal load, isothermally soaking (3–30 min) at the brazing temperature, and slowly cooling the joint to room temperature. Table 6.2 summarizes the C/C/metal joints that we have successfully brazed in flat-back or tubular configurations with Ti, Cu-clad Mo, and Ni-based superalloys (Inconel 625 and Hastelloy). The braze alloys included Ag-, Cu-, Ti-, and Ni-based alloys. Scanning electron microscope (SEM) and energy dispersive spectroscopic (EDS) examinations, microhardness tests, and a limited number of joint strength measurements were done to evaluate the joint integrity, microstructure, strength, and chemical interactions.

Two variations of conventional brazing techniques to join C/C composites bear special mention. In the first, liquid infiltration with braze interlayers is used for joining. For this, transition metals such as Cr and Mo are deposited on the C/C surface via slurry deposition followed by heat treatment to form micrometer-size carbide layers [42]. The

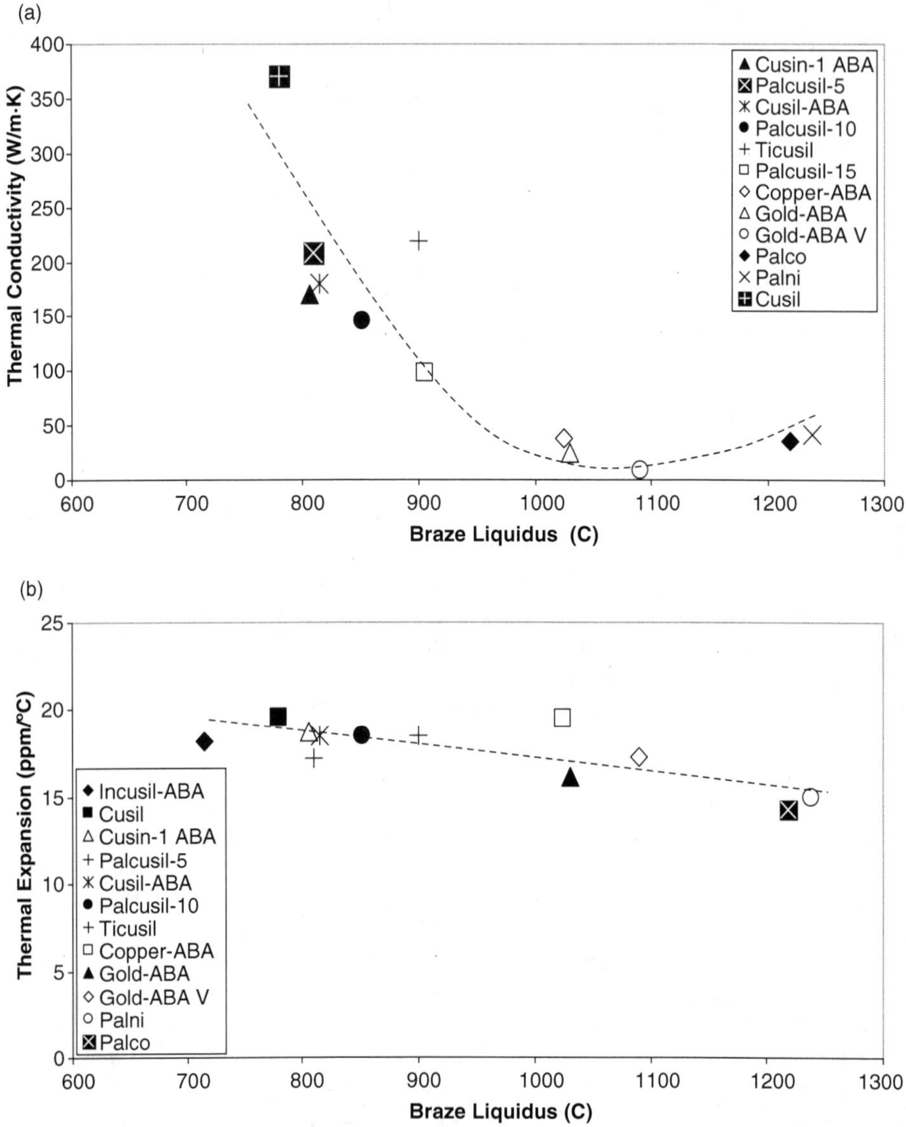

Figure 6.7. (a) Thermal conductivity and (b) thermal expansion of some commercial brazes used to join C/C and ceramics to metals.

TABLE 6.2. Braze Effectiveness in Joining C–C Composites to Metals

Composite	Metallic substrate	Braze	Bonding
C/C[a,f]	Ti	Silcoro-75,[h] Palcusil-15,[h] Cusil	Weak
C/C[b]	Ti	Ticuni, Cu-ABA, Ticusil	Good
C/C[a,f]	Ti and Hastelloy	MBF-20,[h] MBF-30[h]	Good (Ti), fair (Hastelloy)
C/C[c–e]	Ti, Cu-clad Mo,[i] Inconel 625	Ticusil[g]	Good,[d] fair[e]
C/C[c–e]	Cu-clad Mo[i]	Cusil-ABA[g]	Good
C/C[c–e]	Ti and Inconel 625	Cusil-ABA[g]	Good

[a]Polished.
[b]Not polished.
[c]3-D composite (P120 C fiber, CVI C matrix), Goodrich Corp., Santa Fe Springs, CA.
[d]Oriented fiber at the joint (3-D composite).
[e]Nonoriented side at the joint.
[f]T-300 C fibers in resin-derived C matrix, C-CAT, Kennedale, TX.
[g]Braze paste.
[h]Braze foil.
[i]H.C. Starck, Inc., Newton, MA.

carbide-coated C/C composite surface is brought in contact with copper foil or paste and heated. This leads to capillary infiltration of porous carbide layer by Cu, resulting in the formation of a pore-free and crack-free joint. These joined samples can then be directly brazed to other alloys.

The second variation of the brazing approach involves the use of soft and ductile rheocast Cu–Pb alloys as thermal bond layers [43] that are capable of absorbing CTE mismatch-induced residual stresses. Rheocast alloys have a nondendritic, globular microstructure and excellent flow characteristics, and exhibit non-Newtonian behavior. Rheocast alloys are obtained via mechanical or electromagnetic (EM) stirring of partially solidified alloy slurries and prolonged holding of the slurry in the partially solidified state to enable coalescence and globularization of primary solid dendrites. The viscosity of rheocast slurry is shear rate dependent (pseudoplastic) and time dependent at a constant shear rate (thixotropy).

The shear strength of C/C composites joined to themselves and to pure Cu using rheocast Cu–Pb monotectic alloys is, however, quite low (~1.5–3 MPa). The main reason for the low joint strength is the poor wetting characteristics of the Cu–Pb alloys on C/C composite. Contact angle measurements in a hot-stage microscope [43] revealed nonwetting values in the range 101°–104° at 983 K under argon atmosphere. Such large contact angles clearly demonstrate poor spreading characteristics. Conceivably, wettability and joint strength may be improved either by the use of active fillers in Cu–Pb alloys or by judicious surface modification of the C/C composite surface (e.g., deposition of Cr films on C/C using sputtering, vapor deposition, or thermal decomposition).

Methods to join C/C to metals must be cognizant of the role surface coatings on C/C might play. Multilayer, multifunctional coatings are frequently applied to C fibers and C/C composites to serve as diffusion barrier to resist oxidation and corrosion, inhibit thermal degradation, and promote compatibility with various matrices. For example, C/C composites are coated with SiC or Si_3N_4, and they may also contain internal oxidation inhibitors such as boron, titanium, and silicon. At high operating temperatures, these elements form glassy oxide phases within the C/C composite, which seal any CTE mismatch-induced cracks in the external coating. This creates a barrier to the diffusion of oxygen into the porous composite matrix and retards its degradation. An important consideration in joining C/C is whether the external coating should be applied before or after joining. This is especially important to consider for brazed C/C joints because most external coatings are applied at 1273–1673 K, and most brazes will actually melt at these temperatures. In addition, the effect of glass-forming inhibitors added to the composite on the wettability and adhesion of braze must be considered.

JOINT INTEGRITY, MICROSTRUCTURE, AND COMPOSITION

Typical joint microstructures of C/C composites joined to Ti and Cu-clad Mo using a number of brazes for thermal management applications are shown in Figure 6.8. The joint microstructures display intimate contact and sound metallurgical bonding at the composite/metal interface. The contact regions between the composite and braze are generally free of structural imperfections such as interfacial microvoids and cracks. With Ti-containing brazes, the composite/braze interfaces were found to be enriched with Ti. Additionally, some interdiffusion of alloying elements was noted among the C/C, braze, and Ti or Cu-clad Mo. The Cu cladding at the braze/Cu-clad Mo interface remains untransformed because the joining temperatures in most cases were below the melting point of Cu (1359 K). For AgCuTi brazes, the braze matrices generally exhibit the characteristic two-phase eutectic structure with Ag-rich light-gray areas and Cu-rich dark areas. Intermetallics such as AgTi, Ti_2Cu_3, and $TiCu_2$ may also form.

In C/C composites with relatively high porosity content, infiltration of the interfiber regions by Ti-containing brazes to distances on the order of several hundred micrometers in a short (~5 min) span of time may occur (Fig. 6.9). This is because of the excellent wettability of Ti-containing brazes consistent with the sessile-drop wettability test results [26] on Cu–Ti/porous C, in which the sessile drop volume continuously decreased due to the reactive infiltration of open porosity in graphite. In fact, sessile drops of high Ti content (~28%Ti) rapidly and completely disappeared into the graphite substrate. Joint interfaces between resin-derived C/C composite and Cu-clad Mo made using AgCuTi brazes show cracking within the composite. However, as with other systems, Ti segregation occurred at the C–C/braze interface, suggesting possible formation of a Ti-rich interphase, such as TiC_{1-x}, which bonds well to both the carbon and the braze.

Carbon–carbon composites with P120 fibers were joined to porous carbon foam (HTPoco) and Ti using Cusil-ABA braze (Table 6.2) [36]. The braze filled the gaps and pores of the substrates; braze paste exhibited better penetration characteristics than

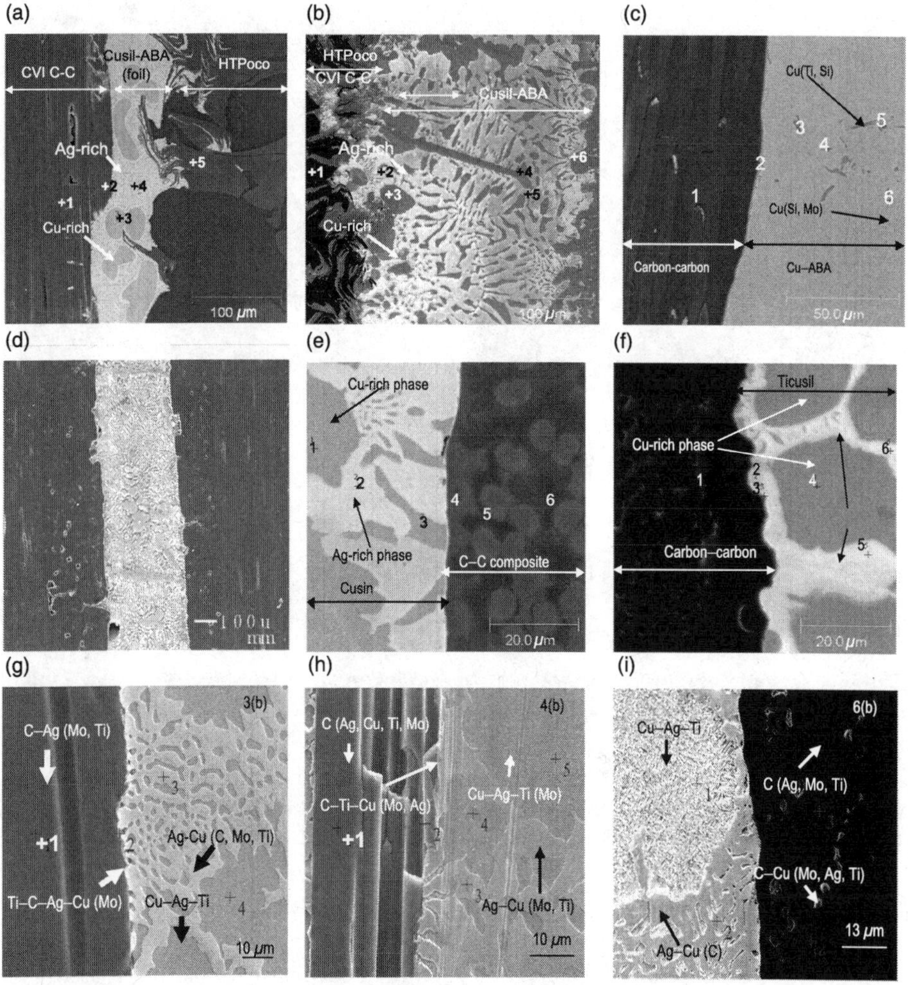

Figure 6.8. C/C brazed to different substrates: (a, b) C/C to Poco foam using Cusil-ABA, (c) C/C to Cu-clad Mo using Cu-ABA, (d) C/C to itself using Cusil-ABA, (e) C/C to Ti using Cusin, (f) C/C to Cu-clad Mo using Ticusil, and (g–i) C/C to Cu-clad Mo using Ticusil or Cusil-ABA.

braze foil. The highly porous foam draws the molten braze away from the interface and leaves an insufficient quantity of braze for joint formation. But the joint microstructures show good bonding devoid of cracks as well as a distinct braze interlayer region. The low-density Poco foam has significant flexibility and can accommodate elastic strains better than dense, solid graphitic carbon, thus preventing cracking. In addition, the foam has a small CTE, which compares favorably with the CTE of the C/C composite (the in-plane CTE of Poco foam is 1.02×10^{-6}/K, and the out-of-pane CTE is -1.07×10^{-6}/K). This is beneficial in reducing residual stress and distortion.

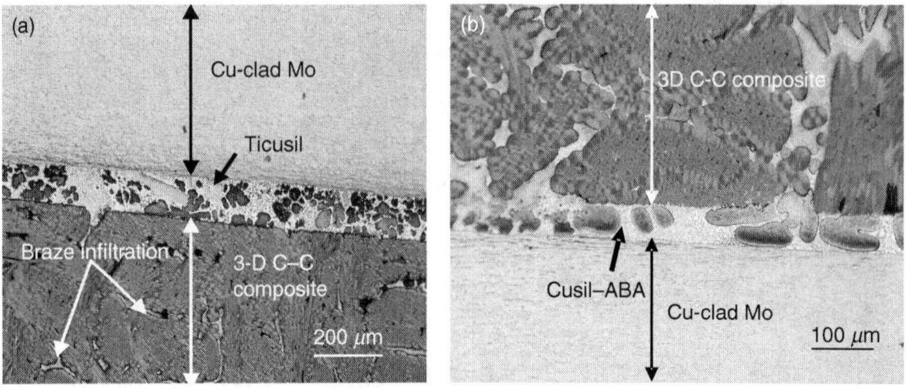

Figure 6.9. Braze infiltration of pores in a CVI C/C composite joined to Cu-clad Mo using (a) Ticusil and (b) Cusil-ABA brazes.

MECHANICAL PROPERTIES OF C–C COMPOSITE/METAL JOINTS

Joint Strength and Fractography

The mechanical strength of the joints between C/C composites and Ti plates or tubes with or without a saddle material (e.g., carbon foam) was characterized as a function of braze type and fiber ply orientation at the mating surface in the joint. For joints involving Ti tubes, a "tube-on-plate" test was developed and implemented [39]. Figure 6.10a shows the average fracture loads and Figure 6.10b shows the strength of the joint structures for different braze compositions with C/C composites that were either oriented with the surface ply fiber tows aligned parallel to the tube axis or perpendicular to the tube axis. Typical fracture surfaces with different fiber orientations are shown in Figure 6.10c. Fracture always occurred within the surface ply of the C/C composite and not within the braze itself indicating good chemical bonding of the braze compositions to the composite.

Two factors definitely contribute to the load-carrying ability of these joints: the bonded area of the braze material and the fiber-tow orientation. The joint structures with the Cu-ABA braze had the highest load-carrying ability because the bonded area of the braze material was largest (Fig. 6.10a). Presumably, this was due to the superior spreading properties of this braze material. In addition, on average, the composite plates with fiber tows in the outer ply aligned perpendicular to the tube axis had higher load-carrying ability than composite plates with outer ply fiber tows aligned parallel to the tube axis for all three braze compositions (Fig. 6.10c). Evidently, the higher load-carrying ability is due to the greater number of tows that are brazed to the Ti tube when fiber tows are oriented perpendicular to the tube axis compared to the parallel orientation.

For the Ti tube/foam/C–C plate sandwich structure, tension and shear tests were performed in the configurations shown in Figure 6.11a, b. For these joints, Ti tubes

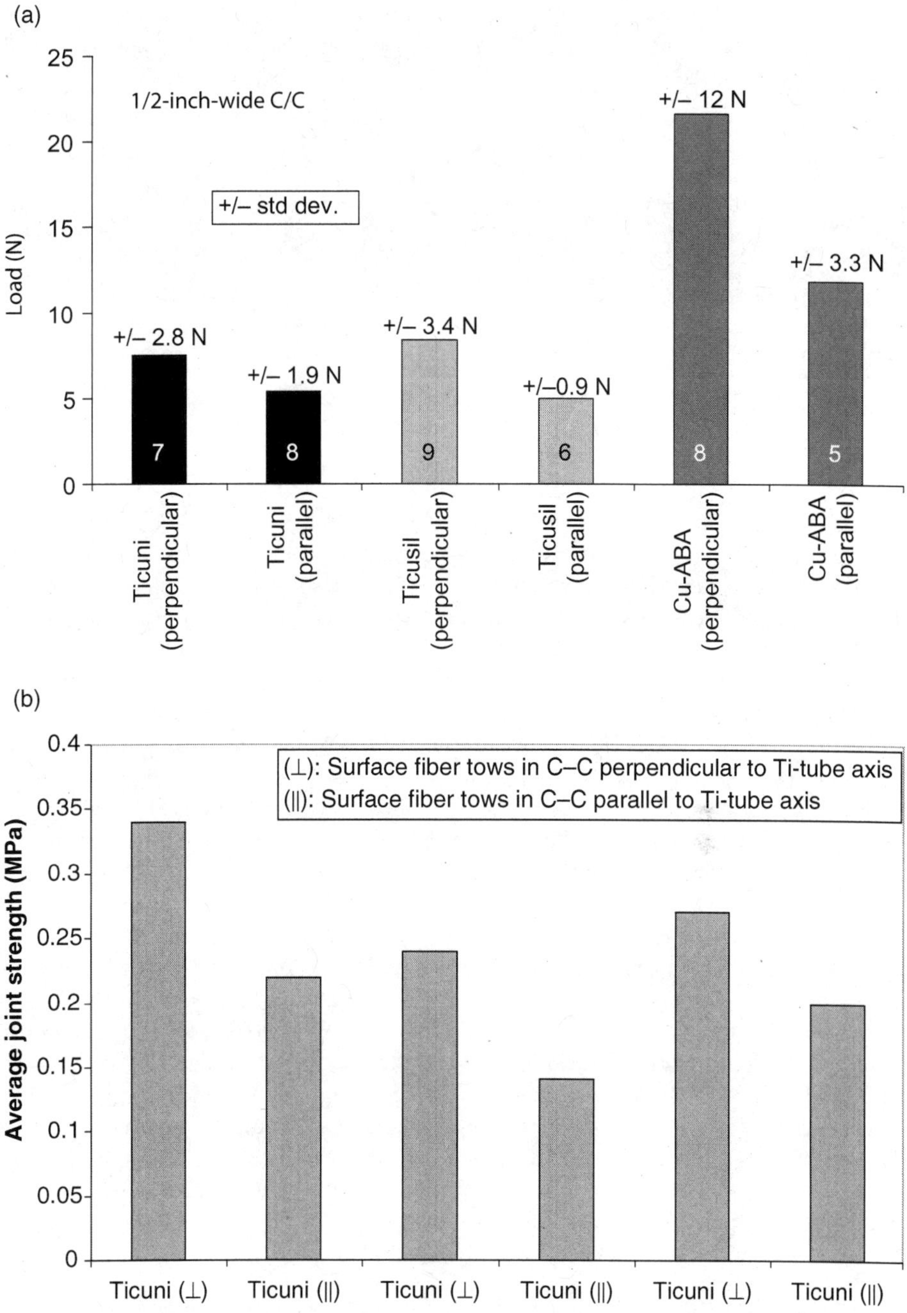

Figure 6.10. (a) Tensile failure loads, (b) joint strength, and (c) typical fracture surfaces of Ticuni and Cu-ABA C/C plate/Ti-tube joint structures (the fiber tows in the surface ply of the composite were oriented perpendicular to the tube axis; the dotted lines show the size of the bonded area for the braze material).

Figure 6.10. (*Continued*)

with an outer diameter of 12.66 mm and a wall thickness of 1.5 mm were joined to a grooved (6.8-mm radii) or flat-sided HTPoco foam that was joined to a P120 C/C or a K1100 five-harness satin C/C composite. The shallow trough HTPoco foam was grooved to a depth of 2.5 mm from the surface, whereas the deep trough HTPoco foam was grooved to a depth of 5 mm from the surface. The reason for joining the tubes to flat and grooved areas was to create different stress states on the braze joints; that is, the shallower the groove, the higher the stress on the tube-to-HTPoco foam joint.

The tube tensile test was used to assess the tensile load-carrying ability of tube-to-plate joints for Ti tubes brazed to C–C plates with a variety of brazes and C/C surface conditions. The P120 C/C–HTPoco foam sandwich was mounted on a rigid block with a 14-mm-wide opening.

A load was applied to the tube via a 6.25-mm steel rod so that the Ti-tube/HTPoco foam joint would be in tension. For tube-to-plate shear stress test, the tube was rigidly fixed to a block using a bolt that locked the tube to the block. A sheath (16-mm inside diameter, 25-mm outside diameter) connected to the crosshead was lowered so as to contact the P120 C/C–HTPoco foam sandwich structure and then was loaded to failure.

The heat-rejection application would favor a grooved approach; however, early on during testing, it became obvious that the HTPoco foam was the weakest member of the structure. In order to increase the stress on the tube-HTPoco brazed joint to promote failure in the joint region, shallow groove and then flat surface specimens were fabricated and tested. The strength of the tube-HTPoco foam joint was greater than 12 MPa in tension or shear. There was no indication that any of the joints were near failure. Thus, failure in tension and shear always occurred in the foam regardless of the type of C/C composite used and whether the Ti tube was brazed to a curved foam plate to maximize the bonded area or a flat foam surface to maximize stress in the joint. Figure 6.11c shows typical failures that occurred in the HTPoco foam.

Microhardness

The microhardness distribution across joints provides an assessment of the combined effects of several different factors, which include distribution of any special phases

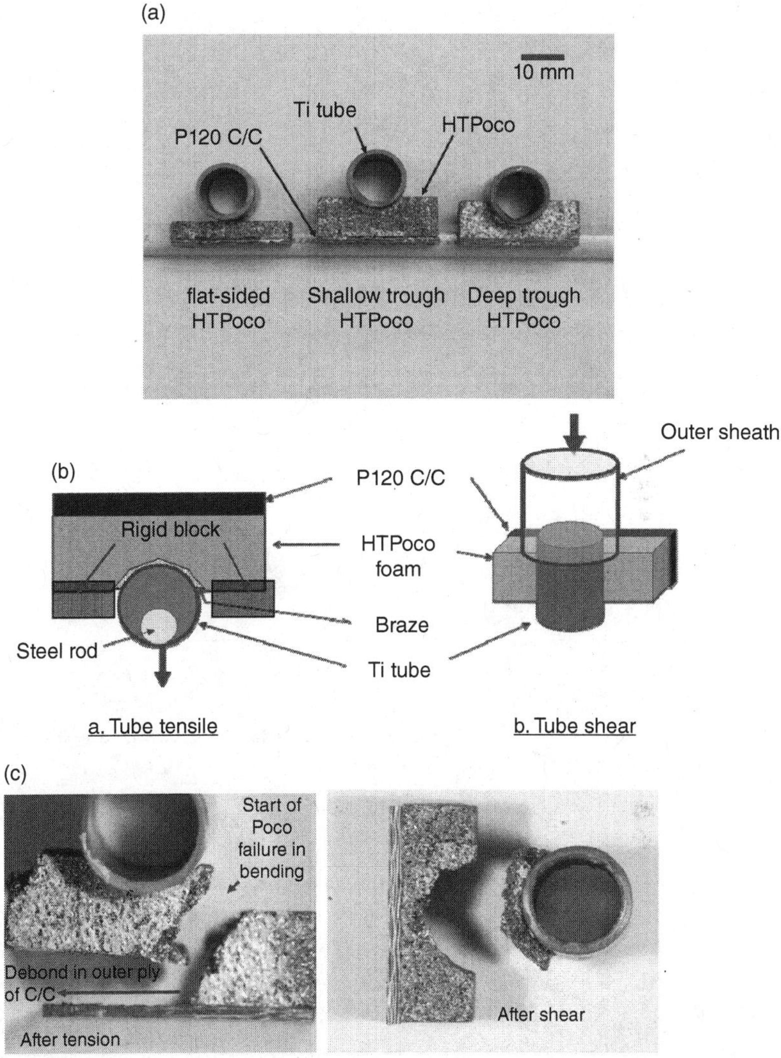

Figure 6.11. (a) Photograph of brazed Ti-tube/HTPoco/P120 C/C composite mechanical test specimens, (b) tensile and shear test procedures, and (c) failed specimens after tube tensile test (left) and tube shear test (right). Both specimens in (c) had shallow grooves.

(e.g., carbides), solid solution hardening, and residual stresses due to CTE mismatch. Large residual stresses due to CTE mismatch yield high hardness in well-bonded joints. The hardness distribution is also sensitive to the measurement path and the distribution of metallurgical phases encountered in the path by the indenter. Whereas hardness distributions may change with heat treatment and service conditions, the microhardness profiles of as-fabricated joints can shed light on mechanical integrity and stress state

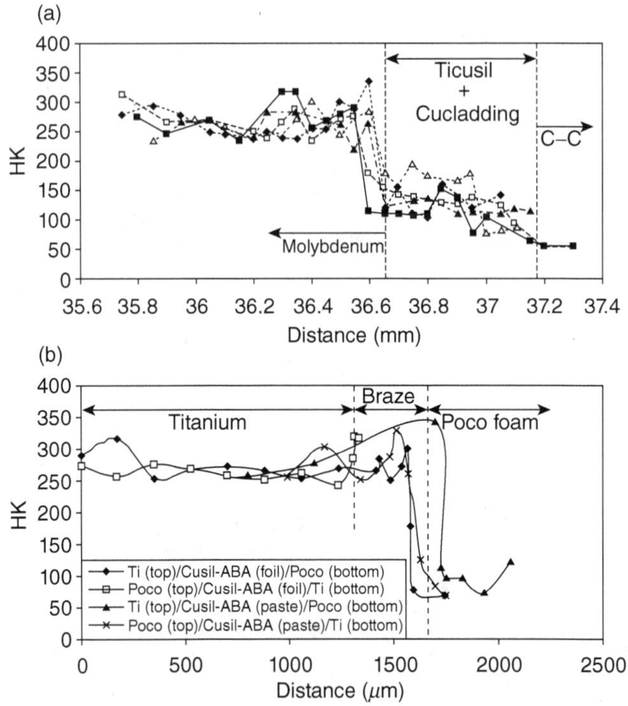

Figure 6.12. Distribution of Knoop microhardness across (a) a C/C composite joined to Cu-clad Mo using Ticusil and (b) Poco foam joined to Ti using Cusil-ABA ([35, 36]).

in the joint. Figure 6.12 shows Knoop microhardness profiles across the interfaces in two types of C/C joints. The measurements were made using a Knoop microindenter under a load of 200 g and loading time of 10 s. The hardness profiles were reproducible and consistent with the expected behavior. Generally, the AgCuTi braze containing higher Ti contents exhibited slightly higher hardness consistent with the greater hardening expected due to Ti.

THERMAL AND THERMOMECHANICAL CONSIDERATIONS

Thermal Expansion Mismatch and Residual Stress

Residual stresses arising from the CTE mismatch and large temperature excursions are detrimental to the integrity of the joint. These stresses may lower the joint strength as revealed in Figure 6.13, which shows the effect of elastic thermal strain, $\Delta\alpha\Delta T$ ($\Delta\alpha$, CTE mismatch; ΔT, temperature interval), which develops during brazing on the shear strength of C/C-to-Ti joints. A very large CTE mismatch might occur in some systems (e.g., Inconel to C/C composite joints), which may increase the propensity for fracture.

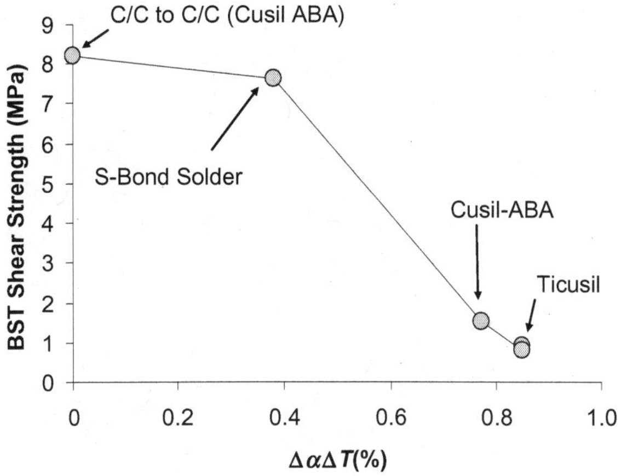

Figure 6.13. C/C/Ti joint strength as a function of thermal strain for different brazes; data for C/C joined to itself are also shown for comparison.

Most metallic brazes also have large CTE values (~15–19 × 10^{-6}/K), but they are usually quite ductile (20–45% elongation). If bond strength is high and the braze is not ductile (e.g., glass interlayer), then the residual stresses could cause joint failure. Judicious use of braze chemistry and application of innovative brazing strategies (e.g., compliant layers, layers with graded CTE, and transient liquid-phase bonding) can be used to reduce residual stresses in joints.

Models of residual stress development in joints show that for well-bonded joints with a metal interlayer, residual stresses are confined to a distance of about three times the thickness of the interlayer. Near the center of the joint far from the free surface, stresses parallel to the interface are tensile in the metal and compressive in the nonmetallic substrate (e.g., C/C or a ceramic) and show a discontinuity (jump) at the interface. Stresses normal to the interface are compressive in the metal and tensile in the nonmetal at the free edge of the interface where these stresses reach their maximum value and may cause fracture. Interlayer plasticity reduces the tensile stress, and ductile interlayers with low yield strength (σ_y) can almost completely eliminate the fracture.

It is well-known that ductile metal interlayers reduce the stresses at the bonded region, and joint strength generally increases with increasing interlayer thickness. For brazed joints of C/C composites to Cu-clad Mo, the CTE of Cu-clad Mo varies with the clad layer thickness (5.6 × 10^{-6}/K to 11.6 × 10^{-6}/K for 0–40% clad layer thickness per side of Mo) and the CTE of C/C composites is ~2.0–4.0 × 10^{-6}/K over 293–2773 K. Thus, CTE mismatch between C/C and Cu-clad Mo is moderate, which is beneficial for the integrity of the joint.

Analytical and numerical models have been developed to estimate the strain energy and fracture stress in well-bonded joints [44–47]. For a small CTE mismatch between a nonmetallic substrate (C) and a metallic substrate (M), but with a large CTE mismatch between the ductile interlayer (I) and the base materials, the elastic strain energy U_{eC}

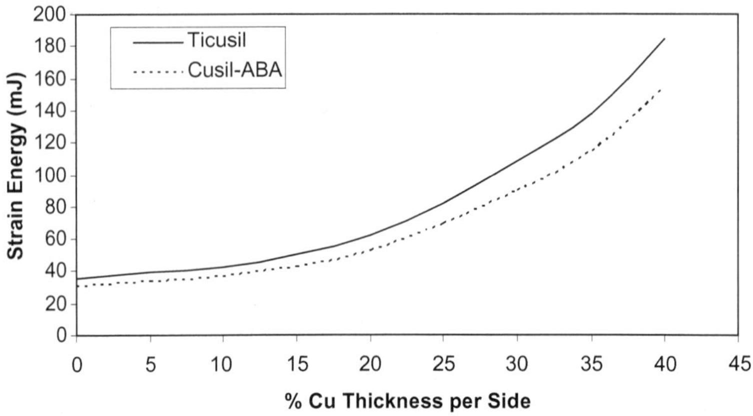

Figure 6.14. Calculated strain energy in the C–C/Cu-clad Mo joint as a function of the percent Cu thickness per side in Cu-clad Mo.

can be calculated using an analytical model presented in Park et al. [47]. The required material properties are yield strength of interlayer (σ_{YI}), radius of the joint, elastic modulus of the nonmetallic substrate (E_C), elastic modulus of the interlayer (E_I), temperature change (ΔT), and the CTE (α) of metal (M), nonmetal (C) and interlayer (I). The calculations are made in terms of two dimensionless parameters Π_I and Φ defined in Kovalev et al. [46], where Π_I is the ratio of the thermal residual strain at the interface to the yield strain of the braze interlayer, and Φ specifies the relative difference in CTEs between the nonmetal (C), braze interlayer (I), and metal substrate (M).

The strain energy in the C–C/Cu-clad Mo joints was computed using generic property data and model of Park et al. [47], and the results are presented in Figure 6.14. These calculations yield U_{eC} as ~38–190 mJ for clad layer thicknesses ranging from 0% to 40%, with U_{eC} increasing at a progressively faster rate with an increase in Cu layer thickness. In addition, less strain energy develops in Cusil-ABA joints than in Ticusil joints. While rigorous numerical models are needed for more reliable calculations, the above results demonstrate the sensitivity of the residual stresses to the clad layer thickness and the properties of braze, metal, and the composite.

Thermal Conduction in Brazed Joints

The thermal conduction in C/C composite-to-Cu-clad Mo joints is important for thermal management applications such as spacecraft radiators and heat exchangers. For 1-D steady-state heat conduction, the joined materials form a series thermal circuit with an effective thermal resistance, $R_{eff} = \Sigma(\Delta x_i/K_i)$, where Δx_i and K_i are the thickness and thermal conductivity, respectively, of the ith layer.

For C/C composite-to-Cu-clad Mo joints created by us, $\Delta x_{C\text{–}C} = \Delta x_{Cu\text{–}Mo} = 0.25 \times 10^{-2}$ m, and $\Delta x_{Ticusil} \sim 100 \times 10^{-6}$ m. For C/C composites, even though $K_{C/C}$ is anisotropic and variable, an average value (~125 W/m-K) can be used for the purpose

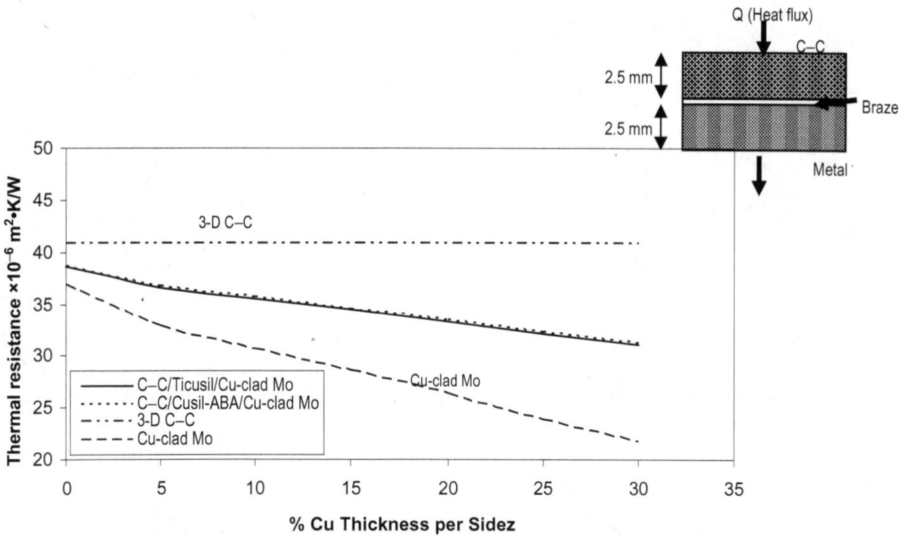

Figure 6.15. Calculated effective thermal resistance of C–C/Cu-clad Mo joint as a function of the percent Cu thickness per side in Cu-clad Mo.

of calculations. The conductivity of Cu-clad Mo, K_{Cu-Mo}, varies with the clad layer thickness and is taken from Harper [41]; K_{Cu-Mo} varies from 138 to 235 W/m-K for 0–30% clad layer thickness. The effective thermal resistance of the joints calculated using the above-mentioned approach is shown in Figure 6.15. R_{eff} varies in the range 31.5 to 38.5 × 10^{-6} m^2·K/W, and there is an insignificant (<1%) difference between Ticusil and Cusil-ABA. Because the difference in the R_{eff} of the joints with the two brazes is insignificant, there may be considerable flexibility in selecting brazes to satisfy other criteria, such as ductility and wetting characteristics, without impairing the thermal conductivity and weight advantages of the joined materials.

Figure 6.15 also compares the R_{eff} values of the joints to the R_{eff} values of C/C and Cu-clad Mo substrates of the same total thickness (5.1 × 10^{-3} m) as the joined materials; the thermal resistance of the C/C block is about 40.8 × 10^{-6} m^2·K/W, and that of Cu-clad Mo substrate decreases with increasing clad layer thickness. The decrease in the conductivity of our joints (Cu : Mo : Cu thickness ratio of 13%:74%:13%) relative to an unjoined Cu-clad Mo substrate is compensated by a 39% decrease in weight. This can be confirmed by estimating the density. The rule-of-mixtures (ROM) density (ρ) of our joints is ~5919 kg/m^3 (with $\rho_{C\text{–}C}$ = 1900 kg/m^3) compared to a density of 9937 kg/m^3 for Cu-clad Mo (ignoring the thin braze interlayer does not introduce any sensible error in the density calculations).

The preceding 1-D steady-state calculations ignore the interface thermal resistance, anisotropy of thermal conduction in the C/C composite, and chemical modification of the interface and constituent properties resulting from diffusion and/or chemical reactions during brazing. In spite of their inherent simplicity, these calculations convincingly point toward potential benefits of joining Cu-clad Mo to C/C composites for thermal management applications.

SUMMARY AND FUTURE PROSPECTS

Carbon–carbon composites have been successfully joined to Ti and Cu-clad Mo for thermal management applications using a number of active braze alloys containing Ti or Cr. Sound metallurgical bonding developed in C–C/Cu-clad Mo joints and C/C/Ti joints, although in some cases, cracking occurred within the composite because of its low interlaminar shear strength. Evidence of chemical interaction particularly interdiffusion of solutes and segregation of Ti at the C/C surface was observed.

The mechanical characterization of the joints indicated that both the area of bonding and the orientation of the C fiber tows in the outer ply of the harness satin weave at the mating surface affected the joint strength. Joint structures with outer fiber tows oriented perpendicular to the axis of Ti tubes had greater load-carrying ability than joint structures with outer fiber tows oriented parallel to the tube axis for all braze compositions. When Ti tubes were joined to C/C composites using a saddle material (Poco foam) in sandwich structures and were tested in tension and shear, failure always occurred in the foam and not at the joint, thus confirming the soundness and integrity of the joints. In addition, microhardness traces across the joints were reproducible and were generally consistent with the expected behavior.

Integrating C/C composites with Ti and particularly with Cu-clad Mo may have significant benefits for thermal management applications because of the possibility to tailor the thermal conductivity and thermal expansion properties. Innovative joining technologies are currently being developed for robust joining of C/C composites to Ti, Cu-clad Mo, and other materials. Technical challenges for scaling up the laboratory-based development should be overcome for implementation in large and complex-shaped components in real systems to be used in aerospace, power generation, and nuclear and transportation industries [48, 49]. Integration of joining technologies for seamless manufacture of components, devices, and systems shall require interrogation and validation of joint design philosophies, design issues, modeling the thermal dynamics, and performance evaluation of prototype and real thermal management components and systems.

Clearly, much additional developmental work is needed on the integration of C/C composites for thermal management applications. Optimum joint designs for thermal applications demand accommodation of a number of factors, and these will need to be addressed. The interplay between stresses and thermal gradients will need to be investigated to design and fabricate joints that possess a high degree of integrity. Joint design specifically for thermal management applications should also be cognizant of the influence of factors such as fiber arrangement, interlaminar shear behavior, role of coatings, oxidation resistance, thermal and mechanical fatigue, stress rupture behavior, and strength and conductivity loss with time under service conditions. The degradation of the thermal performance of joints from undesirable microstructural changes will need to be evaluated. Life prediction analysis for joints used in structures of thermal management systems and devices also will be needed. Resolution of such technical issues and challenges opens up exciting possibilities for continued research on thermal management materials and technologies.

REFERENCES

1. Zweben C. 2007. Advances in high-performance thermal management materials: a review. J Adv Mater 39(1):3–10.
2. Gallagher C, Shearer B, Matijasevic G. 1998. Materials selection issues for high operating temperature (HOT) electronic packaging. In: *Proc. High Temperature Electronic Materials, Devices and Sensors Conference*, IEEE, 180–189.
3. Zweben C. 1998. Advances in composite materials for thermal management in electronic packaging. J Mater (JOM) 50(6):47–51.
4. Zweben C. 1995. The future of advanced composite electronic packaging. In: Chung DDL, editor. *Materials for Electronic Packaging*. Oxford: Butterworth-Heinemann, pp. 127–144.
5. Prasher RS, Shipley J, Prstic S, Koning P, Wang JL. 2003. Thermal resistance of particle laden polymeric thermal interface materials. In: *Proceedings of International Mechanical Engineering Congress and Exposition*, Washington, DC: ASME.
6. Zweben C. 2002. Metal matrix composites, ceramic matrix composites, carbon matrix composites and thermally conductive polymer matrix composites. In: Harper CA, Editor, *Handbook of Plastics, Elastomers and Composites*, 4th edition. New York: McGraw-Hill, pp. 321–344.
7. Thaw C, Zemany J, Zweben C. 1987. Metal matrix composites for microwave packaging components. Electron Packaging Prod August:27–29.
8. Neubauer E, Angerer P, Korb G. 2005. Heat sink materials with tailored properties for thermal management. In: *Proceedings of 28th International Seminar on Electronics Technology*, IEEE, 258–263.
9. Prieto R, Molina JM, Narciso J, Louis E. 2008. Fabrication and properties of graphite flake/metal composites for thermal management applications. Scr Mater 59:11–14.
10. Petitcorps YL, Poueylaud JM, Albingre L, Berdeu B, Lobstein PO, Silvain JF. 1997. Carbon fiber-reinforced Cu-matrix composites, processing routes and properties. Key Eng Mater 127–131:327–334.
11. Weibgarber T, Lefranc G, Schulz-Harder J, Meyer H, Stocker O. 2003. Cu-SiC composites for thermal management application. PM2TEC 2003, *Adv. in Powder Metallurgy & Particulate Materials, Proc. of an International Conference*, Part 6, Las Vegas, June 8–12, Metals Powder Industries Federation (MPIF).
12. Zweben C. 2005. Composite materials and mechanical design. In: Kutz M, editor. *Mechanical Engineers' Handbook, Book 1: Materials and Mechanical Design*, 3rd edition. New York: John Wiley & Sons, Inc.
13. Taylor R. 2000. *Comprehensive Composite Materials*. Boston, MA: Elsevier Science Ltd., pp. 387–426, 4.
14. POCO HTC web site. 2008. http://www.poco.com.
15. Schwartz MM. 1994. *Joining of Composite Matrix Materials*. Materials Park, OH: ASM International.

16. Yalof S. 1987. Joining of carbon-carbon and ceramic-matrix composites. Materials Innovations Lab Presentation, Inter-agency Planning Group Meet, IDA Memo Report M-312, T.F. Kearns (ed.).
17. Valentine PG, Trester PW. 1990. In: *Proc. 15th Conference on Metal-Matrix, Carbon, and Ceramic-Matrix Composites*. NASA CP-3133, Part I, J.D. Buckley (ed.), 39–55.
18. Dadras P, Ngai T. 1991. In: *Proc. of the 5th Conf. Metal-, Carbon- and Ceramic-Matrix Composites*. J.D. Buckley (Ed.), NA*SA CP 3133,* Part I, 25–38.
19. Nicholas MG. 1990. Active metal brazing. In: Nicholas MG, editor. *Joining of Ceramics*. New York: Chapman and Hall, pp. 73–93.
20. Eustathopoulos N, Nicholas MG, Drevet B. 1999. *Wettability at High Temperatures*. Boston: Pergamon, pp. 281–282.
21. Keene BJ. 1993. Review of data for the surface tension of pure metals. Int Mater Revs 38(4):157–192.
22. Standing R, Nicholas M. 1978. The wetting of alumina and vitreous carbon by Cu-Sn-Ti alloys. J Mater Sci 13:1509–1514.
23. Li JG. 1992. Kinetics of wetting and spreading of Cu-Ti alloys on alumina and glassy carbon substrates. J Mater Sci Lett 11:1551–1554.
24. Grigorenko N, Poluyanskaya V, Eustathopoulos N, Naidich Y. 1998. *Interfacial Sci. of Ceram. Joining*. Bellosi A, editor. Boston, MA: Kluwer Academic Publishers, pp. 69–78.
25. Grigorenko N, Poluyanskaya V, Eustathopoulos N, Naidich YV. 1997. In: Eustathopoulos N, Sobczak N, editors. *Proc. 2nd Int. Conf. on High-Temp. Capillarity*. Krakow: Foundry Res. Inst., pp. 27–35.
26. Sobczak N, Sobczak J, Ksiazek M, Radziwill W, Morgiel J. 1997. In: Eustathopoulos N, Sobczak N, editors. *Proc. 2nd Int. Conf. on High-Temp. Capillarity*, Krakow: Foundry Research Institute, pp. 97–98.
27. Whalen JT, Anderson AT. 1976. Wetting of SiC, Si3N4, and carbon by Si and binary Si Alloys. J Am Ceram Soc 34(4):378–383.
28. Humenik M, Jr, Kingery WD. 1954. Metal-ceramic interactions-III. Surface tension and wettability of metal-ceramic systems. J Am Ceram Soc 37(1):18–23.
29. Sobczak N, Asthana R. 2001. The role of wetting phenomenon in the structure and bonding of metal-ceramic interfaces: some recent developments. In: Pandali SG, editor. *Recent Research Developments in Materials Science*, Vol. 2. Trivandrum, India: Research Signpost, pp. 135–160.
30. Sobczak N, Singh M, Asthana R. 2005. High-temperature wettability measurements in ceramic-metal systems: some methodological issues. Curr Opin Solid State Mater Sci 9(4–5):241–253.
31. Asthana R, Singh M, Sobczak N. 2010. Wetting behavior and interfacial microstructure of palladium- and silver-based braze alloys with C–C and SiC–SiC composites. J Mater Sci 45(16):4276–4290.
32. Metals Handbook. 1993. *Welding, Brazing and Soldering*, Vol. 6. Materials Park, OH: ASM International, pp. 117–118.
33. Rabinkin A. 1989. New applications for rapidly solidified brazing foils. Welding J 10:39–46.
34. McKeown SA, LeVasseur RD. 1991. High performance heat sink for surface mount applications. CH 3030-4/91/0000-0153, IEEE, 153–157.

35. Singh M, Asthana R. 2008. Characterization of brazed joints of carbon-carbon composites to Cu-clad-Mo. Compos Sci Tech 68(14):3010–3019.

36. Singh M, Morscher GN, Shpargel TP, Asthana R. 2008. Active metal brazing of titanium to high-conductivity carbon-based sandwich structures. Mater Sci Eng A 498(1–2):31–36.

37. Singh M, Asthana R, Shpargel TP. 2007. Brazing of C–C composites to Cu-clad Mo for thermal management applications. Mater Sci Eng A 452–453:699–704.

38. Morscher GN, Singh M, Shpargel T. 2006. In: Stephens JJ, Weil KS, editors. *Brazing and Soldering* (Proc. 3rd Int. Conf. Brazing and Soldering, San Antonio, TX, April 24–26, 2006). Materials Park, OH: ASM International and AWS, pp. 257–261.

39. Morscher GN, Singh M, Shpargel TP, Asthana R. 2006. A simple test to determine the effectiveness of different braze compositions for joining Ti tubes to C/C composite plates. Mater Sci Eng A 418(1–2):19–24.

40. Singh M, Shpargel TP, Morscher GN, Asthana R. 2005. Active metal brazing and characterization of brazed joints in titanium to carbon-carbon composites. Mater Sci Eng A 412:123–128.

41. Harper CA. 2003. *Electronic Materials and Processes Handbook*. New York: McGraw-Hill.

42. Appendino P, Ferraris M, Casalegno V, Salvo M, Merola M, Grattarola M. 2004. Direct joining of CFC to copper. J Nucl Mater 329–333:1563–1566.

43. Salvo M, Lemoine P, Ferraris M, Montorsi MA, Matera R. 1995. Cu-Pb rheocast alloy as joining materials for CFC composites. J Nucl Mater 226:67–71.

44. Park J-W, Mendez PF, Eager TW. 2002. Strain energy distribution in ceramic-to-metal joints. Acta Mater 50(5):883–899.

45. Park J-W, Eager TW. 2004. Strain energy release in ceramic-to-metal joints with patterned interlayers. Scr Mater 50(4):555–559.

46. Kovalev SP, Miranzo P, Osendi MI. 1998. Finite element simulation of thermal residual stresses in joining ceramics with metal interlayers. J Am Ceram Soc 81(9):2342–2348.

47. Park J-W, Mendez PF, Eagar TW. 2005. Strain energy release in ceramic-to-metal joints by ductile metal interlayers. Scr Mater 53(7):857–861.

48. Singh M, Asthana R. 2008. Advanced joining and integration technologies for ceramic-matrix composite systems. In: Krenkel W, editor. *Ceramic-Matrix Composites: Fiber-Reinforced Ceramics and Their Applications.* Weinheim, Germany: Wiley-VCH, pp. 303–325.

49. Singh M, Asthana R. 2007. Brazing of advanced ceramic composites: issues and challenges. In: Ewsuk K, Nogi K, Waesche R, et al., editor. *Characterization and Control of Interfaces for High Quality Advanced Materials II*, Vol. 198. Ceramic Trans. Hoboken, NJ: John Wiley & Sons, pp. 9–14.

7

CONTACT INTERACTION IN CARBON–METAL SYSTEMS FOR JOINING AND INTEGRATION

V. M. Perevertailo and O. B. Loginova

National Academy of Sciences of Ukraine, Kiev, Ukraine

INTRODUCTION

In many technological processes (crystal growth, production of composite materials, brazing, welding, deposition of coatings, and impregnation), metal melts make contacts with a solid carbon phase. The thermodynamics and kinetics of these processes as well as the properties of the resultant product depend largely on the conditions and properties of the interface, that is, interfacial energy, adhesion, and wettability (Naidich 1972; Naidich et al. 1991a).

Two allotropic carbon modifications (graphite and diamond) have found wide application in engineering. The crystalline structure of hexagonal graphite contains four atoms in a unit cell and is of a layer nature with the lattice parameters $a = 0.246$ nm and $c = 0.671$ nm. Diamond is of a tetrahedral structure; bonds between carbon atoms in the diamond lattice are covalent in character; and the latter is responsible for diamond's unique properties: extremely high hardness and elasticity, and low electrical conductivity. The C–C bond in diamond is the σ-bond, which is caused by the bonds of the sp^3 hybridization between carbon atoms; the C–C bond length is 0.15445 nm.

Diamond is characterized by the perfect cleavage along the (111) plane. The hardness and surface energy of diamond faces having different indices are not equal. The

Ceramic Integration and Joining Technologies: From Macro to Nanoscale, First Edition.
Edited by Mrityunjay Singh, Tatsuki Ohji, Rajiv Asthana, Sanjay Mathur.

calculated surface energies for the (111), (110), and (100) planes are, respectively, 5.3, 6.5, and 9.2 J/m^2. Diamond is stable only in the high-pressure and high-temperature region and, therefore, under the normal pressure at high temperatures, the diamond-to-graphite transition can be observed. The data on the onset of the diamond graphitization are contradictory as the initial diamond graphitization temperature depends on many factors, like dispersion of diamond powders, their surface state, impurities, and medium. In vacuum, noticeable graphitization starts at a temperature above 1100°C (Naidich 1972).

Wetting of carbon materials by metal melts depends mainly on the chemical affinity of a metal for carbon (Naidich 1972; Naidich et al. 1991a). According to the type of interaction between components of the solid and liquid phases, contact systems of carbon with metals can be divided into the following groups:

- Systems with metals nonreactive to carbon. The melts do not wet carbon materials.
- Systems with group VIII metals and the addition of metals nonreactive to the solid phase. The melts wet carbon materials due to the dissolution of the solid phase in the liquid one.
- Systems with carbide-forming metals and the addition of metals nonreactive to carbon. The melts wet due to the formation of chemical compounds in the contact zone.
- Systems with group VIII metals simultaneously containing several components reactive and nonreactive to the solid phase. The melts wet due to the various trends of interaction in the contact zone (the dissolution of the solid phase in the liquid and the formation of chemical compounds).

WETTING OF GRAPHITE AND DIAMOND BY METALS NONREACTIVE TO CARBON

Groups IVb–VIb metals are nonreactive to carbon. They do not wet carbon and their work of adhesion to the surface of the solid phase is low and does not exceed 300 mJ/m^2. As the temperature increases in these systems, the contact angle values and work of adhesion change slightly or even remain constant (Naidich 1972). Anisotropy of diamond crystal properties, which reflects the anisotropy of its surface energy, is revealed in the anisotropy of wetting different diamond faces by metal melts (see Table 7.1). It is seen that the difference in angle of wetting by metal melts between the natural and (111) polished diamond faces is low. Somewhat higher are the wetting angles for a natural face of diamond because of the higher nonuniformity of its surface roughness as compared to the polished surface. The wetting angles increase with the face roughness, which agrees with the Wenzel equation.

The above results are indicative of the anisotropy of wetting different diamond faces by metals nonreactive to carbon. All the metals studied wet the (111) face better than the (100) face. The ratio between the values of the work of adhesion of the metals

TABLE 7.1. Contact Angles θ (degree) and Work of Adhesion W_a (mJ/m^2)

	Roughness	Cu		Ag		Au		Ge		Sn	
Solid face	(R_a) (μm)	θ	W_a	θ	W_a	θ	W_a	θ	W_a	θ	W_a
Face (111) diamond	0.051	158 ± 2	98	146 ± 3	154	158 ± 2	81	156 ± 3	54	149 ± 2	66
	0.010–0.013[a]	137 ± 3	360	126 ± 4	371	150 ± 1	149	132 ± 2	206	135 ± 3	136
	0.010	138 ± 2	344	134 ± 2	275	148 ± 2	169	128 ± 2	240	129 ± 3	172
Face (100) diamond	0.010	147 ± 3	216	142 ± 2	191	157 ± 2	89	152 ± 3	73	137 ± 3	134
MPG-6 graphite	0.100	143 ± 2	344	127 ± 2	358	153 ± 2	121	135 ± 2	101	144 ± 2	84

The metal melts to diamond and graphite at 1100°C (Naidich et al. 1985).
[a]Natural face of diamond.

studied is in the range from 1.4 to 1.9 (1.380 for Sn, 1.44 for Ag, 1.593 for Cu, and 1.912 for Au), which correlates with the ratio between the number of carbon atoms per square centimeter of the diamond crystal surface in the cubic and octahedral structures. The wetting angles of graphite by the metals under study are intermediate between the wetting angles for the (111) and (100) faces of a diamond crystal.

WETTING OF GRAPHITE BY GROUP VIII METALS

In the interaction of group VIII transition metals with carbon, the wetting is due to the dissolution of the solid phase in the liquid phase without the formation of stable carbides. The exception is Fe, for which the Fe_3C carbide is known. For these systems, the carrying out of wetting experiments under the conditions of separate (noncontact) heating, which exclude the presaturation of the alloy with carbon when heated prior to the wetting test, is important. Otherwise, the contact angles are heavily dependent on the experimental conditions and, in the same system, can vary over a wide range, from wetting to nonwetting.

Figure 7.1 gives the dependences of the angles of wetting graphite by the Fe–C ($T = 1550°C$), Co–C ($T = 1550°C$), Ni–C ($T = 1550°C$), Rh–C ($T = 2000°C$), Pd–C ($T = 1560°C$), and Pt–C ($T = 1800°C$) melts on the carbon content (Naidich et al. 1979). Melts of pure metals wet graphite intensively. Carbon addition to the melt before the contact with a solid phase impairs the wetting in the system and decreases the work of adhesion (W_a).

In these systems, the wetting of graphite by the melts is due to the difference in carbon chemical potential ($\Delta\mu$) between the solid and liquid phases and the mass transfer (dissolution) of carbon in the liquid phase, which is preceded by the adsorption of the melt components at the phase boundary. As the carbon content of the melt increases, $\Delta\mu$ decreases and the contact angles increase. The highest degree of the decrease in W_a

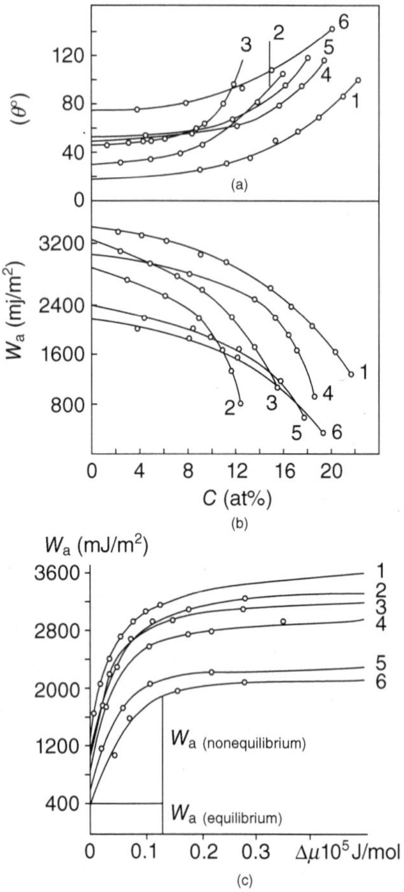

Figure 7.1. (a) Wettability and (b) work of adhesion versus carbon concentration: dependence of W_a of metal to the graphite surface as a function of the difference of carbon chemical potential ($\Delta\mu$) between the solid and liquid phases (c) in the contact systems of graphite–group VIII metals Fe (1), Co (2), Ni (3), Pd (4), Rh (5), and Pt (6).

is achieved with the carbon content of the melt close to the ultimate solubility of carbon at a given temperature (see Fig. 7.1).

In the general case, the work of adhesion of a liquid to a solid surface can be written as (Naidich 1972)

$$W_a = W_{a\,(\mathrm{equilibrium})} + W_{a\,(\mathrm{nonequilibrium})}.$$

The nonequilibrium component of the work of adhesion ($W_{a\,(\mathrm{nonequilibrium})}$) is defined by a chemical interaction (the carbon dissolution process), which occurs due to the inequality in carbon chemical potential between the solid and liquid phases. With the saturation of the melt with carbon, the carbon chemical potential in both phases

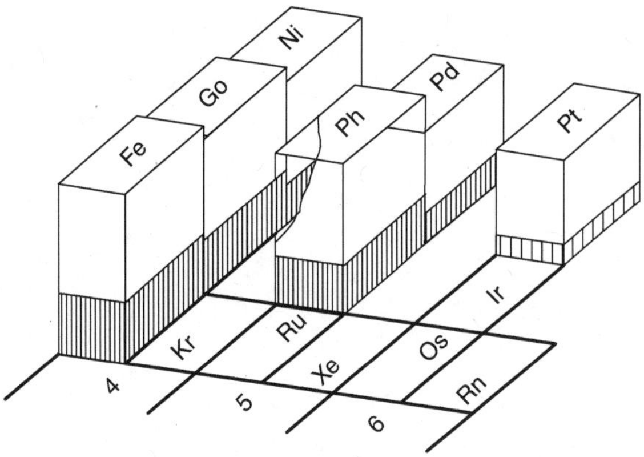

Figure 7.2. Work of adhesion in the graphite–group VIII metals contact systems under the nonequilibrium (the light part) and equilibrium (the shaded part) interaction conditions.

becomes equal, and $W_{a\,(nonequilibrium)}$ tends to zero; that is, the work of adhesion is reduced to the equilibrium contribution, defined by the difference in properties between solid and liquid phases.

Thus, the value of the equilibrium contribution can be isolated from the total work of adhesion. In Figure 7.1, this is done for the platinum–carbon system. Hence, the variation of the wetting in the contact system is possible by monitoring the diffusion of components constituting the solid phase to the bulk melt.

Figure 7.2 gives the diagram of the work of adhesion of group VIII metals to graphite under nonequilibrium (the light part of the bar) and equilibrium (the shaded part) contact conditions. A decrease in the work of adhesion (both total and an equilibrium contribution) is observed according to the arrangement of the chemical element in the periodic table. The data on Rh and Pt are received at higher temperatures. The observed dependence would be more distinct if the wetting experiments were conducted at the same temperature.

The observed dependence of the work of adhesion of metals on the arrangement in the periodic table allows one to predict the work of adhesion to graphite of not-yet-studied metals such as Ru, Ir, and Os. The work of adhesion of these metals to graphite should be within the following ranges (mJ/m^2):

$$(3185)\ W_{a\,Rh} < W_{a\,Ru} < W_{a\,Fe}\,(3620)$$

$$(2190)\ W_{a\,Pt} < W_{a\,Ir} < W_{a\,Rh}\,(3185).$$

Our calculations based on the structural and thermodynamic considerations (Perevertailo and Loginova 1991) verify the above conclusion.

Within the region of the existence of solid solutions in the phase diagram (see Table 7.2), the wetting angles of graphite by the melts do not virtually change (see Fig. 7.1). As the melt composition approaches the eutectic, an abrupt increase of the contact angles is observed and their highest values correspond to the degree of the saturation

TABLE 7.2. Physicochemical Characteristics of Group VIII Metal Melts

Composition of the melt (at%)	T (°C)	Eutectic		Limiting solubility of carbon in the solid solution (at%)	Difference in carbon content between eutectic and solid solution (at%)	θ (degree)	σ_{lg} (mJ/m^2)
		T (°C)	Composition (at%)				
Fe	1550	1153	17.1	8.79	8.21	17	1850
Fe + 21C						99	1505
Co	1550	1319	11.9	4.3	7.6	30	1807
Co + 15C						105	1488
Ni	1550	1314	9.4	3.4	6	48	1797
Ni + 12C						115	1449
Rh	2000	1694	15.5	3.75	11.75	50	1940
Rh + 17C						118	1800
Pd	1560	1504	—	—	—	52	1470
Pd + 18C						121	1295
Pt	1800	1705	16.8	0	16.8	75	1740
Pt + 19C						139	1590

of the melt with carbon at a given temperature. In the Fe–C, Co–C, Ni–C series, the difference between the ultimate carbon content of the solid solution and its content in the eutectic decreases. The minimum difference is for the Ni–C system; in the range under study, the contact angles change most sharply. In this case, the linearity of the dependence of the work of adhesion on the difference between the carbon chemical potentials in the solid and liquid phases occurs at small deviations of the system from the equilibrium state only (see Fig. 7.1); that is, the maximum effect in decreasing the work of adhesion is with the carbon content of the melt close to the carbon's highest possible solubility.

X-ray diffraction analysis of Ni–C melt containing 8 at% carbon has shown that clusters of weakly deformed tetrahedra and octahedra from Ni atoms, carbon chains, and a small amount of closed carbon fragments form (see Figs. 7.3 and 7.4) (Perevertailo et al. 2004). Consideration of more distant carbon atoms shows an elongation of the chains and an increase (growth) of the closed fragments. As the portion of atoms, which form chains, increases regularly (from 23.0% to 56.3%), it can be concluded that carbon chains of different lengths penetrate the nickel matrix and are uniformly distributed throughout the melt. The most likely atomic spacing in chains (0.140 nm; see Fig. 7.3) is close to the corresponding value in planar networks of graphite (0.142 nm), which indicates that covalent bonds between carbon atoms in the melt are partially retained.

Carbon clusters formed in these melts, the bonding energy within which is higher than that between the solvent (nickel) and cluster atoms, are surface active. Clearly, the surface tension of the Ni–C melts of eutectic composition defined from the experimental data turns out to be much lower than for pure nickel (see Table 7.2). The above conclusion is true for other metals of the iron subgroup as well. The nonuniformity of the melt at the nanolevel can also be responsible for some features of isotherms of wetting graphite by carbon-containing melts of group VIII metals. As the melt

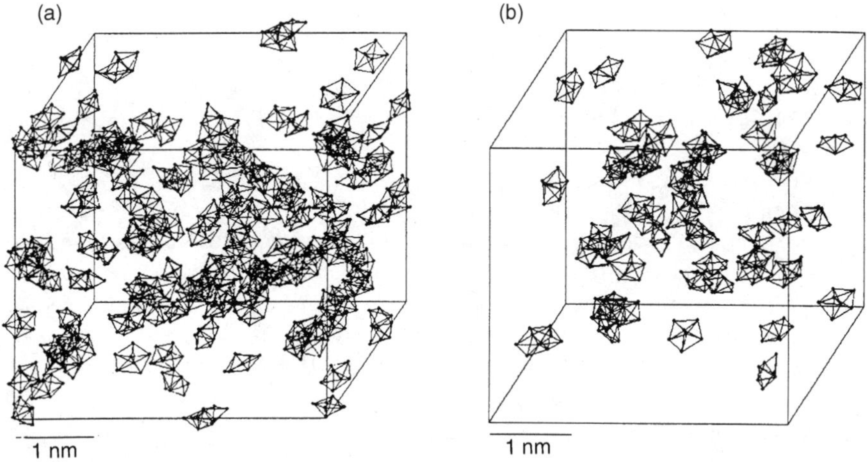

Figure 7.3. Clusters of weakly deformed tetrahedra formed by nickel atoms in (b) liquid Ni and in the (a) 92Ni–8C melt.

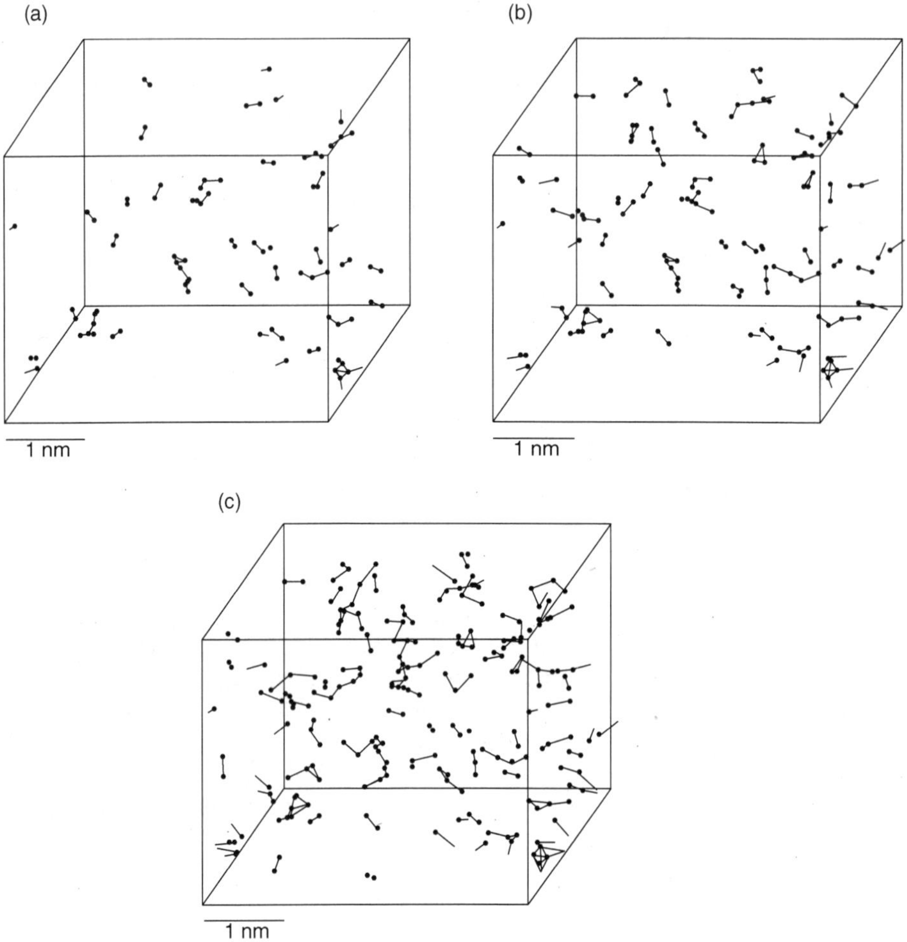

Figure 7.4. Configurations formed by carbon atoms removed from each other at distances of (a) ≤0.20, (b) 0.25, and (c) 0.30 nm.

composition approaches the eutectics, the size and amount of clusters from weakly deformed tetrahedra formed by nickel atoms, which are larger (2.0–2.5 nm) than those of pure liquid nickel (1.3–1.5 nm), are essentially increased (Perevertailo et al. 2004). Evidently, the formation of clusters from nickel atoms in the melt, as well as a decrease in the difference in the chemical potential of carbon between solid and liquid phases, essentially complicates the conditions of adsorption and the mass transfer through the phase boundary, and therefore, the wetting of graphite with the Ni–C melt impairs abruptly (see Fig. 7.1).

Structures of the forming zone of graphite contact interaction with pure and presaturated carbon group VIII metals are similar and correspond completely to the phase diagram (Perevertailo et al. 1998).

CARBIDE-FORMING METALS IN CONTACT WITH CARBON

Pure carbide-forming metals exhibit a considerable chemical affinity for a solid phase and, because of this, intensively wet graphite to form carbides in the contact zone. To obtain reliable experimental data on wetting by this group of metals is rather difficult as they, except for aluminum, have high melting temperatures. Moreover, for groups IV–VII transition metals (as in the case with group VIII metals), during coheating of the drop–substrate composition, a considerable carburization of the metal is possible, which can also affect the contact angle value.

As the studies have shown, the rate of spreading in the contact system will be higher the more intense is the reaction in the contact zone (Naidich et al. 1972). Figure 7.5 gives the spreading rates of silicon, nickel, and carbon-saturated nickel melt over the graphite surface. The intensity of the interaction between the liquid and solid phases decreases in the series Si (the formation of a chemical compound in the contact zone) → Ni (the dissolution of the solid phase in the liquid phase) → Ni–C (the melt of the equilibrium composition). In the systems under study, the coefficient of spreading and the rates of spreading and wetting decrease in this sequence as well (Naidich et al. 1972).

A number of papers report the studies of the aluminum and silicon interaction with carbon, as these metals are widely used in the production of refractory compositions and matrix materials. Investigations on the wetting of various carbon materials by aluminum have shown that at the melting temperature (630°C), aluminum behaves as a nonreactive metal and the contact angles start to decrease only at 1000°C (see Table 7.3). The reason for this is the existence of an oxide film, which is stable up to 1000°C, on the surface of liquid Al.

The structure of the initial carbon material and the density and roughness of the initial surface essentially affect the rate of wetting of the solid-phase surface by liquid aluminum and the contact angle value. In this case, the wetting rate depends on the rate

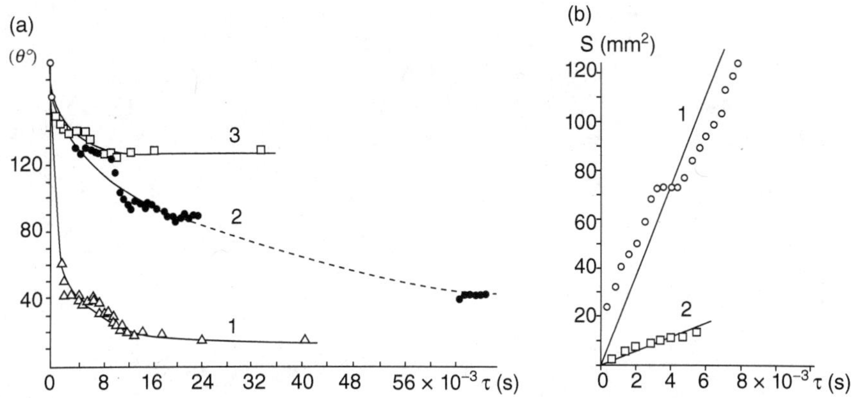

Figure 7.5. Time dependences of (a) the contact angles and of (b) changes of the area of wetting spot at spreading liquid drops over graphite at 1460°C: 1, Si; 2, Ni; and 3, Ni–C.

TABLE 7.3. Contact Angles for Carbide-Forming Metals on Carbon Materials

	Experimental conditions					
Metal	*T* (°C)	Atmosphere, heating	Carbon material	θ (degree)	Interfacial reaction layer	Source
Al	800	Vacuum, coheating	Graphite	150		Naidich (1972)
	1000			75		
	1200			38		
	800		(111) Diamond	150		
	1000		Face	75		
	1000	Vacuum, coheating	MG-1 graphite	137		Bobkovskiy et al. (1972)
	1175			78	Al_4C_3	
	980	Vacuum, coheating	Graphite	110		Landry et al. (1998)
	980		pseudomonocrystal	100		
	820		Pyrolytic carbon	70	Al_4C_3	
	980		Vitreous carbon	50	Al_4C_3	
	1100	Vacuum coheating	MPG-6 graphite	60		Perevertailo et al. (2001)
	1200			15		
Si	1460	Vacuum, separate heating	MPG-6 graphite	15		Naidich et al. (1972)
	1450	Vacuum, coheating	Graphite	0		Naidich (1972)
	1500		(111) Diamond face	0		
	1430	Vacuum, coheating	Vitreous carbon	17	SiC	Kostikov and Varenkov (1981)
	1430		Pyrolytic carbon	21		
	1470	Argon, coheating	(111) Diamond face	0		Kostikov et al. (1974)
	1450	Vacuum, coheating	Electrode graphite	0	SiC	Panasyuk et al. (1982)
			MPG-6 graphite	0		
			Pyrolytic carbon	0		
	1430	Vacuum, coheating	Vitreous carbon	26		Dezellus et al. (2005)
			Monocrystal of graphite	35	SiC	
Cr	1775	Argon, coheating	Graphite	40		Weisweiler and Mahadevan (1972)
			Vitreous carbon	35		
Ti	1740		Graphite	60		
			Vitreous carbon	50		
Ti	1700	Vacuum, coheating	AG-1500 graphite	0		Kostikov and Varenkov (1981)
Zr	1880			0		
V	2500			>0		
Nb	2500			>0		
Mo	2500			>0		
Ta	2750			60		
Mn	1400	Argon, Separate heating	MPG-6 graphite	16	Mn_7C_3, Mn_5C_2	This study
Cr	1700	Argon, coheating		10	Cr_7C_3	
V	1750	Vacuum, coheating		20	V_2C	
Ti	1800	Vacuum, coheating		0	TiC	

of breaking C–C covalent bonds in the structure of the solid-phase surface: For dense carbon materials (pyrolytic and pseudo-monocrystalline graphite), the spreading rate is higher and at 820°C, contact angles are higher than for vitreous carbon (Landry et al. 1998). At 820–980°C in the Al–C contact system, only vitreous carbon contact angles are within 70° and 50°, which corresponds to the angle of wetting of aluminum carbide by aluminum (Landry et al. 1998). According to the data reported in Ferro and Derby (1995), the contact angle for pure aluminum on Al_4C_3 (which has a very high degree of covalence) is 59° at 1100°C. This made it possible to suggest (Landry et al. 1998; Eustathopolos et al. 1999) that carbon-forming metals wet carbon substrates with spreading being controlled by the rate of the formation of a continuous layer of a corresponding carbide at the interface.

Self-bonded silicon carbide is wetted by silica at 1415°C in the helium atmosphere, the contact angle being 36° (Whalen and Anderson 1975). At 1440°C in vacuum, a wetting angle for silica on a SiC monocrystal is 37° (Yupko and Gnesin 1973). Nevertheless, in wetting various graphite materials by Si at 1450°C in vacuum, it has been found that the majority of carbon materials are wetted by silicon to form a smaller contact angle than in the case of silicon carbide (see Table 7.3). As with aluminum, the structure of the initial graphite, its purity, density, and texture essentially affect both the rate of wetting the solid-phase surface by a liquid metal and the contact angle.

Thus, at 1450°C, highly porous electrode graphite interacts actively with liquid Si, the initial contact angle being 53°, which decreases to zero for 35 s. For the MPG-6 fine-grained low-porous graphite, the initial contact angle is 68°, and for 135 s, it also decreases to zero. Contact angles for Si on pyrolytic carbon reach zero for more than 600 s. Pyrolytic carbon differs from other graphite grades in density, and SiC forms uniformly along the solid-phase surface. Silicon spreads over a dense thin layer of silicon carbide distributed uniformly at the interface to form a thin uniform layer. Therefore, the diffusion of Si through the interface is a problem, and the rate of wetting dense graphite decreases abruptly (Panasyuk et al. 1982).

Carbides of transition metals are more metal-like than aluminum and silicon carbides. The majority of carbides of transition metals are substances of variable compositions; the regions of their homogeneity (e.g., TiC and ZrC) are sufficiently wide; and they are capable to dissolve a metal component. Physicochemical properties of carbides formed in the zone of contact with a liquid phase depend heavily on their composition; therefore, with an increase of the deficiency of carbon in carbides, when they are wetted by carbide-forming metals, the contact angle can decrease abruptly and sometimes attain a value near zero (Naidich 1972; Panasyuk et al. 1982).

The examination of the microstructure of the contact interaction zone of Mn, Cr, V, and Ti with graphite on the crystallization has shown that the structure of alloys of the Mn–C system consists of elongated manganese carbide crystals distributed uniformly over the whole volume (see Fig. 7.6a). The contact boundary is rather even, in some places along the Mn_7C_3 grain boundaries; precipitations of the second phase (Mn_5C_2) can be seen as thin lighter interlayers or fine grains. The structure of Cr–C alloys (see Fig. 7.6b) exhibits large Cr_7C_3 crystals distributed uniformly over the whole volume. In parallel with them, the structure shows a large number of fine crystals that

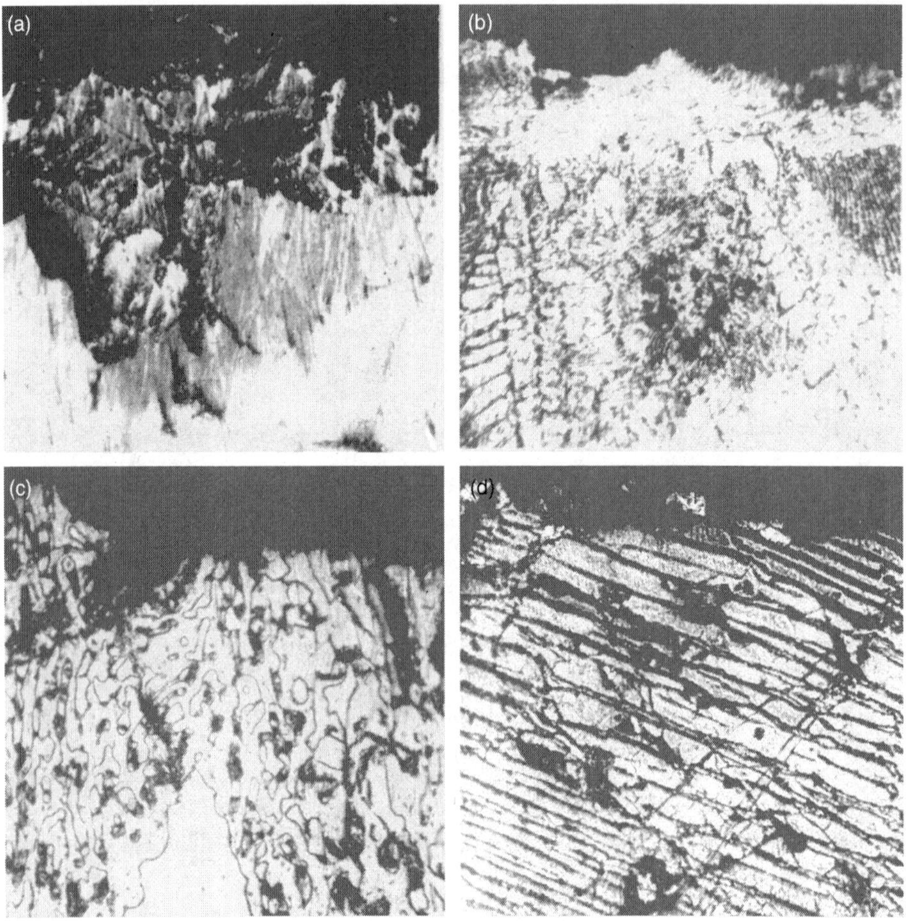

Figure 7.6. Microstructures of the MPG-6 graphite with (a) Mn, (b) Cr, (c) V, and (d) Ti interfaces. 320×.

are of the same structure and elemental composition as Cr_7C_3. The second phase surrounding the Cr_7C_3 crystals is the Cr + Cr_7C_3 eutectics.

At the contact boundary in the V–C system, a V_2C layer forms that consists of a large number of thin, platelike grains located parallel to the graphite surface (see Fig. 7.6c). From this layer deep into the metal part of the alloy, dendritic crystals penetrate. The composition of the latter is identical to V_2C and corresponds to the region of the V_2C carbide homogeneity. The microstructure of alloys of the Ti–C system at the interface with graphite after the crystallization exhibits two layers (see Fig. 7.6d). The graphite substrate is directly in contact with a titanium carbide layer, which is followed by a layer of a solid solution of carbon in titanium. A carbide layer is very dense and consists of intergrown titanium carbide columnar crystals oriented perpendicularly to the interface. The carbide content of the layer as viewed from graphite is close to the stoichiometric composition and decreases slightly with the thickness. The boundary between the carbide layer and the solid solution is mostly even as well; however, in some places, carbide grains penetrating into the upper layer are met. This layer consists of large, equiaxial Ti grains and a low concentration of dissolved carbon, whose quantity is much lower than that, which should correspond to the Ti–C equilibrium phase diagram.

It is evident that the processes occurring in the interaction of graphite with melts of the metals having a high affinity for carbon largely depend on the kinetic factors of the formation and growth of carbide phases. Thus, in case of the Mn saturation with carbon at 1400°C, the alloy composition corresponds to the region of the coexistence of Mn_7C_3 and a liquid phase. During the subsequent cooling at 1340°C, a peritectic transformation to form ε-carbide takes place, which at 1020°C transforms to Mn_5C_2. It is quite possible that the presence of a solid carbide phase in the liquid does not allow it to spread to form zero angles.

In the Ti–C system, the formation of solid phases at 1800°C starts even at low (about 2–3 at%) concentrations of carbon in an alloy. This, in combination with the highest affinity of titanium for carbide, as compared to other elements being studied, gives rise to a rapid formation of titanium carbide in a thin layer adjacent directly to the substrate. This layer practically completely blocks the carbon diffusion to the melt because of the high density of carbide and "absorption" of carbon atoms in completing its composition to the stoichiometric one. This results in the fact that the carbon content of the liquid behind the carbide layer does not attain the equilibrium concentration. Further, as the melt spreads over the graphite surface along the triple line, columnar carbides form continuously; that is why one may consider that titanium spreads over the continuously forming carbide phase layer to zero angles.

Thus, contact angles for the majority of carbide-forming metals on carbon materials are lower than for corresponding carbides (see Table 7.3). The reasons for this can be various rates (that differ by an order of magnitude) of spreading and forming the carbide phase in the contact zone, as well as physicochemical properties of the resulting transition layer, its morphology, and integrity.

WETTING OF GRAPHITE BY NONREACTIVE TO CARBON MELTS WITH THE ADDITION OF CARBIDE-FORMING METALS

The wetting in such systems occurs also due to the carbide formation in the contact zone. In this case, the wetting angle is largely defined by the chemical affinity of the active metal for carbon and its thermodynamic activity in the melt of a nonreactive component. The curves of wetting carbon materials with such alloys look like those of the saturation. Small additions of an active metal heavily decrease the contact angle, while the further increase in the active metal content of the melt affects the angle only slightly (Naidich 1972; Naidich et al. 1991a). Such dependences are of great practical interest as these melts provide a basis for soldering alloys in liquid-phase capillary joining.

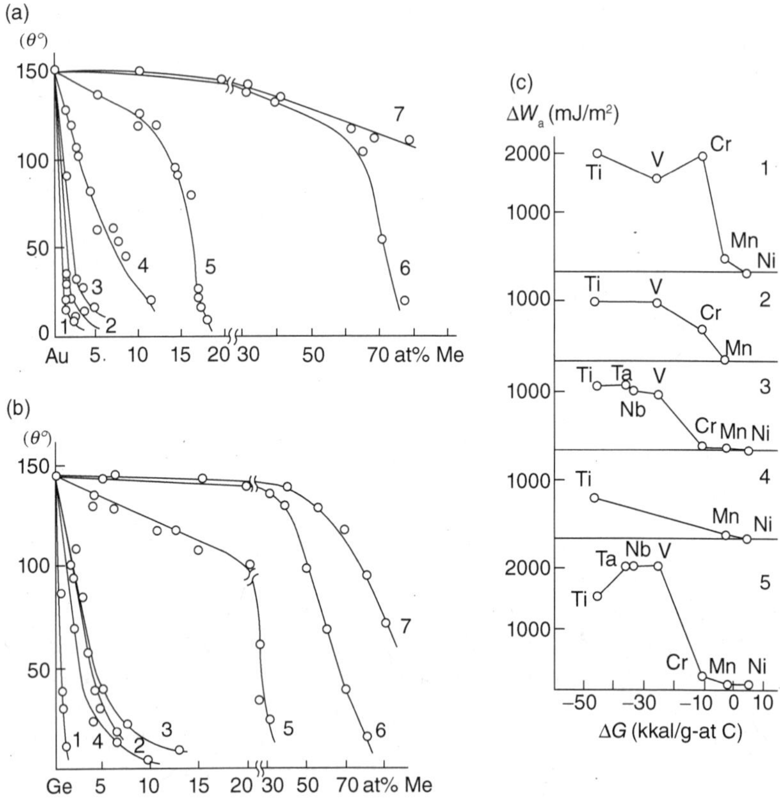

Figure 7.7. Wettability of graphite at 1100°C versus content of Ta (1), Nb (2), V (3), Ti (4), Cr (5), Mn (6), Ni (7) in the (a) Ge- and (b) Au-based melts and dependence of work of adhesion on ΔG for (c) melts based on (1) Cu, (2) Ga, (3) Ge, (4) Sn, and (5) Au, which contain an equal concentration (5 at%) of transition metals.

Pure Au and Ge are nonreactive to the graphite surface, which is due to the weak (physical) forces of the interaction of these metals with carbon; the wetting angles range from 150° to 145° (see Fig. 7.7a). The affinity of metals–additives for carbon increases in the Ni → Mn → Cr → V → Nb → Ta → Ti series; the wetting of graphite by the melts of Au and Ge with the addition of the active metals is improved in the same sequence as the content of the active additive of the melt increases (Naidich et al. 1983).

Alloys of the Au–Ni and Ge–Ni systems do not wet graphite over the whole range of the active component concentrations. Au–Mn and Ge–Mn melts start to wet the carbon surface at considerable concentrations of the active component (above 50 at% Mn), and Au–Cr and Ge–Cr melts at a concentration of Cr above 16 at%. The alloys of Au and Ge with V, Nb, and Na wet graphite and diamond well even at small concentrations of a carbide-forming component (2–5 at%).

Figure 7.7c shows the dependences of the work of adhesion of melts of metals (Cu, Ga, Ge, Sn, and Au) nonreactive to carbon, which contain the same amount (5 at%)

of a transition metal, on the value of isobaric and isothermal potential of the metal–carbon reaction (ΔG). We can see that the Cr activities in alloys having various solvents are different. Its addition to melts of metals nonreactive to carbon results either in a rapid improvement of the wetting in the system and in an increase in the work of adhesion (in alloys with Cu and Ga), or in a slight effect on the interfacial processes (Cr in combination with Ge and Au).

The reason for this is evidently in the fact that the enthalpy of the formation of nontransition metals–chromium compounds (below 2 kcal/g-at) is in a number of cases comparable with the value of the isobaric and isothermal potential of the carbon–chromium reaction; therefore, Cr shows a high adhesion activity in alloys with those metals–solvents, whose formation enthalpy is below ΔG.

The process of wetting in the systems, where a chemical interaction between solid and liquid phases takes place, is based on the phenomenon of the adsorption of the active component at the boundary with the solid phase, which implies the formation of chemical compounds in the contact zone. Metal melts containing metals reactive to the solid-phase substance form intermediate carbide layers with carbon materials at the interface.

For instance, the X-ray spectrum microanalysis of the zone of the contact interaction in the diamond–(Cu–10 at% Cr) melt system has allowed the dependences of the Cu and Cr concentrations at the interface to be found (Naidich et al. 1991a). It has been found that the Cu and Cr diffusion into diamond is absent, and the interface is enriched with Cr to form a layer of carbides having a zonal structure: The layer adjacent to diamond is enriched with carbon to the highest degree and consists of the Cr_3C_2 carbide; this layer is followed by layers of carbides with lower carbon concentrations (Cr_7C_3 and $Cr_{23}C_6$), which completely agrees with carbide phases in the Cr–C phase diagram. The above phenomenon is characteristic of the diffusion saturation of the surface of solids, which is accompanied by chemical interactions and phase transformations observed in binary systems with active components.

In the Cu–Ga–Cr melt–graphite system, carbide phases start to form even at 1.5 at% Cr in a melt. A number of these phases increase with time of holding, and at 1050°C, the resultant transition layer consists of approximately equal amounts of Cr_7C_3 and Cr_3C_2 carbides. Melts of V with Ga when interacting with graphite form a transition layer consisting of VC, V_2C, and a solid solution of carbon in vanadium at the contact boundary. This layer is rather thick, and as the holding time increases, the second carbide content of the layer is slightly increased (Naidich et al. 1991a).

THE INTERRELATION BETWEEN THERMODYNAMIC AND INTERFACIAL ACTIVITIES OF COMPONENTS IN WETTING SOLIDS BY MELTS

This interrelation has been studied rather comprehensively and is well supported for wetting a solid phase by a pure liquid. As to the wetting of solids by multicomponent melts, particularly by those containing additives reactive to the solid-phase substance, this interrelation is more complicated and is less understood.

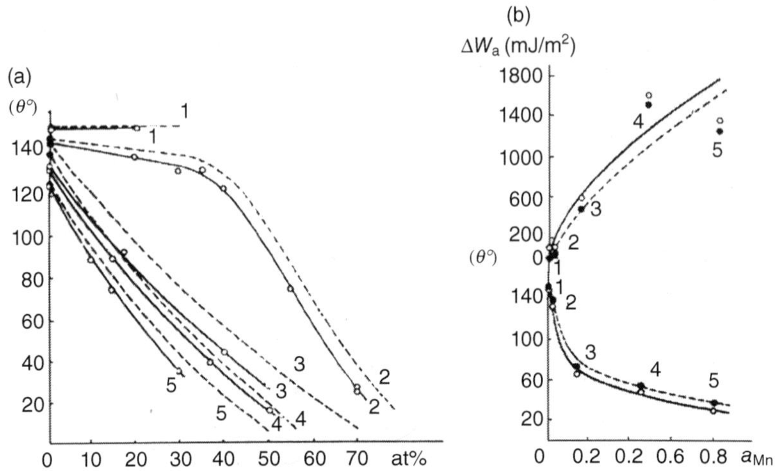

Figure 7.8. (a) Wetting versus Mn concentration and (b) wetting and adhesion versus Mn activity in the alloy in wetting graphite (–) and diamond (—) with (1) Au–Mn, (2) Ge–Mn, (3) Sn–Mn, (4) Cu–Mn, and (5) Ag–Mn melts.

In Au–Mn, Ge–Mn, Sn–Mn, Cu–Mn, and Ag–Mn binary alloys, the Mn activity increases from Au to Ag; the wetting of diamond and graphite by these melts improves in the same sequence (see Fig. 7.8) (Perevertailo et al., 1978). Pure metal solvents (Au, Ag, Sn, Cu, and Ge) do not wet diamond and graphite. The addition of 30 at% Mn to the Cu, Sn, and Ag melts decreases contact angles to 20°–0°. With this concentration of Mn in the Au and Ge melts, the angles of wetting graphite and diamond by these melts do not virtually change, and only an increase of the Mn concentration up to 70–80 at% allows a good wetting in these contact systems to be attained.

Figure 7.8b gives curves of the work of adhesion of various inert metal melts containing equal amounts of Mn (30 at%) to the graphite surface (minus the work of adhesion of a pure solvent). By numerous examples of contact systems, we have demonstrated that there is a correlation between the degree of chemical affinity of an active component for the solid-phase material and the wetting in the contact system and a relationship between adhesive and thermodynamic activities of an active component of the melt in wetting graphite and diamond by various metal melts. The influence of the thermodynamic activity of the melt active component on the melt adhesion to graphite surface is shown in Naidich et al. (1978). The values of the bonding energy of Mn with metal solvents and with graphite differ slightly, which, in our opinion, makes possible this relation to be established. A reason for the absence of the additive relationship between the work of adhesion and the Mn activity in the melts of various nonreactive solvents (see Fig. 7.9) is evidently a disregarded contribution of the process of carbon dissolution in the melt to a change of the melt component activities.

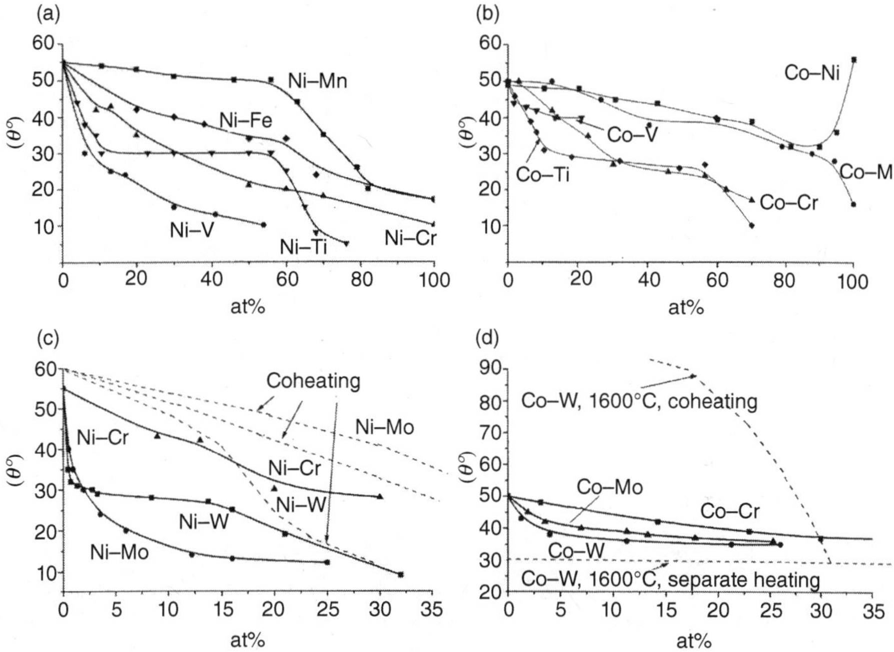

Figure 7.9. Concentration dependences of graphite wettability by Ni- and Co-based melts.

WETTING OF GRAPHITE BY GROUP VIII METAL MELTS WITH THE ADDITION OF REACTIVE AND NONREACTIVE METALS

The contact systems of the similar type exhibit various trends of the interaction in the contact zone when wetting carbon phase by a metal melt: the dissolution of the solid phase in the liquid one and the formation of a chemical compound (carbides and intermetallides). The analysis of the results obtained is most complicated, as the contact angle value depends on the chemical affinity of the melt components for carbon, their activity in the melt, and the difference in carbon chemical potential between the solid and liquid phases. Just this type of interaction is of great practical interest because these systems are the basis for media to grow diamond crystals.

Concentration dependences of contact angles on graphite for nickel and cobalt melts with additions of carbide-forming metals are given in Figure 7.9 (Perevertailo et al. 2006). It is seen that Fe, Mn, Cr, V, Ti, Mo, and W added to a nickel or cobalt melt decrease the contact angle value. An exception is the Ni–Mn system, in which the contact angle does not virtually change up to 60 at% Mn in the melt despite the fact that in this concentration range the carbon solubility in the melt increases from 12 to 15 at% and the melting temperature of eutectic decreases from 1400 to 1260 K (see Table 7.4).

TABLE 7.4. Characteristics of Alloys of the Ni–Mn–C and Co–Mn–C Systems

Composition (at%)			Temperature of termination of the melting of the $L \rightarrow G + \gamma$ eutectic (K)
Ni	Mn	C	
58.0	30.0	12.0	1400
47.0	40.0	13.0	1330
37.0	50.0	13.0	1280
25.0	60.0	15.0	1260
Co	Mn	C	
85.5	5.0	9.5	1500
81.0	10.0	9.0	1420
75.0	15.0	10.0	1400
71.0	17.0	11.0	1390

The low Mn activity in the melt is the reason why the angles of graphite wetting by the Ni–Mn melts do not change in value. At a Mn content of the Ni–Mn system up to 50 at%, the real value of its thermodynamic activity does not exceed one-fifth of the additive value; that is, in this concentration range, considerable negative deviations from the Roault law take place. The bonding energies of manganese with nickel and manganese with carbon differ slightly, which is responsible for unchanged-in-value adhesion activity of Ni–Mn melts containing up to 55 at% Mn toward graphite. As the manganese content of the melt increases above 60 at%, the thermodynamic activity of manganese increases sharply and the wetting in the system improves (see Fig. 7.9).

The Mn activity in the cobalt melt is higher than in the nickel melt, and therefore, the concentration dependences of the contact angles on graphite for Ni–Mn and Co–Mn melts differ in nature (see Fig. 7.9). For the melts of both systems, the wetting angle in the range from 45° to 50° is realized at different manganese concentrations: For the cobalt melt at 20 at% and for the nickel melt at 60 at% though, the carbon solubility in the Co–20 at% Mn is lower (10 at%) than in the Ni–60 at% Mn (15 at%), and the melting temperature of cobalt-containing eutectic is higher (1390 K).

Similar results were previously obtained in wetting graphite by Au–Mn and Ge–Mn melts. The manganese activity in these systems is also characterized by a high negative deviation from the Roault law (at concentrations up to 40–60 at%, it is virtually zero). In this range of Mn concentrations, graphite wetting by the Au–Mn and Ge–Mn melts does not virtually occur (see Fig. 7.10) (Perevertailo et al. 2006).

Au–Ti, Cu–Ti, Ni–Ti, and Co–Ti melts are also characterized by high negative deviations from the ideal solutions. However, the bonding energy of titanium with carbon in titanium-containing systems as opposed to manganese-containing systems is much higher than the bonding energy between atoms of the metals in the melt. A smooth change of the wetting angle with changing titanium content of the melt is observed in the isotherms of graphite wetting by Au–Ti and Cu–Ti melts. At the same time, the

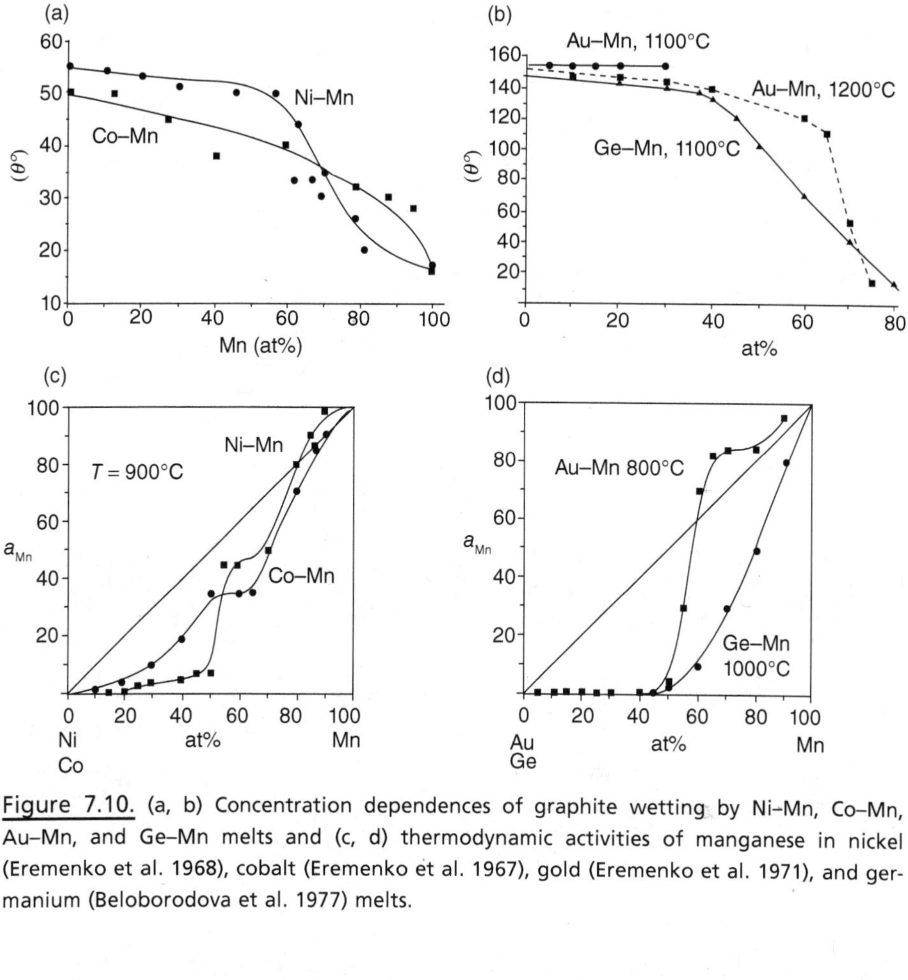

Figure 7.10. (a, b) Concentration dependences of graphite wetting by Ni–Mn, Co–Mn, Au–Mn, and Ge–Mn melts and (c, d) thermodynamic activities of manganese in nickel (Eremenko et al. 1968), cobalt (Eremenko et al. 1967), gold (Eremenko et al. 1971), and germanium (Beloborodova et al. 1977) melts.

wetting angles on graphite for the Ni–Ti and Co–Ti melts do not change in isotherms when the Ti content of the melt ranges from 15 to 55 at% (see Fig. 7.11) (Perevertailo et al. 2006).

The formation of the contact interaction zone in the systems of graphite with nickel melt containing additions of Fe, Mn, Cr, V, Ti, Mo, and W occurs as follows. With a low concentration of the active addition in the melt, a nickel-based solid solution and a small amount of free graphite form. As the concentration of the carbide-forming component in the melt increases, carbide phases of the appropriate compositions appear in the structure of the contact zone.

The critical concentration of the carbide-forming additive, which corresponds to the onset of the formation of carbide phases in the structure of the contact zone, increases in the Ti → V → Cr → Mn (from 4 at% for Ti to 61 at% for Mn) and in the W → Mo → Cr (from 4 at% for W to 31.6 at% for Cr) series, which corresponds in particular to a decrease in the degree of the affinity of metal additive for carbon

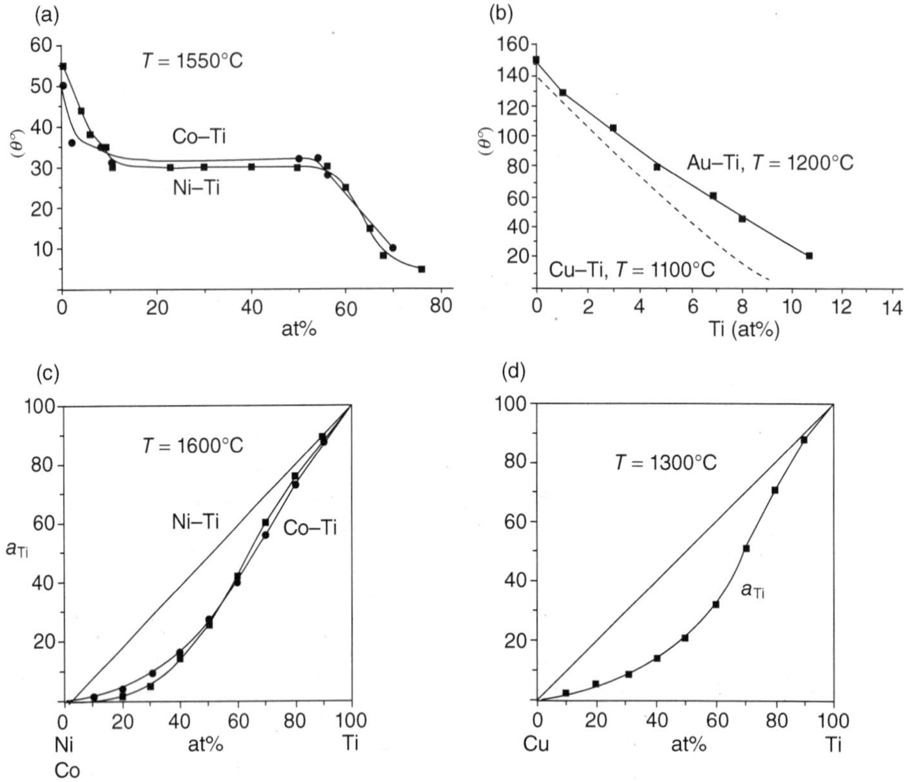

Figure 7.11. Concentration dependences (a, b) of graphite wetting by Ni–Ti, Co–Ti, Au–Ti, Ge–Ti melts and thermodynamic activities (c, d) of titanium in nickel (Turchanin et al. 2001), cobalt (Turchanin and Agraval 2002) and copper (Turchanin et al. 1997) melts.

(Perevertailo et al. 2006). In the Ni–Mn–C system, the Mn_7C_3 carbide is mainly of the hexagonal form, though platelike crystals can also be met (see Fig. 7.12). Crystals of the Cr_3C_2 carbide are also of the platelike form. Carbides of vanadium (VC), titanium (TiC), and tungsten (WC) are characterized by a rectangular form and the Mo_2C carbide has the form of long, thin needles.

At the interfaces in the Ni–V–C, Ni–Ti–C, and Ni–W–C systems, carbide phase particles tend to intergrow. The resultant layer is not always dense; sometimes, the melt penetrates into graphite and near the melt–graphite interfaces, the streaks of the metal that penetrates into graphite are observed. At first, a layer of carbides, which are the richest in carbon, forms. For instance, in the Ni–W–C system, a layer of tungsten carbide forms at the interface with graphite first, and as the tungsten concentration in the melt increases above 30 at%, a layer of the Ni_3W_3C double carbide abuts on it immediately (see Fig. 7.12).

In the concentration range from 15 to 55 at% Ti in the Ni–Ti–C system, a dense thin layer of titanium carbide forms at the interface with graphite, thus stabilizing the

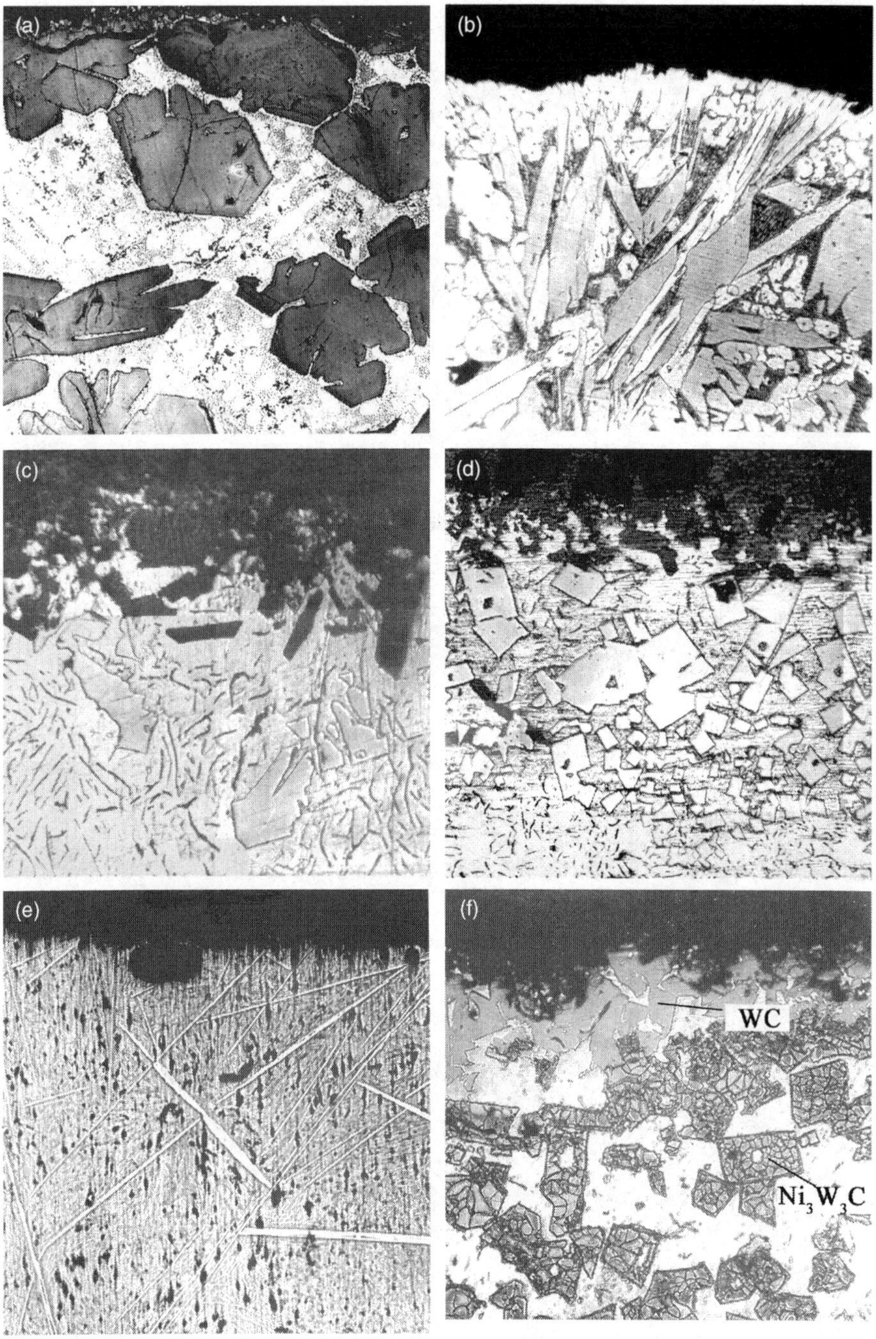

Figure 7.12. Forms of carbides that crystallize from the nickel-based melts: (a) Ni, 81 at% Mn (Mn_7C_3); (b) Ni, 50 at% Cr (Cr_3C_2); (c) Ni, 11.8 at% V (VC); (d) Ni, 12.6 at% Ti (TiC); (e) Ni, 16.2 at% Mo (Mo_2C); (f) Ni, 30.7 at% W (WC, Ni_3W_3C). 320×.

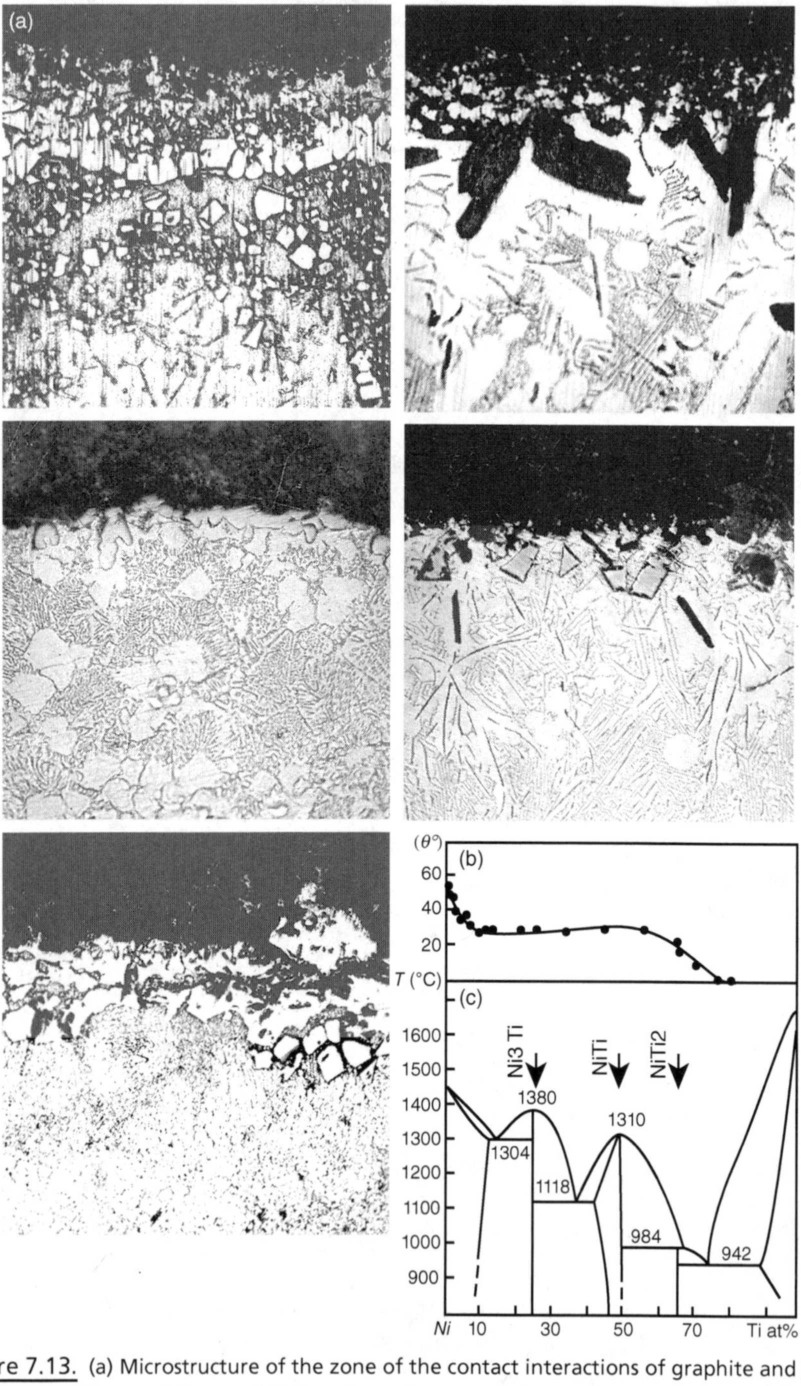

Figure 7.13. (a) Microstructure of the zone of the contact interactions of graphite and Ni–Ti melts, 320×; (b) wettability; and (c) the phase diagram of the Ni–Ti system.

wetting angle values (see Fig. 7.13) (Perevertailo et al. 2006). Something like this takes place evidently in the Co–Ti–C system as well. Under these conditions, the metal melt makes contacts, in fact, not with graphite but with carbide (wetting angles are close to those on titanium carbide for pure nickel or cobalt in value (Panasyuk et al. 1986). The metal part of the sample is easily sheared off (peeled off) from the graphite substrate along the interface, and in this case, no deep erosion of the graphite surface is observed under the drop and the structure of the region adjacent to the interface forms on the basis of the Ni–Ti binary system (see Fig. 7.13). According to the X-ray diffraction analysis, at the Ti concentration of 15–60 at%, the intermetallide of composition Ni_3Ti and titanium carbide TiC are identified in the bulk of the crystallized drop, with no free carbon.

The formation of the dense layer of titanium carbide at the interface in the Ni–Ti and Co–Ti systems is aided by the high enthalpies of mixing and low thermodynamic activity of titanium in nickel and cobalt, which is indicative of the formation of stable complexes based on various kinds of atoms corresponding to nickel intermetallides in the bulk of the melt. As the bonding energy of titanium to nickel or cobalt is higher than that of titanium to carbide, the existence of these complexes essentially complicates the formation of the carbide layer.

An increase in the titanium content of the melt above 60 at% results in an increase of the titanium activity in the melt and in the formation of loose thick layers of titanium carbide at the interface. An intensive penetration of the melt along the grain boundaries deep into graphite is observed at a concentration of 75 at% Ti (see Fig. 7.13). The size of individual carbide grains increases as the temperature of the experiment is much higher than the temperature of the liquid-phase appearance at this relation between nickel and titanium. The melts containing above 60 at% Ti intensively wet graphite, the best wetting being observed for the melt with 75 at% Ti, which corresponds to the lowest-melting eutectic in the Ni–Ti system.

If the formation of the structure of the metal melt–carbon surface interface occurs in full accordance with the phase diagram, the capillary characteristics of the contact system are also affected by the method of carrying out experiments (separate or coheating). In coheating, the material is enriched with carbon even in the course of heating, while in separate heating, the metal melt composition does not change. Therefore, the wetting angles on graphite for Ni–Cr melts containing up to 40 at% Cr on separate heating are lower than on coheating (see Fig. 7.9). A similar picture was observed for the Co–W–C system at 1600°C and for the Ni–Cr–C, Ni–Mo–C, and Ni–W–C systems at 1650°C (Bakun et al. 1992; Zabuga et al. 1992). Some discrepancy in the wetting angle values obtained by us and reported in Bakun et al. (1992) and in Zabuga et al. (1992) may be attributable to different rates of heating the melt (the higher the heating rate, the lower the degree of the alloy carbonization in the solid state). This phenomenon manifests itself most distinctly in the case where the melt contains chemical elements, which intensively dissolve carbon, and slightly in contact angles when the melt contains carbide-forming elements.

As mentioned earlier, the intensity of the interaction of a multicomponent melt with a solid phase, wetting, and adhesion in a contact system are defined, on the one hand, by the chemical affinity of the melt component for the solid-phase substance and,

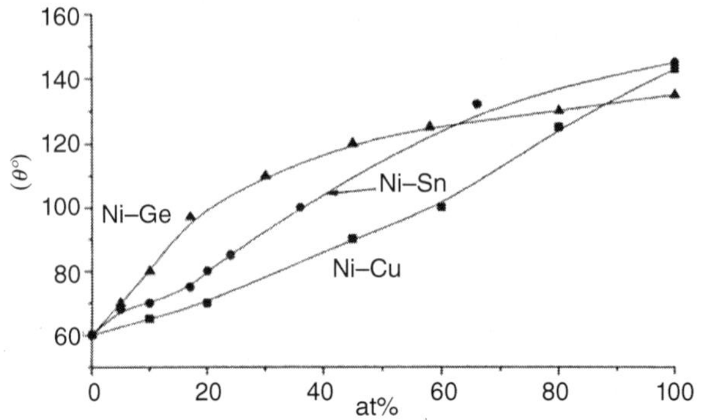

Figure 7.14. Wetting of graphite by Ni–Cu, Ni–Sn, and Ni–Ge melts at $T = 1500°C$.

on the other hand, by the intensity of the interaction of the melt components themselves between each other and their thermodynamic activities. Figure 7.14 shows the concentration dependences of wetting graphite by Ni–Cu, Ni–Sn, and Ni–Ge melts. An addition of Cu, Sn, and Ge to the nickel melt even in small amounts brings about an increase in contact angles. However, in the case of Sn and Ge, the impairment of wetting shows up much more sharply than with Cu. Such a behavior of metal melts in contact with graphite is caused by a number of reasons, the most important of which are different solubility of carbon in a liquid phase and activity of components in the nickel melt. Clearly, Sn and Ge decrease the carbon solubility in nickel melts much more rapidly than Cu does, and the thermodynamic activities of Cu, Sn, and Ge in liquid Ni melts increase in the Ge → Sn → Cu series; the wetting in the contact system changes in the same sequence (see Fig. 7.14).

Very interesting results in wetting should be expected when using a nonreactive melt containing several components reactive to the solid-phase substance. A stronger interaction between a pair of the melt components can make the occurrence of the third component at the interface easier, thanks to a decrease in the energy of its interaction with other atoms of the melt. Clearly, the studies of the wetting of graphite by multicomponent melts based on Ni–Mn, Ni–Cr, Ni–V, and Ni–Ti, in which the relationship between Ni and Mn, Cr, V, and Ti was constant and amounted to 40:60, 85:15, 95:5, and 95:5, respectively (Naidich et al. 1988; Naidich et al. 1991b), showed the following. The affinity of Mn, Cr, V, and Ti for carbon is higher than of Ni, and Cu, Ge, Sn, and Ga are nonreactive to carbon. However, the addition of nonreactive Cu, Ge, Sn, and Ga in certain concentrations improves the wetting of graphite (see Figs. 7.15–7.17). The wetting isotherms exhibit minima, which are not typical for a contact of two-component melts containing reactive (Ni) and nonreactive (Sn, Cu, and Ge) components with a carbon surface (see Fig. 7.14).

There is a rather large region of Cu and Sn concentrations (up to 90 at%) at which a good wetting of graphite by the melt is observed. At the same time, melts containing Ge (50 at%) show an abrupt worsening in wetting graphite. Different mechanisms of the effect of nonreactive additions on the wetting in the system can also be explained

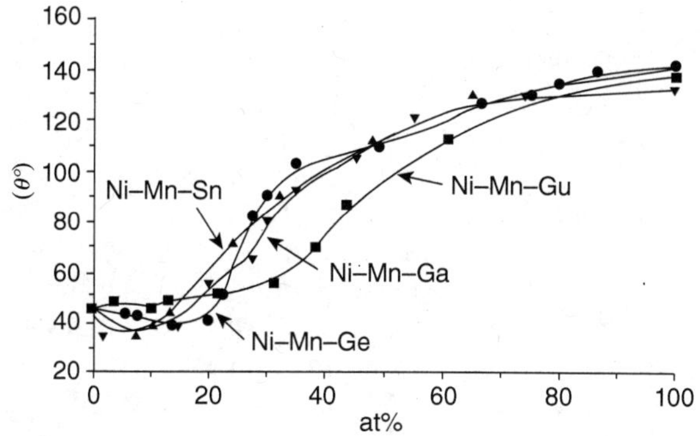

Figure 7.15. Wetting of graphite by Ni–Mn–Ge, Ni–Mn–Sn, Ni–Mn–Cu, and Ni–Mn–Ga melts at $T = 1500°C$.

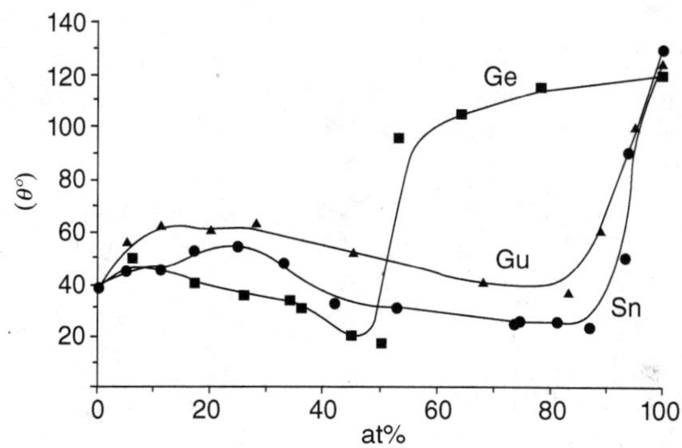

Figure 7.16. Wetting of graphite by Ni–V–Ge, Ni–V–Sn, and Ni–V–Cu melts at $T = 1500°C$.

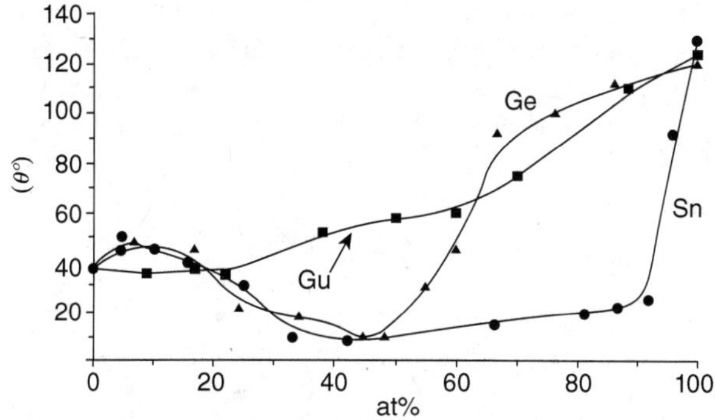

Figure 7.17. Wetting of graphite by Ni–Ti–Ge, Ni–Ti–Sn, and Ni–Ti–Cu melts at $T = 1500°C$.

by a different degree of a decrease of the melt–carbon bonding energy, as Ge and Sn decrease the carbon solubility in the melt much more rapidly than Cu.

It should be taken into account that the graphite wetting by three-component melts is simultaneously affected by two elements reactive to carbon (Ni and Mn, Ni and Cr, Ni and V, or Ni and Ti), and the nonreactive additive interacts with each melt component variously. The analysis of the phase diagrams of binary systems of Ni and Mn, Cr, V, Ti, Cu, Sn, and Ge as well as of thermodynamic characteristics of these system melts has shown that Sn, Ge, Ga, or Cu react with Ni more strongly than with Mn, Cr, V, and Ti. It is likely that the nonreactive additive changes (increases) the activity of Mn, Cr, V, and Ti in the melt, which gives rise to an increasing adhesion activity thereof, caused by an easier access of Mn, Cr, V, and Ti to the solid–liquid interface, as a result of which the interfacial tension decreases and the degree of the graphite wetting by the melt increases (Naidich et al. 1988). An increase in the Ti adhesion activity caused by its easier access to the interface with graphite because of the addition of Ge into the Ni–Ti melt is supported by metallographic analysis (Perevertailo et al. 1992).

The wetting is poorer only in cases that the concentrations of the nonreactive additives become considerable and decrease the percentage of the active metal (Mn, Cr, V, and Ti) both in a unit volume and per unit interfacial area. It should be noted that the worsening of the wetting in the systems under study is observed at various concentrations of nonreactive additives in the melt: 30 at% Ge, Sn, Ga in Ni–Mn; 35 at% Ge in Ni–Cr; 50 at% Cu in Ni–Mn; and 80–90 at% Cu and Sn in Ni–Cr.

Thus, the use of nonreactive additives in multicomponent melts containing several active components can show up in a very unexpected way and in a number of cases improves the wetting in the solid–liquid contact system.

RELATIONSHIP BETWEEN THE PHASE DIAGRAMS, THE INTERFACE STRUCTURE AFTER HARDENING, AND THE TYPE OF WETTING ISOTHERM

The existence of the relationship between the interface structure after hardening and the type of isotherm of graphite wetting by nickel or cobalt melts with the addition of carbide-forming metals allowed us to suggest a scheme that defined the tendency for a wetting variation with a change in the melt composition. A scheme of the formation of the contact interaction zone in the systems with "competing" interaction (the dissolution of the solid phase in the liquid and the formation of a chemical compound) is given in Figure 7.18 (Perevertailo et al. 2006). In the phase diagrams of nickel with carbon and carbide-forming metals, of interest are three equilibrium regions: a solid solution based on the metal or an intermetallide + carbon, a carbide + a solid solution + graphite, and a carbide + a solid solution. Each phase region is characterized by its own concentration dependence and the rate of the wetting angle variation.

In the concentration range corresponding to the solid solution + carbon two-phase equilibrium region, the structure of the contact interaction zone is represented by the nickel-based solid solution and a small amount of free graphite. To this region in the phase diagram, there corresponds section 1 of the wetting isotherm, within which

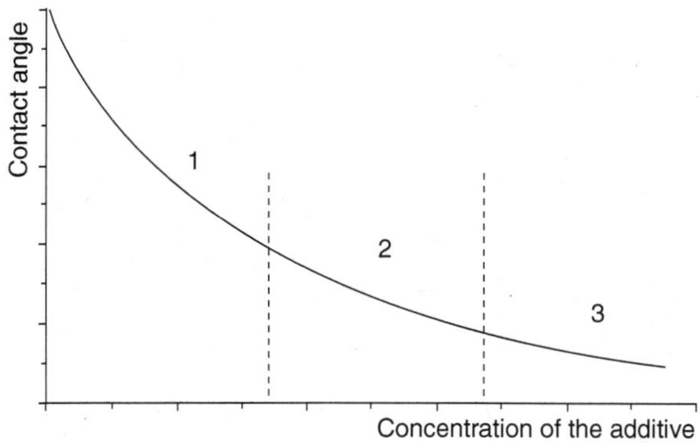

Figure 7.18. Schematic of the formation of the contact interaction zone in the systems with competing interaction of the melt components with the solid phase.

the contact angles change abruptly. The studies of spreading group VIII metals over the graphite surface showed that the rate of spreading was rather high, and it took 10^{-2}–10^{-4} s (see Fig. 7.5). This means that the spreading occurs much faster than the diffusion of the contact system components across the interface. The special calculations with the use of the constants of the dissolution rates (10^{-6}–10^{-5} m/s) (Eremenko and Natanzon 1970) showed that for the time of spreading, the "diffusion deflection" of the solid phase was 10^{-9}–10^{-8} m; that is, at the initial stage of the interaction and the spreading of the melt over the solid-phase surface, the diffusion of the melt components across the interface may be neglected. The wetting angle depends on the experimental conditions (a separate or coheating of the melt and substrate), which is due to the carbonization of the melt in the solid state.

The width of the first region is dictated both by the limiting solubility of the metal additive and carbon in nickel and by the additive activity in the melt. The rate of the melt spreading over the solid-phase surface and the adsorption of the active component on the interface are responsible for the melt enrichment both in carbon and in the active component. A low thermodynamic activity of components in the melt points to a high interaction between the melt components and gives ground to suggest that there is a wide region of solid solution based on nickel (the basic component) in the solid state or that the formation of intermetallides takes place. Thus, the region of the equilibrium between the solid solution and carbon in the phase diagram is characterized by section 1 of the wetting isotherm, within which the interface formation depends on the melt structure and is responsible for the melt spreading over the solid-phase surface and for the adsorption of the active component on the interface.

In the range of concentrations corresponding to the region of three-phase equilibrium (carbide + solid solution based on the metal or intermetallide + graphite), the structure of the contact zone is either a solid solution of the respective carbide-forming metal and carbon in nickel or an intermetallic phase or a mixture of these phases depending on the intensity of the interaction of the additive components and carbon with nickel. The formation of carbide phases in the contact zone structure starts near the phase boundary with graphite, and at the initial stage, these carbides are revealed by metallographic analysis at considerable magnifications (1600×) only. As the concentration of the carbide-forming additive increases, so do the size of carbide phases and the amount of them in the bulk of the melt, while the amount of free carbon decreases. This equilibrium region in the phase diagram has its counterpart in section 2 in the wetting isotherm, which is characterized by a smooth variation of wetting angles.

In the range of concentrations corresponding to the region of two-phase equilibrium in the phase diagram (carbide + solid solution based on the metal or intermetallide), grains of the carbide phase tend to intergrow and to form a layer, which is not always dense. A layer of carbides, which are most rich in carbon, is adjacent immediately to the interface. In section 3 of the isotherm, contact angles change slightly; the rate of spreading of the melt over the solid-phase surface is low (sometimes the spreading takes several hours) and increases with the temperature of the experiment. The formation of dense layers at the interface results in the stabilization of the wetting angle values in the system. Thus, the formation of compounds on the solid-phase surface is characterized by the section of the wetting isotherm, within which the decisive role in the interface formation is played by the diffusion of the contact system components across the interface.

HIGH-PRESSURE EFFECT ON WETTING OF GRAPHITE AND DIAMOND BY METAL MELTS

The development of high-pressure technique has made it possible to study physicochemical characteristics of materials at high pressures. To do this, a procedure of wetting solids by metal melts under high pressures (up to 10 GPa) has been developed (Naidich et al. 1982). The high-pressure cell to study the effect of high pressure on the wetting was designed as applied to solid-phase high-pressure apparatuses (see Fig. 7.19).

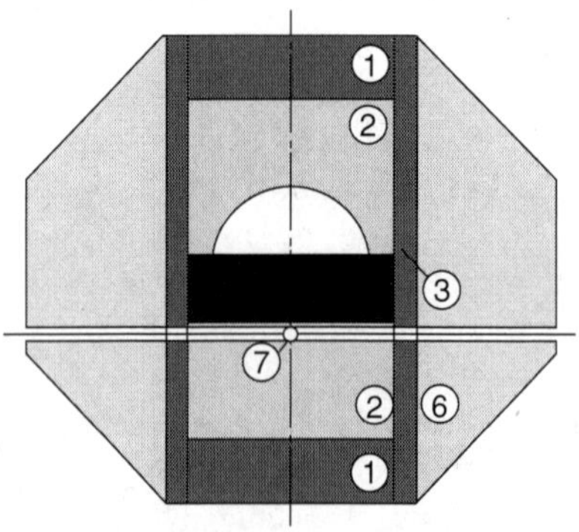

Figure 7.19. High-pressure cell: 1—thermal insulator; 2—flux; 3—graphite heater; 4—metal melt; 5—solid substrate; 6—container; 7—thermocouple.

A substrate with a metal melt sample was placed horizontally in the high-pressure cell under a layer of precompacted flux chosen so that its melting temperature was below the contact melting temperature of the objects under study and it was nonreactive to the container material, substrate, and the melt. As a flux, NaCl was usually used. The choice of NaCl as a pressure medium is motivated by the fact that at a pressure of 4.5 GPa, it melts at 1420°C and ensures free formation of a melt drop and spreading it over the surface of the solid, maintaining the chemical inertness to the liquid and solid phases. After holding at the above temperature and pressure, the cell was allowed to cool under the same pressure. The contact angle was determined from the form of the crystallized drop removed from the high-pressure apparatus with the solid substrate. The measured angle values contain a systematic error caused both by the presence of the flux and the measurements using a cold drop. The accuracy of the contact angle measurements was ±2°–5°. The analysis of the flux effect on the measured angles has shown that it does not essentially affect their value (Naidich et al. 1982).

Original (111) planes of natural diamonds, (100) and (111) planes cut from large crystals of natural diamonds, (001) and (*hk*0) planes of natural graphite, as well as samples of MPG-6 polycrystalline graphite, were used for wetting experiments. The experiments on wetting solids by metal melts at high pressures showed that high pressure (up to 8 GPa) did not essentially affect the wetting of carbon materials by melts of nonreactive to carbon metals and group VIII transition metals (see Table 7.5) (Naidich et al. 1984).

For melts containing carbide-forming metals, a slight improvement in the wetting at high pressures as compared with that in vacuum is probably caused by an increase in the carbide stability at a high pressure. At a pressure of 4.5 GPa, the Ni–Mn and Ni–Cr melts exhibit somewhat lower contact angles on graphite (see Figs. 7.20 and 7.21) than in vacuum, and the minimum in the wetting curve in the Ni–Mn–Ga system (see Fig. 7.22) is observed at a lower Mn concentration in the melt than is observed in a vacuum (Loginova and Perevertailo 2005).

By and large, the processes occurring in the contact system at atmospheric and high pressures are identical, and the contact zones are qualitatively similar in the phase composition and structure. These data are of principal importance, as they allow graph-

TABLE 7.5. Wetting of Graphite and Diamond by Metal Melts at High Pressure

	Graphite				(111) diamond face			
	$P = 4.5$ GPa		$P = 2 \times 10^{-3}$ Pa		$P = 8$ GPa		$P = 2 \times 10^{-3}$ Pa	
Metal	T (°C)	θ (degree)	T (°C)	θ (degree)	T (°C)	θ (degree)	T (°C)	θ (degree)
Cu	1400	145	1100	143	1700	145	1100	137
Ag	1420	149	1100	127	1700	142	1100	126
Au	1420	134	1100	121	1700	126	1100	150
Ni	1400	52	1550	48	2500	21		
Pd	1750	58	1560	52	2500	30		
Pt	1900	61	1800	75	2500	27		

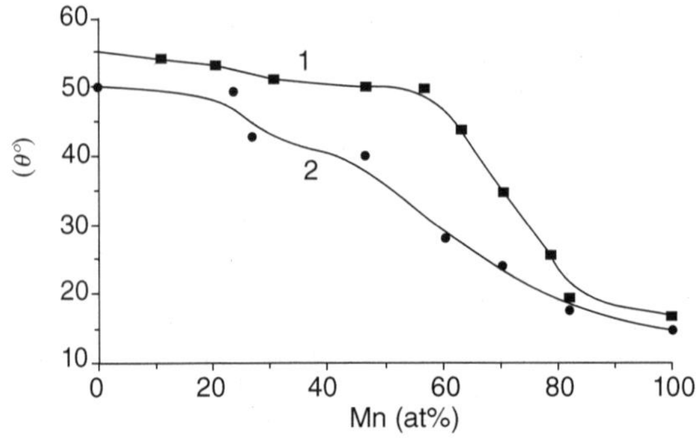

Figure 7.20. Wetting of graphite by Ni–Mn melts at atmospheric pressure (1) and at 4.5 GPa (2).

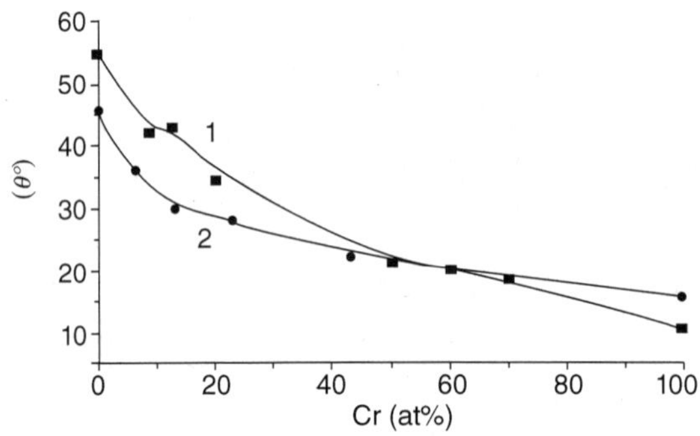

Figure 7.21. Wetting of graphite by Ni–Cr melts at atmospheric pressure (1) and at 4.5 GPa (2).

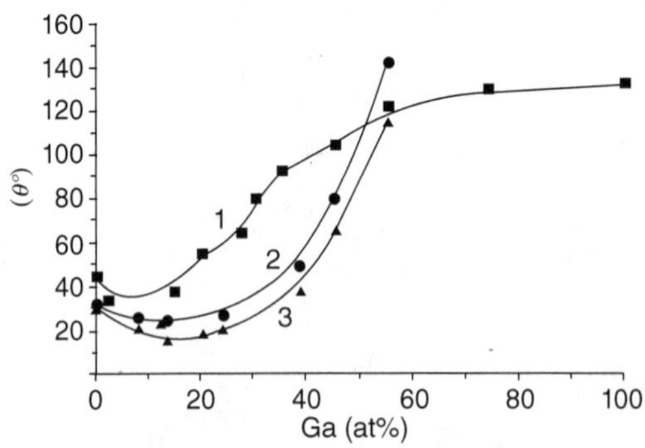

Figure 7.22. Wetting of graphite (1, 2) and diamond (3) by Ni–Mn–Ga melts at (1) atmospheric pressure, and at (2) 4.5 and (3) 7.0 GPa.

ite to be used instead of expensive diamond when developing new growth media to produce diamond crystals, and the contact system capillary characteristics obtained in vacuum or protection media to be used to describe the processes occurring at high pressure.

CONCLUSIONS

The interaction of the metal melt with a solid carbon-containing phase includes several processes that are often very complicated for description. They are spreading, adsorption of the melt active component at the phase boundary, and diffusion of the contact system components through this boundary with a subsequent nucleation and growth of a new phase layer (Saiz et al. 2000). In turn, the interfacial interaction is largely defined by the type of the phase diagram. The phase diagram shows the component ability to form boundary solid solutions and the limiting value of the mutual solubility, as well as the presence of the intermediate phases, their stoichiometric compositions, and regions of homogeneity. In the region of melting, the phase diagram gives the data on the type and melting temperature of the forming intermediate phases, on nonvariant equilibria (eutectic or peritectic), temperature and compositions of nonvariant points, concentration, and temperature limits of the phase regions that include a liquid phase.

If the interaction of a solid and liquid phases runs stationary (at the constant difference in chemical potential $\Delta\mu$ between the components of both phases), a decrease in the interfacial tension is in direct proportion with $\Delta\mu$. It follows that in the systems, where the components of a solid do not completely dissolve in a liquid and the liquid does not absolutely dissolve in a solid phase, the $\sigma_{sl} = f(t)$ curve should show the minimum (t being the time of contact between the phases). If chemical compounds (with the melting points exceeding the contact temperature) form at the interface, only the initial section of this curve is manifested. The further changes in the interfacial tension are sharply hindered as the diffusion of components through the intermediate solid layer is rather slow.

For better understanding of the spreading mechanism in the systems, in which the intermediate phases form, of great importance are the composition, origin, structure, sequence of the phase formation, and growth at the interface, especially at the initial stage of the process. It is necessary to know or make a justified assumption over which surface the basic quantity of a liquid metal is spread, as the intermediate layers forming in the contact zone often differ in properties from the initial surface and, therefore, may be variously wetted. For good wetting of solids of various physicochemical natures by liquid metals, it is necessary that chemical compounds forming at the interface are to a sufficient degree metal-like, dissolve in the basic metal, or are gaseous and removed easily from the contact zone.

The new phase interlayer at the interface can form by different mechanisms and temperatures; the sequence of their formation may not correspond to the coexistence of phases represented in the equilibrium phase diagram. The rate of their formation and the structure depend on thermodynamic and kinetic factors. In spreading, when the supply of a liquid metal to the interfaces is hindered, the coexistence of nonequilibrium phases is a possibility.

The investigation of the features of the contact zone structure formation in the metal–carbon systems is an intermediate element that allows us to relate characteristics typical of the liquid to the state of the system after crystallization, which is described by the diagram of phase equilibria, as the bases for them are the same thermodynamic principles. These investigations will allow metal melts to be classed by the ability to well wet carbon phases and at a high rate, and will provide a possibility to substantiate the choice of the melt compositions for different technological processes.

Nonequilibrium systems are characterized by a heavy temperature dependence of contact angles. A wetting threshold is often observed in graphs. In heating above a certain (threshold) temperature, the contact angle decreases abruptly because of a considerable increase of the work of adhesion. The existence of the wetting threshold is important in terms of application. For instance, to ensure a good wetting of the surfaces of the materials being joined in soldering, the wetting temperature should often much exceed the melting temperature of the solder.

The studies of the processes of wetting and phase interaction under the conditions of competing interaction of the melt components with each other and with a solid phase in contact systems of different physicochemical natures have allowed us to derive criteria for the purposeful control over high-temperature capillary processes. The physicochemical factors that control the wetting of solids by metal melts include the following:

- Chemical affinity of the melt components for the solid-phase substance (ΔG, isobaric–isothermal potential of the reaction of the formation of an intermediate phase at the interface)
- Activity (the effective concentration) of the metal additive in the melt of the basic metal
- Limiting dissolution of the metal additive and solid-phase components, and the interaction products in the basic metal (difference in chemical potentials, $\Delta\mu$, between components of the solid and liquid phases)
- Hysteresis phenomena stemming from the state of the solid-phase surface (porosity, roughness, and chemical nonuniformity) and its orientation (crystalline structure of the solid phase).

The chemical affinity is successfully used in the development of bonds, solders for the capillary joining of solids having different physicochemical natures (Naidich 1972; Naidich et al. 1991). In this case, one or several components reactive to the solid-phase substance (from the row of transition metals) are added to a melt nonreactive to the solid-phase substance. The nonreactive melt is chosen from the row of nontransition metals, most often Cu, Ag, Sn, Au, Ge, Ga, or their melts. The principle of the solder formulation consists in alloying plastic relatively low-melting metals (Cu, Ag, Au, Sn, and their melts) with active metals having a high chemical affinity for the solid-phase components.

However, if the melt components having a high chemical affinity for the solid-phase components are characterized by a low activity in the melt, a good wetting in the system is not possible. The interrelation between the adhesion and thermodynamic activities of the components of metal melts is based on the adsorption of the melt components at the boundary with the solid phase. In common with the type of phase equilibria in the system under consideration, the behavior of atomic interaction in the liquid largely defines the structure of the contact zone being formed. In the case of a low activity of the components, only the redistribution of the melt components in the contact zone is observed, while at a high component activity, transition layers form at the interface.

For systems, in which a solid phase dissolves in a liquid phase, the highest decrease in the work of the melt adhesion to the solid phase is achieved when the concentration of a solid-phase component in the melt is close to its limiting solubility. By varying the value of a nonequilibrium contribution of the work of adhesion, which is defined by a chemical boundary reaction (the dissolution of the solid phase in the liquid one), one can substantially affect the wetting in the system, for example, by controlling the diffusion of the solid-phase components into the melt volume, if this substance in a proper concentration is preadded to the melt.

The best illustration of the effect of the solid-phase crystalline structure on the adhesion and wetting in a contact system is the anisotropy of wetting different crystal faces by metal melts, which indicates the anisotropy of the crystal surface energy. This effect shows up more distinctly in single crystals of complex chemical composition, as the possibilities of controlling the solid–liquid interfacial energy of any solid phase having a complex composition increase. First, the density of various types of atoms on different crystallographic faces differs; second, it is possible to control the adsorption of an impurity on one of the atoms or on all of them; third, in the solid-phase structure, the situation is possible, where the interface on the solid-phase side is represented by atoms of the same type.

Obtaining strong nonmetal/metal joinings and the development of composite materials based on solid phases of different physicochemical natures depend on the proper choice of a solder or binder. Thus, by varying the concentration of an additive, which is active to the solid-phase material (or changing the additive), by selecting the composition of an inactive solvent, and by controlling the process of diffusion between the components of the contact system, it is possible to largely affect the wetting and adhesion of a liquid metal melt to a solid phase and to open up the way to form a given structure of the transition zone. For example, the use of an additive having a high chemical activity and a high thermodynamic activity can ensure the highest work of adhesion of the melt to the solid phase. Under the conditions of an easier adsorption of the active component at the interface, thick transition layers of a new phase can form, which is an important prerequisite for the liquid-phase metallization of superhard material surfaces.

The process of joining superhard materials using adhesion-active melts requires that the concentration of an active additive in the melt should be strictly limited and the diffusion between the components being joined be controlled. Transition layers that form in joining should be of the minimum thickness. Otherwise, a thick layer of a new phase, often a brittle one, would exert a detrimental effect on the brazed tool performance, and intensive diffusion from the solid phase into the bulk melt would cause the structural heterogeneity in the materials being joined, thus reducing their performance. To meet the above requirements, as solders for an active brazing, such contact systems should be used, in which the mutual solubility of components of the solid and liquid phases is minimum.

The above scheme of the formation of the contact interaction zone may be useful in the substantiation of the choice of growth medium compositions for the diamond crystal production. For efficient diamond spontaneous crystallization, metal melts should be used, which are able to well wet carbon phases and at a high rate. In this case, the rates of all the processes governing the formation of the contact interaction zone at the metal melt–carbon surface boundary are the highest and are commensurable with the time of the diamond spontaneous crystallization. These melts, as a rule, should be characterized by a high mutual solubility of contacting phases, low thermodynamic activity of the active additive in the melt of the basic component (a group VIII metal), and moderate affinity of the additive for carbon. It is preferable to grow diamond crystals in systems whose phase diagrams contain the regions of free carbon, and there is no transition to the solid two-phase solution + carbide regions.

Thereby, the following is secured:

- A high rate of the diamond nucleation (in this concentration range, wetting angles change abruptly, providing low values of the interfacial energy at the diamond–growth medium phase boundary, and it is easier for a diamond nucleus to form) at relatively low temperatures of the liquid-phase appearance and technologically justified pressure
- A high growth rate of already formed nuclei as a good wetting of the initial graphite by a metal melt secures the carbon dissolution in the liquid phase at a required rate and in large quantities, and the solid solution, which accumulates carbon, makes for delivery of carbon to the face of the growing crystal.
- The absence of the carbide phase interlayers at the interfaces, which are often nontransparent for carbon diffusion, and the initial building material (carbon), which is not spent for the formation of a competing phase (carbide)

The *d*-metals of groups V–VIII of the periodic table meet these requirements to a large degree. Because of the great affinity for carbon, the *d*-metals of group IV, alkali earth metals, *p*-elements (B, Al, and Si), should be used as additions to growth media in the diamond spontaneous crystallization with caution and in small (up to 2 at%) amounts. The use of such chemical elements in the growth medium composition in large quantities will result in an abrupt increase in the diamond crystallization parameters and in a decrease in the degree of the graphite-to-diamond transformation. This is because the formation of the dense layer of a carbide phase at the graphite interface substantially restricts the diamond growth rate due to the difficulties in the carbon transfer to the melt. In this case, the diamond crystal formation is possible by an increase of the carbon content of the melt through an increase in temperature.

REFERENCES

Bakun OV, Kostrova LI, Nochevkin SA, et al. 1992. Physicochemical features of the contact interface formation in the systems of graphite-nickel melts with group VIa metals. Poroshkovaya Metallurgiya 1:60–64.

Beloborodova EA, Batalin GI, Kazimirov VP. 1977. Thermodynamic properties of germanium–manganese liquid alloys. Zhurnal Fizicheskoi Khimii 51:1800–1801.

Bobkovskiy VN, Kostikov VI, Levin VY, Maurah MA. 1972. Wettability of carbon materials with silica-based melts. In: Eremenko VN, editor. *Wettability and Surface Properties of Melts and Solids (in Russian)*. Kiev: Naukova Dumka, pp. 242–247.

Dezellus O, Jacques S, Hodaj F, Eustathopolos N. 2005. Wetting and infiltration of carbon by liquid silicon. J Mater Sci 40:2307–2311.

Eremenko VN, Lukashenko GN, Sidorko VP. 1967. Thermodynamic properties of alloys of the Mn-Co system. Izvestiya AN SSSR Metally 3:192–196.

Eremenko VN, Lukashenko GN, Sidorko VP. 1968. Thermodynamic properties of alloys of the Ni-Mn system. Izvestiya AN SSSR Metally 2:208–214.

Eremenko VN, Natanzon YV. 1970. Kinetics of the metal dissolution in metal melts under the conditions of the outer problem. Poroshkovaya Metallurgiya 8:39–54.

Eremenko VN, Lukashenko GN, Khmelenko GI. 1971. Thermodynamic properties of gold-manganese alloys. Ukrainsky Khimichesky Zhurnal 37:1206–1211.

Eustathopolos N, Nicholas MG, Drevet B. 1999. *Wettability at High Temperatures*. Amsterdam, NY: Pergamon.

Ferro AC, Derby B. 1995. Wetting behaviour in the Al-Si/SiC system: interface reactions and solubility effects. Acta Metall Mater 43:3061–3073.

Kostikov VI, Varenkov AN. 1981. *Interaction of Metal Melts with Carbon Material (in Russian)*. Moscow: Metallurgiya.

Kostikov VI, Andropov YI, Otopkov PP. 1974. Wetting of diamond with Ni–Si and Cu–Si alloys. In: Eremenko VN, editor. *Adhesion of Melts (in Russian)*. Kiev: Naukova Dumka, pp. 129–131.

Landry K, Kalageropoulou S, Eustathopoulos N. 1998. Wettability of carbon by aluminum alloys. Mater Sci Eng 254-A:99–111.

Loginova OB, Perevertailo VM. 2005. Effect of capillary process on the diamond crystallization from metal melts. In: Novikov NV, editor. *Synthesis, Sintering and Properties of Superhard Materials. Collection of Scientific Works (in Russian)*. Kiev: ISM, pp. 26–38.

Naidich YV. 1972. *Contact Phenomena in Metal Melts (in Russian)*. Kiev: Naukova Dumka.

Naidich YV, Perevertailo VM, Nevodnik GM. 1972. Study of the kinetics of spreading metal melts over solid surfaces. Poroshkovaya Metallurgiya 7:57–61.

Naidich YV, Perevertailo VM, Loginova OB. 1979. Adhesion and wettability of graphite with metals of the VIII group of the periodic system. Izvestiya AN SSSR Metally 4:37–41.

Naidich YV, Perevertailo VM, Shulzhenko AA, Loginova OB, Ivakhnenko SA. 1982. Wettability of the solid phases with the melts at a high pressures. In: Eremenko VN, editor. *Surface Properties of Melts (in Russian)*. Kiev: Naukova Dumka, pp. 189–193.

Naidich YV, Perevertailo VM, Loginova OB. 1983. Adhesion and thermodynamic properties in the graphite–metal melt contact systems. Zhurnal Fizicheskoi Khimii 57:577–580.

Naidich YV, Shulzhenko AA, Andreev AV, Loginova OB, Perevertailo VM. 1984. Wetting of diamond with metal melts in conditions of diamond thermodynamic stability. Dopovidi AN Ukr SSR 6-A:74–79.

Naidich YV, Perevertailo VM, Loginova OB, Polyanskaya ND. 1985. Wettability of different diamond faces with chemically inert to carbon melts. J Superhard Mater 7:19–21.

Naidich YV, Perevertailo VM, Loginova OB, Zabuga VV. 1988. Capillary properties and density of multicomponent nickel alloys in contact with graphite. Izvestiya AN SSSR Metally 3:57–61.

Naidich YV, Perevertailo VM, Lavrinenko IA, Kolesnichenko GA, Zuravlev VS. 1991a. *Surface Properties of Melts and Solids and Their Use in Materials Science (in Russian).* Kiev: Naukova Dumka.

Naidich YV, Perevertailo VM, Loginova OB. 1991b. Wetting of graphite multicomponent alloys based on nickel-titanium and nickel-vanadium metal melts. Adhes Melts Brazing Mater (25):66–69.

Panasyuk AD, Ovchinnikova TA, Beluh AB, Bereza VP. 1982. Wetting of the graphite materials by liquid silica. Adhes Melts Brazing Mater (9):20–25.

Panasyuk AD, Fomenko VS, Glebova GG. 1986. *Resistance of Non-Metallic Materials in the Melts.* Handbook. (in Russian). Kiev: Naukova Dumka.

Perevertailo VM, Loginova OB. 1991. Calculation of work of adhesion to carbon surface of melts of group VIII metals. J Superhard Mater 13:67.

Perevertailo VM, Loginova OB, Naidich YV. 1978. A study of capillary phenomena and wetting diamond single crystals with manganese-containing binary alloys. Adhes Melts Brazing Mater (3):31–33.

Perevertailo VM, Delevi VG, Loginova OB, et al. 1992. Wettability and contact interaction in the graphite-nickel-titanium-germanium system. J Superhard Mater 14:30–32.

Perevertailo VM, Loginova OB, Kulik OG, Turkevich VZ. 1998. Capillary phenomena in the Co-Mn-C system. Fluid Phase Equilibria 150–151:607–614.

Perevertailo VM, Loginova OB, Bagno NG. 2001. Interaction between metal melts and zirconium dioxide. Trans JWRI 30(Special Issue):143–147.

Perevertailo VM, Kazimirov VP, Roik AS, Loginova OB. 2004. Structure of nickel–carbon melt of composition 92Ni–8C. J Superhard Mater 26:43–49.

Perevertailo VM, Loginova OB, Kulik OG. 2006. Formation of the contact interaction zone of nickel alloys with the carbon surface. J Superhard Mater 28:27–38.

Saiz E, Cannon RM, Tomsia AP. 2000. Reactive spreading: adsorption, ridging and compound formation. Acta Mater 48:4449–4462.

Turchanin MA, Agraval PG. 2002. Enthalpies of mixing liquid alloys of titanium, zirconium and hafnium with cobalt. Rasplavy 2:8–16.

Turchanin MA, Belokonenko IV, Agraval PG, et al. 1997. Enthalpies of mixing in liquid alloys of copper and nickel with titanium, zirconium and hafnium. In: *Proceedings of the Second International Conference on Construction and Functional Materials CFM-97*, 68–69.

Turchanin MA, Belokonenko IV, Agraval PG. 2001. Heat of formations of liquid alloys of nickel with IVa-metals. Rasplavy 3:53–60.

Weisweiler W, Mahadevan V. 1972. Wettability characteristics of metal melts on graphite and glasslike carbon substrates and their relation to catalytic graphitization. High Temperature—High Press 4:27–38.

Whalen TJ, Anderson AT. 1975. Wetting of SiC, Si_3N_4 and carbon binary Si alloys. J Am Ceram Soc 58:396–398.

Yupko VL, Gnesin GG. 1973. Contact interaction of SiC with liquid silica. Poroshkovaya Metallurgiya 10:97–101.

Zabuga VV, Taranets NY, Naidich YV. 1992. Wettability in the graphite-Co-W-melt system. Adhes Melts Brazing Mater (28):38–41.

PART III

INTEGRATION ISSUES IN ENERGY GENERATION AND DEVICE FABRICATION

8

INTEGRATION TECHNOLOGIES FOR FERRITES AND POWER INDUCTORS IN CERAMIC CIRCUIT BOARDS

Richard Matz

Siemens Corporate Technology, Munich, Germany

INTRODUCTION

The evolution of a technology from first products to ubiquitous use is commonly described by a specific, simple, and empirical law. For the last half-century, one of the fastest-growing technologies has been transistor integration. Applying Moore's law (Moore 1965; revised in the 1970s), the number of transistors integrated on a semiconductor chip doubles every 18 months. This has fueled an unprecedented growth in the functional range of integrated circuits. Communication technology is another example, albeit less dynamic; since Marconi's first spark gap transmitter experiments in 1895 and the first transatlantic transmission in 1901, the number of voice channels per area has been increasing by a factor of one million every 50 years (Riezenman 2001), mainly due to spatial division multiplexing. For light-emitting diodes (LEDs), a price drop by an order of magnitude and a growth in light efficiency by a factor of 20 per decade is described by Haitz's law (Daewoo Securities 2007).

Thus, the miniaturization of electronic components has reached a high level of maturity and remains a major part of mobile, intelligent, and modular subsystems, from semiconductors to printed circuit boards (PCBs) as well as to assembly and interconnect technologies for passive, that is, nonsemiconductor electronic components. However,

Ceramic Integration and Joining Technologies: From Macro to Nanoscale, First Edition.
Edited by Mrityunjay Singh, Tatsuki Ohji, Rajiv Asthana, Sanjay Mathur.

among the technologies and devices, there is wide variation. Soldered wires with through holes were eliminated from PCBs in the 1970s by surface mounted devices (SMDs). Continuous improvements in fabrication and circuit technology, but also in materials, have allowed considerable size reductions of SMDs since then. This is particularly true for capacitors.

Due to the intricacy of magnetic fields and materials, such miniaturization efforts have been less successful for inductive components. In the microwave region, useful impedances are associated at gigahertz frequencies with relatively small inductances of a few nH; such naturally small components can be integrated on silicon with sufficient self-resonant frequencies and quality factors by creating spatial distance between the coils and the lossy substrate. This is possible by etching away a section of the substrate under the coplanar inductor (Jiang et al. 2000) or even by tilting the conductor windings vertically (Zou et al. 2003). Induced lossy substrate currents can also be efficiently suppressed by etching a comb of high-aspect-ratio trenches into the silicon and overcoating them with a dielectric serving as a new low-loss substrate for the inductor (Rais-Zadeh et al. 2008). The advantage of this approach is that it does not sacrifice mechanical strength and shock resistance.

Alternatively, Si substrate losses have been conquered by placing inductors off-chip on low-loss substrates such as ceramics. As these passive components are an integral part of the board or package and form a functional unit with the active devices, the term "system in package" has been coined. Low-temperature cofired ceramic (LTCC) devices, in particular, exhibit cost advantages over thin-film technologies: Holes are mechanically punched into individual ceramic green tapes and are filled with Ag paste for vertical interconnects; planar conductor traces are formed by screen printing Ag paste onto the tape surface; finally, a multitude of tapes is laminated to form a multilayer board, which is sintered at a relatively low temperature around 900°C. Sintering additives such as various glasses are used to reduce the sintering temperature to below the melting point of Ag, a major constituent of common metallization pastes. LTCC devices have a proven record in the automotive and aviation industries due to their hermeticity and robustness, but also in microwave and wireless applications, where their excellent loss and miniaturization performance are exploited. The latter stems from the fact that distributed circuit components such as directional couplers shrink in size with increasing dielectric constant of the surrounding material. There is pronounced interest in this field to reduce inductor line and space widths by conventional and innovative printing methods to values far below 100 μm (e.g., Lahti et al.2000).

Since higher inductance values require the use of ferrites to enhance the magnetic flux density, the high-temperature processing of ceramic multilayer boards has a basic advantage over low-temperature organic boards to realize such high-end materials in a consistent integrated technology. Magnetic losses are lower in magnetic ceramics than in amorphous magnetic metals at frequencies above a few 100 kHz (Dick et al. 2008). While the wireless communication market exhibited extraordinary growth rates around the turn of the century, initial ferrite developments pursued the adaptation of high-frequency Ba hexaferrite with a relative permeability of up to 20 (Karmazin et al. 2005). Embedded 10-nH inductors in pure ferrite multilayers of NiZnCo and Ba hexaferrite were demonstrated at several gigahertz (Matters-Kammerer et al. 2006).

In power electronics, on the other hand, as, for example, power converters, lamp ballasts, and switched-mode power supplies, current amplitudes are higher and frequencies of operation lower by a few orders of magnitude. Inductors have microhenry size now, and the use of magnetic materials is imperative in order to achieve compact size. The most obvious approach is the integration or embedding of the inductor coil into the PCB, which is then covered by magnetic material. As such, electroplated NiFe layers (Ludwig et al. 2003) and ferrite polymer compound layers (Waffenschmidt 2004) were employed to demonstrate a 4.7-μH/0.5-A inductor for a DC–DC converter and a 60-W transformer, respectively. There is no magnetic core passing vertically through the dielectric PCB in these so-called coreless structures; the deformation of the magnetic field by the magnetic cover layers is sufficient to tolerate the gap imposed by the dielectric board. This reduces the effective permeability but improves the power-handling capability because magnetic saturation occurs at higher current levels than in closed core structures (Billings 2002). A similar inductor with reversed roles was considered by Tada et al. (2002), in which the ferrite is the substrate and the inductor coil is covered by a printed dielectric insulation layer. Very high inductance values were achieved by embedding coils in low-temperature cofired ferrite (LTCF) multilayers. NiZnCu ferrite has been developed for this purpose by several groups, since it can be sintered in air and exhibits high relative permeability μ up to more than 10 MHz. Typical performance data are, for example, permeability of 400 and inductance of several 100 μH at 100 kHz (Barth et al. 2005); permeability of 200–300 and 4.7-μH inductance at 3 MHz (Hahn et al. 2006); permeability of 200 and 10- to 20-nH inductance at 4 MHz (Lim et al. 2007a; Lim et al. 2008). The two latter inductors were used in DC–DC converters at 0.8 and 20 A current load, respectively. Multilayer ferrite or LTCF boards can also serve as substrates for SMDs to obtain a complete electronic circuit; however, the substrate must be shielded from these components and their associated circuitry by a metal layer and a dielectric spacer layer to avoid loss and line impedance distortion (Lim et al. 2007b). The cofiring of dielectric and ferrite ceramic layers would be highly desirable also for two other reasons: Magnetic flux leakage is high and electric insulation is low between the turns of a coil when the turns are immediately surrounded by ferrite; this results from high permeability and low (semiconductor) resistivity, respectively. The difficulty in fabricating a closed ferrite core around a dielectric board lies both in the technology and the thermal mismatch of the materials. Nevertheless, 100-μH inductors of this type were recently made by filling core holes in a dielectric board with NiZnCu ferrite paste and by closing the magnetic loop with top and bottom ferrite tapes (Jao et al. 2007).

While a compromise is usually possible for single inductors between inductance, size, series resistance, and self-resonance frequency on the one hand and fabrication technology and materials on the other, this is incomparably more demanding for transformers, that is, magnetically coupled inductors. There are no compensation techniques available for the reduction of power transfer and the overall deterioration of circuit performance by magnetic leakage. Discrete transformers use a ferrite core with wires wound around it and thus maximize magnetic coupling and minimize parasitic capacitive coupling. This is difficult to achieve in monolithic devices due to the high aspect ratio of the conductive traces. However, at least in radio frequencies, the capacitive coupling can be taken into account in the design stage, when transmission line models are used to construct the transformer. The Guanella and Ruthroff transformers are examples (Post 2007). At lower frequencies, three basic structures may be distinguished as approximations to the discrete transformer: The first structure is the *closed core type*, in which the magnetic loop is closed between ferrite cover plates or multilayers by filling ferrite paste into voids inside the dielectric multilayer board carrying the primary and secondary coils (Lipkes 1999). The second structure is the *dielectric gap type*, based on the finding that even a simple dielectric circuit board of several 100-μm thickness without magnetic feedthroughs suffices to yield considerable flux coupling above 80% and transmitted power of several 10 W (Sato et al. 2001; Waffenschmidt 2004; Yu et al. 2006; Waffenschmidt and Jacobs 2008). This allows the use of a conventional PCB as carrier of the metallization. The third structure is the *embedded type*, in which even the dielectric board is omitted and the coils are placed directly inside the ferrite layers. As there is only a ferrite multilayer now, the embedded transformer raises the least claims in terms of fabrication technology. For the same reason, it is well suited for single-chip components but not for integration with other passive components. To counteract the low reluctance path surrounding each conductor and the associated flux leakage, additional screen-printed dielectric layers were suggested to reduce the effective permeability between the windings (Abel 2001; Wahlers et al. 2002). Embedded NiZn ferrite transformers for the power level below 1 W were made commercially available in this way (Slama 2003).

In view of the state of the art, this chapter describes performance and design rules of individual as well as magnetically coupled inductors using NiZnCu and MnZn ferrite multilayers, which have emerged in our laboratory with support by national cooperation and funding over years of materials and component development. Starting from generic physical relations governing these components, the synthesis of LTCC ferrites and their relevant materials characteristics are addressed in two subsequent sections. Two further sections cover the performance of embedded inductors and transformers. The relation between materials properties and measured device characteristics in terms of small-signal S-parameter analysis and power-handling capability is elucidated for the three transformer structures: the embedded, the core, and the gap types. The analysis supports the conclusion that reduced ferrite volume, flat shape, and parasitics are of limited disadvantage and that integrated ceramic transformers with mixed dielectric–ferrite multilayers should be feasible after a few manageable steps in design and technology development.

DEVICE PHYSICS

In order to understand the chances and limitations of ferrites and their use in multilayer circuit boards, the basic physical laws of magnetic fields and magnetic induction need to be reviewed. The generic building block of a lumped, low-frequency inductor is a circular current filament, for which Figure 8.1 shows the magnetic flux density with details relevant for further discussion. It can be calculated numerically from the Ampère–Laplace law

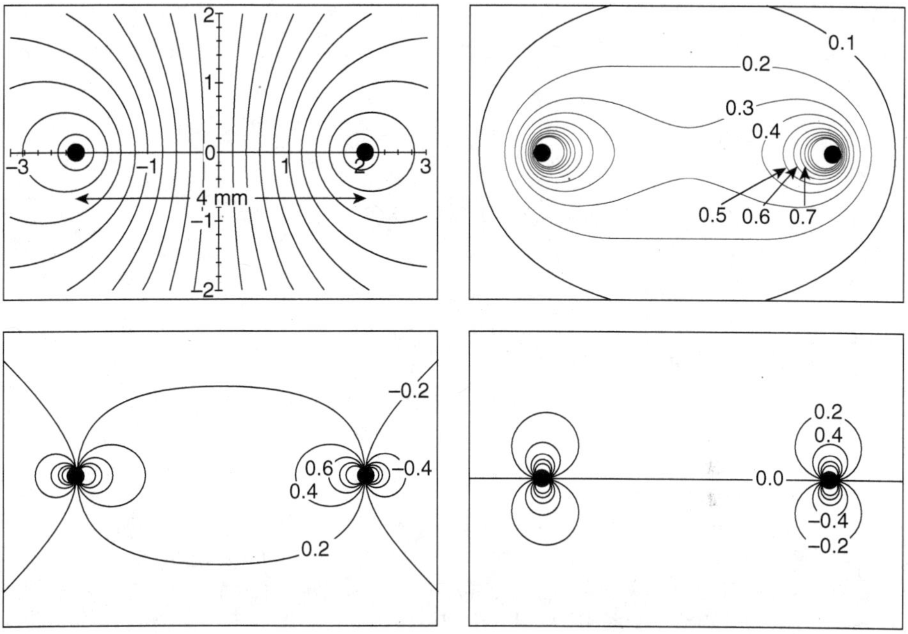

Figure 8.1. Magnetic flux density in the axial plane, created by a circular current filament of 4-mm diameter in air when a current of 1 A flows through it. Top left: magnetic field lines. Top right: contour lines showing the amplitude of flux density in mT. Bottom left: contour lines showing the vertical flux density component in mT. Bottom right: contour lines showing the radial flux density component in millitesla.

$$\vec{B} = \frac{\mu\mu_0 I}{4\pi} \oint \frac{\vec{u}_T \times \vec{u}_r}{r^2} dl, \tag{8.1}$$

where μ and μ_0 are the permeabilities of the surrounding material and vacuum, respectively; I is the current in the filament, and the integral extends along the whole filament. r is the distance between the section dl of the filament and the position at which the field is calculated; u_r is the associated unit vector; and u_T is the unit vector tangential to the filament. The field is directed mainly along the "vertical" symmetry axis of the circle and field lines close in a toroidal shape outside the circle.

The total flux passing through the ring defines the inductance L and the impedance Z of the conductor at frequency $f = \omega/2\pi$:

$$\Phi = \int_{\text{circle}} \vec{B} \cdot d\vec{f} \qquad L = \frac{\Phi}{I} \qquad Z = j\omega L. \tag{8.2}$$

It is evident from Equation 8.1 that the permeability of the embedding material acts like a scalar factor on the field. Magnetic flux, inductance, and impedance therefore increase linearly with permeability. When an identical second circular conductor is placed parallel to the first one at some vertical distance, it is penetrated by a fraction of the magnetic flux in the first ring. This fraction defines the magnetic coupling constant:

$$k = \frac{\int_{\text{circle 2}} \vec{B} \cdot d\vec{f}}{\int_{\text{circle 1}} \vec{B} \cdot d\vec{f}} = \frac{\Phi_2}{\Phi_1} < 1. \tag{8.3}$$

When the primary current is time dependent, it induces a voltage in the secondary coil by electromagnetic induction:

$$V_2 = -\frac{d\Phi_2}{dt}. \tag{8.4}$$

An electromagnetic transformer requires that the magnetic coupling between the primary and the secondary side is brought as close as possible to 100% by proper choice of materials and design. Usually, a ferrite or magnetic metal core is employed for this task. Without such a structured core, a high flux density surrounds the primary conductor in close proximity without passing through the secondary coil. A considerable radial flux component then exists or leaks out above and below the conductor as shown in Figure 8.1. Simple homogeneous embedding of the coils inside a ferrite multilayer is useless since it enhances all inductances but leaves the shape of the magnetic field and the coupling unaffected according to Equations 8.1 and 8.3. The design goal is therefore to find a core structure that is simultaneously efficient and manufacturable.

In a single ideal inductor with multiple, helically arranged turns, each turn adds the same flux through itself and all other turns. The inductance is therefore correlated in a quadratic way to the number of turns. In reality, magnetic flux leakage and reduced coupling lead to a subquadratic relation. An approximate design rule can be derived for conductors of width w carrying a current filament in the center and surrounding a rectangle $a \times b$. To this end, the vector potential $\vec{A}$ created by all inductor turns is integrated along the periphery of the rectangular core to yield the magnetic flux and the inductance:

$$\Phi = L \cdot I = \mu\mu_0 \oint \vec{A} d\vec{l}, \text{ where } \vec{A} = \frac{1}{4\pi} \int \frac{\vec{j}}{r} dV. \tag{8.5}$$

The analytic result for a straight coil with rectangular cross-section and N helical turns separated by a distance z_0 is

$$L = \frac{\Phi}{I} = \frac{\mu\mu_0}{2\pi} \sum_{k,m=1}^{N} [f_1(a,0,k,m) - f_2(a,0,k,m) - f_1(a,b,k,m) + f_2(a,b,k,m)$$
$$+ f_1(b,0,k,m) - f_2(b,0,k,m) - f_1(b,a,k,m) + f_2(b,a,k,m)]$$

where

$$f_1(x_1, x_2, k, m) = \int_{-x_1/2}^{x_1/2} \ln\left[\frac{1}{2}\left(\sqrt{(w+2x_2)^2 + 4(m-k)^2 z_0^2 + (x_1 + w - 2x')^2} + (x_1 + w - 2x')\right) dx'\right]$$

$$f_2(x_1, x_2, k, m) = \int_{-x_1/2}^{x_1/2} \ln\left[\frac{1}{2}\left(\sqrt{(w+2x_2)^2 + 4(m-k)^2 z_0^2 + (x_1 + w + 2x')^2} - (x_1 + w + 2x')\right) dx'\right]. \quad (8.6)$$

Considering a coil with square turns of $2 \times 2\,\text{mm}^2$ and 0.2 mm conductor width as an example, algebraic evaluation of this expression reveals, in agreement with experimental evidence, that the exponent κ relating the number of turns N with the inductance decreases from 2 for narrowly spaced turns to 1.4 at 0.5-mm spacing. This can be described by

$$L \propto N^{\kappa} \text{ with } \kappa = 2 \cdot e^{-0.7\,s}, \quad (8.7)$$

where s is the spacing of turns.

While leakage flux may be compensated in a single inductor coil by adding more turns or conductor length, a transformer basically relies on the magnetic induction between the primary and the secondary coil. Leakage deteriorates device performance in this case and cannot be compensated by any counter measures. Ferrites have therefore been traditionally employed to selectively amplify the coupled flux according to the flux density versus field relation

$$\vec{B} = \mu\mu_0 \vec{H} \quad (8.8)$$

and to suppress the relative amount of uncoupled flux. It is a vital challenge to optimize the properties and the shaping of these materials under the specific conditions of the application and the fabrication technology.

A straightforward solution is the insertion of a ferrite cylinder or core passing through both transformer coils along their common axis. The magnetic field is predominantly parallel along the cylinder surface and hence continuous at the materials interface; the flux B is consequently amplified inside the core. To first approximation, field deformations due to the core are negligible and the flux given by the Ampère–Laplace law can be multiplied by the respective material permeabilities.

As shown in Figure 8.2, the magnetic coupling can reach values well above 95% according to this model. Evidently, the core should ideally have a permeability above 400 and fill 70% of the coil diameter, because larger core diameters will also amplify the radial or leakage flux and the coupling will decrease again. One might expect that such a vertical ferrite core providing axial magnetic coupling between the coils is necessary for sufficient guidance and enhancement of the magnetic flux.

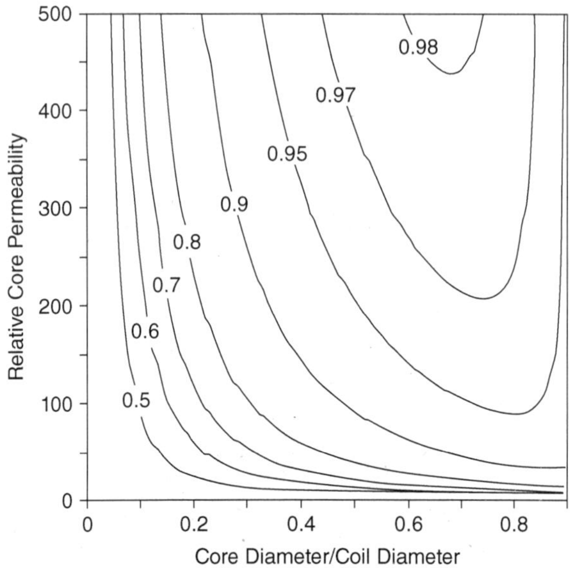

Figure 8.2. Contour lines of the calculated electromagnetic coupling between two circular coils of 4-mm diameter and 100-μm axial distance when a ferrite core of variable permeability and diameter is inserted.

However, vertical cores are difficult to fabricate by tape lamination techniques, and simpler approximate structures need to be devised. Frequently, just ferrite layers attached to the bottom and top of the circuit board have a positive impact on magnetic coupling through field distortion. This immediately results from the well-known boundary conditions that the normal component of the flux density field B and the tangential component of the magnetic field H must be continuous at the interface between the dielectric board (medium 1) and the magnetic cladding layers (medium 2); that is,

$$\begin{aligned} B_{1n} &= B_{2n} \\ H_{1t} &= \frac{B_{1t}}{\mu_0} = \frac{B_{2t}}{\mu\mu_0} = H_{2t}. \end{aligned} \tag{8.9}$$

The angular deviations θ of the flux density field from the vertical direction in the two media are related by

$$\tan(\theta_1) = \frac{B_{1t}}{B_{1n}} = \frac{B_{2t}}{\mu B_{2n}} = \frac{1}{\mu}\tan(\theta_2). \tag{8.10}$$

Since the magnetic field lines are closed without sinks and sources, a field deformation by the ferrite cladding must occur, such that the tan (θ) becomes much smaller in the dielectric and larger in the ferrite. Effectively, the field lines are "diffracted" at

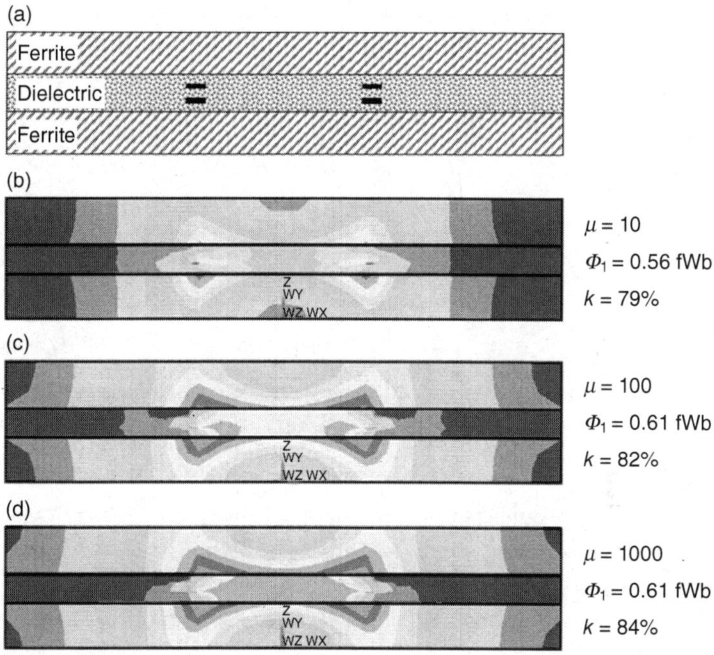

Figure 8.3. Finite element simulation of the magnetic flux density around single-turn transformer coils in a dielectric board covered by ferrite cladding layers. (a) Cross-sectional structure of primary and secondary turns, with 4-mm diameter, 0.26-mm dielectric thickness, and 1-mm ferrite thickness. (b–d) Flux density at 1-A primary current. The maximum is 20 pT in the dark gray regions near the center of the ferrite layers. The legend shows the permeability, primary flux, and coupling coefficient.

the dielectric–ferrite interface according to Equation 8.10, such that a small horizontal component (e.g., 1° azimuth) in the dielectric leads to an almost purely horizontal flux in the ferrite (83° azimuth for $\mu = 500$). The magnetic flux in the dielectric is obviously forced into the vertical direction by the "external" ferrite cladding. It spreads very flat, however, in the ferrite layers. The original oval field lines of Figure 8.1 are therefore distorted into rectangles. While the vertical sections are favorable for the penetration through the coils, the horizontal ones are squeezed into a thin ferrite layer, which is an advantage for making flat devices. Finite element simulations in Figure 8.3 confirm the expected effect of concentration of the flux along the axis in the dielectric material and improved coupling efficiency for relative ferrite permeabilities above 100. Without ferrite, the coupling would be a mere 56% for this structure.

To drive progress in design rules and materials development, a thorough understanding of the correlations between materials properties, device structure, and device performance in an electronic circuit is essential. Measured characteristics of inductors and transformers are therefore traced back to their most basic quantities. Comparing these finally with original simulations and design helps derive a generic set of design rules. To this end, the devices of this chapter are experimentally characterized by

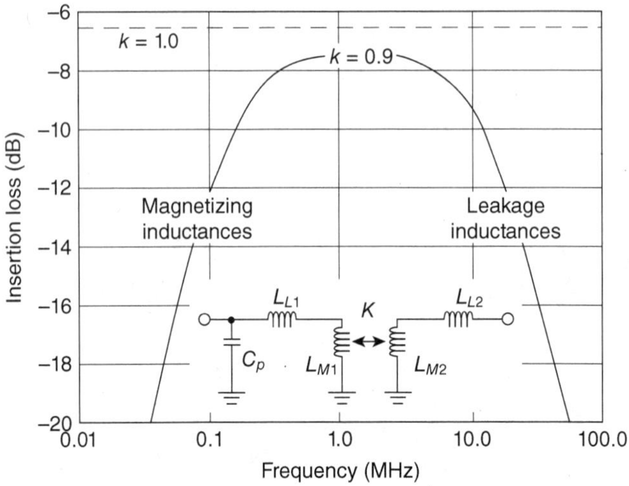

Figure 8.4. Insertion loss (solid line) for a transformer with 16:4 turns, 3-μH primary magnetizing inductance, 90% coupling, and no parasitic capacitance. The deviation from the level of an ideally coupled 16:4 transformer (dashed line) is well resolved.

small-signal scattering parameter measurements and by current–voltage analysis under power conditions. Inductors are analyzed as a single component using an impedance analyzer or are integrated into a low-pass filter. Transformers are generally studied by their insertion loss in a 50 Ω environment and their power transmission capability and efficiency.

Basically, a transformer is described by its impedance matrix:

$$\begin{pmatrix} V_1 \\ V_2 \end{pmatrix} = j\omega \begin{pmatrix} L_1 & M \\ M & L_2 \end{pmatrix} \begin{pmatrix} I_1 \\ I_2 \end{pmatrix}, \tag{8.11}$$

where L_1 and L_2 are the primary and secondary total inductances, which are the sum of the leakage inductance and the magnetizing inductance (see inset in Fig. 8.4). The magnetizing inductances are related to the ratio of turns by

$$L_{M1}/L_{M2} = N_1^2/N_2^2 = 1/N_{21}^2. \tag{8.12}$$

Their product defines the mutual inductance M and the coupling coefficient k through

$$M^2 = L_{M1} \cdot L_{M2} = k^2 L_1 \cdot L_2. \tag{8.13}$$

The leakage inductance will usually be different on the primary and on the secondary side. However, in order to limit the number of fit parameters, the same fraction of leakage is assumed for both sides throughout the present discussion so that

$$L_1 = L_{M1}/k; \quad L_2 = L_{M2}/k. \tag{8.14}$$

Assuming an additional shunt capacitor C_p (see inset in Fig. 8.4) on the primary side to account for stray capacitances between the screen-printed flat conductor traces, Equations 8.11–8.14 yield the insertion loss:

$$|S_{21}|^2 = Z_1^2/(Z_2^2 + Z_3^2) \tag{8.15}$$

with

$$Z_1 = 2\omega L_{M1} N_{21}, \tag{8.16}$$

$$Z_2 = Z_0 + Z_0\omega^2 L_{M1} C_p N_{21}^2/k - \omega^2 L_{M1}^2 N_{21}^2 (1/k^2 - 1)/Z_0, \tag{8.17}$$

$$Z_3 = \omega L_{M1}(1 + N_{21}^2)/k - \omega^3 L_{M1}^2 C_p N_{21}^2 (1/k^2 - 1), \tag{8.18}$$

and

$$Z_0 = 50\ \Omega. \tag{8.19}$$

A fit of Equation 8.15 to the measured insertion loss will later allow accurate determination of the three parameters, k, L_{M1}, and C_p, as well as additional derived quantities. It can be shown that Equation 8.15 is invariant under commutation of primary and secondary ports; that is, the transformer behaves symmetrically.

A typical transmission curve for a transformer with flux leakage is shown in Figure 8.4. Generally, the transmission of a transformer is low at low frequencies. It increases according to

$$|S_{21}| \approx 2N_{21}\omega L_{M1}/Z_0 \tag{8.20}$$

to the level

$$|S_{21}| = \frac{2k}{N_{21} + 1/N_{21}}. \tag{8.21}$$

The low-frequency behavior and this level unambiguously define the magnetizing inductance and the coupling coefficient since the ratio of turns is known. This also defines the high-frequency decrease:

$$|S_{21}| = \frac{Z_0}{\omega L_{M1}} \frac{2k^2}{N_{21}(1 - k^2)}. \tag{8.22}$$

Any additional high-frequency drop in the measurement can be readily attributed to the stray capacitance.

Although the power transmission capacity of a transformer is finally limited by its resistive and magnetic losses, it is valuable to study first the lossless situation again. For purely resistive load R_2 currents and voltages can be expressed by each other according to Equations 8.11–8.14 as

$$I_1 = \frac{(R_2 - j\omega L_2)V_1}{\omega^2 L_1 L_2 (1-k^2) + j\omega R_2 L_1}, \tag{8.23}$$

$$I_2 = \frac{j\omega k\sqrt{L_1 L_2} V_1}{\omega^2 L_1 L_2 (1-k^2) + j\omega R_2 L_1}, \tag{8.24}$$

and

$$V_2 = \frac{j\omega k\sqrt{L_1 L_2} R_2 V_1}{\omega^2 L_1 L_2 (1-k^2) + j\omega R_2 L_1}. \tag{8.25}$$

The voltage and current ratios

$$\left|\frac{V_2}{V_1}\right| = \frac{k}{\sqrt{L_1/L_2 + \omega^2 L_1 L_2 (1-k^2)^2 / R_2^2}} \xrightarrow{\text{open loop}} \frac{N_2}{N_1} \tag{8.26}$$

and

$$\left|\frac{I_2}{I_1}\right| = k\sqrt{\frac{L_1}{L_2}} \frac{1}{\sqrt{1+(R_2/\omega L_2)^2}} \xrightarrow{\text{short circuit}} \frac{N_1}{N_2}. \tag{8.27}$$

collapse into the well-known ratio of turns under open loop and short circuit conditions, respectively. The averaged primary and secondary power is given by

$$\bar{P} = \frac{1}{2}\mathrm{Re}(V \cdot I^*) \tag{8.28}$$

and by inserting Equations 8.23–8.25, it can be shown that both are equal to

$$\bar{P} = \left(\frac{1}{2}\frac{V_1^2}{R_2}\frac{N_2^2}{N_1^2}\right)\frac{k^2}{1+(1-k^2)^2 \dfrac{\omega^2 L_1 L_2}{(R_2 N_1/N_2)^2}}. \tag{8.29}$$

The power transmission efficiency of 100% necessarily follows from our assumption of a lossless device. In spite of this, the transmitted power decreases when the magnetic coupling coefficient k drops below 1, because the associated serial leakage inductances block the power flow through the transformer. Furthermore, the same power can be transmitted with permutated primary and secondary ports, that is, the reciprocal voltage conversion factor, when the load and the primary voltage are changed by the factors $(N_1/N_2)^2$ and N_2/N_1, respectively, where N_1 and N_2 denote the initial primary and secondary numbers of turns.

SYNTHESIS OF FERRITES

NiZn(Cu) and MnZn ferrite are the two magnetic ceramics most widely used for traditional power electronic applications in the 100 kHz to 10 MHz window. They stand out by their high permeability and magnetic saturation, low magnetic loss, and acceptable gyromagnetic resonance frequency in the 1 MHz to 10 MHz range. Particularly, NiZnCu ferrite is well compatible with conventional LTCC technology since it can be sintered in air and ferrite tapes are already commercially available (Slama 2003; Barth et al. 2005; Hahn et al. 2006; Jao et al. 2007; Lim et al. 2007a; Lim et al. 2008). MnZn ferrite exhibits higher permeability and saturation, but at the expense of lower resonance frequency and the need to maintain a well-controlled low-oxygen partial pressure during sintering, which has retarded its adaptation to LTCC (Sato et al. 2001). This chapter describes performance and design rules of individual as well as magnetically coupled LTCC inductors using NiZnCu and MnZn ferrite multilayers, recently developed in a joint industrial and academic effort (NIKOL 2008).

A substantial reduction of the sintering temperature as compared to traditional ferrite fabrication employs extensive formation of the magnetic phase already in the powder and subsequent sintering with sintering aids. In the conventional processing of MnZn ferrite, for example, the powder is calcinated between 800 and 900°C and sintered after the addition of binder and sintering aids in shaped form between 1100 and 1300°C in nitrogen atmosphere with controlled oxygen partial pressure. For LTCC, these calcination and sintering temperatures are interchanged; that is, commercially available ferrite powder is calcinated at high temperature and pressed into pellets or cast into tapes. Sintering proceeds, after binder burnout in air, at only 900°C in nitrogen. To boost sintering activity, however, an additional high-energy milling step is applied to the powder after calcination and prior to mixing with the sintering aids.

The characteristic microstructure of LTCC MnZn ferrite is revealed by combining scanning electron microscopy (SEM, Fig. 8.5) with energy-dispersive X-ray (EDX) analysis. Most of the material consists of the magnetic spinel phase of MnZn ferrite (dark gray) into which a few Fe-rich grains (nonmagnetic hematite) are embedded. The Bi_2O_3-B_2O_3-SiO_2-ZnO (BBSZ) glass used as sintering aid in an amount of a few volume percent segregates at the grain boundaries (white). The porosity is typically around 2 vol%.

MAGNETIC PROPERTIES

The complex initial permeability of a ferrite material can be determined by machining a ring core from a bulk pellet or a multilayer with, for example, inner and outer diameters of $r_1 = 6$ mm and $r_2 = 12$ mm, respectively, and a height of $h = 1$–3 mm, wrapping several turns of copper wire around it to form a toroidal coil, and measuring the single-port impedance with a commercial analyzer (e.g., HP4194A):

$$Z = Z' + jZ'' = R_0 + j\omega L. \tag{8.30}$$

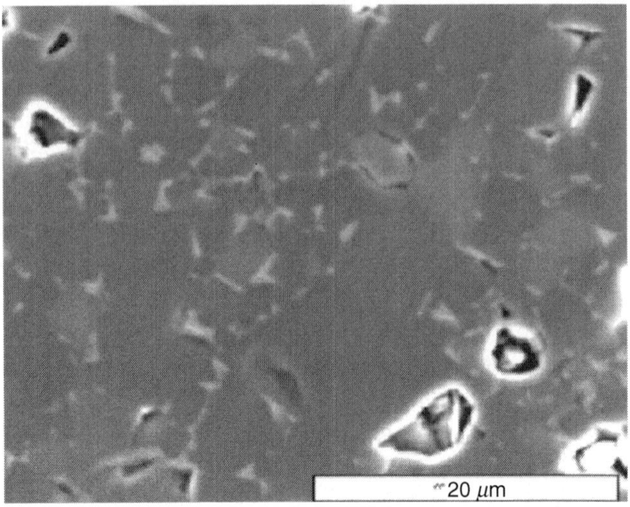

Figure 8.5. Microstructure of MnZn ferrite after sintering at 900°C as seen by scanning electron microscopy (SEM).

R_0 is the wire resistance and L is the complex inductance of the toroidal coil; that is,

$$L = N^2 \mu_0 (\mu' - j\mu'') \frac{h \ln(r_2/r_1)}{2\pi}. \tag{8.31}$$

The complex permeability is given as a function of frequency by

$$\mu' = \frac{2\pi Z''(\omega)}{\omega N^2 \mu_0 h \ln(r_2/r_1)} \quad \text{and} \quad \mu'' = \frac{2\pi (Z'(\omega) - R_0)}{\omega N^2 \mu_0 h \ln(r_2/r_1)}. \tag{8.32}$$

The amplitude permeability and the magnetic loss are measured with identical toroidal coils at larger currents and voltages as a function of the exciting current. To this end, a sinusoidal voltage is applied, for example, by a signal generator driving a power amplifier. The voltage and the current transients are stored on a digital oscilloscope. Despite the nonlinearity of the ferrite, a quick analysis can be limited to the fundamental wave. The effective quantities are then

$$U_{\text{eff}} = U/\sqrt{2} \quad \text{and} \quad I_{\text{eff}} = I/\sqrt{2}. \tag{8.33}$$

The impedance of the coil is given by

$$Z = U(t)/I(t) = R + j\omega L = Z_0 e^{i\varphi}, \text{ where } \quad Z_0 = \frac{U_0}{I_0} = \frac{U_{\text{eff}}}{I_{\text{eff}}}. \tag{8.34}$$

Calculating the total power loss (resistive and magnetic)

$$P = \frac{\omega}{2\pi} \int_{0}^{2\pi/\omega} U(t) \cdot I(t)\,dt = U_{\text{eff}} I_{\text{eff}} \cos\varphi \tag{8.35}$$

yields the effective resistance

$$R = (U_{\text{eff}} / I_{\text{eff}}) \cos\varphi = P / I_{\text{eff}}^{2}. \tag{8.36}$$

The inductance then follows as flux divided by current or

$$L = NA_e B_0 / I_0 = \sqrt{Z_0^{2} - R^{2}} / \omega. \tag{8.37}$$

Using the effective magnetic cross-section A_e and length l_e of the ring core, the amplitudes of flux density and magnetic field are

$$B_0 = I_0 L / NA_e \quad \text{and} \quad H_0 = I_0 N / l_e, \tag{8.38}$$

which finally yields the amplitude permeability $\mu = B_0/\mu_0 H_0$. The instantaneous current and voltage signals during a full period can be used to obtain the nonlinear hysteresis curve through integration of the induction law:

$$B(t) = \frac{1}{NA_e} \int_{0}^{t} U(t')\,dt' \quad \text{and} \quad H(t) = I(t) N / l_e. \tag{8.39}$$

For MnZn ferrite, we find a μ' between 500 and 600 up to 6 MHz, that is about a factor of two below that of commercial, high-temperature sintered material. The resonance frequencies ($\tan\delta = \mu''/\mu' = 1$) of LTCC bulk and tape material are 6 and 15 MHz, respectively, indicating smaller, more mobile magnetic units in the tape. Since grain boundaries impede the domain wall movement, a grain size of at least several micrometers was found to be necessary for good magnetic performance. A glass content above 5 vol% is similarly detrimental. Since the macroscopic specific resistivity is determined by the grain boundary composition, however, the glass phase helps to improve the material's performance: Specific DC resistivities above 1 kΩ·m were found after long (20 h) sintering, while standard material typically exhibits a few Ω·m. Shorter sintering of LTCC material delivers intermediate values between these extremes. At 2.5 MHz, the loss tangent amounts to 0.15 and does not exhibit any pronounced dependence on grain size or glass content. This is attributed to compensation effects as larger grains permit better magnetic interaction and reduce the hysteresis loss, while smaller grains with a higher fraction of insulating glass-filled grain boundaries reduce the eddy current loss.

Ceramic properties are strongly dependent on the details of ceramic processing involving selection of raw powder, powder milling, sintering aids, and the temperature profile for binder burnout and sintering. It is therefore possible to trade in an application-specific way the resonance or cutoff frequency of a material against its low-frequency

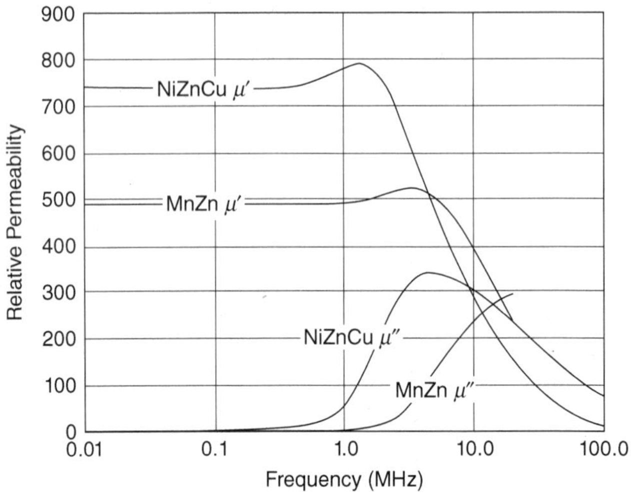

Figure 8.6. Real and imaginary parts of the permeabilities of MnZn and NiZnCu ferrite as measured with multilayer ring cores.

permeability and vice versa. This is exemplified by the NiZnCu ferrite shown in Figure 8.6, which achieves higher permeability at lower resonance frequency and lower magnetic saturation than MnZn ferrite, and hence behaves contrary to the usual performance of high-temperature sintered ferrites.

Magnetic hysteresis and saturation come into play at the larger excitation levels of amplitude permeability and magnetic loss measurements. The total magnetic loss comprises three contributions: hysteresis loss, eddy current loss, and resonance loss. The hysteresis loss increases linearly with frequency and dominates at low frequency. Though leaving this out here for brevity, the shape of the low-frequency hysteresis curve provides useful information on density, mobility, and interaction of magnetic domains in relation to the polycrystalline microstructure. It is found for LTCC ferrites that the permeability and the magnetic saturation are reduced, while the coercive field and the hysteresis loss (=area enclosed) are increased due to the content of sintering additives, foreign phases and smaller grains. Grain growth also seems impeded by the layer interfaces in multilayer samples.

The measured amplitude permeabilities displayed in Figures 8.7 and 8.8 also reflect the S-shaped *B*–*H* relation by an initial increase and a final drop due to saturation. This saturation shifts to a lower level for higher frequencies because the magnetizing contribution of slow-switching domains is lost (Fig. 8.7). At low excitation levels, the amplitude permeability increases considerably with frequency due to resonant behavior. Compared to commercial material (e.g., 3F4 by Ferroxcube) at 2.5 MHz and low flux, the amplitude permeability is not appreciably deteriorated by the modifications in composition and processing, which have been applied to the LTCC material. However, magnetic saturation occurs at a lower flux of about 40 mT. This is attributed to the reduced density of the magnetic phase in LTCC material.

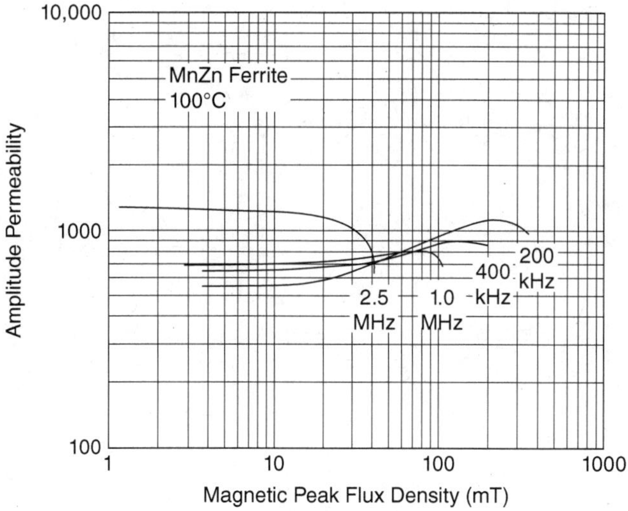

Figure 8.7. Amplitude permeability of MnZn LTCC ferrite measured at 100°C versus flux density and frequency.

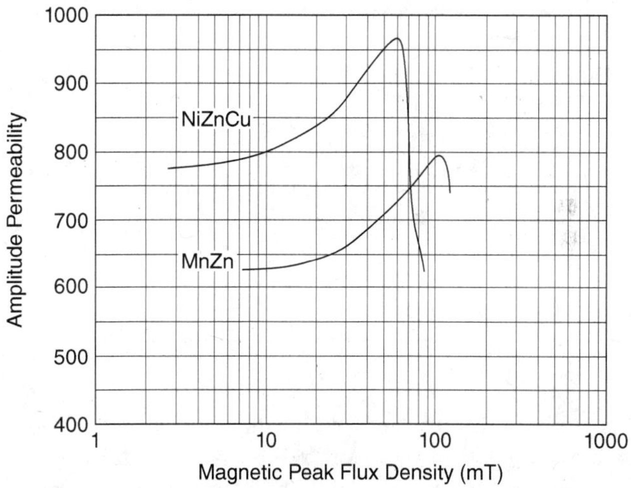

Figure 8.8. Comparison of amplitude permeabilities of specifically developed NiZnCu and MnZn LTCC ferrites at 1 MHz versus flux density. NiZnCu ferrite saturates at lower flux density than MnZn ferrite.

Regarding the volume-specific magnetic loss shown in Figure 8.9, the overall behavior resembles that known from commercial MnZn ferrites. It is determined by hysteresis and eddy current contributions well below the gyromagnetic resonance. Both of them increase quadratically with flux density B but can be separated by analyzing the frequency response since the hysteresis term is linear and the eddy current term is quadratic.

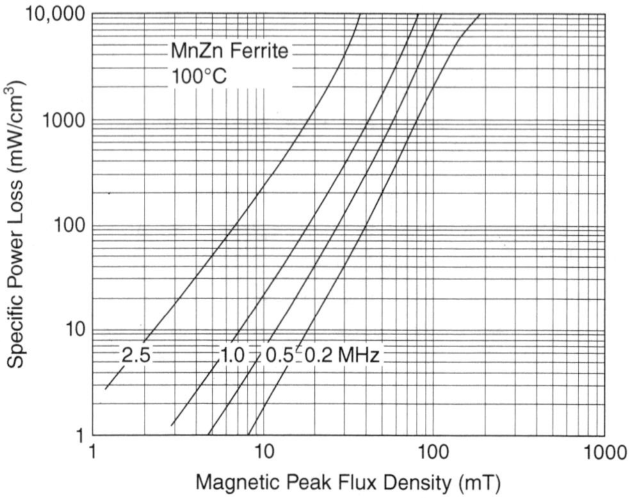

Figure 8.9. Specific power loss of MnZn LTCC ferrite measured at 100°C versus flux density and frequency.

TABLE 8.1. Magnetic Losses in Commercial and LTCCMnZn Ferrite (Measured with Ring Cores)

	Total loss (W/cm^3)	Fractional contribution (%)		
		Eddy currents	Hysteresis	Resonance
High *T* MnZn ferrite	6	30	10	60
LTCC MnZn ferrite	4	50	25	25

In detail, not only lower loss turns out for the LTCC ferrite (Table 8.1) but also a different distribution among the three loss mechanisms. Eddy currents are almost on the same level; hysteresis losses are slightly higher in the LTCC case due to smaller grain size; and resonance losses drop from almost 4 to only 1 W/cm^3 as a consequence of the rather high resonance frequency of 15 MHz. It must be pointed out that the minimization of losses is essentially important for the miniaturization of inductive components; otherwise, they lose efficiency.

As ferrites are being used in power electronic circuits with active and passive components dissipating heat, vital interest exists in their operation at elevated temperature. This will be even more crucial for the higher integration densities in ceramic multilayer circuit boards. The amplitude permeability and the power losses are therefore usually measured at 100°C as shown in the Figures 8.7 and 8.9. Ultimately, however, the magnetic order is destroyed by random thermal effects above the Curie temperature,

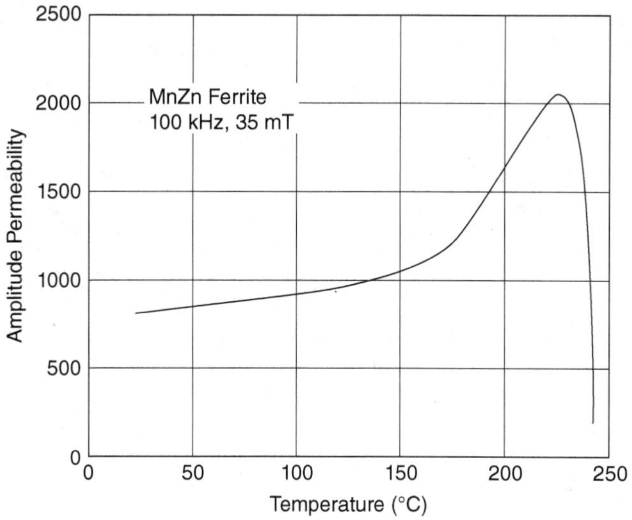

Figure 8.10. Amplitude permeability of MnZn LTCC ferrite as a function of temperature. The Curie temperature is 240°C.

which therefore sets the upper operational limit. The Curie temperature can be determined from either the initial permeability or the amplitude permeability by taking data at different temperatures, for example, by placing the toroidal coil with a thermocouple in an oven and by giving sufficient time for thermal equilibration. The data of Figure 8.10 were taken in this way. The loss of permeability at the Curie temperature of 240°C is clearly distinguishable. Below this point, the permeability increases because the thermal energy helps surmount the barrier of domain alignment and magnetization becomes more efficient.

EMBEDDED POWER INDUCTORS

Toroidal coils provide insight into basic ferrite properties without the need to embed screen-printed conductors. They are therefore well suited for materials optimization in a quick feedback loop. However, they do neither reveal the full potential and limitations of integrated LTCC components nor do they yield comprehensive design rules. By studying single and magnetically coupled inductors, the next two subsections will address such aspects as performance shifts induced by self-heating, magnetic saturation, and field shaping.

We will first look at an embedded inductor as a particularly simple structure that has potential for both application and analysis of the interplay between material properties, materials shaping, and device performance (Matz et al. 2007). The inductance of an embedded conductor line is proportional to the permeability of the surrounding material according to Equations 8.1 and 8.2 as long as a sufficient material thickness is provided. With the ferrite tapes described, it is therefore an easy task to realize inductors of several 10-μH inductances in homogeneous ferrite multilayers. For a power line filter, for example, a single oval turn embedded between five MnZn ferrite bottom and top layers yields a design value of 2.2 μH, when the conductor is 500 μm wide and 20 mm long. We realized it in both ferrite materials by two parallel, screen-printed Ag lines on the top and bottom of a central ferrite layer to enhance the current capacity. Hence, the total layer count was 11 and the sample volume was $11.3 \times 6.5 \times 1.0\,\text{mm}^3$. Two shunt capacitors of $3 \times 3.3\,\mu$F each were surface mounted after firing to form a low-pass filter. The inductor was characterized by an impedance analyzer both as an isolated element by single-port measurement and as part of the filter by measuring the insertion loss.

Figures 8.11 and 8.12 show the characteristic filter transmission and how it is modified by a DC current passing through the filter for the MnZn and the NiZnCu versions, respectively. The insertion loss exhibits a pronounced peak for zero bias current at the cutoff frequency (Pozar 1998)

$$f_c = \frac{1}{\pi\sqrt{L \cdot C}}. \tag{8.40}$$

With the observed frequencies of 70 and 57 kHz, this simple relation reveals inductances of 2.1 and 3.1 μH in good agreement with finite element simulations and the permeabilities of the two ferrites in Figure 8.6. The sharpness of the cutoff peak reflects the conductor DC resistance of 20 mΩ.

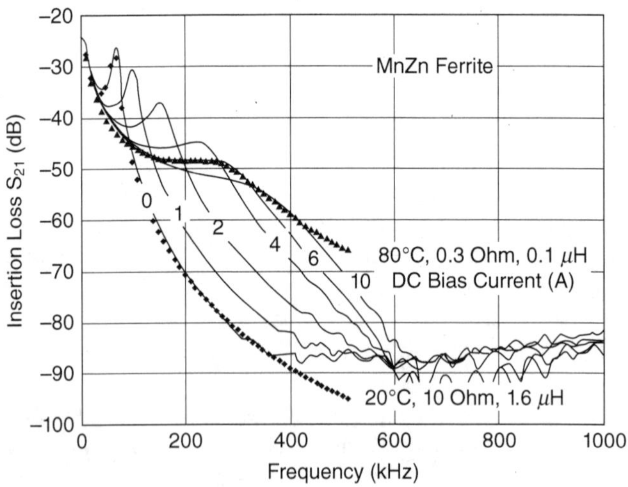

Figure 8.11. The unbiased cutoff peak of the power line filter at 70 kHz reveals an inductance of 2.1 μH of the inductor in a MnZn ferrite multilayer. It decreases under bias due to magnetic saturation and thermal effects.

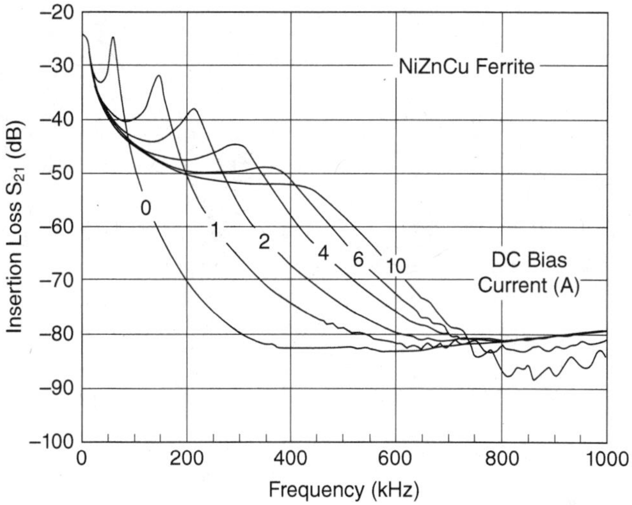

Figure 8.12. The 56-kHz cutoff in NiZnCu ferrite reveals higher permeability and inductance (3.1 μH) than in MnZn ferrite, but has similar saturation and insulation.

Two additional effects occur under the DC bias current: First, the sample heats up to about 100°C at 10A because the conductor loss of 2 W is being dissipated by air convection only. The temperature change has a severe effect on the conductivity of the ferrites far below the Curie temperature, which are well-known semiconductors. Second, the magnetization of the material saturates with increasing bias current; therefore, the slope of the B–H curve around the DC point of operation (the small-signal permeability) decreases. The measured data are therefore well explained by dotted example curves based on an equivalent circuit, which accounts for both the loss of inductance as well as the semiconducting behavior of the ferrite. This can be described by a shunt resistor with a negative temperature coefficient. The model yields a strong reduction from 1.6 μH and 10 Ω bypass at 20°C to only 0.1 μH and 0.3 Ω bypass at 80°C. The smoothing of the cutoff peak is directly related to the current bypassing the inductor through the surrounding ferrite. Hence, a successful integration requires careful balancing between fabrication complexity, thermal management, and operational conditions imposed by the surrounding circuit.

The quantitatively determined decay of inductance by such DC premagnetization is quite useful to derive the effective *B–H* curve of the ferrites. Since the magnetization of the volume elements surrounding the embedded conductor is not homogeneous, finite element simulation (ANSYS) must be employed. At first, the magnetic field of the DC current is calculated. An effective *B–H* curve with a shifted operating point is then assigned to each volume element from a generic curve (based on hysteresis measurements) and the local magnetic field strength. The net AC inductance is finally calculated in a second FEM pass using this spatially modulated shift of the generic curve (Fig. 8.13). Low- and high-field permeabilities near 1000 and around 10 are particularly

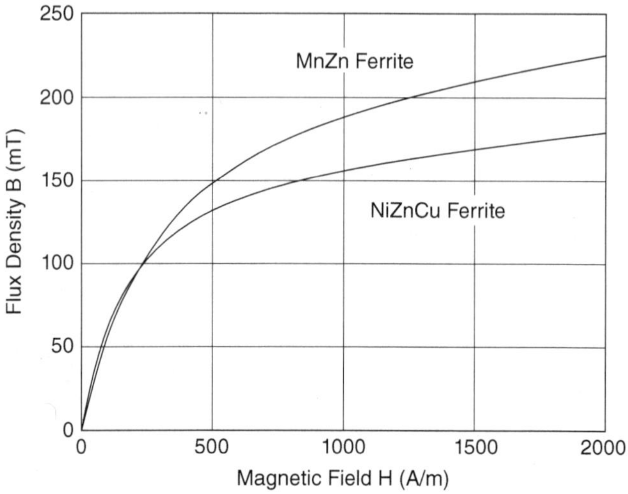

Figure 8.13. Effective B–H curves of NiZnCu and MnZn ferrite surrounding an embedded inductor.

quantified, but also a lower saturation flux density in the NiZnCu case. This correlates with the behavior of ring cores as displayed by Figure 8.8.

So far, we have identified a number of important boundary conditions to be observed for the integration of power electronic inductors. The shrinkage of magnetic cross-sections must be pursued well controlled, if at all possible, since increased flux density causes magnetic saturation effects and disproportionate increase of losses. The residual, thermally dependent conductivity of ferrites enforces dielectric embedding of conductors; this would also help to suppress magnetic leakage to an acceptable level as we saw before. The technology of mixed integration of magnetic and dielectric LTCC tapes is unfortunately highly demanding and just at the beginning. Major obstacles are mismatches in sinter shrinkage, sintering atmospheres (nitrogen for MnZn ferrite and air for dielectrics), and thermal contraction during cooling after firing (10–12 ppm/K in ferrites vs. 6 ppm/K in dielectrics). A suitable dielectric material with a coefficient of thermal expansion (CTE) close to that of ferrites seems to be zirconia, which is not available at present for LTCC.

It is possible, however, to sinter standard dielectric layers outside with ferrite layers inside; in this case, the outer layers contract less and are under compressive stress after cooling. Thus, the bottom and top layers of the 11-layer inductor discussed before were replaced by commercial dielectric tape (e.g., DuPont 951) to provide an insulation layer for SMDs. Leakage currents were indeed considerably reduced, but the cutoff frequency of the filter also shifted from 70 to 180 kHz, indicating reduced values of inductance and permeability (0.3 μH and 70, respectively). The effect is caused by the mechanical strain and band structure deformation in the ferrite (inverse magnetostriction). It can be estimated from elastic theory assuming homogeneous stress distribution in each of the two materials according to

TABLE 8.2. Calculation of Stress in LTCC-Integrated MnZn Ferrite Using Equation 8.41

Quantity		Unit	Material 1 (MnZn ferrite)	Material 2 (dielectric)
Thermal expansion	α	ppm/K	11.5	5.8
Temperature interval	ΔT	K	850	850
Poisson ratio	ν	—	0.17	0.17
Young's modulus	E	GPa	90–150	152
Total thickness	d	μm	810	180
Calculated stress	σ	MPa (10 bar)	160 (tension)	−720 (compression)

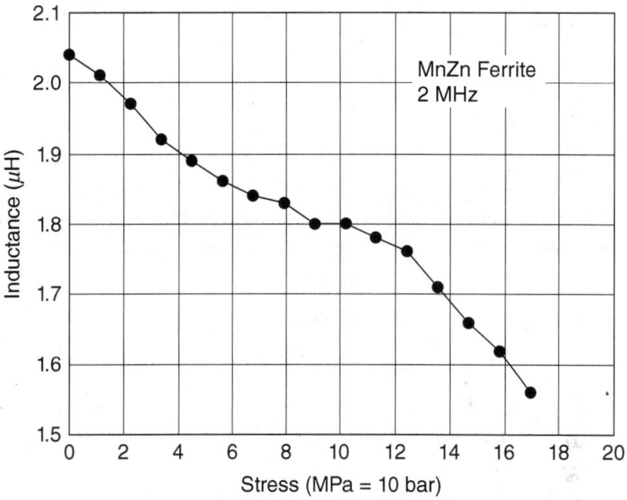

Figure 8.14. Measured inductance loss of an embedded inductor in MnZn ferrite under mechanically applied uniaxial stress normal to the multilayer surface.

$$\sigma_1 d_1 = -\sigma_2 d_2 = \frac{(\alpha_1 - \alpha_2)\Delta T}{(1-\nu_1)/E_1 d_1 + (1-\nu_2)/E_2 d_2}. \tag{8.41}$$

Quantities are defined and evaluated in Table 8.2 using typical elastic properties to yield a tensile stress in the ferrite of 160 MPa. The dielectric material is much thinner and is therefore subject to higher stress. Since the cladding layers are under compressive stress, no cracks develop and the samples remain intact.

The quenching of permeability by lattice deformation allows for simple designs of pressure sensors and variable inductors (variometers) without moving mechanical parts. Figure 8.14 shows the inductance measured by an impedance analyzer when a mechanical workbench exerts increasing uniaxial pressure.

A 25% change occurs for approximately 20 MPa. This compares well with the 90% change under built-in 160 MPa mentioned before. By placing a ferrite-embedded inductor and a piezoelectric multilayer in a rigid box, this extraordinary tuning effect on the inductance can be triggered without any moving mechanical parts. On the other side, the effect constricts mixed integration: From Equation 8.41, it can be derived that the CTE matching must be better than 0.1 ppm/K to limit the stress-related permeability loss to 10% in a structure consisting of a 1-mm-thick dielectric board with 1-mm-thick cofired ferrite layers on both sides. This constitutes an appreciable challenge for materials development.

CERAMIC MULTILAYER TRANSFORMERS

In the absence of a mature cofiring technology for ferrites and dielectrics, three generic structures of PCB transformers shown in Figure 8.15 have emerged over the years. Traditionally, the core type prevails; two pieces of a separately manufactured ferrite core connect through holes in the board. They are formed as E- and I-shapes and are held together by brackets. The core shape follows the field lines around the coils (see Fig. 8.1) and permits the best coupling and flux amplitude for power transmission. The sections of the core reaching through the board, however, may be omitted to simplify the technology; this results in the gap transformer. The flux must now overcome the dielectric gap between the upper and lower ferrite plates, which costs coupling efficiency and flux amplitude. The simplest transformer in view of the traditional LTCC multilayer technology is finally the embedded type; its coils are placed between ferrite layers without appreciable use of dielectric material. It permits the highest flux density, but since this exists preferably near the conductors, the magnetic coupling between the turns is even weaker than in the gap transformer.

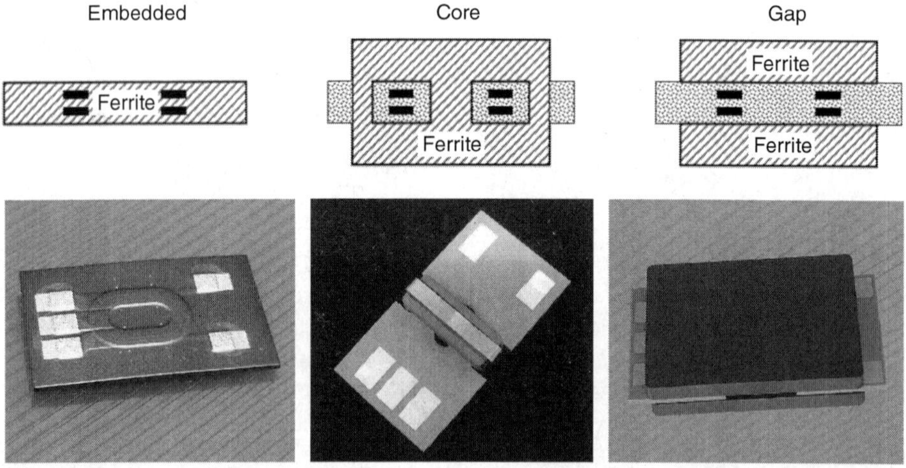

Figure 8.15. Cross-sectional structures of PCB transformers (top) and their realization in LTCC (bottom). Left: embedded type with NiZnCu ferrite (27 × 14 × 1 mm^3). Center: external core type with MnZn ferrite (27 × 14 × 5 mm^3). Right: gap type with MnZn ferrite (30 × 18 × 5 mm^3).

The three types have been fabricated in various versions; typical ones are shown in Figure 8.15. The embedded structure resembles the embedded inductor described before. Due to the lower resistivity of the MnZn ferrite, however, sufficient primary to secondary isolation is achieved only with NiZnCu ferrite material. Ten primary turns are placed on two layers with each layer carrying five spirals. One secondary turn is placed on each adjacent layer on the top and bottom to yield a ratio of 10:2 turns. The primary turns are 300 μm, and the secondary ones are 2.7 mm wide. To improve the current capacity, each metal layer is duplicated on an additional, immediately adjacent ferrite layer. Therefore, eight screen-printed metal levels and nine ferrite layers are used in total.

The core type has an identical layer count, but the turns are embedded in a conventional dielectric multilayer. The ferrite core is attached externally. It comprises two pieces having conventional E- and I-shapes and being formed by mechanical processing from sintered MnZn ferrite multilayers of 2.6- and 1.5-mm thicknesses, respectively. In a first embodiment, 16 primary and 4 secondary turns are placed on the eight metallization layers without duplication. The ferrite core has a $3 \times 5\,\text{mm}^2$ inner cross-sectional area and an overall footprint of $14 \times 5\,\text{mm}^2$ ("E14"). In a second embodiment, the inner layers between the 16 primary turns are replaced by a highly dielectric material with a dielectric constant of 65, while the outer layers are replaced by the corresponding base tape with a standard dielectric constant of 7. This design aims at testing the possibility to monolithically integrate a transformer with capacitors. The gap transformer finally resembles the core type but has 1.5-mm-thick MnZn ferrite multilayer plates only without connection through the board. Further designs have been analyzed and described elsewhere (Matz et al. 2008).

A detailed description of performance should begin with the discussion of the scattering parameter S_{21} or insertion loss. It is measured by taking the amplitudes of voltages on the primary and the secondary ports of the transformer as a function of frequency with a commercial impedance analyzer and then expressing their ratio in decibels (dB). The direction of measurement is irrelevant since transformers behave symmetrically. An illustrative example is the embedded NiZnCu ferrite transformer described before. Its insertion loss (Fig. 8.16) approaches a maximum level of −10.8 dB, which corresponds to 75% magnetic coupling according to Equation 8.21. Even with 100% coupling, a transformer with a different number of primary and secondary turns cannot be matched simultaneously to the 50-Ω environment at both ports and hence would not reach the zero decibel level. $k = 1$ would yield about −8 dB in the present case. Fitting Equation 8.15 to the measured data yields the magnetizing inductance from the low-frequency increase and the coupling coefficient from the maximum level. The resulting leakage inductance then immediately defines the decrease in the MHz range (Equation 8.22). The circuit parameters are summarized in Table 8.3.

Equation 8.15 interestingly accounts for the blocking of the transformer at higher frequencies by its own leakage inductance but does not describe the second maximum at 10 MHz. Since the leakage inductance definitely causes the drop in transmission to −17 dB, the following recovery to −10.8 dB must be attributed to its disappearance. The permeability in fact drops off above 5 MHz (Fig. 8.6). Inclusion of this effect into

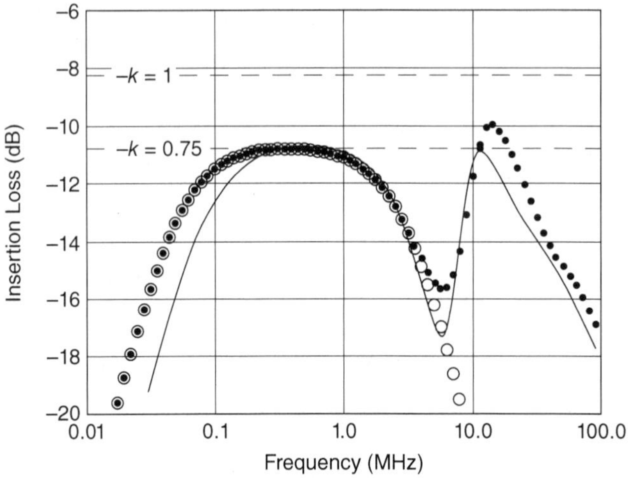

Figure 8.16. A fit (circles) to the measured insertion loss (line) of a NiZnCu-embedded 10/2 transformer reveals the primary magnetizing inductance and the magnetic coupling coefficient in the low-frequency range. The dotted curve additionally accounts for the cutoff in the permeability and for the stray capacitance (Table 8.3).

Equation 8.15 perfectly matches the measured data (dotted curve). However, the permeability obviously has vanished completely at already 10 MHz rather than decaying slowly like in Figure 8.6. The difference can be explained by the different orientation of the magnetic field relative to the multilayer. While the field lines are parallel to the ferrite layers in the toroidal coil of Figure 8.6, they are in a perpendicular plane for buried conductors. The effective permeability is, in this case, very sensitive to the weakening of the interfaces between layers at higher frequency. Vice versa, the slow decay of permeability in toroidal LTCC ferrite coils seems to indicate a combination of a magnetically intact interior of the ferrite layers and presumably "dead" defect-rich interface layers spreading with higher frequency throughout the structure.

An expanded frequency range of operation requires low leakage and perfect magnetic coupling. Core transformers achieve such coupling coefficients above 95%. The example in Figure 8.17 is therefore well suited up to 10 MHz. It would even work above this limit if there was no band stop spike. For the sake of simplicity, the band stop has been excluded from the analytical model (Equation 8.15), but more detailed simulation identifies it as an interaction between parts of the (more inductive) primary coil and the parasitic capacitance between the primary and the secondary coil. In embedded transformers, it is obscured by the leakage inductance. The band stop may be shifted to higher frequency by reducing the parasitic primary-to-secondary capacitance.

The capacitance inside each of these coils is to be considered as well. One might wish to integrate, for example, capacitors with the transformer in the same board. This requires that a few high-K layers run also through the transformer. As a test, Figure 8.18 shows the effect of three high-K layers between the 16 primary turns of a

TABLE 8.3. Circuit Parameters of LTCC Transformers

			Embedded Figure 8.16	Core Figure 8.17	Core Figure 8.18	Gap Figure 8.19
				Low-K board	High-K inside	Low-K board
			NiZnCu ferrite	MnZn ferrite	MnZn ferrite	MnZn ferrite
Size		mm^3	27 × 14 × 1	27 × 14 × 5	27 × 14 × 5	30 × 18 × 5
Ratio of turns	$N_{21} = N_2/N_1$	—	2/10	4/16	4/16	4/16
Coupling coefficient	k	—	0.75	0.97	0.98	0.97
Best insertion loss	Equation 8.21	dB	−10.8	−6.8	−6.7	−6.8
Primary magnetizing inductance	L_{M1}	μH	125	46.4	48.0	10.9
Secondary magnetizing inductance	$L_{M2} = N_{21}^2 \cdot L_{M1}$	μH	5.0	2.9	3.0	0.68
Mutual inductance	$M = \sqrt{L_{M1} L_{M2}}$	μH	25	11.6	12.0	2.7
Primary leakage inductance	$L_{L1} = \left(\frac{1}{k} - 1\right) L_{M1}$	μH	41.67	1.43	0.98	0.34
Secondary leakage inductance	$L_{L2} = \left(\frac{1}{k} - 1\right) L_{M2}$	μH	1.67	0.09	0.06	0.02
Stray capacitance (primary)	C_p	pF	60	40	450	40

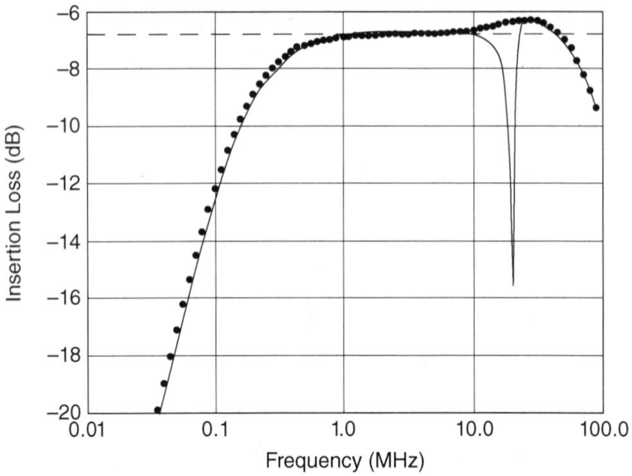

Figure 8.17. In the case of a 16/4 transformer with an external E14 core and coils embedded in a standard dielectric multilayer (ε =7.8), a fit (dots) to the measured insertion loss (line) reveals the primary magnetizing inductance and the magnetic coupling coefficient below 10 MHz as well as a small stray capacitance at higher frequencies (Table 8.3).

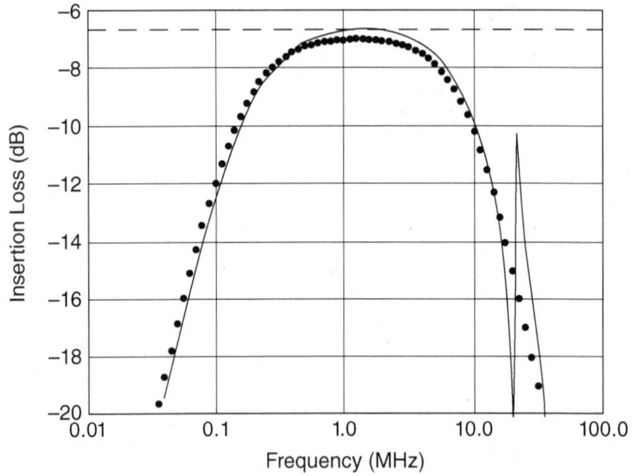

Figure 8.18. In the case of a 16/4 transformer with an external E14 core and a primary coil having high-permittivity layers (ε =65) between its 16 turns, a fit (dots) to the measured curve (line) reveals the primary magnetizing inductance and the magnetic coupling coefficient in the 0.1- to 1.0-MHz range as well as a larger stray capacitance at higher frequencies (Table 8.3).

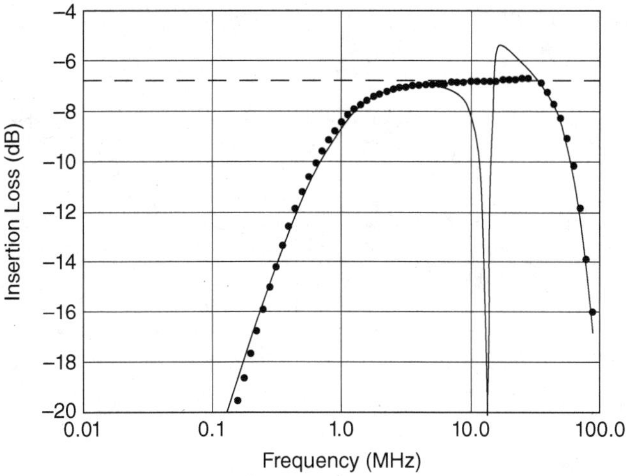

Figure 8.19. For the 16/4 gap transformer with standard dielectric board (ε = 7.8), a fit (dots) to the measured insertion loss (line) reveals good magnetic coupling but weak performance below 1 MHz due to reduced magnetizing inductance (Table 8.3).

transformer being otherwise identical to that of Figure 8.17. The enlarged intracoil capacitance now constitutes a capacitive bypass, which suppresses transmission above 3 MHz. The band stop remains unchanged since we still use the standard dielectric tapes between primary and secondary coils (Fig. 8.18).

As we have demonstrated, LTCC core transformers exhibit a favorable frequency window of operation between 1 and 10 MHz when properly designed. The borders are set by the magnetizing inductance at low frequency and the parasitic capacitance (band stop) at high frequency. By omitting the core and simplifying the structure toward a gap transformer, the window gets narrower because the gap reduces the magnetizing inductance. This is clearly evident from Figure 8.19, where the lower frequency limit is shifted upward by a factor of 3–5. The coil diameter was expanded slightly from a core area of $3 \times 5\,mm^2$ to $3 \times 7\,mm^2$ for compensation, but the enlarged parasitic capacitance moves the band stop from 20 to 13 MHz.

Hence, the analysis of small-signal scattering parameters is generally very useful in the assessment of a transformer design. The nonlinear behavior of the ferrite core material, however, as it might show up in power applications, is not contained. Power transmission measurements are therefore indispensable to identify effects of magnetic saturation and loss and to translate them into adequate design rules. Table 8.4 exemplifies transmission data of the three principal transformer types: embedded, gap, and core. The core type is slightly larger than the one used for Figure 8.17, but the other two are identical to those of Figures 8.16 and 8.19. The embedded transformer was used at a lower frequency of 500 kHz due to its large leakage inductance and handicap at

TABLE 8.4. Power Performance of LTCC Transformers for $N_2 > N_1$ and Small Load Resistors (Actual Load)

				Transmission				
	Size	Turns	Ferrite	Actual load (ohm)	Visible load (ohm)	Frequency (MHz)	Power (W)	Efficiency (%)
Embedded	27 × 14 × 1	10:2	NiZnCu	4	100	0.5	20	69
Gap	30 × 18 × 5	16:4	MnZn	3	50	2.5	110	82
Core	30 × 16 × 5	16:4	MnZn	3	50	2.5	120	90

The load transformed to the primary side is designated as visible load.

2.5 MHz. The transmitted power was limited to only 20 W in this case, while the gap and core types with their better magnetic coupling clearly exceed the 100-W limit. Also, the measured efficiencies agree with the small-signal data above: Since the leakage flux of the embedded case exists completely in the ferrite, this case exhibits the highest losses. The gap type has similarly low leakage as the core type, but its efficiency suffers from the flux concentration near the ferrite–dielectric interface according to the field diffraction effect expressed by Equation 8.10 and Figure 8.3. The specific magnetic losses increase quadratically with flux density in these regions according to Figure 8.9.

According to Equation 8.29, these powers and efficiencies were also achieved with reversed primary and secondary ports of the transformers and appropriately matched load resistors. This is to be expected because all voltages and currents are simply reversed and hence the internal field and loss remain unchanged. The fact that the actual and the visible load also change roles by this operation, however, needs special diligence since this requires new impedance matching between the transformer and the primary power supply. Otherwise, operation of the supply circuit under nonlinear conditions can cause unrecognized higher harmonics and associated high-frequency magnetic losses.

The modulation amplitude of the ferrite along its respective *B–H* curve can be estimated from the transmitted power, for example, for the core transformer, as follows:

$$\overline{P} = \frac{V_2^2}{2R_2} = 120\ \mathrm{W} \Rightarrow V_2 = 27\ \mathrm{V} \Rightarrow$$

$$\Phi = \frac{V_2}{\omega N_2} = 4.3\times10^{-7}\ \mathrm{As} \xrightarrow{3\times7\ \mathrm{mm}^2\ \mathrm{core}} B = 20\ \mathrm{mT}.$$

Compared to the specific losses in Figure 8.9 and the *B–H* curves in Figure 8.13, this moderate amplitude of flux density *B* does not overload the ferrite or operate the device in extreme nonlinearity.

CONCLUSION

Both MnZn and NiZnCu ferrites are well suited for low-temperature cofiring with screen-printed Ag conductors. The modifications in ceramic processing and the sintering additives necessary to achieve the desired sintering behavior reduce the permeability of these materials by about a factor of two as compared to traditional high-temperature sintered material. This nevertheless leaves room for numerous power electronic applications. Due to the restricted grain growth and interface phases, magnetic domains appear more mobile. The resulting higher resonance frequency cuts residual losses without enhancing hysteresis or eddy current losses.

The integrated embedded inductors of this study revealed higher electrical conductivity, lower loss, and higher magnetic saturation in MnZn ferrite than in NiZnCu ferrite, but these properties may be varied in some range by ceramic processing. As semiconductors, both ferrites exhibit a large positive temperature coefficient of electrical conductivity. This must be taken into account when dealing with embedded conductors. The permeability of both materials is furthermore very sensitive to mechanical stress as it occurs in mixed multilayer structures. This permits compact monolithic variometers but has so far inhibited a useful cofiring technology for dielectric and ferrite tapes in LTCC.

The design rules of the three basic transformer structures were investigated by fabrication, characterization, and modeling. The relations governing device performance were studied in the small-signal and power regime. The procedures for the measurement of circuit parameters were derived, and we have particularly seen how the physics of magnetic fields prevents the essential magnetic coupling between the primary and the secondary coil, when these are embedded in homogeneous material, be it dielectric or magnetic. Despite its suitability for multilayer technology, the embedded transformer therefore disqualifies by weak coupling and large leakage inductance. The conventional transformer with an E-shaped ferrite core, on the other side, excels in terms of coupling and transmission efficiency but resists integration into a tape-based process like LTCC. The so-called dielectric gap transformer, which is generally addressed also by organic circuit board designers, could be a useful compromise since it avoids the complex processing of magnetic interconnects through the board while still maintaining high magnetic coupling between the coils. The core and the gap transformers were shown to transmit powers well above 100 W at 2.5 MHz with efficiencies between 80% and 90%.

Hence, LTCC ferrites have reached a proven and mature performance level as a material. The remaining challenges and subjects of ongoing research are related to cost-efficient sintering and shaping of these materials in combination with low-dielectric multilayer boards. MnZn ferrite requires an inert sintering atmosphere, while dielectric ceramic usually need oxygen. Sensitivity to strain derails straightforward integration of magnetic with dielectric tapes, although this would facilitate electric insulation of conductors and intentional shaping of the magnetic flux. Recent progress in ceramic technology and component evaluation has demonstrated the potential of LTCC ferrites. The goal to achieve the smallest possible component size with properly shaped ferrites of sufficient magnetic cross-section will constitute a challenge between materials performance, processing technology, and natural boundary conditions for the years to come.

ACKNOWLEDGMENTS

This work was supported by the German Ministry for Education and Research (BMBF) under grant number 03X4503B. Contributions by the funded consortium, particularly sintered NiZnCu ring cores by J. Töpfer of the University of Applied Sciences in Jena as well as NiZnCu green tapes by S. Barth of the Hermsdorf Institute for Technical Ceramics are appreciated. Furthermore, D. Götsch, R. Karmazin, R. Männer, and A. Schilling of Siemens Corporate Technology have made this comprehensive and consistent view on LTCC ferrites possible by developing MnZn ferrite, by fabricating components, and by analyzing performance.

REFERENCES

Abel DA. 2001. Multi-layer transformer apparatus and method. U.S. Patent 6198374 B1.

Barth S, Bechtold F, Müller E, Mürbe J, Töpfer J. 2005. Low sintering Ni-Cu-Zn ferrite tapes for LTCC integrated inductors. In: *Proceedings of the 1st IMAPS/ACerS International Conference and Exhibition on Ceramic Interconnect and Ceramic Microsystems Technologies (CICMT).*

Billings K. 2002. Why have an air gap? Power Electron Technol 28:52–53.

Daewoo Securities. 2007. *LED: The Dream Light Source of the 21st Century*. September 18.

DickCP, Hirschmann D, Plum T, Knobloch D, De Doncker RW. 2008. Novel high frequency transformer configurations—amorphous metal versus ferrites. In: *Proceedings of the 39th Annual IEEE Power Electronics Specialists Conference (PESC)*, pp. 4264–4269.

Hahn R, Sommer G, Dörr I, Schwerzel S, Reichl H. 2006. Design of integrated inductances based on ferromagnetic LTCC layers. Adv Microelectronics 33:8–16.

Jao JC, Li P, Wang SF. 2007. Characterization of inductor with Ni-Zn-Cu ferrite embedded in B_2O_3-SiO_2 glass. Jpn J Appl Phys 46:5792–5796.

Jiang H, Wang Y, Yeh JLA, Tien NC. 2000. On-chip spiral inductors suspended over deep copper-lined cavities. IEEE Trans Microw Theory Tech 48:2415–2423.

Karmazin R, Dernovsek O, Ilkov N, Wersing W, Roosen A, Hagymasi M. 2005. New LTCC-hexaferrites by using reaction bonded glas ceramics. J Eur Ceram Soc 25:2029–2032.

Lahti M, Lantto V, Leppävuori S. 2000. Planar inductors on an LTCC substrate realized by gravure-offset-printing technique. IEEE Trans Components Packaging Technol 23:606–610.

Lim MH, Liang Z, van Wyk JD. 2007a. Low profile integratable inductor fabricated based on LTCC technology for microprocessor power delivery applications. IEEE Trans Components Packaging Technol 30:170–177.

Lim MH, Dong Y, van Wyk JD, Lee FC, Ngo KDT. 2007b. Shielded LTCC inductor as substrate for power converter. In: *Proceedings of the IEEE 38th Annual Power Electronics Specialists Conference (PESC)*, pp. 1605–1611.

Lim MH, van Wyk JD, Lee FC, Ngo KDT. 2008. A class of ceramic-based chip inductors for hybrid integration in power supplies. IEEE Trans Power Electron 23:1556–1564.

Lipkes Z. 1999. Core and coil structure and method of making the same. U.S. Patent 5 945 902.

Ludwig M, Duffy M, O'Donnell T, McCloskey TP, Ó Mathùna SC. 2003. PCB integrated inductors for low power DC/DC converter. IEEE Trans Power Electron 18:937–945.

Matters-Kammerer M, Mackens U, Reimann K, Pietig R, Hennings D, Schreinemacher B, Mauczok R, Gruhlke S, Martiny C. 2006. Material properties and RF applications of high k and ferrite LTCC ceramics. Microelectronics Reliability 46:134–143.

Matz R, Götsch D, Karmazin R, Männer R, Siessegger B. 2007. Low temperature cofirable MnZn ferrite for power electronic applications. In: *Proceedings of the International Conference on Electroceramics (ICE)*. To be published in J Electroceramics.

Matz R, Götsch D, Goßner T, Karmazin R, Männer R, Siessegger B. 2008. Power inductors in ceramic multilayer circuit boards. In: *Proceedings of the IMAPS/ACerS 4th International Conference and Exhibition on Ceramic Interconnect and Ceramic Microsystems Technologies (CICMT)*, pp. 213–222.

Moore GE. 1965. Cramming more components onto integrated circuits. Electronics 38: 114–117.

NIKOL. 2008. *Neue, reaktionskontrollierte Keramik zur Integration passiver Komponenten in leistungselektronische Module*. Supported by the Bundesministerium fuer Bildung und Forschung (BMBF) under grant no. 03X4503B.

Post JE. 2007. Analysis and design of planar, spiral-shaped, transmission-line transformers. IEEE Trans Adv Packaging 30:104–114.

Pozar DM. 1998. *Microwave Engineering*, 2nd edition. New York: John Wiley & Sons, pp. 431–436.

Rais-Zadeh M, Laskar J, Ayazi F. 2008. High performance inductors on CMOS-grade trenched silicon substrate. IEEE Trans Components Packaging Technol 31:126–134.

Riezenman MJ. 2001. The rebirth of radio. IEEE Spectr 38:62–64.

Sato T, Yokoyama H, Yamasawa K, Toya K, Kobayashi S, Minamisawa T. 2001. Multilayered transformer utilizing Mn-Zn ferrite and its application to a forward-type DC-DC converter. Electr Eng Jpn 135:1–8.

Slama G. 2003. Low-temp co-fired magnetic tape yields high benefits. Power Electron Technol 29:30–34.

Tada N, Tabuchi T, Ikezaki H. 2002. Inductor part and method of producing the same. European Patent ApplicationEP 1 367 611 A1.

Waffenschmidt E. 2004. Design and application of thin, planar magnetic components for embedded passive integrated circuits. In: *Proceedings of the 35th Annual IEEE Power Electronics Specialists Conference (PESC)*, pp. 4546–4552.

Waffenschmidt E, Jacobs J. 2008. Planar resonant multi-output transformer for printed circuit board integration. In: *Proceedings of the 39th Annual Power Electronics Specialists Conference (PESC)*, pp. 4222–4228.

Wahlers RL, Huang CYD, Heinz MR, Feingold AH, Bielawski J, Slama G. 2002. Low profile LTCC transformers. In: *Proceeding of the 2002 Int Symp Microelectronics, IMAPS Int Microelectronics and Packaging Soc*, pp. 76–80.

Yu Q, Wang H, Geng Y, Liu Z. 2006. Research of LTCC NiCuZn transformer prototype. In: *Proceedings of the 6th World Congr on Intelligent Control and Automation*, pp. 5272–5276.

Zou J, Liu C, Trainor DR, Chen J, Schutt-Ainé JE, Chapman PL. 2003. Development of three-dimensional inductors using plastic deformation magnetic assembly (PDMA). IEEE Trans Microw Theory Tech 51:1067–1075.

9

OXIDE THERMOELECTRIC POWER GENERATION

Ryoji Funahashi, Saori Urata, Atsuko Kosuga, and Delphine Flahaut

National Institute of Advanced Industrial Science and Technology (AIST), Osaka, Japan

INTRODUCTION

In view of global energy and environmental problems, research and development have been promoted in the field of thermoelectric power generation as a means of recovering vast amounts of waste heat emitted by automobiles, factories, and similar sources. Waste heat from such the sources offers a high-quality energy source equal to about 70% of total primary energy, but it is difficult to reclaim due to the source amounts being small and widely dispersed. Thermoelectric generation systems offer the only viable method of overcoming these problems, by converting heat energy directly into electrical energy irrespective of source size and without the use of moving parts or the production of environmentally deleterious wastes. The requirements placed on materials needed for this task, however, are not easily satisfied. They not only must possess high conversion efficiency but also must be composed of nontoxic and abundantly available elements having high chemical stability in air even at temperatures of 800–1000 K. Thermoelectric modules are composed of intermetallic compounds, such as Bi_2Te_3, Pb–Te, and Si–Ge. Practical applications of materials such as these have been delayed, however, by problems such as low melting or decomposition temperatures, toxicity or scarcity, and prohibitive cost. Recently, oxide compounds have attracted attention as promising thermoelectric materials because of their potential to overcome the above-mentioned problems [1–8].

Ceramic Integration and Joining Technologies: From Macro to Nanoscale, First Edition.
Edited by Mrityunjay Singh, Tatsuki Ohji, Rajiv Asthana, Sanjay Mathur.

To fabricate practically usable modules, not only thermoelectric materials but also electrodes possessing good thermal and electrical properties must be developed. In this chapter, the use of oxides is reviewed in the progress from material to modules.

THERMOELECTRIC POWER GENERATION

Thermoelectric power generation can be described by the Seebeck effect, discovered in 1821 by T. Seebeck. Voltage is generated in an electrical conductor by applying a temperature differential. This phenomenon is caused by diffusion of thermally activated electrons. One example using this effect is a thermocouple. For this application, the absolute values of the Seebeck coefficient (thermoelectric motive force per 1 K) are not important, but linearity between voltage and temperature difference is. On the other hand, the absolute values of the Seebeck coefficient are nevertheless important factors and should be high for the power application. Since typical thermoelectric materials can generate only a few hundred microvolts per kelvin, it is necessary to construct modules in which the thermoelectric materials are connected in series (Fig. 9.1). Two kinds of materials are needed for modules. In one material, the voltage is higher at the hot side than at the cold side; in the other, it is opposite. These correspond to n- and p-type materials, respectively. In general, the performance of thermoelectric materials is evaluated using thermoelectric figure of merit Z, which is described as follows:

$$Z = S^2/\rho\kappa. \tag{9.1}$$

Here, S, ρ, and κ are the Seebeck coefficient, electrical resistivity, and thermal conductivity, respectively. The better thermoelectric materials have the larger Z values, namely, good materials show large S and small ρ and κ. However, κ is usually large in good electrical conductors such as metals. On the other hand, S values increase with decreasing carrier density. Therefore, the discovery of good thermoelectric materials is a very difficult issue.

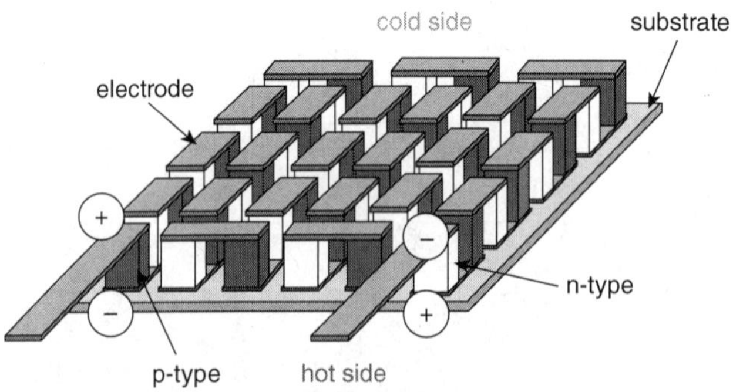

Figure 9.1. Schematic structure of a thermoelectric module.

THERMOELECTRIC OXIDE MATERIALS

p-Type Oxides

It has been generally accepted that the thermoelectric efficiency of oxide compounds is lower than that of the intermetallic compounds. However, a report by Terasaki et al. throws this assumption into question. With the discovery that, in spite of being a highly conductive oxide, $NaCo_2O_4$ single crystals have a high Seebeck coefficient [1], exploration of oxide-based thermoelectric materials began in earnest. Here, thermoelectric properties at high temperature of two Co-based oxide compounds with layered structure, $Ca_3Co_4O_9$ (Co-349) and $Bi_2Sr_2Co_2O_9$ (BC-222), first reported by the authors [3–5], is discussed.

Experimental Procedures. Both Co-349 and BC-222 single-crystalline whiskers were prepared using the glass-annealing method [9]. $CaCO_3$, $SrCO_3$, Bi_2O_3, Co_3O_4, and PbO powders used as starting materials were mixed with cationic compositions of Ca : Sr : Bi : Co : Pb = 1:1:1:2:0 and 1:1:1:2:1 for the Co-349 and BC-222 whiskers, respectively. The mixture was melted in air using an alumina crucible at 1573 K for 30 min. The melt was quenched by insertion between two copper plates to obtain glassy plates, which were heated in a stream of O_2 gas at 1193–1203 K for 300–1000 h to grow the whiskers from their surface (Fig. 9.2). The ribbonlike whiskers measured 1.0–5.0 μm in thickness, 50–200 μm in width, and less than 1.2 mm in length. Measurements of in-plane ρ and S along the whisker length were performed at 300–1000 K and 373–973 K, respectively, in air. ρ was measured using a standard DC four-probe method. S values were calculated from a plot of thermoelectric voltage against temperature differential as measured using a laboratory-designed instrument. Silver paste was used for terminal connection. Because the Co-349 whiskers are very small single crystals, measurement of κ for them is very difficult. The flake-like single crystals with few square millimeter planes can be prepared by a flux method [10]. Thermal diffusivity of the Co-349 phase was measured using these single crystals. On the other hand, BC-222 whiskers have enough dimensions for thermal diffusivity measurement. In-plane κ of

Figure 9.2. Scanning electron microscopic images for (a) Co-349 and (b) BC-222 whiskers.

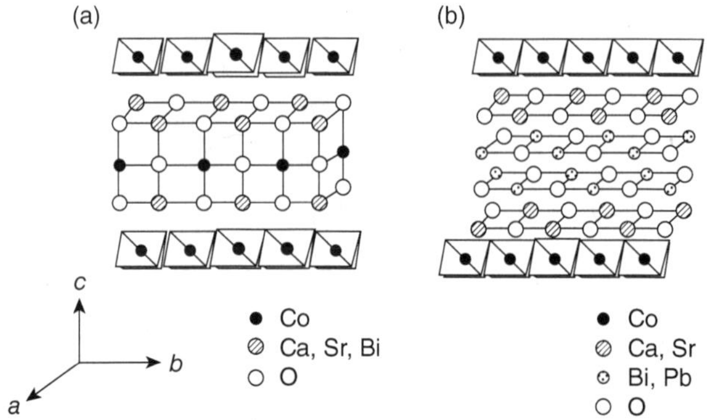

Figure 9.3. Schematic crystal structure of (a) Co-349 and (b) BC-222 phases.

the Co-349 single crystal and the BC-222 whisker was calculated by thermal diffusivity, heat capacity, and density. The thermal diffusivity was measured using an AC calorimetric method [11, 12] and the heat capacity using a differential scanning calorimetry (DSC) method. Density was evaluated using a pycnometer.

Results and Discussion. The crystal structure of the Co-349 and the BC-222 phases has been clarified just after reporting their thermoelectric properties [13–15]. They have layered structure in which two different layers alternate in the direction of the *c*-axis (Fig. 9.3). In one of the layers, a Co atom is surrounded by six O atoms in an octahedral configuration, and the octahedra are edge shared (CoO_2 layers) and form a triangle lattice. In the other layers, Ca, Co, Sr, Bi, Pb, and O form triple rock salt (Ca_2CoO_3) layers and quadruple rock salt ($Bi_2Sr_2O_4$) layers in the Co-349 and the BC-222 phases, respectively, namely, the triangle and quadrangle lattices are stacked alternatively in these phases. As a result, a misfit structure is formed between CoO_2 and rock salt layers. This structure in nanoscale seems to be effective to suppress phonon thermal conductivity.

Figure 9.4 indicates the temperature dependence of S and ρ for both whiskers. S values basically increase with temperature throughout the temperature range measured and are about 200 μV/K at temperatures higher than 873 K. The temperature dependence of ρ shows a semiconducting-like pattern at lower temperatures than 300 and 200 K for the Co-349 and BC-222, respectively. And at higher temperatures than these, metallic-like behavior is observed for both whiskers. Such phenomenon is due to formation of a pseudogap by strong electron correlation [16]. As a result, the high S in spite of low ρ is obtained in CoO_2 layered compounds.

Figure 9.5a shows the temperature dependence of κ for the Co-349 single crystal and the BC-222 whisker. The κ value is very low compared with that of general conducting oxides. A phonon mean free path l_{ph} was roughly evaluated as 0.5 nm using the Debye formula ($\kappa_{ph} = Cvl_{ph}/3$, where κ_{ph} is the lattice thermal conductivity and v is the sound velocity) for the BC-222 phase. κ_{elc} was calculated by the difference of electron thermal conductivity ($\kappa_{ele} = LT/\rho$, $L = 2.45 \times 10^{-8}$ WΩ/K^2) from κ. The l_{ph} value is

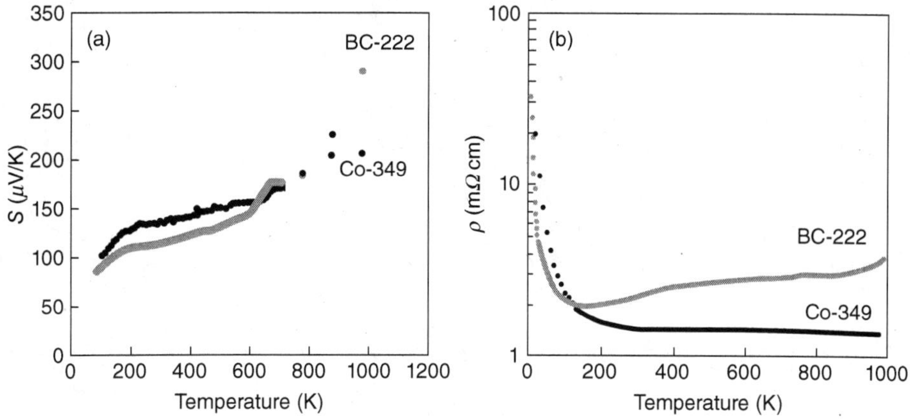

Figure 9.4. Temperature dependence of (a) S and (b) ρ for Co-349 and BC-222 whiskers.

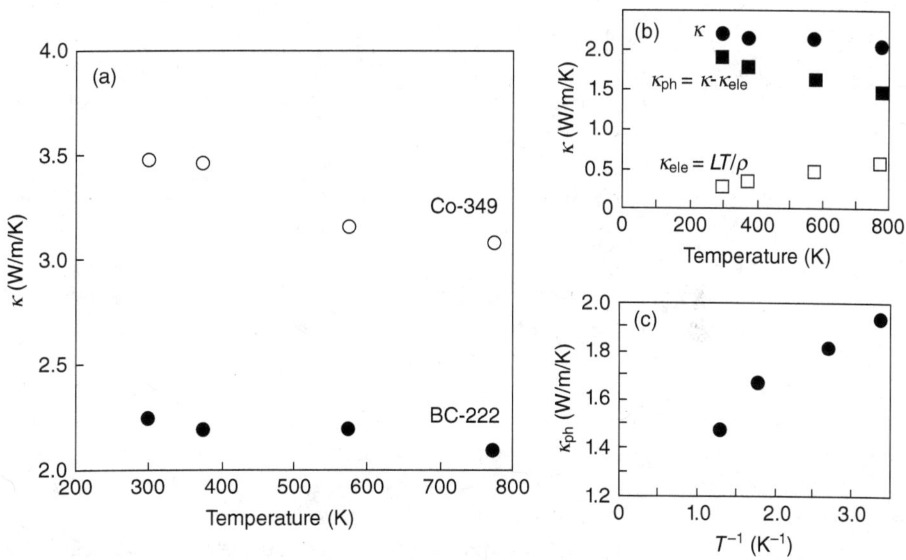

Figure 9.5. Temperature dependence of (a) κ for Co-349 single crystal and BC-222 whisker and (b) calculated κ_{ele} and κ_{ph} and (c) κ_{ph}–T^{-1} plot for BC-222 whisker.

close to the *a*- and *b*-cell parameters of the BC-222 layer. The total κ decreases with increasing temperature because it is dominated by κ_{ph} (Fig. 9.5b). In general systems in which the phonons are scattered by a phonon–phonon interaction, the κ_{ph} is proportional to the inverse of *T*, while in the BC-222 phase, no linearity appears in the κ_{ph} versus T^{-1} plot (Fig. 9.5c). This indicates that low κ is due not only to simple phonon–phonon interaction but also to other complex mechanisms, such as the misfit between the subcells.

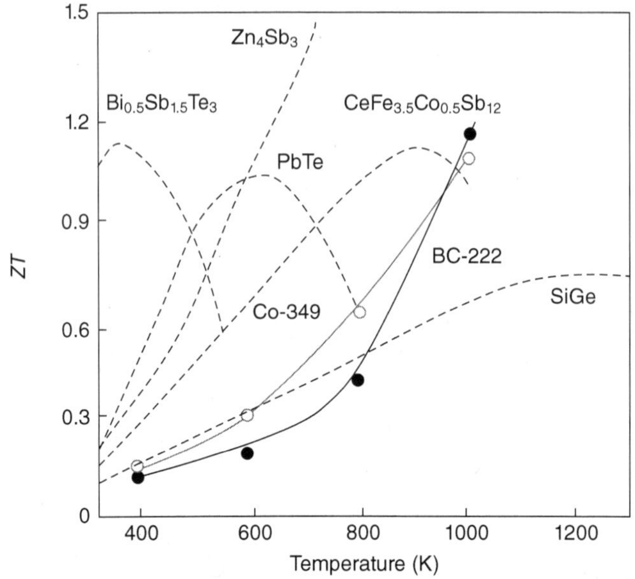

Figure 9.6. Temperature dependence of ZT for Co-349, BC-222, and conventional intermetallic materials.

The *ZT* was calculated using *S*, ρ, and κ (Fig. 9.6). The *ZT* value at 973 K was obtained using κ evaluated by the extrapolation of the κ–*T* line. It is over 1.0 at 973 K in air, which is equivalent to 10% conversion efficiency for both phases. While thermoelectric materials whose *ZT* is comparable to Co-349 and BC-222 phases have been reported [17], they are not only unstable in air at high temperatures but also contain harmful or not abundantly available elements. Moreover, operating temperatures are limited to those below 800 K. The oxide Co-349 and BC-222 compounds, however, contain no elements that are scarce or considered seriously harmful, and are chemically stable in air even at 1000 K. Thermoelectric modules fabricated using these oxides would therefore be low in cost and could be used in air for extended periods. The discovery of CoO_2 layered compounds with high thermoelectric performance at 800–1000 K leads to the expectation that electric power generation using waste heat from automobiles, factories, and similar sources will be realized in the near future.

n-Type Oxides

In order to achieve efficient thermoelectric power generation, n-type oxide materials are indispensable. Although several n-type oxide materials have been reported, some of them are not suitable for high-temperature applications in air because they need oxygen deficiency [18, 19] or include toxic elements [6]. A perovskite $CaMnO_3$ phase, which is one of the strong candidates as the n-type oxide, exhibits a rather high *S* value. ρ is high in no substituted $CaMnO_3$ because of lack of charge carriers. Substitution on a Ca or Mn site has been attempted in considering decrease in ρ. Power factor ($S^2\rho$) of

0.3 mW/m/K^2 can be reached for the $CaMn_{0.96}Nb_{0.4}O_3$ [20] and 0.28 mW/m/K^2 for $Ca_{0.9}Bi_{0.1}MnO_3$ at 1000 K [7]. For these compounds, the S value remains as high as $-100\,\mu$V/K.

For enhancement of ZT values in the $CaMnO_3$ phase, we investigated systematically thermoelectric properties at high temperature in the $CaMnO_3$ substituted by rare earth (Yb, Tb, Nd, and Ho) on the Ca site.

Experimental Procedures. Polycrystalline samples of $Ca_{0.9}A_{0.1}MnO_3$ (A = Yb, Tb, Nd, Ho) were synthesized using a solid-state reaction in air. The compounds starting from stoichiometric mixtures of $CaCO_3$, Yb_2O_3, Tb_2O_3, Ho_2O_3, and Nd_2O_3 were calcined at 1073 K. Then, the powders were heated at 1273 K for 10 h and 1475 K for 12 h with intermediate grinding. Finally, the products were pressed into pellets and were sintered in air at 1573 K for 15 h. The pellets were cooled down to room temperature in the furnace. Moreover, polycrystalline samples of $Ca_{1-x}Yb_xMnO_3$ ($x = 0.05$–0.5) were synthesized using a solid-state reaction in air in order to optimize the Yb content. The compounds starting from stoichiometric mixtures of $CaCO_3$, Mn_2O_3, and Yb_2O_3 were calcined at 1073, 1273, and 1475 K for 12 h with intermediate grinding. Then, the products were pressed into pellets and were sintered in air at 1573 K for 15 h.

X-ray powder diffraction (XRD) analysis was carried out using a Rigaku diffractometer using Cu-Kα radiation. Lattice parameters were obtained from the Rietveld analysis of the X-ray data [21]. The microstructures of the samples were observed by scanning electron microscopy (SEM). The constituent analysis was carried out by using an energy-dispersive X-ray (EDX) spectrometer.

Resistivity measurement was carried out by a standard dc four-probe method 300–1100 K in air. The thermoelectromotive forces (ΔV) and temperature difference (z) were measured at 373–973 K, and the S values were calculated from a plot of ΔV–ΔT. Two Pt-Pt/Rh (R-type) thermocouples were adhered to both ends of the samples using silver paste. For voltage terminals, the Pt wires of the thermocouples were used. Measured S values were reduced by those of Pt wires to obtain the net S values of the samples. Thermal conductivity is obtained from the thermal diffusivity, specific heat capacity, and density. Thermal diffusivity and specific heat were measured by a laser flash method (ULVAC-TC3000V) and DSC (MDSC2910, TA Instruments, Delaware), respectively.

Results and Discussion

THERMOELECTRIC PROPERTY. The temperature dependence of ρ for the samples is represented in Figure 9.7a. The undoped $CaMnO_3$ exhibits a ρ value around 300 mΩcm at room temperature. The rare earth substitution induces a strong decrease of the ρ values of two orders of magnitude according to the increase in charge carriers (electron) and Mn^{3+} content in the Mn^{4+} matrix. Moreover, the conduction mode changes from a semiconducting behavior toward a metallic one. We have also noticed that, for the same Mn^{4+}/Mn^{3+} ratio, ρ decreases with decreasing rare earth ionic radius. Many studies report the influence of the average ionic radius of cation at the A-site in hole-doped $AMnO_3$ on the transport properties. Actually, the average ionic radius $\langle r_A^{3+}\rangle$ determining the Mn–O–Mn bond length and angle strongly affects the overlapping of

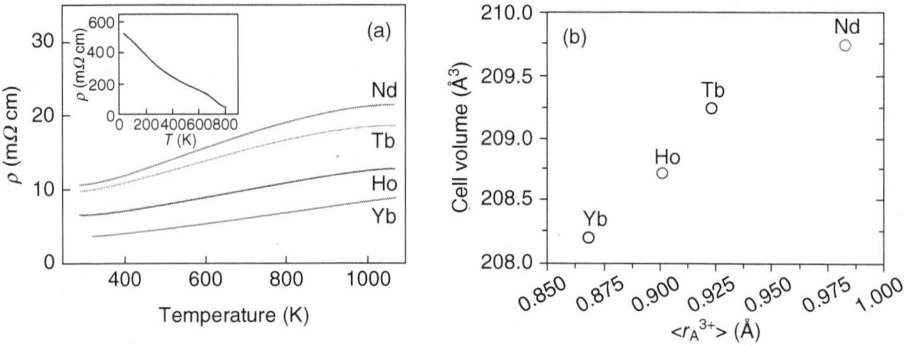

Figure 9.7. Temperature dependence of resistivity ρ (a) and the relationship between radii of rare earth ions $<r_A^{3+}>$ and cell volume (b) for $Ca_{0.9}A_{0.1}MnO_3$ samples. Insert: resistivity versus temperature of $CaMnO_3$.

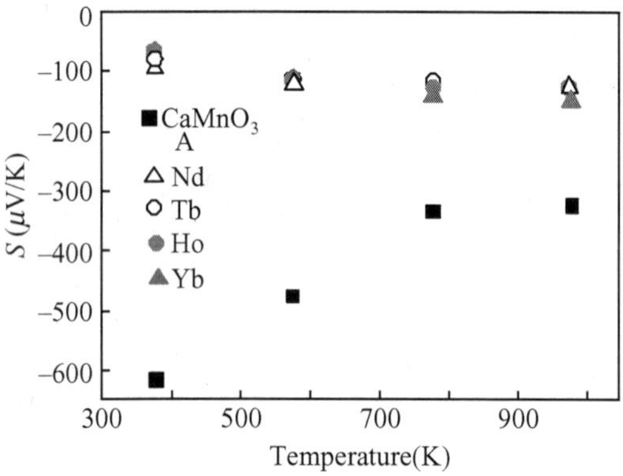

Figure 9.8. Temperature dependence of Seebeck coefficient of $CaMnO_3$ and $Ca_{0.9}A_{0.1}MnO_3$.

Mn and O orbitals. It has been demonstrated that the decrease in $<r_A^{3+}>$ makes ρ low. As shown in Figure 9.7b, cell volumes are smaller with decreasing $<r_A^{3+}>$. Therefore the decrease in ρ by substitution with small $<r_A^{3+}>$ ions is due to the compression of cell volumes.

Figure 9.8 shows the S values versus the temperature for the $CaMnO_3$ and $Ca_{0.9}A_{0.1}MnO_3$. The undoped $CaMnO_3$ shows large absolute values of S, which is decreasing with increasing temperature. This is related to its low carrier content and semiconductor behavior. The rare earth substitution induces a remarkable decrease in the S values, which is in agreement with the increase in charge carrier content. For the substituted compounds, absolute values of S increase with the temperature from 300 to 900 K. While the ρ values depend on $<r_A^{3+}>$, there is no obvious relation between S and $<r_A^{3+}>$. S is decided by only the charge carrier content.

Figure 9.9a demonstrates the temperature dependence of total thermal conductivity κ of samples. The κ was calculated from the following formula: $\kappa = DC_pd$, where D, C_p, and d are the thermal diffusivity, specific heat capacity, and density, respectively.

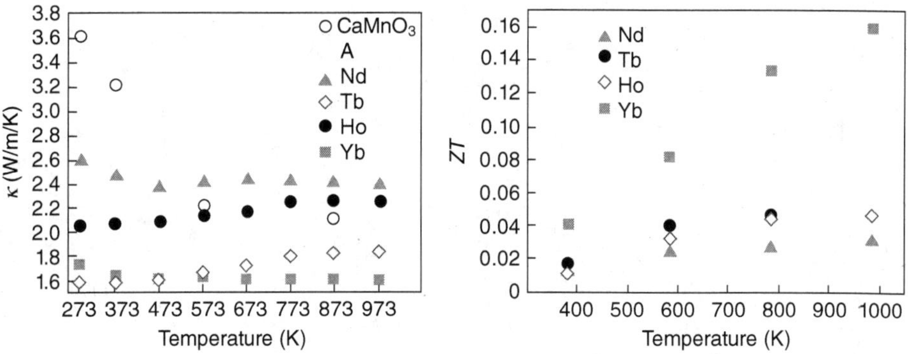

Figure 9.9. Temperature dependence of (a) thermal conductivity and (b) dimensionless figure of merit for $CaMnO_3$ and $Ca_{0.9}A_{0.1}MnO_3$.

The thermal conductivity is constant in the whole range of temperature. As reported by Cong et al. in the $Ca_{1-x}Pr_xMnO_3$ systems, the rare earth substitution decreases κ by the production of phonon scattering [22]. Depending on $<r_A^{3+}>$, the κ values change from 1.6 to 2.4 W/m/K. The thermal conductivity due to electron κ_{ele} is smaller than thermal conductivity due to phonon κ_{ph} by 10 times more in the $Ca_{0.9}A_{0.1}MnO_3$ system. The total κ is dominated mainly by κ_{phon} and tends to decrease by substitution with heavy elements.

The temperature dependence of *ZT* for $Ca_{0.9}A_{0.1}MnO_3$ samples is shown in Figure 9.9b. For all samples, the *ZT* values increase with increasing temperature over the whole measured temperature range. By the fact that *S* values are independent of the rare earth elements and that ρ and κ decrease by substitution with heavy elements, it is expected that higher *ZT* than that reported in previous papers is obtained [7]. The highest *ZT* was reached 0.16 at 1000 K for the $Ca_{0.9}Yb_{0.1}MnO_3$ sample.

As mentioned earlier, Yb substitution is most effective to enhance *ZT* values in the $Ca_{0.9}A_{0.1}MnO_3$ system. We need to optimize the Yb content to improve *ZT* more. $Ca_{1-x}Yb_xMnO_3$ ($x = 0.05$–0.5) samples were prepared and their thermoelectric properties were measured.

Figure 9.10a indicates ρ variations with temperature. The substitution of Yb for Ca in $CaMnO_3$ decreases ρ in a spectacular way due to the creation of Mn^{3+} cations in the Mn^{4+} matrix. The ρ values decrease with increasing Yb content up to $x = 0.15$ and exhibits a minimum (3 mΩcm at room temperature) for $Ca_{0.85}Yb_{0.15}MnO_3$. Temperature dependence of ρ shows metallic-like behavior in $x \leq 0.3$ in the measured temperature regain and metal–insulator transition is observed at $x = 0.5$. Many studies have been reported on the influence of average ionic radii of A-site $<r_A>$ in hole-doped $AMnO_3$ on the transport properties. Actually, the $<r_A>$ determines the Mn–O–Mn bond length and/or angle and affects ρ in the same charge carrier content as mentioned earlier.

The temperature dependence of *S* is shown in Figure 9.10b. The substitution of Yb^{3+} on the Ca^{2+} site increases charge carrier content, which is reflected by a continuous decrease in the absolute *S* values.

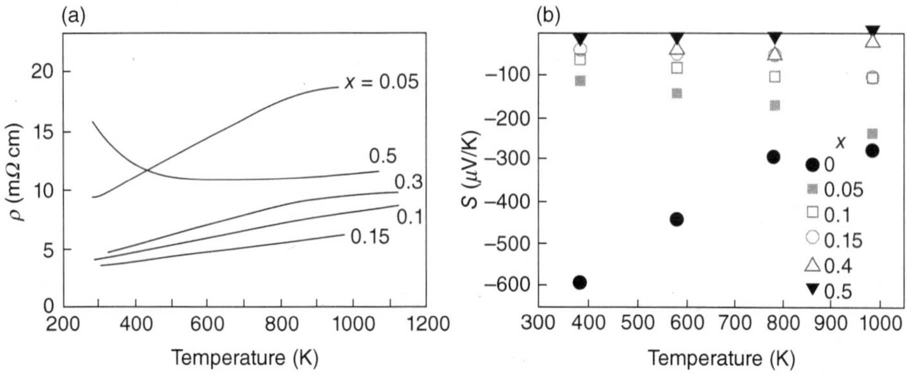

Figure 9.10. Temperature dependence of (a) resistivity and (b) Seebeck coefficient for $CaMn_{1-x}Yb_xO_3$ ($0 \leq x \leq 0.5$).

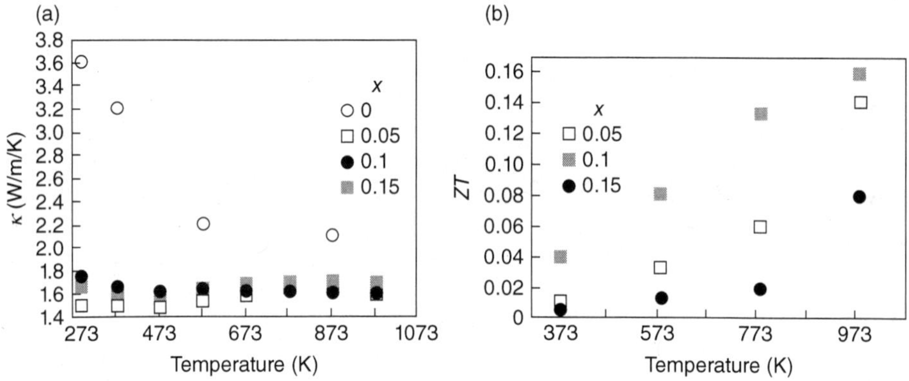

Figure 9.11. Relation between temperature and (a) κ and (b) ZT for $CaMn_{1-x}Yb_xO_3$ ($0 \leq x \leq 0.15$).

The κ values of substituted samples are lower than those of $CaMnO_3$ (Fig. 9.11a). One can claim that the κ_{ele} is negligible compared with κ_{ph}. Although total κ values are suppressed by Yb substitution, there is no obvious dependence on substitution content. The ZT values have been calculated from S, ρ, and κ. The highest ZT of 0.16 is obtained at $x = 0.1$ in the $Ca_{1-x}Yb_xMnO_3$ system.

DEVICE TECHNOLOGY

In order to achieve thermoelectric power application, power generators, namely, thermoelectric modules, are necessary. The modules consist of both p- and n-type devices. The generated power is proportional to the total volume of the thermoelectric devices. Because growth of oxide single crystals takes a long time, the oxide devices must be polycrystalline materials, sintered bodies. As mentioned earlier, since the p-type Co-349

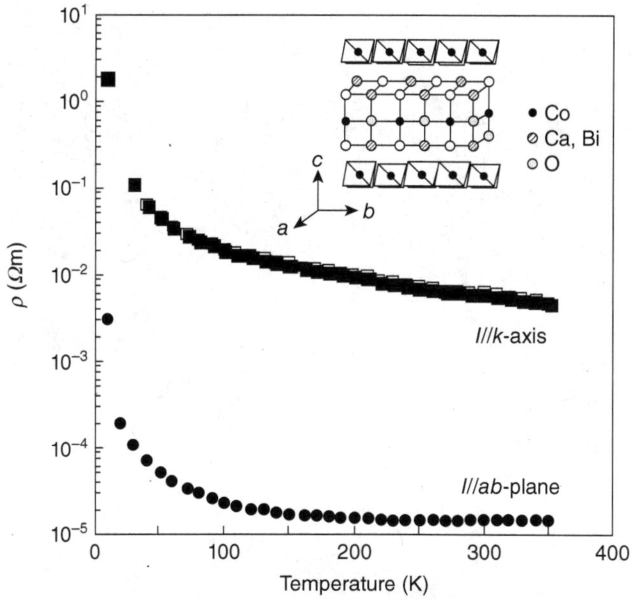

Figure 9.12. Temperature dependence of electrical resistivity ρ in cases of current (I) flowed parallel to the c-axis and to the ab-plane for the Co-349 single crystal.

phase has a layered structure, it shows two-dimensional electrical properties. The ρ values in the direction of the *ab*-plane (in-plane) are lower than in the *c*-axis direction (out-of-plane) by several hundred times (Fig. 9.12) [23]. Therefore, not only densification but also grain alignment is indispensable for high *ZT* in the Co-349 devices.

As shown previously, $CaMnO_3$ substituted Ca and Mn sites are one of good candidates for n-type devices. The bulks of $CaMnO_3$, however, have a serious problem of brittle fracture. The Mn-113 devices were broken completely by heating for power generation in a thermoelectric module composed of Co-349/Mn-113 devices (Fig. 9.13) [24]. In order to fabricate good thermoelectric modules, well-grain aligned Co-349 bulks with low ρ and tough Mn-113 bulks with good mechanical property should be prepared. In this section, the device technology for the preparation of good thermoelectric oxide bulks is discussed.

p-Type Bulk Materials

Good thermoelectric performance at high temperature in air has been reported for single-crystalline whiskers of two such materials: $Ca_3Co_4O_9$ (Co-349) [4] and $Bi_2Sr_2Co_2O_9$ (BC-222) [5, 25]. Though these whiskers are about 1.0–10 mm long, their cross-sectional area has thus far been limited to ca. 5.0×10^{-4} mm^2. Bulk polycrystalline materials are therefore necessary in order to suppress electrical resistance and are indispensable for power generation applications. The study of the Co-349 single crystals thus far has shown a two-dimensional electrical property that originates from their

Figure 9.13. Brittle fracture of a Mn-113 device.

crystal structure (Fig. 9.12). This causes Co-349 bulk materials with no preferred grain alignment to have ρ several times higher than that of the whiskers and, consequently, the *ZT* of the former is lower than the latter [3, 26, 27], that is to say, the arrangement of all Co-349 grains in the crystallographic *ab*-plane is very important. Moreover, the influence of grain boundaries, which have higher ρ than the intragrain regions, should be reduced. Although hot pressing is indeed effective for improved grain alignment and decreased ρ, the *ZT* of hot-pressed materials is 0.3 at most [27], and as such, a different approach is required in the preparation of Co-349 bulk materials for the desired application. Here, the preparation and thermoelectric properties of a Co-349 composite consisting of the Co-349 large single crystals and powder are described.

Experimental Procedures

Co-349 Single Crystals. The large single crystals were synthesized by heating a mixture of $CaCO_3$ and Co_3O_4 powders in $CaCl_2$ and $SrCl_2$ fluxes. $CaCO_3$ and Co_3O_4 powders were weighed in a molar ratio of Ca : Co = 1.0:4.0. $CaCl_2$ and $SrCl_2$ powders were weighed in a molar ratio of Ca : Sr = 1.0:2.5. The $CaCO_3$ and Co_3O_4 powders (58.67 g) and the $CaCl_2$ and $SrCl_2$ powders (141.33 g) were mixed using an agate container and balls for 1 h. The mixed powder was heated at 1173 K for 3 h in an alumina crucible to melt the $CaCl_2$ and $SrCl_2$. Subsequent cooling was carried out at rates of 1 K/h down to 1023 K and 133 K/h down to 623 K. The sample was then cooled to room temperature in the furnace without temperature control. Water-soluble salts were washed away by hot water, and Co_3O_4 and other non-water-soluble remnants on the surface of the single crystals were eliminated by soaking in a diluted nitric acid solution (~2 vol%) for 3 h. After neutralization by washing in water, the single crystals were dried in air at room temperature. The dimensions of the single crystals were 0.5–2.0 mm^2 of the well-grown *ab*-plane and 10–50 μm thick in the *c*-axis direction (Fig. 9.14a).

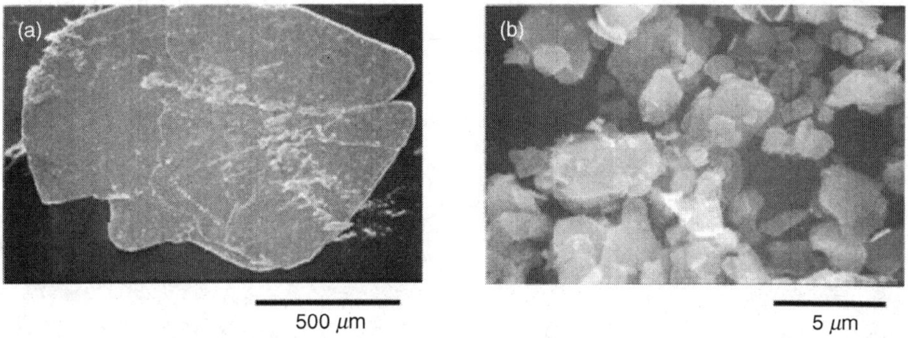

Figure 9.14. Scanning electron microscopic images for Co-349 (a) single crystal and (b) powder.

Co-349 Powder. The powder was prepared using a solid-state reaction and to achieve partial substitution of Ca by Bi because Bi substitution is effective to enhance S and to reduce ρ and κ [26, 27]. The $CaCO_3$, Bi_2O_3, and Co_3O_4 powders were mixed in a cationic ratio of Ca:Bi:Co = 2.7:0.3:4.0 using the agate container and balls for 1 h. The mixed powder was calcined at 1073 K for 10 h in air. The calcined sample was pulverized, palletized, and sintered at 1153 K for 20 h under an oxygen gas flow of 150 cm^3/min. The Co-349 powder was obtained by pulverizing the sintered pellets. The grains of the Co-349 powder have a platelike shape with the well-grown crystallographic *ab*-plane and the thickness direction corresponding to the *c*-axis. The dimensions of particles comprising the powder were about 1.0–5.0 μm^2 and were less than 0.5 μm thick (Fig. 9.14b).

Co-349 Composite. The Co-349 composite was prepared by hot pressing a mixture of Co-349 single crystals and powder. The amount of incorporated single crystals was 20 wt% (20 wt% composite). The single crystals and powder were mixed with an organic binder consisting of methylcellulose, ethanol, and water. The mixture was pressed to align the *ab*-plane of Co-349 single crystals and powder. After drying in a vacuum at room temperature to evaporate the solvent, the platelike Co-349 materials combined in the binder were piled up and pelletized. The pelletized precursor was hot pressed at 1123 K for 20 h under 10 MPa in air. For comparison, a hot-pressed sample without the single crystals (0 wt% bulk) was prepared under the same conditions.

Measurements. Measurements of S and ρ for the 20 wt% composite and the 0 wt% bulk were performed in the hot-pressed plane, which largely corresponds to the *ab*-plane of the Co-349 phase. For the single crystals, these measurements were carried out exclusively in the *ab*-plane. The S values were measured as mentioned earlier at 373–973 K. ρ was measured using a standard dc four-probe method in air from room temperature to 1000 K. In this section, we are discussing electrical conductivity σ ($= \rho^{-1}$). κ measurement for both 20 wt% composite and 0 wt% bulk was carried out using the laser flash method (RIGAKU LF/TCM FA8510B) along the direction parallel

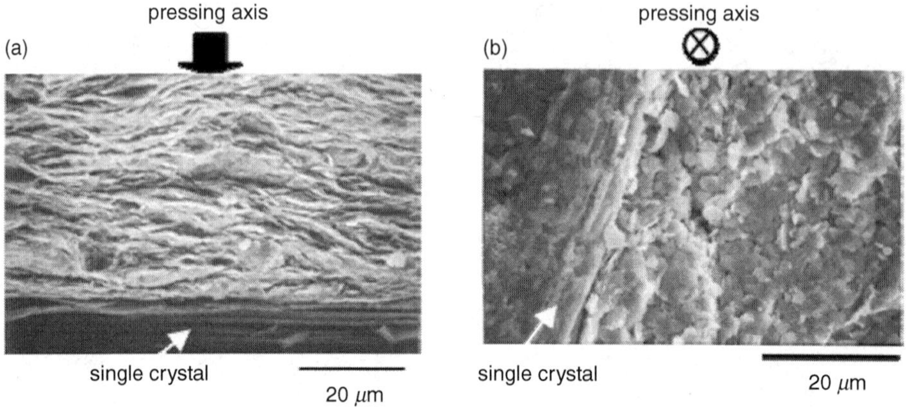

Figure 9.15. Scanning electron microscopic images for (a) cross-sectional area parallel to the hot-pressing axis and (b) surface area perpendicular to the hot-pressing axis of 20 wt% composite.

to the hot-pressing axis, almost perpendicular to the *ab*-plane in a vacuum. The measurement error with this instrument is about ±3% and ±5% for thermal diffusivity and heat capacity, respectively. On the other hand, κ of the single crystal was calculated by thermal diffusivity for the *ab*-plane, heat capacity, and density. The thermal diffusivity was measured using an ac calorimetric method [11, 12] and the heat capacity using a DSC method. In these methods, measurement error is about ±20% and ±1% for thermal diffusivity and heat capacity, respectively.

Results and Discussions

Microstructure. Figure 9.15 shows the microstructure of the 20 wt% composite. A stacking structure composed of platelike grains is observed (Fig. 9.15a), indicating that the well-grown plane, corresponding to the *ab*-plane of the single crystals and the small grains originating from the powder, is well aligned perpendicular to the hot-pressing axis (Fig. 9.15b). Although a quantitative analysis of the grain alignment has yet to be carried out, the alignment of the small grains seems to be improved by the incorporation of the single crystals close to them in the composite, due to the large single crystals behaving as pressure media [28].

Electrical Properties. Positive S values increase with temperature for all samples with almost no difference among them (Fig. 9.16). The in-plane σ values (σ_{in}) of the 20 wt% composite, on the other hand, are higher than those of the 0 wt% bulk as shown in Figure 9.17I, indicating that incorporating the single crystals is effective to increase σ_{in}. A step appears at 550 and 650 K in the σ_{in}–T curves for the bulk samples and the single crystal, respectively. The step at 550 K for the 0 wt% bulk and 20 wt% composite seems not to be due to measuring errors but to the release of oxygen. Thermogravimetric analysis showed a broad decrease in weight at temperatures higher

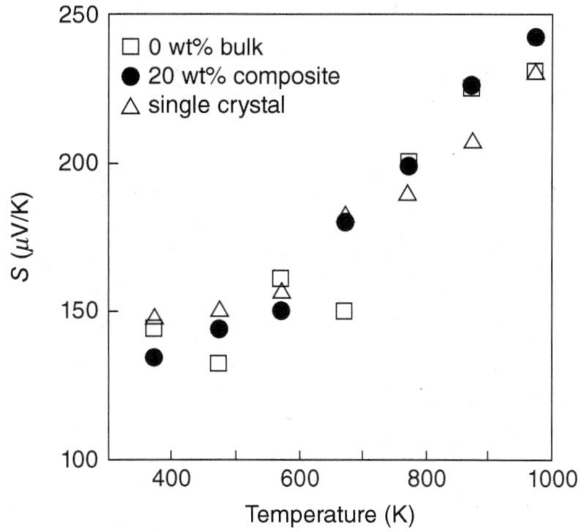

Figure 9.16. Temperature dependence of Seebeck coefficient of 20 wt% composite, 0 wt% bulk, and single crystal.

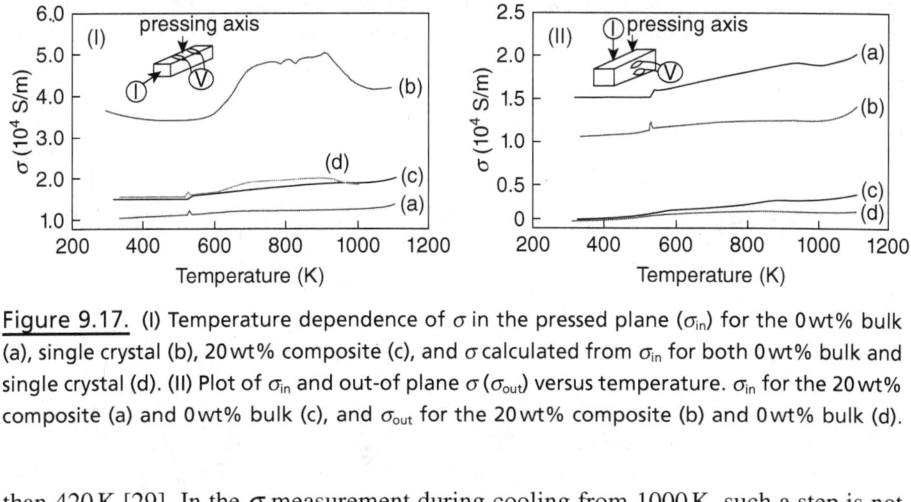

Figure 9.17. (I) Temperature dependence of σ in the pressed plane (σ_{in}) for the 0 wt% bulk (a), single crystal (b), 20 wt% composite (c), and σ calculated from σ_{in} for both 0 wt% bulk and single crystal (d). (II) Plot of σ_{in} and out-of plane σ (σ_{out}) versus temperature. σ_{in} for the 20 wt% composite (a) and 0 wt% bulk (c), and σ_{out} for the 20 wt% composite (b) and 0 wt% bulk (d).

than 420 K [29]. In the σ measurement during cooling from 1000 K, such a step is not observed. As no step appears at 650 K in the 20 wt% composite, it would seem that the current does not flow preferentially through the single crystals. However, the increase in σ due to the incorporation of single crystals indicates that the single crystals in the composite act as bypasses for the current and reduce the influence of the grain boundaries. σ calculated by the summation of 20% and 80% of σ_{in} for the single crystal and 0 wt% bulk, respectively, is also plotted in Figure 9.17I. Actually, the alignment of the small grains seems to be improved by the incorporation of the single crystals in close proximity to them in the composite, and the current bypasses the grain boundaries through the single crystals. The σ_{in} values, however, are almost comparable to the calculated ones based on the supposition of a simple parallel circuit model. To induce

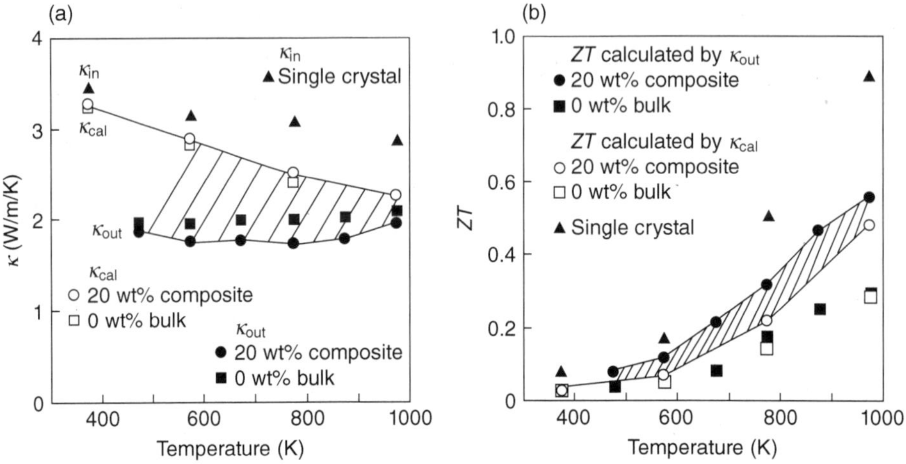

Figure 9.18. (a) Measured κ of the 20 wt% composite, 0 wt% bulk, and single crystal. These values are out-of-plane κ (κ_{out}) and in-plane κ (κ_{in}) for the bulk samples and the single crystal, respectively. κ_{in} for the bulk samples (κ_{cal}) is the calculated one using in-plane κ_{ph} of the single crystal and κ_{ele} for each sample, which is evaluated by Wiedemann–Franz's law. The actual κ_{in} for the 20 wt% composite seems to be in the region indicated by oblique lines. (b) Plot of dimensionless thermoelectric figure of merit ZT versus T. ZT is calculated using measured κ (κ_{out}) for the 20 wt% composite and 0 wt% bulk. The in-plane ZT evaluated using κ_{cal} is also plotted for the 20 wt% composite and 0 wt% bulk.

the above effects expected from the incorporation of large single crystals, the content of single crystals should be increased; a preparation technique for precursor bulks allowing better grain alignment should be developed; and the electrical connection between the small grains and single crystals should be improved. Moreover, to discuss the electrical properties in detail, more composites of different compositions will need to be prepared.

Thermal Conductivity and *ZT*. Figure 9.18a shows the temperature dependence of κ for the 20 wt% composite, 0 wt% bulk, and single crystal. Only the κ for the single crystal corresponds to in-plane values. Considering the anisotropy of σ (Fig. 9.17II), a difference in κ should be found between the in-plane (κ_{in}) and out-of-plane (κ_{out}) values. While the κ_{in} is indispensable for the calculation of the in-plane *ZT*, it has yet to be measured using the laser flash method for the bulk samples. The κ due to electrons, κ_{ele} for the single crystal, can be evaluated by Wiedemann–Franz's law ($\kappa_{ele} = L\sigma T$, $L = 2.45 \times 10^{-8}\,\mathrm{W}\Omega/\mathrm{K}^2$). Using this κ_{ele}, the in-plane κ due to phonon κ_{ph} is calculated for the single crystal and is greater than that for the bulk samples because of the lack of grain boundaries and pores, and the absence of Bi-substitution, which is effective in suppressing the κ_{ph} [26, 27]. For evaluation of the κ_{in} for the bulk samples, their in-plane κ_{ele} evaluated by Wiedemann–Franz's law and the in-plane κ_{ph} for the single crystal are used. Therefore, calculated κ_{in} values (κ_{cal} in Fig. 9.18a) for the bulk samples seem to be higher than the actual κ_{in} values, which seem to be between κ_{out} and κ_{in}. Using κ_{in} for the single crystal and the 0 wt% bulk, κ for the 20 wt% composite is calculated using the weight fractions in the same way as σ. The calculated values are almost the same as actual κ_{in} for the 20 wt% composite.

ZT values are calculated using both κ_{out} and κ_{in} (or κ_{cal}) and increase with temperature (Fig. 9.18b). Actual *ZT* values for the 20 wt% composite seem to be in the region indicated by oblique lines. Although the single crystal displays good thermoelectric performance, *ZT* of the 0 wt% bulk is 0.3 at most. The incorporation of single crystals was thus effective in increasing *ZT* and resulted in *ZT* higher than 0.48 at 973 K in air for the 20 wt% composite. Bergman and Fel reported that considerable enhancement of power factor is often possible with but a modest reduction in *Z* [30]. Particularly, in the composites made up of two thermoelectric components with similar parameters, the power factor is determined almost entirely using the volume fraction of the components. As σ and κ for the 20 wt% composite are comparable to those of the composite calculated by the weight fraction, which almost corresponds to the volume fraction, *Z* values also are determined by the weight fractions.

For the purpose of enhancing the *ZT* of Co-349 bulk samples, the incorporation of large single crystals is effective because of the microstructural improvement and the bypass effect. While *ZT* was successfully enhanced through this study, the full potential of the effects from the incorporation of single crystals has yet to be realized. Further study in order to optimize conditions of preparation, particularly the amount of single crystals and the technique used to align the grains, will be necessary to achieve this. It will be very interesting to discover whether the *ZT* of composites containing single crystals can match or surpass that of single crystals alone.

n-Type Bulk Materials

Although thermoelectric properties of materials composing the modules should be enhanced, high chemical and mechanical durability of the materials, contact resistance, and strength at the junctions are also very important in the practical use of the modules. The polycrystalline sample of $Ca_{0.9}Yb_{0.1}MnO_3$ (Mn-113) was found to exhibit a moderate *ZT* of 0.16 at 973 K in air [31]. We found, however, that the Mn-113 legs in the modules broke after the power generation test due to thermal stress, which was attributed to the large difference of thermal expansion coefficient between the Ag electrode and Mn-113 legs, and the low mechanical strength of the Mn-113 legs [24]. One possible approach to overcome this problem is to enhance the mechanical properties of the Mn-113 sintered ceramics. There are some material designs to strengthen and toughen ceramics by using composite techniques, incorporating particles, whiskers or platelet reinforcement, and precipitation of secondary phases [32, 33]. Among them, ceramic composites having ductile metal dispersion were reported to show excellent mechanical properties such as hardness, Young's modulus, bending strength, and toughness [34, 35]. The mechanical properties of oxide melt composites are discussed here.

Mechanical Properties of Monolithic Bulks

EXPERIMENTAL PROCEDURES. The $Ca_{2.7}Bi_{0.3}Co_4O_9$ (Co-349) and $CaMn_{0.98}Mo_{0.02}O_3$ (Mn (Mo)-113) bulk samples for the investigation of mechanical properties were prepared as follows. The Co-349 powder was prepared by solid-state reaction at 1123 K for 10 h in air. As starting materials, $CaCO_3$, Co_3O_4, and Bi_2O_3 powders were used and mixed thoroughly in the stoichiometric composition. The Co-349 bulks were prepared using a hot-pressing technique. The obtained Co-349 powder was hot pressed for 20 h in air under a uniaxial pressure of 10 MPa at 1123 K to make density and grain alignment high. Preparation of the Mn (Mo)-113 bulks was started using $CaCO_3$, Mn_2O_3, and MoO_3 powders. These powders were mixed well and were treated at 1273 K for 12 h in air. The powder was densified using a cold isostatic pressing (CIP) technique for 3 h under about 150 MPa. After the CIP process, the precursor pellets were sintered at 1473 K 12 h in air. The Mn-113 bulks can be densified well by CIP and sintering under the atmospheric pressure even without hot pressing. For comparison, n-type $La_{0.9}Bi_{0.1}NiO_3$ (Ni-113) bulks were prepared by hot pressing. The bulks were cut into dimensions of 4.0 mm wide, 3.0 mm thick, and 40.0 mm long for a three-point bending test and 5.0 mm wide, 5.0 mm thick, and 10.0 mm long for the thermal expansion coefficient.

Linear thermal expansion coefficient α was measured using a differential dilatometer (Thermo Plus TMA8310, Rigaku, Tokyo) at 323–1073 K in air. A three-point bending test was carried out at room temperature; the loading speed was 0.5 mm/min; and span length was 30.0 mm (Autograph AG-20kNG, Shimadzu, Kyoto) as shown in Figure 9.19. In the case of Co-349 bulks, measurement was performed in the loading direction perpendicular and parallel to the hot-pressing axis.

Bulk density of the samples was measured by the Archimedean method. Powder density was determined by the picnometric method in a He gas atmosphere. Relative density for the oxide bulks was calculated using these densities.

RESULTS AND DISCUSSION. To investigate the cause of the destruction of the Mn (Mo)-113 legs in the modules, α and three-point bending strength σ_f were investigated. The α values for the Co-349 bulks are lower than those for the Mn (Mo)-113 and Ni-113 ones and are closer to the alumina plates (Fig. 9.20a). The thermal expansion coefficient increases with temperature for all samples. The differential between ther-

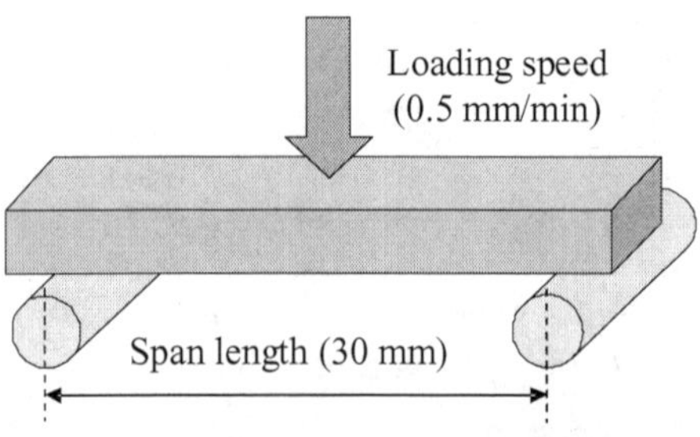

Figure 9.19. Schematic picture of a three-point bending test.

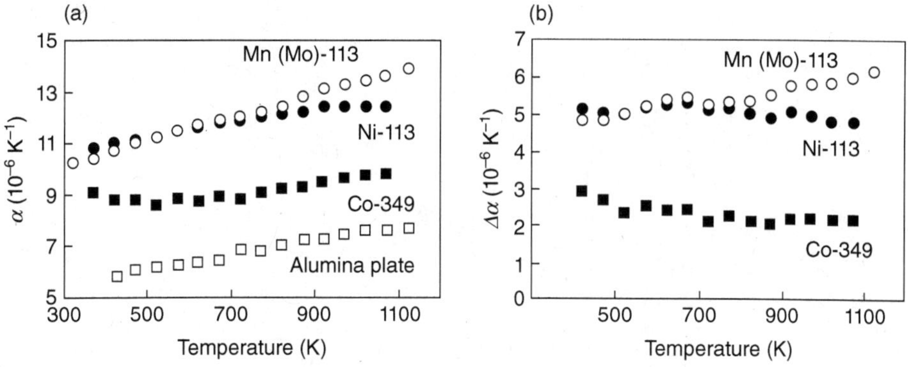

Figure 9.20. Temperature dependence of α for Co-349, Mn (Mo)-113, Ni-113 bulks, and alumina plate of the substrate (a). Temperature dependence of differential of α ($\Delta\alpha$) between thermoelectric oxide bulks and the alumina plate (b).

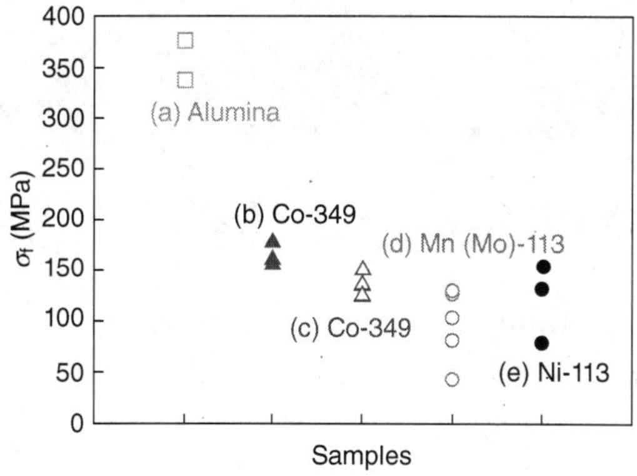

Figure 9.21. Three-point bending strength σ_f of (a) alumina plates, (b) Co-349 (loading direction parallel to the hot-pressing axis), (c) Co-349 (loading direction perpendicular to the hot-pressing axis), (d) Mn (Mo)-113, and (e) Ni-113 bulks.

moelectric oxide bulks and the alumina plate is shown in Figure 9.20b. For the Co-349 and Ni-113 bulks, the differential from the alumina plate tends to decrease with increasing temperature. On the other hand, it increases in the Mn (Mo)-113 bulk. This seems to be one of the reasons for the destruction of Mn-113 legs.

The σ_f of the Co-349, Mn (Mo)-113 bulks, and the alumina plates are shown in Figure 9.21. This strength corresponds to the maximum load in the load–displacement curve. At this load, the initial cracking happens. The alumina substrates show the highest σ_f. Anisotropy of σ_f is observed in the Co-349 bulks. The strength in the case of loading direction parallel to the hot-pressing axis is higher than perpendicular to the hot-pressing axis. σ_f of the Mn (Mo)-113 bulk is lower than that of the Co-349 and Ni-113 bulks. This low σ_f is one of the main reasons for the destruction of the Mn (Mo)-113 legs only.

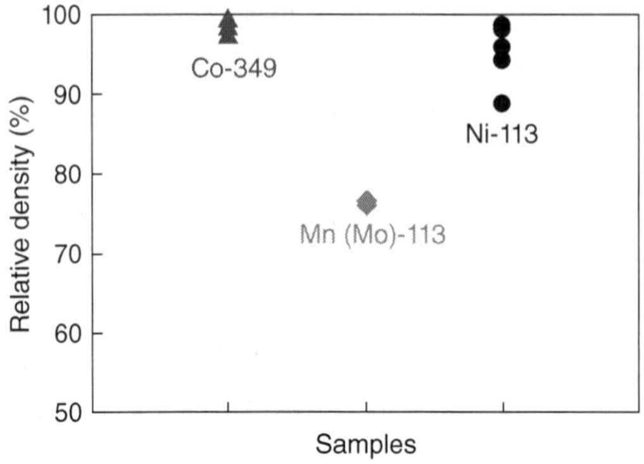

Figure 9.22. Relative density of Co-349, Mn (Mo)-113, and Ni-113 bulks.

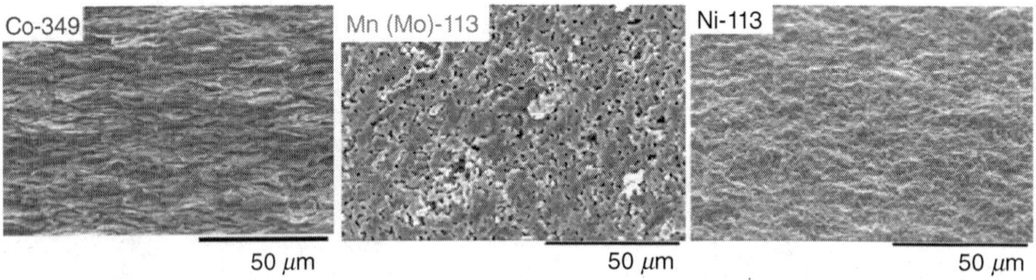

Figure 9.23. Scanning electron microscopic images for Co-349, Mn (Mo)-113, and Ni-113 bulks.

The relative density of the Mn-113 bulks is lower than the other oxide bulks (Fig. 9.22). This low density is a reason for the weak σ_f for the Mn (Mo)-113 bulks. SEM images for the three kinds of oxide bulks are shown in Figure 9.23. Many large pores are observed in the Mn (Mo)-113 bulks clearly. More densification is necessary to enhance σ_f in the Mn (Mo)-113 bulks. In the SEM images, the Co-349 and Ni-113 bulks are seen as "aggregates of small grains," but Mn (Mo)-113 grains are grown much more than the other oxide bulks. Such the microstructure in the Mn (Mo)-113 bulks allows the cracks to run easily. It has not been clear whether cracking happens in the Co-349 and Ni-113 legs in the modules after power generation. Even if it happens, however, the extension of cracks is prevented by the microstructure as seen in Figure 9.23, namely, fracture toughness K_{IC} of these bulks is better than the Mn-113 bulk. σ_f and K_{IC} for the Mn-113 bulk should be improved by densification and microstructure.

Mechanical Properties of $CaMnO_3$/Ag Composites

EXPERIMENTAL PROCEDURES. $Ca_{0.9}Yb_{0.1}MnO_3$ (Mn (Yb)-113) powder was synthesized by a conventional solid-state reaction. The appropriate amounts of $CaCO_3$, Mn_2O_3, and Yb_2O_3 were mixed well and calcined at 1273 K for 15 h and then at 1523 K for 12 h in air with an intermediate grinding. The Mn (Yb)-113/Ag composites were prepared by wet milling various amounts of Ag_2O (0, 5, 10, and 20 wt%) with the Mn (Yb)-113 powder in an agate pot using ethyl alcohol and agate balls for 24 h. After sintering, the composites included 0, 4.7, 9.4, and 18.8 wt% of Ag formed by decomposition of Ag_2O. Mixed slurries were dried and milled for 12 h. Green pellets were prepared under a uniaxial pressure of 40 MPa and then were fired at 1523 K for 2 h in air. The bulk density was calculated based on the weight and dimension of each sample. The powder density was measured using a pycnometer. The crystallographic structure was analyzed by powder XRD at room temperature using Cu-K radiation. The microscopic structure of the composites was observed by SEM observation. The K_{IC} values were evaluated at room temperature by the indentation fracture technique using a Vickers indenter and the relationship was proposed by Niihara [36]. σ_f was measured on the bar-shaped specimens by a three-point bending method at room temperature. The loading speed was 0.1 mm/min and the span length was 10.0 mm.

Mechanical Properties. The Ag content dependence of the relative density for Mn (Yb)-113/Ag composites is shown in Figure 9.24. The powder density increases with increasing Ag content. The relative density of the composites sintered at 1673 K is above 95%. Their brittleness, however, made the composites difficult to form the shapes for the various measurements. Therefore, although the relative density was around 85%, the composites sintered at 1523 K were used for the measurement of mechanical properties.

The XRD patterns of all the composites are shown in Figure 9.25. The starting Ag_2O was completely reduced to metallic Ag after sintering at 1523 K. No secondary phases and solid solutions between Mn (Yb)-113 and Ag were detected. No obvious change in the orthorhombic Mn (Yb)-113 lattice parameters was observed in all the composites. Figure 9.26 shows SEM images of the polished surface for the composites including Ag by 0 (a) and 18.8 wt% (b), respectively. The bright dots in Figure 9.26 correspond to the metallic Ag and are homogeneously dispersed in the $Ca_{0.9}Yb_{0.1}MnO_3$ matrix. The particle size of Ag grew with increasing Ag content. The average particle

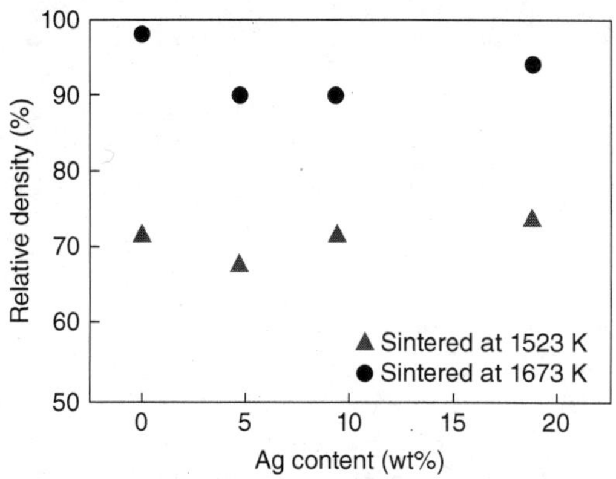

Figure 9.24. Ag content dependence of relative density for $Ca_{0.9}Yb_{0.1}MnO_3$/Ag composites.

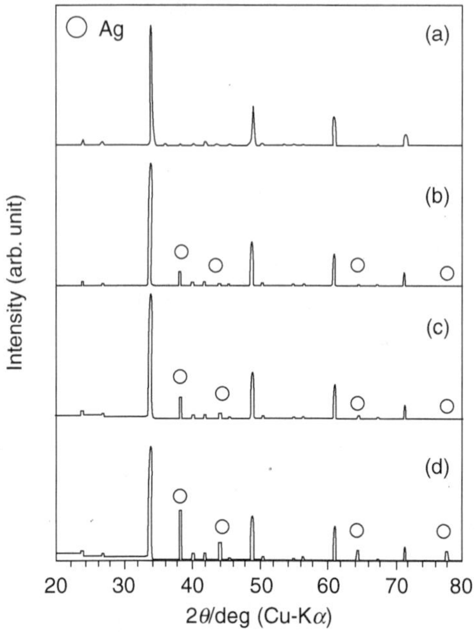

Figure 9.25. X-ray diffraction patterns of monolithic Mn (Yb)-113 (a), 4.7 wt% (b), 9.4 wt% (c), and 18.8 wt% (d) Mn (Yb)-113/Ag composites.

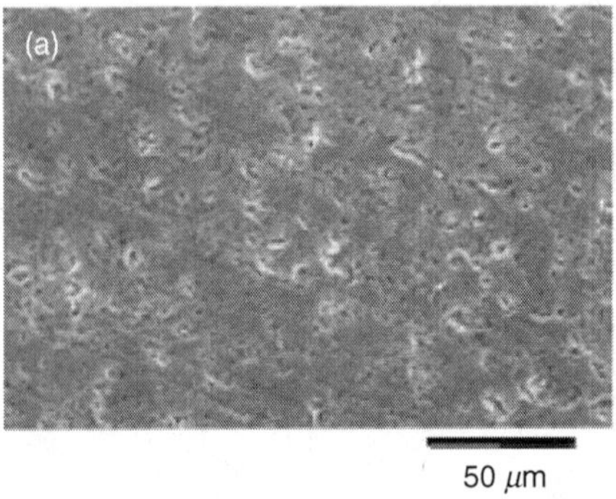

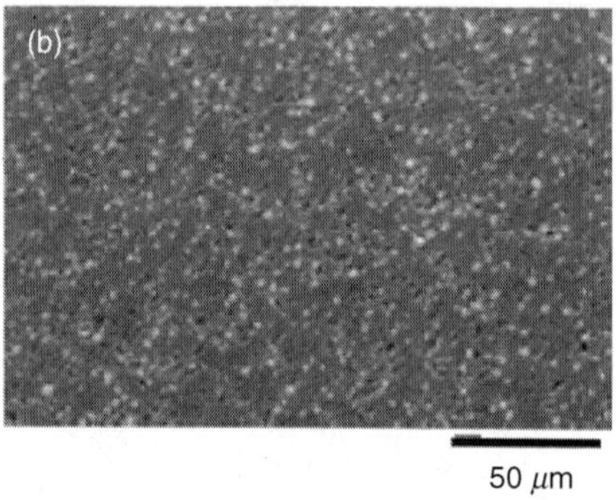

Figure 9.26. Scanning electron microscopic images of monolithic Mn (Yb)-113 (a) and 18.8 wt% Mn (Yb)-113/Ag composite (b).

size of the dispersed Ag was 0.8, 1.6, and 3.9 μm for 4.7, 9.4, 18.8 wt% composites, respectively.

Figure 9.27 shows the Ag content dependence of K_{IC}. The K_{IC} values increase with increasing Ag content and reaches a maximum value of 2.2 MPa m$^{1/2}$ for the 9.4 wt% composite and then slightly decreases at 18.8 wt% of the Ag content.

The improvement mechanism of K_{IC} seems to be the plastic stretching of metallic inclusions by bridging and deflecting the growing crack in ceramic/metal composites [37, 38]. As the crack reached the ceramic/metal interface, the difference in the defor-

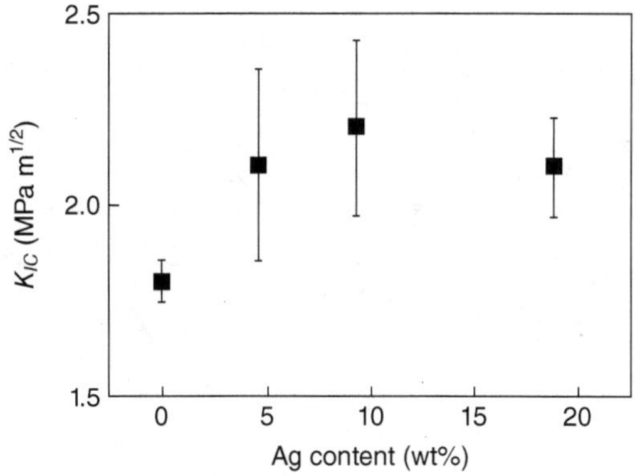

Figure 9.27. Ag content dependence of the fracture toughness KIC for $Ca_{0.9}Yb_{0.1}MnO_3$/Ag composites.

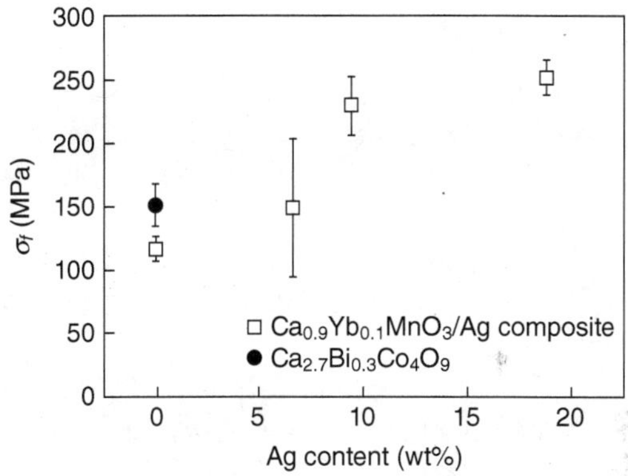

Figure 9.28. Ag content dependence of σf for $Ca_{0.9}Yb_{0.1}MnO_3$/Ag composites.

mation ability between the ductile particles and the brittle matrix causes the crack to be blunted locally. However, if the size of Ag particles becomes larger beyond a critical value, this effect would be small. Thus, the K_{IC} of $Ca_{0.9}Yb_{0.1}MnO_3$/Ag composites had the maximum value due to the Ag particle exceeding the critical size between 9.4 and 18.8 wt% of the Ag content.

Figure 9.28 shows the Ag content dependence of σ_f for the composites. Generally, the σ_f values of brittle materials are related K_{IC} and the length of an initial crack c as indicated by Equation 9.2 [39]:

$$\sigma_f = \frac{1}{Y} \cdot \frac{K_{IC}}{\sqrt{c}}, \tag{9.2}$$

where Y is a dimensionless geometrical parameter. σ_f of 6.6 wt% composite becomes comparable to that of the p-type $Ca_{2.7}Bi_{0.3}Co_4O_9$ bulk and increases by about 25% than that of the monolithic $Ca_{0.9}Yb_{0.1}MnO_3$ bulk. Although the relative density is independent

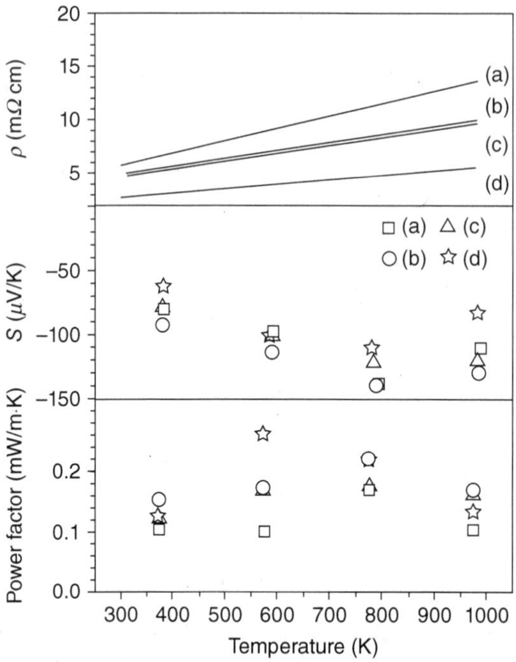

Figure 9.29. Temperature dependence of the electrical resistivity ρ, Seebeck coefficient S, and power factor S^2/ρ for (a) $Ca_{0.9}Yb_{0.1}MnO_3$, (b) 4.7 wt%, (c) 9.4 wt%, and (d) 18.8 wt% $Ca_{0.9}Yb_{0.1}MnO_3$/Ag composites.

of the Ag content, σ_f is increased by the incorporation of Ag. The improvement in σ_f seems to be attributed to the enhancement of K_{IC}.

The power factor of the monolithic $Ca_{0.9}Yb_{0.1}MnO_3$ bulk and $Ca_{0.9}Yb_{0.1}MnO_3$/Ag composites is almost the same because of simultaneous decreases in S and ρ values by the incorporation of Ag (Fig. 9.29).

MODULES

Although fabrication of thermoelectric modules using oxide materials has been reported, their performance is much lower than that expected considering the properties of their starting materials [40, 41]. This is thought to be because the contact resistance at electrodes in which oxide/metal junctions are usually formed is very high, thus severely limiting the magnitude of output power. Moreover, cracking or exfoliation due to the great difference of thermal expansion between oxides and metals is also a serious problem. Given the high temperatures (>773 K) of applications with thermoelectric oxides, conventional materials and methods for constructing electrodes cannot be used [42, 43]. Accordingly, the preparation of electrodes possessing good thermal and electrical properties is considered one of the most important issues in realizing thermoelectric power generation. Focusing on this issue, the construction of thermoelectric unicouples composed of the Co-349 and Mn-113 bulk materials and their properties are discussed first, and then fabrication and power generation of modules are mentioned.

Experimental Procedures

Bulk Materials. Bulk materials of p-type Co-349 was prepared using a hot-pressing technique. The specific compositions are $Ca_{2.7}Bi_{0.3}Co_4O_9$. The Co-349 precursor powder, prepared by solid-state reaction, was hot pressed for 20 h in air under about 10 MPa at 1123 K. Hot-pressed plates were cut to provide a leg element area of 3.7×4.0–$4.5\,mm^2$ and a leg element length of 4.7 mm. The length direction was made perpendicular to the hot-pressing axis for the purpose of inducing a temperature gradient. Bulk materials of n-type Mn-113 were sintered under the atmospheric pressure. In order to dope electrons, a part of Mn was substituted with Mo. The composition is $CaMn_{0.98}Mo_{0.02}O_3$ (Mn (Mo)-113). The Mn (Mo)-113 precursor powder was prepared by solid-state reaction, compacted into pellets and densified using a CIP technique. The pellets were sintered at 1473 K in air and were cut to provide a leg element area of $5.0 \times 5.0\,mm^2$ and a leg element length of 4.5 mm.

Thermoelectric Unicouples. An alumina plate possessing dimensions of 5.0 mm wide, 8.0–14 mm long, and 1.0 mm thick was used as a substrate. In order to achieve electrical conduction, Ag paste was applied to one side of the alumina plate, and an Ag sheet possessing the same width and length with the alumina plate and 50 μm thick was put on the plate. Then, the paste was solidified by heating. The surface of the Ag sheet was polished. Ag paste was mixed with p- and n-type oxide powders to make connections between oxide legs and Ag electrodes possessing high mechanical strength and low contact resistance. The oxide precursor powders were pulverized by ball milling to obtain a grain size smaller than 10 μm then were mixed in varying ratios with the Ag paste. After connecting the legs and substrate, the wet Ag paste was dried at 373 K and solidified by heating at 1123 K for 3 h under 6.4 MPa in air. The Co-349/Mn-113 unicouple is shown in Figure 9.30.

Internal resistance R_I was measured using a standard dc four-probe method for the whole unicouple at room temperature.

Module. A thermoelectric module composed of 108 pairs of p-type Co-349 and n-type Mn (Yb)-113 legs were constructed using Ag electrodes and Ag paste including the p-type powder by 6 wt%. The dimensions of both oxide legs were 7 mm wide, 3.5 mm thick, and 5 mm high. The preparation condition is same with that for the unicouples. There was no substrate on both sides of the module, namely, a skeleton-type module (Fig. 9.31a). Measurement of generated power was carried out using a plate-type electric furnace as shown in Figure 9.31b.

Results and Discussion

Co-349/Mn (Mo)-113 Unicouples. The R_I of Co-349/Mn (Mo)-113 unicouples prepared by Ag paste, including various ratios of Co-349 and Mn (Mo)-113 powders,

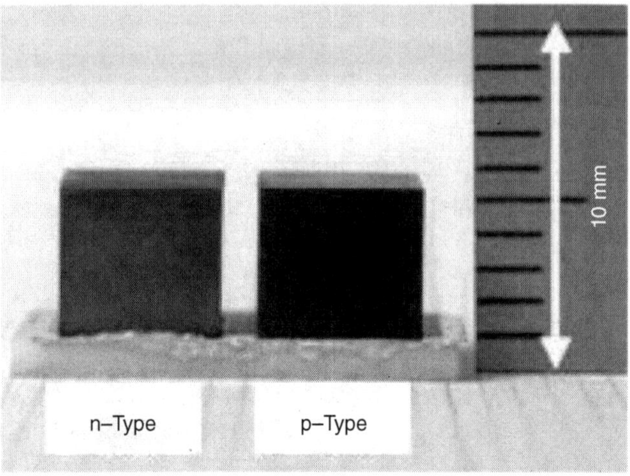

Figure 9.30. Ca-349/Mn (Mo)-113 thermoelectric unicouple.

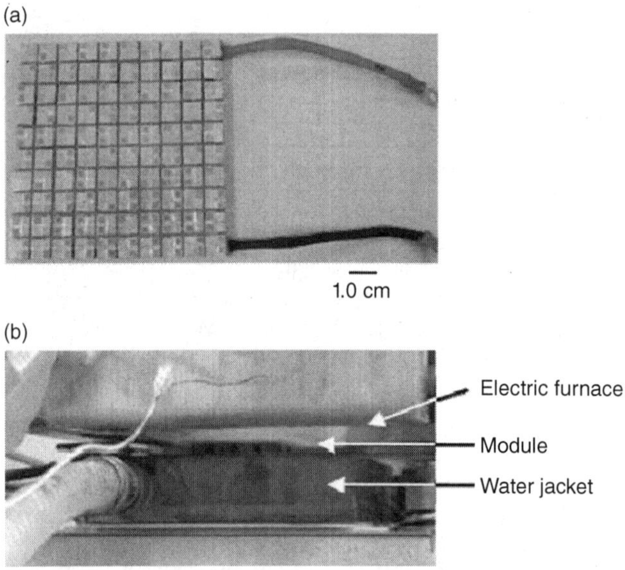

Figure 9.31. Thermoelectric module composed of 108 pairs of $Ca_{2.7}Bi_{0.3}Co_4O_9$ and $Ca_{0.9}Yb_{0.1}Mn_3$ (a) and measurement system for generated power (b).

is shown in Figure 9.32. Incorporation of oxide powders into the Ag paste is effective to reduce R_I in the Co-349/Mn (Mo)-113 unicouples. Although V_O has never been measured yet, it seems to be independent of the ratio of oxide powder included in Ag paste [44]. Because the lowest R_I is observed at 6 wt% of Co-349 powder content, the modules have been prepared using this composite paste.

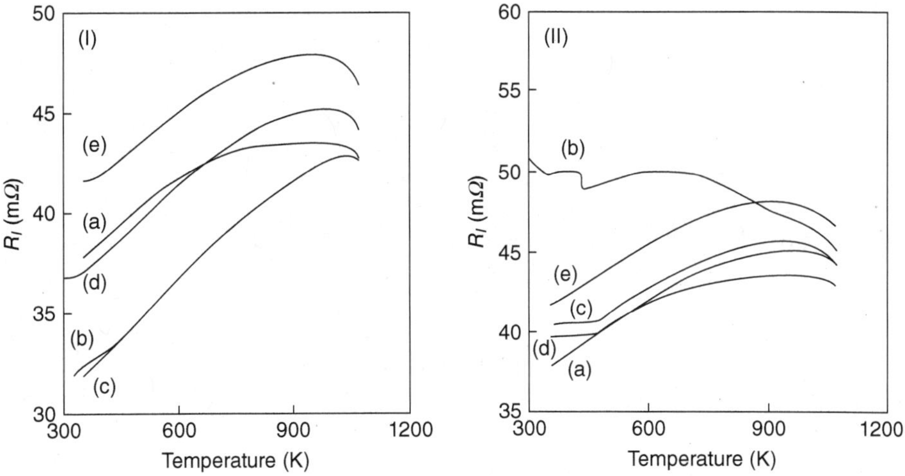

Figure 9.32. Temperature dependence of R_l for Co-349/Mn (Mo)-113 unicouples prepared using Ag paste, including Mn (Mo)-113 (I) and Co-349 power (II). 0 wt% (a), 1.5 wt% (b), 3 wt% (c), 6 wt% (d), and 10 wt% (e) of oxide powders were incorporated.

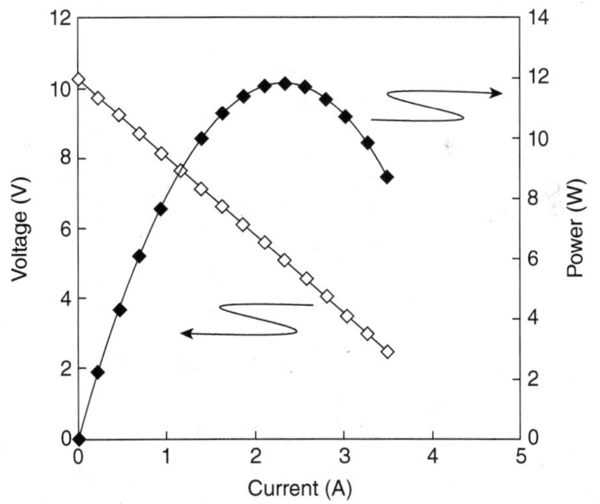

Figure 9.33. Thermoelectric power generation properties for the module consisting with 108 pairs of p-type Co-349 and n-type Mn (Yb)-113 legs.

Figure 9.33 displays generated power using the electric furnace and circulating water. The measurement was carried out in the air. Maximum power reached 12 W at a temperature condition of 1073 and 400 K for the hot-side temperature T_H and the temperature difference between the hot- and cold-side ΔT of 1073 and 400 K, respectively. At present, the power density of the module composed of the Co-349 and Mn (Yb)-113 legs reaches higher than 1.0 kW/m^2.

CONCLUSION

Although the history of research on thermoelectric oxide materials is never long, it has reached the stage for development on application. In order to achieve thermoelectric power generation, applications in which the oxide materials can show their strong points (high temperature, environment, cost, and so on) should be found out. And the durability of thermoelectric systems in the application conditions should be evaluated and improved. Moreover, technologies for heat transfer from heat sources to modules, for cooling the modules, and for power conditioning (DC–AC convertors, etc.) should be developed.

In a material point of view, the thermoelectric conversion efficiency of the oxide materials discussed in this review is insufficient for wide application. Discovery of new materials possessing higher ZT values than the present ones (hopefully, $ZT > 2$) even at low temperature (~373 K) is indispensable. For this purpose, the creation of new thermoelectric materials should be accomplished by new appropriate concepts (phonon glass–electron crystal and quantum effect) and technologies (controlling nanostructure and high-throughput screening technique, etc.).

REFERENCES

1. Terasaki I, Sasago Y, Uchinokura K. 1997. Phys Rev B 56:R12685.
2. Fujita K, Mochida T, Nakamura K. 2001. Jpn J Appl Phys 40:4644.
3. Li S, Funahashi R, Matsubara I, Ueno K, Yamada H. 1999. J Mater Chem 9:1659.
4. Funahashi R, Matsubara I, Ikuta H, Takeuchi T, Mizutani U, Sodeoka S. 2000. Jpn J Appl Phys 39:L1127.
5. Funahashi R, Matsubara I. 2001. Appl Phys Lett 79:362.
6. Yasukawa M, Murayama N. 1997. J Mater Sci Lett 16:1731.
7. Ohtaki M, Koga H, Tokunaga T, Eguchi K, Arai H. 1995. J Solid State Chem 120:105.
8. Koumoto K, Seo WS, Ozawa S. 1997. Appl Phys Lett 71:1475.
9. Funahashi R, Matsubara I, Shikano M. 2001. Chem Mater 13:4473.
10. Shikano M, Funahashi R. 2003. Appl Phys Lett 82:1851.
11. Hatta I, Sasuga Y, Kato R, Maesono A. 1985. Rev Sci Inst 56:1643.
12. Yamane T, Mori Y, Katayama S, Todoki M. 1997. J Appl Phys 82:1153.
13. Masset AC, Michel C, Maignan A, Hervieu M, Toulemonde O, Studer F, Raveau B, Hejtmanek J. 2000. Phys Rev B 62:166.
14. Miyazaki Y, Kudo K, Akoshima M, Ono Y, Koike Y, Kajitani T. 2000. Jpn J Appl Phys 39:L531.
15. Leligny H, Grebille D, Perez O, Masset AC, Hervieu M, Raveau B. 2000. Acta Crystallogr B 56:173.
16. Funahashi R, Matsubara I, Ikuta H, Takeuchi T, Mizurani U. 2001. Mater Trans 42:956.
17. Chen G, Dresselhaus MS, Dresselhaus G, Fleurial J-P, Caillat T. 2003. Int Mat Rev 48:45.
18. Ohta S, Nomura T, Ohta H, Koumoto K. 2005. J Appl Phys 97:034106.
19. Ohtaki M, Tsubota T, Eguchi K, Arai H. 1996. J Appl Phys 79:1816.

20. Xu G, Funahashi R, Matsubara I, Shikano M, Zhou Y. 2002. J Mater Res 17:1092.
21. Rodriguez-Carvajal J. 1993. Physica B 195:55.
22. Cong BC, Tsuji T, Thao PX, Thanh PQ, Yamamura Y. 2004. Physica B 352:18.
23. Funahashi R, Urata S, Sano T, Kitawaki M. 2003. J Mater Res 18:1646.
24. Urata S, Funahashi R, Mihara T, Kosuga A, Sodeoka S, Tanaka T. 2007. Int J Appl Ceram Technol 4:535.
25. Funahashi R, Shikano M. 2002. Appl Phys Lett 81:1459.
26. Li S, Funahashi R, Matsubara I, Ueno K, Sodeoka S, Yamada H. 2000. Chem Mater 12:2424.
27. Xu G, Funahashi R, Shikano M, Matsubara I, Zhou Y. 2002. Appl Phys Lett 80:3760.
28. Funahashi R, Matsubara I, Konishi M, Dimesso L, Umeda M, Ogura T, Yamashita H, Kosaka S, Ono N. 1994. Physica C 235–240:3439.
29. Funahashi R. unpublished data.
30. Bergman DJ, Fel LG. 1999. J Appl Phys 85:8205.
31. Flahaut D, Mihara T, Funahashi F, Nabeshima N, Lee K, Ohta H, Koumoto K. 2006. J Appl Phys 100:084911.
32. Fan C, Rahaman MN. 1992. J Am Ceram Soc 75:2056.
33. Nakada Y, Kimura T. 1997. J Am Ceram Soc 80:401.
34. Sekino T, Nakajima T, Ueda S, Niihara K. 1997. J Am Ceram Soc 80:1139.
35. Nawa M, Yamazaki K, Sekino T, Niihara K. 1996. J Mater Sci 31:2849.
36. Niihara K. 1983. J Mater Sci Lett 2:221.
37. Ashby AF, Blunt FJ, Bannister M. 1989. Acta Metall 37:1847.
38. Flinn B, Ruehle M, Evans AG. 1989. Acta Metall 37:3001.
39. Tuan WH, Wu HH, Yang TJ. 1995. J Mater Sci 30:855.
40. Matsubara I, Funahashi R, Takeuchi T, Sodeoka S, Shimizu T, Ueno K. 2001. Appl Phys Lett 78:3627.
41. Shin W, Murayama N, Ikeda K, Sano S. 2001. J Power Sources 103:80.
42. Matsubara K. 2002. *Proc. 21st International Conference on Thermoelectrics*, pp. 418.
43. Hori Y, Kusano D. 2004. *Proc. of 22nd International Conference on Thermoelectrics*, pp. 602.
44. Funahashi R, Urata S, Mizuno K, Kouuchi T, Mikami M. 2004. Appl Phys Lett 85:1036.

10

INTEGRATION TECHNOLOGIES FOR SOLID OXIDE FUEL CELLS (SOFCS) AND OTHER ELECTROCHEMICAL REACTORS

Yoshinobu Fujishiro,[1] Toshio Suzuki,[1] Toshiro Yamaguchi,[1] Koichi Hamamoto,[1] Masanobu Awano,[1] and Nigel Sammes[2]

[1]*National Institute of Advanced Industrial Science and Technology (AIST), Nagoya, Japan*
[2]*Colorado School of Mines, Golden, Colorado*

INTRODUCTION

In this chapter, we review electrochemical reactors, such as solid oxide fuel cells (SOFCs), from the viewpoint of three-dimensional (3-D) integration and nanostructural control for energy and environmental applications.

Electrochemical ceramic reactors, such as SOFCs, which consist of durable ceramic components (electrolyte and electrodes), with their high efficiency, have a wide range of applications for energy and environmental issues. Electrochemical modules are composed of functional ceramic parts with various features and are occasionally joined even with metallic parts. The manner in which they are assembled affects their electrochemical modulus properties as much as the properties of the unit parts. Hyperstructural control, predominately nano- and microstructural control of the ceramic parts, plays an important role in determining the types of ceramic electrochemical modules for various applications.

3-D functional modules have been used in electrochemical power modules, such as SOFCs, and electrochemical reactors for materials substitution, such as deNOx reactors,

---

Ceramic Integration and Joining Technologies: From Macro to Nanoscale, First Edition.
Edited by Mrityunjay Singh, Tatsuki Ohji, Rajiv Asthana, Sanjay Mathur.

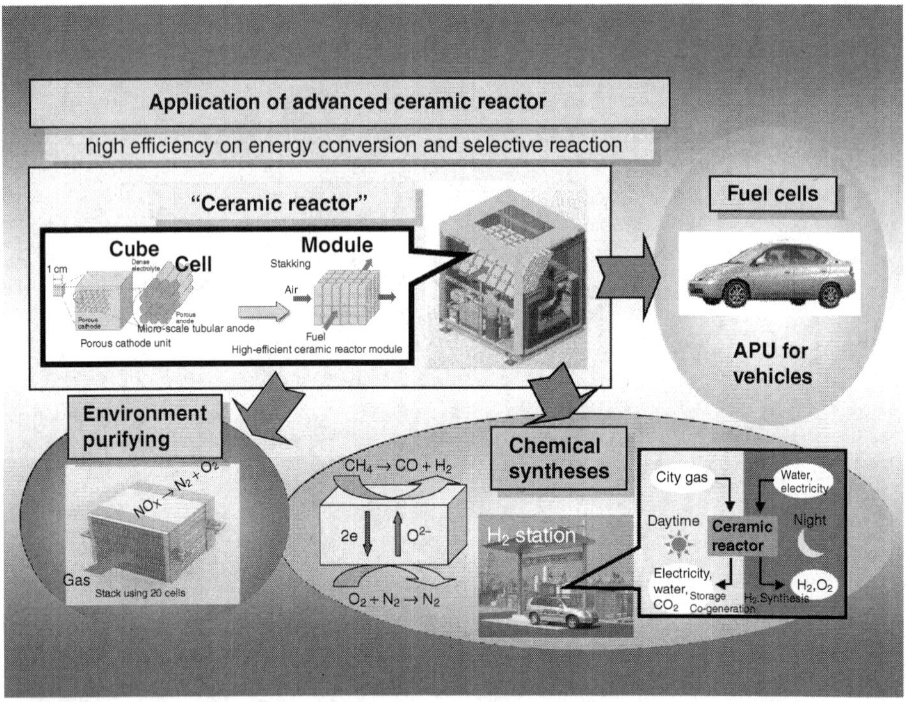

Figure 10.1. Schematic drawing of a three-dimensional assembly of high-performance electrochemical reactors for various applications. APU, auxiliary power unit.

particulate matter (PM) purifiers, and syngas generators. The principal advantage of the electrochemical reactors is the high efficiency of the reactions because of the direct reaction conversion of the ions on the electrodes (Fig. 10.1).

BASIS OF ELECTROCHEMICAL REACTORS

Electrochemical Reactivity

Electrochemical reactivity is a phenomenon that transports electrons through electrodes with a chemical reaction. For example, SOFCs are considered one type of electrochemical energy converter. Since essential electrochemical reactions proceed on the nanometer scale in the electrodes, nanoparticles play an important role in the enhancement of the electrochemical reaction of these electrochemical cells [1, 2]. Another well-known phenomenon concerning nanoscale reactions occurs at the "space charge" layer. The concept is called "nanoionics," newly proposed by J. Maier [3]. It explains the remarkable increase in ionic conductivity by polarization, which is caused by a nanoscale alternation of thin films of different types of ionic conductors such as fluorides and oxides. Nanoscale electrochemical reactions are enhanced by networking of the nanoparticles. Furthermore, there are many possibilities to design and fabricate interfaces of the matrix and nanoparticles for hyperionic conductivity. Electrochemical properties can be described by consideration of the contribution of nanoparticles, especially from the standpoint of chemical reactions with material transportation and energy conversion.

Electrochemical Reactions Influenced by Nanostructural Control

SOFCs are the best-known examples of devices using an electrochemical reaction. The reaction occurs on the electrodes with the electrolyte. Electrochemical reactions are defined as reactions that transport electrons at both electrodes in an electrochemical cell. In the case of SOFCs, the concentration difference of oxygen between the fuel (hydrogen, hydrocarbon, etc.) and air or oxygen promotes the diffusion of oxygen ions through the electrolyte (oxygen ionic conductor) at an elevated temperature, which allows sufficient mobility of the ions.

The reaction in SOFCs is the conversion of chemical to electrical energy. The reverse reaction is the electrodecomposition of water. The most important factors influencing the electrochemical reaction are the mobility of an ion and microstructural control for the enhancement of chemical or electrochemical reactions. Figure 10.2 shows a schematic drawing of an SOFC cathode whereby the gas molecules passing

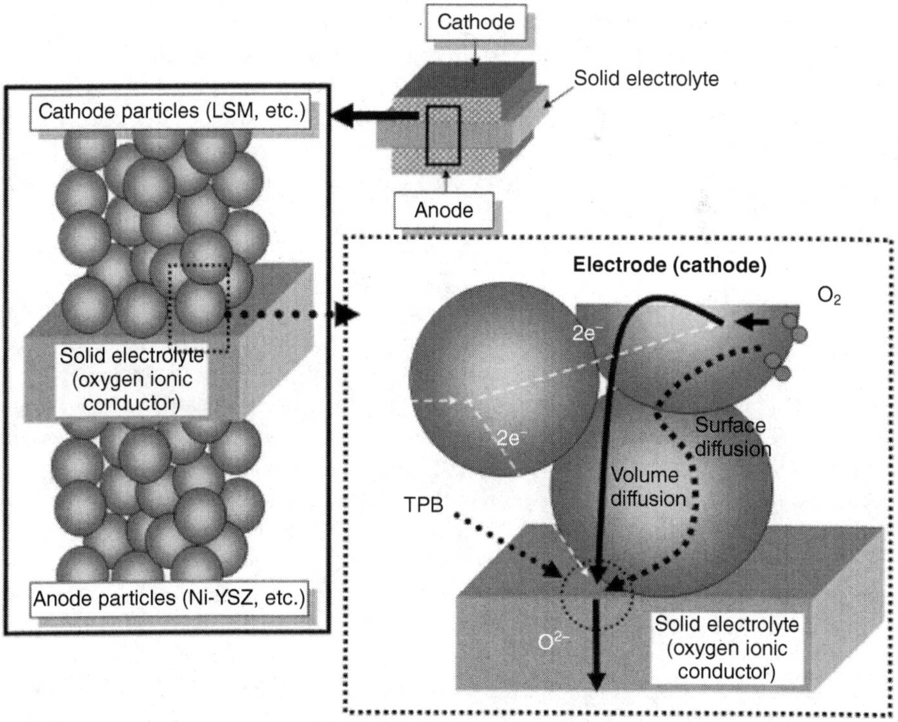

Figure 10.2. Schematics of the electrochemical reaction that converts gas molecules to ions at the three-phase boundary in the cathode with nanoparticles and nanospace.

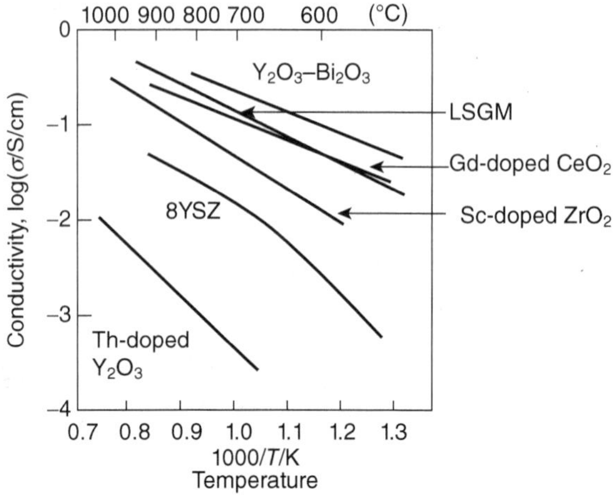

Figure 10.3. Oxygen ionic conductivity of various materials as a function of cell working temperatures.

through the surfaces of the solid particles are ionized by the electrochemical reaction at the three-phase boundary (TPB) site, which is usually distributed at the interface of nanoparticle and surrounding space (gas phase). Another typical example of electrochemical reactions occurring at the interfaces is the photocatalytic reaction, which causes the effective separation of electrons and holes [4].

Controlling the electrical and ionic conductivities and the microstructure of electrodes, to improve the electrochemical reactivity is the main target of the research and development in SOFC applications. Figure 10.3 shows the ionic conductivities of various electrolyte materials as a function of cell working temperature. Yttria-stabilized zirconia (YSZ) has been the most popular electrolyte material because of its relatively high ionic conductivity with good mechanical integrity and phase stability at high working temperatures. Recently, scandium-doped zirconia (scandium-stabilized zirconia [ScSZ]), lanthanum gallate (lanthanum strontium gallium magnesium oxide [LSGM]), and ceria-based doped oxides are expected to be new electrolyte candidates for low-temperature applications. Electrochemical reactions proceed at the interfaces between the electrodes and electrolyte. The electrolyte transports the reaction media (ions). That is why thinner and even electrolyte-supported films are preferred for high-performance cells to produce more electrons. On the other hand, a thick electrode with fine particles is required for having high reaction volume, good ionic connectivity, and electrical conducting grains. Therefore, controlling the structure of various components in the electrochemical cells in the range of nano-, micro-, and macroscale is being studied via processing technology development such as the fabrication of nanocomposite powders.

The concept of "percolation" in the electrodes is also important in improving the performance of electrochemical reactions. It is necessary to maintain a network structure for both ionic and electrical conducting phases such as that required in SOFCs. This network structure includes pore size distribution from the surface of the electrode to the permeable area where the gas molecules are converted to ions.

Control of the Electrochemical Reaction and Its Application to the Development of SOFCs

Reaction at the TPB in an SOFC is the most significant example of applying nanoparticles in the field of electrochemistry. The reaction at the TPB to ionize gas molecules, and its reverse reaction, is affected not only by nanoparticles themselves but also by the nanostructures they form and their surroundings. In this chapter, the nanoscale reaction zone in an SOFC is described in detail from the viewpoint of electrochemistry.

Since fuel cells have high efficiencies for power generation, they can drastically reduce the emission of CO_2, which is considered to be responsible for global warming. Various types of fuel cells, such as polymer electrolyte fuel cells (PEFCs), molten carbonate fuel cells (MCFCs), phosphoric acid fuel cells (PAFCs), and SOFCs, have been developed. Among them, PEFCs have gained the most attention for home and vehicle applications. On the other hand, SOFCs have great potential because they have higher power generation efficiency and utilize ceramic technologies.

Recent trends in SOFC research and development focus on the improvement of reactivity of gas molecules and ions at electrodes and on the enhancement of ionic conductivity of the electrolyte. In the former case, nanoparticles and their surrounding nanomicron-scale area play an important role in their electrochemical reactivity. As mentioned previously, the enhancement of the electrochemical reaction at the TPB point is most important in the development of the SOFC. Recently, the introduction of nanotechnology, with an attempt to apply nanoparticles to construct a nanostructure, has become the focus of SOFC research and development.

Because of the restriction of the electrochemical reaction, electrodes should have good ionic and electrical conductivities to expedite the reaction. Therefore, the electrode materials should exhibit a high catalytic activity for the desired electrochemical reactions. Nanoparticles are expected to have high reactivity as catalysts. Overpotential in the fuel cell reflects the loss of reactivity due to various causes when converting gas molecules to ions. The voltage drops in each cell component as a function of cell operating temperature. Suppressing the overpotential in the SOFC can significantly enhance its electrochemical reactivity. The overpotential is induced by the loss in reactivity between gas molecules and ions at the electrodes, which is primarily due to the polarization losses at the cathode. This is why nanoparticles are used to make electrodes with better performance controlling factors, such as gas diffusion, distribution of nanoreaction zone (TPB), and ionic or electrical conductivity, to reduce the overpotential.

Figure 10.4 shows a schematic reaction of an SOFC cell, with a cathode and an anode in the cell for oxidation and reduction reactions, respectively. The reaction of a gas molecule to an ion is, in principle, reversible. At the cathode, oxygen molecules are adsorbed on the surface of an electrode particle and migrate or transport via a surface diffusion or volume diffusion mechanism, respectively. Oxygen molecules are converted into oxygen ions, with the supplied electrons, after arriving at the TPB, and are transported under a concentration gradation of oxygen between the cathode and the anode through the electrolyte. In the anode, the oxygen ion is converted back to an oxygen molecule, water, or carbon dioxide, depending on the kind of fuel supplied to

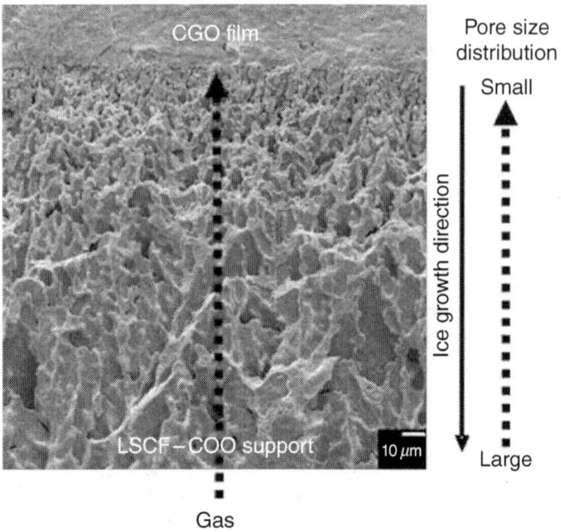

Figure 10.4. An example of controlling the internal structure of SOFC in the range of nano-, micro-, and macroscale, showing uniaxial pore shape and pore size distribution vertical to the cell surface. LSCF, lanthanum strontium calcium iron oxide; CGO, cerium gadolinium oxide.

the anode. The released electrons cause output currents of the SOFC by connecting both electrodes via an applied load. The differences in the ionization energy of the electrodes derived from their composition and microstructure affect the reactivity between gas molecules and ions [5–8].

In the state-of-the-art anode, Ni–YSZ cermet, fine particles are used to produce a large number of active TPB sites. The electrochemical active zone consists of a percolated matrix having ionic and electrical conductors with the following advantages: (1) a high porous surface area with excellent electrochemical conductivity, (2) high electronic conductivity, (3) high stability through prolonged electrochemical reaction time, (4) a good match with the electrolyte with respect to thermal expansion properties, (5) stable catalytic properties for hydrogen–proton conversion reactions, and (6) good sustainability as a material for its supply source and cost-effectiveness for mass production.

Novel applicable fabrication technology for the electrodes has been developed, including screen printing, plasma coating, and electrostatic deposition methods, to layer the electrode materials on the electrolyte [9]. After the reduction of coated Ni–YSZ film, metallic nickel particles are precipitated to form a skeleton structure in the porous composite anode with YSZ particles.

Structural Control of the Electrode-Supported Thin-Film Electrolyte

An important factor for improving the SOFC properties (especially for the power density) is the control of the nanomicron structure in the macroelectrode construction, that is, the hyperstructural control of the electrode to support the thin electrolyte. To reduce the polarization losses at the electrolyte, electrode-supported thin-film electrolytes are believed to be the most effective. The entire electrode should be porous for good gas permeability to maintain high reactivity while dense enough at the coated thin-film electrolyte interface to have high ionic conductivity. It is difficult to accomplish all these requirements with ordinary layering procedures. "Colloidal processing," a novel processing technology, was developed to fabricate unidirectional pore shapes (cylindrical) and size distributions through the electrode with nano- to micron-scale structure at both of its surfaces. This simultaneously allowed for high permeability and high reactivity of the electrode and resulted in an increase of the ionic conductivity of the fuel cell [10].

Ceria-based composite electrodes are a typical example of using nanoparticle-derived microstructures to enhance the properties of SOFCs. The ceria composite is well-known for its high catalytic performance in exhaust gas purification. Ceria has a high oxygen ionic conductivity at relatively lower cell working temperatures because of its nanoscale crystal structure. There have been many attempts to apply ceria compounds to make cells using a chemical processing method such as spin coating. It has been reported that SOFCs could be successully fabricated to perform at working temperatures as low as 500°C [11].

Recent approaches to support the improvement of SOFCs include the use of "real-time," "in situ," and "nanoscale" analyses, such as the electrochemical scanning tunneling microscopy (STM), for the electrochemical cells [12]. In addition, galvanometric measurement in the nanoscale range will further be developed because of its advantage of providing the essential analysis for the electrochemical phenomenon.

As described earlier, 3-D nanostructural control plays an important role in electrochemical reaction. Further applications of the nanoparticles are expected to provide solutions to various energy and environmental problems.

SOFC AND RELATED RESEARCH AND DEVELOPMENT

Ceramic, high-temperature SOFCs are based on the ability of oxide ions to be conducted through a solid at elevated temperatures. Oxide ion conductivity was observed in ZrO_2-Y_2O_3 by Nernst at the end of the eighteenth century. Since the 1960s, many oxide systems have been examined as oxide ion conductors. In the 1980s and 1990s, SOFCs gained attention as a potentially economical, clean, and efficient means of producing electricity in a variety of commercial and industrial applications. After 2000, following the commercialization of the PEFC as a cogeneration system, application of SOFCs has reached the initial field testing stage. For example, ceramic industries such as Kyocera and NGK Insulators have released kilowatts-class SOFC modules for cogeneration systems available for domestic use.

The electrolyte for an SOFC has to meet the following requirements: high ionic conductivity, low electronic conductivity, and chemical and physical stability of preparation in the form of a dense film. To obtain a high-performance SOFC, the

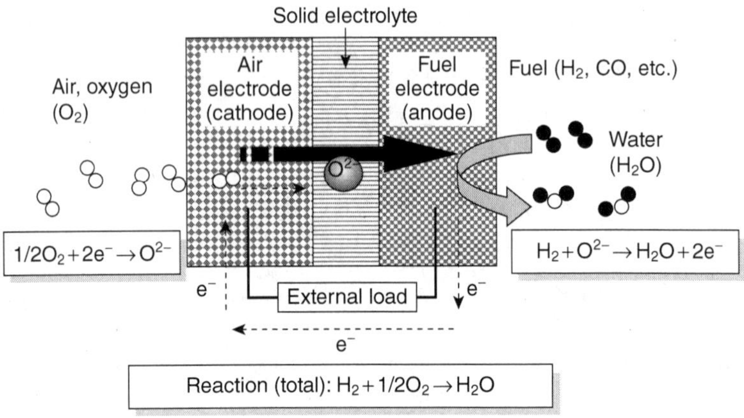

Figure 10.5. Schematic of the principle of solid oxide fuel cells.

contribution of the electrolyte resistance should be suppressed to less than the $0.1\,\Omega\text{cm}^2$ level at the operating temperature, thus resulting in a thin film with the thickness of a typical electrolyte, YSZ, at around hundreds of micrometers to realize high conductivity at the operation temperature. Thin sheets of YSZ electrolytes have been fabricated by advanced ceramic technologies such as tape casting (the "doctor blade" method).

The operating principle of an SOFC is schematically shown in Figure 10.5. When an external load is applied to the cell, oxygen is reduced at the porous air electrode to produce oxide ions. These ions migrate through the solid electrolyte, such as YSZ, to the fuel electrode, and oxide ions react with the fuel, typically H_2 or CO originated from hydrocarbon fuels, to produce H_2O or CO_2. SOFCs of two different configurations are presently under investigation, namely, planar and tubular. Most progress to date has been achieved with the Westinghouse tubular cell, started in the 1970s. In this design, the active cell components fabricated are thin-layered electrodes on a porous Ca–SZ support tube with gas-tight thin layers of electrolyte and interconnect. KW-class SOFC modules containing over 100 cells have been created and delivered for field testing. The system operated continuously over thousands of hours without significant degradation [13].

On the other hand, using these YSZ sheets, planar-type SOFC cells and modules have been developed. Electrodes are coated on both sides of the electrolyte followed by an assembly process, which stacks the individual sheets together. By using glass ceramics, gas-tight sealing of the cells with the interconnect, which is composed of doped $LaCrO_3$ plates or metal alloy plates, affects the power generation property of a module significantly. In the planar design, high-temperature SOFCs can be developed with extremely high current densities, as high as $1\,A/cm^2$, for operation at elevated temperatures over 800–1000°C.

To obtain high-performance SOFCs for output power, stability, and reliability, many technical and materials problems remain to be solved before their real application and commercialization.

MICRO-SOFC DEVELOPMENT

Research Background

Microtubular SOFCs, as described by Kendall and others, are fabricated using YSZ, extruded and fired to form 2-mm-diameter, 200-μm-wall thickness electrolyte tubes that can be painted with nickel/YSZ cermet anodes and strontium-doped lanthanum manganite cathodes, giving excellent thermal shock resistance. J. Mizusaki at Tohoku University and colleagues demonstrated the advantage of a microtubular cell for its thermal shock resistance on quick heating-up and cooling-down tests, and revealed its applicability for use in cars. Another important objective is to apply the tubular bundle for combined heat and power (CHP) generation.

Polymer electrolyte membranes (PEMs) have been extensively studied for portable telecommunications, electronics, medical, and transportation applications. The use of fuel cells in vehicle applications has accelerated in recent years, particularly PEM fuel cell systems, such as those produced by Ballard Power Systems, for application at lower temperatures than that of the SOFC.

Since fuel cells have high efficient power generation performance, they can drastically reduce the generation of CO_2. Various types of fuel cells, such as PEFCs, MCFCs, PAFCs, and SOFCs, have been developed. So far, among these, PEFC has gained attention as a fuel cell for home and vehicle use. SOFCs, on the other hand, have the highest efficiency for power generation (Fig. 10.6). The operating temperatures of conventional SOFCs are high, that is, 800–900°C, and their applications have been limited to large-scale power generation facilities. Thus, to use SOFCs as distributed electric sources for homes, portable electronic devices, and auxiliary power sources for vehicles, the development of SOFCs that could be operated at temperatures of 500–600°C was necessary. To decrease the operating temperature, ceramic electrolyte materials with high ionic conductivity such as doped ceria and lanthanum gallate have already been utilized, and in practice, the operating temperature has been reduced to 500–600°C. However, despite the reduction in operating temperature, thermal distortion remains an obstacle for

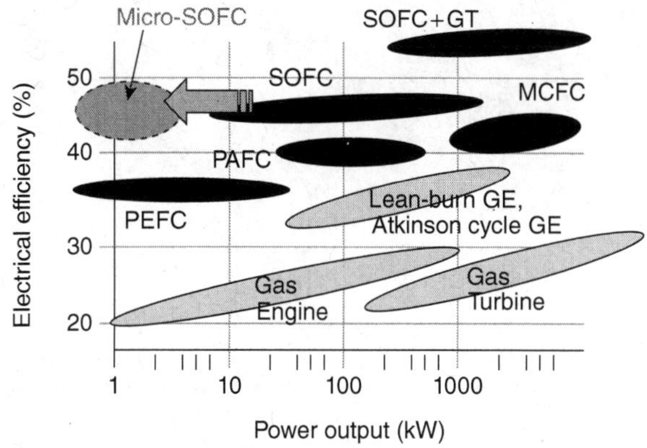

Figure 10.6. Various kinds of fuel cells.

devices that need frequent startup and shutdown operations. Thus, a microtubular cell has been developed, with the goal of reducing the thermal distortion of the cell [14].

Fabrication of Microtubular Cells

A highly efficient microtubular SOFC of millimeters to submillimeters in diameter, which can solve the thermal distortion problem by such miniaturization of the SOFC, has been fabricated. The ceria-based materials that enable the reduction in operating temperature down to 500–600°C are mechanically fragile, and thus their microfabrication has so far been considered to be difficult. However, using an advanced microtube processing technique that enables the control of microstructures and dense film-coating techniques, we have succeeded in the fabrication of microtubular SOFCs. In addition, we have succeeded in optimizing electrode microstructures to greatly enhance fuel reaction efficiency. In cell fabrication, nickel–ceria-based materials and lanthanum cobalt–ceria-based materials are used as the fuel and air electrodes, respectively.

The micro-SOFC that has been fabricated has a tubular structure of approximately 1 cm in length and 0.8–1.6 mm in diameter. When hydrogen gas is introduced into a microtube of 1.6 mm in diameter at 450–570°C, power densities of 0.17–1.0 W/cm^2 have been obtained (Fig. 10.7). This value is one of the highest levels attained so far for SOFCs with ceria-based electrolytes.

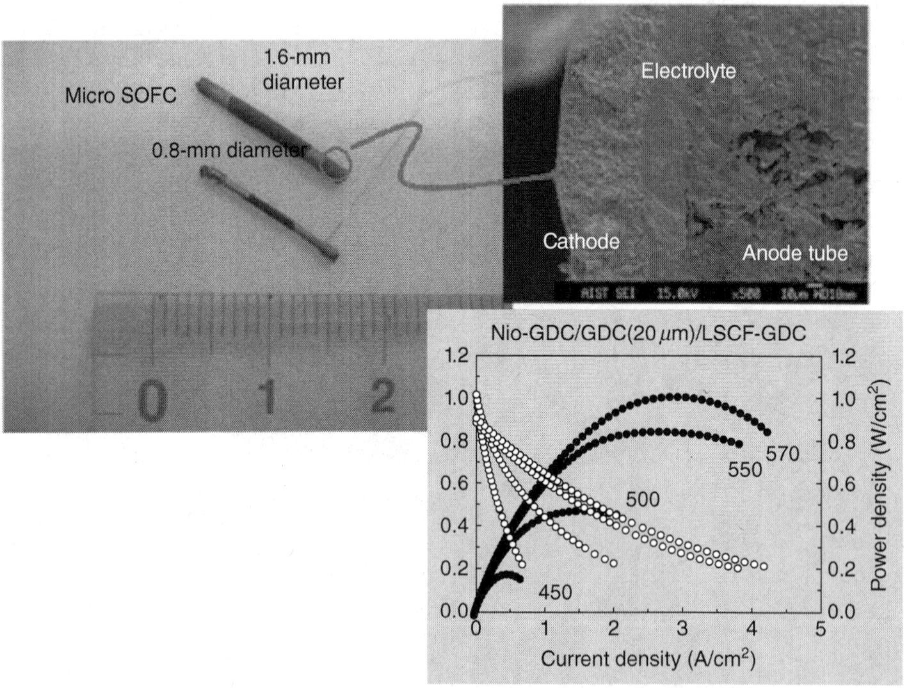

Figure 10.7. Fabrication and high performance of microtubular SOFC using ceria electrolyte.

For the micro-SOFC of 0.8 mm in diameter, a hundred micro-SOFCs can be integrated in each 1 cm^3, and thus power densities of 7 W/cm^3 at 500°C and 15 W/cm^3 at 550°C can, in theory, be expected. This demonstrates that our technique can open the gateway for applications for distributed power sources for homes and portable electronic devices and auxiliary power sources for vehicles [15–18].

Development of a Small, High-Performance Micro-Fuel Cell Bundle

The SOFC bundle uses lanthanum cobalt as the air electrode material that forms part of the SOFC bundle. This makes it possible to establish the technology for producing a small cubic SOFC bundle with a power output of 2 W or more per cubic centimeter at below 600°C. This technology is suitable for mass production because it relies on basic ceramic-forming technology.

The actual micro-SOFC cube bundle is an integrated structure that has a volume of 1 $cm^3$and tubular cells with a diameter of 0.8–2.0 mm. Toho Gas examined the performance of the SOFC bundles by flowing hydrogen into microtubular SOFCs of 2 mm in diameter, operating at 550°C. The tests confirmed that the SOFC cube bundle generated more than 2 W of power (Fig. 10.8). As shown in the figure, the test data shows that a volume of 1 cm^3 can generate more than 2 W of power for a current of 4.5 A at the operating temperature of 550°C.

The tests confirmed that the SOFC bundle produces the highest volumetric power density of any known fuel cell, and does so at a current of 4.5 A and at an operating temperature below 600°C. At the same time, the bundle is the smallest fully operational microtube SOFC cube with passages for fuel and air thus far obtained.

The newly developed micro-SOFC opens the prospect of stacking micro-SOFCs to produce stacks and modules ranging in size from small mobile electric power units of several tens of watts with volumes of several tens of cubic centimeters to auxiliary power units for vehicles or household power units with outputs of several kilowatts and volumes of several thousand cubic centimeters. SOFCs are expected to be applied as distributed power units for household use, as power units for mobile electronics devices and as auxiliary power sources for vehicles [19].

Figure 10.8. High-performance microtubular cell accumulated cube bundle shows output power over 2 W/cm^3.

Development of the Low-Temperature SOFCs and a Compact Module Fabrication

Understanding the correlation between the microstructure of the anode electrode of an SOFC and its electrochemical performance in the tubular design is the key to improving the properties of SOFCs. It was shown that the electrochemical performance of the cell improved when the size of the constituent particles was reduced so as to yield a highly porous microstructure. The SOFC had a power density of greater than $1\,W/cm^2$ at an operating temperature as low as 600°C with a conventional zirconia-based electrolyte, a nickel–cermet anode, and a (La, Sr)(Co, Fe)O_3 cathode material. The effect of the hydrogen fuel flow rate (linear velocity) was also examined for the optimization of operating conditions. Higher linear fuel velocity led to better cell performance for the cell with higher anode porosity. A zirconia-based cell could be used for a low-temperature SOFC system under 600°C just by optimizing the microstructure of the anode electrode and the operating conditions [20, 21] (Fig. 10.9).

Furthermore, realization of high-power small-sized SOFC systems is ongoing by means of the control of the microstructure of the electrodes. The targets of the tubular microcell bundles were systems of several hundred watts to 1 kW; additional technical issues had to be solved in order to realize micro-SOFC modules for small-sized fuel cell systems of several watts to tens of watts. Among these technical issues, optimization of the cell position, air passage, and thermal balance, which are trade-offs, was the most important issue. A highly efficient micro-SOFC module using high-performance tubular SOFCs of submillimeter diameter can function as a ceramic reactor. By optimizing the inner structure and the alignment of micro-SOFC bundles of this module, the power needed for the air supply is minimized, and the thermal balance is easily controlled. The module works with high efficiency even if the supply of air is via natural convection. Fuel is fed to the stack through a fuel feed manifold; the air flows along the module surface and is fed into the stack by natural convection. The microtubular cell adopted in this study is a tube of 0.8-mm diameter that uses ceria-based ceramic

Figure 10.9. Microtubular SOFC shows the highest output power at a low-temperature (600°C) cell working and using zirconia electrolyte.

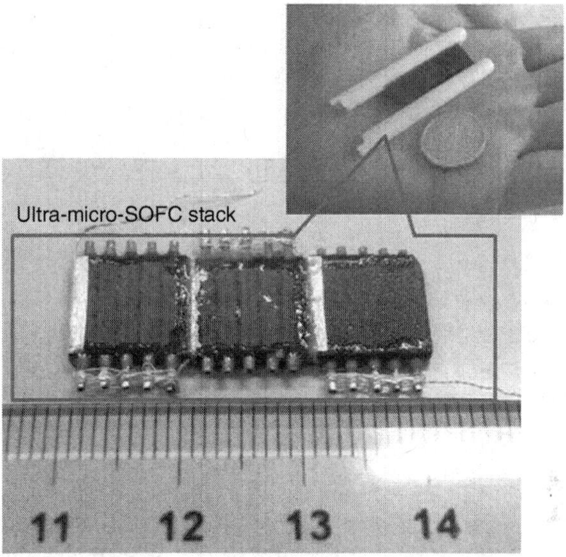

Figure 10.10. An ultra-micro-SOFC stack working without air blow for oxygen supply to the cells.

electrolytes, nickel ceria-based ceramics for the anode, and lanthanum cobalt ceria-based ceramics for the cathode. To form a microbundle, the microtubular cells are integrated in the porous structure made of lanthanum cobalt ceramic. Figure 10.10 shows a stack (0.6 cm^3) in which three 10 × 10 × 2 mm micro-SOFC bundles (made of five integrated microtubular cells) are connected in series. A micro-SOFC module is obtained by attaching a fuel manifold to this stack structure. Hydrogen at a temperature of 450–550°C was input into the module, and a power density of 0.88–2.2 W/cm^3 (450–550°C) was realized. The stack weighs only about 2 g and provides an excellent power output per weight. With the development of the micro-SOFC module, there is a good possibility that portable power sources for small-sized electronic devices can be realized, assuming that systems of tens of watts can be obtained by stacking adequate numbers of the modules to fulfill the requirements of electric output power [22].

Development of 3-D Controlled Micro-SOFCs: Honeycomb Electrochemical Reactors

Fuel cells have attracted attention for high energy-conversion efficiency applicable to on-demand power generation systems. However, until now SOFCs have been applied only to heavy power systems that need to be continuously operated at temperatures of 800–1000°C due to the properties of materials used for cells, components, and stacking structures. For applications to home and automobile use, it is necessary to develop small and highly efficient SOFCs with a power level of several kilowatts with quick start and stop performance.

Honeycomb structured ceramics have used supported substrates of catalysts for the cleanup of exhaust gas from cars, for example, because they have larger areas for gaseous reactions and are structurally stable. If layered structures of fuel cells are constructed on the wall of each regularly arranged channel in the ceramic honeycomb, high cell-packing density SOFC modules can be realized. Such a ceramic honeycomb can be used as an electrode support for cell accumulation by developing a coating process for the dense electrolyte layer on the inner wall of the porous honeycomb substrate [23].

To solve these technical problems, we have developed precise slurry injection techniques for forming layers of various electrolytes and electrodes on the inner surface (wall) of the channels in the porous honeycomb substrates that are made of perovskite-type lanthanum strontium manganese oxide (LSM).

A honeycomb support with a size of 15 × 15 mm of LSM using extrusion molding, as shown on the left of Figure 10.11, has been produced. Using a slurry of ScSZ or ceria-based oxides (gadolinium doped ceria [GDC]), we have coated the surface of 256 submillimeter square channels together using a special jig and heated this at 1300°C after drying. Then, we again coated with a nickel slurry including ceria-based oxides (50 vol% Ni–GDC), followed by a heating process at a temperature higher than 1100°C. As a result, a dense, thin electrolyte layer of 20 μm and a NiO–GDC layer, as an electrode, were formed on the surface of each channel in the LSM honeycomb support. Quick (few-minute) cooling tests were carried out with the honeycomb SOFCs, and no destruction of the cell structure was observed [24, 25].

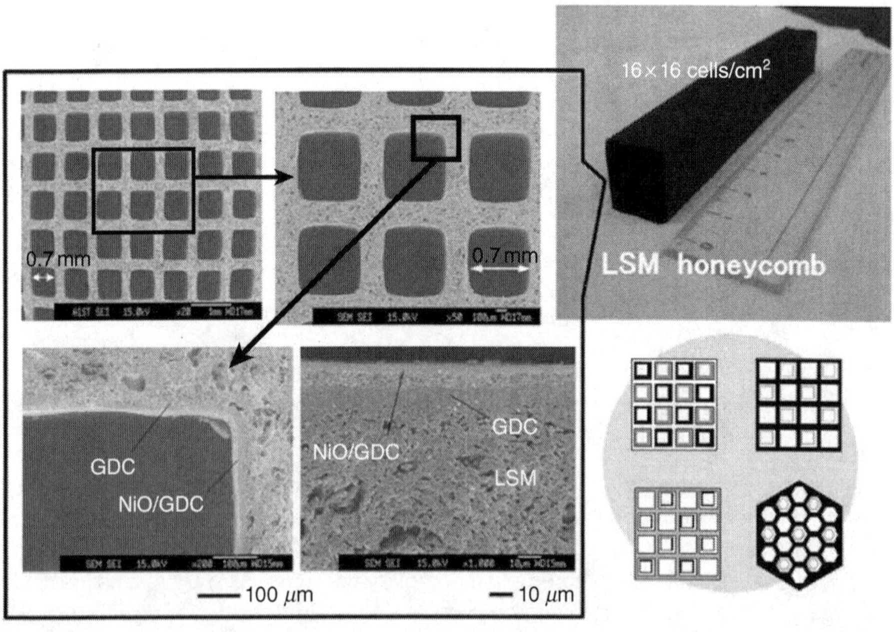

Figure 10.11. Microhoneycomb SOFC involves over 250 cells per cubic centimeter volume.

A power generation test of cells produced using the same coating techniques was then carried out. Using humidified hydrogen gas, a power generation performance of 0.23 W/cm^2 (700°C), which corresponds to the highest level so far attained in the mid-temperature range for cathode-supported SOFCs, was verified [26].

More recently, a honeycomb module has been realized by the assembly of high-performance honeycomb SOFC units connected in parallel and in series to generate output power of tens of watts at high power density reaching 2 W/cm^2 in a very small, hand-sized space. The honeycomb modules are expected for their application as micro-SOFCs and as other electrochemical reactors via their advantages using a conventional manufacturing process for low-cost fabrication.

ELECTROCHEMICAL DE-NO_X REACTOR AND OTHER APPLICATIONS FOR THE CLEAN CAR TECHNOLOGY

Development of High-Performance Electrochemical Reactors

Environment Purifying by Electrochemical Reactors. Essentially, the electrochemical reaction is a directional conversion of materials and energy. High efficiency of the electrochemical reaction is based on the reaction at the electrodes for oxidation and reduction. Use of the ideal hydrogen energy for many applications requires a long time for its realization. For example, fuel cell cars are expected as the ideal zero-emission system, but the substitution of fuel cell cars for gasoline and diesel engine systems will take a long time, 20–30 years or more. Therefore, technologies for exhaust gas purification are likely to be realized only in the next few decades. Electrochemical reactors are expected to be used due to their high efficiency for environmental applications, similar to that of energy generation.

In this section, the details of the development of electrochemical reactors for NOx decomposition, and a novel combined reactor for electrochemical and thermoelectric devices, are described.

High-Performance NO_x Decomposition in Exhaust Gases by Electrochemical Reactors. The reduction of NO_x emission from vehicles has become one of the most important research targets in environmental protection. Air pollution by nitrogen oxides (NO_x) in combustion waste causes serious environmental problems, especially in urban areas. The successful decomposition of NO gas into oxygen and nitrogen without coexisting oxygen was demonstrated over 30 years ago by the development of the "three-way catalyst" (TWC) and other effective catalysts. However, in the presence of oxygen, such as lean-burn conditions, TWCs do not provide sufficient decomposition. The widely used catalytic decomposition of NO_x requires an additional 2–3% of fuel as a reducing reagent, and as a result, there is an increase in the amount of fuel used by the engine. The high efficiency of diesel engines is expected to realize savings in energy consumption even when the exhaust gas includes high partial pressure of oxygen, approaching 10%. This is one reason for the development of high-performance catalysts and other systems workable under high PO_2

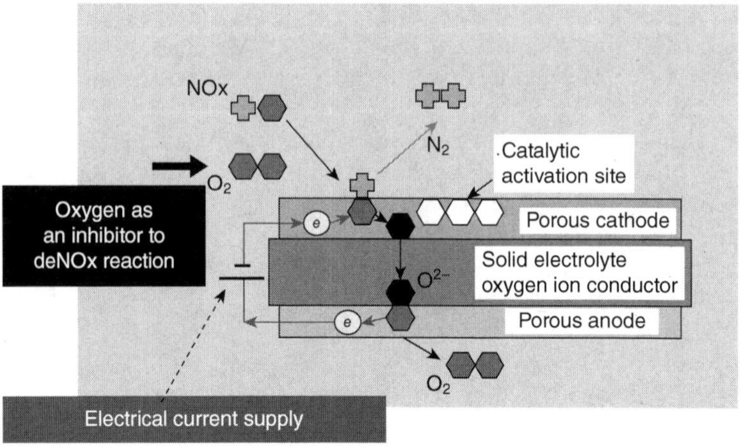

Figure 10.12. Schematic of selective NOx reduction in exhaust gas components.

conditions. Environmental solutions must also be accompanied by energy-saving technologies.

From this viewpoint, the reduction of NO_x emissions can be achieved not only by catalytic NO_x decomposition but also by electrochemical decomposition. The successful decomposition of NO gas into oxygen and nitrogen without coexisting oxygen in a primitive electrochemical cell [27] is shown in Figure 10.12. Recently, many attempts to improve the properties of electrochemical cells operating in the presence of excess oxygen have been carried out by using different catalysts as the cathode material [28]. Unfortunately, in all known cells to date, the coexisting oxygen is decomposed in preference to the NO_x gas, and as a result, the electrical power required for NO_x decomposition is extremely high. The main disadvantage of all previous electrochemical cells is their use of the cathode as both a current collector and a catalyst for NO decomposition.

On applying a voltage to such cells, NO gas is directly reduced at the triple-phase boundary (cathode–electrolite–gas), forming gaseous N_2 and oxygen ions. The oxygen ions are transported through the solid electrolyte from the cathode to the anode, and gaseous O_2 is created at the anode. At the same time, excess O_2 in the combustion exhaust gas is adsorbed and decomposed at the TPB in preference to the NO gas. As a result, due to this unwanted reaction, the additional current associated with the oxygen ions produced far exceeds the current associated with the desired reaction. Pumping of the coexisting oxygen molecules should be suppressed in order to decrease the electrical energy consumption to levels of practically applicable systems.

To solve the problem of effective NO decomposition in the presence of excess oxygen, a new type of electrochemical cell with a functional multilayer electrode has been designed. This new type of electrochemical cell could be represented by an asymmetrical arrangement of the general SOFC-type symmetric cell composed of an electrolyte sandwiched between electrodes. The external voltage applied between the cathode and the anode leads to the polarization of the YSZ (electrolyte) disk and to the generation of a high concentration of oxygen vacancies in the near cathode region. Due to the gradient in the concentration of the oxygen ions, diffusion of oxygen ions from the electrocatalytic electrode to the YSZ disk takes place. For effective cell operation, the cathode should be an electronic and oxygen ionic current conductor with high electronic and ionic conductivity along the cathode plane and from the electrocatalytic electrode to the YSZ solid electrolyte.

In addition to such macro- and microscale arrangements of the cell, applying nanoscale control at the reaction sites is expected to be effective for the acceleration of the electrochemical reaction for the selective absorption and decomposition of NO_x molecules among the predominant oxygen molecules. Interaction between restricted space for gas molecule movement and the electrochemically active surfaces of ceramics to enhance the reduction and oxidation reaction is a possibility. Thus, nanostructural evolution of the electrocatalytic electrode was investigated and correlated with the deNOx properties of the cell. Combining the NO_x reduction accelerator with the oxygen acceptor, and the nanoscale reaction space, promoted an extremely high efficiency of NOx decomposition, even under high PO_2 conditions, as is the case in diesel engine exhaust gas applications [29–34].

Reaction mechanisms are described as follows. The electrochemical reactors for selective NO_x decomposition could be represented by the following asymmetrical reactor arrangement, that is, the functional multilayer electrode–cathode–YSZ–anode. The YSZ disk is used as an oxygen ionic conducting substrate for manufacturing the electrochemical reactor. The functional multilayer electrode on the cathode, in particular, the nanoporous NiO–YSZ electrocatalytic electrode, plays the most important role for the deNO_x properties; that is, the distinguishing feature of such reactors is the artificial nanostructure formed in the NiO/YSZ interface of the electrocatalytic electrode under operation. In this new reactor, the external voltage is applied between the cathode and the anode, and the electrocatalytic electrode covering the cathode is free from the voltage drop of the applied electrical current. This geometry is very important for the deNOx cell operation, which depends not on the ionic conductivity of the circuit, as in SOFC-type reactors, but on the diffusion process of the reacting species. This macromechanism of the cells contributes to a decrease in the cell working temperature from 800 to 500°C, which is close to the expected temperature range for the purification of exhaust gas from diesel engines.

An external electric field leads to the polarization of the YSZ solid electrolyte and the generation of a high concentration of oxygen vacancies inside the 3-D network of YSZ particles in the NiO–YSZ electrocatalytic electrode and to the reduction of the near-surface region of NiO grains into nickel (Fig. 10.13). As a result, the formation of nanosized nickel grains takes place in the near YSZ/NiO interface boundary region. When the external voltage is switched off, the penetration of NO_x and O_2 gases through the nanopores, between the NiO and YSZ grains, leads to the oxidation of the new nanosize nickel grains into NiO grains. Nickel grains with diameters of 10–50 nm are located around the untransformed central part of a primary grain in the figure. In the region with new nickel grains, the pores are located both in the interfacial vicinity and in front of the newly growing nanograins. This phenomenon is explained as an electrochemical reduction reaction occurring at the YSZ/NiO

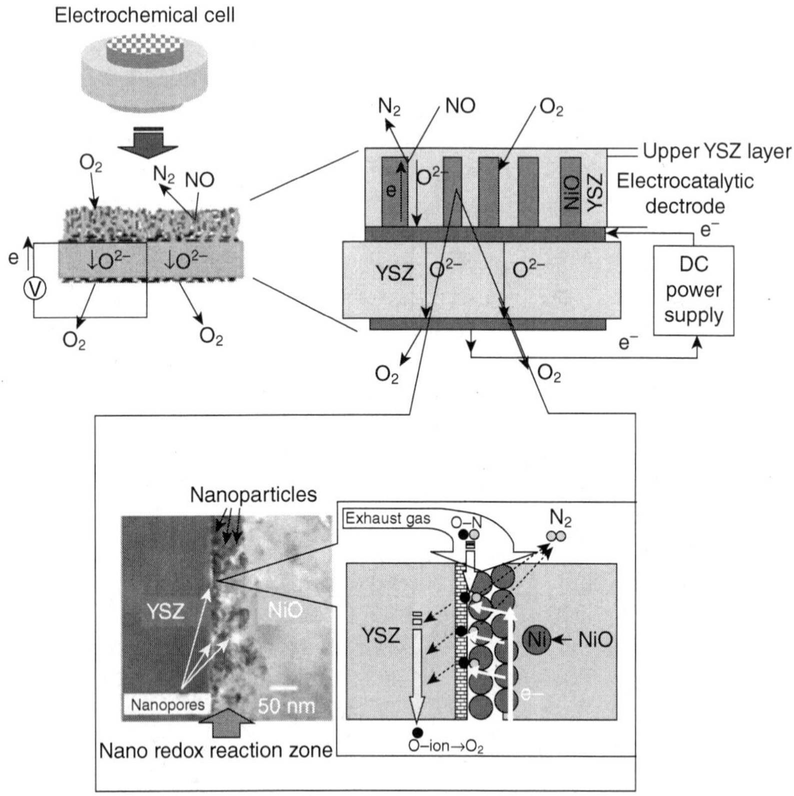

Figure 10.13. Transmission electron microscopy (TEM) image of an interface of NiO and YSZ and a reaction model of selective NOx decomposition.

interface due to the induced electric field in the cell when the voltage is applied between the electrodes.

Nanoparticles were formed by an electrochemical reduction reaction at the interface of the ionic conducting phase and the electrical conducting phase. An applied voltage causes the reduction reaction and recrystallization of NiO to nickel, resulting in decreased volume. Voltage-induced generation and diffusion of the oxygen vacancy accelerates the phase transformation, and Ni nanoparticles are distributed at the interfaces. The strength of the applied electric field affects the thickness of the nanoparticle layer at the interface. An increase in voltage increased the thickness of the nanoparticle distribution layer. After applying voltage in excess, required for removing oxygen, micron-sized pores in the nanoparticle distribution layer were generated. The selectivity of NO_x molecules was depressed drastically by the existence of such large pores. The microstructural construction procedure using this electrochemical reaction was irreversible. Therefore, increasing the efficiency of the selective NO_x decomposition reaction requires an increase in the thickness of the interface layer with nanoparticles and nanopores, which is enabled by the voltage application over a prolonged period.

The distinguishing feature of the microstructure of the YSZ/NiO interface is the 10- to 50-nm nickel grains reversibly produced during the electrochemical procedure and cell operation. It is well-known that the adsorption and decomposition of NO_x gas molecules occur in preference to oxygen gas molecules on nickel grain surfaces. In addition, we should mention that rough surfaces and nanosized nickel grains are much more active for breaking NO chemical bonds than smooth, flat surfaces. Based on the above results, the following reaction mechanism was proposed for NO decomposition on the nanosized nickel grains produced during the reactor operation. The electrochemical reaction described as equation $NO + Ni \rightarrow Ni\text{–}NO$ is followed by the reaction described as equation $2Ni\text{–}NO \rightarrow 2NiO + N_2$.

NO gas molecules are first chemisorbed on nickel. During the second step, the chemisorbed NO decomposes to form N_2, oxidizing nickel to NiO. NO_x gas molecules introduced in the nanoscale spaces have an extremely high probability of interaction with nickel and an oxygen vacancy in YSZ as a counterpart to receive oxygen molecules through this redox reaction.

Oxygen ionic current passed though the network of YSZ particles surrounding the nickel grains removed the oxygen species produced and permitted the reaction to reoccur. The regeneration reaction, of the reduction of NiO to Ni, takes place at the NiO/YSZ interface under the reactor operation, as in the following equation: $NiO + V_O(ZrO_2) + 2e \rightarrow Ni + O^{2}\text{–}(YSZ)$. Therefore, the reduction of NiO grains into nickel grains and the oxidation of nickel grains into NiO take place continuously during reactor operation. As a result the catalytic activity for NO decomposition is independent of the operation time.

In Figure 10.14, it is seen that the efficiency of NO decomposition by newly developed electrochemical reactors far exceeds the efficiency of traditional types of electrochemical cells. Previous results for the application of electrochemical cells to deNO_x were unsuccessful in suppressing the supplied electrical power due to the effect of coexisting oxygen molecules. Precise nanostructural control of the electrocatalytic electrode layer has enabled us to improve the cell properties, with an energy consumption level typically half of that of presently employed catalysts (calculated from the engine power and fuel loss to convert electrical energy).

For electrochemical reactor applications, there are still many problems to be resolved. These include enlargement or multiplication of the number of cells in order to obtain sufficient reaction area, and evaluation and improvement of the cell properties under real gas compositions and flow rates. Stack assembly and module construction will allow this "nanotechnology" to be employed in real applications as a solution to environmental problems. The cell enlargement and stack fabrication to produce a gas purifying module for vehicles has been undertaken. Uniform development of the nanostructure by the electrochemical treatment of the cell has enabled us to operate it with a sufficiently reduced electrical power input and sustainability of the cells. Furthermore, the microstructure of the cells should be optimized for NO_x selective decomposition. Figure 10.15 reveals the importance of nanostructural control suitable for NO_x selective decomposition by forming nanoreaction zones by applied voltage control under moderate conditions.

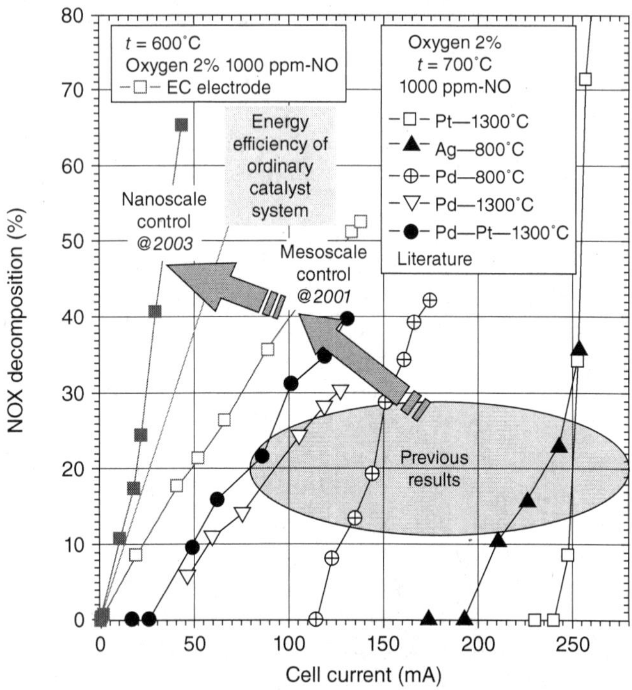

Figure 10.14. Improved deNOx efficiency for applied current by the nanoscale structurally controlled electrochemical cell in comparison with previous results.

The newly developed electrochemical reactors work effectively even at low concentrations of NO_x (300–500 ppm) and at high concentrations of oxygen (10–20% and more) in the exhaust gas. The new type of electrochemical reactors with nanostructured multilayer electrocatalytic electrodes can be used for effective NO decomposition even in the presence of high oxygen concentrations.

Most recently, NOx selective decomposition property at the suppressed temperature around 200–300°C is revealed by introducing an anisotropic structure in nanometer scale as a nanowired electrode (Fig. 10.16). The electrochemical reactor with nanostructured electrodes decomposes NO_x in diesel engine exhaust gas containing high concentrations (about 20%) of oxygen at a lower temperature than 300°C. Furthermore, it improves the fuel efficiency because of the reduction in energy needed to purify exhaust gas. This technology enables the construction of a system that combines air quality conservation with a reduction in CO_2 emission [35–38].

Other problems to be solved, before application can be realized, include durability of the cell under operating conditions involving real exhaust gases such as H_2O, CO, CH, SO_x, improved lifetime durability, and reduction in the cost of cell fabrication. The deNO_x property of the cell before and after introduction of humidity at high temperature indicated good performance of the cell even in 10% humidity at a relevant working temperature. The cell was durable under these conditions, with prolonged lifetime over hundreds of hours in experiments.

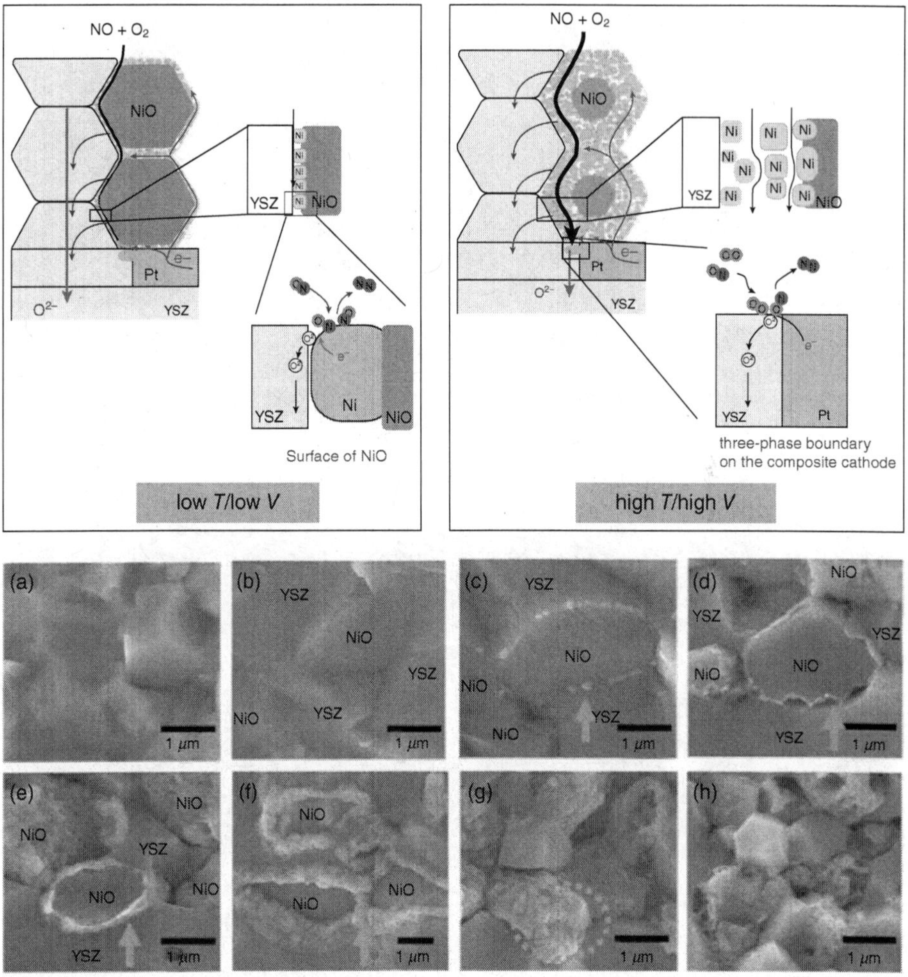

Figure 10.15. Scheme of nanostructural control suitable for NOx selective decomposition by forming a nanoreaction zone by applied voltage control.

Development of Electrochemical Reactors for NO_x/PM Simultaneous Purification

With a view to making use of the oxidation/reduction that takes place simultaneously at the electrodes at both ends of the electrochemical cell, we have made further improvements to the electrodes so that strong oxidation activity will occur due to oxygen radicals on the anode side in addition to NO_x reduction on the cathode side. As a result, the decomposition of PM (a sootlike mass) in the exhaust gas has become possible. Unlike the diesel particulate filter (DPF) that is in common use today, our technology has substantial potential as an energy-saving purification method to remove

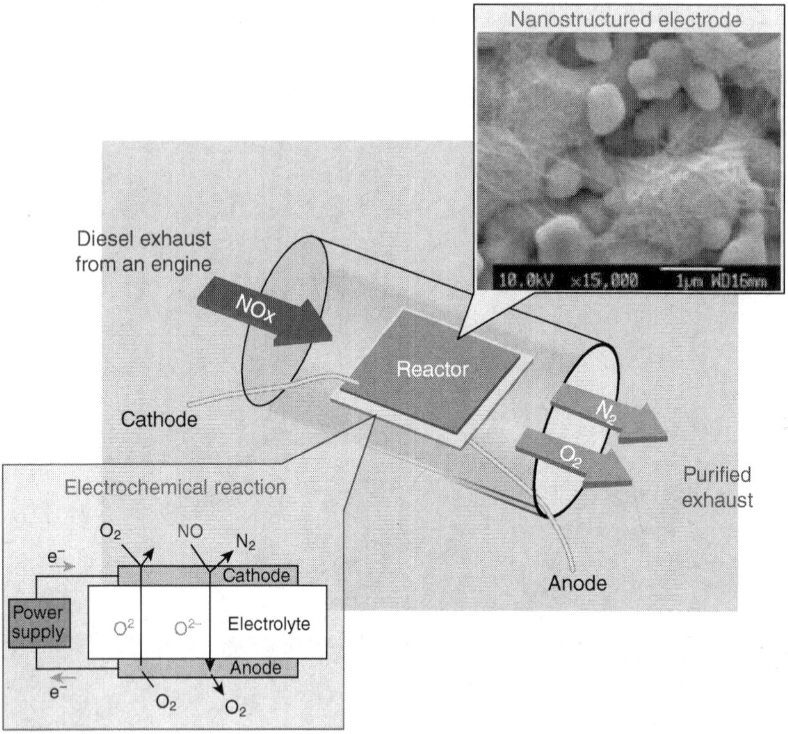

Figure 10.16. Image drawing and a microphotograph of "nanowired electrode" for the low-temperature active NOx decomposition property of the electrochemical reactor.

NO_x and PM simultaneously, independent of the engine operating conditions, and is potentially capable of being activated "on demand" [39] (Fig. 10.17).

Promotion of energy efficiency in the transportation sector, in particular, is an important challenge to attaining CO_2 emission reduction goals under the Kyoto Protocol. A new concept of an electrocatalytic reactor for NO_x decomposition using nanostructural control of an electric conductor (NiO) and an ionic conductor (YSZ) in the electrocatalytic electrode has been proposed. Electrochemically induced formation of nanosize nickel grains surrounded by nanopores, in a NiO/YSZ interface of the electrocatalytic electrode under the reactor operation, resulted in a remarkable improvement of the deNO_x efficiency, under the applied current and voltage. Oxygen ions passing though the network of YSZ particles surrounding the nickel grains repeatedly removed the produced oxygen species. Restricted gas molecular movement in the nanospace at the interfaces accelerated the selective high-performance reaction of NO_x molecules in coexistence with excess oxygen. The electrical power required for NO decomposition in the presence of large amounts of oxygen is reduced to an applicable level [40]. Furthermore, the high efficiency of the gas molecule reaction in the nanoreaction space will be applicable for the improvement of any other application of electrochemical cells, including SOFCs and chemical synthesis reactions.

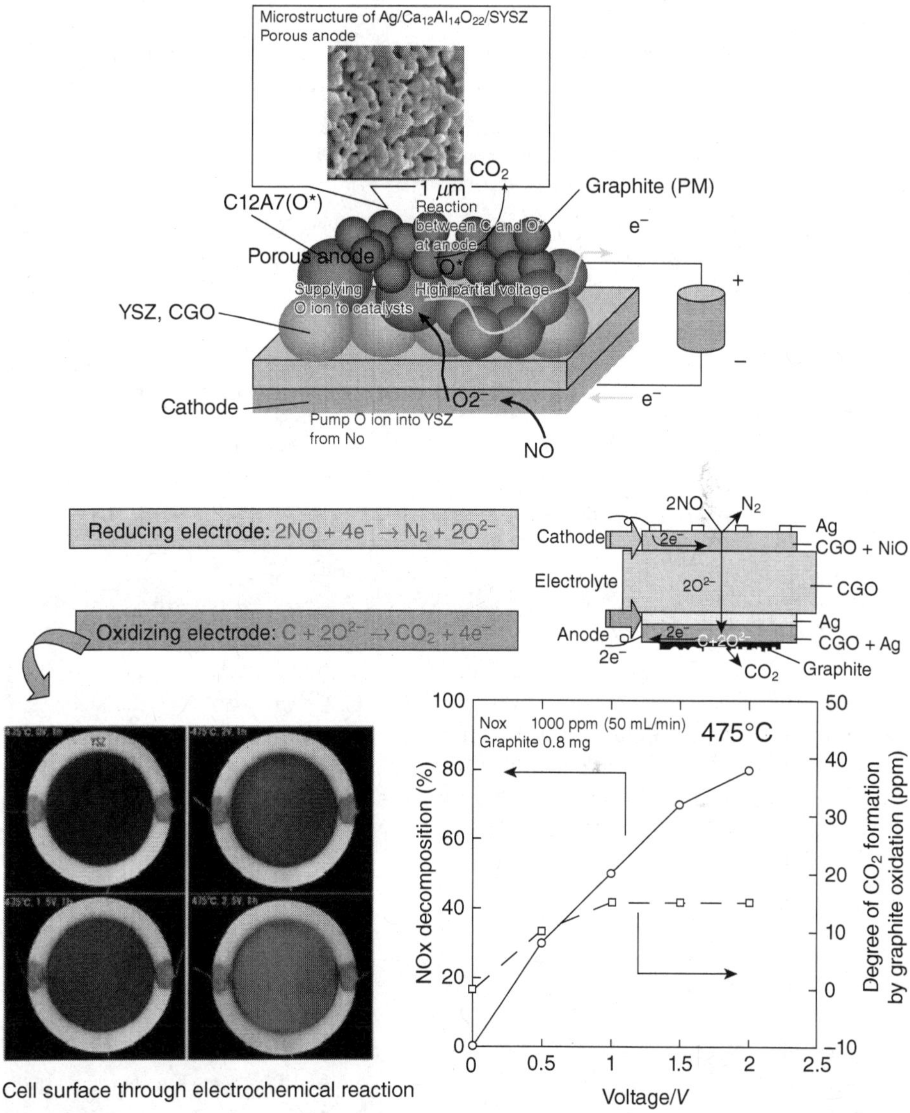

Figure 10.17. Scheme and photographs of anode on increased voltage of the simultaneous NOx/PM removal electrochemical cell.

REFERENCES

1. Steele BCH, Heinzel A. 2001. Materials of fuel-cell technology. Nature 414:345–352.
2. Nakamura O, Goodenough JB. 1982. Solid State Ionics 7:119–124.
3. Sata N, Eberman K, Ebert K, Maier J. 2000. Nature 408:946–949.
4. Adachi M, Murata Y, Takao J, Jiu J. 2004. J Am Chem Soc 126:14943.

5. Ivers TE, Webee A, Herbstritt D. 2001. J Eur Ceram Soc 21:1805–1811.
6. Atkinson A, Barnett S, Gorte RJ, Irvine JTS, McEvoy AJ, Mogensen M, Singhal SC, Vohs J. 2004. Nature 3:17–27.
7. Yokogawa H. 2001. Ceramics 36(7):472–476.
8. Yamaguchi T, Matsumoto M, Matsubara H. 2004. Ceramics 39(4):281–285.
9. Aizawa M. 2001. Ceramics 36(7):493–495.
10. Moon J-W, Hwang H, Awano M. 2002. J Ceram Soc Japan 110(5):479–484.
11. Wang S, Awano M, Maeda K. 2002. J Ceram Soc Japan 110(8):703–709.
12. Yamada T, Batina N, Itaya K. 1995. J Phys Chem 335:204.
13. Sriramulu S, Carlson E, Singh A. 2006. SOFC durability: going the distance. Fuel Cell Rev 3:15–24.
14. Blum L, Meulenberg WA, Nabielek H, Steinberger R. 2005. Worldwide SOFC technology overview and benchmark. Int J Appl Ceram Technol 2(6):482–492.
15. Suzuki T, Funahashi Y, Yamaguchi T, Fujishiro Y, Awano M. 2007. Design and fabrication of light-weighted, sub-millimeter tubular solid oxide fuel cells. Electrochem Solid State Lett 10(8):177–179. 2) Suzuki T, Funahashi Y, Yamaguchi T, Fujishiro Y, Awano M. 2007. Anode supported micro tubular SOFCs for advanced ceramic reactor system. J Power Sources 171:92–95.
16. Suzuki T, Yamaguchi T, Fujishiro Y, Awano M. 2007. Current collecting efficiency of micro tubular SOFCs. J Power Sources 163:737–742.
17. Suzuki T, Awano M, Petrovsky V, Anderson HU, Jasinski P. 2006. Composite (La, Sr) MnO3—YSZ Cathode for SOFC. Solid State Ionics 177(19–25):2071–2074.
18. Suzuki T, Funahashi Y, Yamaguchi T, Fujishiro Y, Awano M. 2008. Fabrication of needle-type micro SOFCs for micro power devices. Electrochem Comm 10:1563–1566.
19. Suzuki T, Funahashi Y, Yamaguchi T, Fujishiro Y, Awano M. 2008. Cube-type micro SOFC stacks using sub-millimeter tubular SOFCs. J Power Sources 183:544–550.
20. Suzuki T, Funahashi Y, Yamaguchi T, Fujishiro Y, Awano M. 2008. Fabrication and characterization of micro tubular SOFCs with multi-layered electrolyte. Electrochem Solid State Lett 11:(6):B87–B90.
21. Sammes N, Roy B, Galloway K, Suzuki T, Awano M, Serincan F. 2009. A study of GDC-based micro tubular SOFC. Mater Sci Forum 638–42:1152–1157.
22. Suzuki T, Funahashi Y, Yamaguchi T, Fujishiro Y, Awano M. 2008. Development of cube-type SOFC stacks using anode-supported tubular cells. J Power Sources 5:68–74.
23. Yamaguchi T, Shimizu S, Suzuki T, Fujishiro Y, Awano M. 2008. Development and evaluation of a novel cathode-supported SOFC having a honeycomb structure. Electrochem Solid State Lett 11(7):G117–G121.
24. Yamaguchi T, Shimizu S, Suzuki T, Fujishiro Y, Awano M. 2008. Fabrication and characterization of high performance cathode supported small-scale SOFC for intermediate temperature operation. Electrochem Commun 10:1381–1383.
25. El-Toni AM, Yamaguchi T, Shimizu S, Suzuki T, Fujishiro Y, Awano M. 2008. Development of a dense electrolyte thin film by the ink-jet printing technique for a porous LSM substrate. J Am Ceram Soc 91(1):] 346–349.
26. Yamaguchi T, Shimizu S, Suzuki T, Fujishiro Y, Awano M. 2008. Evaluation of micro LSM supported GDC/ScSZ bilayer electrolyte with LSM-GDC activation layer for IT-SOFCs. J Electrochem Soc 155:B423–B426.

27. Pancharatnam S, Huggins RA, Mason DM. 1975. J Electrochm Soc 122(7):869–875.
28. Washman ED, Jayaweera P, Krishnan G, Sanjurjo A. 2000. Solid State Ionics 136–137:775–782.
29. Hamamoto K, Hiramatsu T, Shiono O, Katayama S, Fujishiro Y, Bredikhin S, Awano M. 2004. J Ceram Soc Jpn 112:S1071–S1074.
30. Fujishiro Y, Hamamoto K, Awano M. 2004. J Mater Sci Mater Electron 15:769–773.
31. Awano M, Bredikhin S, Aronin A, Abrosimova G, Katayama S, Hiramatsu T. 2004. Solid State Ionics 175:605–608.
32. Awano M, Fujishiro Y, Hamamoto K, Katayama S, Bredikhin S. 2004. Int J Appl Ceram Technol 1:277–286.
33. Awano M et al. 2004. Japanese Patent No. 3626971; Chemical Reactor.
34. Awano M et al. 2004. US Patent US6818107B2.
35. Hamamoto K, Fujishiro Y, Awano M. 2006. Intermediate temperature electrochemical reactor for NO_X decomposition. J Electrochem Soc 153(11):D167–D170.
36. Hamamoto K, Fujishiro Y, Awano M. 2007. Reduction and re-oxidation reaction of catalytic layer in the electrochemical cell for NO_X decomposition. J Electrochem Soc 154(9):F172–F175.
37. Hamamoto K, Fujishiro Y, Awano M. 2008. Gas sensing property of the electrochemical cell with a multilayer catalytic electrode. Solid State Ionics 179:1648–1651.
38. Hamamoto K, Fujishiro Y, Awano M. 2008. Low temperature NOx decomposition using electrochemical reactor. J Electrochem Soc 155:E109–E111.
39. Hamamoto K, Fujishiro Y, Awano M. 2006. Simultaneous removal of nitrogen oxides and diesel soot particulate in nano-structured electrochemical reactor. Solid State Ionics 177:2297–2300.
40. Bredikhin S, Hamamoto K, Fujishiro Y, Awano M. 2008. Electrochemical reactors for NO decomposition. Basic aspects and a Future. Ionics 13(3):285–299.

11

INTEGRATION TECHNOLOGIES FOR SENSORS

Woosuck Shin, Maiko Nishibori, and Ichiro Matsubara

National Institute of Advanced Industrial Science and Technology (AIST), Nagoya, Japan

INTRODUCTION

For the development of chemical microsystems, the integration of functional materials, mostly ceramic oxides, into the Si technology is the most interesting issue to improve the transducer characteristics of a sensor or actuator. The recent use of microsystem technologies in hotplate-type micro gas sensors became fundamental for the use of pulse heating, transient analysis, self-cleaning processes, as well as topics in microelectronics such as low power consumption, integration, and reliability [1]. However, the functional materials cannot always be obtained by the standard complementary metal oxide semiconductor (CMOS) technology—sputtering and CVD. In these cases, transducer materials are obtained separately and novel technologies for the implantation of such active layers on microsystems are required.

In the case of micro gas sensors, at least a part of the scientific and technical community considers the advantages of some typical chemical routes such as traditional precipitation or sol–gel [2], which lead to sensitive, controlled, and mass-produced powders, instead of thin films using the vacuum system [3]. Among these chemical

Ceramic Integration and Joining Technologies: From Macro to Nanoscale, First Edition.
Edited by Mrityunjay Singh, Tatsuki Ohji, Rajiv Asthana, Sanjay Mathur.

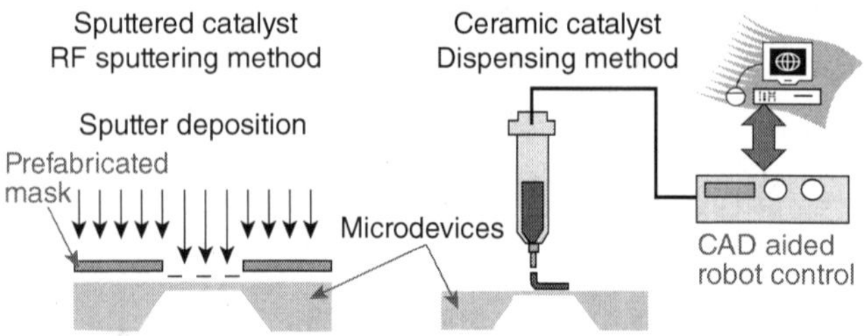

Figure 11.1. Ceramic integration on the membrane. Thin-film process versus direct deposition method. CAD, computer-aided design.

routes, powders of functional materials are prepared to obtain the required functionality, although this increases processing time and costs related to deposition equipment and handling. Sensor or actuator devices have also been fabricated by dispensing the pretreated ceramic materials on the microdevices using the so-called microprinting technology, which is described elsewhere [4, 5]. Furthermore, the deposition of functional films, either by classical screen printing or by more sophisticated drop deposition techniques, such as the ink-jet system, is performed after combining the functional material with organic carriers. It has been recently demonstrated that the dispensing technique can be successfully employed for the preparation of a ceramic catalyst combustor with nanoparticles for gas-sensing applications [6]. This is a straightforward method for fast fabrication, without further high-temperature annealing.

Briefly, the microceramic patterns are fabricated on a double side polished silicon wafer. Si_3N_4 or SiO2 dielectric layers have been used as a thin dielectric membrane, as shown in Figure 11.1. The area of the membrane is typically 1 mm, whereas the heated catalyst area is about 0.4 mm. In this area, thick paste drops are implanted, and the thermal and mechanical behaviors are predicted for micromachined substrates following implantation.

The large extra thermal mass introduced by microdropping and microdrying is expected to affect the thermal distribution as well as the thermal transient, increasing both the thermal time response and the power consumption of the hotplate. The issues of microsensor fabrication are discussed in detail in a previous review paper on the relationship between the SnO_2 sensing layer and the thin microhotplate [7].

The authors have developed thermoelectric (TE) microdevices combined with ceramic combustor patterns. The device consists not only of a microhotplate but also a TE transducer structure on the thin membrane. On this micromachined thin dielectric surface, a paste containing the ceramic powders of Pt and alumina is dispensed through a nozzle with an air pressure driving system to the membrane. For this, an improved micromanipulator, a microinjector, and a visualization system are required. As a result, the microdrops of ceramic catalyst have been deposited on the silicon devices like those in Figure 11.2.

Figure 11.2. Thermoelectric microdevices with ceramic combustor integrated in the membrane and or free standing parts (redrawn from reference).

MICRODISPENSING METHOD

Ink-jet and Dispenser

Controlling the rheological parameters of the paste is most important for the fabrication of the catalyst pattern, whose dimension changes the performance of the device. The process parameters of paste dispensing are also strongly dependent on the rheological properties of paste. Furthermore, a better understanding of the rheological properties of the paste will lead us to a more advanced technology of 3-D forming, such as that reviewed in References 8 and 9.

Speaking of the rule of thumb of the paste rheology, the viscosity of the catalyst pastes has been increased with the concentration of powder in the dispersant of terpineol or by adding ethyl-cellulous polymer. In this study, the viscoelastic properties of the paste have been investigated quantitatively and discussed for better control of dispensing fine ceramic patterns. We have changed the ceramic particle mixture of the dispersant by adding polymers. We have examined the viscoelasticity of paste and discussed how it affects the pattern formation characteristics. The microjet is a computer-automated device for precision printing of liquids or particular slurries, as shown in Figure 11.3. The system uses a computer-driven *x-y* stage for printing, where the print pattern is defined by CAD file. The files are easily modified, which permits on-site designing. This is a great merit over the other techniques such as screen printing or photolithography. These techniques need new masks or screens for changing the design.

The dimension of the fine feature is mostly determined by the nozzle dimension and liquid dispensing speed. Important parameters are the contact angle between the substrate and the dispensed volume of the liquid per time or moving velocity of the nozzle head. The ink-jet system typically uses a 0.05-mm nozzle. The dispenser nozzle size varies much widely from several micrometers to 2 mm. The finest nozzle of the dispenser for micrometer level patterns has an inner diameter of 1 μm and requires

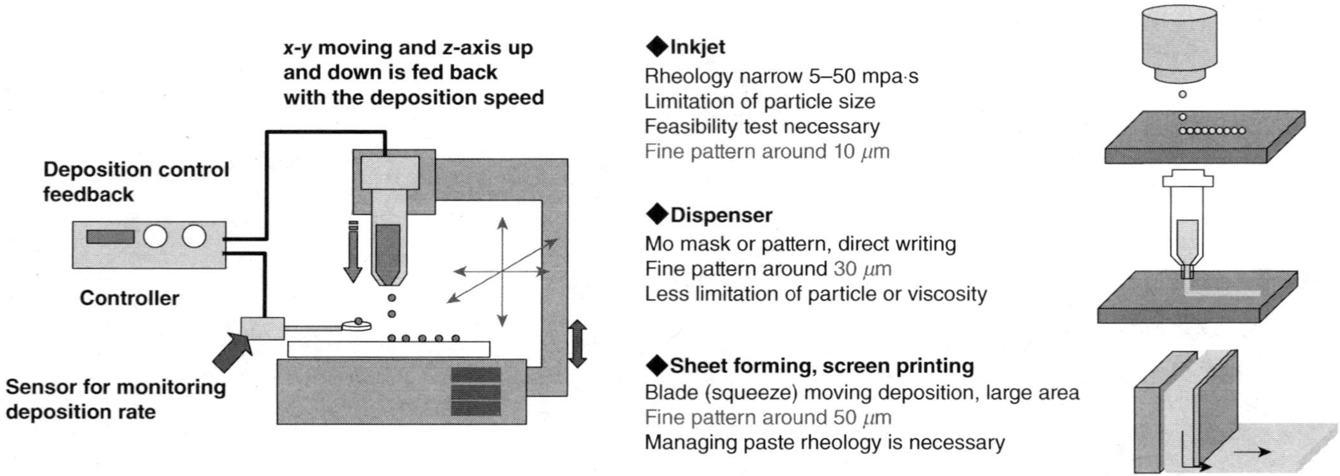

Figure 11.3. Typical micro-ink-jet and microdispenser systems.

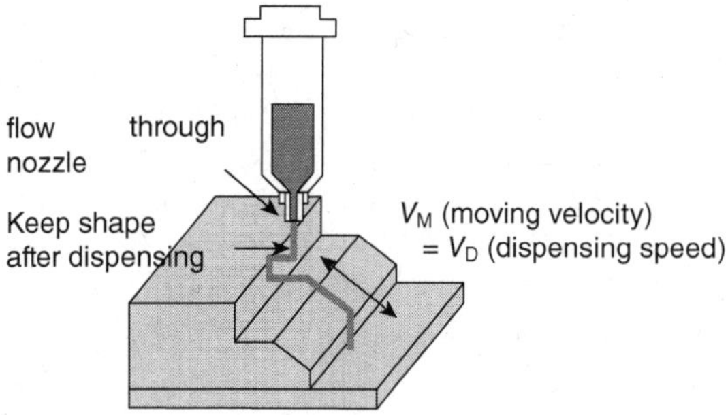

Figure 11.4. Dispenser tip motions for the x-y plane and three dimension. The moving speed of the tip is balanced to the dispensing speed to keep the same width or diameter of the deposited pattern.

well-dispersed colloidal suspensions of the nanolevel to print out through this tip. The paste is loaded into a syringe and the air pressure dispenser causes the paste to move down. Some systems screw the paste into the pump block assembly, to eliminate air bubbles in the paste, which is important for a uniform and reproducible process.

Suitability for 3-D Direct Writing

Some dispenser systems use force feedback control on the nozzle tip to control printing and to maintain a constant force on the tip, to enable conformal writing by electromagnet or air blowing. This control is important to the control of the print thickness and the ability to print over variations in the topography (i.e., height) of the workpiece. The system is also inherently capable of laying down multiple materials in a single layer, which cannot be done in conventional tape-casting techniques or in screen printing (Fig. 11.4).

The direct writing or dispensing system needs *x-y* alignment functions and the nozzle tip and work distance control functions, which are fairly expensive; therefore, the system is not directly suitable for a continuous mass processing of large-size structures, in contrast to other techniques, but it is ideally suited for step-and-repeat procedures, which are appropriate for multilayer, multimaterial electronic components. Many commercial dispensing systems are produced by Asymtek, Musashi Engineering, Inc., Sciperio, and Micropen, and differ from each other in their degrees of resolution, vision control systems, processing flexibility, and materials options.

Rheological Characteristics of Ceramic Pastes

In the production of thick ceramic pastes, the flow of the paste should be controlled for the fine patterns on the workpieces. Screen printing formulations are designed to achieve metal conductor patterns with the resolutions below 50 μm (typically 120 μm) with high reproducibility. Below the optimized viscosity, the deposited pastes spread out and lose resolution. If the paste is rigid, the dispensing controllability becomes worse. The microdispensing system uses the nozzles for depositing material, and there are important parameters for each nozzle cross-section, dispensing speed or pressure, and pastes. There are several characteristics in the ceramic pastes used in the dispensing system, and many of them are similar to the pastes developed for screen printing. The rheological properties of the pastes are tailored to achieve better control and finer resolution. The theoretical and phenomenological understanding of the paste rheology would lead to high-resolution and high-reliability microceramic pattern fabrication.

Paste Rheology

The dispenser can be used to fabricate a ceramic pattern using the paste with a wide range of rheological property, but the understanding of paste flow during the dispensing is essential for a reliable and reproducible process. The flow behavior of terpineol and catalyst particle paste mixtures reflects the profound structural changes in the paste. Typically, as the viscosity decreased sharply with the shear rate, in the absence of polymers, the pastes exhibited a so-called strong shear thinning behavior characteristic of a flocculated system. This can be expressed by Equation 11.1:

$$\eta = \frac{\tau_o}{\dot{\gamma}} + k\dot{\gamma}^{n-1}, \tag{11.1a}$$

or in terms of shear stress,

$$\tau = \tau_o + k\dot{\gamma}^n, \tag{11.1b}$$

where, η is viscosity, τ is stress, τ_o is yield stress to initiate flow (in the case of a solid-like paste), $\dot{\gamma}$ is shear rate, and k is the consistency index [10]. For Newtonian liquid, the power term of parameter n is 1 and τ_o is 0, but the paste with the particles became nonlinear and the power term of the shear rate became negative (Fig. 11.5).

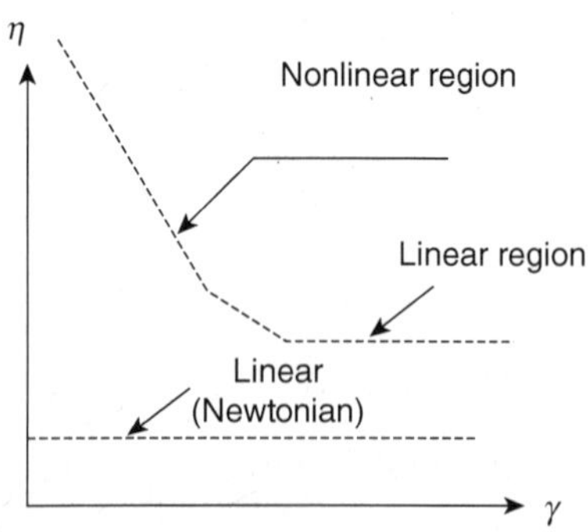

Figure 11.5. Schematics of paste rheology and shear rate dependence of viscosity. The shear thinning is the sharp decrease of viscosity with shear rate in the nonlinear region.

During dispensing, the paste experiences varying shear stresses and rates, which affect the size and shape of the patterns. However, the dispenser operation can only control the parameters of air pressure inside the syringe and pressing time. Other parameters are decided by the needle used for dispensing. As shown earlier, the paste in this study is not Newtonian flow, and checking the linearity of the dispensing condition is a kind of requisite step for understanding the dispensing process.

Deposition Rate Monitoring

Regardless of the ink types, the piezoelectric ink-jet printing reproducibly formed from sub- to 10 pL drops for a typical deposition process. This drop size seems popular, considering the higher reproducibility for small-scale pattern fabrication. For the dispensing process, the dispensing speed rather than the drop size of the ink is a more important parameter, and it typically varies from 0.01 to 1 nL/s, mainly depending on the nozzle diameter.

In most cases, controlling the dispensing volume is the most critical issue for fine pattern fabrication and also for product-level fabrication with controlled quality level. Although the popular method of monitoring the ink-jet-printed volume is to measure millions of drops by microbalance, it is a very time-consuming process, and the condition could change after the long dispensing. Imagine the 1 μg drop, which can be measurable by microbalance, will be obtained by injections of 1 million picoliter drops. If the droplets were injected at a frequency of 200 Hz, then one should wait for 5000 s or 80 min to monitor the drop volume.

It is also possible to monitor the dispensed liquid mass by a quartz crystal microbalance (QCM). This novel idea of in situ monitoring of dispensing volume for this direct writing technique has been demonstrated [11]. As the variations of the frequency changes of the QCM sensor are well correlated with the deposited mass of the droplet of the ink-jet as well as the dispenser, we can predict the dispensing volume with proper calibration of this sensor.

Figure 11.6 shows the drop mass monitoring for ink-jet of *i*-propanol. Using the simple equation of $\Delta f(\text{Hz}) = 1153\Delta m$ (ng), the change of the mass, Δm, is plotted on the right-side y-axis. From this result, the single droplet can be calculated to be 0.25 ng. Adding this drop monitoring system is expected to increase the feedback controllability of the ink-jet or dispenser system, as shown in Figure 11.3.

DEVICE FABRICATION

Hotplate-Type Microdevice

Sensors were obtained from dispensing the oxide material on silicon micromachined substrates (microhotplates). An example of this microdispensing or dropping technology has been reported earlier [5]. The devices were fabricated on double side polished p-type silicon. Si_3N_4 has been used as a dielectric membrane. The resistor heater was polysilicon and Pt was used as electrode. Sputtered electrodes patterned with liftoff

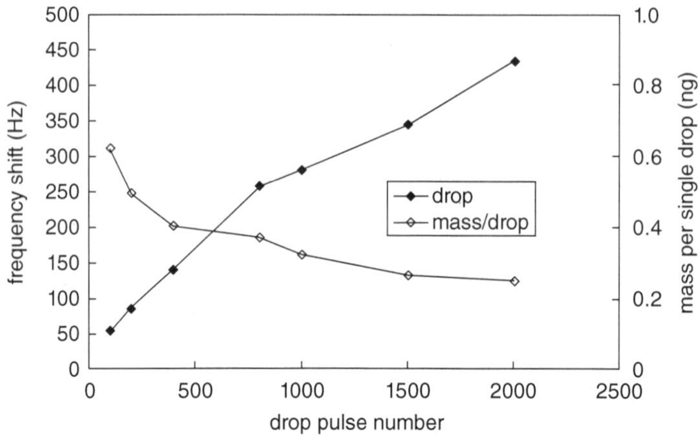

Figure 11.6. In situ quantitative monitoring of the ink-jet droplets using a QCM sensor device, showing the frequency shift of QCM was plotted with the number of drop pulses [11].

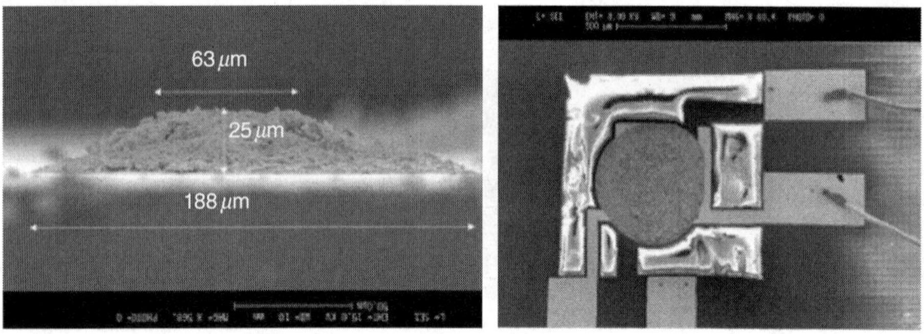

Figure 11.7. SEM image of a cross-section drop, left, and top view, right, of the sensing material deposited on the membrane of the Si microdevice [5].

have been designed with interdigit-pattern layouts, and the change of the resistance of the sensing materials of SnO_2 was read by this Pt electrode. The typical area of the membrane is $1000 \times 1000\,\mu m$, whereas the heated area is $400 \times 400\,\mu m$.

Micro-Gas Sensor Device Using SnO_2

SnO_2 nanopowders were obtained by irradiation with a microwave energy source of a saturated tin chloride solution in an organic solvent. In situ catalytic additives were introduced in chloride form in the tin chloride precursor solution, as previously reported [12]. On this micromachined surface, a paste containing the microwave SnO_2 powders is transferred from the capillary system to the membrane. For this, improved micromanipulator, microinjector, and visualization systems are required. As a result, microdrops are obtained, as shown in Figure 11.7.

The material was deposited on the micromachined membranes by means of a controlled system formed from a pneumatic microinjection part connected to a micropipette and mounted onto a micromanipulator equipped with an optical microscope. This simple system allows a fast deposition of different pastes on the silicon wafers mounted on the transister outline (TO) package after the cleaning of the used microinjection components. Finally, the deposited sensitive layer is fired to remove the organic solvent used in the paste preparation in order to obtain good adhesion. The control of the paste parameters and those of the microdropping deposition technology has a high influence on the characteristic deposited layer, such as porosity, density, and drop diameter.

TE Gas Sensor Device Using Combustor [13]

The pattern layout of the μ-thermoelectric hydrogen sensor (THS) is shown in Figure 11.8. The substrate size was $5 \times 5\,mm^2$ with a single-membrane structure. The double side polished silicon wafer ([100] p-type, Shinetsu Inc., Tokyo) of 350-nm thickness was used as substrate. The wafers were thermally oxidized to form an insulating layer of SiO_2, and then nitride film was deposited by low-pressure chemical vapor deposition to form multilayers of 250-nm Si_3N_4/80-nm SiO_2 on both the front and backside. The membrane was heated up using two meanders of Pt microheater by a single-current mode. The hot point of the catalyst on the membrane with low thermal conductivity minimizing the conduction heat losses is patterned at the center of the membrane. The cold junction is also located on the same membrane whose configuration leads to a very stable voltage signal even in the unpredictable atmosphere.

A sintered target of 1 at% boron-doped $Si_{0.8}Ge_{0.2}$ (p-type) with a purity of 99.997% was used for the radio frequency (RF) magnetron sputtering deposition of a SiGe thin film. The Pt microheater and electrode lines were fabricated by the lift-off technique. The wet-etching windows were made by a nitride layer patterned by reactive ion etching (RIE), and aqueous KOH solution etching was performed to make the membrane structures. The final step was to cover the top surface of the active area by the insulating layer of plasma enhanced (PE) CVD SiO_2 and then to deposit the ceramic catalyst.

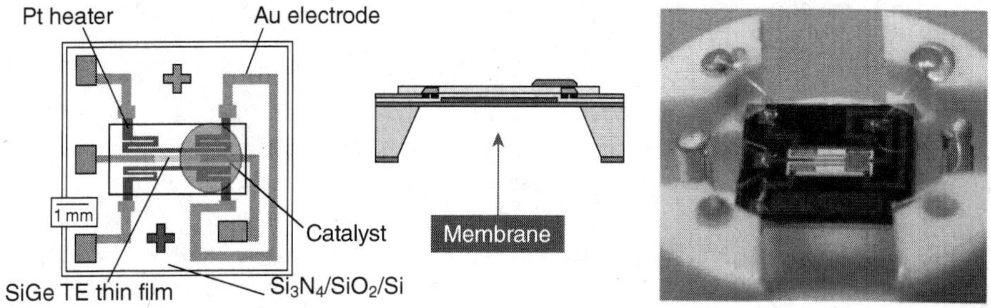

Figure 11.8. Pattern layout of the μ-THS.

Ceramic Catalyst Deposition

The Pt/alumina catalyst was prepared by the impregnation of a commercial alumina support with an aqueous solution of platinum(IV) chloride pentahydrate. The alumina powder was alpha-Al_2O_3 of 100-nm size, and the organic vehicle was a conventional dispersant, a blend of terpineol and ethyl cellulose. As the best catalytic activity was found to be the 40 wt% Pt/alumina catalyst by the previous research [14], we have deposited the catalyst of the 40 wt% Pt/alumina. The slurry was kept under stirring at 100°C for 30 min and was dried at 120°C for 2 h in air. A submillimeter-size drop of the paste was deposited on one side of the membrane, the hot side (as indicated in Fig. 11.8), of the μ-THS by a dispenser system with CAD robot operation (Musashi Engineering, Inc., Mitaka, Japan). After this integration of the ceramic paste on the thin membrane of the μ-THS, it was fired to remove the organic vehicle in air at 300°C, and then the chip was packaged as shown in Figure 11.8. This shows the μ-THS with a thin dielectric membrane is transparent and the silhouette of heater meander is shown.

The ceramic catalyst system discussed in this chapter is not usual, as reported in other reports. Normally, used catalyst noble metal, Pt or Au, contents are around 4 wt% and are well below 10 wt%, but the optimized Pt content in the hydrogen combustor catalyst system is 40 wt%. This big difference in the metal content results in extremely high activity, which is a very essential function for the saturate and a complete catalytic reaction for specific applications such as sensors and microdevices.

Besides the chemical composition of the catalyst, the viscosity of the paste is an important parameter for the deposition process. The viscosity determines not only the achieved size of the drop and the final thickness of the thick film but also affects the dispensing conditions, air pressure and needles. By changing the mixing ratio of particles and vehicles, the viscosity of the paste was controlled, and finally, the thickness of the paste was changed. With the concentration of the powder, the viscosity of the paste increased, leading, as a result, to a thicker catalyst film. The dispensing condition of the paste was tuned to shape the same diameter, prohibiting spreading out as much as possible.

Figure 11.9 shows the shear rate dependence of the viscosity of the three different pastes. By mixing the catalyst powder and the organic vehicle with the weight ratios 1:2, 1:4, and 1:8, the pastes were prepared. The viscosity of the ceramic paste (mixing powder and vehicle in weight) showed clear nonlinear behavior; that is, viscosity decreased with the shear rate and the paste flows as liquid, Newtonian, at higher shear rate.

This linear flow was useful for dispensing a small catalyst pattern on the microdevices, and we consider that the polymer of ethyl cellulose could play a role to inhibit the flocculation of the particle, making the paste easy-flow. If someone wants to prepare a thicker catalyst, a non-Newtonian paste should be used. Because the paste should flow at a high shear rate and stop at a low shear rate; that is, the paste must have a clear yield stress. This yield stress behavior is important for the paste to run out through the narrow nozzle and can be quantitatively evaluated by a conventional rheology technique [8, 13]. For the different particle concentrations, usually, the particle concentration increases both the viscosity and the shear yield stress and paste elasticity.

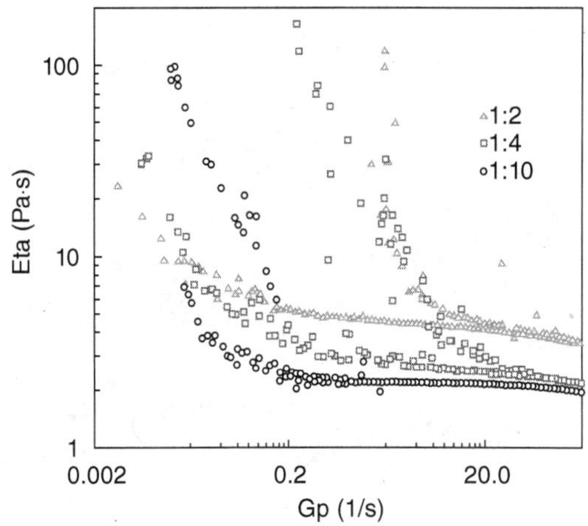

Figure 11.9. Shear rate dependence of the viscosity of the ceramic paste 40 wt% Pt/alumina catalyst [13].

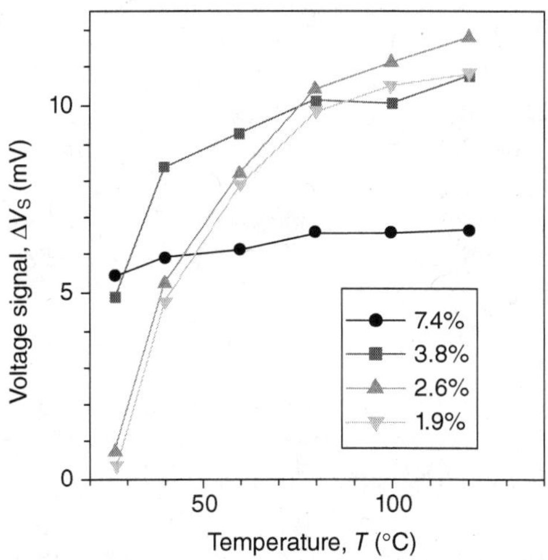

Figure 11.10. The optimized catalyst particle concentration in the paste was 3.8 wt%, with the sensor performances of the higher voltage output at operating temperature 100°C and the better temperature independence [13].

SENSOR PERFORMANCE

Catalyst Size and Thickness

Figure 11.10 shows the output voltage signal, V_S, for 1% hydrogen in air of the μ-THS with the catalyst of different thickness, which was controlled by changing the

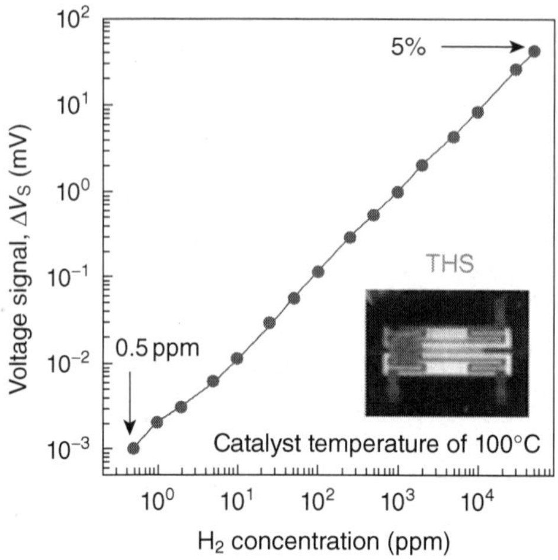

Figure 11.11. Detection of wide range of hydrogen can be demonstrated using ceramic catalyst.

concentration of particles [13]. For the 1% hydrogen gas in air, ΔV_S was about 10 mV at around 100°C. Keeping the catalyst temperature as high as 100°C ensures the stable catalytic combustion and stable sensor operation, prohibiting the unwanted effect of moisture and temperature dependence of the ΔV_S. The temperature dependence of the combustion was also related with the thickness of the catalyst. The thicker was advantageous over the thinner in the sense of low-temperature dependency.

The effect of the catalyst size on catalytic activity has been investigated previously by changing the catalysts with different diameters were prepared on the μ-THS, and the optimized size for the catalyst was found to be just smaller than the membrane width [15]. Through further optimization, including new design of the TE device and the catalyst film thickness, we have demonstrated that the μ-THS could detect a very wide range of hydrogen in air [16] (Fig. 11.11).

Long-Term Stability of the Ceramic Catalyst

The degradation of the catalyst for long-term operation is also an important problem for real application. From these reasons, for better reliability and temperature independence, a thicker catalyst combustor film is preferred. We have studied how the stability of the sensor with the catalyst thickness and have confirmed this again. Furthermore, we have found that the aging process after the fabrication of the catalyst combustor is very essential for long-term stability. A deeper study on the microstructure changes of the nanosize Pt in the catalyst has been reported [17]. Self-heating by the microheater in the THS device has enhanced the stability of the thick ceramic catalyst integrated in the microdevices to prohibit Pt diffusion and sintering, the so-called anchor effect, as shown in Figure 11.12.

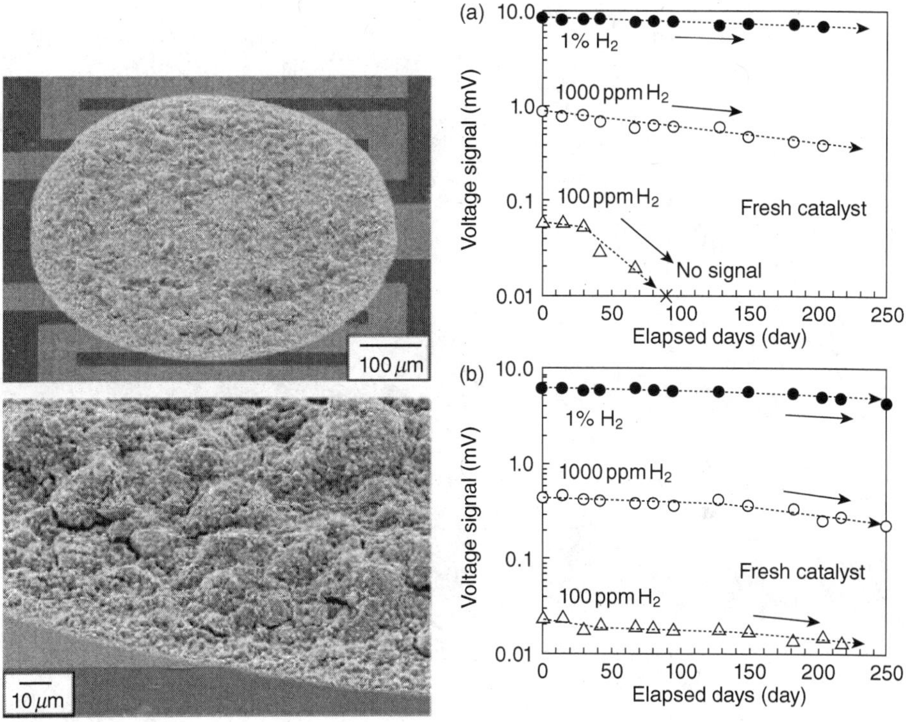

Figure 11.12. Thick ceramic catalyst is a promising candidate for robust use showing long-term stability [17].

Catalyst for Thermopile Device

The increasing use of portable electronics has driven research in the area of portable electrical generators. Chemical fuels, for example, hydrocarbons, when reacted with air, release 10–100 times more energy per unit mass than is released from state-of-the-art batteries. Recent technologies of portable power generators, which use chemical fuel, include micro fuel cells, microturbines, and micro-TE power generation using catalytic combustors. The advantage of TE generators over other power generators is their simplicity. This simplicity and the absence of moving parts make TE generators good targets for miniaturization. Schaevitz et al. have demonstrated the conversion of heat from the catalytic combustion of hydrogen into electrical power using a micro-TE generator with a Pt thin-film combustor [18]. The micro-TE generator operated at a temperature of 500°C and produced a voltage of 7 V.

We have succeeded in getting rid of this heater structure by the integration of the ceramic catalyst on the microdevices [19]. Figure 11.13 shows a schematic of the thermopile design, and we have prepared an optical microscope photo of the thermopile. The thermopile consists of 20 couples of B-doped SiGe/Au. The thermocouple legs are aligned from the center of the membrane (hot junctions) to the edge of the silicon (cold

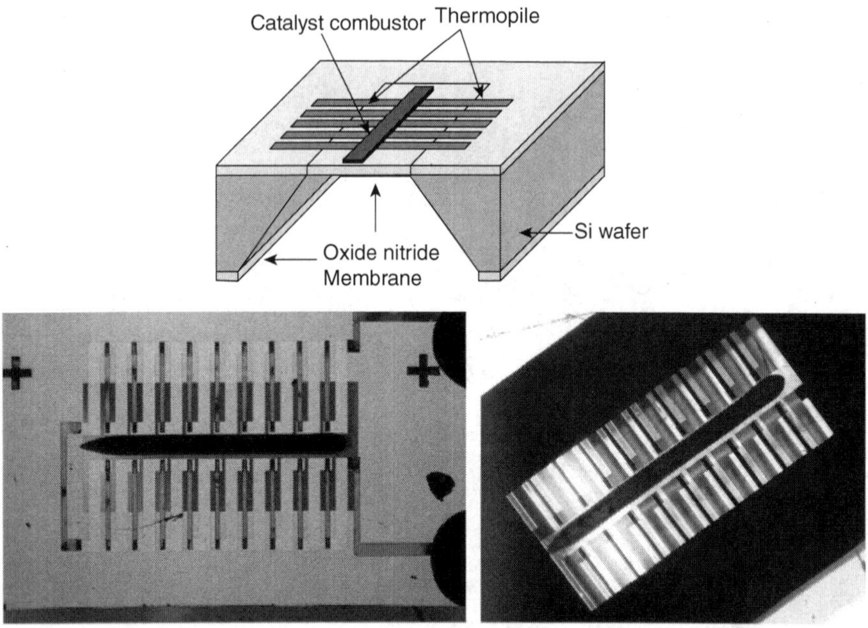

Figure 11.13. Thermopile generator device with ceramic catalytic combustor patterns.

junctions). The thick-film ceramic catalyst pattern is on the center of the membrane, aligned at the hot end of the thermopile. Without heater, at room temperature, the catalytic combustor burns the low-concentration hydrogen, providing thermal energy to the system to create a temperature gradient across the thermopile part. This thermopile structure was also applied to the sensor device [20], in which the heater structure is included for reliable gas sensing even at humid condition.

SUMMARY

Both new catalyst system and new integration method for microdevices are necessary. The ceramic catalyst system discussed in this chapter is not the usual kind as those reported in other studies. The usual ceramic catalysts have noble metals, Pt or Au contents around 4 wt%, or well below 10 wt%, although the optimized Pt content in the hydrogen combustor catalyst system is 40 wt%. This big difference in the metal content results in extremely high activity, which is a very essential function for the saturated and perfect catalytic reaction for specific applications such as sensors and microdevices.

In the microdevice applications, there is only a small chance and only a short path for the chemical reaction to progress, and the catalyst should be as active as possible. Furthermore, most microdevices are fabricated by Si wafer technology and their working temperature is relatively low, typically below 125°C. To make the catalyst fully active at this low temperature, high metal content is adopted and can be used without sintering (metal grain growth in catalyst at high temperature). We have challenged the Au–transition metal oxide system for CO gas sensing [21] and now we are developing the Pt system with metal alloying. The integration method is, of course, also a key technology in this field, as listed in a recent paper, for the ceramic catalyst [22]. What we want to address in the end of this chapter is that some more interesting work is waiting to be done on the integration process.

After treatment of the functional ceramic materials, for instance, the aging process, as introduced here, seems to be also a very unique and interesting field of study. The in situ self-poling process as integrated in the microdevices can be easily done by applying even very small electrical or magnetic fields, with the help of its microsize effect. On the other hand, restive oxides such as tin oxide and microheaters can induce an unwanted effect that makes uniform heating difficult because the microheater is too close to the oxide thick films. These new functions need to be integrated in the same microdevices.

REFERENCES

1. Gardner JW, Varadan VK, Awadelkarim OO. 2001. *Microsensors MEMS and Smart Devices*. New York: Wiley.
2. Dieguez A, Romano-Rodriguez A, Morante R, Kappler J, Barsan N, Göpel W. 1999. Nanoparticle engineering for gas sensor optimisation: improved sol–gel fabricated nanocrystalline SnO_2 thick film gas sensor for NO_2 detection by calcination, catalytic metal introduction and grinding treatments. Sens Actuators B 60:125.
3. Semancik S, Cavicchi RE, Wheeler MC, Tiffany JE, Poirier GE, Walton RM, Suehle JS, Panchapakesan B, DeVoe DL. 2001. Microhotplate platforms for chemical sensor research. Sens Actuators B 77(1–2):579–591.
4. Cirera A, Cabot A, Cornet A, Morante JR. 2001. CO–CH_4 selectivity enhancement by in situ Pd-catalysed microwave SnO2 nanoparticles for gas detectors using active filter. Sens Actuators B 78(1–3):151–160.
5. Cerda J, Cirera A, Vila A, Cornet A, Morante JR. 2001. Deposition on micromachined silicon substrates of gas sensitive layers obtained by a wet chemical route: a CO/CH_4 high performance sensor. Thin Solid Films 391:265.
6. Shin W, Tajima K, Choi Y, Izu N, Matsubara I, Murayama N. 2006. Planar catalytic combuster film for thermoelectric hydrogen sensor. Sens Actuators, B 108:455–460. 2005.
7. Simon I, Barsan N, Bauer M, Weimar U. 2001. Micromachined metal oxide gas sensors: opportunities to improve sensor performance. Sens Actuators B 73:1–26.
8. Lewis J. 2003. Direct writing in three dimensions. *Materials Today* 32–39.
9. Li Q, Lewis J. 2003. Nanoparticle inks for directly assembled three-dimensional periodic structures. Adv Mater 15:1639–1643.
10. Reed JS. 1995. Rheology of saturated system. In: Reed JS, *Principles of Ceramic Processing*, 2nd edition. New York: John Wiley & Son Inc., pp. 277–305.
11. Shin W, Nishibori M, Itoh T, Izu N, Matsubara I. 2008. Monitoring of dispersed fluid with the QCM for the better control of inkjet or dispenser system. J Ceram Soc Japan 116: 459–61.

12. Puigcorbe J, Cirera A, Cerda J, Folch J, Cornet A, Morante JR. 2002. Microdeposition of microwave obtained nanoscaled SnO_2 powders for gas sensing microsystems. Sens Actuators B 84:60–65.
13. Shin W, Nishibori M, Tajima K, Houlet LF, Choi Y, Izu N, Murayama N, Matsubara I. 2006. Integration of ceramic catalyst on micro-thermoelectric gas sensor. Sens Actuators, B 118:238–291.
14. Choi Y, Tajima K, Sawaguchi N, Shin W, Izu N, Matsubara I, Murayama N. 2005. Planar catalytic combustor application for gas sensing. Appl Catalyst A 287:19–24.
15. Shin W, Tajima K, Choi Y, Izu N, Matsubara I, Murayama N. 2006. Micro thermoelectric device with ceramic combustor. Sens Actuators, A 130–131:411–418.
16. Nishibori M, Shin W, Houlet L, Tajima K, Izu N, Itoh T, Murayama N, Matsubara I. 2006. New structural design of micro-thermoelectric sensor for wide range hydrogen detection. J Ceram Soc Japan 114:853–856.
17. Nishibori M, Shin W, Houlet L, Tajima K, Izu N, Itoh T, Matsubara I. 2008. Long-term stability of Pt/alumina catalyst combustors for micro-gas sensor application. J Eur Ceramic Soc 28:2183–2190.
18. Schaevitz SB, Franz AJ, Jensen KF, Schmidt MA. 2001. *Proc. 11th Int. Conf. Solid-State Sensors and Actuators*, p. 30.
19. Nishibori M, Shin W, Houlet L, Tajima K, Izu N, Matsubara I, Murayama N. 2006. Catalyst combustors with B-Doped SiGe/Au thermopile for micro-power-generation. Jpn J Appl Phys 45:L1130–L1132.
20. Houlet L, Shin W, Tajima K, Nishibori M, Izu N, Itoh T, Matsubara I. 2008. Thermopile sensor-devices for the catalytic detection of hydrogen gas. Sens Actuators,B 130:200–206.
21. Nishibori M, Tajima K, Shin W, Izu N, Itoh T, Matsubara I. 2006. CO oxidation catalyst of Au-TiO2 on the thermoelectric gas sensor. J Ceram Soc Japan 115:34–41.
22. Meille V. 2006. Review on methods to deposit catalysts on structured surfaces. Appl Catalysis A 315:1–17. (no detail information but long list).

12

ON-CHIP INTEGRATION OF FUNCTIONAL HYBRID MATERIALS AND COMPONENTS IN NANOPHOTONICS AND OPTOELECTRONICS

Talha Erdem and Hilmi Volkan Demir

Bilkent University, Ankara, Turkey

Incorporation of custom-designed nanomaterial into nanophotonic devices and components in optoelectronic systems enables the realization of optical functionalities favorably controlled with external optical and electrical effects. To this end, innovative nanophotonic devices and optoelectronic systems are developed using multiple combinations of nanostructures (epitaxially grown, chemically synthesized, deposited, plated, etc.) that are embedded in hybrid architectures and on-chip integration of components for a variety of applications including light generation, displays, modulation, sensing, imaging, and communications in a wide spectral range from ultraviolet (UV) to visible to infrared.

Such integration of functional materials and components in optoelectronics requires the use of different methods developed for chip-scale integration at the range of micrometers to nanometers, including monolithic integration, hybrid integration, layer-by-layer assembly, and directed assemblies (electrical, optical, mechanical, etc.). This chapter reviews these state-of-the-art integration approaches currently used in optoelectronics.

Ceramic Integration and Joining Technologies: From Macro to Nanoscale, First Edition.
Edited by Mrityunjay Singh, Tatsuki Ohji, Rajiv Asthana, Sanjay Mathur.

Additionally, the chapter summarizes those innovative integration approaches that are presently being investigated for cutting-edge optoelectronics and nanophotonics.

MONOLITHIC INTEGRATION TECHNIQUES

Optoelectronic components used in current communication systems are typically based on discrete components. These components with different functions are generally combined with each other using fiber splices to provide a specified function. Although the approach of packaging each component individually and then connecting them through fiber connections is a common practice, it manifests some disadvantages. One of them is the great difficulty in efficiently coupling light to and from each discrete chip, which is one of the main sources of optical loss at the system level. In addition, this approach also places a burden at the system implementation level due to cost issues.

To address these problems, monolithic integration may provide possible solutions [1]. For example, using the on-chip integration method, device-to-device coupling problem can be completely eliminated, conveniently leading to a significant decrease in packaging costs and sizes. Additionally, elimination of possible mechanical movements in the structure and possible reduction of driving currents may reduce the power consumption of such integrated chips and may increase the reliability of the full system.

For monolithic integration to become useful and meaningful, a list of some requirements should be fulfilled [2]:

1. The operation of one component should not be affected by the operation of the other components. In other words, each component should function as if it is discrete.
2. Trade-off between fabrication difficulty and performance optimization should be carefully considered because a careless design may increase the overall cost and/or may decrease the device performance.
3. The integration method should not be costly and time-consuming. There are several monolithic integration schemes.

Butt Joint Growth

Butt joint growth is based on the removal of a specific part of the epitaxial region (e.g., waveguide/multiple quantum well [MQW] region and, subsequently, the nonplanar selective regrowth to fill in the removed part (e.g., with desired band edge and waveguide architecture). The main advantage of this integration technique is its high versatility [3]. An exemplary cross-section of butt joint grown epitaxy is shown in Figure 12.1a. In this case, for example, for a laser diode, the flexibility of this process comes from the possibility to grow a quantum well active region for a maximized modal gain. Also, allowing each growth to lead to a desired absorption band edge is an additional strength of this approach. In spite of the ability to separately optimize individual components on the chip [4], there is also some difficulty that arises with the requirement to match the growth thicknesses and to achieve the desired composition to avoid reflection and

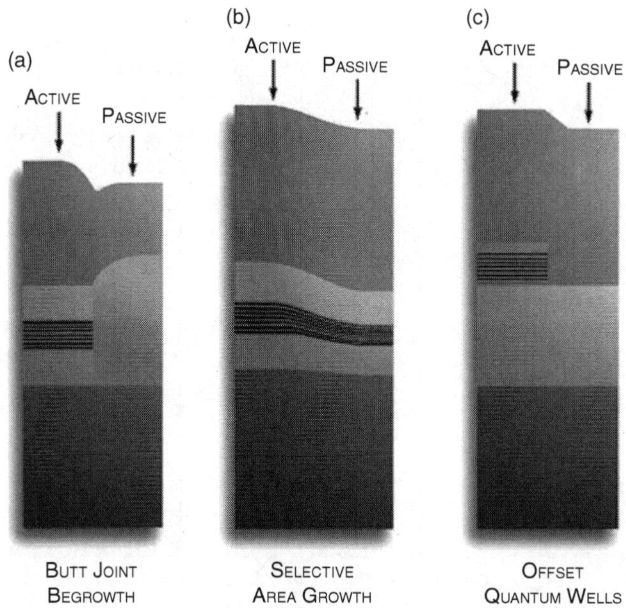

Figure 12.1. (a) Butt joint growth, (b) selective area growth, and (c) offset quantum well techniques showing cross-sectional active and passive sections perpendicular to the growth direction [4].

loss at the interfaces [3]. Increasing the number of desired band edges and waveguide architectures increases the complexity since each new band edge and structure requires an additional regrowth [4].

Selective Area Growth (SAG)

The SAG technique can be used effectively to provide multiple band edges across the epitaxial wafer in a single growth step [5]. This method involves patterning of a dielectric mask on the epiwafer, which is followed by the epitaxial growth (e.g., metal organic chemical vapor deposition [MOCVD] growth). This dielectric mask limits the growth. After butt joint growth, additional regrowth for the upper cladding is performed following the removal of the dielectric mask (Fig. 12.1b). For example, in the case of making a waveguide that integrates a laser diode with a modulator diode along the length of the waveguide, the centered multiple quantum well (c-MQW) active regions in the waveguide are grown on a single chip both for maximum gain in the laser diode section and for maximum electroabsorption in the quantum well electroabsorption modulator (QW EAM). As this technique is based on the contrast in the surface kinetics of the growth elements on the semiconductor and dielectric, the reactor conditions should be controlled very carefully. However, one problem is that the surface diffusion around the transition region might be on the order of tens of micrometers. Also, thickness variation may cause a nonideal optical mode overlap with the MQW region [4].

Offset Quantum Wells

This method involves the growth of the MQW active region on a passive bulk waveguide. This technique allows the nonabsorbing waveguide section to be left for the selective etching of MQWs because the gain is not a concern in these etched regions. After building the active and passive regions, the upper cladding is regrown in this technique as well (Fig. 12.1c) [6]. The strength of this technique lies in its simplicity. However, allowing only for two band edges at most (one from MQW and one from the waveguide) is a drawback of this method [4].

The offset quantum well method can be improved with the addition of a wider band gap quantum well in the passive waveguide [6], known as dual offset quantum wells, shown in Figure 12.2a. This improvement provides more efficient electroabsorption and phase modulators by using a quantum-confined Stark effect resulting in a more abrupt absorption edge [2].

Quantum Well Intermixing (QWI)

Another method for the monolithic integration is the QWI technique. In this method, the metastable nature of gradient at the interfaces is used. Diffusion tendency under compositional/concentrational gradients is the driving force of this process. The diffusion rate can be adjusted/increased using catalysts. QWI process can be made selective by using a lithographically defined catalyst patterning. Some of the techniques used for

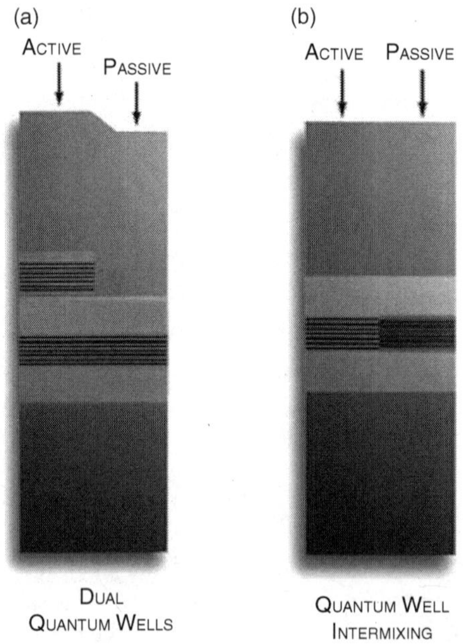

Figure 12.2. Schematics for (a) dual quantum wells and (b) quantum well intermixing methods [4].

QWI are impurity-induced disordering (IID) [7], impurity-free vacancy-enhanced disordering (IFVD) [8], photoabsorption-induced disordering (PAID) [9], and implantation-enhanced interdiffusion [10]. A schematic representation of the process is given in Figure 12.2b. QWI does not change the average composition; however, it changes the compositional profile slightly. Therefore, reasonable index continuity is observed at the interface between adjacent parts. This property of the QWI method avoids parasitic reflections, which would otherwise decrease the performance of the device [4].

Coldren and coworkers used the implant-enhanced interdiffusion technique to build up a 10-Gb/s widely tunable transmitter. This technique involves the diffusion of point defects created during ion implantation into an InP implant buffer layer grown on an MQW active region [2]. Good spatial resolution and controllability (using annealing time, temperature, and implant dose) are the strengths of this method [11].

The IID method relies on the use of impurities, generally dopants, to change the equilibrium defect concentration, which depends on the Fermi level to enhance self-diffusion into the crystal resulting from intermixing [7]. A disadvantage of this method is the adverse effects of dopants on the electrical properties of the device [12].

The PAID technique is based on the heating of the MQW structure through laser irradiation to provide intermixing [12]. The drawback of this technique is its poor spatial resolution, although the material quality is very high [9].

The IFVD method relies on the postgrowth introduction of vacancies that diffuse through the lattice while exchanging places with adjacent atoms as a result of a thermal process. This diffusion mechanism is the driving force of the intermixing for this technique. In spite of its advantage of simplicity, high annealing temperatures to adjust the quantum well are a disadvantage of this method [9].

Multigrowth Monolithic Integration

Sabnis et al. proposed and demonstrated an improved version of multistep selective area growth (MSAG) monolithic integration technique [13]. This approach mainly consists of five steps:

1. Proper MSAG surface preparation and device isolation are maintained by stopping the incorporation of buried epitaxial etch. Thanks to these layers, it is possible to integrate different epitaxial layers with clean and planar surfaces before epitaxial regrowths. Sequential surface preparation for MSAG can be realized by burying multiple etch-stop layers, one for every MSAG process.
2. Different epitaxial structures are combined by an MSAG process.
3. Device mesas in the SAG are constructed by an MSAG epitaxy etching while using selective wet etches and shallow dry etches.
4. Planarization and passivation processes are applied.
5. Another planarization and passivation process is applied to enable multilevel metal routing for maintaining the desired circuit architecture. After the wafer planarization, interconnection of metal from level to level is still needed. This requirement is fulfilled by using vias and plug metallization. Another approach

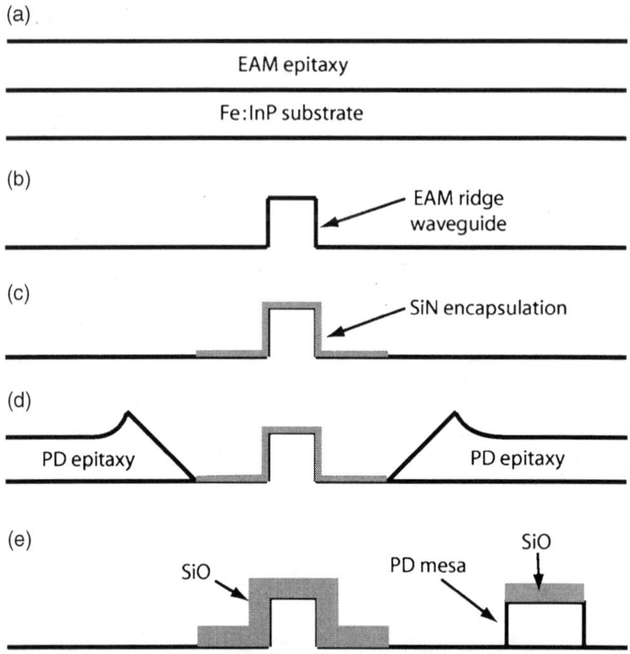

Figure 12.3. MSAG process for the monolithic integration of a photodetector (PD) and an electroabsorption modulator (EAM): (a) epitaxial growth of EAM, (b) ridge etching of EAM, (c) SiN mask definition for SAG, (d) PD SAG, and (e) PD mesa etch [13].

is to etch the planarization material such that a slope is constructed at the interface between layers where the metal connection is desired.

These steps are shown in Figure 12.3.

Surface Passivation and Planarization

During the selective removal of the epitaxially grown material by vertical etching, the sidewalls of the device are typically left exposed. These regions are chemically active and may introduce surface states, which further results in current leakage and stability problems. Sidewall passivation and planarization of the passivation material are therefore required procedures for the vertically etched III–V semiconductor devices [14]. The sidewall passivation suppresses the leakage current, whereas the planarization of the passivation material is needed for the metal interconnection and device integration.

In the etch-back planarization method [15], which is a conventional approach, the planarization of the entire wafer is difficult to seal and passivate device sidewalls. To solve this problem, Demir et al. proposed an approach to realize self-aligning planarization and passivation. In this method, the passivation is achieved by the use of a passivation polymer to passivate the sidewalls across the epitaxial wafer.

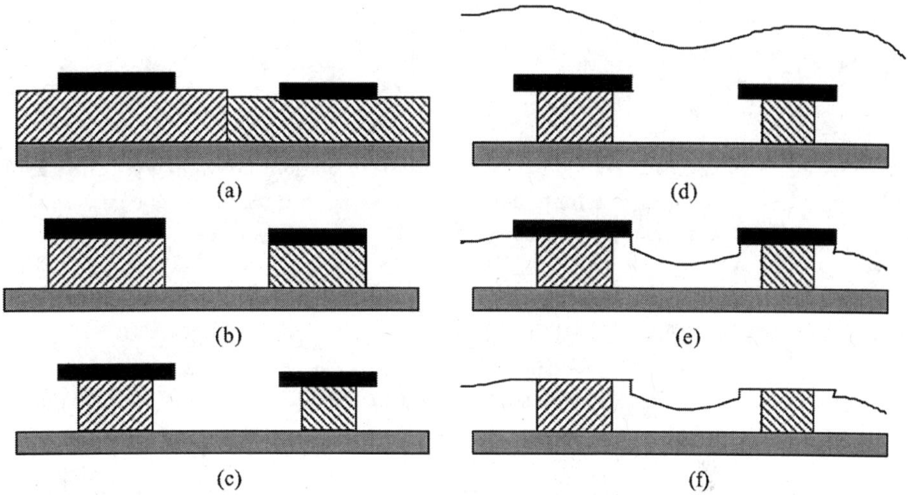

Figure 12.4. Illustration of self-planarizing passivation sequence: (a) The hard mask on the epitaxial wafer is defined; (b) device areas are patterned by vertical etching; (c) the device materials underlying the hard mask perimeters are removed for the undercut creation; (d) the device areas are sealed; (e) the cured polymer is etched back; and (f) the hard mask is removed [14].

The process can be summarized in Figure 12.4 as follows: (1) A dielectric mask is defined on the wafer, which is maintained in the subsequent steps; (2) by vertically etching the unmasked parts of the wafer, the semiconductor device area, which is masked, is defined; (3) the sidewalls of the semiconductor mesa are selectively etched horizontally to form an undercut beneath the hard mask to protect the device as a roof; (4) the sidewalls of the device layers under the hard mask are sealed and passivated with a spin-on polymer; (5) the cured polymer past the top level of the hard mask is etched back with an anisotropic dry etch; and (6) the hard mask is removed in order to construct a passivation structure with the polymer around the device perimeter. In this work, benzocyclobutene (BCB) is used as the passivation polymer, while silicon nitride or silicon oxide is used as the hard mask [16].

Via and Trench Building for Metal Contacts

In III–V semiconductor materials, vias and trenches are used for metal contacting [17]. In the work of Zheng et al., the construction of these structures involves the use of a sacrificial layer, which is made of either a SiO_2 dielectric hard mask layer on the device layers or a semiconductor layer grown onto the device epitaxial layer, both before patterning. The construction steps follow: (1) formation of a hard mask using conventional dielectric thin-film deposition, (2) etching of the semiconductor by reactive ion etching using the hard mask, (3) spinning of a polymer layer and back etching within the hard mask for every device in the wafer, and (4) removal of the hard mask by wet etching to have a via structure. This procedure is illustrated in Figure 12.5.

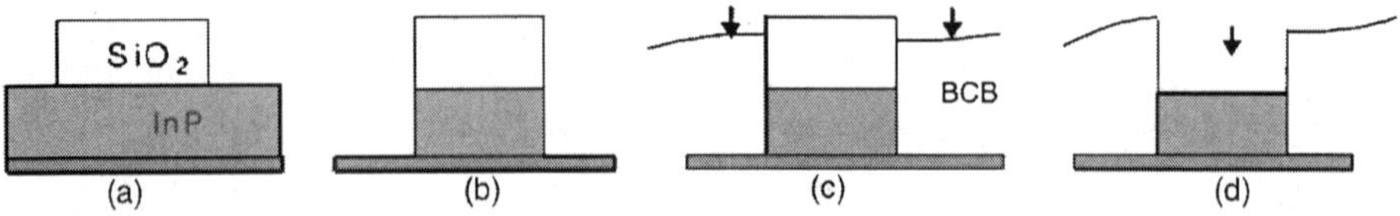

Figure 12.5. Schematic illustration of via and trench formation: (a) formation of the hard mask, (b) etching the semiconductor to define the active region, (c) spinning and back etching of the polymer, and (d) removal of the hard mask to obtain via and trench opening [17].

NANOFABRICATION TECHNIQUES

Nanoscale material design and nanoscale fabrication techniques are in general grouped as top-down and bottom-up approaches [18]. The top-down method involves various techniques of lithography to pattern nanoscale structures. This approach may make use of serial and parallel techniques to pattern features. On the other hand, the bottom-up method is based on the interactions between molecules and colloidal particles to construct discrete nanoscale structures in two and three dimensions.

Photolithography and scanning beam lithography are considered in the conventional top-down approach. Their limitations are their high costs, difficulty in accessing the required facilities, and lack of applicability in a wide range of problems. Less conventional approaches also cover both top-down and bottom-up methods. Molding, embossing, and printing can be classified as top-down methods; on the other hand, scanning probe lithography (SPL), edge lithography, and self-assembly can be interpreted as bridge methods between the top-down and bottom-up approaches.

In general, conventional techniques suffer high cost and low throughput, and they are typically restricted to planar fabrication in semiconductor materials and are incompatible with many problems in nonstandard fabrication. Additionally, these techniques expose substrates to corrosive etchants, cause high-energy radiation, and require high temperatures. These are the driving force for the investigation of new nanofabrication methods.

Photolithography

Photolithography is the chosen method for the manufacturing of microelectronic devices by the industry. In photolithographic systems, mask aligners shine collimated light trough a quartz plate that supports a patterned chromium coating and expose a photoresist thin film coated on a planar substrate. The photoresist is an organic material that becomes insoluble or more soluble as a result of chemical changes when exposed to high-energy (short wavelength) light [19–21]. The photoresist exposed to light is soaked in a solvent that dissolves exposed or unexposed regions. This process thus allows for transferring the pattern of the starting mask to the exposed and developed photoresist film on the substrate. This patterning also functions as a mask for the substrate during subsequent steps (Fig. 12.6). A modern photolithography system can pattern with a resolution of 27 nm [22]. In 2005 semiconductor nanofabrication productions, photolithography is used

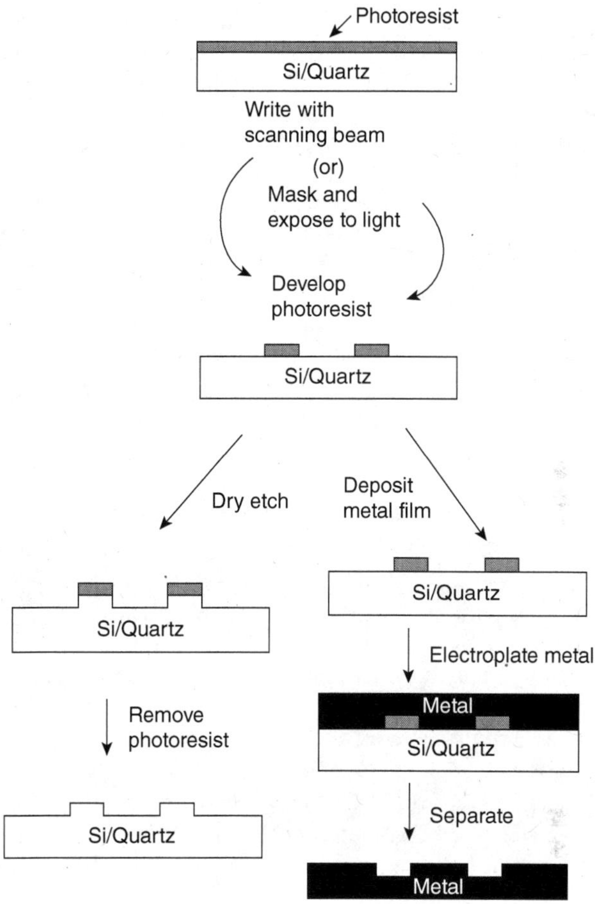

Figure 12.6. Schematic illustration of fabrication using photolithography and electroplating in hard materials [18].

to pattern 37-nm-wide features with 193-nm wavelength light. Attempts to decrease these sizes require optical proximity correction (OPC) and phase-shift mask technology; however, all of these modifications increase the cost significantly.

Immersion lithography is proposed as a potential lithography method for features with sub-50-nm resolution exposing light at 193 nm [23]. This lithography technique is the analogous of immersion microscopy that is widely used in biology [24]. In this technology, the imaging resolution is improved by the increased refractive index of the medium between the imaging lens and the imaging plane [24]. This method requires contacting the photoresist-coated substrate and mask with water (or another solvent) and using complex, aspheric, catadioptric lenses. Another solution toward patterning smaller structures using photolithography might be decreasing the wavelength of exposing light. This, however, requires new types of photoresists, new light sources, and new types of optics.

The main drawback of the photolithography lies on the time and cost requirements to fabricate the photomask. On the other hand, interferometric lithography provides the production of simple patterns without using a photomask [25, 26]. This method is based on the constructive and destructive interference of multiple laser beams at the surface of photoresist. Although no photomask is needed for this method, pattern projection is restricted only for regularly spaced arrays.

Scanning Beam Lithography

Scanning beam lithography is a process that is often used for the production of photomasks for projection lithography instead of actual device fabrication. Patterning dense arrays of sub-20-nm features on an area of 1 cm^2 takes ca. 24 h [18]. Therefore, scanning beam lithography is restricted to small areas or to areas with low densities.

Scanning beam lithography is a slow process compared to photolithography. However, this technique also allows for the fabrication of arbitrary patterns. Scanning beam lithography has three classes: (1) scanned laser beams with ca. 250-nm resolution (the least expensive), (2) focused electron beams with sub-50-nm resolution, and (3) focused ion beam (FIB) lithography systems with sub-50-nm resolution (often used in research).

High-resolution patterning comes with some trade-offs. An increase in the resolution can be achieved only by decreasing the diameter of the particle beam. As a result, the beam current also decreases, which results in an increased processing time, required to pattern the entire surface.

An FIB enables writing patterns into a photoresist or directly onto the substrate [27]. Using this technique, materials can be selectively removed through ion bombardment, or patterns can be created additively by localized chemical vapor deposition or by ion deposition [28–30]. FIB lithography enables patterning features in a semiconductor with resolution down to 20 nm and with lateral dimensions down to 5 nm [31].

SPL

SPL is a promising tool for manipulation and imagination with an atomic-scale resolution [32]. However, this tool requires improvements for use in large-area manufacturing. Scanning tunneling microscopy (STM) (Fig. 12.7A), atomic force microscopy (AFM), and near-field scanning optical microscopy (NSOM) are some of the most important SPL techniques. Although atomic-scale manipulations on surfaces are possible using these techniques, there is still a long way to go in order to make these techniques practical [18]. In general, the operation of SPL can be described as the localized modification of surface by oxidation or by material transfer using a sharp probe in contact with the surface [18, 33, 34].

Serial Patterning of Surfaces

Dip-pen nanolithography (DPN) provides methods to dispose nanoparticles or other molecules selectively onto a surface [34]. Inking AFM tips with a solution of the materials to be transferred enables positioning the material onto the substrate when the

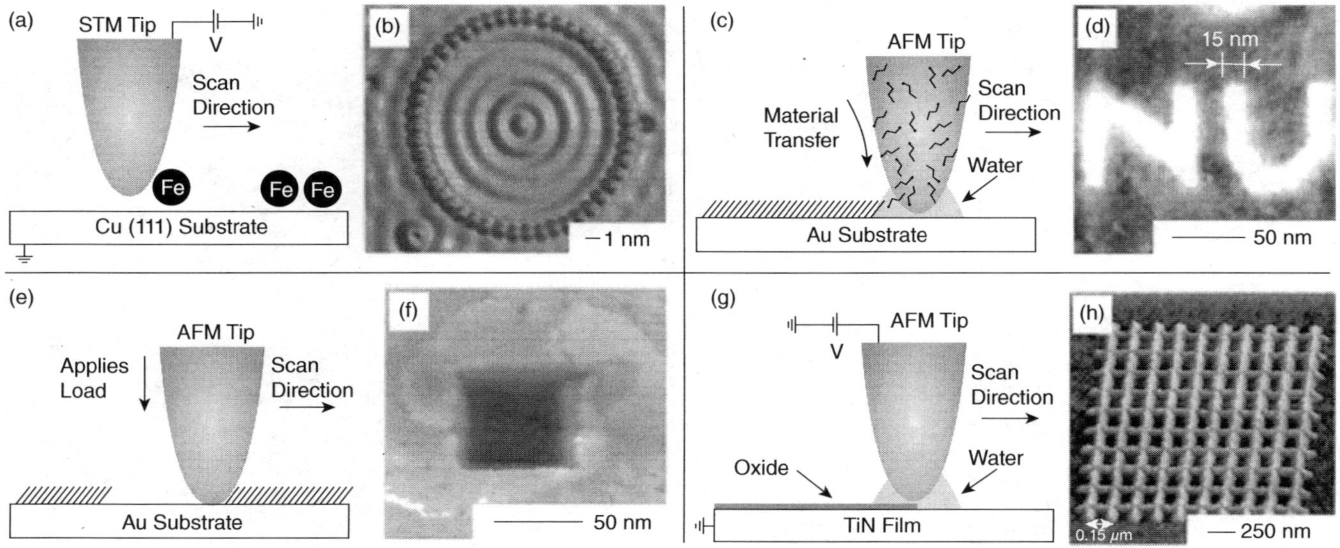

Figure 12.7. Schematic illustrations of four approaches to SPL and patterns produced using (A) STM [35] (B) to make a quantum corral of a 48-atom Fe ring formed on Cu enclosing a defect-free region [198]; (C) dip-pen lithography (DPN) [35] (D) to pattern features down to 15 nm [199]; (E) nanoshaving [35] to remove regions of SAMs (F) to pattern a square hole within octadecane thiolate SAMs on Au [200], and (G) scanning electrochemical oxidation [35] to selectively oxidize a surface (H) to build ca. 50-nm-wide lines of oxide written on TiN [38].

probe scans the surface (Fig. 12.7C) [35]. Using DPN 15-nm patterns is achievable (Fig. 12.7D). This mechanism cannot be understood perfectly; however, several theories have been proposed. According to one theory, a water layer between the tip and surface makes the material transfer possible. According to another one, the material transfer is possible due to the solid–solid interactions. Humidity, chemical interaction with the substrate, radius of curvature of the probe, and linear velocity of the probe determine the spreading of the ink.

Another approach for the patterning of the surfaces is the selective removal of material from the surface. This removal operation is performed by mechanical abrasion using a scanned AFM probe [36]. The probe displaces a thin film when the contact load between the tip and the substrate is higher than the displacement threshold of the surface layer. This process is called "nanoshaving," and it is used for the patterned removal of the self-assembled monolayers (SAMs) (Fig. 12.7E, F) [20]. Material abrasion can also be performed after multiple scans when the contact load of the AFM tip is still below the displacement threshold of the material [37]. Nonuniformities in the topography, however, can cause irregular patterns. A surface can be chemically modified using SPL. This operation can be made using a conductive AFM or STM tip. In Figure 12.7G, H, a TiN substrate patterned by localized electric field-induced oxidation using a conductive AFM probe is shown [38].

Parallel Patterning of Surfaces

Although SPL methods provide flexible patterning possibilities, the use of a single tip makes these methods impractical for mass production. Using multiple probes that work in parallel may remedy this problem [35, 39]. A two-dimensional array of independently addressed AFM probe is developed, and this concept is known as "millipede" (Fig. 12.8) [39]. In this approach, deflection in the vertical direction and resistive heating can also be provided.

Figure 12.8. Millipede approach example [39].

Edge Lithography

In edge lithography, topographic edges are used in the fabrication of nanoscale structures [35, 40, 41]. By using this method, arrays of structures that are smaller than 100 nm can be produced. Edge lithography can be grouped in two techniques: (1) edge lithography using topography-directed pattern transfer and (2) cleaved or cut edge nanostructures.

Pattern Transfer Directed by Topography. Decoration of step edges: At the step edges of single-crystalline surfaces, metals and other materials are deposited selectively with success [42, 43]. This deposition procedure results in the generation of continuous nanowires.

Electrodeposition or etching at the edge defined defects in SAMs: Selective removal and deposition of materials are the other ways of patterning nanostructures. [40, 44, 45] An example might be that sharp metal corners within topographically patterned metal substrates avoiding well-ordered SAM formation expose underlying metal at the edges (Fig. 12.9A) [45]. Selective etching of this metal makes it possible to transfer the outline of the patterned metallic topography onto the underlying film.

For example, SAMs can be formed on silver, but not on titanium, when the patterned substrate is put into an alkanethiol solution. The exposed Ti/TiO_2 layer builds a ~5-nm gap in the SAM. This gap may serve as a nanoelectrode upon which metal can be electrochemically deposited (Fig. 12.9B) [31]. A curved or flat substrate can be produced while using an adhesive to remove these structures. Incident light can be polarized using the arrays of Cu lines on an adhesive (Fig. 12.9C) [45].

Controlling undercutting at edges: After the deposition of a thin film, topographic features can be undercut in a controlled fashion. This technique also provides the fabrication of nanostructures (Fig. 12.9D) [44]. For example, isotropic wet etching of a thin metal film on a silicon or CaF_2 substrate covered with a photoresist provides patterned structures following the undercutting of the photoresist [44]. Trenches at the edges of the pattern can be observed by lifting off the photoresist (Fig. 12.9E, F) [44].

Phase-shifting photolithography: Controlling the phase of the exposing light, which is used to create narrow constructive and destructive interference, allows for significant modifications in the intensity profile when the light meets the vertical edges of a transparent, topographically patterned substrate. This phenomenon is used to create dark and bright spots of incident light onto the photoresist. Optimally, the light should have a phase shift of π at the photoresist–mask interface.

Cleaved or Cut Edge Nanostructures. Various techniques to grow thin films with a careful control of thickness between 1 and 50 nm enable additional following methods for the fabrication of nanostructures.

Edges Produced by Fracture. Simple quantum structures can be patterned using the cross-sections of multilayer films produced by molecular beam epitaxy (MBE) as templates [41]. For example, field-effect transistors (FETs) with a 20-nm gate length are produced using this procedure. In this example, alternating

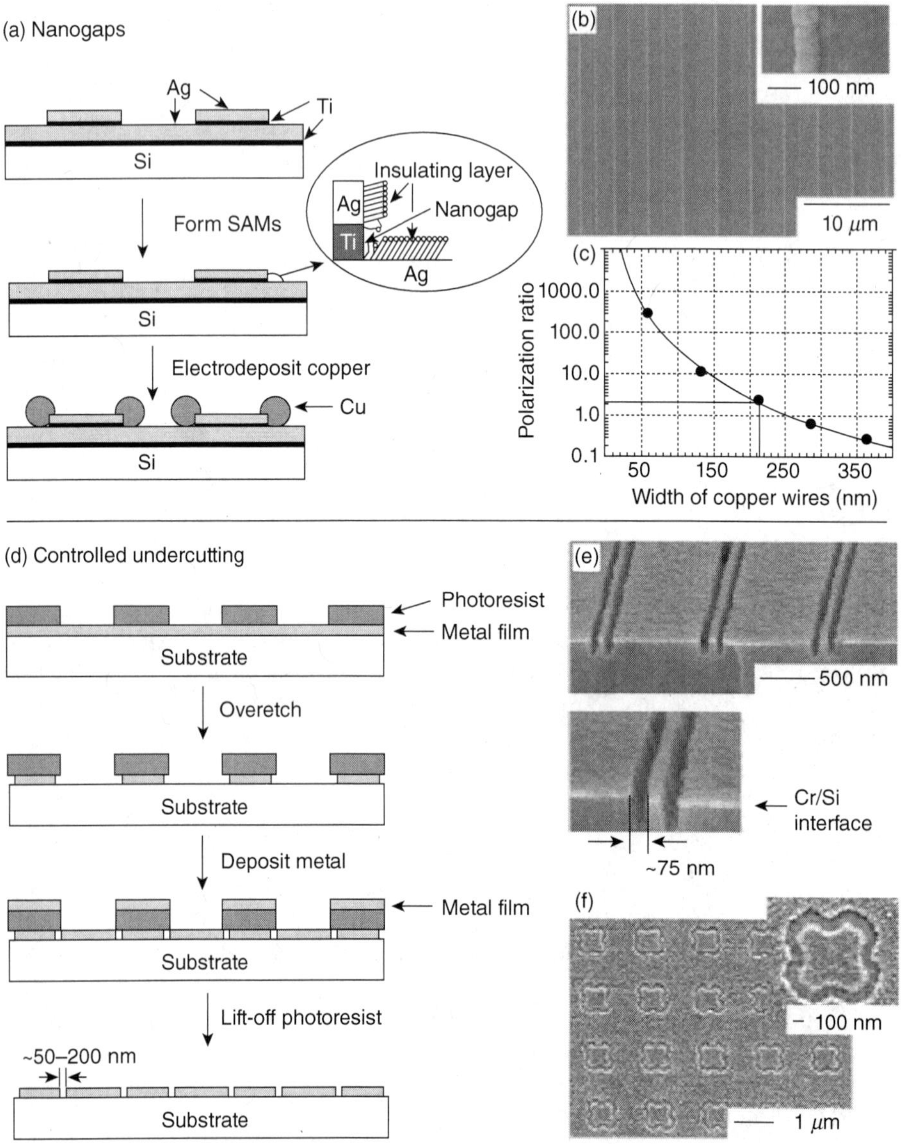

Figure 12.9. (A) Production of nanogaps by engineering defects in Ag SAMs, (B) scanning electron microscope (SEM) image of 70-nm-wide lines of Cu selectively electrodeposited on these nanogaps [45], (C) polarization ratio of the electric component of light as a function of line width [45], (D) illustration of controlling undercutting, (E) cross-section of ~75-nm-wide trenches in silicon [44], and (F) frequency-selective surface of 100-nm-wide trenches in Al [44].

layers of AlGaAs and GaAs, which are grown using MBE, are used (Fig. 12.10A, B) [41]. Parallel arrays of nanowires can be built using selectively etched, multilayered MBE-grown GaAs/AlGaAs substrates as the physical templates with narrowly spaced groves for patterning by physical vapor deposition (PVD). The widths of the wires are determined by the GaAs layers, and the spacing between wires is controlled by AlGaAs layers [46].

Edges Produced by Sectioning with a Microtome. A nanostructured edge can be built up by embedding a thin film in a matrix and by taking its cross-section. One approach to do this is based on removing the matrix material by repeated sliding against a rough surface [47, 48]. This technique can be extended using a microtome to section a polymer-encapsulated patterned metal film (Fig. 12.10C, D) [49, 50].

Edges Positioned by Reorientation. Tipping periodic arrays of posts onto one side can be used to pattern large-scaled areas (~cm^2) (Fig. 12.10E) [51]. This process enables one to generate regular arrays of features in nanoscale lateral dimensions in plane (Fig. 12.10F, G).

Soft Lithography

Soft lithography is a collection of techniques that involve organic (soft) materials to provide replication and pattern transfer on a wide range of length scales (from nanometer to centimeter) [41]. The term soft is used for the stamp and/or substrate composed of materials including polymers or SAMs. The pattern transfer is achieved generally by first fabricating a topographical patterned master then molding this master to build up a patterned stamp, and finally generating a replica of the original template (Fig. 12.11). Conventional methods are good enough for the fabrication of masters, whereas the rest of the process needs new techniques. Soft lithography can be described in three steps: replica molding, embossing, and microcontact printing.

Replica Molding. Replica molding transfers topographical features from a rigid or elastomeric mold into another material by solidifying a liquid in contact with the original pattern [52]. Step-and-flash lithography technique (Fig. 12.12A), developed by Willson and colleagues, involves the use of a transparent master, for example, quartz [53]. After contacting this mold with a low-viscosity, photocurable polymer, the polymer is held under UV light. With this method, features as small as 30 nm can be produced [54]. The strengths of this approach are the rapid cyclic time (<5 min/replication) and the ability of optical mold alignment with features on the underlying surface [53, 55]. However, the difficulty of patterning nanostructures on nonplanar surfaces replicating isolated features caused by air at the mold–polymer interface is the weakness of this technique. Replica molding can also be achieved using deformable templates, for example, polydimethlysiloxane (PDMS). Figure 12.12B makes an illustrative summary of this procedure. Another way for the replica molding is the microtransfer molding, in which the liquid prepolymer is put into the molding regions. After cleaning the excess polymer using a flat edge (a doctor blade), the mold is placed in contact with the rigid substrate. Finally, the prepolymer is cured (Fig. 12.12C). This technique is used for

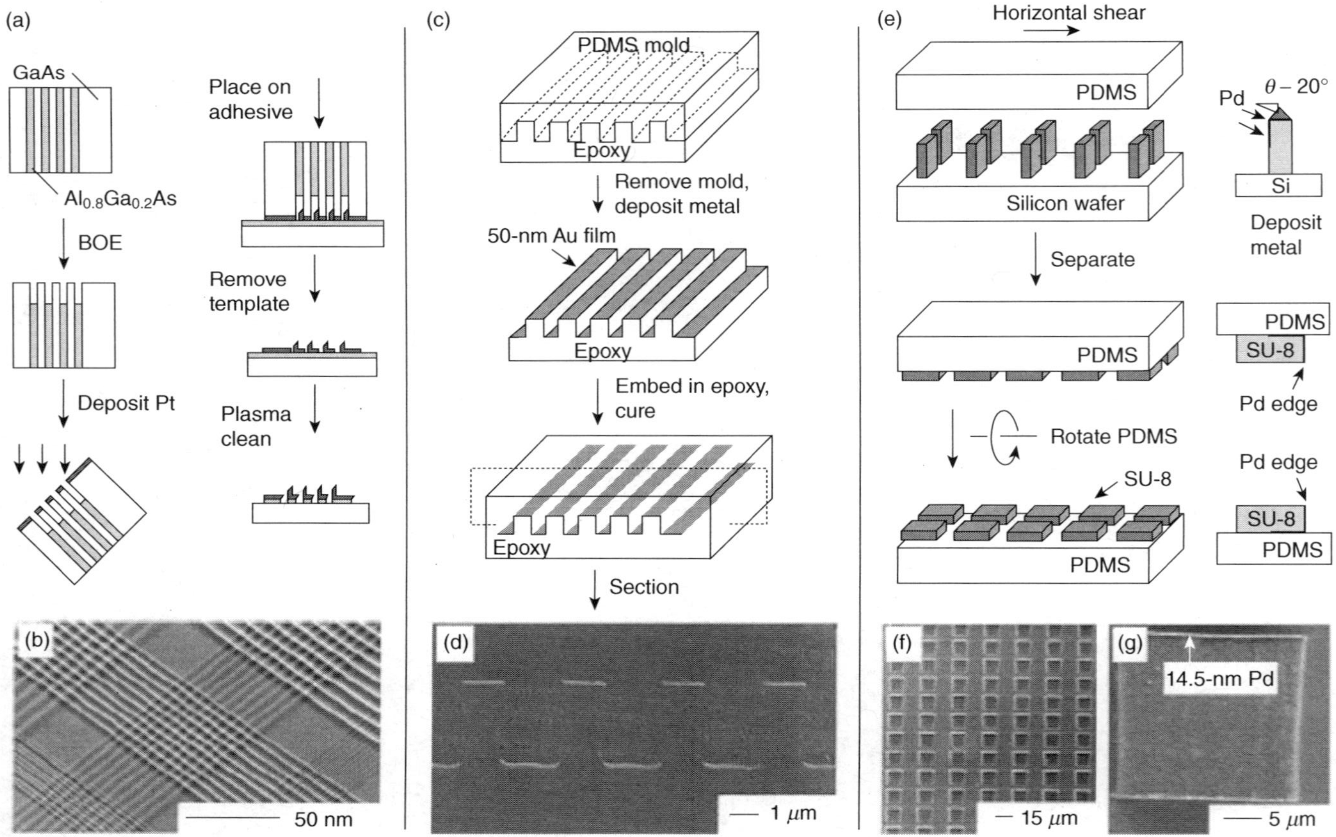

Figure 12.10. (A) A schematic illustration and (B) SEM images of Pt nanowire arrays [41], (C) illustration of the fabrication of a patterned array of epoxy-embedded conducting metal edges, (D) SEM image of these metal edges [41], (E) schematic illustration of microdomino array uniformly collapsed after the application of a horizontal shear (PDMS) [42], (F) SEM images of these structures, and (G) a ~15-nm-wide edge of Pd on a collapsed post [42]. BOE, buffered oxide etch.

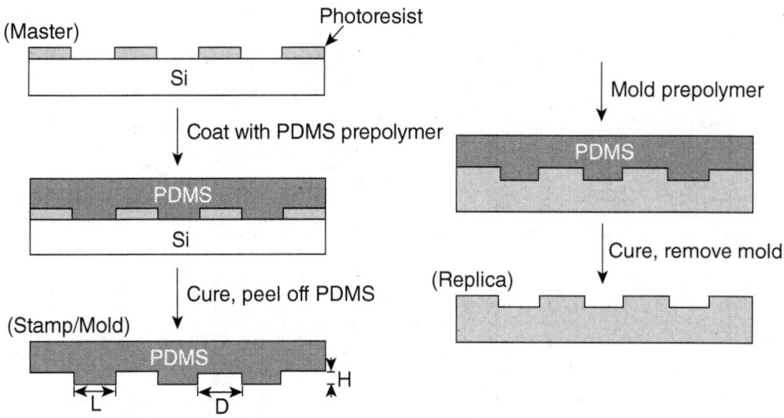

Figure 12.11. Schematic illustration of soft lithography [35].

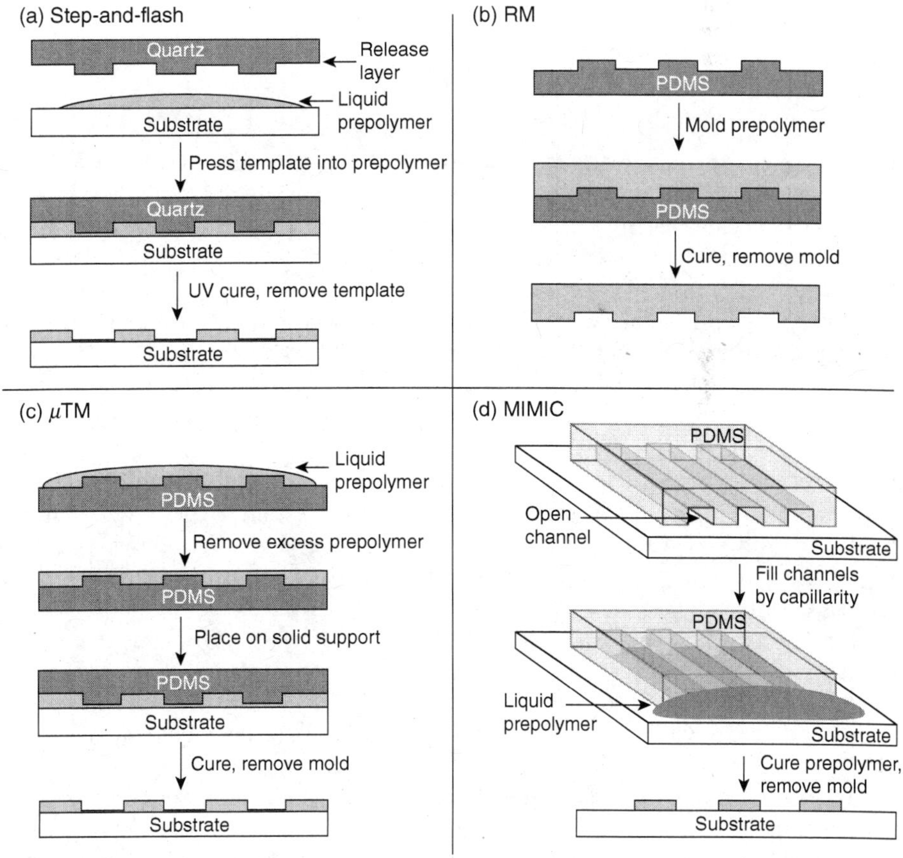

Figure 12.12. Schematic illustrations of molding techniques: (A) step and flash, (B) replica molding (RM), (C) microtransfer molding (μTM), and (D) micromolding in capillaries (MIMIC) [35].

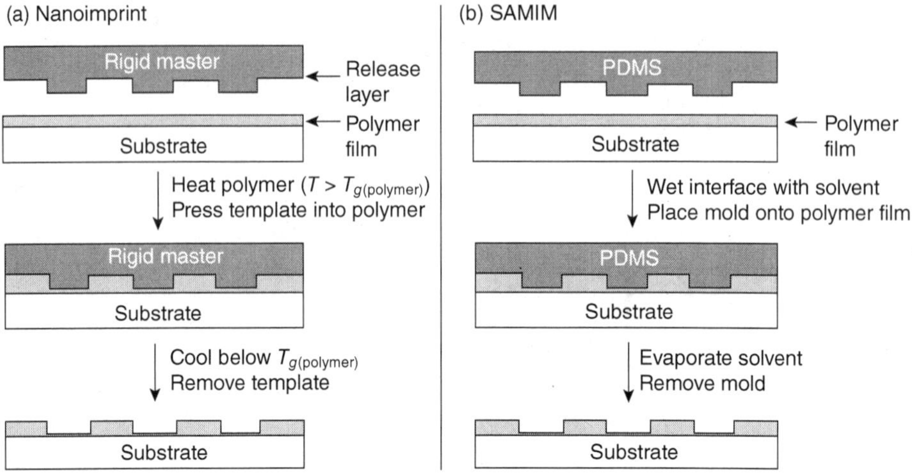

Figure 12.13. Illustration of embossing procedures: (A) nanoimprint lithography and (B) solvent-assisted micromolding (SAMIM) [35].

patterning planar or curved surfaces [56]. Micromolding in capillaries is also another way of molding [57]. This technique involves the use of capillaries to fill the channels of a topographically patterned stamp with a liquid. The precursor material is cured within the capillaries and the mold is removed (Fig. 12.12D).

Embossing and Nanoimprinting. Embossing techniques are used for manufacturing compact disks and digital versatile disks (DVDs). The modifications of these techniques can also be extended to nanofabrication with a resolution below 50 nm [58]. Nanoimprint lithography is used to transfer a pattern from a rigid mold into a deformable material coated on a rigid surface (Fig. 12.13A). In general, the material is first heated under pressure. The mold is later removed after cooling the material below its glass transition temperature. A structure that is as small as 10 nm can be replicated using imprinting with aspect ratios 10:1 [59]. In this technique, rigid mold and spin-cast polymer film must be positioned in parallel, and the applied pressure should be uniform. These are the main drawbacks of this approach.

Embossing can also be applied using soft mold. Using PDMS mold solvent-assisted micromolding (SAMIM) takes place with an appropriate solvent instead of using rigid mold, high temperatures, or pressures (Fig. 12.13B). After placing the polymer into the mold, evaporation is completed and, finally, the mold is removed.

Microcontact Printing. The general procedure for microcontact printing involves contacting the surface of a metal, metaloxide, or semiconductor with a topographically patterned PDMS stamp, wetted with a solution of an alkanethiol or other molecules that can form SAMs for a few seconds. An ordered monolayer is rapidly formed at the points of contact. The strengths of microcontact imprinting for fabricating nanoscale structures are that the minimum feature size on the pattern depends on the physical dimensions of the step (instead of optical diffraction), that the stamp provides mechanical flexibility, and that various types of materials can be used as inks on different materials. Blurring of pattern by lateral diffusion of the ink, the deformation in the pattern reflecting deformations in the stamp, and high number of defects in the film are, on the other hand, the disadvantages of this technique [60].

GENERAL SELF-ASSEMBLY TECHNIQUES

Self-assembly is defined as the self-organization of different components into regular structures without human intervention. Self-assembly processes can be observed and used in various fields of technology and nature. For nanotechnology, self-assembly appears to be one of the fundamental and most promising techniques to construct nanostructures. Self-assembly can be classified mainly into two groups: static self-assembly and dynamic self-assembly.

Static self-assembly is based on the systems that are in equilibrium and that do not dissipate energy (Fig. 12.14) [61]. Molecular crystals and globular proteins are among the examples [62, 63]. Structures formed by static self-assembly may require energy during their building process; however, once they are formed, the structure is stable and therefore no energy is further needed to hold the system stable. Most of the research projects focus on this type of self-assembly.

In self-assembly, the components should be able to change their positions with respect to each other so that they can balance attractive and repulsive forces in their steady-state position. In molecular scale, these forces are often observed as the results of van der Waals, electrostatic, hydrophobic, and hydrophilic interactions. The forces observed in the self-assembly of larger structures include gravitational force, external electromagnetic forces, and magnetic, capillary, and entropic forces. Since the movement of particles is critical for the self-assembly processes, the interactions occur often in liquid media or on smooth surfaces.

Although researches are mainly focused on static self-assembly, dynamic self-assembly also has a significant importance (Fig. 12.15) [60]. The main essence of dynamic self-assembly is to mimic the reactions and structures that enable the continuity of life in nature. In living cells, many self-assembly processes are dynamic. When the energy flow stops, these cells die, showing the importance of these processes.

Dynamic self-assembly can also be observed in nonliving environments and systems. Oscillating reactions in solution and on the surface of catalysts, Rayleigh–Bernard convection cells are some examples for dynamic self-assembly.

Templated Self-Assembly

Templating self-assembly makes the creation of patterns into a self-assembled structure possible [20]. Additionally, it is used for increasing the order of the self-assembled structure. Surface topography, electric and magnetic fields, or shear forces can be used to direct self-assembly. Top-down methods to direct the bottom-up assembly of molecules and other particles are often preferred in the templated self-assembly approach [64].

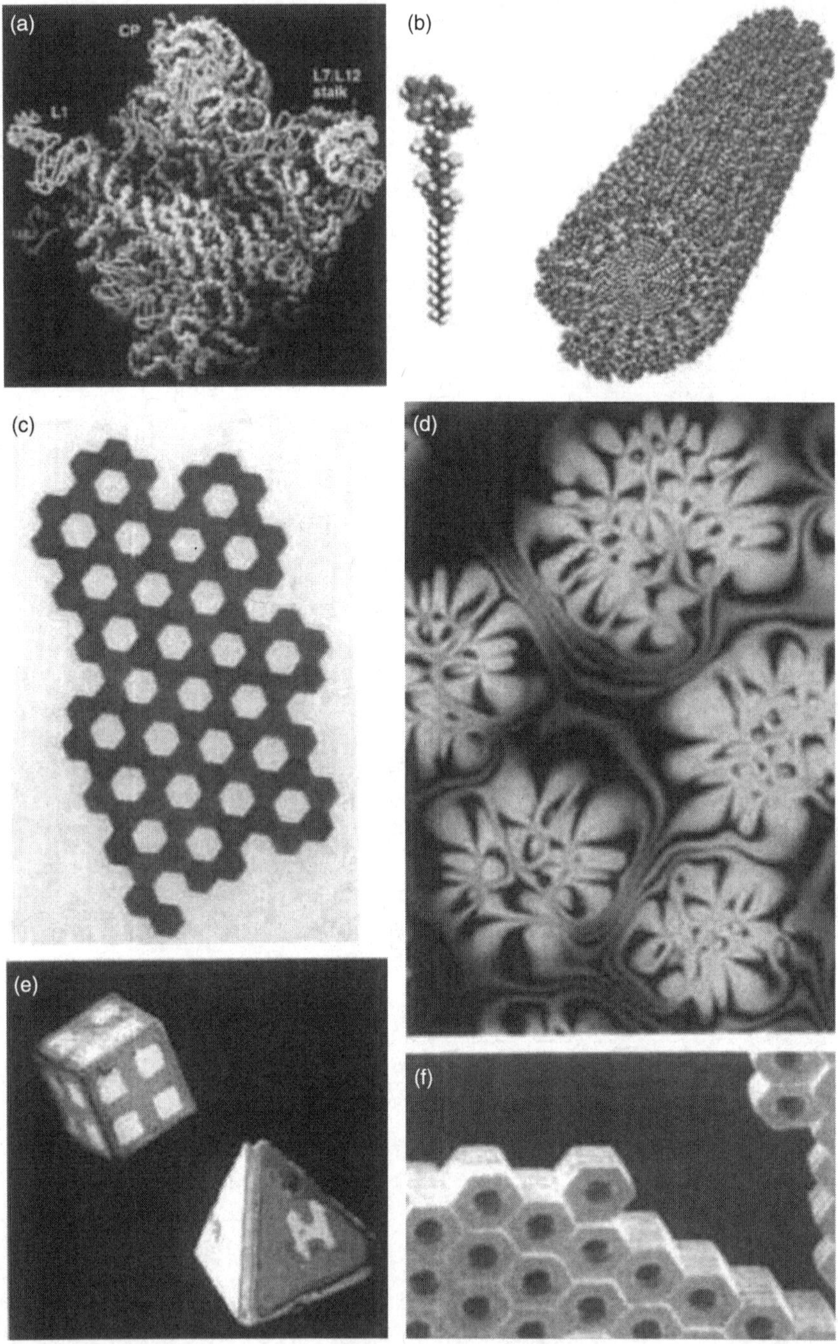

Figure 12.14. Examples of static self-assembly: (A) crystal structure of a ribosome [201], (B) self-assembled peptide-amphiphile nanofibers [95], (C) an array of millimeter-sized polymeric plates [61], (D) thin film of a nematic liquid crystal on an isotropic substrate [61], (E) micrometer-sized metallic polyhedra folded from planar substrates [202], and (F) a three-dimensional aggregate of micrometer plates [203]. (D) Courtesy of O. Lavrentovich.

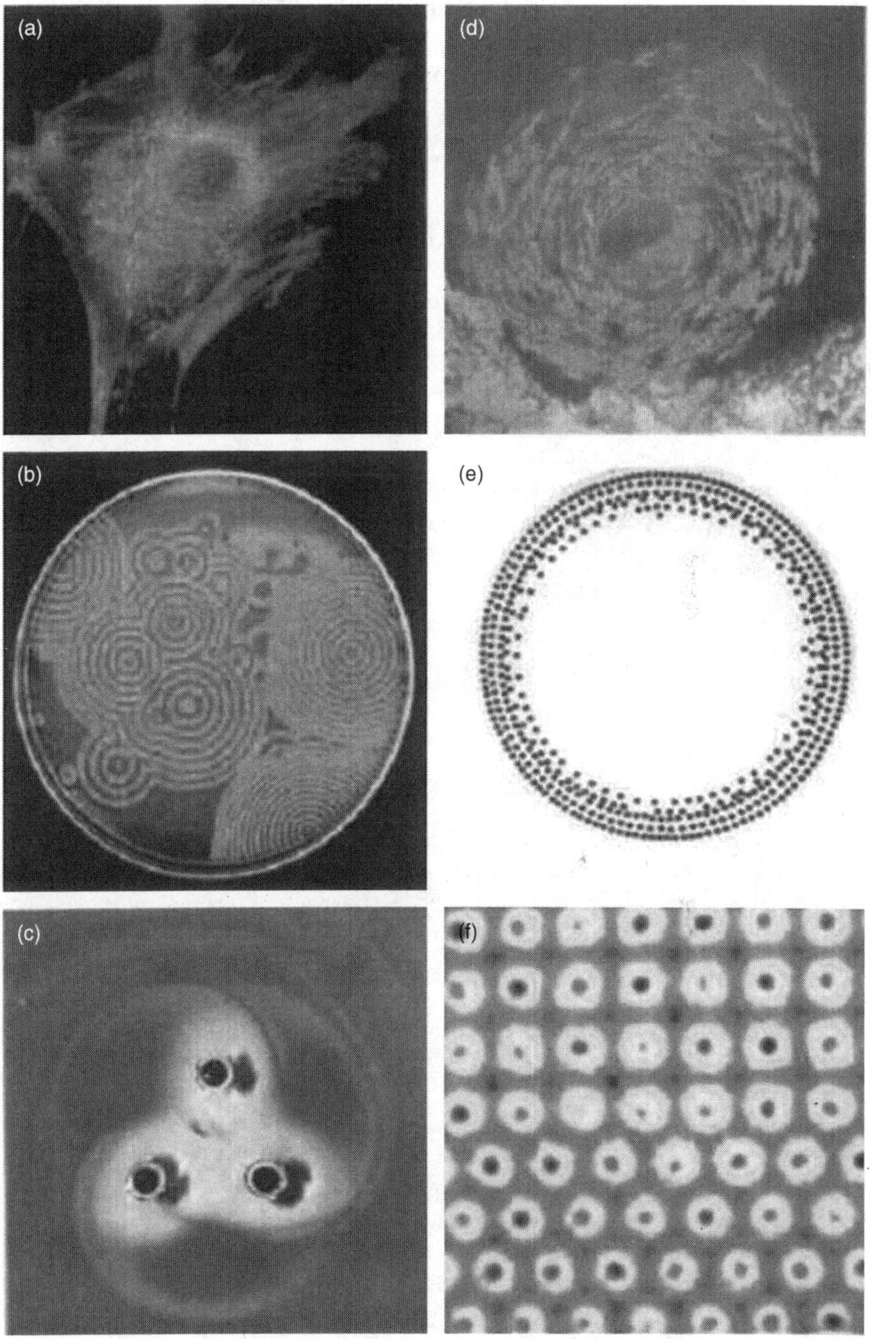

Figure 12.15. Examples of dynamic self-assembly: (A) an optical micrograph of a cell [204], (B) reaction–diffusion waves in Belousov–Zabatinski reaction [61], (C) a simple aggregate of a magnetized disk [205], (D) a school of fish [61], (E) concentric rings by charged metallic beads [61], and (F) convection cells above a metallic support [61].

Templating from Molecules. Some organic nanostructures, which are not totally functional, can be used as templates to mask the deposition of metal or to guide the deposition of metal nanoparticles and nanowires [65, 66]. Patterning block copolymers can be achieved by the combined usage of top-down and bottom-up approaches. Extreme ultraviolet (EUV) interferometric lithography can be used to pattern oxidized SAMs or random copolymers on silica [67]. When a block of copolymer is made to self-assemble, the patterned SAM has the role of a template that guides the phase separation of the polymer. Moreover, regular copolymer structures can also be formed by annealing a block copolymer film confined by physical boundaries [68]. Low level of defects, ability to control and pattern the phase-segregated regions, and controlled drift in the pattern over nonlocal dimensions are the advantages of this approach compared to nontemplated self-assembly.

Templating from Polymers. The modification of colloidal particle surfaces can be realized by using charged polymers and polyelectrodes [69]. Electrostatic attraction between charged surface and charged particles suffices for the adsorption. In most of the cases, polymers have excess charge; therefore, they change/reverse the charge on the surface [70]. This electrostatic change on the surface allows for the formation of another polymer layer, which has the oppositely signed charge of the first polymer layer. As a result, a layer-by-layer assembly process is observed. Colloidal particles are also used to template the self-assembly of nanoparticles that have diameters of >100 nm in general [71].

Templating Using External Forces. Nanospheres with similar sizes and shapes can be employed in the production of thin films of ordered lattices by using their tendency to self-assemble. In order to direct the self-assembly of nanoparticles and nanorods with different structures, electric [72], magnetic [73], shear forces [74], and additional spatial constraints[75] are used. Compared to nanospheres, the self-assembly of nanowires and nanorods is more difficult because these nanostructures have anisotropic forms [76]. Self-assembled nanowires are observed in partially ordered, small domains.

According to another classification of templated self-assembly, there are three main approaches; these are patterned chemical modifications of the surface, patterned charge, and patterned topography [77]. Preferential adsorption of particles on to the selected regions can be observed using the first approach, that is, patterned chemical modifications of the surface of the substrate [78]. The second approach can be realized by using FIB radiation [79], microcontact printing of alkanethiols on gold [80], selective adsorption of polyelectrolytes [81], and electrostatic microcontact printing [82]. Well-ordered high-quality lattices could not be observed using these methods. In order to confine and control colloidal assemblies, patterned topography of the substrate, like wells [83], lithographically patterned reliefs [84], or microfluidic channels[85] are used in the literature. The ratio of the size of the particle to the size of the patterned feature affects the formation of the resulting lattices [86]. Also, shape-selective reliefs, physical boundaries, and external fields are effective on the form of the lattice. An example of such a process is illustrated in Figure 12.16 [77].

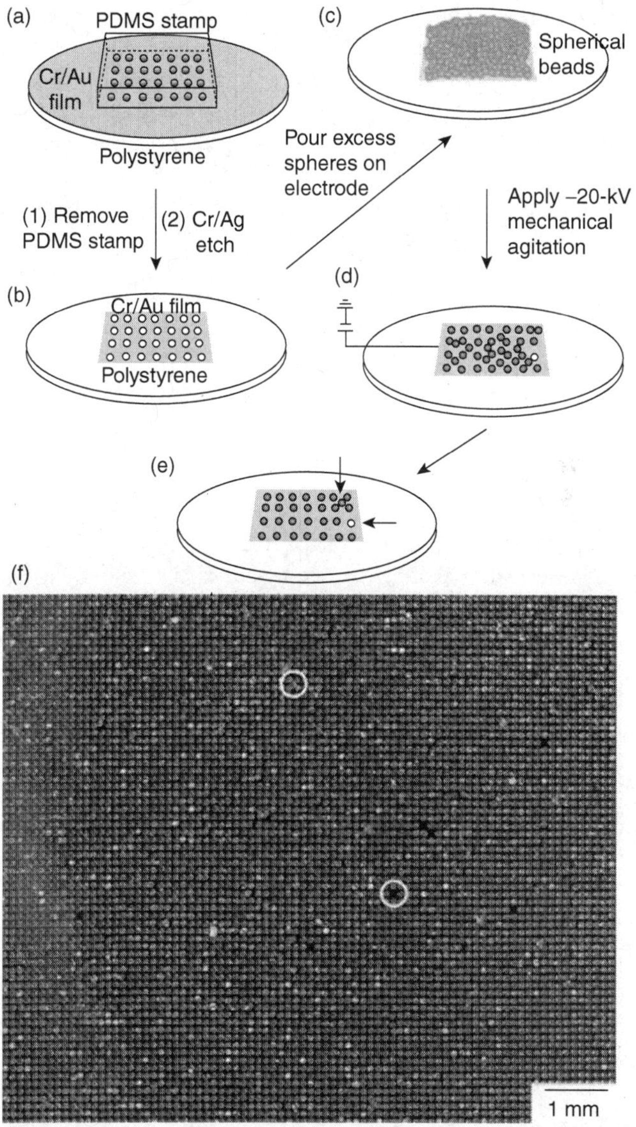

Figure 12.16. Schematic representation of template-directed self-assembly microspheres on a patterned gold electrode using an applied electric field: (a, b) microcontact printing and etching building a templated electrode; (c, d) microspheres self-assembled over the templated electrodes under –20-kV voltage applied to gold electrode; (e) ideally, one microsphere remained on each window and no microspheres on the gold surface; and (f) a typical large-area square matrix of microspheres [77].

Inorganic Templates. Alumina nanoholes and inorganic nanowires are successfully used in the templated self-assembly of nanoparticles [87]. These provide better toughness and prevent changes in the structural characteristics of nanoparticles [88]. Wirelike or rodlike nanostructures are assembled by using alumina nanoholes as templates, which are created by electrochemical etching. Anodic porous aluminas serve as nanocells that are used by building various nanostructured arrays, like magnetic nanorod arrays for longitudinal magnetic data storage, semiconductor particles for optical devices [89], and electroluminescence display devices [90]. These arrays can be used in assembling large-area arrays of parallel wires.

Nanoparticle assembly can be realized both inside and outside these nanowires/rods. As an example, the work of Ajayan and Iijima can be shown, in which they arrange Pd nanoparticles into the carbon nanotubes using capillary forces [91]. Nanoparticles can be assembled outside of these nanotemplates by direct deposition of material (physical or chemical deposition), sol–gel approach, or modifications of the surface functionalities [92].

Biological Templates. Motivation for the use of self-assembly in nanofabrication comes from the observation that many biological structures are assembled by the molecular-level self-assembly [93]. DNA strands [94], peptides [95], and viruses[96] are the possible templates on which a significant amount of research work currently continues. Covalent and noncovalent interactions are in charge during the assembly of nanostructures on these templates. Direct deposition or in situ growth of nanoparticles on these templates can be realized.

Chemically Assisted Assembly

Highly ordered nanoparticle assemblies or patterns can be created using a chemically assisted self-assembly technique [87, 97]. This technique is based on the covalent and noncovalent interactions of a nanoparticle surface protecting group. This approach is widely preferred both in solution and on surfaces. Chemically assisted self-assembly of nanoparticles can be applied by electrostatic layer-by-layer assembly, chemical templating, and SAM methods.

Hydrogen Bonding. Hydrogen bonding is one of the most investigated interaction types for the assembly of nanoparticles together with electrostatic interactions. Based on the hydrogen bonding phenomena, the exchange cross-linking precipitation route for nanoparticle assembly on surfaces is developed. Zirbs et al. showed, for example, the highly selective assembly of barbituric acid-stabilized Au nanoparticles onto Au substrates covered with Hamiltonian-type receptors (Fig. 12.17) [98].

Additionally, different biomolecular linkers are also used for the assembly of nanoparticles based on hydrogen bonding [99]. The chemical characteristics of these molecules are very critical in order to realize the assembly of nanoparticles such as antibody–antigen interactions [100].

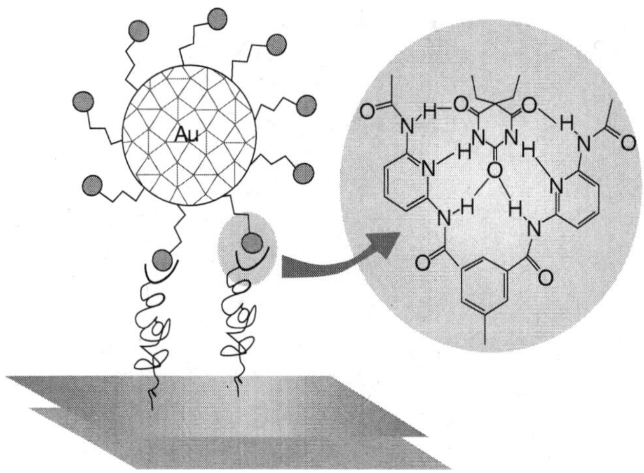

Figure 12.17. Directed self-assembly of Au nanoparticles via multiple hydrogen bonding [98].

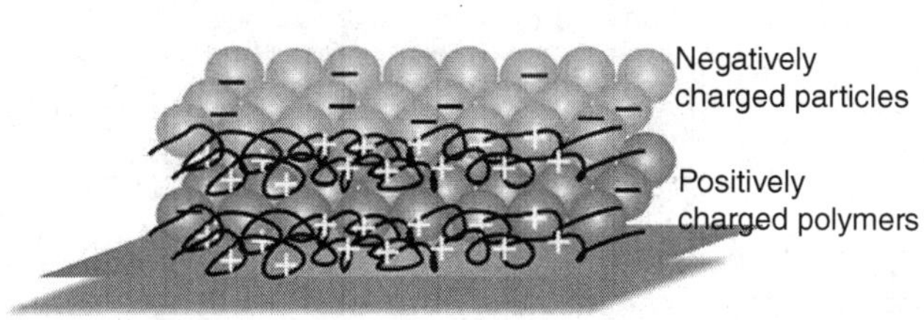

Figure 12.18. Electrostatic assembly of negatively charged particles on positively charged polymers [87].

Electrostatic Assembly. Simple and flexible layer-by-layer assembled nanostructures can be created using electrostatic interactions (Fig. 12.18) [87]. Using this approach, different materials with various optical and electronic properties are structured. The layer-by-layer assembly is first declared by Decher using polyelectroytes [69], whereas Iler reported it for macroscale colloids [101]. However, the use of this technique for nanoparticle assembly is reported by Kotov et al. for the first time [102].

Entropic, van der Waals, steric, dipolar forces and electric charges on the sterically charged nanoparticles are effective in the determination of the stoichiometry of the structures. In general, the nanoparticles are negatively charged, whereas the polymer has a positive charge.

Layer-by-layer assembly is also possible by the epitaxial growth of layers. This technique is especially preferred in the production of different optoelectronic devices, for example, light-emitting diodes. Different layers of thin films are grown epitaxially while using MOCVD. Demir and coworkers hybridized nanocrystals, which are excited by near UV or blue light coming from the diode, on these LEDs to produce white light [72, 103–105]. By using this method, it is possible to precisely control the color of the emitted light on the chromaticity diagram, its correlated color temperature, and its color rendering index.

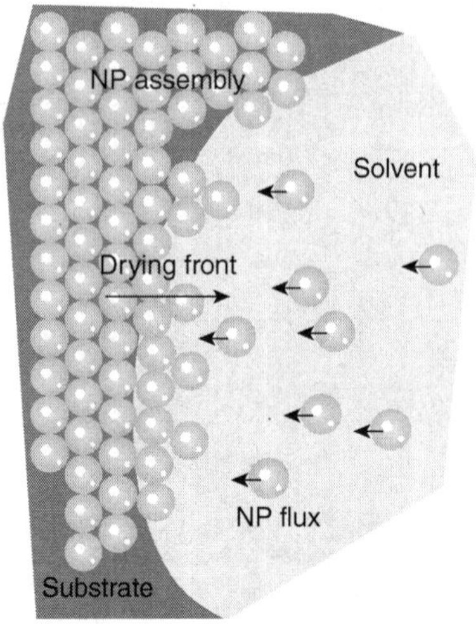

Figure 12.19. Evaporation-induced assembly of nanoparticles (NPs) on surfaces [87].

Drying-Mediated (Evaporation-Induced) Self-Assembly

This method of self-assembly is one of the simplest assembly methods of nanoparticles. Passivated nanoparticles, which are slowly and irreversibly evaporated, assemble in highly ordered patterns on the surfaces (Fig. 12.19) [87]. When the evaporation process is slowed down, the weak attraction forces between nanoparticles in the solution start to be important. According to Denkov et al., the first step of nucleus formation and generation of attractive forces between the particles in the solution results in the second step of the colloidal crystallization [106]. The explanation of the aggregation process is also based on the thermodynamics [107]. The interparticle attractions cannot greatly exceed the thermal energies; therefore, equilibrium structures can be observed depending on the temperature and concentration.

Often the nanoparticles are thought to be uniformly suspended in the solution. The particle concentration increases right underneath of the liquid surface if the diffusion rate of nanoparticles in the liquid is slower than the evaporation speed of the liquid surface. This process often causes the self-assembly of a two-dimensional monolayer at the surface. It is believed that the surface tension is the effective force to hold the assembled structure. Continuation of the evaporation results in the formation of another monolayer on the first layer. If the evaporating region lies on a hydrophilic surface, the dispersed nanoparticles in the solution are positioned toward the drying region by convection. These nanoparticles are dispersed there as the drying front gets thinner.

Continuing the drying process of the nanoparticle solution on surfaces causes the formation of concentric rings with high ordering of quantum dots or other nanoparticles [108]. The rings and the spokes are nanometers in height, submicrons to a few microns in width, and millimeters in length.

Magnetically, Optically, or Electrically Oriented Self-Assembly

The nanoparticles can be aligned and assembled by using magnetic, electric, or optical fields [87]. Electric and magnetic fields lead to the formation of nanoparticle monolayers with increased symmetry [109]. Strong interparticle interactions caused by strong magnetic fields make magnetic nanoparticles assemble in one-, two-, or three-dimensional structures. In general, the ordering of nanoparticles is effected by the external magnetic fields [110]. According to the direction of magnetic field, nanostructures can be grown in a specified direction [111].

Like magnetic fields, electric fields can also be used in the assembly of nanoparticles. In the work of Hermanson et al., micrometer-long wires are created by assembling simple colloidal metallic particles suspended in water under applied electric field [112]. The assembly process relies on the mobility and interactions of particles related to the alternating electric fields (dielectrophoresis [DEP]). In the presence of an alternating electric field, the manipulation and assembly of particles are observed without the interference of the electro-osmotic and electrochemical effects existing in the direct current system. Microwire formation can be described as a collective effect in which the nanoparticles gather at the end of the tip to extend the wire in the direction of the field gradient. The growth of the wire can be effected by controlling the homogeneity of the electric field.

In addition to the magnetic field- and electric field-directed self-assembly, the nanocrystals can also be assembled under the illumination of light [113]. When the surface-bound amino groups are organized to yield thiol ends, the semiconductor and metal nanocrystals can be assembled using light.

Interfacial Self-Assembly

Another method to assemble nanoparticles for large-scale applications is the interfacial assembly (Fig. 12.20) [114]. For example, in Pickering emulsions, large particles of $>1\,\mu m$ stabilize emulsions via adsorption on the liquid–liquid interface [115]. In assembling nanoparticles at the fluid interfaces, a similar approach can be used. This process is highly dependent on the temperature fluctuations and interfacial energy. Lin et al. reported that this assembly process is related to the minimization of the Helmholtz free energy [116]. The particles escape as a result of the thermal activation because of the weak energy associated with the placement of nanoparticles.

Nanoparticles can be easily assembled as densely packed, disordered monolayers. The nanoparticle adsorption depends highly on the size of the particles. Nanoparticle surface modifications change the interfacial energy. In general, small particles assemble more weakly on to the interface in comparison with larger particles. Size-dependent particle exchange is also observed, with larger nanoparticles changing places with the smaller ones at a rate related to their adsorption energies [117]. In general, important factors in these processes are the dipole moment, small positive charge, and directional hydrophobic attractions.

In another method, based on hydrophobic interactions, surface-protected nanoparticles are gathered on the water surface. Here, a monolayer formation is observed at

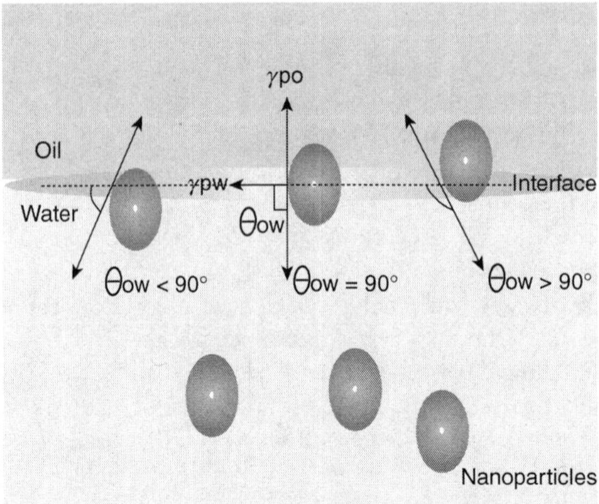

Figure 12.20. Interfacial self-assembly is controlled by the tension of two phases and the contact angle. Energy balance relationship is as follows: $\Delta E = -\pi r^2/\gamma_{OW} \times (\gamma_{OW} - [\gamma_{PW} - \gamma_{PO}])^2$. γ, interfacial energy; P, particle; W, water; O, oil; r, effective NP radius [206].

the water–air interface. Compressing this monolayer slowly, the nanoparticles are transferred to substrates via horizontal or vertical liftoff.

Shape-Selective Assembly

Nonspherical nanoparticles with varying shapes like tetra-/octahedral or cubic can be produced [87]. With these nanoparticles having different shapes, interparticle forces play a critical role in their assembly. As a result of the different polarity between nanoparticles of different shapes, dipole moments are generated. Using this phenomenon, anisotropic structures can be constructed. Utilizing this anisotropy in adsorption and desorption, the assembly of nanoparticles is observed in different dimensions [118]. The driving force of this assembly procedure is the energy difference between crystallographic planes. Since different structures show different affinities toward different inorganic materials, asymmetric self-assembly of nanoparticles can be achieved. More complex structures can be created by mixing the nanoparticles with different phases and sizes [119]. As a result of the mixture of the solutions of differently sized nanoparticles, phase separation is also observed.

Preparation of nanostructures in controlled arrays requires an understanding of the mechanism of the assembly. After understanding the mechanism, new models need to be found in order to understand possible behaviors of the system according to the changes in the parameters of the system. The hierarchical self-assembly of nanoparticles model, developed by Banin and coworkers, is one of the models that provide understanding on the nanostructure construction [120–122]. This model is based on stochastic calculations that simulate the position changes of self-assembling nanoparticles.

SAMS

SAMs are themselves nanoscale structures that provide some useful controllable properties [123]. Their building process relies on the free energy change of the materials. Metal and metal oxide surfaces have the tendency to adsorb organic materials since the free energy between the surface and ambient environment decreases as a result of this adsorption process [124]. The adsorbed materials may change the stability of the nanostructures on the surface; additionally, they may serve as physical or electrostatic barriers to avoid aggregation.

SAMs are organic structures that are built by the adsorption of molecules from the solution or gas phase onto the solid or liquid surfaces. After adsorption, crystalline and semicrystalline structures can be observed on the surface. Molecules or ligands forming SAMs have head groups that have a strong affinity for a specific substrate. SAMs typically have a thickness of about 1–3 nm [123]. SAMs can be laterally patterned with 10- to 100-nm dimensions while using most of the conventional nanofabrication tools and methods. SAMs have numerous advantages in nanotechnology, some of which are the following: (1) The preparation of SAMs is easy; there is no need for special environments or techniques; (2) SAMs can be built up on objects independent of their size and can be used to alter the chemical and/or physical functionalities of these objects; (3) SAMs can affect the electric and optical properties of metallic structures in relation with the external environment; and (4) SAMs make connections between molecular-level structures and macroscopic interfacial phenomena, like wetting and friction.

SAM Substrate Types

A substrate is defined here as the physical object supporting the surface on which the SAM is to be built. Substrates can be in the form of planar surfaces or even nanostructures that are highly curved. The type of the SAM (and thus its preparation method) is application specific [123]. For example, polycrystalline materials can be used in applications including etch resists, in templates for crystallization, and in model surfaces for biological studies. In order to measure electron transport through organic molecules, single crystals or polycrystalline materials are preferred.

Thin films of metals supported on silicon wafers, glass, mica, or plastic substrates are the most frequently used planar substrates for SAMs. Gold, silver, copper, palladium, platinum, nickel, and their alloys are the materials with which thin films can be conveniently constructed easily using PVD methods.

Among these materials, gold is the most frequently used one since gold can form very good SAMs and it is a well-known material [123]. Additionally, Au can be obtained easily in the form of thin film and colloid. Another advantage of Au is its appropriateness for lithographic processes, although it is an expensive material. Its low affinity for oxidation under melting temperature makes gold favorable for nanofabrication procedures as well.

Thin Films on Glass or Silicon by PVD. A thin primer or adhesion layer of titanium, chromium, or nickel (1–5 nm) and a layer of noble metal (10–200 nm) are the building blocks of a thin film deposited onto a silicon wafer or glass support [123]. The primer serves as the improvement layer for the adhesion of metals that do not form oxides on substrates with an oxidized surface.

Metals with high melting points form smaller grains than metals with lower melting points. The size differences of the grains affect the properties of the resulting SAMs, which determine their applications. Polycrystalline films with smallest possible grains are used in microcontact printing and etching, whereas films with larger grain sizes are used as an insulation barrier against electrochemical processes or biased electron transport.

Thin Films on Mica. The films are usually produced by thermally evaporating gold at a rate of ~0.1–0.2 nm/s on to a heated (400–650°C) sample of mica. By using this method, thin films with grain sizes of ~1000 nm with flat terraces of ~100 nm in width can be constructed. Using the template stripping method, it is possible to build surfaces with a roughness of <1 nm [125]. This technique is based on gluing a glass slide (or another substrate) onto the gold film that is deposited on mica. Finally, the gold film is removed from the mica to expose the surface initially in direct contact with mica.

Electroless Deposition of Thin Films. Electroless deposition is defined as the process in which chemical reduction of metal salts is used [126]. The advantages of this approach are that it does not require vacuum processing devices, that commercially available chemicals are used in the process, and that there is no need of conductive electrodes in this technique. Therefore, deposition of films onto the nonconducting films can be realized. The electroless methods can be used only with solutions. This is advantageous for the thin-film deposition of nanostructures, including colloids and nanopores.

Underpotential Deposition. For the modification of thin-film surfaces, the underpotential deposition technique is used. This is an electrochemical approach to form a submonolayer coverage of one metal onto another one. This layer grows epitaxially by adopting the ordering of the undersurface [127]. This metal layer varies the physical and optical properties of a SAM.

Mechanisms of Assembly from Gas and Solution

In order to control the assembly process of SAMs, the mechanisms of the assembly should be understood. These processes basically depend on kinetic and thermodynamic factors [123]. For example, for thiolates, though the mechanisms cannot be understood completely, it is certain that the SAM formation is strongly related to the energetics of the metal–sulfur bonds and noncovalent lateral interactions of organic groups. Maximization the attractive lateral forces (van der Waals, hydrogen bonding) results with the assembly of the organic layer.

Organization of SAM thiolates on gold from the gas phase involves complex growth kinetics related to the intermediacy of low-coverage phases. During the SAM assembly of alkanethiols on gold from the gas phase, it is being thought that a precursor occurs, which may involve chemisorbed thiolates with noncrystalline geometry.

Since the solution environment is relatively complex, the mechanisms of SAM configuration could not be understood in details. It is believed that the assembly from solution obeys a kinetic model like the Langmuir adsorption model [128, 129]. Although it is believed that the evolution pathway of the assembly process in solution is similar to the process in gaseous phase, it could not be demonstrated perfectly.

Patterning of SAMs

There are numerous ways of patterning SAMs, including e-beam and X-ray lithography, atomic beam lithography, patterning by gradient formation, ink-jet printing, and orthogonal self-assembly.

Photolithography. When the SAMs of alkanethiols are exposed to UV light using a pattern of apertures, for example, in a chromium film, the photooxidation process of SAMs can be observed in the UV light-illuminated regions [130]. The species, which are photooxidized, can be removed from the substrate using a polar solvent, for example, water or ethanol, for rinsing operation. The resolution of the procedure and produced structures depends on the capability of the optical devices that are used. For UV exposure, the resolution limit is ~0.3 μm [130], and the minimum required time for the exposure is 15 min. Using a phase mask, arrays of lines with 100 nm in ca. 1 min can be produced [131]. Using laser beams, thermal desorption on the SAM can be realized [132], which can be further used to pattern SAMs. Although its exposure time is as short as 0.1 s, the feature size remains to be as large as 20 μm.

E-Beam and X-Ray Lithography. SAMs can also be patterned by using electron beams [133]. The electrons with low energies cause chemical changes in SAMs such as cleavage of bonds, formation of bonds, cross-linking of adjacent molecules, fragmentation of molecules, and conformational disorder [134]. These chemically changed regions are observed to have better resistance against etchants. By using this property, SAM features can be patterned. As another high-energy radiation source to pattern SAMs, X-rays can be used as well [135]. It is observed that X-rays cause similar chemical changes to electron beams on SAMs [136].

Atomic Beam Lithography. SAM patterning can be realized by using rare gases as well. Neutral rare gases that are excited into metastable states can be used to damage SAMs [137]. As a result of the collision of the metastable atom and SAM, energy is released and the metastable atom returns to its ground state. During this process, organic material is locally ionized [138]; here, conformational disorder is observed [139].

Gradient Formation. In order to produce uniform lateral gradients of SAMs, which consist of one or two molecular components, there are several methods, including (1) controlled immersion of one thiol and subsequent immersion in a second thiol (Fig. 12.21) [140], (2) diffusion of two thiols from opposite ends of a gold substrate [141],

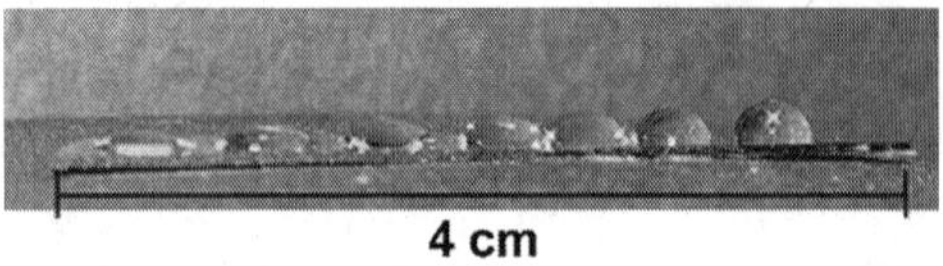

Figure 12.21. Optical photograph demonstrating a gradient in hydrophilicity [140].

(3) electrochemical desorption of thiols from dynamic or potential gradients [142], and (4) gray-scale photolithography on photosensitive SAMs [143].

Ink-Jet Printing. Commercial ink-jet printers are used to deposit nanoliter volumes of solutions with organic dyes onto surfaces like paper. This technology can be used to deposit solutions on metals to construct patterns of SAMs with 100-μm feature size [144].

Orthogonal Self-Assembly. Generation of substrates, which consist of two or more materials, can be considered as an alternative approach for SAM patterning. These generated substrates then form SAMs having affinities toward certain materials.

Current Applications of SAMs in Nanofabrication from Integration Point of View

In the formation of nanoscale features and objects, templated synthesis is a promising approach [123]. Using interesting physical, electrical, and optical properties of these structures, a variety of devices including sensors, selective filters, and probes for biological applications can be designed. Since these structures have smaller sizes in comparison with the cells, a large range of biological applications exists. Alkanethiols can be used to add new functionalities to the SAMs and metals like gold, on which SAMs can form.

In the work of Martin [145], length-controllable nanowires with different metal compositions are demonstrated by using the electrodeposition of metals within mesoporous carbonate or alumina membranes. By using these rods, the orthogonal functionalization of different metallic sections with different SAMs can be realized. Additionally, it is shown that by using charged cysteine SAMs, the ion flux through nanopores can be controlled [146]. By changing the pH of the solution, the membrane permits either cations or anions. It is also possible to make size selection.

Arrays of metallic nanostructures can be used in cellular automata [147], arrays of biomolecules, cell sorting, and information storage. The block copolymer micelle nanolithography technique can be used in the formation of gold nanoparticles in a close-packed hexagonal lattice. The distance between these dots is dependent on the molecular weight and the linear composition of the copolymer [148]. This procedure includes, first, the preparation of a single layer of adsorbed micelles by removing a glass slide from a block copolymer micellar solution, which is also in coordination with Au(III) salt. The self-assembly of micelles is observed in a hexagonal lattice on the surface of the slide. After hydrogen plasma treatment, a hexagonal array of gold nanoparticles is produced. More complex nanoparticles can be produced using the focused e-beam technique [149].

Colloidal crystals of hexagonal close-packed silica or polystyrene (PS) spheres are used as mask in the nanosphere lithography technique for material deposition [150]. Silver nanoparticle arrays with triangular shape and regular size and spacing can be constructed on various surfaces using this technique.

Metallic half shells are created using e-beam deposition of thin layers of metal onto colloidal arrays. Dissolving colloidal spheres results in free metallic half shells that have nanometer-sized thicknesses.

SAMs can be used to stabilize nanoparticles during their formation. However, a range of functional groups are observed between the nanoparticle and the environment at the interface. Polyvalent interactions of biological tools and platforms are studied by using these functional groups.

ASSEMBLY OF NANOCRYSTALS

Epitaxially Grown, Self-Organized Solid-State Quantum Dots

Two-dimensional planes of semiconductors can be built by using MBE and metal organic vapor-phase epitaxy (MOVPE or MOCVD) methods with nanoscale precision [151]. These techniques can also be applied to assemble quantum dots on semiconductor surfaces by using the principle of lattice mismatch between alternating layers [152]. In this method, a thin layer is epitaxially grown on a substrate with a different lattice constant, such that occurring strain can lead the self-organization of quantum dots. The size and distribution of the quantum dots are dependent on the temperature, substrate angle, flux ratio, lattice mismatch, and growth rate [153, 154].

Self-Assembly of Colloidal Quantum Dots

By setting the surface chemistry of colloidal nanocrystal quantum dots appropriately, ordered self-assembled nanocrystal superlattices in two and three dimensions can be achieved [155].The interactions between these nanocrystals is mostly based on the van der Waals force between them [151].

FePt magnetic nanocrystals can be aligned using the thermal annealing approach. These nanocrystals build up a face-centered tetragonal superlattice. These structures hold promise for data storage applications [156].

The biological systems are also widely investigated for the self-assembly of nanocrystals and nanowires. Gold nanocrystals are assembled using DNA strands [157]. In the mixture of gold nanocrystals and DNA strands, gold nanocrystals form mixtures of dimer (two nanocrystals) and trimer aggregates (three nanocrystals).

Simultaneous reduction of $Pt(acac)_2$ and thermodecomposition of $Co_2(CO)_8$ result in $CoPt_3$ magnetic nanocrystal formation in the presence of organic molecules terminating the growth of Co–Pt bulk phases [158]. Because these nanocrystals have similar sizes and uniform spherical shapes, they can be easily self-organized into two- or three-dimensional superlattices. As a result of slow evaporation, the carrier

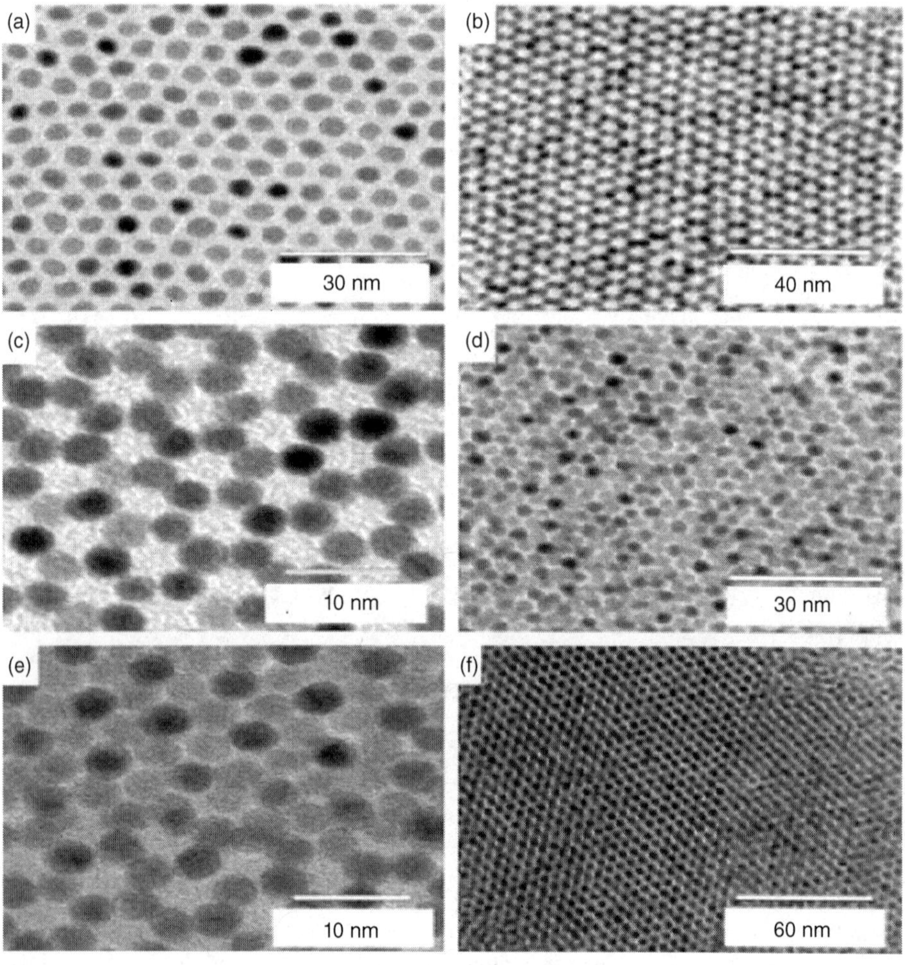

Figure 12.22. TEM images of (a) a 4.8-nm $CoPt_3$ nanoparticle monolayer, (b, c) two layers of 3.6- and 4.0-nm $CoPt_3$ nanoparticles, (d) three layers of 4.0-nm $CoPt_3$ nanoparticles, and (f) more than five layers of 4.5-nm $CoPt_3$ nanoparticles [158].

solvent including colloidal nanocrystals are self-assembled. Several kinds of superstructures are found depending on the solvent evaporation conditions and particle size (Fig. 12.22a).

In case the nanocrystals build more than one monolayer, the nanocrystals in the second monolayer take position between the ones in the first monolayer (Fig. 12.22b, c). The nanocrystals in the third layer are observed to have classical cubic close packing (CCP) (Fig. 12.22d, e).

From monodisperse nanocrystals, the formation of planar superlattices followed [159]. Au or Ag and Au nanoparticles form regular AB_2 and AB alloy superstructures depending on the amount of each species and the ratio of particle diameters.

Nanocrystals that are colloidally synthesized need to be uniform in composition, size, shape, and surface chemistry. The use of thiols as capping agents for II–VI semiconductor nanocrystal synthesis results in molecular-like structures with exact composition [160]. This synthesis operation consists of some consecutive stages: nucleation from initially homogeneous solution, growth of the previously formed nuclei, isolation of particles from the reaction mixture after reaching the desired size, and postpreparative size fractionation.

Nanoparticle Distribution Control by Polymers (According to Aggregation States)

Stabilized Nanoparticles (sNPs) on Microcapsules with Nonuniform Distribution. After the mixture of precoated PS and aqueous suspension of sNPs, they are adsorbed on films due to their positive charge [161]. Using this method, layers from one to six sNP layers are built. This approach makes it possible to adjust the control of the deposited nanoparticle concentration by changing the volume or the concentration of the dipping solution. Although the amount of sNPs can be controlled, the distribution is always observed to be nonuniform.

sNPs on Microcapsules with Uniform Distribution. In the above described method, there is no control over the mechanism of interaction between nanoparticles and the polyelectrolyte surface [161]. The mixture of sNP and identically charged polyelectrolyte leads to a competition between nanoparticles. As a result of this competition, a uniform distribution of nanoparticles is obtained.

Nonstabilized Nanoparticles (nsNPs) on Microcapsules with Nonuniform Distribution. The mixture of precoated PS cores and the suspension of nsNPs result in the adsorption of nanoparticles on films since they are negatively charged [161]. Similar to the sNP case, the amount of nanoparticles can be tuned; however, their distribution becomes nonuniform.

nsNPs on Microcapsules with Uniform Distribution. Premixing the negatively charged nsNP with opposite charged polyelectrolytes forms stable suspensions [161]. It is observed that the polyelectrolytes stabilize the nanoparticles and push them apart upon the adsorption. The result is nonaggregated, uniform distributions of nsNP on the capsules.

Two- and Three-Dimensional Arrays of Monodisperse Nanocrystals Formed by Self-Assembly

The self-assembly process relies on the process of spontaneous formation of ordered arrays from monodisperse particles [160].This effect is caused by the dispersive attraction of nanoparticles due to van der Waals forces [155, 162]. While placing a drop of colloidal solution of monodisperse nanoparticles on a suitable substrate and permitting slow evaporation, self-assembly of two- and three-dimensional arrays can be observed.

The particle spacing in the arrays can be adjusted using different ligands at the synthesis process or by postpreparative exchange [155].As surface coverage increases, nanocrystals tend to form three-dimensional arrays instead of two-dimensional arrays by getting positioned between the nanoparticles of the layer at the top.

Selectivity in Adsorption of Semiconductor Nanocrystals onto a Self-Organized Pattern

Selectivity in the adsorption of semiconductor nanocrystals onto an organic self-organized pattern has a time-dependent behavior [163]. Construction of potential energy landscape on the surface helps to understand the behavior of the delivered solvent. There are two consecutive stages during the self-assembly of patterns: (1) the molecular exchange delivered solvent and lipid molecules to construct the adsorption sites for nanocrystals and (2) the adsorption of nanocrystals onto the adsorption sites because of the interaction between the nanocrystals and the substrate.

Rogach et al. studied the time dependence of nanocrystal deposition controllability [163]. In their experiment, CdSe nanocrystals were dripped in 1-phenyloctane on a patterned surface. After some time, the droplet was taken out using a pipet or a filter paper. This procedure was applied for several different time intervals. If the wetting time remained less than 10 min, the majority of the nanocrystals tended to position in the channels of the patterned surface (Fig. 12.23a). Increasing the waiting/wetting time to 20 min caused the nanocrystals not just to fill in the channels of the patterned surface but also to assemble on the stripes (Fig. 12.23b). Increasing the wetting time even further led to more nanocrystals accumulating on top of the stripes. In Figure 12.23c, a schematic illustration is given for the time dependence of the nanocrystal deposition on wetting time.

In this procedure, the solvent should be carefully selected. The solvent should be chosen to be appropriate for the nanocrystal dispersion, while the selected solvent should not dissolve the selected patterned surface. The surface coverage is modeled by

$$C = -A \cdot \exp(-t/\tau) + C_0,$$

where C is the surface coverage, A is the front factor, τ is the lifetime, and C_0 is the surface coverage at infinity. The rate constants are different for the different regions of patterned surface.

Au Nanocrystal/DNA Conjugates

There are a variety of biological polymers that can be used to couple nanocrystals. Among these biological structures, DNA is particularly investigated because of its inherent programmability [164]. The Watson–Crick base pairing of an olgionucleotide with 12 bases is thermally stable at room temperature. Twelve base pairs correspond to a length of approximately 4 nm, which contain a sufficient number of pairs for the placement of nanoparticles.

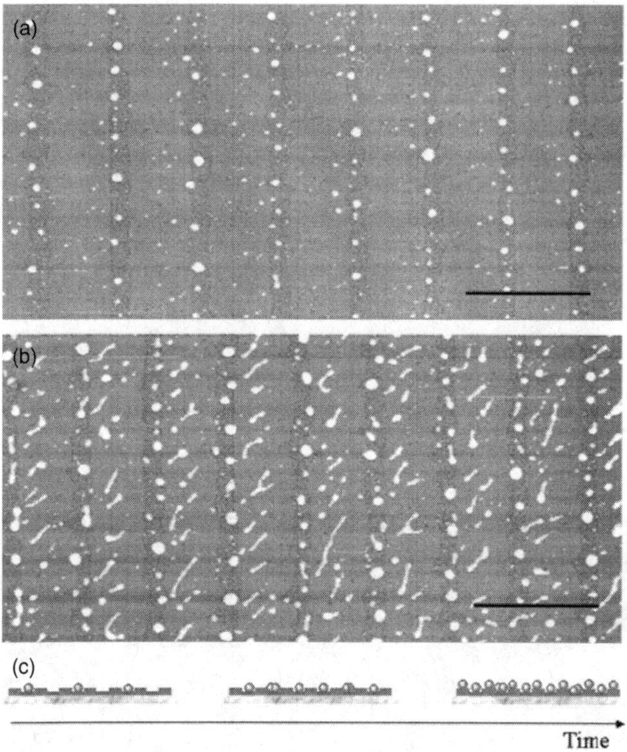

Figure 12.23. AFM images of CdSe nanocrystals on the self-organized lipid patterns show the distribution of nanocrystals with (a) 5-min wetting time and (b) 20-min wetting time. (c) A representative schematic illustration of CdSe nanocrystals depending on wetting time [163].

In order to make use of these properties of DNA/nanocrystal conjugates, it is necessary to have control over the number of nanocrystals conjugating with DNA. Although there is no complete control over this issue, statistical conjugation results are obtained. The average number of DNA strands per particle can be controlled by the adjustment of the DNA:Au ratio.

Entrapment of Nanocrystals in Sol–Gel-Derived Composite Hydrophobic Silica Spheres

An example for the entrapment procedure is given as follows [165]. First, two solutions are prepared. One of the solutions consists of ethanol, ammonium hydroxide, and surfactant Tween 80. The second solution is prepared using tetraethyl orthosilicate (TEOS), various amounts of PS, and toluene solution of the coated nanocrystals. Following the addition of the hydrophobic solution to the hydrophilic solution, the system is stirred overnight. The formed spheres are taken out from the solvent following the centrifugation. The aggregate particles are separated by sonification. Various conditions for specific cases are summarized in Table 12.1.

TABLE 12.1. Entrapment Conditions for Various Nanocrystals in Silica/Polystyrene Spheres [165]

Nanocrystal type and size (nm)	NC amt (mg)	PS amt (mg)	Size of spheres(nm)
CdSe rods, 24.5 × 4.9	20	30	250
CdSe dots, 3.5	40	40	500
CdSe dots, 6.0	20	50	780
CdSe rods, 11.0 × 3.0	30	60	1000
CdSe rods, 15.0 × 3.8	35	30	300
PbSe dots, 10.0	50	40	500
InAs/ZnSe CS, 4.3	20	35	400
InAs/ZnSe CS, 6.3	20	35	400
Au dots, 6.0	30	45	750

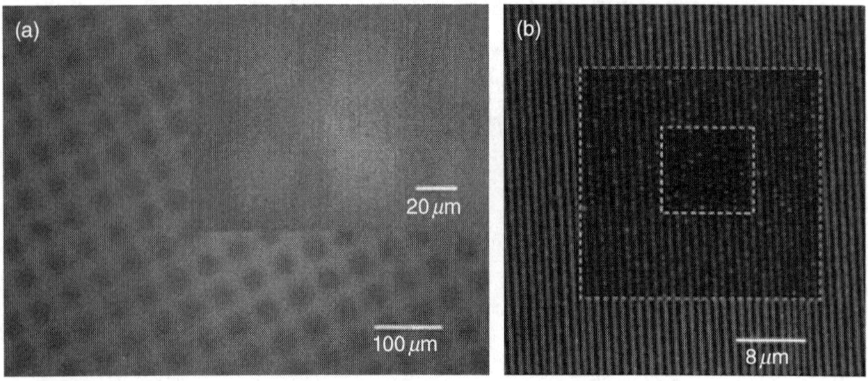

Figure 12.24. Hierarchical luminescence patterns: (a) fluorescence image and (b) confocal laser scanning microscopy (CLSM) image [166].

Hierarchical Luminescence Patterning by Multiscaled Self-Assembly

An alternative method for the nanofabrication hierarchical luminescence patterning is proposed by Chen et al. [166]. Instead of e-beam lithography (EBL) and photolithography, this method offers a self-assembly solution to the nanofabrication. This method involves first generation of two levels of hierarchical luminescence patterning. In the third stage of the process, light is exposed to the samples using a shadow mask (Fig. 12.24a, b).

This approach for nanofabrication offers a relatively inexpensive process; it is also highly controllable, high yielding, and easy to implement. This method also offers the ability to adjust the overall spectrum.

Advantages of Hybrid Nanocomposite Materials with Organic and Inorganic Components

Nanocomposite materials can be classified into two groups [167]: The first class is the inorganic material-embedded polymeric matrix [168], whereas the second class is the organic polymer-embedded inorganic templates [169]. In these approaches, the useful properties of organic components such as good process ability and low density of the polymeric component are combined with the useful properties of inorganic materials, such as their high mechanical durability and well-defined optoelectronic properties.

ALIGNMENT OF NANORODS USING DC ELECTRIC FIELD

Nanostructures including nanorods and nanowires that have significantly different lengths in different dimensions have attracted a great level of attention in scientific research [170, 171]. These structures have the potential to be the fundamental parts of new-generation optical, electrical, and optoelectronic devices [172].

One of the techniques for nanowire production is template-directed assembly. As templates the linear pores and channels in porous materials are used as templates [170]. In the work of Hornyak et al., the production of gold nanowires is reported, while the gold nanocrystals are directed into the pores of an alumina or polymer membrane [173]. Nanowire production also followed using block copolymers as templates [174]. Additionally, further research continues on the production of nanowire by using DNA [175] and viruses [176] as templates. As another template-directed approach, patterned substrates can also be employed in nanowire production. As Yin and Xia [177], and Zhang et al. [178] demonstrated, nanowire production can be realized by a catalyst-mediated phase separation approach [179] and seed-mediated growth in solution [180]. Pacholski et al. successfully demonstrated the production of nanowires by the self-assembly of nanocrystals [181]. In another method, it is proposed to form nanowires from Ag nanocrystal (NC) suspension at the air–water interface and DEP. Moreover, nanowires with different shapes (e.g., linear, zigzag, helical, centipedes, and rings) are produced by the attachment of nanocrystals in different orientations [182]. In addition, micrometer-long nanowires are formed while using the dipole–dipole interactions between spherical nanocrystals [183]. Finally, Rakovich et al. reported the self-assembly of nanocrystals into nanowire networks in phosphate-buffered aqueous solutions [170].

Nanorod Alignment

The deposition of the nanorods from the solution results in an isotropic spatial distribution of nanostructures [184]. However, this situation is not desirable for many applications because the nanostructures are often desired to have shape- and size-dependent optical, electrical, and physical properties. On the other hand, slow evaporation of nanorods from solution leads to slightly ordered structures. This orientation occurs in the liquid crystalline phase. In this phase, the use of electric field can work for the alignment of nanorods as well.

Electric Field Assembly of Perpendicularly Oriented Nanorod Superlattices

In the work of Ryan et al., perpendicularly oriented II–IV semiconductor nanorod superlattices are formed using DC electric field and slow evaporation of solvent [185]. In this work, CdS nanorods are generated by sulfur/tri-n-octylphosphine solution injection at high temperature into hot cadmium oxide/surfactant mixtures [182]. The DC electric field is applied to the CdS nanorods in toluene that are between two electrodes (Fig. 12.25). The strength of the applied electric field is 1 V/μm. Near saturation of the entire assembly in toluene ensures gentle evaporation. As the evaporation continues, the diameter of the droplets gets smaller, resulting from the formation of nanorods into a planar supercrystal. It is observed that the slower the evaporation is, the higher the degree of spatial order of the nanorods is.

In the absence of the electric field, the nanorods in the outer layers tend to align parallel to the surface, whereas the nanorods at the center are observed to align perpendicular to the substrate. When the electric field is applied, the nanorods tend to position perpendicular to the substrate. Since the electric field is responsible for the rotation of nanorods around their axis to align them perpendicular to the substrate, the evaporation process results with two-dimensional arrays of semiconductor nanorods. Figure 12.26a, b shows the alignment of nanorods in the absence of an electric field, while Figure 12.26c shows a transmission electron microscopy (TEM) image of nanorods aligned under an electric field. Figure 12.26d shows the hexagonal nanorod structures; this alignment is observed when the total growth rate is slow. The evaporation process of the solvent yields densely packed nanorods.

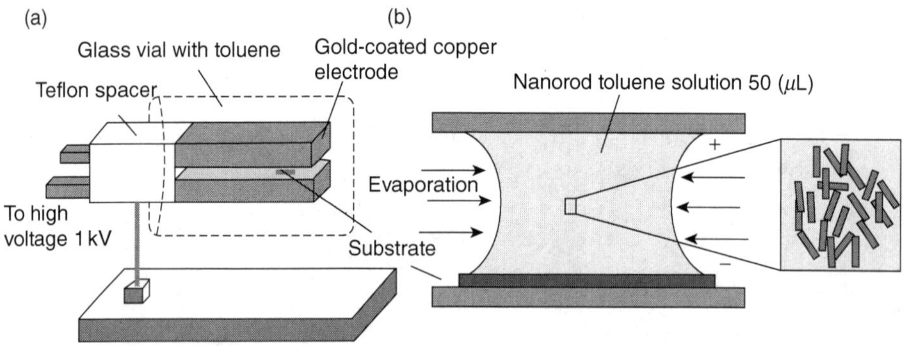

Figure 12.25. Schematic illustration of the electric field assembly of nanorod alignment: (a) a three-dimensional sketch and (b) a two-dimensional view of the meniscus. The substrate is positioned between the electrodes, and 50 μL of the nanorod toluene solution is deposited between the electrodes such that a meniscus is generated [185].

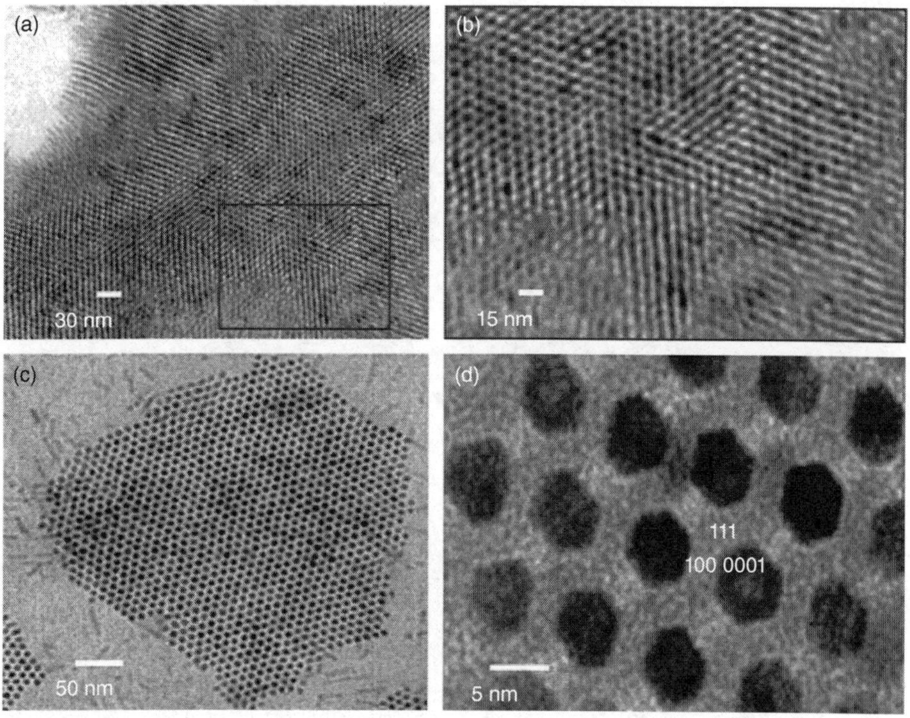

Figure 12.26. (a, b) TEM image of nanorods in the absence of an electric field, (c) TEM image of nanorods aligned under an electric field of 1 V/μm, (d) azimuthal alignment of nanorods [185].

ASSEMBLY OF NANOSTRUCTURES USING DEP AND OPTOELECTRONIC TWEEZERS (OETS)

DEP

To build up devices, high aspect ratio nanostructures, for example, nanowires and nanorods, can be aligned using electric field guiding. By using this method, precise alignment of these nanostructures is achieved across electrodes [186]. Different examples include rod-shaped nanostructures [176, 187] and multiwalled carbon nanotubes [181]. This method provides fast nanowire assembly at room temperature.

The dielectrophoretic forces[188] stem from the induced dipole moment in a non-uniform electric field [189]. This force can be formalized as $\mathbf{f}_{DEP} = (\mathbf{p} \cdot \nabla)\mathbf{E}$, where **p** stands for the dipole moment and **E** for the electric field. For alternating electric fields, the dielectrophoretic force can be approximated by replacing **p** with the effective dipole moment (EDM) of the particle [189]. However, this approximation works only when the particle size is much smaller than the characteristic length of the electric field, which should not be disturbed by the presence of the particle. Therefore, this approximation cannot be used for the objects that are larger than or comparable with the gap between

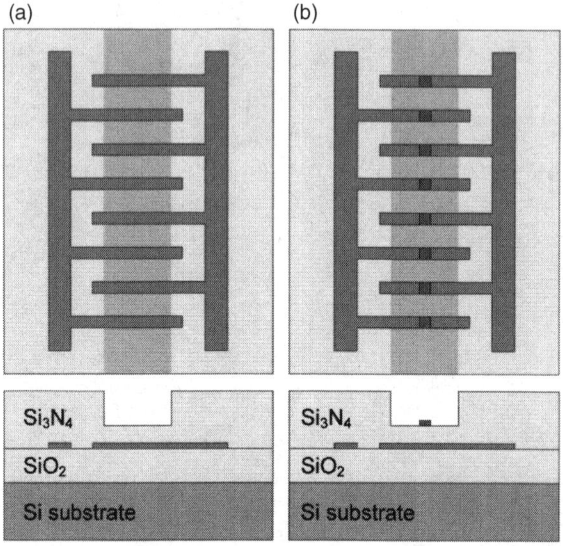

Figure 12.27. Electrode structures used in the electric field guided assembly. (a) Ti/Au electrode with thinned silicon nitride coverage and (b) the same electrode with the addition of top electrodes for the improvement of alignment precision [176].

opposing electrodes. Additionally, it is also problematic when the interaction of multiple particles is considered. To provide a solution for the dielectrophoretic force acting on high aspect ratio nanowires near microelectrodes under electric field, the resulting force and electric field should be computed numerically using the Maxwell stress tensor method [190].

The electric field guided assembly of nanowires in a fluid can be realized using the immersed electrokinetic finite element method (IEFEM). This method involves the movement of independent solid meshes on top of a fixed background fluid using an electric field.

The assembly of nanowires is studied and realized using an electric field by electrically isolated nanowires (Fig. 12.27a) [176]. The metal electrodes are built using 50-nm Ti/150-nm Au liftoff on a silicon dioxide substrate. The finger width and the separation are 3 and 5 μm, respectively. The protection of the electrodes is succeeded by 500 nm of silicon nitride to avoid short circuiting of the electrodes to the metallic nanowires.

The assembly procedure follows: The nanowire suspension is diluted and drop-cast onto the sample, which is biased with AC voltages between $V_{rms} = 5$ V and $V_{rms} = 70$ V. These voltages generate an electric field between 10,000 and 140,000 V/cm given the electrode architecture of the sample. Using this setup, it is observed that the alignment of nanowires is orthogonal to the electrode fingers in the high field regions covered with thin silicon nitride (Fig. 12.28a, b). Another observation is that the alignment occurs if the voltage is greater than $V_{rms} = 25$ V, and nanowire distribution is random. Additionally, it is found out that the relationship between applied voltage and alignment time is inversely proportional.

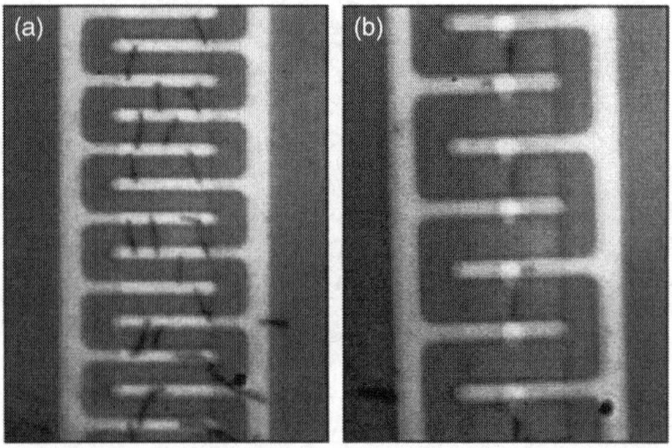

Figure 12.28. Optical microscope image of Au nanowires aligned by the application of (a) 30V at 1kHz with upper field electrodes and (b) 20V at 1kHz with upper field electrodes [176].

It is thought that the random alignment of nanowires is caused by the electric field change between the electrodes following the positioning of a nanowire. The strength of the dielectrophoretic forces is found out to be related to the voltage and time dependencies of the alignment. For low voltages, DEP force remains insufficient, which cannot align the nanowires. Better alignment of nanowires is observed by increasing the applied voltage.

In order to facilitate alignment at certain locations, additional metal electrodes can be deposited on top of the thin silicon nitride layer (Fig. 12.28b). Capacitive coupling between upper and lower electrodes improves the strength of the electric field, which results in better alignment at the periphery of top electrodes.

The frequency dependency of the nanowire alignment process is also investigated. It is found out that the alignment time decreases as the frequency increases. This suggests that the polarization of the dielectric medium around the nanowires is effective in the alignment process at low frequencies. A theory for this phenomenon suggests that at low frequencies, the polar molecules in the dielectric medium can shield the charge separation on the nanowire [191]. However, at high frequencies, the polarized molecules cannot position, such that they shield the charge separation and avoid the torque on the nanowire. Therefore, a strong alignment force can be observed at high frequencies.

The dielectrophoretic force on a spherical particle is given by $\mathbf{F}_{\text{dep}} = 2\pi r^3 \varepsilon_m \text{Re}(K)\nabla \mathbf{E}^2$, where r is the radius of the spherical particle, ε_m is the permittivity of the media, K is the Clausus–Mossotti factor, and E is the electric field. This K factor is a function of permittivity and conductivity of the particle. For the ellipsoid objects with radius r and length l, the complex particle permittivity ε_p^* and medium permittivity ε_m^*, the dielectrophoretic force is given as follows [189]:

$$F_{\text{DEP}} = \frac{\pi r^2 l}{3} \varepsilon_m \operatorname{Re}\left\{ \frac{\varepsilon_p^* - \varepsilon_m^*}{\varepsilon_m^*} \right\} \nabla |\vec{E}|^2.$$

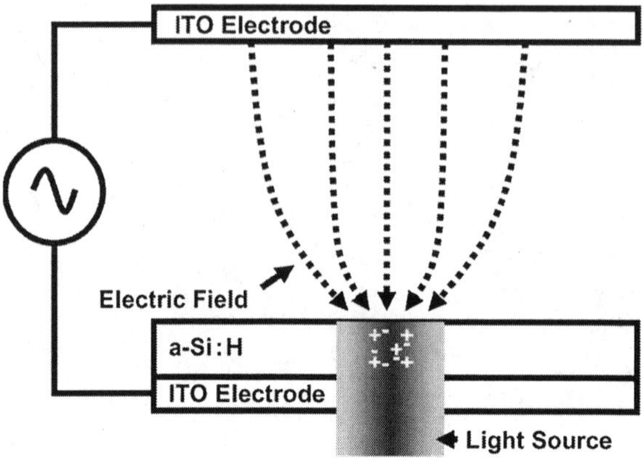

Figure 12.29. Schematics of optoelectronic tweezers [208]. ITO, indium tin oxide.

OETs

The OET method is a combined approach using two different well-known approaches. One approach is the use of optical tweezers to align structures and the other is DEP. The optical tweezer method is developed in the Bell Laboratories by Arthur Ashkin [192]. Using this method, artificial assemblies of cells, colloids, and other objects in three dimensions can be produced by independently controlled tweezers [193]. The main drawback of this method is the weakness of the forces generated by the optical interactions [194]. Increasing the light intensity can be offered as a solution; however, this increase is limited due to the limitations of laser power.

Wu et al. overcome this problem by replacing Ashkin's light forces with dielectrophoretic forces. In the DEP approach, metal electrodes are needed for the establishment of the electric field, which introduces important limitations. Wu et al. removed these limitations by replacing light with electrodes. In other words, optical signals play the role of virtual electrodes formed in a layer of photoconductor (Fig. 12.29) [195].

NANOSKYVING: A NEW METHOD TO PRODUCE ARRAYS OF NANOSTRUCTURES

Nanoskyving uses the deposition of thin films on flat or topographically patterned polymeric substrates using physical vapor methods and subsequent sectioning these films with an ultramicrotome. This technique allows for the fabrication of nanostructures and their arrays for the edges of the thin films [196].

For the nanoskyving method, the first step is to obtain an epoxy substrate after curing an epoxy prepolymer on a flat or topographically patterned PDMS stamp. As the second step, thin films (with nanometer thickness) are deposited onto this epoxy structure. After the generation of the epoxy block containing an embedded thin film,

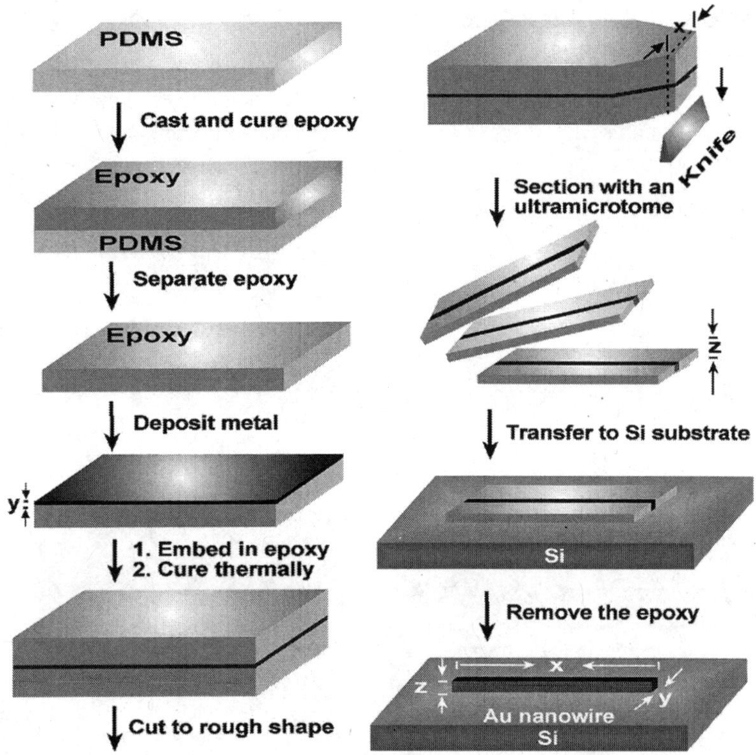

Figure 12.30. Fabrication of metal nanowires as an example of nanoskyving [196].

an ultramicrotome is used to cut this block in slabs that are as thin as 30 nm. After this sectioning operation, the slabs typically float on the surface of water filling the sample collecting reservoir of the microtome (water avoids damage). The removal of epoxy is carried out by oxygen plasma after the epoxy sections are transferred onto a solid substrate (Fig. 12.30).

Nanoskyving builds nanometer-sized structures by cutting thin films deposited on a master structure fabricated in epoxy. Nanoskyving can be thought of as engineering two combined techniques: (1) deposition of the thin film on the master from vapor phase (using e-beam evaporation or sputtering) and (2) sectioning the patterned block by master using a microtome for the production of two-dimensional nanoscale structures. The third dimension is determined by the master topography.

Since the master production is not difficult, nanoskyving does not have an important restriction for the first step in its application. The availability of ultramicrotomes makes nanoskyving implementable for the production of nanostructures that are problematic to produce using conventional techniques such as EUV, X-ray photolithography, EBL, and FIB. In addition, ultramicrotomes are cheaper than clean rooms, e-beam writers, and related equipments [18, 35, 123].

Fabrication of Complex Nanostructures Using Topographically Patterned Substrates

Although Figure 12.30 illustrates a simple structure to fabricate nanosized materials [196], complex nanostructures can also be produced using nanoskyving by choosing the topography of the substrate and selecting the orientation of the sectioning process (Fig. 12.31A). In Figure 12.31B, C, nanostructure arrays with complex patterning are shown as an example.

In nanoskyving, the sliced parts need to be collected before they are transferred onto the substrate layer. Because of its convenience, wet sectioning is the most commonly used sectioning technique. The floating sections/slices are collected and transferred using several techniques. One technique relies on the removal of the slabs using a collection loop, in which a meniscus of water supports the sample. The transfer of the slices can also be achieved by touching the loop to the substrate using the capil-

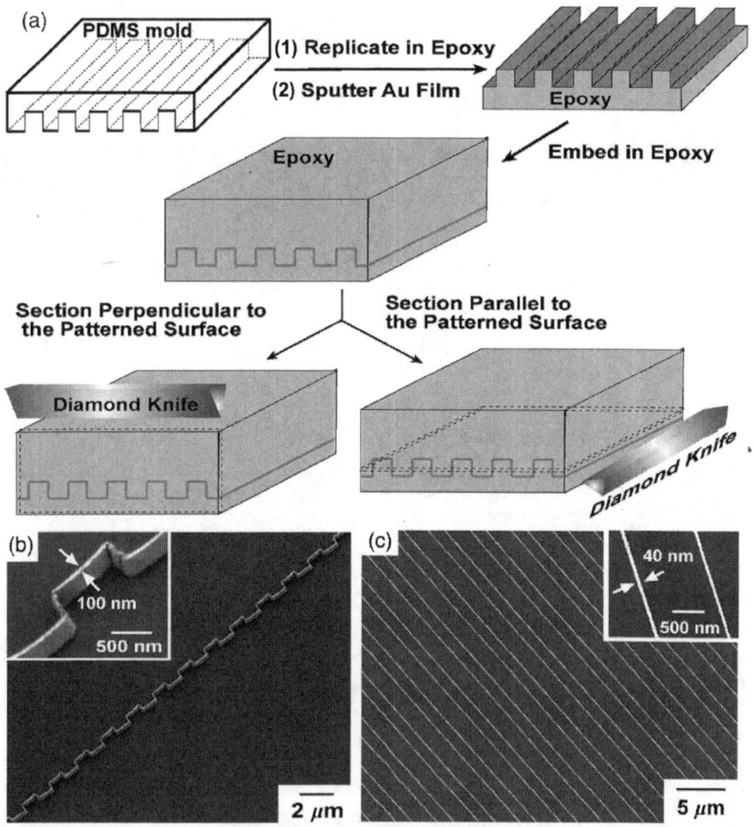

Figure 12.31. (A) Schematic illustration of complex nanostructure production and (B, C) structures obtained by different orientations of slicing, perpendicular and parallel, respectively [196].

lary action. Another method is the collection of slabs by submerging the substrate directly below the floating epoxy slices such that the slices are assembled on the substrate.

Nanoskyving provides simple and inexpensive ways of production of nanostructures in relatively large areas (~mm^2). However, this area still remains small in comparison with the areas patterned by photolithography and nanoimprint lithography, which are on the order of ~500 cm^2 [197]. Nanostructures with higher aspect ratios can be easily produced using nanoskyving, whereas this is a problem in the other methods. Because of the ability to manipulate the slices, the nanomaterials can be easily positioned and assembled on planar or curved regions. In summary, nanoskyving offers substitution of exposure in photolithography, writing in EBL and SPL, and printing/molding in soft and imprint lithography with sectioning by microtome of nanoskyving. However, nanoskyving does not allow the fabrication of connected nanostructures. Therefore, using this technique for integrated circuits seems very difficult. Another drawback of nanoskyving is the lack of stability in the patterns to generate slices that can be registered precisely.

REFERENCES

1. Skogen EJ, Raring JW, Morrison GB, Wang CS, Lal V, Masanovic ML, Coldren LA. 2005. IEEE J Sel Topics Quantum Electron 11:343.
2. Coldren LA, Raring J, Sysak M, Barton J, Johansson L. 2006. Proceedings of Indium Phosphide and Related Materials.
3. Binsma J, Thijs P, VanDongen T, Jansen E, Staring A, Van Den Hoven G, Tiemeijer L. 1997. IEICE Trans Electron E80-C:675.
4. Raring JW, Sysak MN, Tauke-Pedretti A, Dummer M, Skogen EJ, Barton JS, DenBaars SP, Coldren LA. 2006. Proc SPIE 6126:61260H.
5. Aoki M, Suzuki M, Sano H, Kawano T, Ido T, Taniwatari T, Uomi K, Takai A. 1993. IEEE J Quantum Electron 29:2088.
6. Mason B, Fish G, DenBaars S, Coldren L. 1998. IEEE Photon Technol Lett 10:1211.
7. Deppe D, Holonyak N, Jr. 1988. J Appl Phys 64:93.
8. Si SK, Yeo DH, Yoon KH, Kim SJ. 1998. IEEE J Sel Topics Quantum Electron 4:619.
9. Qui B, Bryce A, De La Rue R, Marsh J. 1998. IEEE Photon Technol Lett 10:769.
10. Charbonneau S, Poole P, Feng Y, Aers G, Dion M, Davies M, Goldberg R, Mitchell I. 1995. Appl Phys Lett 67:2954.
11. Skogen EJ, Raring J, Barton J, DenBaars S, Coldren LA. 2003. IEEE J Sel Topics Quantum Electron 9:1183.
12. Skogen EJ, Barton JS, Denbaars SP, Coldren LA. 2002. IEEE J Sel Topics Quantum Electron 8:863–869.
13. Sabnis VA, Demir HV, Fidaner O, Zheng J-F, Harris JS, Jr., Miller DAB, Li N, Wu T-C, Chen H-T, Houng Y-M. 2005. IEEE J Sel Topics Quantum Electron 11:1255.
14. Demir HV, Zheng J-F, Sabnis VA, Fidaner O, Hanberg J, Harris JS Jr., Miller DAB. 2005. IEEE Trans Semiconductor Manufacturing 18:182.
15. So Y-H, Garrou P, Im J-H, Scheck DM. 2001. Chemical Innovation 31:40.

16. Processing procedures for dry-etch cyclotene advanced electronics resins. Dow Chemicals, Tech Rep.
17. Zheng J-F, Demir HV, Sabnis VA, Fidaner O, Harris JS, Jr., Miller DAB. 2006. J Vac Sci Technol B 24:1117.
18. Gates BD, Xu Q, Stewart M, Ryan D, Willson CG, Whitesides GM. 2005. Chem Rev 105:1171.
19. Stewart MD, Patterson K, Somervell MH, Willson CG. 2005. J Phys Org Chem 13:767.
20. Willson CG, Trinque BC. 2003. J Photopolym Sci Technol 16:621.
21. Ito H. 2003. J Polym Sci A 41:3863.
22. http://www.asml.com/asml/show.do?ctx=41905&rid=41906 (accessed May 19, 2011).
23. Switkes M, Kunz RR, Rothschild M, Sinta RF, Yeung M, Baek SY. 2003. J Vac Sci Technol B 21:2794.
24. Bradbury S. 1998. *Introduction to light microscopy*. Oxford, UK: BIOS Scientific Publishers Ltd.
25. Zaidi SH, Brueck SR. 1993. J Vac Sci Technol B 11:658.
26. Bozler CO, Harris CT, Rabe S, Rathman DD, Hollis MA, Smith HI. 1994. J Vac Sci Technol B 12:629.
27. Gamo K. 1996. Microelectron Eng 32:159.
28. Morita T, Kometani R, Watanabe K, Kanda K, Haruyama Y, Hoshino T, Kondo K, Kaito T, Ichihashi T, Fujita J, Ishida M, Ochiai Y, Tajima T, Matsui S. 2003. J Vac Sci Technol B 21:2737.
29. Prewett PD, Gentili M, Maggiora R, Mastrogiacomo L, Watson JG, Turner GS, Brown GW, Plumb D, Leonard Q, Cerrina F. 1992. Microelectron Eng 17:249.
30. Blauner PG, Ro JS, Butt Y, Thompson CV, Melngailis J. 1989. Mater Res Soc Symp Proc 129:483.
31. Kubena RL. 1989. Mater Res Soc Symp Proc 129:483.
32. Ginger DS, Zhang H, Mirkin CA. 2004. Angew Chem Int Ed 43:305.
33. Nyffenegger RM, Penner RM. 1997. Chem Rev 97:1195.
34. Kraemer S, Fuierer RR, Gorman CB. 2003. Chem Rev 103:4367.
35. Gates BD, Xu Q, Love JC, Wolfe DB, Whitesides GM. 2004. Annu Rev Mater Res 34:339.
36. Jung TA, Moser A, Hug HJ, Brodbeck D, Hofer R, Hidber HR, Schwarz UD. 1992. Ultramicroscopy 42:1446.
37. Miyake S, Watanabe S, Miyazawa H, Murakawa M, Kaneko R, Miyamoto T. 1994. Appl Phys Lett 65:3206.
38. Gwo S, Yeh CL, Chen PF, Chou YC, Chen TT. 1999. Appl Phys Lett 74:1090.
39. Vettiger P, Despont M, Drechsler U, Haberle W, Lutwyche MI, Rothuizen HE, Stutz R, Widmer R, Binnig GK. 2000. IBM J Res Dev 44:323.
40. Aizenberg J, Black AJ, Whitesides GM. 1998. Nature 394:868.
41. Pfeiffer L, West KW, Stormer HL, Eisenstein JP, Baldwin KW, Gershoni D, Spector J. 1999. Appl Phys Lett 56:1697.
42. Jun T, Schlitter R, Gimzewski JK, Himpsel FJ. 1995. Appl Phys A 61:467.
43. Himpsel FJ, Jung T, Kirakosian A, Lin JL, Petrovykh DY, Rauscher H, Viernow J. 1999. MRS Bull 24:20.
44. Love JC, Paul KE, Whitesides GM. 2001. Adv Mater 13:604.

45. Yang H, Love JC, Arias F, Whitesides GM. 2002. Chem Mater 14:1385.
46. Melosh NA, Boukai A, Diana F, Gerardot B, Badolato A, Petroff PM, Heath JR. 2003. Science 300:112.
47. Morris RB, Franta DJ, White HS. 1087. J Phys Chem 91:3559.
48. Campbell JK, Sun L, Crooks RM. 1999. J Am Chem Soc 121:3779.
49. Glauert A. 1974. Ultramicroscopy Pt. 2:213.
50. Xu Q, Gates B, Whitesides GM. 2004. J Am Chem Soc 126:1332.
51. Gates B, Xu Q, Thalladi VR, Cao T, Knickerbocker T, Whitesides GM. 2004. Angew Chem Int Ed Engl 43:2780.
52. Haverkorn van Rijsewijk HC, Legierse PEJ, Thomas GE. 1982. Philips Technol Rev 40:287.
53. Colburn M, Johnson S, Stewart M, Damle S, Bailey T, Choi B, Wedlake M, Michaelson T, Sreenivasan SV, Ekerdt JG, Willson CG. 1999. Proc SPIE Int Soc Opt Eng 3676:379.
54. Resnick DJ, Dauksher WJ, Mancini DP, Nordquist KJ, Ainley ES, Gehoski KA, Baker JH, Bailey TC, Choi BJ, Johnson SC, Sreenivasan SV, Ekerdt JG, Willson CG. 2002. Proc SPIE Int Soc Opt Eng 4688:205.
55. Sotomayor Torres CM, Zankovych S, Seekamp J, Kam AP, Cedeno CC, Hoffmann T, Ahopelto J, Reuther F, Pfeiffer K, Bleidiessel G, Gruetzner G, Maximov NV, Heidari B. 2003. Mater Sci Eng 77:596.
56. Zhao XM, Xia Y, Whitesides GM. 1996. Adv Mater 8:837.
57. Kim E, Xia Y, Whitesides GM. 1995. Nature 376:581.
58. Chou SY, Krauss PR, Renstrom PJ. 1995. Appl Phys Lett 67:3114.
59. Chou SY, Krauss PR, Zhang W, Guo L, Zhuang L. 1997. J Vac Sci Technol B 15:2897.
60. Zhao XM, Wilbur JL, Whitesides GM. 1996. Langmuir 12:3257.
61. Whitesides GM, Grzybowski B. 2002. Science 295:2418.
62. Desiraju GR. 1989. *Crystal engineering: the design of organic solids*. Amsterdam: Elsevier.
63. Isaacs L, Chin DN, Bowden N, Xia Y, Whitesides GM, Reinhoudt DN. 1999. In: Reinhoudt DN, ed. *Perspectives in supramolecular chemistry,* New York: VCH, pp. 1–46.
64. Lindsey JS. 1991. New J Chem 15:153.
65. Deng Z, Mao C. 2004. Angew Chem Int Ed 43:4068.
66. Yang J, Mayer M, Kriebel JK, Garstecki P, Whitesides GM. 2004. Angew Chem Int Ed 43:1555.
67. Edwards EW, Montague MF, Solak HH, Hawker CJ, Nealey PF. 2004. Adv Mater 16:1315.
68. Sundrani D, Darling SB, Sibener SJ. 2004. Nano Lett 4:273.
69. Decher G. 1997. Science 277:1232.
70. Afsharrad T, Bailey AI, Luckham PF, Macnaughtan W, Chapman D. 1987. Colloids Surf 25:263.
71. Wang DY, Mohwald HJ. 2004. Mater Chem 14:459.
72. Harnack O, Pacholski C, Weller H, Yasuda A, Wessels JM. 2003. Nano Lett 3:1097.
73. Love JC, Urbach AR, Prentiss MG, Whitesides GM. 2003. J Am Chem Soc 125:12696.
74. Gourdon D, Yasa M, Alig ARG, Li Y, Safinya CR, Israelachvili JN. 2004. Adv Funct Mater 14:238.
75. Nagle L, Ryan D, Cobbe S, Fitzmaurice D. 2003. Nano Lett 3:51.

76. Xia YN, Yang PD, Sun YG, Wu YY, Mayers B, Gates B, Yin YD, Kim F, Yan YQ. 2003. Adv Mater 15:353.
77. Winklemann A, Gates BD, McCarty LS, Whitesides GM. 2005. Adv Mater 17:1507.
78. Fan FQ, Stebe KJ. 2003. Langmuir 19:5179.
79. Fudoizi H, Kobayashi M, Shinya N. 2002. Langmuir 13:7648.
80. Tien J, Terfort A, Whitesides GM. 1997. Langmuir 13:5349.
81. Chen KM, Jiang XP, Kimerling LC, Hammond PT. 2000. Langmuir 16:7825.
82. Jacobs HO, Whitesides GM. 2001. Science 291:1763.
83. Yin YD, Xia YN. 2002. Adv Mater 14:605.
84. Cui Y, Bjork MT, Liddle JA, Sonnichsen C, Boussert B, Alivisatos AP. 2004. Nano Lett 4:1093.
85. Kumacheva E, Garstecki P, Wu HK, Whitesides GM. 2003. Phys Rev Lett 91:128301.
86. Yin YD, Lu Y, Gates B, Xia YN. 2001. J Am Chem Soc 123:8718.
87. Kinge S, Crego-Calama M, Reinhoudt DN. 2008. ChemPhyschem 9:20.
88. Ugarte D, Stoeckli T, Bonard JM, Chatalain A, De Heer WA. 1998. Appl Phys A 67:101.
89. Saito M, Kirihara M, Taniguchi T, Miyagi M. 1989. Appl Phys Lett 55:607.
90. Majetich SA, Yin Y. 1999. Science 284:470.
91. Ajayan PM, Iijima S. 1993. Nature 361:333.
92. Yin LW, Bando Y, Zhu YC, Golberg D, Li MS. 2004. Adv Mater 16:929.
93. Niemeyer CM. 2001. Angew Chem 113:4254.
94. Warner MG, Hutchison JE. 2003. Nat Mater 2:272.
95. Hartgerink JD, Beniash E, Stupp S. 2001. Science 294:1684.
96. Mao C, Solis DJ, Reiss BD, Kottmann ST, Sweeney RY, Hayhurst A, Georgiou G, Iverson B, Belcher AM. 2004. Science 303:213.
97. Chung SW, Ginger DS, Morales MW, Zhang ZF, Chandrasekhar V, Ratner MA, Mirkin CA. 2005. Small 1:64.
98. Zirbs R, Kienberger F, Hinterdorfer P, Binder W. 2005. Langmuir 21:8414.
99. Katz E, Willner I. 2004. Angew Chem 116:6166.
100. Niemeyer C. 2000. Curr Opin Chem Biol 4:609.
101. Iler RK. 1966. J Colloid Interface Sci 21:569.
102. Kotov A, De'ka'ny I, Fendler JH. 1995. J Phys Chem 99:13065.
103. Nizamoglu S, Zengin G, Demir HV. 2008. Appl Physics Lett 92:031102.
104. Huyal IO, Ozel T, Koldemir U, Nizamoglu S, Tuncel D, Demir HV. 2008. Optics Express 16:1115.
105. Demir HV, Nizamoglu S, Ozel T, Mutlugun E, Huyal IO, Sari E, Holder E, Tian N. 2007. New J Phys 9:362.
106. Denkov ND, Velev OD, Kralchevsky PA, Ivanov IB, Yoshimura H, Nagayama K. 1993. Nature 361:26.
107. Gelbart WM, Sear RP, Heath JR, Chaney S. 1999. Faraday Discuss 112:299.
108. Deegan RD, Bakajin O, Dupont TF, Huber G, Nagel SR, Witten TA. 1997. Nature 389:827.
109. Giersig M, Hilgendorff M. 1999. J Phys D 32:L111.
110. Laoste D, TC Lubensky. 2001. Phys Rev E 64:41506.

111. Chushkin Y, Chitu L, Luby S, Majkova E, Satka A, Holy V, Ivan J, Giersig M, Hilgendorff M, Metzger T, Konovalov O. 2005. Mater Res Soc Symp Proc 877E(S6):18.
112. Hermanson KD, Lumsdon SO, Williams JP, Kaler EW, Velev OD. 2001. Science 294:1082.
113. Vossmeyer T, Delonno E, Heath JR. 1997. Angew Chem 109:1123.
114. Binder WH. 2005. Angew Chem 117:5300.
115. Kralchevski PP, Ivanov IB, Ananthapadmanabhan KP, Lips A. 2005. Langmuir 21:50.
116. Lin Y, Skaff H, Ermick T, Dinsmore AD, Russell TP. 2003. Science 299:226.
117. Hamley IW. 2003. Nanotechnology 14:R 39.
118. Tang ZY, Wang Y, Sun K, Kotov NA. 2005. Adv Mater 17:358.
119. Ikkale O, ten Brinke G. 2002. Science 295:2407.
120. Kletenik-Edelman O, Ploshnik E, Salant A, Shenhar R, Banin U, Rabani E. 2008. J Phys Chem C 112:4498.
121. Grousson M, Tarjus G, Viot P. 2000. Phys Rev E 62:7781.
122. TenWolde PR, Sun SX, Chandler D. 2001. Phys Rev E 65:0011201.
123. Love JC, Estroff LA, Kriebel JK, Nuzzo RG, Whitesides GM. 2005. Chem Rev 105:1103.
124. Adamson AW, Gast AP. 1997. *Physical Chemistry of Surfaces*, 6th edition. New York: Wiley-Interscience.
125. Herner M, Wagner P, Semenza G. 1993. Surf Sci 291:39.
126. Baudrand D. 2000. Plat Surf Finish 87:42.
127. Herrero E, Buller LJ, Abruna HD. 2001. Chem Rev 101:1897.
128. Schreiber F. 2000. Prog Surf Sci 65:151.
129. Chen SH, Frank CW. 1990. ACS Symp Serv 447:160.
130. Behm JM, Lykke KR, Pellin MJ, Leggett GJ. 2002. J Am Chem Soc 124:2414.
131. Friebel S, Aizenberg J, Abad S, Wiltzius P. 2000. Appl Phys Lett 77:2406.
132. Shadnam MR, Kirkwood SE, Fedosejevs R, Amirfazli A. 2004. Langmuir 20:2667.
133. Sondag-Huethorst JAM, van Helleputte HRJ, Fokkink LG. 1994. J Appl Phys Lett 64:285.
134. Golzhauser A, Geyer W, Stadler V, Eck W, Grunze M, Edinger K, Weimann T, Hinze PJ. 2000. Vac Sci Technol B 18:3414.
135. Klauser R, Hong IH, Wang SC, Zharnikov M, Paul A, Goelzhaeuser A, Terfort A, Chuang TJ. 2003. J Phys Chem B 107:13133.
136. Heister K, Zharnikov M, Grunze M, Johansson LSO, Ulman A. 2001. Langmuir 17:8.
137. Berggren KK, Bard A, Wilbur JL, Gillaspy JD, Helg AG, McClelland JJ, Rolston SL, Phillips WD, Prentiss M. 1995. Science 269:1255.
138. Harada Y, Masuda S, Ozaki H. 1997. Chem Rev 97:1897.
139. Chabinyc ML, Love JC, Thywissen JH, Cervelli F, Prentiss MG, Whitesides GM. 2003. Langmuir 19:2201.
140. Morgenthaler S, Lee S, Zuercher S, Spencer ND. 2003. Langmuir 19:10459.
141. Liedberg B, Tengvall P. 1995. Langmuir 11:3821.
142. Plummer ST, Bohn PW. 2002. Langmuir 18:4142.
143. Herbert CB, McLernon TL, Hypolite CL, Adams DN, Pikus L, Huang CC, Fields GB, Letourneau PC, Distefano MD, Hu W-S. 1997. Chem Biol 4:731.
144. Bietsch A, Hegner M, Lang HP, Gerber C. 2004. Langmuir 20:5119.
145. Martin CR. 1994. Science 266:1961.

146. Lee SB, Martin CR. 2001. Anal Chem 73:768.
147. Lieberman M, Chellamma S, Varughese B, Wang Y, Lent C, Bernstein GH, Snider G, Peiris FC. 2002. Ann N Y Acad Sci 960:225.
148. Kastle G, Boyen HG, Weigl F, Lengl G, Herzog T, Ziemann P, Riethmuller S, Mayer O, Hartmann C, Spatz JP, Moller M, Ozawa M, Banhart F, Garnier MG, Oelhafen P. 2003. Adv Funct Mater 13:853.
149. Glass R, Arnold M, Blummel J, Kuller A, Moller M, Spatz JP. 2003. Adv Funct Mater 13:569.
150. Haynes CL, Van Duyne RP. 2001. J Phys Chem B 105:5599.
151. Parviz BA, Ryan D, Whitesides GM. 2003. IEEE Trans Adv Packaging 26(3):233–241.
152. Notzel R. 1996. Semicond Sci Technol 11:1365.
153. Solomon GS, Trezza JA, Harris JS. 1995. Appl Phys Lett 66:3161.
154. Solomon GS, Trezza JA, Harris JS. 1995. Appl Phys Lett 66:991.
155. Murray CB, Kagan CR, Bawendi MG. 2000. Annu Rev Mater Sci 30:545.
156. Sun SH, Murray CB, Weller D, Folks L, Moser A. 2000. Science 287:1989.
157. Alivisatos AP, Johnsson KP, Peng XG, TE Wilson, Loweth CJ, Bruchez MP, Schultz PG. 1996. Nature 382:609.
158. Shevchenko EV, Talapin DV, Rogach AL, Kornowski A, Haase M, Weller H. 2002. J Am Chem Soc 124:11480.
159. Kiely CJ, Fink J, Brust M, Bethell D, Schiffrin DJ. 1998. Nature 396:444.
160. Rogach AL, Talapin DV, Shevhenko EV, Kornowski A, Haase M, Weller H. 2002. Adv Fun Mater 12(10):653.
161. Skirtach AG, Dejugnat C, Braun D, Susha AS, Rogach AL, Sukhorukov GB. 2007. J Phys Chem C 111:555.
162. Collier CP, Vossmyer T, Heath JR. 1998. Annu Rev Chem 49:371.
163. Chen X, Hirtz M, Rogach AL, Talapin DV, Fuchs H, Chi L. 2007. Nano Lett 7:3483.
164. Zanchet D, Micheel CM, Parak WJ, Gerion D, Alivisatos AP. 2001. Nano Lett 1:32.
165. Mokari T, Sertchook H, Aharoni A, Ebenstein Y, Avnir D, Banin U. 2005. Chem Mater 17:258.
166. Chen X, Rogach AL, Talapin DV, Fuchs H, Chi L. 2006. J Am Chem Soc 128:9592.
167. Holder E, Tessler N, Rogach AL. 2008. J Mater Chem 18:1064.
168. Gangopadhyay R, De A. 2000. Chem Mater 12:2064.
169. Oriakhi CO. 2000. J Chem Educ 77:1138.
170. Rakovich YP, Volkov Y, Sapra S, Susha AS, Döblinger M, Donegan JF, Rogach AL. 2007. J Phys Chem C 111:18927.
171. Markovich G, Collier CP, Heath JR. 1998. Phys Rev Lett 80:3807.
172. Murray CB, Kagan CR, Bawendi MG. 1995. Science 270:1335.
173. Hornyak GL, Patrissi CJ, Martin CR. 1997. J Phys Chem B 101:1548–1555.
174. Glass R, Arnold M, Blummel J, Kuller A, Moller M, Spatz JP. 2003. Adv. Funct. Mater. 13:569–575.
175. Dujardin E, Hsin L-B, Wang CRC, Mann S. 2001. Chem Commun 1264–1265.
176. Mao C, Christine EF, Hayhurst A, Sweeney R, Qi J, Georgiou G, Iverson B, Belcher AM. 2003. PNAS 100:6947–6951.
177. Yin Y, Xia Y. 2003. J Am Chem Soc 125:2048–2049.

178. Zhang L, Gaponik N, Müller J, Plate U, Weller H, Erker G, Fuchs H, Rogach AL, Chi L. 2005. Small 1:524–527.
179. Hu J, Odom TW, Lieber CM. 1999. Acc Chem Res 32:435–445.
180. Jana NR, Gearheart L, Murphy CJ. 2001. Chem Commun 617.
181. Pacholski C, Karnowski A, Weller H. 2002. Angew Chem Int Ed 41:1188–1191.
182. Cho K-S, Talapin DV, Gaschler W, Murray CB. 2005. J Am Chem Soc 127:7140–7147.
183. Tang Z, Kotov NA, Giersig M. 2002. Science 297:237–240.
184. Scher EC, Manna L, Alivisatos AP. 2003. Phil Trans R Soc Lond A 361:241.
185. Ryan KM, Mastroianni A, Stancil KA, Liu H, Alivisatos AP. 2006. Nano Lett 6:1479.
186. Liu Y, Chung J-H, Liu WK, Ruoff RS. 2006. J Phys Chem B 110:14098.
187. Chen XQ, Saito T, Yamada H, Matsushige K. 2001. Appl Phys Lett 78:3714.
188. Green NG, Morgan H. 2003. *AC Electrokinetics-Colloids and Nanoparticles*. Hertfordshire: Research Studies Press.
189. Jones TB. 1995. *Electromechanics of Particles*. New York: Cambridge University Press.
190. Wang X, Wang XB, Cascoyne PRC. 1997. J Electrost 39:277.
191. Bockris JOM, Reddy AKN. 1970. *Modern electrochemistry*.
192. Ashkin A, Dziedzic JM, Bjorkholm JE, Chu S. 1986. Opt Lett 11:288.
193. Grier DG. 2003. Nature 424:810.
194. Rogers JA. 2008. Nature Photonics 2:69.
195. Chiou PY, Ohta AT, Wu MC. 2005. Nature 436:370.
196. Xu Q, Rioux RM, Dickey MD, Whitesides GM. 2008. Acc Chem Res A41:1566.
197. Li NH, Wu W, Chou SY. 2006. Nano Lett 6:2626.
198. Crommie MF, Lutz CP, Eigler DM. 1993. Science 262:218.
199. Hong S, Zhu J, Mirkin CA. 1999. Science 288:1808.
200. Liu GY, Xu S, Qian Y. 2000. Acc Chem Res 33:457.
201. Ban N, Nissen P, Hansen J, Moore PB, Steitz TA. 2000. Science 289:905.
202. Gracias DH, Kavthekar V, Love JC, Paul KE, Whitesides GM. 2002. Adv Mater 14:235.
203. Clark TD, Tien J, Duffy DC, Paul KE, Whitesides GM. 2001. J Am Chem Soc 123:7677.
204. Whitesides GM, Ismagilov RF. 1999. Science 284:89.
205. Ismagilov RF, Rosmarin D, Gracias DH, Strook AD, Whitesides GM. 2001. Appl Phys Lett 79:439.
206. Zhang H, Edwards EW, Wang D, Möhwald H. 2006. Phys Chem Chem Phys 8:3288.
207. Valley JK, Jamshidi A, Ohta AT, Hsu H-Y, Wu MC. 2008. J Microelectromech Syst 17:342.

13

INTEGRATION OF MULTIFUNCTIONAL PROPERTIES IN THERMAL BARRIER COATINGS BY CHEMICAL VAPOR DEPOSITION

Takashi Goto

Tohoku University, Sendai, Japan

INTRODUCTION

A long lifetime under high-temperature operating conditions is of fundamental importance for gas turbines to achieve high efficiencies and low energy consumption. The combustion temperature of gas turbines in the 1960s and 1970s was below the melting point of metal components, around 1000°C. Since then, the operation temperature of gas turbines has been rising year by year, currently reaching over 1500°C, mainly due to the development of air cooling and thermal barrier coating (TBC) systems [1]. Turbine blades are directly exposed to high-temperature gases, and therefore, high refractoriness with excellent mechanical properties, particularly toughness and creep strength, is an essential requirement for practical applications. Although the metal components of blades (metal substrate), mainly nickel-based superalloys, have been significantly improved from conventionally cast polycrystalline to advanced directionally solidified single-crystalline materials, the advancement of TBC is another crucial issue for the development of gas turbines.

TBC should be characterized by high refractoriness, thermal stability, and good adhesion to the metal substrate, and yttria stabilized zirconia (YSZ) is now the most common coating material due to its high strength, ductility, and a relatively large

Ceramic Integration and Joining Technologies: From Macro to Nanoscale, First Edition.
Edited by Mrityunjay Singh, Tatsuki Ohji, Rajiv Asthana, Sanjay Mathur.

thermal expansion coefficient. However, despite the closeness of thermal expansion coefficients between YSZ coating and metal substrates, a slight difference of thermal expansion causes a significant thermal stress at the interface during severe heat cycles, yielding cracking around the YSZ/substrate boundary and finally resulting in fatal delamination after prolonged operation. Bond coat layers, typically MCrAlY (M : Ni, Co, etc.), should be precoated on the metal substrate to reduce the thermal stress and also to protect the metal surface from oxidation. To maximize the protection of the bond coat layer, the surface temperature should be lowered by the thermally insulative TBC layer. Since ceramics have different characteristics from metals, TBC requires sophisticated integration between metallurgy and ceramic technology.

As the microstructure of the YSZ coating significantly affects stress relaxation and thermal conductivity, control of the microstructure is of primary importance to improve the performance of TBC [2]. The microstructure control in TBC also requires an integrated understanding of various aspects such as crystal nucleation, anisotropic columnar grain growth, and nanopore alignment. Various kinds of microstructures have been fabricated using atmospheric plasma spray (APS) and electron-beam physical vapor deposition (EBPVD); however, a new method of YSZ coating should be sought to achieve further development of TBC. Chemical vapor deposition (CVD) can be a candidate process for TBC due to its many advantages, in particular, well-adherent conformal coverage. Conventional CVD uses thermal energy for chemical reactions, resulting in heterogeneous deposition reactions on the substrate surface. Lasers have often been employed as an auxiliary energy source to accelerate chemical reactions to deposit thin films, nanodots, and whiskers. By integrating recent advanced laser technology and conventional CVD, a tremendously high-speed deposition process can be implemented, which would be applicable to many thick coatings.

This chapter briefly reviews the processes for depositing TBC [3], including CVD, and introduces a new high-speed coating process of laser CVD for a new generation of TBC processes.

TBC PROCESS

Since TBC should be able to endure a severe temperature gradient of about several hundred degrees, the thickness of TBC should be at least several hundred micrometers to accommodate high thermal stress, and therefore high-speed deposition processes, particularly APS and EBPVD, have been traditionally employed for TBC. Figures 13.1 and 13.2 present a schematic of the APS setup and a typical cross-sectional microstructure of an APS coating, respectively [4]. In an APS process, YSZ source powder is supplied to a plasma torch and is instantly melted/vaporized and sprayed onto a metal substrate. Although APS has wide-ranging versatile applications due to its simple and economical setup, APS coatings often contain unmelted particles, voids, and cracks that constitute a characteristic splat texture. The laminar nature of an APS coating might cause delamination by transversely extending cracks, sometimes resulting in a catastrophic failure of TBC. On the other hand, in an EBPVD process, a strong electron beam is emitted to a YSZ sintered body rod to form YSZ vapor, which is deposited on

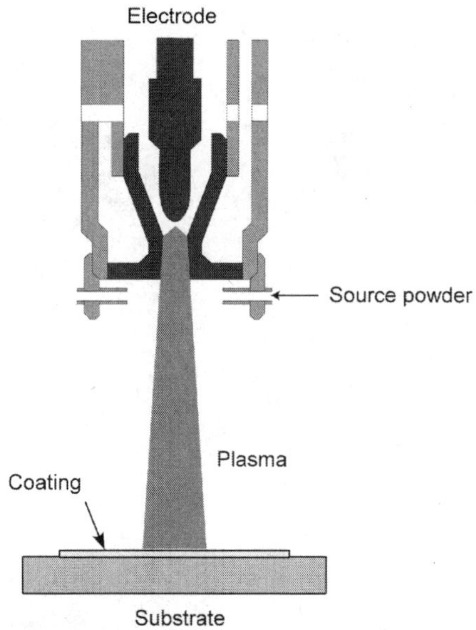

Figure 13.1. Schematic of a plasma spray apparatus.

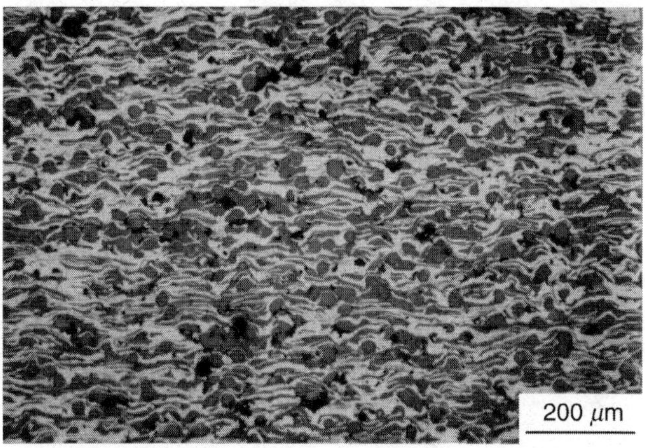

Figure 13.2. Typical cross-sectional microstructure of YSZ coating prepared by plasma spray.

a metal substrate, resulting in the formation of a more advanced YSZ coating for TBC. Figures 13.3 and 13.4 are, respectively, a schematic of the EBPVD setup and of a typical cross-sectional texture of YSZ coating [1]. An elongated columnar texture is particularly effective for the relaxation of thermal stress at the YSZ/substrate interface, mainly owing to nanoscale gaps at the columnar grain boundary. Such texture yields vertical cracks preventing the delamination of the whole TBC coating. Figure 13.5 depicts the

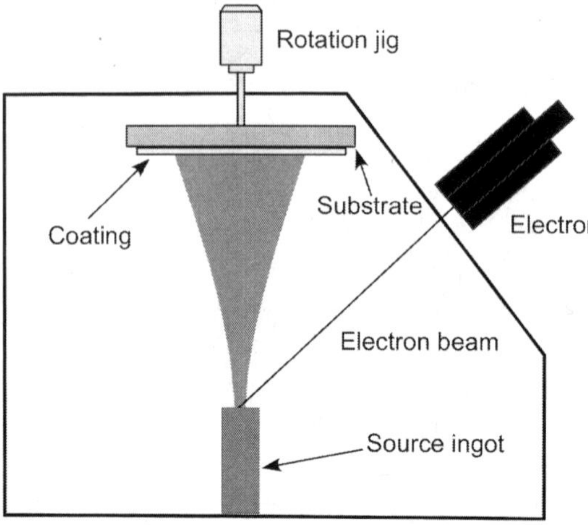

Figure 13.3. Schematic of EBPVD apparatus.

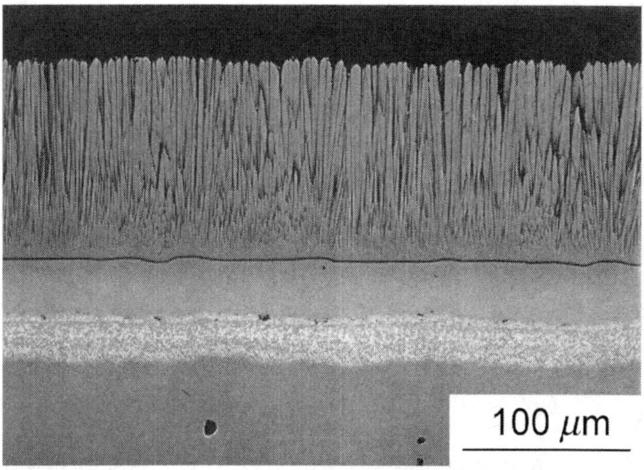

Figure 13.4. Typical cross-sectional microstructure of YSZ coating prepared by EBPVD.

nanoscale texture of YSZ coating prepared by EBPVD [5]. Each grain contains a large number of nanopores, which can effectively reduce the thermal conductivity to less than 2 W/m/K, half of the bulk of the YSZ sintered body. The formation mechanism of nanopores and columnar grains in EBPVD can be explained by a "shadowing effect," as illustrated in Figure 13.6 [6]. During the EBPVD process, a highly concentrated YSZ vapor is produced by a strong electron beam, and the vapor then arrives at the substrate surface. Since the EBPVD process is a highly directional line-of-sight process, YSZ vapor cannot reach behind the predeposited grains, inevitably forming a large number of nanopores inside the coating. Although the highly porous nanoscale structure is

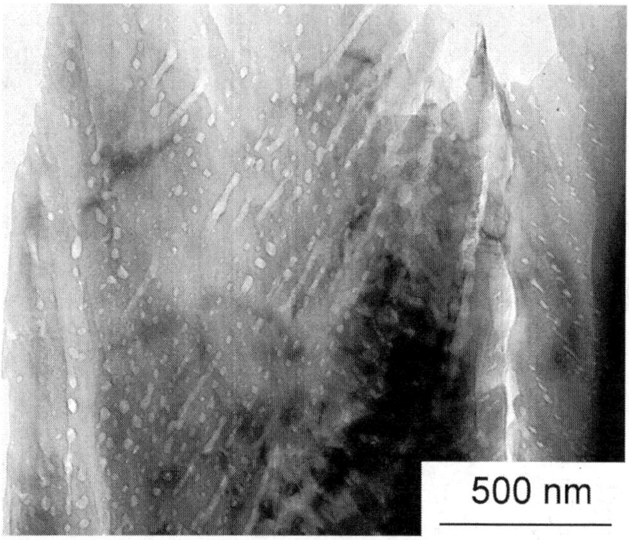

Figure 13.5. Nanoscale texture of YSZ coating prepared by EBPVD.

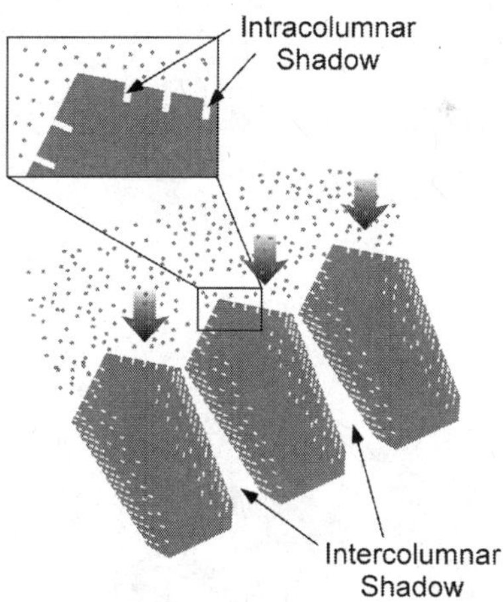

Figure 13.6. Formation mechanism of nanopores and columnar grains in EBPVD (shadowing effect).

advantageous for TBC, nanopores and a complicated surface morphology could be thermally unstable, as they are prone to shrinkage during the high-temperature operation and finally may result in the delamination of the entire coating.

HIGH-SPEED COATING BY CONVENTIONAL CVD

CVD, applied in many industrial fields for such things as semiconductor devices and cutting tools, is a coating process providing highly pure and dense films with excellent conformal coverage [7]. Figure 13.7 briefly demonstrates the relationship between deposition rates and applicable processes ranging from surface treatment to LSI (large-scale integrated circuit). Ion beam epitaxy (IBE) and molecular beam epitaxy (MBE) have been used for preparing thin films at a submicrometer level with a finely controlled nanostructure owing to their low deposition rates of around 0.1–1.0 μm/h. Since CVD can be adapted for various deposition conditions, including temperature, pressure, source gas ratio, and various types of precursors, a wide range of morphologies and micro-/nanotextures can be fabricated by CVD. However, it is commonly understood that CVD is a rather low-deposition rate process with rates ranging from one to a few tens of micrometers per hour and is therefore employed for preparing thin films several micrometers in thickness. Figure 13.8 demonstrates a general trend of morphology as functions of typical deposition parameters, deposition temperature (T_{dep}), and supersaturation [8]. Generally, at low T_{dep} and high supersaturation, CVD deposits become powdery due to homogeneous reactions in the gas phase. With increasing T_{dep} and decreasing supersaturation, the morphology changes from amorphous to fine crystals, columnar grains, platelike crystals, and finally to epitaxial films. The columnar texture essentially required for TBC is common in CVD.

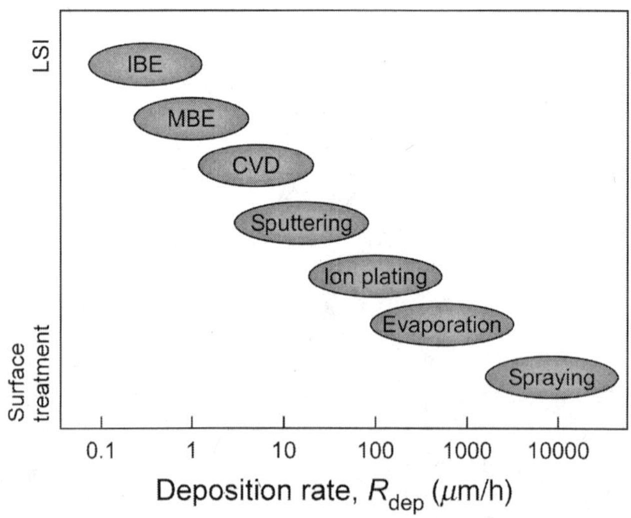

Figure 13.7. Relationship between deposition rates and applicable processes.

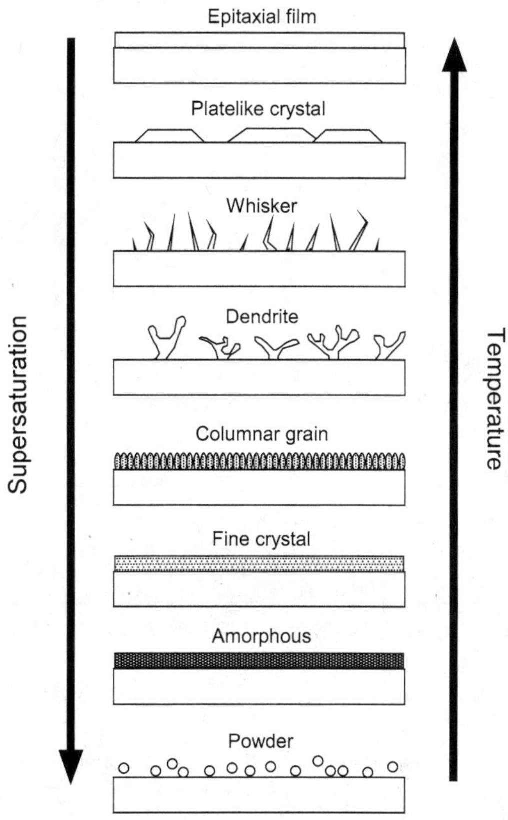

Figure 13.8. General trend of morphology as functions of deposition temperature (T_{dep}) and supersaturation in CVD.

Although the deposition rates of common CVD have been rather low or moderately slow, thick films of several materials such as SiC [9] and Si_3N_4 [10] have been prepared at exceptionally high deposition rates of more than 1.0 mm/h by the present author. The deposition rate of CVD is closely associated with its kinetic process. Figure 13.9 depicts the elemental kinetic processes of CVD for depositing films [8]. The film formation occurs through a sequence of steps: mainly, diffusion of source gas in a main gas stream, diffusion in a gas boundary layer near the substrate, and the absorption, reaction, and desorption of chemical species on the substrate. Since chemical reactions and crystal growth occur at atomic/molecular levels, a macroscopically rough substrate surface can be well covered at a nanometer scale. Even the backside of a substrate can be coated because the gases can arrive there by diffusion. This non-line-of-sight nature of CVD is in contrast to that of the APS and EBPVD processes, and is advantageous for TBC.

YSZ films have been widely prepared by CVD, particularly for oxide ion conduction and buffer layers for oxide high-temperature superconductor films. Figure 13.10 summarizes the deposition rates of YSZ films prepared by conventional CVD [11–15].

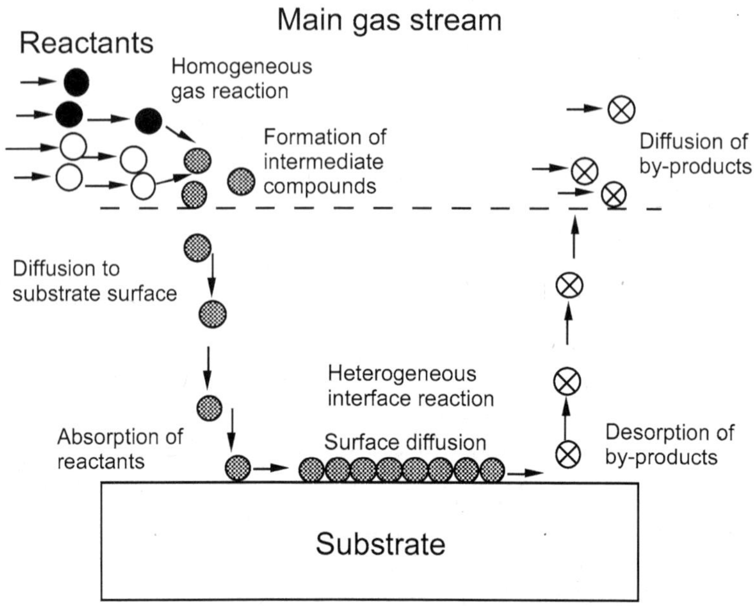

Figure 13.9. Elemental kinetic processes of CVD for depositing films.

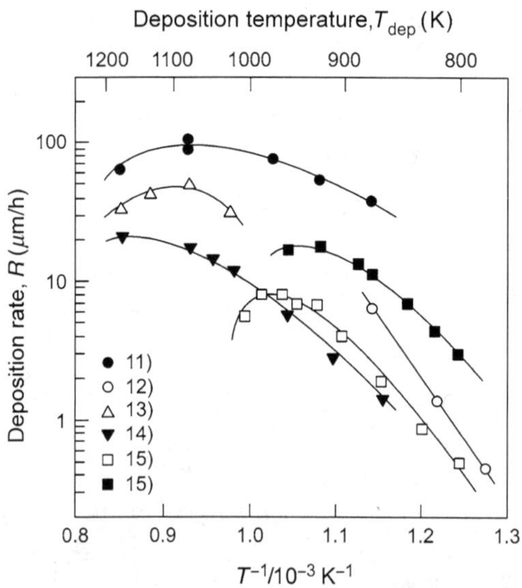

Figure 13.10. Deposition rates of YSZ films as a function of deposition temperature (T_{dep}) in conventional CVD.

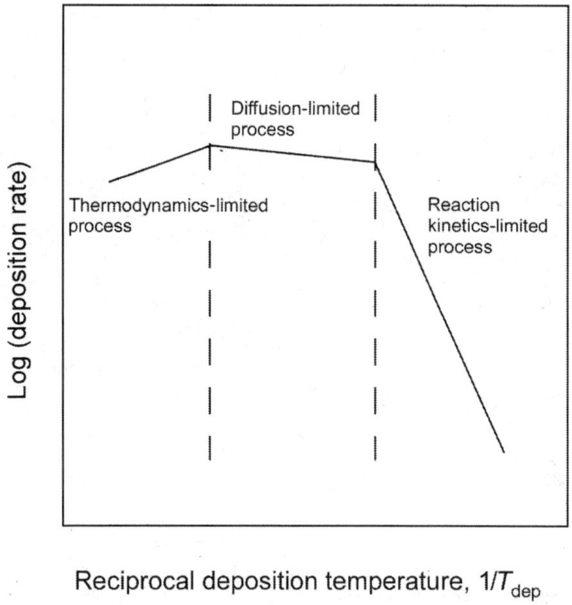

Figure 13.11. General trend of deposition rate as a function of deposition temperature (T_{dep}).

The deposition rates of CVD YSZ films are generally less than several micrometers per hour, and the films have a total thickness of below 10–20 μm when halide gases ($ZrCl_4$ and YCl_3) or various metal organic (MO) prescursors such as acetylacetonate (acac), methylhepfadionato (thd) and dipivaloylmethanato (dpm) are used.

While CVD can provide coatings characterized by good adherence and conformal coverage, a significant increase in the deposition rates of CVD for its application to TBC would be needed. The deposition rate would not be enhanced simply by increasing T_{dep} or the source gas supply, as it is controlled by the slowest process among many elemental processes depicted in Figure 13.9. Figure 13.11 demonstrates a general trend of deposition rate as a function of T_{dep} [8]. In a low T_{dep} region, chemical reaction rates on the surface could be the lowest step in general, at which the deposition rate exponentially increases with T_{dep}. At further higher T_{dep}, the chemical reaction rates become sufficiently high, and then the deposition rate should be controlled by a source gas supply, that is, a diffusion process in a gas phase. Under the diffusion-limited process, the deposition rate would slightly increase with T_{dep} and would increase with increasing source gas flux. However, the deposition rate cannot be further increased at a higher source gas supply because premature reactions may occur in a gas phase, resulting in the formation of powder or intermediate compounds. At further higher T_{dep}, the deposition rate may be decreased by powder formation in the gas phase or by a decrease of driving force of chemical reactions (i.e., a decrease of $-\Delta G°$, Gibbs free energy) [16]. In any case, the rate-controlling step in CVD is a thermally activated process, and thus conventional CVD is generally termed thermal CVD.

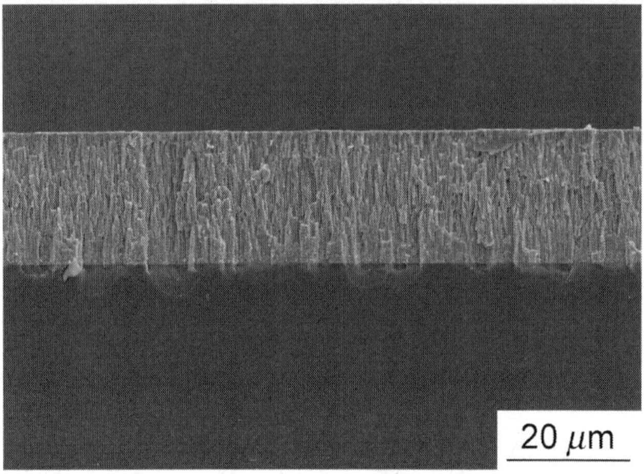

Figure 13.12. Cross-sectional microstructure of YSZ coating prepared by thermal CVD.

The application of CVD to TBC has not been widely conducted, and several modifications of the CVD process are needed mainly, solving the problems of how to supply a large amount of precursors to the deposition zone and how to uniformly heat a substrate (turbine blade) with a complex shape. Wahl et al. have prepared YSZ coatings for TBC at a high deposition rate of 50 μm/h by using $Zr(thd)_4$ and $Y(thd)_3$ precursors [12]. The present author's group has also employed thermal CVD for TBC and further increased the deposition rate of YSZ coatings up to 102 μm/h by modifying a cold wall-type CVD setup and optimizing deposition conditions using $Zr(dpm)_4$ and $Y(dpm)_3$ precursors [11]. Figure 13.12 depicts a cross-sectional microstructure of YSZ coating prepared by the present author's group. A well-grown columnar texture with a (100) orientation can be identified, in which a large number of nanopores are contained in columnar grains, as shown in Figure 13.13 [12]. The shape of nanopores by the thermal CVD is slightly angular, whereas that by EBPVD is almost spherical or an elongated spheroid. The formation mechanism of nanopores in the thermal CVD might have been different from that in EBPVD. Figure 13.14 is a schematic diagram of the nanopore formation mechanism in the thermal CVD [17]. It is generally known that the morphology and microstructure of coatings by thermal CVD may be affected by the rate-controlling step. Under the reaction-limited step in a low T_{dep} region, nucleation sites for grain growth would be kinks or steps, that is, the lowest nucleation energy sites, and the grains would grow upward along the crystal surface from the bottom to the top surface, as depicted in Figure 13.14a. As a result, the coatings prepared under such conditions might be dense without forming nanopores. On the other hand, under the diffusion-limited step in a high T_{dep} region, nucleation may be initiated at sites of the highest source gas concentration, where the chemical reaction rates are sufficiently high due to high T_{dep}. Therefore, grain growth would start from the top surface and proceed downward to inside the coating, as depicted in Figure 13.14b [18]. Each grain may collide at the bottom, resulting in the formation of nanopores, and the inner surface of

Figure 13.13. Nanopores in columnar grains of YSZ coating prepared by thermal CVD.

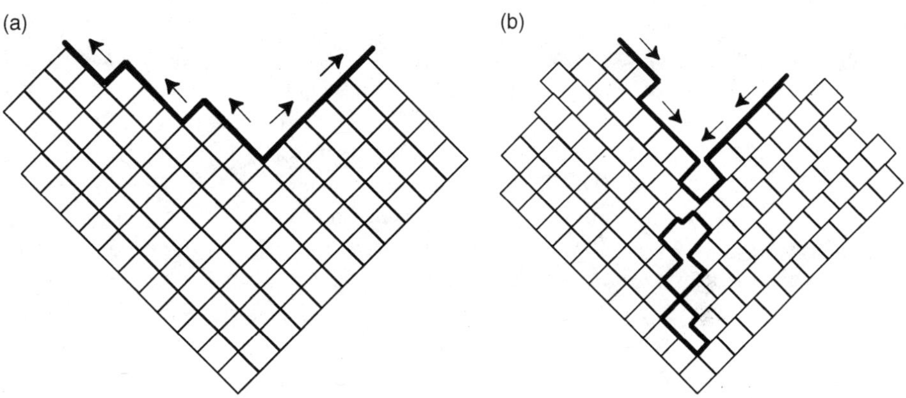

Figure 13.14. Schematic diagram of the nanopore formation mechanism in thermal CVD: (a) reaction-limited condition and (b) diffusion-limited condition.

nanopores might be surrounded by specific crystal planes, as observed in an angular shape. These angular nanopores in the YSZ coating formed by thermal CVD would be crystallographically more stable and more durable than the spherical nanopores formed by EBPVD at high temperatures. Such characteristics of nanopores in the YSZ coatings with low thermal conductivity are advantageous for application to TBC. CVD YSZ coatings have been found to have a low thermal conductivity of 0.8 W/m/K, exhibiting excellent heat cycle resistance and oxidation resistance [19].

Nonequilibrium cold plasma typically produced by a radio frequency or microwave electromagnetic field can provide a highly reactive environment for CVD by forming ionizing species; this technique is termed plasma-enhanced chemical vapor deposition (PECVD). Many kinds of film have been prepared by PECVD, mostly for application

Figure 13.15. Cross-sectional microstructure of YSZ coating prepared by PECVD.

in the fabrication of semiconductor thin-film devices. On the other hand, Préauchat and Drawin have utilized PECVD for preparing YSZ coatings to apply to TBC [20]. They prepared YSZ coatings at a deposition rate of 250 μm/h with a thickness of 200 μm by using $ZrCl_4$ and $Y(thd)_3$ precursors. Figure 13.15 shows a cross-sectional microstructure of the YSZ coating by PECVD. Well-developed columnar grains with a (100) oriented quasi-tetragonal (t') phase were observed, and excellent thermal shock resistance with low thermal conductivity (1.6 W/m/K) was reported. PECVD has high potential for obtaining high deposition rates by modifying the experimental setup and source supply. Figure 13.16 shows a microjet plasma (a type of high-temperature plasma) CVD apparatus where ionized source gases were blown onto a substrate zone by using a plasma torch [21]. A high deposition rate of 120 μm/s (432 mm/h) was reported for preparing SiO_2 film by using hexamethyldisiloxane.

HIGH-SPEED COATING BY LASER CVD

Lasers are an auxiliary energy source applicable to CVD, hence the name "laser CVD" [22]. Laser technology has been intensively applied to material processing since the 1970s, and many processes such as laser welding, laser ablation, and laser annealing have been developed. A laser has two kinds of energy: heat and light. Thus, laser CVD is generally categorized into two types: pyrolytic and photolytic [23]. In pyrolytic laser CVD, a CO_2 laser (wavelength, $\lambda = 9 \sim 11\,\mu$m) has been generally used as a heat source. By focusing a laser beam, a small area can be heated, causing thermally activated chemical reactions (pyrolytic reactions). By moving a substrate or by scanning with a laser beam, microscale three-dimensionally structured deposits or direct patterned deposits have been fabricated [24]. In pyrolytic laser CVD, TiN [25] and TiB_2 [26] coatings with a thickness up to 10 μm have been prepared on carbon fibers and

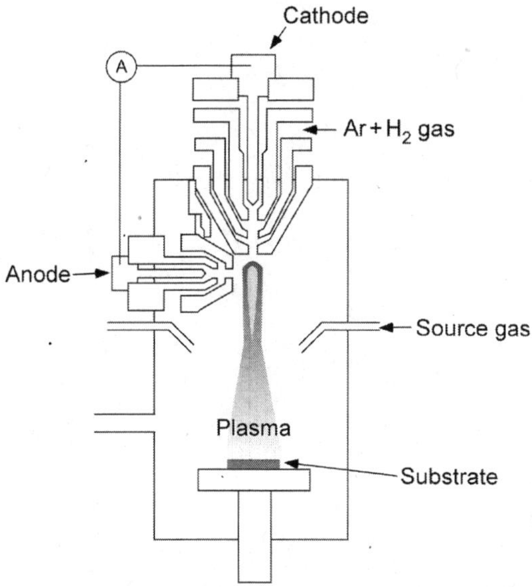

Figure 13.16. Microjet plasma CVD apparatus.

plates by using a CO_2 laser beam with a diameter of 3.5–9.0 mm. In photolytic laser CVD, on the other hand, a high-energy ultraviolet laser, typically an Excimer laser, has been employed to enhance photochemical reactions, decomposing source precursors directly without heating substrates. Since the ultraviolet laser scarcely affects the impurity profile of semiconductor substrates, photolytic laser CVD has been widely employed to fabricate various thin films. In both pyrolytic and photolytic laser CVD, however, deposits have always been small in scale, either thin films, dots, or whiskers; no thick or wide-area coatings have been prepared by conventional laser CVD.

Due to the recent developments in laser technology, high-power lasers are readily available for material processing. The present author has developed a laser CVD process by which thick coatings on complicated-shaped substrates at high deposition rates are created, a process particularly applicable to TBC [27]. A distinct feature of this laser CVD is plasma formation around a deposition zone. Source gases are ionized, similar to the "plasma plume" in laser ablation. Figure 13.17 is a schematic of the laser CVD. The CVD chamber is made of water-cooled stainless steel. A substrate is preheated on a hotplate at 300–500°C, and O_2 gas and precursor vapors are separately introduced into the chamber through a double-tube nozzle. A laser beam (with power at most 250 W) is slightly expanded to 20 mm in diameter, where the substrate temperature increases to around 150–250°C by the laser radiation in the case of an Al_2O_3 substrate. The temperature of the AlN substrate can be promptly increased by the laser radiation close to its melting point at around 2200°C. The temperature increase in a substrate depends on the characteristics of the substrate; that is, the substrate is either laser absorbable or reflective. After the introduction of source gases, a "plasma" forms around the deposition zone by the interaction between the laser and the source gases.

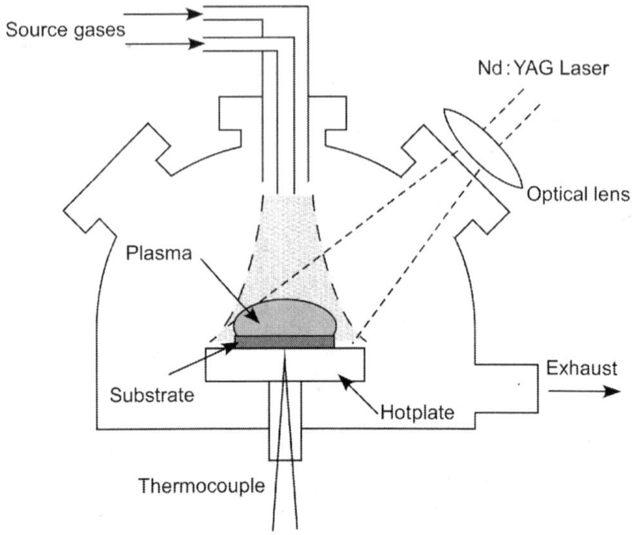

Figure 13.17. Schematic of laser CVD (LCVD).

Figure 13.18 is a schematic of a feature of the laser CVD. The plasma can be characterized by usual plasma diagnosis using a Langmuir probe and emission spectroscopy. The plasma formation causes a significant increase in the substrate temperature up to 800–1000°C, emitting a specific bright light. The emission spectrum is similar to that of a Planck distribution with an electron temperature of approximately 4000°C [28]. Figure 13.19 shows a practical application of the laser CVD to YSZ coating on a gas turbine blade [29]. The laser CVD is not a line-of-sight process because the deposition reaction that occurs is mainly assisted by the plasma around the deposition zone. Figure 13.20 presents a cross-section of YSZ coating prepared by the laser CVD. The YSZ coating consists of (100) oriented columnar grains as observed in EBPVD. By changing the deposition conditions, a pebblelike or a cauliflower-like texture can be obtained. The highest deposition rate of YSZ coating in the laser CVD is 660 μm/h by using $Zr(dpm)_4$ and $Y(dpm)_3$ precursors [30]. A large number of nanopores are observed in the columnar grains, and the highest porosity of the YSZ coating is about 30%. A low thermal conductivity is strongly required by TBC. Figure 13.21 depicts the relationship between the deposition rate and the thermal conductivity of YSZ coating. The thermal conductivity decreases with increasing deposition rate and becomes almost constant (0.6 W/m/K) at a deposition rate in excess of 40 μm/h, implying that the higher the deposition rate, the greater the number of nanopores. The columnar morphology and low thermal conductivity of YSZ coating should be maintained during the high-temperature operation for TBC.

Figure 13.22 shows the morphology change of a laser CVD YSZ coating after high-temperature heat treatment. Even after heat treatment at 1400°C for 100 h, the

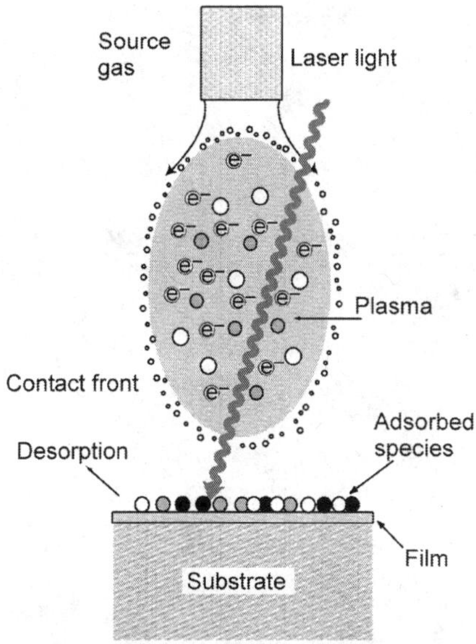

Figure 13.18. Schematic of feature of laser CVD.

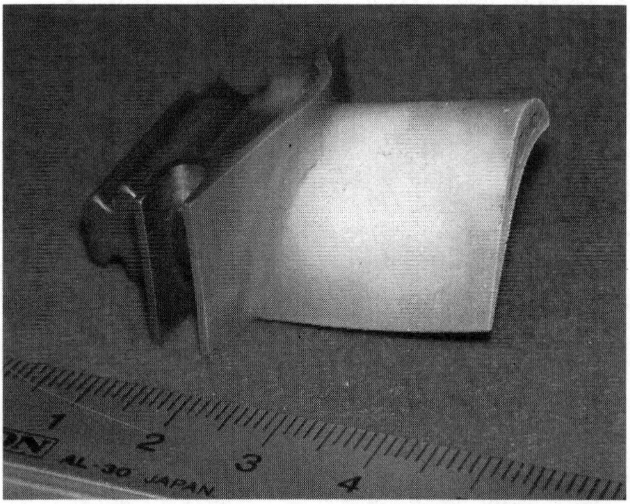

Figure 13.19. Practical application of laser CVD to YSZ coating on a gas turbine blade.

faceted surface and the columnar cross-section retained their original shapes. After 1600°C for 100h, the morphology was deformed by sintering and by the growth of grains, where the crystal structure of YSZ coating was transformed from tetragonal to monoclinic, accompanying the decrease in the number of nanopores. Although the operating temperature of a gas turbine is at most around 1500°C, the surface

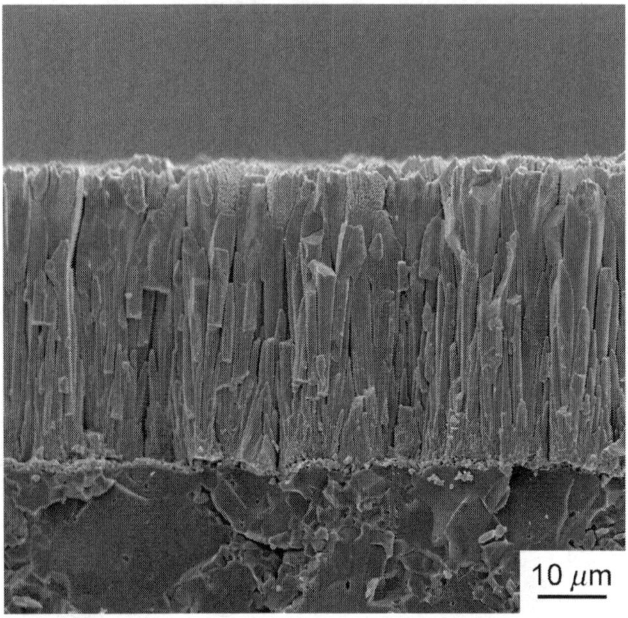

Figure 13.20. Cross-section of YSZ coating prepared by laser CVD.

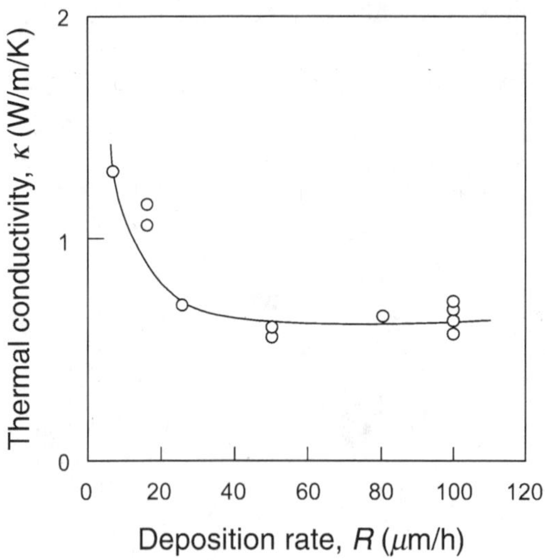

Figure 13.21. Relationship between deposition rate and thermal conductivity of YSZ coating prepared by LCVD.

Figure 13.22. Morphology change of laser CVD YSZ coating after high-temperature heat treatment: (a) surface, 1400°C, 100 h; (b) surface, 1600°C, 100 h, (c) cross-section, 1400°C, 100 h, (d) cross-section, 1600°C, 100 h.

temperature of YSZ coating should be far less than 1500°C due to an air-cooling system. Therefore, the YSZ coating by the laser CVD has sufficient thermal stability under the condition of practical applications.

The laser CVD can be used to fabricate a wide range of oxide films, such as TiO_2, Y_2O_3, Al_2O_3, and SiO_2 by changing source gas precursors [31]. TiO_2 films can be prepared by using Ti-iso-propoxide (Ti(O-n-Bu)$_4$) at a deposition rate of 2300 μm/h, in which anatase- or rutile-type TiO_2 films in a single phase are prepared by changing deposition conditions. Y_2O_3 films are prepared at 270 μm/h by using Y(dpm)$_3$ and exhibit excellent antiplasma etching properties [32]. α-type Al_2O_3 films can be prepared by the laser CVD at a low T_{dep} and at a high deposition rate. A thermally stable and high hardness α-type Al_2O_3 coating is required for cutting tools. The deposition temperature of α-Al_2O_3 coating in conventional thermal CVD is over 1000°C, whereas it is around 800°C by the laser CVD at a deposition rate of 1200 μm/h by using

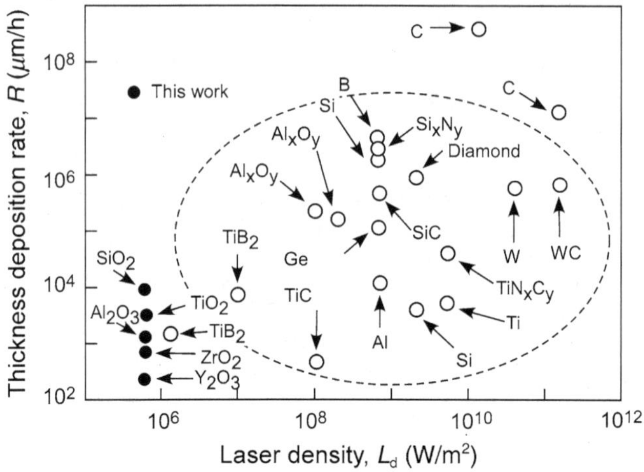

Figure 13.23. Relationship between one-dimensional (linear) deposition rate and laser density in laser CVD.

$Al(acac)_3$. The deposition rate of SiO_2 film is high at 27.5 mm/h by using tetraetoxysilane ($[C_2H_5O]_4Si$).

In conventional laser CVD, many materials such as silicon, carbon, and diamond in shapes of nanodots or whiskers have been prepared at significantly high deposition rates. Figure 13.23 summarizes the relationship between the one-dimensional (linear) deposition rate (increase in thickness or length of deposits per time) and laser density [33]. The linear deposition rates in the literature are about 10^2–10^5 times greater than those of the present laser CVD. However, the laser beam sizes have been focused to around several hundred micrometers to several millimeters at most in conventional laser CVD. As previously shown in Figure 13.11, the deposition rate can be determined by the rate-controlling step, either chemical reaction limited or diffusion rate limited. Since the chemical reactions are sufficiently accelerated by laser, it might be possible to determine the deposition rate by gas transport (diffusion process) in a gas phase in laser CVD. Because the deposition area is localized by a focused laser beam, source gases can easily access the deposition area. Therefore, it has been generally understood that high-speed deposition in laser CVD can be attained only for deposits in small areas, particularly almost one-dimensional fiberlike deposits. Thick wide-area coatings at high deposition rates have never been reported by laser CVD [23]. On the contrary, with the present laser CVD, it is possible to prepare thick, wide-area coatings at high deposition rates. The plasma formation might have been closely associated with the high deposition rates. The laser CVD may be a kind of "laser-induced plasma CVD."

The volume deposition rate can be calculated by multiplying the linear deposition rate by the laser beam sizes. Figure 13.24 illustrates the relationship between the volume deposition rate and the laser density. The volume deposition rates of the present laser CVD are the highest level among those of conventional laser CVD. Laser CVD

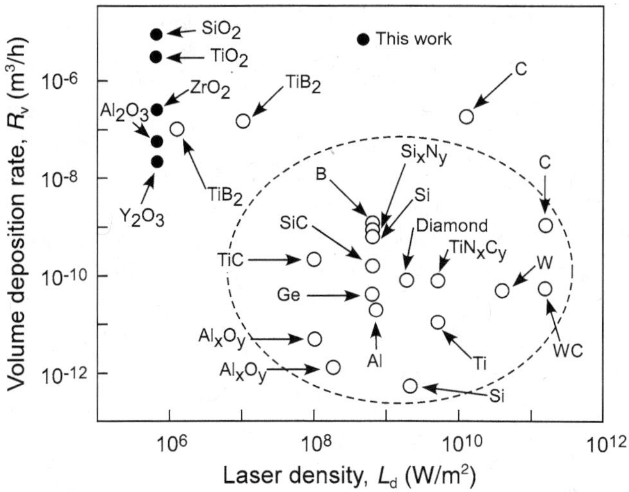

Figure 13.24. Relationship between volume deposition rate and laser density in laser CVD.

represents a technically superior approach to integrating TBCs in the manufacture of turbine blades.

SUMMARY

TBC is a key issue in improving the performance of gas turbines. Although TBC has been traditionally fabricated by APS and EBPVD, another coating process, CVD, is a possible candidate process that provides adherent coating with excellent coverage owing to its non-line-of-sight nature. To apply CVD to TBC, a significant increase in deposition rate is crucial. An auxiliary energy-assisted CVD, in particular, laser CVD, is promising for the next-generation TBC process to produce an integrated microstructure consisting of nanopores and anisotropic columnar grains. Laser CVD, integrating conventional thermal CVD, and recent laser technology appear to be promising for a wide range of applications of thick coatings in many engineering fields.

REFERENCES

1. Clarke DR, Levi CG. 2003. Materials design for the next generation thermal barrier coatings. Annu Rev Mater Res 33:383–417.
2. Xu H, Guo H, Gong S. 2008. Thermal barrier coatings. In: Gao W, Li Z, editors. *Development in High-Temperature Corrosion and Protection of Materials*. Cambridge: Woodhead, pp. 476–489.
3. Goto T, Kimura T. 2003. Development of high-speed CVD process for thermal barrier coating. Bull Ceram Soc Jpn 38:369–373.

4. Kammer PA. 1996. *Flame Spray*. Handbook of Thin Film Technology, London: IOP Pub. A4.1, p. 1.
5. Li TJ, Levi CG, Wadley HNG, Evans AG. 2001. Distributed porosity as a control parameter for oxide thermal barrier made by physical vapor deposition. J Am Ceram Soc 84:2937–2946.
6. Yamaguchi N, Wada K, Kimura K, Matubara H. 2003. J Ceram Soc Jpn 11:883.
7. Blocher JM, Jr. 1966. *Vapor Deposition*. New York: John Wiley & Sons.
8. Goto T. 2005. Chemical processes. In: the Chemical Society of Japan, editor. *Inorganic Compounds, the Fifth Series of Experimental Chemistry*. Tokyo: Maruzen, pp. 4–7.
9. Hirai T, Goto T, Kaji T. 1983. Preparation of silicon carbide by chemical vapor deposition. J Ceram Soc Jpn 91:502–509.
10. Hirai T, Niihara K, Goto T. 1977. Rapid chemical vapour-deposition of Si_3N_4. J Mater Sci 12:631–632.
11. Tu R, Kimura T, Goto T. 2002. Rapid synthesis of yttria-partially-stabilized zirconia films by metal-organic chemical vapor deposition. Mater Trans 43:2354–2356.
12. Whal G, Nemetz W, Giannozzi M, Rushworth S, Baxter D, Archer N, Cernuschi F, Boyle N. 2001. Chemical vapor deposition of TBC: an alternative process for gas turbine components. Trans ASME 123:520–524.
13. Akiyama Y, Sato T, Imaishi N. 1995. Reaction analysis for ZrO_2 and Y_2O_3 thin film growth by low-pressure metalorganic chemical vapor deposition using β-diketonate complexes. J Crys Growth 147:130–146.
14. Bourhila N, Felten F, Senateur JP, Schuster F, Madar R, Abrutis A. 1997. Deposition and characterization of ZrO_2 and yttria-stabilized ZrO_2 films using injection-LPCVD. Proc. 14th Intern. Conf. and EUROCVD-11, Electrochem. Soc. Proc. Vol. 97–25, M. D. Allendorf C. Bernard (Eds.), pp. 417–424.
15. Pulver M, Nemetz W, Wahl G. 2000. CVD of ZrO_2, Al_2O_3 and Y_2O_3 from metalorganic compounds in different reactors. Surf Coat Technol 125:400–406.
16. Bryant WA. 1997. The fundamental of chemical vapour deposition. J Mater Sci 12:1285–1306.
17. Goto T, Kimura T, Tu R. 2004. High-speed deposition of nano-pore dispersed zirconia by CVD and improvement of thermal barrier performance. Power Powder Met 51:821–828.
18. Holman WR, Huegel FJ. 1974. Interrelationships between process parameters, structure, and properties of CVD tungsten and tungsten-rhenium alloys. J Vac Sci Technol 11:701–708.
19. Tu R, Goto T. 2005. Thermal cycle resistance of yttria stabilized zirconia coatings prepared by MO-CVD. Mater Trans 46:1318–1323.
20. Préauchat B, Drawin S. 2001. Properties of PECVD-deposited thermal barrier coatings. Surf Coat Technol 142–144:835–842.
21. Silmy K, Hollander A, Dillmann A, Thomel J. 2005. Micro-jet plasma CVD with HMDSO/O_2. Surf Coat Technol 200:368–371.
22. Bauerle D. 2000. *Laser Processing and Chemistry*. Berlin: Springer, pp. 337–360.
23. Duty C, Jean D, Lackey WJ. 2001. Laser chemical vapour deposition: materials, modelling, and process control. Inter Mater Rev 46:271–287.
24. Bauerle D. 2000. *Laser Processing and Chemistry*. Berlin: Springer, pp. 375–395.
25. Hopfe V, Tehel A, Baier A, Scharsig J. 1992. IR-laser CVD of TiB_2, TiC_x and TiC_xN_y coatings on carbon fibers. Appl Surf Sci 54:78–83.

26. Elders J, wan Voorst JDW. 1993. Laser-induced CVD of titanium diboride and the influence of atomic hydrogen. Appl Surf Sci 69:267–271.

27. Garcia JRV, Goto T. 2003. Thermal barrier coatings produced by chemical vapor deposition. Sci Technol Adv Mater 4:397–402.

28. Miyazaki H, Kimura T, Goto T. 2003. Acceleration of deposition rates in a chemical vapor deposition process by laser irradiation. Jpn J Appl Phys 42:L316–L318.

29. Goto T. 2006. High-speed deposition of oxide films by laser CVD. Ceramic Data Book 2006 34:104–106.

30. Kimura T, Goto T. 2003. Rapid synthesis of yttria-stabilized zirconia films by laser chemical vapor deposition. Mater Trans 44:421–424.

31. Goto T. 2004. High-speed deposition of zirconia films by laser-induced plasma CVD. Solid State Ionics 172:225–229.

32. Goto T, Banal R, Kimura T. 2007. Morphology and preffered orientation of Y_2O_3 film prepared by high-speed laser CVD. Surf Coat Technol 201:5776–5781.

33. Goto T, Kimura T. 2006. High-speed oxide coating by laser chemical vapor deposition and their nano-structure. Thin Solid Films 515:46–52.

14

THE CHANGING PHYSICS IN METAL INTERCONNECT RELIABILITY

Cher Ming Tan and Yuejin Hou

Nanyang Technological University, Singapore

Interconnection is essential to any network. For electrical network, interconnections are needed to connect different electrical functional blocks so that electrical signals can flow in between the blocks for information processing. Interconnections also provide paths for energy supply to the blocks. Therefore, interconnections in electrical networks are crucial and must be good electrical conductors.

When the concept of integration began in the early 1950s for electronics, interconnections remained important to connect all the transistors and passive components in a single integrated chip, and one of the first few failure mechanism that occurred in the first integrated chip was interconnect-related failure (Blech and Sello 1966). As integration technology advances, the interconnect technology also advances with a continuous shrinking in dimension, but its reliability remains a key determining factor to the reliability of an integrated system (Srinvasan et al. 2004).

In this chapter, we will present an overview of the various failure mechanisms of an electrical interconnection and focus on the most important failure mechanisms to showcase the changing underlying physics of the failure mechanisms as we go from micro- to nanoscale. This understanding of the changing physics is essential for the design-for-reliability (DFR) approach in integration system design and development.

Ceramic Integration and Joining Technologies: From Macro to Nanoscale, First Edition.
Edited by Mrityunjay Singh, Tatsuki Ohji, Rajiv Asthana, Sanjay Mathur.

The DFR approach is increasingly important in today's global competitive market (Crowe and Feinberg 2001).

BRIEF OVERVIEW OF INTERCONNECTION FAILURES

For integrated electronic chips, the interconnection is mainly metal, and Al, Cu, and Sn alloy solders are the most commonly used metals for interconnections. The common failure mechanisms for these metals are electromigration (EM), stress-induced voiding (SIV), stress-induced hillock, whisker formation, and corrosion.

Corrosion

There are basically three different types of corrosion that can occur in metals, namely, electrochemical corrosion, galvanic corrosion, and gas corrosion.

Electrochemical corrosion occurs due to the different oxidation potentials of the adjacent metal films and the presence of moisture. This difference in oxidation potential is the result of externally applied bias. If catalysts such as chlorides are present, the corrosion rate will be greatly increased, leading to corrosion through the formation of stable corrosion products, which may not contain chloride itself. An example of a corrosion product is lead carbonate.

Chloride ions play a key role as a catalyst in many corrosion mechanisms occurring in the semiconductor, microelectronics, and electronics industries. Only traces of chlorides are needed to promote and accelerate corrosion reaction.

In the absence of external bias, the dissimilar regions within an interconnect will also possess different oxidation potentials. This dissimilar region could be the different metals or different amounts of impurity in the metal. In this case, the anodic and cathodic processes are localized. The electron current will flow between the anode and cathode through the metal, and in the electrolyte, the movement of cations is toward the cathodic region and the anions toward the anode. As a result, the anode will suffer metal loss (cations).

Even in the complete absence of moisture on a metal surface, corrosion can still occur due to the presence of chemically active gases. An example is the exposure of Ag film to S_8 saturated vapors.

For a detailed discussion of corrosion, readers are referred to Giacomo (1997).

Whisker Formation

Whiskers formed on a metal film surface can cause a short circuit or an increased leakage current. Figure 14.1 shows the whisker formation between the tin-plated connector pins in field application as an example.

Tin whisker is commonly observed in electronic products. It was discovered in 1951 by Compton et al. (1951) but was found to be mitigated with the addition of Pb in the solder in 1954 by Koonce and Arnold (Koonce and Arnold 1954). Figure 14.2 shows an example of tin whisker (C. Handwerker, pers. comm.).

Figure 14.1. Tin-plated connector pins after 10 years (courtesy of NASA's Goddard Space Flight Center through Prof. C. Handwerker, Purdue University) (http://nepp.nasa.gov/whisker/index.html and http://www.klabs.org/richcontent/General_Application_Notes/tin_whiskers_ak.doc).

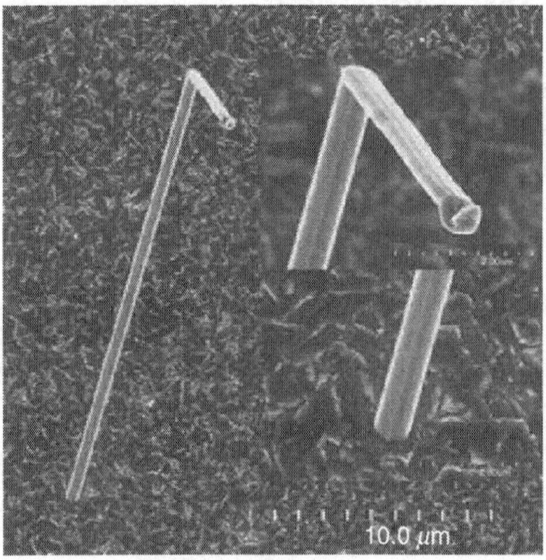

Figure 14.2. An example of kinked filaments of tin whisker (courtesy of Prof C. Handwerker, Purdue University).

However, the shrinkage of the size of solder today renders EM to occur in the solder. This EM will activate the formation of tin whisker even in leaded solder as observed recently by Fan-Yi et al. (2007).

Besides solders, Al interconnects can also form whisker mainly due to the results of EM (Mori et al. 1997). Hinode et al. (1996) found that Al whisker formation is preceded by the following steps:

1. Nonuniform grain growth during annealing (formation of local low-stress positions)
2. Diffusion driven by stress gradients (along grain boundaries and lattices)
3. Dislocation climb from Frank–Read source
4. Dislocation glide to extrude aluminum
5. Step formation on the film surface by breaking through the oxide layer
6. Rearrangement of side surfaces to reduce the surface energy.

After their thorough studies of whiskers on Al, they concluded the following (Hinode et al. 1996):

1. Whiskers nucleated in a relatively narrow temperature range (230–300°C) in the sample films, and this temperature range corresponds to the start of the plastic deformation of the overall film.
2. Whisker formation is caused by thermal stress that results from a thermal expansion mismatch between the film and the substrate.
3. Most whiskers were straight (though sometimes kinked at an angle of 120°), rod-shaped, and had plane side surfaces.
4. The whiskers were mostly single crystals with their bottom part within the film, and the bottom part was a larger grain than the surrounding grains.
5. Whiskers tend to grow 20°–55° from the perpendicular direction. This may be related to the slip plane directions in the highly prefer-oriented (111) films.
6. Growth only occurs at the bottom of a whisker.
7. An oxidizing atmosphere suppresses nucleation of the whiskers.
8. A higher heating rate decreases the whisker number density and length.
9. Aluminum films contained many impurities, such as oxygen and nitrogen, can suppress grain growth, causing localization of stress relief, resulting in whisker formation.

Takatsuija et al. (1999) also found that whisker generation is promoted as the exposure times in the temperature range of 230–300°C becomes longer. Furthermore, most whiskers grow from (110)-oriented and (211)-oriented grains, and that no whiskers grow from (111)-oriented grains.

Hillock Formation

Thermally induced Al hillocks cause manufacturing and reliability problems in integrated devices. These thermal hillocks interfere with the uniformity of subsequent spin-on materials such as photoresist, and can create pinholes in passivation on the top metal line rendering the metal surface susceptible for corrosion or electrical breakdown. Thermal hillocks have both lateral and vertical dimensions on the order of 1 μm (Pico and Bonifield 1991).

Iwamura et al. (1994) found that whiskers are variants of hillocks. The formation of both hillocks and whiskers is observed at the same time during heating, and their sizes increase with increasing temperature. With further heating, the bottom of a whisker changes its form to a hillock.

In a similar way, tin alloy solders will also form a hillock as shown in Figure 14.3. In fact, it is found that interconnect materials can generally form either hillock and/or whisker after prolonged operation as a result of stress relaxation as shown in Table 14.1 (C. Handwerker, pers. comm.). They are highly dependent on the microstructure of the thin film and on the nature of the stress in the film.

SIV

SIV is also called stress migration (SM), and it is a phenomenon of metal voiding in conductor lines that are under tension in the absence of electrical current. The voids can totally sever a conductor line to cause circuit failures.

Stress-induced voids can be classified into two major categories: a wedge-shaped and a slit-shaped void. The wedge-shaped voids form at the edges of conductor lines and tend to peak in lines of 2- to 3-μm width. Slitlike voids are not easy to observe but

Figure 14.3. Examples of solder hillock (courtesy of Prof. C. Handwerker, Purdue University).

TABLE 14.1. Formation of Hillocks/Whiskers by Interconnect Material (C. Handwerker, Pers. Comm.)

Material	Formation of hillock only?	Formation of whisker only?	Formation of both hillock and whiskers?
Pb	Y	N	N
Sn	N	N	Y
Al	N	N	Y
Au	Y	N	N
Cu	Y	N	N
Zn	N	Y	N
Pd	Y	N	N
Pt	Y	N	N
Cd	N	Y	N

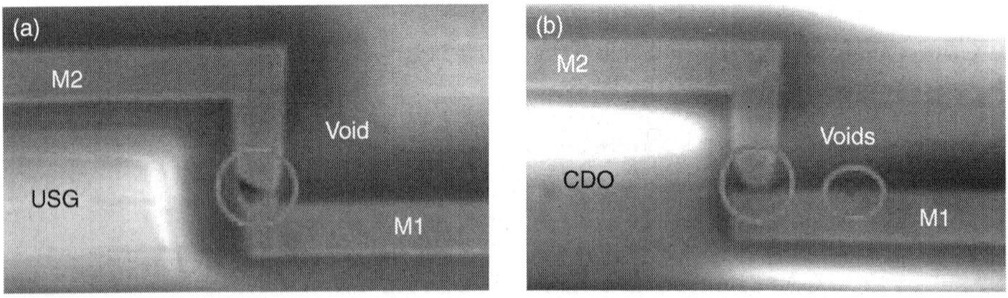

Figure 14.4. Stress-induced voiding in ULSI interconnections with (a) USG and (b) CDO as intermetal dielectrics (courtesy of Prof. Subodh Mhaisalkar, Nanyang Technological University, Singapore).

nearly all the open-conductor failures are caused by them (Chang and Sze 1996). Figure 14.4 shows examples of SIV in ultra-large scale integration (ULSI) interconnection with undoped silicate glass (USG) and carbon-doped oxide (CDO) as intermetal dielectrics, respectively.

In contrast to EM where Black's equation is widely accepted and employed for the closed-form lifetime estimation, closed-form lifetime formulation for SIV is still under debate (McPherson and Dunn 1987; JEDEC 2000; Fischer and Zitzelsberger 2001; Ogawa et al. 2002). McPherson et al. proposed a power-law creep model as the mechanism for stress relaxation where the thermal-induced stress gradient served as the driving force for vacancy movement (McPherson and Dunn 1987). Further improvement was done by Ogawa et al. by introducing the "active diffusion volume" concept, and only those vacancies within the active diffusion volume were able to participate in the SIV process (Ogawa et al. 2002). Fischer et al. also derived a phenomenological equation to describe the SIV lifetime (Fischer and Zitzelsberger 2001). However, lattice relaxation due to strain energy relief near the void region during the SIV process has not been considered in the above-mentioned models.

In fact, the contribution by the strain energy relief is higher than the contribution by the vacancy diffusion in a typical SIV process. Korhonen et al. suggested that the relaxation of stress will be far more through the lattice volume than through the total vacancy volume (Korhonen et al. 1993). Taking aluminum interconnection as an example, a typical reported vacancy concentration in the interconnection is within 0.01–0.1% (Carling et al. 2003). Hence, the void volume fraction due to vacancy clustering will be within 0.01–0.1% by assuming all the vacancies are able to participate in the clustering process. On the other hand, the void volume fraction due to the complete strain energy relaxation is 3 $(\alpha_{metal} - \alpha_{surr})\cdot\Delta T$, where α_{metal} and α_{surr} are the coefficients of thermal expansion (CTEs) for the metallization and the surrounding materials, respectively. ΔT is the difference between the stress-free temperature (SFT) and the test temperature. Using typical values of 24×10^{-6}/K for $(\alpha_{metal} - \alpha_{surr})$ and 200 K for ΔT for the Cu/SiO2 interconnect system (Suo 2003), the void volume fraction is evaluated to be 1.5% for strain energy relief, which is at least 10 times larger than that contributed by the vacancy clustering. In other words, the severity of SIV will be underestimated if the contributions from lattice relaxation are ignored in the void-forming process.

Based on the fact that SIV is one of the means to release the stored strain energy of an interconnection through mass redistribution, Tan et al. (Tan and Hou 2007) proposed a lifetime model from the energy perspective, and besides the thermal stress, they also considered metallization grain structures, interconnect passivation integrity, interlayer dielectric (ILD) confinement effect to interconnection, and interconnect dimensions in the model.

According to the diffusion pathways available for stress relaxation volume growth, they classified the interconnect SIV process into three categories as linear, square, and cubic as shown in Figure 14.5a–c. Figure 14.5a shows the stress relaxation volume in the form of cuboids expanding along the line direction. This represents the case of SIV in very narrow Cu interconnections where stress along the width direction is relaxed in a very short time. Figure 14.5b demonstrates the case when the stress relaxation volume is in the shape of a half cylinder, expanding in both length and width directions. Figure 14.5c depicts the case when the grain boundary diffusion is as important as the interface diffusion.

The time to failure (TTF) of the SIV for the cases of linear, square, and cubic, respectively, is given as follows (Tan and Hou 2007):

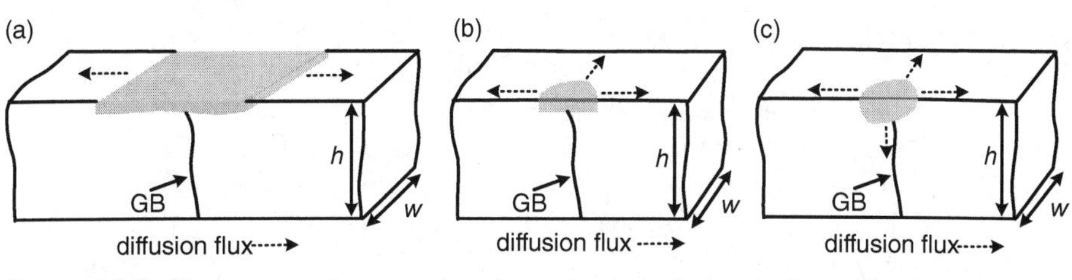

Figure 14.5. Three cases of stress relaxation volume evolution indicated by the gray area. The relationship between the stress relaxation volume and diffusion length is categorized as (a) linear, (b) square, and (c) cubic. GB, grain boundary.

$$t_f = C_1 \frac{\gamma^2 T}{B^3 D_{I0}(T_0 - T)^4} e^{\frac{E_a}{k_B T}} \quad \forall \quad \text{linear}, \tag{14.1}$$

$$t_f = C_2 \frac{\gamma T}{B^2 D_{I0}(T_0 - T)^2} e^{\frac{E_a}{k_B T}} \quad \forall \quad \text{square}, \tag{14.2}$$

and

$$t_f = C_3 \frac{\gamma^{2/3} T}{B^{5/3}\left(3D_{I0}D_{G0}^{1/2} + D_{G0}^{3/2}\right)(T_0 - T)^{4/3}} e^{\frac{E_a}{k_B T}} \quad \forall \quad \text{cubic}, \tag{14.3}$$

where C_1 to C_3 are constants given as

$$C_1 = \frac{4k_B S^2}{81\Omega(\alpha_{Cu} - \alpha_{surr})^2 W^2 d^2}, \tag{14.4}$$

$$C_2 = \frac{2k_B S}{9\pi\Omega(\alpha_{Cu} - \alpha_{surr})^2 d}, \tag{14.5}$$

and

$$C_1 = \frac{(4S)^{2/3} R_B}{(3\pi)^{2/3}\Omega(\alpha_{Cu} - \alpha_{surr})^{4/3}}. \tag{14.6}$$

Here, k_B is the Boltzmann constant, B is the effective bulk modulus, γ is the interface energy at Cu/cap layer per unit area, S is the free surface area of a fatal void for the interconnect, Ω is the atom volume, T_0 is the SFT, E_a is the activation energy of atomic diffusion, d is the interfacial layer thickness, w is the interconnection width, D_{I0} is the prefactor of the atomic diffusivity along the interface layer, and D_{G0} is the prefactor of the atomic diffusivity along the grain boundary.

From Equations 14.4–14.6, it can be seen that C_1, C_2, and C_3 are dependent on the void surface area at failure, the interfacial layer thickness of the cap layer/Cu interconnection, and the interconnection width.

From Equations 14.1–14.3, one can see that the SIV lifetime is strongly dependent on the passivation condition as good passivation leads to higher interface energy γ. This is in agreement with the experimental work by Lloyd et al. (2005). Equations 14.1–14.3 also imply that interconnections with lower effective bulk modulus have larger SIV lifetimes, which agrees with the work presented by Suo (2003). The SIV lifetime is also shown to be inversely proportional to the diffusivities in the dominant diffusion paths from Equations 14.1–14.3. The temperature exponent of the SIV lifetime is consistent with the commonly observed values from one to four by McPherson and Dunn (1987) and Fischer and Zitzelsberger (2001).

By grouping the temperature-independent terms as a constant, C, the generalized expression for the SIV TTF can be expressed as

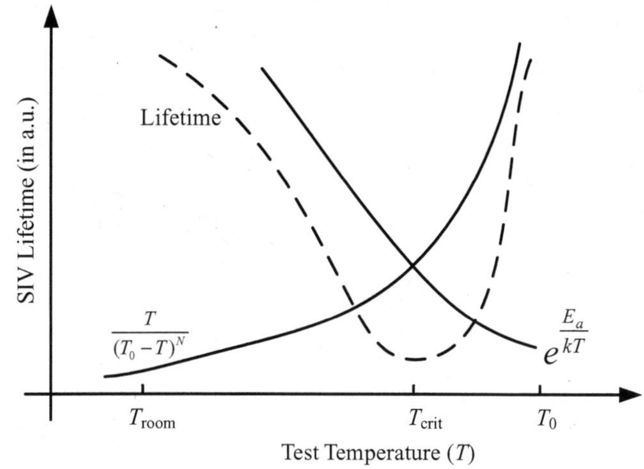

Figure 14.6. SIV lifetime as a function of temperature. As temperature drops from stress-free temperature $T0$, the strain energy increases, but the atomic diffusivity decreases, resulting in an inverted bell shape for the SIV lifetime.

$$t_f = C\frac{T}{(T_0 - T)^N} e^{\frac{E_a}{k_B T}}, \tag{14.7}$$

where N is the temperature exponent, which depends on the particular geometries and microstructures of the interconnections. Equation 14.7 implies that the failure time approaches infinity when the SIV test is carried out at either the SFT (T_0) or the absolute 0 K, and there is a critical temperature, T_{crit}, at which the SIV lifetime is the lowest. This is shown in Figure 14.6. This T_{crit} can be determined by differentiating Equation 14.7 with respect to T, which gives

$$T_{crit} = \frac{2E_a T_0}{\sqrt{(E_a - k_B T_0)^2 + 4E_a T_0 k_B N} + E_a + k_B T_0}. \tag{14.8}$$

Equation 14.8 reveals that T_{crit} is dependent on the activation energy, temperature exponent, as well as SFT. Unlike the SIV lifetime, T_{crit} is independent of the passivation condition γ and confinement effect B. The value of T_{crit} is found to agree well with experimental data from Ogawa et al. (2002), Fischer and Zitzelsberger (2001), and An and Ferreira (2006), as shown in Figure 14.7 (Hou and Tan 2008).

EM

EM is a phenomenon of mass transport in metal film when the film is stressed with high electrical current density (~10^5 A/cm^2). Due to the small cross-sectional area of the interconnections (or thin films) used in microelectronic devices, the current density at normal operating condition is extremely high and, as a consequence, the EM-induced

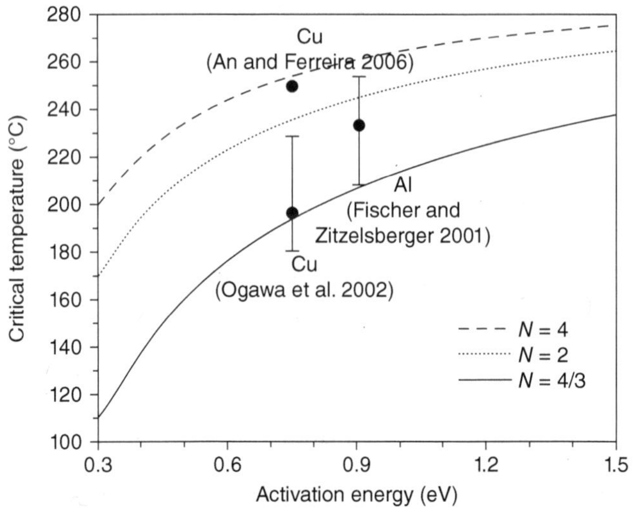

Figure 14.7. Critical temperature profile as a function of the activation energy and temperature exponent. Stress-free temperature is set at 300°C.

mass transport may occur through diffusion. This significant mass transport during EM renders accumulation of vacancies or atoms, creating voids or hillocks in the interconnection. The void formation in the interconnection results in open circuit or increased line resistance, which can cause circuit functional (or parametric) failure. The hillock formation in the interconnection results in a short circuit between adjacent interconnects. The EM-induced mass transport is accelerated when the temperature of an interconnection is increased.

The first reported work on EM was presented by Fiks in 1959 (Fiks 1959). In 1961, Huntington et al. (Huntington and Grone 1961) concluded that thermally activated metal ion becomes essentially free to lattice and is acted upon by two opposing forces in a metal. These forces are called (1) direct force and (2) "electron wind" force. The direct force is the force experienced by the activated positive metal ion in the opposite direction to the electron flow due to the application of electric field. On the other hand, electron wind force (EWF) is the force experienced by metal ion in the direction of electron flow due to the momentum exchange between the moving electrons with the ion.

In the late 1960s, EM threatened the existence of integrated circuits (ICs) (Blech and Sello 1966). Electron moves with very high speed, and higher momentum is transferred to metal ions when the current density is high in the conductor. As a result of higher momentum transfer to ion, EWF dominates and causes an appreciable mass transport in the direction of electron flow. The resultant force of the direct and EWFs was termed as "electron wind force" for simplicity or simply "wind force" in the ballistic model of EM developed by Huntington and Fiks (Fiks 1959; Huntington and Grone 1961; Huntington 1975; Verbruggen et al. 1986; Duryea and Huntington 1988). By the end of the 1960s, EM degradation for many metals, including Au, Cu, Al, and

Ag, and metal alloys were also studied (Fiks 1959; Huntington and Grone 1961; Blech and Sello 1966; Hartman and Blair 1969; Kang et al. 1969; Patil and Huntington 1969; Huntington 1975; Verbruggen et al. 1986; Duryea and Huntington 1988).

In the late 1960s, Black carried out a series of EM experiments on Al-based metallization systematically (Black 1969). He formulated a semiempirical equation for the median TTF of a group of identical conductors that underwent the same EM stressing as follows:

$$t_{50} = Aj^{-2} \exp\left(\frac{E_a}{k_B T}\right), \tag{14.9}$$

where

- t_{50} = median TTF,
- A = a constant, which contains a factor involving the cross-sectional area of the film,
- j = current density in the conductor,
- Ea = activation energy,
- k_B = Boltzman's constant, and
- T = conductor temperature.

This equation is known as the Black equation and is used widely in the EM study of VLSI interconnects.

To be more exact, atomic mass transport cannot induce failure in metallization unless there is a nonzero atomic flux divergence (AFD). Thus, for a given section of interconnect, if the AFD, $\nabla J = 0$, the material entering into and flowing out from the section is exactly the same. On the other hand, when $\nabla J \neq 0$, mass accumulation (i.e., when $\nabla J < 0$) or depletion (i.e., when $\nabla J > 0$) occurs, and this results in EM failure.

Roy et al. (Roy and Tan 2006) derived that the failure time is inversely proportional to the AFD. Hence, it is highly desirable to reduce the AFD in metallization so as to improve the interconnect EM lifetime.

There are many causes by which AFD can occur in metallization. One of the major causes of AFD in polycrystalline interconnection was reported to be the nonuniform grain structure and their orientation in the metallization (Attardo et al. 1971; d'Heurle and Rosenberg 1973; Harrison 1988; Marcoux et al. 1989; Rosenberg and Ohring 1971; Schoen 1980; Nikawa 1981). A well-known structural inhomogeneity was found at the junction of three grains, the so-called triple point (Tan and Arijit 2007).

Another cause of AFD is the local grain size variation (Tan and Arijit 2007). Significant mass accumulation can occur at the boundary from small to large grains, whereas a corresponding mass depletion proceeds at the boundary from large to small grains. Hence, a uniform distribution of grain size is desired to reduce EM-induced damage.

The important microstructure parameters are grain size distribution characteristics (e.g., average grain size and standard deviation in grain size distribution), grain orientation (e.g., [111] and [200]), grain boundary misorientation angle (Tan and Arijit 2007), and grain boundary inclination angle with respect to the electron flow direction. The increase in the average grain size with respect to the line width was shown to improve the EM lifetime. Hence, the metallization microstructure is one of the prime factors in governing EM characteristics, and this will be discussed in a later section of this work.

The presence of minute discontinuity in structure or material or defects in the metallization causes AFD and enhances the EM degradation failure process. The obvious causes of AFD in metallization are at the ends of the conductors where a perfectly blocking boundary (to atomic flux) exists. AFD can also result from nonuniform distribution of current, temperature, or stress fields in the metallization at the EM test condition. The structural changes in the direction of flux such as that in a tapered structure, meander structure, and via structure obviously cause AFD in the metallization during an EM test.

For details of EM physics, readers can refer to Tan and Arijit (2007).

The above-mentioned failure mechanisms in metal interconnects are mainly dependent on their microstructures as well as the driving forces for the failure mechanisms. As the size of interconnects reduces as we move from macro- to nanoscale, their microstructures and some of the driving forces will change. In this chapter, we choose line interconnect EM as a first example to illustrate the changing physics of interconnect reliability, since hillocks and whiskers in the interconnects can also be induced by the EM as mentioned earlier.

THE CHANGING PHYSICS OF EM

As the feature size of Cu interconnects is scaling down into nanometer range, Cu interconnects are more prone to EM and SM failures, not only because they are subjected to more severe use conditions due to increased current density but also because the critical volume for the fatal void is smaller (Michael et al. 2003a). Since the interconnect dimensions are approaching the length of the mean free path (MFP) of electrons in Cu, size effects are also becoming important. This is manifested in the increase of the resistivity of nanointerconnects (Schindler et al. 2005) and the size dependency of their elasticity (Sharma et al. 2004). Also, new failure mechanisms start appearing with the reduced barrier layer thickness and the implementation of low-k interlevel dielectrics (ILD) in Cu nanointerconnects.

The Increase of Metal Resistivity Under Size Effect

Size Effect on Resistivity. Two kinds of size effect can be observed in thin films, namely, the quantum size effect and the classical size effect. The quantum size effect arises when the thickness of the film is comparable to the de Broglie wavelength of electron, which is a few nanometers for Cu and Al. Such a dimension is beyond the current limit of lithographic technologies, and hence it is not relevant in this discussion.

In the classical size effect, increases in the resistivity and changes in the material properties of the film are observed when the thickness of the metallic film is commensurate with the MFP of electrons in the film, which is usually a few tens of nanometers. For nanointerconnects, it is this size effect that is becoming important.

The classical size effect is the result of the increased scattering at the surfaces/interfaces as well as grain boundaries as the interconnect dimensions of thin films are scaled down. Several physical models have been proposed to characterize this size effect quantitatively in order to predict the resistivity of Cu interconnect under deep submicron regime. In the Fuchs and Sondheimer (FS) model (Fuchs 1938; Sondheimer 1952), the increase of the interconnect resistivity due to the surface/interface scattering is described by the specularity coefficient P, which represents a fraction of the incident electrons that scattered specularly at the surface/interface.

Later on, Mayadas and Shatzkes (MS) (MS model) (Mayadas and Shatzkes 1970) observed that electron scattering at grain boundaries also contributed to the increase of the film resistivity in addition to the surface/interface scattering because the typical grain size is smaller than the MFP for the nanointerconnects. In the MS model, an additional parameter R is introduced as an empirical specularity parameter. Both the models have been extensively tested against the experimental data for thin films (Vancea et al. 1987; Vancea 1989), and large variations in the reported P and R values are observed due probably to the inaccuracy in the dimensional measurement of the Cu damascene interconnects (Shimada et al. 2006).

From the previous discussion, it can be easily understood that the increase in the resistivity of an interconnect due to the size effect is strongly determined by the ratio between the MFP (λ_{MPF}) and the line width (W_{line}). At room temperature, the λ_{MPF} of the conduction electrons in Al is 14 nm, which is much smaller than that of Cu (45 nm), making it much less prone to the size effect. Steilesberger et al. showed the simulation results of the size effect for both Al and Cu lines using the combined FS and MS models (Steinlesberger et al. 2004), and they showed that the resistivity for both Cu and Al lines increases sharply as their line widths are scaled down below 100 nm, implying a strong size effect for interconnects below 100 nm. In fact, Engelhardt et al. (2002) demonstrated experimentally that the resistivity obtained for a 50-nm Cu nanointerconnect was almost three times higher than its bulk value.

The increase in the resistivity of the interconnect due to the size effect will increase the Joule heating effect as the current flows through it, thus increasing the line temperature, which will degrade its reliability adversely. This will be elaborated later.

Size Effect on Temperature Coefficient of Resistivity (TCR). The temperature dependence of the resistivity for metal interconnects can be derived as (Steinhogl et al. 2002):

$$\left(\frac{d\rho}{dT}\right)_{\text{nanostructure}} = \rho_0(T)\times TCR_0(T), \tag{14.10}$$

where ρ_0 is the resistivity of the bulk material (for Cu, $\rho_0 = 1.7\,\mu\Omega\cdot$cm at 20°C) and TCR is the corresponding TCR. Schindler et al. (2005) measured the room temperature TCR of Cu interconnects of 0.15 μm thick and with different line widths. They observed that the TCR decreases with line width when the scaling is below 100 nm.

In order to explain their results, we consider the resistivity of Cu interconnects. According to the Matthiessen rule, the electrical resistance of a conductor consists of two contributions as follows:

$$\rho(T) = \rho_{\text{phonon}}(T) + \rho_{\text{defect}}. \tag{14.11}$$

The temperature-dependent part of the resistivity, $\rho_{\text{phonon}}(T)$, is caused by electron scattering by phonons. The temperature-independent contribution ρ_{defect} is due to crystal imperfections, which comprise defects and electron scattering at the internal and external surfaces/interfaces. It is assumed that the size effect contribution to the resistivity does not depend on the temperature since the scattering sources are static defects (Steinhogl et al. 2005), and this is indeed proven experimentally (Steinhogl et al. 2002). Hence, the size effect of nanointerconnects affects only the defect part in Equation 14.11.

From Equations 14.10 and 14.11, we have

$$TCR_0(T) = \frac{1}{\rho_{\text{phonon}}(T) + \rho_{\text{defect}}} \cdot \frac{d\rho_{\text{phonon}}(T)}{dT}. \tag{14.12}$$

At a particular temperature, $d\rho_{\text{phonon}}(T)/dT$ should be independent of line width since it is solely contributed by phonon scattering. For an interconnect with a narrower line width, the ρ_{defect} in the denominator of Equation 14.12 is larger due to the size effect as explained earlier, making the corresponding TCR smaller. Therefore, TCR is decreasing with shrinking line widths as observed by Schindler et al. (2005).

For a given temperature, Equation 14.12 can be modified as

$$\rho_{\text{nano}} = \frac{TCR_{\text{bulk}}}{TCR_{\text{nano}}} \rho_{\text{bulk}}. \tag{14.13}$$

For bulk Cu at 20°C, $\rho_0 = 1.7\,\mu\Omega\cdot\text{cm}$ and TCR = 0.004/K (Tan and Roy 2006). With a TCR for an 80-nm line around 0.0025/K as obtained from Schindler et al., the calculated resistivity using Equation 14.13 is approximately $2.72\,\mu\Omega\cdot\text{cm}$, which is close to the typical computed value of $2.7\,\mu\Omega\cdot\text{cm}$ (Shimada et al. 2006) and the experimentally measured value of $2.8\,\mu\Omega\cdot\text{cm}$ (Steinhogl et al. 2005).

Implications of the Size Effect on EM Reliability. According to the International Technology Roadmap for Semiconductor Report, 2006 update (ITRS, 2006), the maximum current density in the interconnects will reach a level of 5.15 MA/cm^2 by 2010. The maximum current density allowed in interconnects is limited by two factors: Joule heating and EWF. Both are strongly dependent on the size effect as illustrated next.

The increased resistivity results in more severe Joule heating. With the heat dissipation included in the interconnect system, we perform finite element analysis for the Cu single line structure shown in Figure 14.8. The line width is scaled from 1 μm down to 0.03 μm while keeping a fixed aspect ratio of 2. The input current density is

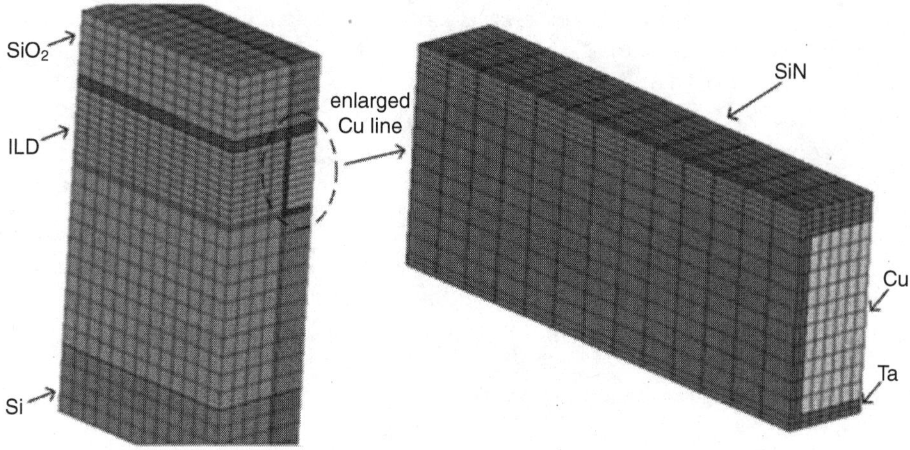

Figure 14.8. Finite element model of the structure for Joule heating effect calculation.

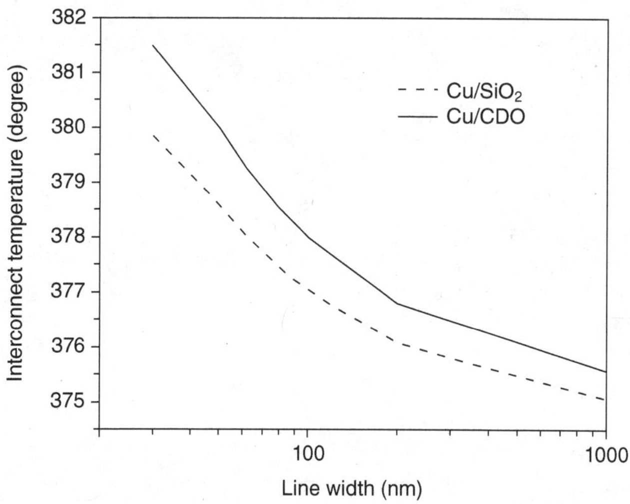

Figure 14.9. Simulation results for the Joule heating effect. The substrate is fixed at 573 K.

1 MA/cm^2. To include the size effect in the analysis, Cu resistivity variation according to the resistivity model shown in Steinlesberger et al. (2004) is considered. The test temperature at the substrate bottom is fixed at 300°C.

Coupled field electrical-thermal analysis is employed for the analysis. It can be seen that the interconnect temperature rises sharply as the Cu interconnect width shrinks below 100 nm due to the size effect as shown in Figure 14.9. It is noteworthy that Figure 14.9 is a conservative estimation since the current density applied in the simulation is the same across all line widths. In fact, the current density increases dramatically with line width scaling, further deteriorating the Joule heating effect. Besides, the temperature increment by Joule heating is larger for Cu/CDO interconnects as compared with its oxide counterpart, and this is attributed to the lower thermal conductivity of low-k material (CDO).

The EWF per atom, f_e, is proportional to the electric current density j, given as (Hu et al. 1999)

$$f = Z^*_{epj}, \tag{14.14}$$

where e is the fundamental electron charge and Z^* is the effective charge number. The Nernst–Einstein equation relates the EWF to atomic flux J, which is the number of atoms crossing a unit area per unit time as follows (Korhonen et al. 1993):

$$J = \frac{Z^* De\rho j}{\Omega k_B T} = \frac{Z^* DeE}{\Omega k_B T}, \tag{14.15}$$

where $D = D_0 \exp(-E_a/k_B T)$ is the self-diffusivity of interconnect atoms, E_a is the activation energy, Ω is the atomic volume of the metal, T is the local temperature of the interconnect, k_B is the Boltzmann constant, and E is the applied electrical field. It is noted that D in Equation 14.15 represents an effective diffusivity of metal atoms along interconnects because the EM mass transport can occur through different diffusion paths.

As indicated by Equation 14.15, the atomic flux is proportional to the product of EWF (f_e) and the diffusivity (D). With the continuous scaling of the Cu interconnect, the dominant diffusion path changes from grain boundary to Cu/cap interface (Shao et al. 2007), and a much lower value of Z^* is reported for surface diffusion as compared with the grain boundary diffusion (Sorbello 1996). Rosenberg et al. (2000) reported the values of Z^* of −14 and −0.1 for grain boundary and surface diffusion respectively. As a result, even though the Cu resistivity increases under the size effect, the decrease of Z^* for nanointerconnect is much larger than the increase in resistivity, rendering a reduction in the EWF for nanointerconnects at a given current density (Rosenberg et al. 2000).

However, the reduction in the EWF does not imply a better EM reliability for nanointerconnect. This is due to the following reasons:

1. As the dominant diffusion path changes to the Cu/cap layer (Shao et al. 2007) for Cu interconnects, and the surface-to-volume ratio increases with line downscaling, the fractions of atoms diffusing along the interface are higher for nanointerconnects. For Cu interconnects, surface diffusivity can be a few orders higher than that of grain boundary and lattice due to its low activation energy (Tu 2003). As a result, a higher effective diffusivity for Cu atoms is observed and higher atomic flux is found in nanointerconnects according to Equation 14.15.
2. The larger temperature increment by Joule heating, as shown in Figure 14.9, and with an activation energy of 1 eV for Cu/cap layer interface diffusion, the surface diffusivity for an 80-nm line is 27% higher than that of the 1000-nm

line for a Cu/CDO interconnect under the same current density. Thus, the mass transport along the Cu/cap interface is expected to be more severe as the line width shrink. The 27% higher value of surface diffusivity is computed using $D = D_0 \exp(-E_a/k_BT)$ with $E_a = 1\,\text{eV}$, $T = 376$ and 379°C for the 1-μm and 80-nm lines, respectively, as obtained from Figure 14.9.

With the above-mentioned reasons, the EM reliability degradation will be more severe for nanointerconnection even if the current density remains the same.

From the previous discussions, it is therefore possible to enhance the EM lifetime of nanointerconnect by reducing the atomic flux. By adopting the various means of suppressing the diffusivity in the dominant diffusion path in interconnects, current density limitations for future technology nodes may then be only limited by the Joule heating alone (Lloyd et al. 2004). And to reduce the Joule heating and EWF, a better thermal management and/or better interconnect design is necessary in order to maintain the same EM reliability for nanointerconnects. Large-grained Cu interconnects and ultrathin barrier layers are essential in reducing the resistivity of nanointerconnects in future ULSIs. In fact, Koike and Wada (2005) and Usui et al. (2005) have demonstrated an improved EM performance for Cu interconnect by using the process of self-forming diffusion layer they have developed. The self-forming Mn oxide barrier layer is only about 2 nm thick, and the resistivity of the Cu line decreases significantly.

Barrier Layer Scaling

According to ITRS 2006 (ITRS, 2006), barrier layer thickness has to be scaled down to 3.3 nm by 2010 to reduce interconnect resistivity. The barrier layer is an integrated part of the Cu damascene interconnects, providing the mechanical confinement of the interconnect to sustain EM mass transport. It also serves as a base layer for Cu seed deposition and electroplating, where processing defects in the barrier can induce early failure, which degrades EM reliability (Michael et al. 2003b).

In the recent work by Pyun et al. (2005), it was found that with decreasing Ta barrier thickness, Blech product was reduced due to weaker structural confinement. The experiment was carried out at three different barrier layer thicknesses of 7.5, 10.0, and 17.5 nm for Cu/methylsilsesquioxane (MSQ) interconnects. They showed that the Blech product is reduced more than expected when the barrier thickness is reduced to 7.5 nm due probably to the presence of the defects in the barrier layer. Bimodal failure distribution was also found in interconnects with barrier thicknesses of 7.5 and 10 nm due to Cu out diffusion through the thin barrier layer under EM-induced stress.

Schindler et al. (2005) also correlated the TTF with the barrier layer thickness in their EM experiments. It was found that a thinner barrier layer thickness led to shorter TTF. One possible mechanism of this dependence of TTF on barrier layer thickness is that a thinner barrier layer yields a weaker adhesion between the dielectrics and Cu, resulting in an easier diffusion of Cu atoms along the Cu/barrier interface, and thus worsens its EM performance (Schindler et al. 2005). Another possible reason is that a thinner barrier layer leads to lower effective bulk modulus as shown in the work by

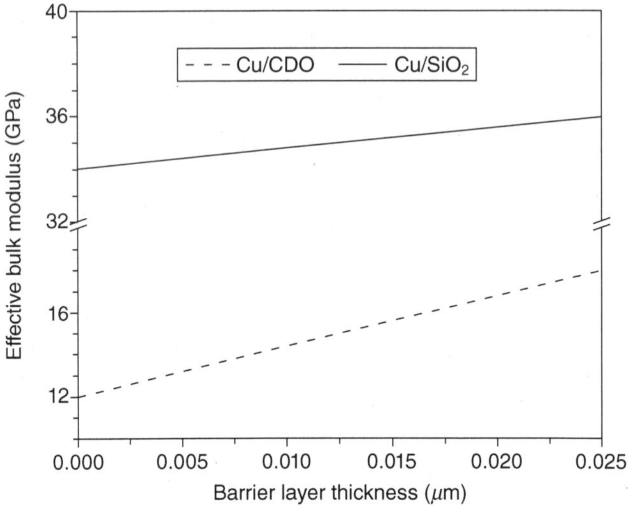

Figure 14.10. Simulation results bulk of effective modulus for both Cu/SiO_2 and Cu/low-k interconnects.

Pyun et al. (2005), and this weaker confinement decreases the backflow stress, resulting in an increase in the net drift velocity of atoms and a reduction of the EM lifetime.

Finite element analysis is employed to investigate the effective bulk modulus scaling with the barrier layer thickness for both Cu/SiO_2 and Cu/CDO interconnects. The line width and thickness for the finite element model are 0.1 and 0.2 μm, respectively. The barrier layer thicknesses are from 0.025 to 0.001 μm. The detailed calculation of the effective bulk modulus can be found in the work by Hou and Tan (2007).

The calculation result is shown in Figure 14.10. The effective bulk modulus for Cu/low-k is much lower than that of Cu/SiO_2 due to the lower mechanical strength of the low-k materials. For barrier layer thickness from 0.025 down to 0.001 μm, the effective bulk modulus only drops by 5% for Cu/SiO_2 interconnects as compared with a drop of 33% for Cu/low-k interconnects. This indicates that the confinement effect for Cu/SiO_2 interconnects is less prone to the barrier layer scaling than its low-k counterpart, which can be explained by a much higher mechanical strength of SiO_2 (~71 GPa) as compared with that of CDO (~16 GPa).

Based on the previous analysis, the scaling of the barrier layer is more critical for low-k interconnects. The implication of the thinner barrier layer thickness is a lower Blech product and a reduced EM lifetime (Pyun et al. 2005).

Also, with the thinner barrier layer, the chance for Cu atoms to outdiffuse through the barrier layer increases due to the degraded integrity of the barrier layer, resulting in a reduction in the time-dependent dielectric breakdown (TDDB) lifetime of Cu interconnects as demonstrated by Tokei et al. (2004). Lloyd et al. (2006) also found that the TDDB lifetime is indeed dominated by the diffusion of Cu at low test voltages.

With the implementation of low-k interlayer dielectrics, the lower breakdown strength of the dielectrics may further shorten the TDDB lifetime in Cu nanointerconnects.

The Impact of Scaling on Diffusion Paths

In metal interconnects, metal atoms can diffuse along several diffusion paths. The effective Z^*D from Equation 14.15 resulting from all the diffusion paths for a polycrystalline line with width w and thickness h can be written as

$$Z^*D = Z_B^*D_B + Z_G^*D_G\delta_G \frac{1}{d}(1-d/w) + Z_I^*D_I\delta_I\left(\frac{2}{w}+\frac{1}{h}\right) + Z_S^*D_S\delta_S\frac{1}{h}, \tag{14.16}$$

where the subscripts B, G, I, and S signify the diffusion path along the bulk, the grain boundary, the Cu/barrier interface, and the Cu/cap interface, respectively; δ_G, δ_I, and δ_S are the width of the respective individual diffusion paths; and d is the average grain size. In the following,

$$\delta_G \frac{1}{d}(1-d/w), \quad \delta_I\left(\frac{2}{w}+\frac{1}{h}\right), \quad \text{and} \quad \delta_S\frac{1}{h}$$

are the fractions of atoms diffusing along the grain boundary, the Cu/barrier interface, and the Cu/cap interface, respectively.

As the line width of Cu interconnects is scaling down, the microstructures migrate from the polycrystalline structure to the bamboo structure. Hu et al. (1999) observed bamboo, near-bamboo, and polycrystalline structures in interconnects with line widths between 0.15–0.5 μm, 0.75–1 μm, and 2–10 μm, respectively. It is widely accepted that the Cu/cap interface layer is the most significant diffusion path for bamboo structures (Hu et al. 1999; Shao et al. 2007). As interconnects continue to scale down, Wang et al. (2004) showed that current crowding in future narrow interconnects cannot be neglected as it can enhance EM-induced mass transport along otherwise less critical diffusion paths such as the Cu/barrier interface. This is indeed observed by Roy et al. (2006).

For bamboo Cu interconnects where the dominant diffusion path is along the Cu/cap interface, effective Z^*D can be rewritten as

$$Z^*D = Z_S^*D_S\delta_S\frac{1}{h}. \tag{14.17}$$

Similarly, when the dominant diffusion path is along the Cu/liner interface, effective Z^*D can be rewritten as

$$Z^*D = Z_I^*D_I\delta_I\left(\frac{2}{w}+\frac{1}{h}\right). \tag{14.18}$$

From Equation 14.17, one can see that when the Cu/cap interface is the dominant diffusion path, scaling the line width alone does not result in a decreased EM lifetime. However, Chang et al. (2007) proved that the above statement is true only when the EM failure is nucleation limited. In the case where the dominant diffusion path is along the Cu/liner interface, scaling both the line width and thickness will result in an increased $Z^{*}D$ as shown in Equation 14.18, hence a decreased EM lifetime.

In reality, the scaling is done through both the line width and thickness, and hence the ratio of the diffusion volume to the total line volume increases due to the increase in the surface-to-volume ratio. This will result in a decreased EM lifetime (Hu et al. 2002; Bruynseraede et al. 2005). Furthermore, the reduction in the critical void volume for a narrower interconnect can also lead to an even shorter EM lifetime.

The ratio of the total diffusion volume to the total line volume is estimated for different line widths with a constant aspect ratio of two. For simplicity, the interface thickness is taken as 0.5 nm (Hu et al. 1999), and the line is assumed to be infinitely long. The calculation results are shown in Figure 14.11. It can be seen that the ratio of the total diffusion volume over the total line volume increases dramatically for line width below 100 nm, and therefore the interface diffusion will be more prominent.

Experimentally, Michael et al. (2003a) reported that for narrow interconnects, the EM lifetime decreased with increasing surface-to-volume ratio, and the EM lifetime did not decrease with line width linearly. Instead, the EM lifetime decreased more rapidly when the line width was less than 0.2 μm. This is probably due to a process-induced defect at the interfaces and/or the extremely small critical void volume. The same finding was also reported by Chang et al. (2007).

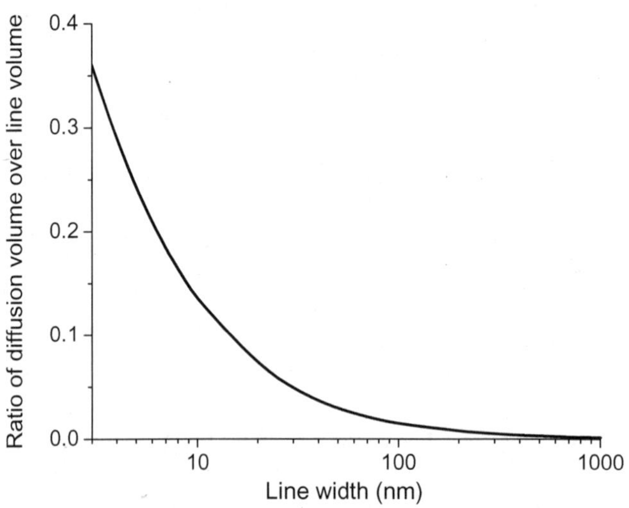

Figure 14.11. The ratio of total diffusion volume to line volume. The line is assumed to be infinitely long with an aspect ratio of two.

Evolvement of Various Driving Forces

From the work reported by Tan et al. (2005), temperature gradient and stress gradient contribute to the EM failure, and their effects on EM are more than just mere modifications of the flux divergence as is done in the conventional diffusion path approach. With the continuous scaling of interconnects, the dominant diffusion path changes from grain boundary to Cu/cap interface, and therefore EWF also reduces as mentioned earlier. On the other hand, the thermomechanical stress computed is found to be larger for nanointerconnects (Tan et al. 2007a).

Hence, it is necessary to reevaluate the strength of the various driving forces present during EM using finite element analysis.

The atomic flux due to the EWF can be expressed as (Black 1974)

$$\vec{J}_A = \frac{N}{k_B T} Z^* e\rho D_0 \exp\left(-\frac{E_a}{k_B T}\right) \cdot \vec{j}. \tag{14.19}$$

Similarly, the atomic flux due to temperature gradient-induced driving force (TGIDF) and thermomechanical stress gradient-induced driving force (SGIDF) are shown in Equations 14.20 and 14.21, respectively (Shewmon 1963; Ross 1991):

$$\vec{J}_{th} = -\frac{NQ^* D_0}{k_B T^2} \exp\left(-\frac{E_a}{k_B T}\right) \cdot \nabla T \tag{14.20}$$

$$\vec{J}_S = \frac{N\Omega D_0}{k_B T} \exp\left(-\frac{E_a}{k_B T}\right) \cdot \nabla \sigma_H. \tag{14.21}$$

Therefore, the ratio of atomic fluxes between TGIDF and EWF can be derived as

$$r_{21} = \frac{J_{th}}{J_A} = \frac{Q \cdot \nabla T}{Z^* e\rho jT}, \tag{14.22}$$

and the ratio of atomic fluxes between SM and EWF can be expressed as

$$r_{31} = \frac{J_S}{J_A} = \frac{\nabla \sigma_H}{NZ^* e\rho j}. \tag{14.23}$$

In this work, we employ finite element analysis to numerically calculate the ratios as expressed in Equations 14.22–14.23 for the Cu line–via structure as shown in Figure 14.12. By virtue of the symmetry of the structure, only half of the model is simulated. The Ta diffusion barrier layer and SiN cap layer are also included as shown in Figure 14.12. The line thickness is fixed at $0.35\,\mu$m, while the line width is scaled from 1 to $0.05\,\mu$m. Via diameter is scaled together with the line width, while the via height is fixed at $0.8\,\mu$m. Different ILDs, namely, the SiO_2 and CDO, are also used for comparison. The test structure is subjected to a typical current density of $1\,MA/cm^2$ and the SFT is set at 400°C. The simulation results for r_{21} and r_{31} are plotted in Figures 14.13 and 14.14, respectively.

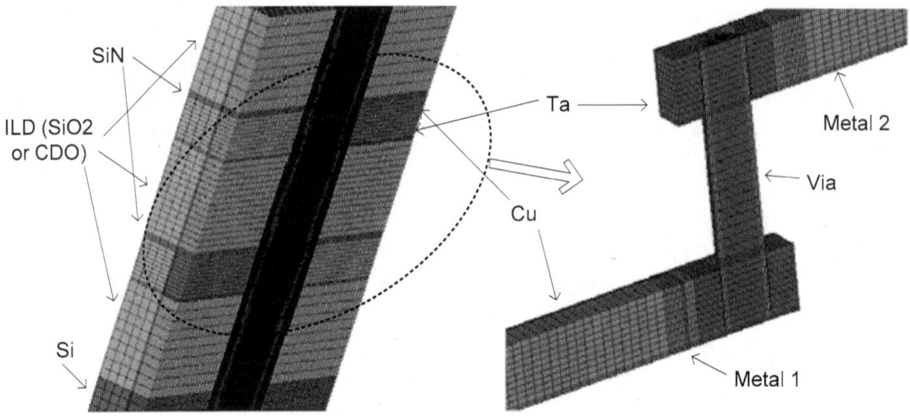

Figure 14.12. Finite element model of line–via structure for driving force evaluations.

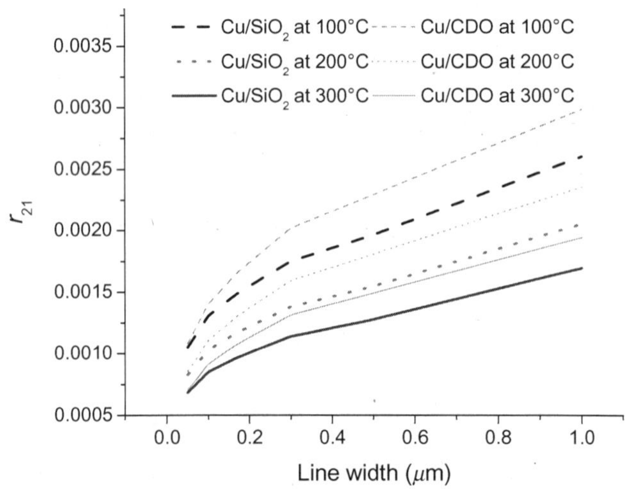

Figure 14.13. The ratio of temperature gradient-induced driving force to electron wind force.

The simulated temperature gradient is found to be between 0.002 and 0.01 K/μm in the line region for line width between 0.05 and 1 μm. However, at the inner corner of the via bottom, the temperature gradient can reach up to a level of 0.1 K/μm due to the additional structural induced temperature gradient. Figure 14.13 reveals that r_{21} is larger for the Cu/low-k interconnects as compared with the Cu/SiO_2-based interconnects for a given line width. This can be explained by the poor thermal conductivity of the low-k materials, and hence a larger temperature gradient in the interconnect. Besides, we also found that the temperature gradient decreases with decreasing line width, indicating a smaller contribution from TGIDF for narrower interconnects.

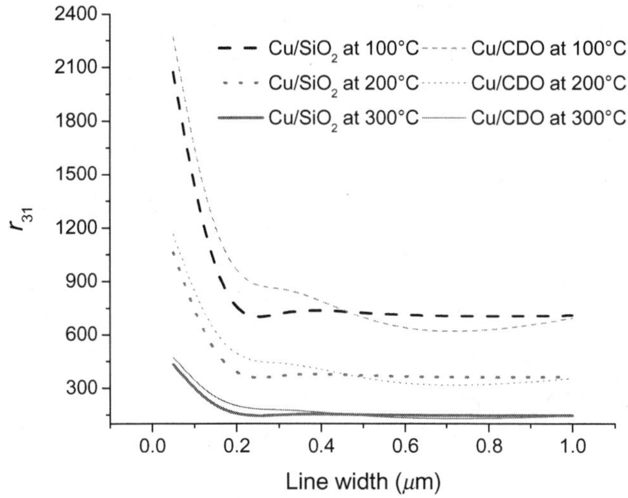

Figure 14.14. The ratio of thermomechanical stress gradient-induced driving force over electron wind force.

Therefore, for interconnects with narrower line widths, the TGIDF is smaller. EWF is two to three orders higher than TGIDF as shown in Figure 14.13, implying TGIDF is unlikely to be a dominant driving force for EM failure in nanointerconnects.

Figure 14.14 shows the computed r_{31} and reveals that the stress gradient increases with decreasing line width, indicating a higher thermomechanical stress gradient for nanointerconnects. The stress gradient is also higher for low-k interconnects, which is consistent with the work reported by Gan et al. (2006). The SGIDF is two to three orders higher than EWF as shown in Figure 14.14.

The importance of high thermomechanical stress on EM reliability was reported by Tan et al. (Tan and Roy 2006). Later on, Li et al. (2007) showed that vacancies in interconnect tend to cluster at the intersections of the grain boundary and the passivation layer due to the thermomechanical stress in Al interconnects. Recently, preexisting microvoids were found in both wide and narrow Cu interconnects under the thermomechanical stress (Chang et al. 2007). These microvoids can be swept by the EWF and collapse into a fatal void at the cathode end. In other words, SGIDF is likely to be of increasing concern in Cu nanointerconnects.

EM Failure Statistics

Traditionally, lognormal distribution is used to analyze EM failure data for both Al- and Cu-based interconnects (Pierce and Brusius 1997; Li et al. 2006). The proof of the appropriateness of the lognormal distribution is as follows (Tan et al. 2007b).

Denoting the interconnect resistance by R, we can model the gradual increase in resistance as

$$R_i = (1+\delta_i)\cdot R_{i-1}, \tag{14.24}$$

where R_i and R_{i-1} represent consecutive discrete resistance values and δ_i is a small quantity indicating the gradual infinitesimal increase in the resistance value. As time progresses from $t = t_0$ to t_N, the resistance value increases from its initial value R_0 to R_N.

Based on Equation 14.24 and using a recursive approach, the resistance value at failure, denoted by R_N, can be expressed as

$$R_N = \prod_{i=1}^{N} (1+\delta_i) \cdot R_0. \tag{14.25}$$

Taking the natural logarithm of Equation 14.25, the expression for $\ln(R_N)$ is expressed by

$$\ln(R_N) = \sum_{i=1}^{N} \ln(1+\delta_i) + \ln(R_0). \tag{14.26}$$

For gradual degradation in resistance, $\delta_i << 1$. Using the Taylor series expansion, $\ln(1 + \delta_i) \sim \delta_i$. The terms of second and higher order in the Taylor series are ignored as δ_i is very small. Hence, Equation 14.26 may be rewritten as

$$\ln(R_N) \approx \sum_{i=1}^{N} \delta_i + \ln(R_0). \tag{14.27}$$

Since δ_i is a random variable, the resistance value R_N is also a random variable. As the degradation in resistance is gradual, the number of terms in the δ summation is large. In such a case, the statistical distribution of the summation term can be approximated by a normal distribution by means of the "central limit theorem" (Tobias and Trindade 1995). Thus, $\ln(R_N)$ follows a normal distribution, and R_N follows the lognormal distribution. In other words, the EM degradation process is lognormal in nature.

However, for deep submicron interconnects, the resistance degradation becomes abrupt. One example of the resistance change profile in nanointerconnects for a 100-nm interconnect is shown in Figure 14.15 (Tan et al. 2007b). In this case, the change in resistance for a narrow interconnect is a combination of gradual and catastrophic degradation, rendering the use of the lognormal distribution model as inappropriate, as shown next.

The degradation trends observed in Figure 14.15, which consist of "gradual" and "catastrophic" regions of the resistance degradation, can be described by Equations 14.28 and 14.29, respectively, as follows:

$$R_i = (1+\delta_i) \cdot R_{i-1}; 1 \le i \le M \tag{14.28}$$

$$R_i = (1+\Delta_i) \cdot R_{i-1}; (M+1) < i \le N, \tag{14.29}$$

where δ and Δ represent the corresponding increment in resistance values in the gradual and catastrophic regions, respectively, as time progresses, and they are both random variables with $\delta << 1$ and $\Delta >> 1$.

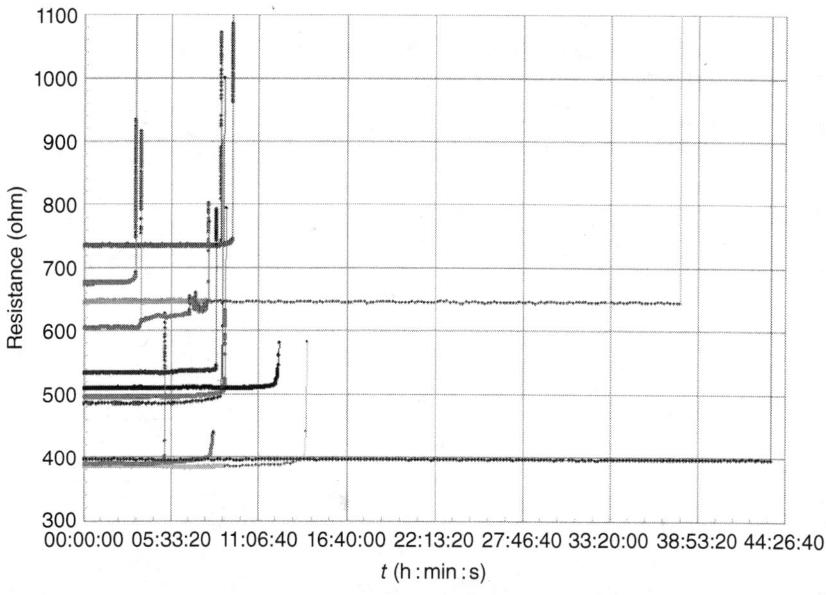

Figure 14.15. Resistance degradation plot for the EM failure process in deep submicron copper line interconnects. The line width of this structure is 100 nm, and the test conditions include a current density, $j = 3.5\,\text{MA/cm}^2$, and temperature, $T = 325°\text{C}$ (Tan et al. 2007b).

Similar to Equation 14.25, with Equations 14.28 and 14.29, the resistance value at failure can be expressed as

$$R_N = \prod_{i=1}^{M}(1+\delta_i)\cdot\prod_{i=M+1}^{N}(1+\Delta_i)\cdot R_0. \tag{14.30}$$

Taking the natural logarithm of Equation 14.30, the expression for $\ln(R_N)$ is

$$\begin{aligned}\ln(R_N) &= \sum_{i=1}^{M}\ln(1+\delta_i)+\sum_{i=M+1}^{N}\ln(1+\Delta_i)+\ln(R_0)\\ &\approx \sum_{i=1}^{M}\delta_i+\sum_{i=M+1}^{N}\ln(1+\Delta_i)+\ln(R_0)\end{aligned} \tag{14.31}$$

While Taylor series approximation can be used for the term $\ln(1+\delta_i)$ as $\delta_i \ll 1$, this approximation is inapplicable for the second logarithmic summation term in Equation 14.31 because $\Delta_i \gg 1$. Therefore, the resulting statistic for $\ln(R_N)$ is now the sum of a normal distribution and an additional unknown statistical terms represented by the second summation term. As a result, the overall statistical nature of $\ln(R_N)$ does not follow a normal distribution, and the lognormal distribution cannot be applied to analyze such EM failure data.

TABLE 14.2. Comparing the Data Fitting Accuracy of the Gamma Model versus the Lognormal Model for M1 and M2 Test Structures with Different Via Configurations

Data set #	Test type	Line width (μm)	Gamma LSR	Lognormal LSR
1	M1	0.28	0.0057	0.0065
2	M2	0.28	0.0041	0.0063
3	M1	1.0	0.0041	0.0039
4	M2	1.0	0.0048	0.0033

LSR, least square residue.

TABLE 14.3. Comparison of the B1 Life, B10 Life, and t_{50} for Narrow and Wide Cu EM Lines with Respect to the Gamma and Lognormal Distributions

Statistical distribution	Test structure	B1 life (h) t_{50} (h)	B10 life (h)	t_{50} (h)
Gamma	Narrow	6.80×10^6 3.58×10^7	1.58×10^7	3.58×10^7
Lognormal	Narrow	8.90×10^6 3.40×10^7	1.63×10^7	3.40×10^7
Gamma	Wide	7.50×10^5 1.04×10^7	3.05×10^6	1.04×10^7
Lognormal	Wide	1.45×10^6 9.38×10^6	3.35×10^6	9.38×10^6

To derive a suitable distribution and to model the resistance degradation pattern as shown by Figure 14.15, Tan et al. (2007b) discretized the range of resistance values into a finite number of distinct states and a multistate Markov model is subsequently developed. They concluded that the statistical nature of the overall EM degradation process closely resembles the gamma multimodal statistics.

Comparison of the lognormal and gamma distribution on EM test data revealed that a better fit is indeed observed for the narrow line as shown in Table 14.2. The implication of the use of different statistical distributions for extrapolating to normal operating conditions is shown in Table 14.3, and one can see the significant difference in the extrapolated results for the narrow line on the B1 life, that is, the time for 1% early failure.

THE CHANGING PHYSICS OF SOLDER JOINT FAILURE

Our next example for the change in failure physics in metal interconnect is on the tin alloy solder, an essential interconnect between chip and package. The volume of solder reduces as we go from macro- to microscale in order to meet the severe requirements of future electronic packaging. This reduction in the solder volume renders a change in the failure physics as follows.

Due to the continuous downscaling of the size of the solder joint to accommodate higher input–output counts, the solder bumps must carry a much higher current density. Therefore, EM phenomena become a serious threat to the reliability of solder bumps (Liu et al. 2006). It is found that EM degradation of the larger volume joints is different from that of an ultrasmall flip chip. Instead of severe Kirkendall voiding in the larger volume joint, there are changes in the microstructure of the solder during the EM of small joints (Tu et al. 2003), and its most important degradation mechanism is the formation of intermetallic phases in the solder and phase separation (Meier et al. 2007).

When tin-containing solder comes into contact with the copper pad surface, Cu–Sn intermetallics are formed (Tu et al. 2003). Although the Cu consumption was much higher for the solder with higher solder volume, the mean thickness of the intermetallics is smaller (Sharif et al. 2004). On the other hand, as the solder volume decreases, the larger thickness of the intermetallic can turn the majority part of the solder into intermetallic. Since the intermetallics are brittle in nature and the thermal expansion is very different from that of solder (Tu et al. 1997), excessive formation of intermetallics is expected to impair the reliability of the joint.

As the size of the solder joint is generally larger than the chip interconnect, there exists a transition region near the intersection of the chip interconnect and the solder joint where the current density changes abruptly in the solder joint. Simulation of current distribution in the solder bump of the flip chip structure showed that the current density in the current crowd region is much higher than the averaged current density inside the solder bump (Yi-Shao and Chin-Li 2006).

A finite element model for solder bump and its current density distribution are shown in Figure 14.16a, b, respectively. One can see that the solder bump temperature can be very much larger than the ambient oven temperature under the Joule heating effect, leading to the thermomigration (TM) failure in the solder bump (Ye et al. 2003) or the local melt of the solder bump (Tsai et al. 2006). The Joule heating problem is

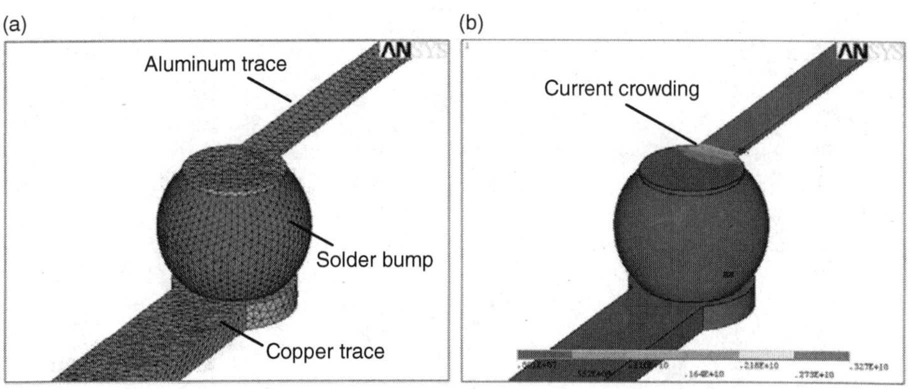

Figure 14.16. (a) Finite element model for solder bump. (b) Current density distribution.

more severe with a smaller dimension of the solder joint because the conductive heat transfer through the solder joint is lesser owing to the small surface area, making the heat dissipation less efficient.

Besides EM and TM, mechanical failures such as fatigue are also critical for solder joint reliability. Chen et al. (2006) studied the effects of solder joint structure and shape on the solder fatigue lifetime through finite element simulations. They found that the convex-shaped solder with larger height and smaller pitch has a smaller maximum equivalent plastic strain range, which leads to longer fatigue life. They also reported that a larger solder joint was associated with longer fatigue lifetime, which can be explained by a longer crack propagation time for a larger solder ball (Chaparala et al. 2005).

CONCLUSIONS

In this chapter, we have looked at the different failure mechanisms associated with interconnections in integrated chip and electronic packaging/systems. For the integrated chips, the present main interconnection is either Al or Cu, and their failure mechanism is EM. As we proceed from macro- to nanoscale, the underlying physics of EM is changing as described in this chapter. Such a change requires a different approach to the improvement of chip interconnect reliability.

For chip assembly, solder is the main interconnection, and again we show the changing physics of solder failures as we go from macro- to microscale in the solder volume.

It is worth noting that the changes in the failure physics described in this chapter assume no change in the materials for interconnects. As we know, interconnect materials will be changing as technology advances, and this change in material will bring additional changes in failure physics in addition to that mentioned in this chapter, which is beyond the scope of this work.

REFERENCES

An JH, Ferreira PJ. 2006. In situ transmission electron microscopy observations of 1.8 μm and 180 nm Cu interconnects under thermal stresses. Appl Phys Lett 89:151919.

Attardo MJ, Rutledge R, Jack RC. 1971. Statistical metallurgical model for electromigration failure in aluminum thin flim conductors. J Appl Phys 42:4343–4349.

Black JR. 1969. Electromigration—a brief survey and some recent results. IEEE Trans Electron Devices 16:338–347.

Black JR. 1974. *Proc. Conf. 12th IEEE/IRPS*. Las Vegas, NV, pp. 142–149.

Blech IA, Sello H. 1966. A study of failure mechanisms in silicon planar epitaxial transistors. Phys Fail Electron 5:496–505.

Bruynseraede C, Tokei Z, Iacopi F, Beyer GP, Michelon J, Maex K. 2005. *Proc. Conf. 43rd IEEE/IRPS*, pp. 7–17.

Carling KM, Wahnstrom G, Mattsson TR, Sandberg N, Grimvall G. 2003. Vacancy concentration in Al from combined first-principles and model potential calculations. Phys Rev B 67: 054101.

Chang CY, Sze SM, editors. 1996. *ULSI Technology*. Singapore: McGraw-Hill.

Chang CW, Thompson CV, Gan CL, Pey KL, Choi WK, Lim YK. 2007. Effects of microvoids on the linewidth dependence of electromigration failure of dual-damascene copper interconnects. Appl Phys Lett 90:193505.

Chaparala SC, Roggeman BD, Pitarresi JM, Sammakia BG, Jackson J, Griffin G, McHugh T. 2005. Effect of geometry and temperature cycle on the reliability of WLCSP solder joints. IEEE Trans Compon Pack Technol 28:441–448.

Chen RS, Tseng SC, Wan CS. 2006. Effects of solder joint structure and shape on thermal reliability of plastic ball grid array package. Int J Advd Manuf Technol 27:677–687.

Compton KG, Mendizza A, Arnold SM. 1951. Filamentary growths on metal surfaces: whiskers. Corrosion 7:327–334.

Crowe D, Feinberg A, editors. 2001. *Design for Reliability*. Boca Raton, FL: CRC Press LLC.

d'Heurle FM, Rosenberg R. 1973. *Physics of Thin Films*. Hass G, et al., editors. New York: Academic Press, pp. 257–310.

Duryea TW, Huntington HB. 1988. The driving force for electromigration of an atom adsorbed on simple metal surface. Surf Sci 199:261–281.

Engelhardt M, Schindler G, Steinhogl W, Steinlesberger G. 2002. Electrical assessment of copper damascene interconnects down to sub-50 nm feature sizes. Microelectron Eng 64:409–416.

Fan-Yi O, Kai C, Tu KN, Yi-Shao L. 2007. Effect of current crowding on whisker growth at the anode in flip chip solder joints. Appl Phys Lett 91:231919–231911.

Fiks VB. 1959. On the mechanism of the mobility of ions in metals. Sov Phys Solid State 1:14.

Fischer AH, Zitzelsberger AE. 2001. *Proc. Conf. 39th IEEE/IRPS*, Orlando, FL. New York: IEEE, pp. 334–340.

Fuchs K. 1938. The conductivity of thin metallic films according to the electron theory of metals. Proc Cambridge Philos Soc 34:100.

Gan ZH, Shao W, Mhaisalkar SG, Chen Z, Gusak A. 2006. *Stress-Induced Phenomena in Metallization: Eighth International Workshop*. New York: AIP, pp. 269–274.

Giacomo GD. 1997. *Reliability of Electronic Packages and Semiconductor Devices*. New York: McGraw-Hill.

Harrison JW. 1988. A simulation model for electromigration in fine line metallization of integrated circuits due to repetitive pulse currents. IEEE Trans Electron Devices 35: 2170–2179.

Hartman TE, Blair JC. 1969. Electromigration in thin gold films. IEEE Trans Electron Devices 16:407–410.

Hinode K, Homma Y, Sasaki Y. 1996. Whiskers grown on aluminum thin films during heat treatments. J Vac Sci Technol A 14:2570–2576.

Hou Y, Tan CM. 2007. *14th International Symposium on the Physics and Failure Analysis of Integrated Circuits*. Bangalore, India: IEEE, p. 65.

Hou Y, Tan CM. 2008. Stress-induced voiding study in integrated circuit interconnects. Semicond Sci Technol 23:075023.

Hu C-K, Rosenberg R, Lee KY. 1999. Electromigration path in Cu thin-film lines. Appl Phys Lett 74:2945.

Hu C-K, Gignac L, Rosenberg R, Liniger E, Rubino J, Sambucetti C, Domenicucci A, Chen X, Stamper AK. 2002. Comparison of Cu electromigration lifetime in Cu interconnects coated with various caps. Appl Phys Lett 81:1782.

Huntington HB. 1975. Effect of driving forces on atom motion. Thin Solid Films 25:265–280.

Huntington HB, Grone AR. 1961. Current-induced marker motion in gold wires. J Phys Chem Solids 20:76–87.

ITRS (International Technology Roadmap for Semiconductors). 2006. http://www.itrs.net/Links/2006Update/2006UpdateFinal.htm.

Iwamura E, Ohnishi T, Yoshikawa K, Itayama K. 1994. In-situ scanning electron microscope observation of hillock and whisker growth on Al-Ta alloy films for interconnections of thin film transistor- liquid crystal displays. J Vac Sci Technol A 12:2922–2924.

JEDEC. 2000. Constant temperature aging to characterize aluminum interconnect metallization for stress-induced voiding. JEDEC Publication No. 139, pp. 1–19.

Kang KD, Burgess RR, Coleman MG, Keil JG. 1969. A Cr-Ag-Au metallization system. IEEE Trans Electron Devices 16:356–360.

Koike J, Wada M. 2005. Self-forming diffusion barrier layer in Cu–Mn alloy metallization. Appl Phys Lett 87:041911–041913.

Koonce SE, Arnold SM. 1954. Metal whiskers. J Appl Phys 25:134–135.

Korhonen MA, Black RD, Li C-Y. 1993. Stress evolution due to electromigration in confined metal lines. J Appl Phys 73:3790–3799.

Li B, Christiansen C, Gill J, Sullivan T, Yashchin E, Filippi R. 2006. Threshold electromigration failure time and its statistics for Cu interconnects. J Appl Phys 100:114516.

Li W, Tan CM, Hou Y. 2007. Dynamic simulation of electromigration in polycrystalline interconnect thin film using combined Monte Carlo algorithm and finite element modeling. J Appl Phys 101:104314.

Liu CY, Lin K, Chuang YC, Wang SJ. 2006. Study of electromigration-induced Cu consumption in the flip-chip Sn/Cu solder bumps. J Appl Phys 100:83702–83701.

Lloyd JR, Lane MR, Liu X-H, Liniger E, Shaw TM, Hu C-K, Rosenberg R. 2004. Reliability challenges with ultra-low k interlevel dielectrics. Microelectron Reliab 44:1835–1841.

Lloyd JR, Lane MW, Liniger EG, Hu C-K, Shaw TM, Rosenberg R. 2005. Electromigration and adhesion. IEEE Trans Device Mater Reliab 5:113–118.

Lloyd JR, Murray CE, Ponoth S, Cohen S, Liniger E. 2006. The effect of Cu diffusion on the TDDB behavoir in a low-k interlevel dielectrics. Microelectron Reliab 46:1643–1647.

Marcoux PJ, Merchant PP, Naroditsky V, Rehder WD. 1989. A new 2D simulation model of electromigration. Hewlett-Packard J 40(3):79–84.

Mayadas AF, Shatzkes M. 1970. Electrical resistivity model for polycrystalline films: the case of arbitrary reflection at external surfaces. Phys Rev B 1:1382–1389.

McPherson JW, Dunn CF. 1987. A model for stress-induced metal notching and voiding in very large-scale-integrated Al–Si (1%) metallization. J Vac Sci Technol 5:1321–1325.

Meier K, Roellig M, Wiese S, Goette C, Deml U, Wolter K-J. 2007. *Electromigration in Large Volume Solder Joints*. London: IEEE.

Michael NL, Kim C, Gillespie P, Augur R. 2003a. Electromigration failure in ultra-fine copper interconnects. J Electron Mater 32:988–993.

Michael NL, Kim C, Gillespie P, Augur R. 2003b. Mechanism of reliability failure in Cu interconnects with ultralow-k materials. Appl Phys Lett 83:1959.

Mori H, Okabayashi H, Komatsu M. 1997. Electromigration in layered Al lines studied by in-situ ultra-high voltage electron microscopy. Thin Solid Films 300:25–29.

Nikawa K. 1981. Monte Carlo calculations based on the generalized electromigration failure model. *IEEE IRPS*, pp 175–181.

Ogawa ET, McPherson JW, Rosal JA, Dickerson KJ, Chiu T-C, Tsung LY, Jain MK, Bonifield TD, Ondrusek JC, Mckee WR. 2002. In: *Proc. Conf. 40th IEEE/IRPS Dallas*, Texas. New York: IEEE, pp. 312–331.

Patil HR, Huntington HB. 1969. Electromigration and associated void formation in sliver. J Phys Chem Solids 31:463–474.

Pico CA, Bonifield TD. 1991. Thermal hillocks on half-micron aluminum lines: the next generation reliability issue? In: *IEEE VMIC Conference*, pp. 256–257.

Pierce DJ, Brusius PJ. 1997. Electromigration: a review. Microelectron Reliab 37:1053–1072.

Pyun JW, Lu X, Chung J, Yoon S, Ho PS, Henis N, Neuman K, Smith L, Pfeifer K. 2005. Effect of barrier thickness on electromigration reliability of Cu/porous low k interconnects. Appl Phys Lett 87:061907.

Rosenberg R, Ohring M. 1971. Void formation and growth during electromigration in thin films. J Appl Phys 42:5671–5679.

Rosenberg R, Edelstein DC, Hu C-K, Rodbell KP. 2000. Copper metallization for high performance silicon technology. Annu Rev Mater Sci 30:229–262.

Ross CA. 1991. Stress and electromigration in thin film metallization. Mat Res Soc Proc 225:35–46.

Roy A, Tan CM. 2006. Experimental investigation on the impact of stress free temperature on the electromigration performance of copper dual damascene submicron interconnect. Microelectron Reliab 46:1652–1656.

Roy A, Kumar R, Tan CM, Wong KS, Tung C-H. 2006. Electromigration in damascene copper interconnects of line width down to 100nm. Semicond Sci Technol 21:1369–1372.

Schindler G, Penka S, Steinlesberger G, Traving M, Steinhogl W, Engelhardt M. 2005. Reliability studies of narrow Cu lines. Microelectron Eng 82:645–649.

Schoen JM. 1980. Monte Carlo calculations of structure-induced electromigration failure. J Appl Phys 51:513–521.

Shao W, Mhaisalkar SG, Sritharan T, Vairagar AV, Engelmann HJ, Aubel O, Zschech E, Gusak AM, Tu KN. 2007. Direct evidence of Cu/cap/liner edge being the dominant electromigration path in dual damascene Cu interconnects. Appl Phys Lett 90:052106.

Sharif A, Chan YC, Islam RA. 2004. Effect of volume in interfacial reaction between eutectic Sn-Pb solder and Cu metallization in microelectronic packaging. Mater Sci Eng B 106:120–125.

Sharma P, Ganti S, Ardebili H, Alizadeh A. 2004. On the scaling of thermal stresses in passivated nanointerconnects. J Appl Phys 95:2763–2769.

Shewmon PG. 1963. McGraw-Hill Series in Mat. Sci. and Eng. p. 189.

Shimada M, Moriyama M, Ito K, Tsukimoto S, Murakami M. 2006. Electrical resistivity of polycrystalline Cu interconnects with nano-scale linewidth. J Vac Sci Technol 24:190–194.

Sondheimer EH. 1952. Adv Phys 1:1–42.

Sorbello RS. 1996. Advanced metallization for future ULSI. In Proceedings of the Materials Research Society Symposium, vol. 427, pp. 73–81.

Srinvasan J, Adve SV, Bose P, Rivers JA. 2004. The impact of technology scaling on lifetime reliability. *International Conference on Dependable Systems and Networks*, IEEE Computer Society.

Steinhogl W, Schindler G, Steinlesberger G, Engelhardt M. 2002. Size-dependent resistivity of metallic wires in the mesoscopic range. Phys Rev B 66:075414.

Steinhogl W, Schindler G, Steinlesberger G, Traving M, Engelhardt M. 2005. Comprehensive study of the resistivity of copper wires with lateral dimensions of 100 nm and smaller. J Appl Phys 97:023706.

Steinlesberger G, Schindler G, Engelhardt M, Steinhogl W, Traving W. 2004. Interconnect technology conference, 2004. *Proceedings of the IEEE 2004 International.* New York: IEEE, pp. 51–53.

Suo Z. 2003. Reliability of interconnect structures. Interfacial Nanoscale Failure 8:265–235.

Takatsuija H, Tsujimoto K, Kuroda K, Saka H. 1999. Relationship of grain nanostructure and orientation in whisker growth on aluminum thin films on glass substrates. Thin Solid Films 343–344:41–464.

Tan CM, Arijit R. 2007. Electromigration in ULSI interconnects. Mater Sci Eng R 58:1–75.

Tan CM, Hou Y. 2007. Lifetime modeling for stress-induced voiding in integrated circuit interconnections. Appl Phys Lett 91:061904.

Tan CM, Roy A. 2006. Investigation of the effect of temperature and stress gradients on accelerated EM test for Cu narrow interconnects. Thin Solid Films 504:288–293.

Tan CM, Roy A, Vairagar AV, Krishnamoorthy A, Mhaisalkar SG. 2005. Current crowding effect on copper dual damascene via bottom failure for ULSI applications. IEEE Trans Device Mater Reliab 5:198–205.

Tan CM, Hou Y, Li W. 2007a. Revisit to the finite element modeling of electromigration for narrow interconnects. J Appl Phys 102:033705.

Tan CM, Raghavan N, Arijit R. 2007b. Application of gamma distribution in electromigration for submicron interconnects. J Appl Phys 102:103703.

Tobias PA, Trindade DC. 1995. *Applied Reliability*, 2nd edition. New York: Van Nostrand Reinhold.

Tokei Z, Patz M, Schmidt M, Iacopi F, Demuynck S, Maex K. 2004. Correlation between barrier integrity and TDDB performance of copper porous low-k interconnects. Microelectron Eng 76:70–75.

Tsai CM, Lin YL, Tsai JY, Yi-Shao L, Kao CR. 2006. Local melting induced by electromigration in flip-chip solder joints. J Electron Mater 35:1005–1009.

Tu KN. 2003. Recent advances on electromigration in very-large-scale-integration of interconnects. J Appl Phys 94:5451.

Tu PL, Chan YC, Lai JKL. 1997. Effect of intermetallic compounds on the thermal fatigue of surface mount solder joints. IEEE Trans Compon Pack Manuf Technol B 20:87–93.

Tu KN, Gusak AM, Li M. 2003. Physics and materials challenges for lead-free solders. J Appl Phys 93:1335–1353.

Usui T, Nasu H, Koike J, Wada M, Takahashi S, Shimizu N, Nishikawa T, Yoshimaru A, Shibata H. 2005. *Low Resistive and Highly Reliable Cu Dual-Damascene Interconnect Technology Using Self-Formed MnSixOy Barrier Layer.* Burlingame, CA: IEEE, pp. 188–190.

Vancea J. 1989. Erratum: unconventional features of free electrons in polycrystalline metal films. Int J Mod Phys B 3:1455–1501.

Vancea J, Reiss G, Hoffmann H. 1987. Mean-free-path concept in polycrystalline metals. Phys Rev B 35:6435–6437.

Verbruggen AH, Griessen R, Groot DGd. 1986. Electromigration of hydrogen in vanadium, niobium and tantalum. J Phys F Met Phys 16:557–575.

Wang H, Bruynseraede C, Maex K. 2004. Impact of current crowding on electromigration-induced mass transport. Appl Phys Lett 84:517.

Ye H, Basaran C, Hopkins D. 2003. Thermomigration in Pb–Sn solder joints under joule heating during electric current stressing. Appl Phys Lett 82:1045–1047.

Yi-Shao L, Chin-Li K. 2006. Characteristics of current crowding in flip-chip solder bumps. Microelectron Reliab 46:915–922.

15

INTEGRATION ISSUES OF BARIUM STRONTIUM TITANATE THIN FILM FOR TUNABLE MICROWAVE APPLICATIONS

Ashok Kumar,[1,3] Supriya Ketkar,[2,3]
and Venkataraman Gurumurthy[2,3]

[1]*Mechanical Engineering Department, University of South Florida, Tampa, Florida*
[2]*Electrical Engineering Department, University of South Florida, Tampa, Florida*
[3]*Nanotechnology Research and Education Center, University of South Florida, Tampa, Florida*

INTRODUCTION

Small size, low cost, and high field-dependent permittivity are desirable characteristics in the current and next-generation microwave applications. These requirements impose significant challenges on existing technologies and create the necessity for new materials and technologies. Thin-film ferroelectrics are receiving increased attention because of their potential in producing tunable radio frequency (RF) and microwave circuits. RF and microwave applications benefit greatly from the use of tunable devices and circuits. With components that can be tuned over a broad range, filters can be made to tune over multiple frequency bands of operation; impedance matching networks can be adjusted for amplifier power level or antenna

Ceramic Integration and Joining Technologies: From Macro to Nanoscale, First Edition.
Edited by Mrityunjay Singh, Tatsuki Ohji, Rajiv Asthana, Sanjay Mathur.

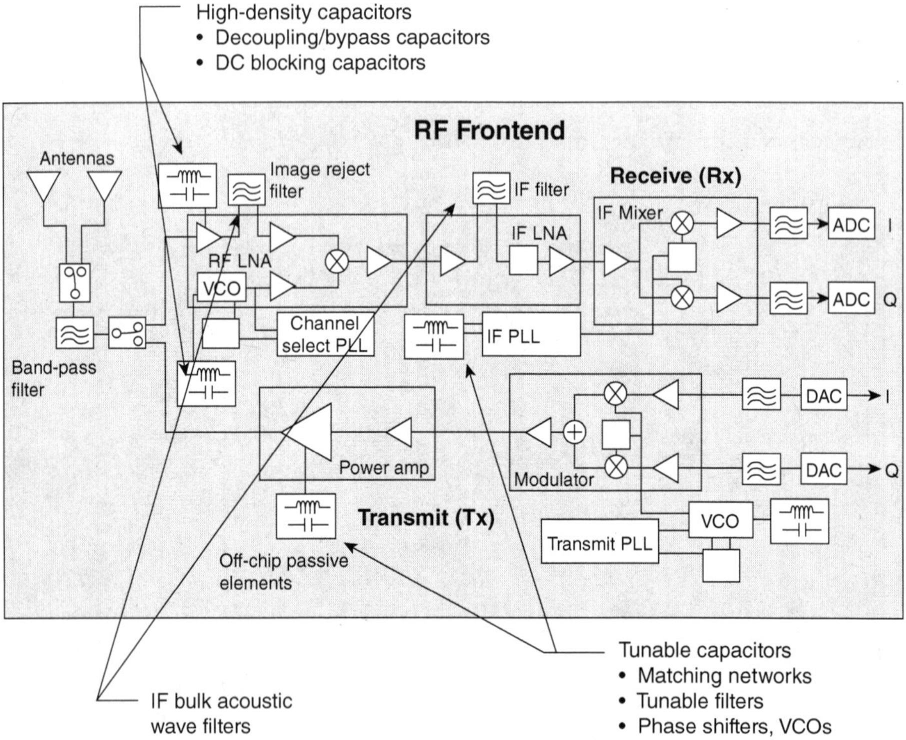

Figure 15.1. Tunable capacitors in the RF front end. ADC, analog to digital converter; DAC, digital to analog converter; VCO, voltage-controlled oscillator; PLL, phase-locked loop; LNA, low noise amplifier.

impedance optimization. In general, tunability in these circuits gives the designer an additional tool in meeting the stringent frequency and power requirements of wireless communication systems, even in the changing operating environment inherent in these systems. In addition, tunable circuit responses can be made to compensate for the deleterious effects of aging and temperature changes in sensitive RF circuits. Figure 15.1 shows the different types of tunable capacitors used in RF front ends. As the communications spectrum becomes more crowded, the demands on battery life, size, and weight constraints intensify, and cost pressures increase, in which case technologies that enable such adaptability become increasingly important.

Barium strontium titanate (BST) is a ferroelectric (FE) material that exhibits an electric field-dependent dielectric constant. Capacitors fabricated with BST as the dielectric material therefore have capacitance values that can be adjusted by applying a DC voltage across them. Because of its voltage tunability and the high value of dielectric constant, high power-handling capability, fast tuning response, and ease of integration with other thin-film components, devices based on thin-film BST present an attractive technology for the design of tunable circuits.

BST DEVICE TECHNOLOGY FOR TUNABLE MICROWAVE APPLICATIONS

BST and microelectromechanical systems (MEMS) technologies are the current competing technologies for low-cost tunable circuits. The cost reduction is achieved by using inexpensive substrates and employing high-volume deposition technologies. Table 15.1 displays the competing technologies available for making tunable capacitors. While the BST varactors are cheaper to fabricate and to process, the MEMS counterparts offer excellent power-handling capabilities.

Application of BST varactors in tunable microwave devices is manifold and requires precise film optimization and capacitor design. BST thin films offer a low-cost scalable varactor technology, and it is the most widely researched material for the current generation of tunable RF and Microwave components and systems. The various applications for which BST is being currently researched for are shown in Figure 15.2.

As illustrated in Figure 15.3, the high dielectric constant of BST makes it suitable for small area capacitors, which in turn make smaller device packages possible [1].

Some of the other applications are described in detail in the following subsections:

TABLE 15.1. Comparison of Competing Technologies for Making Tunable Circuits

Properties	Semiconductor	MEMS	BST
Tunability (high Q)	Good (2–3:1)	Low (<1.5:1.0)	Good (2–3:1)
RF loss	Moderate (Q < 60 typ)	Very Good (Q < 200)	Moderate (Q < 100 typ)
Control voltage (V)	<10	<60	<5–30
Tuning speed	Fast 1–5 ns	Slow >5 μs	Fast <30 ns
Power-handling capability	Poor	Excellent	Trades with control voltage

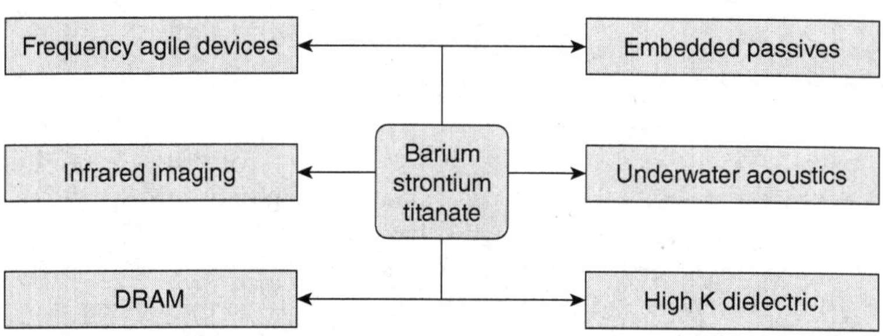

Figure 15.2. Prospective areas of application of barium strontium titanate.

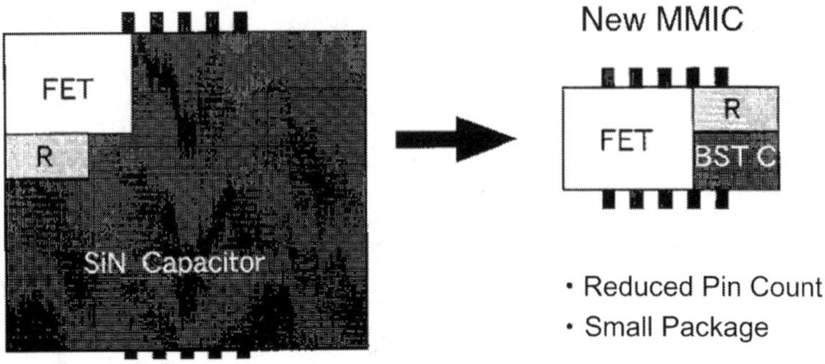

Figure 15.3. Reduction in device packages using BST capacitors [1]. MMIC, monolithic microwave integrated circuit.

1. *Tunable Filters.* Tunable filters are widely used in the receiver front end, which has potential applications in most military applications and satellite communication systems. Most of the current generation of tunable filters rely on mechanical tuning or semiconductor-based varactors. The disadvantages of mechanical tuning are low tuning speed and large size. Semiconductor varactors are much faster and have low power-handling capabilities. BST varactors have the potential to overcome these difficulties and can be used in low-pass and band-pass tunable filters (refer to Fig. 15.1).
2. *Phase Shifters.* Phase shifters are the critical elements for electronically scanned phased array antennas. Phased array antennas are used in fighter aircraft radar and certain commercial systems such as cellular telephone base stations. Phase shifters in an electronically scanned array antenna allow the antenna beam to be steered in the desired direction without physically repositioning the antenna. When a signal is output from an amplifier, the amplification process generates certain undesired mixing products that introduce noise into the system. These unwanted signals need to be filtered to have a cleanly amplified signal. Phase shifters provide an interesting and elegant way to perform this function. As phase shifters possess the ability to change the phase of a signal, it is possible to sample the undesired harmonics and to change their phase by 180°. Signals that are 180° out of phase cancel each other out. Figure 15.4 displays the design of a BST-based phase shifter [2]
3. *Tunable Antennas.* Microstrip antennas are widely used in wireless communication systems because they are lightweight, compact, conformable to planar and nonplanar surfaces, simple, and inexpensive to manufacture using modern printed circuit technology. However, the main disadvantage of microstrip antennas is their narrow operating bandwidth. In situations when there is no need for high instantaneous bandwidth, like frequency hopping in cell phone systems, one can improve the operating frequency range of antennas by making them tunable. Microstrip antennas with multiple-varactor loading at the radiating edges are designed, fabricated, and measured. This type of tuning can achieve more tunability than using multidielectric layers with BST in between.

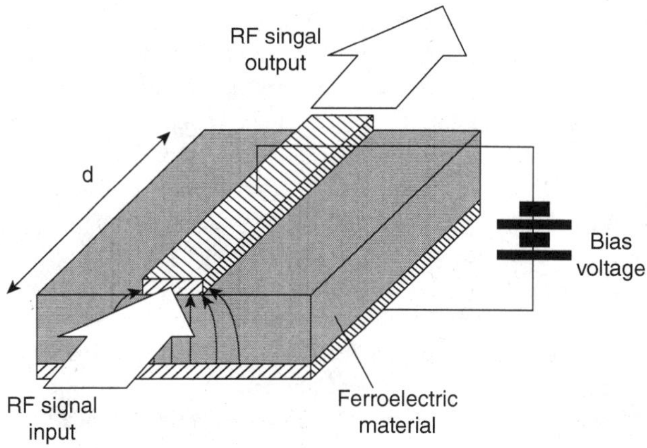

Figure 15.4. Structure of a BST microstrip-based phase shifter [2].

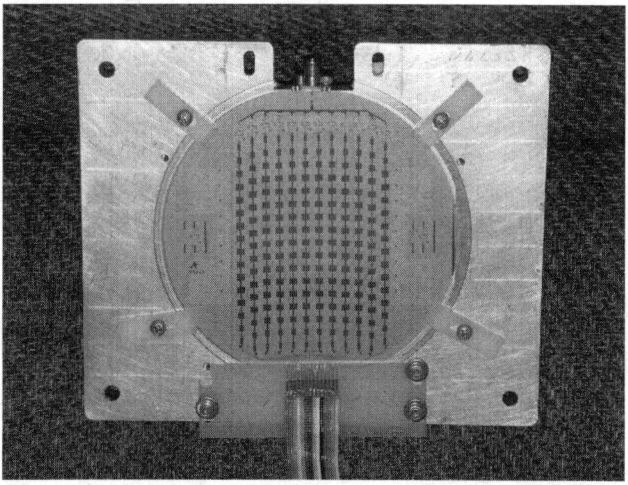

Figure 15.5. Integrated electronically scanned antenna mounted in test fixture [3].

Figure 15.5 depicts an integrated electronically scanned antenna (antenna mounted in test fixture) [3]

BST: STRUCTURE AND PROPERTIES

In certain crystalline minerals, the crystals become electrically polarized when subjected to mechanical forces. Tension and compression generate voltages of opposite polarity, which is proportional to the applied force. Conversely, if the crystals were exposed to an electric field, it lengthens and shortens according to the polarity of the field and is proportional to the strength of the field. These behaviors are labeled the piezoelectric and inverse piezoelectric effects, respectively. For the piezoelectric effect to occur, the absence of the center of symmetry or the inversion point is necessary. As a result of a unique polar axis, the crystals of this group, when uniformly heated, generate an additional electric charge due to spontaneous polarization parallel to the polar axis in addition to piezoelectric effect-related charge creation. This effect is known as pyroelectricity. In some pyroelectric crystals, an additional property is observed wherein an external electric field can reverse the spontaneous polarization of the crystals over some temperature range. The response of polarization to an external electric field manifests itself as a hysteresis loop. The materials that exhibit this property are called ferroelectrics. All ferroelectrics are pyrolectric and piezoelectric [4–6].

These materials generally are physically strong and chemically inert. The composition, shape, and dimensions of a ceramic material can be tailored to meet the requlirements of a specific application [7]. The widespread use of FE ceramics started with the use of barium titanate ($BaTiO_3$)-based ceramics in capacitor applications and piezoelectric transducer devices. Since then, many other FE ceramics including lead titanate ($PbTiO_3$), lead zirconate titanate (PZT), lead lanthanum zirconium titanate (PLZT), and relaxor ferroelectrics like lead magnesium niobate (PMN) have been developed and utilized for a variety of applications. The biggest use of FE ceramics has been in areas such as dielectric ceramics for capacitor applications, FE thin films for nonvolatile memories, piezoelectric materials for medical ultrasound imaging and actuators, and electro-optic materials for data storage and displays [8]. $Ba_{1-x}Sr_xTiO_3$ is a continuous solid solution between two traditional ferroelectrics—barium titanate ($BaTiO_3$) and strontium titanate ($SrTiO_3$). The Curie temperature of BST decreases linearly with increasing Sr concentration at a rate of 3.4°C/mol% Sr. Therefore, BST's Curie temperature and hence its properties can be tailored over a wide range to meet the requirements of different applications [1]. This kind of flexibility in properties, and hence the prospective application range, is absent in conventional FE ceramics.

In the following section, the crystal structure, phase transition behavior, polarization behavior, the microstructure of BST, and the influence these properties have on its electric performance are discussed.

Crystal Structure

An FE ceramic is an arrangement of perovskite (oxide having the same crystalline structure as the mineral, $CaTiO_3$) crystals, each consisting of tetravalent metal ions (usually titanium or zirconium), in a lattice of large divalent metal ions (usually lead or barium), and usually O^{2-} ions. The resulting crystal structure is in general called the perovskite ABO_3 structure, shown in Figure 15.6, where A represents the large cations located at the corners of the unit cell, B represents the smaller cations located at the body center, and O is usually the oxygen atoms positioned at the face centers, though halides like fluorine and chlorine can also take its place.

The perovskite structures are chemically very flexible; that is, many different cations can be substituted with transition metal ions on both the A and B sites without drastically changing the overall structure. The properties of the final structures are

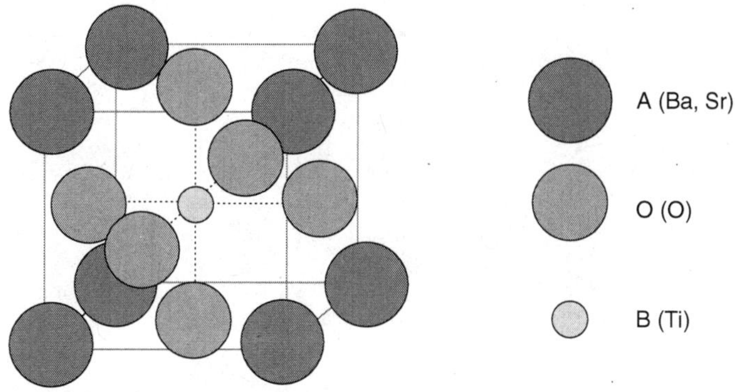

Figure 15.6. The perovskite crystal structure of BST in cubic phase with no externally applied electric field [9].

completely different from one another depending upon the metal ion that has been used for substitution.

Thus, the composition, shape, and dimensions of perovskite structures and hence their electronic properties can be tailored over a wide range. This is a major reason of application of perovskite FE ceramics over a broad spectrum of applications [7]. BST belongs to the family of ABO_3 perovskites. BST crystals exist in both cubic and tetragonal symmetry depending upon the (Ba/Sr) ratio and the corresponding Curie temperature of the chemical composition. In BST, Ba^{2+} and Sr^{2+} occupy the corners of the cube and they are shared between eight adjacent cells. This gives a total of 0.5 barium atoms and 0.5 strontium atoms per cell. Six oxygen atoms occupy the face-centered position on the six sides, and each is shared between two adjacent cells, which makes it three oxygen atoms per cell. Ti^{4+}, which is the smallest of the atoms in the crystal, is located in the center of the cubic cell. It is relatively free to move and the spontaneous polarization of BST is mainly attributed to the movement of the Ti atom [9, 10]. The crystal structure of BST in its cubic phase is shown in Figure 15.6.

Phase Transitions

FE materials undergo a transformation from a paraelectric (PE) phase to an FE phase when cooled down to a certain temperature known as the Curie temperature (T_c). The schematic of FE phase transition and the change in crystal structure below and above Curie temperature is represented in Figure 15.7 [7, 11].

As depicted in Figure 15.7, in the FE phase, the dielectric constant increases as the temperature increases. While in the PE phase, the dielectric constant decreases with an increase in temperature, obeying the Curie–Weiss law [12]. Below the Curie temperature, the position of smaller tetravalent metal ion and the octahedral structure changes from cubic to tetragonal symmetry, with the smaller tetravalent metal ion in an off-center position corresponding to a permanent dipole. In the FE phase, adjoining dipoles align themselves in a particular direction forming regions of local alignment

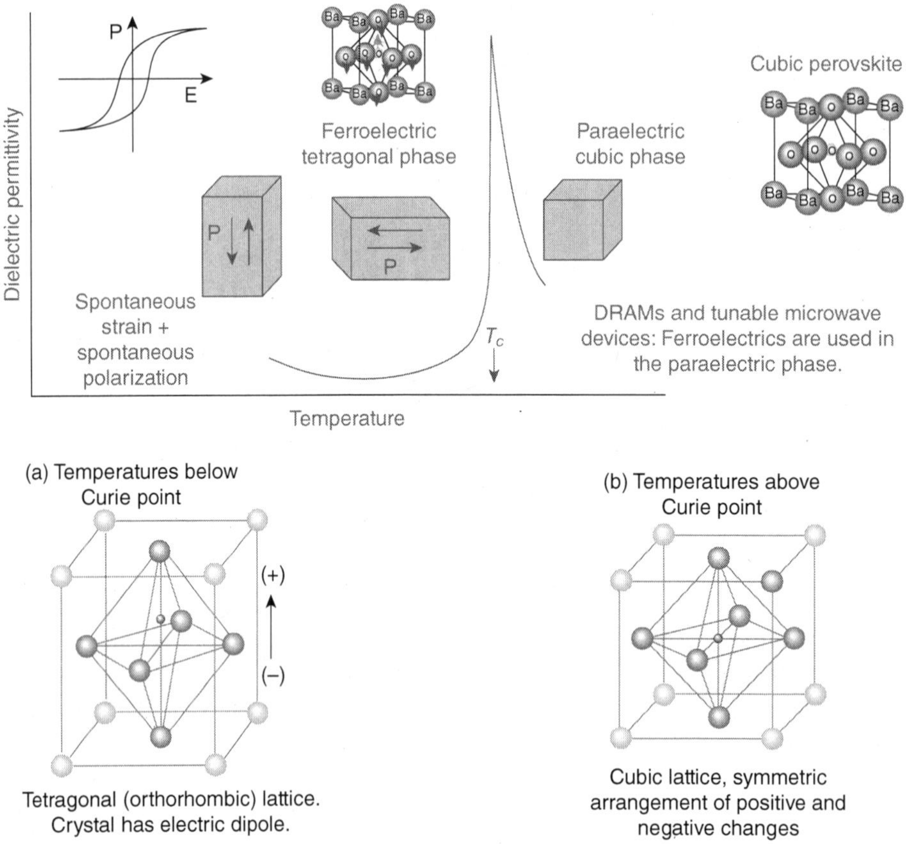

Figure 15.7. Phase transitions and corresponding changes in the crystal structure of perovskite ferroelectric materials [7, 11].

called domains. The alignment gives a net dipole moment to the domain and thus a net polarization called as spontaneous polarization (P_s). Domains are formed due to stray electric fields which are created by noncompensated polarization charges. However, the direction of polarization in the neighboring domains is not the same; hence, the FE material displays no overall polarization unless an electric field is applied to it. The domains are separated by domain walls and a particular domain pattern is a result of stresses created at the Curie point, uncompensated surface charges, and physical imperfections (vacancies, dislocations, and dopants). 3-D piezoresponse force microscopy (PFM) is used to visualize domains in thin films. Piezoresponse images of epitaxial PZT film are shown in Figure 15.8 [13].

It is also important to note that in most ABO_3, perovskites have different stable phases at different temperatures. For example, barium titanate has four stable phases, which, in the order of transformation, are tetragonal, orthorhombic, rhombhoehedral, and cubic phases, respectively. Above the Curie temperature, thermal energy or an electric field is sufficient to induce movement of the metal atoms randomly from one

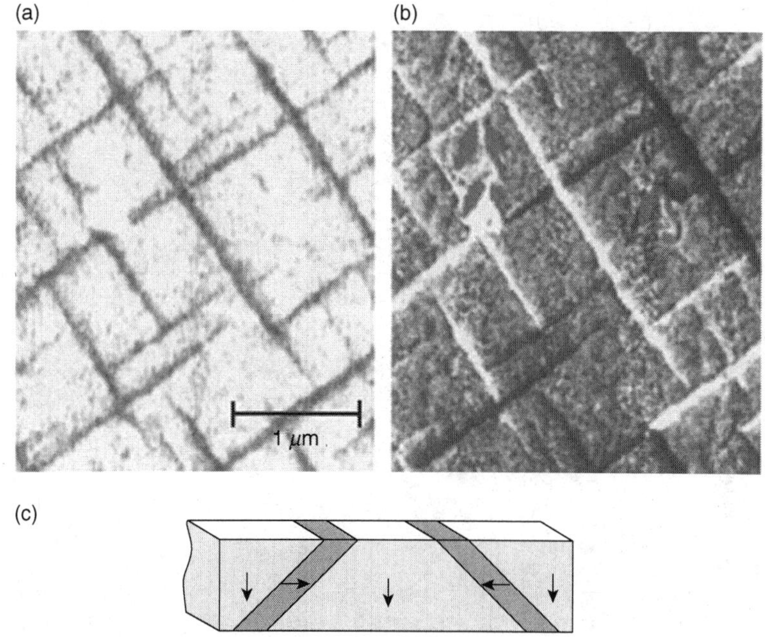

Figure 15.8. PFM images of an epitaxial PZT film. (a) Domains with vertical (out-of-plane) polarization. (b) Domains with horizontal (in-plane) polarization. (c) Schematic showing the different directions of orientation of dipoles in different domains [13].

position to another, and so there is no fixed asymmetry. The open octahedral site allows the tetravalent metal atom to develop a large induced dipole moment when an electric field is applied, which is the main reason why perovskite FE materials have high permittivity in the PE phase. In the FE phase, the strong hysteresis behavior observed makes the FE material suitable for nonvolatile memory applications and piezoelectric applications. Above the Curie temperature (in the PE phase), the material no longer has spontaneous polarization. However, the dielectric permittivity still remains high, and hence the FE material is highly suitable for dynamic random access memory (DRAM) and tunable microwave applications.

The polarization of a dielectric is due to the redistribution of charges inside the dielectric when an electric field is applied to it. From a macroscopic point of view, the dielectric has no net charges if no electric field is applied to it. If one considers a boundary perpendicular to the direction of polarization, there is a net positive or negative charge on the surface, which is not compensated by the charges inherent to the dielectric. Therefore, the polarization appears on the surface of the dielectric. Thus, polarization of a dielectric produces a macroscopic electric field, which is determined by these surface charges. Due to the different behavior of its parent FE materials $BaTiO_3$ and $SrTiO_3$, the phase transition behavior of BST evolves with composition. The variation of Curie temperature and lattice parameters (defined as the spacing between unit cells) of BST with barium composition is shown in Figure 15.9 [14].

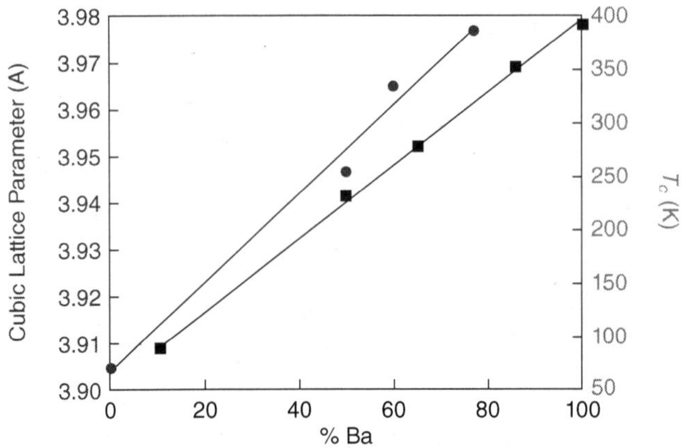

Figure 15.9. Variations of Curie temperature and lattice parameters of BST with change in barium percentage composition [14].

$BaTiO_3$ ($T_c = 393\,K$) is a prototype FE, which undergoes a sequence of crystal structure transitions as it switches from the FE to the PE phase [15]. $SrTiO_3$ is an incipient FE; that is, it exhibits quantum ferroelectricity, which means that though it undergoes phase transition at very low temperatures ($T_c = 105\,K$), the quantum effect prevents a complete transition to the PE phase. Therefore, $SrTiO_3$ exhibits FE behavior well above its Curie temperature or PE effect well below the Curie temperature [16, 17]. BST undergoes structural changes as it undergoes phase transitions like $BaTiO_3$ for Ba concentrations varying from 20% to 100%. As the composition of BST comes close to pure $SrTiO_3$, its behavior is complicated by the competition between prototype FE transitions and quantum FE transition-related instabilities.

The variations of dielectric constants of BST with temperature for different concentrations of barium and strontium are given in Figure 15.10 [18]. It suggests that BST with $x = 0.5$ is ideal for room temperature tunable microwave applications. This is because for this composition, the relative dielectric permittivity is high and phase transitions occur at temperatures well below room temperature. Hence, a good trade-off is obtained between dielectric permittivity and losses [19].

The phase transition behavior of BST also varies with thickness. The important occurrence that has been observed in the phase transition properties of BST with variation in thicknesses is that the variation of dielectric constant with temperature becomes less and less pronounced with decrease in thickness. At thicknesses in the range of 100 nm, the dielectric constant shows almost no variation with temperature. Thus, it would be advantageous to use thin BST films for applications that require stability in device performance with variations in temperature [1, 20]. The comparison of variation of dielectric constant with temperature for ceramic and thin-film $Ba_{0.7}Sr_{0.3}TiO_3$ is shown in Figure 15.11a [1].

The sharp peak in permittivity that is observed in the phase transition region for thick films is suppressed more and more as the film thickness becomes lesser and lesser. The phase transition region gets spread over a wider range of temperature (diffuse phase

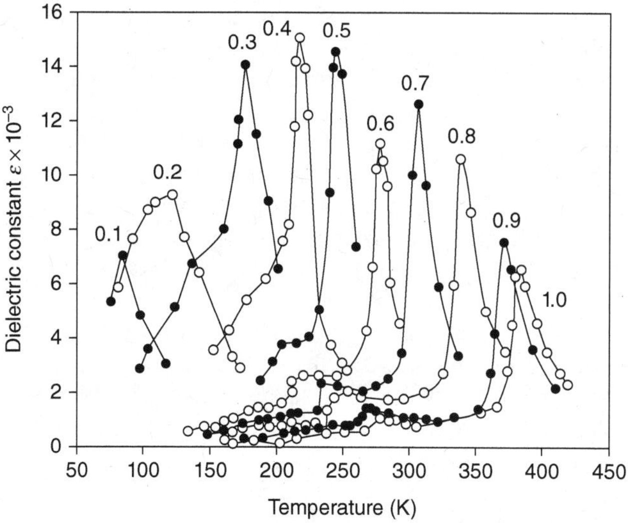

Figure 15.10. Variations of dielectric constants of BST for different concentrations of barium and strontium [18].

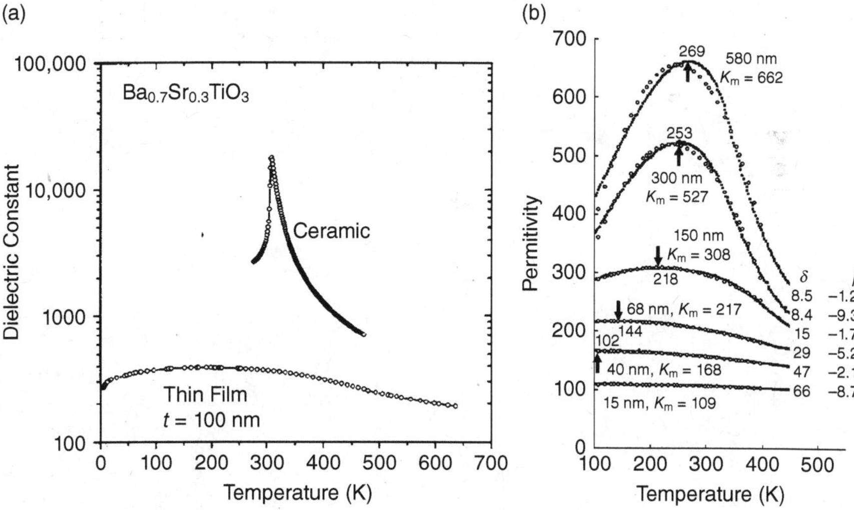

Figure 15.11. Phase transition behavior of $Ba_{0.7}Sr_{0.3}TiO_3$. (a) Comparison of variation of dielectric constant with temperature between ceramic and thin-film $Ba_{0.7}Sr_{0.3}TiO_3$ [1]. (b) Difference in phase transition behavior with decrease in film thickness (K_m is the maximum permittivity value) [21].

transitions). Hence, characterization of BST thin films becomes more complicated as FE behavior might be observed in the PE region and vice versa [21]. The origin of this variation in properties of ferroelectrics with decrease in thickness has still not been well understood. In one of the theories, it has been suggested that though the "apparent" Curie temperature obtained from Curie–Weiss plots from as-measured data suggests a diffuseness in phase transitions accompanied by a shift in Curie temperatures, the "inherent" Curie temperature does not change with thickness [22]. The variation of permittivity with temperature for six different film thicknesses of $Ba_{0.7}Sr_{0.3}TiO_3$ is shown in Figure 15.11b. It has also been reported that the Curie temperature of BST thin films fabricated on various substrates increases due to the internal stress of the films arising from the difference in thermal coefficient of expansion (TCE) and/or difference in the lattice parameter between the film and the substrate [16, 23].

$Ba_{1-x}Sr_xTiO_3$ films with $x < 0.7$ show little hysteresis behavior at room temperature and above. A very weak hysteresis behavior is observed for films with x values ranging from 0.7 and 0.9 as seen from Figure 15.12 [24], which shows the results of polarization–electric field (P–E) measurements done on $Ba_{1-x}Sr_xTiO_3$ thin films at room temperature.

Polarization

There are different types of polarization contributing to dielectric response depending upon the structure of molecules constituting a particular material [13, 15, 25, 26]:

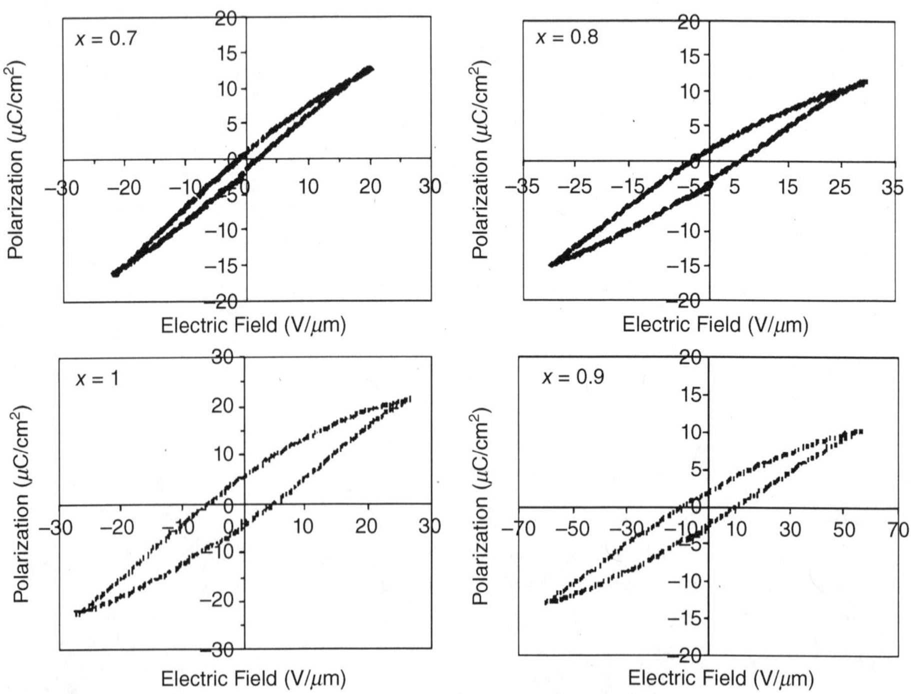

Figure 15.12. P–E hysteresis loops for $Ba_xSr_{1-x}TiO_3$ thin films at room temperature [24].

- Electronic polarization
- Ionic polarization
- Dipolar or molecular polarization
- Space-charge or interfacial or Maxwell–Wagner polarization
- Orientation polarization
- Domain wall polarization.

The general prediction of mean-field models is that near the surface of an FE thin film, the polarization normal to the surface should continuously decrease from the bulk value [27]. It has been theoretically concluded that there exists a surface-modified layer in a thin film when the FE material is in the PE phase. The polarizability of material near the surface is reduced relative to a bulk FE. The effect of the reduced polarizability is the reduced overall capacitance of the film. The film behavior suggests a total suppression of the ionic contributions to the polarization due to this "dead layer" in the surface [28].

The domain structure of thin films is also affected by growth stresses, lattice mismatches, and thermal expansion mismatch stresses imposed by the substrate [29]. In the case of perovskite materials like BST, tensile stress at the transformation temperature pulls the polarization into the plane of the film, while compressive stress leads to polarization being directed normal to the surface. Tensile stress leads to high dielectric constants and low remnant polarizations, and compressive stresses lead to the opposite [30–32]. Tensile stresses are advantageous for acoustic applications as the hysteresis losses decrease. Since varactors are operated in the PE phase and thin films are generally used, the polarization and hence the effects of stresses are negligible. Many FE films undergo a slow polarization when a voltage is applied to the capacitor and the current resulting from the polarization follows an inverse power law over many decades in time [33–35]. This is known as the Curie–von Schweidler law and it has an exponent of close to unity. Discharging also follows the same pattern but with a current of opposite polarity. The polarization on charging and discharging is directly related to capacitance and dielectric losses. In BST thin films, dielectric relaxation is a dominant cause of dielectric loss. The origin of the relaxation currents is poorly understood, and several mechanisms to explain these currents have been put forward in literature [29, 36–39]. Figure 15.13 [29] shows the thickness dependence of relaxation currents observed in BST thin films for thicknesses varying from 8 to 100 nm, keeping the electric field constant at 167 KV/cm.

Polarization and Frequency

When an alternating field is introduced, the polarizability of the dielectric material is selectively excited by different polarization mechanisms at different frequencies. The contribution of different polarization mechanisms emerges when the dielectric response of a material is studied across a wide frequency range. The contribution of a particular polarization mechanism at a particular frequency range depends upon the mass of the particle that initiates it. For example, at radio frequencies, a point is reached where the

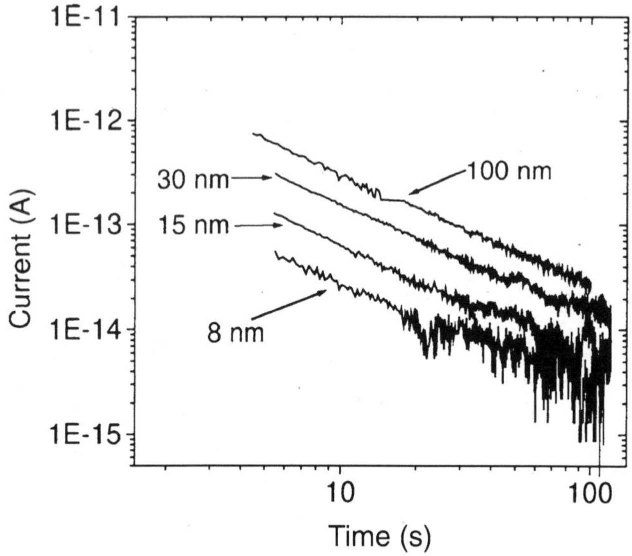

Figure 15.13. Thickness dependence of relaxation currents observed in metal organic chemical vapor deposited BST thin films [29].

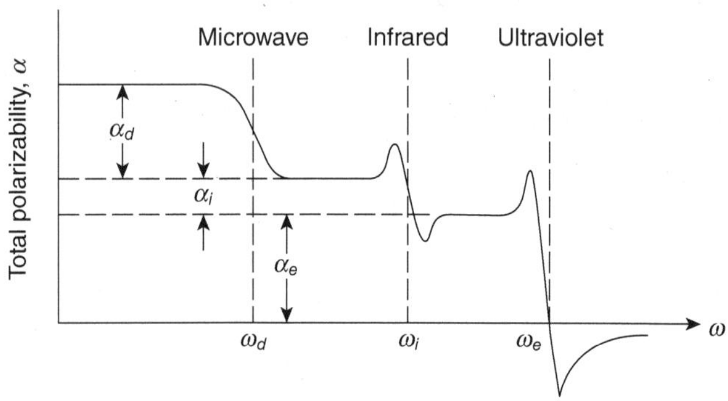

Figure 15.14. Variation of the total polarizability of a dielectric material with frequency [15].

molecules of the material (molecular polarization) can no longer rotate fast enough to contribute to the net polarization. As a result, the polarizability of the material reduces in magnitude as frequency increases. At a frequency range above the infrared region, only the electronic polarization remains effective because it is due to electrons that are still able to follow the field at high frequencies. At still higher frequencies, even the electronic contribution to polarizability vanishes. The variation of polarizability due to extrinsic contributions with frequency is shown in Figure 15.14 [15], where α_e, α_i, and α_d stand for electronic, ionic, and dipolar polarizations, which occur up to frequencies

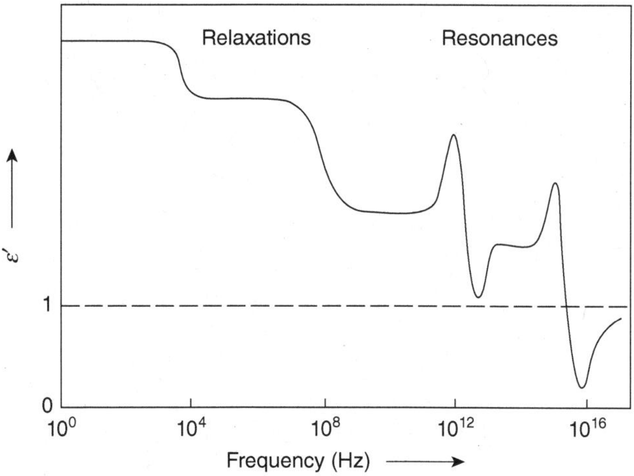

Figure 15.15. Frequency dependence of the real part of relative dielectric permittivity [13].

ω_e, ω_i, and ω_d, respectively. The frequencies ω_i and ω_d depend on the material under study, but their orders of magnitude remain in the regions indicated in the figure [40].

In Figure 15.14, relaxation behavior is observed in a frequency range extending from $\omega = 0$ to $\omega = \omega_d$, which is due to dipolar (molecular) polarization and internal mechanisms, namely, orientation, domain wall, and space-charge polarizations. A resonance behavior is observed from the end of the microwave band, which is due to ionic and electronic polarization. The resonance behavior leads to an increase in both the real part of permittivity and the loss tangent [25]. Again, since relative permittivity is directly related to polarizability, the variation of dielectric constant (ε') with frequency follows a similar pattern as shown in Figure 15.15 [13].

Effect of Electric Field on the Behavior of FE Materials

Large amplitudes of externally applied field lead to strong nonlinearity in FE materials. These significant nonlinearities and electric field tunable permittivities open up a broad field of applications for FE materials. Therefore, it is important to understand the effects of large amplitudes of externally applied fields on FE films. When capacitance–voltage (CV) measurements are done on FE films at frequencies of interest, a small AC ripple propagates through the film along with the applied DC bias. The small increase in permittivity when bias is initially applied is due to the movement of domain walls whose motion was inhibited by defects in the FE film (a phenomenon called domain wall pinning). But with further application of bias, the domain walls become immobile, and extrinsic contributions to particle movements are nullified due to elastic constraints [9, 41–44]. Therefore, at high applied fields, the AC signal is mainly probing the intrinsic contributions to dielectric response.

As the DC bias is decreased, the extrinsic contributions again start influencing the dielectric behavior and the contributions reach a maximum point at zero bias. These

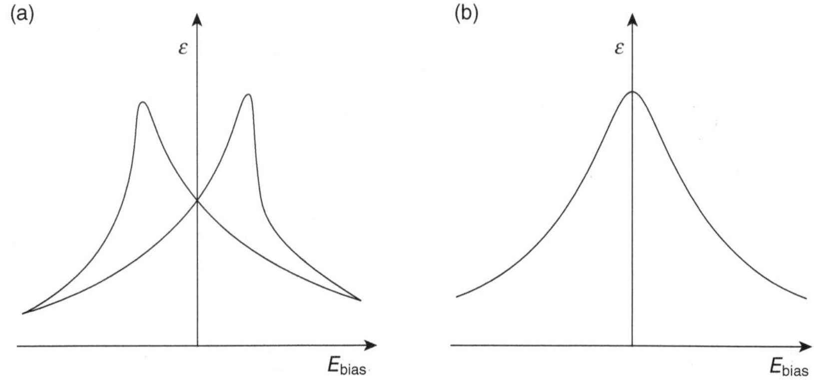

Figure 15.16. Variation of permittivity with applied bias for a ferroelectric material in the (a) ferroelectric and (b) paraelectric phases [13].

particle movements result in dielectric losses as described in previous sections; therefore, maximum permittivity accompanied by maximum losses are observed at zero bias. Conversely, minimum permittivity and losses are observed at maximum bias value. Thus, the variation of capacitance (or dielectric permittivity) of an FE film with positive and negative bias voltages assumes a characteristic bell-shaped curve [42]. A clear distinction is observed in the CV characteristics of FE films in the FE phase and PE phase. In the FE phase, a "butterfly loop" is obtained theoretically with two distinct peaks in capacitance values. These peaks correspond to different domain orientations formed on a film in the FE phase [42, 43]. In the PE phase, no butterfly loop is observed due to absence of hysteresis behavior. The CV characteristics of an FE film in FE and PE phases, respectively, are depicted in Figure 15.16 [13].

An FE material for microwave applications is usually operated in the PE stage, close to Curie temperature to ensure high tunability and low loss tangent [9]. It is inevitable that a large electric field has to be applied to obtain large tunability in FE thin films. For example, a metal–insulator–metal (MIM)-type capacitor with an FE material as a dielectric requires electric fields in the range of 1000 KV/cm to obtain high tunability [16, 45]. The quality factor (Q-factor) is used to characterize the losses in lumped circuit elements. At low frequencies, the capacitances of BST in the PE phase decreases monotonically with applied bias. The increase in magnitude of externally applied bias further restricts the extrinsic polarization mechanisms in the material, hence the steadily decreasing permittivities (or capacitances) with increasing frequencies and electric fields. Typical capacitance–frequency curves for 1-V increments in applied bias voltages for $Ba_{0.5}Sr_{0.5}TiO_3$ thin films are shown in Figure 15.17 [46].

This bias-induced resonance at high frequencies observed in BST and strontium titanate has an adverse residual effect on the loss properties of the device. It is extremely important to understand and characterize these resonances in order to obtain reliable capacitors for tunable microwave applications [16, 46]. Figure 15.18 shows the bias-induced resonant dips observed in a BST-based MIM capacitor with increasing bias values [47].

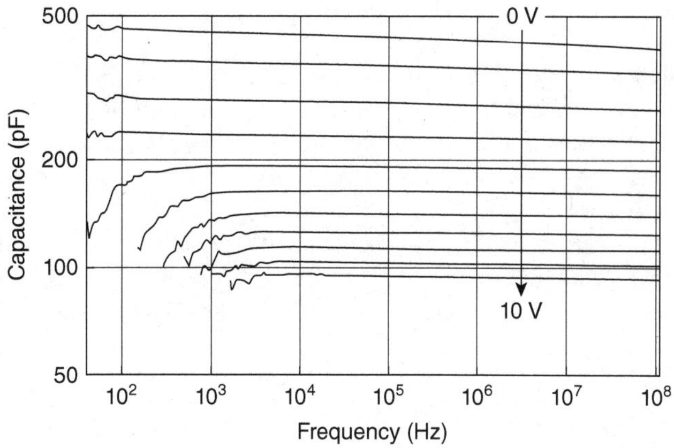

Figure 15.17. Variation of capacitance of $Ba_{0.5}Sr_{0.5}TiO_3$ thin films with frequency for 1-V increments in bias [46].

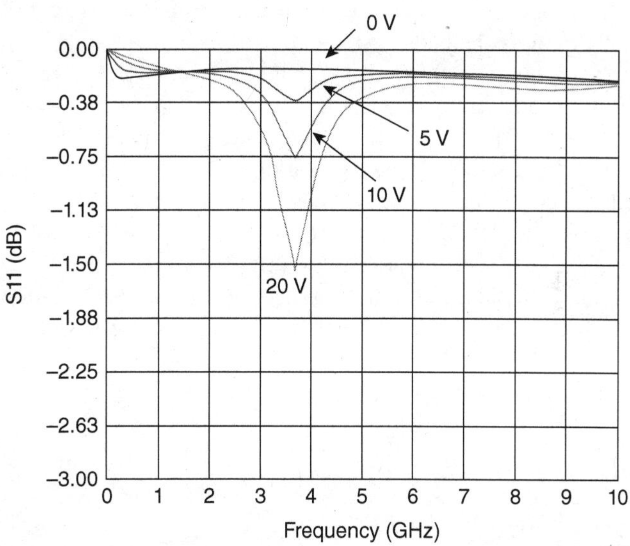

Figure 15.18. Bias-induced resonant dips observed in a BST-based MIM capacitor from 0–20 V in 5-V increments [46].

A time-dependent decrease in resistance under applied DC bias is observed in both bulk and thin-film titanate dielectrics [29, 47, 48]. Waser et al. [49] have suggested that resistance degradation occurs due to migration of charged oxygen vacancies. Resistance degradation is quantified by measurement of the characteristic time required for the leakage current density to increase by an order of magnitude from its steady-state value. Resistance degradation is reported as a big factor in BST thin films [47, 48, 50].

Microstructure and Point Defect Chemistry

The microstructure of the thin film influences many properties of thin films. It is extremely important to keep the microstructure of the film constant while studying the properties of a thin film. For FE thin films, it becomes even more important because of the "size effects" (huge variation in properties from bulk to thin films) whose origins and effects are yet not completely understood [22, 29]. The crystallinity of BST thin films affects its dielectric properties. The X-ray diffraction (XRD) spectrum of a sintered BST target is shown in Figure 15.19 [51]. York et al. [52] have reported that (100) orientation is more preferable and leads to smoother films. The substrate and the deposition parameters play a major role in deciding the crystallinity of thin films. Usually, polycrystalline films are obtained on silicon. Magnesium oxide (MgO) and lanthanum aluminate ($LaAlO_3$) are preferred substrates for obtaining single-crystalline and epitaxial BST films [53].

A film with good texture reduces the variability in the performance of devices fabricated in the same wafer. Finally, it is important to note that crystallinity does not contribute to the size effects in FE thin films and that films with low crystallinity display subdued dielectric properties irrespective of the deposition method used [29]. The grain size of thin films strongly depends on the deposition method, and in general, FE thin films with grain sizes below 1 μm show a steady decrease in dielectric permittivity due to existence of a low permittivity layer in the grain boundaries. The grain size decreases with an increase in strontium content, and the dielectric permittivity decreases with a decrease in grain size [24, 29]. The existence of a low permittivity layer at the grain boundaries is the main reason for the decrease in dielectric permittivity [29]. In polycrystalline BST thin films, the grain boundaries play a major role in decreasing dielectric permittivity. The ([Ba + Sr]/Ti) ratio has a strong effect on the dielectric behavior of BST thin films. The maximum value of the dielectric constant is found when the ([Ba + Sr]/Ti) ratio is equal to unity. When the amount of excess Ti in the film increases,

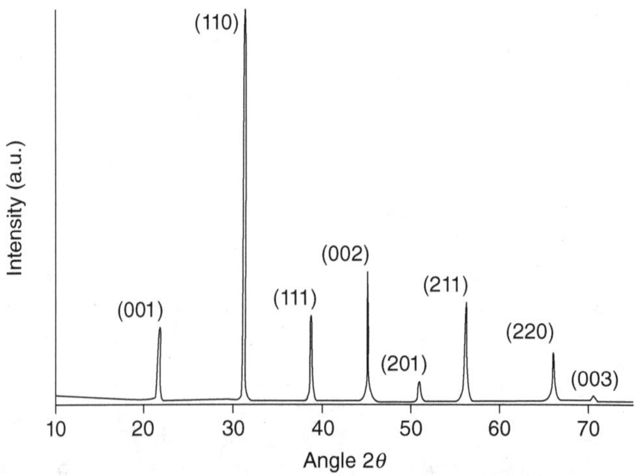

Figure 15.19. X-ray diffraction spectrum of a sintered BST target [51].

the relative permittivity monotonically decreases [54]. In films with a columnar structure, grain boundaries have a negligible effect on the overall permittivity as they lie perpendicular to the electrode. In fine-grain films, they have a marked effect (depending on the grain size) as they lie parallel to the electrode layer [29].

The behavior of BST-based capacitors under high electric fields depends on the migration of ionic defects. Hence, a greater knowledge of the concentration and behavior of mobile point defects is necessary for accurate modeling. The point defect chemistry of thin films is strongly influenced by the presence of nearby interfaces and interdiffusion of ions across the interfaces. Analysis of the nature, volume, and behavior of various ionic defects is complicated, and many direct and indirect methods have been put forward [55, 56].

BST VARACTOR TECHNOLOGY

The two most widely reported device geometries for BST varactors are the interdigital (IDT) structure and the MIM structure, which are shown in Figure 15.20a, b [57].

The IDT structures are easier to fabricate as they require only a single-step metallization, while the fabrication of MIM structures require at least a three-step process. IDT structures can be processed at high temperatures as they are fabricated directly on the substrate, while in MIM structures, the chemical stability of the bottom electrode decides the processing temperature. IDT structures suffer from reduced tunability due to large fringing electric fields in the air, while MIM structures offer higher tunabilities as the fields are better confined between the top and bottom electrodes. Typical operating voltages of MIM structures are <30 V, while IDT structure operating voltages are in the range of 100 V. Figure 15.21a shows the process flow for the fabrication of a BST (1) parallel plate and (2) IDT capacitors [58].

The MIM structures are ideal for evaluating the dielectric properties of BST thin films due to the direct relation between the measured capacitance and the relative dielectric permittivity. FE devices are fast, small, and lightweight, and, because they work using an electric field, have low power consumption. The range of tuning is quite large and devices are relatively simple in nature. Thus, they offer the following advantages:

- Faster operation speeds
- Lower power requirements
- High-yield, low-cost units
- Improved space efficiency
- Large operational bandwidth: 1 MHz to >20 GHz.

BST thin films generally attain a crystalline phase when deposited at temperatures around 450–700°C in an oxygen environment. Therefore, for MIM capacitor applications, the bottom electrode layer should have good thermal stability and oxidation resistance [1]. Platinum (Pt) is primarily chosen as the standard electrode material

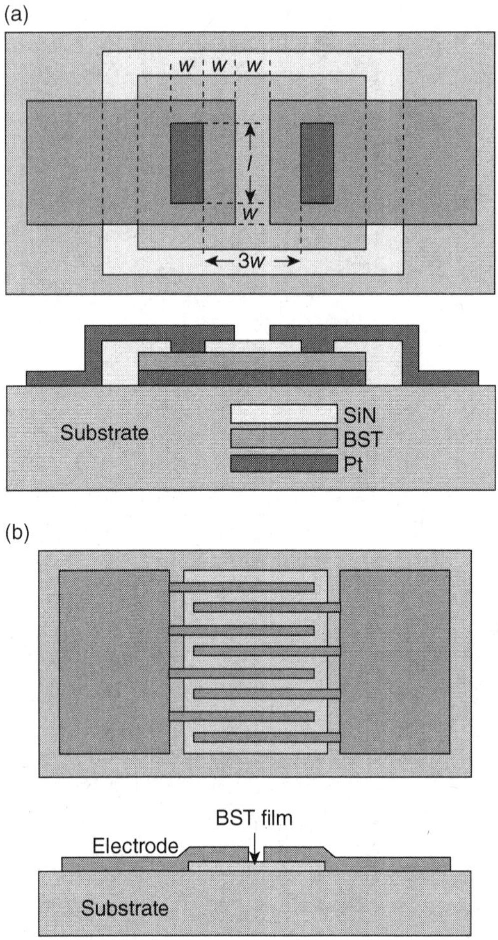

Figure 15.20. Two different configurations of BST varactors: (a) parallel plate capacitor and (b) interdigital capacitor [57].

mainly because of its chemical inertness in a high-temperature oxidizing ambient. It has been reported that better quality of platinum bottom electrode translates into better overall device performance. Typically, a silicon dioxide (SiO_2) diffusion barrier layer is grown on silicon substrate before depositing Pt to prevent diffusion of platinum through the substrate [1, 19].

The SiO_2 layer gets decomposed due to diffusion of Pt into the Si/SiO_2 interface, which leads to the formation of volatile SiO [59]. Titanium (Ti) layer is sputtered to act as an adhesion layer between platinum and SiO_2 and to prevent the decomposition of the SiO_2 layer. However, when depositions are done in an oxygen ambient at high temperatures, it was observed that there was interdiffusion between the elemental titanium and platinum. This interdiffusion phenomenon leads to defect formation in the platinum layer, which results in the degradation of device performance, depending upon

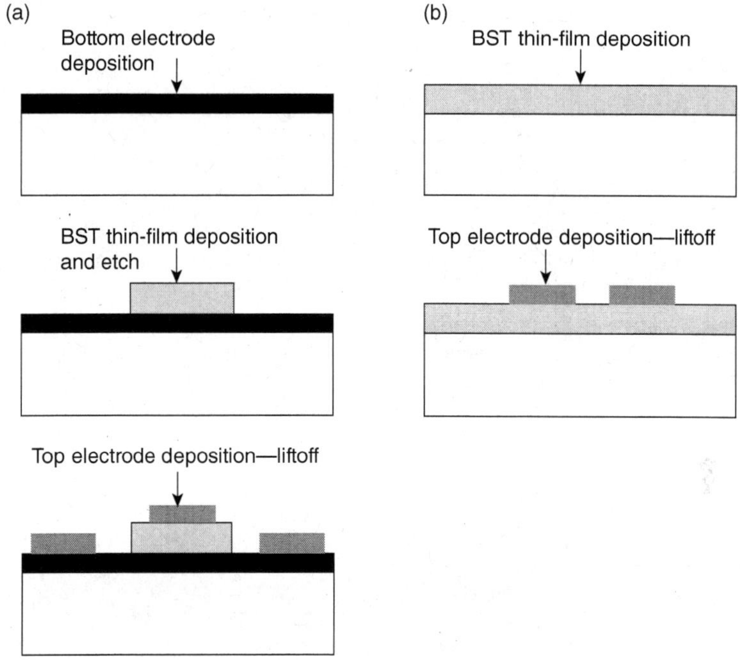

Figure 15.21. Process flow for the fabrication of a BST (a) interdigital capacitors and (b) parallel plate [58].

the thickness of the adhesion layer [60–64]. Tombak et al. have reported that above 300-MHz electrode losses play a principal role in degrading the quality factor of BST thin-film capacitors [65]. Therefore, it is very important to obtain a stable defect-free bottom electrode stack to achieve superior device performance. The integration issues of BST thin films for tunable microwave circuits will be dealt with in detail later in this chapter.

DEPOSITION TECHNOLOGIES FOR BST THIN FILMS

The deposition process used to prepare the film is one of the major parameters that decide the material properties of BST thin films. A careful study of the influence of growth parameters on material properties and correlation with electrical properties is imperative. There are four main deposition methods used for obtaining BST thin films for device applications. They are the following:

- Chemical solution deposition (CSD) or sol–gel deposition
- Pulsed laser deposition (PLD)
- RF magnetron sputtering
- Metal organic chemical vapor deposition (MOCVD).

TABLE 15.2. Comparison of the Two Capacitor Configurations

Configuration of capacitor	Parallel plate capacitors	Interdigital capacitors
Advantages	Very small fringing capacitance Maximum utilization of film tunability Lower control voltages (15–30 V) Power handling adjustable at expense of control voltage	Avoids problem of growth on bottom electrode Easy to fabricate/integrate High power-handling capability
Disadvantages	Bottom electrode hard to process Fabrication/integration process is more complicated	Large fringing capacitance reduces tunability Large finger spacing results in lower applied control/tuning field High control voltages (~100 V)

TABLE 15.3. Different Deposition Methods and Their Advantages and Disadvantages

Method	Advantages	Disadvantages	Integration issues
Chemical solution deposition (CSD)	Simple and rapid Low capital investment Film stoichiometry	Repeatability Scalability Morphology	Less compatible with standard microfabrication processes
Pulsed laser deposition (PLD)	Good film quality Film stoichiometry	Formation of particulates Scalability Low volume deposition	Manufacturing compatibility issue
RF magnetron sputtering	Uniformity Scalability High volume deposition	Film stoichiometry Defect concentration Residual stresses	Less due to highly developed manufacturing expertise
Metal organic chemical vapor deposition (MOCVD)	Flexibility Scalability High volume deposition	Lack of suitable precursors Control Complexity	Less due to highly developed manufacturing expertise

The advantages and disadvantages of the deposition methods are listed in Table 15.3 [9].

CSD

The typical procedure used to produce sol–gel solutions for thin-film deposition is provided in Figure 15.22. Specifically, equal molar amounts of barium and strontium

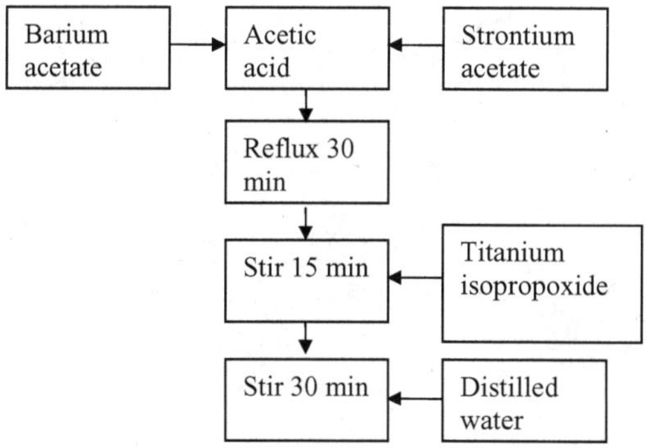

Figure 15.22. Procedures for sol–gel solution preparation and thin-film deposition [66].

acetates are combined with acetic acid, and the solution is refluxed. After refluxing, titanium isopropoxide is added so as to achieve a Ba : Sr : Ti molar ratio of 1:1:2 in the solution mixture. The solution is hydrated and stirred for 2h with filtered deionized water at a water : titanium isopropoxide molar ratio of 1000:1. This solution is immediately used for sol–gel deposition via the dip-coating method. The substrates, with patterned or continuous metal bilayers, are immersed in the sol–gel solution and withdrawn at a rate of 1–3 mm/s and subsequently annealed at 50°C for 30 min in 1 atm of wet oxygen. Multiple dip coatings (i.e., twice and thrice) are done by repeating the above-mentioned procedure. Upon completion of dip coating, the samples are slowly heated to 350°C in air to burn volatile organic components in the sol–gel coating. Densification and crystallization of the thin films is carried out at 750°C also in air [66].

PLD

PLD is a thin-film physical vapor deposition technique where a high power nanosecond-range pulsed laser beam is focused inside a vacuum chamber to strike a rotating target of the material that is to be deposited. The laser pulse with a high energy of typically 2–5 J/cm^2 leads to an evaporation of the target surface and the immediate formation of plasma, containing energetic neutral atoms, ions, and molecules. This so-called plume reaches the substrate surface with an energy of 0.1 to >10.0 eV, depending on the pressure of the background gas environment. This process can occur in ultrahigh vacuum or in the presence of a background gas, such as oxygen, which is commonly used when depositing oxides to fully oxygenate the deposited films [66]. A typical PLD system is depicted in Figure 15.23 [67], along with the image of a typical system. As a result of the short, high energetic laser pulses, the evaporated material is not in the thermodynamic equilibrium, and the relative amount of different compounds in the plume corresponds to the target composition even for strongly differing melting points. Therefore, in contrast to most other deposition techniques, the desired thin-film stoichiometry is easily achieved by PLD [68].

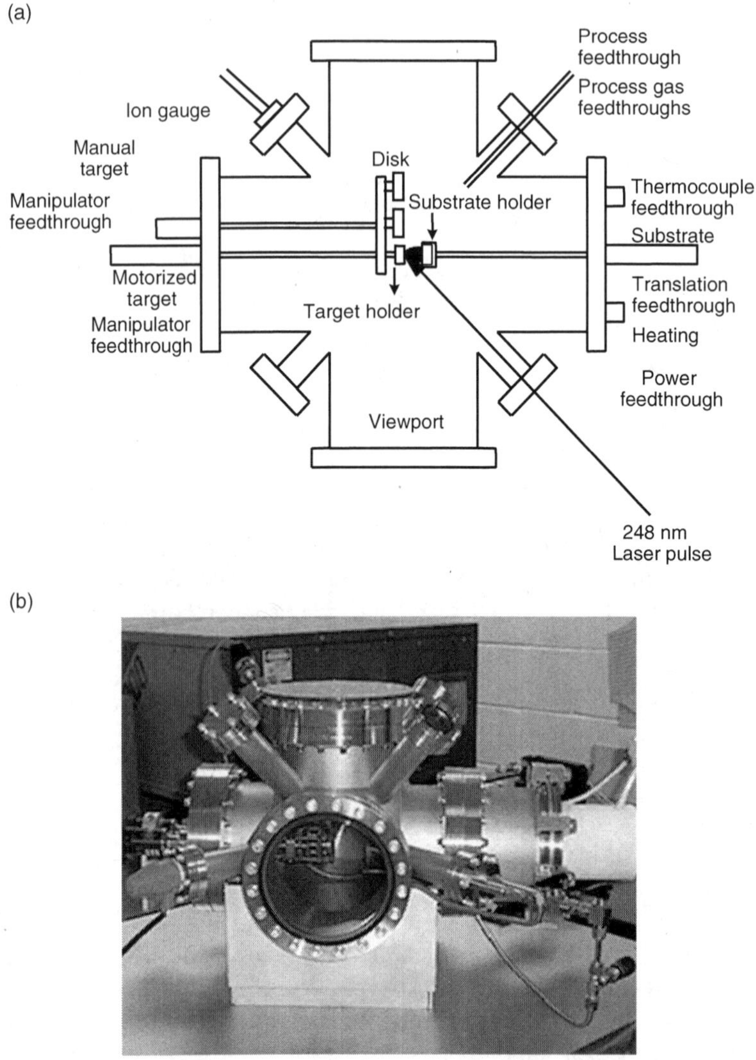

Figure 15.23. (a) Typical PLD system thin-film deposition [67] and (b) image of PLD system available in the Nanomaterials Research Lab at the University of South Florida, Tampa, FL.

RF Magnetron Sputtering

Sputtering is a physical process whereby atoms and ions in a solid target material are ejected into the gas phase due to bombardment of the material by energetic ions. The sputter gas is inert, and in most sputtering systems, argon is used as the sputtering gas. Sputtering is done at pressures in the order of 10^{-6} Torr or more inside specially designed chambers. Inside the chamber, the target material to be sputtered is bonded onto one circular electrode, which acts as cathode and is placed in front of the substrate to be coated. Gas is introduced and maintained at a suitable pressure in the chamber, which is ionized by a suitable voltage applied between the two electrodes.

As the target is subjected to a negative electrical potential compared to the plasma body, the positively charged ions in the plasma will be accelerated toward the object. At the target surface, a number of processes and combinations of them can occur. As shown in Figure 15.24 [69], in Figure 15.24a, the arriving ion is adsorbed onto the surface and a chemical or physical bonding is created. The incoming species may also transfer some of its momentum to one of the surface atoms (Fig. 15.24b) and cause it to move around on the surface (displacement). If the energy of the incoming ion is sufficiently high, it may start to remove some of the surface atoms (sputtering, shown in Fig. 15.24c), and if the energy is increased even further, the incoming ion can be implanted (Fig. 15.24d). In addition, there can also be a number of combinations of these processes. Sputtering is initiated when the energy of the incoming atom or ion is of the order of 100 eV, while implantation occurs at ~1 keV.

Dielectric or insulating materials like BST are sputtered by using RF power instead of DC biasing to prevent charge accumulation at the surface of the target and results in electrical discharging or arcing, wherein macromolecules of target species are transported to the substrate and form poor-quality films. Ionization of the gas in RF sputtering is caused by the electrons, which are more mobile than ions and hence respond and oscillate with the applied frequency. The deposition rate, however, is reduced in RF sputtering due to reduced work done per cycle. BST depositions are generally done at temperatures in the range of 450–700°C to obtain crystalline films. To account for the loss of oxygen atoms at high temperatures, oxygen gas is introduced in the chamber along with argon. The argon/oxygen ratio is one of the important deposition parameters that decide the quality of the deposited films. The other important parameters are RF power, substrate temperature, deposition pressure, and distance between target and substrate. Since high-temperature depositions in oxidizing conditions adversely affect the stability of the interfaces between the substrate, bottom electrode, and BST, it is important to keep the temperature as low as possible. It has been found that the losses in the deposited film reduce with increased growth rates [1]. But too much increase in RF power could lead to the formation of defective crystals in the film during growth due to heavy bombardment [29].

MOCVD

MOCVD is a chemical vapor deposition process that uses metal organic precursors to produce high-purity, high-performance thin films. Atoms that you desire to be in your film are combined with complex organic gas molecules and passed over a hot substrate. The heat breaks up the molecules and deposits the desired atoms on the surface, layer by layer. The properties of the film can be tailored by varying the composition of the gas. It can grow high-quality semiconductor layers (as thin as a millionth of a millimeter), and the crystal structure of these layers is perfectly aligned with that of the substrate [70]. Figure 15.25 shows a typical MOCVD setup.

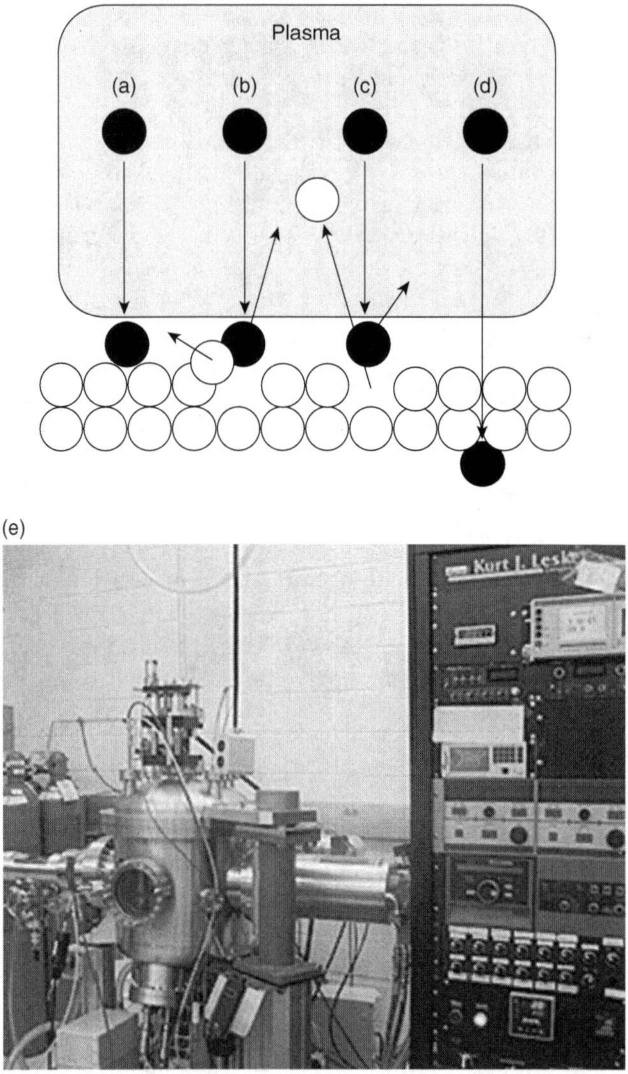

Figure 15.24. Processes initiated by bombardment of ions onto the sputtering target: (a) adsorption, (b) displacement (momentum transfer), (c) removal of surface atom (sputtering or etching), (d) ion implantation [69], and (e) RF magnetron sputtering tool (model-CMS 18, Kurt J. Lesker, Pittsburgh, PA), available in the Nanomaterials Research Lab at University of South Florida, Tampa, FL.

PROPERTIES INFLUENCING INTEGRATION ISSUES IN BST THIN FILMS

Commercial electronic circuits are stringent with respect to voltage requirements, and BST-based parallel plate structures require low-bias voltages to provide the desired

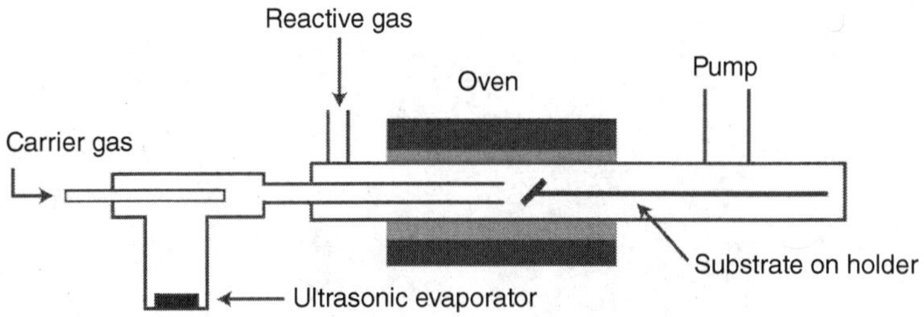

Figure 15.25. Typical MOCVD setup.

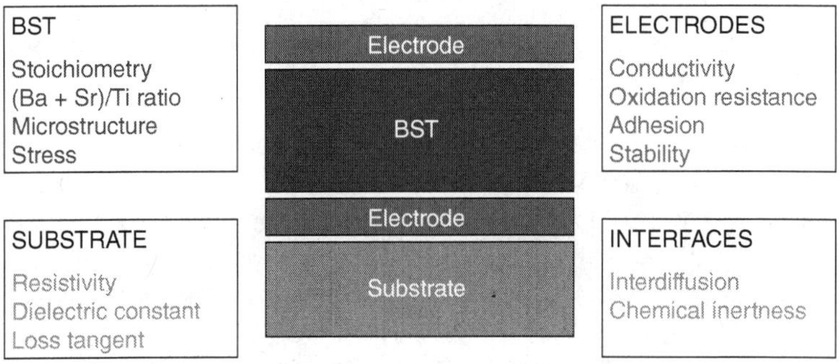

Figure 15.26. Material issues to be considered in a BST-based MIM structure [1].

tenability, which makes them more flexible and compatible for many circuit applications. The important material issues to be considered in BST-based MIM structures are summarized in Figure 15.26 [1].

Since BST deposition is done at relatively high temperatures ranging from 450 to 700°C in the presence of oxygen gas, it becomes important to use a very stable and oxidation-resistant metal as the bottom electrode. Hence, noble earth metals like platinum (Pt), ruthenium (Ru), and iridium (Ir) or their conducting oxides are preferred for use as bottom electrodes. Typically, a silicon dioxide (SiO_2) diffusion barrier layer is grown on silicon substrate before depositing Pt to prevent diffusion of platinum through the substrate during BST deposition. Additionally, an adhesion layer is required for platinum and development of a stable adhesion layer is a major challenge as interdiffusion between layers is a major issue with BST-based MIM structures, as BST deposition happens at high temperatures in oxidizing conditions. A very thin titanium (Ti) or titanium dioxide (TiO_2) adhesion layer is usually deposited before depositing platinum. Figure 15.27 shows the schematic of stacked BST metal insulator metal structures.

Parameters that have been identified to affect the dielectric properties of BST film capacitors are processing methods, film composition, crystalline structure, microstructure, surface morphology, film thickness, electrode materials, and interdiffusion. The main goal is to optimize all the parameters to achieve high tunability and low dielectric loss with the optimized circuit design.

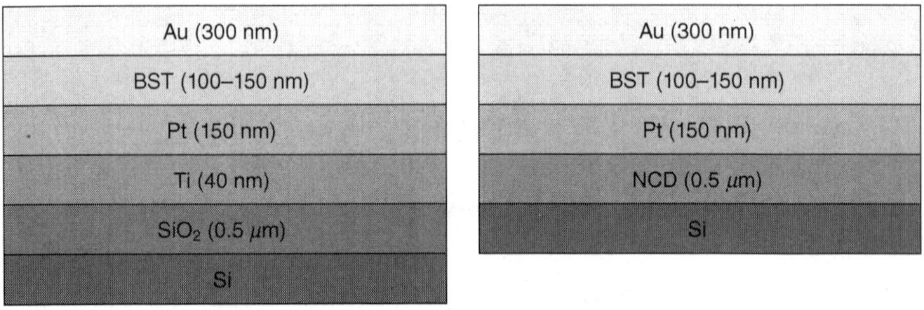

Figure 15.27. Schematic of stacked BST metal insulator metal structures.

1. *Processing Methods.* A variety of deposition techniques, including RF magnetron sputtering, MOCVD, PLD, and sol–gel, have been used to synthesize BST thin films [52, 58, 71, 72]. The various techniques used to fabricate BST thin films are designed to produce specific microstructure and dielectric properties. It is essential to obtain the lowest possible process temperature to comply with other technologies and to minimize postdeposition thermal treatments under low oxygen partial pressure to maintain the resistance of the films. While maintaining precise microscopic control of the stoichiometry, it is also essential to achieve good step coverage and large-area deposition. For example, Ezhilvalavan and Tseng [73] have reported that deposited BST dielectric requires high-temperature annealing around 750°C in an O_2 ambient in order to achieve crystalline structure. During high-temperature annealing in O_2 ambient, a considerable amount of oxygen penetrates storage Pt electrode, resulting in oxidizing polysilicon at the interface between plugged polysilicon and BST film. Therefore, the barrier layer to block the oxygen penetration is needed between Pt electrode and plugged polysilicon. The barrier layer should have resistance to oxygen penetration at high annealing temperature.
2. *Film Composition.* The dielectric constant can be varied by changing the Ba/Sr ratio [74–77]. It is shown that dielectric constant increases with a decrease in Sr content. It has been conferred that maximum dielectric constant is obtained at room temperature when the (Ba + Sr)/Ti ratio is 1:1. At extreme titanium excess, an amorphous phase is found between grains. Longer reaction time resulted in films with more strontium ratio. Strontium is more readily incorporated into the BST films, relative to barium [78]. The preferred incorporation of strontium into BST films under hydrothermal conditions can be explained by the larger thermodynamic driving forces of formation of $SrTiO_3$ than that of $BaTiO_3$ [79]. In another report by Zonghui Zhou et al., the results showed that any sample with a different Ba/Sr ratio contained two phases—the perovskite phase and the tungsten bronze phase. The dielectric constants of BST basically decreased, while the tanδ increased with the increase in Sr ratio. Both dielectric

constant and tanδ decreased with the applied frequency increased for the same Sr value [80].

3. *Crystalline Structure.* Crystalline BST films are usually obtained at relatively high substrate temperatures. During BST film deposition at high temperatures, interlayers and specific grain structures are formed that cause a decrease in dielectric constant and high leakage current. An alternative approach is to grow amorphous BST films at low temperatures and then to crystallize them in a postannealing process. The dielectric constant of polycrystalline film is found to be lower than the epitaxial film [81]. Kim et al. investigated the crystallinity dependence of the microwave dielectric losses in (Ba, Sr)TiO_3 thin films [82]. The thin-film crystallinity was varied by controlling the sputter-deposition temperatures on a Pt/Si substrate. The microwave dielectric losses were seen to increase from 0.0024 ± 0.0018 to 0.0102 ± 0.0017 with an increase in crystallinity. Raman spectra confirmed the relation between the deteriorated dielectric losses and symmetry-breaking defects inducing microscopic polar regions above the Curie temperature of the bulk $(Ba_{0.43}Sr_{0.57})TiO_3$.
4. *Microstructure.* The grain size profoundly affects the dielectric properties of BST thin films. The grain size decreases and grain boundary increases when the thickness of the film reduces. Due to the increased grain boundaries in thinner films, it has a higher leakage current density and a lower dielectric constant compared to thicker films. It has been reported that increased dielectric constant and increased temperature variation is observed with increased grain size [83]. By increasing the sintering temperature of the samples, there was a decrease in the height of the peak of dissipation in the FE region. This corresponded to a reduced loss as sintered density increased. In general, the peak loss value is less than 0.015 when the sintered density is more than 90% [84].
5. *Surface Morphology.* Increasing the $O_2/(O_2 + Ar)$ ratio during the film deposition process using RF sputtering increases the surface roughness of BST thin films [76]. Figure 15.28 shows the two-dimensional surface plots of the BST thin films deposited at 90/10 and 70/30 Ar/O_2 ratio. The BST films deposited at 90/10 Ar/O_2 have a surface roughness of 3.4 nm, while the BST films deposited at 70/30 Ar/O_2 ratio have a surface roughness of 7 nm. The surface roughness increases with the oxygen content as the cation mobility on the growth surface reduces with higher oxygen pressures. Also, sputtered BST thin films have a much lower surface roughness than pulsed laser deposited BST thin films [58]. The leakage current characteristics of BST thin films are strongly related to the surface roughness of BST thin film [85]. The surface roughness of BST thin film is dependent on bottom electrode materials.
6. *Film Thickness.* BST thin films behave differently than bulk BST [86]. The dielectric constant of thin-film BST is much smaller than bulk BST. It is shown that the dielectric constant decreases with a decrease in thickness of BST thin film [87]. The thickness dependence of the dielectric constant varies with the substrate temperature in connection with the grain size effect. The thickness

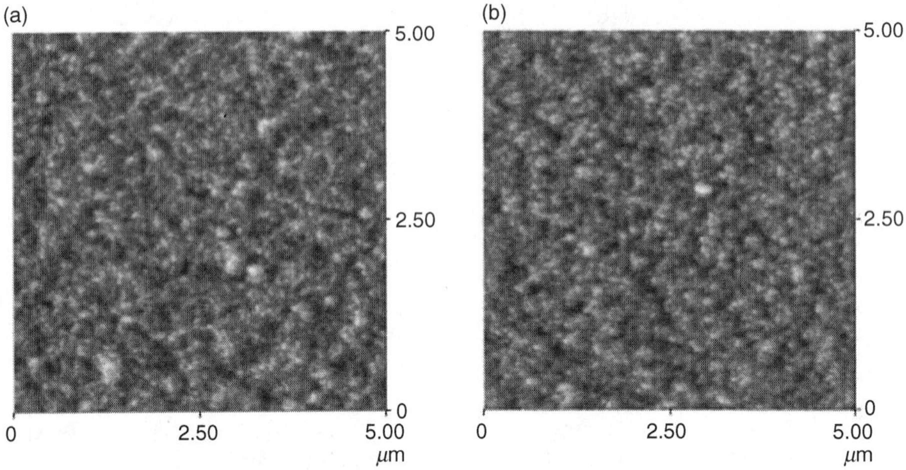

Figure 15.28. Surface view of BST films sputtered at Ar/O_2 ratios (a) 90/10 and (b) 70/30 [58].

dependence of permittivity is explained in the Schottky barrier model. In this model, variation in apparent capacitance with bias is explained via a voltage-dependent interfacial layer and capacitance in series with the capacitance of the bulk of the film, whose permittivity is taken to be bias dependent.

7. *Electrode Materials.* In the application of BST films as a dielectric, electrode materials must meet certain requirements, such as high conductivity, sufficient resistance against oxidation, good adhesion to BST, and interfacial smoothness to reduce leakage current and to increase capacitance. In addition, factors such as grain size distribution, crystalline orientation, interface, and surface structures significantly influence the performance of electrode materials. Metal electrodes, such as Pt, Ir, and Ru, and conducting oxide electrodes, like RuO_2, IrO_2, and $SrRuO_3$, have been investigated for BST varactors [88–91]. Among these, platinum is promising due to its good oxidation resistant at higher temperatures and high conductivity. Platinum also has some drawbacks like hillock formation at higher temperatures, poor adhesion, and difficulty with patterning.
8. *Interdiffusion Effects in BST Device Structures.* Since BST depositions are usually done at high temperatures in the presence of oxygen, both the structural defects in the different layers and the temperature gradients that exist at different locations contribute to interdiffusions between different layers, which lead to the formation of defects in the $Pt/Ti/SiO_2$ layers, which are extremely detrimental to the performance of the devices. The Si/SiO_2 layer undergoes decomposition at high temperatures in the presence of oxygen due to contamination from metals like Ti and Pt. The surface defects in the Si/SiO_2 act as nucleation sites for void formation when the deposited metals getter at these defects, leading to decomposition of the oxide layer. Metals can form precipi-

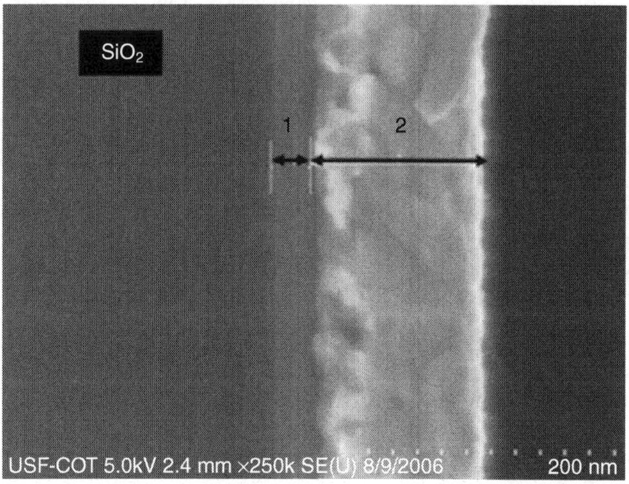

Figure 15.29. Silicon dioxide, titanium, and platinum layers on Si (100) wafer [87].

tates, generate dislocations, or enhance the roughness of the Si/SiO_2 interface [59, 92].

The Ti layer is deposited to act as an adhesion layer and also to prevent the formation of platinum silicide at high temperatures [93]. A cross-sectional image of SiO_2, Ti, and Pt layers on a Si wafer is shown in Figure 15.29. The layer marked "1" is the Ti layer, and the layer marked "2" is the Pt layer. Both the layers were sputtered at room temperature and hence show no defects due to interdiffusion. At high temperatures in an oxygen ambient, Ti diffuses and oxidizes in the Pt layer, which leads to hillock formation in the platinum layer, which degrades the performance of FE devices and increases the probability of device shorting. The formation of hillocks in Pt–Ti electrodes has been widely reported [60–64]. Hillocks are believed to result from the volume expansion in the Pt layer due to the formation of TiO_2 in the Pt layer. This volume expansion gives rise to the compressive stress of the Pt layer, which is believed to be the cause for the formation of the hillocks. It is also important to note that the shorting of capacitors is more due to the size of the hillocks than the distribution of hillocks. The size and distribution of the hillocks depend on the thicknesses of Ti and Pt layers and the amount of oxygen used during the deposition of BST [26, 62]. The use of TiO_2 as the adhesion layer instead of Ti also poses adhesion issues, which leads to process-induced delamination of the Pt layer [60, 94]. Thus, SiO_2/Ti/Pt/BST- and SiO_2/TiO_2/Pt/BST-based MIM structures suffer from severe interdiffusion and delamination problems, which affect the reliability and yield of the resulting devices. Tombak et al. [65] have reported that above 300-MHz electrode losses play a principal role in degrading the quality factor of BST thin-film capacitors. Therefore, it is very important to obtain a stable defect-free platinum bottom electrode layer to achieve superior device performance. Platinum deposited on a sapphire substrate at high temperatures has been employed for BST-based varactor fabrication [1].

Interdiffusion Solution: Nanocrystalline Diamond (NCD)/Pt/BST Structure

Chemical vapor deposited (CVD) diamond, also known as NCD, would make an excellent material for high-frequency applications, and prior research work has shown that diamond films deposited on high-resistive silicon substrates displayed very low RF losses comparable with the performance of thermally grown SiO_2 on high-resistive silicon substrate from DC to 65 GHz [95]. Very stable electrodes can be obtained by depositing metals on diamond films as they are chemically inert and highly corrosion resistant [91]. Diamond also has the highest thermal conductivity among all materials, which means that it can act as an effective heat sink that can reduce the detrimental effects that the electromigration phenomenon has on the metal layers [96]. Thus, it exhibits excellent properties for application as a stable diffusion barrier layer for FE thin film-based MIM devices. Figure 15.30 shows the scanning electron micrograph (SEM) of the interface of a NCD/Pt/BST structure. NCD acts as a barrier layer between Pt and Si to counter the interdiffusion of Pt into Si [87].

As seen from the image, a clear and stable interface exists between all the layers. One disadvantage with using diamond films is that the surface roughness of the diamond films is relatively higher than SiO_2. But for initial performance analysis, devices with large active areas were used to circumvent the roughness effects. Figure 15.31 [98] shows the variation of the real part of impedance (resistance) with frequency for the Pt film on NCD and SiO_2 diffusion barrier layers, respectively. As seen in Figure 15.31, it was found that the platinum film on NCD displayed less resistive losses than the one

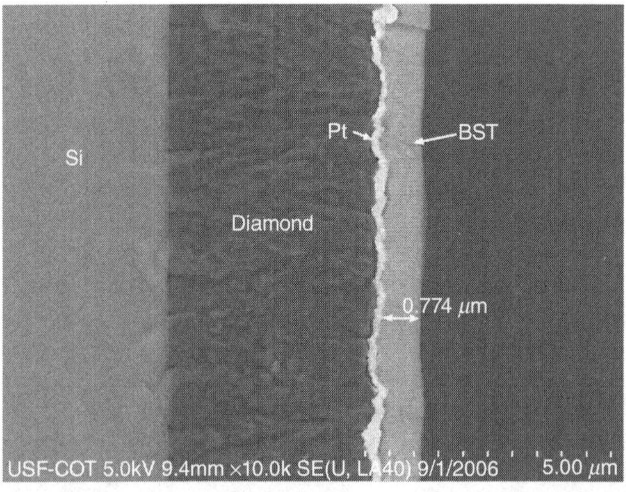

Figure 15.30. Nanocrystalline diamond, platinum, and BST layers on Si (100) wafer [87].

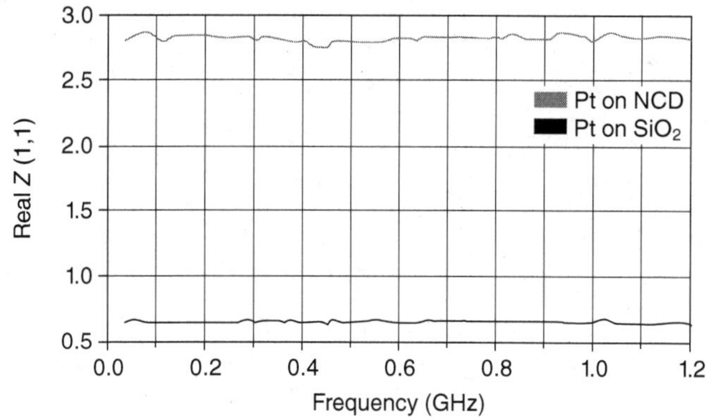

Figure 15.31. Variation of the real part of impedance (resistance) with frequency for the Pt film on NCD and SiO_2 diffusion barrier layers [98].

on SiO_2, which indicates lesser mass transport along the platinum layer and a reduced contribution to device losses [97].

Mass Transportation Effects in BST Device Structures. Mass transport phenomena have a distinct influence on thin-film formation, properties, and performance. It leads to instabilities in the functioning of components and devices due to effects like decrease in conductivity, short or open circuiting of conductors, lack of adhesion, and generation of stress. Such problems frequently surface when two adjacent layers of materials are chemically reactive. The mass transport phenomena are aided largely by lattice sites that are unoccupied, interstitial, or on the surfaces, which result from marginal properties of materials. In polycrystalline thin films, a large proportion of atomic defect combinations are associated with nonlattice sites like grain boundaries, dislocations, surfaces, and interfaces relative to lattice sites. This is because the loosely bound atoms at these nonlattice sites attract different point defect populations and are more mobile than lattice atoms. A more important reason for high diffusivity through nonlattice sites is their low activation energies. Therefore, they act as diffusion paths that short-circuit the lattice. Grain boundary diffusion is believed to account for a major amount of mass transportation. In thin films with small grains, the high density of boundaries allows rapid and widespread diffusion, leading to serious reliability problems.

Another important mass transport phenomenon to be considered is electromigration, which is the migration of metal atoms along metals carrying large current densities. In thin films, the heat generated due to high current densities is conducted away to an extent by the substrate, which acts as a massive heat sink. In thin films, electromigration manifests itself by void and hillock formation, cracked dielectric film overlayers, and grain boundary grooving. This leads to film damage either due to accumulation or depletion of atoms. Film damage is caused by either accumulation or depletion of atoms, which depends on whether the major mass transport is structurally induced or induced by a temperature gradient. If more atoms enter such a junction of grains than leave it, a mass pileup or formation of hillocks is expected. Voids are formed when

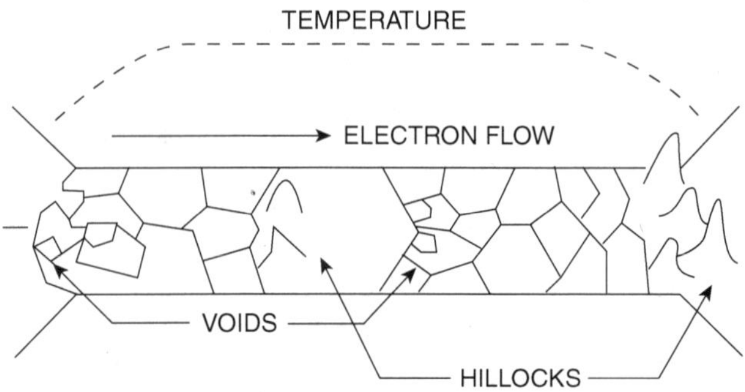

Figure 15.32. Model showing electromigration-related mass transport arising due to a non-uniform grain structure and temperature gradients [97].

mass depletion occurs. To prevent these defects, a uniform grain distribution is desired. In device structures, electromigration frequently occurs in the presence of a temperature gradient that develops at locations of poor film adhesion, in regions of different thermal conductivities, such as metal–dielectric contacts. Figure 15.32 shows how hillocks and voids occur in thin films due to mass transport, which in turn arises due to the grain structure and temperature variations [97].

Varactor Breakdown and Tunability. Dielectric breakdown occurs due to either one of four mechanisms, namely, thermal (due to ionic conduction with increase in temperature), avalanche (at high electric fields, accelerated free electrons strike atoms of the material and free more electrons, which in turn create more free electrons, and the process continues until the material breaks down), discharge (large electric field arc across pores of the material leading to erosion and more arcing until the material breaks down), and electrolytic (conduction paths created over time due to ionic and/or environmental conditions). The breakdown mechanism in ferroelectrics is avalanche breakdown [99]. The breakdown strength of a dielectric is not related to its dielectric properties [100]. The breakdown fields are the same above and below the Curie temperature, in the FE and PE phases, and have nothing to do with domains. The breakdown field depends mainly upon the thickness of the dielectric layer, the type of electrodes used, and temperature [101]. For the BST varactors fabricated on SiO_2, thicker films give higher tunabilities (at the expense of increased-bias voltages) and display higher breakdown fields as indicated by Acikel [1]. It has been suggested that both tunability and breakdown fields of BST varactors are thickness dependant and that geometry effects can be ignored for large-area MIM varactors as tunability of the varactor is directly proportional to BST's tenability [46].

SUMMARY

BST has been one of the most researched electronic materials in the recent past due to its pertinent application areas in DRAMs and RF and microwave circuits. A thorough review of its material properties has been presented in this chapter to address issues related to tunability and integration. The essential factors contributing to integration issues are interdiffusion in the substrate layers and formation of voids and hillocks. Platinum is known to be the most preferred electrode of choice for BST in R&D as it is nonreactive to BST and forms an interface, which has favorable electrical properties. However, Pt has several integration issues such as poor adhesion, difficulty in dry-etching, and diffusion of BST–Pt layers. To circumvent this integration issue, a thin-film layer of NCD can be used as a diffusion barrier layer. Thus, to conclude, efforts are in progress to achieve high-quality, tunable BST thin-film structures in RF and microwave circuits.

ACKNOWLEDGMENTS

Part of the research work was supported from NSF ECCS grant #0901779. We would also like to thank Dr. Tom Weller and Dr. Andrew Hoff for their research inputs.

REFERENCES

1. Acikel B. 2002. *High Performance Barium Strontium Titanate Varactor Technology for Low Cost Circuit Applications*. Santa Barbara: University of California.
2. Lancaster MJ, Powell J, Porch A. 1998. Thin-film ferroelectric microwave devices. Supercond Sci Technol 11:1323.
3. Shi S, Purden J, Jin J, York RA. 2005. A 24 GHz wafer scale electronically scanned antenna using BST phase shifters for collision avoidance systems. IEEE Antennas Propagation Soc Int Symp 1B:84.
4. Jaffe B, Cook WR, Jr., Jaffe H. 1971. *Piezoelectric Ceramics*. New York: Academic Press.
5. Gallego-Juarez JA. 1989. Piezoelectric ceramics and ultrasonic transducers. J Phys E 22:804.
6. Jordan TL, Ounaies Z. 2001. *Piezoelectric Ceramics Characterization*. Hampton, VA: NASA Langley Research Center.
7. Arnau A. 2008. *Piezoelectric Transducers and Applications*, 2nd edition. New York: Springer.
8. Safari A, Panda RK, Janas VF. 2000. *Ferroelectric Ceramics: Processing, Properties and Applications*. Piscataway, NJ, USA: Rutgers University.
9. Ghosh D. 2005. Tunable microwave devices using BST and base metal electrodes. PhD thesis. North Carolina State University.
10. Mason WP. 1948. Electrostrictive effect in barium titanate ceramics. Phys Rev 74:1134.
11. Stemmer S. 2005. *Tutorial overview of tunable dielectrics*. IEEE MTT-S International Microwave Symposium, Long Beach, CA.
12. Jona F, Shirane G. 1993. *Ferroelectric Crystals*. New York: Dover.
13. Bottger U. 2005. Dielectric properties of polar oxides. In: Waser R, Bottger U, Tiedke S, editors. *Polar Oxides: Properties, Characterization, and Imaging*. Hoboken, NJ: Wiley.

14. Suherman PM, Jackson TJ, Lancaster MJ. 2007. Comparison of techniques for microwave characterization of BST thin films. IEEE Trans Microwave Theory Tech 55(2):397.
15. Tsymbal EY, Pettifor DG. 1999. The influence of impurities within the barrier on tunneling magnetoresistance. J Appl Phys 85:5801.
16. Morito K, Iwazaki Y, Suzuki T, Fujimoto M. 2003. Electric field induced piezoelectric resonance in the micrometer to millimeter waveband in a thin film $SrTiO_3$ capacitor. J Appl Phys 94:5199.
17. Muller KA, Burkard H. 1979. $SrTiO_3$ an iintrinsic quantum paraelectric below 4K. Phys Rev B 19:3593.
18. Smolenskii GA, Isupov VA. 1954. Segnetoelektricheskie svoistva tverdykh rastvorov stannata bariya V titanate bariya. Zhurnal Tekhnicheskoi Fiziki 24:1375.
19. Bubb DM, Horwitz JS, Qadri SB, Kirchoefer SW, Hubert C, Levy J. 2004. $(Ba,Sr)TiO_3$ thin films grown by pulsed laser deposition with low dielectric loss at microwave frequencies. Appl Phys A 79:99.
20. Shaw TM, Suo Z, Huang M, Liniger E, Laibowitz RB, Baniecki JD. 1999. The effect of stress on the dielectric properties of barium strontium titanate thin films. Appl Phys Lett 75:2129.
21. Parker CB, Maria JP, Kingon AI. 2002. Temperature and thickness dependent permittivity of $(Ba, Sr)TiO_3$ thin films. Appl Phys Lett 81:340.
22. Lookman A, Bowman RM, Gregg JM, Scott JF, Dawber M, Reudiger A. 2004. Understanding thickness effects of thin film capacitors. Integr Ferroelectr 61:51.
23. Ban ZG, Alpay SP. 2002. Phase diagrams and dielectric response of epitaxial barium strontium titanate films: a theoretical analysis. J Appl Phys 91:9288.
24. Adikary SU, Chan HLW. 2002. Ferroelectric and dielectric properties of sol–gel derived $BaxSr1\text{-}xTiO_3$ thin films. Thin Solid Films 424:70.
25. Goodge AC. 1997. *Lens Focused Microwave Reflectometry Concepts for Ceramic Coating Characterization*. Charlottesville: University of Virginia.
26. Botsco R, McMaster R. 1986. *Electromagnetic testing*. Vol. 4: The American Society for Nondestructive Testing.
27. Kretschmer R, Binder K. 1973. Surface effects on phase transitions in ferroelectrics and dipolar magnets. Phys Rev B 30:1065.
28. Zhou C, Newns DM. 1997. Intrinsic dead layer effect and the performance of ferroelectric thin film capacitors. J Appl Phys 82:3081.
29. Shaw TM, Trolier-McKinstry S, McIntyre PC. 2000. The properties of ferroelectric films at small dimensions. Annu Rev Mater Sci 30:263.
30. Tuttle BA, Garino TJ, Voight JA, Headley TJ, Dimos D, Eatough MO. 1995. *Science and Technology of Electroceramic Thin Films*, Auciello RWO, editor. New York: Springer, p. 117.
31. Kholkin AL, Calzada ML, Ramos P, Mendiola J, Setter N. 1996. Piezoelectric properties of Ca modified $PbTiO_3$ thin films. Appl Phys Lett 69:3602.
32. Trolier-McKinstry S, Shepard JF, Jr., Lacey JL, Su T, Zavala G, Fendler J. 1998. Piezoelectricity in ferroelectric thin films: domains and stress issues. Ferroelectrics 206:381.
33. Schumacher M, Deitz GW, Waser R. 1995. Dielectric relaxation of perovskite—type oxide thin films. Integr Ferroelectr 10:231.

34. Streiffer SK, Basceri C, Kingon AI, Lipa S, Bilodeau S, Carl R, Van Buskirk PC. Dielectric behavior of CVD (Ba,Sr)TiO_3 thin films on Pt/Si. *Presented at Materials Research Society Symposium*, 1996.

35. Baniecki JD, Laibowitz RB, Shaw TM, Duncombe PR, Neumayer DA, Kotecki DE, Shen H, Ma QY. 1998. Dielectric relaxation of $Ba_{0.7}Sr_{0.3}TiO_3$ thin films from 1 mHz to 20 GHz. Appl Phys Lett 72:498.

36. Fukuda Y, Numata K, Aoki K, Nishimura A. 1996. Origin of dielectric relaxation observed for Ba0.5Sr0.5TiO3 thin-film capacitor. Jpn J Appl Phys 35:5178.

37. Fukuda Y, Haneda Y, Sakaguchi H, Numata K, Aoki K, Nishimura A. 1997. Dielectric properties of (Ba, Sr)TiO_3 thin films and their correlation with oxygen vacancy density. Jpn J Appl Phys 36:L1514.

38. Schumacher M, Waser R. 1998. Curie—Von Schweidler behaviour observed in ferroelectric thin films and comparison to superparaelectric thin film materials. Integr Ferroelectr 22:109.

39. Baniecki JD, Laibowitz RB, Shaw TM, Duncombe PR, Neumayer DA. 1999. Low temperature hydrogen induced degradation of (Ba, Sr) TiO_3 thin film capacitors. *MRS Proceedings*.

40. Hudak O. 2005. Landau theory of 180 domain walls in $BaTiO_3$ type ferroelectric particles: microcomposite materials. Condens Matter 2:1.

41. Moulson AJ, Herbert JM. 1997. *Electroceramics: Materials, Properties, Applications*. New York: Chapman and Hall.

42. Damajanovic D. 1998. Ferroelectric, dielectric, and piezoelectric properties of ferroelectric thin films and ceramics. Rep Prog Phys 61:1267.

43. Waser R. 2003. *Nanoelectronics and Information Technology*. Germany: Wiley-VCH.

44. Losego MD. 2005. *The chemical solution deposition of lead zirconium titanate (PZT) thin films directly on copper surfaces*. MS: North Carolina State University.

45. Im J, Auciello O, Baumann PK, Streiffer K, Kaufman DY, Krauss AR. 2000. Composition-control of magnetron-sputter-deposited $(Ba_xSr_{1-x})Ti_{1+y}O_{3+z}$ thin films for voltage tunable devices. Appl Phys Lett 76:625.

46. York R. 2009. Tunable dielectrics for RF circuits. In: Michael S, Palmer WD, editors. *Multifunctional Adaptive Microwave Circuits and Systems*. Raleigh, NC: SciTech Publishing.

47. Basceri C, Lash SE, Parker CB, Streiffer SK, Kingon AI, Grossmann M, Hoffmann S, Schumacher L, Waser R, Bilodeau S, Carl R, van Buskirk PC, Summerfelt SR. 1998. An important failure mechanism in MOCVD (Ba,Sr)TiO3 thin films: resistance degradation. *Presented at Materials Research Society*.

48. Numata K, Fukuda Y, Aoki K, Nishimura A. 1995. Analysis of the Resistance Degradation of $SrTiO_3$ and BaxSr(1 − x)TiO3 Thin Films. Jpn J Appl Phys 34:5245.

49. Waser R, Baiatu T, Hardtl KH. 1990. DC electrical degradation of perovskite-type titanates. J Am Ceram Soc 73:1645.

50. Grossmann M, Hoffmann S, Gusowski S, Waser R, Streiffer SK, Basceri C, Parker CB, Lash SE, Kingon AI. 1998. Resistance degradation behavior of $Ba_{0.7}Sr_{0.3}TiO_3$ Thin films compared to mechanisms found in titanate ceramics and single crystals. Integr Ferroelectr 22:83.

51. Delage T, Champeaux C, Catherinot A, Seaux JF, Madrangeas V, Cros D. 2004. High-K BST films deposited on MgO by PLD with and without buffer-layer. Thin Solid Films 453:279.
52. Padmini P, Taylor TR, Lefervre MR, Nagra AS, York RA, Speck JS. 1999. Realization of high tunability barium strontium titanate thin films by RF magnetron sputtering. Appl Phys Lett 75:3186.
53. Chang W, Horwitz JS, Carter AC, Pond JM, Kirchoefer SW, Gilmore CM, Chrisey DB. 1999. The effect of annealing on the microwave properties of $Ba_{0.5}Sr_{0.5}TiO_3$ thin films. Appl Phys Lett 74:1033.
54. Stemmer S, Streiffer SK, Browning ND, Basceri C, Kingon AI. 2000. Grain boundaries in barium strontium titanate thin films: structure, chemistry and influence on electronic properties. Interface Sci 8:209.
55. Stemmer S, Streiffer SK, Browning ND, Kingon AI. 1999. Accommodation of nonstoichiometry in (100) fiber-textured $(Ba_xSr_{1-x})Ti_{1+y}O_{3+z}$ thin films grown by chemical vapor deposition. Appl Phys Lett 74:2432.
56. Pompe XGW, Suo Z, Speck JS. 1993. Elastic energy release due to domain formation in the strained epitaxy of ferroelectric and ferroelastic films. J Appl Phys 74:6012.
57. York B, Nagra A, Erker E, Taylor T, Periaswamy P, Speck J, Auciello O, Streiffer S. 2003. *Microwave integrated circuits using thin-film BST*. Progress report prepared by DARPA and Argonne National Labs.
58. Manavalan SG. 2004. *Structural and Electrical Properties of Barium Strontium Titanate Thin Films for Tunable Microwave Applications*. Tampa: University of South Florida.
59. Boubekeur H, Mikolajick T, Palmer W, Hopfner J, Frey L, Ryssel H. 2002. Platinum contamination issues in ferroelectric memories. J Appl Phys 92:3257.
60. York R, Nagra A, Erker E, Taylor T, Periaswamy P, Speck J, Streiffer S, Auciello O. 2000. Microwave integrated circuits using thin-film BST. IEEE Int Symp Appl Ferroelectr 1:195.
61. Wang K, Yao K, Chua SJ. 2005. Titanium diffusion and residual stress of platinum thin films on Ti/SiO2/Si substrate. J Appl Phys 98:013538–013531.
62. Nam HJ, Choi DK, Lee WJ. 2000. Formation of hillocks in Pt Ti electrodes and their effects on short phenomena of PZT films deposited by reactive sputtering. Thin Solid Films 371:264.
63. Al-Shareef HN, Dimos D, Tuttle BA, Raymond MV. 1997. Metallization schemes for dielectric thin film capacitors. J Mater Res 12:347.
64. Matsui Y, Hiratani M, Kumagai Y, Miura H, Fujisaki Y. 1998. Thermal stability of Pt bottom electrodes for ferroelectric capacitors. Jpn J Appl Phys 37:L465.
65. Tombak A, Maria JP, Ayguavives F, Jin Z, Stauf GT, Kingon AI, Mortazawi A. 2002. Tunable barium strontium titanate thin film capacitors for RF and microwave applications. IEEE Microw Wireless Compon Lett 12:3.
66. Coda ME, Du H, Bhandarkar S, Johnson DW, Jr. 1997. Selected-area sol-gel deposition of barium strontium titanate thin films on thermally oxidized silicon through mediation of self-assembled monolayers. Supramol Sci 4:43.
67. Vadlamani S. 2003. *Synthesis and characterization of wide band gap semiconductor thin films*. Master's Thesis, University of South Florida, Tampa, FL.
68. Changa W, Gilmore CM, Kim WJ, Pond JM, Kirchoefer SW, Qadri SB, Chirsey DB, Horwitzb JS. 2000. Influence of strain on microwave dielectric properties of (Ba,Sr)TiO_3 thin films. J Appl Phys 87:3044.

69. Böhlmark J. 2005. *Fundamentals of High Power Impulse Magnetron Sputtering*. Sweden: Linkoping University.

70. Park JH. 1998. *Chemical Vapor Deposition*, 1st edition. Chicago, IL: ASM International.

71. Regnery S, Ehrhart P, Szot K, Waser R, Ding Y, Jia CL, Schumacher M, McEntee T. 2003. MOCVD of (Ba,Sr)TiO_3: nucleation and growth. Integr Ferroelectr 57:1175.

72. Jain M, Majumder SB, Katiyar RS, Bhalla AS. 2004. Structural and dielectric properties of heterostructured BST thin films by sol-gel technique. Thin Solid Films 447:537.

73. Ezhilvalavan S, Tseng T-Y. 2000. Progress in the developments of (Ba,Sr)TiO_3 (BST) thin films for gigabit era DRAMs. Mater Chem Phys 65:227.

74. Tian HY, Chan HLW, Choy CL, No K. 2003. The effects of composition gradients of $Ba_xSr_{1-x}TiO_3$ thin films on their microstructures, dielectric and optical properties. Mater Sci Eng B 103(3):246.

75. Lu SG, Zhu XH, Mak CL, Wong KH, Chan HLW, Choy CL. 2003. High tunability in compositionally graded epitaxial barium strontium titanate thin films by pulsed-laser deposition. Appl Phys Lett 82(17):2877.

76. Lu SG, Zhu XH, Mak CL, Wong KH, Chan HLW, Choy CL. 2003. Compositionally graded epitaxial barium strontium titanate thin films derived from pulsed laser deposition. Mater Chem Phys 79:164.

77. Lee SJ, Moon SE, Ryu HC, Kwak MH, Kim YT, Han SK. 2003. Microwave properties of compositionally graded (Ba, Sr)TiO_3 thin films according to the direction of the composition gradient for tunable microwave applications. Appl Phys Lett 82(13):2133.

78. Zhao J, Wang X, Li L, Wang X, Li Y. 2008. Stoichiometry control and structure evolution in hydrothermally derived (Ba,Sr)TiO_3 films. Ceram Int 34:1223.

79. Kajiyoshi K, Yoshimura M, Hamaji Y, Tomono K, Kasanami T. 1996. Growth of (Ba, Sr) TiO_3 thin films by the hydrothermal-electrochemical method and effect of oxygen evolution on their microstructure. J Mater Res 11:169.

80. Zhou Z, Cheng X, Du P, Chang J, Huang S, Lu L, Ye Z. 2007. Effect of Sr/Ba ratio on dielectric properties of BSTN ceramics. Key Eng Mater 336–338:125.

81. Goux L, Gervais M, Catherinot A, Champeaux C, Sabary F. 2002. Crystalline and electrical properties of pulsed laser deposited BST on platinized silicon substrates. J Non-Crystalline Solids 303:194.

82. Kim TG, Oh J, Moon T, Kim Y, Park B, Lee YT, Nam S. 2003. Effect of crystallinity on the dielectric loss of sputter-deposited (Ba,Sr)TiO_3 thin films in the microwave range. J Mater Res 18:682.

83. Cukauskas EJ, Kirchoefer SW, Chang W. 2002. Morphology and dielectric properties of $Ba_{0.5}Sr_{0.5}TiO_3$ thin films on annealed (100) MgO. J Cryst Growth 236:239.

84. Su B, Holmes JE, Cheng BL, Button TW. 2002. Processing effects on the microstructure and dielectric properties of barium strontium titanate (BST) ceramics. J Electroceramics 9:111.

85. Tsai MS, Sun SC, Tseng TY. 1997. Effect of oxygen to argon ratio on properties of (Ba,Sr) TiO_3 thin films prepared by radio-frequency magnetron sputtering. Appl Phys 82(7):3482.

86. Parker CB, Maria JP, Kingon AI. 2002. Temperature and thickness dependent permittivity of (Ba,Sr)TiO_3 thin films. Appl Phys Lett 81(2):340.

87. Gurumurthy V. 2007. *Barium strontium titanate films for tunable microwave and acoustic wave applications*. Master's Thesis, University of South Florida, Tampa, FL 33613.

88. Yan F, Wang Y, Chan HLW, Choy CL. 2003. Ferroelectric properties of $(Ba_{0.5}Sr_{0.5})TiO_3$/ $Pb(Zr_{0.52}Ti_{0.48})O_3$/$(Ba_{0.}\ 5Sr_{0.5})$ TiO_3 thin films with platinum electrodes. Appl Phys Lett 82(24):4325.

89. Stauf GT, Ragaglia C, Roeder JF, Vestyck D, Maria JP, Ayguavives V, Kingon A, Mortazawi A, Tombak A. 2001. Thick electrodes for high frequency high Q tunable ferroelectric thin film varactors. Integr Ferroelectr 39:321.

90. Tarutani M, Yamauka M, Takenaga T, Kuroiwa T, Horikawa T. 2002. Improved fabrication process for Ru/BST/Ru capacitor by liquid source chemical vapor deposition. Thin Solid Films 409(1):8.

91. Cha SY, Jang BT, Lee HC. 1999. Effects of Ir electrodes on the dielectric constants of $Ba_{0.5}Sr_{0.5}TiO_3$ films. Jpn J Appl Phys 38:L49.

92. Liehr M, Dallaporta H, Lewis JE. 1988. Defect formation in SiO2/Si(100) by metal diffusion and reaction. Appl Phys Lett 53:589.

93. Narayana GS, Misra NK. 1990. X-ray studies on growth, thermal vibrations and internal stress in thin condensed films of indium. Indian J Phys A 64A:151.

94. Taylor TR. 2003. *Stoichiometry and Thermal Mismatch Effects on Sputtered Barium Strontium Titanate Thin Films.* Santa Barbara: University of California.

95. Chee J, Karru R, Fisher TS, Peroulis D. 2005. *DC-65GHz characterization of nanocrystalline diamond leaky film for reliable RF MEMS switches.* IEEE EGAAS 2005. pp. 581.

96. Beuille C, Dutarde E, Schneider H, Castex MC, Lefeuvre E, Achard J, Silva F. 2002. Characterization of metal-diamond-silicon associations for active power electronics applications. *IEEE 33rd Annual Power Electronics Specialists Conference*, Vol. 4, pp. 1764.

97. Ohring M. 1992. *The Material Science of Thin Films.* San Diego: Academic Press.

98. Gurumurhty V, Jeediguntha S, Baylis S, Spangnol P, Bumgarner J, Kumnar A, Weller T. 2008. Structural and electrical properties of nanocrystalline diamond based barium strontium titanate varactors. Ferroelectrics 377:75.

99. Scott JF. 1999. Device physics of ferroelectric thin film memories. Jpn J Appl Phys 38:2272.

100. Ban ZG, Alpay SP. 2002. Phase diagrams and dielectric response of epitaxial barium strontium titanate films: a theoretical analysis. J Appl Phys 91(11):9288.

101. Scott JF. 2000. *Ferroelectric Memories.* Germany: Springer Verlag.

102. York R, Nagra A, Erker E, Taylor T, Periaswamy P, Speck J, Streiffer S, Auciello O. 2000. Microwave integrated circuits using thin-film BST. *Presented at International Society of Applied Ferroelectrics (ISAF), Hawaii.*

16

AEROSOL DEPOSITION (AD) INTEGRATION TECHNIQUES AND THEIR APPLICATION TO MICRODEVICES

Jun Akedo

National Institute of Advanced Industrial Science and Technology (AIST), Ibaraki, Japan

INTRODUCTION

The manufacturing of ceramics usually requires sintering at temperatures higher than 1000°C, which makes it difficult to compound or to integrate ceramics with low melting point, metals, glass, or plastics. This is a serious problem for upgrading electroceramic components and optical components. For example, an effective and reliable technology is required to fabricate actuators made of piezoelectric thick films for optical scanners, micromotors [1, 2], scanning force microscopy [3], micropumps, ultrasonic mixers [4, 5], and micromanipulators for medical applications in microelectromechanical systems (MEMS). Improvements in technological processes will be very important in the near future to produce. For example, ink-jet printer heads that will reduce the printing time or flapper actuators that will control the high-speed positioning of recording heads for high-density storage drives [6]. For these applications, using piezoelectric or electrostrictive materials with large strain and high-speed response often require dense and thick micropatterned films, with thickness exceeding 5 μm in many cases [2, 7, 8]. However, thick layers produced by conventional thin- or thick-film methods usually have cracks and may easily peel from the substrates. With these methods, it is also

Ceramic Integration and Joining Technologies: From Macro to Nanoscale, First Edition.
Edited by Mrityunjay Singh, Tatsuki Ohji, Rajiv Asthana, Sanjay Mathur.

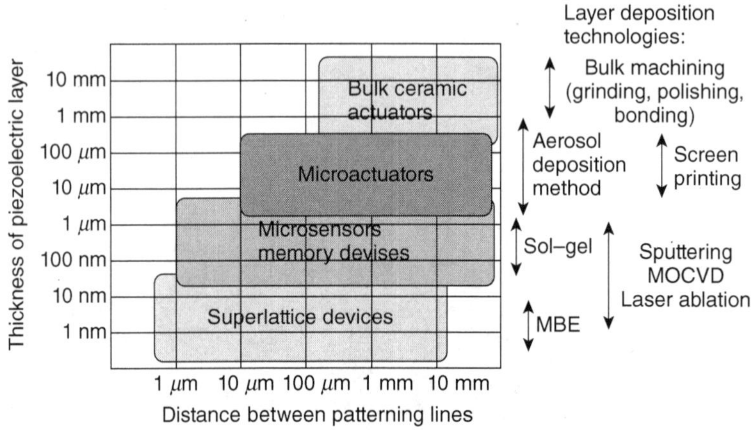

Figure 16.1. Fabrication method for PZT layer in various thickness ranges. MBE, molecular beam epitaxy.

difficult to produce complex material compositions, and fabrication can be time-consuming and costly.

Figure 16.1 summarizes the conventional methods used to fabricate PZT films on substrates and the applications of these films. There are many reports on the fabrication of lead–zirconate–titanate ($Pb[Zr_{0.52},Ti_{0.48}]O_3$) (PZT) films of thicknesses ranging from 0.08 to 5 μm by sol–gel [9, 10], sputtering [11, 12], metal organic chemical vapor deposition (MOCVD) [13], pulse laser ablation [14], electron beam evaporation [15], and ion beam deposition [16]. In these methods, dense PZT films can be formed and oriented on a $Pt/Ti/SiO_2/Si$ substrate. However, fabrication of PZT films with a thickness of over 1–3 μm using these methods requires much attention, as there are many problems to overcome regarding film quality or deposition. Hydrothermal synthesis [17, 18] has the advantage of a low process temperature at 150°C, and the poling procedure is not required. However, the surface roughness and the density of the films are poor. PZT thick films fabricated by the screen-printing method [19, 20] have low density, and PZT/Pt/Si structures can be damaged because of the long firing time at temperatures higher than 800°C. An improved screen-printing method with low-temperature sintering and a high-resistance electrode has been reported [21], but the piezoelectric properties of films produced by this method were not reported. For sputtered and sol–gel-derived PZT thin films, a postdeposition rapid thermal annealing treatment [22] was introduced to reduce the damage to the substrate or structure and to improve the electrical properties. However, for thick films with a thickness of over 1 μm, this process was not so effective. The etching of thick ceramic films by plasma etching [23], inductively coupled plasma etching [24, 25], or reactive ion etching [26] is also difficult. For bulk PZT adhered to a silicon membrane, it is difficult to ensure an adequate mechanical and electrical coupling between the films and to assemble a complex structure. Thus, it can be concluded that, by conventional methods, fine patterning of thick (over 1 μm) PZT films on silicon-based substrates is still difficult to realize.

To reduce costs and fabrication time and to avoid damages on circuitry already present on the substrate, it is very important to have a high-speed deposition rate, a low process temperature, and fine patterning. Until now, a number of studies have aimed at reducing the sintering temperature for the purpose of reducing energy consumption and at implementing innovative functional components through the integration with metal or glass materials. Several deposition methods based on the principle of particle collisions have already been investigated. Recently, the cold spray method (CSM) for metal materials has attracted much attention. However, this method is not successful for ceramic materials.

For these reasons, a new deposition technique based on the collision adhesion of fine particles for fabrication and micropatterning of thick ceramic layers has been developed, named the aerosol deposition (AD) method. AD [27–31] is a novel and a very attractive coating method for ceramic integration. Submicrometer ceramic powder is mixed with a carrier gas to form an aerosol flow, ejected through a micro-orifice nozzle, and deposited onto a substrate in the deposition chamber kept under vacuum during the deposition. Special attention is paid to the character of primary powder particles and the deposition conditions. Using this method, ceramic layers of Pb(Zr,Ti)O_3, α-Al_2O_3, Y_2O_3, YSZ, AlN, MgB_2, and other ceramic materials with nanocrystalline structure, high transparency, high hardness, and high breakdown voltage were formed at room temperature (RT). It is suggested that during impact of the fine particles with the substrate, a part of the particle's kinetic energy is converted into a bonding energy between the substrate and the fine particles and between the fine particles themselves. In the fabrication of thick films with a thickness range of 1–100 μm, the AD method has many advantages in comparison with conventional thin- or thick-film methods or thermal spray methods.

AD METHOD

The AD method is based on shock-loading solidification due to the impact of ultrafine ceramic particles with a surface. [27, 28] First, particles are mixed with a gas to generate an aerosol. This aerosol is ejected through a nozzle at low pressure and is impacted onto a substrate to form a thin/thick film. During the impact with the substrate, part of the particle's kinetic energy is converted into thermal energy, causing an increase in temperature at the point of impact and promoting bonding between the substrate and the particles and also between multiple particles. However, the detailed mechanism occurring during the particle impingement with a surface in the AD method has not yet been fully clarified.

Figure 16.2 shows the deposition apparatus and Table 16.1 gives the deposition conditions for the AD method. The AD apparatus consists of two vacuum chambers connected by a tube. The first chamber is an aerosol-generation chamber and the second chamber is a deposition chamber. The deposition chamber is used for the formation and patterning of films. The aerosol-generation chamber has a carrier gas system and a vibration system to mix the powder with the carrier gas. The aerosol generated in this chamber is delivered to the deposition chamber by a pressure difference between the

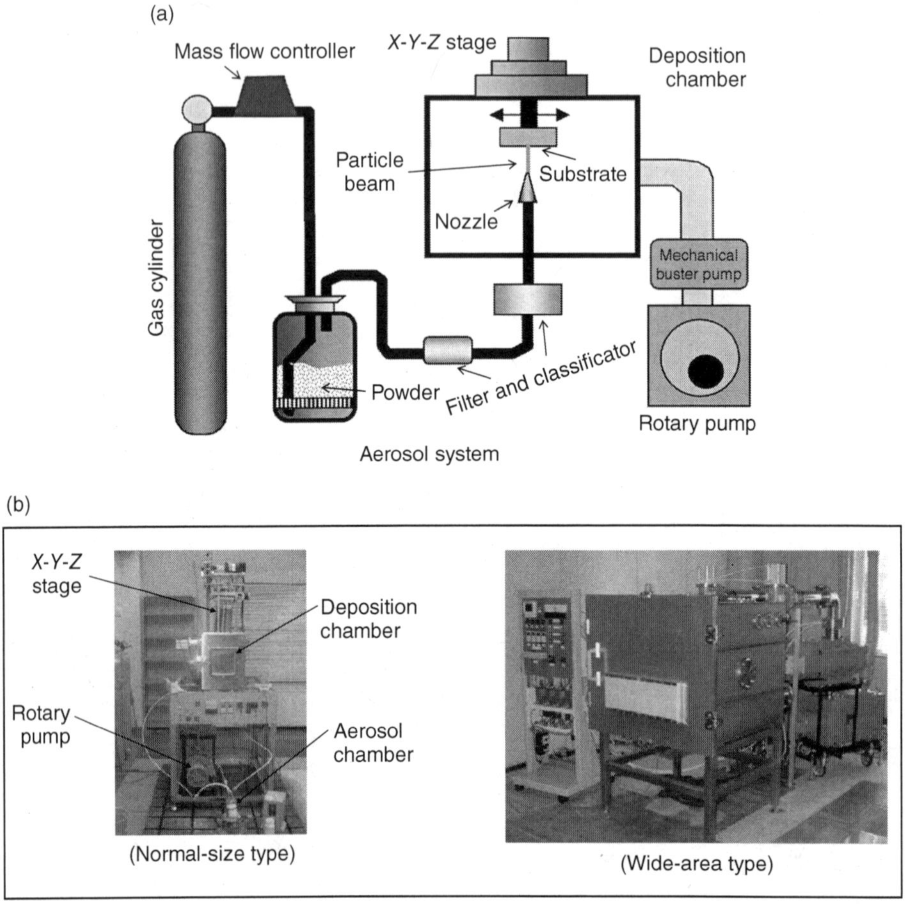

Figure 16.2. (a) Schematic and (b) prototype photograph of an aerosol deposition (AD) apparatus.

TABLE 16.1. Typical Deposition Conditions (Experimental Parameters)

Pressure in deposition chamber	0.05–2.0 kPa
Pressure in aerosol chamber	10–80 kPa
Size of nozzle orifice	$5.0 \times 0.3\,mm^2$, $10.0 \times 0.4\,mm^2$
Accelerating gas	He, N_2, air
Consumption of accelerating gas	1–10 L/min
Maintained substrate temperature during deposition	300 K
Scanning area (area of deposition)	$40 \times 40\,mm^2$, $400 \times 400\,mm^2$
Scanning speed of the nozzle motion along the substrate	0.125–10 mm/s
Distance between the nozzle and the substrate	1–40 mm

two chambers. The deposition chamber contains a nozzle, a substrate holder with a heating system, and a mask alignment system used for making patterned films. A rotary vacuum pump coupled to a mechanical booster pump is used to vacuum this chamber to a pressure of about 50–2000 Pa during deposition. Figure 16.2b shows various-sized deposition machines.

Sintered ceramic powders with a particle size range of about 0.08–2.0 μm are typically used as the deposition particles. After suspension in the carrier gas to form an aerosol, the aerosol is accelerated to several hundred meters per second through a rectangular orifice with a width less than 1 mm. Not all of the aerosol particles will be deposited onto the substrate. The ratio of deposited to nondeposited particles strongly depends on particle size and the degree of aggregation of the particles. To form films with an acceptable density and material properties, particles with a certain size and morphology must be used. To generate a jet of particles with acceptable size and morphology, a deagglomeration device and a filter are used between the aerosol-generation and deposition chambers to break the particles apart as much as possible and to select particles in a prescribed size range.

ROOM TEMPERATURE IMPACT CONSOLIDATION (RTIC)

Consolidation of Ceramic Powders at RT

For the AD method, high-speed formation of ceramic films at RT with high density and high transparency is possible by optimizing the particle diameter and deposition conditions. The result is a process that yields an acceptable consolidation at RT without the need for a thermal treatment. This process is called RTIC [28, 29, 32].

Figures 16.3 and 16.4 show micrographs of typical RTIC ceramic layers, a cross-sectional scanning electron microscope (SEM) image of an as-deposited α-Al_2O_3 layer at RT, comparison of its X-ray diffraction (XRD) profile with the profiles obtained for the raw powder and bulk material, and transmission electron microscope (TEM) images of the as-deposited layer and starting powder. The formation of thick films of over 100 μm having high transparency was confirmed, as shown in Figure 16.3a, b. Using a multinozzle system, the coating area could be enlarged as shown in Figure 16.3c. The aerosol deposited films showed a relatively good uniformity of thickness of 1.4%. The surface roughness (Ra) was less than 100 nm. The deposited film had a high density and randomly oriented polycrystalline nanostructure with crystal grains smaller than 20 nm in length. TEM and electron diffraction imaging did not show either amorphous layers or heterostructures at the boundary of the crystal grains. XRD profiles confirmed that the spectral phases of the α-Al_2O_3 particles were retained in the deposited layer. However, a broadening of the spectral peaks and a slight shift in their angle positions were observed. The change between the spectra of the raw particles and the deposited layer is due to the reduction of the film crystal size and/or the distortion during the deposition. Clear lattice images in crystal grains less than 10 nm were observed, as well as uniform microstructures at the boundary between the substrate and the deposited layer. For α-Al_2O_3 layers deposited at RT, the film density was greater than 95% of the theoretical density, and Vickers hardness was 1600 HV [27]. Such α-Al_2O_3 layers are

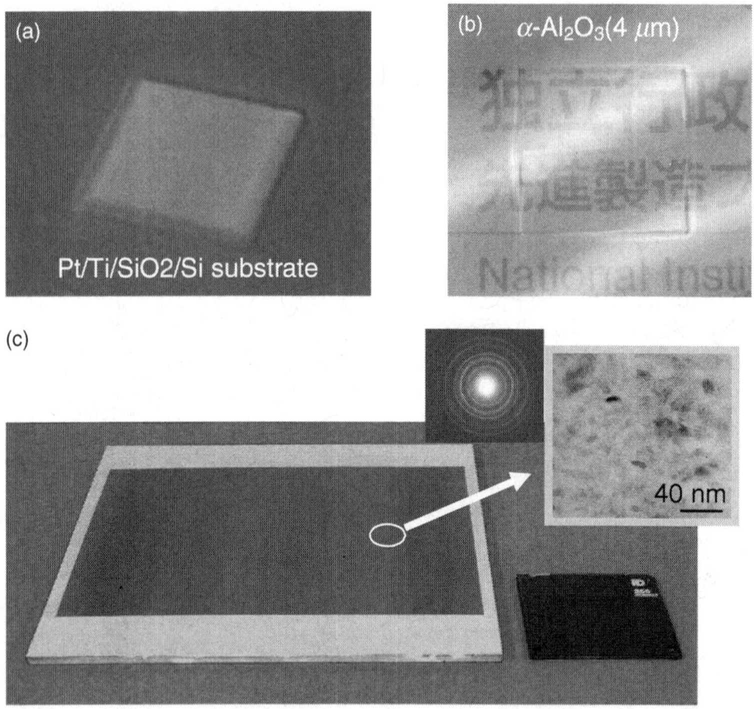

Figure 16.3. Photograph of AD-ceramic layer deposited at room temperature using the RTIC phenomenon. (a) Thick ceramic layer: over 500 mm; (b) transparent ceramic layer: 99% at 400–900 nm; (c) wide-area coating: $200 \times 200\,mm^2$.

acceptable for use as abrasion-resistant coatings [30]. The crystal grain size of the as-deposited films was smaller than that of the starting particles because the starting particles break down during collisions.

Table 16.2 shows the relationship between micro-Vickers hardness, the crystallite size, and particle impact velocity in various AD deposited materials. The layer hardness increased with increasing particle impact velocity and sometimes was higher than that of the bulk material, which was sintered at a high temperature. Critical particle velocities for an acceptable RTIC ranged from 150 to 500 m/s, and the velocity needed to create films with acceptable hardness tended to increase with increasing sintering temperature for a particular ceramic material.

Although distortions were included inside the films, a rise of the substrate temperature due to particle collisions was not observed. Consolidation of ceramic powders was possible at RT without the use of binders. Even if the formation of thin films using RTIC does not require sintering, it produces high-density films. Acceptable RT deposition was observed not only for oxide materials such as α-Al_2O_3, lead zirconate titanate(PZT: $Pb[Zr_{52},Ti_{48}]O_3$), and Ni-Zn—Fe_3O_4, but also for nonoxide materials such as AlN and MgB_2. In either case, particles with diameters greater than 80 nm are needed to make films having an acceptable hardness.

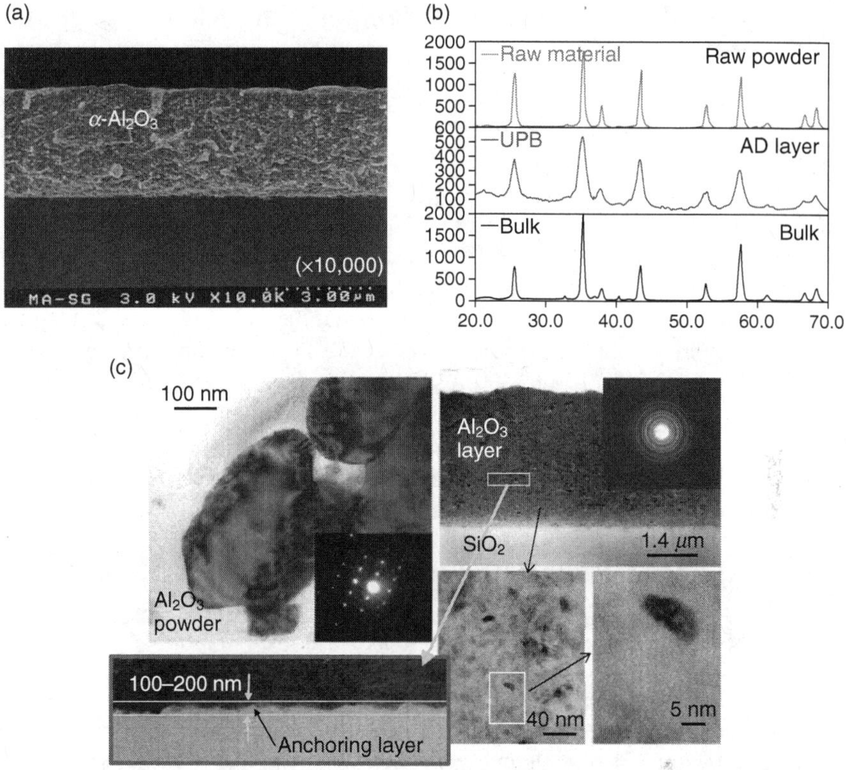

Figure 16.4. Microstructure of an α-Al_2O_3 layer deposited at room temperature by the AD method. (a) Cross-sectional SEM image of an as-deposited layer at RT, (b) comparison of XRD profiles, and (c) TEM images of the starting powder and the as-deposited layer at RT.

TABLE 16.2. Mechanical Properties of As-Deposited AD Layers

Material	Hardness (HV) Layer (deposited at RT)	Hardness (HV) Bulk[a]	Average crystallite size of the layer (nm)	Particle velocity at the collision (m/s)
Oxide				
α-Al_2O_3	1200–2100	1900 ± 100	13–20	150–500
PZT	400–550	350 ± 50	10–30	100–300
(Ni, Zn)Fe_2O_3	700–750	1040 ± 80	5–20	250–600
Nonoxide				
AlN	1100–1470	1180 ± 90	5–15	200–600
MgB_2	700	—	5–20	300–550

[a]Bulk sample was prepared from the same starting powder as layers by the conventional sintering procedure (at a temperature over 1200°C).

Impact Particle Velocity and Local Temperature Increase During the AD Process

The particle's velocity during the AD process was measured by a time-of-flight method, in which some part of the particle flow was mechanically cut from the total flow and deposited onto a moving substrate. The deflection of the deposited pattern from the center axis, the geometrical dimensions, and the moving substrate speed provide data to determine the particle flow velocity [31]. The merit of this method is that the impact particle velocity is directly measured. This is a more accurate method than the conventional measurement using a high-speed camera, which can only estimate particle flow velocity. Figure 16.5 shows the relationship between the impact particle velocity and gas consumption in controlling the particle velocity. The particle impact velocities were estimated as varying from 150 to 500 m/s. The critical velocity for RTIC of α-Al_2O_3 was 150 m/s. This implies a very small kinetic energy compared with that obtained in the conventional shock compaction process. Using the values measured for particle impact velocity, the local rise in temperature and the shock pressure at the point of impact between the particle and the substrate were simulated by finite element method (FEM) computational simulations using AUTODYNs-2D (CRC Solutions Co., Tokyo, Japan) with the Johnson–Holmquist material model [29, 33, 34]. In the simulation, the particle diameter was set to 0.3 μm, which was the same as the average diameter of the starting α-Al_2O_3 powder. Bulk material constants of α-Al_2O_3 [35] were used for these simulations. The maximum local temperature rise and the shock pressure at the point of impact during layer formation do not exceed 500°C and 2.5 GPa, as shown in Figure 16.6a, b. This local increase in temperature is too small to induce any ceramic sintering. Consolidation features of ceramic material fabricated using the AD method are funda-

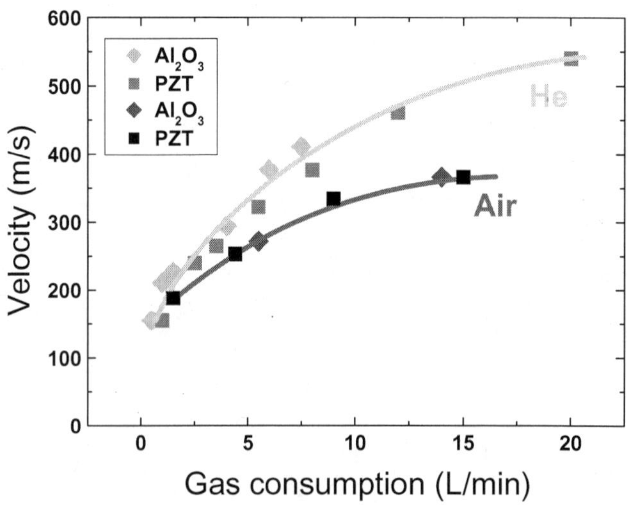

Figure 16.5. Relationship between impact particle velocity and gas consumption to control the particle velocity for different gases (nozzle of 10 × 0.4 mm²).

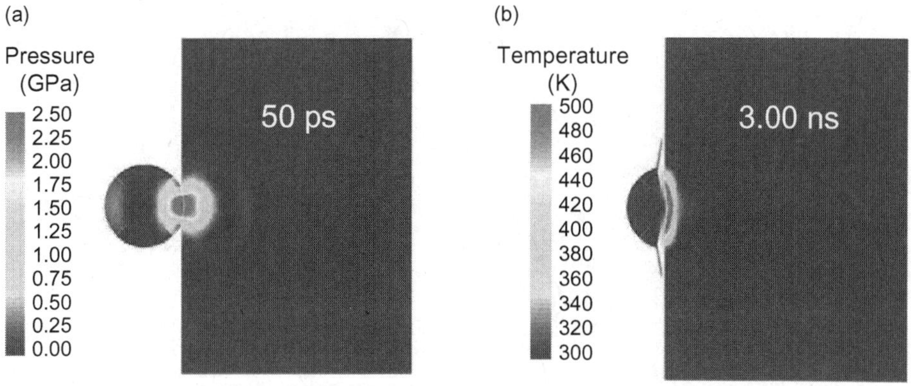

Figure 16.6. FEM simulation of the local rise in temperature and shock pressure during impact of particle with the substrate for AD. The impact particle velocity was 300 m/s [29].

mentally similar to those in the case of the shock compaction method. However, the local rise in temperature and the compact pressure for brittle materials such as ceramics and glasses were less than 1000°C and 10 GPa, which are typical for conventional shock wave synthesis [36, 37].

Densification Mechanism of Ceramic Layers in the AD Process

From the simulation and experimental results, it may be concluded that partial or total melting of particles does not occur during collision. Then, how can the reduction of the crystallite size for the starting particle and the production of dense structures during deposition be explained?

There are two possible explanations:

1. Only small particles less than 20 nm in the jet flow will actually take part at the deposition to form the layer.
2. Fracture and deformation of starting particles are obtained during the impact with the substrate to fill the gaps between the deposited particles.

To clarify this, a mixed aerosol of α-Al_2O_3 and PZT powders was deposited to form a composite layer. The cross-section of this layer was observed using high-resolution TEM to investigate the densification and bonding mechanism of ceramic particles. Figure 16.7a, b shows TEM images of the α-Al_2O_3/PZT composite layers. Selected area diffraction (SAD) for both starting particles indicates that these particles have a single crystal structure. Black and white regions indicate PZT and α-Al_2O_3, respectively, because lead is much heavier than aluminum and the electron absorption for each element depends on their mass. A laminar structure along the substrate plane was observed only in the cross-section TEM image. Inside black and white regions, electron diffraction patterns with an 80-nm electron beam irradiating diameter indicate

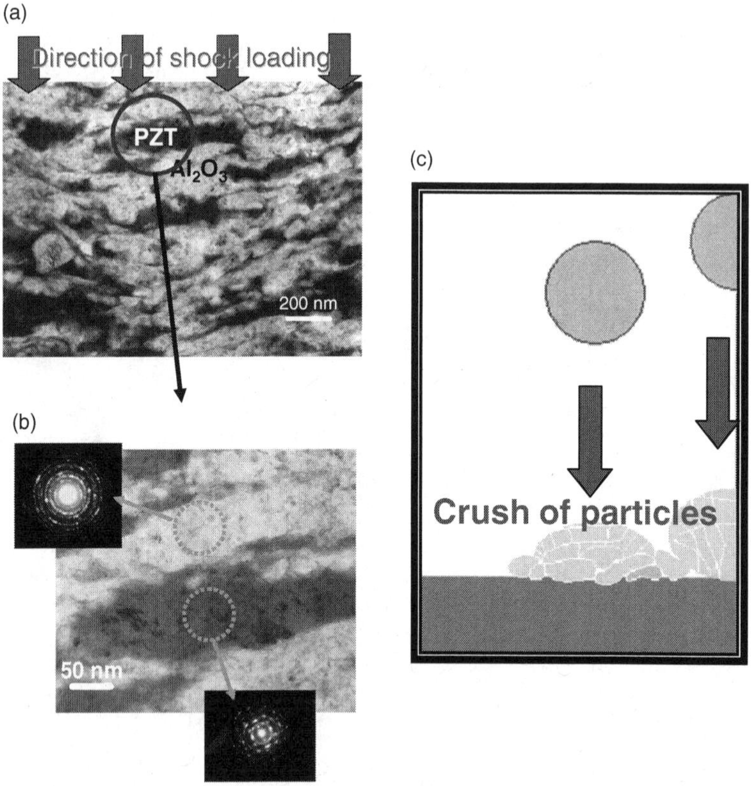

Figure 16.7. TEM image α-Al_2O_3/PZT composite layer and densification mechanism. (a) Cross-sectional. TEM image, (b) zoom-up image, and (c) densification image [29].

a typical net pattern image, as shown in Figure 16.7b, and randomly oriented small crystallites less than 20 nm were observed in a zoomed image. These crystallites were almost the same in size as those in the as-deposited α-Al_2O_3 films (shown in Fig. 16.4c). The volume of the black part, assuming an oval shape, was close to that of starting particles of PZT. The fracture toughness of the α-Al_2O_3 starting particle was estimated as 3.0–6.5 GPa by a special nanoindentation system [38]. This value was close to that obtained for maximum impact pressure in the simulation, meaning that the α-Al_2O_3 starting particle is most likely to be fractured during impact. From these results, it was concluded that the dense structure occurs due the reduction of crystallite size by fracture or plastic deformation at RT during the AD process. This explanation for densification of ceramic particles during the AD process is different from that of conventional shock compaction. The plastic deformation of ceramic particles at RT can be assumed not only for α-Al_2O_3 and may be a general phenomenon because AD layers for a large variety of ceramic materials could be easily formed.

By comparison, the CSM [39, 40] for metal coating requires particle velocities over 500 m/s to fabricate high-density layers, being much higher than the ones required for the AD method. Also, melting points of ceramic materials are higher than those of metal materials, and ceramics are more difficult to deform. Therefore, the bonding mechanism in the case of AD cannot be explained as for the case of the CSM. To date, the bonding mechanism between the fine particles themselves and between the substrate and the fine particles has not been clarified. There are still some questions to be answered regarding AD. Does the generation of a clean and active layer of starting particles make the particle bonding at a low temperature possible? Is some kind of chemical reaction induced during impact? To answer these questions, more studies are needed on the bonding mechanism of ceramic particles during AD.

Influence of Carrier Gas

If the carrier gas and pressure in the deposition chamber are properly selected to reduce the electrical discharge during the particle deposition, transparent PZT and alumina films with transmission efficiencies of 60–90% at wavelengths ranging from 400 to 800 nm can be formed at RT [29, 32]. This can be achieved because during deposition, the size of the defects (pores and grain boundaries) in the starting particles is reduced, being smaller than those found in sintered, bulk materials, and because the particle grain size is smaller than the wavelength of visible light. Carrier gas and gas flow seem to have an effect on the optical properties of AD films. Plasma illuminations have been observed during collision of fine particles with the substrate due to the fracture of brittle materials [41]. Photographs of PZT films deposited at RT using He or N_2 as carrier gas and the emission spectra at the point of impact are shown in Reference 32. The color of the deposited films is influenced by the carrier gas. Considering previous reports [28, 29], the film color change is explained by discharging during the deposition. Such discharging induces defects in the film and the drastic decrease of the film's transmittance. When helium gas was used, if consumption was increased, plasma discharging was increased, and thus the film color became even darker. The results of the emission investigation will be reported in detail later.

DEPOSITION PROPERTIES AND FILM PATTERNING

Deposition Ratio and Influence of Starting Powder Properties

High deposition rates can be achieved easily with the AD method because the source material is already in particle form. These deposition rates are at least 30 times higher than other conventional thin-film formation methods. Therefore, the AD method is an attractive manufacturing process due to high throughput. The particle velocity and concentration in the aerosol jet at the nozzle increases with increasing carrier gas flow rate, resulting in increased deposition efficiency. Starting particle properties, such as the average particle size and the size distribution, as well as mechanical and surface properties, can also strongly affect the deposition efficiency [42].

The deposition rates for PZT AD layers using powders subjected to various milling procedures are shown in Figure 16.8. It can be seen that, by increasing the milling time, the deposition rate of the PZT layer significantly increased and reached a maximum of

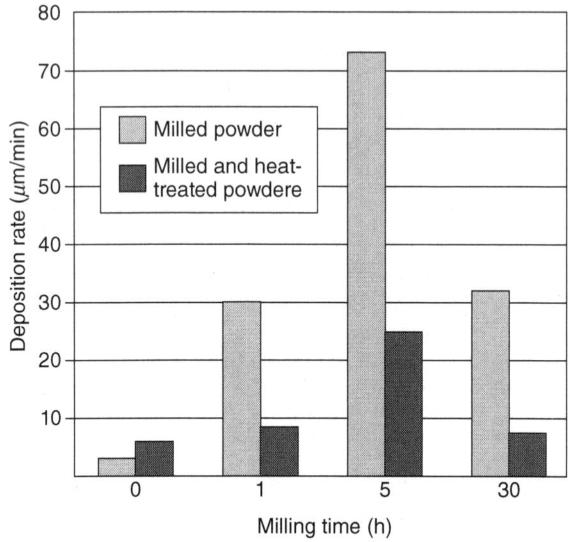

Figure 16.8. Deposition rate for PZT film formation at room temperature using powder milled for different duration times with (black bar)/without (gray bar) heat-treatment procedure at 800°C for 4h in air [42].

73 μm/min for a 5-mm^2 deposition area when powder was milled for 5h. This value is 30 times higher than that for a starting powder without the milling procedure. An interesting fact is that the deposition rate decreased as a result of further milling to 30h. It is assumed that particle surface properties (e.g., surface activation, defects, and gas absorption) will change by long-time milling, making it less probable that they will be deposited in the same conditions. This milling procedure is applicable to control the porosity of ceramic layers deposited by the AD method.

The starting powder particle size and shape strongly influence the RTIC phenomenon in the AD method. If spherical α-Al_2O_3 ultrafine particles with an average particle size of around 50nm were used, after AD, the films have a pressed-like structure and the RTIC phenomenon could not be observed even if the ejecting particle velocity from the nozzle was over 400m/s and particle size was very small (Fig. 16.9a). In contrast, if nonspherical α-Al_2O_3 fine powder with an average particle size of around 1 μm was used, the deposited particles on the substrate were consolidated at RT and the RTIC phenomenon was observed even for particle velocities of around 200m/s, as shown in Figure 16.9b. As a result, high-density and transparent ceramic layers were obtained. These results are explained by the aerodynamic properties of the particle jet flow near the substrate. If particle size and weight are too small, the particle follows the carrier gas flow, as shown schematically in Figure 16.10. Therefore, the particle velocity normal to the substrate is largely decreased and is not sufficiently high to obtain the RTIC phenomenon. A more detailed investigation about the particles' aerodynamic properties in the AD method still needs to be conducted.

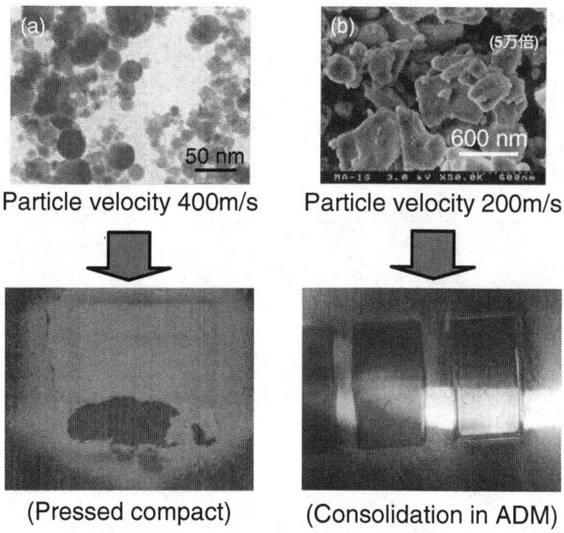

Figure 16.9. Influence of starting particle diameter for the RTIC phenomenon on the AD method. (a) Average diameter: 50 nm. (b) Average diameter: 700 nm.

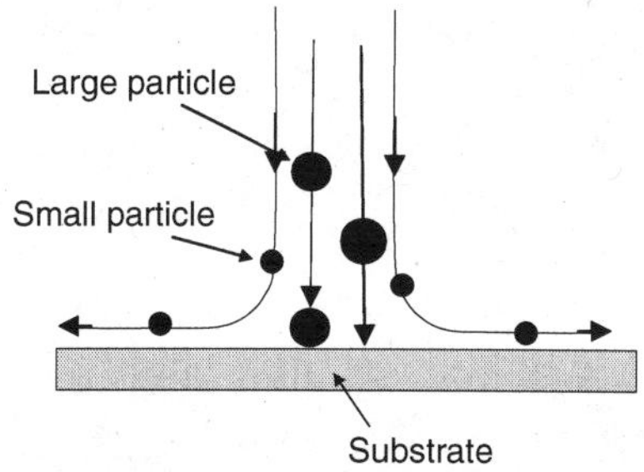

Figure 16.10. Particle trajectories in an aerosol jet flow near the substrate on the AD method.

An advantage of the AD method over conventional thin-film and thermal spray coating methods is that substrate surface does not need precleaning to achieve good deposition. During the initial deposition stage, the particles impacting the substrate will act as cleaning agents in a similar way as in sandblasting processes. Surface contaminants such as dirt and oils are removed by the initial particle collisions. The deposition automatically begins when the surface becomes sufficiently clean. The film adhesion strength to glass and metal substrates may be in excess of 30 MPa because an anchoring layer having a thickness of about 100–200 nm was formed in the interface between the substrate and the deposited layer. To obtain maximum adhesive strength, a substrate with suitable hardness and elasticity is needed to allow the formation of the anchoring layer. A substrate that is very soft will be etched by the particle jet flow and the deposition will not occur. On the other hand, when a substrate with high hardness is used, the adhesion strength between the deposited layer and the substrate is weak and the film may easily peel off.

Patterning Properties of Ceramic Film

Ceramic film patterning can be achieved by using a mask deposition method, in which a particle jet is impacted onto the substrate through a defined pattern mask that contains openings with a width of at least 50 μm. In this case, it is important to consider the aerosol jet flow in the deposition chamber and through the mask orifices. If the pressure in the deposition chamber is not sufficiently low, the particle jet is scattered by the edge of the openings in the mask and the resulting mask pattern is not preserved on the substrate.

The effect of the ceramic particles sprayed onto a substrate changes from deposition to erosion [43], as shown in Figure 16.11, depending on the particle diameter, velocity, and angle of incidence β of the particle jet to the substrate. In the figure, β_c is the angle of incidence when the deposition rate and the abrasion rate have the same value, and the normalized thickness is the thickness of the PZT layer after AD relative to the initial value of PZT thickness. These factors also influence the layer density and surface roughness. However, to achieve acceptable patterns through a mask with acceptable detail, the angle of incidence of the particle jet must be kept within a specified range. This is because the angle of incidence affects the flow patterns on the downstream side of the mask, which in turn affects the deposition efficiency and the

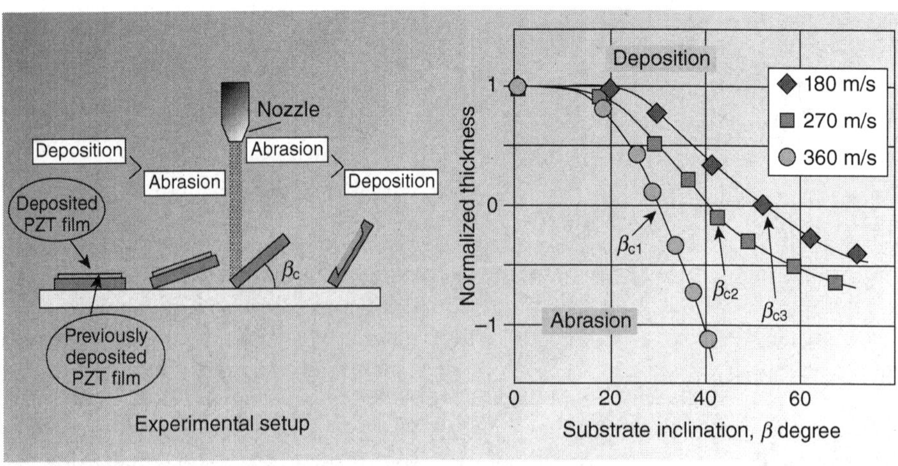

Figure 16.11. Particle jet flow angle of incidence effect to deposition properties [43].

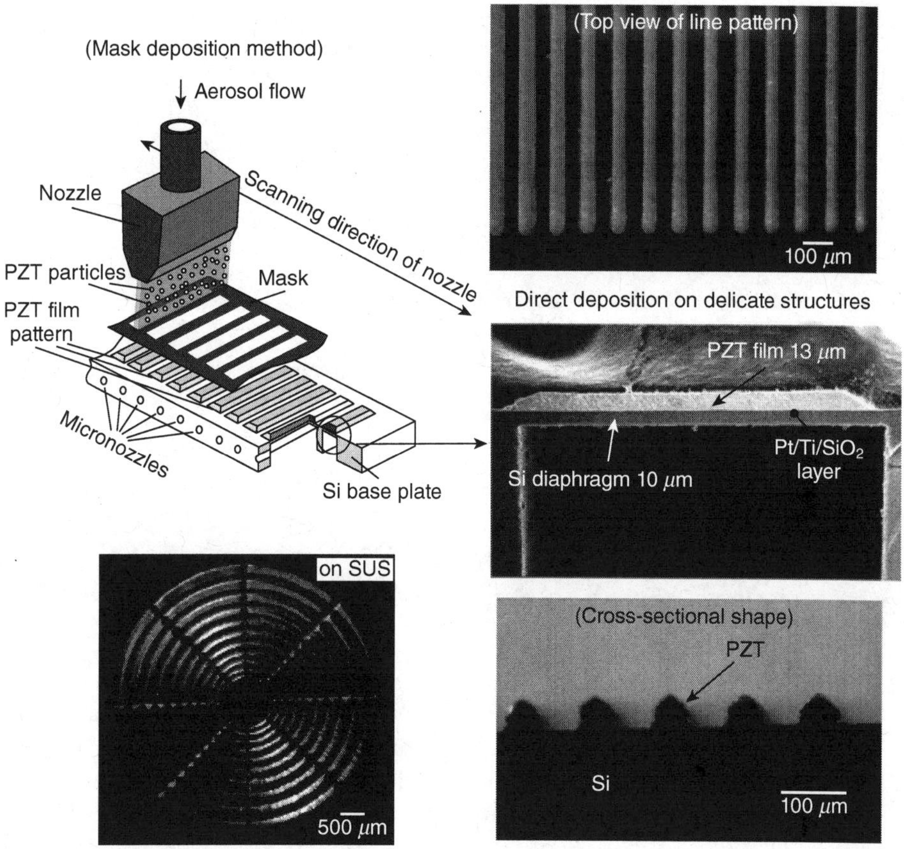

Figure 16.12. Patterning properties of thick ceramic layers on the AD method using the mask deposition method SUS, stainless steel.

degree to which the mask profile is distorted. Figure 16.12 shows a thick, patterned PZT layer deposited under optimum deposition conditions onto silicon, stainless steel, and Pt/Si substrates [44]. A ceramic microstructure with a 50-μm line width and an aspect ratio (line height/line width) greater than 1 can be patterned by controlling the substrate heating temperature and the starting particle properties. The AD method is useful for making piezoelectric films greater than 10 μm thick for applications such as ultrasonic devices. However, a pattern width less than 50 μm was difficult to obtain. Recently, the production of ceramic fine patterns using the AD method and the lift-off process was tried. The hardness and thickness of the photoresist layer were chosen carefully. As a result, a minimum pattern width less than 10 μm for a 2-μm-thick PZT and α-Al_2O_3 layers was obtained, as shown in Figure 16.13.

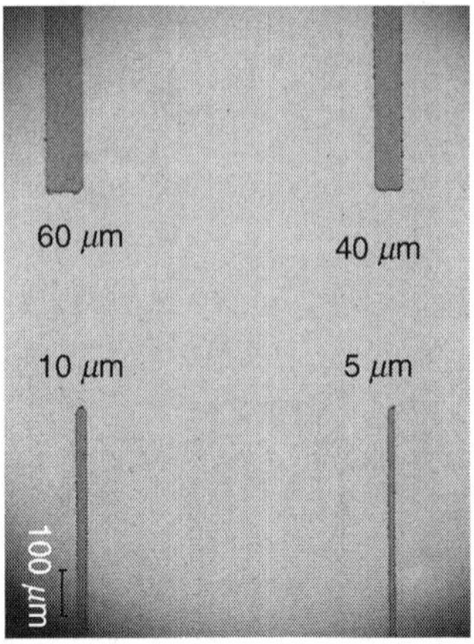

Figure 16.13. Fine patterning of ceramic layers deposited by the AD method using the lift-off process with photoresist.

OTHER SIMILAR METHODS AND COMPARISON WITH THE AD METHOD

Coating Process Based on Collision of Solid-State Particles

Several deposition methods based on the principle of solid-state particle collision have already been investigated. From previous studies, particle deposition with the AD method strongly depends on the particle diameter and velocity. Figure 16.14 shows a comparison between the AD method and other deposition methods based on the collision of solid-state particles according to the particle diameter and velocity. The deposition methods can be classified into two groups, depending on the type of particle acceleration: (1) acceleration by electric field (electrostatic particle impact deposition [EPID] method [45] and macron beam method [46]) and (2) acceleration by gas transportation (CSM [47], gas deposition [GD] method [48], and AD method). For all of these methods, the formation of thin/thick films is based on the acceleration of small particles and their impact with the substrate. The EPID process was the precursor of the coating technologies using electrical field acceleration of ultrafine particles and was originally developed by Ide et al. [49]. The GD method [50] is a coating technology of ultrafine particles via acceleration by mixing with high-speed carrier gas flow and was originally developed in ERATO Hayashi UFP Project in Japan. The CSM [39, 40] was developed by Anatolii Papyrin at the Institute of Theoretical and Applied Mechanics of the Russian Academy of Sciences, for producing thick metallic coatings. The hypersonic plasma particle deposition (HPPD) method was originally developed at the University of Minnesota [51] for Si, SiC, and ceramic thick coatings. The supersonic cluster beam deposition (SCBD) method [52] was not included here because the cluster size was about 10^3 atoms and the collision phenomenon was slightly different from that in collision of fine or ultrafine particles.

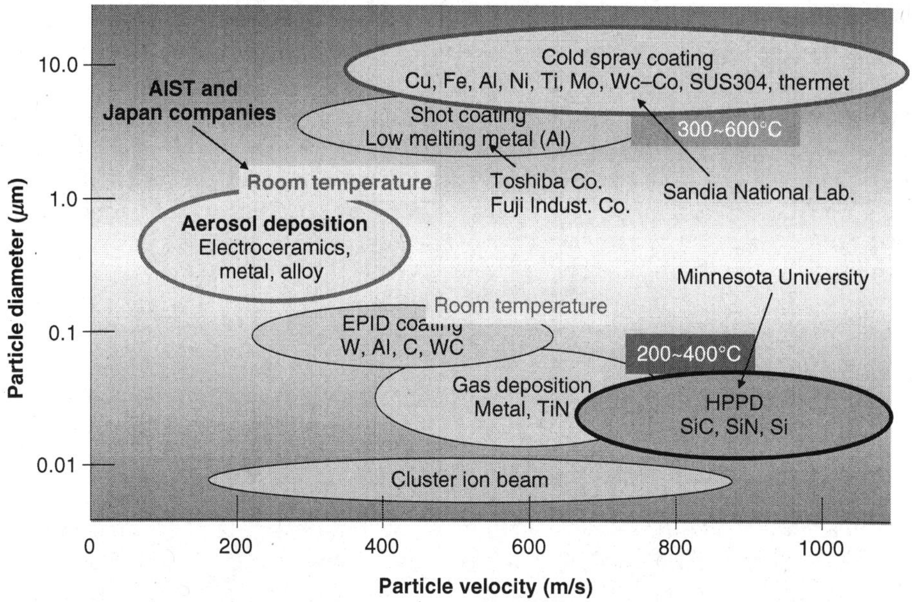

Figure 16.14. Comparison between AD method and other methods based on collision of solid-state particles.

Generally, in these coating methods, the kinetic energy of the fine particles is released at the impact during the very short collision time, which is less than 1 μs, resulting in an increase of the surface temperature of particles at the contact area above the melting point, thereby helping the particle to bond to the surface [49, 50]. However, it has been observed that the particle bonding is almost realized in solid-state form. The operating principles of these methods are completely different from those of the conventional thermal spray method, which requires melting or partial melting of the starting particles to make particles bond. In addition, it is not clear whether the bonding state between particles in fine particle layers formed by each technique is the same. There are few reports about the mechanism of energy conversion during a particle collision with a substrate, but we suspect that some differences in the deposition mechanisms between these methods exist.

The EPID method is more appropriate for conductive material due to the necessity of charging up the particles. Thick-film formation (over 1 μm) with the EPID method has not been reported yet [49]. It is believed that ultrafine particles could be embedded partially in the substrate at the time of collision, and a mixed layer consisting of particles and the substrate materials is formed. It is also believed that layer growth for this method is low because of weak bonding between the jet particles. For the GD method, metallic and ceramic ultrafine particles having a diameter less than 100 nm and a highly activated surface are used. These ultrafine particles are formed by the condensation of an evaporated metal under high-pressure conditions in a vacuum chamber.

In the HPPD method, active ultrafine particles are also used, and these particles are produced under high-pressure conditions after the condensation from the gas phase in the nozzle. Deposition properties in the HPPD method and the SCBD method show the same tendency as that in the GD method, as reported in References 51 and 52. The deposition efficiency of the EPID method seems to be very low. On the other hand, the GD, CSM, and HPPD methods have high deposition rates. In the case of CSM, large-sized particles with a diameter greater than 10 μm are accelerated by hot carrier gas at 300–600°C and sprayed on a substrate under atmospheric condition using an ultrasonic (de Laval) nozzle. This method is very similar to the AD method and the conventional thermal spray coating process, but producing ceramic coatings has not been successful so far.

Comparison Between the AD Method and Other Methods

The GD method enables the formation of metallic nanocrystalline layers at RT. However, the layer density is low, ranging between 50% and 80% of that of the bulk material. To obtain a high electrical conductivity comparable to that of the bulk material, the grain growth of the deposited layer using a thermal heat treatment such as substrate heating or heating of carrier gas is necessary [50]. In addition, a high-density layer structure and high adhesion with a substrate have not been obtained for impact velocities less than 500 m/s [31]. With AD and CSM, highly dense and hard layers having a thickness of several millimeters can be obtained with relative ease. In addition, these methods seem to be very practical methods due to the simplicity of the device structure. However, for CSM, the formation of ceramic layers has not been reported at all, though there were many reports for the formation of metal and metal alloy layers. In addition, for low melting point materials such as aluminum, nickel, or copper, the particle impact velocity to obtain the formation of a coating, known as critical velocity, was very high, ranging between 500 and 700 m/s [39, 47]. In contrast, high-temperature melting materials such as α-Al_2O_3 can be deposited at RT by the AD method. The particle impact velocity is low and ranges between 150 and 400 m/s, as shown in Figure 16.4. The deposition properties are different from those of the CSM. For the AD method, the kinetic energy, particle diameter, and velocity of the particles are smaller than those in the CSM. These results can be explained by the fact that particle impact velocity on the CSM is reduced by the reflection of the jet flow and by the resistance of the air layer near the substrate in atmospheric conditions. As a result, impact velocity and pressure in the CSM are insufficient to produce the RTIC phenomenon, though the particle velocity just after ejection from the nozzle is very high. It is believed that a critical particle mass is necessary to overcome the slowdown resistance by the atmospheric layer near the substrate. Therefore, no reports on the deposition of small particles having a diameter less than 5 μm with the CSM could be found. With the AD method, it is possible to deposit not only metallic materials but also ceramic materials. It is believed that AD and CSM share the RTIC phenomenon for solid-state particles, as both deposition conditions and microstructures are comparable in the reported papers. However, the AD method is different from the CSM because RTIC conditions for these two methods are very different.

Presently, the deposition mechanisms due to the collision of solid-state fine particles are not well understood. A more detailed study is necessary.

ELECTRICAL PROPERTIES OF AD FILMS

Fundamentally, the above-mentioned coating methods are based on the shock-loading consolidation of solid-state particles with or without thermal or plasma energy assistance. The microstructures of the films produced by these methods are completely different from those of the films made by typical spray coating methods because the primary powders used are not melted before impact with the substrate and particles are directly bonded in their solid state. The expected features for these methods are a dense film formation at a low temperature and the retention of the primary material composition and the crystal structure. However, there is no report about applications of these methods to electrical and functional materials. If the crystal structure of the ceramic particles is retained during deposition, the temperature of the process might be decreased, and these methods may then be applied to electroceramic materials.

AD layers deposited at RT generally have high electrical insulation and breakdown characteristics that exceed the corresponding properties of the bulk material. For example, the electrical breakdown of α-Al_2O_3 and Y_2O_3 exceeded 3 MV/cm, and for PZT, it was found to exceed 500 kV/cm [27, 53]. The volume resistivity, dielectric constant, and dielectric loss of the α-Al_2O_3 layer formed by the AD method were $1.5 \times 10^{15}\,\Omega$cm, 9.8 at 1 kHz, and 0.2%, respectively [54]. Those values are almost the same as those of the bulk material. Such electrical characteristics can be useful for developing devices such as electrostatic chucks [55], electrical insulation layers with good thermal conductivity for high-power electric devices.

Although PZT layers deposited at RT exhibit piezoelectric and ferroelectric behaviors, their properties are unacceptable for practical applications because of the structural defects introduced and the reduction of the crystallite size during deposition. By postannealing in air at temperatures ranging from 500 to 600°C, a grain growth of fine crystals and a defect recovery in AD layers were observed, which dramatically improved the ferroelectric properties. The dielectric constant (ε) and the piezoelectric constant (d_{31}) of postannealed layers formed at 600°C were 800~1200 and −100 pm/V [56], respectively, which is comparable to the values obtained with conventional thin-film formation methods. Moreover, the electrical breakdown (<1 MV/cm) and Young's modulus (>80 GPa) of the AD films exceeded those obtained with conventional thick-film formation technologies. By postannealing at temperatures up to 850°C, a remnant polarization (Pr) of 38 μC/cm^2 and a coercive field strength (Ec) of 30 kV/cm were obtained in the PZT films [57, 58]. Even without adding any special additives to the fed particles or using special procedures, compared with conventional screen-printing methods, the AD method permits a 300–400°C reduction of the process temperature. The most important characteristic of the AD method is that, compared with the properties of conventional screen-printed thick films, the as-deposited film has a more dense and crystallized structure that is close to the bulk material.

DEVICE APPLICATION

Antiplasma Corrosion Components Using AD–Yttrium Oxide Film

Figure 16.15 shows the cross-sectional SEM and TEM images of the yttrium oxide films fabricated by the AD method, and Table 16.3 shows the electrical and mechanical properties of the films. Conventionally, to sinter yttrium oxide requires the hot isostatic pressing (HIP) process with a temperature in excess of 1700°C. However, the AD method not only attained the dense consolidation at RT but also yielded an AD film with a greater electrical insulating property and mechanical strength than the bulk material. Figure 16.16 shows the test result of the antiplasma corrosion. Even before exposure to plasma (Fig. 16.15, upper images), the AD yttrium oxide film had better surface smoothness than the bulk yttrium oxide, and for the film produced by thermal spray coating, after exposure to plasma (Fig. 16.15 lower images), significant differences were found between the AD yttrium oxide films and the films produced by the other processes; that is, the AD yttrium oxide film was free from pores, although noticeable pores were found in the yttrium oxide samples fabricated by the other processes. This result demonstrates that the AD method considerably improved the antiplasma corrosion and the surface smoothness of yttrium oxide. With the AD method, we achieved the antiplasma corrosion, wear resistance, and reduction of adsorption gas and particle emission that are required of the next-generation semiconductor manufacturing industries. Figure 16.17 shows the large-area ($50 \times 50\,cm^2$) AD yttrium oxide films and the electrostatic chuck produced by the AD method.

The use of the AD method has gained popularity since 2007 in the coating technology of antiplasma corrosion films for next-generation semiconductor manufacturing

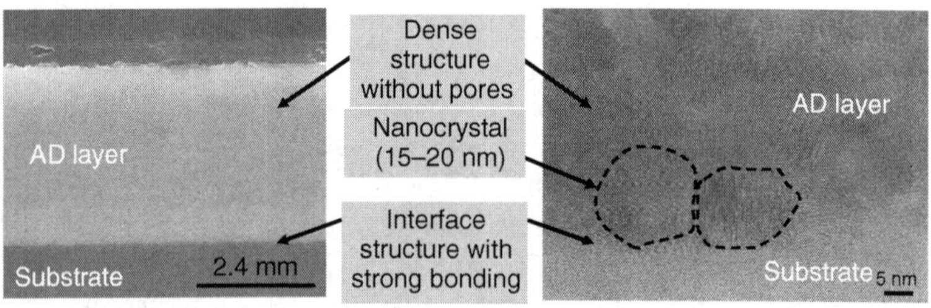

Figure 16.15. The fine structure of yttrium oxide films formed by the AD method at room temperature.

TABLE 16.3. The Electrical and Mechanical Properties of AD–Yttrium Oxide Films

Volume resistance (RT)	$>10^{14}\,\Omega cm$
Breakdown voltage	$150\,V/\mu m$
Vickers hardness	9.2 GPa
Adhesion force	80 MPa

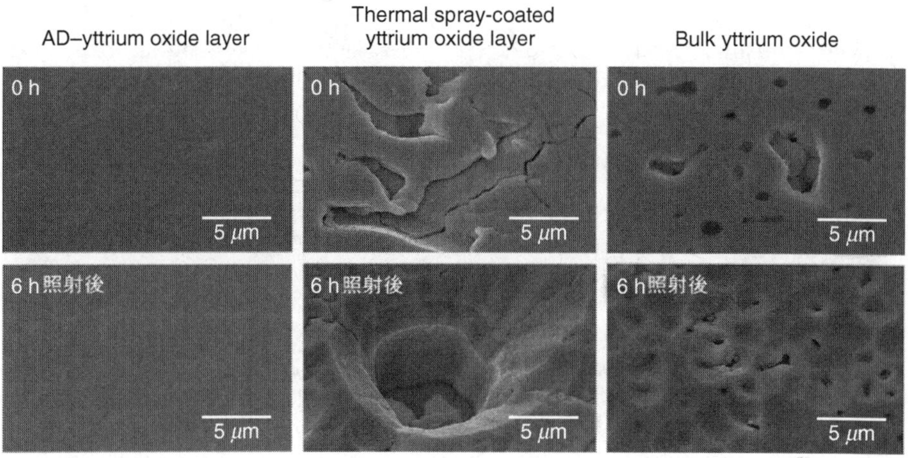

Figure 16.16. The surface morphology of yttrium oxide films produced by various methods before and after being exposed to plasma.

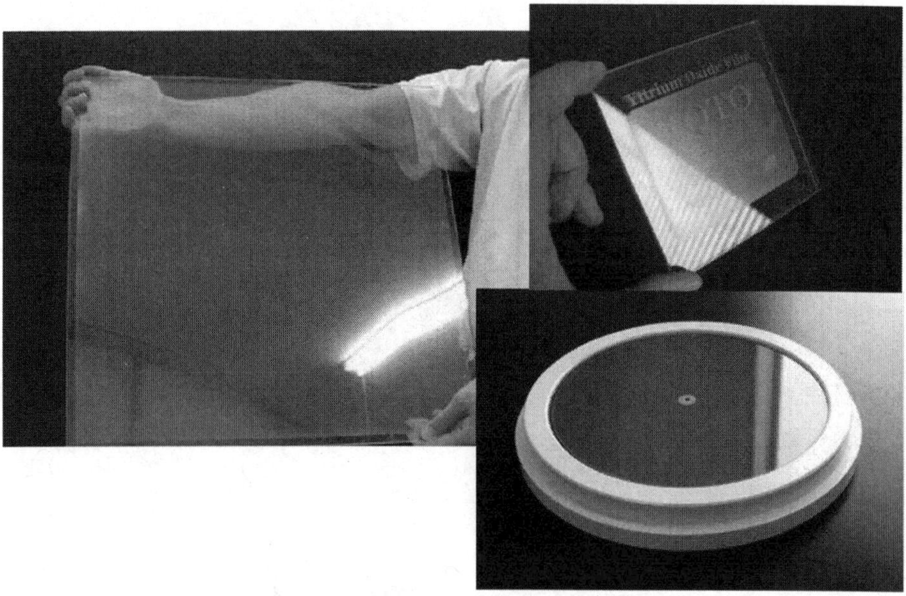

Figure 16.17. The large-area (50 × 50 cm^2) AD yttrium oxide films and the electrostatic chuck produced by the AD method.

facilities and the electrostatic chucks by TOTO Co. Ltd. The success of the development is due to the distinct feature of the AD method that can form high-performance ceramic films for the antiplasma corrosion and the wear resistance at low cost.

Piezoelectric Device Applications

Microdevice applications such as MEMS optical scanners, ink-jet heads, and ultrasonic motors have already been developed using piezoelectric materials. Using the AD method, a high-performance optical microscanner with a scanning speed at a resonance frequency over 30 kHz and a scan angle (peak-to-peak value) greater than 30° under atmospheric conditions was successfully fabricated by the deposition of the piezoelectric materials at a high rate onto the scanner structure fabricated by Si micromachining or mechanical punching, as shown in Figure 16.18 [59, 60]. This optical scanner with high scanning speed is expected to be a key component for various types of sensors for the next generation of projection display devices. Further research is still under way targeting the realization of 3-D displays, holographic data storage, and the development

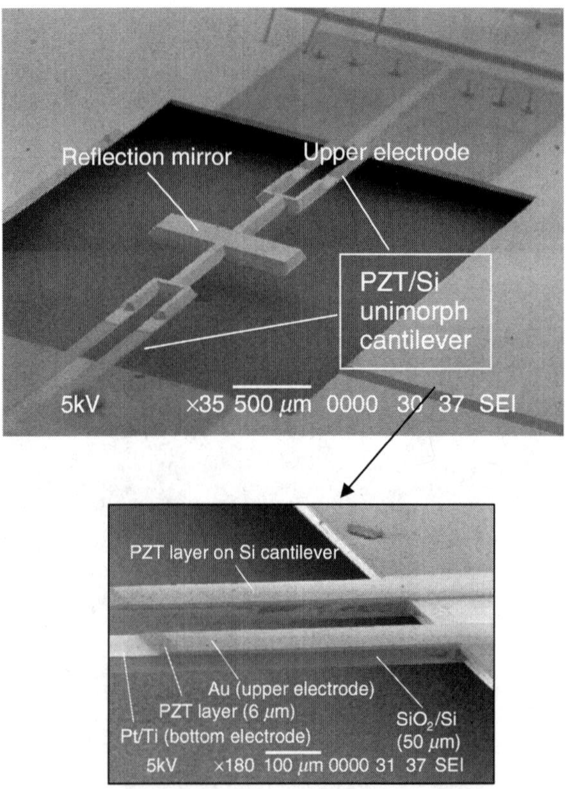

Figure 16.18. Optical microscanner driven with a PZT thick layer deposited on a Si MEMS structure by the AD method. Scanning speed: over 30 kHz; scanning angle: over 30°.

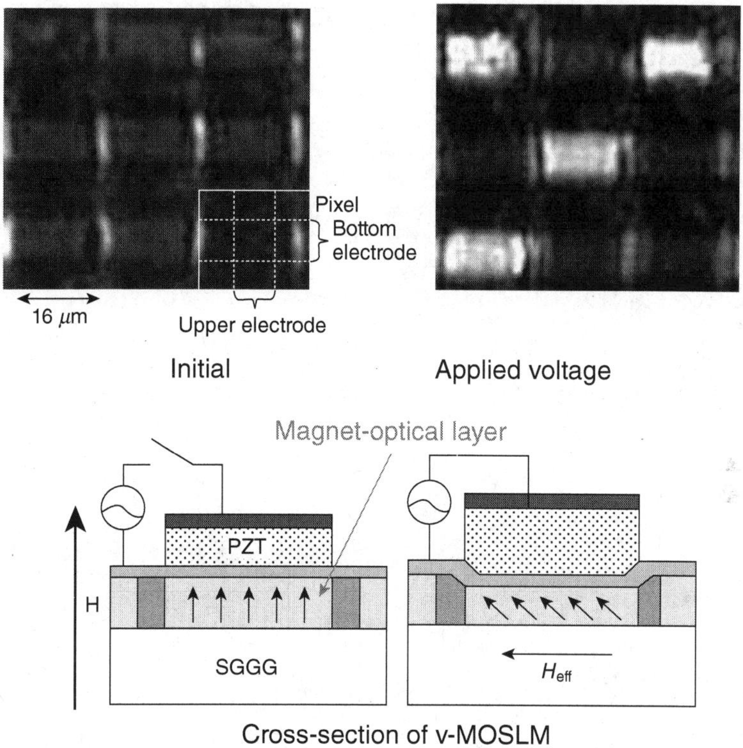

Figure 16.19. PZT-based magneto-optic spatial light modulators driven with a PZT thick layer deposited on a Bi-YIG single crystal structure by the AD method [61]. YIG, yttrium iron garnet.

of fast-response spatial light modulators, aiming to replace liquid crystal technology. A prototype PZT-based magneto-optic spatial light modulator (PZT-MOSLM), shown in Figure 16.19, has been fabricated by incorporating a piezoelectric thick film into magneto-optic materials [61]. In magneto-optic spatial light modulators (MOSLMs), the high switching speed results from the fact that the pixel switching is achieved by switching the direction of magnetization, up or down, within 1 ns. A novel MOSLM driven by an electric field instead of a magnetic field was first achieved by using the piezoelectric effect of AD-PZT thick films, which reduces the anisotropic energy of the structured garnet film. The pixels could be easily switched in the presence of a small external bias field. As a result, the power consumption of such an MOSLM was drastically reduced, being 10 times smaller than that of a conventional current-driven one. Successful pixel switching at 20 MHz has already been achieved with an 8-V drive voltage. Tube-type ultrasonic micromotors, shown in Figure 16.20, were also fabricated as the prototype for the application of the AD method [62]. In this device, a 10-μm PZT thick layer was deposited on a stainless steel tube having a 2-mm diameter by using another advantage of AD method, which consists in successful deposition onto

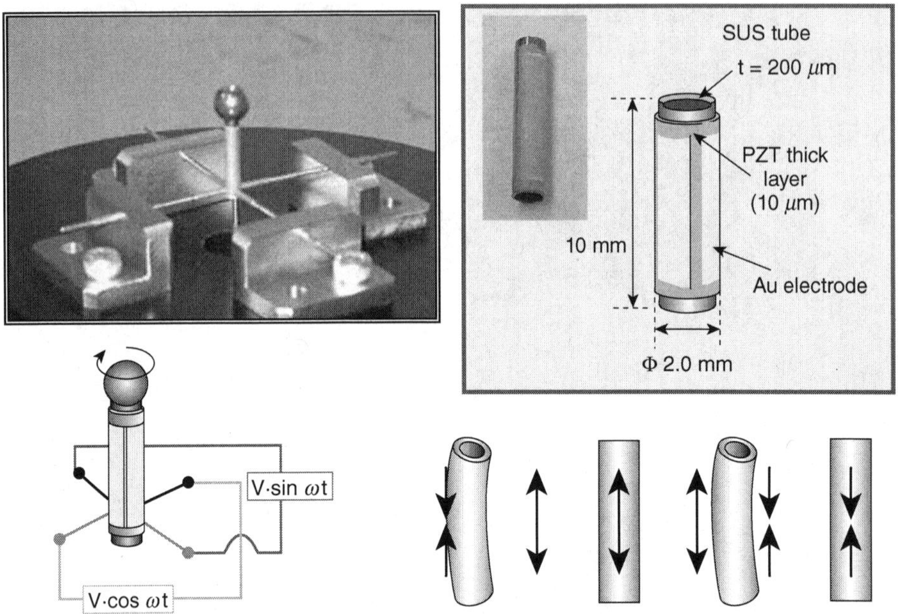

Figure 16.20. Tube-type ultrasonic motor driven with a PZT thick layer deposited on a SUS tube by the AD method.

curved surfaces. The rotation speed of this ultrasonic motor ranged between 1200 and 1500 rpm for a 7- to 15-V drive voltage.

High-Frequency Device Applications

With increasing central processing unit (CPU) speeds and higher communication frequencies, the surface mounting technology has reached its limit in the development of high-frequency devices in the gigahertz band. The present technologies of low-temperature cofired ceramic (LTCC) and ceramics/polymer composites are not necessarily satisfying the demands because of the degradation of material properties. To tackle the problem, a highly accurate fine-scale integration of the dielectric, magnetic, and metallic materials is required, and further miniaturization and higher-performance devices are needed. An embedded capacitor with a multilayer structure in an FR-4 printed circuit board, shown in Figure 16.21, has been developed by depositing $BaTiO_3$ ferroelectric materials onto a copper substrate using the AD method [63, 64]. A capacitance of over 300 nF/cm^2 was achieved, 10 times larger than the competitive technology of ceramics/polymer composite films. Thus, this technology has yielded the highest performance for a capacitor fabricated at process temperatures less than 300°C.

Additionally, electromagnetic interference (EMI) wave absorbers and microwave imaging sensors are currently under development. EMI suppression properties of Fe–ferrite film prepared by the AD method on a polyimide film were studied [65]. The

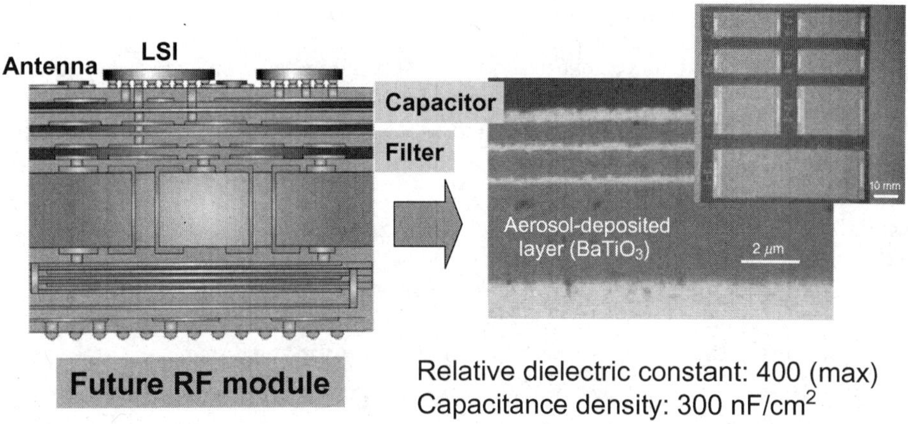

Figure 16.21. Formation of a thin-film capacitor in FR4 resign printed circuit boards at room temperature using the AD method [63]. LSI, large-scale integration.

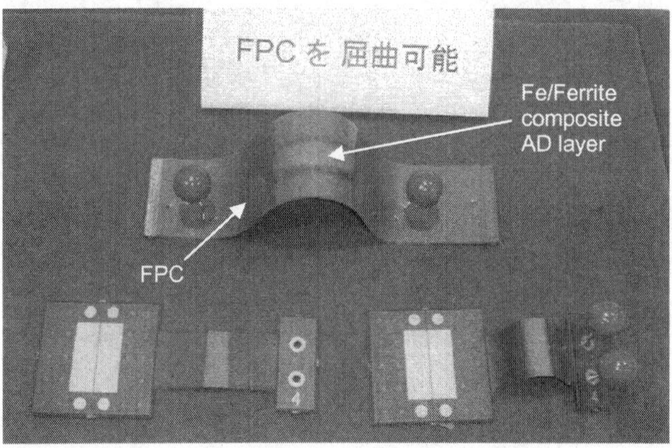

Figure 16.22. EMI suppression layer (Fe/Ferrite) deposited on an FPC at room temperature by the AD method.

deposition rates of Fe–ferrite composite films increased proportional to the iron ratio of Fe–ferrite powder before deposition. The Fe–ferrite composite film with Fe : ferrite = 8:2 (weight ratio) showed a remarkable EMI mitigation characteristic for microstrip line structures in the range of 2.5–10.0 GHz. Fe–ferrite composite films were also applied by the AD method on a flexible printed circuit (FPC), as shown in Figure 16.22. The FPC was connected to a transmitter board and a receiver board. Compared with an FPC without Fe–ferrite composite films, the FPC with such a film could suppress electric field intensity by about 10 dB around 1.2 GHz in the far field, which was the resonance frequency of the FPC [66].

Optical Device Applications

With the anticipated requirement of ultra-high-speed integrated optical circuits to deal with the need for high-capacity information processes, the development of an ultra-high-speed optical modulator has been studied. Using the AD method, lead lanthanum zirconium titanate (PLZT) electro-optical materials have been successfully deposited onto a glass substrate at temperatures 100°C lower than the conventional processing temperature. A transparent film was successfully obtained (Fig. 16.23) having an electro-optical constant (rc) of 102–168 pm/V that is two times larger than that of conventional thin films and five to six times larger than that of single-crystal $LiNbO_3$, showing the highest performance found so far [66, 67]. A Fabry–Perot-type optical modulator using this film was also fabricated. Recently, a microscopic electro-optical field probe [68] was fabricated by directly depositing a lead zirconate-titanate ($Pb[Zr_{0.3}Ti_{0.7}]O_3$) layer onto an optical fiber edge with a 125-μm diameter using the AD method, as shown in Figure 16.24. A radio frequency (RF) electro-optical signal was successfully measured over a microstrip line. The capability for gigahertz-range fields was also shown. This device has great potential for detailed electrical characterization in the microscopic regions of high-performance electronic products such as the interconnecting parts between LSI packages and printed circuit boards and spaces among different LSI chips in a package. In the future, the development of a high-speed optical modulator with a low driving voltage using a ceramic film is intended for applications in a wide variety of areas, such as the miniaturization of network equipment and high-speed computer data transfer.

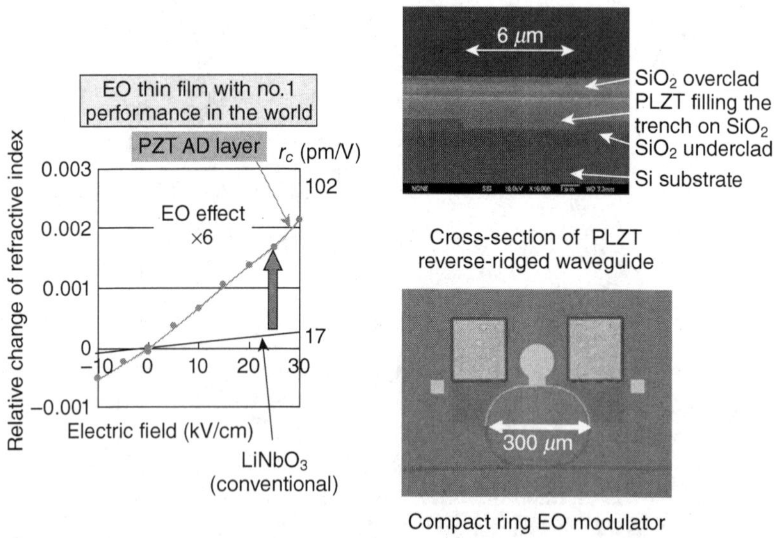

Figure 16.23. Electro-optical layer directly deposited on a Si structure for Si nanophotonic devices (ultra-high-speed integrated optical circuits at a 10-GHz range) by the AD method, and the EO constant (rc) of this layer [66]. EO, electro-optic.

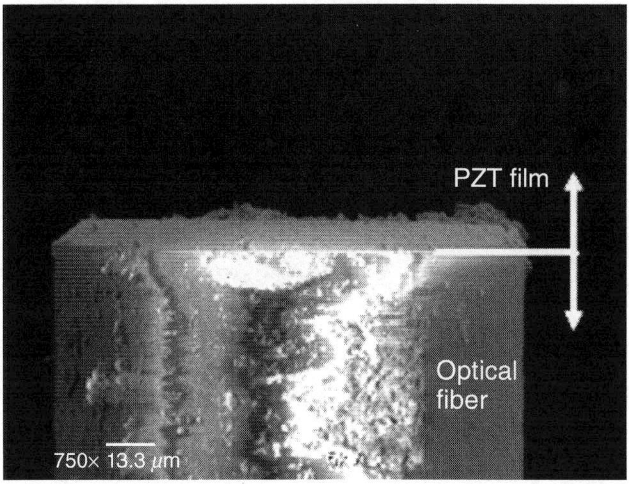

Figure 16.24. SEM image of an optical fiber edge that was taken from the side after the PZT film was deposited on a fiber facet [68].

SUMMARY

In this chapter, the high potential of the AD technique on the fabrication of electroceramic thick films for micro device applications has been highlighted. By comparison with conventional thin-film technology and thermal spray coating processes, the AD method shows important advantages, such as (1) high deposition rate, (2) low process temperature, (3) dense film formation at a low temperature, and (4) retention of primary material composition and crystal structure in the deposited films due to the RTIC process.

For the AD method, the following topics have been discussed:

1. High-speed formation of ceramic films at RT with high densification (over 95%) and high transparency (60–90%) is possible by deposition conditions and/or ceramic particle diameter optimization. Film formation is possible without thermal or plasma energy assistance due to a process called RTIC. The influence of particle impact velocity, the carrier gas, and starting powder properties on the deposition rate has been analyzed in an attempt to offer an explanation for the deposition mechanism. The deposition is explained by particle–substrate and particle–particle bonding due to the pressure and rise in temperature at the point of impact and particle breakdown during collisions.
2. The electrical properties of AD films were investigated along with the effects of a heat treatment on the deposited films for α-Al_2O_3 and PZT. The AD α-Al_2O_3 films generally have high electrical insulation and electrical breakdown characteristics that exceed the values for the bulk material. Also, the dielectric and piezoelectric constants of the thermally treated (600°C) PZT are comparable

with that obtained by conventional coating methods, but electrical breakdown and Young's modulus have higher values than those for conventional film coating techniques.

3. The application of the AD method to the fabrication of electronic components by the integration of thick and thin films has been introduced. Examples of piezoelectric, high frequency, and optical devices fabricated using the AD method were shown, emphasizing the advantages offered by this technique.

The AD technique offers the advantage of RT solid-state deposition on a wide variety of substrate materials.

More research is needed to clarify the deposition and recovery by annealing mechanisms and to understand the consolidation and particle–substrate and particle–particle bonding mechanisms.

ACKNOWLEDGMENTS

The author is very grateful to New Energy and Industrial Technology Development Organization (NEDO) project members for their collaboration. This research was supported in part by the NEDO project on "Nanostructure Forming for Advanced Ceramic Integration Technology in Japan: Nanotechnology Program."

REFERENCES

1. Muralt P, Kholkin A, Kohli M, Maeder T, Brooks KG, Luthier R. 1995. Fabrication and characterization of PZT thin films for micromotors. Integr Ferroelectr 11:213–220.
2. Akamine S, Albrecht TR, Zdeblick MJ, Quate CF. 1990. A planar process for microfabrication of a scanning tunneling microscope. Sens Actuators A 23:964–970.
3. Massood T-A. 1998. *Microactuators: Electrical, Magnetic, Thermal, Optical, Mechanical, Chemical and Smart Structures*. Tuller HL, editor. Norwell, MA: Kluwer Academic Publishers.
4. Manz A, Graber N, Widmer HM. 1990. Miniaturized total chemical analysis systems: a novel concept for chemical sensing. Sens Actuators B 1:244–248. Bergveld P. 1994. *Proc. Micro Total Analysis Systems Workshop μ-TAS'94*, 1.
5. Koganezawa S, Uematsu Y, Yamada T, Nakano H, Inoue J, Suzuki T. 1999. Dual-stage actuator system for magnetic disk drives using a shear mode piezoelectric microactuator. IEEE Trans Magn 35(2):988–992.
6. Akedo J, Lebedev M. 2002. Ceramics coating technology based on impact adhesion phenomenon with ultrafine particles: aerosol deposition method for high-speed coating at low temperature. Materia 41(7):459–466 (in Japanese).
7. Sayer M, Sreenivas K. 1990. Ceramic thin films: fabrication and applications. Science 247:1056–1060.
8. Abothu IR, Ito Y, Poosanaas P, Kalpat S, Komarneni S, Uchino K. 1999. Sol-gel processing of piezoelectric thin films. Ferroelectrics 232:191–195.

9. Chen HD, Udayakumar KR, Gaskey CJ, Cross LE, Bernstein JJ, Niles LC. 1996. Fabrication and electrical properties of lead zirconate titanate thick films. J Am Ceram Soc 79: 2189–2192.
10. Luginbuhl P, Racine G-A, Lerch P, Romanowicz B, Brooks KG, de Rooij NF, Renaud P, Setter N. 1996. Piezoelectric cantilever beams actuated by PZT sol-gel thin film. Sens Actuators A 54:530–535.
11. Watanabe S, Fujiu T, Fujii T. 1995. Effect of poling on piezoelectric properties of lead zirconate titanate thin films formed by sputtering. Appl Phys Lett 66:1481–1483.
12. Sakata M, Wakabayashi S, Ikeda M, Goto H, Takeuchi M, Yada T. 1995. Pb-based ferroelectric thin film actuator for optical applications. Microsystem Technologies 2:26–31.
13. Sakashita Y, Ono T, Segawa H, Tominaga K, Okada M. 1991. Preparation and electrical properties of MOCVD-deposited PZT thin films. J Appl Phys 69:8352–8357.
14. Kidoh H, Ogawa T, Morimoto A, Shimizu T. 1991. Ferroelectric properties of lead-zirconate-titanate films prepared by laser ablation. Appl Phys Lett 58(25):2910–2912.
15. Oikawa M, Toda K. 1976. Preparation of $Pb(Zr,Ti)O_3$ thin films by an electron beam evaporation technique. Appl Phys Lett 29(8):491–492.
16. Castellano RN, Feinstein LG. 1979. Ion-beam deposition of thin films of ferroelectric lead zirconate titanate (PZT). J Appl Phys 50(6):4406–4411.
17. Ohba Y, Miyauchi M, Tsurumi T, Daimon M. 1993. Analysis of bending displacement of lead zirconate titanate thin film synthesized by hydrothermal method. Jpn J Appl Phys 32: 4095–4098.
18. Shimomura K, Tsurumi T, Ohba Y, Daimon M. 1997. Preparation of lead zirconate titanate thin film by hydrothermal method. Jpn J Appl Phys 30:2174–2177.
19. Morten B, De Cicco G, Gandolfi A, Tonelli C. 1992. PZT-based thick films and the development of a piezoelectric pressure sensor. Hybrid Circuits 28:25–28.
20. Chen HD, Udayakumar KR, Cross LE, Bernstein JJ, Niles LC. 1995. Dielectric, ferroelectric, and piezoelectric properties of lead zirconate titanate thick films on silicon substrates. J Appl Phys 77:3349–3353.
21. Akiyama Y, Yamanaka K, Fujisawa E, Kowata Y. 1999. Development of lead zirconate titanate family thick films on various substrates. Jpn J Appl Phys 38:5524–5527.
22. Vasant Kumar CVR, Sayer M, Pascual R, Amm DT, Wu Z. 1991. Lead zirconate titanate films by rapid thermal processing. Appl Phys Lett 58(11):1161–1163.
23. Poor MR, Fledderman CB. 1991. Measurements of etch rate and film stoichiometry variations during plasma etching of lead-lanthanum-zirconium-titanate thin films. J Appl Phys 70:3385–3387.
24. Chung CW. 1998. Reactive ion etching of $Pb(Zr_xTi_{1-x})O_3$ thin films in an inductively coupled plasma. J Vac Sci Technol B 16:1894–1900.
25. Li X, Abe T, Esashi M. 2000. Deep reactive ion etching of Pyrex glass. *Proc. IEEE MEMS'2000*, p. 271, Miyazaki, Japan.
26. Saito K, Choi JH, Fukuda T, Ohue M. 1992. Reactive ion etching of sputtered $PbZr_{1-x}Ti_xO_3$ thin films. Jpn J Appl Phys 31:L1260–L1262.
27. Akedo J. 2004. Aerosol deposition method for fabrication of nano crystal ceramic layer: novel ceramics coating with collision of fine powder at room temperature. Mater Sci Form 449–452:43–48.
28. Akedo J, Lebedev M. 1999. Microstructure and electrical properties of lead zirconate titanate $(Pb(Zr_{0.52}/Ti_{0.48})O_3)$ thick film deposited with aerosol deposition method. Jpn J Appl Phys 38(9B):5397–5401.

29. Akedo J. 2006. Aerosol deposition of ceramic thick films at room temperature -densification mechanism of ceramic layer. J Am Ceram Soc 89(6):1834–1839.
30. Kiyohara M, Tsujimichi Y, Mori K, Hatono H, Migita J, Kusunoki T, Minami N, Lebedev M, Akedo J. 2002. *Proc. of 15th Ceram. Soc. Jpn. Autumn Symp.*, 228.
31. Levedev M, Akedo J, Mori K, Eiju T. 2000. Simple self-selective method of velocity measurement for particles in impact-based deposition. J Vac Sci Technol A 18(2):563–566.
32. Akedo J, Lebedev M. 2001. Influence of carrier gas conditions on electrical and optical properties of Pb(Zr,Ti)O3 thin films prepared by aerosol deposition method. Jpn J Appl Phys 40:5528–5532.
33. Johnson GR, Holmquist TJ. 1994. An improved computational constitutive model for brittle materials. In: Schmidt SC, Shaner JW, Samara GA, Ross M, editors. *High Pressure Science and Technology—1993*, Vol. 2, New York: AIP Press, pp. 981–984.
34. Holmquist TJ, Johnson GR, Grady DE, Lopatin CM, Hertel ES. 1995. High strain rate properties and constitutive modeling of glass. *Proceedings of 15th International Symposium on Ballistics*, pp. 237–244, TB31, Jerusalem, Israel, May 21–24.
35. Anderson CE, Johnson GR, Holmquist TJ. 1995. Ballistic experiments and computations of confined 99.5% AL2O3 ceramic tiles. *Proceedings of 15th International Symposium on Ballistics*, pp. 65–72, G6, Jerusalem, Israel, May 21–24.
36. Hirai H, Kondo K. 1994. Shock-compacted Si_3N_4 nanocrystalline ceramics: mechanisms of consolidation and of transition from α- to β-form. J Am Ceram Soc 77:487–492.
37. Petrovic JJ, Olinger BW, Roof RB. 1985. Explosive shock loading on alpha-Si_3N_4 powder. J Mater Sci 20:391–398.
38. Yoshida M, Ogiso H, Nakano S, Akedo J. 2005. Compression test system for a single submicron particle. Rev Sci Instrum 76:e-093905-1–5.
39. Dykhuizen RC, Smith MF, Gilmore DL, Neiser RA, Jiang X, Sampath S. 1999. Impact of high velocity cold spray particle. J Therm Spray Technol 8:559–564.
40. Vlcek J, Huber H, Voggenreiter H. 2001. Kinetic powder compaction applying the cold spray process—a study on parameter. *Proc. of ITSC2001*, 417.
41. Scudiero L, Dickinson JT, Enomoto Y. 1998. The electrification of flowing gases by mechanical abrasion of mineral surfaces. Phys Chem Miner 25:566–573.
42. Akedo J, Lebedev M. 2002. Powder preparation for lead zirconate titanate thick films in aerosol deposition method. Jpn J Appl Phys 41:6980–6984.
43. Lebedev M, Akedo J. 2000. Patterning properties of lead zirconate titanate (PZT) thick films made by aerosol deposition. IEEJ Trans Sens Micromach 120-E(12):600–601.
44. Akedo J. 2000. Study on rapid micro-structuring using jet-molding: present status and structuring subjects toward HARMST. Microsystem Technologies 6(11):205–209.
45. Ide T, Mori Y, Ikawa N, Yagi H. 1991. The film formation method using hypervelocity microparticle impact by electrostatic acceleration (1st report): electrodynamic behavior of ultrafine particles during film growth. J Jpn Soc Prec Eng 57:122–127 (in Japanese).
46. Fukuzawa F. 1991. Micro particle ion beam (macron beam). Oyo Buturi 60(7):720–721 (in Japanese).
47. Alkimov P, Kosarev VF, Papyrin AN. 1990. A method of cold gas-dynamic deposition. Dokl Akad Nauk SSSR 315(5):1062–1065.
48. Hayashi C, Kashu S, Oda M, Naruse F. 1993. The use of nanoparticles as coatings. Mater Sci Eng A 163:157–161.

49. Ide T, Mori Y, Konda I, Ikawa N, Yagi H. 1991. The film formation method using hypervelocity microparticle impact by electrostatic acceleration (2st report): preparation of diamond-like carbon film. J Jpn Soc Prec Eng 57:143–148 (in Japanese).

50. Hayahi C. 1987. Ultrafine particles. J Vac Sci Technol A 5(4):1375–1384.

51. Rao NP, Tymiak N, Blum J, Neuman A, Lee HJ, Girshick SL, McMurry PH, Heberlein J. 1998. Hypersonic plasma particle deposition of nanostructured silicon and silicon carbide. J Aerosol Sci 29:707–720.

52. Barborini E, Piseri P, Podesta A, Milani P. 2000. Cluster beam microfabrication of pattern of three-dimensional nanostructured objects. Appl Phys Lett 77:1059–1061.

53. Akedo J, Kiyohara M. 2006. Nanostructuring and shock compaction using fine particle beam—aerosol deposition for forming of nanocrystal layer and powder techology. J Soc Powder Technol 43(5):376–384 (in Japanese).

54. Nam S-M, Momotani M, Mori N, Kakemoto H, Wada S, Akedo J, Tsurumi T. 2004. Microstrip band pass filter of GHz region employing aerosol-deposited alumina thick films. Integr Ferroelectr 66:301–310.

55. Akedo J. 2005. Aerosol deposition for coating of transparent and high resistive ceramic layer. Metal (AGNE Gijutsu Center) 75(3):16–23 (in Japanese).

56. Akedo J, Lebedev M. 2000. Piezoelectric properties and poling effect of $Pb(Zr,Ti)O_3$ thick films prepared for microactuators by aerosol deposition. Appl Phys Lett 77(11): 1710–1712.

57. Akedo J, Lebedev M. 2002. Effects of annealing and poling conditions on piezoelectric properties of $Pb(Zr_{0.52},Ti_{0.48})O_3$ thick films formed by aerosol deposition method. J Cryst Growth 235:397–402.

58. Kawakami Y, Akedo J. 2005. Annealing effect on $0.5Pb(Ni_{1/3}Nb_{2/3})O_3$–$0.5Pb(Zr_{0.3}Ti_{0.7})O_3$ thick film deposited by aerosol deposition method. Jpn J Appl Phys 44:6934–6937.

59. Asai N, Matsuda R, Watanabe M, Takayama H, Yamada S, Mase A, Shikida M, Sato K, Lebedev M, Akedo J. 2003. A novel high resolution optical scanner actuated by aerosol deposited PZT films. *Proc. of MEMS 2003*, Kyoto, Japan, pp. 247–250.

60. Akedo J, Lebedev M, Sato H, Park JH. 2005. High-speed optical microscanner driven with resonation of lamb waves using $Pb(Zr,Ti)O_3$ thick films formed by aerosol deposition. Jpn J Appl Phys 44:7072–7077.

61. Takagi H, Mizoguchi M, Park JH, Nishimura K, Uchida H, M Lebedev, Akedo J, Inoue M. 2004. PZT-driven micromagnetic optical devices. Mat Res Soc Symp Proc 785: D6.10.1–D6.10.6.

62. Nanotech 2007 NEDO project. 2007. *Nanostructure forming for advanced ceramic integration technology in Japan: nanotechnology program*. Pamphlet, 2/21.

63. Imanaka Y, Akedo J. 2004. Integrated RF module produced by aerosol deposition method. Proceedings of the 54th Electronic Components and Technology Conference (ECTC), pp. 1614–1620.

64. Nam S-M, Yabe H, Kakemoto H, Wada S, Tsurumi T, Akedo J. 2004. Low temperature fabrication of $BaTiO_3$ thick films by aerosol deposition method and their electric properties. Tran MRS Jpn 29(4):1215–1218.

65. Sugimoto S, Haga K, Nakada M, Kagotani T, Inomata K, Akedo J. 2005. Magnetic properties of $Fe/(NiZnCu)Fe_2O_4$ composite films prepared by aerosol deposition method. *INTERMAG 2005*, Nagoya Japan, April 4–8, EC-04.

66. Nakada M, Ohashi K, Akedo J. 2005. Optical and electro-optical properties of $Pb(Zr,Ti)O_3$ and $(Pb,La)(Zr,Ti)O_3$ films prepared by aerosol deposition method. J Cryst Growth 275:e1275–e1280.

67. Nakada M, Ohashi K, Lebedev M, Akedo J. 2005. Electro-optic properties of $Pb(Zr_{1-x}Ti_x)O_3$ ($x = 0$, 0.3, 0.6) films prepared by aerosol deposition. Jpn J Appl Phys 44:L1088–L1090.

68. Iwanami M, Nakada M, Tsuda H, Ohashi K, Akedo J. 2007. Ultra small electro-optic field probe fabricated by aerosol deposition. IEICE Electro Express 4(2):26–32.

PART IV

NANO- AND BIOINTEGRATION

17

ADVANCES IN NANOINTEGRATION METHODOLOGIES: PATTERNING, POSITIONING, AND SELF-ASSEMBLY

Yoshitake Masuda[1] and Kunihito Koumoto[2]

[1]*National Institute of Advanced Industrial Science and Technology (AIST), Nagoya, Japan*
[2]*Nagoya University, Graduate School of Engineering, Nagoya, Japan*

INTRODUCTION

Metal oxides have recently been fabricated in solutions without high-temperature sintering in order to reduce energy consumption and to allow application for various substrates having low heat resistance. The fabrication of ceramic thin films from solutions has been encouraged by the development of environment-friendly chemistry such as green and sustainable chemistry [1–6], bioinspired materials chemistry [7], biomimetic materials chemistry [7], soft-solution processing [8–10], soft chemistry (*chimie douce* in French) [11], liquid-phase deposition [7, 12], chemical bath deposition (CBD) [12, 13], electroless deposition (ED) with catalyst [12, 13], successive ion layer adsorption and reaction (SILAR) [12, 13], sol–gel process [14, 15], hydrothermal reaction [16], electrodeposition [17, 18], and so on. Solution processing of ceramics allows us to prepare ceramic thin films on the surface of solids such as substrates, particles, and fibers. Ceramic nano/microstructures can also be fabricated by applying these solution systems to electronic or photonic devices.

Ceramic Integration and Joining Technologies: From Macro to Nanoscale, First Edition.
Edited by Mrityunjay Singh, Tatsuki Ohji, Rajiv Asthana, Sanjay Mathur.

Many kinds of lithography or patterning techniques have been developed to prepare patterns of thin films, for instance, X-ray/electronbeam lithography and photolithography [19], microcontact printing [20, 21], wet etching [22], ink-jet printing [23], embossing [24, 25], slip-pressing [26], charge-based printing [27], micromolding [28], and cold welding [29]. However, etching or lift-off processes are required in many of these methods, which causes degradation of performance, increases waste and energy consumption, and makes the process complicated. Additionally, etching or lift-off processes cannot be applied to corrosion-resistant ceramics. The deposition of thin films only on desired areas of a substrate is thus required for the pattering of ceramic thin films.

Additionally, nanointegration (NI) and liquid-phase patterning (LPP) of colloidal crystals have attracted much attention for future devices such as photonic crystals, dye-sensitized solar cells, molecular sensors, gas sensors, and nanoparticle devices. Colloidal crystals and particulate films have been thus fabricated by various methods. Natural opal is an assembly of SiO_2 particles and is prepared in nature by the process of self-assembly. Particles are assembled to face-centered cubic (FCC) (FCC lattice, FCC structure) or hexagonal close-packed (HCP) (HCP lattice, HCP structure) to form a closed pack structure. Nano/microparticle device fabrication can be learned from self-assembly in nature.

Patterning of assembled particles was realized by several methods [21, 23, 30–33]. Micropatterns of randomly deposited particles were fabricated by gas-jet [30, 31] and ink-jet [23, 32] printing methods or by electrostatic interaction [21]. Aizenberg et al. [21] supported substrates micropatterned with anionic and cationic regions in a particle suspension for 2 min. The substrates were then removed and the excess suspension was rinsed. The observation of the drying process of this wetted substrate showed that the particles within each negatively charged region moved toward the center of the circles and formed dense clusters. They fabricated micropatterns of randomly deposited particles and dots using electrostatic interactions and a drying process. Micropatterns of particle assemblies with high arrangement accuracy have also been realized by using templates such as micromolds or grooves. Kim et al. [34] fabricated micropatterns of the FCC (or HCP) structure from polymer particles (200 nmϕ) using a micromold. Colloidal solution was flowed into the crevice between a micromold with patterned grooves and a flat substrate to form a micropattern of close-packed particle assemblies along the grooves. Ozin et al.[35] added a drop of an aqueous dispersion of SiO_2 particles (840 nmϕ) at the crevice between the patterned silicon wafer having a pattern of etched V-shaped grooves and the flat block to fabricate the patterns of FCC colloidal crystal parallel lines. A particle suspension was also spread on the patterned wafer by spin coating to form (100)-oriented FCC colloidal crystals. Xia et al. [36] fabricated a two-dimensional (2-D) array of particle assemblies using a fluidic cell and a patterned template that has a 2-D array of cylindrical holes or trenches. Spherical particles were packed into each hole or trench by capillary, gravitational, and electrostatic forces. These methods realized micropatterns of close-packed particle assemblies having high arrangement accuracy. However, independent patterns cannot be formed by the micromold method because colloidal solutions were inserted through continuous molds. Additionally, preprocessing of substrates is necessary to have micromolds or grooves, and further progress is required to fabricate various complicated structures for application to photonic or electronic devices.

Here, we propose NI and LPP of ceramic thin films and particle assemblies [37]. Nano/micropatterns of metal oxide thin films and colloidal crystals were fabricated on self-assembled monolayers (SAMs). Many kinds of self-assemblies were effectively utilized in these processes.

NI OF CERAMICS

SAM Preparation for Patterning of Metal Oxides

SAM can modify the surface of solids such as a substrate, particles, or fibers with various functional groups, and the molecular recognition of functional groups of SAM was the key technique for LPP. Patterned SAMs were used as templates in our LPP processes to deposit ceramic thin films on desired areas of substrates.

An Si wafer (p-type Si [100], NK Platz Co., Tokyo, Japan) was sonicated in water, ethanol, or acetone for 10 min, respectively, and exposed for 15 min to ultraviolet (UV) light (184.9 and 253.7 nm) (low-pressure mercury lamp 200 W, PL21-200, 15 mW/cm^2 for 254 nm, SEN Lights Co, Osaka, Japan) to clean the surface. The UV light that we used in a recent study (PL21-200) has stronger power than that we used in former studies (NL-UV253, Nippon Laser & Electronics Lab, Nagoya, Japan). The octadecyltrichlorosilane (OTS, $C_{18}H_{37}SiCl_3$)-SAM or aminopropyltrimethoxysilane (APTS, $H_2NC_3H_5Si(OCH_3)_3$) were prepared by immersing the Si substrate in an anhydrous toluene (Aldrich Chemical Co., Tokyo, Japan) solution containing 1 vol% OTS (Acros Organics, Tokyo, Japan) for 15 min or APTS (TCI) for 2 h under a N_2 atmosphere (Fig. 17.1). The substrate with the SAM was baked at 120°C for 5 min to remove residual solvent and to promote chemisorption of the SAM. The control of preparation

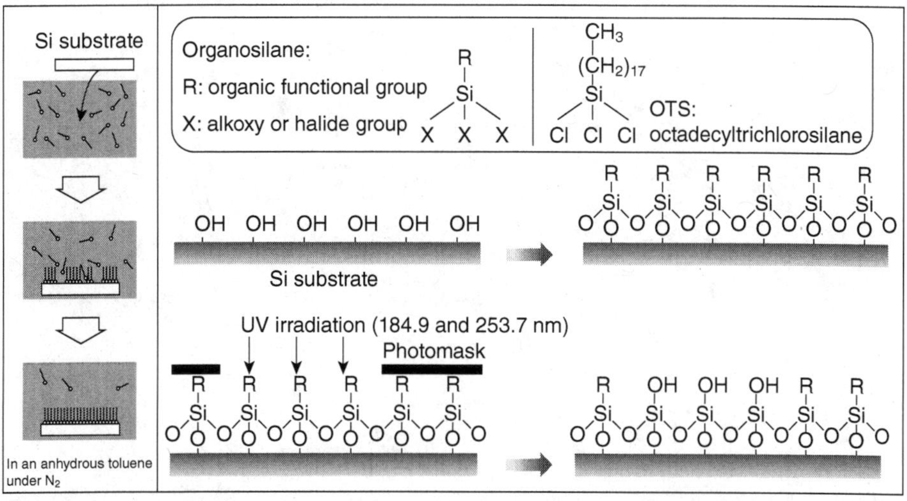

Figure 17.1. Conceptual process for the fabrication of a self-assembled monolayer.

conditions such as humidity is very important to fabricate organically modified surfaces, which realize site-selective deposition.

The SAMs on the silicon substrates were exposed for 15 min to UV light through a photomask to be used as a template for micropatterning of ZnO crystals. UV-irradiated regions became hydrophilic due to silanol group formation, while the nonirradiated part remained unchanged. Formation of the SAMs and the modification to silanol groups by UV irradiation were verified using the static water drop contact angle (θw) (a contact angle meter CA-D, Kyowa Interface Science Co., Saitama, Japan) and X-ray photoelectron spectroscopy (XPS) (ESCALAB 210, VG Scientific, Tokyo, Japan). The X-ray source (MgKα, 1253.6 eV) was operated at 15 kV and 18 mA, and the analysis chamber pressure was $1–3 \times 10^{-7}$ Pa. The initially deposited OTS-SAM or APTS-SAM showed a static water contact angle (WCA) of 105° or 63°, but the UV-irradiated surface of SAM was wetted completely (contact angle < 5°) (Fig. 17.2). The spectrum peak corresponding to the N 1s binding energy centered at 399.5 eV was observed for the surface of the APTS-treated Si substrate on which APTS-SAM was formed; however, it was not detected from the surface after UV irradiation. These experiments show the decomposition and removal of SAMs from the surface of substrates.

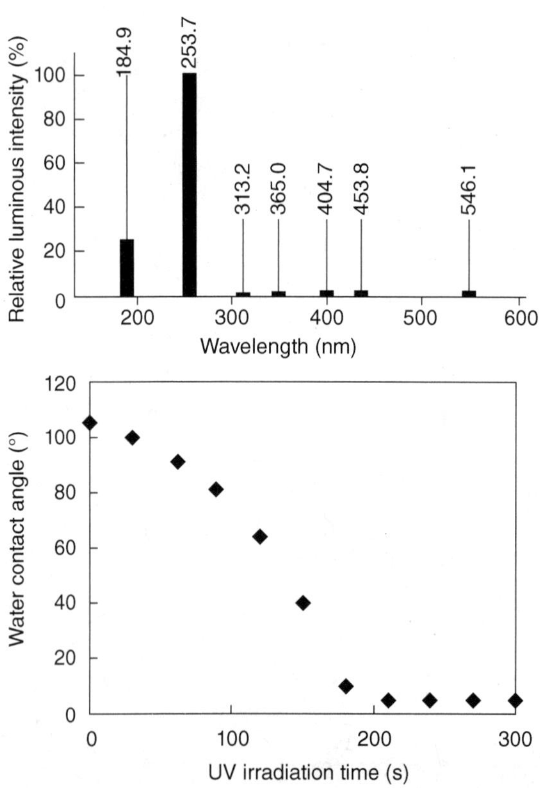

Figure 17.2. Relative luminous intensity of a UV lamp and water drop contact angle of OTS-SAMs as a function of UV-irradiation time.

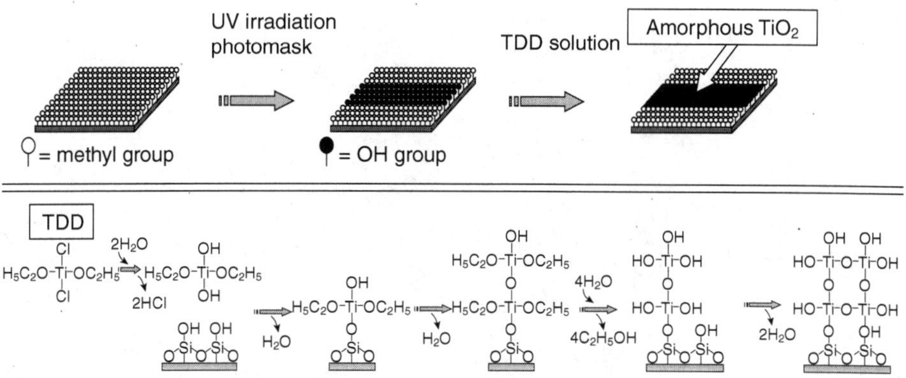

Figure 17.3. Conceptual process for the selective deposition of amorphous TiO_2 thin film using a self-assembled monolayer.

LPP of Amorphous TiO_2 Thin Films [38–41]

Patterned OTS-SAM was immersed in an anhydrous toluene (99.8%, water < 0.002%; Aldrich Chemical Co., Inc.) solution containing 0.1 M titanium dichloride diethoxide (TDD) for 30 min under an N_2 atmosphere using a glove box (Fig. 17.3) [38–41]. All glassware was dried in a dry box at 50°C before use. The estimated partial pressure of H_2O in an N_2 atmosphere is below 0.1 hPa. Chlorine atoms of TDD react with H_2O and change into OH, which further react with silanol groups of SAM resulting in the formation of Ti–O–Si bonds [42]. The ethoxy group, OC_2H_5, of TDD is hydrolyzed into hydroxyl groups, which are further condensed to form Ti–O–Ti bonds [42]. The thickness of films can be easily controlled by varying the soaking time. After SAM substrates had been rinsed with toluene and preserved in air, thin films appeared on the silanol surfaces of OTS-SAM but were not observed on octadecyl surfaces [38] (Fig. 17.4). A micropattern of amorphous TiO_2 thin films was thus fabricated on a patterned OTS-SAM. Line width measurements at 15 equally spaced points on each line indicated an average printed line width of 23.3 μm. Line edge roughness, as measured by the standard deviation of the line width, was ~0.5 μm, representing a ~2.1% variation (i.e., 0.5/23.2) in the nominal line width [38]. X-ray diffraction (XRD) measurements (Rigaku RU-200, Rigaku Co., Tokyo, Japan) with CuKα radiation (40 kV, 30 mA) for as-deposited thin films showed that they were composed of amorphous phases. The ratio of oxygen to titanium was evaluated after 20 min of Ar^+ ion sputtering to avoid the influence of the contaminated layer on the surface. The 1s peak of O can be deconvoluted into two curves (ratio of 529.7 eV [films] and 531.3 eV [silicon oxide] is 1.0:0.22). The ratio of oxygen to titanium was estimated to be 2.2:1.0. Small amounts of chlorine and carbon were also detected (Ti : O : Cl : C = 1.0:2.2:0.17:0.37) [38].

LPP of Crystalline Anatase TiO_2 Thin Films Using a Seed Layer [43]

The concept of LPP using a seed layer, which accelerates the deposition of thin films, was proposed [43]. The deposition process of anatase TiO_2 from an aqueous solution

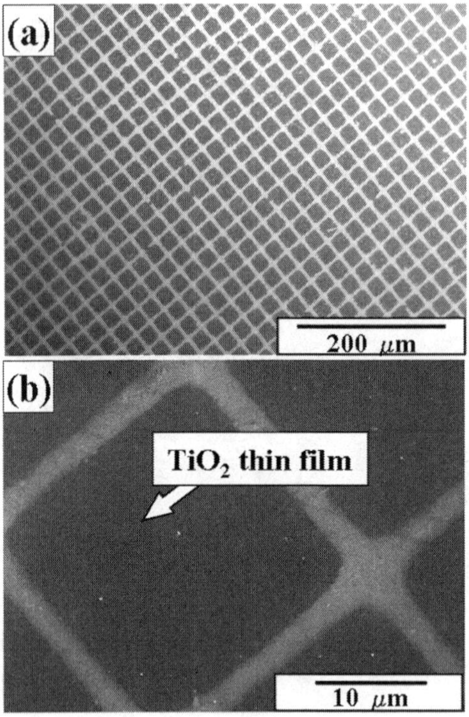

Figure 17.4. SEM of (a) a micropattern of amorphous TiO_2 thin films and (b) magnified area of (a).

was evaluated in detail using a quartz crystal microbalance (QCM), and it was found that the nucleation and initial growth of anatase TiO_2 were accelerated on amorphous TiO_2 thin films compared to silanol, amino, phenyl, or octadecyl groups. Amorphous TiO_2 thin films were deposited on silanol regions of a patterned OTS-SAM (Fig. 17.5a, b) from a TDD solution (Fig. 17.5c, d). This substrate was immersed in an aqueous solution containing a Ti precursor at pH 1.5 for 1 h to be used as a template for LPP. Anatase TiO_2 was selectively deposited on amorphous TiO_2 regions to form thin films. Consequently, a micropattern of anatase TiO_2 thin film, which had high feature edge acuity was successfully fabricated in an aqueous solution (Fig. 17.5e, f). The center of the anatase TiO_2 thin film region was 61 nm higher than the octadecyl regions, and the thickness of the anatase TiO_2 thin film was estimated to be 36 nm considering the thickness of amorphous TiO_2 thin film (27 nm) [38] and OTS molecules (2.4 nm) [43]. This result is similar to that estimated by QCM measurement (36 nm). The atomic force microscopy (AFM) image showed the film roughness to be 3.7 nm (horizontal distance between measurement points: 6.0 μm), which is less than that of amorphous TiO_2 thin film (root mean square [RMS]: 9.7 nm, 27 nm thick, horizontal distance between measurement points: 6.0 μm) [38]. Additionally, the roughness of the octadecyl group regions was shown to be 0.63 nm (horizontal distance between measurement points:

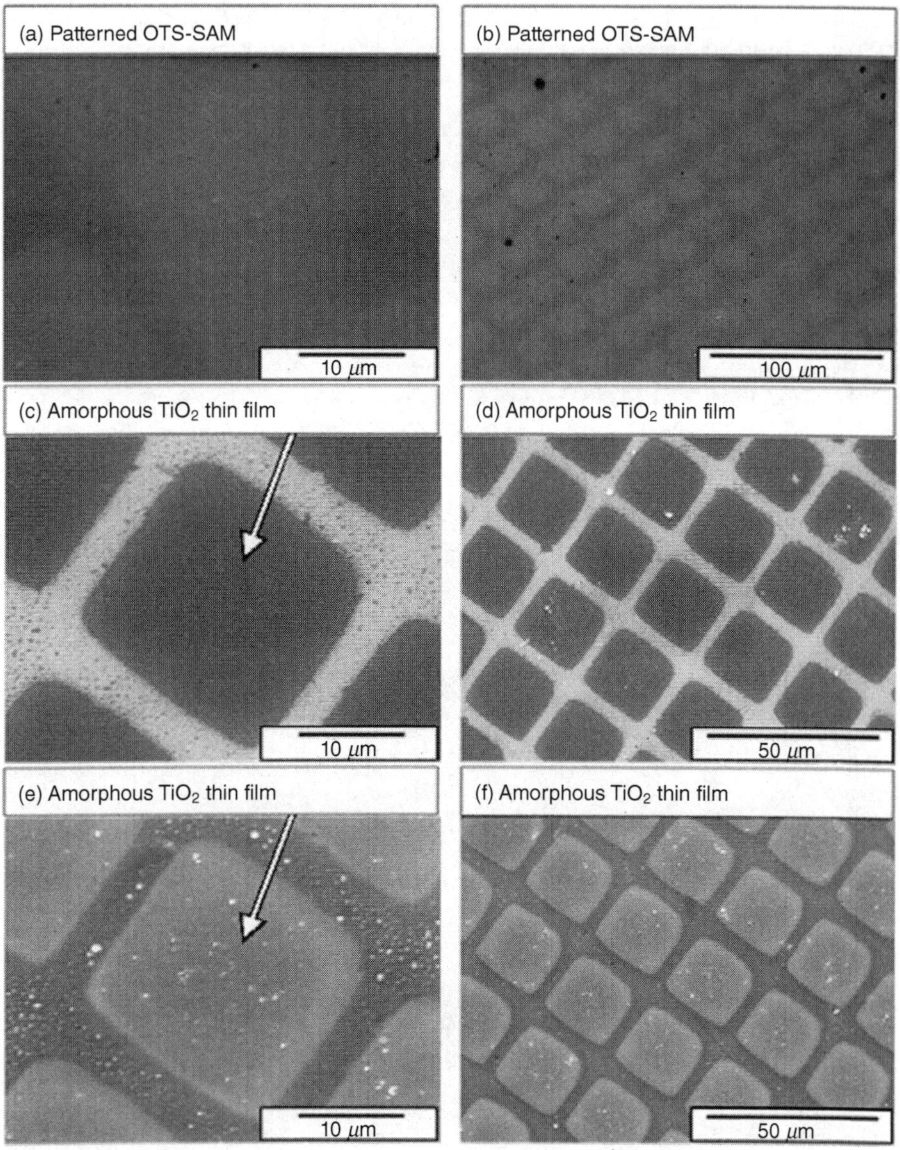

Figure 17.5. SEMs of (a, b) a patterned OTS-SAM, (c, d) a micropattern of amorphous TiO_2 thin films, and (e, f) a micropattern of anatase TiO_2 thin films deposited at pH 1.5.

1.8 μm) [43]. This study showed the good performance of the LPP process using a seed layer and the importance of quantitative analysis of the deposition process.

LPP of Crystalline Anatase TiO_2 Thin Films Using Site-Selective Elimination [44]

Ammonium hexafluorotitanate ($[NH_4]_2TiF_6$) (purity: 96%, 1.031 g) and boric acid (H_3BO_3) (purity: 99.5%, 0.932 g) were dissolved separately in deionized water (50°C, 50 mL) [44]. An appropriate amount of HCl was added to the boric acid solution to control pH, and ammonium hexafluorotitanate solution was added. Solutions (100 mL) with 0, 0.1, or 0.6 mL of HCl showed pH 3.8, 2.8, or 1.5, respectively. Supersaturation of solution can be changed by pH value as discussed in Reference 55. TiO_2 thin films can be formed fast by the deposition of homogeneously nucleated particles at a high pH condition such as pH 3.88, and uniform films can be obtained slowly by heterogeneous nucleation at a low pH condition. SAMs were immersed in the solution (100 mL) containing 0.05 M $(NH_4)_2TiF_6$ and 0.15 M (H_3BO_3) at pH 1.5, 2.8, or 3.8 and were kept at 50°C for 4 h to deposit anatase TiO_2. Ultrasonication was carried out during the immersion period. Deposition of TiO_2 proceeded by the following mechanism:

$$TiF_6^{2-} + 2H_2O \rightleftarrows TiO_2 + 4H^+ + 6F^- \quad (17.1)$$

$$BO_3^{3-} + 4F^- + 6H^+ \rightarrow BF_4^- + 3H_2O. \quad (17.2)$$

Equation 17.1 is described in detail by the following two equations:

$$TiF_6^{2-} \xrightarrow{nOH^-} TiF_{6-n}(OH)_n^{2-} + nF^- \xrightarrow{(6-n)OH^-} Ti(OH)_6^{2-} + 6F^- \quad (17.3)$$

$$Ti(OH)_6^{2-} \rightarrow TiO_2 + 2H_2O + 2OH^- \quad (17.4)$$

Fluorinated titanium complex ions gradually change into titanium hydroxide complex ions in an aqueous solution as shown in Equation 17.3. An increase in F^- concentration displaces Equations 17.1 and 17.3 to the left; however, produced F^- can be scavenged by H_3BO_3 (BO_3^{3-}) as shown in Equation 17.2 to displace Equations 17.1 and 17.3 to the right. Anatase TiO_2 was formed from titanium hydroxide complex ions ($Ti(OH)_6^{2-}$) in Equation 17.4, and thus the supersaturation degree and the deposition rate of TiO_2 depend on the concentration of titanium hydroxide complex ions. The high concentration of H^+ displaces the equilibrium to the left in Equation 17.1, and the low concentration of OH^-, which is replaced with F^- ions, suppresses ligand exchange in Equation 17.3 and decreases the concentration of titanium hydroxide complex ions at low pH, such as pH 1.5. The solution actually remained clear at pH 1.5, showing its low degree of supersaturation. On the other hand, the solution at high pH such as pH 2.8 or 3.8 became turbid because of homogeneously nucleated anatase TiO_2 particles caused by a high degree of supersaturation. Anatase TiO_2 thin film was formed by heterogeneous nucleation in the solution at pH 1.5, while the film was formed by heterogeneous nucleation and deposition of homogeneously nucleated particles at pH 2.8 or 3.8.

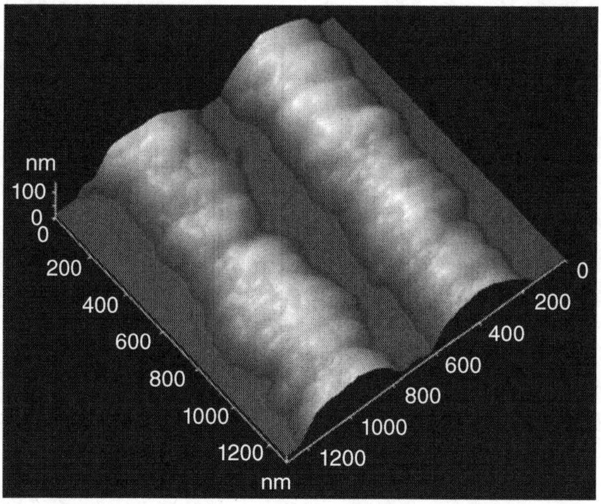

Figure 17.6. AFM image of a nanopattern of anatase TiO_2 fabricated by site-selective elimination.

After having been immersed in the solution with ultrasonic treatment, the substrates were rinsed with distilled water. Thin films were observed on the silanol group regions to form nano/microscaled patterns at pH 3.88, 2.8, or 1.5. The films were deposited for 4 h from the solution without the addition of HCl. Thin films were observed as being dark in an optical micrograph. Separated parallel lines 200–400 nm in width at 100- to 200-nm intervals were successfully fabricated with this method. The length of the separated parallel lines reached more than 100 μm. A cross-section of the lines was shown as a semicircle, and the thickness of the center of the lines was estimated to be about 100 nm by AFM observation (Fig. 17.6). The feature edge acuity of the pattern was higher than that of the pattern fabricated by our lift-off process or by the site-selective immersion method [45]. Site-selective deposition was realized at any pH condition, such as pH 3.88, 2.8, or 1.5. Patterns that have higher feature edge acuity can be obtained at low pH conditions because films were formed slowly without the deposition of homogeneously nucleated particles.

For the adhesion of TiO_2 films to silanol groups, the pH of the deposition solution is critical. Pizem et al. [13] reported that adherent TiO_2 films formed from solutions similar to those used here at pH = 3.9 but that the films were less adherent at pH = 2.9. They related this difference in adherence to an increased electrostatic attraction of TiO_2 to the oxidized surface of silicon at the higher pH. In our experiment, there also were less adherence of TiO_2 at pH 2.8 and 1.5 than that at pH 3.5 due to low supersaturation shown in Equation 17.1 and low electrostatic attraction [13].

However, some depositions were observed on octadecyl group regions in scanning electron micrographs (SEMs). One probable cause is that pinholes and other defects in the films provide at least some degree of access of water to underlying unreacted OH groups in the OTS films. Once exposed to the solution, these sites can act as nucleation points for TiO_2 growth. Because the depositions are performed at elevated temperatures, it is likely that pinholes and defects will continually open and close on the OTS film surfaces due to thermal motions of alkyl chains in the films. The TiO_2 precursors formed in these defects would act as points for eventual growth of TiO_2 over the entire SAM-covered region. This would provide a weakly bound TiO_2 film on the OTS film regions due to the limited number of connections to the underlying silanol sites in the film regions. In fact, Sagiv and others have shown that macroscopic defects induced in alkylsiloxane films can readily be accessed by solution species [46–48]. More recently, Dressick and coworkers demonstrated that solvent accessibility to underlying substrates in aromatic siloxane films is also important [49–51] and may be an even greater factor in controlling the properties of those films, which may account for our previous selectivity observation using phenylsiloxane films, as shown in the lift-off process.

In the lift-off process [52], thin films were formed on the entire area of patterned SAM that has silanol group regions and phenyl (or octadecyl) group regions. After being dried, the substrate was sonicated in water to lift off thin films on phenyl (or octadecyl) group regions selectively. Thin films on phenyl (or octadecyl) group regions were peeled off along the cracks that formed during the drying process. Thin films on phenyl (or octadecyl) group regions without cracks were not peeled off because depositions strongly connected to each other to form solid timber (monolith). The lift-off along cracks decreased the feature edge acuity of the pattern in this method. Thin films were formed on silanol group regions selectively and site-selective deposition was realized with our newly developed method. This resulted in high feature edge acuity of the patterns compared to our previous works [52].

Additionally, the micropattern of thin films was also fabricated by the site-selective immersion method [45]. A solution containing a Ti precursor contacted the hydrophilic regions during the experiment and briefly came into contact with the hydrophobic regions. The solution on the hydrophilic surface was replaced with a fresh solution by continuous movement of bubbles. Thus, TiO_2 was deposited and a thin film was grown on the hydrophilic regions selectively. This technique can be applied for the formation of many kinds of films from any solution and to fabricate micropatterns for many kinds of thin film because the technique creates the difference in contact time of the solution between hydrophilic and hydrophobic regions. However, it is difficult to form a solution layer on nanoscaled hydrophilic regions selectively and to replace it with a fresh solution by continuous movement of bubbles while avoiding contact of the solution on hydrophobic regions. This prevents fabrication of nanoscaled pattern with this method. On the other hand, site-selective deposition was realized in the solution with our newly developed method using the difference of adhesive strength of depositions to substrates. Heterogeneously nucleated deposition and homogeneously nucleated particles and/or clusters can be removed from octadecyl group regions even if these regions are designed in nanoscale order, in which depositions are smaller. This allowed us to realize high feature edge acuity of the patterns compared to site-selective immersion [43, 45].

The distribution of elements on the surface of the substrates was evaluated by energy-dispersive X-ray (EDX) analysis (EDAX Falcon, EDAX Co., Tokyo, Japan), which is built into SEM. Titanium was detected from thin films selectively and oxygen was detected mainly from silanol group regions by EDX. Other elements, except for silicon from the substrate, were not observed from the thin film and substrate by EDX.

Oxygen was detected from not only the deposited thin film but also from the natural oxide layer (amorphous SiO_2 layer) formed on all surface areas of a silicon substrate. These observations showed a predominant deposition of titanium oxide on silanol group regions.

The deposited thin films were also investigated using an X-ray diffractometer (RAD-C, Rigaku) with CuKα radiation (40kV, 30mA) and Ni filter plus a graphite monochromator. Thin films deposited at pH 3.8 for 4h showed an XRD pattern of anatase-type TiO_2 having an orientation similar to that of films deposited in the solution at pH 1.5 or 2.8 [45]. The diffraction from parallel to c-plane such as (004) was observed as being strong compared to that of the randomly oriented powder diffraction pattern. Pizem et al. [13] postulated that the commonly observed [001] orientation of anatase films could be due to the slight polarity of the planes parallel to the [001] axis, unlike other low-index planes of this structure such as {100}, (110), and (210). The orientation and crystal growth mechanism are further discussed in a separate article [53].

Thin films were further evaluated by XPS (ESCA-3200, Shimazu Corporation, Kyoto, Japan; 1×10^{-5} Pa). The X-ray source (MgKα, 1253.6eV) was operated at 8kV and 30mA. The spectral peaks corresponding to Ti 2p (458.7eV) were observed from thin films deposited on the silanol region. This binding energy is higher than that of Ti metal (454.0eV), TiC (454.6eV), TiO (455.0eV), TiN (455.7eV), and Ti_2O_3 (456.7eV), and similar to that of TiO_2 (458.4–458.7eV) [54–56]. This suggests that the titanium atoms in thin films are positively charged relative to that of titanium metal by the formation of direct bonds with oxygen. On the other hand, this spectrum was not observed from octadecyl group regions. The O 1s spectrum was observed from the silanol regions and was divided into O 1s (530.2eV) and O 1s (532.3eV). O 1s (532.3eV) can be assigned to the silicon oxide layer on the surface of the silicon wafer (532.0eV [55]), whereas the binding energy of O 1s (530.2eV) is similar to that of TiO_2 (529.9eV [56] and 530.1eV [55]) as observed by Shin et al. [57]. This shows that oxygen is negatively charged compared with neutral oxygen molecules (531.0eV), possibly through the formation of chemical bonds with Ti. The ratio of titanium to oxygen was estimated from the Ti $2p_{3/2}$ (458.7eV) spectrum and O 1s (530.2eV) spectrum to be Ti : O = 1.0:2.0.

LPP of Magnetite Particulate Thin Films Using a Pd Catalyst [58, 59]

A catalyst solution containing Na_2PdCl_4 (0.38mM) and NaCl (0.01M) in a 0.01M 2-morpholinoethane sulfonate pH 5 aqueous buffer was prepared [58, 59]. The details of preparation of this solution are described in Reference 60. Hydrolyzed Pd colloids were formed in this solution [61]. The patterned APTS-SAM was immersed into the colloidal dispersion of catalyst at 25°C for 30min and catalyzed APTS-SAM was rinsed with water.

Catalyzed SAM was immersed in an aqueous solution containing iron(III) nitrate (0.0025M) and dimethylamine borane (DMAB) (0.03M) and was kept at 80°C using a water bath for 30min to deposit magnetite particulate thin film [62]. DMAB was used to reduce nitrate ions, giving rise to OH– ions and hence raising the solution pH to precipitate Fe_3O_4.

A black-colored iron oxide film was selectively deposited onto regions of the APTS-SAM that had not been exposed to UV radiation following application of the Pd catalyst dispersion. Figure 17.7 shows an optical microscope image and a SEM image of as-deposited films, respectively. The black contrast represents a deposited film in an optical microscope image, whereas the white contrast shows deposited films in SEM images. The EDX mapping images shown in Figure 17.7c indicate the films deposited on the amino surface regions, showing mapping images consisting mainly of iron and oxygen. The thickness of the films was easily controlled in the range from several tens of nanometers to several micrometers by the change of immersion period. The XRD pattern of the thin film deposited on the whole surface of the APTS-SAM clearly indicates that it is a magnetite (Fe_3O_4) film composed of randomly oriented crystallites of about 20 nm in diameter, which was evaluated using the Scherer equation. These evaluations show the successful fabrication of a micropattern of crystalline magnetite films in an aqueous solution using a patterned APTS-SAM and Pd colloid catalysts adsorbed on amino group ($-NH_2$) regions of a SAM.

LPP and Morphology Control of Crystalline ZnO [63]

Zinc acetate ($Zn(CH_3COO)_2$, Kishida Chemical Co., Osaka, Japan) was dissolved into water to be 15 mM at 50°C, and ammonia (28% solution, Kishida) was then added to be 30, 60, or 90 mM ($[NH_3]/[Zn] = 2.0$, 4.0 or 6.0) with stirring as complexing agent [63]. These solutions showed pH = 7.04, 7.50, or 8.93, respectively. Zinc ions reacted with ammonium ions (NH_4^+) formed from ammonia to form tetraamine zinc(II) [64] $[Zn(NH_3)_4][2+]$. ZnO was crystallized from the reaction between $[Zn(NH_3)_4][2+]$ and OH^-. The solution became clouded shortly after adding ammonia due to homogeneous nucleation of ZnO crystals. The morphology of ZnO crystals was controlled by the ratio of ammonia to zinc acetate, that is, the supersaturation degree for crystallization. Patterned OTS-SAMs were immersed downward into the solution containing zinc acetate (15 mM) and ammonia (30 mM) as complexing agents ($[NH_3]/[Zn] = 2.0$) at 50°C for 3 h (Fig. 17.8).

ZnO crystals having long hexagonal cylinder shape were homogeneously nucleated to make the solution turbid shortly after adding ammonia. Crystals showed sharp hexagonal facets of about 100 nm in diameter and larger than 500 nm in length. The morphology indicated high crystallinity of ZnO nanoparticles. The nanoparticles were deposited and further grown on hydrophobic octadecyl group regions of a patterned SAM selectively. Consequently, a micropattern of light-emitting ZnO crystals was successfully fabricated in an aqueous solution without Pd catalyst. ZnO crystals were also deposited on hydrophobic regions of patterned SAMs such as DTS-SAM, HTS-SAM, PTS-SAM, MTS-SAM, PTCS-SAM or APTS-SAM. This showed that the method is highly versatile and offers good potential for the fabrication of devices.

ZnO crystals were deposited on hydrophobic SAM regions such as OTS-SAM, APTS-SAM, or other SAMs rather than hydrophilic silanol regions. The zeta potential of ZnO crystals deposited on a silicon substrate was measured to be 10 mV at pH 8.1, and ZnO crystals should thus have a positive zeta potential not less than 10 mV in the solution at pH 7.04. The SAM of OTS, silanol, and APTS showed zeta potentials of

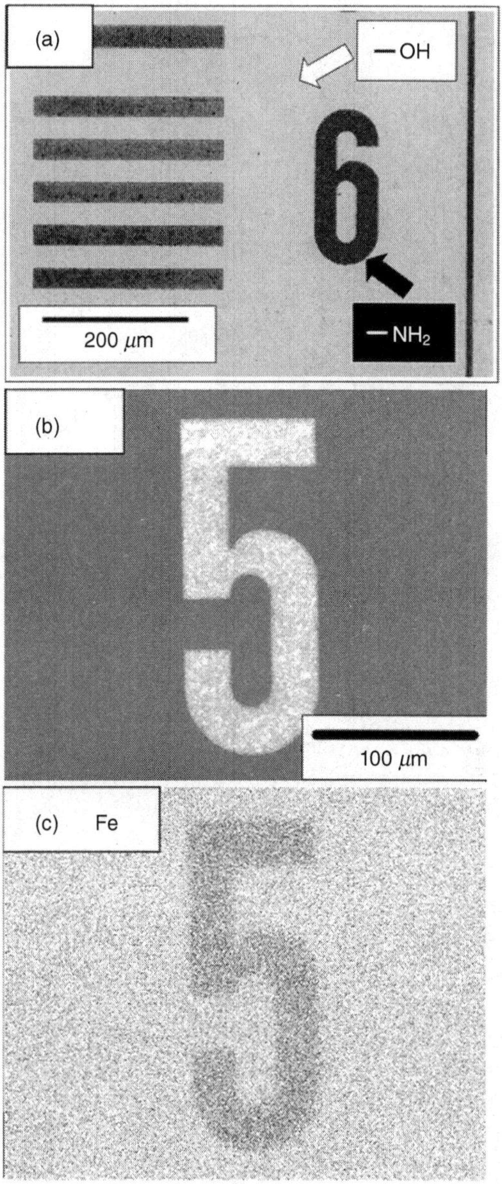

Figure 17.7. (a) Optical microscope image, (b) SEM image, and (c) characteristic X-ray images (Fe) of a micropattern of crystalline Fe_3O_4.

−3.0, −38.2, or +22.0 mV, respectively. ZnO having positive zeta potential should be deposited on silanol regions having a negative zeta potential rather than other SAMs if the site-selective deposition was caused only by electrostatic interactions. The site-selective deposition of ZnO crystals would be caused not only by electrostatic

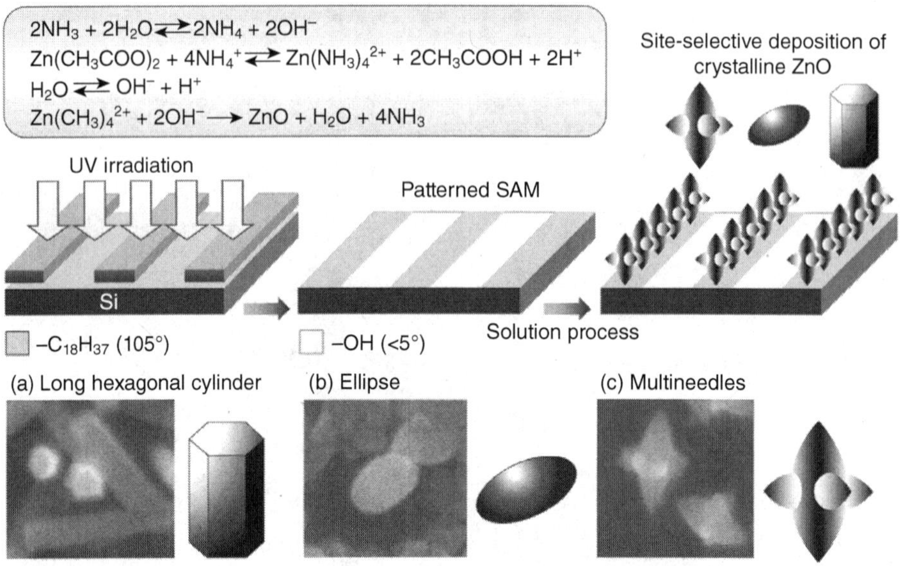

Figure 17.8. Conceptual process for the self-assembly patterning of light-emitting crystalline ZnO nanoparticles in an aqueous solution.

interactions as shown by the relation of zeta potentials. ZnO crystals having long hexagonal cylinder shapes were deposited on a hydrophilic silicon substrate to evaluate the surface of crystals. The substrate covered with many deposited ZnO crystals exhibited high WCA (140°). The deposited ZnO crystals were found from the experiment to have hydrophobic surfaces. Surfaces of naked ZnO crystals would be hydrophilic because of surface hydroxyl groups, and they would become hydrophobic by being covered with organic molecules having hydrophobic functional groups. CH_3COO^- ions coming from $Zn(CH_3COO)_2$ might be adsorbed to ZnO crystal surfaces by the interaction between Zn and $-COO^-$ to cover the surface with hydrophobic $-CH_3$ groups, and some of $Zn(CH_3COO)_2$ would exist in the surface layer of ZnO crystals. Additionally, deposited ZnO crystals having long hexagonal cylinder shape became hydrophilic (<10°), and their zeta potential shifted positively by UV irradiation in air. ZnO crystals deposited on a silicon substrate showed zeta potentials of 10 mV at pH 8.1, 0 mV at pH 8.8, and −15 mV at pH 9.2, while they shifted to 20 mV at pH 8.1, 10 mV at pH 8.8, and 7 mV at pH 9.2 by UV irradiation. The decomposition of CH_3COO^- ions and the breakage of the bond between CH_3COO and Zn would be caused by light excitation, ozone, and active oxygen by UV irradiation in air. This finding suggests that organic molecules, such as CH_3COO^- ions, which show negative zeta potential and can be removed by UV irradiation, would be absorbed onto the surfaces of ZnO crystals. Furthermore, ZnO crystals were confirmed to deposit on a hydrophobic polyethylene terephthalate surface rather than on a hydrophilic polyethylene terephthalate surface modified by UV irradiation in the same solution. Additionally, organic molecule was reported to adsorb to growing ZnO crystals, in which poly(ethylene oxide)-block-poly(methylacrylic acid) (PEO-b-PMAA) was adsorbed preferentially to the {0001} face of ZnO to retard crystal growth perpendicular to this face [65]. Consequently, site-selective deposition was achieved by the effective molecular recognition caused by a combination of the forces composed mainly of hydrophobic interactions between functional groups of SAMs and ZnO crystal surfaces.

Patterned SAMs were also immersed into the solution containing zinc acetate (15 mM) and ammonia (60 mM or 90 mM) as complexing agent ($[NH_3]/[Zn] = 4.0$ or 6.0) for 3 h. ZnO crystals having ellipse or multineedle shapes (two large needles and four small needles) were homogeneously nucleated to make the solution turbid shortly after adding ammonia. Nucleation and deposition of ZnO crystals were accelerated by the addition of ammonia. Each ZnO crystal was about 500 nm in size. The crystals were deposited and further grown on hydrophobic regions of patterned SAMs selectively. Micropatterns of light-emitting ZnO crystals having ellipse or multineedle shapes were fabricated on patterned SAMs such as OTS-SAM, dodecyltrichlorosilane (DTS)-SAM, hexyltrichlorosilane (HTS)-SAM, propyltrichlorosilane (PTS)-SAM, methyltrichlorosilane (MTS)-SAM, phenyltrichlorosilane (PTCS)-SAM, or APTS-SAM in aqueous solutions.

XRD spectra of ZnO crystals having ellipse or multineedle shapes showed dominant peaks corresponding to ZnO (0002) planes revealing that ZnO crystals were deposited with a high degree of orientation of their *c*-axes perpendicular to the substrate. Enhanced (0002) and (10-10) peaks from ZnO crystals having long hexagonal cylinder shapes showed that crystals were deposited to make (0002) or (10-10) planes parallel to the substrate. Crystals having high crystallinity and high purity with no additional phase were shown to be prepared in an aqueous solution with precise control of their morphologies without the use of a Pd catalyst. The aqueous solution system showed high ability for fabricating nano/microdevices composed of crystalline materials. ZnO crystals are well-known to grow along the *c*-axis. The orientations evaluated from XRD patterns were consistent with SEM observations and were shown to be controlled precisely by the solution conditions.

Photoluminescence properties of ZnO crystal patterns were further evaluated. Micropatterns of ZnO crystals were observed by an optical microscope (Fig. 17.9), and strong visible luminescence from ZnO crystals excited by a 330- to 385-nm light was observed by a photoluminescence microscope. ZnO crystals showed strong UV luminescence (around 390 nm) attributed to band-edge luminescence and visible-light luminescence caused from oxygen vacancy (450–600 nm) [66, 67]. All of the crystals showed photoluminescence due to high purity and high crystallinity with optimal oxygen vacancy, and this caused the bright visible photoluminescence image. Luminescence properties can be controlled by changing the crystalline morphologies. ZnO crystals deposited from an aqueous solution were shown to have high visible light-emitting properties.

LPP of Y_2O_3: Eu Thin Films [68]

The patterned APTS-SAM was immersed in an aqueous solution containing $Y(NO_3)_3{\cdot}6H_2O$ (4 mM), $Eu(NO_3)_3$ $6H_2O$ (0.4 mM), and NH_2CONH_2 (50 mM) at 25°C [68]. The solution was heated to 77°C gradually as shown in Figure 17.10 since urea

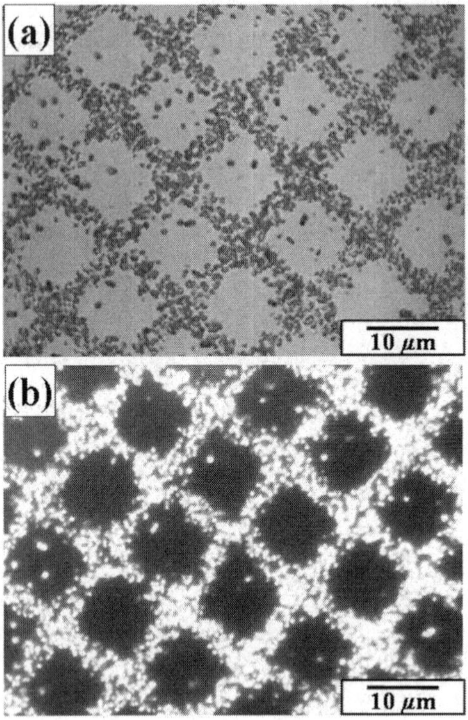

Figure 17.9. (a) Optical microscope image and (b) photoluminescence image of patterned ZnO particles under white light or UV light (330–385 nm).

(NH_2CONH_2) decomposes to form ammonium ions (NH_4^+) above 70°C (Equation 17.1). The decomposition of urea at elevated temperature plays an essential role in the deposition of yttrium oxide. The aqueous solution of urea yields ammonium ions and cyanate ions (OCN^-) at temperatures above 70°C [69] (Equation 17.1). Cyanate ions react rapidly according to Equation 17.2. Yttrium ions are weakly hydrolyzed [70, 71] in water to $YOH(H_2O)_n^{2+}$ (Equation 17.3). The resulting release of protons (H^+) and/or hydronium ions (H_3O^+) accelerates urea decomposition (Equation 17.2). The precipitation of the amorphous basic yttrium carbonate ($Y(OH)CO_3{\cdot}xH_2O$, $x = 1$) can take place through the reaction in Equation 17.4 [72, 73]. The controlled release of cyanate ions by urea decomposition causes deposition of basic yttrium carbonate once the critical supersaturation in terms of reacting component is achieved. Since the decomposition of urea is quite slow, the amount needed to reach supersaturation within a given period of time must be considerably higher than the stoichiometric amount of yttrium ions, as revealed by previous studies of lanthanide compounds [74].

The temperature of the solution increased gradually and reached 77°C in about 80 min. The solution was kept at ~77°C during deposition. The pH of the solution increased from 5.2 to 5.8 in about 90 min and then gradually decreased to 5.6. Temperature and pH increased for the initial 90 min and became stable after 90 min.

(1) Decomposition of urea (NH_2CONH_2)

$$NH_2\text{-}CO\text{-}NH_2 \rightleftarrows NH_4 + OCN^-$$

$$OCN^- + 2H^+ + H_2O \longrightarrow OC_2 + NH_4^+$$

(2) Deposition of amorphous basic yttrium carbonate ($Y(OH)CO_3 \cdot XH_2O$)

$$Y(NO_3)_3 \cdot 6H_2O \longrightarrow [YOH(H_2O)_n]^{2+} + 3NO_3^- + H^+ + (5\text{-}n)H_2O$$

$$[YOH(H_2O)_n]^{2+} + CO_2 + H_2O \rightleftarrows Y(OH)CO_3 \cdot H_2O + 2H^+ + (n-1)H_2O$$

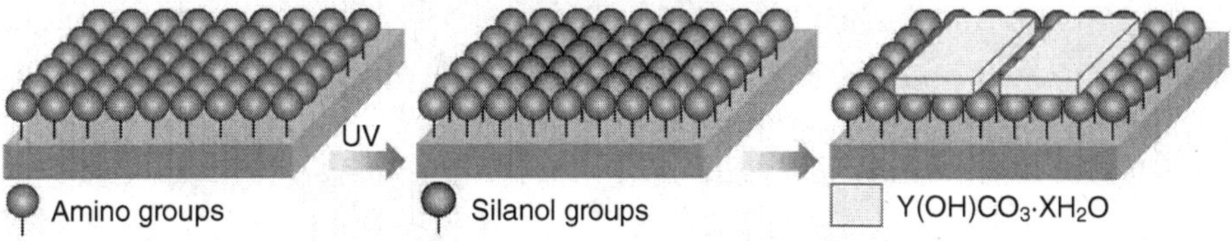

$Y(NO_3)_3 \cdot 6X_2O$	4 mM	NH_2CONH_2 50 mM
$Eu(NO_3)_3 \cdot 6X_2O$	0.4 mM	at 77°C

Figure 17.10. Conceptual process for site-selective deposition of visible light-emitting Y_2O_3:Eu thin films using a self-assembled monolayer.

The average size of particles homogeneously nucleated in the solution at 100 min was about 227 nm and increased to 262 nm at 150 min, 282 nm at 180 min, 310 nm at 210 min, and 323 nm at 240 min. Particles nucleated and grew after the solution temperature exceeded 70°C because urea decomposes above 70°C to form carbonate ions [69], which causes deposition of basic yttrium carbonate [70–73]. The particles grew rapidly at the beginning of the growth period and then their growth rate decreased exponentially. The decrease in growth rate was caused by the decrease of supersaturation degree influenced by a decrease in solution concentration.

Yttrium carbonate films were observed to deposit on amino regions of a patterned SAM after the immersion in an aqueous solution (Fig. 17.11). Deposits showed white contrast, while silanol regions without deposition showed black contrast in SEM observation. Narrow lines of depositions having 10- to 50-μm width were successfully fabricated in an aqueous solution. Patterned APTS-SAM showed high ability for site-selective deposition of yttrium carbonate in solution systems.

Yttrium carbonate films were also deposited on the hydrophobic octadecyl surface of OTS-SAM having a WCA of 116° and as-purchased silicon wafer having a WCA of about 20°–50°, which was kept in a plastic case in air. On the other hand, the films were not deposited on a UV-irradiated silicon wafer having a WCA < 5°. The superhydrophilic surface of WCA < 5° suppressed film deposition, whereas the hydrophobic surface and medium surface of WCA > 20°–30° accelerated film deposition possibly because of hydrophobic interaction between deposition and substrate surface. This is consistent with a former study [32]. Yttrium carbonate was deposited both on bare single-crystal Si wafers and on Si wafers coated with sulfonate-functionalized organic SAMs.

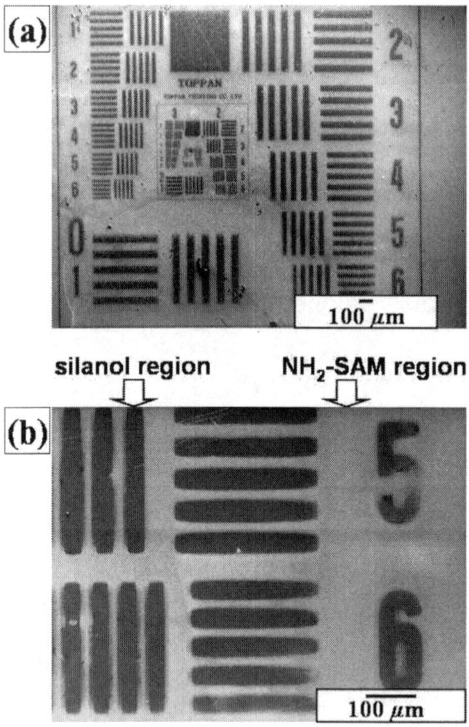

Figure 17.11. (a) SEM of patterned Y_2O_3:Eu thin films and (b) magnified area of (a).

Yttrium, europium, oxygen, and carbon were observed from as-deposited thin films on amino regions, while silicon and oxygen were detected from noncovered silanol regions by EDX. The molecular ratio of yttrium to europium was determined to be 100:8. It was close to that of $Y(NO_3)_3{\cdot}6H_2O$ to $Eu(NO_3)_3{\cdot}6H_2O$, that is, 100:10, in the solution because the chemistry of $Eu(NO_3)_3$ is similar to that of $Y(NO_3)_3$ to incorporate europium in the precipitation. The content of europium was in the range we had expected. Y_2O_3:Eu with atomic ratio Y:Eu = 100:~8 was reported to have strong photoluminescence [75, 76]. Carbon was detected from yttrium carbonate. Silicon and oxygen were detected from silicon wafer covered with a natural oxide layer (amorphous SiO_2).

Amino regions were covered with thin films composed of many large particles (about 100–300 nm in diameter) and very high roughness (RMS: 25.6 nm). Silanol regions, on the other hand, showed only nanosized small particles (about 10–50 nm in diameter) and very low roughness (RMS: 1.7 nm). The high site selectivity of deposition and the big difference in surface morphology and roughness were clearly shown by AFM observation. The thickness of the films was estimated from AFM scans across deposited and undeposited regions of the substrate. It increased with immersion time after 45 min (0 nm at 45 min, 60 nm at 70 min, and 100 nm at 90 min). The average growth rate

(70 nm/h = 100/90 min) was higher than that previously reported (2 nm/h = 35 nm/15 h) [73]. An amorphous yttrium basic carbonate film was deposited at 80°C from aqueous solutions of $YNO_3 \cdot 5H_2O$ and urea on Si wafers coated with sulfonate-functionalized organic SAMs in previous studies. The thickness was then evaluated by transmission electron microscopy (TEM) after the treatment with ultrasonication for half an hour in distilled water. The difference of growth rate was caused mainly by the difference of the substrate treatment by ultrasonication. Additionally, the thickness of our film was smaller than the particle size in the solution (227 nm at 100 min). Heterogeneous nucleation and attachment of initial particles of yttrium carbonate occurred without the attachment of aggregated large particles. The yttrium carbonate was then grown on the substrate to form a film of 100-nm thickness after immersion for 90 min. The particles of about 100 nm in height were removed by ultrasonication for 30 min, and the film of several nanometers in height remained as reported [73].

Yttrium was not detected by XPS from the substrate immersed for 45 min; however, it was clearly observed from that immersed for 90 min. This indicates that the deposition began between 45 and 90 min after immersion. The solution temperature reached 70°C in ~45 min, and then the solution began to decompose and release carbonate ions, causing the deposition of basic yttrium carbonate. The deposition mechanism evaluated by XPS is consistent with the change of solution temperature, the decomposition temperature of urea, and the chemical reaction of this system. The binding energy of the Y 3d5/2 spectrum from the deposition (158.2 eV) was higher than that of metal yttrium (155.8 eV) [77]. The spectrum shifted to lower binding energy (156.7 eV) after annealing at 800°C in air for 1 h and is similar to that of Y_2O_3 (157.0 eV) [78]. The binding energies of Y 3d5/2 spectra in as-deposited films and annealed films were higher than that of metal yttrium possibly due to the chemical bonds formed between yttrium ions and oxygen ions. The chemical shift of Y 3d5/2 binding energy by annealing is consistent with crystallization of as-deposited films to crystalline Y_2O_3. C 1s spectra were detected at 289.7 and 284.6 eV from as-deposited films. The C 1s spectrum at 289.7 eV then disappeared by the annealing. C 1s at 284.6 eV was assigned to surface contamination, and C 1s at 289.7 eV was detected from as-deposited yttrium carbonate. The disappearance of C 1s at 289.7 eV is consistent with the phase transition from yttrium carbonate to Y_2O_3.

As-deposited film was shown to be an amorphous phase by XRD measurement. The film showed no diffraction peak after annealing at 400°C for 1 h; however, it showed 222, 400, and 440 diffraction peaks of crystalline cubic Y_2O_3 [79] without any additional phase after annealing at 600°C for 1 h, and the intensities of diffraction peaks increased further by annealing at 800°C for 1 h. The film was shown to be a polycrystalline Y_2O_3 film constructed from randomly deposited Y_2O_3 particles without crystal axis orientation. The crystal structure model and diffraction pattern of Y_2O_3 were calculated from the crystal structure data of the Inorganic Crystal Structure Database (ICSD) 23811. The crystallization by annealing confirmed from XRD measurement is consistent with XPS evaluation.

We attempted to remove Y_2O_3 films from the silicon substrate by debonding with scotch tape or by ultrasonication for 5 min in water. However, the films maintained their bonds with the substrate, indicating that strong adhesion had formed between films and substrate.

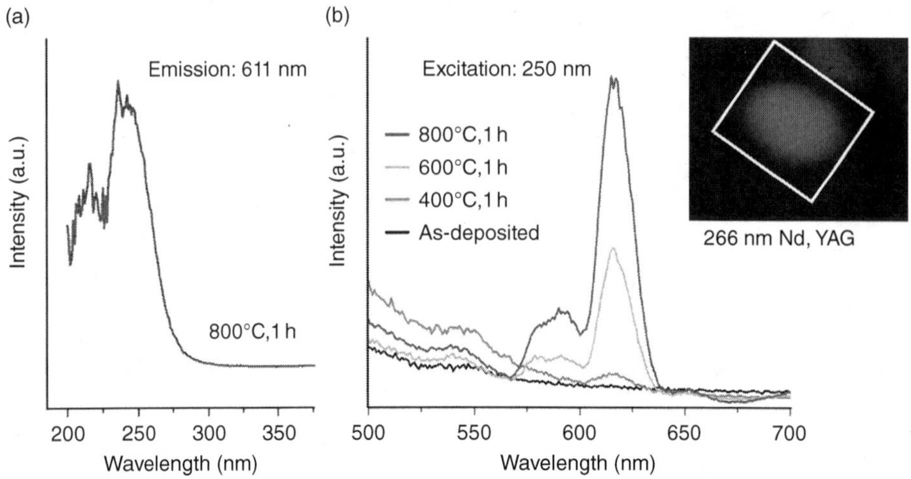

Figure 17.12. (a) Fluorescence excitation spectrum (emission: 611 nm) for Y_2O_3:Eu thin film after annealing at 800°C for 1 h. (b) Fluorescence emission spectra (excitation: 250 nm) for Y_2O_3:Eu thin films before and after annealing at 400, 600, or 800°C for 1 h. Inset: photoluminescence image for Y_2O_3:Eu thin film annealed at 800°C for 1 h (excitation: 266 nm).

The thin film annealed at 800°C for 1 h, that is, crystalline Y_2O_3:Eu thin film, was shown to be excited by 230–250 nm (center: 243 nm) and to emit red light photoluminescence centered at 611 nm in the fluorescence excitation spectrum (Fig. 17.12a). Neither the as-deposited film nor the film annealed at 400°C for 1 h showed photoluminescence; on the other hand, the films annealed at 600 or 800°C for 1 h emitted light centered at 617 nm by 250 nm in fluorescence emission spectra (Fig. 17.12b). The fluorescence intensity of the film annealed at 800°C was stronger than that of the film annealed at 600°C. Fluorescence intensity increased by the phase transformation from amorphous yttrium carbonate to yttrium oxide and crystal growth by the heat treatments, and is consistent with the crystallization observed by XRD [32]. The spectra are described by the well-known $^5D_0$–^7F_J line emissions ($J = 0, 1, 2, \ldots$) of the Eu^{3+} ion with the strongest emission for $J = 2$ at 612 nm. The thin film annealed at 800°C produced visible red light photoluminescence by excitation from Nd:YAG laser (266 nm) (Fig. 17.12, inset). The white square shows the edges of the Y_2O_3:Eu thin film, and the red color shows visible red emission from the irradiated area on the substrate.

Summary

We proposed a novel concept LPP to fabricate 2-D ceramic patterns in an aqueous solution without the etching process. Nano/micropatterns of ceramics such as TiO_2, Fe_3O_4, and ZnO were successfully fabricated on a patterned SAM. This result is a step toward fabricating nano/microstructures having high performance for next-generation ceramic devices.

NI OF PARTICLE ASSEMBLIES

Patterning of Colloidal Crystals in Liquids [80–82]

Patterning of Colloidal Crystals in Liquids. Novel processes to realize low-dimensional arrangement of SiO_2 particles were proposed [80–82]. Particle wires and a pattern of the close-packed particle monolayer were fabricated in the solution at room temperature. SAMs were formed on Si substrates and were modified to be suitable for templates of precise arrangement. Particles were arranged precisely in the desired positions in the solution using well-controlled electrostatic interactions and chemical bond formation between particles and substrates.

SAM Preparation for Patterning of Colloidal Crystals. OTS-SAM was prepared by immersing the Si substrate in an anhydrous toluene solution containing 1 vol% OTS for 5 min under a N_2 atmosphere. SAMs were exposed for 2 h to UV light (184.9 nm) through a photomask. The UV-irradiated regions became hydrophilic due to Si–OH group formation, while the nonirradiated part remained unchanged; that is, it was composed of hydrophobic octadecyl groups, which gave rise to patterned OTS-SAM. To check successful film formation and functional group change, water drop contact angles were measured for irradiated and nonirradiated surfaces. Initially deposited OTS-SAM had a WCA of 96°, while the UV-irradiated SAM surface was saturated (contact angle < 5°).

The patterned OTS-SAM was immersed in a toluene solution containing 1 vol% APTS in air for 1 h. APTS molecules combined to silanol groups of SAM and hence, octadecyl/amino group patterned SAM was fabricated. OTS-SAM exhibited a WCA of 96°, while that of the amino surface was 28°. These observations indicated successful fabrication of octadecyl/amino group patterned SAM.

Surface Modification of SiO2 Particles. Silica particles (1 $\mu m\phi$, HIPRESICA UF, UNK, Tokyo, Japan) were immersed in a dicyclohexyl and were sonicated for 10 min under a N_2 atmosphere for good dispersion. One volume percent of trichlorocyanoethylsilane (TCES) was added to the dicyclohexyl solution under a N_2 atmosphere, and the solution was stirred gently for 30 min in order to chemisorb TCES onto the SiO_2 particle surfaces. SiO_2 particles with TCES were centrifuged several times to remove unreacted TCES using dicyclohexyl.

The SiO_2 particles with TCES were further dispersed in a tetrahydrofuran solution containing potassium tert-butoxide (t-BuOK) and 18-crown 6-ether for 48 h under an ambient atmosphere to oxidize the CN groups to carboxyl groups. The solution was centrifuged several times using distilled water to remove t-BuOK, 18-crown 6-ether, and a tetrahydrofuran. SiO_2 particles modified with carboxyl groups were thus obtained.

SiO_2 particles covered by silanol groups or carboxyl groups were arranged selectively in silanol regions or amino regions of SAM using interactions between particles and SAMs. Zeta potentials of SiO_2 particles that have silanol groups and SiO_2 particles modified by carboxyl groups were measured (Zetasizer 3000HSA, Malvern Instruments Co., Tokyo, Japan) as shown in Figure 17.13. Zeta potentials measured in aqueous

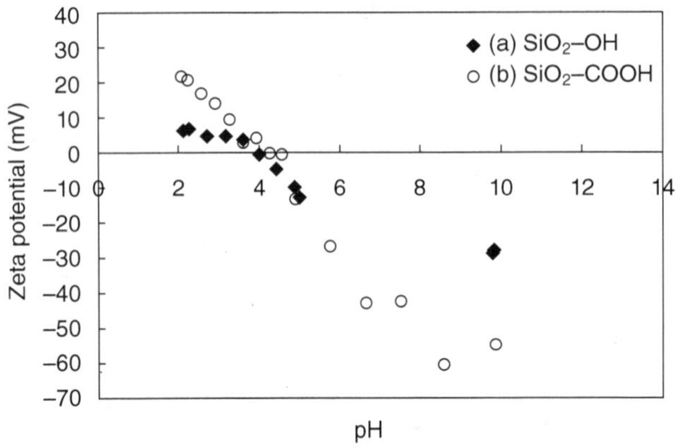

Figure 17.13. Zeta potential of (a) SiO_2 particles and (b) SiO_2 particles modified with carboxyl groups.

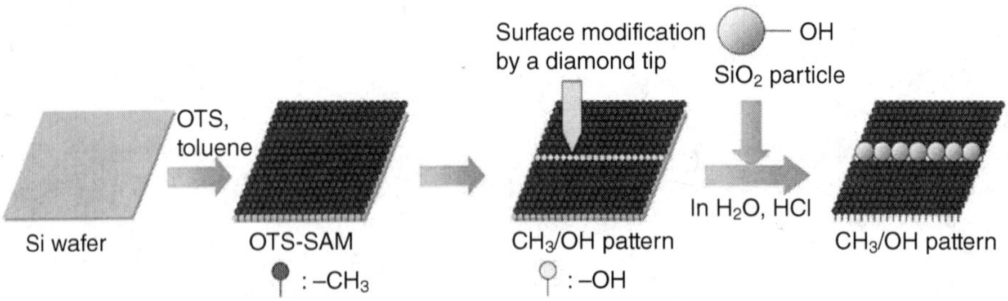

Figure 17.14. Conceptual process for the fabrication of a particle wire on a patterned SAM modified by a diamond tip.

solutions (pH = 7.0) for the surface of silicon substrate covered with silanol groups, phenyl groups (PTCS), and amino groups (APTS) are −38.23, +0.63, and +22.0 mV, respectively.

Fabrication of Particle Wires Employing Selective Arrangement Process. OTS-SAM was modified by a diamond tip to form a line of silanol groups of approximately 0.5-μm width (Fig. 17.14). The diamond tip was contacted to OTS-SAM surface lightly and was traced with low contact pressure in order to modify the SAM surface. The surface modified by a diamond tip, that is, the white area in Figure 17.14, corresponds to silanol groups, showed low contact angle (<5°). This modified region was shown to be white compared to the OTS-SAM region in a SEM (S-3000N, Hitachi, Tokyo, Japan). Octadecyl groups were broken mechanically by contact pressure with the diamond tip, and they possibly changed into silanol groups. The diamond tip was used to avoid contamination from a metal tip and the influence of a chemical reaction between the tip and the SAM.

Patterned SAMs were immersed in the aqueous solution containing SiO_2 particles and a hydrochloric acid as a condenser and rinsed in water, and were observed by SEM.

Figure 17.15. SEMs of (a) a single-particle wire, (b) a double-particle wire, (c) a triple-particle wire, and (d) a curved double-particle wire.

SiO_2 particles were observed on lines of silanol groups selectively indicating particles were successfully arranged well (Fig. 17.15a).

Because particles were not easily removed by sonication, it was judged that iloxane bonds had been formed by condensation of silanol groups between particles and a SAM.

It is clearly seen that the accuracy of particle arrangement has been improved compared to our former experiments [81, 82]. A double- and a triple-particle wire were likewise fabricated on wide silanol group regions with about 1.4 and 2.2 μm in width, respectively (Fig. 17.15b, c). The double-particle wire that has a triangular lattice also demonstrates high arrangement accuracy, though there is a defect in arrangement between the seventh particle from the left and the eighth particle. Additionally, a curved double-particle wire was fabricated on curved region of silanol groups. Curved double-particle wires have not been reported previously, and they may have useful applications for an optical waveguide.

The accuracy of particle arrangement was evaluated from Figure 17.15a. Center position ($[x_i, y_i]$ μm) of each particle was plotted to estimate the standard deviation (Fig. 17.16). The bottom left corner of Figure 17.15a was set to be the origin of the x-y coordinate. The approximated straight line ($f(x)$) and its slope (θ) are represented as follows:

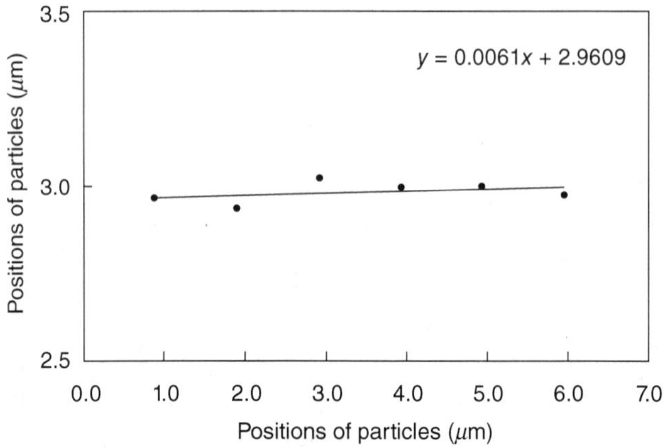

Figure 17.16. Positions of particles in Figure 17.3a showing the accuracy of particle arrangement.

$$f(x) = 0.0061x + 2.9609 \tag{17.5}$$

$$\cos\theta = 0.9940. \tag{17.6}$$

Standard deviation from the approximated straight line is described by the expression

$$S\ (\text{standard deviation}) = \frac{\left[\sum_i \{\cos\theta \cdot (f(x_i) - y_i)\}^2\right]^{1/2}}{n-1}, \tag{17.7}$$

where n is the number of particles ($n = 6$). We obtained $S = 0.0126$. The accuracy of particle arrangement in Figure 17.15b, c was estimated by the same manner. Standard deviation of seven particles from left in an upper particle line and a bottom particle line in Figure 17.15b were estimated to be $S = 5.66 \times 10^{-3}$ and $S = 3.84 \times 10^{-3}$. And the standard deviations of an upper particle line, a middle particle line, and a bottom particle line in Figure 17.15c were estimated to be $S = 8.11 \times 10^{-4}$, 8.27×10^{-3}, and 2.30×10^{-2}, respectively.

Precise Arrangement of Particles on Small-Area Silanol Sites Modified by AFM Lithography. OTS-SAM was modified to silanol groups by an AFM (atomic force microscope, Nanoscope E, Digital Instruments, Tokyo, Japan) to control position of arrangement accurately (Fig. 17.17). A source measure unit (SMU Model 236, Keithley, Tokyo, Japan) was installed in the AFM in order to control the electric current passing through the probe and the SAM. The SAM was biased positively, and the AFM probe was scanned with constant current mode (50 nA), and the scanned area was used as a template for arrangement. Scanning area (100 × 100 nm) was set smaller than the diameter of the particles (500 nmϕ) to facilitate the precise arrangement of particles.

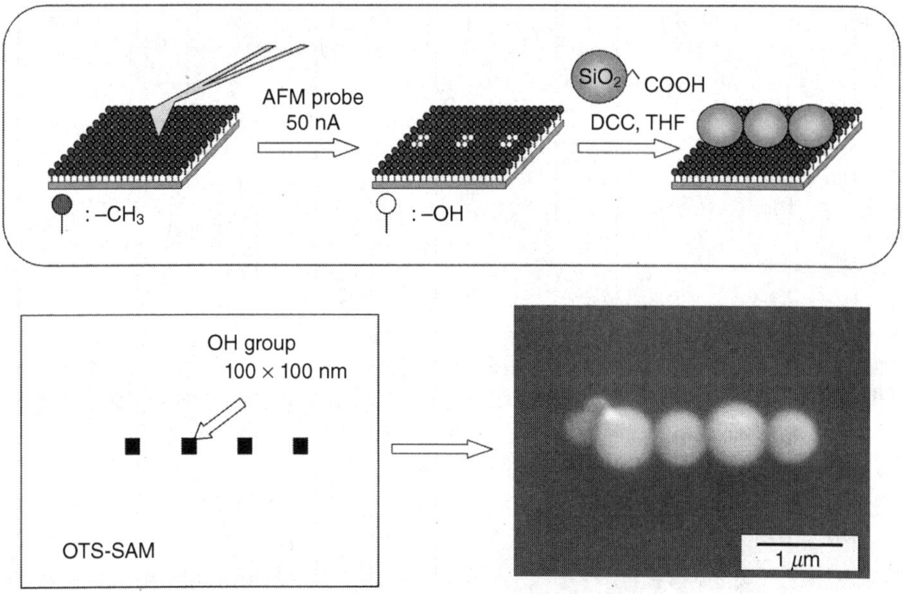

Figure 17.17. Conceptual process and SEM of the particle arrangement on a patterned SAM modified by AFM lithography. DCC, dicyclohexylcarbodiimide; THF, tetrahydrofuran.

SiO_2 particles (500 nmϕ, powder, SO-E2, Admatechs Co., Aichi, Japan) modified with carboxyl groups were sonicated for 10 min in tetrahydrofuran or dichloromethane, and this solution was refrigerated to –20°C for 1 h. *N,N′*-dicyclohexylcarbodiimide was added to this solution as a condenser to form ester bonds between carboxyl groups of SiO_2 particles and silanol groups of a SAM. Modified OTS-SAM was then immersed in this solution for 2 h. The temperature of the solution was increased slowly to 25°C and was kept for 2 h. After having been rinsed in water, a SAM was observed by a SEM. Particles were arranged in silanol regions and a line of particle was fabricated (Fig. 17.17). 2-D arrangement with required features can easily be realized with this technique, though it takes a long time to modify a SAM with an AFM probe. Particles were not removed easily from a SAM by sonication, indicating that ester bonds were formed by condensation. The accuracy of particle arrangement in Figure 17.17 was estimated to be $S = 1.17 \times 10^{-2}$. This might be decreased by decreasing the dimension of each silanol region.

In order to verify the formation of ester bonds between carboxyl groups and silanol groups, bromopropionic acid, whose molecule has a carboxyl group at one end and a bromo group at the other, was reacted with silanol groups of a Si substrate using the same reaction scheme as we used to attach SiO_2 particles to silanol groups. After having been sonicated in acetone for 5 min, the substrate surface was analyzed by XPS (ESCALAB 210, VG Scientific Ltd., 1-3 × 10^{-7} Pa, measurement area; 3 × 4 mm). The X-ray source (MgKα, 1253.6 eV) was operated at 15 kV and 18 mA. The spectrum corresponding to Br 3d binding energy centering at 74.35 eV was observed. Although the observed binding energy is higher than that of KBr, this chemical shift must have been caused by carbon atoms neighboring bromine atoms. Since bromo groups cannot react directly with silanol groups under the present conditions, the XPS result firmly indicates that carboxyl groups of bromopropionic acid reacted with silanol groups to form possible ester bonds.

Patterning of Close-Packed Particle Monolayers. SiO_2 particles modified with carboxyl groups were dispersed in water. Octadecyl/amino group patterned SAM was then immersed in the solution for several minutes (Fig. 17.18). The substrate was rinsed with water and observed with a SEM. SiO_2 particles were observed in silanol regions selectively, forming a close-packed monoparticle layer (Fig. 17.19a). Boundaries between the monoparticle layer and otcadecyl region are clearly observable, and a few particles are observed in the octadecyl region. SiO_2 particles modified with carboxyl groups are charged negative, and amino groups of SAM are charged positive in water. Accordingly, particles are attracted to amino groups and form a monoparticle layer. Particles in the solution did not adhere to the monoparticle layer since both the particles and the monoparticle layer have negative charges and repel

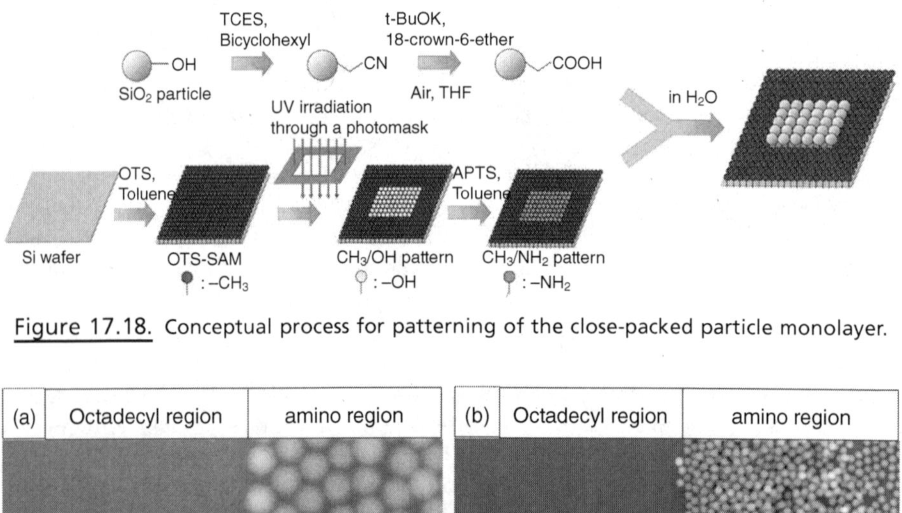

Figure 17.18. Conceptual process for patterning of the close-packed particle monolayer.

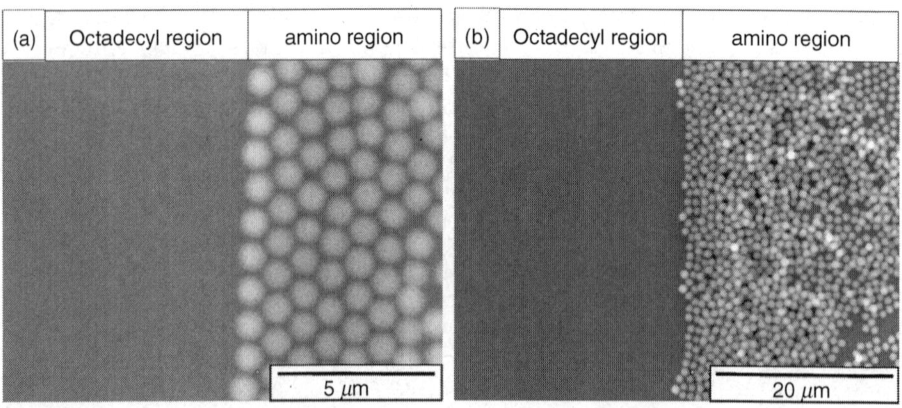

Figure 17.19. SEMs of (a) a close-packed particle monolayer of SiO_2 particles modified with carboxyl groups formed in the amino region and (b) randomly deposited particles in the amino region.

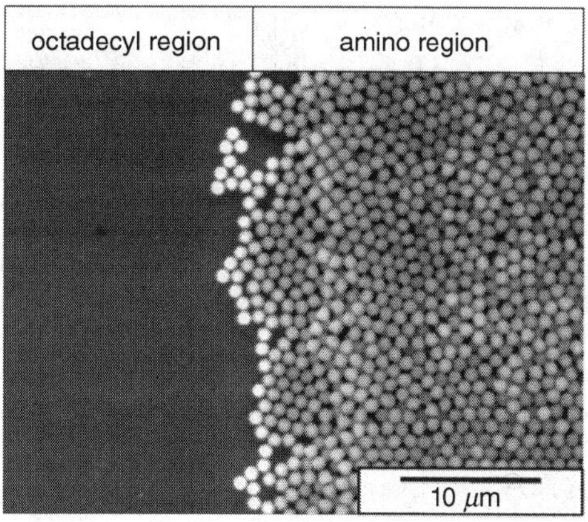

Figure 17.20. SEM of randomly deposited SiO_2 particles in the amino region.

each other. Particles were also deposited randomly in some areas (Fig. 17.19b), and this suggests that it is difficult to obtain the pattern of the close-packed particle monolayer in a large area.

Additionally, octadecyl/amino group patterned SAM was immersed in the solution containing nonmodified SiO_2 particles for several minutes (Fig. 17.20). While SiO_2 particles were observed in the silanol region predominantly, the feature edge acuity of the pattern was lower than that of the pattern in which SiO_2 particles modified with carboxyl groups was used. This demonstrates the applicability of surface modification of SiO_2 particles with carboxyl groups. Furthermore, octadecyl/silanol group patterned SAMs were immersed in solutions containing SiO_2 particles modified with carboxyl groups or nonmodified SiO_2 particles for several minutes, respectively. The particles were not adhered to either octadecyl groups or silanol groups. This means that the difference in surface potential between SiO_2 particles modified with carboxyl groups and amino groups of SAM accelerate the adhesion of particles to the amino groups. The particles were attracted and adhered to amino groups predominantly by electrostatic interactions between particles and SAMs, and chemical bonds were not formed because pure water was used as a solution with no condensation agent. The surface of SiO_2 particles modified with carboxyl groups and amino SAM must have changed into $-COO^-$ and $-NH_3^+$, respectively, in water to attract each other.

Dry Patterning of Colloidal Crystals and 2-D Arrays

Self-Assembly Patterning of Colloidal Crystals by Dry Patterning [83]. Particle wires were fabricated through self-assembly on hydrophilic regions of SAMs [83]. A SAM of OTS was formed on a silicon substrate and was modified by UV irradiation to create a pattern of hydrophobic octadecyl and hydrophilic silanol groups. Ethanol- or water-containing particles (550 or 800 nmϕ) were dropped onto a patterned SAM. The solution was separated into two droplets with a liquid bridge between the droplets along the hydrophilic regions of a patterned SAM. The droplets and the liquid bridge were used as a mold for the fabrication of a 2-D pattern of colloid crystals. A particle wire was formed between two droplets, and colloid crystals such as an opal structure were formed at both ends of the particle wire after drying the solution. The particle wires constructed from a close-packed structure or a non-close-packed structure, that is, square lattice, were fabricated through self-assembly at room temperature using this method.

The UV-irradiated regions became hydrophilic due to the formation of Si–OH groups, while the nonirradiated part remained unchanged; that is, it was composed of hydrophobic octadecyl groups, which gave rise to patterned OTS-SAM. To confirm successful film formation and functional group change, water drop contact angles were measured for irradiated and nonirradiated surfaces. Initially deposited OTS-SAM had a WCA of 96°, while the UV-irradiated SAM surface was saturated (contact angle < 5°). This observation indicated successful fabrication of SAM patterned with octadecyl/silanol groups (Fig. 17.21). Polystyrene particles in water (150 μL) (550-nm φ particle or 820-nm φ carboxylated particle, 10 wt%, dispersed in water; Seradyn Co., Indianapolis, IN) were further dispersed in ethanol (3 mL) or water (3 mL) and were poured onto a patterned OTS-SAM. The contact angles of the ethanol solution or water solution measured 10–20° or 96° on the OTS-SAM, respectively, while they were saturated

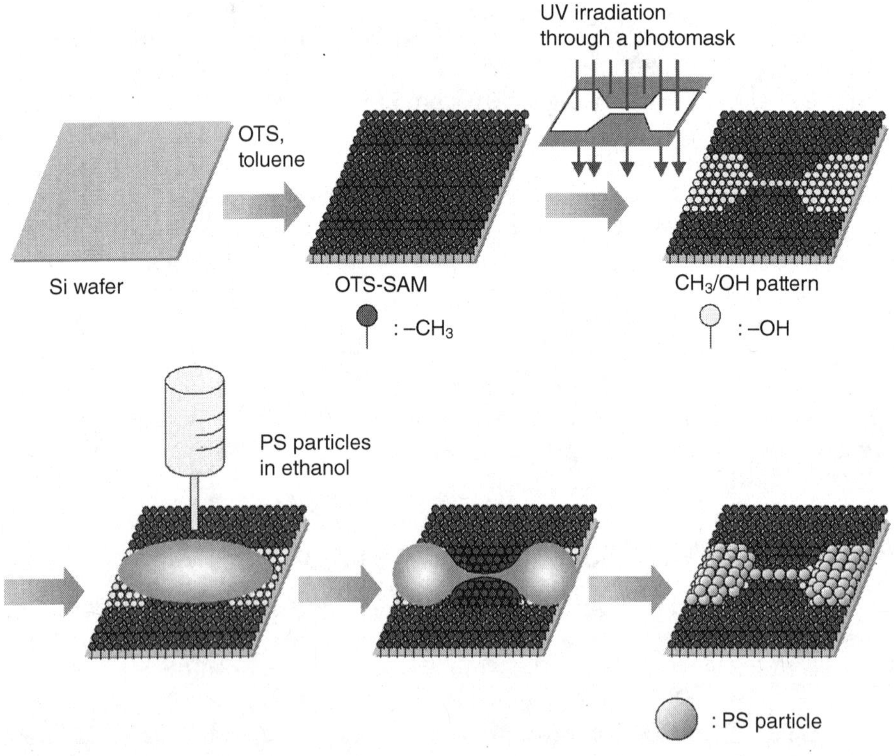

Figure 17.21. Conceptual process for the fabrication of a particle wire using a patterned SAM and a liquid bridge. PS, polystyrene.

(contact angle < 5°) on silanol groups. The droplets were observed to separate into two drops and a bridge of solution was formed on a silanol line. The droplets and the liquid bridge were used as a mold for the fabrication of a 2-D pattern of colloid crystals. After the evaporation of the solution, substrates were observed by a scanning electron microscope (S-3000N, Hitachi, Ltd.).

The dispersibility of particles in the solution is very important for particle assembly, and high dispersibility is necessary to form a close-packed structure. The zeta potentials of particles dispersed in the solutions were examined by direct measurement of electrophoretic mobility using an electrophoretic light scattering spectrometer (Zetasizer 3000HS, Malvern Instruments Co., Ltd.). The zeta potentials of polystyrene particles (550 nmφ) in water, carboxylated particles (820 nmφ) in water, and polystyrene and carboxylated particles in ethanol were determined to be −38.3, −50.2, −53.9, and −44.0 mV, respectively. Surface modification by carboxyl groups decreased the negative zeta potential in both solutions. Furthermore, particles in ethanol had slightly low negative zeta potentials compared with those in water; that is, the particles were slightly well-dispersed compared with those in water.

Polystyrene particles in water were poured onto a patterned OTS-SAM (Fig. 17.21) and were observed after the evaporation of the water. The water at the liquid bridge evaporated in about 24 h to form a particle wire, and droplets at the two ends completely evaporated in about 48 h. In this manner, particle wires constructed from a close-packed structure, that is, triangular lattice, were produced from the water solution (Fig. 17.22a). The middle of a particle wire was narrower than its end (Fig. 17.22b, c). The width of the particle wire does not depend on the width of the silanol line but rather on the interfacial tensions between solution and substrate, solution and atmosphere, and atmosphere and substrate. The silanol line was not used to decide the width of the particle wire but rather the position of the liquid bridge and particle wire. Close-packed

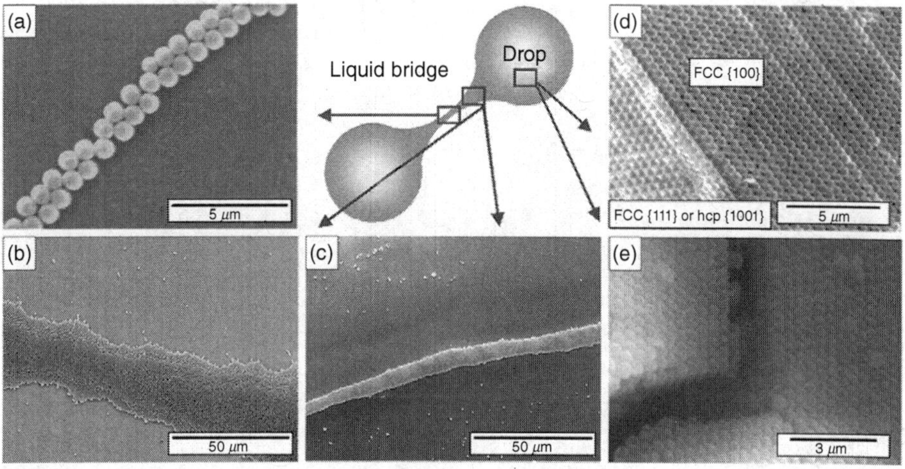

Figure 17.22. SEMs of particle structures fabricated from an aqueous solution containing microparticles using a liquid bridge. (a–c) Particle wires constructed from a triangular lattice (close-packed structure) and (d, e) a close-packed 3-D structure.

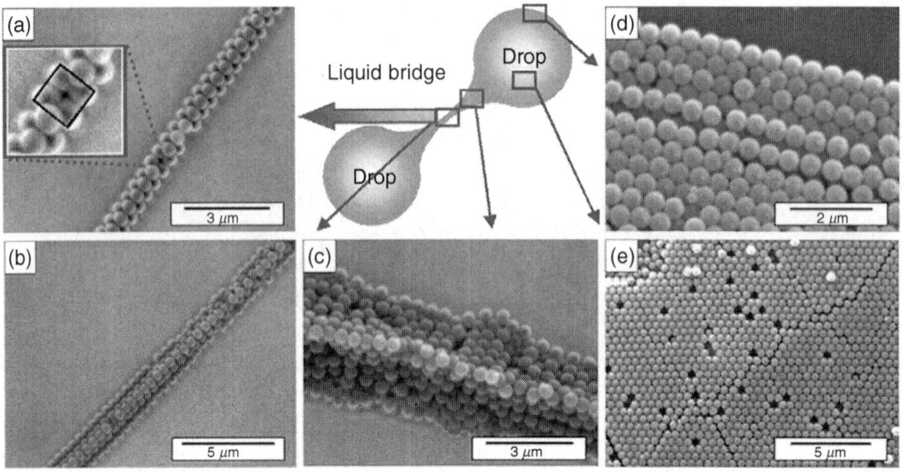

Figure 17.23. SEMs of particle structures fabricated from an ethanol solution containing microparticles using a liquid bridge. (a, b) Particle wires constructed from square lattice, (c) particles deposited on the edge of a silanol line, and (d, e) close-packed 3-D structures.

structures were also formed on large silanol regions (Fig. 17.22d, e). The right-hand area of Figure 17.22d can be regarded as the {100} plane of the FCC structure, and the left-hand area can be regarded as the {111} plane of the FCC structure or the {0001} plane of the HCP structure. The close-packed structure was thus considered to be an FCC structure.

Figure 17.23 shows particle wires and 3-D structures fabricated from ethanol solution containing polystyrene particles. The ethanol at the liquid bridge evaporated in about 1 min to form a particle wire, and droplets at the two ends evaporated in about 20 min. The liquid bridge of ethanol evaporated faster than that of water for several reasons. The saturated vapor pressure of ethanol (59 mmHg [0.078 atm] at 25°C) is higher than that of water (24 mmHg [0.031 atm] at 25°C), explaining the difference in the evaporation rate of the two droplets. The ratio in evaporation rate of the ethanol liquid bridge to the ethanol droplets is higher than that of the water liquid bridge to the water droplets. This can be explained as follows: Water has high surface tension (71.8×10^{-3} N/m at 25°C) compared with ethanol (22.0×10^{-3} N/m at 25°C). Ethanol existed along patterned hydrophilic regions with a small meniscus at the angle between droplets and a liquid bridge. However, water formed a large meniscus at the angle between droplets and a liquid bridge, causing a wide line width of water liquid bridge compared with ethanol on our patterned surfaces. Additionally, the water liquid bridge was higher than that of ethanol due to high surface tension. These made the cross-section area of water larger than that of ethanol. The thick liquid bridge evaporated slowly because of its large volume and low vapor pressure calculated from the Kelvin equation, in which the smaller convex liquid surface gives rise to higher internal pressure and faster evaporation rate. Furthermore, the solution at the droplets flowed into a liquid bridge, and this further complicated the evaporation mechanism.

A narrow particle wire was formed at the center of the liquid bridge (Fig. 17.23a, b), and wide wires were formed at the edges of the liquid bridge (Fig. 17.23c) along the silanol line. The particle wire in Figure 17.23a is not a close-packed structure and is constructed from a square lattice, which is a metastable phase compared with a close-packed structure. High dispersibility of particles is necessary to form a close-packed structure in the solution. However, the particles dispersed well in ethanol and the dispersibility of particles in ethanol is similar to that in water. This shows that the non-close-packed structure was caused not only by the influence of dispersibility but also by many other factors. Movement and rearrangement of deposited particles is necessary to construct a close-packed structure. However, the ethanol evaporated quickly and suppressed the movement of particles by liquid bridge force. Additionally, adhesion between particles and a substrate and cohesion between particles probably caused moderate suppression of the rearrangement of particles. Factors such as evaporation rate, interaction force between particles, and interaction force between particles and a substrate were important in the packing process. Close-packed 3-D structures were also formed on large silanol regions (Fig. 17.23d, e), and they contained many defects (Fig. 17.23e). The ethanol evaporated so quickly that the particles did not rearrange well to form a close-packed structure during the evaporation of ethanol. This is one of the factors of forming a loosely packed structure. To directly evaluate the effect of the evaporation rate, a similar experiment using ethanol was conducted in a small airtight container with small pinholes to allow the ethanol to evaporate slowly. The size and number of pin holes were adjusted for ethanol at the liquid bridge to evaporate in about 24 h. The particle wire constructed from a close-packed structure was formed after about 24 h, and droplets at both ends were dried after about 48 h. The close-packed 3-D structures were formed in large silanol regions. The number of defects was smaller than that formed from the ethanol solution with a shorter time and was similar to that formed from water. The humidity in the container was close to 100%, and the saturated vapor pressure of ethanol was 59 mmHg (0.078 atm) at 25°C. This showed that particles dispersed well in ethanol and the interaction between particles and a substrate was sufficiently weak to produce a close-packed structure in the drying process of 24 h. Additionally, the results showed that not only the interaction force between particles and that between particles and a substrate but also the evaporation rate needs to be controlled to fabricate particle wires.

The accuracy of particle arrangement was evaluated from Figure 17.23a as calculated in a recent work [80]. The center position ($[x_i, y_i]$ μm) of each particle in an upper layer was plotted to estimate the standard deviation. The bottom left corner of Figure 17.23a was set to be the origin of the x-y coordinate.

The approximated straight line ($f(x)$) and its slope (θ) are represented as follows:

$$f(x) = 1.3965x - 5.3344 \tag{17.8}$$

$$\cos\theta = 0.5822. \tag{17.9}$$

Standard deviation from the approximated straight line is described by the expression

$$S\text{ (standard deviation)} = \frac{\left[\sum_i \{\cos\theta \cdot (f(x_i) - y_i)\}^2\right]^{1/2}}{n-1}, \qquad (17.10)$$

where n is the number of particles ($n = 19$). The accuracy of the particle arrangement in Figure 17.23a was estimated to be $S = 1.63 \times 10^{-3}$. This is lower than that of the particle arrangement obtained in our previous work [80].

Particle wires were fabricated on hydrophilic regions of a patterned SAM. Ethanol- or water-containing particles were separated into two droplets with a liquid bridge between the droplets along hydrophilic regions of a patterned SAM. Particle wires constructed from a close-packed structure or a non-close-packed structure were then formed through self-assembly between two droplets after drying of the solution.

Self-Assembly 2-D Array of Colloidal Crystal Wires by Dry Patterning [84, 85]. An orderly array of particle wires constructed from a close-packed colloidal crystal was fabricated without the preparation of patterned templates [84, 85]. A substrate was immersed vertically into a SiO_2 colloidal solution, and the liquid surface moved downward upon evaporation of the solution. Particles formed a mono/multiparticle layer, which was cut by the periodic drop-off of the solution. The orderly array of particle wires was successfully fabricated, showing the suitability of the self-assembly process for the fabrication of nano/microstructures constructed from nano/microparticles or blocks. The mechanism of the assembly process and control of thickness, width, and interval of particle wires were further discussed. Moreover, an array of particle wires, constructed not from close-packed FCC (or HCP) structure but from two kinds of particles, was realized to fabricate an array of particle wires with NaCl structure by this self-assembly process.

OTS-SAM was prepared by immersing the Si substrate in an anhydrous toluene solution containing 1 vol% OTS for 5 min under an N_2 atmosphere. The contact angles of the ethanol solution or water solution measured 10–20° or 96° on the OTS-SAM, respectively. The water solution showed a contact angle of 10°–20° on a silicon wafer that was kept in air.

The OTS-SAM was immersed in ethanol solution (80 mL) containing SiO_2 particles (1000 nmϕ, 10 mg). The bottom of the solution was heated at 70°C and a condenser tube was kept at the top of the solution to cool it. The temperature difference between the top and bottom of the solution was controlled so as to stir and move particles by convection. The surface of the solution moved on the OTS-SAM surface upon evaporation of the ethanol. Particles began to assemble at the surface of the solution and the particle layer was fabricated by the movement of the solution surface (Fig. 17.24). Further evaporation of the solution caused separation of the particle layer and the solution surface because particles were not sufficiently supplied from the solution. The liquid surface then dropped off and the particle layer separated from the solution surface. The next particle layer was formed by the same procedure. Consequently,

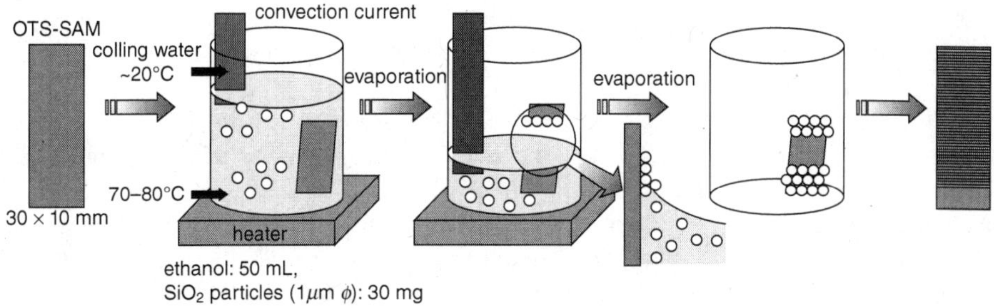

Figure 17.24. Schematic for the self-assembly process to fabricate an orderly array of particle wires constructed from a close-packed structure.

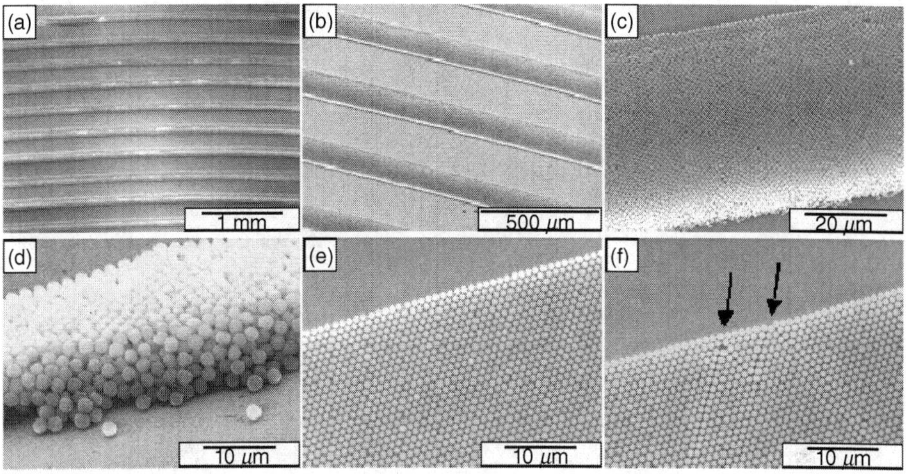

Figure 17.25. SEMs of the array of particle wires constructed from (a–d) the multiparticle layer or (e–f) the monoparticle layer, (d) bottom side of a particle wire, (f) a monoparticle wire having two defects on the upper side.

separated particle wires, that is, an array of particle wires, were successfully fabricated by our newly developed method.

After having been immersed in the solution, which evaporates quickly, the substrates were observed using a scanning electron microscope (S-3000N, Hitachi, Ltd.), an optical microscope (BX51WI Microscope, Olympus Optical Co., Tokyo, Japan) with a digital camera (DP50, 5.8 megapixels, Olympus Optical Co., Ltd.) and a computer for capturing data, and a digital video camera recorder (DCR-TRV 50, Sony Corporation, Tokyo, Japan) with an optical magnifying glass.

The width and interval of particle wires fabricated from the ethanol solution (80 mL) containing SiO_2 particles (10 mg) at 70°C were shown to be about 150 and 200 μm, respectively (Fig. 17.25a–d). Particle wires were constructed from a close-packed particle structure and their upper side showed high feature edge acuity (Fig. 17.25c). The array of particles finished suddenly as shown on the bottom side of the particle wires (Fig. 17.25c–d). These observations suggest that particle wires were formed from the upper side and were cut by the drop-off of solution, and are consistent with the procedure shown in Figure 17.24.

Particle wires were formed on the OTS-SAM from the ethanol solution (80 mL) containing a small amount of SiO_2 particles (1000 nmϕ, 1 mg) to fabricate thin wires constructed from a monoparticle layer (Fig. 17.25e–f). Particles were slowly supplied to particle wires from the solution. The number of particle layers was shown to be controlled by the change of particle number in the solution. The upper side of the particle wire in Figure 17.25f has two defects, shown by arrows. The influence on the arrangement below the defects shows that particles were constructed from the upper side to form a close-packed structure. In addition, the disorder disappeared at the middle of the particle wire (Fig. 17.25f), which shows that the formation process of the particle wire has self-recovery ability.

Figure 17.14 shows the formation process of the particle wire. The upper particle line was formed (Fig. 17.26a) first and particles were supplied gradually from the solution to form a close-packed structure (Fig. 17.26b) to prepare a wide monoparticle wire (Fig. 17.26c). High dispersibility of particles in the solution and the effective meniscus force allowed us to prepare a highly ordered close-packed structure.

Particle wires showed iridescent diffraction (Fig. 17.27) caused by the high regularity of the particle array shown in Figure 17.25. The diffracted wave number was changed drastically by the diffraction angle.

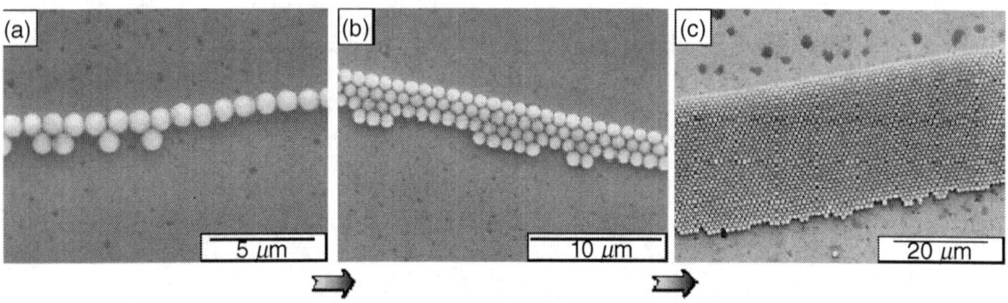

Figure 17.26. SEMs of the formation process of a particle wire.

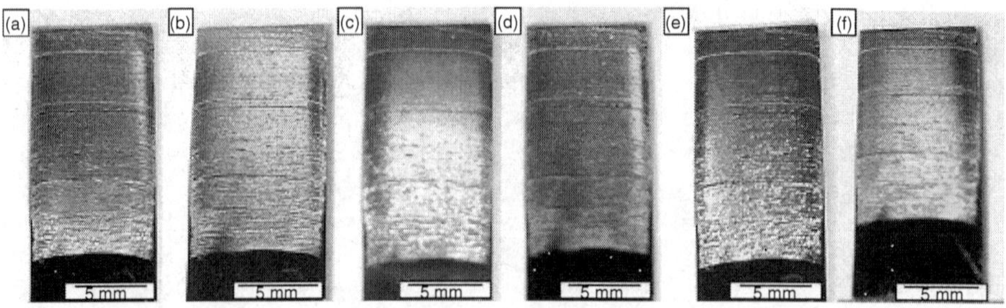

Figure 17.27. Photographs of particle wires taken from different directions: (a) front view and (b–f) slightly cross shot of the same sample.

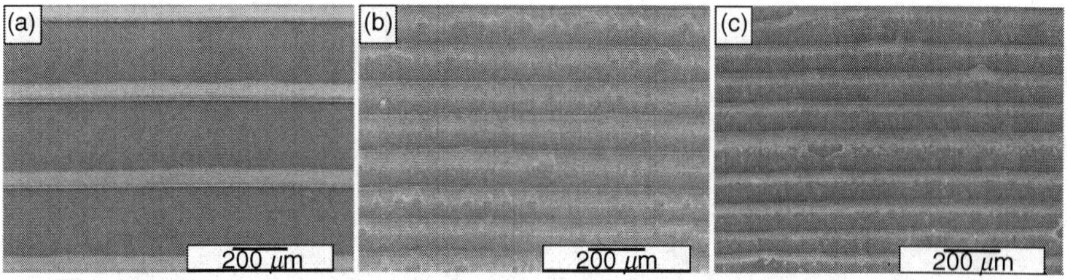

Figure 17.28. SEMs of the array of particle wires constructed at (a) 70°C, (b) 60°C, or (c) 50°C.

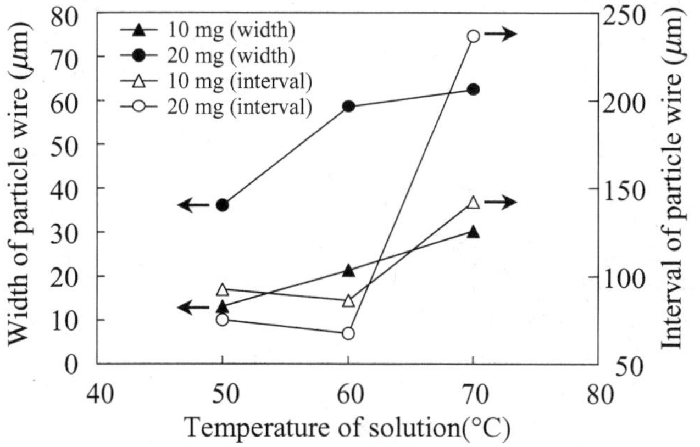

Figure 17.29. Width and interval of particle wires as functions of the temperature of the solution.

A close-packed structure was easily obtained by the use of meniscus force compared to the site-selective deposition in the solution [80–82]. A 2-D ordered array can be fabricated without the preparation of a template, although templates are required for the liquid mold method [83].

Control of the interval of particle wires was achieved by a change in solution temperature (Figs. 17.28 and 17.29). The interval of a particle wire fabricated from the ethanol solution (80 mL) containing SiO_2 particles (20 mg) at 70°C (Fig. 17.28a) was about three times that of particle wires fabricated at 60°C (Fig. 17.28b) or 50°C (Figs. 17.28c and 17.29). The same tendency was confirmed in the dilute solution system (80 mL) containing SiO_2 particles (10 mg) (Fig. 17.29). Wide interval was caused by the high descent speed of the solution surface at high temperature. The regularity of particle wires at 70°C was much higher than that at 60 or 50°C, in which the movement of particles caused by convection and the descent speed of the solution surface were low compared to those at 70°C.

The width of the particle wires was also controlled by the change in the solution temperature or particle concentration in the solution (Fig. 17.29). The width of particle wires was increased by the increase of solution temperature in both particle concentrations (10 or 20 mg), and the width at high particle concentration (20 mg) was about two

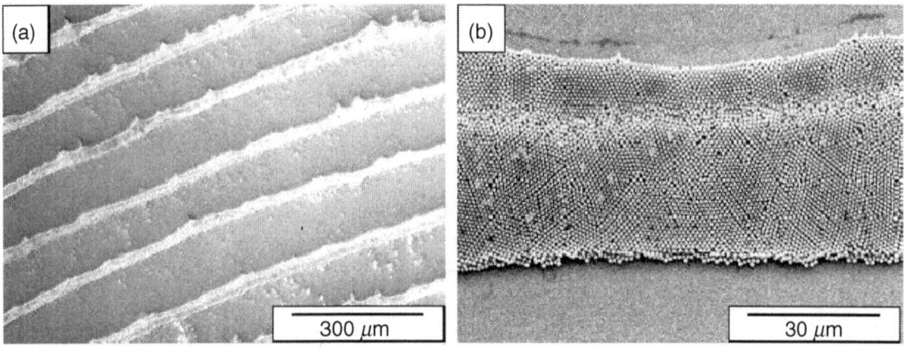

Figure 17.30. SEMs of the array of particle wires fabricated from water solution.

times that at low particle concentration (10 mg). High descent speed of the solution surface and high movement of particles caused by convection were brought about at high temperature. These factors probably cause the change in width; however, further precise control and clarification of the mechanism are necessary. The descent speed of the solution surface would be controlled by the control of the lift speed of a substrate instead of control of the evaporation speed. Further improvement of the process is required for application to future devices.

Self-assembly of particle wires was realized using water instead of ethanol without SAM (Fig. 17.30). The particle wires also showed iridescent diffraction and the diffracted wave number was changed drastically by the diffraction angle. We have reported that a particle layer was formed on the whole area of OTS-SAM by the use of water, which evaporates slowly compared to ethanol. Fast movement of the solution surface caused by quick evaporation or lift of the substrate is probably necessary for drop-off of the solution surface, which allows the fabrication of separated particle wires. The condenser tube, which is used to cool the solution surface, was removed to evaporate water more quickly. This makes the descent speed of the water surface fast enough to separate the particle wires. Additionally, silicon wafer, which was kept in air to show a contact angle of 10°–20° to water, was used instead of OTS-SAM. Because the contact angle of the water on the OTS-SAM (96°) was higher than that of ethanol (10–20°), the shape of the water surface was not suitable for making an array of particle wires. These improvements allowed us to fabricate an array of particle wires using water without SAM. This process is environmentally friendly compared with the process using ethanol with OTS-SAM; however, the regularity and the feature edge acuity of particle wires were low compared to the process shown in Figure 17.25, and thus further improvement is required in this system.

The array of particle wires constructed not from FCC (or HCP) but from NaCl structure (rock salt structure), in which each ion is 6-coordinate and has a local octahedral geometry, was realized by the use of two kinds of particles (SiO_2 modified with carboxyl groups: 2000 nmϕ, SiO_2 modified with amino groups: 1000 nmϕ; ratio of particle radii is 0.5). SiO_2 particles 2000 nmϕ in diameter and SiO_2 particles 1000 nmϕ in diameter were modified to have carboxyl groups or amino groups on their surfaces, respectively. The zeta potential of SiO_2 particles modified with carboxyl groups was measured to be −40 mV in an aqueous solution at pH 7.0 [8] by direct measurement of electrophoretic mobility using an electrophoretic light-scattering spectrometer (Zetasizer 3000HS, Malvern Instruments Co., Ltd.). On the other hand, APTS (3-aminopropyltrimethoxysilane)-SAM was prepared by the immersion of silicon wafer in an anhydrous toluene solution containing 1 vol% APTS for 1 h in air to measure the zeta potential instead of SiO_2 particles modified with amino groups. Amino groups of APTS-SAM were measured to be +22.0 mV in aqueous solutions (pH = 7.0) by an electrophoretic light-scattering spectrophotometer (ELS-8000, Otsuka Electronics Co., Osaka, Japan). Electrostatic interaction between SiO_2 particles modified with carboxyl groups and SiO_2 particles modified with amino groups was utilized for self-assembly of these particles to form ionic crystal, such as NaCl structure. Although the ratio of ion radii is 0.611 ($=r^+(Na^+)/r^-(Cl^-)$) in NaCl, cations can be in contact with anions in the range of 0.414–0.732 in the ratio of ion radii to form a NaCl structure. Figure 17.31 shows the upper side of the particle wire and is the same arrangement as the (100) face of NaCl structure. Each particle can be 6-coordinate when the same particle layer stacks on this layer with a slide of half lattice constant (sum of each particle radii). Small spherical particles (cations in NaCl structure) are shown as square or triangular in SEMs because electrons from the SEM electron gun flowed from large particles into small particles to show the contacted area as white. Two kinds of dispersed particles would be adhered on the substrate and rearranged well to form a 2-D layer. Particles were also deposited randomly in some areas, and this suggests that it is difficult to obtain a NaCl structure in a large area. In addition, the NaCl structure was prepared from two kinds of SiO_2 particles without surface modification (SiO_2: 2000 nmϕ, SiO_2: 1000 nmϕ; ratio of particle radii is 0.5), though the regularity of particle array was slightly lower than that prepared from SiO_2 particles with surface modification. A NaCl-type arrangement of bimodal SiO_2 particles can be formed thermodynamically without electrostatic interaction because the (100) face of NaCl structure constructed from large particles (x nmϕ) and small particles (($\sqrt{2}-1)x$ nmϕ) was a close-packed structure. This process should be improved to prepare a NaCl structure without the use of surface modification of particles to apply for various kinds of particles. Additionally, the interaction between

Figure 17.31. SEM of the array of particle wires constructed from NaCl structure.

particles, substrates and liquid and the behavior of particles should be controlled well to form a particle assembly having high regularity. Particles should be dispersed well in the solution to avoid aggregation, and attractive interaction between particles and a substrate should be decreased enough to accelerate rearrangement of particles on the substrate. Appropriate movement of particles accelerates the formation of a particle assembly; however, excess convection suppresses the assembly of particles. This process has many factors and should be improved to fabricate particle assembly having desired structure and high regularity. The fabrication of an array of particle wire constructed from a NaCl structure shows the high ability of the self-assembly process to realize future devices such as photonic crystal, in which various structures are required to effectively utilize photonic band gap (PBG).

Particle wires constructed from a close-packed multiparticle layer or monoparticle layer were fabricated without patterned templates. The array shows iridescent diffraction caused by high regularity of the particle array. The mechanism of the process was discussed and the control of the thickness, width, and interval of particle wires was realized by change of solution temperature and concentration of colloidal solution. Furthermore, an array of particle wires constructed from an ionic crystal such as NaCl structure was fabricated using electrostatic interaction between particles showing positive zeta potential and particles showing negative zeta potential. This shows the suitability of the self-assembly process for creating future devices such as photonic crystals.

Patterning of Colloidal Crystals and Spherical Assemblies by the Two-Solution Method

Self-Assembly Patterning of Colloidal Crystals by the Two-Solution Method [86]. Desired patterns of colloidal crystals having high feature edge acuity and high regularity were fabricated by the two-solution method [86]. A micropattern of colloidal methanol prepared on a SAM in hexane was used as a mold for particle patterning, and slow dissolution of methanol into hexane caused shrinkage of molds to form micropatterns of close-packed SiO_2 particle assemblies. This result is a step toward the realization of nano/microperiodic structures for next-generation photonic devices by a self-assembly process.

Silicon substrate was immersed into toluene solution containing 1 vol% OTS molecules under nitrogen atmosphere for 5 min to prepare a hydrophobic OTS-SAM. OTS-SAM was irradiated by UV light (PL21-200, SEN Lights Co.; 18 mW/cm^2, distance from a lamp: 30 mm, 24°C, humidity: 73%, air flow: 0.52 m^3/min, 100 V, 320 W) through a photomask for 10 min. UV irradiation modified hydrophobic octadecyl groups to hydrophilic silanol groups forming a pattern of octadecyl and silanol regions. Patterned OTS-SAM having hydrophobic octadecyl regions and hydrophilic silanol regions was used as a template for patterning of the colloidal solution.

SiO_2 particles (1 μm in diameter) (0.002–0.2 mg) were thoroughly dispersed in methanol (20 μL) and were dropped on a patterned OTS-SAM (Fig. 17.32). The solution was lightly repelled by hydrophobic regions and mainly exists on hydrophilic

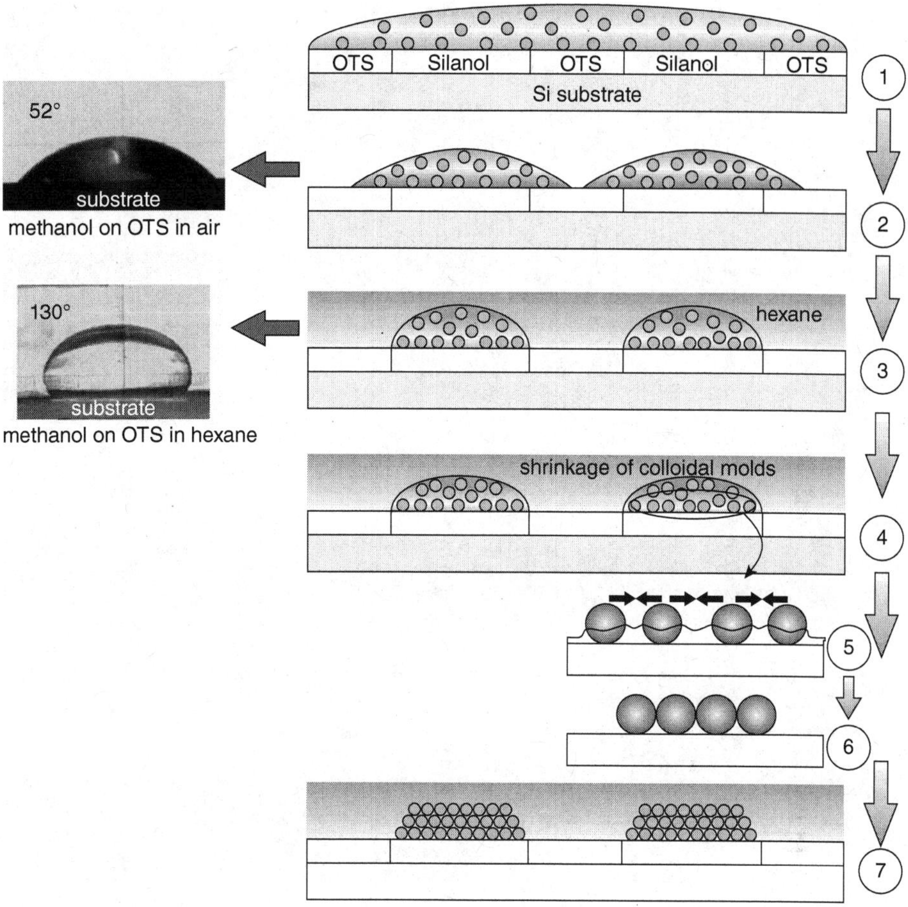

Figure 17.32. Conceptual process for the two-solution self-assembly method to fabricate patterned colloidal photonic crystals.

silanol regions. The substrate was then immersed into hexane and was carefully swung to remove the residual solution. The solution was repelled well by octadecyl regions in hexane. The contact angle of the methanol solution on OTS-SAM was confirmed to increase from 51.6° in air to 129.5° in hexane (Fig. 17.32), indicating that the methanol solution tends to exist on silanol regions selectively.

Methanol was then gradually dissolved into hexane to shrink the colloidal solution mold containing particles. The shrinkage of the mold increased the concentration of particles in the solution. The particles then attracted each other by meniscus force to form a close-packed FCC or HCP (hexagonal closest packing) structure during the drying process of methanol. Close-packed structures were thus formed on hydrophilic silanol regions selectively (Fig. 17.33a–h).

2-D micropatterns of multiparticle layers (Fig. 17.33a, b), partially double-particle layers (Fig. 17.21c, d), or single-particle layers (Fig. 17.21e–h) were successfully

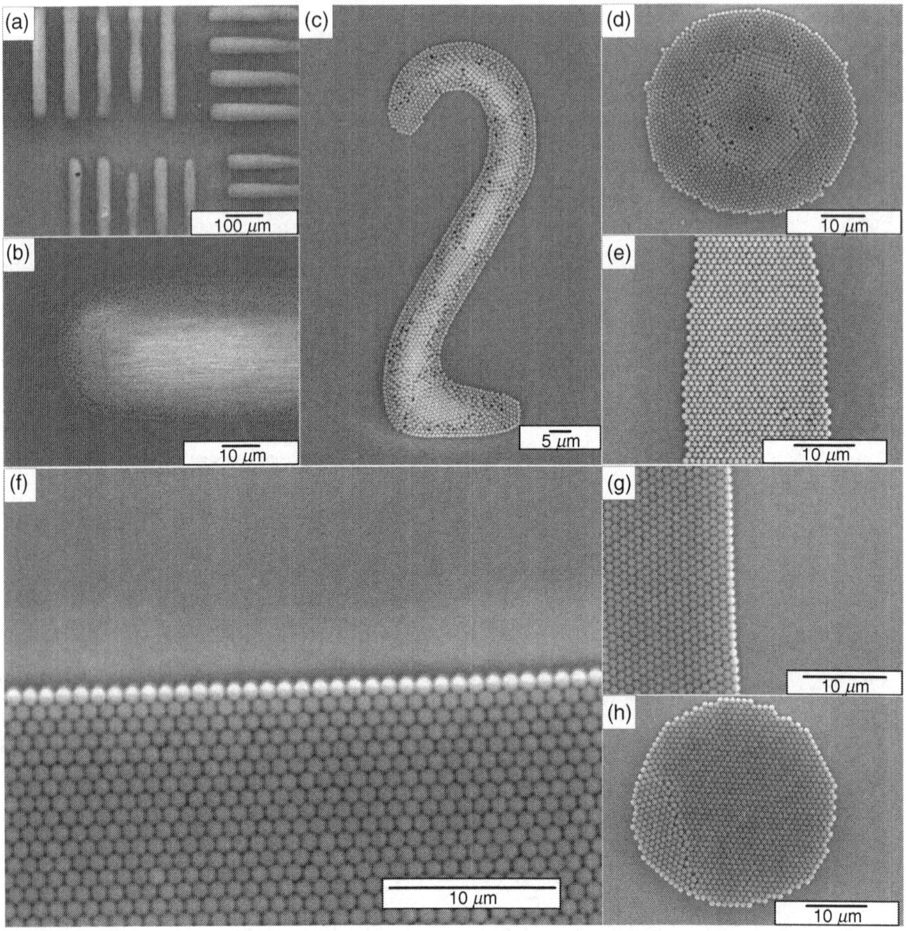

Figure 17.33. SEMs of patterned colloidal photonic crystals constructed from (a, b) multi-particle layers, (c, d) partially double-particle layers, or (e–h) single-particle layers. Image (b) is a magnified area of (a).

fabricated by changing the particle concentration in the solution and the solution volume per unit of hydrophilic area (SiO_2/methanol: 0.2 mg/20 μL for Fig. 17.33a, b, 0.02 mg/20 μL for Fig. 14.33c, d, and 0.002 mg/20 μL for Fig. 17.33e–h). 2-D patterns of colloidal crystals with high regularity in particle assembly have not been prepared by our processes as previously reported [80–83, 85]. The feature edge acuity of patterns and regularity in particle assembly [80–83, 85] presented here are clearly higher than those previously reported [80–83, 85]. Figure 17.33b shows a magnified area of the patterns constructed from thick particle layers (Fig. 17.21a). The edge of patterns (Fig. 17.33a–h) show high feature edge acuity due to close-packing induced by meniscus force in the drying process. The core area of the particle circle (Figs. 17.33d and 17.34) was a double-particle layer of close-packed hexagonal lattice, that is, the arrangement

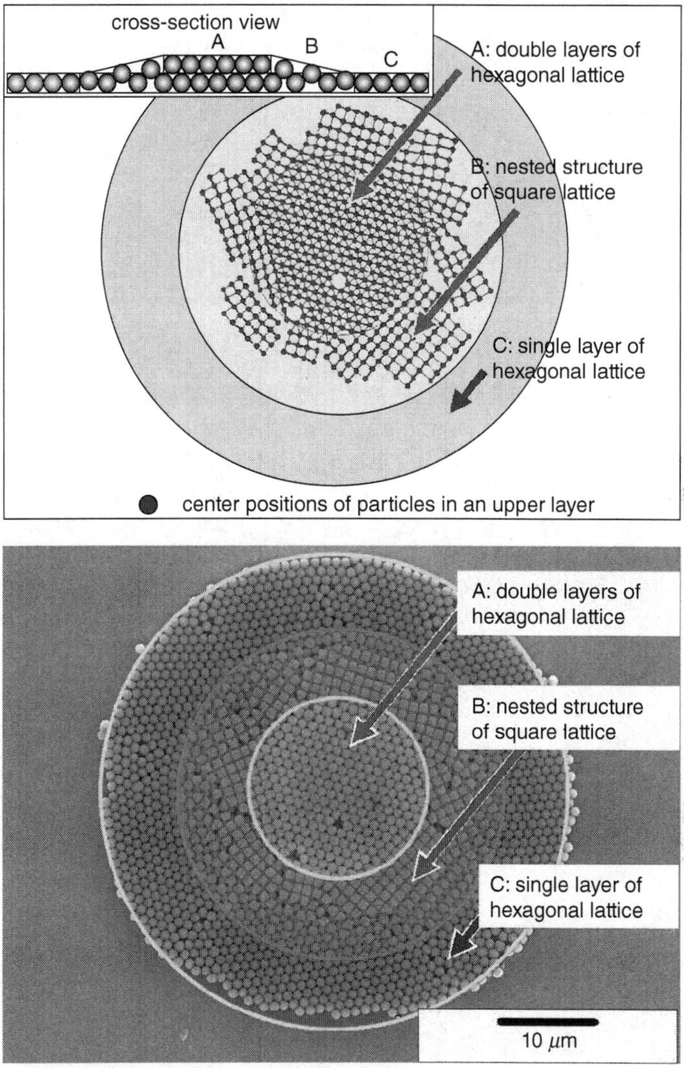

Figure 17.34. Particle arrangement view (top) prepared from the SEM (bottom) of patterned colloidal photonic crystals constructed from partially double-particle layers. Center positions of particles in an upper layer show hexagonal or square lattice arrangement regions. The cross-sectional view shows the difference in height caused by the difference in particle arrangement modes. The three colored areas show regions of (a) double layers of hexagonal lattice, (b) a nested structure of square lattice, and (c) a single layer of hexagonal lattice.

of FCC {111}, and the outer shell of the circle was a single-particle layer of hexagonal lattice (FCC {111}). The boundary area of these two flat terraces, that is, the inner shell, was constructed from a nested structure of square lattice, that is, the arrangement of FCC {100}, to form a gentle slope between the core double layer and the outer shell single layer. The lattice constant of the square lattice at the inner shell increased gradually with distance from the core of the circle to form a gentle slope. The difference in height was caused by the assembling process and the shape of the liquid mold in which the center is higher than the outside. The particle arrangement in the patterns constructed from a single-particle layer (Fig. 17.33e–h) was assigned to the arrangement in FCC {111}, which is a close-packed structure. The border line of particle layer in Figure 17.33e showed a different shape from that in Figure 17.33f, g because the border line of particle layer in Figure 17.33f, g was FCC <1, 1, 0> and that in Figure 17.33e was FCC <1/2, 1/2, 1>, which is orthogonal to FCC <1, 1, 0>. There were far fewer defects in the particle patterns than in our former processes [80–83, 85] because of the effective meniscus force. The standard deviation for the edge of the pattern constructed from a single-particle layer (Fig. 17.33f) was calculated in the same manner as we have reported previously [83]. The center position ($[x_i, y_i]$ μm) of each particle at the edge was plotted to estimate the standard deviation. The particle at the far left in the edge line of Fig. 17.33f was set to be the origin of the *x-y* coordinate.

The approximated straight line ($f(x)$) and its slope (θ) are represented as follows:

$$f(x) = 0.0273x - 0.09 \tag{17.11}$$

$$\cos\theta = 0.9996. \tag{17.12}$$

The standard deviation from the approximated straight line is given by the expression

$$S\,(\text{standard deviation}) = \sqrt{U} = \frac{\left[\sum_i \{\cos\theta\cdot(f(x_i) - y_i)\}^2\right]^{1/2}}{n-1}, \tag{17.13}$$

where n is the number of particles ($n = 35$). Unbiased variance (U) was used because the number of particles (n) is smaller than universe. The accuracy of the particle arrangement in Figure 17.33f was estimated to be $S = 8.75 \times 10^{-3}$. This is lower than that of the pattern constructed from a single-particle layer prepared in the solution using chemical reactions ($S = 3.89 \times 10^{-2}$, Fig. 17.7a in a previous report [80]). The standard deviation for the edge of the pattern constructed from a single-particle layer (Fig. 17.33h) was calculated in the same manner. Distance (r_i) from center of a circle ($[x_o, y_o]$ μm) to each particle ($[x_i, y_i]$ μm) at the edge and its average ($\bar{r}$) are presented as follows:

$$r_i = ((x_i - x_o)^2 + (y_i - y_o)^2)^{1/2}, \tag{17.14}$$

$$\bar{r} = \frac{\sum_{i=1}^{n}\left((x_i - x_o)^2 + (y_i - y_o)^2\right)^{1/2}}{n}, \tag{17.15}$$

and

$$x_o = 21.9\ \mu\text{m}, \quad y_o = 17.7\ \mu\text{m}, \quad \bar{r} = 16.53\ \mu\text{m}.$$

The standard deviation from the approximated circle is given by the expression

$$S\ (\text{standard deviation}) = \sqrt{V} = \left[\frac{\sum_{i=1}^{n}\left\{\left((x_i - x_o)^2 + (y_i - y_o)^2\right)^{1/2} - \bar{r}\right\}^2}{n}\right]^{1/2}, \tag{17.16}$$

where n is the number of particles ($n = 112$). Variance (V) was used instead of unbiased variance because the number of particles (n) is the same as universe, that is, all particles at the edge. The accuracy of the particle arrangement in Figure 17.33h was estimated to be $S = 3.89 \times 10^{-1}$. This is higher than that in Fig. 17.33f because a perfect circle cannot be constructed from a small number of particles, which were packed in hexagonal arrangement.

The assembly process can be assumed from the details of structures and defects. Particle circles (Fig. 17.33d, h) showed no defects at their core, implying that particles were probably assembled from the core of the particle circle and not from the outer shell. The particle circle (Figs. 17.33d and 17.34) would be formed not layer by layer, and the upper layer at the core was also formed before the outer shell was assembled, since the first layer at the boundary area was not a close-packed assembly. A close-packed hexagonal lattice would be formed in the lower layer as shown in the single-particle circle (Fig. 17.33h) in the case wherein particles were assembled layer by layer. Other particle patterns (Fig. 17.33a–c) would be assembled from the core area in the same manner.

The methanol solution containing particles was dropped onto a patterned OTS-SAM and was dried in air without immersion into hexane for comparison. However, a particle pattern only with low feature edge acuity was obtained. This indicates that the contact angle of methanol solution on OTS-SAM, that is, the shape of the methanol solution mold, is important for particle patterning. Additionally, SiO_2 particles were dispersed in water instead of methanol, and a patterned OTS-SAM covered with the water colloidal solution was immersed into hexane, but the particle pattern was not obtained because the water solution cannot form a micropattern of the mold on hydrophilic regions selectively due to its high contact angle. These results suggest that the combination of methanol, hexane, OTS, and silanol allowed us to fabricate micropatterns of particle assembly having high feature edge acuity and high regularity.

This two-solution self-assembly method successfully produced precise 2-D patterns of colloidal photonic crystals, that is, close-packed FCC or HCP particle assemblies having a photonic band gap, with short fabrication period and low energy. Various complicated patterns of colloidal crystals having high feature edge acuity and high regularity in particle assembly were fabricated, and the number of particle stacking layers, that is, thickness of colloidal crystals, was demonstrated to be controlled. This process has the advantage of the drying process in which meniscus force can be utilized for close-packed particle assembly, and also the advantage of the static solution process in which particle assembling processes can be controlled well. The low dissolution rate of methanol into hexane and the appropriate contact angle of methanol on a substrate achieved by the combination of methanol, hexane, OTS, and silanol allowed us to utilize the solution mold for particle patterning. Additionally, high dispersibility of particles and high repulsion force between particles and substrate suppressed the aggregation of particles, and the meniscus force between particles was effectively utilized to form a close-packed structure. The quality of SAM, shape, and shrinkage rate of solution molds also greatly influence the feature edge acuity and the regularity of the particle assembly. The newly developed method achieved much higher regularity in particle assembly and feature edge acuity of the pattern than those previously reported [80–83, 85]. The result is a step toward the realization of nano/microperiodic structures for next-generation photonic devices by the self-assembly process.

Self-Assembly Patterning of Spherical Colloidal Crystals by the Two-Solution Method [87]. Micropatterns of spherical particle assemblies were fabricated by the two-solution method [87]. Hydrophilic regions of a patterned SAM were covered with methanol solution containing SiO_2 particles and immersed in decalin to control the shape of droplets and gradually to dissolve the methanol into decalin. Interfacing of methanol/decalin and shrinkage of methanol droplets were utilized to obtain meniscus force to form spherical particle assemblies; additionally, its static solution system allowed precise control of the conditions. Particles were assembled to form spherical shapes on hydrophilic regions of a SAM, and consequently, micropatterns of spherical particle assemblies were successfully fabricated through self-assembly. This patterned two-solution process has the advantages of being both a drying process having meniscus force and a static solution process having high controllability.

An Si wafer (p-type Si [100], NK Platz Co., Ltd.) was sonicated in water, ethanol, or acetone for 10 min, respectively, and was exposed for 2 h to UV light (184.9 nm) (low-pressure mercury lamp, NL-UV253, Nippon Laser & Electronics Lab) to clean the surface. The OTS-SAM or heptadecafluoro-1,1,2,2-tetrahydrodecyltrichlorosilane (HFDTS)-SAM [45] was prepared by immersing the Si substrate in an anhydrous toluene (Aldrich Chemical Co., Inc.) solution containing 1 vol% OTS (Acros Organics) or HFDTS (Lancaster Synthesis Ltd., Windham, NH) for 5 min under an N_2 atmosphere (Fig. 17.35). The substrate with the SAM was baked at 120°C for 5 min to remove residual solvent and to promote chemisorption of the SAM.

The HFDTS-SAM on the silicon substrate was exposed for 2 h to UV light through a photomask such as a mesh for transmission electron microscopy (Okenshoji Co.,

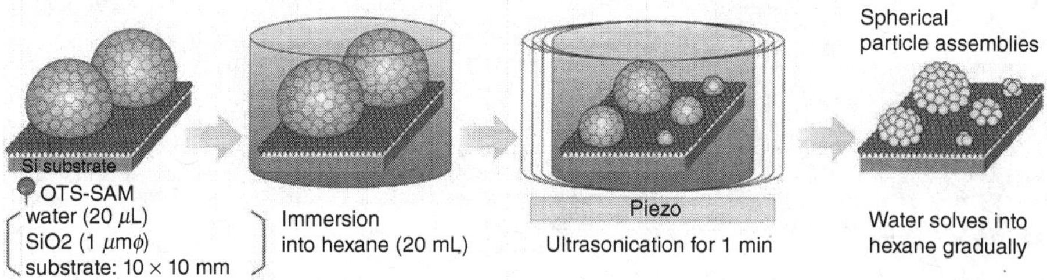

Figure 17.35. Conceptual process for fabrication of spherical particle assemblies.

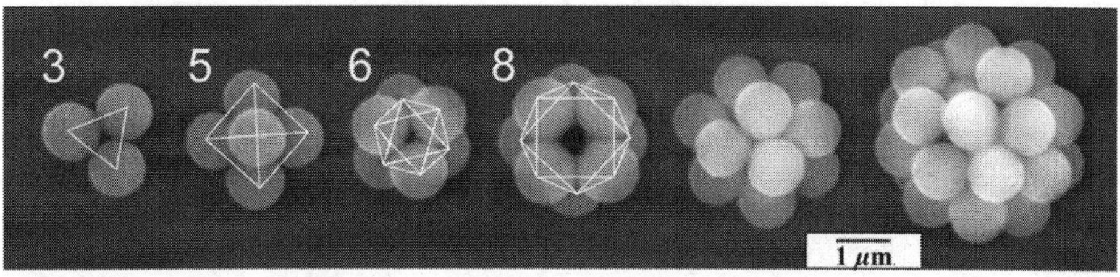

Figure 17.36. SEMs of spherical particle assemblies with different diameters. Particle assemblies constructed from a small number of particles such as 3, 5, 6, 8, and so on.

Tokyo, Japan) to be used as a template for micropatterning of spherical particle assemblies. UV-irradiated regions became hydrophilic due to silanol group formation, while the nonirradiated part remained unchanged. Formation of the SAMs and the modification to silanol groups by UV irradiation were verified using the water drop contact angle (θw). The initially deposited OTS-SAM or HFDTS-SAM showed a WCA of 105° or 112°, but the UV-irradiated surface of SAM was wetted completely (contact angle < 5°).

SiO_2 particles (1.13 $\mu m\phi$, Hipresica UF N3N, CV: 3.57%, specific gravity: $1.8 \pm 0.1\,g/cm^3$, Ube-Nitto Kasei Co., Tokyo, Japan) (0.2 mg, 1.5×10^{27} particles) were thoroughly dispersed in water (20 μL) and dropped on a hydrophobic OTS-SAM (Fig. 17.35). The OTS-SAM with droplets was then immersed in hexane (20 mL, solubility of hexane in water at 20°C: 0.0013 g/100 mL, specific gravity: 0.7) and ultrasonicated for 1 min. Large water droplets containing SiO_2 particles were separated into many small emulsions that kept them spherical on hydrophobic OTS-SAM. Water in the emulsions was gradually extracted to hexane to reduce the size of emulsions forming spherical particle assemblies [88]. After having been immersed for 12 h, spherical particle assemblies with different diameters were observed on OTS-SAM; it was also observed that the assemblies were constructed from various numbers of particles such as 3, 5, 6, 8, or many particles (Figs. 17.36–17.38). Quantities of 3, 5, 6, or 8 particles were assembled into triangular, pyramidal, octahedral, or decahedral particle clusters, respectively. The number of particles in spheres can be controlled by the change of emulsion size or particle concentration in water [88]. The spherical shape of particle assemblies was caused by the high contact angle of water emulsion on hydrophobic OTS-SAM in hexane. Consequently, various sizes of spherical particle assemblies can be prepared using this emulsion process.

Figure 17.37. Spherical particle assembly and tilted images.

The particles would be strongly bound to the droplet interfaces by surface tension in emulsions [89]. In the formation process of small clusters of particles (Fig. 17.36), the water in emulsions dissolved into hexane to reduce droplet size, and this restricted the area in which particles could exist [88]. Particles touched together by the reduction of emulsion size and formed a spherically packed assembly. Deformation of the interface then led to the rearrangement that formed close-packed particle assemblies. The clusters were formed using emulsions not dispersed in the solution but adsorbed on the flat substrate, and thus some of the clusters showed imperfect symmetry such as pyramidal, octahedral, or decahedral, which have a large flat face touching the substrate.

Additionally, large spherical particle assemblies (e.g., ~57 μmϕ in Fig. 17.38) were prepared from large methanol emulsions (~100–300 μmϕ) in decalin (decahydronaphthalene) without the use of ultrasonication, which makes emulsions smaller. Many linear disclinations, that is, grain boundaries, were formed on the surface to reduce elastic strain energy because a close-packed triangular particle lattice composed of a particle surrounded by six particles in plane cannot cover a spherical surface [90]. Linear disclinations were composed of an alternative arrangement of the particle surrounded by 5 particles in plane (red, this particle can be assumed to have a charge of +1) and the particle surrounded by 7 particles in plane (yellow, charge −1). Both ends of a linear disclination were particles surrounded by 5 particles in plane to make the total charge of each linear disclination +1 as observed in previous report[90]. The minimum number of linear disclinations can be calculated for the large sphere (~57 μmϕ) (Fig. 17.38) to be 22 (N: minimum number of linear disclinations, $N = 12 + 0.41r/d = 22$; r: radius of sphere, ~28.5 μm; d: mean particle spacing, 1.13 μm) from geometric calculations [90, 91]. The surface area in a purple circle can also be calculated to be about 10.5% of the total surface area of the sphere from the formula (surface area ratio: S_2/S_1, surface area of the sphere: $S_1 = 4\pi r^2$, surface area in a purple circle: $S_2 = 2\pi r(r - r_1)$,radius of the purple circle shown in Fig. 17.38 $\sqrt{r^2 - r_1^2} = 17.5$ μm). The minimum number of linear disclinations on the surface area in a circle can therefore be estimated to about 2.3(n: minimum number of linear disclinations in a purple circle, $n = N \cdot S_2/S_1$). However, many linear disclinations were observed in a SEM (Fig. 17.38, bottom). It is suggested that our assembly method can be further optimized to reduce linear disclinations.

Furthermore, the upper sides of spherical particle assemblies were removed using a manipulator installed in an optical microscope (BX51WI Microscope, Olympus Optical Co., Ltd.) to evaluate the packing structure. Particle assemblies have a densely

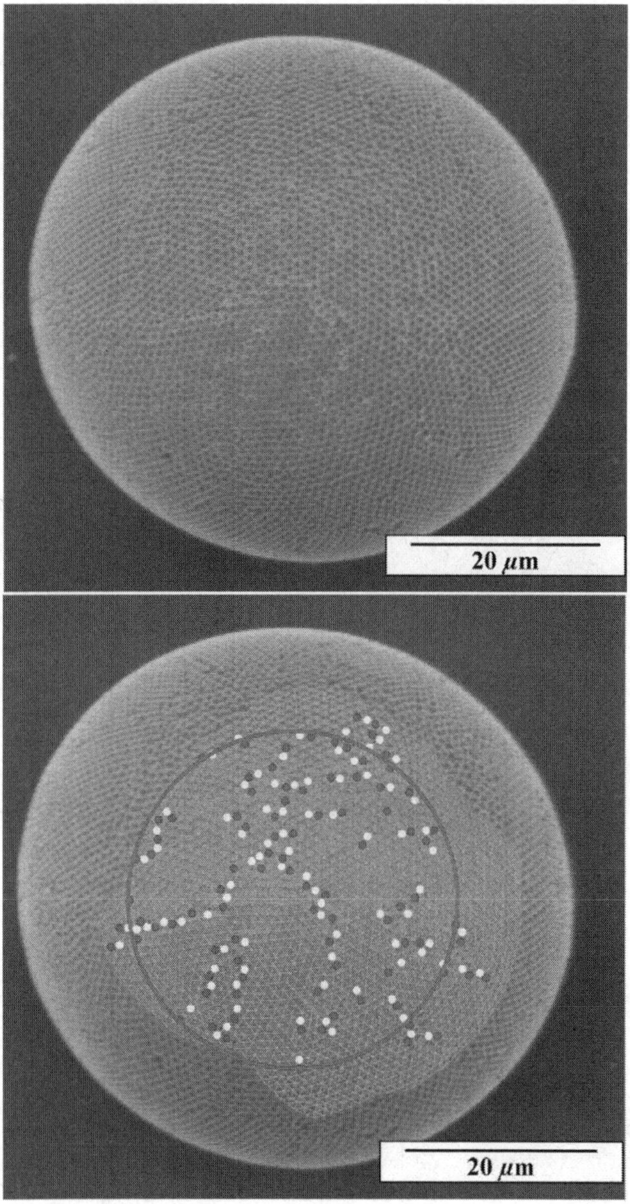

Figure 17.38. Top: a large spherical particle assembly. Bottom: linear disclinations on a large spherical particle assembly.

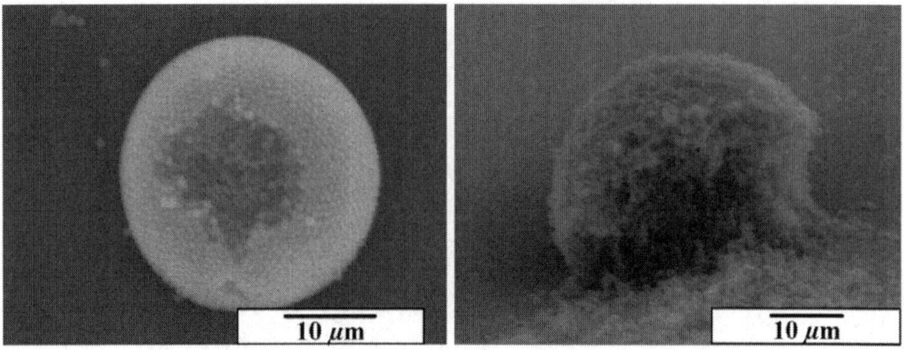

Figure 17.39. SEMs showing the inside of spherical particle assemblies formed from methanol emulsions in decalin.

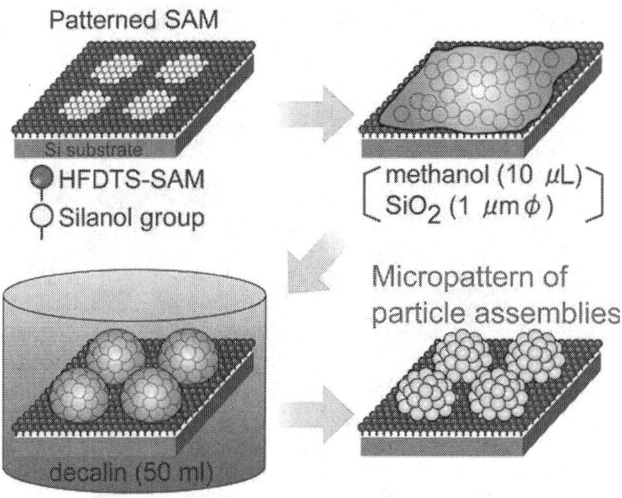

Figure 17.40. Conceptual process of micropattern of spherical particle assemblies.

packed structure (Fig. 17.39). Particles would be completely rearranged to form a densely packed structure during the reduction of emulsion size due to the high dispersibility of particles in emulsions.

SiO_2 particles (1.13 $\mu m\phi$, 1 g/L) were thoroughly dispersed in methanol solution (10 μL) and were dropped onto a patterned HFDTS-SAM [45] having hydrophobic HFDTS-SAM regions and hydrophilic silanol regions, photopatterned using a mesh for transmission electron microscopy as a photomask (Fig. 17.40). The patterned SAM covered with the solution was then carefully immersed in decalin so as not to remove the solution because the density of methanol (0.79) is lower than that of decalin (0.88), causing methanol to float on decalin. The patterned SAM was then gently vibrated to remove additional methanol solution and to assist the movement of droplets to silanol regions. The methanol solution was selectively contacted on hydrophilic regions to

Figure 17.41. SEMs of micropattern of spherical particle assemblies. Top left: Micropattern of spherical particle assemblies. Bottom left: Magnified area of top left. Top right: Tilted micropattern of spherical particle assemblies. Bottom right: Magnified area of top right.

form a micropattern of the solution, which became clearer after immersion for a few hours. The methanol solution containing particles formed a spherical shape because of the surface interaction between methanol, decalin and the surface of a SAM, and the buoyant force of methanol in decalin. Methanol in emulsions was gradually dissolved into an exterior decalin phase to form particle assemblies. After having been immersed for 12 h, particle assemblies having a dome shape were formed at the center of each silanol region (Fig. 17.41). The distance between the centers of each spherical particle assembly was the same as the distance between holes of a mesh. The diameter of spherical particle assemblies was about 18 $\mu m\phi$. It was smaller than that of hydrophilic regions (about 100 μm) due to low particle concentration in methanol and the shape of methanol droplets on a substrate. This indicates that the arrangement regularity of spherical particle assemblies can be improved more by the use of the photomask having small holes to decide positions of each droplet precisely. Some extra particles, that is, noise particles, were also deposited on hydrophobic regions. The process should be further optimized to control many factors such as volume of methanol solution on a substrate, quality of a SAM, or aggregation of particles in the solution to avoid noise particles. Particle assemblies were shown to have a densely packed structure by destructive inspection using a manipulator. Consequently, the dot array of spherical particle assemblies was successfully fabricated by this self-assembly process.

Micropatterns of particle assemblies were fabricated without the use of a template, having microstructures such as molds or grooves. A 2-D array of spherical particle assemblies was fabricated by self-assembly with this method. Interfacing between two solutions and shrinkage of droplets were utilized to obtain a meniscus force to form spherical particle assemblies, and additionally, its static solution system allowed precise control of the conditions. These showed the high ability of self-assembly processes to prepare microstructures constructed from colloidal crystals. Further investigations of the solution–solution, solution–SAMs, and solution–particle interfaces and the behavior of particles and solutions would allow us to develop this two-solution system to prepare desirable particle assembly structures.

Summary

Self-assembly patterning of colloidal crystals was realized by LPP, dry patterning, and two-solution method. Micropatterns of colloidal crystals were fabricated under environmentally friendly conditions using SAMs. They can be applied to photonic crystals, dye-sensitized solar cells, molecular sensors, gas sensors, nanoparticle devices, and so on. These showed the high ability of SAMs and self-assembly processes.

REFERENCES

1. http://www.epa.gov/greenchemistry/whats_gc.html.
2. Anastas PT, Warner JC. 1998. *Green Chemistry: Theory and Practice*. New York: Oxford University Press.
3. http://www.meti.go.jp/policy/chemistry/.
4. Green Chemistry Workshop, Ministry of Economy. 1999. Trade and Industry Japan.
5. http://www.rsc.org/is/journals/current/green/greenpub.htm.
6. http://www.gdch.de/.
7. Mann S. 1996. *Biomimetic Materials Chemistry*. Weinheim: VCH Publishers.
8. Yoshimura M, Livage J. 2000. MRS Bull 25:12–13.
9. Yoshimura M, Suchanek WL, Byrappa K. 2000. MRS Bull 25:17–25.
10. Yoshimura M. 1998. J Mater Res 13:796–802.
11. Figlarz M. 1988. Chem Scr 28:3–7.
12. Niesen TP, DeGuire MR. 2001. J Electroceram 6:169–207.
13. Pizem H, Sukenik CN, Sampathkumaran U, McIlwain AK, De Guire MR. 2002. Chem Mater 14:2476–2485.
14. Lev O, Wu Z, Bharathi S, Glezer V, Modestov A, Gun J, Rabinovich L, Sampath S. 1997. Chem Mater 9:2354–2375.
15. Klein L. 1994. *Sol-Gel Optics: Processing and Applications*. Boston: Kluwer Acadmic Publishers.
16. Byrappa K, Yoshimura M. 2000. *Handbook of Hydrothermal Technology*. Park Ridge, NJ: LLC/Noyes Publications.
17. Yoshimura M, Suchanek W. 1997. Solid State Ionics 98:197–208.
18. Yoshimura M, Suchanek W, Han KS. 1999. J Mater Chem 9:77–82.
19. Cho YR, Lee JH, Song YH, Kang SY, Hwang CS, Jung MY, Kim DH, Lee SK, Uhm HS, Cho KI. 2001. Mater Sci Eng B 79:128–132.

20. Jeon NL, Clem PG, Nuzzo RG, Payne DA. 1995. J Materials Res 10:2996–2999.
21. Aizenberg J, Braun PV, Wiltzius P. 2000. Phys Rev Lett 84:2997–3000.
22. Nashimoto K, Haga K, Watanabe M, Nakamura S, Osakabe E. 1999. Appl Phys Lett 75: 1054–1056.
23. Mott M, Song JH, Evans JRG. 1999. J Am Ceram Soc 82:1653–1658.
24. Stutzmann N, Tervoort TA, Bastiaansen CWM, Feldman K, Smith P. 2000. Adv Mater 12: 557–562.
25. Matsuda A, Matsuno Y, Tatsumisago M, Minami TJ. 1998. Am Ceram Soc 81:2849–2852.
26. Bauer W, Ritzhaupt-Kleissl HJ, Hausselt J. 1999. Ceram Inter 25:201–205.
27. Jacobs HO, Whitesides GM. 2001. Science 291:1763–1766.
28. Kim E, Xia YN, Whitesides GM. 1995. Nature 376:581–584.
29. Kim C, Burrows PE, Forrest SR. 2000. Science 288:831–833.
30. Hayashi C. 1985. Oyobutsuri (in Jpn) 54:687.
31. Fuchida E. 2000. Kinzoku (in Jpn) 70:443.
32. Song JH, Edirisinghe MJ, Evans JRG. 1999. J Am Ceram Soc 82:3374–3380.
33. Vlasov YA, Bo XZ, Sturm JC, Norris DJ. 2001. Nature 414:289–293.
34. Kim E, Xia Y, Whitesides GM. 1996. Adv Mater 8:245–247.
35. Ozin GA, Yang SM. 2001. Adv Funct Mater 11:95–104.
36. Xia YN, Yin YD, Lu Y, McLellan J. 2003. Adv Funct Mater 13:907–918.
37. Masuda Y. http://staff.aist.go.jp/masuda-y/index.html.
38. Masuda Y, Sugiyama T, Lin H, Seo WS, Koumoto K. 2001. Thin Solid Films 382: 153–157.
39. Masuda Y, Jinbo Y, Yonezawa T, Koumoto K. 2002. Chem Mater 14:1236–1241.
40. Masuda Y, Wang DJ, Yonezawa T, Koumoto K. 2002. Key Eng Mater 228–229:125–130.
41. Wang DJ, Masuda Y, Seo WS, Koumoto K. 2002. Key Eng Mater 214:163–168.
42. Lee LH. 1968. J Colloid Interface Sci 27:751–760.
43. Masuda Y, Ieda S, Koumoto K. 2003. Langmuir 19:4415–4419.
44. Masuda Y, Saito N, Hoffmann R, De Guire MR, Koumoto K. 2003. Sci Tech Adv Mater 4:461–467.
45. Masuda Y, Sugiyama T, Koumoto K. 2002. J Mater Chem 12:2643–2647.
46. Gun J, Sagiv J. 1986. J Colloid Interface Sci 112:457–472.
47. McGovern ME, Kallury KMR, Thompson M. 1994. Langmuir 1994(10):3607–3614.
48. Duchet J, Chabert B, Chapel JP, Gerard JF, Chovelon JM, Jaffrezic-Renault N. 1997. Langmuir 13:2271–2278.
49. Dressick WJ, Chen M-S, Brandow SL. 2000. J Amer Chem Soc 122:982–983.
50. Dressick WJ, Chen M-S, Brandow SL, Rhee KW, Shirey LM, Perkins FK. 2001. Appl Phys Lett 78:676–678.
51. Dressick WJ, Nealey PF, Brandow SL. 2001. Proc SPIE 4343:294–305.
52. Koumoto K, Seo S, Sugiyama T, Seo WS, Dressick WJ. 1999. Chem Mater 11:2305.
53. Masuda Y, Sugiyama T, Seo WS, Koumoto K. 2003. Chem Mater 15:2469–2476.
54. Collins RJ, Shin H, De Guire MR, Heuer AH, Sukenik CN. 1996. Appl Phys Lett 69: 860–862.
55. Huang D, Xiao ZD, Gu JH, Huang NP, Tuan C-W. 1997. Thin Solid Films 305:110–115.

56. Zhang F, Mao Y, Zheng Z, Chen Y, Liu X, Jin S. 1997. Thin Solid Films 310:29–33.
57. Shin H, Collins RJ, DeGuire MR, Heuer AH, Sukenik CN. 1995. J Mater Res 10:699–703.
58. Nakanishi T, Masuda Y, Koumoto K. 2004. Chem Mater 16:3484–3488.
59. Nakanishi T, Masuda Y, Koumoto K. 2005. J Cryst Growth 284:176–183.
60. Brandow SL, Dressick WJ, Marrian CRK, Chow GM, Calvert JM. 1995. J Electrochem Soc 142:2233–2243.
61. Dressick WJ, Dulcey CS, Georger JH, Calabrese GS, Calvert JM. 1994. J Electrochem Soc 141:210–220.
62. Izaki M, Shinoura O. 2001. Adv Mater 13:142–145.
63. Masuda Y, Kinoshita N, Sato F, Koumoto K. 2006. Cryst Growth Des 6:75–78.
64. Call RL, Jaber NK, Seshan K, Whyte JR. 1980. Solar Energy Mater 2:373–380.
65. Oner M, Norwig J, Meyer WH, Wegner G. 1998. Chem Mater 10:460–463.
66. Wu XL, Siu GG, Fu CL, Ong HC. 2001. Appl Phys Lett 78:2285–2287.
67. Kang JS, Kang HS, Pang SS, Shim ES, Lee SY. 2003. Thin Solid Films 443:5–8.
68. Masuda Y, Yamagishi M, Koumoto K. 2007. Chem Mater 19:1002–1008.
69. Shaw WHR, Bordeaux JJ. 1955. J Am Chem Soc 77:4729–4733.
70. Ryabchikov DE, Ryabukhin VA. 1970. *Analytical Chemistry of Yttrium and the Lanthanide Elements*. Ann Arbor, MI: Humphrey Science.
71. Baes CF, Mesmer RE. 1976. *The Hydrolysis of Captions*. New York: Wiley.
72. Aiken B, Hsu WP, Matijevic E. 1988. J Am Ceram Soc 71:845–853.
73. Agarwal M, DeGuire MR, Heuer AH. 1997. Appl Phys Lett 71:891–893.
74. Matijevic E, Hsu WP. 1987. J Colloid Interface Sci 118:506–523.
75. Sharma PK, Jilavi MH, Nass R, Schmidt H. 1999. J Lumin 82:187–193.
76. Kwaka MG, Parkb JH, Shon SH. 2004. Solid State Commun 130:199–201.
77. Fuggle JC, Martensson N. 1980. J Electron Spectrosc Relat Phenom 21:275.
78. Wagner CD. 1990. *Practical Surface Analysis*, vol. 1, 2nd edition. New York: John Wiley.
79. Paton MG, Maslen EN. 1967. Acta Crystallographica 1:1948–1923.
80. Masuda Y, Itoh M, Yonezawa T, Koumoto K. 2002. Langmuir 18:4155–4159.
81. Masuda Y, Seo WS, Koumoto K. 2000. Jpn J Appl Phys 39:4596–4600.
82. Masuda Y, Seo WS, Koumoto K. 2001. Thin Solid Films 382:183–189.
83. Masuda Y, Tomimoto K, Koumoto K. 2003. Langmuir 19:5179–5183.
84. Masuda Y, Itoh T, Itoh M, Koumoto K. 2004. Langmuir 20:5588–5592.
85. Masuda Y, Itoh M, Koumoto K. 2003. Chem Lett 32:1016–1017.
86. Masuda Y, Itoh T, Koumoto K. 2005. Langmuir 21:4478–4481.
87. Masuda Y, Itoh T, Koumoto K. 2005. Adv Mater 17:841–845.
88. Manoharan VN, Elsesser MT, Pine DJ. 2003. Science 301:483–487.
89. Levine S, Bowen BD, Partridge SJ. 1989. Colloids Surfaces 38:325–343.
90. Bausch AR, Bowick MJ, Cacciouto A, Dinsmore AD, Hsu MF, Nelson DR, Nikolaides MG, Travesset A, Weitz DA. 2003. Science 299:1716–1718.
91. Bowick MJ, Nelson DR, Travesset A. 2000. Phys Rev B 62:8738–8751.

18

INTEGRATION OF NANOWIRES IN NEW DEVICES AND CIRCUIT ARCHITECTURES: RECENT DEVELOPMENTS AND CHALLENGES

F. Hernández-Ramírez,[1] J. D. Prades,[1] A. Romano-Rodriguez,[1] S. Barth,[2] H. Shen,[3] and S. Mathur[3]

[1]*University of Barcelona, Barcelona, Spain*
[2]*University College Cork, Cork, Ireland*
[3]*University of Cologne, Cologne, Germany*

INTRODUCTION

Quasi-one-dimensional (1-D) nanostructures offer size- and shape-dependent functional behaviors attributed to confinement effects (Xia et al. 2003; Comini et al. 2009), which make them excellent building blocks in sensors, switches, memory elements, electrodes, and many other electronic components (Yasuda et al. 2001; Cui et al. 2001a; Law et al. 2002; Law et al. 2004a, b; Yang 2005; Kuchibhatla et al. 2007; Tian et al. 2007). 1-D semiconductor nanomaterials have been established as attractive device elements for circuits and interconnects with potential to reach much higher device densities (Xia et al. 2003; Chaudhry et al. 2007; Comini et al. 2009) when compared to conventional semiconductor technology. With size reduction, the electronic excitations in semiconductor materials shift to higher energies (band-gap broadening), and the oscillator strength is concentrated into just a few discrete transitions. Functional properties of anisotropic nanostructures such as electrical, optical, and chemical behaviors can thus be modulated, exploiting quantum confinement of conduction electrons

Ceramic Integration and Joining Technologies: From Macro to Nanoscale, First Edition.
Edited by Mrityunjay Singh, Tatsuki Ohji, Rajiv Asthana, Sanjay Mathur.

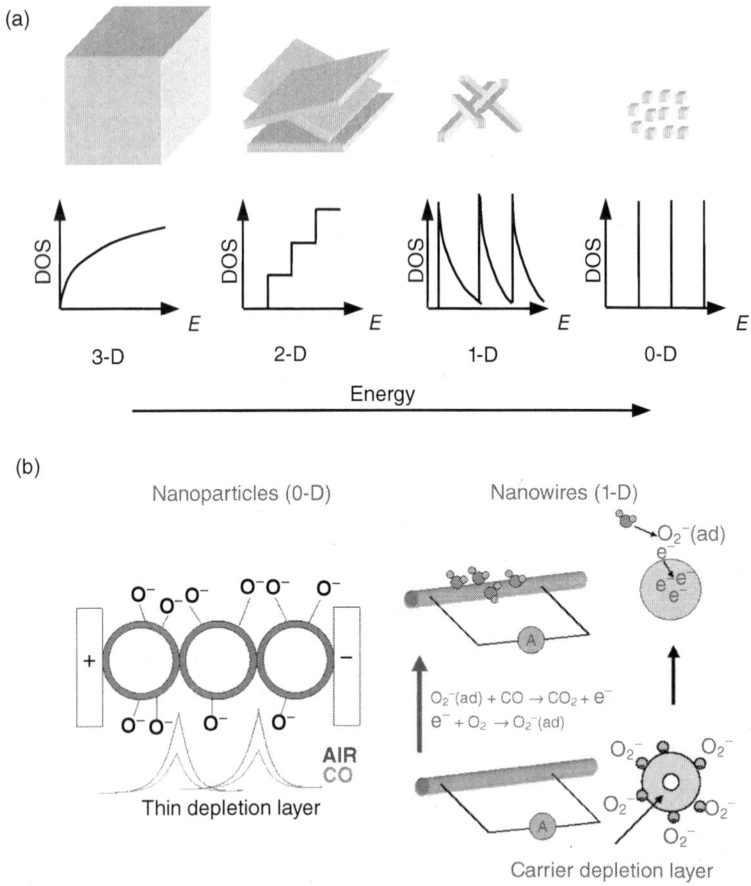

Figure 18.1. (a) Forms of nanomatter and corresponding density of states (DOS) with reducing dimensions. (b) Carrier transport behaviors in 0-D and 1-D systems.

in reduced dimensions. In comparison to nanoparticles, wirelike 1-D nanostructures have mixed discrete/continued density of states (DOS), which influence the carrier transport behavior known for bulk materials (Fig. 18.1a). Since the diameter of nanowires (NWs) can be controlled during growth, more efficient control over chemical and physical properties is feasible in NWs, whereas only averaged properties can be realized for nanoparticulate materials. Although nanoparticles possess larger surface/volume ratios, advantageous for transduction properties and for the detection of gas molecules, photons, and biomolecules, the electronic mechanism that is charge carrier movement is more effective in 1-D systems due to higher crystallinity, defined conduction channel, and absence of grain boundaries (GBs) that are generally detrimental for electron transport due to scattering and carrier trapping phenomena (Comini et al. 2009).

Semiconductor NWs are effective "probes" to detect any kind of molecular species inducing a chemical stimulus upon adsorption on the NW surface. In addition, the wirelike geometry of NWs is suitable for integration in two- and four-probing configurations in which the electrical signals, such as variation of current, can be measured. Furthermore, 1-D nanostructures are mainly single crystalline with defined facets and an atomically precise surface structure, which enable high sensitivity, reliable sensing mechanism, and mechanical stability. Synthetic nanotechnology including both "top-down" and "bottom-up" principles demonstrates the potential of rational synthesis of nanomaterials in nanoparticles (zero-dimensional) and films (two-dimensional) forms.

The development of nanodevices demands not only controlled synthesis and growth of NWs but also novel concepts for handling, manipulation, and integration of NWs in devices. Technical challenges related to rapid and reproducible fabrication of interconnects at the nanoscale represent an underdeveloped domain (Wang et al. 2001; Wang 2003; Patolsky et al. 2007). The steps taken to circumvent these difficulties and to demonstrate the final potential of nanoelectronics are surveyed in this chapter, providing a unifying framework for the basic physical and electrical principles that determine the transport behavior of NWs and highlight the newborn discipline of nanoelectronics and efforts devoted to obtain a deeper insight into the electrical behavior of NWs and to study the role of nanocontacts in their final performance as functional blocks (Hernandez-Ramirez et al. 2006; Zhang et al. 2007; Hernandez-Ramirez et al. 2007a, b). These two points are crucial for the formation of NW architectures and for interfacing it in the macroscopic world.

A general overview of recently fabricated microelectronic devices such as field-effect transistors (FETs) (Bryllert et al. 2006; Goldberger et al. 2006; Xiang et al. 2006) and other design strategies allowing a large-scale and massively parallel assembly of NWs in complex architectures (Javey et al. 2007; Lu and Lieber 2007) reveals the potential and manifold benefits in electronics, computers, information technology, medicine, and energy storage (Yasuda et al. 2001; Cui et al. 2001a, b; Law et al. 2002; Wang 2003; Law et al. 2004a, b; Yang 2005; Kuchibhatla et al. 2007; Tian et al. 2007; Wang 2008a, b) (Fig. 18.2).

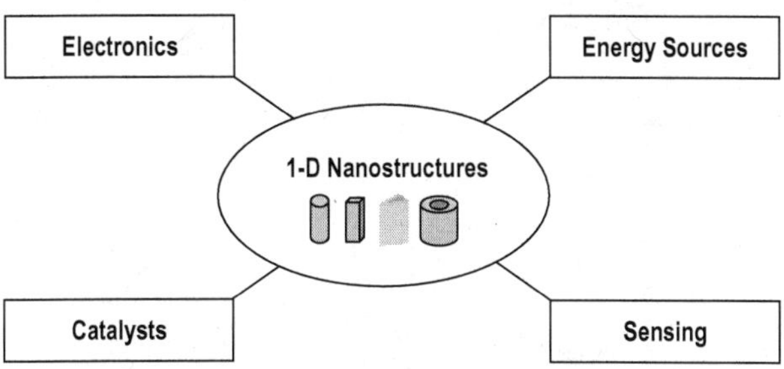

Figure 18.2. Potential applications of one-dimensional inorganic nanostructures.

1-D NANOSCALE BUILDING BLOCKS: SYNTHESIS AND GROWTH MECHANISM

Synthetic Approaches

During the last decade, several approaches including both top-down and bottom-up concepts for the synthesis of 1-D nanostructures were described in detail (Table 18.1); however, generic methodologies suitable for a large variety of materials are limited. This subsection is devoted to common strategies used to obtain high inorganic aspect ratio nanostructures.

Crystal Structure Governed Nucleation in Solution. Several materials show preferential growth of certain (kinetically favored) facets under appropriate experimental conditions such as educt, concentration, pH, and ionic strength of solution (Vayssieres et al. 2001). The method relies on the crystal structure and crystal chemistry of NW materials and is intrinsically limited, but it provides access to less expensive, high-yield and potential large-area coverage on surfaces. Certain materials such as metal oxides (ZnO, SnO_2, FeO_x, etc.) exhibit an aqueous-phase nucleation of NW potential from a commercial point of view.

Templated Synthesis. In template-based approaches, the actual skeleton is replicated either by filling or covering a host structure (Fig. 18.3). The most popular representatives of structure-directing *hard templates* with predefined pores are anodic alumina oxide (AAO) and polymeric membranes, which can be filled with a solution or covered with a thin layer of the desired material by solvent (Brumlik and Martin 1991) or atomic layer deposition (ALD) techniques (Bachmann et al. 2007). The pore filling can be achieved by capillary force, chemical vapor infiltration (CVI) of molten material, sol–gel, or electrolytic reactions. Advantages of the pore-filling techniques

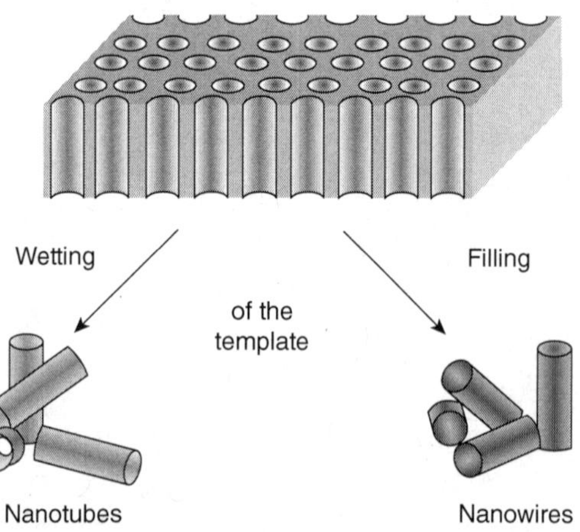

Figure 18.3. Scheme of template-based synthesis of 1-D nanostructures.

TABLE 18.1. Summary of 1-D Nanostructures Synthesized Using Different Methods

Material	Morphology	Growth method	Applications	References
Si	Nanowires	Metal organic CVD method	Transistor	Wagner and Ellis (1964)
		Chemical beam epitaxy		Liu et al. (1999)
Ge	Nanowires	Metal organic CVD method	Photodetector	Bootsma and Gassen (1971)
		Pulsed laser deposition		Morales and Lieber (1998)
Bi	Nanowires	Electrochemical deposition	FET device	Cornelius et al. (2005)
Ag	Nanowires	Electrochemical deposition	FET device	Hu et al. (2004)
Au	Nanowires	Solution-phase synthesis	Nanodevice	Wang and Sun (2009)
Co	Nanowires	Pulsed electrodeposition technique	Nanodevice	Huang et al. (2009)
Se	Nanowires	Solution-phase method	Resistor	Gates et al. (2002)
Te	Nanowires	Vapor-phase method	Resistor	Furuta et al. (1975)
Si–SiGe	Nanowires	Laser-ablation method	FET device	Wu et al. (2002)
Si–Ge	Nanowires	Metal organic CVD method	FET device	Lauhon et al. (2002)
Ge-Si	Nanowires	Metal organic CVD method	Transistor	Lauhon et al. (2002)
NiSi–Si	Nanowires	Solid-phase reaction method	Transistor	Wu et al. (2004a, b)
ZnO	Nanowires	Vapor–solid method	FET device Resistor Nanosensor Photodetector	Yang et al. (2002)
		Vapor–liquid–solid method		Wang et al. (2002)
		AAO template-assisted method		Li et al. (2000)
		Microemulsion method		H Zhang et al. (2004a, b)
		Template-free solution method		H Zhang et al. (2005a)
	Nanobelts	Vapor–solid method		Pan et al. (2001)
		Vapor–liquid–solid method		Ding et al. (2004a, b)
		Hydrothermal method		XY Zhang et al. (2004c)
	Nanorods	Template-free aqueous method		Vayssieres (2003)
		Vapor–liquid–solid method		Gao and Wang (2004)
		Vapor–solid method		Wang (2008b)
		Pulsed laser ablation without catalyst		Hartanto et al. (2004)
	Nanotubes	Vapor–solid method		Xing et al. (2003)
		Vapor–liquid–solid method		Kong and Wang (2003a)
		Solution method		Sun et al. (2005)

(*Continued*)

TABLE 18.1. (*Continued*)

Material	Morphology	Growth method	Applications	References
In_2O_3	Nanowires	Catalyst-assisted laser-ablation method	FET device Nanosensor	Li et al. (2003a)
		Vapor–solid method		Peng et al. (2002b)
		Vapor–liquid–solid method		J Zhang et al. (2003)
		AAO-templated solution method		Cao et al. (2003)
	Nanobelts	Vapor–solid method		Kong and Wang (2003b)
	Nanotubes	Thermal evaporation method		Li et al. (2003b)
		Solvothermal method		Chen et al. (2006a, b)
MgO	Nanowires	Vapor–solid method	FET device	Yin et al. (2002)
		Vapor–liquid–solid method		Kim and Shim (2006)
		Catalyst-assisted laser-ablation method		Nagashima et al. (2007)
SnO_2	Nanowires	Catalyst-assisted laser-ablation method	Resistor Nanosensor Photonic device	Z Liu et al. (2003)
		Vapor–solid method		Luo et al. (2006)
		Vapor–liquid–solid method		Chen et al. (2003a, b)
		Solution method		Wang et al. (2003)
	Nanobelts	Thermal oxidation		Sun et al. (2003)
		Vapor–solid method		Duan et al. (2005)
		Laser-ablation method		Hu et al. (2003)
	Nanorods	Microemulsion method		Liu et al. (2001)
		Hydrothermal method		DF Zhang et al. (2003)
		Solution methodl		Cheng et al. (2004)
		Vapor–liquid–solid method		He et al. (2006)
	Nanotubes	Template hydrothermal method		Liu and Zeng (2004)
		Aqueous solution method		Du et al. (2008)
		Microemulsion method		Wang et al. (2008)

Ga_2O_3	Nanowires	Thermal evaporation method	Resistor Photonic device	Zhang et al. (1999a, b)
		Catalyst-assisted arc discharge method		Choi et al. (2000)
		Laser-ablation method		Hu et al. (2002)
		Catalyst-assisted vapor method		Chang and Wu (2003)
	Nanobelts	Vapor–solid method		Geng et al. (2003)
		Vapor–liquid–solid method		Zhang et al. (2005b)
	Nanotubes	Vapor–solid method		Gong et al. (2008)
V_2O_5	Nanowires	Hydrothermal method	Resistor	Zhou et al. (2008)
	Nanobelts	Vapor–solid method		Chan et al. (2007)
TiO_2	Nanotubes	AAO-templated solution method	Photonic device	Imai et al. (1999)
		Solution method		Liu et al. (2002)
	Nanowires	Hydrothermal method		Zhang et al. (2002)
		Vapor–solid method		Wu et al. (2005)
WO_3	Nanowires	SBA-15 templated solution method	Resistor	Zhu et al. (2003a, b)
		Hydrothermal method		Baek and Baek (2007)
		Thermal evaporation method		Song et al. (2007)
	Nanobelts	Vapor–solid method		Li et al. (2003c)
ZrO_2	Nanowires	AAO-templated solution method	Photonic device	Cao et al. (2004)
	Nanorods	Precursor thermal decomposition method		Li et al. (2003a)
	Nanotubes	AAO-templated solution method		Dae et al. (2008)
Nb_2O_5	Nanobelts	Precursor thermal decomposition method	FET device	Wei et al. (2008)
	Nanowires	Vapor–liquid–solid method		Varghese et al. (2008)
	Nanotubes	Precursor thermal decomposition method		Kobayashi et al. (2007)
Fe_3O_4	Nanotubes	MgO-templated pulsed laser deposition	Bi-device Resistor	Liu et al. (2005a, b)
	Nanowires	Magnetic field-induced hydrothermal method		Wang et al. (2004a, b)

(*Continued*)

TABLE 18.1. *(Continued)*

Material	Morphology	Growth method	Applications	References
IrO_2	Nanotubes	Metal organic CVD method	Photonic device	Chen et al. (2004)
	Nanowires	Metal organic CVD method		Chen et al. (2006a, b)
Ta_2O_5	Nanotubes	Precursor thermal decomposition method	FET device	Kobayashi et al. (2007)
MoO_3	Nanotubes	Hydrothermal method	FET device Nanodevice	Hu et al. (2008)
		Carbon nanotube templated method		Satishkumar et al. (1997)
		Vapor–solid method		Li and Bando (2002)
	Nanowires	Thermal evaporation method		Zhou et al. (2003)
		Solution method		Qi et al. (2008)
MnO_2	Nanowires	Hydrothermal method	FET device Nanodevice	Wang and Li (2002)
		SBA-15 templated synthesis		Imperor-Clerc et al. (2004)
	Nanotubes	Hydrothermal method		Zheng et al. (2005)
Fe_2O_3	Nanowires	Thermal oxidation method	Nanosensor Resistor Nanodevice	Fu et al. (2001)
		Hydrothermal method		Xiong et al. (2003a, b)
	Nanobelts	Thermal oxidation method		Wen et al. (2005)
NiO	Nanowires	Wet chemical route	Resistor Photonic device	Xu et al. (2003)
		AAO-templated sol–gel method		Yang et al. (2005)
	Nanotubes	AAO-assisted solution method		Shi et al. (2008)
Co_3O_4	Nanowires	Thermal oxidation method	Resistor Nanodevice	Dong et al. (2007)
		Hydrothermal method		Zhang et al. (2008)
	Nanotubes	Carbon nanotube templated method		Du et al. (2007)
		Solution method		Lou et al. (2008)
Cu_2O	Nanowires	Solid-state reduction method	Resistor	Wang et al. (2002)
		Surfactant-assisted solution method		Xiong et al. (2003a, b)
		Hydrothermal method		Tan et al. (2007)
CuO	Nanowires	Thermal oxidation method	Resistor Nanodevice	Jiang et al. (2002)
		AAO-templated deposition method		Ko et al. (2006)
		Solution method		Du et al. (2004)
	Nanobelts	Solution method		Du et al. (2004)

CdO	Nanowires	AAO-assisted electrochemical deposition	Nanosensor Photonic device	Peng et al. (2002d)
		Chemical bath deposition method		Dhawale et al. (2008)
	Nanoneedles	Vapor–liquid–solid method		Liu et al. (2003)
Al_2O_3	Nanotubes	Pulse anodization method	Resistor Nanodevice	Lee et al. (2008a, b)
		Thermal evaporation method		Li et al. (2005a, b)
		Surfactant-assisted solution method		Kuang et al. (2003)
		Carbon nanotube-assisted growth		Ogihara et al. (2006)
	Nanowires	Vapor–solid method		Peng et al. (2002c)
SiO_2	Nanowires	Vapor–liquid–solid method	Transistor	Huang et al. (2000)
LiV_3O_8	Nanorods	Hydrothermal method	Resistor	Liu et al. (2007)
MnV_2O_6	Nanorods	Hydrothermal method	Resistor	Lei et al. (2007)
$ZnAl_2O_4$	Nanorods	Polycarbonate membrane template	Resistor	Thomas et al. (2006)
$BaTiO_3$	Nanowires	Hydrothermal method	Nanodevice	Bao et al. (2009)
$CoFe_2O_4$	Nanowires	Mild chemical method	Nanodevice	Ersen et al. (2008)
$MnFe_2O_4$	Nanorods	Surfactant-free hydrothermal method	Nanodevice	Zhen et al. (2008)
$LiFePO_4$	Nanowires	SBA-15 templated synthesis	Batteries	Lim et al. (2008)
$ZnFe_2O_4$	Nanorods	AAO-templated deposition method	Resistor	Jung et al. (2005)
GaAs	Nanowires	Pulsed laser deposition	Resistor Photonic device	Ohlsson et al. (2002)
		Chemical beam epitaxy		Lee et al. (1999)
		Vapor–liquid–solid method		Gudiksen et al. (2002)
GaP	Nanowires	Pulsed laser deposition	Resistor	Duan and Lieber (2000a)
Bi_2S_3	Nanorods	Assisted hydrothermal method	Nanodevice	Zhu and Liu (2009)
InP	Nanowires	Pulsed laser deposition	FET device	Ohlsson et al. (2002)
InAs	Nanowires	Chemical beam epitaxy	FET device	Dick et al. (2005)
		Vapor–liquid–solid method		Duan and Lieber (2000b)
In_2Se_3	Nanowires	Vapor–liquid–solid method	Nanodevice	Sun et al. (2006)
GaN	Nanowires	Laser-assisted catalytic growth	Photonic device	Ohlsson et al. (2002)
		Magnatron sputtering		Ai et al. (2007)

(*Continued*)

TABLE 18.1. (*Continued*)

Material	Morphology	Growth method	Applications	References
CdSe	Nanowires	Pulsed laser deposition	Nanodevice	Ohlsson et al. (2002)
CdS	Nanowires	Laser-assisted catalytic growth	Nanodevice	Cui et al. (2001b)
Cd_3P_2	Nanowires	Situ nanorod template method	Nanodevice	Shen et al. (2006)
Cd_3As_2	Nanowires	Thermal evaporation	Nanodevice	Omari et al. (2008)
SiGe	Nanowires	Pulsed laser deposition	Transistor	Ohlsson et al. (2002)
SiC	Nanowires	Anodization and electrodeposition processes	Transistor	Barrelet et al. (2003)
Si_3N_4	Nanowires	Anodization and electrodeposition processes	Transistor	Chan et al. (2007)
$TiSi_2$	Nanowires	CVD	Nanosensor	Zhou et al. (2009)
ZnS	Nanowires	Nanocluster-catalyzed vapor–liquid–solid	FET device	Wu et al. (2002)
ZnSe	Nanowires	Vapor–liquid–solid method	FET device	Asoh et al. (2003)
Zn_3P_2	Nanowires	Electrochemical deposition	FET device	Choi et al. (2003)
GaP–GaAs	Nanowires	Alternating laser-ablation method	Nanodevice	Gudiksen et al. (2002)
ZnSe–CdSe	Nanowires	Atomic layer deposition	Nanodevice	Solanki et al. (2002)
ZnO–ZnMgO	Nanowires	Catalyst-free metal organic vapor-phase epitaxy	Nanodevice Resistor	Park et al. (2003)
InAs–GaAs	Nanowires	Catalyst-free metal organic vapor-phase epitaxy	Photonic device	Panev et al. (2003)
InAs–InP	Nanowires	Catalyst-free metal organic vapor-phase epitaxy	Photonic device	Poole et al. (2003)
GaN–AlGaN	Nanowires	Preferential etching	Nanodevice	Goldberger et al. (2003)
Ge-SiC_xN_y	Nanocable	CVD	Nanodevice	Mathur et al. (2007)
SnO_2–M_2O_3	Nanowires	Thermal evaporation method	Nanosensor	Li et al. (2007)
SnO_2–VO_x	Nanowires	CVD	Nanosensor	Mathur et al. (2007)
SnO_2–Fe_3O_4	Nanowires	CVD	Nanodevice	Mathur et al. (2008)

are the flexibility to form several material classes and the solubility of the alumina backbone in concentrated basic solution or phosphoric acid, which enables the release of as-grown nanostructures (Martin 1991). Also it is possible to construct AAO with hyperbranched structures, which allows creating hierarchical structures (Modi et al. 2003). However, one of the drawbacks of this method is obviously the predominantly polycrystalline nature of the resulting material. The filling factor of the pores is usually quite high as there are no favored specific sites in the template. On the other hand, positive replication, which means coating of the outermost surface of already 1-D features, such as carbon nanotube (CNT) or noncarbon structures as oxides, is also a commonly applied methodology. Hereby, the backbone can act purely as removable skeleton or contribute to the hybrid structure or act as second reagent for solid-phase reactions (Fan et al. 2006a, b, c).

Soft templates are usually surfactant-based mesophasic structures, which are formed by self-assembly of polar molecules at critical micelle concentrations. Chemical reactions within this dual-phase system allow, similar to the vesicular micelles, synthesis of diverse networks and assemblies, respectively. The surface facet-dependent binding strengths of surfactants facilitate the formation of anisotropic nanostructures with specific growth directions. This was demonstrated for silver wires, which were seeded by Pt nanoparticles, but the shape of the NWs was predominantly governed by kinetically controlled synthesis using a polymer (Sun et al. 2002). The polymer surfactant has different residence times and binding affinity on specific crystal planes and enhances thereby the predominate growth of specific planes of face-centered cubic silver.

Catalyst-Assisted Growth in Liquids. NW growth in a liquid can also be catalyzed by metal colloids following either a solution–liquid–solid (SLS) (Trentler et al. 1995) or a supercritical fluid–liquid–solid (SFLS) mechanism (Holmes et al. 2000). In the SLS process, metal particles (Bi, Sn, In, Bi/Au, etc.) with low melting temperatures are used as growth seeds, which act as nucleation centers for decomposed metal organic species and induce the nucleation of single-crystalline wires. The chemical interaction between metal catalyst and NW-forming species is similar to the vapor–liquid–solid (VLS) mechanism (Section 2.2). Similar catalytic reactions are expected in the SFLS growth, where the main differences are higher temperature (ca. 500–550°C), high pressures in the range of 200–300 bar, and solvents (benzene, toluene, hexane, and CO_2). The major benefits of these procedures are narrow size distributions and the high yield of thin wires (4–10 nm) (Holmes et al. 2000).

Catalyst-Assisted Vapor Deposition. The most commonly used growth initiator in vapor-phase synthesis of 1-D structures is gold, while other metals also show similar properties (VLS growth mode: Fig. 18.4) (Wagner and Ellis 1964). However, several vapor-phase techniques are viable for the synthesis of anisotropic nanostructures including thermal and pulsed laser-induced evaporation, chemical vapor deposition (CVD) and molecular beam epitaxy (MBE). Thermal evaporation generally requires high temperatures in order to break strong chemical bonds in materials with high melting points (Xia et al. 2003). Pulsed laser deposition (PLD) is a versatile technique

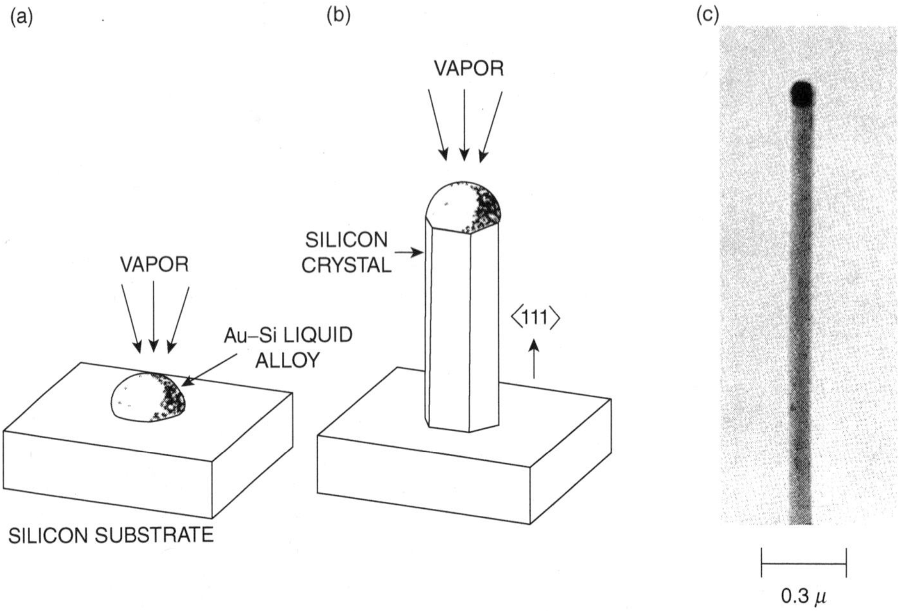

Figure 18.4. Schematic illustration of the growth of a silicon crystal by VLS mechanism with (a) the initial liquid droplet on the substrate and (b) the growing crystal with the liquid droplet on the tip. (c) TEM image of a Si NW leading to the development of the VLS theory. Reprinted from Wagner and Ellis (1964).

in which the solid-phase material is evaporated through high-energy laser pulses (Morales and Lieber 1998). It provides access to a large variety of materials and compositions. MBE is based on the thermal evaporation of the phase-forming elements under high-vacuum conditions, which react on the surface to form the desired crystal lattice (Xia et al. 2003). CVD techniques offer easy variation of deposition parameters. The chemical reaction could be thermolysis, hydrolysis, oxidation, reduction, disproportionation, nitration, and carboration, which depends on the precursor species used. Once the gaseous species are in proximity to the substrate, they could either adsorb directly on the catalyst particle or on the surface itself, and diffusion processes as well as concentration of the adsorbates (supersaturation) are responsible for the growth of a solid phase at the catalyst–nanoparticle surface interface (Wagner and Ellis 1964).

Catalyst-Free Growth Modes. Vapor transport synthesis is comparable to a resublimation process and leads to elongated and beltlike structures in most of the cases, which show high crystallinity and defined side facets without the need for any external catalysts. The nucleation step could be induced by defects, which will be present at the growth front and indicated by a nonsmooth growth front (Dai et al. 2003). Therefore, the formed structures exhibit strong variation in the two confined directions, depending on the interplay of crystal chemistry and surface energies, which leads to formation of nanobelts, springs, and other interesting morphological features (Wang 2003). The morphology depends on several factors, such as substrate temperature, carrier gas flow, and composition of starting materials. However, in some reactions involving disproportionation reaction, signs of a VLS growth mode (presence of a metallic seed at the tip of anisotropic nanostructures) are observed, which shows the lower activation energy barrier for catalyst-assisted growth (Xia et al. 2003). Continuous delivery of gaseous species by physical vapor deposition (resublimation or PLD) at appropriated conditions leads to the formation of NWs covered with an oxide shell. The growth front is expected to be a semiliquid suboxide that promotes the growth and is usually thinner than the formed shell. Finally, screw dislocations were described in literature to promote the formation of elongated crystals (Sears 1955).

Top-Down Approach/Chemical Etching. The top-down approach often uses the traditional microfabrication methods such as photolithography, which are used to cut, mill, and shape materials into the desired shape and order. In comparison to conventional methods, the chemical etching technique is a more effective method to synthesize 1-D nanostructures. Due to the facet-dependent etching rates and site-selective etching induced by metallic coating, 1-D nanostructures can be obtained under control of size, dimension, and position (Peng et al. 2002a). The critical issue in this approach is how the features of mask can be transferred to the surface of materials, especially in the case of high-density arrays with small dimension scales.

Growth Mechanism

In the section on "Synthetic Approaches," we have briefly discussed the various approaches to synthesize 1-D nanostructures; however, the evidences to support the hypothesis of different growth mechanisms are missing in literatures (Xia et al. 2003; Kolasinki 2006; Comini et al. 2009). Most of the published data are based on the possible mechanisms supported by crystallographic and morphological observations made after the growth had occurred. The real-time investigation on the growth of 1-D nanostructures is possible in gas-phase synthesis but difficult in the solution-based methods and demands sophisticated setups for in situ experiments in electron microscope chambers to study the growth mechanism in order to gain a general physical picture of NW growth.

A common technique for the growth of NWs, namely, VLS mechanism proposed by Wagner and Ellis (1964), involves three growth regimes: (1) A nanoscaled metallic particle forms an eutectic alloy (liquid) with the wire material (supplied as vapor); (2) the vapor of the source material is further absorbed by the liquid catalyst until supersaturation occurs; (3) the vapor atoms diffuse and condense at the liquid/solid interface, which will be pushed forward to form an NW.

Wu et al. have reported the first real-time observation of semiconductor NW growth in an in situ high-temperature transmission electron microscope (TEM, JEOL CX200), which unambiguously demonstrated the validity of the catalyst-assisted growth mechanism at nanometer scale (Wu and Yang 2001). The growth stages postulated for VLS mechanism, namely, metal alloying (Fig. 18.5b, c), crystal nucleation (Fig. 18.5d, e), and axial growth (Fig. 5f), were evidently observed in this experiment. In situ observation of wire nucleation/growth at nanometer scale also showed that the diameter of a

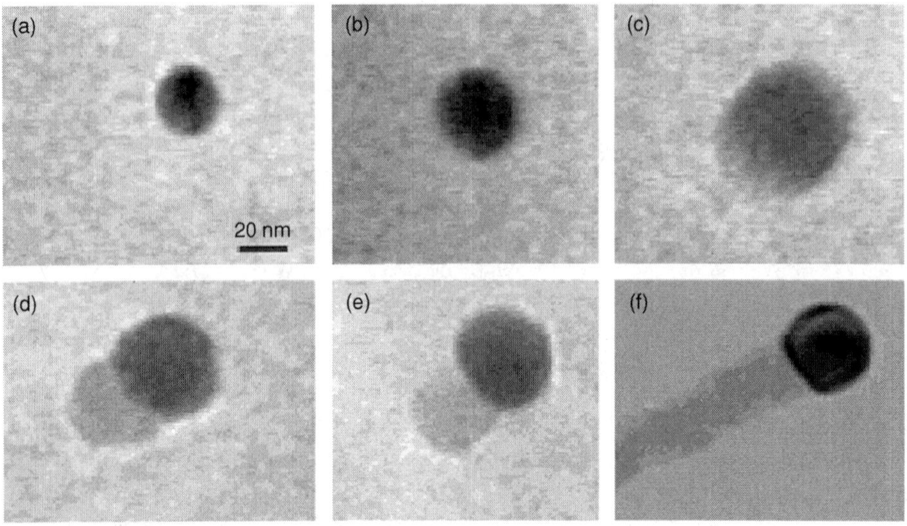

Figure 18.5. In situ TEM images recorded during the growth process of Ge nanowire. (a) Au nanoclusters in solid state at 500°C; (b) alloying initiates at 800°C; at this stage, Au exists in mostly solid state; (c) liquid Au/Ge alloy; (d) the nucleation of Ge nanocrystal on the alloy surface; (e) Ge nanocrystal elongates with further Ge condensation; (f) eventually forming a nanowire. Reprinted from Wu and Yang (2001).

Ge NW is dependent on the size of the Ge/Au alloy droplet instead of the size of Au particles used as growth seeds.

The observed growth process reveals following aspects of NW formation:

1. *Alloying Process (Fig. 18.5a–c).* At the beginning of the process, Au particles remain in the solid state. With an increasing amount of Ge vapor condensation and dissolution on Au particles, Ge and Au form an alloy, which is usually in a liquid state. As the size of the alloy droplets increases, the elemental contrast decreases because of the dilution of the heavy metal Au with the lighter element Ge. This alloying process can be depicted as an isothermal line in the Au–Ge phase diagram, which moves from point A to point B (Fig. 18.6).
2. *Nucleation (Fig. 18.5d, e).* Once the composition of the alloy supersedes the second liquidus line (i.e., the liquid–solid transition), the nucleation of solid Ge is initiated within the supersaturated alloy leading to a liquid/solid interface. The crystallographic orientation of nucleated solid on the liquid/solid interface is dictated by the surface energy considerations, whereby facets with the lowest surface energies show preferential growth. The formation of a catalyst/NW interface is rather complex and depends on growth condition and the nature of materials.
3. *Axial Growth (Fig. 18.5d–f).* Once the solid Ge nanocrystals nucleate at the liquid/solid interface, further dissolution of Ge vapor in liquid alloy and con-

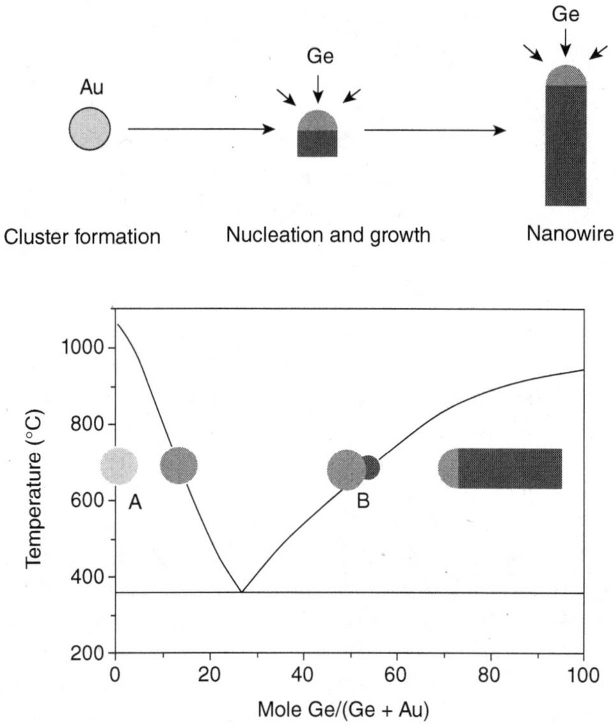

Figure 18.6. VLS process and the isothermal behavior in a Au–Ge phase diagram.

tinuing crystallization lead to the growth of NWs. Although solid Ge can nucleate on any position on the surface of an alloy catalyst, the Ge species preferentially diffuse at the existing solid/liquid interface, primarily due to the fact that less energy is required by the adatoms to be adsorbed on the low-energy facets, when compared to the energies of secondary nucleation of Ge from a Au/Ge alloy. Consequently, secondary nucleation events are generally suppressed, and no new solid/liquid interface is created, which is evident in the upward (axial) movement of the growth front.

The growth of Si NWs from a AuSi eutectic liquid examined using ultrahigh-vacuum transmission electron microscopy revealed the different stages of VLS growth mode in real time (Kim et al. 2008). The physical model suggests that nucleation of solid Si is heterogeneous and occurs usually at the edge of the Au/Si droplet. The data showed that after silicon precursor was introduced ($t = 0$ s), a certain time elapsed (incubation time) before the solid Si nuclei appeared. Each nucleus initially grew very rapidly, but after a few seconds, the growth process equilibrated and slowed down. Since the time of nucleation depends on the diffusion kinetics of Si in Au/Si alloy, the larger droplets can have a longer incubation time in comparison to that of smaller droplets. Further, the crystal structure and roughness of the substrate surface influence the thermodynamic behavior of Au and Au/Si droplets, such as wetting behavior and mobility of precursor fragments on the surface.

In the conventional VLS picture, the catalyst does not change during growth and the sidewalls of the NW consist of smooth facets. Many growth experiments have shown that this assumption is not valid because tapered NWs and NWs with rough-faceted sidewalls are commonly obtained despite careful control over the experimental parameters. In situ TEM studies (Hannon et al. 2006; Kodambaka et al. 2006a) showed that surface migration of catalyst particles (Au) will strongly alter the length, shape, and sidewall properties of silicon NWs in Au-catalyzed VLS growth. In the case of droplets present on the tips of NWs, it was found that (1) Au wets the surface of the substrate, the sidewalls of NWs, and the catalytic tips; (2) growth on the sidewalls of NWs is slow due to the decreased density of Au nuclei. Thermodynamically driven diffusion of Au from smaller droplets to the larger ones (Ostwald ripening) leads to the change in diameter of NWs and the formation of tapered NWs (some NWs become thicker, whereas others become thinner). Ostwald ripening of the droplets via curvature-dependent chemical potential is possibly the cause of larger droplets growing at the expense of smaller ones (Fig. 18.7).

In order to fabricate long and uniform-diameter NWs, the introduction of surface-active species has been suggested. For instance, oxygen residues present during the growth (Kodambaka et al. 2006b) may affect the growth of NWs: (1) The adsorbed oxygen is sufficiently mobile on the substrate surface and thus inhibits the diffusion of Au on the surface; (2) Au agglomeration kinetics may be modified and may be suppressed by oxygen. The measured growth rates for NWs and sidewalls showed the later growth rate to be ~100 times slower than the growth rate in axial direction.

Investigation on the eptaxial growth of Au-assisted axial heterostructure NWs composed of group IV and III–V materials (Dick et al. 2007) such as InAs-InP, GaAs-GaP, GaAs-InP, InAs-InP, InAs-GaP, and InAs-AlAs revealed two distinct growth behaviors:

1. *A–B Straight Growth.* The growth of NW B continues in the same crystal direction of NW A when the second material B is introduced.
2. *Kinked Growth.* The NW B tends to kink or wrap around A and grows backward. It is noted that even kinked wires can have an epitaxial growth.

The fabrication of axial 1-D heterostructures is facilitated by their narrow diameter, which allows for efficient relaxation on the nanoscale and the formation of epitaxial structures even for large lattice-mismatched materials. In the simple VLS model, the growth front to form NW A via the interface between A and wire/Au alloy means that nucleation of A occurs via layer-by-layer growth (Frank–van der Merwe) at the interface. In the case of branched heterostructures, the island formation of material B on NW A results in the parallel growth of B on NW A, whereas the layer formation of material B on NW A tends to straight and vertical growth of B on NW A (Fig. 18.8).

The growth rate with solid Au is slower than that with liquid Au, indicating that vapor–solid–solid (VSS) growth mode is less likely in catalyzed growth of NWs. Instead of Au catalyst, Si NWs were successfully synthesized using aluminum as cata-

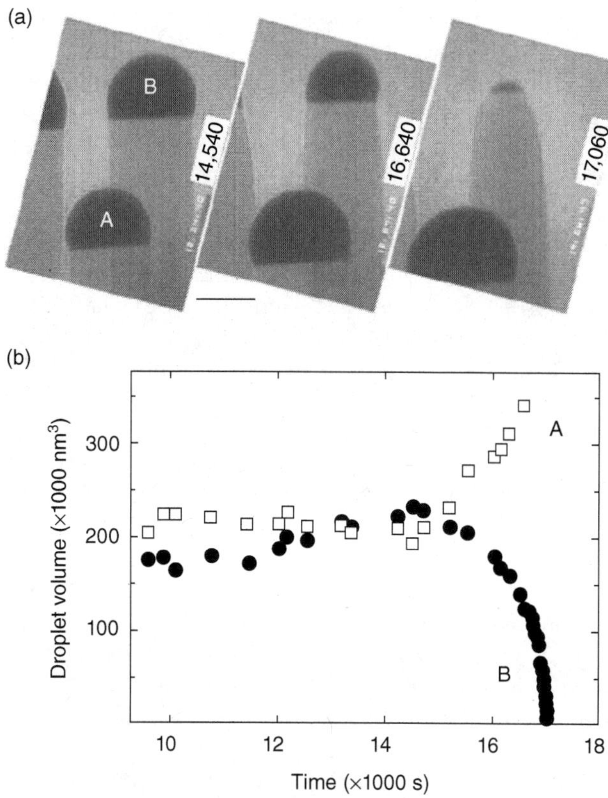

Figure 18.7. In situ TEM images showing the growth of droplets A and B during the growth of Si nanowires. Reprinted from Hannon et al. (2006).

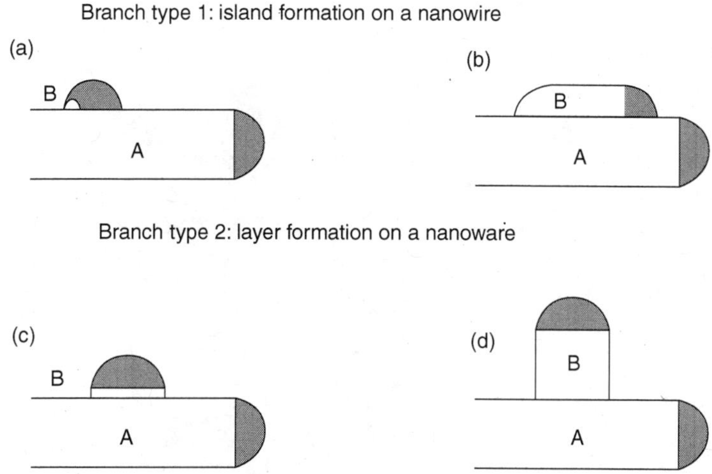

Figure 18.8. Schematic illustration of the formation of various types of branched heterostructures. Redrawn from Dick et al. (2007).

lyst at 450°C below the eutectic point of an Al–Si alloy (~577°C) (Wang et al. 2006). The replacement of Au with other metals is based on the fact that gold traps electrons and holes in Si and possess a serious contamination problem for Si complementary metal oxide semiconductor (CMOS) processing (Wang et al. 2006). The use of Cu as an optional catalyst material to grow Si NWs showed that the structure of Si NWs can be switched from diamond to the wurtzite phase along the growth direction (Arbiol et al. 2007). In comparison to the Au–Si system, the Cu–Si phase diagram is rather complex resulting in the formation of Cu_5Si followed by Cu_3Si in η, η', and η'' forms, whereby the three Cu_3Si solid phases have distinct crystalline structures that strongly influence the final structure and morphology of the Si NWs.

The adsorption–nucleation–growth process observed in the formation of NWs is dependent on the partial pressure of the source material, catalyst–NW material alloying, diffusion in a liquid or solid alloy particle, and crystallization of the solid NWs. Any fluctuation in the above processing parameters can result in the morphological variation of final NWs. For instance, the growth rate of Si NW was reduced by an order of magnitude upon the addition of a low concentration of trimethyl antimony (Me_3Sb) during the growth process (Nimmatoori et al. 2009). The Sb atoms can segregate on the catalyst surface and in the interface, which can reduce the alloying and thus the growth rate of NWs (Fig. 18.9).

The physical as well as functional properties of undoped NWs, such as electrical or optical properties, can be modified to match the practical applications by the incorporation of adequate dopants. However, the synthesis of doped NWs is challenging

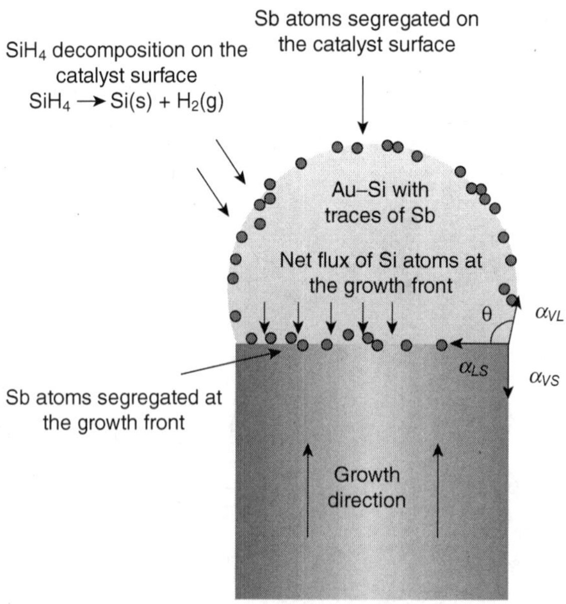

Figure 18.9. Suppression of VLS growth of Si nanowires by Sb addition. Redrawn from Nimmatoori et al. (2009).

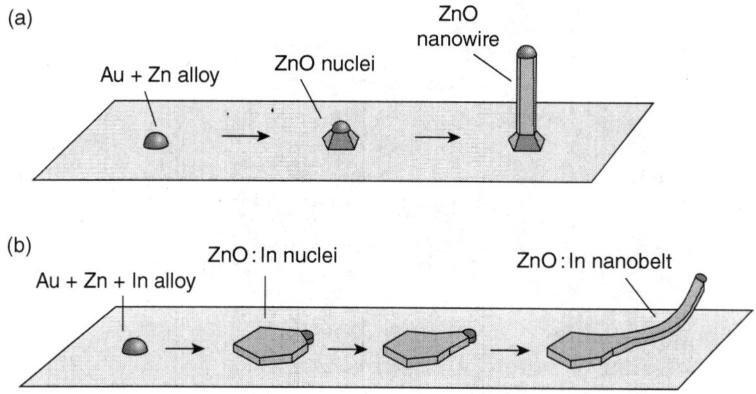

Figure 18.10. Proposed growth processes for nanowire (a) and nanobelt (b). Reprinted from Fan et al. (2006b).

because the synthetic pathway and growth mechanistic aspects can be totally different. Fan et al. have studied the change of morphology and crystallinity of Au-catalyzed ZnO nanostructures by introducing indium during the growth (Fan et al. 2006b), which showed that ZnO nanostructures could be switched from *c*-axial NWs to *a*-axial nanobelts by indium doping (Fig. 18.10).

Semiconducting 1-D Materials

Semiconductors are widely used in functional devices, which include a large variety of material classes. This section presents reliable strategies for the synthesis of 1-D elemental, III/V, and oxide semiconductors and describes some examples of heterostructures, which are important for the alteration of physical and functional properties of 1-D nanomaterials.

Silicon. Silicon NWs are commonly synthesized by catalyzed growth techniques (Wagner and Ellis 1964). The most popular method, based on the number of literature reports, is the CVD using gaseous sources, such as silane (SiH_4) (Bootsma and Gassen 1971; Wu et al. 2004a), disilane (R_2SiH_2) (Hannon et al. 2006), and silicon tetrachloride ($SiCl_4$) (Wagner and Ellis 1964; Givargizov and Sheftal 1971; Zhang et al. 2001). Enhanced effectiveness of the precursor decomposition could be achieved by plasma-enhanced CVD techniques, which allow a precracking and weakening of chemical bonds (Hofmann et al. 2003). In addition, evaporation techniques of pure silicon sources by PLD (Morales and Lieber 1998; Zhang et al. 1998), by MBE (Liu et al. 1999), or by thermal evaporation (Feng et al. 2000) have also been widely practiced. Thermal evaporation of SiO followed by its disproportionation in the presence of a catalyst was reported to form Si/SiO_2 core–shell NWs (Zhang et al. 1999a, b; Kolb et al. 2004). The successfully tested metallic catalysts, which promote VLS growth of

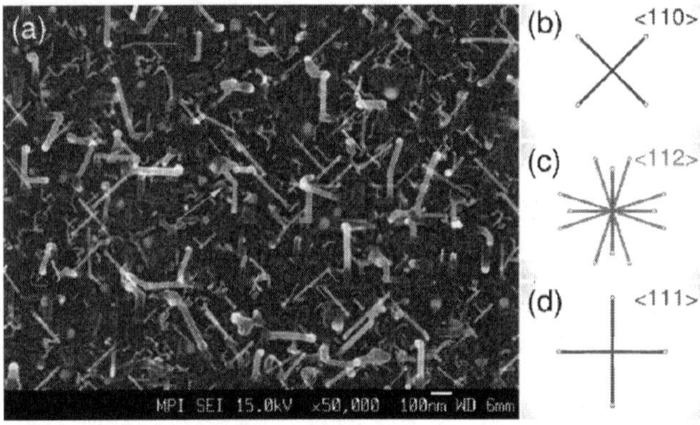

Figure 18.11. (a) Scanning electron micrograph (SEM) image of diameter-dependent growth direction of Si NWs on Si(100) with schematic illustration of crystallographic orientations of wires (b–d). Reprinted with permission from Schmidt et al. (2005b).

silicon NWs, include Ni, Cu, Pd, Ag, Pt, Al, Ga, In, Zn, and Fe (Wagner and Ellis 1964; Givargizov and Sheftal 1971; Morales and Lieber 1998, Yu et al. 2000, Wang et al. 2006). Further, a solid metal-induced VSS growth is reported for Ti and Dy (Schmidt et al. 2005a). The common growth direction for VLS-grown Si NWs with diameters > 50 nm is <111>; however a change of the preferential growth direction with diminishing radial dimensions is described, which results in NWs oriented in <112> and <110> directions as shown in Figure 18.11 (Schmidt et al. 2005b).

Solution-based techniques were also successfully applied for synthesizing silicon NWs via SFLS and SLS mechanisms. Au growth seeds in supercritical organic solvents are able to induce the formation of single-crystalline NWs in the range of 5–10 nm (Holmes et al. 2000). Moreover, solid-phase seeding of silicon wires is also described for several materials including Co, Ni, Mn, CuS, and MnPt in supercritical toluene (Tuan et al. 2006). Recently, Si NWs were seeded by Au and Bi nanoparticles in a high boiling solvent such as octacosane ($C_{28}H_{58}$) or squalane ($C_{30}H_{62}$) via decomposition of trisilane to follow an SLS growth mode (Heitsch et al. 2008). Electrolytic deposition of silicon using $SiCl_4$ in ionic liquids allows filling of membrane templates to form amorphous silicon NWs under mild conditions (Al-Salman et al. 2008). The synthesis of high aspect silicon nanostructures through chemical etching, allowing fabrication of aligned arrays in plane to the substrate as well as in different angles out of plane, remains a popular technique majorly due to its compatibility (Peng et al. 2002b; Huang et al. 2008).

Germanium. In general, most of the techniques described for the growth of silicon NWs can also be applied to obtain 1-D Ge nanostructures by choosing appropriate Ge sources. The common growth procedures include catalyst-assisted chemical vapor-phase deposition by use of Au and sources such as germane GeH_4 (Bootsma and Gassen 1971; Miyamoto and Hirata 1975) and chemical vapor transport of solids (Wu and Yang 2000), and laser or thermally assisted evaporation of metallic germanium

(Morales and Lieber 1998; Gu et al. 2001). Interestingly, indium-catalyzed Ge NWs showed preferential <110> growth direction on insulators, while generally <111> oriented wires are reported by Au-promoted growth (Yu et al. 2006). Germanium aryls also serve as efficient precursors for the formation of NWs, which grow via a base growth model (Mathur et al. 2004) or via Ni-catalyzed growth (Mathur and Barth 2008). Core–shell Ge/SiC_xN_y NWs formed at high temperatures from molecular sources showed a self-catalytic growth process (Mathur et al. 2007). Solution-based techniques include the filling of membranes by supercritical inclusion (Ziegler et al. 2004) and electrolysis (Al-Salman et al. 2008), reduction of germanium chloride by sodium at elevated temperatures and pressure (Heath and LeGoues 1993), and SFLS techniques (Hanrath and Korgel 2002). Besides the usually used alkyl hydrides, also germanium aryloxides (Gerung et al. 2006) showed potential to act as precursors to 1-D germanium nanostructures via SFLS, the major feature of the synthesis being the clearage of Ge–O bonds to form elemental germanium.

Gallium Arsenide. Gallium arsenide (GaAs) is one of the most promising III/V semiconductors for high-speed digital, high-frequency microwave, and electro-optical device applications (Adachi 1985). However, gallium arsenide can either crystallize in a zincblende (cubic) or wurtzite (hexagonal) structure (polymorphs). Thus, a common problem of stacking faults in the growth of these nanostructures occurs, which should be controlled to ensure the functional performance of the as-synthesized materials. The most popular growth technique for GaAs is the metal organic vapor-phase epitaxy (MOVPE), whereby trialkyl gallium and arsine or alkyl arsines are used as precursors (Persson et al. 2004). Catalyst-free growth of NWs under similar conditions was achieved by the patterning of a GaAs substrate with SiO_2, which enabled the growth of NWs on the uncovered areas (Noborisaka et al. 2005). Physical vapor growth on gold seeds was described by laser ablation of GaAs substrates (Duan and Lieber 2000a) and by evaporation of the elements in MBE (Dubrovskii et al. 2005). MBE allows the growth of NWs on silicon oxide-coated GaAs due to self-nucleation of NWs on Ga droplets. Even oriented growth could be realized using very thin oxide coatings, whereby the reaction of Ga with the oxide led to the formation of microcracks, which allowed contact with the III/V substrate (Colombo et al. 2008).

GaAs NW formation by SFLS synthesis also requires two separate sources such as $As(SiMe_3)_3$, $Ga(tBu)_3$, and Au colloidal catalysts (Davidson et al. 2004), whereas NWs were also obtained using similar precursors with In catalyst particles in the SLS method (Yu and Buhro 2003). Chemical, anisotropic etching of large GaAs substrates with H_3PO_4, H_2O_2, and H_2O in combination with lithography allows predefinition of horizontal arrays, which can be transferred onto plastic substrates (Sun and Rogers 2004).

Zinc Oxide. Among 1-D oxides, zinc oxide is the most studied material due to its optical and electrical properties. Given the large number of potential applications demonstrated for anisotropic ZnO nanostructures, such as optical switches, sensors, piezo response, photovoltaics, varistors, and transparent conducting oxides (Minne et al. 1995; Kong et al. 2003). In contrast to the elemental and III/V semiconductors,

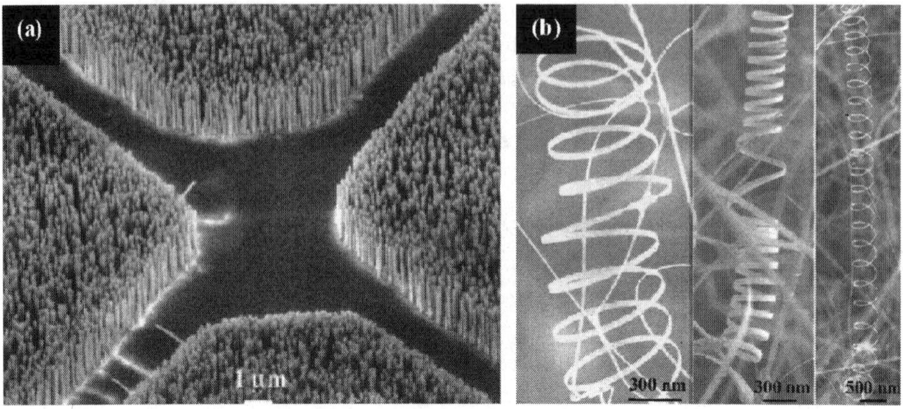

Figure 18.12. (a) Aligned ZnO nanowires obtained by carbothermal reaction on sapphire (reprinted with permission from Huang et al. 2001a, b) and (b) SEM images of ZnO nanobelts with helical structure. Reprinted with permission from Kong et al. (2003).

metal oxide nanostructures can be synthesized by a large variety of methods. Metal organic CVD of ZnO is described by the reaction of oxygen with dimethylzinc in the presence of gold to promote VLS growth (Park et al. 2002). Physical methods to obtain NWs such as Au-supported growth via PLD (Zhang et al. 2005a, b, c) and carbothermal reaction (Huang et al. 2001a, b) were also successfully demonstrated (Fig. 18.12a). Thermal evaporation in the absence of metal catalysts allowed the formation of very interesting 1-D structures such as nanobelts (Pan et al. 2001), rings (Kong et al. 2004), and nanosprings as shown in Figure 18.12b (Gao et al. 2005). The formation of nanorings was attributed to the polarity of ZnO lattice, whereas the nanosprings demonstrate switching from a belt to a superlattice structure. Carbothermal reduction of ZnO onto porous aluminum oxide resulted in the condensation of metallic zinc, which acts as seed for ZnO NW nucleation (Jie et al. 2004). Electrochemical deposition from zinc nitrate aqueous solution into an AAO membrane was described to produce single-crystalline zinc oxide nanorods (Zheng et al. 2002), whereas deposition from zinc sulfate and boric acid produced metallic zinc wires, which were oxidized in a second step to ZnO (Li et al. 2000). Large areas of small oriented ZnO nanorods on glass substrates were described by thermal decomposition of methenamine in the presence of low concentrations of zinc nitrate in H_2O (Vayssieres 2003), the key parameter being the variation of ionic strength.

Tin Oxide. Tin oxide is an important material for gas sensors, catalysis, and conducting oxides, which has led to several physical and chemical syntheses of SnO_2 NWs (Dai et al. 2002; Comini et al. 2009). Thermal evaporation of SnO or SnO_2 can be used in the formation of nanobelts in high quantity (Dai et al. 2001). In addition, VS synthesis by SnO evaporation under pulsed flow conditions led to the formation of segmented wires (Lilach et al. 2005). Further, NWs have been obtained by a self-catalyzed growth mechanism via carbothermal synthesis (Wang et al. 2005). The laser ablation of pure tin in oxidizing Ar/O_2 atmosphere (Liu et al. 2003a, b) and the oxidation of electrodeposited tin wires within structure-directing AAO membranes (Zheng et al. 2001) or through oxidation of tin vapors at elevated temperatures (Ma et al. 2004) allowed the formation of NWs. The catalyst-assisted molecule-based CVD growth is a generic approach toward controlled synthesis of metal oxide nanostructures and had been demonstrated for SnO_2 as well (Mathur et al. 2005). Large surfaces covered with tin oxide nanorods could be achieved by an aqueous approach controlling the ionic strength and pH of precursor solutions (Vayssieres and Graetzel 2004).

Heterostructures. A natural progression from single-component materials to complex structures and compositions is the next step for the development of smart materials. The combination of material classes such as metals, semiconductors, and polymers has led to unique properties and functionalities that are not accessible in the single-component materials. In this subsection, we will focus on complex materials with nanoscale heterostructures (Fig. 18.11). The strategies to grow and control these structures are similar to single-component materials and often include multiple synthetic steps.

Semiconductor heterostructures were described by several groups using a sequential VLS-based growth of different materials. The MBE technique allowed the synthesis of Si/Ge/Si segmented NWs by alternating source evaporation cycles (Zakharov et al. 2006). Si/SiGe superlattices were grown by the continuous deposition of Si via decomposition of silicon chloride and the sequential laser ablation of a Ge target, which resulted in the silicon–germanium intermixing within the catalyst particle (Wu et al. 2002). Segmented heterostructures of III–V semiconductor wires are generally synthesized via CVD techniques by switching the gas-phase supply of either the group III or group V precursor. An abrupt change in the segment composition of an InAs/InP wire is obtained via a sequence where growth is interrupted as the indium source (TMIn) is switched off, followed by a change of the group V sources (Bjork et al. 2002). The sharpness and crystalline structure are shown in Figure 18.13. Metal-to-semiconductor axially grown heterostructures (Ag/Si) can be realized by a combinatorial approach including the electrochemical deposition of Ag and growth of the Si segments through decomposition of $SiCl_4$ in H_2 atmosphere (Luo and Zhu 2006). Solid-state reaction is used for the formation of NiSi/Si axial heterostructures via a site-selective physical vapor deposition of Ni onto Si NWs through masking techniques and subsequent annealing. The atomically sharp interface of these structures opens up the possibility of integrating both active devices and high-performance interconnects from a single nanoscale building block (Wu et al. 2004b). Material manipulation at the nanometer scale was shown by electrochemical dip-pen nanolithography via a site-selective conversion of postgrown GaN NWs into Ga_2O_3 underneath an atomic force microscope (AFM) tip (Maynor et al. 2004).

Radial or core–shell heterostructures are generally formed in a two- or multistep procedure where either the growth parameters or the synthetic method are changed. An intrinsic silicon (i-Si) core and a p-type Si (p-Si) shell were fabricated by altering the axial and radial growth modes during CVD. After the VLS growth of the Si core, a shell was formed with diborane added as a p-type dopant by lowering the synthesis

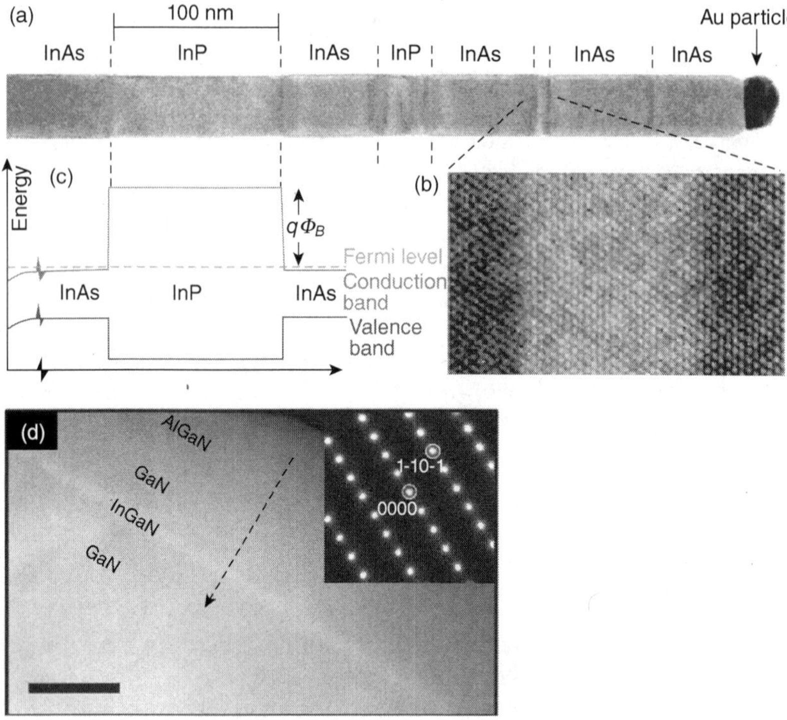

Figure 18.13. (a) TEM image of InP segments (100, 25, 8, and 1.5 nm) inside an InAs nanowire. (b) Magnification of the 8-nm segment region showing crystalline perfection and the interface abruptness on the level of monolayers. (c) Simulated band structure diagram of the InAs/InP heterostructures. Reprinted with permission from Bjork et al. (2002). (d) A dark field scanning TEM image of a cross-sectional GaN/$In_xGa_{1-x}N$/GaN/AlGaN core–multishell nanowire. Scale bar: 50 nm. Reprinted with permission from Qian et al. (2005).

temperature (Lauhon et al. 2002). The same procedure was used to fabricate Ge/Si core–shell NWs. The amorphous shell could be crystallized by thermal annealing in both cases, which resulted in epitaxial coaxial structures. Similar procedures were described for several III/V combinations, for example, a core–multishell GaN/InGaN/GaN/AlGaN structure is shown in Figure 18.13d (Qian et al. 2005). In addition, non-epitaxial growth of coaxial structures such as SnO_2/Fe_3O_4 (Mathur et al. 2008) and Si/CdS (Hayden et al. 2005) has been recently reported. Vertical FETs could also be seen as core–shell structures with wrapped dielectric materials; this kind of features will be explained in detail in the section Development of Nanodevice Architectures.

A further degree of complicity of heterostructures is hierarchical or branched structures, described on the basis of their morphologies as nanotrees, nanobrushes, nanoflowers, and so on. Controlled branching events are demonstrated for GaP/GaAsP hierarchical structures, which were formed in a two-stage VLS mode MOVPE growth process (Dick et al. 2004). The GaP trunks were seeded by Au particles and were grown on oriented gallium phosphide substrates. In a second step, additional growth seeds were deposited followed by the growth of the branches. ZnO branches on several backbone materials, such as SiC, GaP, GaP, and CNTs, were described by In- and Ga-induced seeding of ZnO, which was formed by oxidation of the metallic Zn vapor source (Bae et al. 2004). For metal oxides, merely a few such heterostructures such as In_2O_3/Ga_2O_3 (Xu et al. 2007) and SnO_2/V_2O_5 (Mathur and Barth 2007) were reported. Both of the described examples were achieved via vapor-phase routes; while the first one was grown in a one-step synthesis, the second system required a two-step process.

STRUCTURE–PROPERTY CHARACTERIZATION AND RELATIONSHIP

Investigation of 1-D Structures

Since the composition and morphology of simple and complex 1-D nanostructures are important to determine the chemical and physical properties, the following section will discuss the recent development of the state-of-the-art analytical methods applied on the structural studies of 1-D nanostructures.

Axially Constructed Heterostructures. The A/B/A/B superlattice 1-D structures are synthesized following the general concept, in which the same catalyst is used to grow an NW of a A/B/A/B superlattice structure by modulating the reactants during the growth.

Figure 18.14a shows the TEM image and composition profile of InAs/InP 1-D heterostructures (Bjork et al. 2002). The contrast between different compositions is small due to the small difference in the atomic weight of the constituting elements. The contrast between different segments can be enhanced by Fourier transform and inverse Fourier transform techniques, in which the original TEM image is modified by superposing an additional image obtained from masks around the split 200 reflection in reciprocal space, followed by an inverse Fourier transform (Fig. 18.14b, c). The TEM picture can be further modified to visualize the InAs and InP lattice spacings with different color codes. For instance, the lattice spacing in a 40-nm diameter nanowhisker of InAs containing three segments of InP (ca. 1.5, 8.0, and 25.0 nm in thickness) was clearly illustrated by pseudocoloration of the TEM image (Fig. 18.14d).

The InAs/InP heterostructure displays an atomically smooth interface between InAs and InP zones, which can be shown in high-resolution TEM (Fig. 18.11b). The heterostructure leads to modulation of individual band gaps, which influences the overall transport behavior of conducting electrons (Fig. 18.11c).

A similar concept has been applied for the fabrication of Si/Ge 1-D heterostructures (Wu et al. 2002). Due to the relatively large difference of atomic weight between Si and Ge, the mass contrast can be clearly observed in TEM image, whereas the periodic variation of Si and Ge content is supported by the line profile of energy-dispersive X-ray (EDX) analysis (Fig. 18.15).

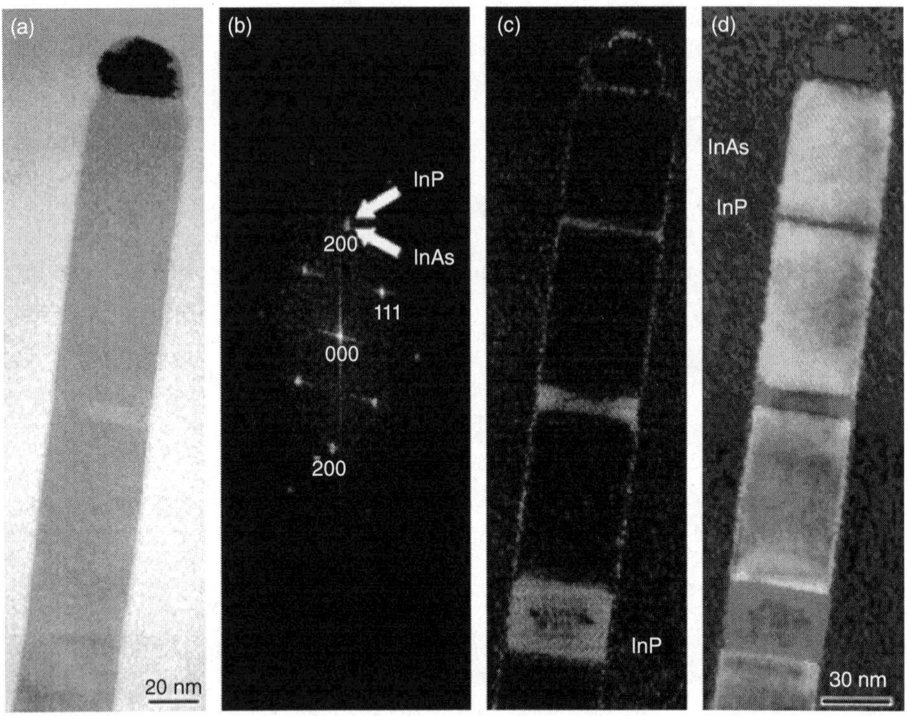

Figure 18.14. TEM analysis of InAs/InP heterostructures. Reprinted from Bjork et al. (2002).

Radically Controlled Heterostructures. The previous reports have shown (Lauhon et al. 2002) that the understanding of the growth mechanism of homogeneous NW structures allows the fabrication of radically modulated heterostructures. In the case of axial heterostructures, one or more heterojunctions are created within the NW, whereas core–shell structures generally exhibit a single interface. The EDX elemental mapping and line scans are powerful tools to identify the composition variation in nanometer scale as shown for Ge/Si heterostructures (Fig. 18.16).

The single-step synthesis of 1-D Ge/SiC_xN_y core–shell nanocables was achieved by CVD of the molecular precursor $[Ge\{N(SiMe_3)_2\}_2]$ (Mathur et al. 2007). The EDX analysis showed that the shell contained Si, C, and N elements, whereas the X-ray photoelectron spectroscopy (XPS) analysis suggested the phase to be possibly SiC_xN_y based on the analysis of the chemical shifts of Si2p, N1s, and C1s XPS spectra (Fig. 18.17).

Three-dimensional (3-D) single-crystalline branched NW heterostructures, where the backbones and branches are assembled with ZnS and CdS phases, respectively (Jung et al. 2005), were obtained by sequential seeding of gold nanocluster catalysts during the growth process. The lattice-resolved HRTEM images of the junction regions of the ZnS (Fig. 18.18D) and CdS (Fig. 18.18E) branched NWs clearly suggest that these branched NWs have a single-crystalline structure without structural defects

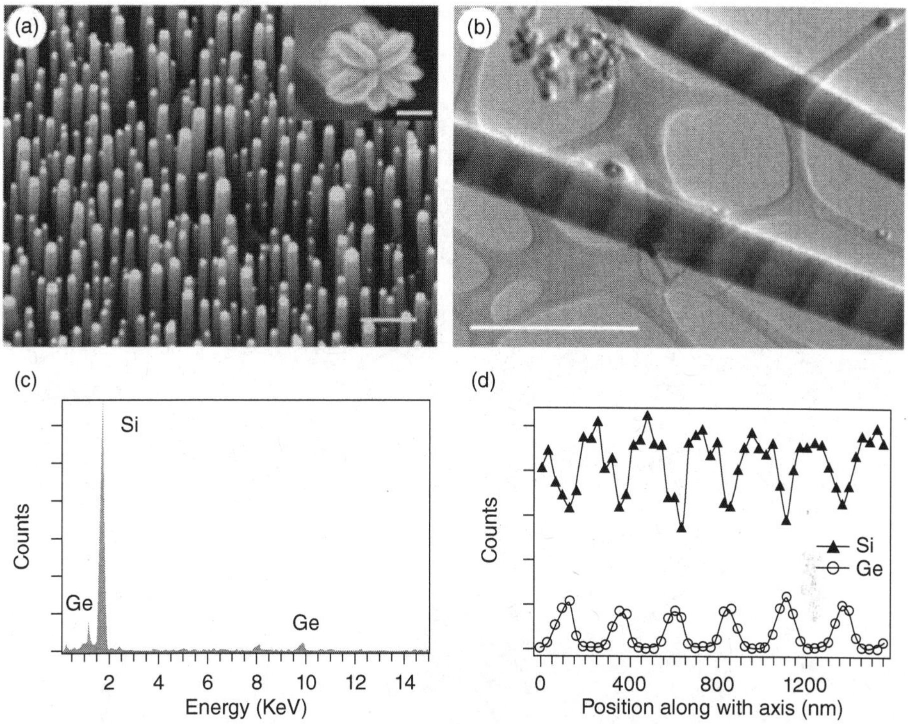

Figure 18.15. (a, b) Electron microscope images and (c, d) elemental profiles in Si/SiGe superlattice nanowires. Reprinted from Wu et al. (2002).

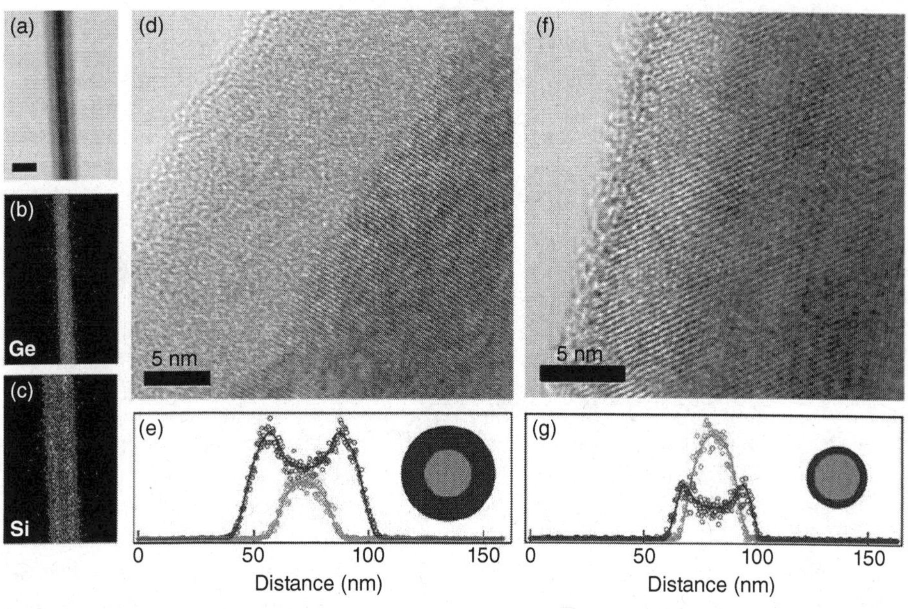

Figure 18.16. Structural analysis of Ge/Si core–shell structures. Reprinted from Lauhon et al. (2002).

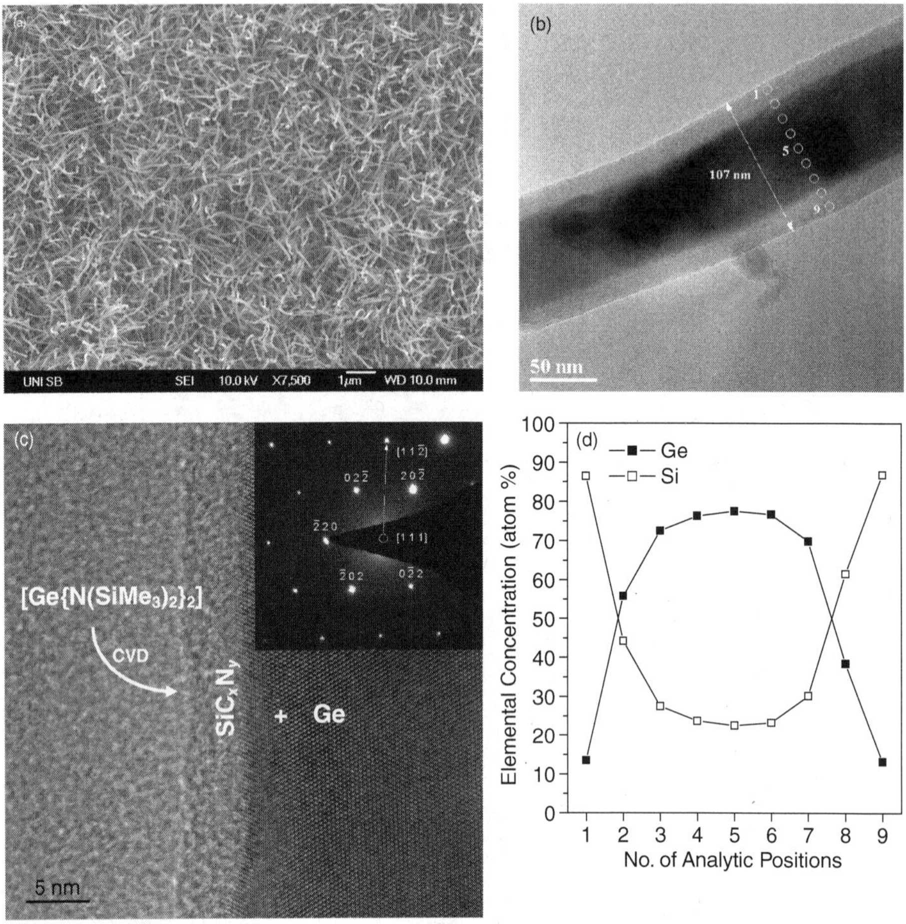

Figure 18.17. Structural analysis of Ge/SiC_xN_y core–shell structures. Reprinted from Mathur et al. (2007).

throughout the branch and the backbone, and the junction possesses a clean interface between the backbone and the branch. Although ZnS and CdS structures exhibit a large lattice mismatch, the heterostructure may undergo structural relaxation during its growth to overcome the lattice mismatch by forming structures of intermediate chemical compositions such as $Cd_{1-X}Zn_XS$. Transmission electron microscopy studies have proven that the growth of heterostructure branches occurred epitaxially from the backbone while maintaining single-crystalline structure.

Xiong et al. have observed coherent twinning phenomena in GaP and InP NWs grown by the VLS mechanism (Xiong et al. 2006). The coherence takes place over distances along the wire axis of ~0.5–1.0 μm. They have observed that the twin correlates with [1$\bar{1}$1] faceting on the surface of the NWs, which is possibly governed by

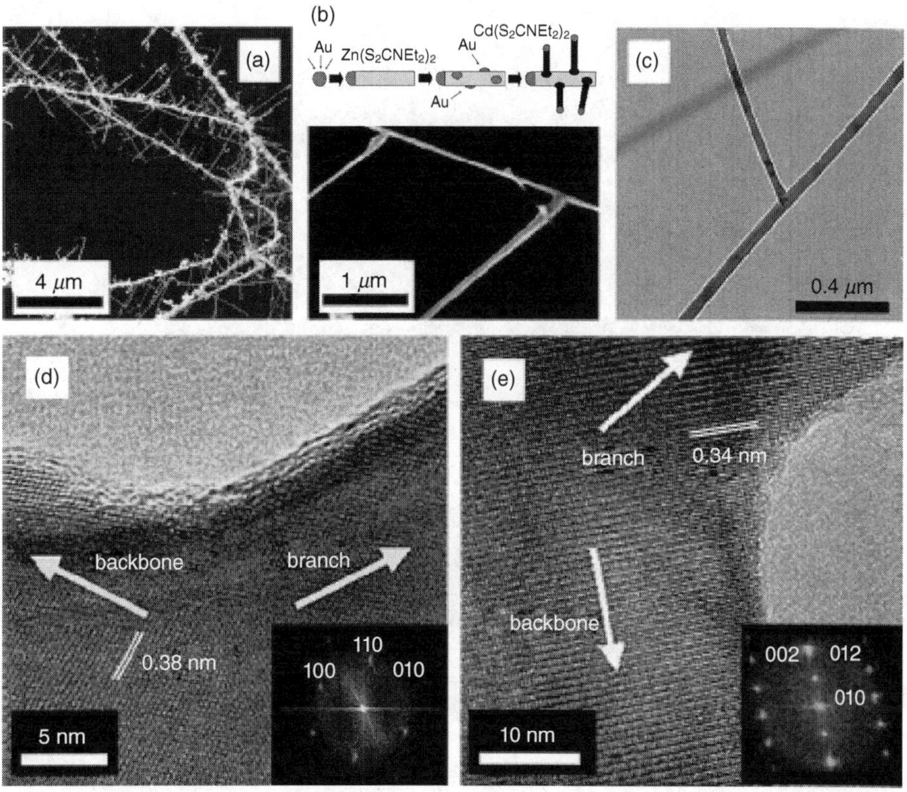

Figure 18.18. Electron microscopic images of branched CdS/ZnS nanowires. Reprinted from Jung et al. (2005).

the surface energy. The quasi-periodic structural modulation along the wire axis may be attributed to a regenerative oscillation of the semiconductor concentration in the alloy droplet during the growth process (Fig. 18.19).

Combined scanning transmission electron microscopy and electron energy loss spectroscopy (STEM-EELS) can provide the data correlating the microscopic details such as crystallinity, surface chemistry, and optical and electrical properties. The EELS spectrum contains information from low-energy (less than 25 eV) interband transitions and plasma excitations to higher-energy (tens to hundreds of electronvolts) ionization of core electrons to the conduction band. The technique is highly useful in providing both structural information and nanoscaled spatial resolution at the individual nanostructure.

Korgel et al. have used STEM-EELS to examine single-crystalline Ge NWs and the size dependence of the Ge 3d core electron ionization edges to determine changes in the conduction band edge and band structure (Hanrath and Korgel 2004). The Ge 3d spectra with spatial resolution can deliver the information about the change of electronic structures due to the surface modification. Figure 18.20 shows the Ge 3d ionization

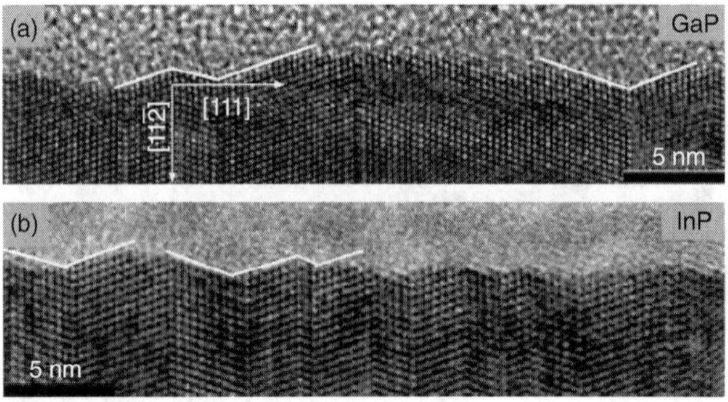

Figure 18.19. The twinning correlated surfaces of (a) GaP and (b) InP nanowires. Reprinted from Xiong et al. (2006).

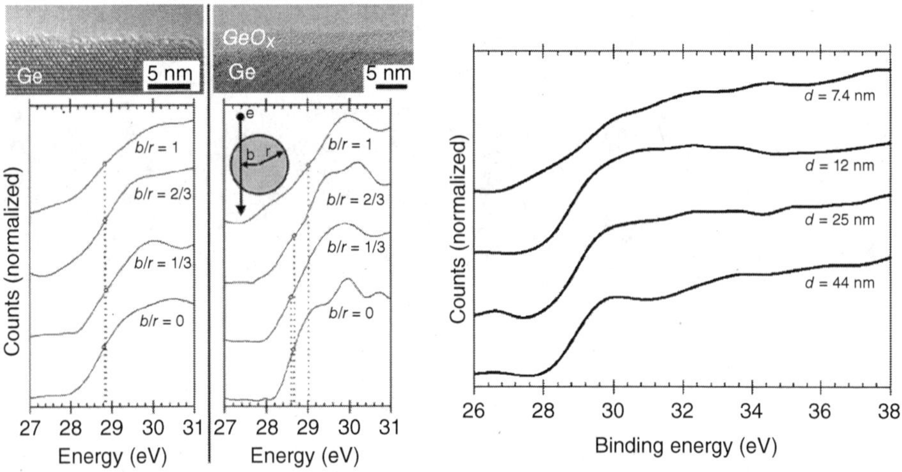

Figure 18.20. Surface and size-dependent EELS spectra of Ge nanowires. Reprinted from Hanrath and Korgel (2004).

edge, which is very sensitive to surface chemistry; the oxidized NWs with poor electrical passivation exhibit up to a 0.3-eV shift in edge inflection point when the probe is positioned near the surface. When the diameter of Ge NW is smaller than 25 nm (exciton Bohr radius of Ge), the Ge 3d ionization edge also shifts to higher energy with significant changes in the peak fine structure.

Size- and Shape-Dependent Physical Properties

Synthetic methods for growing NWs have made significant progress in recent years, enabling the production of 1-D nanostructures with controlled size, single-crystalline quality (Yang 2005; Fan et al. 2006c), and complex structures such as coaxial and linear heterostructured NWs (Xiang et al. 2006; Liang et al. 2007; Tian et al. 2007; Moore et al. 2008) (Fig. 18.21). The synthesis of 1-D materials paves the way to develop multifunctional devices, which make use of their high anisotropic geometry and size-dependent physical properties. It was shown that NWs exhibit unique size-dependent electrical transport properties (Yang 2005), high optical efficiency (Sham et al. 2004),

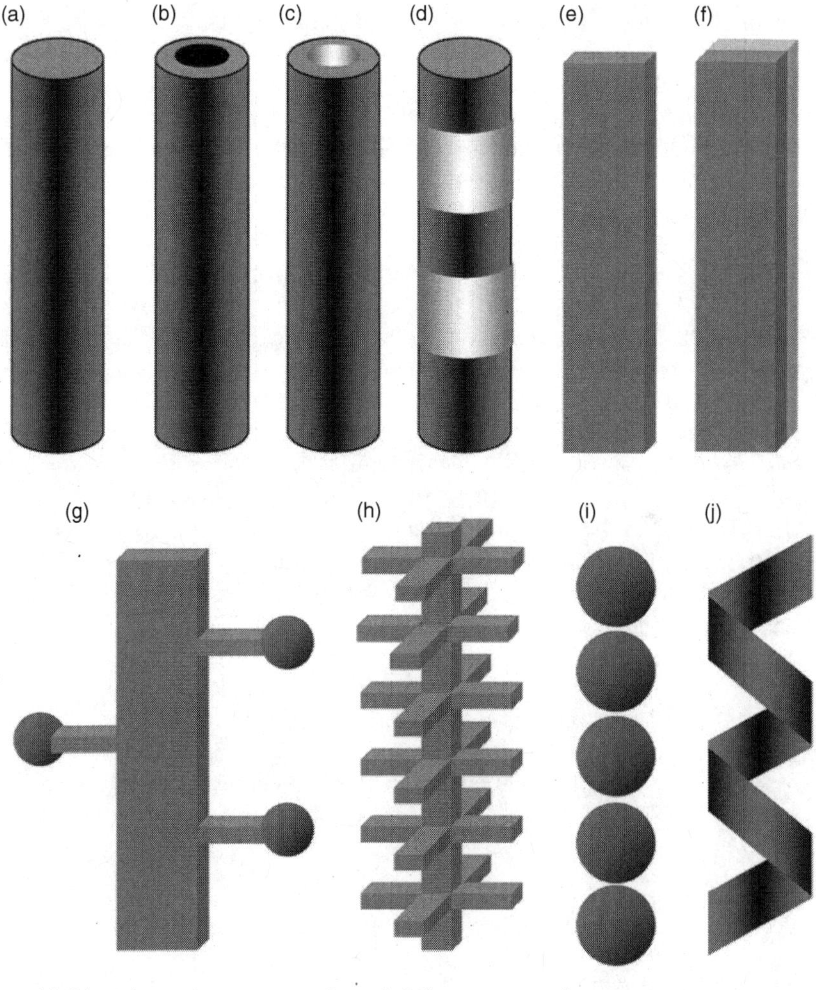

Figure 18.21. Schematic representation of different types of quasi-one-dimensional nanostructures. (A) Nanowires and nanorods; (B) core–shell structures with metallic core, semiconductor, or metal oxide shell; (C) nanotubes/nanopipes and hollow nanorods; (D) segmented heterostructures; (E) nanobelts/nanoribbons; (F) nanotapes; (G) dendrites; (H) hierarchical nanostructures; (I) nanosphere assembly; (J) nanosprings. Reprinted from Kuchibhatla et al. (2007).

enhanced thermoelectric efficiency compared to conventional Peltier elements (Lin et al. 2000), and better mechanical (Rubio-Bollinger et al. 2001; Cuenot et al. 2004) and electromechanical behaviors (Zhao et al. 2004a, b). In this section, the most significant physical properties of NWs are briefly reviewed.

Electrical Transport. The possibility to obtain single-crystalline NWs free from structural defects has been successfully demonstrated for a number of compositions (Xia et al. 2003; Yang 2005; Fan et al. 2006c). The studies on intrinsic electrical properties of the NWs allow to develop a physical model to study the carrier transport behaviors in quasi-1-D systems (Yasuda et al. 2001; Cui et al. 2001a, b; Law et al. 2002; Law et al. 2004a, b; Kuchibhatla et al. 2007). Although the preliminary results showed many attractive application prospects (Wang 2003), the fabrication of cost-effective and multifunctional nanodevices still remains a challenging avenue. The charge carrier mobility μ is an important parameter for device fabrication; however, the mobility is strongly influenced by the main source of carrier scattering, namely, the surface states of NWs (Fan et al. 2004). The mobility is decreased with the decreasing diameter of NWs (Chang et al. 2004, 2006, 2007) (Table 18.2). To circumvent this problem, surface passivation of NWs with polymers (Park et al. 2004; He et al. 2007; Hong et al. 2008; Prades et al. 2008c) or isolating materials such as Si_3N_4 (Chang et al. 2006) has shown to be the viable solution to restrict the detrimental influence of surface states and thus to restore high mobility values (Fig. 18.22) essential for any possible device applications.

Note that band-gap of semiconductor NWs is influenced by lateral dimensions (Mathur et al. 2005). This size effect becomes more significant for ultrathin NWs ($r < 20$ nm) since the electrons are confined and separated in different band edges (Zheng et al. 2006), which offers promising possibilities to design new devices for innovative applications. However, the application of the effect in device fabrication is technically challenging due to the difficult manipulation of NWs on a nanometer scale (Hernandez-Ramirez et al. 2006). Recent reports have shown that the charge carriers are confined in both coaxial and longitudinal heterostructured NWs (Xiang et al. 2006; Liang et al. 2007; Tian et al. 2007; Moore et al. 2008). Additionally, NWs may find possibilities in many other applications; for example, electrons can be easily extracted out from NWs tips by an external electric field (field emitters). It was demonstrated that semiconductor NWs (e.g., GaN, AlN, CdS, and ZnO) grow vertically on conductive

TABLE 18.2. Comparison of the Geometrical and Electrical Transport Properties of ZnO Nanowires

Length (μm)	Diameter (nm)	Surface/volume	Mobility (cm^2/Vs)	Conductivity (S/m)
1	18	2.22×10^9	5.41	119.25
3	190	2.11×10^7	30.22	8.81

The thinner the nanowire, the lower the mobility as a result of the strong influence of the surface. Data were taken from Chang et al. (2007).

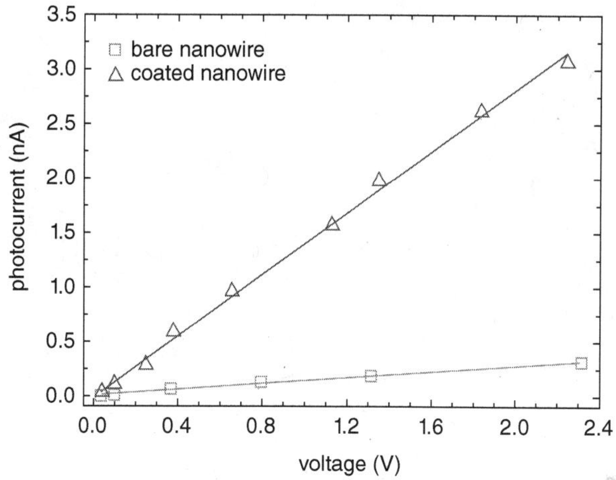

Figure 18.22. Mobility improvement in an individual ZnO nanowire after passivation of its surface with poly(methyl methacrylate) (PMMA). This is demonstrated by the systematic increase in the photocurrent under UV illumination which is proportional to the carrier mobility. Data taken from Prades et al. (2008c).

substrates, which show strong low-threshold and stable field-emission currents (Tseng et al. 2003; Zhu et al. 2003a, b; Tang et al. 2004; Li et al. 2005a, b; Liu et al. 2005a, b; Shi et al. 2005). This effect could be used in highly efficient illumination displays by fabricating arrays of NW cathodes (Gangloff et al. 2004).

Thermal Transport. Phonon transport in 1-D nanostructures is significantly impeded since boundary scattering and reduced phonon group velocity are stemmed by confinement effects (Hochbaum et al. 2008). Theoretical models of heat conduction in semiconducting NWs predict a large reduction of thermal conductivity of NWs with only a few tens of nanometer diameter (up to 90% compared to bulk) (Chen et al. 2008). On one hand, this size-dependent effect is an issue that needs to be addressed in the miniaturization process in the semiconductor industry; on the other hand, this tunable physical property is useful for thermoelectric applications. Thermoelectric materials are characterized by a figure of merit defined by

$$\mathrm{ZT} = \alpha^2 T/[\rho(\kappa_p + \kappa_e)], \tag{18.1}$$

where α is the Seebeck coefficient, T is the absolute temperature, ρ is the electronic resistivity, κ_p is the lattice thermal conductivity, and κ_e is the electronic thermal conductivity. It can be asserted that thermoelectric effects in NWs will be improved if phonon transport is hindered. It can be achieved by tailoring the diameters, compositions, and carrier concentrations (Hicks and Dresselhaus, 1993; Lin and Dresselhaus 2003). Yang et al. (Li et al. 2003b, c) have shown the possibility to reach Seebeck coefficients and electrical resistivity values in Si NWs (ca. 50 nm) similar to that of doped bulk Si. The Si NWs exhibited a 100-fold reduction of thermal conductivity, yielding a value of ZT = 0.6 at room temperature. Although bulk Si is a poor thermoelectric material (ZT ~ 0.01 compared to Bi, Te, Pb, Sb, and Ag with ZT > 1), the arrays of Si NW are promising architectures for thermoelectric applications. It was also reported that quantum confinement effect reduces the heat conductance in nanomaterials, which could be used for new cooling concepts based on the NW system (Chen et al. 2003a, b).

It is noted that the superlattice NWs may provide better performances by the combination of sharp periodic band offsets, which can guarantee high quantum confinement, low phonon scattering at interfaces, and good electrical conductivity (Li et al. 2003b, b). This configuration could be achieved by stacking different compounds in linear heterostructures or by inducing planar defects/phase changes perpendicular to the NW axis (Morral et al. 2007). In this direction, superlattice wires were compared to conventional Si NWs (Li et al. 2003b, c), showing that thermal conductivity of the former is up to five times lower than that of Si NWs at 300 K. Nevertheless, future studies are needed to guide an accurate engineering of these complex structures.

Optical Properties. The ability to manipulate light pulses within submicron volumes is crucial to realize the integrated light-based devices such as optical computers. A versatile approach is to assemble photonic circuits from a collection of NWs that assume different functions (i.e., light creation, guiding, and detection). NWs exhibit several features including inherent one dimensionality, low surface roughness, and the power to operate above and below the diffraction limit, which make them suitable photonic building blocks (Yang 2005). The toolbox of NW elements includes transistors, light-emitting diodes, lasers, and photodetectors (Kind et al. 2002). Nevertheless, NW-based photonics still require a development of waveguides, which are able to integrate these elements in a cost-effective and flexible process. This would pave a way toward complex devices such as logic gates (Tian et al. 2007).

Surface effects also cause detrimental effects on NW optical performances (Prades et al. 2008c). For instance, they act as recombination centers for minority carriers, degrading the final response of optoelectronic prototypes (Lauhon et al. 2004). Again, this problem is circumvented if NWs are passivated with a dielectric shell, which forms a high-quality optical cavity around the NW, namely, as the gain medium (Lauhon et al. 2002; An et al. 2004; Han et al. 2004; Qian et al. 2004). Note that thin NWs (radius close to the exciton Bohr radius) exhibit quantum confinement effects with strong consequences for their optical properties. This point was demonstrated for NWs of III–V semiconductors (Bhunia et al. 2003), ZnO (Park et al. 2003; Wang et al. 2004a, b), as well as for Si (Ma et al. 2003; Zhao et al. 2004a, b). Here, it should be pointed out that silicon is currently the most important material in microelectronic industry, but it has poor optical emission because of its indirect band-gap transitions (O'Mara et al. 1990). For this reason, Si/Ge heterostructured NWs are considered as a promising alternative to overcome this fundamental bottleneck, since it is expected that these new nanomaterials will exhibit much higher luminescence efficiency than bulk Si because their band structure may be modified so that radiative recombination processes are enhanced (Giovine et al. 2001). Therefore, new optoelectronic devices (group of IV semiconductors) could be developed, allowing the combination of both electrical and optical functionalities on the same integrated circuit.

The potential of semiconductor NWs to develop light detectors was demonstrated as well (Kind et al. 2002; Soci et al. 2007). Nevertheless, a critical analysis of photoconduction mechanisms in NWs indicates that high performance photodetectors cannot be achieved with NWs. Their high aspect ratio dramatically reduces the volume where photons are absorbed, leading to weak electrical responses (Prades et al. 2008a, b, c).

Yang et al. demonstrated that NWs with flat-end facets can be used as optical resonance cavities to generate coherent light at the nanoscale (Johnson et al. 2001). Well-faceted NWs support axial Fabry–Perot waveguide modes separated by

$$\Delta\lambda = \lambda^2/[2Ln(\lambda)], \tag{18.2}$$

where L is the cavity length and $n(\lambda)$ is the group index of refraction. This behavior is originated by the large diffraction losses suffered by light in transverse paths. According to experiments (Johnson et al. 2002), diffraction avoids smaller wires from lasing since photoluminescence is lost rather than the surrounding radiation field. Far-field imaging indicates that photoluminescence and lasing emission are localized at the ends of NWs, suggesting strong waveguiding behavior that is fully consistent with axial Fabry–Perot modes. Room temperature UV lasing has been observed in ZnO NW systems (Huang et al. 2001a, b; Johnson et al. 2002), and similar effects were also observed in other wide band-gap NWs, such as GaN (Johnson et al. 2002), CdS (Duan et al. 2003; Agarwal et al. 2005), and ZnS (Ding et al. 2004a, b) (Fig. 18.23). Finally, it should be noted that the electroluminescence of n-ZnO NW arrays vertically grown on p-GaN substrate was reported by Park et al. (Park and Yi 2004) with a maximum emission at yellow (560 nm), which is normally assigned to defects in the structure.

Mechanical Properties. NWs are usually long and strong enough to be pushed, bent, and shaped with commercial micro/nanomanipulators (Law et al. 2004b). Free-standing NWs can be repeatedly curved into loops of radii as small as 5 μm without degradation or without being broken. This behavior of NWs is remarkable for ceramic that are brittle in bulk form. The extraordinary length, flexibility, and strength of these nanostructures enable their controlled manipulation and the accurate positioning and optical linking to other NW-based elements in order to form optical networks and complex device components.

Furthermore, all these outstanding mechanical properties have important consequences in their electrical transport properties. For instance, Yang et al. reported that Si NWs exhibit giant piezoresistance effects, which could have significant applications in NW-based flexible electronics, as well as in nanoelectromechanical systems (NEMSs) (He et al. 2006). Wang et al. recently demonstrated that it is possible to convert mechanical energy into an electrical current using piezoelectric ZnO NW arrays. This mechanism of energy harvesting has been called piezotronics (Wang 2008a). In this experimental approach, aligned NWs are deflected with a conductive AFM tip operated

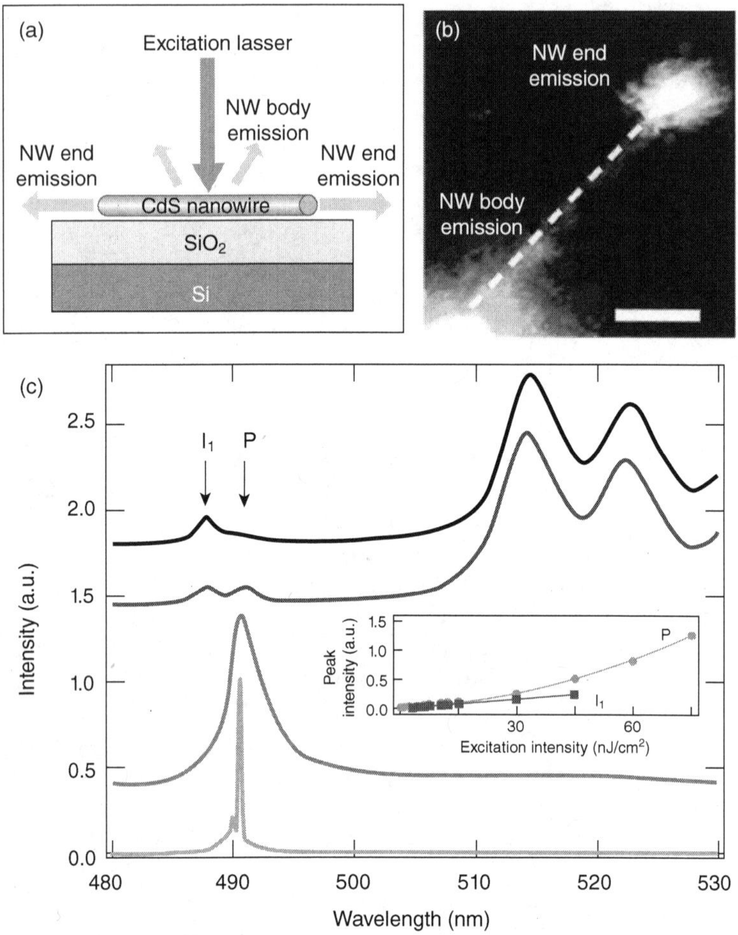

Figure 18.23. (A) Sketch of optical experiments performed with a single nanowire. (B) Photoluminescence image shows luminescence from both the excitation area and the end of a CdS nanowire (scale bar: 5 mm). (C) Photoluminescence spectra of CdS nanowire end emission recorded at 4.2 K with excitation powers of 0.6, 1.5, 30, and 240 nJ/cm^2 (from top to bottom, respectively). Inset shows peak intensity of I_1 (squares) and P (circles) bands as a function of incident laser power. Reprinted from Agarwal et al. (2005).

in contact mode (Fig. 18.24) (Song et al. 2007). The coupling between piezoelectric and semiconducting properties in zinc oxide creates a strain field-and-charge separation across the NW as a result of the induced bending. Rectifying characteristics of the Schottky barriers formed between the metal tip and NWs lead to a measurable electrical current. At present, the efficiency of these piezoelectric power generators is estimated to range from 17% to 30%, exhibiting high potential to convert mechanical, vibrational, and/or hydraulic energy into electricity for powering autonomous nanodevices.

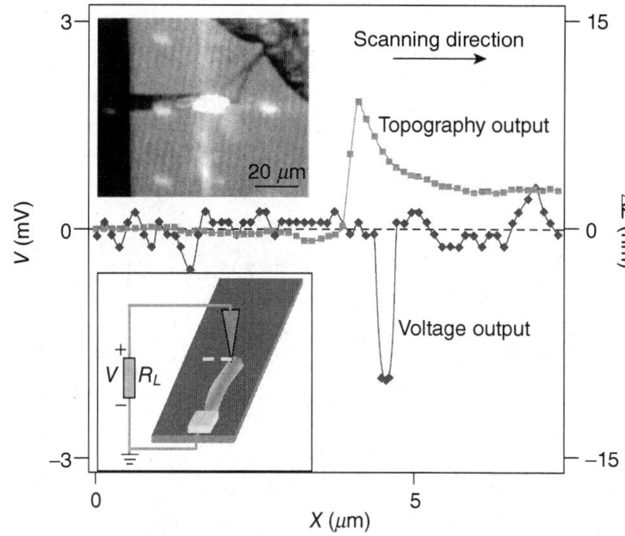

Figure 18.24. In situ observation of piezoelectric potential of a one-end-free ZnO wire/belt using an AFM tip. The ZnO is clean without oligomer coating. The top inset is a snapshot optical image of an AFM tip that is deflecting the wire/belt. The bottom inset is a schematic experimental setup. The curves are the aligned plot of the output voltage (diamonds) over the external load and the topography profile (squares). Reprinted from Song et al. (2007).

DEVELOPMENT OF NANODEVICE ARCHITECTURES

In this section, the winding way toward real devices is briefly surveyed, stressing the most important breakthroughs reached up to now. The first part focuses on the attempts to fabricate real devices based on a limited number of NWs, and the second one makes a short introduction to recent works, which are faced with the key issue of large-scale integration. Nevertheless, authors expect that readers will find helpful contents and acknowledge the effort to capture brand new results illustrative of the present state of the art.

FET Device Fabrication

Since a complete review of NW-based device fabrication is beyond the scope of this section, only a few examples will be presented here, mainly devices based on single or ensemble NWs.

FETs are the type of devices on which the largest amount of the efforts of NW community is devoted. In comparison to conventional microelectronic technology, CMOS, the simple device's design, and a uniform NW as active material simplify the development of electronic systems based on NWs (Fig. 18.25). However, some efforts have led to other electronic devices, like p–n homo- and heterojunctions, and p–i–n heterostructures.

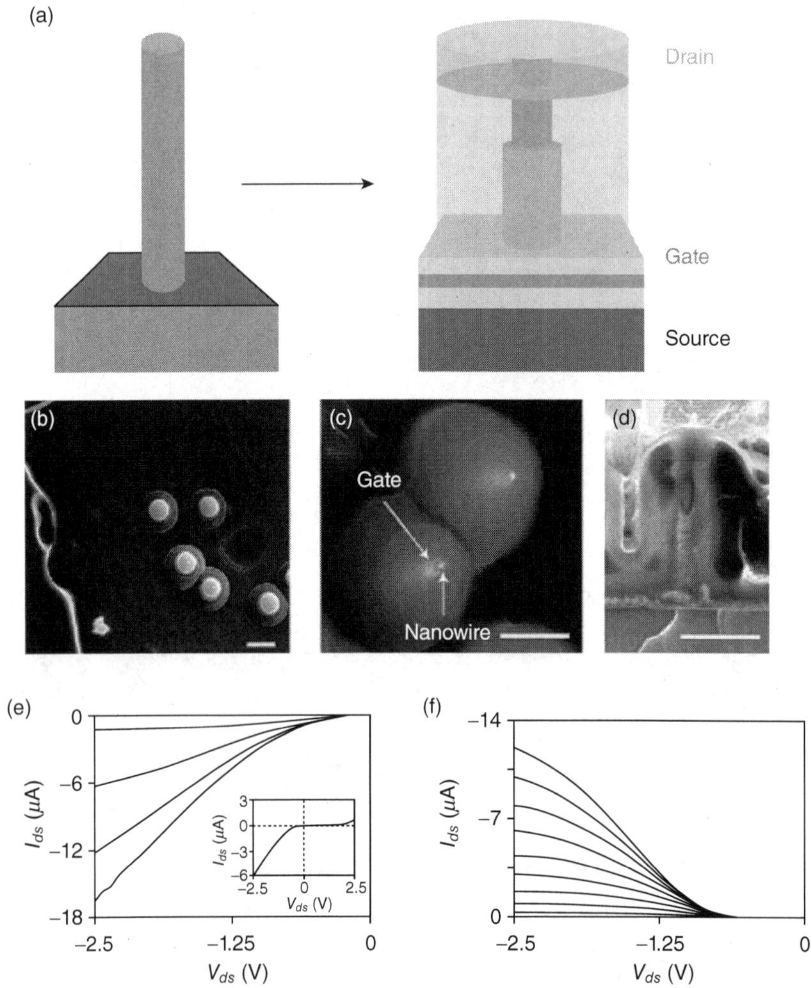

Figure 18.25. Vertically integrated nanowire field-effect transistor (VINFET) device configuration. (A) Schematic illustration of the device fabricated from vertical Si nanowires. (B) Top-view SEM image of the device (scale bar: 2 µm). (C) Top-view SEM image of the midsection of the VINFET device (scale bar: 1 µm). (D) Cross-sectional pseudocolor SEM image of the VINFET device (scale bar: 500 nm). (E) I_{ds} versus V_{ds} at V_{gs} = −2.5, −2.0, −1.5, −1.0, and −0.5 V from bottom to top. Inset shows the I_{ds} versus V_{ds} spectrum at V_{gs} = −1.5 V. (F) I_{ds} versus V_{gs} with V_{ds} ranging from −2.5 to −0.25 V, from top to bottom, respectively. Reprinted from Goldberger et al. (2006).

The simplest way to fabricate such a device can be described as the following: (1) removing NWs from substrates on which they have been grown, (2) sonicating their suspension in a liquid to reduce their agglomeration, and (3) dispersing a drop of this solution over a substrate. Typically, a Si wafer with a relatively thin silicon oxide overlayer is employed. After locating the NWs, drain and source contacts are fabricated by e-beam lithography, employing the highly doped substrate as the back-gate contact. Using this method, FET devices from several classical semiconductors like Si (Cui et al. 2003), Ge (Greytak et al. 2004), GaN (Huang et al. 2002), ZnO (Choi et al. 2008), and In_2O_3 (Jo et al. 2008) have been fabricated. Most of these devices have shown improved electrical properties as compared to their thin-film equivalent devices, showing the potential of such structures in device fabrication.

The NW-based device architectures, on a first look, appear complex for the single-device fabrication but offer potential to simplify the integration of the NWs in larger systems. Such approaches consist of the fabrication of a vertical transistor around an as-grown NW with a concentric gate around it (so-called wrap-gated), conveniently isolated, and has been carried out on Si (Schmidt et al. 2006), ZnO (Ng et al. 2004), and InAs materials (Bryllert et al. 2006). This has been performed without removing the NWs from the substrate and, even, by employing the catalyst used for the growth as an interface material between the NW and the drain contact deposited after the isolation and growth of NWs. In the second approach, this contact is suspended and air-bridged. The advantages of such architectures are evident, such as the serial fabrication of FET devices in view of their integration into larger systems involving tens or thousands of devices.

New degrees of freedom in the development of electronic devices can be obtained when replacing the homogeneous NWs by heterogeneous NWs. Here, various routes have been envisaged according to the use of axially or radially variating heterostructures. For example, in Si–core/Ge–shell NWs (Xiang et al. 2006), the formation of 1-D hole gas and the HfO_2 or ZrO_2 high-k dielectric gate isolator gave rise to a high-transconductance value, a quality factor of the FETs, which has been obtained as four times larger than state-of-the-art metal oxide semiconductor field effect transistor (MOSFET) devices. Furthermore, these devices have shown potential for operation up to 2 THz (Hu et al. 2008). On the other hand, axially doped Si NWs applied in the development of a classical planar FET as well as a wrap-gated FET have been demonstrated (Ho et al. 2008). The latter shows a more controlled device switching. Furthermore, other axially variating NW heterostructures, namely, an InAs/InP axial NW grown by MBE (Nilsson et al. 2008), have been used for the fabricaton of a single-electron transistor (SET). In this case, air was employed as gate isolator, as the NW was suspended at only 15 nm above the gate electrode. It was shown for the first time that such SET devices could be operated in the radio frequency range.

Recently, branched InAs NWs have been synthesized and three-terminal devices have been fabricated on them (Suyatin et al. 2008), presenting tunable nonlinear electrical behavior. The authors have demonstrated the ability of such devices for frequency mixing, multiplication, and phase-difference detection of electronic signals at room temperature. This opens new ranges of applications of devices based on NWs.

Axial and radial heterostructures of NWs have also been used for the fabrication of other devices. For example, a solar cell based on Si p–i–n radial NW was fabricated by contacting the core p-type and the shell n-type (Tian et al. 2007). Although the open-circuit voltage was relatively low, the photovoltaic efficiency can reach in the range of 3.4%. An important pointer is to use such NW-based solar cells as power supply for other NW-based electronic devices, like nanosensors and nanologic circuits. The same group presented p–i–n and 2 p–i–n configurations as solar cell (Kempa et al. 2008). However, in this case, the tandem cell was realized by modulating the doping of Si backbone during the NW growth.

Additional device versatility can be obtained by using crossed NWs as the building blocks. This was first proposed by Duan et al. (2001) using p- and n-doped InP NWs. Two crossed NWs, contacted on both ends, have been demonstrated as conduction channels so that current can flow from one NW to the other once the crossed NWs are in contact mode. The contact of NWs with same doping exhibits an ohmic behavior, whereas the rectifying behavior has been observed in crossed NWs with n and p doping. This situation is even further complicated when combining two different materials with different dopings, like p-Si/n-GaN (Huang et al. 2002). Finally, by crossing one p-Si NW with 3 n-GaN NWs, a NOR logic gate has been fabricated and demonstrated.

Crossing NWs have also been employed to develop other optoelectronic devices apart from solar cells. Emission from crossing p- and n-type InP NWs has been measured (Duan et al. 2001), giving rise to single NW-based nanoLED. This result has been further reported on other NW materials. More recently, lasing and waveguiding effects on CdS or GaN NWs have been reported (Greytak et al. 2005).

Integration of NW Elements into Complex Nanodevice Architectures

Controlled synthesis and manipulation of NWs have made great progress in recent years, enabling their fundamental characterization and preliminary integration in simple proof-of-concept devices, as shown in the former sections. Nevertheless, the integration of NWs in consumer-class electronics based on CMOS technology remains yet a hard-to-achieve objective due to important growth and engineering issues, such as the lack of control during fabrication, compatibility limitations, high cost, and inability to meet industrial standards. For all these reasons, it can be asserted that researchers and developers are still on their primary steps with large-scale integration of NWs in complex architecture devices. Nevertheless, intensive efforts are currently on the way to overcome these limitations.

Toward Real Demonstrators Based on NWs. Design and fabrication of portable platforms with consumer-class electronics for monitoring the electrical properties of NWs have been demonstrated (Hernandez-Ramirez et al. 2007b). These devices, which are more affordable and easier to operate than traditional lab equipments, can be combined with different NWs in order to obtain different functionalities. This has enabled the fabrication of low-cost photodetectors and gas sensors with long-term stability (Hernandez-Ramirez et al. 2007b; Prades et al. 2008a, b, c). It is well-known that metal oxide surfaces need to be heated (activated) to maximize their sensing response to specific target species (Diéguez et al. 2000). The combination of both bottom-up and top-down fabrication techniques to integrate NWs in microhotplates has been extremely useful to obtain setups being able to modulate their effective temperature and thus, to activate the NW response toward gas molecules (Sysoev et al. 2006; Hernandez-Ramirez et al. 2007b). In parallel, other research efforts are focusing their attention to obtain more complex systems based on arrays of NWs, which could enhance the selectivity of final devices (Sysoev et al. 2006). The response behaviors of NW contacts in parallel are simultaneously monitored and recorded. Later, experimental data processed by pattern recognition software (i.e., fuzzy logic) can determine the accurate composition of external stimuli (Sysoev et al. 2007). Although all these studies are still in a preliminary step, Kolmakov et al. (Sysoev et al. 2006; Sysoev et al. 2007) demonstrated that they could pave the way toward the first electronic nose (e-nose) based on NWs. The main advantage of these devices might be a significant energy-saving device (Prades et al. 2008b) (Fig. 18.26).

Massive Integration of NWs. Detection and quantification of biological and chemical species are central to many areas of health care and life sciences, ranging from diagnosing the disease to screening new drug molecules. Semiconductor NWs configured as electronic devices have emerged as a general platform for ultrasensitive direct electrical detection of biological and chemical species (Patolsky et al. 2006, 2007). The most conventional bottom-up strategies start with (1) the growth of uniform, single-crystalline NWs, (2) subsequent isolation of the NWs as stable suspensions, then (3) nanofabrication techniques are used to produce devices based on individual or few NWs (Hernandez-Ramirez et al. 2009). Finally, recent works demonstrated (Zhou et al. 2005; Prades et al. 2008c) that chemical modifications on the NW surfaces with functional receptors or passivating layers are an effective method to tailor the performance of the resulting devices.

The integration of NWs in the fabrication of nanodevices is still technologically challenging; however, several approaches have been promoted, although the difficulties of manipulation and assembly of NWs on the nanometer scale need to be solved. Present photolithographic techniques have witnessed a continuing price increase with decreasing feature size into nanometer dimensions. In view of these limitations, it has been suggested that future device integration may be based on alternative approaches, such as assembly of 1-D nanostructures. Therefore, research has focused on developing new chemical and physical assembly techniques to control the placement of these 1-D nanostructures with the precision required to form well-ordered electrical circuits. The assembling methods can be divided into three classes: assembly using macroscopic field forces (electric or magnetic field), microfluidic assisted template integration, and chemical assembly (Langmuir–Blodgett [LB]) based on surface functionalization on nanostructures and substrates (Huang et al. 2001a,b; Whang et al. 2003; Zhang et al. 2004a, b, c; Englander et al. 2005). The field-assisted technologies can assemble NWs by controlling the direction and intensity of the electric or magnetic field, while the microfluidic method represents a general assembly strategy for any kind of 1-D nanostructures. NWs can be aligned with controlled nanometer- to micrometer-scale range using the LB technique and can be transferred to planar substrates in a layer-by-layer process to form parallel and crossed NW structures.

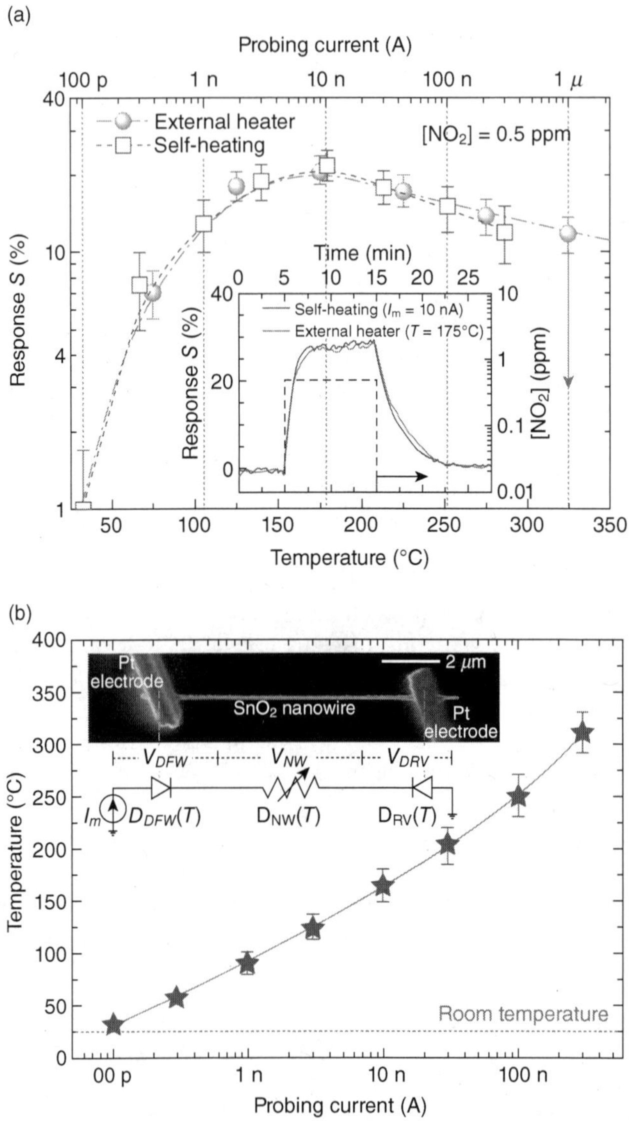

Figure 18.26. Response of a SnO_2 nanowire operated in self-heating mode and with external microheater to $[NO_2] = 0.5$ ppm: (a) comparison of the responses obtained with both measuring methodologies. The similarities to the calibration curve of the external heater were used to roughly estimate the effective temperature. (b) Estimated temperature of the devices at different I_m (r_{nw}~35 nm). Inset: SEM image of a SnO_2 nanowire connected to two Pt microelectrodes fabricated with focused ion beam (FIB). The equivalent circuit of this structure corresponds to two back-to-back diodes (D_{FW} and D_{RV}) in series with the nanowire resistance (R_{NW}). Reprinted from Prades et al. (2008b).

Alternatively, top-down approaches also offer attractive possibilities for 3-D multifunctional electronics based on the layer-by-layer assembly of NW building blocks (Fig. 18.27). Lieber et al. demonstrated that it is possible to fabricate high-performance FETs based on stacked layers of Ge/Si core–shell NWs. This approach enables developing cost-effective fabrication strategies (like printing techniques) with novel functionalities that go beyond the silicon wafer-based electronics (flexible plastic substrates) (Javey et al. 2007).

Despite the recent advances in nanolithography techniques, these techniques are only suitable for the fabrication of devices for prototyping and academic purposes. To extend the use of NWs for low-cost and large-scale fabrication processes, new concepts for large-scale applications of self-assembling techniques must be developed. In this direction, the first steps have been made by self-aligning of 1-D metal oxide nanostructures using dielectrophoretic techniques (Kumar et al. 2005; Suehiro et al. 2006; Seo et al. 2006a, b; Lee et al. 2008a, b).

Dielectrophoresis is an attractive alternative for the positioning and alignment of NWs due to its low cost, simplicity, and flexibility (Duan et al. 2001; Kumar et al. 2005; Suehiro et al. 2006; Seo et al. 2006a, b; Lee et al. 2008a, b). This method is based on the well-known forces that appear when dielectrically polarized materials are applied by a nonuniform electric field (Pohl 1951; Pohl 1978). It has been tested for different nanomaterials like single-walled (Lucci et al. 2005; Seo et al. 2006a, b; Stokes and Khondaker 2008) and multiwalled CNTs (Tung et al. 2007), polymeric (Dan et al. 2007), metal (Smith et al. 2000; Boote and Evans 2005), semiconductor NWs (Duan et al. 2001; Seo et al. 2006a, b), and metal oxide NWs (Suehiro et al. 2006; Seo et al. 2006a, b; Lee et al. 2008a, b).

Dielectrophoresis can be applied to the fabrication of a new generation of nanodevices, and it can be easily combined with other techniques like e-beam or conventional photolithography. Using the assembling technique, NWs cannot only be aligned but can also be positioned at any desired electrodes, and thus the time to fabricate a device is significantly reduced. This feature can be applied in the fabrication of the simplest electronic elements, like rectifying junctions (Lee et al. 2007) and transistors (Stokes and Khondaker 2008), paving the way for the development of novel electronic devices in the future.

SUMMARY

This review has briefly described the recent development and progress of 1-D nanostrcutures in the aspects of basic materials science, advanced characterization, as well as the perspective of practical applications. Further, intrinsic and technological limitations related to the integration of device fabrication are also discussed. Semiconductor NWs with on-chip microscale components by combining the bottom-up NW synthesis with top-down methods are summarized. With this rapid development, more approaches to the growth of NWs, technological advances to a more controlled formation of NWs, new innovative methods for the characterization of microstructures, and physical

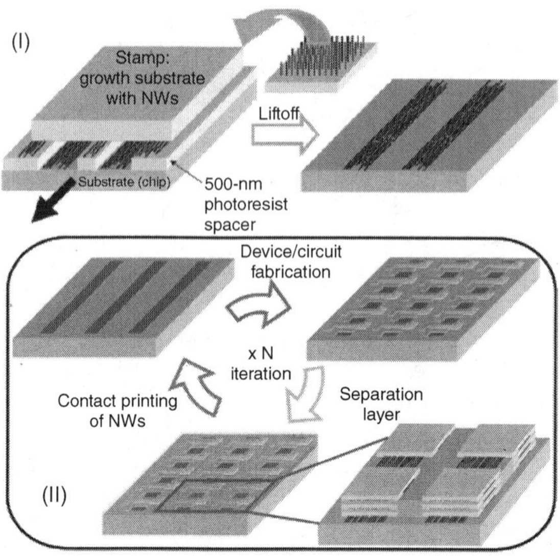

(I)
Stamp: growth substrate with NWs
Liftoff
Substrate (chip)
500-nm photoresist spacer
Device/circuit fabrication
x N iteration
Contact printing of NWs
Separation layer
(II)

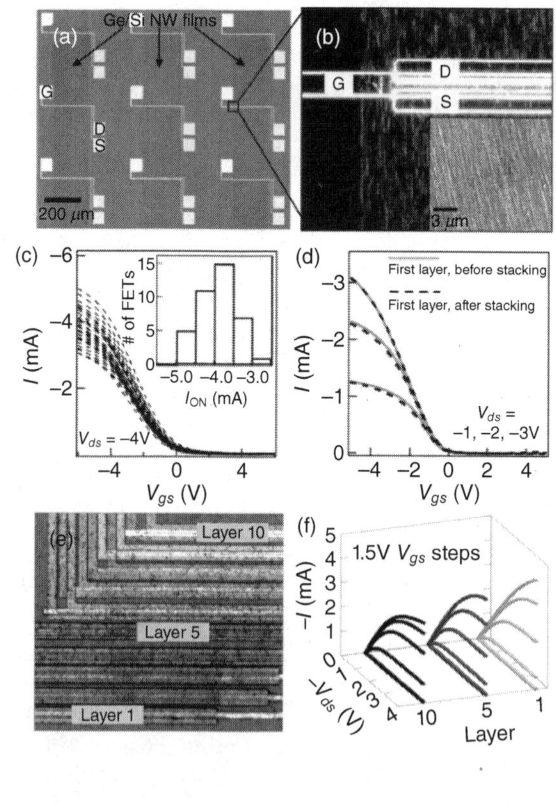

(a)
Ge/Si NW films
G
D
S
200 μm
(b)
G
D
S
3 μm
(c)
I (mA)
# of FETs
I_{ON} (mA)
V_{ds} = −4V
V_{gs} (V)
(d)
First layer, before stacking
First layer, after stacking
I (mA)
V_{ds} = −1, −2, −3V
V_{gs} (V)
(e)
Layer 10
Layer 5
Layer 1
(f)
1.5V V_{gs} steps
−I (mA)
$-V_{ds}$ (V)
Layer

Figure 18.27. An overview of 3-D nanowire circuit integration. (I) Contact printing of nanowires from growth substrate to prepatterned substrate. Nanowires are grown with random orientation and are aligned by sheer forces during the printing process. (II) The 3-D nanowire circuit is fabricated by iteration of contact printing, device fabrication, and separation layer deposition steps. 3-D nanowire FETs: (a) Optical microscopic image shows the array of nanowire FETs. (b) TEM image demonstrates a parallel array of nanowires is aligned between source (S), drain (D), and top-gate (G) electrodes. (c) I-V_{gs} characteristics of a FET device composed of 40 nanowires. (d) Transfer characteristics of the first-layer nanowire FET before and after second-layer fabrication. (e) Optical image of 10 layers of Ge/Si nanowire FETs. (f) Current versus drain–source voltage characteristicsfor nanowire FETs from layers 1, 5, and 10. Reprinted from Javey et al. (2007).

properties as well as the innovative approaches toward the use of NWs in electronics, photonics, and sensor applications have made progress.

In comparison to nanoparticles, NWs have inherited not only the large surface-to-bulk ratio but also the amazing length-to-width ratio. The wirelike NWs cannot only bridge the classic quantum world due to the partial limitation of dimension sizes but can also accumulate the variation of charge potential from a sensitive surface to influence the transport behaviors of charge carriers. Its future is all visionary promise.

It has been realized that the integration of 1-D materials in nanodevices has encountered two different barriers (Fig. 18.28): knowledge and engineering techniques. Although the proof-of-concept devices have been demonstrated in various scientific reports, the knowledge about control over dimensions, doping, growth mechanism, surface modification, and assembly are far away from the well-developed level.

The materials engineering of 1-D building blocks will require more developing time due to the similar facts: (1) Si CMOS technology has steadily grown in importance for the past 50 years until it is now the dominant logic technology in the electronics industry; (2) a carbon nanotube (CNT) is a cylindrical structure of rolled-up sheet of graphite. The materials have been discovered since 1990; however, scientists are still struggling for the controlled synthesis because its physical properties (metallic or semiconducting) depend on how the graphite rolls into the cylinder form. It is evident that the knowledge and engineering barriers can be overcome only by the long-term interdisciplinary collaboration between chemists, physicists, and materials scientists. Since research efforts in NW production and assembly are continuing across the world, we believe that it is just a matter of time before someone comes up with a viable way to mass assemble and to produce NWs for NW-based nanodevices.

In the practical fabrication process, the connection of NWs should be electrically ohmic and mechanically robust, whereas the chemical state of NWs should be stable. Instead of the "first bottom-up synthesis and second top-down fabrication" approaches, the controlled placement of NWs can be achieved by various assembly techniques. Direct integration of NW growth into device fabrication will remarkedly simplify the process and allow the access of individual NWs in a parallel fashion. This assembly approaches (Integration of NW Elements into Complex Nanodevice Architectures)

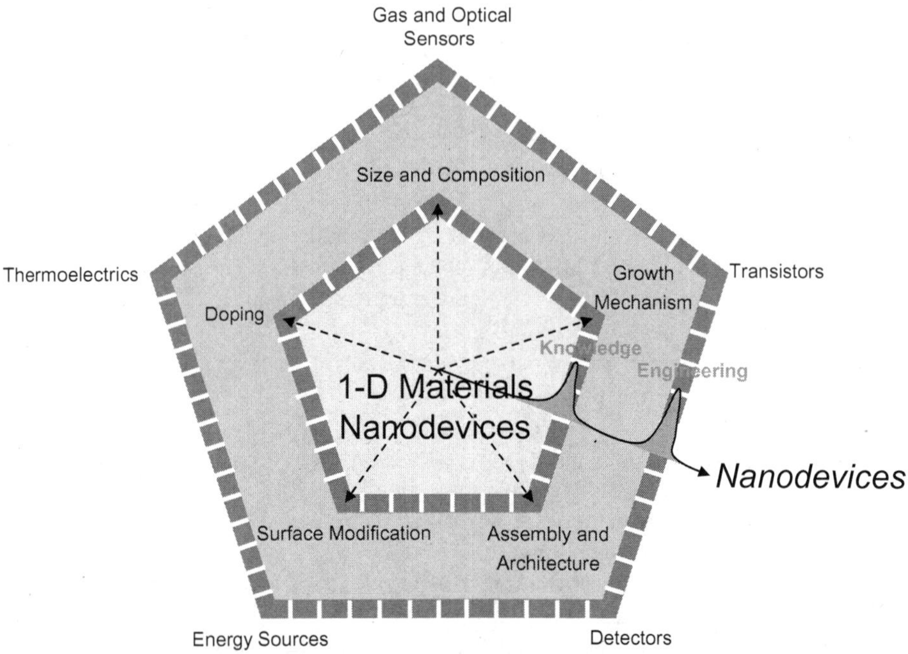

Figure 18.28. Integration of 1-D nanostructures in new devices and circuit architectures: recent developments and challenges.

can potentially eliminate the postgrowth processing steps and solve the "position registry" problems for NW-based electronics. It is noted that targeted development of chemical methods, achieving inorganic nanostructures by controlled and/or self-organized growth, need greater attention. A better understanding of underlying knowledge and engineering barriers of NW synthesis and the process of NW integration requires

- New strategies for chemically driven synthesis of 1-D nanostructures (Fig. 18.29)
- General chemical principles for designing morphology and size in the final material
- Control of surface modification to tune the surface behaviors of NWs
- Combinatorial methods to explore the sensing mechanism of NW-based devices.
- Computational methods applicable to design and simulate the device systems
- Proof of concept for the commercial viability of NW-based devices.

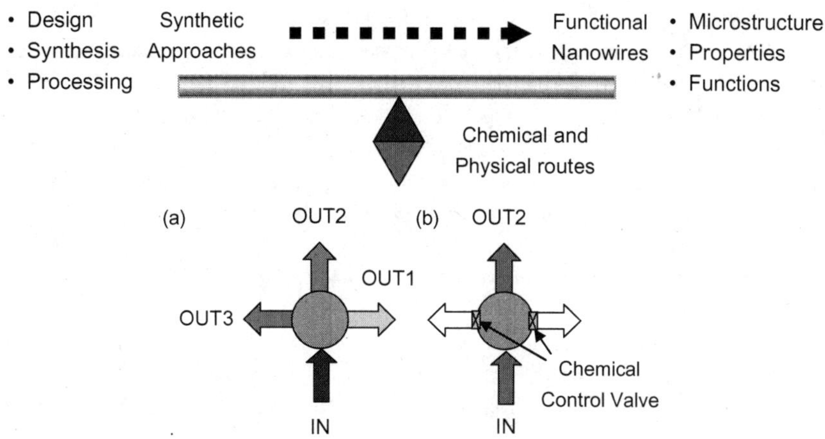

Figure 18.29. The technical road map of chemical and physical approaches to functional inorganic nanowires.

ACKNOWLEDGMENTS

Thanks are due to the German Science Foundation (DFG), the Federal Ministry of Education and Research (BMBF), and the University of Cologne for supporting this work.

REFERENCES

Adachi S. 1985. GaAs, AlAs, and $Al_xGa_{1-x}As$ material parameters for use in research and device applications. J Appl Phys 58:R1–R129.

Agarwal A, Barrelet CJ, Lieber CM. 2005. Lasing in single cadmium sulfide nanowire optical cavities. Nano Lett 5:917–920.

Ai Y, Xue C, Sun C, Sun L, Zhuang H, Wang F, Li H, Chen J. 2007. Synthesis of GaN nanowires through Ga_2O_3 films' reaction with ammonia. Mater Lett 61:2833–2836.

Al-Salman R, Mallet J, Molinari M, Fricoteaux P, Martineau F, Troyon M, Zein El Abedin S, Endres F. 2008. Template assisted electrodeposition of germanium and silicon nanowires in an ionic liquid. Phys Chem Chem Phys 10:6233–6237.

An SJ, Park WI, Yi GC, Kim YJ, Kang HB, Kim M. 2004. Heteroepitaxal fabrication and structural characterizations of ultrafine GaN/ZnO coaxial nanorod heterostructures. Appl Phys Lett 84:3612–3614.

Arbiol J, Kalache B, Cabarrocas PR, Morante JR, Morral AF. 2007. Influence of Cu as a catalyst on the properties of silicon nanowires synthesized by the vapor-solid-solid mechanism. Nanotechnology 18:305606.

Asoh H, Ono S, Hirose T, Nakao M, Masuda H. 2003. Growth of anodic porous alumina with square cells. Electrochim Acta 48:3171–3174.

Bachmann J, Jing J, Knez M, Barth S, Shen H, Mathur S, Gosele U, Nielsch K. 2007. Ordered iron oxide nanotube arrays of controlled geometry and tunable magnetism by atomic layer deposition. J Am Chem Soc 129:9554–9555.

Bae SY, Seo HW, Choi HC, Park J. 2004. Heterostructures of ZnO nanorods with various one-dimensional nanostructures. J Phys Chem B 108:12318–12326.

Baek Y, Yong K. 2007. Controlled growth and characterization of tungsten oxide nanowires using thermal evaporation of WO_3 powder. J Phys Chem C 111:1213–1218.

Bao NZ, Shen LM, Gupta A, Tatarenko A, Srinivasan G, Yanagisawa K. 2009. Size-Controlled one- dimensional monocrystalline $BaTiO_3$ nanostructures. Appl Phys Lett 94:253109.

Barrelet CJ, Wu Y, Bell DC, Lieber CM. 2003. Synthesis of CdS and ZnS nanowires using single-source molecular precursors. J Am Chem Soc 125:11498–11499.

Bhunia S, Kawamura T, Watanabe Y, Fujikawa S, Tokushima K. 2003. Metalorganic vapor-phase epitaxial growth and characterization of vertical InP nanowires. Appl Phys Lett 83: 3371–3373.

Bjork MT, Ohlsson BJ, Sass T, Persson AI, Thelander C, Magnusson MH, Deppert K, Wallenberg LR, Samuelson L. 2002. One-dimensional steeplechase for electrons realized. Nano Lett 2:87–89.

Boote JJ, Evans SD. 2005. Dielectrophoretic manipulation and electrical characterization of gold nanowires. Nanotechnology 16:1500–1505.

Bootsma GA, Gassen HJ. 1971. A quantitative study on the growth of silicon whiskers from silane and germanium whiskers from germane. J Cryst Growth 10:223–234.

Brumlik CJ, Martin CR. 1991. Template synthesis of metal microtubules. J Am Chem Soc 113: 3174–3175.

Bryllert T, Wernersson LE, Löwgren T, Samuelson L. 2006. Vertical wrap-gated nanowire transistors. Nanotechnology 17:S227–S230.

Cao HQ, Qiu XQ, Liang Y, Zhu Q, Zhao M. 2003. Room-temperature ultraviolet-emitting In_2O_3 nanowires. Appl Phys Lett 83:761–763.

Cao H, Qiu X, Luo B, Liang Y, Zhang Y, Tan R, Zhao M, Zhu Q. 2004. Synthesis and room-temperature ultraviolet photoluminescence properties of zirconia nanowires. Adv Funct Mater 14:243–246.

Chan CK, Peng H, Twesten RD, Jarausch K, Zhang XF, Cui Y. 2007. Fast, completely reversible Li insertion in vanadium pentoxide nanoribbons. Nano Lett 7:490–495.

Chang KW, Wu JJ. 2003. Catalytic growth and characterization of Ga_2O_3 nanowires. Appl Phys A 76:629–632.

Chang PC, Fan Z, Wang D, Tseng WY, Chiou WA, Hong J, Lu JG. 2004. ZnO nanowires synthesized by vapor trapping CVD method. Chem Mater 16:5133–5137.

Chang PC, Fan T, Chien C, Stichtenoth D, Ronning C, Lu JG. 2006. High-performance ZnO nanowire field effect transistors. Appl Phys Lett 89:133113.

Chang PC, Chien CJ, Stichtenoth D, Ronning C, Lu JG. 2007. Finite size effect in ZnO nanowires. Appl Phys Lett 90:113101.

Chaudhry A, Ramamurthi V, Fong E, Islam MS. 2007. Ultra-low contact resistance of epitaxially interfaced bridged silicon nanowires. Nano Lett 7:1536–1541.

Chen G, Dresselhaus MS, Dresselhaus G, Fleurial JP, Caillat T. 2003a. Recent developments in thermoelectric materials. Int Mater Rev 48:45–66.

Chen Y, Cui X, Zhang K, Pan D, Zhang S, Wang B, Hou JG. 2003b. Bulk-quantity synthesis and self-catalytic VLS growth of SnO_2 nanowires by lower-temperature evaporation. Chem Phys Lett 369:16–20.

Chen RS, Huang YS, Tsai DS, Chattopadhyay S, Wu CT, Lan ZH, Chen KH. 2004. Growth of well aligned IrO_2 nanotubes on $LiTaO_3$(012) substrate. Chem Mater 16:2457–2462.

Chen C, Chen D, Jiao X, Qang C. 2006. Ultrathin corundum-type In_2O_3 nanotubes derived from orthorhombic InOOH: synthesis and formation mechanism. Chem Commun 44:4632–4634.

Chen YL, Hsu CC, Song YH, Chi Y, Carty AJ, Peng SM, Lee GH. 2006. Iridium metal thin films and patterned IrO2 nanowires deposited using iridium(I) carbonyl precursors. Chem Vapor Dep 12:442–447.

Chen R, Hochbaum AI, Murphy P, Moore J, Yang PD, Majumdar A. 2008. Thermal conductance of thin silicon nanowires. Phys Rev Lett 101:105501.

Cheng B, Russell JM, Shi WS, Zhang L, Samulski ET. 2004. Large-scale, solution-phase growth of single-crystalline SnO_2 nanorods. J Am Chem Soc 126:5972–5973.

Choi YC, Kim WS, Park YS, Lee SM, Bae DJ, Lee YH, Park GS, Choi WB, Lee NS, Kim JM. 2000. Catalytic growth of b-Ga2O3 nanowires by arc discharge. Adv Mater 12:746–750.

Choi J, Wehrspohn R, Gösele U. 2003. Moiré pattern formation on porous alumina arrays using nanoimprint lithography. Adv Mater 15:1531–1534.

Choi JH, Khan DY, Myoung JM. 2008. Fabrication and characterization of ZnO nanowire transistors with organic polymer as a dielectric layer. Solid State Commun 143:126–130.

Colombo C, Spirkoska D, Frimmer M, Abstreiter G, Morral AFI. 2008. Ga-assisted catalyst-free growth mechanism of GaAs nanowires by molecular beam epitaxy. Phys Rev B 77:155326.

Comini E, Baratto C, Faglia G, Ferroni M, Vomiero A, Sberveglieri G. 2009. Quasi-one dimensional metal oxide semiconductors: preparation, characterization and application as chemical sensors. Prog Mater Sci 54:1–67.

Cornelius TW, Brötz J, Chtanko N, Dobrev D, Miehe G, Neumann R, Molares MET. 2005. Controlled fabrication of poly- and single-crystalline bismuth nanowires. Nanotechnology 16:S246–S249.

Cuenot S, Fretigny C, Champagne SD, Nysten B. 2004. Surface tension effect on the mechanical properties of nanomaterials measured by atomic force microscopy. Phys Rev B 69:165410.

Cui Y, Wei QQ, Park HK, Lieber CM. 2001a. Nanowire nanosensors for highly sensitive and selective detection of biological and chemical species. Science 293:1289–1292.

Cui Y, Lauhon LJ, Gudiksen MS, Wang J, Lieber CM. 2001b. Diameter-controlled synthesis of single-crystal silicon nanowires. Appl Phys Lett 78:2214–2216.

Cui Y, Zhong Z, Wang D, Wang WU, Lieber CM. 2003. High performance silicon nanowire field effect transistors. Nano Lett 3:149–152.

Dae C, Kim S, Ahn B, Kim J, Sung M, Shin H. 2008. Template-directed gas-phase fabrication of oxide nanotubes. J Mater Chem 18:1362–1367.

Dai ZR, Pan ZW, Wang ZL. 2001. Ultra-long single crystalline nanoribbons of tin oxide. Solid State Commun 118:351–354.

Dai ZR, Gole JL, Stout JD, Wang ZL. 2002. Tin oxide nanowires, nanoribbons, and nanotubes. J Phys Chem B 106:1274–1279.

Dai ZR, Pan ZW, Wang ZL. 2003. Novel nanostructures of functional oxides synthesized by thermal evaporation. Adv Funct Mater 13:9–24.

Dan Y, Cao Y, Mallouk TE, Johnson AT, Evoy S. 2007. Dielectrophoretically assembled polymer nanowires for gas sensing. Sensor Actuat B-Chem 125:55–59.

Davidson FM III, Schricker AD, Wiaceck RJ, Korgel BA. 2004. Supercritical fluid-liquid-solid synthesis of gallium arsenide nanowires seeded by alkanethiol-stabilized gold nanocrystals. Adv Mater 16:646–649.

Dhawale DS, More AM, Latthe SS, Rajpure KY, Lokhande CD. 2008. Room temperature synthesis and characterization of CdO nanowires by chemical bath deposition (CBD) method. Appl Surf Sci 254:3269–3273.

Dick KA, Deppert K, Larsson MW, Martensson T, Seifert W, Wallenberg LR, Samuelson L. 2004. Synthesis of branched "nanotrees" by controlled seeding of multiple branching events. Nat Mater 3:380–384.

Dick KA, Deppert K, Martensson T, Mandl B, Samuelson L, Seifert W. 2005. Failure of the vapor–liquid–solid mechanism in Au-assisted MOVPE growth of InAs nanowires. Nano Let 5:761–764.

Dick KA, Kodambaka S, Reuter MC, Deppert K, Samuelson L, Seifert W, Wallenberg LR, Ross FM. 2007. The morphology of axial and branched nanowire heterostructures. Nano Lett 7:1817–1822.

Diéguez A, Vilà A, Cabot A, Romano-Rodríguez A, Morante JR, Kappler J, Barsan N, Weimar U, Göpel W. 2000. Influence on the gas sensor performances of the metal chemical states introduced by impregnation of calcinated SnO_2 sol–gel nanocrystals. Sensor Actuat B-Chem 68:94–99.

Ding JX, Zapien JA, Chen WW, Lifshitz Y, Lee ST, Meng XM. 2004a. Lasing in ZnS nanowires grown on anodic aluminum oxide templates. Appl Phys Lett 85:2361–2363.

Ding Y, Gao PX, Wang ZL. 2004b. Catalyst-nanostructure interfacial lattice mismatch in determining the shape of VLS grown nanowires and nanobelts: a case of Sn/ZnO. J Am Chem Soc 126:2066–2072.

Dong Z, Fu YY, Han Q, Xu Y, Zhang H. 2007. Synthesis and physical properties of Co_3O_4 nanowires. J Phys Chem C 111:18475–18478.

Du GH, Van Tendeloo G. 2004. $Cu(OH)_2$ nanowires, CuO nanowires and CuO nanobelts. Chem Phys Lett 393:64–69.

Du N, Zhang H, Chen B, Wu J, Ma X, Liu Z, Zhang Y, Yang D, Huang X, Tu J. 2007. Porous Co_3O_4 nanotubes derived from $Co_4(CO)_{12}$ clusters on carbon nanotube templates: a highly efficient material For Li-Battery applications. Adv Mater 19:4505–4509.

Du N, Zhang H, Chen B, Ma X, Yang DR. 2008. One-pot, large-scale synthesis of SnO_2 nanotubes at room temperature. Chem Commun 14:3028–3030.

Duan X, Lieber CM. 2000a. General synthesis of compound semiconductor nanowires. Adv Mater 12:298–302.

Duan X, Lieber CM. 2000b. Laser-assisted catalytic growth of single crystal GaN nanowires. J Am Chem Soc 122:188–189.

Duan X, Huang Y, Cui Y, Wang J, Lieber CM. 2001. Indium phosphide nanowires as building blocks for nanoscale electronic and optoelectronic devices. Nature 409:66–69.

Duan X, Huang H, Agrawal R, Lieber CM. 2003. Single-nanowire electrically driven lasers. Nature 421:241–244.

Duan J, Yang S, Liu H, Gong J, Huang H, Zhao X, Zhang R, Du Y. 2005. Single crystal SnO_2 zigzag nanobelts. J Am Chem Soc 107:6180–6181.

Dubrovskii VG, Cirlin GE, Soshnikov IP, Tonkikh AA, Sibirev NV, Samsonenko YB, Ustinov VM. 2005. Diffusion-induced growth of GaAs nanowhiskers during molecular beam epitaxy: theory and experiment. Phys Rev B 71:205325.

Englander O, Christensen D, Kim J, Lin L, Morris S. 2005. Electric-field assisted growth and self-assembly of intrinsic silicon nanowires. Nano Lett 5:705–708.

Ersen O, Begin S, Houlle M, Amadou J, Janowska I, Greneche JM, Crucifix C, Pham-Huu C. 2008. Microstructural investigation of magnetic $CoFe_2O_4$ nanowires inside carbon nanotubes by electron tomography. Nano Lett 8:1033–1040.

Fan ZY, Wang DW, Chang PC, Tseng WY, Lu JG. 2004. ZnO nanowire field-effect transistor and oxygen sensing property. Appl Phys Lett 85:5923–5925.

Fan HJ, Knez M, Scholz R, Nielsch K, Pippel E, Hesse D, Zacharias M, Gosele U. 2006a. Monocrystalline spinel nanotube fabrication based on the Kirkendall effect. Nat Mater 5:627–631.

Fan HJ, Fuhrmann B, Scholz R, Himcinschi C, Berger A, Leipner H, Dadgar A, Krost A, Christiansen S, Gösele U, Zacharias M. 2006b. Vapour-transport-deposition growth of ZnO nanostructures: switch between c-axial wires and a-axial belts by indium doping. Nachtechnology 17:S231–S239.

Fan HJ, Werner P, Zacharias M. 2006c. Semiconductor nanowires: from self-organization to patterned growth. Small 2:700–717.

Feng SQ, Yu DP, Zhang HZ, Bai ZG, Ding Y. 2000. The growth mechanism of silicon nanowires and their quantum confnement effect. J Cryst Growth 209:513–517.

Fu Y, Chen J, Zhang H. 2001. Synthesis of Fe_2O_3 nanowires by oxidation of iron. Chem Phys Lett 350:491–494.

Furuta N, Ohasi Y, Itinose H, Igarashi Y. 1975. Kinetics of vapor-grown tellurium whiskers. Jpn J Appl Phys 14:929–934.

Gangloff L, Minoux E, Teo KBK, Vincent P, Semet VT, Binh VT, Yang MH, Bu IYY, Lacerda RG, Pirio G, Schnell JP, Pribat D, Hasko DG, Amaratunga GAJ, Milne WI, Legagneux P. 2004. Self-aligned, gated arrays of individual nanotube and nanowire emitters. Nano Lett 4:1575–1579.

Gao PX, Wang ZL. 2004. Substrate atomic-termination induced anisotropic growth of ZnO nanowires/nanorods by VLS process. J Phys Chem B 108:7534–7537.

Gao PX, Ding Y, Mai WJ, Hughes WL, Lao CS, Wang ZL. 2005. Conversion of zinc oxide nanobelts into superlattice-structured nanohelices. Science 309:1700–1704.

Gates B, Mayers B, Cattle B, Xia Y. 2002. Synthesis and characterization of uniform nanowires of trigonal selenium. Adv Funct Mater 12:219–227.

Geng B, Zhang LD, Meng GW, Xie T, Peng X, Lin Y. 2003. Large-scale synthesis and photoluminescence of single-crystalline β-Ga_2O_3 nanobelts. J Cryst Growth 259:291–295.

Gerung H, Boyle TJ, Tribby LJ, Bunge SD, Brinker CJ, Han SM. 2006. Solution synthesis of germanium nanowires using a Ge^2+ alkoxide precursor. J Am Chem Soc 128:5244–5250.

Giovine E, Notargiacomo A, Gaspare LD, Palange E, Evangelisti F, Leoni R, Castellano G, Torrioli G, Foglietti V. 2001. Investigation of SiGe-heterostructure nanowires. 2001. Nanotechnology 12:132–135.

Givargizov EI, Sheftal NN. 1971. Morphology of silicon whiskers grown by the VLS-technique. J Cryst Growth 9:326–329.

Goldberger J, He RR, Zhang YF, Lee SW, Yan HQ, Choi HJ, Yang PD. 2003. Single-crystal gallium nitride nanotubes. Nature 422:599–602.

Goldberger J, Hochbaum AI, Fan R, Yang PD. 2006. Silicon vertically integrated nanowire field effect transistors. Nano Lett 6:973–977.

Gong N, Lu M, Wang CY, Chen Y, Chen LJ. 2008. Au(Si)-filled β-Ga_2O_3 nanotubes as wide range high temperature nanothermometers. Appl Phys Lett 92:073101.

Greytak AB, Lauhon LJ, Gudiksen MK, Lieber CM. 2004. Growth and transport properties of complementary germanium nanowire field-effect transistors. Appl Phys Lett 84:4176–4178.

Greytak AB, Barrelet CJ, Li Y, Lieber CM. 2005. Semiconductor nanowire laser and nanowire waveguide electro-optic modulators. Appl Phys Lett 87:151103.

Gu G, Burghard M, Kim GT, Dusberg GS, Chiu PW, Krstic V, Roth S, Han WQ. 2001. Growth and electrical transport of germanium nanowires. J Appl Phys 90:5747–5751.

Gudiksen MS, Lauhon LJ, Wang J, Smith DC, Lieber CM. 2002. Growth of nanowire superlattice structures for nanoscale photonics and electronics. Nature 415:617–620.

Han S, Li C, Liu Z, Lei B, Zhang D, Jin W, Liu X, Tang T, Zhou C. 2004. Transition metal oxide core–shell nanowires: generic synthesis and transport studies. Nano Lett 4:1241–1246.

Hannon JB, Kodambaka S, Ross FM, Tromp RM. 2006. The infuence of surface migration of gold on the growth of silicon nanowires. Nature 440:69–71.

Hanrath T, Korgel BA. 2002. Nucleation and growth of germanium nanowires seeded by organic monolayer-coated gold nanocrystals. J Am Chem Soc 124:1424–1429.

Hanrath T, Korgel BA. 2004. A Comprehensive study of electron energy losses in Ge nanowires. Nano Lett 4:1455–1461.

Hartanto AB, Ning X, Nakata Y, Okada T. 2004. Growth mechanism of ZnO nanorods from nanoparticles formed in a laser ablation plume. Appl Phys A 78:299–301.

Hayden O, Greytak AB, Bell DC. 2005. Core-shell nanowire light-emitting. diodes Adv Mater 17:701–703.

He JH, Wu TH, Hsin CL, Li KM, Chen LJ, Chueh YL, Chou LJ, Wang ZL. 2006. Beaklike SnO_2 nanorods with strong photoluminescent and field-emission properties. Small 2:116–120.

He JH, Lin YH, McConney ME, Tsuruk VV, Wang ZL, Bao G. 2007. Enhancing UV photoconductivity of ZnO nanobelt by polyacrylonitrile functionalization. J Appl Phys 102:084303.

Heath JR, LeGoues FK. 1993. A liquid solution synthesis of single crystal germanium quantum wires. Chem Phys Lett 208:263–268.

Heitsch AT, Fanfair DD, Tuan HY, Korgel BA. 2008. Solution-liquid-solid (SLS) growth of silicon nanowires. J Am Chem Soc 130:5436–5437.

Hernandez-Ramirez F, Tarancon A, Casals O, Rodríguez J, Romano-Rodriguez A, Morante JR, Barth S, Mathur S, Choi TY, Poulikakos D, Callegari V, Nellen PM. 2006. Fabrication and electrical characterization of circuits based on individual tin oxide nanowires. Nanotechnology 17:5577–5583.

Hernandez-Ramirez F, Tarancon A, Casals O, Pellicer E, Rodríguez J, Romano-Rodriguez A, Morante JR, Barth S, Mathur S. 2007a. Electrical properties of individual tin oxide nanowires contacted to platinum electrodes. Phys Rev B 76:085429–085421.

Hernandez-Ramirez F, Prades JD, Tarancon A, Barth S, Casals O, Jimenez-Diaz R, Pellicer E, Rodríguez J, Juli MA, Romano-Rodriguez A, Morante JR, Mathur S, Helwig A, Spannhake J, Mueller G. 2007b. Portable microsensors based on individual SnO_2 nanowires. Nanotechnology 18:495501.

Hernandez-Ramirez F, Prades JD, Tarancon A, Barth S, Casals O, Jimenez-Diaz R, Pellicer E, Rodriguez J, Morante JR, Juli MA, Mathur S, Romano-Rodriguez A. 2008. Insight into the role of oxygen difussion into the sensing mechanisms of SnO_2 nanowires. Adv Funct Mater 8:2990–2994.

Hernandez-Ramirez F, Prades JD, Barth S, Romano-Rodriguez A, Mathur S, Tarancón A, Casals O, Jimenez-Diaz R, Rodriguez J, Pellicer E, Juli MA, Andreu T, Estrade S, Rossinyol E, Morante JR. 2009. Fabrication of nanodevices based on individual SnO_2 nanowires and their electrical characterization in metal oxide nanostructures and their applications. In: Umar A, editor. *Metal Oxide Nanostructures and Their Applications*. New York: American Scientific Publishers.

Hicks LD, Dresselhaus MS. 1993. Effect of quantum-well structures on the thermoelectric figure of merit. Phys Rev B 47:12727–12731.

Ho T, Wang Y, Eichfeld S, Lew KK, Liu B, Mohney SE, Redwing JM, Mayer TS. 2008. In situ axially doped n-channel silicon nanowire field-effect transistors. Nano Lett 8:4359–4364.

Hochbaum AI, Chen R, Delgado RD, Liang W, Garnett EC, Najarian M, Majumdar A, Yang PD. 2008. Enhanced thermoelectric performance of rough silicon nanowires. Nature 451: 163–164.

Hofmann S, Ducati C, Neill RJ, Piscanec S, Ferrari AC, Geng J, Dunin-Borkowski RE, Robertson J. 2003. Gold catalyzed growth of silicon nanowires by plasma enhanced chemical vapor deposition. J Appl Phys 94:6005–6012.

Holmes JD, Johnston KP, Doty RC, Korgel BA. 2000. Control of thickness and orientation of solution-grown silicon nanowires. Science 207:1471–1473.

Hong WK, Kimm BJ, Kim TW, Jo G, Song S, Kwon SS, Yoon A, Stach EA, Lee T. 2008. Electrical properties of ZnO nanowire field effect transistors by surface passivation. Colloids Surf A 313–314:378–382.

Hu J, Li Q, Meng X, Lee CS, Lee ST. 2002. Synthesis of β-Ga_2O_3 nanowires by laser ablation. J Phys Chem B 106:9536–9539.

Hu J, Bando Y, Liu Q, Golberg D. 2003. Laser-ablation growth and optical properties of wide and long single-crystal SnO_2. Adv Funct Mater 13:493–496.

Hu ZA, Xu T, Liu RJ, Li HL. 2004. Template preparation of high-density, and large-area Ag nanowire array by acetaldehyde reduction. Mater Sci Eng A 371:236–240.

Hu Y, Xiang J, Liang G, Yan H, Lieber CM. 2008. Sub-100 nanometer cannel length Ge/Si nanowire transistors with potential for 2 THz switching speed. Nano Lett 8:925–930.

Huang MH, Wu Y, Feick H, Webber E, Yang P. 2000. Catalytic growth of zinc oxide nanowires by vapor transport. Adv Mater 13:113–116.

Huang MH, Mao S, Feick H, Yan HQ, Wu YY, Kind H, Weber E, Russo R, Yang PD. 2001a. Room-temperature ultraviolet nanowire nanolasers. Science 292:1897–1899.

Huang Y, Duan X, Wei Q, Lieber CM. 2001b. Directed assembly of one-dimensional nanostructures into functional networks. Science 291:630–633.

Huang Y, Duan X, Cui Y, Lieber CM. 2002. Gallium nitride nanowire nanodevices. Nano Lett 2:101–104.

Huang Y, Zhang X, Reiche M, Liu L, Lee W, Shimizu T, Senz S, Gosele U. 2008. Extended arrays of vertically aligned sub-10 nm diameter [100] Si nanowires by metal-assisted chemical etching. Nano Lett 8:3046–3051.

Huang XH, Li GH, Dou XC, Li L. 2009. Magnetic properties of single crystalline Co nanowire arrays with different diameters and orientations. J Appl Phys 105:084306.

Imai H, Takei Y, Shimizu K, Matsuda M, Hirashima H. 1999. Direct preparation of anatase TiO_2 nanotubes in porous alumina membranes. J Mater Chem 9:2971–2972.

Imperor-Clerc M, Bazin D, Appay MD, Beaunier P, Davidson A. 2004. Crystallization of β-MnO_2 nanowires in the pores of SBA-15 silicas: in-situ investigation using synchrotron radiation. Chem Mater 16:1813–1821.

Javey A, Nam SW, Friedman RS, Yan H, Lieber CM. 2007. Layer-by-layer assembly of nanowires for three-dimensional, multifunctional electronics. Nano Lett 7:773–777.

Jiang XC, Herricks T, Xia YN. 2002. CuO nanowires can be synthesized by heating copper substrates in air. Nano Lett 2:1333–1338.

Jie J, Wang G, Wang Q, Chen Y, Han X, Wang X, Hou JG. 2004. Synthesis and characterization of aligned ZnO nanorods on porous aluminium oxide template. J Phys Chem B 108:11976–11980.

Jo G, Hong WK, Maeng J, Kim TW, Wang G, Yoon A, Kwon SS, Song S, Lee T. 2008. Structural and electrical characterization of intrinsic n-type In_2O_3 nanowire. Colloid Surface A 313-314:308–311.

Johnson J, Choi HJ, Knutsen KP, Schaller RD, Yang PD, Saykally RJ. 2002. Single gallium nitride nanowire lasers. Nat Mater 1:106–110.

Johnson J, Yan H, Schaller R, Saykally R, Yang PD. 2001. Single nanowire lasers. J Phys Chem B 105:11387–11390.

Jung JS, Jung YK, Kim EM, Min SH, Jun JH, Malkinski LM, Barnakov Y, Spinu L, Stokes K. 2005. Synthesis and magnetic characterization of $ZnFe_2O_4$ nanostructure in AAO template. IEEE Tran Magn 41:3403–3405.

Kempa TH, Tian B, Kim DR, Hu J, Zheng X, Lieber CM. 2008. Single and tandem axial p-i-n nanowire photovoltaic devices. Nano Lett 8:3456–3460.

Kim HW, Shim SH. 2006. Growth of MgO nanowires assisted by the annealing treatment of Au-coated substrates. Chem Phys Lett 422:165–169.

Kim BJ, Tersoff J, Kodambaka S, Reuter MC, Stach EA, Ross FM. 2008. Kinetic of individual nucleation events observed in nanoscale vapor-liquid-solid growth. Science 322:1070–1073.

Kind H, Yan H, Messer B, Law M, Yang PD. 2002. Nanowire ultraviolet photodetectors and optical switches. Adv Mater 14:158–160.

Ko E, Choi J, Okamoto K, Tak Y, Lee J. 2006. Cu_2O nanowires in an alumina template: electrochemical conditions for the synthesis and photoluminescence characteristics. ChemPhysChem 7:1505–1509.

Kobayashi Y, Hata H, Salama M, Mallouk TE. 2007. Scrolled sheet precursor route to niobium and tantalum oxide nanotubes. Nano Lett 7:2142–2145.

Kodambaka S, Tersoff J, Reuter MC, Ross FM. 2006a. Diameter-independent kinetics in the vapor-liquid-solid growth of Si nanowires. Phys Rev Lett 96:096105–096101.

Kodambaka S, Honnon JB, Tromp RM, Ross FM. 2006b. Control of Si nanowire growth of oxygen. Nano Lett 6:1292–1296.

Kolasinki KW. 2006. Catalytic growth of nanowires: vapor-liquid-solid, vapor-solid-solid, solution-liquid-solid and solid-liquid-solid growth. Curr Opin Solid State Mater Sci 10:182–191.

Kolb FM, Hofmeister H, Scholz R, Zacharias M, Gosele U, Ma DD, Lee ST. 2004. Analysis of silicon nanowires grown by combining SiO evaporation with the VLS mechanism. J Electrochem Soc 151:G472–G475.

Kong XY, Wang ZL. 2003a. Spontaneous polarization-induced nanohelixes, nanosprings, and nanorings of piezoelectric nanobelts. Nano Lett 3:1625–1631.

Kong XY, Wang ZL. 2003b. Structures of indium oxide nanobelts. Solid State Commun 128:1–4.

Kong X, Sun XM, Li XL, Li YD. 2003. Catalytic growth of ZnO nanotubes. Mater Chem Phys 82:997–1001.

Kong XY, Ding Y, Yang R, Wang ZL. 2004. Single-crystal nanorings formed by epitaxial self-coiling of polar nanobelts. Science 303:1348–1351.

Kuang DB, Fang YP, Liu HQ, Frommen C, Fenske D. 2003. Fabrication of boehmite AlOOH and γ-Al_2O_3 nanotubes via a soft solution route. J Mater Chem 13:660–662.

Kuchibhatla SVNT, Karaoti AS, Bera D, Seal S. 2007. One dimensional nanostructured materials. Prog Mater Sci 52:699–913.

Kumar S, Rajaraman S, Gerhardt RA, Wang ZL, Hesketh PJ. 2005. Tin oxide nanosensor fabrication using AC dielectrophoretic manipulation of nanobelts. Electrochim Acta 51:943–951.

Lauhon LJ, Gudiksen MS, Wang CL, Lieber CM. 2002. Epitaxial core-shell and core-multishell nanowire heterostructures. Nature 420:57–61.

Lauhon LJ, Gudiksen MS, Lieber CM. 2004. Semiconductor nanowire heterostructures. Trans R Soc Lond Ser A 362:1247–1260.

Law M, Kind H, Messer B, Kim F, Yang PD. 2002. Photochemical sensing of NO_2 with SnO_2 nanoribbon nanosensors at room temperature. Angew Chem Int Ed 41:2405–2408.

Law M, Goldberger J, Yang PD. 2004a. Semiconductor nanowires and nanotubes. Ann Rev Mater Res 34:83–122.

Law M, Sirbuly D, Johnson J, Goldberger J, Saykally R, Yang PD. 2004b. Nanoribbon waveguides for subwavelength photonics integration. Science 305:1269–1273.

Lee ST, Wang N, Zhang YF, Tang YH. 1999. Oxide-assisted semiconductor nanowire growth. MRS Bull 24:36–42.

Lee Y, Kim TH, Suh DI, Park JE, Kim JH, Youn CJ, Ahn BK, Lee SK. 2007. An electrical characterization of a hetero-junction nanowire (NW) PN diode (n-GaN NW/p-Si) formed by dielectrophoresis alignment. Physica E 36:194–198.

Lee SY, Umar A, Suh DI, Park JE, Hahn YB, Ahn JY, Lee SK. 2008a. The synthesis of ZnO nanowires and their subsequent use in high-current field-effect transistors formed by dielectrophoresis alignment. Physica E 40:866–872.

Lee W, Scholz R, Gosele U. 2008b. A continuous process for structurally well-defined Al_2O_3 nanotubes based on pulse anodization of aluminum. Nano Lett 8:2155–2160.

Lei S, Tang K, Jin Y, Chen C. 2007. Preparation of aligned MnV_2O_6 nanorods and their anodic performance for lithium secondary battery use. Nanotechnology 18:175605.

Li Y, Bando Y. 2002. Quasi-aligned MoO_3 nanotubes grown on Ta substrate. Chem Phys Lett 364:484–488.

Li Y, Meng GW, Zhang LD, Phillipp F. 2000. Ordered semiconductor ZnO nanowire arrays and their photoluminescence properties. Appl Phys Lett 76:2011–2013.

Li C, Zhang D, Han S, Liu X, Tang T, Zhou C. 2003a. Diameter-controlled growth of single-crystalline In_2O_3 nanowires and their electronic properties. Adv Mater 15:143–146.

Li D, Wu Y, Fan R, Yang PD, Majumdar A. 2003b. Thermal conductivity of Si/SiGe superlattice nanowires. Appl Phys Lett 83:3186–3188.

Li D, Wu Y, Kim P, Shi L, Yang PD, Majumdar A. 2003c. Thermal conductivity of individual silicon nanowires. Appl Phys Lett 83:2934–2936.

Li SQ, Liang YX, Wang TH. 2005a. Electric-field-aligned vertical growth and field emission properties of In_2O_3 nanowires. Appl Phys Lett 87:143104.

Li YB, Bando Y, Golberg D. 2005b. Single-crystalline α-Al_2O_3 nanotubes converted from Al_4O_4C nanowires. Adv Mater 17:1401–1405.

Li Y, Qian F, Xiang J, Lieber CM. 2006. Nanowire electronic and optoelectronic devices. Mater Today 9:18–27.

Li Y, Xu G, Zhu YL, Ma XL, Cheng HM. 2007. SnO_2/M_2O_3 one-dimensional nano-core-shell structures: synthesis, characterization and photoluminescence properties. Solid State Commun 142:441–444.

Liang G, Xiang J, Kharche N, Klimeck G, Lieber CM, Lundstrom M. 2007. Performance analysis of a Ge/Si core/shell nanowire field-effect transistor. Nano Lett 7:642–646.

Lilach Y, Zhang JP, Moskovits M, Kolmakov A. 2005. Encoding morphology in oxide nanostructures during their growth. Nano Lett 5:2019–2022.

Lim SY, Yoon CS, Cho JP. 2008. Synthesis of nanowire and hollow $LiFePO_4$ cathodes for high-performance lithium batteries. Chem Mater 20:4560–4564.

Lin YM, Dresselhaus MS. 2003. Thermoelectric properties of superlattice nanowires. Phys Rev B 68:075304.

Lin YM, Sun X, Dresselhaus MS. 2000. Theoretical investigation of thermoelectric transport properties of cylindrical Bi nanowires. Phys Rev B 62:4610–4623.

Liu B, Zeng HC. 2004. Salt-assisted deposition of SnO_2 on α-MoO_3 nanorods and fabrication of polycrystalline SnO_2 nanotubes. J Phys Chem B 108:5867–5874.

Liu JL, Cai SJ, Jin GL, Thomas SG, Wang KL. 1999. Growth of Si whiskers on Au/Si(111) substrate by gas source molecular beam epitaxy (MBE). J Cryst Growth 200:106–111.

Liu YK, Zheng CL, Wang WZ, Yin CR, Wang GH. 2001. Synthesis and characterization of rutile SnO2 nanorods. Adv Mater 13:1883–1887.

Liu S, Gan L, Liu L, Zhang W, Zeng H. 2002. Synthesis of single-crystalline TiO_2 nanotubes. Chem Mater 14:1391–1397.

Liu X, Li C, Han S, Han J, Zhou C. 2003a. Synthesis and electronic transport studies of CdO nanoneedles. Appl Phys Lett 82:1950–1952.

Liu ZQ, Zhang DH, Han S, Li C, Tang T, Jin W, Liu XL, Lei B, Zhou CW. 2003b. Laser ablation synthesis and electron transport studies of tin oxide nanowires. Adv Mater 15:1754–1456.

Liu BD, Bando Y, Tang CC, Xu FF, Golberg D. 2005a. Excellent field-emission properties of P-doped GaN nanowires. J Phys Chem B 109:21521- 21524.

Liu ZQ, Zhang DH, Han S, Li C, Lei B, Lu W, Fang J, Zhou C. 2005b. Single crystalline magnetite nanotubes. J Am Chem Soc 127:6–7.

Liu H, Yang H, Huang T. 2007. Synthesis, structure and electrochemical properties of one-dimensional nanometer materials LiV3O8. Mater Sci Eng B 143:60–63.

Lou XW, Deng D, Lee JY, Feng J, Archer LA. 2008. Self-supported formation of needlelike Co_3O_4 nanotubes and their application as lithium-ion battery electrodes. Adv Mater 20:258–262.

Lu W, Lieber CM. 2007. Nanoelectronics from the bottom up. Nat Mater 6:841–850.

Lucci M, Regoliosi R, Reale A, Carlo AD, Orlanducci S, Tamburri E, Terranova ML, Lugli P, Natale CD, D'Amico A, Paolesse R. 2005. Gas sensing using single wall carbon nanotubes ordered with dielectrophoresis. Sensor Actuat B-Chem 111:181–186.

Luo J, Zhu J. 2006. Arrays of one-dimensional metal/silicon and metal/carbon nanotube heterojunctions. Nanotechnology 17:S262–S270.

Luo S, Chu PK, Liu W, Zhang M, Lin CL. 2006. Origin of low-temperature photoluminescence from SnO_2 nanowires fabricated by thermal evaporation and annealed in different ambients. Appl Phys Lett 88:183112.

Ma DDD, Lee CS, Au FCK, Tong SY, Lee ST. 2003. Small-diameter silicon nanowire surfaces. Science 299:1874–1877.

Ma YJ, Zhou F, Lu L, Zhang Z. 2004. Low-temperature transport properties of individual SnO_2 nanowires. Solid State Commun 130:313–316.

Martin CR. 1991. Template synthesis of polymeric and metal microtubules. Adv Mater 3:457–459.

Mathur S, Barth S. 2007. Molecule-based chemical vapor growth of aligned SnO_2 nanowires and branched SnO_2/V_2O_5 heterostructures. Small 3:2070–2075.

Mathur S, Barth S. 2008. One-dimensional semiconductor nanostructures: growth, characterization and device applications. Z Phys Chem 222:307–317.

Mathur S, Shen H, Sivakov V, Werner U. 2004. Germanium nanowires and core-shell nanostructures by chemical vapor deposition of $[Ge(C_5H_5)_2]$. Chem Mater 16:2449–2456.

Mathur S, Barth S, Shen H, Pyun JC, Werner U. 2005. Size-dependent photo-conductance in SnO_2 nanowires. Small 1:713–717.

Mathur S, Shen H, Donia N, Rugamer T, Sivakov V, Werner U. 2007. One-step chemical vapor growth of Ge/SiC_xN_y nanocables. J Am Chem Soc 129:9746–9752.

Mathur S, Barth S, Werner U, Hernandez-Ramirez F, Romano-Rodriguez A. 2008. Chemical vapor growth of one-dimensional magnetite nanostructures. Adv Mater 20:1550–1554.

Maynor BW, Li JY, Lu CG, Liu J. 2004. Site-specific fabrication of nanoscale heterostructures: local chemical modification of GaN nanowires using electrochemical dip-pen nanolithography. J Am Chem Soc 126:6409–6413.

Minne SC, Manalis SR, Quate CF. 1995. Parallel atomic force microscopy using cantilevers with integrated piezoresistive sensors and integrated piezoelectric actuators. Appl Phys Lett 67:3918–3920.

Miyamoto Y, Hirata M. 1975. Growth of new form germanium whiskers. Japan J Appl Phys 14:1419–1420.

Modi A, Koratkar N, Lass E, Wei BQ, Ajayan PM. 2003. Miniaturized gas ionization sensors using carbon nanotubes. Nature 424:171–174.

Moore D, Morber JR, Snyder RL, Wang ZL. 2008. Growth of ultralong ZnS/SiO_2 core-shell nanowires by volume and surface diffusion VLS process. J Phys Chem C 112:2895–2903.

Morales AM, Lieber CM. 1998. A laser ablation method for the synthesis of crystalline semiconductor nanowires. Science 279:208–211.

Morral AFI, Arbiol J, Prades JD, Cirera A, Morante JR. 2007. Synthesis of silicon nanowires with wurtzite crystalline structure by using standard chemical vapor deposition. Adv Mater 19:1347–1351.

Nagashima K, Yanagida T, Tanaka H, Kawai T. 2007. Epitaxial growth of MgO nanowires by pulsed laser deposition. J Appl Phys 101:124304.

Ng HT, Han J, Yamada T, Nguyen P, Chen YP, Meyyappan M. 2004. Single crystal nanowire vertical surround-gate field-effect transistor. Nano Lett 4:1247–1252.

Nilsson HA, Duty T, Abay S, Wilson C, Wagner JB, Thelander C, Delsing P, Samuelson L. 2008. A radio frequency single-electron transistor based on an InAs/InP heterostructure nanowire. Nano Lett 8:872–875.

Nimmatoori P, Zhang Q, Dickey E, Redwing JM. 2009. Suppression of the vapor-liquid-solid growth of silicon nanowires by antimony addition. Nanotechnology 20:025607.

Noborisaka J, Motohisa J, Fukui T. 2005. Catalyst-free growth of GaAs nanowires by selective-area metalorganic vapor-phase epitaxy. Appl Phys Lett 86:213102.

Ogihara H, Sadakane M, Nodasaka Y, Ueda W. 2006. Shape-controlled synthesis of ZrO_2, Al_2O_3, and SiO_2 nanotubes using carbon nanofibers as templates. Chem Mater 18:4981–4983.

Ohlsson BJ, Björk MT, Persson AI, Thelander C, Wallenberg LR, Magnusson MH, Deppert K, Samuelson L. 2002. Growth and characterization of GaAs and InAs nano-whiskers and InAs/GaAs heterostructures. Physica E 13:1126–1130.

O'Mara WC, Haber RB, Hunt LP. 1990. *Handbook of Semiconductor Silicon Technology*. New York: William Andrew.

Omari M, Kouklin N, Lu G, Chen J, Gajdardziska-Josifovska M. 2008. Fabrication of Cd_3As_2 nanowires by direct vapor–solid growth, and their infrared absorption properties. Nanotechnology 19:105301.

Pan ZW, Dai ZR, Wang ZL. 2001. Nanobelts of semiconducting oxides. Science 291:1947–1949.

Panev N, Persson AI, Skold N, Samuelson L. 2003. Sharp exciton emission from single InAs quantum dots in GaAs nanowires. Appl Phys Lett 83:2238–2240.

Park WI, Yi GC. 2004. Electroluminescence in n-ZnO nanorod arrays vertically grown on p-GaN. Adv Mater 16:87–90.

Park WI, Kim DH, Jung SW, Yi GC. 2002. Metalorganic vapor-phase epitaxial growth of vertically well-aligned ZnO nanorods. Appl Phys Lett 80:4232–4234.

Park WI, Kim JS, Yi GC, Bae MH, Lee HJ. 2004. Fabrication and electrical characteristics of high-performance ZnO nanorod field-effect transistors. Appl Phys Lett 85:5052–5054.

Park WI, Yi GC, Kim M, Pennycook SJ. 2003. Quantum confinement observed in ZnO/ZnMgO nanorod heterostructures. Adv Mater 15:526–529.

Patolsky F, Zheng G, Lieber CM. 2006. Fabrication of silicon nanowire devices for ultrasensitive, label-free, real-time detection of biological and chemical species. Nat Protoc 1:1711–1724.

Patolsky F, Timko BP, Zheng G, Lieber CM. 2007. Nanowire-based nanoelectronic devices in the life sciences. MRS Bull 32:142–149.

Peng XS, Meng GW, Zhang J, Wang XF, Wang YW, Wang CZ, Zhang LD. 2002a. Synthesis and photoluminescence of single-crystalline In_2O_3 nanowires. J Mater Chem 12:1602–1605.

Peng XS, Wang XF, Wang YW, Wang CZ, Meng GW, Zhang LD. 2002b. Novel method synthesis of CdO nanowires. J Phys D-App Phys 35:101–104.

Peng XS, Zhang LD, Meng GW, Wang XF, Wang YW, Wang CZ, Wu GS. 2002c. Photoluminescence and infrared properties of α-Al_2O_3 nanowires and nanobelts. J Phys Chem B 106:11163–11167.

Peng KQ, Yan YJ, Gao SP, Zhu J. 2002d. Synthesis of large-area silicon nanowire arrays via self-assembling nanoelectrochemistry. Adv Mater 14:1164–1167.

Persson AI, Larsson MW, Stenstrom S, Ohlsson BJ, Samuelson L, Wallenberg LR. 2004. Solid-phase diffusion mechanism for GaAs nanowire growth. Nat Mater 3:677–681.

Pohl PA. 1951. The motion and precipitation of suspensoids in divergent electric fields. J Appl Phys 22:869–871.

Pohl HA. 1978. *Dielectrophoresis*. London: Cambridge University Press.

Poole PJ, Lefebvre J, Fraser J. 2003. Spatially controlled, nanoparticle-free growth of InP nanowires. Appl Phys Lett 83:2055–2057.

Prades JD, Jimenez-Diaz R, Hernandez-Ramirez F, Fernandez-Romero L, Andreu T, Cirera A, Romano-Rodriguez A, Cornet A, Morante JR, Barth S, Mathur S. 2008a. Toward a systematic undrestanding of photodetectors based on individual metal oxide nanowires. J Phys Chem C 112:14639–14644.

Prades JD, Jimenez-Diaz R, Hernandez-Ramirez F, Barth S, Cirera A, Romano-Rodriguez A, Mathur S, Morante JR. 2008b. Ultralow power consumption gas sensors based on self-heated individual nanowires. Appl Phys Lett 93:123110.

Prades JD, Hernandez-Ramirez F, Jimenez-Diaz R, Manzanares M, Andreu T, Cirera A, Romano-Rodriguez A, Morante JR. 2008c. The effects of electron–hole separation on the photoconductivity of individual metal oxide nanowires. Nanotechnology 19:465501.

Qi B, Ni X, Li D, Zheng H. 2008. A facile non-hydrothermal fabrication of uniform α-MoO_3 nanowires in high yield. Chem Lett 37:336–340.

Qian F, Li Y, Gradec S, Wang D, Barrelet CJ, Lieber CM. 2004. Gallium nitride-based nanowire radial heterostructures for nanophotonics. Nano Lett 4:1975–1979.

Qian F, Gradecak S, Li Y, Lieber CM. 2005. Core/multishell nanowire heterostructures as multicolor, high-efficiency light-emitting diodes. Nano Lett 5:2287–2291.

Rubio-Bollinger G, Bahn SR, Agrait N, Jacobsen KW, Vieira S. 2001. Mechanical properties and formation mechanisms of a wire of single gold atoms. Phys Rev Lett 87:026101.

Satishkumar BC, Govindaraj A, Vogl EM, Basumallick L, Rao CNR. 1997. Oxide nanotubes prepared using carbon nanotubes as templates. J Mater Res 12:604–606.

Schmidt V, Senz S, Gosele U. 2005a. UHV chemical vapour deposition of silicon nanowires. Z Metallkd 96:427–428.

Schmidt V, Senz S, Gosele U. 2005b. Diameter-dependent growth direction of epitaxial silicon nanowires. Nano Lett 5:931–935.

Schmidt V, Riel H, Senz S, Karg S, Riess W, Gösele U. 2006. Realization of a silicon nanowire vertical surround-gate field-effect transistor. Small 2:85–88.

Sears GW. 1955. A growth mechanism for mercury whiskers. Acta Metall 3:361–366.

Seo HW, Han CS, Hwang SO, Park J. 2006a. Dielectrophoretic assembly and characterization of individually suspended Ag, GaN, SnO_2 and Ga_2O_3 nanowires. Nanotechnology 17: 3388–3393.

Seo HW, Han CS, Jang WS, Park J. 2006b. Synthesis of gold nanoparticles using N,N-dimethylacetoacetamide: size and shape control by the reaction temperature. Curr Appl Phys 6:216–218.

Sham TK, Naftel SJ, Kim PSG, Sammynaiken R, Tang YH, Coulthard I, Moewes A, Freeland JW, Hu YF, Lee ST. 2004. Electronic structure and optical properties of silicon nanowires: a study using x-ray excited optical luminescence and x-ray emission spectroscopy. Phys Rev B 70:045313.

Shen GZ, Bando Y, Ye C, Yuan X, Sekiguchi T, Golberg D. 2006. Single-crystal nanotubes of II_3-V_2 semiconductors. Angew Chem Int Ed 45:7568–7572.

Shi SC, Chen CF, Chattopadhyay S, Chen KH, Chen LC. 2005. Field emission from quasi-aligned aluminum nitride nanotips. Appl Phys Lett 87:073109.

Shi CS, Wang GQ, Zhao NQ, Du XW, Li JJ. 2008. NiO nanotubes assembled in pores of porous anodic alumina and their optical absorption properties. Chem Phys Lett 454:75–79.

Smith PA, Nordquist CD, Jackson TN, Mayer TS, Martin BR, Mbindyo J, Mallouk TE. 2000. Electric-field assisted assembly and alignment of metallic nanowires. Appl Phys Lett 77:1399–1401.

Soci C, Zhang A, Xiang B, Dayeh SA, Aplin DPR, Park J, Bao XY, Lo YH, Wang D. 2007. ZnO Nanowire UV photodetectors with high internal gain. Nano Lett 7:1003–1009.

Solanki R, Huo J, Freeouf JL, Miner B. 2002. Atomic layer deposition of ZnSe/CdSe superlattice nanowires. Appl Phys Lett 81:3864–3866.

Song XC, Zheng YF, Yang E, Wang Y. 2007. Large-scale hydrothermal synthesis of WO_3 nanowires in the presence of K_2SO_4. Mater Lett 61:3904–3908.

Stokes P, Khondaker SI. 2008. Local-gated single-walled carbon nanotube field effect transistors assembled by AC dielectrophoresis. Nanotechnology 19:175202–175201.

Suehiro J, Nakagawa N, Hidaka S, Ueda M, Imasaka K, Higashihata M, Okada T, Hara M. 2006. Dielectrophoretic fabrication and characterization of a ZnO nanowire-based UV photosensor. Nanotechnology 17:2567–2573.

Sun Y, Rogers JA. 2004. Fabricating semiconductor nano/microwires and transfer printing ordered arrays of them onto plastic substrates. Nano Lett 4:1953–1959.

Sun Y, Gates B, Mayers B, Xia Y. 2002. Crystalline silver nanowires by soft solution processing. Nano Lett 2:165–168.

Sun S, Meng GW, Zhang G, Gao T, Geng B, Zhang LD, Zuo J. 2003. Raman scattering study of rutile SnO_2 nanobelts synthesized by thermal evaporation of Sn powders. Chem Phys Lett 376:103–107.

Sun Y, Fuge GM, Fox NA, Riley DJ, Ashfold MNR. 2005. Synthesis of aligned arrays of ultrathin ZnO nanotubes on a Si wafer coated with a thin ZnO film. Adv Mater 17:2477–2481.

Sun XH, Yu B, Ng G, Nguyen TD, Meyyappan M. 2006. III-VI compound semiconductor indium selenide (In_2Se_3) nanowires: synthesis and characterization. Appl Phys Lett 89:233121.

Suyatin DB, Sun J, Fuhrer A, Wallin A, Fröberg LE, Karlsson LS, Maximov I, Wallenberg LR, Samuelson L, Xu HQ. 2008. Electrical properties of self-assembled branched InAs nanowire junctions. Nano Lett 8:1100–1104.

Sysoev VV, Button BK, Wepsiec K, Dmitriev S, Kolmakov A. 2006. Toward the Nanoscopic "Electronic Nose": hydrogen vs carbon monoxide discrimination with an array of individual metal oxide nano- and mesowire sensors. Nano Lett 6:1584–1588.

Sysoev VV, Goschnick J, Schneider T, Strelcov E, Kolmakov A. 2007. A gradient microarray electronic nose based on percolating SnO_2 nanowire sensing elements. Nano Lett 7: 3182–3188.

Tan YW, Xue XY, Peng Q, Zhao H, Wang TH, Li YD. 2007. Controllable fabrication and electrical performance of single crystalline Cu_2O nanowires with high aspect ratios. Nano Lett 7: 3723–3728.

Tang Q, Chen XH, Li T, Zhao AW, Qian QT, Yu DP, Yu WC. 2004. Template-free growth of vertically aligned CdS nanowire array exhibiting good field emission property. Chem Lett 33: 1088–1089.

Thomas A, Premlal B, Eswaramoorthy M. 2006. Synthesis of mesoporous Zn-Al spinel oxide nanorods with membrane like morphology. Mater Res Bull 41:1008–1014.

Tian B, Zheng X, Kempa TJ, Fang Y, Yu N, Yu G, Huang J, Lieber CM. 2007. Coaxial silicon nanowires as solar cells and nanoelectronic power sources. Nature 449:885–890.

Trentler TJ, Hickmann KM, Goel SC, Viano AM, Gibbons PC, Buhro WE. 1995. Solution-liquid-solid growth of crystalline III-V semiconductors- an analogy to vapor-liquid-solid growth. Science 270:1791–1794.

Tseng YK, Huang CJ, Cheng HM, Lin IN, Liu KS, Chen IC. 2003. Characterization and field-emission properties of needle-like zinc oxide nanowires grown vertically on conductive zinc oxide films. Adv Funct Mater 13:811–814.

Tuan HY, Lee DC, Korgel BA. 2006. Nanocrystal-mediated crystallization of silicon and germanium nanowires in organic solvents: the role of catalysis and solid-phase seeding. Angew Chem Int Ed 45:5184–5187.

Tung S, Rokadia H, Li WJ. 2007. A micro shear stress sensor based on laterally aligned carbon nanotubes. Sensor Actuat A-Phys 133:431–438.

Varghese B, Haur SC, Lim CT. 2008. Nb_2O_5 nanowires as efficient electron field emitters. J Phys Chem C 112:10008–10012.

Vayssieres L. 2003. Growth of arrayed nanorods and nanowires of ZnO from aqueous solutions. Adv Mater 15:464–466.

Vayssieres L, Graetzel M. 2004. Highly ordered SnO_2 nanorod arrays from controlled aqueous growth. Angew Chem Int Ed 43:3666–3670.

Vayssieres L, Beermann N, Lindquist SE, Hagfeld A. 2001. Controlled aqueous chemical growth of oriented three-dimensional crystalline nanorod arrays: application to iron(III) oxides. Chem Mater 13:133–235.

Wagner RS, Ellis WC. 1964. Vapor-liquid-solid mechanism of single crystal growth. Appl Phys Lett 4:89–90.

Wang ZL. 2008a. Towards self-powered nanosystems: from nanogenerators to nanopiezotronics. Adv Funct Mater 18:1–15.

Wang ZL. 2008b. Oxide nanobelts and nanowires—growth, properties and applications. J Nanosci Nanotechnol 8:27–55.

Wang X, Li YD. 2002. Selected-control hydrothermal synthesis of α- and β-MnO_2 single crystal nanowires. J Am Chem Soc 124:2880–2881.

Wang C, Sun SH. 2009. Facile synthesis of ultrathin and single-crystalline Au nanowires. Chem An Asian J 4:1028–1034.

Wang J, Gudiksen MS, Duan X, Cui Y, Lieber CM. 2001. Highly polarized photoluminescence and photodetection from single indium phosphide nanowires. Science 293:1455–1457.

Wang WZ, Wang GH, Wang XS, Zhan YJ, Liu YK, Zheng CL. 2002. Synthesis and characterization of Cu_2O nanowires by a novel reduction route. Adv Mater 14:67–69.

Wang YW, Zhang LD, Wang GZ, Peng XS, Chu ZQ, Liang CH. 2002. Catalytic growth of semiconducting zinc oxide nanowires and their photoluminescence properties. J Cryst Growth 234:171–175.

Wang YL, Jiang XC, Xia Y. 2003. A solution-phase, precursor route to polycrystalline SnO_2 nanowires that can be used for gas sensing under ambient conditions. J Am Chem Soc 125:16176–16177.

Wang J, Chen Q, Zeng C, Hou B. 2004a. Magnetic-field-induced growth of single-crystalline Fe_3O_4 nanowires. Adv Mater 16:137–140.

Wang X, Ding Y, Summers CJ, Wang ZL. 2004b. Large-scale synthesis of six-nanometer-wide ZnO nanobelts. J Phys Chem B 108:8773–8777.

Wang B, Yang YH, Wang CX, Yang GW. 2005. Nanostructures and self-catalyzed growth of SnO_2. J Appl Phys 98:073520–073521.

Wang YW, Schmidt V, Senz S, Gosele U. 2006. Epitaxial growth of silicon nanowires using an aluminium catalyst. Nat Nanotechnol 3:186–189.

Wang N, Cao X, Guo L. 2008. Facile one-pot solution phase synthesis of SnO_2 nanotubes. J Phys Chem C 112:12616–12622.

Wei M, Qi ZM, Ichihara M, Zhou H. 2008. Synthesis of single-crystal niobium pentoxide nanobelts. Acta Mater 56:2488–2494.

Wen X, Wang S, Ding Y, Wang ZL, Yang S. 2005. Controlled growth of large-area, uniform, vertically aligned arrays of α-Fe_2O_3 nanobelts and nanowires. J Phys Chem B 109:215–220.

Whang D, Jin S, Wu Y, Lieber CM. 2003. Large-scale hierarchical organization of nanowire arrays for integrated nanosystems. Nano Lett 3:1255–1259.

Wu YY, Yang PD. 2000. Germanium nanowire growth via simple vapor transport. Chem Mater 12:605–607.

Wu YY, Yang PD. 2001. Direct observation of vapor-liquid-solid nanowire growth. J Am Chem Soc 123:3165–3166.

Wu YY, Fan R, Yang PD. 2002. Block-by-block growth of single-crystalline Si/SiGe superlattice nanowires. Nano Lett 2:83–86.

Wu Y, Cui Y, Huynh L, Barrelet CJ, Bell DC, Lieber CM. 2004a. Controlled growth and structures of molecular-scale silicon nanowires. Nano Lett 4:433–436.

Wu Y, Xiang J, Yang C, Lu W, Lieber CM. 2004b. Single-crystal metallic nanowires and metal/semiconductor nanowire heterostructures. Nature 430:61–65.

Wu JM, Shih HC, Wu WT. 2005. Electron field emission from single crystalline TiO_2 nanowires prepared by thermal evaporation. Chem Phys Lett 413:490–494.

Xia YN, Yang PD, Sun YG, Wu YY, Mayers B, Gates B, Yin YD, Kim F, Yan HQ. 2003. One-dimensional nanostructures: synthesis, characterization and applications. Adv Mater 15: 353–389.

Xiang J, Lu W, Hu Y, Wu Y, Yan H, Lieber CM. 2006. Ge/Si nanowire heterostructure has high-performance field-effect transistors. Nature 441:489–493.

Xing YJ, Xi ZH, Xue ZQ, Zhang XD, Song JH, Wang RM, Xu J, Song Y, Zhang SL, Yu DP. 2003. Optical properties of the ZnO nanotubes synthesized via vapor phase growth. Appl Phys Lett 83:1689–1691.

Xiong Y, Xie Y, Li Z, Zhang R, Yang J, Wu C. 2003a. Complexing-reagent assisted synthesis of α-Fe and γ-Fe_2O_3 nanowires under mild conditions. New J Chem 27:588–590.

Xiong YJ, Li ZQ, Zhang R, Xie Y, Yang J, Wu CZ. 2003b. From complex chains to 1D metal oxides: a novel strategy to Cu_2O nanowires. J Phys Chem B 107:3697–3702.

Xiong Q, Wang J, and Eklund PC. 2006. Coherent twinning phenomena: towards twinning superlattices in III–V semiconducting nanowires. Nano Lett 6:2736–2742.

Xu CK, Hong KQ, Liu S, Wang GH, Zhao XN. 2003. A novel wet chemical route to NiO nanowires. J Cryst Growth 255:308–312.

Xu L, Su Y, Li S, Chen YQ, Zhou QT, Yin S, Feng Y. 2007. Self-assembly and hierarchical organization of Ga_2O_3/In_2O_3 nanostructures. J Phys Chem B 111:760–766.

Yang PD. 2005. The chemistry and physics of semiconductor nanowires. MRS Bull 30:85–91.

Yang PD, Yan HQ, Mao S, Russo R, Johnson J, Saykally R, Morris N, Pham J, He RR, Cho HJ. 2002. Controlled Growth of ZnO Nanowires and Their Optical Properties. Adv Funct Mater 12:323–331.

Yang Q, Sha J, Ma XY, Yang DR. 2005. Synthesis of NiO nanowires by a sol-gel. Process Mater Lett 59:1967–1970.

Yasuda R, Noji H, Yoshida M, Kinosita K, Itoh H. 2001. Resolution of distinct rotational substeps by submillisecond kinetic analysis of F1-ATPase. Nature 410:898–904.

Yin Y, Zhang G, Xia YN. 2002. Synthesis and characterization of MgO nanowires through a vapor-phase precursor method. Adv Funct Mater 12:293–298.

Yu H, Buhro WE. 2003. Solution-liquid-solid growth of soluble GaAs nanowires. Adv Mater 15: 416–419.

Yu JY, Chung SW, Heath JR. 2000. Silicon nanowires: preparation, device fabrication, and transport properties. J Phys Chem B 104:11864–11870.

Yu B, Sun XH, Calebotta GA, Dholakia GR, Meyyappan M. 2006. One-dimensional germanium nanowires for future electronics. J Clust Sci 17:579–597.

Zakharov ND, Werner P, Gerth G, Schubert L, Sokolov L, Gosele U. 2006. Growth phenomena of Si and Si/Ge nanowires on Si(111) by molecular beam epitaxy. J Cryst Growth 290: 6–10.

Zhang YF, Tang YH, Wang N, Yu DP, Lee CS, Bello I, Lee ST. 1998. Silicon nanowires prepared by laser ablation at high temperatures. Appl Phys Lett 72:1835–1837.

Zhang HZ, Kong YC, Wang YZ, Du X, Bai ZG, Wang JJ, Yu DP, Ding Y, Hang QL, Feng SQ. 1999a. Ga_2O_3 nanowires prepared by physical evaporation. Solid State Commun 109: 677–682.

Zhang YF, Tang YF, Wang N, Lee CS, Bello I, Lee ST. 1999b. One-dimensional growth mechanism of crystalline silicon nanowires. J Cryst Growth 197:136–140.

Zhang YJ, Zhang Q, Wang NL, Yan YJ, Zhou HH, Zhu J. 2001. Synthesis of thin Si whiskers (nanowires) using $SiCl_4$. J Cryst Growth 226:185–191.

Zhang Y, Li G, Jin Y, Zhang Y, Zhang J, Zhang LD. 2002. Hydrothermal synthesis and photoluminescence of TiO_2 nanowires. Chem Phys Lett 365:300–304.

Zhang DF, Sun LD, Yin JL, Yan CH. 2003. Low-temperature fabrication of highly crystalline SnO2 nanorods. Adv Mater 15:1022–1025.

Zhang J, Qing X, Jiang FH, Dai ZH. 2003. A route to Ag-catalyzed growth of the semiconducting In_2O_3 nanowires. Chem Phys Lett 371:311–316.

Zhang H, Yang D, Ji YJ, Ma XY, Xu J, Que DL. 2004a. Low temperature synthesis of flowerlike ZnO nanostructures by cetyltrimethylammonium bromide-assisted hydrothermal process. J Phys Chem B 108:3955–3958.

Zhang H, Boussaad S, Ly N, Tao N. 2004b. Magnetic-field-assisted assembly of metal/polymer/metal junction sensors. Appl Phys Lett 84:133–135.

Zhang XY, Dai JY, Ong HC, Wang N, Chan HLW, Choy CL. 2004c. Hydrothermal synthesis of oriented ZnO nanobelts and their temperature dependent photoluminescence. Chem Phys Lett 393:17–21.

Zhang H, Yang DR, Ma XY, Que DL. 2005a. Synthesis and field emission characteristics of bilayered ZnO nanorod array prepared by chemical reaction. J Phys Chem B 109: 17055–17059.

Zhang J, Jiang FH, Yang Y, Li J. 2005b. Catalyst-assisted vapor-liquid-solid growth of single-crystal Ga_2O_3 nanobelts. J Phys Chem B 109:13143–13147.

Zhang YF, Russo RE, Mao SS. 2005c. Femtosecond laser assisted growth of ZnO nanowires. Appl Phys Lett 87:133115–133111.

Zhang Z, Yao K, Liu Y, Jin C, Liang X, Chen Q, Peng LM. 2007. Quantitative analysis of current-voltage characteristics of semiconducting nanowires: decoupling of contact effects. Adv Funct Mater 17:2478–2489.

Zhang H, Wu J, Zhai C, Ma X, Du N, Tu J, Yang DR. 2008. From cobalt nitrate carbonate hydroxide hydrate nanowires to porous Co_3O_4 nanorods for high performance lithium-ion battery electrodes. Nanotechnology 19:035711.

Zhao MH, Wang ZL, Mao SX. 2004a. Piezoelectric characterization of individual zinc oxide nanobelt probed by piezoresponse force microscope. Nano Lett 4:587–590.

Zhao XY, Wei CM, Yang L, Chou MY. 2004b. Quantum confinement and electronic properties of silicon nanowires. Phys Rev Lett 92:236805.

Zhen L, He K, Xu CY, Shao WZ. 2008. Synthesis and characterization of single-crystalline $MnFe_2O_4$ nanorods via a surfactant-free hydrothermal route. J Magnet Magnet Mater 320: 2672–2675.

Zheng MJ, Li GH, Zhang XY, Huang SY, Lei Y, Zhang LD. 2001. Fabrication and structural characterization of large-scale uniform SnO_2 nanowire array embedded in anodic alumina membrane. Chem Mater 13:3859–3861.

Zheng MJ, Zhang LD, Li GH, Shen WZ. 2002. Fabrication and optical properties of large scale uniform zinc oxide nanowire arrays by one-step electrochemical deposition technique. Chem Phys Lett 363:123–128.

Zheng D, Sun SX, Fan W, Yu H, Fan C, Cao G, Yin Z, Song X. 2005. One-step preparation of single-crystalline β-MnO_2 nanotubes. J Phys Chem B 109:16439–16443.

Zheng XG, Li Q, Zhao JP, Chen D, Zhao B, Yang YJ, Zhang LC. 2006. Photoconductive ultraviolet detectors based on ZnO films. Appl Surf Sci 253:2264–2267.

Zhou J, Deng S, Xu NS, Chen J, She JC. 2003. Synthesis and field-emission properties of aligned MoO_3 nanowires. Appl Phys Lett 83:2653–2655.

Zhou C, Gundersen MA, Curreli M, Thompson ME, Li C, Lei B, Sun Y. 2005. Selective functionalization of In_2O_3 nanowire mat devices for biosensing applications. J Am Chem Soc 127: 6920–6921.

Zhou F, Zhao X, Liu YM, Yuan CQ, Li L. 2008. Synthesis of millimeter-range orthorhombic V_2O_5 nanowires and impact of thermodynamic and kinetic properties of the oxidant on the synthetic process. Eur J Inorg Chem 16:2506–2509.

Zhou S, Liu XH, Lin YJ, Wang DW. 2009. Rational synthesis and structural characterizations of complex $TiSi_2$ nanostructures. Chem Mater 21:1023–1027.

Zhu GQ, Liu P. 2009. Low-temperature urea-assisted hydrothermal synthesis of Bi_2S_3 nanostructures with different morphologies. Crys Resea Techn 44:713–720.

Zhu K, He H, Xie S, Zhang X, Zhou W, Jin S, Yue B. 2003a. Crystalline WO_3 nanowires synthesized by templating method. Chem Phys Lett 377:317–321.

Zhu YW, Zhang HZ, Sun XC, Feng SQ, Xu J, Zhao Q, Xiang B, Wang RM, Yu DP. 2003b. Efficient field emission from ZnO nanoneedle arrays. Appl Phys Lett 83:144.

Ziegler K, Ryan KM, Rice R, Crowley T, Erts D, Olin H, Patterson J, Spalding TR, Holmes JD, Morris MA. 2004. The synthesis of matrices of embedded semiconducting nanowires. Faraday Discuss 125:311–326.

19

INTEGRATING DIAMOND-LIKE CARBON INTO NANOSTRUCTURE DESIGNS (FABRICATING MICROSCALE AND NANOSCALE ARCHITECTURES OF DIAMOND-LIKE CARBON FILMS)

Xijun Li and Daniel H. C. Chua

National University of Singapore, Singapore

BASICS OF MICROMECHANICAL AND NANOMECHANICAL DEVICES

A mechanical device utilizes the mechanical motion or distortion of a solid or liquid component to excite or sense its environment, thus performing as an actuator or a sensor. One of the simplest mechanical devices is a cantilever as illustrated in Figure 19.1. With one of its ends fixed on a pad, the cantilever can bend in the x–z plane. Its motion can be well described (Genta 1999) by

$$\rho A \frac{\partial^2 u_x(z,t)}{\partial^2 t} + EI_y \frac{\partial^4 u_x(z,t)}{\partial^4 z} = 0, \tag{19.1}$$

where $u_x(z, t)$ is the movement of the cantilever along the x-direction at position z and at time t. The origin of the coordinates is set on the pad; E, A, ρ, and I_y are Young's

Ceramic Integration and Joining Technologies: From Macro to Nanoscale, First Edition.
Edited by Mrityunjay Singh, Tatsuki Ohji, Rajiv Asthana, Sanjay Mathur.

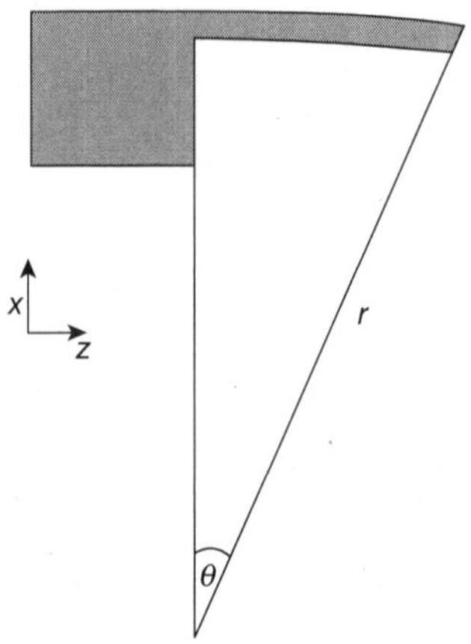

Figure 19.1. Schematics of a cantilever.

TABLE 19.1. Digitalization of a Silicon-Based Cantilever

	t = 200 nm			t = 350 nm		
	l = 1 μm	l = 4 μm	l = 8 μm	l = 1 μm	l = 4 μm	l = 8 μm
Frequency (MHz)	1650	103	28	2887	180	87

modulus, the cross-section area, density, and moment of inertia of the cantilever, respectively. For a cantilever with a rectangular cross-section, the moment of inertia can be expressed as

$$I_y = \frac{wt^3}{12}. \tag{19.2}$$

When the movement $u_x(z, t)$ of the cantilever is much smaller than the radius r illustrated in Figure 19.1 and the cantilever's sizes in the x- (thickness t) and y- (width w) directions are much smaller than its length l along the z-axis, with no stress or force applied in the x–y plane, the resonant motion can be described by

$$\omega_i = \frac{\beta_i^2}{l^2}\sqrt{\frac{EI_y}{\rho A}}. \tag{19.3}$$

This resonant motion can be excited with an intrinsic mode; β_i represents the mode factor of the vibration, and, for a cantilever, its first four values are 3.5, 22.0, 61.7, and 121.0, respectively.

For such a nanometer-scale rectangular cantilever fabricated in silicon, Table 19.1 lists the corresponding fundamental resonance frequencies of different cantilever sizes.

Clearly, as the dimensions of such a cantilever are reduced, the resonance frequency is in the gigahertz range. This is important for applications in wireless telecommunications. The current drawback of electronic resonator devices working in the same frequency range is that they have a quality factor of less than 100, while nanomechanical resonator devices can have a quality factor greater than 10,000 (De Los Santos 1999).

From Equations 19.2 and 19.3, it is clear that the resonance frequencies of a cantilever with a rectangular cross-section satisfy

$$\varpi \propto \frac{t}{l^2}\sqrt{\frac{E}{\rho}}. \tag{19.4}$$

In the preceding descriptions, the cantilever device is too simple to include the damping process taking place in the mechanical motions. To analyze the effect of damping on the physics of mechanical devices, Roukes and Cleland (Cleland and Roukes 1999) developed a harmonic-like solution for cantilevers. By incorporating Coulomb damping (proportional to displacement) into the effective Young's modulus as

$$E_{\text{eff}} = E\left(1+i\frac{1}{Q_{\text{coul}}}\right) \tag{19.5}$$

and viscous damping into the friction force as

$$F_{\text{friction}} = \frac{m\varpi_0}{Q_{\text{vis}}}\frac{dx(t)}{dt}, \tag{19.6}$$

where $x(t)$ is the displacement, they found that for large Q, the cantilever displacement can be effectively expressed as

$$x_{\max}(\varpi) \propto \frac{1}{\sqrt{4\left(1-\frac{\varpi}{\varpi_0}\right)^2+\left(\frac{1}{Q_{\text{total}}}\right)^2}}, \tag{19.7}$$

where Q_{tot} satisfies

$$\frac{1}{Q_{\text{tot}}} = \frac{1}{Q_{\text{coul}}}+\frac{1}{Q_{\text{vis}}} \tag{19.8}$$

and Q_{tot} is the total damping effect on the mechanical vibration. From Equation 19.7, it is obvious that the damping effect degrades the frequency selectivity of the mechanical vibrator from the ideal ω_0 to a frequency band $\Delta\omega$ around ω_0. The Q_{tot} is sometimes known as the quality factor of the mechanical device.

In a real mechanical device, both the resonance frequency and the quality factor are needed to characterize the device. Both of these factors can be applied to sensors independently. From Equation 19.4, it clearly shows that the factors affecting the

TABLE 19.2. A Comparison of the Various Properties of Diamond-Like Carbon Films, Amorphous Silicon, and Microcrystalline Diamond Films

Properties	DLC[a]	a-Si	Diamond
Maximum density (g/cm^3)	~3.4	~2.2	~3.5
Hardness (GPa)	~60	~10	110
Young's modulus (GPa)	~800	~5.9	>1000
Optical band gap (eV)	>3.5	1.8	5.4

[a]The DLC in this case refers to nonhydrogenated diamond-like films with high sp^3 content (>80%), sometimes also known as tetrahedral amorphous carbon or amorphous diamond films.

resonance frequency are not dependent on the dimensions of the microstructure; the material utilized plays a critical role. By properly selecting an appropriate material to build micromechanical and nanomechanical devices, a high-frequency range can also be obtained. Applications of these high-frequency devices can extend from wireless communication to highly sensitive biosensors or chemical sensors, for sensing small mass objects such as cells or even a single biomolecule (Anderson et al. 2007). To have the highest possible range of working frequency for an increase in sensitivity, materials with a higher Young's modulus and low mass density are desirable. Table 19.2 compares some of the known materials for biosensor applications. Both diamond and diamond-like carbon (DLC) are better candidates for fabricating such high-frequency sensors. However, the process of depositing large-area films is much more costly and complicated for diamond films than for DLC films. Figure 19.2a shows a microcantilever fabricated using a 1-μm-thick DLC film, and Figure 19.2b shows the simulated vibrational frequencies of both a polysilicon and the DLC cantilever. Experimental results are consistent with the simulation (Chua et al. 2003; Chua et al. 2004a). This demonstrates that DLC has significant advantages over silicon at similar dimensions when applied to mechanical devices for sensors.

As a mechanical device is reduced from micrometer to nanometer scales, the damping effect becomes much more serious, which will degrade device performance. For example, current silicon microstructures may face a higher failure rate when their micron-size dimensions are further miniaturized into the nanometer scale. One possible cause is surface stiction, arising from the unbonded residual surface hydrogen bond. On the other hand, DLC is highly hydrophobic with a contact angle as high as 81° (Han 1999). As such, the stiction issue for DLC-based nanomechanical devices may pose a less serious problem than for silicon devices. In addition, the hardness of DLC materials is much higher than that of silicon and other noble metals. It is expected that DLC materials are potential candidates for microelectromechanical system (MEMS) or nanoelectromechanical system (NEMS) applications.

PREPARATION AND PROPERTIES OF DLC

DLC is a form of carbon material with a structure similar to that of diamond and in which most of the carbon atoms are bonded in sp^3 hybridization. It has been widely

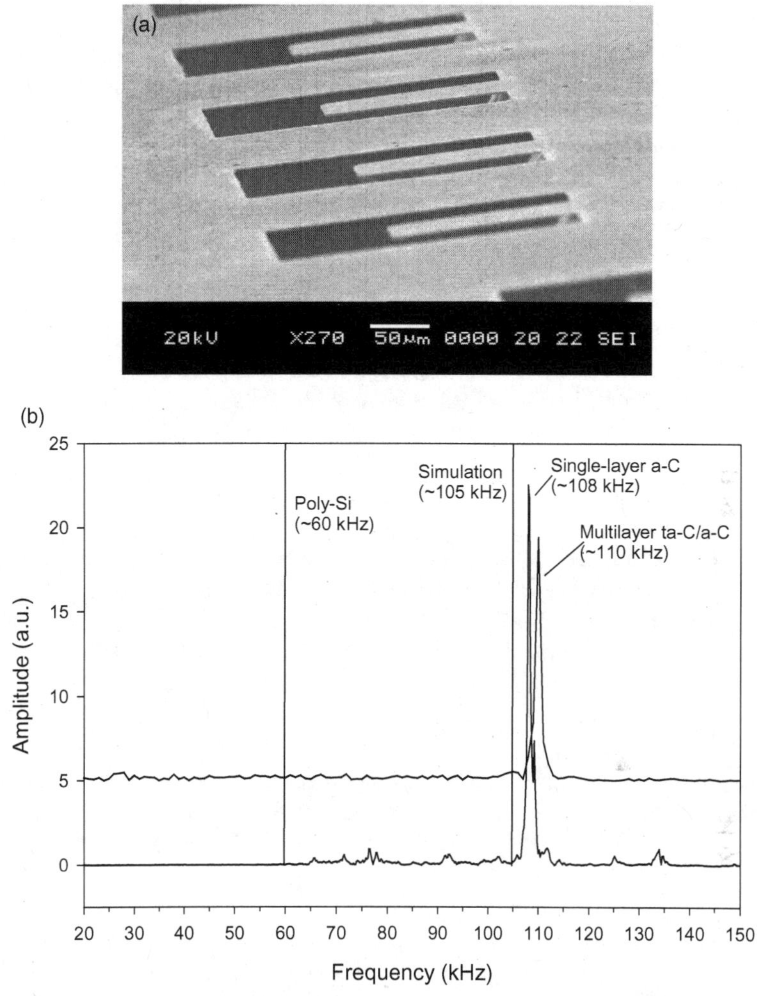

Figure 19.2. (a) A SEM image of a fabricated cantilever of DLC and (b) measured first-mode resonance frequency of two different types of diamond-like carbon cantilevers, a single-layer 1.5-μm-thick PIII a-C and a multilayer 1.5-μm-thick ta-C/a-C fabricated microcantilevers. The simulation results and the resonance from a poly-Si beam are also shown for comparison.

studied since 2000 in the past decade and is also sometimes referred to as amorphous diamond (a-D), tetrahedral amorphous carbon (ta-C), or diamond-like amorphous carbon (DL-aC). There are several review papers on the synthesis of DLC films (Funada et al. 1994; Milne et al. 1996; Martin and Bendavid 2001; Saw et al. 2000; Robertson 2002), which address the film growth methods, such as pulsed laser deposition (PLD) (Milne et al. 1996; Martin and Bendavid 2001; Robertson 2002), filtered cathodic vacuum arc (FCVA) (Milne et al. 1996; Martin and Bendavid 2001; Robertson 2002), radio frequency (RF) magnetron sputtering (Saw et al. 2000), and mass-selected ion-beam (MSIB) techniques (Funada et al. 1994; Milne et al. 1996; Martin and Bendavid 2001; Robertson 2002). DLC films deposited by these techniques tend to have little hydrogen (hence the term "nonhydrogenated") and to be amorphous with a high sp^3 (>80%) content. Films grown by these methods are very smooth and are sometimes considered to be atomically flat (Casiraghi et al. 2005). Through these different techniques, DLC films can be easily deposited on the surface of many materials. In this chapter, we only concentrate on DLC films deposited using the FCVA technique because of the ease of deposition as well as its ability to deposit on large-area substrates.

Due to the high fraction of sp^3 bonding, DLC films fabricated with FCVA have further been shown to have high abrasion resistance, an ultrasmooth surface, and a low coefficient of friction (COF). These films have found several niche applications, including protective coating materials for tribological tools such as razor blades (Hauert 2004; Casiraghi et al. 2005). In addition to its excellent wear resistance, DLC is a very hard material with a hardness close to diamond (Martin et al. 1988). Likewise, it is also chemically inert, making it a very effective high-performance material for harsh environmental applications.

Some of the engineering applications for these DLC films include MEMS and electron emission devices. For the former, DLC films deposited using FCVA techniques give a high sp^3 content of greater than 80%, with a density of ~3.2 g/cm^3 and Young's modulus as high as 800 GPa (see Table 19.2). These properties show that for two microresonator devices having similar dimensions but are made of different materials, a higher working frequency response will be obtained from one made from DLC material compared with other standard materials such as silicon or silicon nitride. In addition, DLC films have been shown to be hydrophobic (described further in Section 19.3) and wear resistant, which is a further advantage for DLC-based MEMS/NEMS moving assemblies. Some silicon MEMS devices have close contact with each other (e.g., wheels), which will create wear and tear as well as stiction problems. This limits the lifetime of the moving components of silicon-based MEMS devices. A comprehensive review of MEMS applications can be found in Luo et al. (2007) and in Peiner et al. (2007). In this section, we discuss the synthesis and properties of DLC films.

DLC Film: Synthesis

DLC Films Produced by FCVA Technique. The FCVA technique is among the most popular and widely used tools for the deposition of DLC films. A schematic diagram of the FCVA system is given in Figure 19.3. Several reviews give more information on the history and development of FCVA as well as various thin-film applications (Milne et al. 1996; Robertson 2002).

The process of depositing DLC films using FCVA usually requires a pure carbon target. The target is loaded onto a cathode, which is loaded in the arc chamber (see Fig. 19.3). It is triggered by an anode striker where an arcing process occurs after the chambers are evacuated to a vacuum pressure of 10^{-5} Pa or better. A high current (whether direct current or pulse) is applied to the target through the striker. Both carbon ions and micro/macrocarbon particles are generated during the arcing. A magnetic field gener-

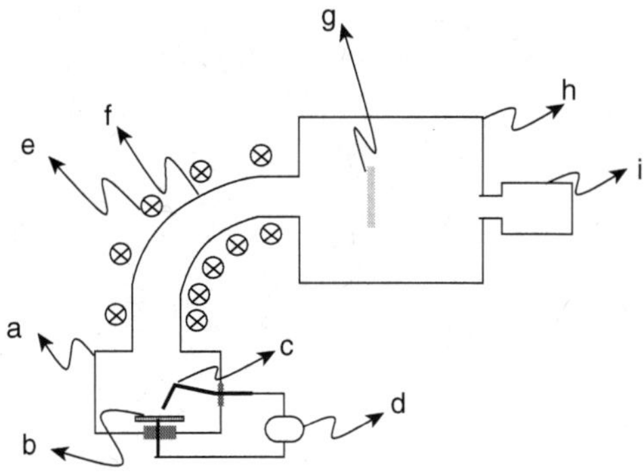

Figure 19.3. Schematics of an FCVA: a, arc chamber; b, target on cathode; c, anode striker; d, arc power supply; e, solenoid; f, filter bend; g, substrate; f, substrate chamber; i, vacuum pump.

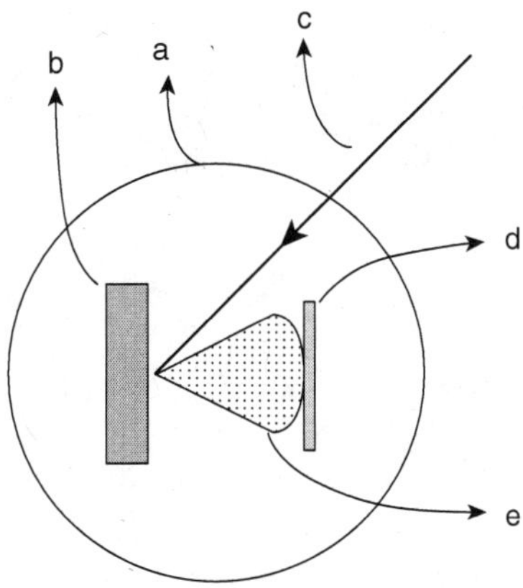

Figure 19.4. Schematics of a PLD system: a, vacuum chamber; b, target; c, pulsed laser beam; d, substrate; e, laser plume/ablation from the target.

ated by a solenoid directs the carbon ions around a bend onto the substrate in another vacuum chamber where the thin film is deposited. As the microparticles and macroparticles follow a line-of-sight direction, they are usually filtered out by the filter bend.

DLC Films by Laser Deposition. Since the invention of epitaxial growth of high-temperature superconductor material using PLD (Eason 2007), this thin-film growth technique has become popular. In applying PLD to growing a DLC film, a pulsed laser beam is focused onto a rotating carbon-based target. Each pulse of the laser beam has a power density high enough to ablate and ionize the target material due to the absorption of the laser beam by the target material. In a high vacuum condition, the ionized material (plasma) beam can be directed to a substrate where DLC film growth is desired. A simple schematic diagram of a PLD system is shown in Figure 19.4.

To deposit a DLC film with high sp^3 fraction, the laser wavelength, substrate temperature, and vacuum condition need to be well controlled. Voevodin et al. (1997) found that if the substrate temperature is higher than 150°C, a graphitelike rather than a DLC film will be obtained. In our DLC growth experiments, a KrF laser beam line at a wavelength was applied to ablate the desired target in a background vacuum of 3×10^{-6} Torr. The pulse width is 25 ns and the pulse energy is 200 mJ with a repetition rate of 20 kHz. The obtained DLC film has an sp^3 fraction higher than 70%, and this fraction is very sensitive to metal doping (Foong et al. 2010).

DLC Film: Material Properties

This section briefly describes various characterization techniques used to determine the material and physical properties of DLC films. Common properties include the composition, sp^3 bonding fraction, surface morphology, and microstructural and various mechanical properties, such as Young's modulus, hardness, and COF. These properties are needed as part of the nanomechanical device design and system improvement. Common materials characterization techniques, such as scanning electron microscopy (SEM), transmission electron microscopy (TEM), Raman spectroscopy, X-ray photoelectron spectroscopy (XPS), and atomic force microscopy (AFM), and mechanical properties testing tools, such as the microscratch test and nanoindenter, can be used to study the film.

Determining the sp^3 Content. The sp^3 and sp^2 bonding fractions in a DLC film determines the main material properties that lead to important applications. The main techniques used to quantify the sp^2 and sp^3 bonding fraction in the film are electron energy loss spectroscopy (EELS) (Schwan et al. 1996), XPS, and Raman spectroscopy (Tay et al. 1999). Other techniques such as energy-dispersive X-ray (EDX) scattering (Egerton 1996) can be used, but that is limited to elemental composition. Usually, EDX is incorporated in a scanning electron microscope, while EELS is incorporated in a transmission electron microscope.

For EELS in a TEM, a sample is exposed to a beam of high-energy electrons. Some of the electrons will undergo inelastic scattering, which means that they lose energy and have their paths slightly deflected. The energy loss will be carefully measured using an electron spectrometer and interpreted in terms of what caused the energy loss. Inelastic interactions give rise to phonon excitations, interband and intraband transitions, plasmon excitations, inner-shell ionizations, and so on. The inner-shell ionizations are most particularly useful for detecting the elemental components of a material. For example, in the case of carbon, a large number of electrons are absorbed at 285 eV because this is the energy required to remove an inner-shell electron from a carbon atom. In addition to chemical composition, EELS is capable of measuring chemical bonding, valence- and conduction-band electronic properties, surface properties, and element-specific pair distance distribution functions (Egerton 1996). One advantage of EELS lies in its ability to clearly detect the differences between diamond, graphite, and amorphous carbon (a-C) through the anti-π bonding orbital. An example of the EELS spectrum of a high-sp^3-content DLC is shown in Figure 19.5. Analysis of the EELS

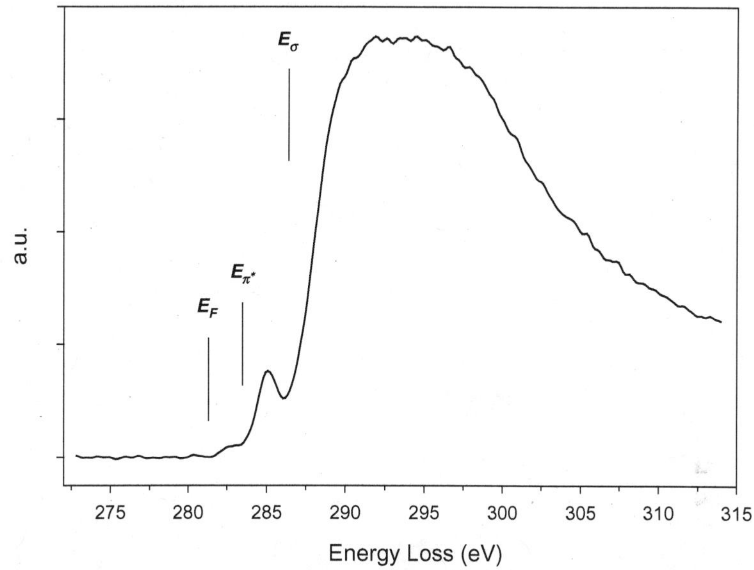

Figure 19.5. EELS spectrum of a diamond-like carbon sample. The low intensity of the 1s-π^* peak confirms that the DLC film has high sp^3 content.

data was done by taking the area of the 1s-π^* peak and comparing it with the total area of the carbon K-edge (Berger et al. 1998; Chua et al. 2004b). The peak at 285 eV is due to electronic excitations from the ground state 1s core level to the vacant π^*-like states. Excitations to higher-lying σ^* states occur above 290 eV. To estimate the sp^3 fraction, it is assumed that the ratio of the integrated areas under the energy windows around the π^* peak and around the σ^* edge is proportional to N_π/N_σ. This gives the fraction of sp^3 bonding of the films and shows the sp^3 fraction to be approximately 85% or above. In general, X-ray absorption near-edge spectroscopy (XANES) gives information similar to that of EELS and they are comparable to each other.

Raman spectroscopy is based on the Raman effect, which occurs when monochromatic light incident into a material specimen scatters at frequencies that are different from the incident frequency. This phenomenon results from an interaction between the incident photons and the vibrational energy levels of the molecules of the target specimen. For DLC films, the Raman spectra usually can be fitted with two Gaussian-like peaks (Ferrari and Robertson 2000) as shown in Figure 19.6a. The positions and line widths of the deconvolved D and G bands give an approximation for the sp^3 content in the thin films. The G peak was assigned to ~1550/cm, while the D peak was assigned to 1330/cm. The Raman spectra were virtually the same for all the films deposited under different conditions. The D peak was nearly negligible for all the samples, indicating that a low degree of sp^2 clustering and no short sp^2 chains existed in the films. The sp^3 fraction can be quantified by

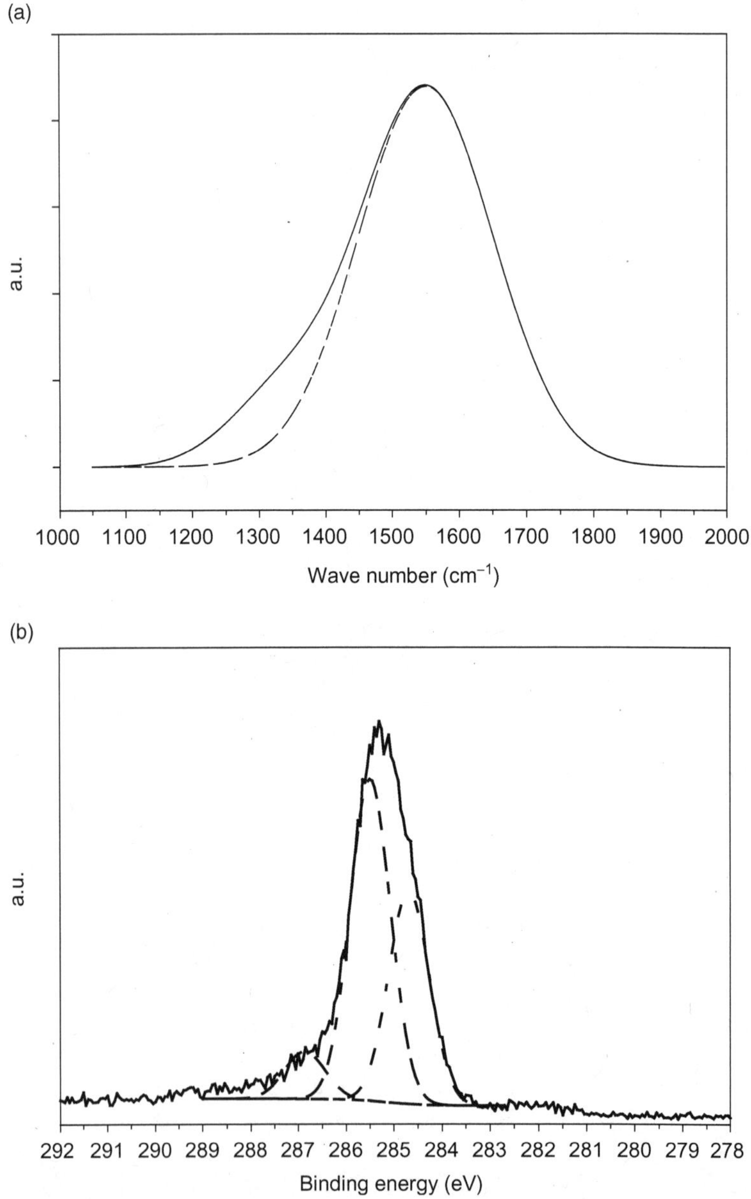

Figure 19.6. Raman shift (a) and deconvolved XPS (b) of a diamond-like carbon sample.

$$\mathrm{sp^3 fraction} = 0.934 - 0.6 \times \frac{I_D}{I_G}, \tag{19.9}$$

where I_D and I_G are the deconvolved areas of the D and G bands, respectively.

XPS is among the most essential techniques for detecting the elements present as well as the bonding states on the surface of a material. This method uses X-rays to eject core–shell electrons, and the core–shell energy will shift depending on the chemical state of the atom. In Figure 19.6b, the XPS spectrum of a DLC film is shown and deconvolved according to its sp^3/sp^2 content. It has further been shown that the ratio of peak areas at 285.39 and 284.29 eV, which corresponds to the sp^3-to-sp^2 ratio, can be used to determine the sp^3 fraction (Tay et al. 1999).

Various researchers have shown that there is a good correlation between all three of these techniques, XPS, Raman, and EELS (Tay et al. 1999), which indicates that they can analyze any DLC film and give a good indication of the sp^3 content. In the example given previously, Figures 19.5 and 19.6 show DLC films with an sp^3 fraction higher than 80%.

Intrinsic Stress of DLC Films. The deposition process to obtain high-sp^3-content DLC film, deposited by either FCVA or PLD, also gives the film very high intrinsic residual stress. This stress can be as high as 10 GPa (Zhang et al. 1999; Chua et al. 2004b) and can limit the film thickness to a hundred nanometers or so when growing on a silicon substrate. In the fabrication of nanoarchitectures of DLC, it is essential to have stress-free or nearly stress-free films. Without controlling the residual stress, the film thickness on a silicon substrate is limited to only several tens of nanometers, which is not thick enough for designing a practical nanoarchitecture.

There are several ways to control the residual stress of a DLC film on a substrate. Various ways have been proposed and demonstrated, such as by doping the target carbon material with metals (Hsieh et al. 1999; Wang et al. 2000; Chiu et al. 2005; Weng et al. 2006) or rare earth metal oxides (Zhang et al. 2008a, b; Foong et al. 2010) or by applying the proper voltage (Hsu et al. 1998; Zhang et al. 2002) to the substrate (substrate biasing) during arc deposition. It is also known that using metal-doped targets for cathodic arc deposition usually gives a nanocomposite film of metal nanoparticles dispersed in a DLC medium instead of a pure DLC film.

When substrate biasing is used during the film deposition process, there are two ways of applying voltage to the substrate to reduce the residual stress of DLC film on a silicon substrate. One way is to apply a short pulse of high voltage (a few thousand volts) to the substrate during the arcing process. This method increases the energy of the carbon ions; this, in turn, increases their penetration depth for a subsurface growth mechanism or a subplantation model (Davis et al. 1998). This technique is sometimes called plasma ion immersion implantation (PIII). This process has been shown to modify the material properties of the DLC film (Ager et al. 1999; Sheeja et al. 2001). The other way is to apply a DC voltage to the substrate during the arc growth of the DLC film. Usually, voltages as low as 300 VDC are sufficient to grow a DLC film nearly free of residual stress by this method. However, this method also reduces the sp^3 content of the film. Though the exact stress suppression mechanism is not well-known, it was found that a high DC voltage could change the residual stress of DLC from compressive stress at 0V to tensile stress at voltages over 300V (Li and Chua 2008).

DLC with low compressive stress can be grown to a thickness of 2 μm in 30 min, which is one of the requirements in microfabrication and nanofabrication processes. In our research, we did not find evidence of layer growth (Uhlmann et al. 1998) in our thick DLC films. The film thicknesses obtained using the DC biasing technique can satisfy the requirements of fabricating both micromechanical and nanomechanical devices.

Surface Morphology and Surface Energy. When DLC films with the desired sp^3 fraction are applied in nanomechanical and micromechanical devices, surface morphology is another highly critical parameter that determines the device's performance. The surface topography and roughness can be studied using AFM, with a noncontact or tapping mode at a proper scan frequency being used for nondestructive testing. The contact mode is not desirable as it may introduce scratches onto the sample surface and distort the topography. As the films are generally very smooth, the atomic force microscope has to be suspended on an antivibration swing to enhance the tracking. Good tracking is important to ensure that accurate images of the surface are captured, as illustrated in Figure 19.7.

The root mean square (RMS) roughness of the films was recorded. A typical surface morphology of DLC over a surface area of 1 μm^2 was measured with AFM, as shown in Figure 19.7. The surface roughness of the film is within 0.2 nm or less depending on instrumental limitations. "Pure" DLC films have been reported to possess an RMS roughness of 0.12 nm or less. The value of 0.08 nm obtained was found to be comparable to that reported in literature. However, the incorporation of metals into a DLC film makes it rougher than the pure DLC films. An increase in the metal content generally leads to an increase in the RMS roughness of deposited films. It is believed that this trend could be explained by using the thermal spike model, which describes the in situ deposition condition at the surface of the as-growing film when two different ions impinge onto the surface (Zhang 2004). During deposition, an infinitesimally small area with a very high temperature was created when ions from the plasma impinged on it. This temperature could only be understood in a local, time-dependent sense; the size of the area was determined by the collision cascades, which were known to quench in some picoseconds. When biasing was added to the substrate, the surface was observed to have an increase of roughness due to the higher energy of the ions impinging onto the film. The atoms impinged into the film were not able to move within such a short thermal spike time.

Shi et al. (1996, 2000) conducted studies to correlate the surface morphology properties with the sp^3 content of DLC films and concluded that there was an inverse relationship between the RMS roughness of the surface and the sp^3 content of the films. In other words, a higher RMS roughness generally tends to correspond to lower sp^3 content. As with the previous results from Raman spectroscopy, XPS, and nanoindentation, the incorporation of metals does decrease the sp^3 content.

The surface energy of the films is another important factor due to the fact that it can give some clues to the surface properties of the thin film. The correlation between

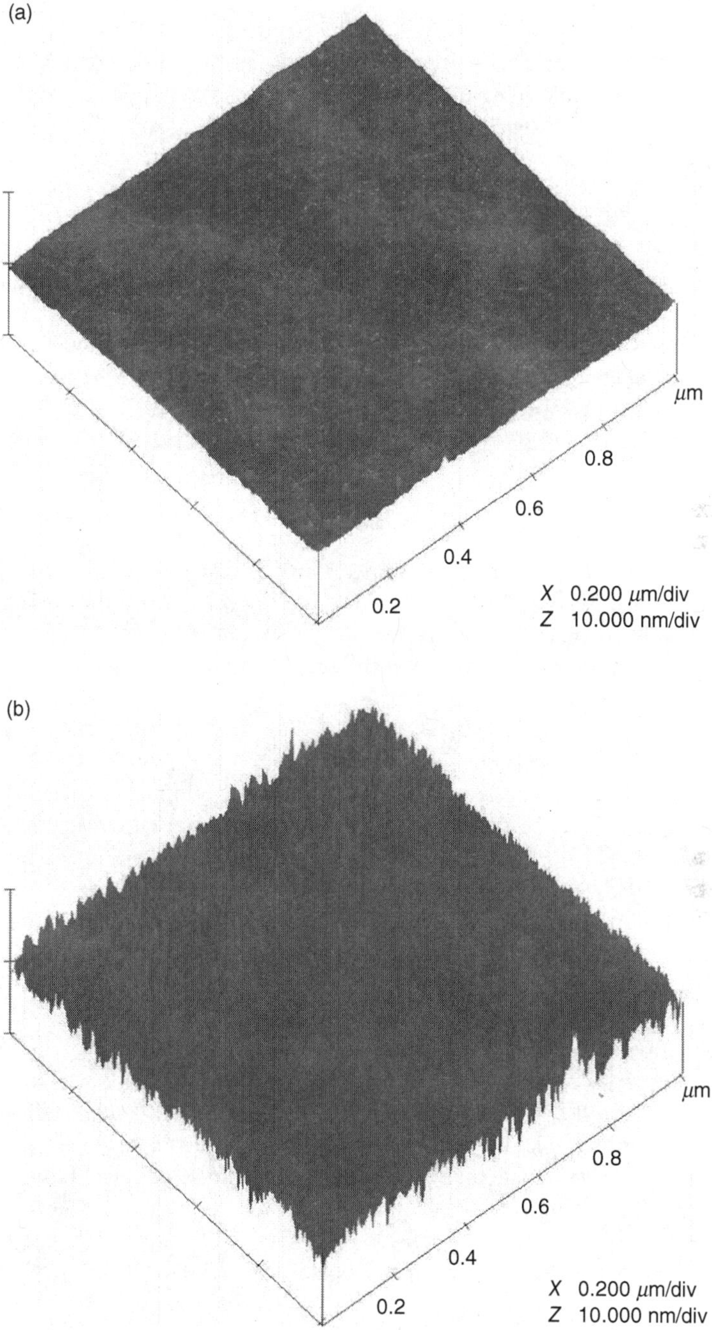

Figure 19.7. Surface morphology of (a) a near-atomically smooth high-sp^3-content DLC film deposited at floating voltage and (b) a 1-μm-thick DLC film, with 50% sp^3 content, deposited at −800 V. The roughness is negligible.

contact angle and surface free energy of a solid surface and a liquid was provided by Young for a drop of liquid lying on a solid surface. Young's equation (Young 1805, Wu 1982) is

$$\gamma_{sv} = \gamma_{sl} + \gamma_{lv} \cos\theta, \tag{19.10}$$

where γ_{sv} and γ_{lv} are the solid and liquid surface energy, respectively; γ_{sl} is the solid–liquid interfacial energy; and θ is the contact angle. The contact angle θ and the liquid surface energy γ_{lv} are measurable parameters. To determine the surface energy of the solid through measurement of the contact angle, various theoretical statements describing the interfacial energy γ_{sl} as $\gamma_{sl} = f(\gamma_{sv}, \gamma_{lv})$ have been used. In this work, the harmonic mean model was used (Weiler et al. 1994):

$$(1+\cos\theta)\gamma_{lv} = 4\left(\frac{\gamma_{sv}^{d}\gamma_{lv}^{d}}{\gamma_{sv}^{d}+\gamma_{lv}^{d}} + \frac{\gamma_{sv}^{p}\gamma_{lv}^{p}}{\gamma_{sv}^{p}+\gamma_{lv}^{p}}\right). \tag{19.11}$$

Here $\gamma_{lv} = \gamma_{lv}^{d} + \gamma_{lv}^{p}$ and the superscripts d and p refer to the dispersive and polar components, respectively. When two liquids are used, and the values of γ_{lv}^{d} and γ_{lv}^{p} for these liquids are known, the dispersive and polar components of the solid surface energy (γ_{sv}^{d} and γ_{sv}^{p}) can be obtained by solving the two simultaneous equations. The surface energy is a combination these two separate components, the dispersive and polar components.

In general, the contact angle for DLC films can range from 75° to a high contact angle of 80°, with the theoretical limit for a-C films reported to be ~81° (Han 1999). It has been further reported that the contact angle for typical DLC films deposited by FCVA is around 77° depending on deposition conditions (Chua et al. 2004b). The surface energy of these DLC films has been reported to be as low as ~4.0×10^{-6} N/m using multiple liquid testing (Pivin et al. 1991). It was noted that the dispersive component tends not to change; hence, the main component that affects the surface energy is the polar component, which can vary from 1.3×10^{-6} to 0.9×10^{-6} N/m. The polar component comprises three different factors: (1) intermolecular forces, (2) permanent and induced dipoles, and (3) hydrogen bonding, whereas the dispersive component of the surface energy is simply due to instantaneous dipole moments. As FCVA-grown DLC films contain little to no hydrogen, the polar component is due to the change of the sp^2 carbon formation and dangling bonds in the a-C network. As such, the process conditions required to obtain high-sp^3-content DLC films probably require a highly dense plasma; this, in turn, will inevitably affect the surface energy as confinement within the same unit volume may cause a significant reduction in the dangling bonds or other defect sites (i.e., a decrease in surface energy; Chua et al. 2004b) which does give rise to a higher contact angle with water (hydrophobic).

Mechanical Property Measurement. There are several ways to measure the value of Young's modulus of these films. One conventional technique is by nanoindentation. Another technique is by laser-induced surface acoustic wave (L-SAW). This laser-acoustic technique uses short pulses of a laser to generate wideband surface wave impulses. The surface wave impulse is detected by a wideband piezoelectric transducer. Both specimen and transducer are fixed to a translation stage that moves perpendicularly to the laser beam position. The surface acoustic waveform is detected at different distances between the laser focus line and the transducer. The experimental dispersion curve is fitted with the theoretical curve to obtain the film parameters. The fit yields the value of the Young modulus of the film (Weiler et al. 1994; Schneider et al. 1997).

The density ρ of the film can be indirectly deduced from Young's modulus E, as reported by an empirical method:

$$\rho = 1.79\,(\mathrm{g/cm^3})\,(1 + E/780\,[\mathrm{GPa}] - (E/1620\,[\mathrm{GPa}])^2). \qquad (19.12)$$

Using this relation, the density of the thin film could be determined. This equation was derived using some test samples for which the density of the films was determined using X-ray reflectivity (XRR), Rutherford backscattering spectrometry (RBS), and EELS (Schneider et al. 1997).

In the case of high-sp^3-content DLC films, a high of ~800 GPa can be obtained. Using the empirical method described in Equation 19.12, the density of the film can reach a high of ~3.4 g/cm^3. This estimated value was very close to the current reported values for polycrystalline diamond, ~3.5 g/cm^3. As such, it had been observed that DLC films possibly are very compact and stiffer, with high sp^3 bond content.

The dynamic friction coefficient is shown in Figure 19.8. The COF for DLC films is generally very low, with an average value of less than 0.1. This value is very close to the COF of graphitic (soft) films. During the measurement where there is direct mechanical contact on the surface of the DLC films, the force (and possibly localized heat) causes the sp^3 bonds to become graphitized. This effect reduces the friction of the

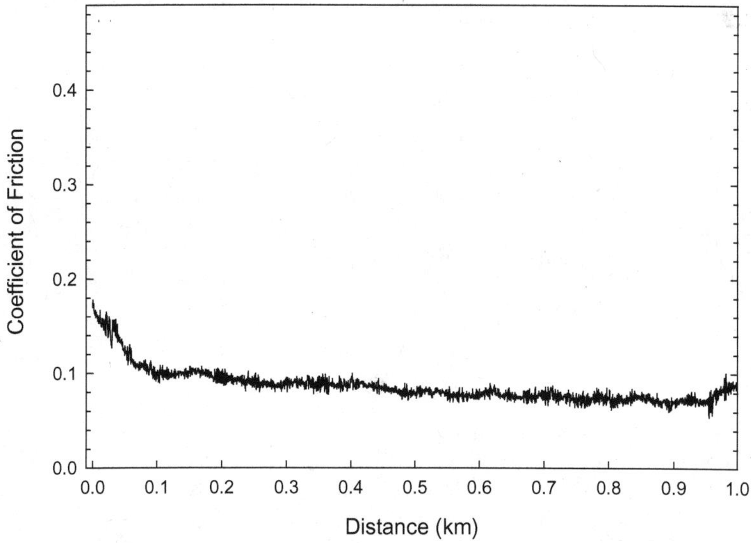

Figure 19.8. Coefficient of friction of diamond-like carbon measured by a tribometer.

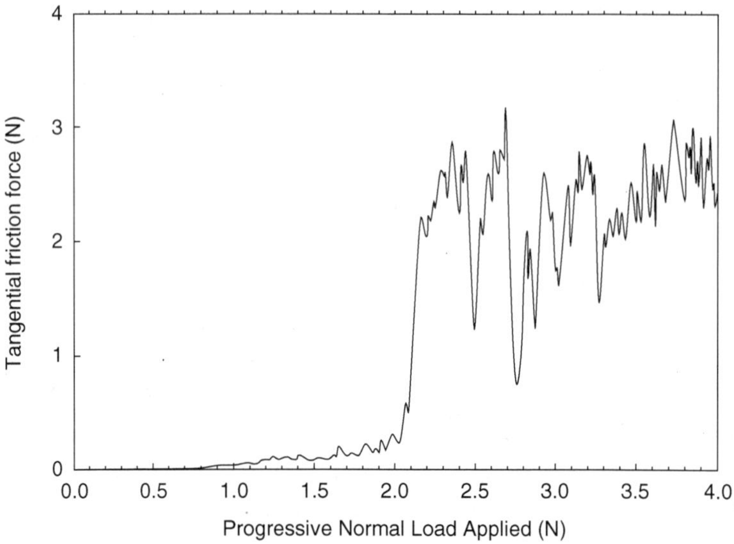

Figure 19.9. Scratch test results of a typical high-sp^3-content diamond-like carbon film on silicon substrate. The tests were performed at room temperature, in ambient air.

surface layer, wherein Figure 19.8 shows a further decrease in the COF as the sliding distance increases.

Microscratch tests are usually carried out under a progressively increasing load from 0 to 5 N. Figure 19.9 shows the result of a typical scratch test for deposited DLC films where a critical load of 2.1 N was obtained (Chua et al. 2004b). However, the adhesion strength is heavily dependent on the type of substrate used. Where silicon was used, there have been reports that a thin silicon carbide interface had formed due to the high-energy carbon ions impinging onto the silicon substrate, which enhances the interface strength. Further, DLC film has a tendency to form carbide bonds with metals, thus making it a very suitable material for wear coating on metals.

The wear rate of high-sp^3-content DLC films can be calculated using the relation in Equation 19.13. Wear rate (k) is given by

$$k = V/(W \times S), \tag{19.13}$$

where V is the wear volume (in cubic millimeters), W is the normal applied load (in newtons), and S is the sliding distance (in meters). The wear rate was approximately 1.2 (±0.5) × 10^{-8} mm^3/N·m for DLC thin films. For a thin 30-nm film, high-quality DLC films can withstand slide cycles as much as 20,000 times (~0.2 km) against a sapphire ball under a 10-N applied load in normal ambient air conditions.

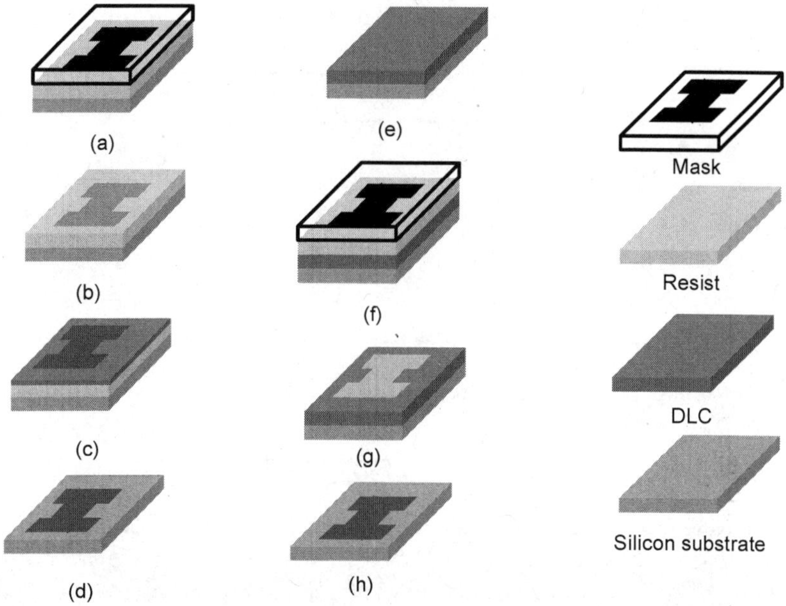

Figure 19.10. Schematic diagram of top-down (a–d) and bottom-up (e–h) processes for fabricating DLC-based nanoarchitectures.

DLC MECHANICAL DEVICES: FABRICATION AND PERFORMANCE

To fabricate DLC-based nanoarchitectures for practical applications, there are two main technological routes. One is to build the architecture directly during the growth process, in which the architecture is defined by the material growth with little or no postdeposition treatment. This route is usually called the bottom-up approach. The other is to start with an as-deposited material and, by using well-developed technology from microelectronics, to subtract the undesired portions for target architectures at the nanoscale level. This route is called the top-down approach. These two routes are described in Figure 19.10. An example of each route, with direct applications in DLC, is given in this section. The bottom-up approach is described first, followed by a maskless top-down approach. For simplicity, only two-dimensional microscale DLC architecture or micromechanical resonators will be described in this chapter. Details of other micro/nanoarchitecture fabrications with the bottom-up approach can be found in Li and Chua (2008). A top-down approach needing a lithographic mask will not be described here, but a similar example can be found in Tsai et al. (2002).

DLC Micromechanical Device Fabricated by Patterned Growth and Liftoff

The growth and fabrication of the cantilevers follow the mask-based bottom-up route, as illustrated in Figure 19.10. A negative resist on silicon is exposed by a mask with a

TABLE 19.3. Materials Properties of PIII Deposited a-C Thick Film

Physical properties	a-C
Young's modulus, E	300 GPa
Density, ρ	2.33 g/cm^3
Hardness measurement	~40 GPa
Coefficient of friction	0.08
Wear rate	~10^{-9} mm^3/Nm
Critical load	1.71 N
Smoothness, R_{RMS}	~0.75

cantilever pattern. After developing, the patterned substrate will be used as a material growth template. After the DLC with an appropriate thickness has been deposited, a lift-off process is applied to remove both the resist and the DLC on the resist. The remaining pattern of the DLC (e.g., a microcantilever) will be left. Two different types of DLC cantilevers were fabricated based on a 1.5-μm-thick PIII single-layer DLC film (also called an a-C film) and another on a 1.5-μm-thick high-sp^3/low-sp^3 multilayer film. The PIII process is described elsewhere in Bilek et al. (2002). The sp^3 content is highly dependent on the ion energy, and for carbon films deposited under high biasing (−300 VDC or more), the sp^3 content decreases, but in addition, the residual stress decreases as well, thus allowing the formation of thick a-C films (Ager et al. 1999; Sheeja et al. 2001). These a-C films have a content of approximately 50% sp^3 and 50% sp^2. The advantage of having a multilayer film is that the lower sp^3 a-C films are present to decrease the residual stress, while the high-sp^3 DLC films give the strength and toughness.

The various properties of the DLC have been described in detail as shown in Table 19.3. These material properties are essential for the simulation and modeling tests. Inaccurate values will give rise to unreliable and dubious results. Unfortunately, it was not possible to model multilayer films due to their complexity. The only approximation available was that most of its properties should lie in between a "pure and thick" DLC film (see Table 19.2) and the PIII a-C film.

Vibratory Response of Microcantilevers. A custom-designed high-vacuum vibrometer was used to measure the resonance frequency of the cantilever beam. A vibration stage with an external piezoelectric driver and lock-in amplifier was used. This stage had a built-in heater that allowed in situ annealing of the attached films while studying the frequency response of the sample.

Figure 19.11 shows the SEM image of the PIII DLC microcantilever; the simulation of the resonance of the cantilever is also shown (Fig. 19.11b).

The experimental results for the first fundamental (mechanical) vibration mode were in very good agreement with the simulation results, at ~108 kHz, compared with poly-Si at ~60 kHz. One possible reason the experiments gave a slightly higher frequency compared with the simulation and modeling might be due to the experimental

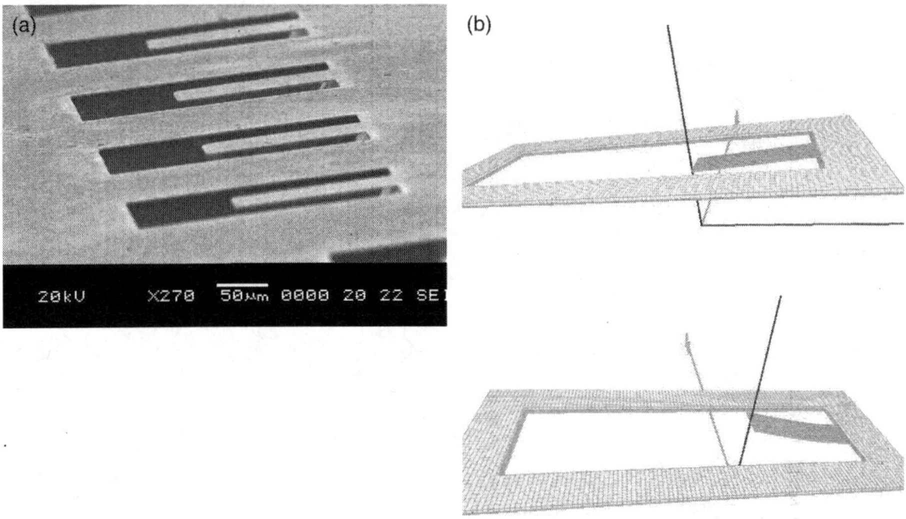

Figure 19.11. (a) SEM image of a-C microcantilevers prepared at −5 kV, 600 Hz, and 25 μs, dimensions 150 × 50 μm, and (b) an actual beam resonance deflection as predicted by the simulation, but with 20× magnification along the *z*-axis.

environment. The results were obtained in vacuum, in the absence of various air friction and damping forces that would be encountered if it were tested in air.

Figure 19.12 shows the resonance frequency of both types of DLC cantilevers. A poly-Si cantilever of the same dimensions is also shown for comparison with those of the simulation results. At the same dimensions, PIII DLC had nearly double the first-mode resonance frequency of poly-Si. For high-frequency applications, it is easier to fabricate PIII DLC or multilayer DLC MEMS resonators than to further redesign poly-Si resonators using smaller dimensions, which stretch the lithography technology. The reduction in dimensions may also reduce the Q-factor significantly. Hence, this experiment demonstrated that PIII DLC and multilayer DLC microstructures are definitely superior to poly-Si microstructures in MEMS resonator applications. In addition, the low friction, low-wear properties, and good chemical inertness made the carbon structures very appealing especially in harsh environment applications.

The quality factor Q of a resonant system is considered to be high if the system loses little energy over time compared to the energy in the system. As such, it is defined as the total energy stored in the structure divided by the sum of energy losses from the vibrating element per cycle. The Q-factor can also be calculated from the amplitude–frequency spectrum of the vibration by taking the resonance frequency f_R divided by the frequency bandwidth Δf_R at the 3-dB amplitude points (Equation 19.8). In a microcantilever, Q is

$$Q = \frac{f_R}{\Delta f_R}. \tag{19.14}$$

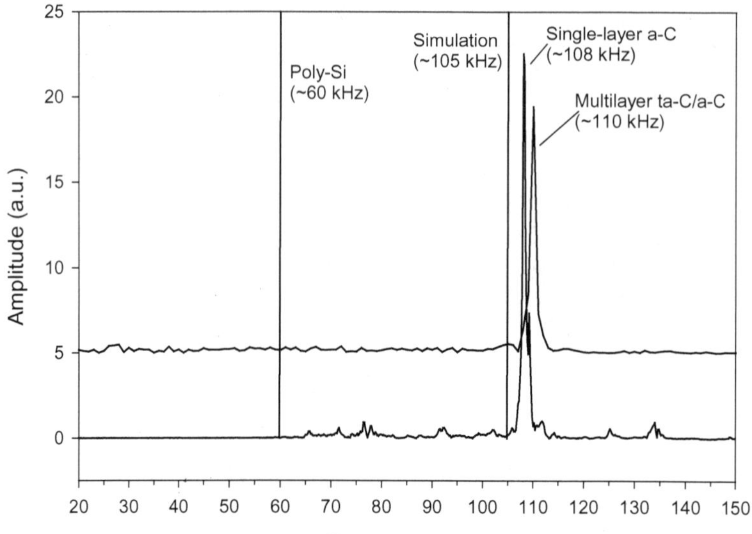

Figure 19.12. Measured first-mode resonance frequency of two different types of diamond-like carbon cantilevers: a single-layer 1.5-μm-thick a-C and a multilayer 1.5-μm-thick ta-C/a-C fabricated microcantilever. The simulation results and the resonance from a poly-Si beam are also shown for comparison.

A high Q-factor is important not only because it simplifies the feedback control electronics, minimizes the perturbing effect of the drive electronics, and gives a high resolution but also because it implies a very low unwanted mechanical coupling to the external world, which gives high accuracy and long-term stability. In the PIII DLC cantilevers fabricated, a Q-factor of ~1020 was obtained; for the multilayer DLC cantilevers, a Q-factor of ~1200 was obtained. This increase could be explained by the harder and stiffer DLC layers in the multilayer structure.

These results further show that not only did micromechanical resonators fabricated from ultrapure, single-crystal semiconductor materials have extremely high-quality factors but a high Q-factor also was demonstrated from an amorphous-material cantilever structure. This value exceeds typical electrical resonators and is higher than poly-Si and crystalline Si cantilevers of the same dimensions measured at room temperature.

Vibratory Response of Multilayer Cantilevers Under Annealing Conditions. In this section, the effects of thermal annealing on the vibratory response of the DLC cantilevers were studied. The samples used were the multilayer DLC cantilevers. The resonance frequency was taken in situ. The results are shown in Figure 19.13.

There is a slight increase in the resonance frequency during annealing. Judging from the SEM image, this could be attributed to the increase in the downward curvature

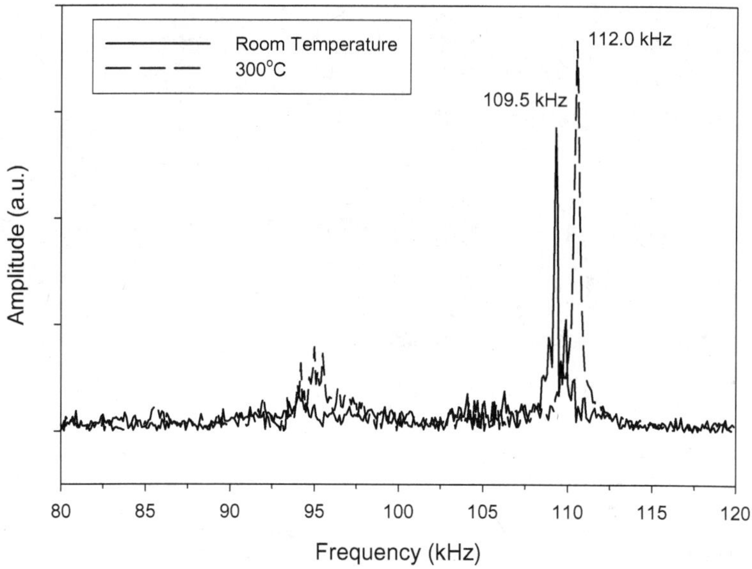

Figure 19.13. Resonance frequency of the cantilever at room temperature and at 300°C.

of the cantilevers. This increase in curvature is due to the increase in the intrinsic stress of the films. It was also observed that the behavior of multilayer DLC films and DLC films differ under annealing conditions.

DLC films undergo stress reduction upon thermal annealing, while a-C films undergo an increase in intrinsic stress due to a graphitization process. However, the multilayer films had an overall increase in intrinsic stress from a superposition of the residual stress of DLC and a-C films (Fig. 19.14). An increase in the intrinsic stress caused an increase in the resonance frequency. Coupled with a high Q-factor, this indicates that there are potential applications for these resonators as temperature sensors or temperature-dependent circuit breakers.

DLC Micro/Nanomechanical Device Fabricated by Focused Ion Beam (FIB) Milling

FIB milling is widely used in the field of material science (Yao 2007; Mayer et al. 2007) as a tool for preparing high-quality TEM sample lamellae. This technique utilizes a FIB to mill material rather than to deposit it as in a FIB–chemical vapor deposition (CVD) method. FIBs can also be applied to fabricating DLC nanoarchitectures by using their milling ability, as illustrated in Figure 19.15. By scanning a highly focused ion beam on a DLC thin film or a DLC thin film covered by a metal film, which is used to reduce the surface charge effect to provide better fabrication resolution, DLC nanoarchitectures as fine as the focusing limit of a FIB facility can be fabricated. The coating metal film may be Cr, which can be easily removed after FIB milling.

Figure 19.14. SEM image of multilayer ta-C/a-C cantilevers (a) as received at room temperature and (b) after thermal annealing at 300°C. A distinct downward curvature could be observed.

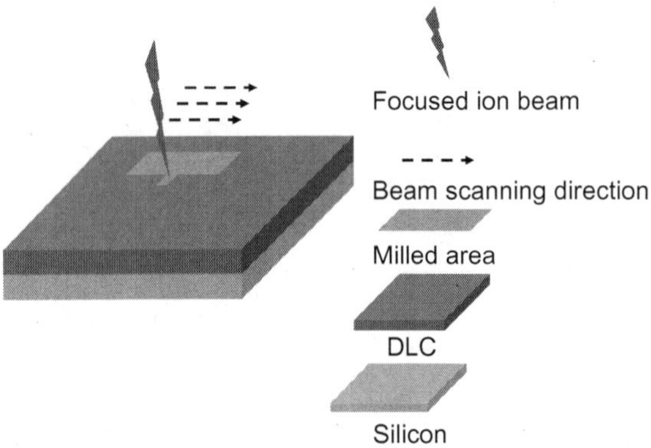

Figure 19.15. Schematics of FIB fabrication of a nanoarchitecture in diamond-like carbon.

One of the most striking properties is that FIB milling ability is almost material independent for the DLC-based materials. It can be applied to any DLC thin film, doped or not. Fabrication rates for variants of the same material are almost the same. Another property is that FIB milling is a maskless process, thus reducing the fabrication period and cost and at the same time adding more design flexibility. As such, devices can be easily fabricated with the help of design software, thus making design, prototyping, and demonstration of new ideas very efficient and economical.

An example of a nanocantilever fabricated by direct FIB milling is illustrated in Figure 19.16. A FEI Quanta 300 dual-beam system was applied to an uncoated DLC film. With carefully controlled fabrication parameters (such as beam current, exposure time, and beam focus condition), high-quality cantilevers can be fabricated at submicrometer scale, as illustrated in Figure 19.16. The milling rate at one of the fabricating beam currents is illustrated in Figure 19.17. On average, the milling depth in DLC by

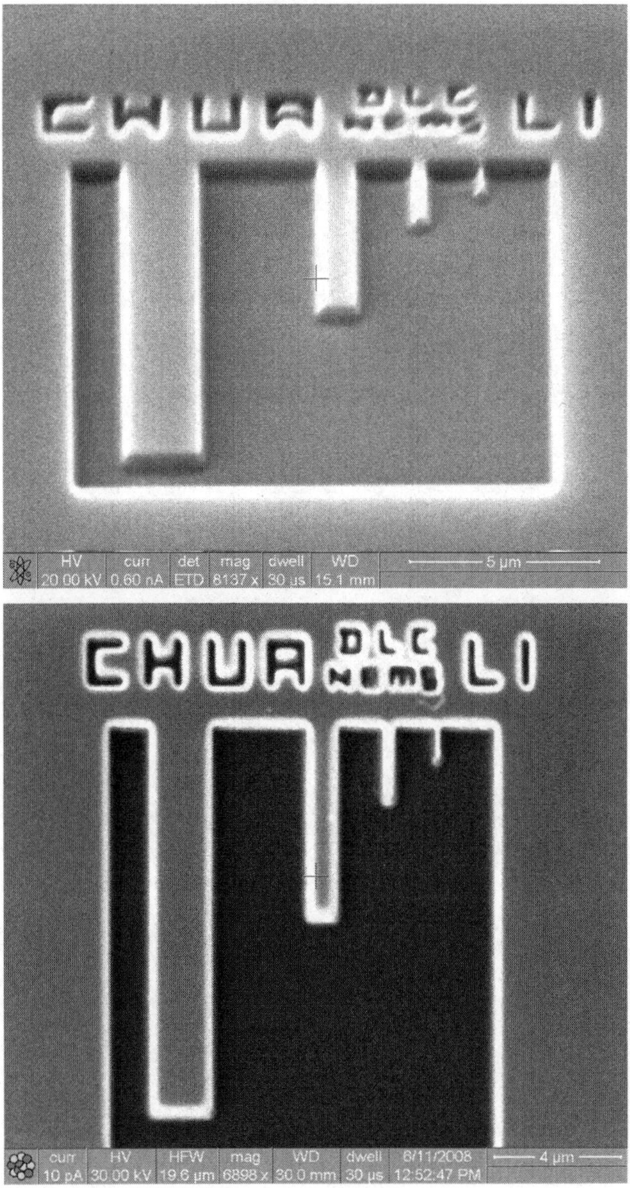

Figure 19.16. Diamond-like carbon cantilevers resting on a silicon substrate. SEM image with a tilt angle of 50° (left) and scanning ion image without tilt (right).

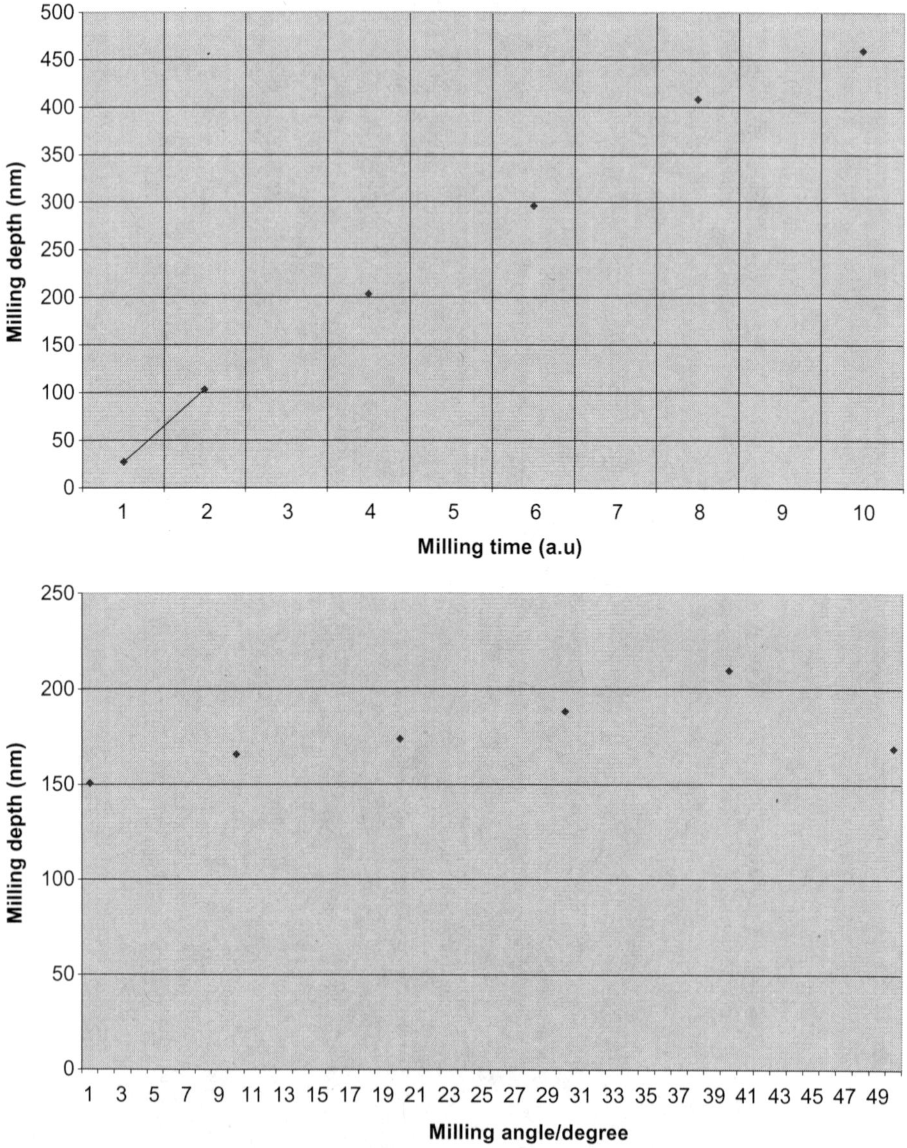

Figure 19.17. Milling depth versus milling time (left) and milling rate versus milling angle (right).

a gallium ion beam is almost proportional to the milling time. The milling rate of DLC is about 40% of that of silicon under the same milling condition due to the high hardness and inertness of DLC films. Though milling at a tile angle of 40° gives a higher milling rate, this rate is not strongly dependent on this milling angle.

DLC Nanoarchitecture Fabricated with FIB-Assisted CVD

FIB-assisted growth of DLC from an organic precursor is another type of ion beam-assisted CVD technology. Due to the tiny beam diameter of a FIB, direct-growth DLC nanoarchitectures are possible. As the beam has excellent control in the x–y plane (two-dimensional mode) while the sample stage can move independently along the z-axis, a fully three-dimensional (3-D) nanoarchitecture can be fabricated. The main fabrication process is illustrated by Figure 19.18 (Matsui 2006). First, the ion beam is focused on the sample (which is held on a stage) while a metal nozzle delivers gusts of carbon-rich organic gas, usually methane (CH_4) or phenanthrene (C_4H_{10}), onto the same area. Once a carbon-rich organic molecule meets the energetic ions near the focal point, it is decomposed into volatile gases such as hydrogen and carbon ions, which will be deposited on the sample surface. The momentum of the energetic beam moving toward the sample surface is partially transferred to the carbon atoms from the decomposed organic molecules; this deflects the carbon ions' or atoms' initial direction of motion onto the sample surface, and the DLC nanoarchitecture begins to nucleate and grow.

One example of a vertical DLC nanoarchitecture grown by this method is illustrated in Figure 19.19, from which we can see that DLC nanopillars (a one-dimensional DLC structure) can be well fabricated with dimension down to less than 100 nm. A FIB with beam width less than 10 nm can be well controlled; thus, this method has the potential to fabricate very fine nanoarchitectures. The same figure shows a scanning electron microscope modified to measure mechanical properties. By using this unique

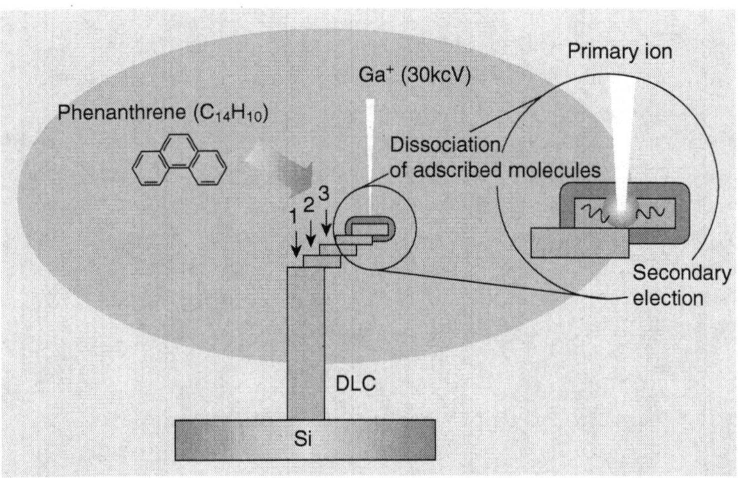

Figure 19.18. Schematics of diamond-like carbon deposition by focused ion beam.

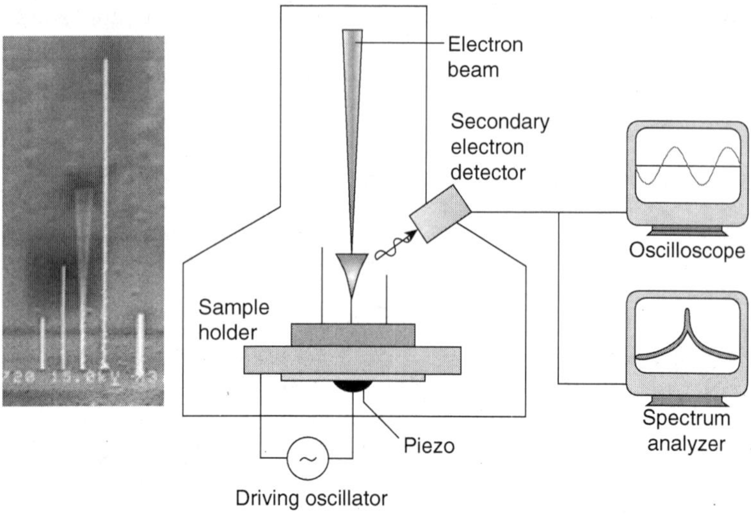

Figure 19.19. Bottom-up nanomechanical resonator and its motion-detection method by SEM.

system, the mechanical property of DLC pillars depending on the aspect ratio of the pillar can be verified.

The ion beam used by Matsui (2006) was composed of gallium ions, which are not volatile. Hence, the simultaneous deposition of DLC nanoarchitecture will also incorporate gallium metal into the bulk of the films and microstructures. Usually, the deposited gallium is at the center of a pillar, though it may be mixed or alloyed. This metal incorporation may affect the total mechanical and electrical properties of the fabricated nanoarchitectures.

As FIB systems have very fine controllability over full space, true 3-D nanoarchitectures of DLC can be well fabricated. It is interesting that complex 3-D nanoarchitectures with capacitors, inductors, and resistors woven together can be easily fabricated using this method (Matsui 2006).

The bottom-up approach of fabricating DLC nanoarchitectures by FIB-based methods is a serial process, which limits the technology to laboratory demonstrations rather than use in industrial production. For DLC nanoarchitectures that have stringent requirements on material and physical properties, removing residual gallium is a key issue for the FIB technique. To solve this, other ion sources, especially those that are volatile, such as argon ions, can be used. In this case, the fabricated DLC nanoarchitecture can be well free of metal contamination.

THE FUTURE OF DLC MICRO/NANOARCHITECTURES

The field of DLC-based microarchitectures and nanoarchitectures is not well explored with respect to practical applications. Many designs and processes are still not fully understood, for example, the etching process of DLC films for nanoarchitectures. Likewise, the optimal nanoarchitecture design for obtaining a high vibration frequency is not clear either.

For many components and devices based on these nanoarchitectures, DLC material is among the best candidates for resistance against degradation in humid conditions and other harsh environmental conditions. For chemical and biological applications, vibrating devices based on the cantilever architecture may not be efficient solutions, as current materials of silicon and silicon nitride may suffer serious degradation of performance due to surface stiction from water or other liquids. Microarchitectures and nanoarchitectures developed from DLC may have better vibration performance in these severe environments due to its hydrophobic properties. It has been reported that DLC material is biocompatible, and there have been some studies of its application to biosensor systems. We can also expect that DLC nanoarchitectures can be developed into other nanofluidic systems for virtual single-molecular sensing and detection. Yet, the combination of both bottom-up and top-down methods can provide more sophisticated nanoarchitectures and better performance for DLC-based mechanical devices.

ACKNOWLEDGMENT

The authors acknowledge National University of Singapore WBS: R-284-000-046-123 for financial support.

REFERENCES

Ager JW, Anders S, Anders A, Wei B, Yao XY, Brown IG, Bhatia CS, Komvopoulos K. 1999. Ion implantation post-processing of amorphous carbon films. Diamond Relat Mater 8: 451–456.

Anderson H, Jonsson M, Vestling L, Lindberg U, Aastrup T. 2007. Quartz crystal microbalance design I. Experimental study of sensor response and performance. Sens Actuators B 123: 27–34.

Berger SD, McKenzie DR, Martin PJ. 1998. EELS analysis of vacuum arc–deposited diamond-like films. Phil Mag Lett 57:285–290.

Bilek MMM, McKenzie DR, Tarrant RN, Lim SHM, McCulloch DG. 2002. Plasma-based ion implantation utilizing a cathodic arc plasma. Surf Coat Technol 156:136–142.

Casiraghi C, Ferrari AC, Robertson J. 2005. The smoothness of tetrahedral amorphous carbon. Diamond Relat Mater 14:913–920.

Chiu MC, Hsieh WP, Ho WY, Wang DY, Shieu FS. 2005. Thermal stability of Cr-doped diamond-like carbon films synthesized by cathodic arc evaporation. Thin Solid Films 476:258–264.

Chua DHC, Milne WI, Yu LJ, Sheeja D, Tay BK. 2003. Fabrication and simulation of amorphous carbon cantilever structures. Mat Res Soc Symp Proc 773:N3.2.1–N3.2.8.

Chua DHC, Milne WI, Sheeja D, Tay BK, Schneider D. 2004a. Fabrication of diamond-like amorphous carbon cantilever resonators. 2004. J Vac Sci Technol B 22:2680–2684.

Chua DHC, Teo KBK, Tsai TH, Milne WI, Sheeja D, Tay BK, Schneider D. 2004b. Correlation of surface, mechanical and microproperties of tetrahedral amorphous carbon films deposited under different magnetic confinement conditions. Appl Surf Sci 221:455–458.

Cleland AN, Roukes ML. 1999. External control of dissipation in a nanometer-scale radiofrequency mechanical resonator. Sens Actuators 72:256–261.

Davis CA, Amaratunga GAJ, Knowles KM. 1998. Growth mechanism and cross-sectional structure of tetrahedral amorphous carbon thin films. Phys Rev Lett 80:3280–3282.

De Los Santos HJ. 1999. Introduction to Microelectromechanical Microwave Systems. Chapter 4. Norwood MA: Artech House Publishers.

Eason R. 2007. *Pulsed Laser Deposition of Thin Films: Applications-Led Growth Functional Materials*. Hoboken, NJ: Wiley-InterScience.

Egerton RF. 1996. *Electron Energy-Loss Spectroscopy in the Electron Microscope*. New York: Plenum Press.

Ferrari AC, Robertson J. 2000. Interpretation of Raman spectra of disordered and amorphous carbon. Phys Rev B 61:14095–14107.

Foong YM, Hsieh J, Li X, Chua DHC. 2010. The study on the effect of erbium on diamond-like carbon deposited by pulsed laser deposition technique. J Appl Phys 106:064904.

Funada Y, Awazu K, Shimamura K. 1994. Diamond-like carbon thin-film formation by ion-beam-assisted deposition. Surface & Coatings Technol 66:514–518.

Genta G. 1999. *Vibration of Structure and Machines: Practical Aspects*. New York: Springer-Verlag.

Han H. 1999. Ultra-thin tetrahedral amorphous carbon film as slider overcoat for high areal density magnetic recording. Surf Coat Technol 120–121:579.

Hauert R. 2004. An overview on the tribological behavior of diamond-like carbon in technical and medical applications. Tribol Int 37:991–1003.

Hsieh WP, Wang DY, Shieu FS. 1999. Characterization of the Ti-doped diamond-like carbon coatings on a type 304 stainless steel. J Vac Sci Technol 17:1053–1058.

Hsu CY, Chen LY, Hong FCN. 1998. Properties of diamond-like carbon films deposited by ion plating with a pulsed substrate bias. Diamond Relat Mater 7:884–891.

Li X, Chua DHC. 2008. Diamond-like carbon: A new material base for nano-architectures. Cosmos 4(2):203–234.

Luo J, Fu YQ, Le HR, Williams JA, Spearing SM, Milne WI. 2007. Diamond-like carbon MEMs. J Micromech Microeng 17:S146–S163.

Martin PJ, Bendavid A. 2001. Review of the filtered vacuum arc process and material deposition. Thin Solid Films 394:1–15.

Martin PJ, Filipczuk SW, Netterfield RP, Field JS, Whitenall DF, McKenzie DR. 1988. Structure and hardness of diamond-like carbon films prepared by arc evaporation. J Mater Sci Lett 7:410–412.

Matsui S. 2006. Three dimensional nanostructure fabrication focused ion beam chemical vapour deposition. In: Bhushan B, editor. *Handbook of Nanotechnologies*. Chapter 6. New York: Spinger-Verlag, pp. 179–195.

Mayer J, Giannuzzi LA, Kamino T, Michael J. 2007. TEM sample preparation and FIB-induced damage. MRS Bull 32:400–407.

Milne WI, Robertson J, Amaratunga GAJ, editors. 1996. *Diamond Like Carbon*. World Scientific.

Peiner E, Tibrewala A, Bandorf R, Luthje H, Doering L, Limmer W. 2007. Diamond-like carbon for MEMs. J Micromech Microeng 17:S83–S90.

Pivin JC, Allouard M, Rotureau G. 1991. Determination of the optimal energy-range for obtaining diamond-like films by ion implantation. Surf Coat Technol 47:433–444.

Robertson J. 2002. Diamond-like amorphous carbon. Mater Sci Engin R 37:129–281.

Saw KG, Idrus RM, Ibrahim K. 2000. Diamond-like amorphous carbon film by d.c. magnetron sputtering. J Mater Sci Lett 19:735–737.

Schneider D, Schwarz T, Scheibe HJ, Panzner M. 1997. Non-destructive evaluation of diamond and diamond-like carbon films by laser induced surface acoustic waves. Thin Solid Films 295:107–116.

Schwan J, Ulrich S, Batori V, Ehrhardt H, Silva SRP. 1996. Raman spectroscopy on amorphous carbon films. J Appl Phys 80:440–447.

Sheeja D, Tay BK, Lau SP, Shi X. 2001. Tribological properties and adhesive strength of DLC coatings prepared under different substrate bias voltages. Wear 249:433–439.

Shi X, Tay BK, Tan HS, Zhong L, Tu YQ, Silva SRP, Milne WI. 1996. Properties of carbon ion deposited tetrahedral amorphous carbon films as a function of ion energy. J Appl Phys 79:7234–7240.

Shi X, Tay BK, Lau SP. 2000. The double bend filtered cathodic arc technology and its applications. Int J Mod Phys B 14:136–153.

Tay BK, Shi X, Tan SH, Chua DHC. 1999. Investigation of tetrahedral amorphous carbon films using X-ray photoelectron and Raman spectroscopy. Surf Interface Anal 28:231–234.

Tsai JTH, Teo KBK, Milne WI. 2002. Approach for a self-assembled thin film edge field emitter. J Vac Sci Technol B 20(1):1–4.

Uhlmann S, Frauenheim T, Lifshitz Y. 1998. Molecular-dynamics study of the fundamental processes involved in subplantation of diamondlike carbon. Phys Rev Lett 81:641–644.

Voevodin AA, Donley MS, Zabinski JS. 1997. Pulsed Laser deposition of diamond-like carbon wear protective coatings: a review. Surf Coat Technol 92:42–49.

Wang DY, Weng KW, Hwang SY. 2000. Study on metal doped diamond-like carbon films synthesized by cathodic arc evaporation. Diamond Relat Mater 9:1762–1766.

Weiler M, Sattel S, Jung K, Ehrhardt H, Veerasamy VS, Robertson J. 1994. Highly tetrahedral, diamond-like amorphous hydrogenated carbon prepared from a plasma beam source. Appl Phys Lett 64:2797–2799.

Weng KW, Chen YC, Lin TN, Wang DY. 2006. Metal-doped diamond carbon film synthesized by filter-arc deposition. Thin Solid Films 515:1053–1957.

Wu S. 1982. *Polymer Interface and Adhesion.* New York: Marcel Dekker.

Yao N, editor. 2007. *Focused Ion Beam Systems: Basics and Applications.* Cambridge: Cambridge University Press.

Young T. 1805. An essay on the cohesion of fluids. Philos Trans R Soc London 9:225.

Zhang P. 2004. Fabrication and characterisation of composite diamond-like carbon films. PhD Thesis, Nanyang Technological University, 2004.

Zhang ZY, Lu XC, Luo JB, Liu Y, Zhang CH. 2008a. Mechanical properties of La_2O_3 doped diamond-like carbon films. Surf Coat Technol 202:1621–1627.

Zhang ZY, Lu XC, Luo JB. 2008b. Unusual stress behaviour of CeO_2-doped diamond-like carbon nanofilms. Philos Mag Lett 88:145–151.

Zhang W, Tanaka A, Wazumi K, Koga Y. 2002. Structural, mechanical and tribological properties of diamond-like carbon films prepared under different substrate bias voltage. Diamond Relat Mater 11:1837–1844.

Zhang S, Xie H, Zeng XT, Jing P. 1999. Residual stress characterization of diamond-like carbon coatings by X-ray diffraction method. Surf Coat Technol 122:219–224.

20

SYNTHESIS, PROPERTIES, INTEGRATION, AND APPLICATIONS OF VERTICALLY ALIGNED CERAMIC NANOSTRUCTURES

D. Pliszka,[1,4] S. Sundarrajan,[1] and S. Ramakrishna[1,2,3]

[1]*Nanoscience and Nanotechnology Initiative, National University of Singapore, Singapore*
[2]*Department of Mechanical Engineering, National University of Singapore, Singapore*
[3]*Division of Bioengineering, National University of Singapore, Singapore*
[4]*Institute of Physics, Pomeranian University in Slupsk, Slupsk, Poland*

INTRODUCTION

Nanostructures are defined as having at least one of the dimensions in nanometer range (1–1000 nm). One-dimensional (1-D) nanostructured materials such as tubes, wires, belts, and rods have attracted much attention in the last decade due to their unique fundamental properties and potential superior performance compared to their bulk counterparts (Xia et al. 2003) and possible applications in the fabrication of nanoscale devices. In the development of such high-performance devices, it is crucial to create highly oriented and ordered arrays of nanostructures since the anisotropy of their properties could be important. One of the most striking examples of nanostructure applications is in microelectronics, where smaller structures mean higher performance owing to an increased number of transistors within an integrated circuit, and more components per chip, both of which reduce size and lower power consumption. Also, information

Ceramic Integration and Joining Technologies: From Macro to Nanoscale, First Edition.
Edited by Mrityunjay Singh, Tatsuki Ohji, Rajiv Asthana, Sanjay Mathur.

storage techniques shall benefit by the introduction of single-nanometer size structures.

Various chemical, electrochemical, and physical deposition techniques such as chemical vapor deposition, pulsed laser deposition, sol–gel methods, hydrothermal solution synthesis, and electrochemical deposition in porous membranes have been reported to create oriented nanostructures.

Selection of the possible materials used for nanopatterning of vertically aligned nanowires offers a wide range including polymers by nanoimprinting techniques, metals, and pure semiconductor materials like silicon and germanium. Among them, growing interest is directed to ceramic materials. Their unique physical properties allow for numerous possible applications.

Ceramic nanowires due to their unique properties have found numerous potential applications, such as ultraviolet (UV) nanolasers (Huang et al. 2001), field-effect transistors (Cha et al. 2006), solar cell electrodes (Law et al. 2005), nanogenerators (Wang et al. 2007), information storage (Luo et al. 2005), optoelectronic devices (Kang and Kim 2006), solar cells (Beermann et al. 2000), humidity sensors (Kuang et al. 2007), and chemical sensors (Chen et al. 2008; Li et al. 2008).

1-D nanowire structures are an ideal candidate to explore properties, which may be inaccessible or at least hardly achieved in larger scales. For example, discreteness of electron states comes into play when the Coulomb energy associated with the addition of a single electron becomes larger than the thermal energy.

From a fundamental physics point of view, the low-dimensional nanowire structure is an ideal platform to probe properties, which may be inaccessible or hard to achieve in larger devices, due to the reduced device size and ideal material properties. As a result, the electrical and optical properties in these nanoscale devices are determined not exclusively by the material composition but can also be tailored by specific device geometry.

In this chapter, we present synthesis methods of vertically aligned ceramic nanowires. Various physicochemical properties of such structures are discussed and their applications are also presented.

VERTICALLY ALIGNED CERAMIC NANOSTRUCTURE SYNTHESIS METHODS

Non-template-Assisted Methods

Since the crystalline quality of 1-D nanostructures is essential for their performance, the most commonly used methods to synthesize such structures are based on crystalline structure growth. However, in spite of the fact that crystallization has been studied for several centuries, there is a relative lack of quantitative knowledge about this process. It is also not trivial to obtain strict control over the process to synthesize precisely determined nanostructures. In the development of the crystal synthesis for generation of nanostructures, the most important issues are simultaneous control over nanostructure dimensions, morphology, and uniformity. Basically, there are two main types of

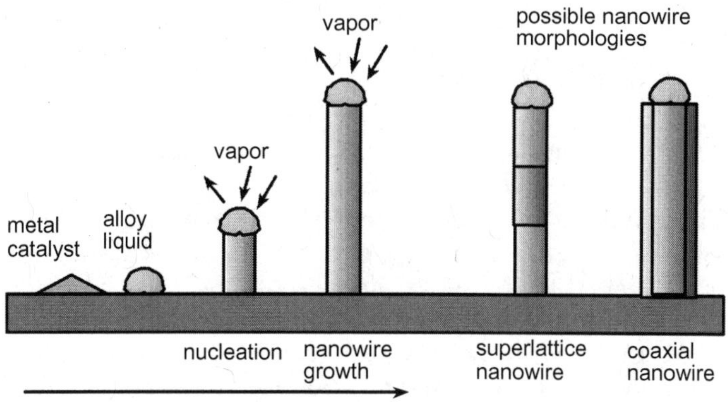

Figure 20.1. Synthesis of nanowires by the VLS method.

the process presented next: vapor–liquid–solid (VLS) and vapor–solid (VS). Additionally, we discuss solution growth crystallization method.

VLS Process. The 1-D growth of nanowires was demonstrated in the 1960s by Wagner and Ellis (1964). At that time, the sizes of such structures were in the range of micrometers. A theory of growth of such structures was developed on the basis of the so-called VLS mechanism. The VLS crystal synthesis process can be divided into the following steps: the formation of a small liquid droplet, and then the alloying, nucleation, and growth of the nanowire (Fig. 20.1).

Growth is started on a clean, defect-free surface of a substrate. In many cases, semiconductor wafers are used, which are important in the integration of the structures with electronic devices; however, other materials such as sapphire crystal or glasses can also be applied. Initially, nanosize metal clusters (diameter below 100 nm) are deposited on the surface to serve as a catalyst for the nanowire growth. Different techniques can be applied for arranging the catalyst particles on the substrate including other nanopatterning techniques. The as-prepared substrates are placed in a reaction chamber, heated until the clusters melt and form liquid droplets. Frequently, this is achieved by dissolving the semiconductor material to form an alloy with a reduced melting temperature as compared to the pure metal used. A suitable temperature in the range of 300–1100°C is chosen according to the binary-phase diagram (diagram showing equilibrium conditions between thermodynamically distinguished phases) between the metal and the target material. For epitaxial growth of nanowires, direct contact of the droplet with the crystalline substrate must be assured. In the second step, a gas containing the growth material flows through the reaction chamber. Since the surface of the liquid droplet has a much larger sticking coefficient than the solid substrate, the precursor atoms prefer to deposit on the surface of the liquid and form an alloy. Surface absorption of precursor atoms into the liquid droplet leads to supersaturation of the semiconductor component in the catalyst droplet. As a consequence, crystal growth occurs at the solid–liquid interface by precipitation and nanowire growth commences. The nanowire diameter is determined mainly by the size of the droplet. The growth rate depends on the supersaturation of the semiconductor component, which can be influenced by the concentration of the precursor vapor and the substrate temperature. The droplet in most cases remains at the tip of the nanowire in the course of the subsequent growth.

In 1992, Hitachi applied the VLS mechanism using gold droplets as catalysts for the growth of III–V nanowires, for example, GaAs and InGaAs (Haraguchi et al. 1992). Since that time, the doping of wires has been widely demonstrated, which proved the general possibility for the formation of p–n junctions within the wires and, hence, for nanowire-based light-emitting diodes (LEDs). The Yang group (Huang et al. 2001) investigated the formation of micropatterned nanowires of ZnO by prepatterning catalytic Au. The group showed for the first time the lasting effect of ZnO nanowires (Huang et al. 2001), which sparked extensive interest in the fabrication of well-aligned and highly ordered ZnO nanowires or rods. In the case of the VLS synthesis method, a problem is the possible influence of gold incorporated in the semiconductor on the electrical properties of nanowires. For example, gold incorporation is known to result in deep-level defects near the midgap of Si, which drastically reduces the minority carrier lifetime and generally should be avoided in the context of active devices.

VS Process. Nanowires can also be grown without extra metal catalysts by thermally evaporating a suitable source material near its melting point and then by depositing at cooler temperatures (Dai et al. 2003). Such a self-organization process, which does not involve liquid droplets as the catalyst, is referred to as a "VS" mechanism.

In many cases, the growth mechanism works in an analogous way to the VLS process, differing in that here, one component of the gaseous atoms of source materials might play the role of the catalyst itself (Liu et al. 1995). Common examples are the formation of ZnO and Ga_2O_3 nanowires. Either the precursor gas is decomposed due to a high-temperature reaction or by the pure metal (Zn and Ga) powder evaporation under a suitable flow gas atmosphere (Fan et al. 2004). Zn and Ga are characterized by low melting points and sublimation temperatures, which are comparable to the temperatures in the reaction chamber. At places with inhomogeneous areas or defects, or at the cooler parts of the reaction chamber walls, segregation at the surface of the substrate is very probable. In the case of ZnO when oxygen is added to the reaction chamber, the liquid droplets solidify quickly by oxidation and the formation of ZnO wires can be observed. A similar behavior has been reported for the formation of GaN nanowires on Si substrates. In the VS growth mode, control of the nanowire diameter is accomplished mainly by changing the evaporation and collection temperatures, as well as the vapor pressure. However, fabrication of various hierarchical semiconductor nanostructures through the VS growth mode has been reported in the literature (Dai et al. 2003; Wang et al. 2004c), but no strict control of the spatial arrangement has been achieved so far.

Solution Growth. The synthesis of nanocrystals (NCs) by solution methods is well-known for II–VI materials (Murray et al. 1993). The formation of CdSe nanorods with an aspect ratio of 30 was reported by, for example, by Manna et al. (2000), who also investigated the factors influencing shape control. Recently, there have been an increasing number of reports on the solution growth of nanowires. Holmes et al. (2000) used a supercritical fluid solution-phase approach to the growth of Si nanowires, including diameter control by using dodecanethiol-capped Au NCs. The growth of ZnO nanowires onto various substrates by the use of thermal decomposition of methylamine and zinc nitrate in aqueous solution was reported by several groups (Vayssieres et al. 2001; Greene et al. 2003). For example, a multistep, seeded growth technique may allow control of the sequential nucleation and growth, leading to complex nanostructures composed of hierarchical nanorods. Overall, solution methods provide little control over the area density, and also the nanowires have smaller aspect ratios than those grown by vapor-phase techniques. Moreover, obtained nanowires display much worse vertical alignment, although they can exhibit some texture. For example, in case of ZnO, the growth and alignment direction is mainly along the *c*-axis, which is crucial to the UV lasing performance of ZnO nanowire arrays. The main advantage of such solution-based methods might be the possibility to create nanowire arrays at low temperatures on large scales at low cost, and on various (even flexible) substrates (Choy et al. 2003).

Template-Assisted Methods

Template-assisted synthesis is a common and straightforward approach to obtain well-aligned 1-D nanoscale structures like nanorods or nanowires. Mostly used are porous templates with nanopores ordered perpendicularly to the template surface. In this approach, a template serves as a scaffold on which various materials are generated and shaped in nanostructures with morphology complementary to the template. This method was pioneered by Martin (1994). Generally, two types of porous membranes are commonly used in nanostructure synthesis. One is polymer films with track-etched channels, alumina films containing anodically etched pores, and block copolymers with vertically self-aligned cylinders. The second method is a template method in which nanosphere lithography is used as template (spots) for nanowire growth using methods discussed in the previous paragraph.

Nanosphere Lithography. The idea of nanosphere lithography is deposition and self-organization of submicrometer spheres into a monolayer with a hexagonal closed-packed structure (Kosiorek et al. 2004). Typically, silica or polystyrene spheres are used, which are commercially available with well-defined size distributions. The deposited single layer of the spheres on a substrate can be used as a lithography shadow mask and for the creation of the template for nanostructures built on the substrate surface (Fig. 20.2). Upon creation of the template masking layer, the microspheres are removed by dissolving them (e.g., using acetone for polystyrene and hydrofluoric acid for SiO_2 spheres).

Upon removal, a pattern of triangular metal catalyst islands is obtained. By this approach, it is possible to produce hexagonal arrays of more than 1 cm^2 with defect-free single domains of up to 100×100 mm (Kempa et al. 2003). Not only are hexagonal patterns available but also other patterns are possible to realize by deposition, such as more than one nanosphere layers or a combination of nanospheres with different shapes

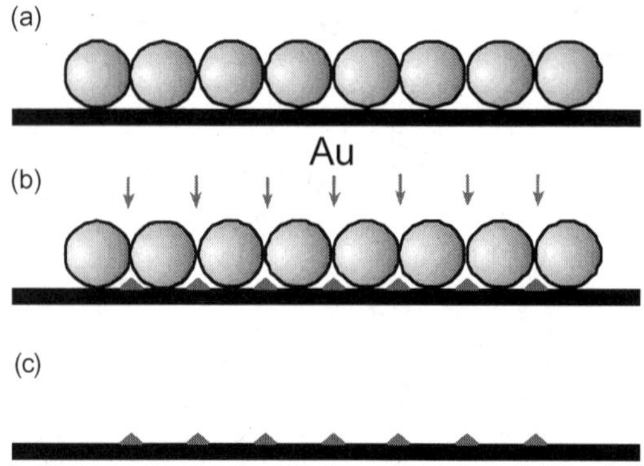

Figure 20.2. Nanosphere lithography technique of nanowire template manufacturing. Technological steps include (a) deposition of the microsphere nanolayer, (b) evaporation of the catalyst material, and (c) removal of nanospheres and further technological process for nanowire growth.

(Banerjee et al. 2005). Nanosphere lithography offers a simple, cost-effective, and high-efficiency lithographical approach, which has been widely utilized for the fabrication of nanostructures, including semiconductor nanowire arrays. Since the size of the initial metal catalyst islands is much larger than the nanowire diameters (50–150 nm), a cluster of nanowires is formed in each lattice site, forming an interconnected nanowire bunch. Nanosphere lithography can be treated as the intermediate method connecting template and non-template-assisted nanowire synthesis methods since, basically, it is a step before the proper nanowire synthesis process.

Alumina Porous Templates. Nanowires may be grown by electrochemical deposition methods by the use of the templates such as porous alumina—anodic aluminum oxide (AAO). Electrochemical oxidation of aluminum (oxidation) was discovered in 1920, allowing the fabrication of disordered porous alumina with pore diameters in the submicrometer range. Since the work performed by Masuda and Fukuda (1995) AAO nanopore arrays have been extensively used in nanotechnology. In their work, it was observed that ordered domains of hexagonally distributed Al_2O_3 pores were established under certain conditions. Free-standing AAO membranes were formed and used also as metal deposition masks. Such membranes should be reasonably thick because of the stability problems with the fragile porous alumina material. Also, shadowing effects can occur if they are used directly for a masking procedure leading to incompletely filled channels. The application of AAO membranes as lithography shadow masks for Au deposition was performed, for example, by Wu et al. (2002) for molecular beam epitaxial growth of GaAs nanowires on a GaAs (111) substrate. The free-standing thin AAO templates (150–200 nm) were transferred directly onto the substrates via van der Waals bonding. The obtained nanowires have shown a much narrower diameter distribution than those grown by using a gold film as a catalyst (discussed previously). Thin AAO membranes are difficult to handle due to their stiffness and transparency, which is their major drawback and limits the pattern transfer to small areas only. Recently, a new technique was developed to transfer the AAO pore arrays into a metal membrane (Lee et al. 2005a). This was achieved by evaporating a thin metal film on top of the AAO followed by electrochemical deposition of metal onto the inner walls of the pores. Afterward, the AAO is chemically dissolved, resulting in a free-standing metal membrane that is mechanically stable and can be used as a macroscopically sized mask for various applications (Lee et al. 2005a). Such metal membranes were used for large-scale nanopatterned growth of nanowires (Fan et al. 2005a). A control of the nanowire diameter can be realized by using mask membranes with different pore sizes, which can be controlled by the parameters of the electrochemical Au deposition onto the membrane. A drawback of this method is that nanowires fabricated in this way are still not perfect in terms of diameter distribution and lattice vacancies. These shortcomings can be avoided by using, for example, pulsed laser deposition instead of thermal evaporation and by further optimization of the growth conditions.

Electrochemical Deposition. Electrochemical deposition with use of template masks is a simple method to obtain nanowires by electrolysis. The template used for nanowire generation is attached to the cathode. The anode is placed in the deposition solution parallel to the cathode. Both are interconnected by deposition solution. With the electric field applied, cations diffuse toward the cathode where they are reduced, resulting in the growth of nanowires inside the pores of the template. After finishing the deposition, free-standing nanowire arrays can be obtained by dissolution of the template membrane as described in an earlier section. Nanowire length is determined by the length of the template membrane pores and may be tuned by etching the obtained nanowires after membrane removal (Fig. 20.3). This method allows for the fabrication of nanowires using a wide range of materials including metals, conductive polymers, and metal oxide materials (Martin 1996). Subsequent electrochemical deposition of different materials is a way to obtain multisegmented nanowires (Lee et al. 2005b).

The key issue for semiconductor nanowires fabricated by electrochemical deposition is crystalline quality. In most cases, the nanowires are not fully epitaxially grown, and they have either an amorphous or polycrystalline structure. They consist of a number of defects, which might affect their technical application, especially in optics.

Block Copolymer Templates. Thin films of diblock copolymers can form large-scale patterns with a periodicity in the nanometer range. The resulting self-organized structures are strongly influenced by the choice of the two different components of the block polymers. The microphase structures generally observed in such systems consist of periodic arrangements of lamellae, cylinders, or spheres. When one polymer material is selectively removed by organic dissolution, a porous film is obtained. The size and periodicity of such pore structures is the result of the length of the block copolymer chains and is typically in the range of some tens of nanometers. In the work by Park et al. (1997), holes with a diameter of around 20 nm and a separation of 40 nm were produced over a large-scale area. The porous polymer thin film can be used as a lithography mask to create metal nanodot arrays, or as an etching mask to produce high-density cavities on the substrate. A useful application is that, when the

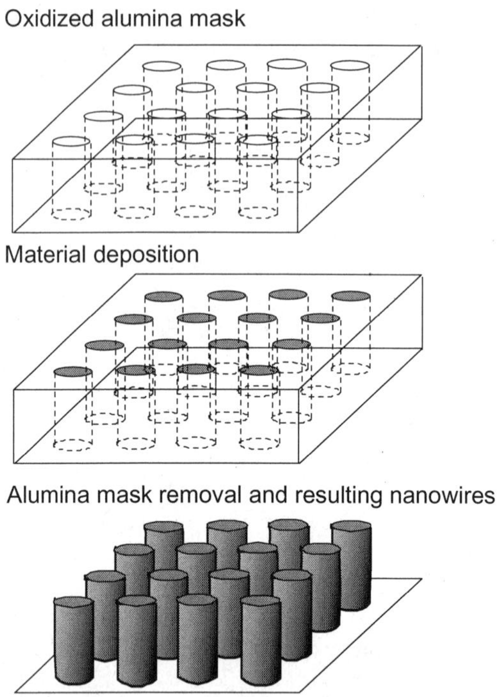

Figure 20.3. Alumina nanomask using production of nanowires.

polymer film is prepared on a silicon nitride film, the pore structure can be transferred into the nitride layer or even to the Si substrate underneath. Subsequently, the substrates could be used as promising growth templates for high-density nanowire arrays if the holes are homogeneously decorated with a metal. Another way of applying diblock copolymers was demonstrated by Glass et al. (2003). Here, self-assembly of block copolymer micelles were combined with e-beam lithography for the deposition of metal nanoparticles. Such a method is capable of generating gold nanodots or lines with a large variety of rather complex patterns. The pattern dimension and geometry is controlled by combining the self-assembly of the block copolymer micelles with the prestructures formed by e-beam or photolithography. Gold nanodots 5 and 6 nm in size were arranged in two differently spaced patterns (57 or 73 nm). It can be envisaged that block copolymer micelle nanolithography may offer great potential for growing spatially separated arrays of very small-sized nanowires.

In contrast to spot patterns from spherical microdomains, cylindrical microdomains provide high aspect ratio features with vertical sidewalls that are available for further processing. Morphology of the copolymer depends on the volume fractions of the blocks and also on the thickness of the copolymer layer, and the size of the channels depends on molecular weight. Due to thermodynamic reasons, control over domain orientation is quite a challenging task. If the thickness of the block copolymer layer is too small and not sufficient to maintain the horizontal alignment of cylinders, it is possible to obtain vertically aligned structures (Fitzgerald et al. 2007). However, using such thin films requires procedures to overcome the preference of at least one of the blocks to wet a substrate and to force the domain structure to be parallel to the substrate. One approach is use of the neutral substrate, which prefers neither of the blocks of copolymer. Usually, this goal is attained by the deposition of random copolymer brushes on the substrate, but also other approaches like hydrogen passivation are used. Besides, these thermodynamically stable approaches, other methods are also used to get vertically aligned cylinders. One is kinetic trapping of the copolymer blocks in nonequilibrium state by the use, for example, of an external electric field.

PROPERTIES OF 1-D NANOSTRUCTURES

Among 1-D nanostructures, nanowires have attracted a lot of attention in nanotechnology research (structures with at least one of the dimensions lesser than 1000 nm are termed as 1-D nanostructures). Nanowire sizes are in micrometer scale and nanometer scale in lengths and diameters, respectively. Nanowires are fabricated using a wide variety of materials including metal oxides, metals, metal alloys, carbon, and polymers. Syntheses of nanowires by various methods are presented in the previous section.

Recently, research on semiconductor nanowires has received increased attention because of the unique and fascinating optical, electrical, mechanical, and thermoelectrical properties of such nanowires and their potential technological applications. The optical, electrical, and mechanical properties of metal oxide nanostructures are highlighted here.

NCs

In bulk metals, the separation between the valence band (VB) and the conduction band (CB) vanishes and the electrons are highly delocalized over a large space, giving metals their conducting properties (Fig. 20.4a). When the size of the metal decreases, its electronic motion is confined; the separation between the VB and CB becomes comparable to or larger than thermal energy, and the metal becomes a semiconductor (Fig. 20.4b). Further increase in the confinement leads to more energy separation and the material becomes an insulator. In semiconductors, excitation involves the separation of electrons and holes, that is, charge carriers by a distance known as Bohr radius. The minimum energy required to separate the charge carriers is known as band-gap energy. When the size of the semiconductor (such as NCs) becomes comparable to, or smaller than, the Bohr radius, the charge carriers are forced to confine in space. This restricted confinement requires additional energy and thereby leads to an increase in the band-gap energy (Fig. 20.4c).

The band gap is dependent on the size of the NC, and it is inversely proportional to crystal size. Figure 20.4d illustrates the difference in absorption between bulk semiconductor (continuous absorption, shown in black solid line) and NCs (a discrete absorption, shown in colored bars).

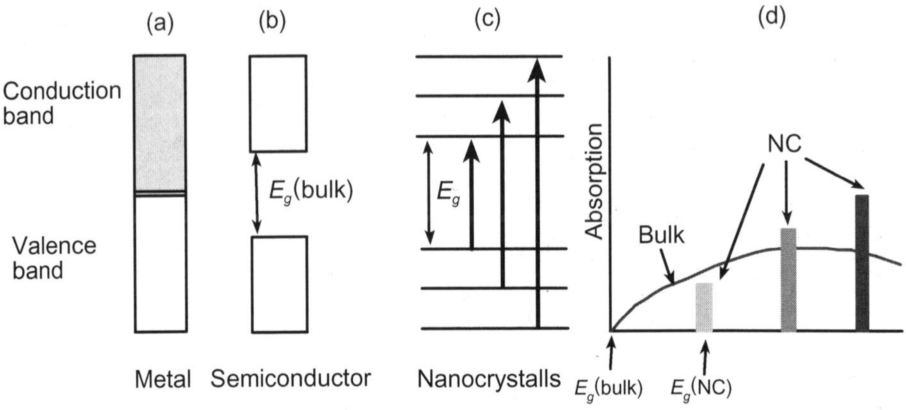

Figure 20.4. Schematic representation of band gaps in (a) metal, (b) semiconductor, and (c) nanocrystals. (d) Optical absorption of bulk and nanocrystals.

It is well-known that bulk materials' electrical properties such as dielectric and ferroelectric properties are intrinsic, which are independent of the size and shape of the materials. However, the properties of nanoparticles can vary, especially at sizes below several tens of nanometers. The change in physical properties of materials with particle size is termed as size effect. The most popular ferroelectric materials are $BaTiO_3$ and $Pb(ZrTi)O_3$ as dielectric and piezoelectric, respectively. These materials with a particle size larger than the critical size exhibit ferroelectricity, and the crystal symmetry changes from tetragonal to cubic at this critical size at room temperature. The reported critical size of $BaTiO_3$ was in the range from 20 to 200 nm and that of $PbTiO_3$ was in the range from 7 to 11 nm.

Effect of Dimensionality of Nanoparticles and Nanowires

According to a simple effective mass approximation model, the increase in band gap of quantum dots (QDs) and quantum wires (QWs) from the bulk value is $\Delta E_g = 2h^2\zeta^2/m^*d^2$, where $1/m^* = 1/m_e^* + 1/m_h^*$ (m_e^* and m_h^* are the electron's and the hole's effective masses, respectively) and d is the diameter. The ζ values for spherical QDs and cylindrical QWs are $\zeta = \pi$ (zero-point spherical Bessel function) and $\zeta = 2.4048$(zero-point cylindrical Bessel function), respectively. The ratio of increase in band gap between these two systems should be $\Delta E_g^{\text{wire}}/\Delta E_g^{\text{dot}} = 0.586$. It has been concluded that the band-gap ratio of QW and QD are material dependent, 0.564 for Si and 0.73 for CdS. A comparative study of the effect of dimensionality on confinement inferred that the quantum confinement observed in the InP wires is weaker than in InP QDs. QWs sometimes behave as a string of QDs (Yu et al. 2003). According to the particle in-box model, ΔE_g (increase in band gap above the bulk value) should depend linearly on $1/d^2$. However, the band-gap shifts in InP QDs ($\Delta E_g \sim 1/d^{1.35}$) and wires ($\Delta E_g \sim 1/\text{d}^{1.45}$) did not follow this model.

Optical Property in Metallic Nanowires and Nanoparticles. 1-D Au and Ag nanorods show that their surface plasmon resonance (SPR) properties are different from their zero-dimensional (0-D) counterparts. 1-D nanorods exhibit two modes due to transverse and longitudinal excitons, in which the former is fixed (520 nm for Au and 410 nm for Ag), whereas the later one can be fine-tuned from visible to near-infrared (NIR) by controlling aspect ratios (El-Sayed 2001).

Optical Property of Metal Oxide Nanowires with Metal Wires. Chen et al. applied the alumina template nanowires in combination with Ni metallic wire to fine-tune the optical property of nanowires. The thickness of the alumina layer and the length of the metal nanowires were varied by changing the anodization voltages (20 and 100 A/m^2) during electrochemical deposition. The color of the metal plate changed from purple to maroon as the observable angle changed from 15° to 55° for the alumina template prepared at 20 A/m^2, whereas the color changed from yellow to green (for 100-A/m^2 deposited one) for the incident angles of 15° and 55°, respectively. These interesting optical properties resulted from the interference between incident light and reflected light from the Ni metallic wires in the alumina template (Chen et al. 2007).

Positional Control of Nanowires into Architectures. Semiconducting nanowires, such as ZnO and Si, are used as building blocks in nanoelectronics and nanophotonics. Specific positioning of these nanowires in three dimensions (3-D) is necessary to build functional circuits, which is very difficult to achieve due to the scattering force experienced by the nanowires. van der Horst et al. have shown the 3-D trapping of ZnO- and silica-coated Si nanowires in the plane perpendicular to the beam axis away from the surface. This would be highly useful in the production of nanoelectronics and nanophotonics devices (van der Horst et al. 2007). This may solve the problems related to the application of these nanowires into various architectures.

Optical and Electrical Properties of ZnO Nanorods and Nanowires. Fabrication of single-crystalline ZnO nanowires (formed at 400°C, mean diameter 20 nm and length range from 0.2 to 1.5 μm) and ZnO nanorods (formed at 600°C, 100- to 500-nm length, diameter range from 50 to 200 nm) has been reported by Liang et al. by controlling the oxidation temperature of Zn precursors under 5 Pa. Photoluminescence (PL) spectra of ZnO nanowires show a considerably weak peak at 380 nm due to a single ionized oxygen vacancy in the ZnO and strong green emission peak at 497 nm due to exciton emission. However, the PL spectrum of the ZnO nanorods exhibits a strong UV emission at 380 nm, which indicates that they are defect free with high optical quality (Liang et al. 2008).

Electrical transport studies of ZnO nanowires confirm the n-type behavior. The field-effect mobility (field-effect transistor [FET]) is in the range of 20–100 $cm^2/V*s$, and after surface treatment, it is increased to more than 4000 cm^2/s. This mobility value is much higher than the reported mobility for Si nanowire (1350 $cm^2/V/s$) (Chang et al. 2006).

Hong et al. studied the electrical properties of single ZnO nanorods with diameters of 100 nm grown on GaN surface. By using the thermionic field emission model of

Padovani and Stratton, carrier concentration and electron mobility were estimated with the values of $9.2 \times 10^{16}/cm^3$ and $33.82\,cm^2/V/s$, respectively (Quang and Jin 2008).

Comparison of Nanofibers with Other Nanostructures and Films. Usually in nanowires, depending on the chosen preparation conditions, both single-crystalline and polycrystalline materials are observed, whereas in ceramic nanofibers, predominantly polycrystalline materials are produced by electrospinning. However, we have recently reported the fabrication of single-crystalline $PbZr_{0.2}Ti_{0.8}O_3$ of 70 nm in diameter by the combination of electrospinning and sol–gel techniques (Zhou et al. 2007). The nanowires exhibit an ultrahigh piezoelectric strain of about 4.2%, which is more than six times larger than that of the thin film (0.6%) and five times larger than the (0-D) single-crystal $BaTiO_3$.The high-strain behavior in nanowires arises from the reversible switching of ferroelastic domains.

We have recently reported that the adhesion difficulties of TiO_2 nanofibers for solar cell applications can be overcome by TiO_2 nanorods. It involves two steps. First, TiO_2 nanofibers were fabricated by the combination of electrospinning and sol–gel techniques and then were subsequently sintered. These nanofibers were mechanically ground to produce TiO_2 nanorods followed by electrospraying to fabricate electrodes. An energy conversion efficiency of 5.8% was achieved by Ramakrishna et al. (Fujihara et al. 2007).

Ceramic nanofibers provide high surface area-to-volume ratio and high porosity, and hence these have been studied for sensor applications by many authors. Wang et al. fabricated WO_3 nanofibers and found that the response time toward NH_3 is much faster than that reported by other methods (films and nanowires) (Wang et al. 2006).

Physical Properties of 1-D Metal Oxides

This section will cover optical and electrical properties of some of the metal oxides, such as ZnO, TiO_2, In_2O_3, Ga_2O_3, SnO_2, Fe_2O_3, CuO, CdO, and V_2O_5. The electrical property values of these metal oxides are tabulated in Table 20.1. Synthesis of metal oxide nanorods and nanowires such as Al_2O_3, CoO, Co_3O_4, Cu_2O, Eu_2O_3, GeO_2, IrO_2, MgO, MnO_2 (β and δ), Mn_3O_4, MoO_2, MoO_3, NiO, PbO_2, RuO_2, Sb_2O_3, and WO_3 is reported, which is not covered in this review due to space constraints. Most of the above metal oxides are also synthesized with nanotubes, nanobelts, and nanoribbon morphologies, which is not included here.

ZnO. ZnO, a direct band-gap (3.37 eV at room temperature) semiconductor with large exciton binding energy (60 meV) and a promising material, is investigated for applications to optoelectronic devices operating in the blue and UV region. ZnO has a noncentrosymmetric wurtzite structure. It is naturally an n-type semiconductor material because of the presence of intrinsic defects (oxygen vacancies and Zn interstitials). It has been suggested that this n-type behavior is due to hydrogen impurity introduction (Cox et al. 2001). Doping of metal oxides with selective elements is one of the effective ways to adjust their electrical, optical, and other properties for practical applications. The enhancement of n-type conductivity into ZnO nanostructures has been achieved

TABLE 20.1. Physical Properties of Selected 1-D Metal Oxides

System	Band gap (eV)	Photoluminescence (emission range [nm]) Band edge	Defect state	Electrical property Carrier concentration (cm^{-1})	Field-effect mobility ($cm^2/V/s$)
ZnO[a]	3.37	380	520	9.2×10^{16} (Le et al. 2008)	20–200 (4000),[b] (Chang et al. 2005) (33.82) (Le et al. 2008)
TiO_2	2.9 rutile[c] 3.2 anatase[d]	380 (Wu et al. 2006)		2×10^{18}[e] (Sysoev et al. 2006)	10^{-2} (Hendry et al. 2006)
SnO_2	3.6	254 (not observed)	570	1.58×10^8	40
In_2O_3	3.75		416, 435	2.3×10^7	98.1 (Zhang et al. 2003)
Ga_2O_3	4.9		414, 432 (Chang et al. 2005)	5.3×10^8	3.5×10^{-3} (Chang et al. 2005)
Fe_2O_3	2.2			8.9×10^{19}	3.5×10^{-1} (Fan et al. 2005b)
CuO	1.2				
CdO	2.28			1.29×10^{20}[f]	
V_2O_5	2.3				[g]

[a]For ZnO nanorods.
[b]After surface treatment.
[c,d]Band gaps reported are for TiO_2 nanoparticles.
[e,f]Units in cubic centimeter.
[g]Conductivity: 0.5 S/cm.

easily by proper doping with group III (Al, Ga, and In) and group IV (Sn) elements, whereas p-type ZnO has been fabricated by doping with group V elements (N, P, As, and Sb). However, there is difficulty in achieving reliable and reproducible p-type conductivity due to compensation of dopant effect by defects (oxygen vacancies or Zn interstitials) and dopant solubility.

The optical properties of 1-D ZnO nanostructures have been extensively studied for their potential applications in optoelectronic devices due to efficient excitonic emission, which occurs at room temperature. In PL spectra of ZnO nanowires with diameters of 100, 50, and 25 nm, band-edge emission at around 380 nm and defect-related green emission at ~520 nm were observed (Yang et al. 2002). The progressive increase of the green emission with a decrease of nanowire diameter suggests that the fraction of oxygen vacancies is higher in the thinner wires because of the increase in the surface area-to-volume ratio. Further reduction in diameter leads to quantum size effect, which exhibits a blue shift of band-edge emission from 387 nm (in 200 nm diameter) to 373 nm (in 6 nm diameter). It has been observed in ZnO nanorods with a diameter of 1.1 nm (less than the bulk exciton Bohr radius, aB, 2.34 nm) that the exciton binding energy is significantly enhanced due to 1-D confinement (Gu et al. 2004). Two absorption spectra at 3.53 eV (in bulk, it is 3.30 eV) and 4.59 eV have been observed, which are attributed to the ground and excited electronic states, respectively. Kong et al. reported an interesting phenomenon of ZnO nanoring formation by self-coiling of nanobelts to minimize the energy contributed by polar charges (Zn^{2+} and O^{2-} charges), surface area, and elastic deformation (Kong et al. 2004).

Optical Property Change by Doping. The doping of ZnO nanowires with 15% concentration of Ga, In, and Sn was carried out by Bae et al. (2005). The absorption and PL of Sn-doped nanowires shift to the lower-energy region than those of In- and Ga-doped nanowires due to the larger charge density of Sn. Doping of ZnO nanowire with Mg (MgO has a band gap of 7.7 eV) changes both optical property of the nanowire and crystal structure (from wurtzite structure in ZnO to cubic). The PL spectra of $Zn_{0.84}Mg_{0.16}O$ (diameter of 30 nm) exhibited a weak emission peak at 384.4 nm (due to ZnO near band-gap emission) with a broad and stronger intense emission at 495.8 nm, which can be assigned to the transition between photoexcited holes and singly ionized oxygen vacancy (Zhi et al. 2005).

***In_2O_3*.** In_2O_3 has a body-centered cubic structure and a wide band gap of 3.75 eV. Due to this wide band gap, it has high optical transparency. Hence, it has been widely used as liquid crystal displays, solar cells, and so on. The PL spectrum of In_2O_3 nanowire when excited at 260 nm emits stable and high intensity blue light with peaks at 416 and 435 nm (Wu et al. 2003) The presence of oxygen defects has been confirmed by X-ray photoelectron spectroscopy (XPS); thus, it is believed that the intensive blue light emission is attributed to oxygen vacancies and indium–oxygen centers. These two peaks may have originated from a radiative recombination of two varieties of trapped excitons (Wu et al. 2003).

Conductance of the device fabricated by nanowire is found to be 7.08×10^{-7} S at V_{ds} =0 V (at 290 K). From the I–V_g curve, an electron mobility of 98.1 $cm^2/V*s$ and a carrier concentration of $2.3 \times 10^7/cm$ were calculated (Zhang et al. 2003). A power-law dependence of In_2O_3 nanowire FET at a large positive gate bias was observed by Liu et al., which is the evidence of 1-D Luttinger liquid behavior (Liu et al. 2005).

***TiO_2*.** TiO_2 is an n-type semiconductor, which has been widely used in solar energy conversion, degradation of environmental pollutants, sensors, lithium ion storage system, and so on. The synthesized nanowires appear in rutile and anatase crystal structures. The PL spectrum of single-crystalline rutile TiO_2 nanowires, when excited at 254 nm, emits a peak at 380 nm (3.26 eV), which results from free exciton emission (Wu et al. 2006). It is to be noted here that this behavior is in contrast to bulk TiO_2, which does not show any emission peak as noticed by some groups. Cathodoluminescence spectra show similar behavior as that of bulk TiO_2 (Wu et al. 2005). Dong et al. observed a decrease in band-gap energy (2.67–3.00 eV) by doping with C, N, and S with TiO_2, which can be ascribed to N and S codoping by mixing the N 2p and S 3p states with O 2p states (Dong et al. 2008). The terahertz mobilities of single-crystalline rutile and porous TiO_2 nanoparticles have been reported by Hendry et al. (Table 20.1). The reduced mobility in the nanoparticles is due to charge trapping at defect sites, and the intrinsic reduction in mobility caused by porous geometry induced path restrictions (Hendry et al. 2006)..

***SnO$_2$*.** SnO_2 is an n-type semiconductor, which has a rutile structure. It has also been reported that some of them have orthorhombic crystal structures. In the PL spectrum, a strong yellow emission band at 570 nm is observed, which originates from surface defect states, corresponding to oxygen vacancies or tin interstitials (Calestani et al. 2005). However, no band-edge absorption is observed for nanowires at 300, which is usually observed for nanoparticles (Calestani et al. 2005).

***Ga$_2$O$_3$*.** Ga_2O_3 has a wide band gap of 4.9 eV and, because of a large band gap, possesses excellent thermal and chemical stabilities. It is a suitable material for the applications of high-temperature oxygen sensor and UV-transparent conductive material. In PL spectra (26-nm diameter and average length of 20 μm), a broadband blue-green emission (two peak positions at 414 and 432 nm) is attributed to the recombination of an electron and a hole originating from donors (due to oxygen vacancies) and acceptors (due to gallium–oxygen vacancy pairs). These donor–acceptor pairs can form trapped excitons resulting in the blue-green emission (Chang et al. 2005). Wu et al. reported that the Ga_2O_3 nanowires with a diameter of 60 nm upon excitation at 378 nm emit only one peak at 446 nm (Wu et al. 2000).

Ga_2O_3 has formidable insulating properties at low temperature, whereas it exhibits n-type semiconducting properties at high temperature. Electron transport measurements show that the Zn-doped Ga_2O_3 nanowires exhibit p-type semiconducting behavior with a significant enhancement of conductivity. The estimated carrier concentration and hole mobility values for the Zn-doped Ga_2O_3 are tabulated in Table 20.1 (Chang et al. 2005)

***Fe$_2$O$_3$*.** α-Fe_2O_3 ($E_g = 2.2$ eV) has been widely used for catalysts, gas sensors, and other applications. Fan et al. has studied pure α–Fe_2O_3 nanobelts as FETs. The as-grown nanobelts possess an n-type behavior. After doping with Zn ions at a lower temperature, four orders of magnitude higher conductivity (17.2 $[\Omega cm]^{-1}$) than as-grown nanobelts was estimated. The electron concentration and mobility of the doped Fe_2O_3 are estimated to be $8.9 \times 10^{19}/cm^3$ and $3.2 \times 10^{-1}\,cm^2/V{*}s$, respectively. It was observed that there was a morphology transition from nanobelt to nanowire when the synthesis temperature was increased from 400 to 800°C. The transition from n-type to p-type was achieved by doping Zn^{2+} at higher temperature (700°C), in which p-type was caused by the substitution of Fe^{3+} by Zn^{2+} (Fan et al. 2005b). Synthesis of rhombohedral Fe_2O_3 nanowires by Fu et al. (2001) and vertically aligned nanowires at 800°C (15- to 75-nm diameter and 10- to 20-μm lengths) and nanobelts at 700°C (5–10 nm in thickness, 30–300 nm in width, and 5–50 μm in length) by Wen et al. (Wen et al. 2005) has been reported in the literature.

***CdO*.** CdO is an n-type semiconductor with a direct band gap of 2.28 eV and an indirect band gap of 0.55 eV. In films, the electron mobility of CdO is found to be greater than 200 cm^2/V/s (Coutts et al. 2000). High career mobility is essential for transparent conducting oxides in electro-optical applications. Single-crystalline 1-D nano-needle-shaped CdO, having a cubic crystal structure, has been synthesized through CVD process by Liu et al. (2003). The resistivity and electron concentration of the CdO nanoneedles are estimated to be $2.25 \times 10^{-4}\,\Omega$cm and 1.29×10^{20}/cm^3, respectively.

***V_2O_5*.** Vanadium pentoxide (E_g = 2.3 eV) exhibits interesting properties for various applications such as electrochemical, photographic industry, and antistatic coatings. It has been estimated that the conductivity of one individual V_2O_5 nanowire is around 0.5 S/cm at room temperature (Muster et al. 2000). Takahashi et al. have grown single-crystalline V_2O_5 nanorods by template growth. The nanorod array electrodes having approximately five times higher current density and energy storage density than sol–gel-derived film was reported (Takahashi et al. 2004). V_2O_5 nanowires were investigated to study the Li–ion diffusion for battery applications by Patrissi and Martin (1999). It was observed that at 200°C, the nanostructured electrode delivered three times the capacity of the thin film. Apart from the above metal oxides, preparation and applications of other oxides such as CuO and Fe_3O_4 have been reported in the literature (Xue et al. 2005), (Jiang et al. 2002).

***Binary Oxide Nanowires*.** 1-D binary oxide nanostructures were prepared by the combination of two materials to obtain the novel physical properties of these two into one nanostructure. It is expected that optical and electrical properties of binary oxide materials can be fine-tuned by the combination of two materials with different band gaps. Among them, Zn_2SnO_4 and In_2O_3/SnO_2 are studied with an aim to explore further functional nanodevices (Chen et al. 2005; Wang et al. 2005). PL spectra of In_2O_3/SnO_2 nanowires emit in the wide range, that is, UV to green emission. Other mixed nanorods such as strontium niobate and $BaTiO_3$ were grown by the sol–gel electrophoresis method (Limmer et al. 2002).

Mechanical Properties of 1-D Nanostructures

Nanostructured materials show promise for applications in a variety of fields because of their high strength-to-weight ratios when compared to their bulk form. The yield stress and hardness of the material increase with decreasing grain size. The increase in yield stress for smaller grain sizes is given by the Hall–Petch equation

$$\sigma_y = \sigma_o + kd^{-1/2},$$

where σ_o is the friction stress in the absence of grain boundaries, k is a constant, and d is the grain size. The yield stress increases as grain size decreases. Due to dislocation pileup in small-size samples that contain fewer dislocations, the stress at the tip of the pileup decreases, thereby resulting in the requirement of increased applied stress to generate dislocations in the adjacent grain. However, when the grain size approaches even smaller values (below 10 nm for metals), an opposite behavior is observed, which is known as the inverse Hall–Petch effect. Many models, such as dislocation-based, diffusion-based, grain boundary-shearing, and two phase-based models, were proposed to explain this abnormal behavior.

As for 1-D nanostructured materials, namely, SiC nanorods and multiwalled nanotubes, it has been reported that they exhibit greater mechanical strengths than micrometer–diameter materials. This effect is attributed to a reduction in the number of defects per unit length (compared to large structures) that leads to mechanical failure (Wong et al. 1997). A Young's modulus of 660 GPa was estimated for these SiC nanorods (21.5 nm).

APPLICATIONS OF NANOWIRES AND INTEGRATION WITH THE DEVICES

The growth mechanisms used in various nanowire fabrication processes are still far from understood, and various nanolithography approaches are under development to realize a better control of the spatial arrangement of the nanowires. Nevertheless, investigations on multifunctional applications of semiconductor nanowires have already started. Due to their high anisotropy in structure, such 1-D structures are distinguished from their bulk counterparts in various aspects of their properties, as has been demonstrated by theory and numerous experiments. Especially single-crystal 1-D semiconductor architectures have a great importance in applications requiring a large surface area, morphological control, and superior charge transport. It was shown that nanowires might have unique size-dependent mechanical properties (Rubio-Bollinger et al. 2001), an increased luminescence efficiency (Sham et al. 2004), a reduced threshold for laser operation (Huang et al. 2001), an enhanced electromechanical response (Zhao et al. 2004), and an enhanced thermoelectric efficiency compared to conventional Peltier elements (Lin et al. 2000). Application of nanowires in advanced nanoscale devices will require high reproducibility in the synthesis of highly regular nanowire arrays directly on the chosen device substrate, as well as new and innovative processes for device integration and electrical contacts. Such processes should retain the structural integrity of the nanowires and conserve their properties of interest while being within the constraints of a technology. Although numerous applications of 1-D nanostructures are explored, due to space constraints, only some of their important applications are discussed here.

FET

Aligned nanowires have found applications in FETs. In this concept, a current flows through a semiconductor nanowire and the respective ends work as a source and drain for the current. A defined area of the wire is covered by an insulating material where a gate voltage can be applied. For example, the transistor gate can be wrapped around the vertically aligned nanowire, and in such case, the substrate acts as the drain. Such configuration allows better electrostatic gate control of the conducting channel and can drive a higher current per device area than it is possible in conventional transistor architecture (Wang et al. 2004a). From a technological point of view such configuration, is favored over structures parallel to the substrate. This concept was demonstrated by Ng et al. (2004), who reported a vertical surround-gate field-effect transistor (VSG-FET) based on a single ZnO nanowire with a 30-nm diameter, grown on a SiC substrate. Still, the application superiority of such structures over classic FET arrangement in the integrated circuits has to be proved experimentally, but demonstrations provide a proof of concept for the processing techniques. Further development will open up the way for the industrial use of vertically aligned FETs. Due to the high aspect ratio of nanowires, electrons are easily transported and extracted out from nanowire ends by an electric field. It has been demonstrated by Shi et al. (2005) that vertically grown semiconductor nanowires on conductive substrates have strong, low threshold and stable field-emission currents. Therefore, like carbon nanotubes, they can find an application in field emission displays and microelectron source devices. An advantage of such vertically aligned structures is ease of fabrication on a large scale by use of nanoimprint or interference lithography.

Optoelectronic Devices

Optical properties of semiconductor nanowires are attracting much attention due to their promising applications as building blocks for nanoscale optoelectronic detectors, LEDs, lasers, optical waveguides, and nonlinear optical frequency converters. Particularly, the integration of light sources with complementary metal oxide semiconductor (CMOS) devices shall be a revolution in information technology, allowing optical computers to be built. In this concept, the electrical signals carrying the information in conventional circuits are converted into optical pulses, which are transferred between components much faster than the electrical signal. Such optical signals, nowadays, are formed by LEDs or lasers based on III–V materials and can also be emitted by nanowires. Electroluminescence of n-doped ZnO nanowire arrays vertically aligned on p-doped GaN substrate was reported by Park et al. (Park and Yi 2004) with a peak position in the yellow (560 nm) range of the emission normally assigned to defects. Regarding the laser applications of semiconducting nanowires, it has been shown that a ZnO nanowire can operate as a natural optical resonator cavity and gain medium, so that lasing can be observed from vertically aligned nanowires when they are grown to a suitable thickness and length (Huang et al. 2001). For nanoscale optical device applications, heterostructural semiconductor nanowires with modulated compositions can have an advantage over nanowires of pure composition. The existence of surface states leads to deleterious defects, which can degrade the performance of the optoelectronic devices by acting as recombination centers for minority carriers (Lauhon et al. 2004). Such negative action can be avoided, immediately after core definition, by passivation and by forming core–shell nanowires. Such a dielectric shell acts as a high-quality optical cavity around a core acting as gain medium.

Semiconductor nanowires of sufficiently small diameter (i.e., close to the corresponding exciton Bohr radius) shall show quantum confinement-related effects, which can have a strong influence on their optical properties (Ma et al. 2003; Park et al. 2003). Silicon is the most important microelectronic material, but it has poor optical emission properties due to its indirect gap transition. Therefore, new (group IV) optoelectronic devices seem to be attractive. This shall allow the combination of electrical and optical functions on the same Si-based chip. It was also reported that quantum effects could reduce the heat conductance, which could be used for new cooling concepts based on nanowires (Chen et al. 2003). Presently, it is too early to judge in which way these properties might find use in future technology.

Sensors

One of the most important applications of nanowires lies in the field of biotechnology. One of the possible functional concepts is nanowire-based sensors. Nanowire nanosensors, due their high active area and ease of the charge transfer, have been suggested for highly sensitive and selective detection of biological and chemical species (Cui et al. 2001). This approach can offer the possibility to detect the selected attachment of single molecules. However, wide application in the field of human medicine is still speculative and under discussion; fast development of applications is expected within the near future. An important factor to be kept in mind is the potential environmental risks associated with nanowires, although this risk should be minimal for semiconductor nanowires embedded in circuits and potentially in the human body. Semiconducting metal oxide sensors have been extensively studied due to their simplicity and sensitivity to ambient conditions. Zinc oxide (ZnO), an n-type metal oxide semiconductor sensing material with a wide band gap, has been extensively studied because of its high chemical stability, low cost, and good flexibility in fabrication. It showed gas sensitivity to such gases as NH_3, O_3, NO_2, CO, H_2, ethanol, and other analytes (Rao and Rao 1999; Koshizaki and Oyama 2000; Xu et al. 2000). Research efforts in the last few years have focused on the improvement of the gas sensitivity as well as selectivity and to decrease the working temperature. However, most of the studies are focused on sintered particles or thin film-based devices, and it has become clear that the sensing properties of 1-D metal oxide nanomaterials are superior. Several research groups have shown that devices based on 1-D nanostructures have great potential to overcome fundamental constraints of traditional semiconductor oxide sensors such as low sensitivity, poor stability, and high working temperature (Kolmakov et al. 2003). The sensing action of a ZnO nanorod sensor is based on reaction with oxygen species chemisorbed on the semiconductor with analyte and release of electrons to the CB. The result is a decrease of the O_2^-, and O^- ion number and increase of electrons in the CB. This leads to an increase in conductivity of ZnO and generation of an electrical signal. Since aligned zinc oxide nanorods have large surface areas, more oxygen absorption events can occur, and sensing response as well as its speed can be superior to the sensors based on 0-D ZnO particles. Another important issue is the stability of the sensor. Experiments have shown that long-term durability of nanowire-based sensors is good.

Nanogenerators

Development of nanodevices creates the need for a power supply to drive them. Usually, nanostructure-based devices require very low power, which can be provided by an external power source such as battery. However, reliance of the power source

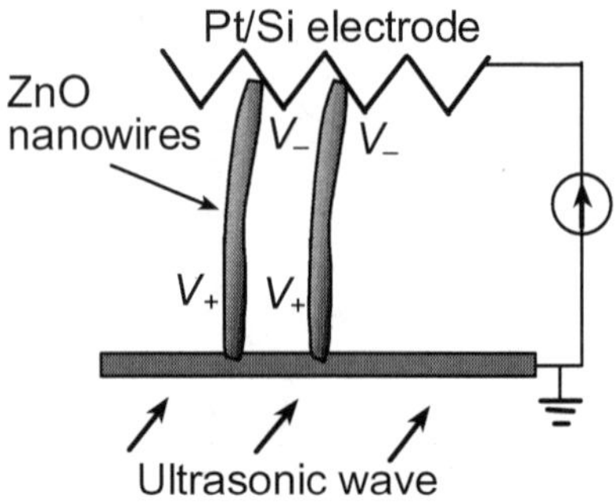

Figure 20.5. Schematic of a nanogenerator based on ceramic nanowires.

may present a limitation for these systems. An interesting novel approach is a nanowire generator driven by an ultrasonic wave, which produces direct current output (Wang et al. 2007). It was fabricated with vertically aligned ZnO nanowire arrays placed under a Pt-plated silicon zigzag-shaped electrode with a small gap between them. The wave drives the electrode up and down, creating vibrations and bending of nanowires. The piezoelectric process converts mechanical energy into electricity. The zigzag electrode works as an array of parallel integrated metal collectors that simultaneously and continuously generate, collect, and transfer output electricity from the nanowires (Fig. 20.5). This approach is mobile, cost-effective, and offers wide possibilities of powering nanodevices like in vivo biosensors, wireless and remote sensors, and nanorobots; additionally, it has also established the basis for building zero-power pressure sensors.

This is explained in detail as follows. An array of aligned ZnO nanowires was covered by a zigzag Si electrode coated with Pt. The Pt coating not only enhanced the conductivity of the electrode but also created a Schottky contact at the interface with ZnO. The nanowires were grown on either GaN substrates or sapphire substrates that were covered by a thin layer of ZnO film (Wang et al. 2004b), which served as a common electrode for directly connecting the nanowires with an external circuit. The output generated by the nanogenerator is continuous and reasonably stable due to the fact that the resonance frequency of the nanowires is much larger than the nanogenerator frequency (about 80 times). With the switching off of the ultrasonic wave, the current disappears, appearing again upon switching on the wave. Current output is in nanoampere range, but it can be increased by the connection of the devices in series and is sufficient for powering up nanodevices.

One of the limitations of the above concept is contact between nanowires and the zigzag electrode. In fact a large majority of nanowires do not produce electricity due to their nonuniformity in height and spatial distribution on the substrate surface. Overcoming this problem will yield higher current output. Besides, some of the nanowires being directly pushed by a zigzag electrode block other nanowires from reaching the contact with the electrode to produce electricity. These are mainly technical difficulties, which shall be overcome by optimized design and processing.

Solar Cells

Dye-sensitized solar cells (DSSCs) have generated much interest because of their potential to be an alternative to p–n junction solar cells based on silicon wafers. The attractiveness of DSSC arises from the fact that they are more cost-effective than classic solar cells. In the conventional solar cells, the semiconductor is responsible both for light absorption and conversion to the current and also for charge transfer, while in DSSCs, the functions are separately controlled by photosensitizing dye and semiconductor, respectively (Fig. 20.6). Usually, anodes in dye-synthesized solar cells are made of thick films (about 10 μm) of TiO_2 or, less often, SnO_2 or ZnO nanparticles (Chappel and Zaban 2002; Matsubara et al. 2003).

Films are deposited as a paste and then are sintered to produce electrical continuity. Thick nanparticle films provide a large internal surface area (characterized by the roughness factor, which is defined as total film area per substrate unit area) important for light-harvesting efficiency, but as a drawback, they show trap-limited diffusion for electron transport limiting efficiency. Another approach is the use of nonordered nanorods. In this case, different techniques are used to produce nanorods, for example, electrospinning technique (Fujihara et al. 2007; Jose and Kumar 2008). A strong limitation of the above-mentioned approaches is trapping of charge carriers instead of their transportation to the collector (diffusivity for electrons in case the nanoparticles are several orders smaller than single crystals).

An attractive alternative is the use of an ordered nanowire photoanode, which, first of all, shall have a large surface area for dye absorption, at least comparable to that of nanoparticle film. Moreover, nanowires shall have aspect ratios no less than 20 due to the small efficiency of the shorter nanowires. The nanowire films are good electrical conductors along the direction of the wire axes. Two-point electrical measurements of the arrays on fluorine-doped tin oxide (FTO) substrates show barrier-free contacts between the nanowire and the substrate. Since particles or nanowire films have similar surfaces, the basic difference is the kinetics of charge transfer at the dye semiconductor interface. As a basic material for anode nanowire for DSSCs, ZnO shows poorer chemical stability and suffers from limited photoconversion efficiences, apporoximately 1%

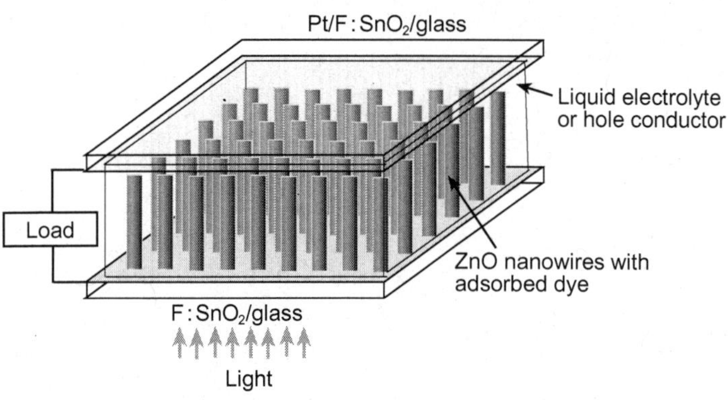

Figure 20.6. Schematic of DSSC based on nanowires.

when compared to TiO_2. DSSCs fabricated using the nanowire arrays demonstrate very encouraging photoconversion efficiencies, 5.02% for a 2- to 3-μm-long nanowire array when using N719 dye (Feng et al. 2008).

Fuel Cells

Demand for clean energy attracted much research attention to fuel cells. Among them, the solid oxide fuel cell (SOFC) seems to be an environmentally friendly power generation device that directly converts chemical energy into electrical energy. Typically, SOFC is built of a porous cathode and anode separated by an electrolyte. In fact, not only the individual functionality of materials requirements must be fulfilled but also, the mechanical and chemical compatibilities of the components for the optimum fuel cell performance must be achieved. Since the electrical power output and efficiency of the cell are often sufficient, long-term durability remains as a main challenge. The problems associated with durability largely result from the thermal expansion coefficient mismatch between the cathode and electrolyte (Mogensen et al. 2000). To overcome these constraints, a composite material is used for the cathode, which is known to give better performance than single-phase materials (Kenjo and Nishiya 1992). Park et al. (Park and Kim 2007) reported work on porous composite cathodes consisting of ZnO single-crystalline nanowire and yttria-stabilized zirconia (YSZ) particles. In this combination, YSZ works as electrolyte. ZnO was selected due to its chemical neutrality, structural stability, and ease to enhance electronic conductivity by doping. It also has a thermal expansion coefficient close to the Y-doped ZrO_2. The porous composite cathode was composed of 0.5 mol% Al-doped ZnO nanowire and YSZ particles. In this configuration, concerns regarding mechanical stability between the cathode and the electrolyte are significantly reduced but are not completely eliminated because YSZ is the main component of both electrolyte and cathode. It is confirmed that ZnO is chemically compatible with YSZ under the conditions in which the SOFC operates. Our study shows that, with size and material optimization, the porous ZnO nanowire/YSZ composite cathode can be a viable alternative to conventional cathodes used in SOFCs. This study can be extended to the fabrication of other composite cathodes employing the various types of chemically stable as well as electronically conducting nanostructures available at present (Park and Kim 2007).

REFERENCES

Bae SY, Na CW, Kang JH, Park J. 2005. Comparative structure and optical properties of Ga-, In-, and Sn-doped ZnO nanowires synthesized via thermal evaporation. J Phys Chem B 109:2526–2531.

Banerjee D, Rybczynski J, Huang JY, Wang DZ, Kempa K, Ren ZF. 2005. Large hexagonal arrays of aligned ZnO nanorods. Appl Phys A-Mater Sci Process 80:749–752.

Beermann N, Vayssieres L, Lindquist SE, Hagfeldt A. 2000. Photoelectrochemical studies of oriented nanorod thin films of hematite. J Electrochem Soc 147:2456–2461.

Calestani D, Zha M, Zappettini A, Lazzarini L, Salviati G, Zanotti L, Sberveglieri G, 2005. Structural and optical study of SnO_2 nanobelts and nanowires. Mater Sci Eng C-Biomimetic Supramol Syst 25:625–630.

Cha SN, Jang JE, Choi Y, Amaratunga GAJ, Ho GW, Welland ME, Hasko DG, Kang DJ, Kim JM. 2006. High performance ZnO nanowire field effect transistor using self-aligned nanogap gate electrodes. Applied Physics Letters 89:3.

Chappel S, Zaban A. 2002. Nanoporous SnO_2 electrodes for dye-sensitized solar cells: improved cell performance by the synthesis of 18 nm SnO_2 colloids. Sol Energy Mater Sol Cells 71: 141–152.

Chang PC, Fan ZY, Tseng WY, Rajagopal A, Lu JG. 2005. Beta-Ga_2O_3 nanowires: synthesis, characterization, and p-channel field-effect transistor. Appl Phys Lett 87:3.

Chang PC, Fan Z, Chien CJ, Stichtenoth D, Ronning C, Lu JG. 2006. High-performance ZnO nanowire field effect transistors. Appl Phys Lett 89:3.

Chen G, Dresselhaus MS, Dresselhaus G, Fleurial JP, Caillat T. 2003. Recent developments in thermoelectric materials. Int Mater Rev 48:45–66.

Chen HY, Wang JX, Yu HC, Yang HX, Xie SS, Li JQ. 2005. Transmission electron microscopy study of pseudoperiodically twinned Zn_2SnO_4 nanowires. J Phys Chem B 109:2573–2577.

Chen HM, Hsin CF, Liu RS, Hu SF, Huang CY. 2007. Controlling optical properties of aluminum oxide using electrochemical deposition. J Electrochem Soc 154:K11–K14.

Chen JJ, Wang K, Hartman L, Zhou WL. 2008. H2S detection by vertically aligned CuO nanowire array sensors. J Phys Chem C 112:16017–16021.

Choy JH, Jang ES, Won JH, Chung JH, Jang DJ, Kim YW. 2003. Soft solution route to directionally grown ZnO nanorod arrays on Si wafer; room-temperature ultraviolet laser. Adv Mater 15:1911.

Coutts TJ, Young DL, Li X, Mulligan WP, Wu X. 2000. Search for improved transparent conducting oxides: a fundamental investigation of CdO, Cd_2SnO_4, and Zn_2SnO_4. J Vac Sci Technol A-Vac Surf Films 18:2646–2660.

Cox SFJ, Davis EA, Cottrell SP, King PJC, Lord JS, Gil JM, Alberto HV, Vilao RC, Duarte JP, de Campos NA, Weidinger A, Lichti RL, Irvine SJC. 2001. Experimental confirmation of the predicted shallow donor hydrogen state in zinc oxide. Phys Rev Lett 86:2601–2604.

Cui Y, Wei QQ, Park HK, Lieber CM. 2001. Nanowire nanosensors for highly sensitive and selective detection of biological and chemical species. Science 293:1289–1292.

Dai ZR, Pan ZW, Wang ZL. 2003. Novel nanostructures of functional oxides synthesized by thermal evaporation. Adv Funct Mater 13:9–24.

Dong F, Zhao WR, Wu ZB. 2008. Characterization and photocatalytic activities of C, N and S co-doped TiO_2 with 1D nanostructure prepared by the nano-confinement effect. Nanotechnology 19:10.

El-Sayed MA. 2001. Some interesting properties of metals confined in time and nanometer space of different shapes. Accounts Chem Res 34:257–264.

Fan HJ, Bertram F, Dadgar A, Christen J, Krost A, Zacharias M. 2004. Self-assembly of ZnO nanowires and the spatial resolved characterization of their luminescence. Nanotechnology 15:1401–1404.

Fan HJ, Lee W, Scholz R, Dadgar A, Krost A, Nielsch K, Zacharias M. 2005a. Arrays of vertically aligned and hexagonally arranged ZnO nanowires: a new template-directed approach. Nanotechnology 16:913–917.

Fan ZY, Wen XG, Yang SH, Lu JG. 2005b. Controlled p- and n-type doping of Fe_2O_3 nanobelt field effect transistors. Appl Phys Lett 87:3.

Feng XJ, Shankar K, Varghese OK, Paulose M, Latempa TJ, Grimes CA. 2008. Vertically aligned single crystal TiO_2 nanowire arrays grown directly on transparent conducting oxide coated glass: synthesis details and applications. Nano Lett 8:3781–3786.

Fitzgerald TG, Borsetto F, O'Callaghan JM, Kosmala B, Holmes JD, Morris MA. 2007. Polymer nanostructures in sub-micron lithographically defined channels: film-thickness effects on structural alignment of a small feature size polystyrene-polyisoprene-polystyrene block copolymer. Soft Matter 3:916–921.

Fu YY, Chen J, Zhang H. 2001. Synthesis of Fe_2O_3 nanowires by oxidation of iron. Chem Phys Lett 350:491–494.

Fujihara K, Kumar A, Jose R, Ramakrishna S, Uchida S. 2007. Spray deposition of electrospun TiO_2 nanorods for dye-sensitized solar cell. Nanotechnology 18:5.

Glass R, Moller M, Spatz JP. 2003. Block copolymer micelle nanolithography. Nanotechnology 14:1153–1160.

Greene LE, Law M, Goldberger J, Kim F, Johnson JC, Zhang YF, Saykally RJ, Yang PD. 2003. Low-temperature wafer-scale production of ZnO nanowire arrays. Angew Chem-Int Edit 42:3031–3034.

Gu Y, Kuskovsky IL, Yin M, O'Brien S, Neumark GF. 2004. Quantum confinement in ZnO nanorods. Appl Phys Lett 85:3833–3835.

Haraguchi K, Katsuyama T, Hiruma K, Ogawa K. 1992. GAAS p-n-junction formed in quantum wire crystals. Appl Phys Lett 60:745–747.

Hendry E, Koeberg M, O'Regan B, Bonn M. 2006. Local field effects on electron transport in nanostructured TiO_2 revealed by terahertz spectroscopy. Nano Lett 6:755–759.

Holmes JD, Johnston KP, Doty RC, Korgel BA. 2000. Control of thickness and orientation of solution-grown silicon nanowires. Science 287:1471–1473.

Huang MH, Mao S, Feick H, Yan HQ, Wu YY, Kind H, Weber E, Russo R, Yang PD. 2001. Room-temperature ultraviolet nanowire nanolasers. Science 292:1897–1899.

Jiang XC, Herricks T, Xia YN. 2002. CuO nanowires can be synthesized by heating copper substrates in air. Nano Lett 2:1333–1338.

Jose R, Kumar A, Thavasi V, Ramakrishna S. 2008. Conversion efficiency versus sensitizer for electrospun TiO2 nanorod electrodes in dye-sensitized solar cells. Nanotechnology 19:7.

Kang Y, Kim D. 2006. Well-aligned CdS nanorod conjugated polymer solar cells. Sol Energy Mater Sol Cells 90:166–174.

Kempa K, Kimball B, Rybczynski J, Huang ZP, Wu PF, Steeves D, Sennett M, Giersig M, Rao D, Carnahan DL, Wang DZ, Lao JY, Li WZ, Ren ZF. 2003. Photonic crystals based on periodic arrays of aligned carbon nanotubes. Nano Lett 3:13–18.

Kenjo T, Nishiya M. 1992. $LaMnO_3$ air cathodes containing ZrO_2 electrolyte for high-tmperature solid oxide fuel-cells. Solid State Ion 57:295–302.

Kolmakov A, Zhang YX, Cheng GS, Moskovits M. 2003. Detection of CO and O^{-2} using tin oxide nanowire sensors. Adv Mater 15:997.

Kong XY, Ding Y, Yang R, Wang ZL. 2004. Single-crystal nanorings formed by epitaxial self-coiling of polar nanobelts. Science 303:1348–1351.

Koshizaki N, Oyama T. 2000. Sensing characteristics of ZnO-based NO_x sensor. Sens Actuator B-Chem 66:119–121.

Kosiorek A, Kandulski W, Chudzinski P, Kempa K, Giersig M. 2004. Shadow nanosphere lithography: simulation and experiment. Nano Lett 4:1359–1363.

Kuang Q, Lao CS, Wang ZL, Xie ZX, Zheng LS. 2007. High-sensitivity humidity sensor based on a single SnO_2 nanowire. J Am Chem Soc 129:6070.

Lauhon LJ, Gudiksen MS, Lieber CM. 2004. Semiconductor nanowire heterostructures. Philos Trans R Soc Lond Ser A-Math Phys Eng Sci 362:1247–1260.

Law M, Greene LE, Johnson JC, Saykally R, Yang PD. 2005. Nanowire dye-sensitized solar cells. Nat Mater 4:455–459.

Le HQ, Tripathy S, Chua SJ. 2008. The influence of nitrogen plasma treatment on the lattice vibrational properties of hydrothermally grown ZnO nanorods. Applied Physics Letters 92:3.

Lee W, Alexe M, Nielsch K, Gosele U. 2005a. Metal membranes with hierarchically organized nanotube arrays. Chem Mat 17:3325–3327.

Lee W, Scholz R, Niesch K, Gosele U. 2005b. A template-based electrochemical method for the synthesis of multisegmented metallic nanotubes. Angew Chem-Int Edit 44:6050–6054.

Li Z, Wu MH, Liu TB, Jiao Z, Zhao B. 2008. Preparation of TiO_2 nanowire gas nanosensor by AFM anode oxidation. Ultramicroscopy 108:1334–1337.

Liang HQ, Pan LZ, Liu ZJ. 2008. Synthesis and photoluminescence properties of ZnO nanowires and nanorods by thermal oxidation of Zn precursors. Mater Lett 62:1797–1800.

Limmer SJ, Seraji S, Wu Y, Chou TP, Nguyen C, Cao GZ. 2002. Template-based growth of various oxide nanorods by sol-gel electrophoresis. Adv Funct Mater 12:59–64.

Lin YM, Sun XZ, Dresselhaus MS. 2000. Theoretical investigation of thermoelectric transport properties of cylindrical Bi nanowires. Phys Rev B 62:4610–4623.

Liu WG, Kong LB, Zhang LY, Yao X. 1995. Study of the surface-layer of lead titanate thin-film by X-ray diffraction. Solid State Commun 93:653–657.

Liu X, Li C, Han S, Han J, Zhou CW. 2003. Synthesis and electronic transport studies of CdO nanoneedles. Appl Phys Lett 82:1950–1952.

Liu F, Bao M, Wang KL, Li C, Lei B, Zhou C. 2005. One-dimensional transport of In_2O_3 nanowires. Appl Phys Lett 86:3.

Luo HM, Wang DH, He JB, Lu YF. 2005. Magnetic cobalt nanowire thin films. J Phys Chem B 109:1919–1922.

Ma DDD, Lee CS, Au FCK, Tong SY, Lee ST. 2003. Small-diameter silicon nanowire surfaces. Science 299:1874–1877.

Manna L, Scher EC, Alivisatos AP. 2000. Synthesis of soluble and processable rod-, arrow-, teardrop-, and tetrapod-shaped CdSe nanocrystals. J Am Chem Soc 122:12700–12706.

Martin CR. 1994. Nanomaterials—a membrane-based synthetic approach. Science 266:1961–1966.

Martin CR. 1996. Membrane-based synthesis of nanomaterials. Chem Mat 8:1739–1746.

Masuda H, Fukuda K. 1995. Ordered metal nanohole arrays made by a 2-step replication of honeycomb structures of anodic alumina. Science 268:1466–1468.

Matsubara K, Fons P, Iwata K, Yamada A, Sakurai K, Tampo H, Niki S 2003. ZnO transparent conducting films deposited by pulsed laser deposition for solar cell applications. Thin Solid Films 431:369–372.

Mogensen M, Primdahl S, Jorgensen MJ, Bagger C. 2000. Composite electrodes in solid oxide fuel cells and similar solid state devices. J Electroceram 5:141–152.

Murray CB, Norris DJ, Bawendi MG. 1993. Synthesis and characterization of nearly monodisperse CdE (E = S, Se, Te) semiconductor nanocrystallites. J Am Chem Soc 115:8706–8715.

Muster J, Kim GT, Krstic V, Park JG, Park YW, Roth S, Burghard M. 2000. Electrical transport through individual vanadium pentoxide nanowires. Adv Mater 12:420.

Ng HT, Han J, Yamada T, Nguyen P, Chen YP, Meyyappan M. 2004. Single crystal nanowire vertical surround-gate field-effect transistor. Nano Lett 4:1247–1252.

Park HJ, Kim S. 2007. Electrochemical characteristics of ZnO-nanowire/yttria-stabilized zirconia composite as a cathode for SOFCs. Electrochem Solid State Lett 10:B187–B190.

Park WI, Yi GC. 2004. Electroluminescence in n-ZnO nanorod arrays vertically grown on p-GaN. Adv Mater 16:87.

Park M, Harrison C, Chaikin PM, Register RA, Adamson DH. 1997. Block copolymer lithography: periodic arrays of similar to 10(11) holes in 1 square centimeter. Science 276: 1401–1404.

Park WI, Yi GC, Kim M, Pennycook SJ. 2003. Quantum confinement observed in ZnO/ZnMgO nanorod heterostructures. Adv Mater 15:526–529.

Patrissi CJ, Martin CR. 1999. Sol-gel-based template synthesis and Li-insertion rate performance of nanostructured vanadium pentoxide. J Electrochem Soc 146:3176–3180.

Quang LH, Jin CS. 2008. Greatly enhanced optical properties of ZnO nanorods grown on GaN in aqueous solution by using two-step treatment method. J Electrochem Soc 155: K105–K109.

Rao GST, Rao DT. 1999. Gas sensitivity of ZnO based thick film sensor to NH_3 at room temperature. Sens Actuator B-Chem 55:166–169.

Rubio-Bollinger G, Bahn SR, Agrait N, Jacobsen KW, Vieira S. 2001. Mechanical properties and formation mechanisms of a wire of single gold atoms. Phys Rev Lett 8702:4.

Sham TK, Naftel SJ, Kim PSG, Sammynaiken R, Tang YH, Coulthard I, Moewes A, Freeland JW, Hu YF, Lee ST. 2004. Electronic structure and optical properties of silicon nanowires: a study using x-ray excited optical luminescence and x-ray emission spectroscopy. Phys Rev B 70:8.

Shi SC, Chen CF, Hsu GM, Hwang JS, Chattopadhyay S, Lan ZH, Chen KH, Chen LC. 2005. Reduced temperature-quenching of photoluminescence from indium nitride nanotips grown by metalorganic chemical vapor deposition. Appl Phys Lett 87:3.

Sysoev VV, Button BK, Wepsiec K, Dmitriev S, Kolmakov A. 2006. Toward the nanoscopic "electronic nose": hydrogen versus carbon monoxide discrimination with an array of individual metal oxide nano- and mesowire sensors. Nano Lett 6:1584–1588.

Takahashi K, Limmer SJ, Wang Y, Cao GZ. 2004. Synthesis and electrochemical properties of single-crystal V_2O_5 nanorod arrays by template-based electrodeposition. J Phys Chem B 108:9795–9800.

van der Horst A, Campbell AI, van Vugt LK, Vanmaekelbergh DAM, Dogterom M, van Blaaderen A. 2007. Manipulating metal-oxide nanowires using counter-propagating optical line tweezers. Opt Express 15:11629–11639.

Vayssieres L, Keis K, Lindquist SE, Hagfeldt A. 2001. Purpose-built anisotropic metal oxide material: 3D highly oriented microrod array of ZnO. J Phys Chem B 105:3350–3352.

Wagner RS, Ellis WC. 1964. Vapor-liquid-solid mechanism of single crystal growth. Appl Phys Lett 4:89–90.

Wang J, Polizzi E, Lundstrom M. 2004a. A three-dimensional quantum simulation of silicon nanowire transistors with the effective-mass approximation. J Appl Phys 96:2192–2203.

Wang XD, Summers CJ, Wang ZL. 2004b. Large-scale hexagonal-patterned growth of aligned ZnO nanorods for nano-optoelectronics and nanosensor arrays. Nano Lett 4:423–426.

Wang ZL, Kong XY, Ding Y, Gao PX, Hughes WL, Yang RS, Zhang Y. 2004c. Semiconducting and piezoelectric oxide nanostructures induced by polar surfaces. Adv Funct Mater 14:943–956.

Wang JX, Chen HY, Gao Y, Liu DF, Song L, Zhang ZX, Zhao XW, Dou XY, Luo SD, Zhou WY, Wang G, Xie SS. 2005. Synthesis and characterization of In_2O_3/SnO_2 hetero-junction beaded nanowires. J Cryst Growth 284:73–79.

Wang G, Ji Y, Huang XR, Yang XQ, Gouma PI, Dudley M. 2006. Fabrication and characterization of polycrystalline WO_3 nanofibers and their application for ammonia sensing. J Phys Chem B 110:23777–23782.

Wang XD, Song JH, Liu J, Wang ZL. 2007. Direct-current nanogenerator driven by ultrasonic waves. Science 316:102–105.

Wen XG, Wang SH, Ding Y, Wang ZL, Yang SH. 2005. Controlled growth of large-area, uniform, vertically aligned arrays of alpha-Fe_2O_3 nanobelts and nanowires. J Phys Chem B 109: 215–220.

Wong EW, Sheehan PE, Lieber CM. 1997. Nanobeam mechanics: elasticity, strength, and toughness of nanorods and nanotubes. Science 277:1971–1975.

Wu XC, Song WH, Huang WD, Pu MH, Zhao B, Sun YP, Du JJ. 2000. Crystalline gallium oxide nanowires: intensive blue light emitters. Chem Phys Lett 328:5–9.

Wu ZH, Mei XY, Kim D, Blumin M, Ruda HE. 2002. Growth of Au-catalyzed ordered GaAs nanowire arrays by molecular-beam epitaxy. Appl Phys Lett 81:5177–5179.

Wu XC, Hong JM, Han ZJ, Tao YR. 2003. Fabrication and photoluminescence characteristics of single crystalline In_2O_3 nanowires. Chem Phys Lett 373:28–32.

Wu JM, Shih HC, Wu WT, Tseng YK, Chen IC. 2005. Thermal evaporation growth and the luminescence property of TiO_2 nanowires. J Cryst Growth 281:384–390.

Wu JM, Shih HC, Wu WT. 2006. Formation and photoluminescence of single-crystalline rutile TiO_2 nanowires synthesized by thermal evaporation. Nanotechnology 17:105–109.

Xia YN, Yang PD, Sun YG, Wu YY, Mayers B, Gates B, Yin YD, Kim F, Yan YQ. 2003. One-dimensional nanostructures: synthesis, characterization, and applications. Adv Mater 15:353–389.

Xu JQ, Pan QY, Shun YA, Tian ZZ. 2000. Grain size control and gas sensing properties of ZnO gas sensor. Sens Actuator B-Chem 66:277–279.

Xue DS, Zhang LY, Gui AB, Xu XF. 2005. Fe_3O_4 nanowire arrays synthesized in AAO templates. Appl Phys A-Mater Sci Process 80:439–442.

Yang PD, Yan HQ, Mao S, Russo R, Johnson J, Saykally R, Morris N, Pham J, He RR, Choi HJ. 2002. Controlled growth of ZnO nanowires and their optical properties. Adv Funct Mater 12:323–331.

Yu H, Li JB, Loomis RA, Wang LW, Buhro WE. 2003. Two- versus three-dimensional quantum confinement in indium phosphide wires and dots. Nat Mater 2:517–520.

Zhang DH, Li C, Han S, Liu XL, Tang T, Jin W, Zhou CW. 2003. Electronic transport studies of single-crystalline In2O3 nanowires. Appl Phys Lett 82:112–114.

Zhao MH, Wang ZL, Mao SX. 2004. Piezoelectric characterization of individual zinc oxide nanobelt probed by piezoresponse force microscope. Nano Lett 4:587–590.

Zhi MJ, Zhu LP, Ye ZZ, Wang FZ, Zhao BH. 2005. Preparation and properties of ternary ZnMgO nanowires. J Phys Chem B 109:23930–23934.

Zhou ZH, Gao XS, Wang J, Fujihara K, Ramakrishna S, Nagarajan V. 2007. Giant strain in $PbZr_{0.2}Ti_{0.8}O_3$ nanowires. Appl Phys Lett 90:052902.

21

NANOINTEGRATION BASED ON THIN-FILM TECHNOLOGY

C. Jin,[1] W. Wei,[2] R. Aggarwal,[2] and R. J. Narayan[1]

[1]*University of North Carolina, Chapel Hill, North Carolina*
[2]*North Carolina State University, Raleigh, North Carolina*

INTRODUCTION

One focus of research in the field of nanoscience and nanotechnology is on an area known as nanointegration [1–4]. Nanointegration refers to the process by which devices and other functional nanoscale structures can be created either using nanostructured or conventional materials. Major challenges in nanointegration include development of well-controlled nanostructures and assembly of these structures into larger assemblies, which can be used as practical devices. Both conventional and nanoscale processing technologies, including lithography, thin-film growth, nanoimprint lithography, and dip-pen nanolithography (DPN), have been used to obtain nanointegration [5–13]. Thin-film growth is commonly used to fabricate nanoscale structures and devices. As a result, thin-film growth is considered to be one of the most promising approaches for achieving nanointegration. Nanointegration based on thin-film technology can be easily adapted to industrial environments as this technology is widely used in the semiconductor industry. In this chapter, advances in thin-film technology pertinent to nanointegration are reviewed. Fabrication of ordered nanostructures containing nanoparticles and/or nanowires is discussed in detail.

Ceramic Integration and Joining Technologies: From Macro to Nanoscale, First Edition.
Edited by Mrityunjay Singh, Tatsuki Ohji, Rajiv Asthana, Sanjay Mathur.

Nanointegration often requires two- or three-dimensional ordering of nanoparticles or nanowires. Two-dimensional ordering refers to the formation of patterned arrays of nanoparticles, nanowires, or other structures in two dimensions. Three-dimensional ordering of nanostructures involves the formation of periodic features in three dimensions [14–16]. Self-assembly and self-organization are terms that are commonly used in order to describe ordering phenomena in nanostructures. The usage of these two terms causes confusion within the nanoscience community. As discussed by Teichert [15], self-assembly involves the formation of a nanostructure containing a narrow size distribution. If the self-assembled nanostructure displays the tendency to form ordered structures, the process is referred to as self-organization. The ordering processes for nanostructures using thin-film growth techniques can be classified into spontaneous processes and forced processes. Spontaneous ordering can occur because of long-range elastic interactions [14–16]. Forced ordering processes typically involve the constraint of nucleation and growth on prepatterned templates or on preprepared screens.

SPONTANEOUS ORDERING OF NANOSTRUCTURES

Spontaneous ordering, such as the formation of periodic surface domains due to strain field relaxation, is a phenomenon that is observed in many classes of materials. In recent years, spontaneous formation of ordered nanoscale surface structures has attracted extensive attention due to the possibility of developing dislocation-free nanostructured materials for device applications.

Several classes of spontaneously ordered nanostructures are commonly described, including (1) periodically faceted surfaces, (2) surface structures containing planar domains, (3) ordered arrays of three-dimensional coherent islands, (4) multisheet arrays of two-dimensional islands, and (5) multisheet arrays of three-dimensional islands [14]. Relaxation processes involving surface stresses may be used to form ordered surface nanostructures. Surface stress can cause instability in surface morphology and minimization of surface energy, which may result in an alteration in surface morphology. These phenomena can be utilized in order to form ordered nanostructures through iteration of process parameters such as surface orientation, temperature, and deposition rate.

Several types of nanostructures can be spontaneously formed on a crystal surface, such as (1) periodically faceted surfaces, (2) planar domains, and (3) ordered arrays of coherently strained islands [14]. The primary driving force for the ordering of these structures is long-range elastic interaction. Ordered structures can be formed either on a crystal surface or on an epilayer of a heteroepitaxial system (e.g., Ge_xSi_{1-x}/Si) [14–17]. Intrinsic surface stress plays an important role in the growth of an ordered nanostructure on a crystal surface. Surface atoms encounter different environments than their bulk counterparts. As a result, surface atoms tend to adopt a different lattice parameter from that of bulk atoms. Surface atoms are either stretched or compressed when they adopt the lattice parameter of bulk atoms, which results in the formation of intrinsic surface stress. The surface of a crystal is unstable and exhibits a higher surface free energy than the bulk material due to the existence of surface stress. Minimization of intrinsic surface energy leads to the formation of periodic faceted surfaces and domain structures on the surface. Mechanisms for the formation of these nanostructures have been categorized by Shchukin and Bimberg based on the theory of equilibrium surface shape, capillarity of solid surfaces, and relaxation of intrinsic surface stress [14].

Faceting structures have been observed on the surfaces of several materials, including silicon, gallium arsenide, and platinum [18–20]. For example, Hibino et al. reported the formation of periodic faceting on a vicinal Si(111) surface that was misoriented 10° to the [112] direction [18]. A vicinal surface is a surface that is created by cutting a material at a small inclined angle with respect to the crystallographic planes. Slender (111) facets with a seven-by-seven structure were observed using high-temperature scanning tunneling microscopy (STM) when the sample was cooled from above the one-by-one to seven-by-seven transition temperature. Periodic faceting structures may be used for the fabrication of complex nanostructures, including nanowire arrays [14].

Several studies on spontaneous ordering have examined heteroepitaxial systems, such as Ge_xSi_{1-x} [15, 16, 20–23]. Three growth modes are usually observed in heteroepitaxial systems. Growth modes include two-dimensional growth (Frank–van der Merwe), three-dimensional growth (Volmer–Weber [VW]), and two-dimensional growth followed by three-dimensional growth (Stranski–Krastanow [SK]). Both silicon and germanium exhibit a diamond cubic structure. There is 4.2% lattice misfit when germanium is epitaxially grown on silicon; the lattice constants for silicon and germanium are 5.431 and 5.646 Å, respectively. Depending on the amount of germanium (x), the lattice constant of Ge_xSi_{1-x} can vary between 5.431 and 5.646 Å. SK growth is commonly observed for heteroepitaxial growth of germanium and Ge_xSi_{1-x} on silicon substrates. The epilayer grows in a pesudomorphological manner in the initial stage, such that the Ge_xSi_{1-x} epilayer maintains the same lattice constant as the substrate. During growth of the wetting layer, lattice mismatch strain is accommodated by elastic deformation of the epilayer. As a result, the cubic lattice of Ge_xSi_{1-x} is strained to a tetragonal one, in which the lattice parameter perpendicular to the growth plane is stretched and the in-plane lattice parameter assumes the lattice parameter of the silicon substrate. Pesudomorphological growth continues until the film reaches a critical thickness, h_c, after which dislocations nucleate and enable relaxation of the strain. Several processes that involve relaxation of the misfit strain lead to different surface morphologies, resulting in the formation of different ordered nanostructures [16]. In general, the relaxation of the misfit strain in SK growth can be classified into three stages: (1) the growth of a pseudomorphological wetting layer, (2) the formation of three-dimensional coherent islands, and (3) the formation of misfit dislocation relaxed three-dimensional islands. Mismatch strain is accommodated during wetting layer growth by tetragonal stretching of the cubic lattice of the epilayer. The elastic strain energy of the wetting layer increases in films with larger thicknesses. The epilayer becomes unstable as the elastic strain energy increases; a long wavelength corrugation of the surface forms as a result. This phenomenon is referred to as Asaro–Tiller–Grinfield (ATG) instability since the presence of nonhydrostatic stress causes the planar surface of a solid to be unstable. Surface corrugation is observed due to stress-driven surface diffusion [24–26]. As the elastic strain energy increases, another stage of strain relaxation takes place, which involves the formation of coherent three-dimensional islands [14–16]. Three-dimensional coherent islands form on top of the wetting layer. The islands minimize the mismatch strain energy by adopting the bulk lattice constant in the upper portion of the islands. As the elastic strain energy continues to increase, strain relaxation results from the formation of misfit dislocations along the epilayer–substrate interface.

Two types of surface morphology changes have been observed on the surface of the wetting layer prior to the formation of three-dimensional coherent islands. The first type of change in surface morphology is the relaxation of the lattice mismatch strain in the wetting layer. The second type of change in surface morphology is the formation of strain-induced step bunching. The formation of self-organized surface nanostructures due to strain relaxation processes has been described in theoretical as well as experimental studies [14–16]. For example, Teichert et al. have investigated the formation of large-scale arrays of surface nanostructures using strain relaxation processes [23]. Their work demonstrates that relaxation of elastic strain can be utilized to produce large-scale ordered nanostructures in heteroepitaxial systems under appropriate deposition parameters. $Si_{0.55}Ge_{0.45}$ epilayers 2.5 nm thick were grown on 100-nm-thick silicon buffer layers using solid-source molecular beam epitaxy (MBE) at 550°C. Silicon (001) wafers were used as the substrate material. The vicinal silicon (001) substrates had different azimuthal miscut angle (ϕ) values and polar miscut angle (θ) values. The polar miscut angle is related to the average distance of substrate steps, and the azimuthal miscut angle is related to the overall deviation of the step direction. Figure 21.1 demonstrates the relationship between the substrate miscut angle and the surface ripple structures in the $Si_{0.55}Ge_{0.45}$ epilayers. Distinct one-dimensional ripple patterns were observed in all of the $Si_{0.55}Ge_{0.45}$ films; orientation of the ripples was observed along the edges of the substrate steps. The average ripple period became smaller as the miscut angle θ increased from 1 μm ($\theta = 0.17°$) to 70 nm ($\theta = 2°$). Their work suggested that stress-induced step bunching of preexisting steps on the Si(001) substrate was responsible for the formation of surface ripples in the $Si_{0.55}Ge_{0.45}$ film. Their work also showed that the direction and period of the ripples in the film were dependent on substrate vicinality. Formation of self-organized three-dimensional islands was also reported. A granular structure was observed when the germanium content in the 2.5 nm $Si_{1-x}Ge_x$ epilayer was increased to $x = 0.75$. Three-dimensional islands with faceted {105} planes were superimposed on the surface ripples. Self-organized "hut"-type three-dimensional islands were observed on the vicinal silicon substrate for miscut angles of $\theta = 2°$ and $\theta = 45°$ (Fig. 21.2). Islands were arranged in lines, which were oriented parallel to the [010] direction.

Large-scale ordered nanoparticles may be fabricated through coherent S growth. Spontaneously ordered nanostructured templates or substrates play a significant role in these self-organization processes. For example, Sekiguchi et al. deposited a germanium epilayer on a vicinal Si(111) surface with a miscut angle of $\theta = 1°$ toward the [112] direction and an azimuthal angle of $\phi = 4°$ [27]. The vicinal surface contained well-defined atomic step edges. A germanium epilayer was subsequently deposited at a substrate temperature of 250°C using MBE. STM revealed that germanium nanoclusters had formed, which were self-aligned along the upper edges of the steps (Fig. 21.3).

Several researchers have observed self-organization of nanostructures on strain-modulated templates. When an epilayer is grown on a template or substrate with non-uniform surface strain distribution, nucleation sites for epilayer growth are modulated according to the strain distribution of the buried template. If the template has a regular

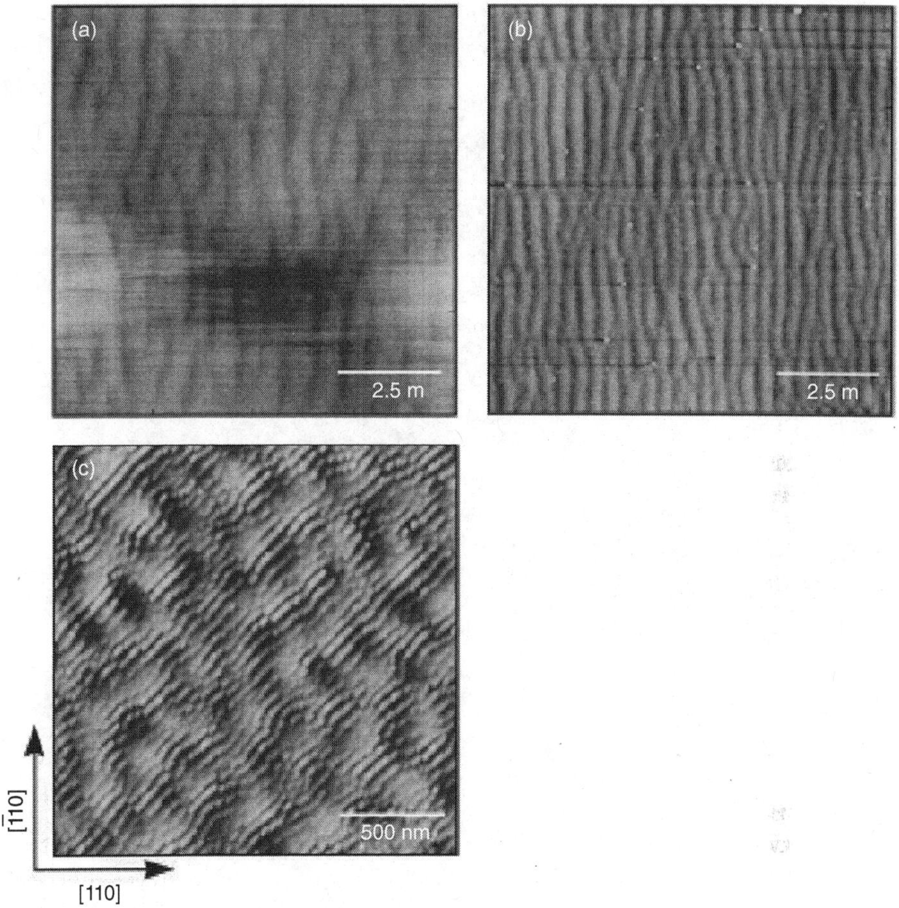

Figure 21.1. (a–c) The effect of substrate miscut on surface ripple structures for $Si_{0.55}Ge_{0:45}$ films grown on vicinal Si(001) substrates. Miscut angles: (a) $\theta = 0.17°$, $\phi = 0°$; (b) $\theta = 0.4°$, $\phi = 0°$; (c) $\theta = 2°$, $\phi = 45°$. Image size: (a, b) $10 \times 10\,\mu m$, (c) $2.5 \times 2.5\,\mu m$. With kind permission from Springer Science + Business Media: Teichert C, Bean J, Lagally M. 1998. Self-organized nanostructures in $Si_{1-}xGe_x$ films on Si(001). Appl Phys A 67:675–685.

strain distribution, preferential nucleation sites will enable formation of an ordered, self-organized structure. For example, Berbezier and Ronda described the use of a strain-modulated buried template layer for the growth of three-dimensional germanium islands [16]. The template layer was obtained by depositing a 10-nm-thick $Ge_{0.65}Si_{0.25}$ film on a vicinal Si(001) substrate. Figure 21.4a shows an atomic force microscopy (AFM) image of a germanium film grown on a square-based periodic corrugation template. The template was formed on a 1.5° off, vicinal (001) Si substrate. In this structure, germanium islands were organized into two-dimensional arrays with a narrow size distribution. It was observed that germanium islands nucleated and decorated the

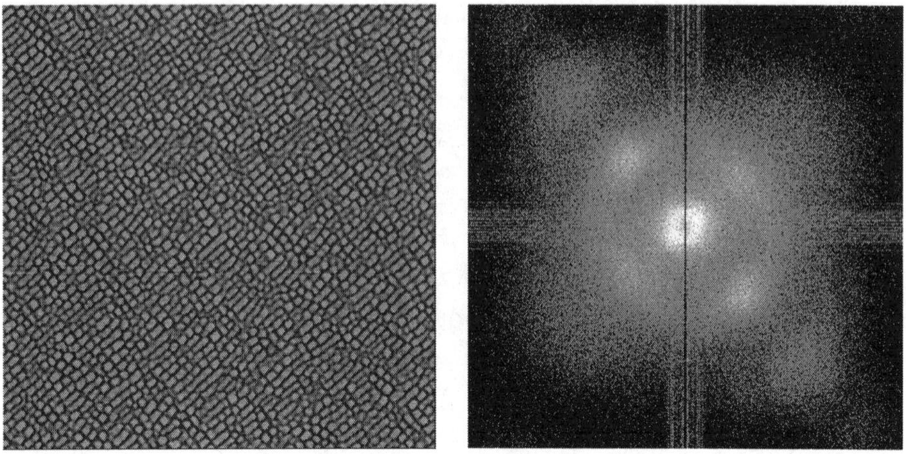

Figure 21.2. Surface morphology of a 2.5-nm $Si_{0.25}Ge_{0.75}$ film grown on a vicinal Si(001) substrate with miscut angles of $\theta = 2°$, $\phi = 45°$. (a) $2.5 \times 2.5\,\mu m$ atomic force microscope image showing prism-like, three-dimensional island chains, in which the long axes of the prism bases lie parallel to the [010] direction of the Si(001) substrate. (b) Two-dimensional power spectrum of the image shown in (a). With kind permission from Springer Science + Business Media: Teichert C, Bean J, Lagally M. 1998. Self-organized nanostructures in $Si_{1-x}Ge_x$ films on Si(001). Appl Phys A 67:675–685.

uppermost surfaces of the periodic mounds. The role of the ordered template in germanium island self-organization was demonstrated by germanium film deposition on an unpatterned silicon surface; in this case, only randomly distributed germanium islands were obtained. Figure 21.4b, c contain AFM images of germanium films that were deposited on template layers, which were prepared on 10° off (001) Si substrates. In this case, the template surface contained self-organized nanowires; germanium islands also nucleated on the uppermost surfaces of the modulations.

The strain effect may also be used to produce nanostructures, including three-dimensional arrays of nanodots and wires, which exhibit vertical correlation [14–16, 28–33]. Vertical correlation is attributed to vertical propagation of the strain field by buried nanostructures. Shchukin and Bimberg reviewed theoretical studies involving strain field-assisted formation of nanostructures and proposed two types of growth modes, which describe the spatial relationship between growing islands and previously grown (i.e., buried) islands [14]. Correlated growth is a growth mode in which the growing islands are situated directly above islands in the buried layer. Anticorrelated growth is a growth mode in which the growing islands are situated in between two buried islands. Vertical growth has been reported for several epitaxial systems, including $In_xGa_{1-x}As$/GaAs [30], InAs/GaAs [31], Ge_xSi_{1-x} [32, 33], and Ge/Si. Figure 21.5 shows a cross-sectional transmission electron microscopy image of 10 Ge/Si bilayers that was obtained by Thanh et al.; in this figure, the average island height is ~6–7 nm [33]. This figure demonstrates that germanium islands exhibit correlated growth. Xie

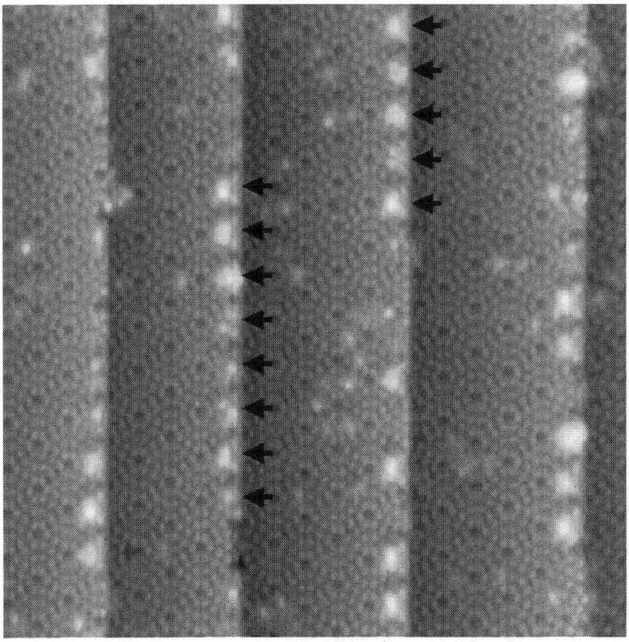

Figure 21.3. Scanning tunneling microscopy image of a germanium film deposited at 250°C on a vicinal Si(111) substrate. The film was subsequently annealed at the same temperature for 24 min ($\theta = 1.0°$, $\phi = 4°$). Germanium nanoclusters were aligned along the upper edges of the steps. Image size: 46.1 × 38.1 nm. Reprinted with permission from Sekiguchi T, Yoshida S, Itoh K, Myslivecek J, Voigtländer B. 2007. One-dimensional ordering of Ge nanoclusters along atomically straight steps of Si(111). Appl Phys Lett 90:013108–013111, American Institute of Physics.

et al. demonstrated self-organized growth of coherent InAs islands, which were separated by a GaAs spacer layer [28]. Their work showed that the InAs/GaAs multilayer structure exhibited self-organized growth along a vertical direction. Vertical correlation has been observed in nonepitaxial multilayered structures as well as multilayered heteroepitaxial systems. For example, Jin et al. reported vertical correlation growth of gold nanoparticles separated by an amorphous alumina matrix [34]. In their work, correlated growth and anticorrelated growth were observed (Fig. 21.6). At relatively low temperatures, round gold nanoparticles demonstrated correlated growth in the amorphous alumina matrix. Anticorrelated growth was observed at higher temperatures.

Misfit dislocation networks may have significant effects on the surface morphology that is observed in heteroepitaxial systems [16]. For example, the formation of surface undulation, surface displacement, and dislocation nucleation has been attributed to misfit dislocation networks [35–40]. The interaction between misfit dislocations and surface morphology implies that nanoislands may be organized on a surface using a strain field that is produced by buried arrays of misfit dislocations. Häusler et al.

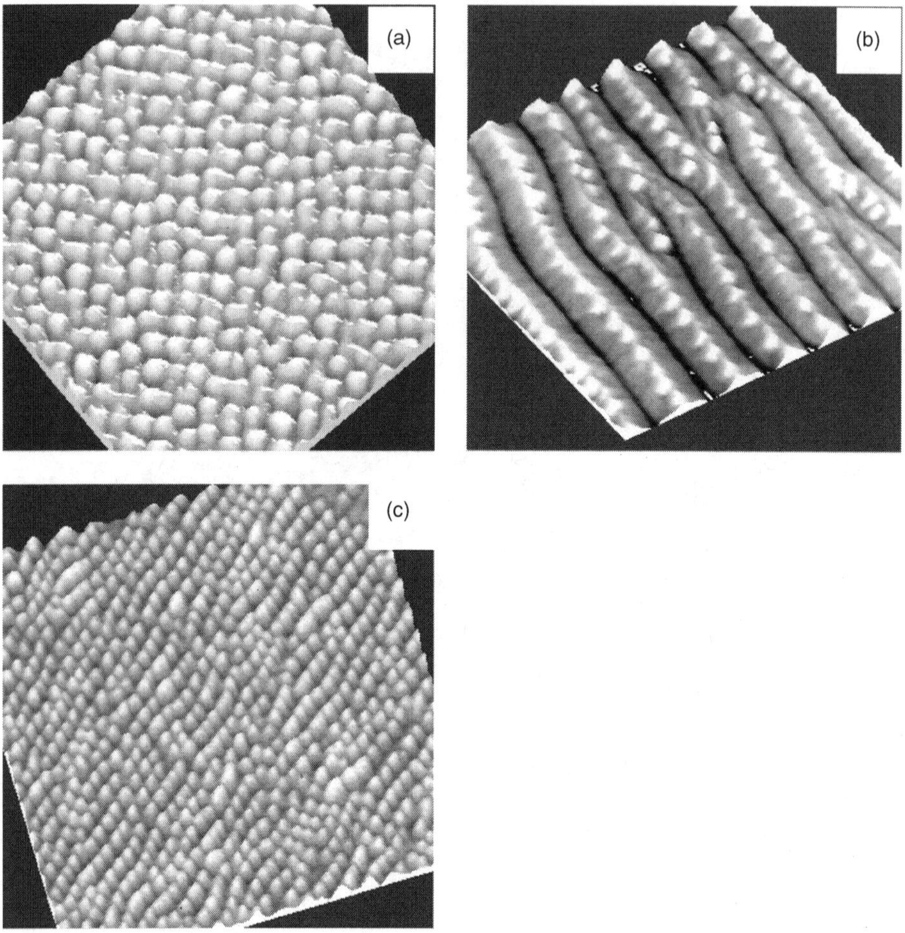

Figure 21.4. Atomic force microscopy images of a germanium epilayer that was deposited on a $Si_{0.65}Ge_{0.35}$ template layer. (a) 1.5° off Si(001), (b) 10° off Si(001), and (c) 10° off Si(001) substrates are shown. Scan size: 5 × 5 μm. Reprinted from Berbezier I, Ronda A. 2009. SiGe nanostructures. Surf Sci Rep 64, pp. 47–98, with permission from Elsevier.

described the growth of an InP/InGaP/GaAs heterostructure using solid-source MBE, in which ordered InP islands formed along the <011> directions of an InGaP buffer layer that was deposited on a GaAs(100) substrate [36]. Shiryaev et al. proposed a mechanism for manipulating the spacing of nanoislands using buried misfit dislocations [37]. Buried arrays of misfit dislocations were created by growing relaxed buffer layers using MBE in a $Si_{1-x}Ge_x$/Si heteroepitaxial system. Germanium was deposited after relaxation of the $Si_{1-x}Ge_x$ buffer layer; the relationship between surface morphology, surface strain, and growth of three-dimensional germanium islands was observed. Figure 21.7 contains AFM images of germanium deposited on $Si_{1-x}Ge_x$/Si, Si-capped

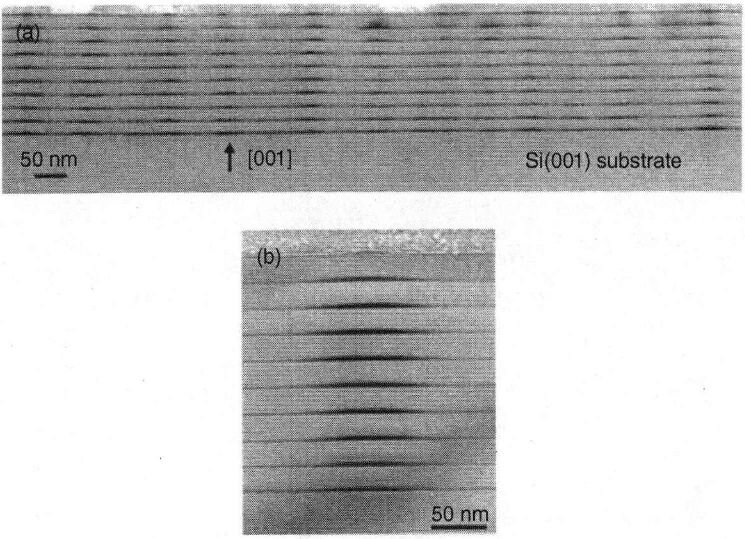

Figure 21.5. (a) Cross-sectional transmission electron microscopy image of 10 germanium/silicon bilayers. (b) A high-resolution image of a germanium island column along the growth direction. Reprinted with permission from Thanh V, Boucaud P, Débarre D, Zheng Y, Bouchier D, and Lourtioz JM. 1999. Vertically self-organized Ge/Si(001) quantum dots in multilayer structures. Phys Rev B 60:5851–5857, American Physical Society.

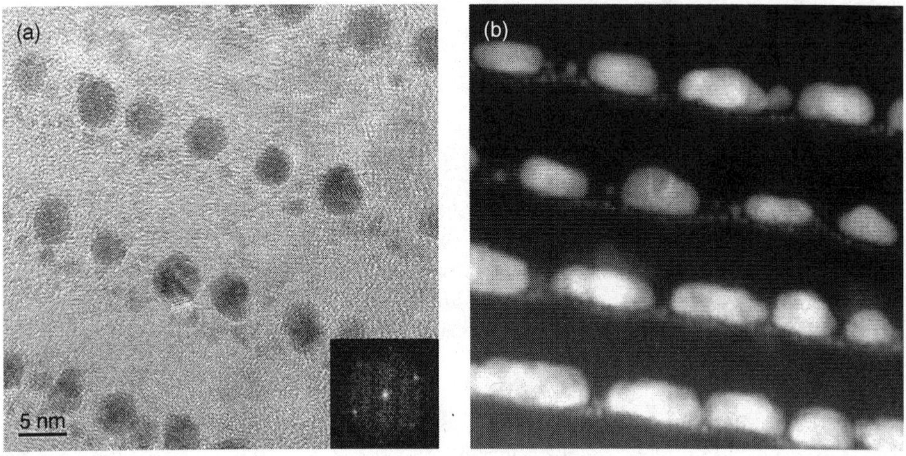

Figure 21.6. (A) Transmission electron microscopy bright field image obtained from gold nanoparticles that were grown at room temperature. The image demonstrates correlative growth of gold nanoparticles along the growth direction. (B) Transmission electron microscopy 500-K bright-field Z-contrast image obtained from gold nanoparticles that were grown at 320°C, showing anticorrelative growth. Reprinted with permission from Reference 34.

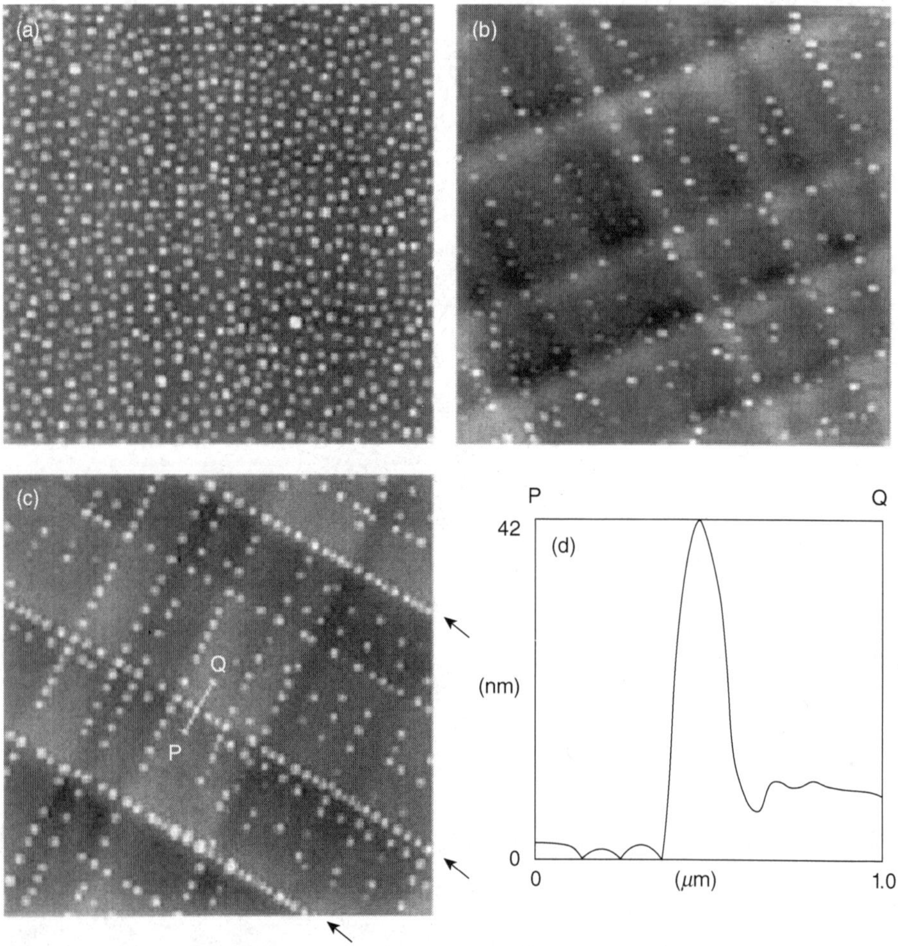

Figure 21.7. Atomic force microscopy image of 1.2-nm-thick germanium films deposited on the surface of a (a) bare Si(001) substrate, a (b) prestructured Si_{1}–xGe_{x}/Si substrate, and a (c) Si-capped/Si_{1}–xGe_{x}/Si substrate. (d) Height profile through an island shown in (c) along the line identified as P–Q. Reprinted with permission from Shiryaev S, Jensen F, Hansen J, Petersen J, Larsen A. 1997. Nanoscale structuring by misfit dislocations in Si_{1}–xGe_{x}/Si epitaxial systems. Phys Rev Lett 78:503–506, American Physical Society.

$Si_{1-x}Ge_x$/Si, and virgin Si(100) substrates. Randomly distributed germanium islands were observed on the Si(100) substrate; on the other hand, ordered germanium islands were observed on the $Si_{1-x}Ge_x$/Si and Si-capped $Si_{1-x}Ge_x$/Si substrates. Hu et al. developed a theoretical model that describes the formation of self-organized, three-dimensional germanium islands on buried misfit dislocation networks; in this model, the film surface is represented by a set of meshed mesoscopic cells [41]. Self-organized three-dimensional germanium islands may be formed by pretreating surfaces with ion sputtering, lithography, or other methods [16, 42, 43]. Strain-related growth of self-organized nanoislands is an appealing approach for integrating nanostructures with devices because it enables nanodots or islands to be fabricated on large wafer areas with relatively low defect densities. It should be noted that the development of self-organized nanostructures necessitates precise control of the processing variables, including substrate, buffer layer, and film processing parameters; strain in the heteroepitaxial system is another important consideration.

SELF-ORGANIZATION WITH TEMPLATE OR SCREENING METHODS

Ordered nanostructures have been also fabricated using prepatterned substrates or screening methods. For example, several investigators have examined the growth of ordered nanowires, which are one-dimensional rodlike structures that exhibit diameters of ~100–200 nm. Due to quantum confinement of electrons within nanometer-scale features, nanowires exhibit physical properties that are not observed in thin films or bulk materials. Nanowires fabricated from Si [44], GaAs [45], ZnO [46], and other semiconductors are of particular interest because these structures may be used in p–n junctions as well as in other nanoscale electronic devices. In contrast to conventional "top-down" approaches, innovative "bottom-up" approaches have been developed for fabricating nanowire-functionalized devices. In bottom-up approaches, nanodevices are directly fabricated at the desired locations with appropriate sizes and other properties, thereby avoiding multistep processes such as thin-film deposition, photolithography, and chemical etching. Achieving precise control over nanowire growth at desired locations remains a major challenge. The vapor–liquid–solid (VLS) mechanism appears to be a very promising method for the controlled growth of nanowires. This method involves the use of catalytic metal seeds in order to initiate nanowire growth; chemical-based and lithography-based approaches have been developed for two-dimensional patterning of metal seeds. Functional devices may be developed by patterning of seeds followed by growth of one-dimensional nanowires; this is one example of "nanointegration" [47]. In this section, nanowire growth using the VLS growth mechanism will be introduced. Selective patterning techniques, which are closely related to lithography-based techniques, will also be discussed. In addition, the use of nanowires for fabrication of nanodevices will be described.

VLS GROWTH

The VLS mechanism is commonly used to fabricate nanowires from a variety of materials. For example, Wagner and Ellis described growth of micrometer-sized silicon whiskers with gold catalyst particles using this mechanism [48]. A typical VLS process involves two steps: (1) In the first step, a supersaturated liquid phase eutectic alloy is formed; (2) in the second step, nucleation and growth of the crystalline solid phase takes place. Small particles of the catalyst material are first deposited on the substrate, which serve as templates for nanowire growth. The diameter of the nanowire is primarily determined by the size of the catalyst particle. Metals such as gold are commonly used as catalysts. Other materials may also be used as catalysts; for example, nickel oxide (NiO) is commonly used for growth of gallium nitride nanowires [49]. The catalyst particles form liquid droplets of eutectic alloy by dissolving gaseous reactants at temperatures greater than the euctectic point. When gold catalyst is used, the euctectic point is 363°C (Au : Si ratio = 4:1). The gaseous reactants are continuously consumed by the eutectic alloy in order to reach a supersaturation state. In the second step, nucleation of nanowires occurs at the solid/liquid interface with precipitates from the supersaturated liquid droplet. Nanowire growth is enabled by material that is continuously supplied during cycles of supersaturation and precipitation; the growth rate is mainly dependent on supersaturation conditions. Gaseous reactants may be supplied using various crystal growth techniques, including laser ablation, chemical vapor deposition, and thermal evaporation. The vapor–solid mechanism is similar to the VLS mechanism; this mechanism is also frequently used for the growth of nanowires due to its technological accessibility. Vapor generated by the techniques previously mentioned is transported and deposited on the substrate, which is maintained at a relatively low temperature. The vapor–solid mechanism is also known as a self-catalytic mechanism since the intermediate species that are generated from the starting materials can serve as catalysts for nanowire growth.

PATTERNING TECHNIQUES

One of the challenges related to the bottom-up growth of nanostuctures involves control over the nanostructure position. For nanostructures grown by the VLS mechanism, the nanostructure position is primarily dependent on the catalyst pattern. Various techniques have been utilized to pattern catalysts for the growth of nanostructures. These techniques include novel nanolithography techniques, such as nanoimprint lithography, nanosphere lithography, and dip-pen lithography, as well as conventional lithography techniques, such as e-beam lithography and photolithography.

PHOTOLITHOGRAPHY AND ELECTRON-BEAM LITHOGRAPHY

Conventional lithography techniques are commonly used in the microelectronics industry. Patterning of catalyst materials for the growth of nanostructures can be easily adapted from conventional lithography techniques. For example, photolithography methods (e.g., phase-shift photolithography) are suitable for large-area patterning of catalyst materials with good control over size and geometry [50]. It should be noted that the feature size of a particular photolithography technique is restricted by the Rayleigh diffraction limit or other limitations. The feature size may in turn affect the ability to obtain consistent growth of nanostructures. For example, Greyson et al. demonstrated that zinc oxide nanowires grown on the relatively large-sized gold patterns (feature size ~200 nm) demonstrated poorly controlled orientations and shapes [50]. The electron-beam lithography technique has been used to pattern features (e.g., gold nanoarrays for nanowire growth) in the sub-100-nm range [51]. This technique usually provides very good control over the formation of patterns in two dimensions; however, the use of the conventional electron-beam lithography technique also faces several limitations. For example, conventional electron-beam lithography equipment has a resolution limit of 30 nm and, as such, cannot create features with sizes less than 30 nm. In addition, electron-beam lithography is a serial process that is not commonly used for creating patterns over large areas. Furthermore, it is a time-consuming and relatively costly process.

NANOLITHOGRAPHY

Novel nanolithography-based techniques have recently been developed, which, in many cases, correspond with conventional lithography techniques. These nanolithography-based techniques may enable limitations associated with conventional lithography techniques, such as formation of small features with high throughput, to be overcome. Nanosphere lithography has emerged as a technique that is capable of achieving patterning over an area of $1\,cm^2$ with defect-free domains up to $100 \times 100\,\mu m^2$ [52]. In nanosphere lithography, a layer of nanometer-sized spheres (e.g., silica and polystyrene) is coated on a substrate using a self-organization process. This layer is used as a shadow mask; a catalyst can subsequently be deposited on the substrate through the holes between the spheres. After removing the shadow mask layer, a honeycomb pattern remains on the substrate. Figure 21.8 contains scanning electron micrographs that demonstrate each step involved in the nanosphere lithography of patterned zinc oxide nanorods. The size of triangle-shaped catalyst islands formed using nanosphere lithography can be controlled by altering the nanosphere diameter. The sixfold honeycomb pattern can be further modified by using multilayers of nanospheres or by using self-organization of nanospheres with different sizes [53]. It should be noted that a limited number of patterns may be generated using nanosphere lithography. The absence of the capability to generate arbitrary patterns is one of the main limitations of nanosphere lithography.

Nanoimprint lithography is another method that has been used for creating patterns of nanostructures; several advances have been made in this process since it was first demonstrated by Chou et al. in 1996 [54]. Nanoimprint lithography may be used to fabricate nanostructures exhibiting feature sizes that are comparable or smaller than those produced by e-beam lithography at lower cost and with higher throughput. In-plane grown ZnO nanowires with feature sizes as small as 12 nm have been achieved using this technique [55]. The nanoimprint lithography process involves two steps, namely, imprinting and etching. In the imprinting step, a stamp/mold containing the desired nanostructures is imprinted with pressure onto a substrate that is coated with a resist film, which is commonly made up of a printable polymer. The substrate is preheated to the printing temperature, which is above the glass temperature of the printable polymer. The stamp and substrate are placed in physical contact for a certain amount of time before the substrate is cooled. The stamp is removed from the resist film after substrate temperature is lowered to the glass temperature. The stamp used in this process is usually fabricated by techniques such as dry etching and e-beam lithography;

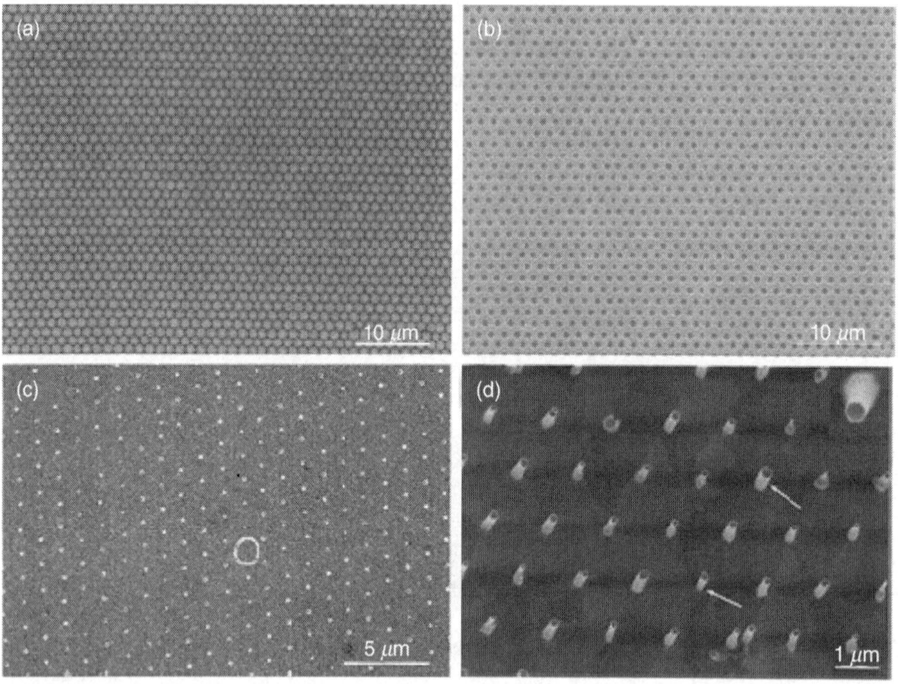

Figure 21.8. Scanning electron microscopy images of each step in the nanosphere lithography process. (a) A layer of polystyrene microspheres was deposited on a gold-coated c-plane sapphire substrate. (b) The polystyrene microsphere layer was imaged after oxygen plasma etching. (c) A gold particle pattern was created after removal of the polystyrene microsphere layer. (d) A zinc oxide nanorod array was grown on the patterned gold particles. Reprinted with permission from Liu DF, Xiang YJ, Wu XC, Zhang ZX, Liu LF, Song L, Zhao XW, Luo SD, Ma WJ, Shen J, Zhou WY, Wang G, Wang CY, Xie SS. 2006. Periodic ZnO nanorod arrays defined by polystyrene microsphere self-assembled monolayers. Nano Lett 6:2375, American Chemical Society.

this structured may be reused. In the next step, the resist film, with imprinted nanostructures, is subjected to an etching process in order to transfer the nanostructures onto the substrate. A thin metal layer may be deposited for use as an etching mask. Mårtensson et al. have demonstrated the use of nanoimprint lithography for the growth of patterned semiconductor nanowires [56]. In this study, InP nanowires were patterned on an InP substrate in order to create a nanophotonic crystal structure; well-ordered InP nanoarrays prepared using this technique are shown in Figure 21.9. The nanowires in this structure are ~1.5 μm long and ~290 nm in diameter. Nanoimprint lithography may be further modified with other techniques (e.g., colloid-based techniques) in order to create complex patterns [57].

DPN is a scanning probe microscope (SPM)-based technique that provides control over the shape, size, and chemical composition of patterned structures in the sub-100-nm

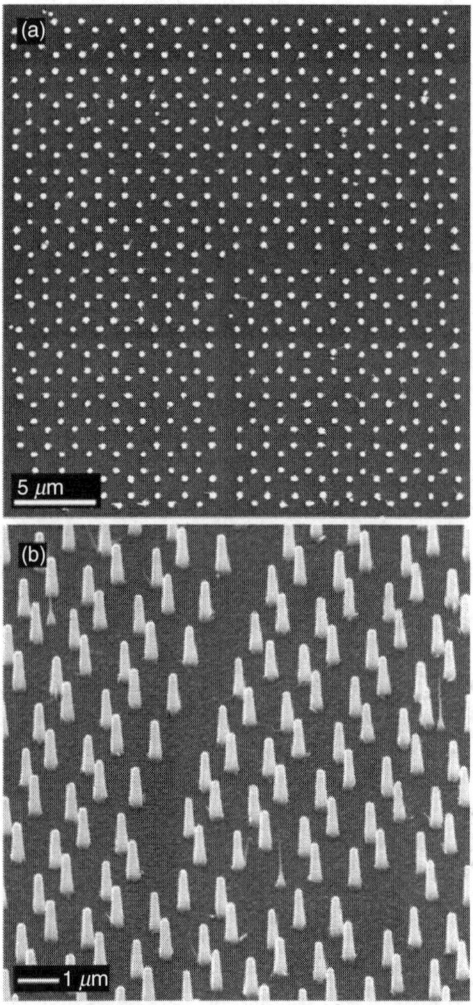

Figure 21.9. Scanning electron micrographs of InP nanowire arrays (nanowire length ~1.5 μm, nanowire diameter ~290 nm) grown on substrates that were patterned using nanoimprint lithography. (a) Top view of an InP nanowire array. (b) InP nanowire array at a 45° viewing angle. Reprinted with permission from Martensson T, Carlberg P, Borgstro M, Montelius L, Seifert W, Samuelson L. 2004. Nanowire arrays defined by nanoimprint lithography. Nano Lett 4:699, American Chemical Society.

regime [58]. The SPM probe/pen is inked with a chemical solution and is subsequently used to deliver the chemical solution by "writing" directly on the substrate. The DPN process may be used to manipulate catalytic materials. For example, Li et al. have demonstrated the use of DPN for patterning catalytic nickel oxide (NiO) nanodots, which enabled patterned growth of gallium nitride nanowires [49]. The NiO nanodot array was generated by coating the probe in nickel nitrate solution; the probe was used for patterning nickel nitrate solution at defined locations on the substrate. The nickel nitrate decomposed to nickel oxide upon heating the substrate. Their work indicated that the number of nanowires obtained from each catalyst island depended on the island size. It was also noted that the size of the DPN-patterned island was influenced by humidity during the DPN process; lower humidity during the DPN process was associated with smaller islands and fewer nanowires. The size of the patterns prepared using DPN may also be controlled by altering the scanning speed of the DPN probe.

Porous alumina membranes may also be used for patterning nanowires. These structures are fabricated by electrochemical oxidation of aluminum, and exhibit high porosities, pore diameters in submicrometer range, and narrow pore size variations [59]. The pores may entirely penetrate the alumina membrane. Porous alumina membranes can be directly used as shadow masks; for example, nanowires can be grown in the pore channels [60]. Porous alumina membranes can also be used in order to create catalyst patterns, which can subsequently be used for nanowire growth [61].

NANOWIRE ELECTRONICS

A reduction in the dimensions of electronic devices is a major focus of research activity in the microelectronics industry. For example, the development of devices containing semiconductor nanowires may enable the dimensions of several ubiquitous electronic devices to be reduced. Various devices based on semiconductor nanowires have been demonstrated, including field-emission displays [62], lasers [63], field-effect transistors [51], and sensors [64]. The field-effect transistor is of particular interest because this device is compatible with widely used complementary metal oxide semiconductor (CMOS) technology. Nanowire-based field-effect transistors with several architectures have been developed [64–66]. For example, wrap-gated field-effect transistors were fabricated by surrounding vertically grown nanowires individually with dielectric material. A gate metal layer was subsequently deposited for use as an electrode. Electrical contacts at the nanowire ends served as source and drain. A schematic of a vertical wrap-gated nanowire transistor, a scanning electron micrograph of a nanowire transistor, and a scanning electron micrograph of an integrated device are shown in Figure 21.10. The fabrication of a vertical field-effect transistor involves more complicated fabrication methods than a planar field-effect transistor laid out on a substrate; however, the vertical field-effect transistor holds significant promise because a high number density of devices can be placed in a vertical layout. The superiority of vertical field-effect transistor technology over conventional CMOS technology still needs to be experimentally demonstrated.

SUMMARY

Numerous advances in nanointegration, specifically the fabrication of ordered nanostructures based on thin-film technology, have been described over the last several

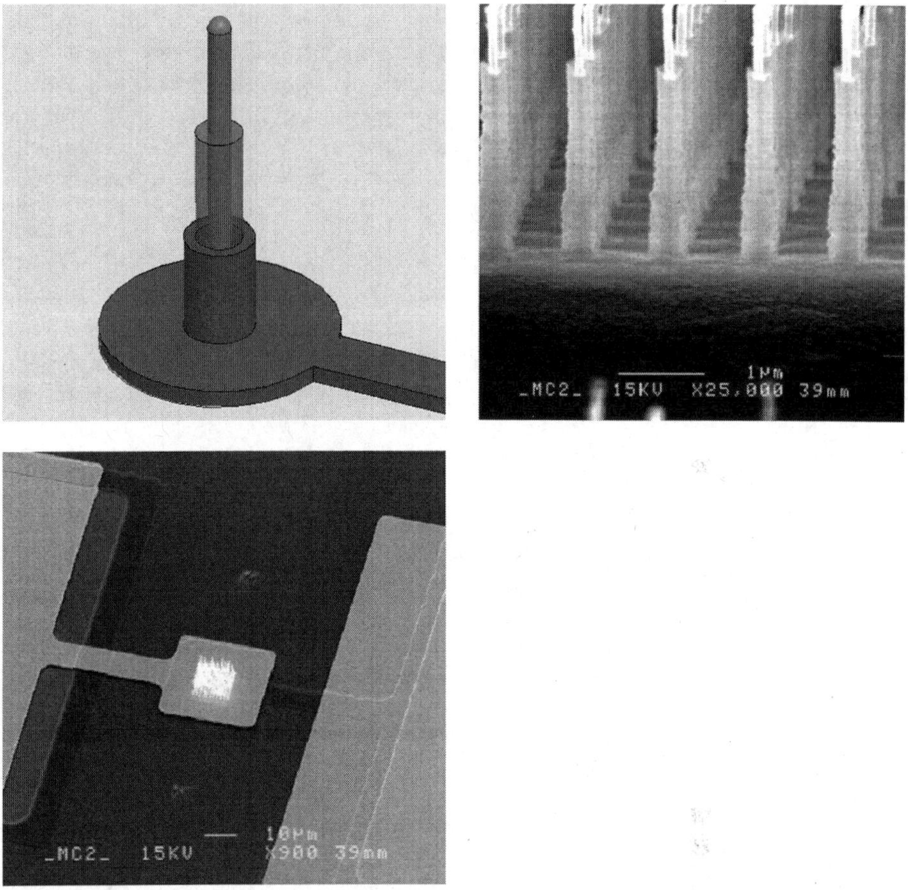

Figure 21.10. Vertical wrap-gated nanowire transistors. (a) Schematic of a nanowire transistor. (b) Scanning electron micrograph of a nanowire transistor after gate material deposition. (c) Scanning electron micrograph of a nanowire transistor device. With kind permission from Bryllert T, Wernersson L, Lowgren T, Samuelson L. 2006. Vertical wrap-gated nanowire transistors. Nanotechnology 17:227, IOP Publishing Ltd.

years. These developments suggest that thin-film growth technology is a very powerful tool for developing ordered nanostructures. Since thin-film technology is commonly used in the microelectronics industry, nanointegration based on thin-film technology can be easily adapted for use in a conventional manufacturing environment. Two methods for processing ordered nanostructures were described. The first method is based on strain engineering during thin-film growth, and the second method involves the use of prepatterned templates and screens. Spontaneous or strain-related growth of ordered nanostructures offers the unique possibility of developing dislocation-free nanostructured devices. In addition, this approach is suitable for the development of nanostructured devices on conventional wafers and other large-scale substrates.

Prepatterned or screened growth of ordered nanostructures is also an important method for integrating nanostructures with functional devices. For example, ordered nanowire arrays and nanowire-based devices may be developed using the VLS growth mechanism. This catalyst-assisted growth mechanism provides very good control over the growth of ordered nanowires. Recent advances suggest that thin film-based nanointegration will play a leading role in the development of next-generation nanostructured devices for optoelectronic, biomedical, and other advanced applications.

REFERENCES

1. Takeuchi K, Tajima Y. 2001. Nano-integration: an ingenuity driven approach in nanotechnology. RIKEN Rev 38:3–6.
2. Han H, Ji R, Park Y, Lee S, Rhun G, Alexe M, Nielsch K, Hesse D, Gösele U, Baik S. 2009. Wafer-scale arrays of epitaxial ferroelectric nanodiscs and nanorings. Nanotechnology 20: 015301.
3. Fisslthaler E, Blumel A, Landfester K, Scherf U, List E. 2008. Printing functional nanostructures: a novel route towards nanostructuring of organic electronic devices via soft embossing, inkjet printing and colloidal self-assembly of semiconducting polymer nanospheres. Soft Matter 4:2448–2453.
4. Heo K, Kim C, Jo M, Hong S. 2009. Massive integration of inorganic nanowire-based structures on solid substrates for device applications. J Mater Chem 19:901–908.
5. Kubo T, Parker J, Hillmyer M, Leighton C. 2007. Characterization of pattern transfer in the fabrication of magnetic nanostructure arrays by block copolymer lithography. Appl Phys Lett 90:233113–2331135.
6. Chen J, Shi J, Decanini D, Cambril E, Chen Y, Haghiri-Gosnet A. 2009. Gold nanohole arrays for biochemical sensing fabricated by soft UV nanoimprint lithography. Microelectronic Engineering 86:632–635.
7. Kostovski G, White D, Mitchell A, Austin M, Stoddart P. 2009. Nanoimprinted optical fibres: biotemplated nanostructures for SERS sensing. Biosens Bioelectron 24:1531–1535.
8. Mo Y, Bai M. 2009. Fabrication and adhesion of biomimetic nanotextures fabricated by local oxidation nanolithography. J Colloid Interface Sci 333:304–309.
9. Kumar D, Zhou H, Nath T, Kvit A, Narayan J. 2001. Self-assembled epitaxial and polycrystalline magnetic nickel nanocrystallites. Appl Phys Lett 79:2817–2819.
10. Cheng A, Tzeng Y, Zhou Y, Park M, Wu T, Shannon C, Wang D, Lee W. 2008. Thermal chemical vapor deposition growth of zinc oxide nanostructures for dye-sensitized solar cell fabrication. Appl Phys Lett 92:092113–092115.
11. Li S, Gan C, Cai H, Yuan C, Guo J, Lee P, Ma J. 2007. Enhanced photoluminescence of ZnO/Er_2O_3 core-shell structure nanorods synthesized by pulsed laser deposition. Appl Phys Lett 90:263106–263108.
12. Nelson B, King W, Laracuente A, Sheehan P, Whitman L. 2006. Direct deposition of continuous metal nanostructures by thermal dip-pen nanolithography. Appl Phys Lett 88: 033104–033106.
13. Belaubre P, Guirardel M, Garcia G, Pourciel J, Leberre V, Dagkessamanskaia A, Trévisiol E, François J, Bergaud C. 2003. Fabrication of biological microarrays using microcantilevers. Appl Phys Lett 82:3122–3124.

14. Shchukin V, Bimberg D. 1999. Spontaneous ordering of nanostructures on crystal surfaces. Rev Mod Phys 71:1125–1171.
15. Teichert C. 2002. Self-organization of nanostructures in semiconductor heteroepitaxy. Phys Rep 365:335–432.
16. Berbezier I, Ronda A. 2009. SiGe nanostructures. Surf Sci Rep 64:47–98.
17. Baski A, Whitman L. 1995. Quasiperiodic nanoscale faceting of high-index Si surfaces. Phys Rev Lett 74:956–959.
18. Hibino H, Fukuda T, Suzuki M, Hommo Y, Sato T, Iwatsuki M, Miki K, Tokumoto H. 1993. High-temperature scanning-tunneling-microscopy observation of phase transitions and reconstruction on a vicinal Si(111) surface. Phys Rev B 47:13027–13030.
19. Kasu M, Kobayashi N. 1993. Equilibrium multiatomic step structure of GaAs(001) vicinal surfaces grown by metalorganic chemical vapor deposition. Appl Phys Lett 62:1262–1264.
20. Watson GM, Gibbs D, Zehner DM, Yoon M, Mochrie SGJ. 1993. Faceting transformations of the stepped Pt(001) surface. Phys Rev Lett 71:3166–3369.
21. Golovin A, Levine M, Savina T, Davis S. 2004. Faceting instability in the presence of wetting interactions: a mechanism for the formation of quantum dots. Phys Rev B 70: 235342–235353.
22. Viernow J, Lin J, Petrovykh D, Leibsle F, Men F, Himpsel F. 1998. Regular step arrays on silicon. Appl Phys Lett 72:948–950.
23. Teichert C, Bean J, Lagally M. 1998. Self-organized nanostructures in Si1−xGex films on Si(0 0 1). Appl Phys A 67:675–685.
24. Asaro R, Tiller W. 1972. Interface morphology development during stress corrosion cracking. I. Via surface diffusion. Metall Trans A3:1789–1796.
25. Srolovitz DJ. 1989. On the stability of surfaces of stressed solids. Acta Metall 37: 621–625.
26. Grinfeld MA. 1986. Instability of the separation boundary between a non-hydrostatically stressed elastic body and a melt. Sov Phys Dokl 31:831–834.
27. Sekiguchi T, Yoshida S, Itoh K, Myslivecek J, Voigtländer B. 2007. One-dimensional ordering of Ge nanoclusters along atomically straight steps of Si(111). Appl Phys Lett 90: 013108–013111.
28. Xie Q, Madhukar A, Chen P, Kobayashi N. 1995. Vertically self-organized InAs quantum box islands on GaAs(100). Phys Rev Lett 75:2542–2545.
29. Springholz G, Pinczolits M, Holy V, Zerlauth S, Vavra I, Bauer G. 2001. Vertical and lateral ordering in self-organized quantum dot superlattices. Physica E 9:149–163.
30. Guha S, Madhukar A, Rajkumar K. 1990. Onset of incoherency and defect introduction in the initial stages of molecular beam epitaxical growth of highly strained $In_xGa_{1-x}As$ on GaAs(100). Appl Phys Lett 57:2110–2112.
31. Solomon G, Trezza J, Marshall A, Harris J Jr. 1996. Vertically aligned and electronically coupled growth induced InAs islands in GaAs. Phys Rev Lett 76:952–955.
32. Schittenhelm P, Engel C, Findeis F, Abstreiter G. 1998. Self-assembled Ge dots: growth, characterization, ordering, and applications. J Vac Sci Technol B 16:1575–1581.
33. Thanh V, Boucaud P, Débarre D, Zheng Y, Bouchier D, Lourtioz JM. 1999. Vertically self-organized Ge/Si(001) quantum dots in multilayer structures. J Phys Rev B 60:5851–5857.
34. Jin C, Zhou H, Wei W, Narayan R. 2006. Three-dimensional self-organization of crystalline gold nanoparticles in amorphous alumina. Appl Phys Lett 89:261103–261105.

35. Teichert C, Hofer C, Lyutovich K, Bauer M, Kasper E. 2000. Interplay of dislocation network and island arrangement in SiGe films grown on Si(001). Thin Solid Films 380:25–28.
36. Häusler K, Eberl K, Noll F, Trampert A. 1996. Strong alignment of self-assembling InP quantum dots. Phys Rev B 54:4913–4918.
37. Shiryaev S, Jensen F, Hansen J, Petersen J, Larsen A. 1997. Nanoscale structuring by misfit dislocations in $Si_{1-x}Ge_x$/Si epitaxial systems. Phys Rev Lett 78:503–506.
38. Jesson D, Chen K, Pennycook S, Thundat T, Warmack R. 1995. Crack-like sources of dislocation nucleation and multiplication in thin films. Science 268:1161–1163.
39. Cullis A, Robbins D, Barnett S, Pidduk A. 1994. Growth ripples upon strained SiGe epitaxial layers on Si and misfit dislocation interactions. J Vac Sci Technol A 12:1924–1931.
40. Häusler K, Noll F, Eberl K. 1996. Ordering of nanoscale InP islands on strain-modulated InGaP buffer layers. Solid State Electron 40:803–806.
41. Hu Q, Ghoniem N, Walgraef D. 2007. Influence of substrate-mediated interactions on the self-organization of adatom clusters. Phys Rev B 75:075405–075416.
42. Hofer C, Abermann S, Teichert C, Bobek T, Kurz H, Lyutovich K, Kasper E. 2004. Ion bombardment induced morphology modifications on self-organized semiconductor surfaces. Nucl Instrum Methods Phys Res B 216:178–184.
43. Vescan L, Grimm K, Goryll M, Holländer B. 2000. Ordered nucleation of Ge islands along high index planes on Si. Mater Sci Eng B 69:324–328.
44. Morales AM, Lieber CM. 1998. A laser ablation method for the synthesis of crystalline semiconductor nanowires. Science 279:208–211.
45. Haraguchi K, Katsuyama T, Hiruma K, Ogawa K. 1992. GaAs p-n junction formed in quantum wire crystals. Appl Phys Lett 60:745–747.
46. Huang MH, Wu Y, Feick H, Tran N, Weber E, Yang P. 2001. Catalytic growth of zinc oxide nanowires by vapor transport. Adv Mater 13(2):113–116.
47. Takeuchi K, Tajima Y. 2001. Nano-integration: an ingenuity driven approach in nanotechnology. RIKEN Rev 38:3–6.
48. Wagner RS, Ellis WC. 1964. Vapor-liquid-solid mechanism of single crystal growth. Appl Phys Lett 4:89–90.
49. Li J, Lu C, Maynor B, Huang S, Liu J. 2004. Controlled growth of long GaN Nanowires from catalyst patterns fabricated by "dip-pen" nanolithographic techniques. Chem Mater 16:1633.
50. Greyson EC, Babayan Y, Odom TW. 2004. directed growth of ordered arrays of small-diameter ZnO nanowires. Adv Mater 16:1348–1352.
51. Ng HT, Han J, Yamada T, Nguyen P, Chen YP, Meyyappan M. 2004. Single crystal nanowire vertical surround-gate field-effect transistor. Nano Lett 4:1247–1252.
52. Liu D, Xiang Y, Wu X, Zhang Z, Liu L, Song L, Zhao X, Luo S, Ma W, Shen J, Zhou W, Wang G, Wang C, Xie S. 2006. Periodic ZnO nanorod arrays defined by polystyrene microsphere self-assembled monolayers. Nano Lett 6:2375.
53. Rybczynski J, Banerjee D, Kosiorek A, Giersig M, Ren ZF. 2004. Formation of super arrays of periodic nanoparticles and aligned ZnO nanorods—simulation and experiments. Nano Lett 4:2037–2040.
54. Chou SY, Krauss PR, Renstrom PJ. 1996. Imprint lithography with 25-nanometer resolution. Science 272:85–87.
55. Nikoobakht B, Michaels CA, Stranicka SJ, Vaudin MD. 2004. Horizontal growth and in situ assembly of oriented zinc oxide nanowires. Appl Phys Lett 85:3244–3246.

56. Mårtensson T, Carlberg P, Borgstrom M, Montelius L, Seifert W, Samuelson L. 2004. Nanowire arrays defined by nanoimprint lithography. Nano Lett 4:699–702.
57. Hochbaum AI, Fan R, He R, Yang P. 2005. Controlled growth of Si nanowire arrays for device integration. Nano Lett 5:457–460.
58. Salaita K, Wang Y, Mirkin CA. 2007. Applications of dip-pen nanolithography. Nat Nanotechnol 2:145–155.
59. Masuda H, Fukuda F. 1995. Ordered metal nanohole arrays made by a two-step replication of honeycomb structures of anodic alumina. Science 268:1466–1468.
60. Zhang X, Zhang L, Meng G, Li G, Jin-Phillipp N, Phillipp F. 2001. Synthesis of ordered single crystal silicon nanowire arrays. Adv Mater 13:1238–1241.
61. Fan HJ, Lee W, Scholz R, Dadgar A, Krost A, Nielsch K, Zacharias M. 2005. Arrays of vertically aligned and hexagonally arranged ZnO nanowires: a new template-directed approach. Nanotechnology 16:913–917.
62. Lee CJ, Lee TJ, Lyu SC, Zhang Y, Ruh H, Lee HJ. 2002. Field emission from well-aligned zinc oxide nanowires grown at low temperature. Appl Phys Lett 81:3648–3650.
63. Huang M, Mao H,S, Feick H, Yan H, Wu Y, Kind H, Weber E, Russo R, Yang P. 2001. Room-temperature ultraviolet nanowire nanolasers. Science 292:1897–1899.
64. Liu A. 2008. Towards development of chemosensors and biosensors with metal-oxide-based nanowires or nanotubes. Biosens Bioelectron 24:167–177.
65. Bryllert T, Wernersson L, Lowgren T, Samuelson L. 2006. Vertical wrap-gated nanowire transistors. Nanotechnology 17:s227–s230.
66. Suk S, Lee S, Kim S, Yoon E, Kim M, Li M, Oh C, Yeo K, Kim S, Shin D, Lee K, Park H, Han J, Park C, Park J, Kim D, Park D, Ryu B. 2005. High performance 5 nm radius twin silicon nanowire MOSFET (TSNWFET): fabrication on bulk Si wafer, characteristics, and reliability. IEEE International Electron Devices Meeting 2005, Technical Digest:735–738.

22

MASS-MANUFACTURABLE NANOWIRE INTEGRATION: CHALLENGES AND RECENT DEVELOPMENTS

Ataur Sarkar and M. Saif Islam

University of California, Davis, California

INTRODUCTION

The ongoing exponential increase in the demand for high-performance computing with the ever-increasing amount of data being transferred among end users, for high-speed audiovisual communication systems, and for even larger data storage capacity, greatly challenges the performance of the existing complementary metal oxide semiconductor (CMOS) integration techniques. The scientific difficulty of continued electronic device density scaling through the transformative "top-down" approach has motivated worldwide efforts to explore alternatives. Massively parallel synthetic "bottom-up" techniques offer the potential for high-density integration of nanoscale devices beyond the capability of conventional technologies with a small fraction of the present-day fabrication cost. The bottom-up technology has significantly advanced and surpassed the incapability of patterning. Using this technique, functional nanostructures can be assembled from chemically synthesized, well-defined nanoscale building blocks similar to the way proteins and biomolecules in nature are used to build complex and functional biological systems. Researchers have been able to implement the bottom-up technology and synthesize a number of one-dimensional (1-D) nanowires with precisely controlled and tunable chemical composition, structure, size, morphology, and electrical properties

Ceramic Integration and Joining Technologies: From Macro to Nanoscale, First Edition.
Edited by Mrityunjay Singh, Tatsuki Ohji, Rajiv Asthana, Sanjay Mathur.

in both homostructure and heterostructure systems (Yazawa et al. 1991; Trentler et al. 1995; Zhang et al. 1998; Kong et al. 1999; Wu and Yang 2000; Cui and Lieber 2001; Huang et al. 2001; Islam et al. 2004; Kamins et al. 2004). Semiconductor nanowires, in particular, have displayed many novel electronic, optical, and chemical properties. Interesting characteristics such as high surface-to-volume ratio, quantum confinement, and simple and low-cost synthesis processes are opening new frontiers in novel electronic and photonic devices. A number of functional nanowire devices ranging from p–n junction diodes to lasers, photovoltaics, and complex logic gates have been successfully prototyped. It is envisioned that quantum properties of the nanowires will contribute to the development of exciting applications and novel device concepts in the near future.

The concept of nanowire synthesis was introduced through the pioneering work of Wagner and Ellis in 1964, which demonstrated micron-sized 1-D whisker structures (Wagner and Ellis 1964). Hiruma and coworkers at Hitachi Labs significantly advanced the work in nanowire synthesis through their well-regarded work almost a decade ago (Yazawa et al. 1991), and their approach was further advanced by several groups to the level where nanoelectronic and photonic devices were realized. Despite significant progress in nanowire synthesis and many promising single-device demonstrations, applications of nanowires have been stalled by our inability to controllably incorporate them within integrated circuits (ICs). While the progress of nanowire synthesis and potential applications is formidable, the real challenge remains in interfacing and interconnecting nanowire devices with other circuit elements to make them useful for their envisioned applications, ranging from chemical and biological sensors to future computing. Successful integration of nanowires and nanotubes requires reproducible interfacing and interconnecting techniques to take advantage of their unique charge transport properties. Highly ohmic contacts between the nanostructures and outer circuitry are essential to realize efficient nanoscale devices and circuits. In this chapter, we review the common approaches to controllably align and interface nanowires to the macroworld, and address the challenges and issues in realizing a mass-manufacturable nanowire integration platform for the integration of future photonic devices and circuits.

NANOWIRE FABRICATION

A nanowire by definition is a structure having a diameter in the range of tens of nanometers and a length of a few micrometers. They can be synthesized out of semiconductors, conductors, and alloys. Nanowires are grown on semiconducting or insulating substrates that can be single crystalline or amorphous. It is important to note that a substrate's crystalline orientation has a major influence on nanowire growth. In general, we need catalyst particles and precursor gases that contain the constituent materials of the nanowires to be synthesized. Usually, nanowires are synthesized using the thermal evaporation of metallic compounds inside a chemical vapor deposition (CVD) reactor. The growth process is mostly catalytic, where metal catalysts such as gold or nickel are widely used. Nanowire growth has three distinct phases: vapor, liquid, and solid, known as the vapor–liquid–solid (VLS) method of nanowire growth.

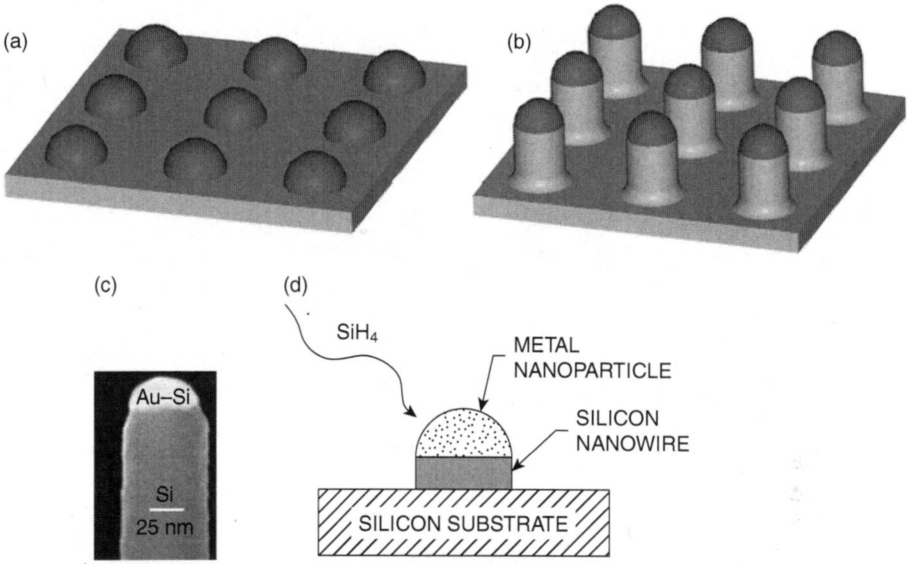

Figure 22.1. Metal-catalyzed Si nanowire growth mechanism with chemical vapor deposition technique. (a) Small metal nanoparticles accelerate the decomposition of a silicon-containing gas; the Si atoms precipitate between the nanoparticle and the substrate, forming a column of silicon as shown in (b). (c) A gold (Au)-catalyzed nanowire with the Au–Si nanoparticle at the tip. (d) Schematic of the VLS technique.

1-D nanowire characteristics are heavily dependent on its crystallization as they evolve from the VLS phases. Nanowire nucleation starts when the concentration of atoms or molecules becomes sufficiently high around a catalyst particle and combines into a small cluster that serves as a seed for further nucleation to form a bigger cluster. This process continues and a nanowire is formed. The metal catalyst acts as a high-energy site for the decomposition of a precursor gas containing the nanowire materials. With the right temperature and pressure, the constituent material atoms deposit onto the catalyst surface (Givargizov 1975; Lauhon et al. 2004) and thus initiate the nucleation. As the growth continues, the nanowire materials are adsorbed into or around the catalyst particles and are deposited to the substrate at the catalyst–substrate interface. The growth site thus forces the nanowire to grow orthogonal to the underlying substrate. Nanowire growth using the catalyst-assisted VLS process is shown in Figure 22.1.

The growth temperature is governed by the type of metal catalyst used (Sze 1981), and hence it should be judiciously chosen. Gold is widely used with silicon nanowire because of the low-temperature growth. The low eutectic temperature of ~363°C of the Au–Si pair with a mole ratio of 15–85% makes gold the first choice when selecting the catalyst. The low-temperature growth of nanowire using a nickel catalyst has been also demonstrated by several research groups through the "solid-phase" process (Kamins et al. 2001; Tuan et al. 2005). Although the solid-phase growth process is not well understood, the use of nickel as a low-temperature catalyst is becoming popular

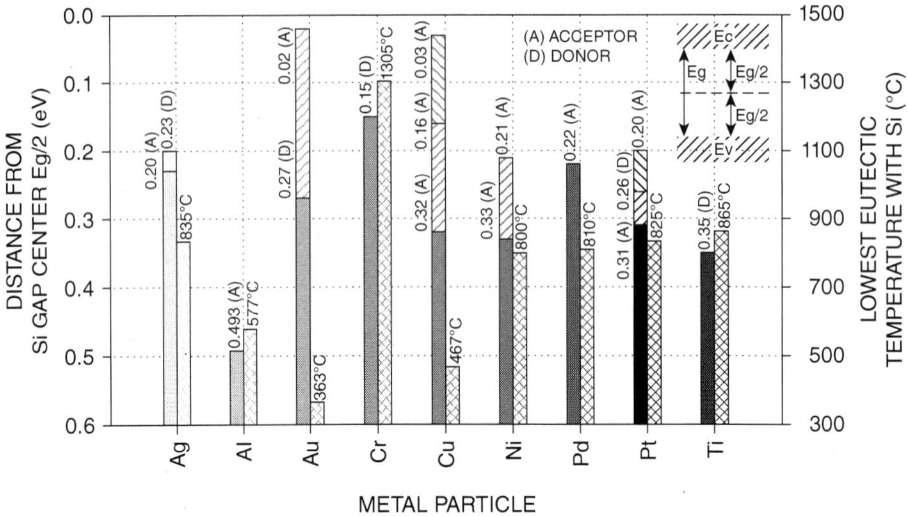

Figure 22.2. Graph of metal atoms mixed with silicon atoms. The left bars show possible ionization energy levels that can be introduced when a metal is mixed with silicon. The right bars show the lowest recorded liquid eutectic temperature when mixed with silicon (Sze 1981).

in multicatalyst growth processes for the conventional semiconductor business. However, metal catalysts can impart impurities to the nanowire and have a performance-degrading effect on nanowire devices. As an example, the gold catalyst can act as midband ionization energy in silicon (Sze 1981). Figure 22.2 plots the distance between the midgap center and the trap sites caused by a metal catalyst. Some metals such as gold and copper contribute to a high level of generation–recombination currents if the metals are present in the semiconductor even if the level of metal is almost undetectable. However, as Figure 22.3 shows, using gold as a catalyst in growing GaAs nanowires is better suited compared with silicon in terms of ionization energy levels, as gold acts as an acceptor level in GaAs.

The heteroepitaxial growth of III–V compound semiconductors onto silicon substrates has been a long sought goal because the integration of high-performance III–V materials with current silicon technology could open many new applications in opto-electronics and photonics. Local growth on the nanometer-scale area allows fabrication of highly mismatched 1-D semiconductor heterostructures using the VLS method. Conventional planar growth of lattice mismatched layers leads to a deformation of the crystal structure and an accumulation of the strain in the growing layer, followed by a transition to the three-dimensional growth mode or by a generation of the misfit dislocation network that propagates through the layer along the growth direction. If the growth area is reduced to the scale where the accumulated lateral strain can be accommodated, the growth of a defect-free structure is feasible. In addition, along the length of the nanowire, a gradual change of the lateral lattice constant may be possible, starting with the strained lateral lattice constant of the substrate and relaxing within a few monolayers to the unstrained lattice constant of the nanowire material. Successful

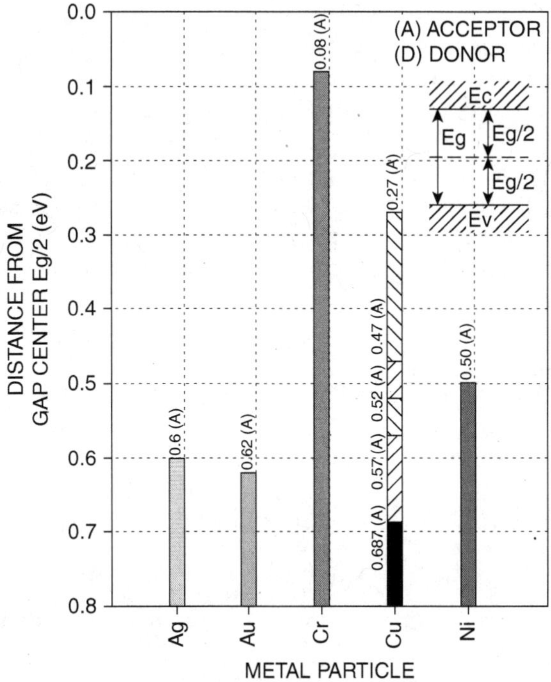

Figure 22.3. Graph of metal atoms mixed with gallium arsenide atoms. The bars show possible ionization energy levels that can be introduced when the metal is mixed with the semiconductor (Sze 1981).

realization of a few tens of nanometer length defect-free InAs nanowhiskers on GaAs substrate as well as InP/InAs/InP nanowires with atomically sharp heterointerfaces was demonstrated (Bjork et al. 2002). Recently, we also grew InP nanowires on silicon substrates (Yi et al. 2005, 2006; Sarkar et al. 2008) as well as quartz surfaces (Sarkar et al. 2008). These results show that lattice mismatch can be accommodated in the nanowire structures because of their small cross-section and, thus, address the long-standing challenge of integrating high-performance III–V semiconductors with mainstream silicon technology.

NANOWIRE ALIGNMENT AND POSITIONING

After nanowire synthesis, the characteristic properties must be evaluated before application to any actual devices or circuits. It is essential to have electrical access to both ends of the nanowire. The technique of parallel interfacing with individual electrical contact to a large number of nanowires is the first requirement to realize novel and efficient nanoelectronic and nanophotonic devices with quantum effects. Another important issue alongside the contact making technique is the proper positioning of the nanowires, which can be performed in in-situ or in postgrowth settings. Generally, nanowires grown via the bottom-up technique are removed from the substrates, dispersed into appropriate solvents, and deposited onto required substrates to orient them in a particular way. Electric field-assisted assemblies (Diehl et al. 2001; Duan et al. 2001), microfluidic channels (Messer et al. 2000; Huang et al. 2001), and Langmuir–Blodgett (LB) assemblies (Whang et al. 2003a, b; Jin et al. 2004) are some examples of postgrowth alignment and positioning techniques. Once positioning is done, (metal) electrodes are deposited on the nanowire ends to connect them to the micro/macroworld. In the following subsections, some useful nanowire aligning and positioning techniques are briefly reviewed.

Microfluidic Channel Assembly

In this method, nanowires dispersed in a solution are injected into a microfluidic channel assembly and nanowires are oriented by their interaction with the channel sidewalls. A demonstration using fluid assembly to control nanowire separation and position was made by Huang et al. in 2001. Figure 22.4 represents the crossed nanowire

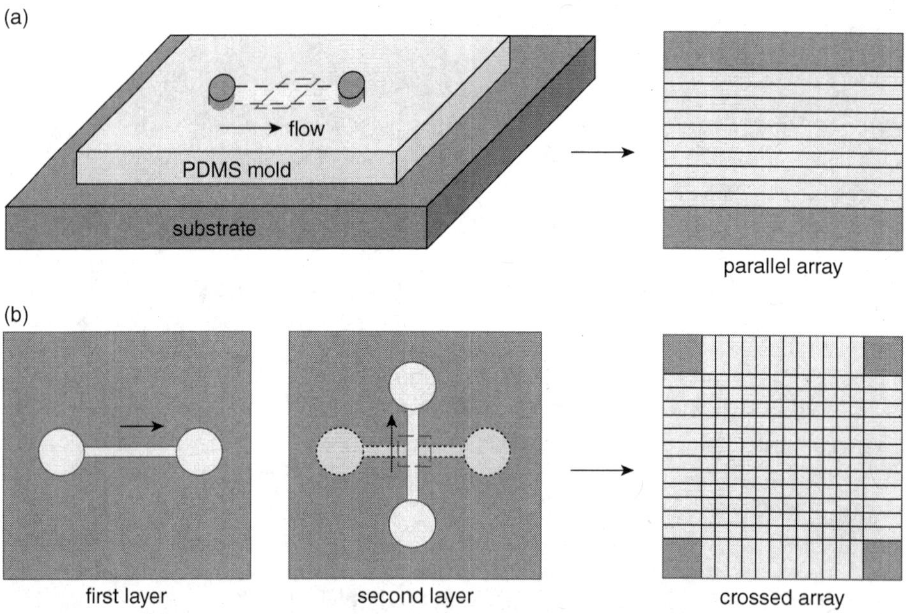

Figure 22.4. Schematic of fluidic channel structures for flow assembly (Huang, Duan et al. 2001). (a) A channel formed when the poly-di-methyl-siloxane (PDMS) mold was brought in contact with a flat substrate. Nanowire (NW) assembly was carried out by flowing an NW suspension inside the channel with a controlled flow rate for a set duration. Parallel arrays of NWs are observed in the flow direction on the substrate when the PDMS mold is removed. (b) Multiple crossed NW arrays can be obtained by changing the flow direction sequentially in a layer-by-layer assembly process.

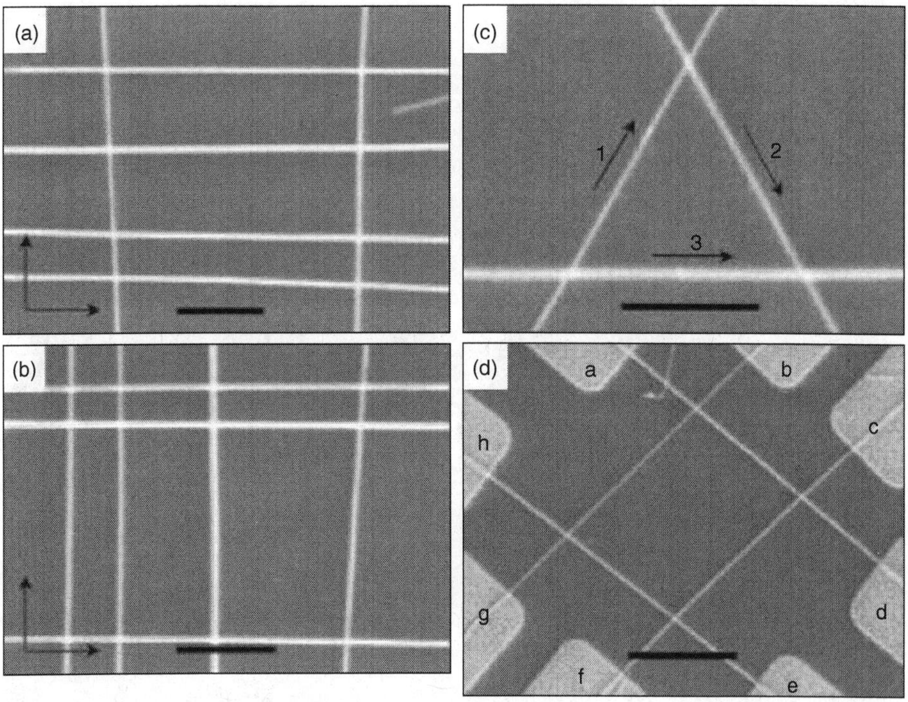

Figure 22.5. (A, B) SEM images of crossed arrays of InP nanowires obtained in a two-step assembly process with orthogonal flow directions for the sequential steps. (C) An equilateral triangle of GaP nanowires obtained in a three-step assembly process, with 60° angles between flow directions. (D) A typical 2 × 2 cross array made by sequential assembly of n-type InP nanowires with orthogonal flows. Ni/In/Au contact electrodes, which were deposited by thermal evaporation, were patterned by e-beam lithography. Flow directions are highlighted by arrows in the images (Huang et al. 2001).

arrays arranged with layer-by-layer assembly of nanowires with different flow directions for sequential steps. Complex geometrical structures can be realized by controlling the angle between the flow directions in sequential steps, as shown in Figure 22.5. One advantage of this layer-by-layer method is that homo- and heterojunction at the cross points can be formed by altering the nanowire composition for each sequence.

Dielectrophoretic Assembly

This method employs an electric or magnetic field to align and position nanowires on a substrate using dielectrophoretic forces to pull the polarizable nanowires toward the high-field region. Under the influence of electric field, the nanowires align parallel to the field direction between two electrodes (Duan et al. 2001). Nanowires coated with a ferromagnetic material, such as nickel or cobalt, can be aligned using magnetic fields

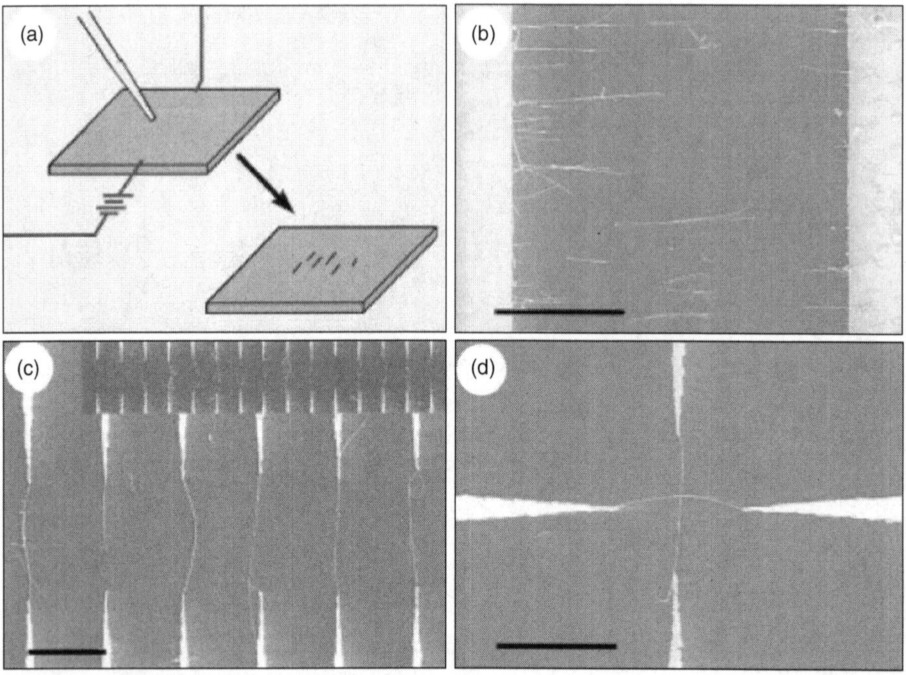

Figure 22.6. Parallel and orthogonal assembly of nanowires with electric fields (Duan et al. 2001). (a) Schematic view of alignment by electric field. The electrodes are biased at 50 ± 100 V after a drop of nanowire solution is deposited on the substrate. (b) Parallel array of nanowires aligned between two parallel electrodes. The nanowires were suspended in chlorobenzene and were aligned using an applied bias of 100 V. (c) Spatially positioned parallel array of nanowires obtained following electric field assembly using a bias of 80 V. Inset: 15 pairs of parallel electrodes with individual nanowires bridging each diametrically opposed electrode pair. (d) Crossed nanowire junction obtained using layer-by-layer alignment with the electric field applied in orthogonal directions in the two assembly steps. The applied bias in both steps was 80 V. Scale bars in b–d: 10 mm.

(Bentley et al. 2004). Electric field assembled parallel and crossbar arrays of nanowires are shown in Figure 22.6 (Duan et al. 2001).

Langmuir-Blodgett Assembly

In this method, a compressive force is applied at a liquid–gas or liquid–liquid interface (Whang et al. 2003a, b). First, a hydrophobic surfactant is used to modify the nanowire surface with the hydrophobic surfactant head facing water. The nanowire is lifted to the water–air interface by the repulsive force between the hydrophobic head and the water molecules. An external force is applied to squeeze the nanowires and to increase the surface pressure, which slowly reaches a critical point. At the critical point, the nanowires get oriented perpendicular to the direction of the applied pressure in the LB

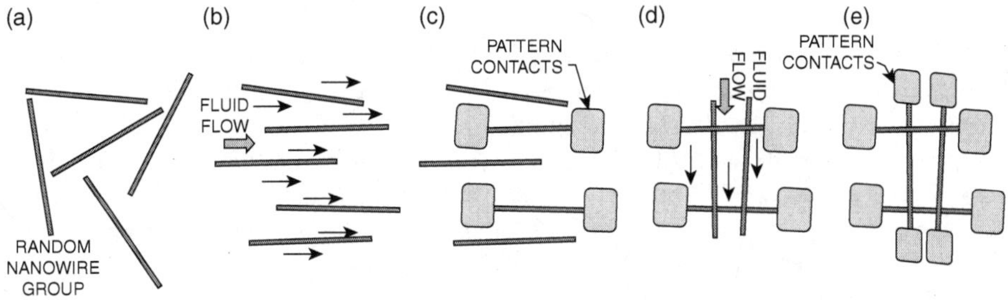

Figure 22.7. An example of the Langmuir–Blodgett technique for producing connections to nanowires. (a) A large group of unattached, randomly dispersed nanowires on a substrate. (b) The nanowires are moved and oriented to a single direction by a flowing fluid. (c) Where devices are to be made, selective contact fabrication is performed at both ends of the nanowires. (d) A second set of nanowires is then exposed to a flowing fluid where the direction of the fluid is orthogonal to the patterned nanowires. These nanowires are intended to be carried and oriented by the flowing fluid and eventually rest on top of the underlying nanowire devices that have fabricated contacts. (e) After patterning contacts on the second set of nanowires, a device having multiple contacts is made (Whang et al. 2003a, b).

trough. This results in an aligned nanowire array on the water surface, which can then be transferred to another substrate for further processing. Figure 22.7 shows ordered hierarchical structures by layer-by-layer transfer of nanowires on a planar substrate using the LB technique, first demonstrated by Whang et al. (2003a, b).

The aforementioned techniques of aligning and positioning the nanowires have not been perfected and need further development. Application of solvents and surfactants to nanowires can foreseeably alter the nanowire properties that will eventually affect the device performance.

NANOWIRE INTERCONNECTION

Integration of 1-D nanowire structures into electronic devices and circuits continues to be an ongoing challenge in the process of developing bottom-up synthesis technology for future ultra-high-density electronic and photonic systems. Nanowires of semiconductors, oxides, nitrides, carbides, and chalcogenides have been generated by numerous groups employing various strategies. There is a growing interest in utilizing 1-D nanoscale wires as memory and logic devices as well as for interconnecting devices and circuits so that the remarkable trend of miniaturization of CMOS fabricated using the top-down approach can be further continued using the bottom-up paradigm. Although a great deal of progress in nanowire synthesis and many promising single-device demonstrations have been made in the last decade, the application to a wide range of applications has been stalled by our inability to successfully incorporate them within ICs. The research community has been sequentially connecting electrodes to individual nanowires for device physics studies since the birth of this exciting area of research. However, a massively parallel and manufacturable interfacing technique is crucial for reproducible fabrication of dense and low-cost nanodevice arrays. A good understanding and control of nanowire growth conditions, dimensions, orientation, and position as well as physical, electronic, and optical characteristics are important preconditions for the reliable design of nanosystems based on 1-D nanowires. Exploration and advancing these important issues require novel fabrication methods, innovative structures, and an in-depth understanding of underlying physics of growth mechanism.

While the forgoing techniques are representative of the progress made toward controlled assembly of nanowires, mass-manufacturable and reproducible techniques need to be addressed. The obstacle of the existing techniques to realizing mass-manufacturable functional nanowire-based devices is the extensive lithography needed to fabricate the electrodes for the assembly of multiple nanowire device structures. In the sections that follow, we summarize different approaches to contact nanowires and make a comparison among some common techniques.

Methods to Contact Nanowires

Among numerous approaches (Langford et al. 2006; Stern et al. 2006) to contact nanowires, the most common technique is electron-beam lithography (EBL), in which isolated electrodes are deposited on top of the nanowires. Atomic force microscopy (AFM) employs manipulating and positioning the wires onto prepatterned electrodes (Avouris et al. 1999), and scanning electron microscopy (SEM) uses in-situ micromanipulators (Kim et al. 2003) to contact the nanowires. Other alternatives such as focused ion beam (FIB)- or focused electron beam (FEB)-assisted deposition are popular for some of the processing steps. FIB has nanomilling capability and has been used to sputter etch the native oxide on bismuth wires prior to deposition of the platinum contact. Contact to nanowires has also been made by dispersing them in a solution and then drop-casting or spin-coating this solution on prepatterned electrodes (Bachtold et al. 1998a, b). In Figure 22.8, carbon nanotubes are contacted to deposited platinum electrodes. Other reported methods of making contacts to nanowires are conducting AFMs (Durkan and Welland 2000a, b), controlled nanowire growth between electrodes (Oon and Thong 2004), and dielectrophoresis (Boote and Evans 2005).

Performance of the Commonly Used Methods to Contact Nanowires

The most common techniques to make electrical contacts to nanowires are EBL, ion and electron beam-assisted deposition, and in-situ micromanipulation. The poor nanowire/metal interface in EBL leading to a Schottky barrier contributes to a high contact resistance of nanowire devices (Stern et al. 2006). Annealing is often performed to make contacts ohmic. EBL-contacted nanowire contact resistance is sensitive to the nanowire surface chemistry before metallization because of the electrode's wrapping around the nanowire surface. Contact resistance less than $100\,\Omega$ with EBL-deposited contacts has been reported (Langford et al. 2006). In the FIB method of contacting

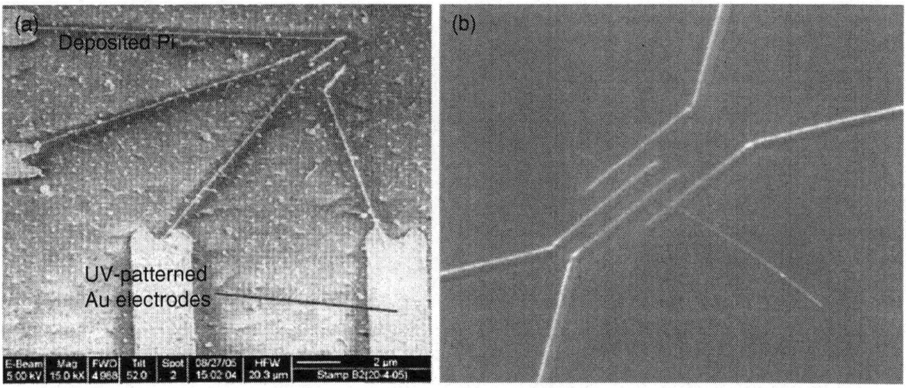

Figure 22.8. (a) Secondary electron (SE) image of Pt-deposited electrodes onto a carbon nanotube (Langford et al. 2006). (b) A single nanotube connected to four 80-nm wide tungsten wires using FIB (Ebbesen et al. 1996).

nanowires, the contact resistance is independent of the wire surface chemistry since metal deposition involves ion milling to the core of the nanowire only. The FIB-assisted contact resistance was observed in the range of 10^4–$10^6\,\Omega$ (Zhao et al. 2001; Bandaru et al. 2005; Langford et al. 2006). The micromanipulation method realizes contact resistance within the range of a few hundred ohms for metal nanowires and 10^3–$10^6\,\Omega$ for carbon nanotubes (Langford et al. 2006).

The aforementioned contact-making techniques are based on a metal–semiconductor contact that enables electrical access to the nanowire ends. All of these techniques involve a large amount of postgrowth processing in which nanowires are separated from their growth substrate and then contacts are forged to the wire placed on a separate insulating substrate. The main drawback of the aforementioned techniques is that they are not suitable for forging contacts in a parallel mass-manufacturable platform. To successfully make contacts to nanowires, the major challenge is to devise a platform that efficiently enables reproducible and parallel contact making to multiple devices in a single shot.

BRIDGED NANOWIRES

As discussed in the previous sections, the major drawback in implementing nanowire-based devices and systems in a massively parallel and reproducible manner is the lack of a suitable interconnecting technique that would enable contacting a large number of nanowires in a single step. There is an urgent need for an efficient and cost-effective nanowire integration technique compatible with current IC processing methods. A massively parallel and mass-manufacturable interfacing technique with reproducible fabrication capability at industry level is thus crucial to develop in order to produce dense and low-cost nanodevice arrays with high throughput to exploit the potential benefits of nanoscale devices.

A novel epitaxial bridging technique for interfacing silicon (Islam et al. 2005) and InP (Yi et al. 2006) nanowires between silicon electrodes was first proposed by Islam et al. with highly linear ohmic contacts due to the epitaxial connection between the nanowires and the electrodes (Sharma et al. 2005). Similar techniques for interfacing silicon nanowires were reported by Yang (He et al. 2005) and Lin et al. (Englander et al. 2003). The advantage of the above epitaxial integration schemes is that these are in-situ techniques, and nanowire deterioration during postprocessing can be totally avoided. Thus, these are highly efficient and suitable for mass-manufacturable production of nanowire-based devices, with a major improvement in the cost/performance ratio.

Nanobridging Between Two Vertical Planes

The nanobridging method is implemented by growing nanowires between a set of electrically isolated prefabricated electrodes of desired material. Nanowire growth initiates from one electrode and finishes on the opposite electrode, making firm contact and thus bridging the electrically isolated electrodes. Nanowire growth is influenced by the crystallographic orientations of the vertical sidewalls of the electrodes; nanowires preferentially grow perpendicular to (111) surfaces. Hence, lateral nanowire growth is observed with (111) sidewalls. The anisotropic wet chemical etching of a (110)-oriented silicon wafer is a well-known technique of forming vertical (111) planes for the vertical electrode planes. The electrodes can be patterned using conventional lithography. Once the electrodes are patterned, a very thin layer of the desired catalyst is deposited on the walls of the electrodes and is annealed to form isolated nanoparticles of the catalyst material, and the structure is then exposed to a precursor gas in a CVD reactor. The decomposition of the precursor gases is slow in the area without a catalyst but is greatly accelerated on the catalyst particles. Considering the case of silicon nanowires, silicon atoms from silicon-containing gases diffuse through or around the catalyst nanoparticles and precipitate on the substrate under the catalyst particle, and thus push the nanoparticle away from the surface. The nanostructure formed between the substrate and the catalyst particle in this process is the nanowire of interest.

As the growth continues and the nanowire reaches the opposite sidewall, it touches and then is "welded" to the sidewall by continued catalyzed decomposition. The welded connection is mechanically strong and possesses a low electrical resistance. With this connection to the opposite electrode sidewall, the formation of nanobridges is completed. The concept was demonstrated by forming a trench bound by two vertical surfaces in a (110)-oriented silicon wafer using conventional lithography, as shown in Figure 22.9.

Thermally grown silicon dioxide, an etch-resistant layer, was used to cover the wafer, and reactive ion etching with CHF_3 and argon gases was used to form an opening in the oxide layer. Using an aqueous solution containing KOH (44% KOH–H_2O at 110°C for 1 min), which has different etching rates at different crystal orientations, the silicon was etched through the opening. With (110)-oriented silicon, the vertical etching rate is much greater than the lateral etching rate (by as much as 100:1), leaving vertical, (111)-oriented silicon surfaces bounding the trench. The trenches patterned in this work are approximately 4–8 μm deep and 2–15 μm wide.

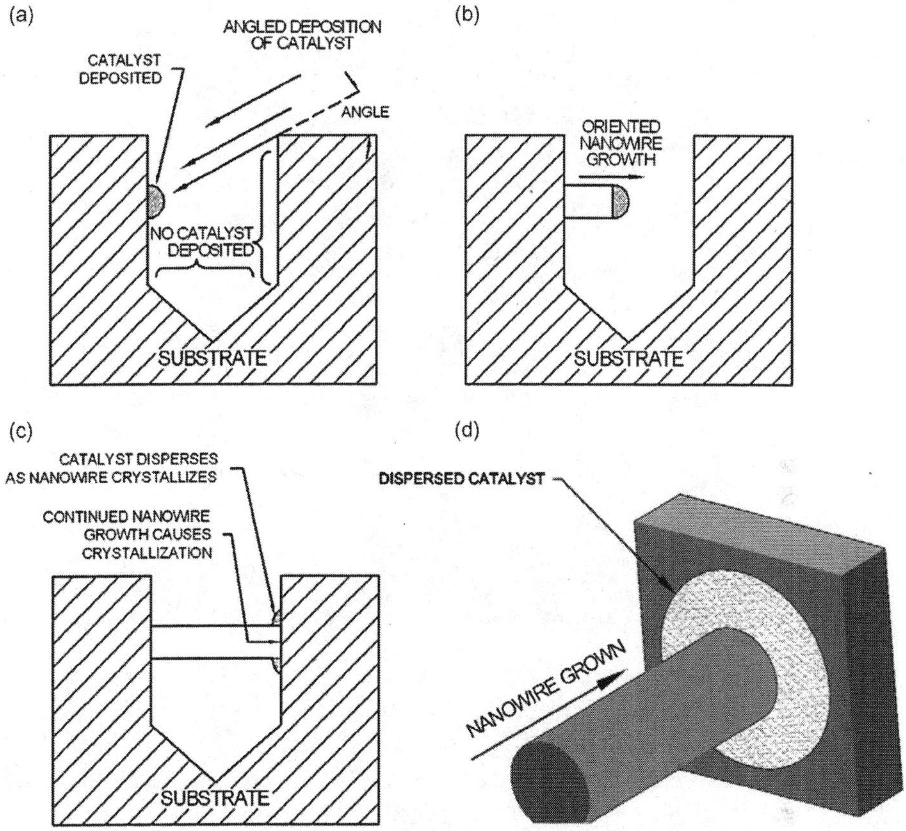

Figure 22.9. Nanobridging technique. (a) Angled deposition into a trench deposits the catalyst onto the growth sidewall. Note the catalyst deposition does not deposit onto the opposite sidewall or onto the trench floor. (b) Catalyst grown nanowire produces oriented nanowire directional growth. (c, d) Continued nanowire growth causes the nanowire to grow onto the opposite sidewall. As the nanowire continues to grow, the catalyst disperses away from the nanowire connection as the nanowire crystallizes to the opposite sidewall.

Metal catalyst nanoparticles, gold or titanium, on the order of 1 nm thick, were deposited by electron-beam evaporation onto the vertical (111)-oriented surfaces of the etched grooves. By angled evaporation, the catalyst was deposited on only one sidewall of the trench so that the nucleating and impinging ends of the nanowires could be studied separately. For the given geometry of the trenches, the deposition angle was so selected that no catalyst particles were deposited on the bottoms of trenches narrower than 8 μm. The samples were then annealed in hydrogen ambient inside a CVD to form Au–Si alloy nanoparticles. Nanowires were grown at ~640°C by introducing a gas mixture of SiH_4 and HCl into the hydrogen ambient.

Figure 22.10 represents Au-catalyzed silicon nanowires grown on (111) vertical sidewalls. The 8-μm-long nanowires extend only partially across a 15-μm-wide trench

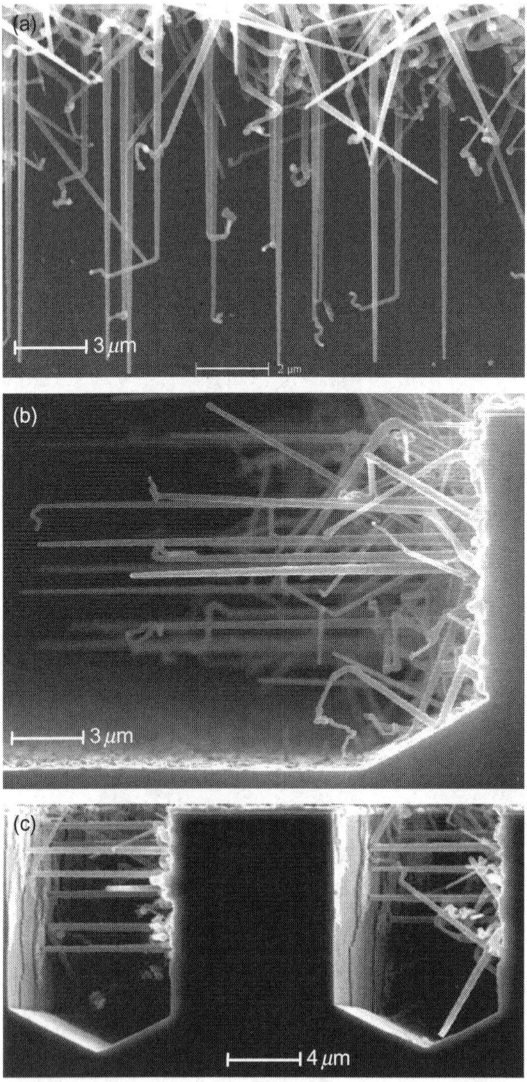

Figure 22.10. (a) SEM micrograph of the top view of laterally grown nanowires on a vertical (111)-oriented Si plane. (b) Cross-section SEM micrograph of lateral epitaxial nanowire growth from (111)-oriented sidewall surface plane at the right of the diagram into a 15-μm-wide trench. (c) "Bridging" across 6-μm-wide trenches and connecting to opposing sidewalls. The nanowires grow from right to left in these images.

(Fig. 22.10a, b). Some 90% of the nanowires have a length of ~8 μm, with the longest nanowire at ~10 μm. Electrical and thermal isolation between the heavily doped electrodes is ensured by means of a silicon-on-insulator (SOI) structure, as shown in Figure 22.11.

Nanocolonnades Between Two Horizontal Planes

Vertically oriented nanowires contacted at the ends with horizontal planes are termed "nanocolonnades." This novel concept of making contacts to vertically oriented nanowires was first demonstrated by Islam et al. using a CVD-assisted growth technique, illustrated in Figure 22.12. This approach has a high potential for enabling design of nanowire lasers with precise cavity length and electrical pumping capability.

To integrate nanocolonnades, two doped epitaxial layers of semiconductor materials separated by an insulating layer are grown and a window was opened in the upper doped layer using conventional lithography and dry etching. A selective isotropic wet etchant is then applied through the opening to remove the intermediate insulating layer to form an "awning"-shaped structure. Using the metal catalysts as discussed in the

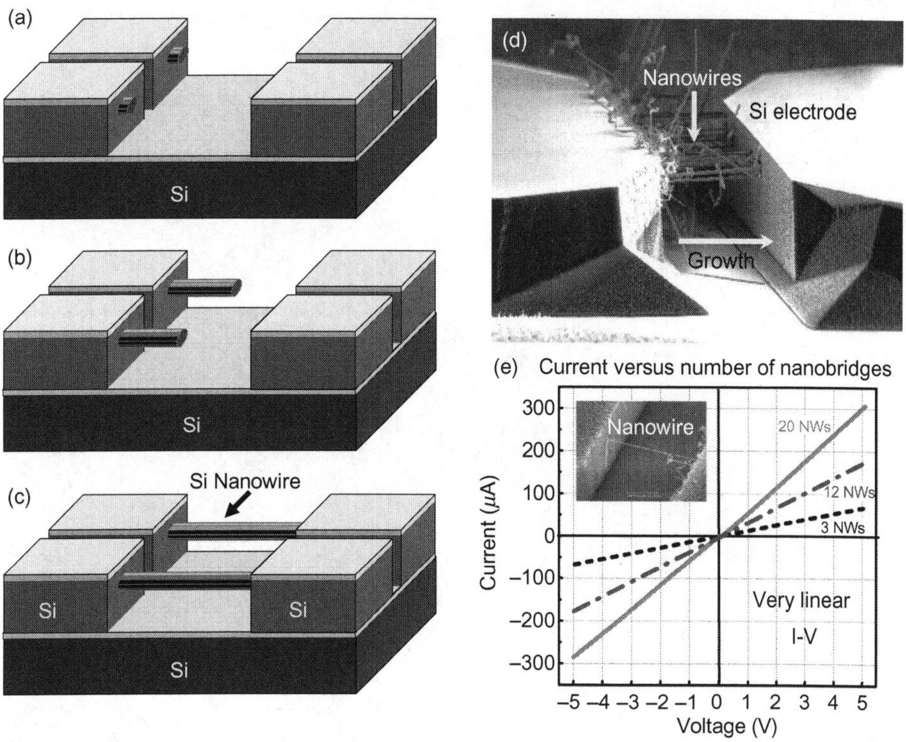

Figure 22.11. (a–c) Schematics depicting the nanobridge synthesis method. (d) SEM micrograph of the first Si nanobridges formed between two vertical (111)-oriented Si surfaces. (e) Linear current voltage characteristics of nanobridges.

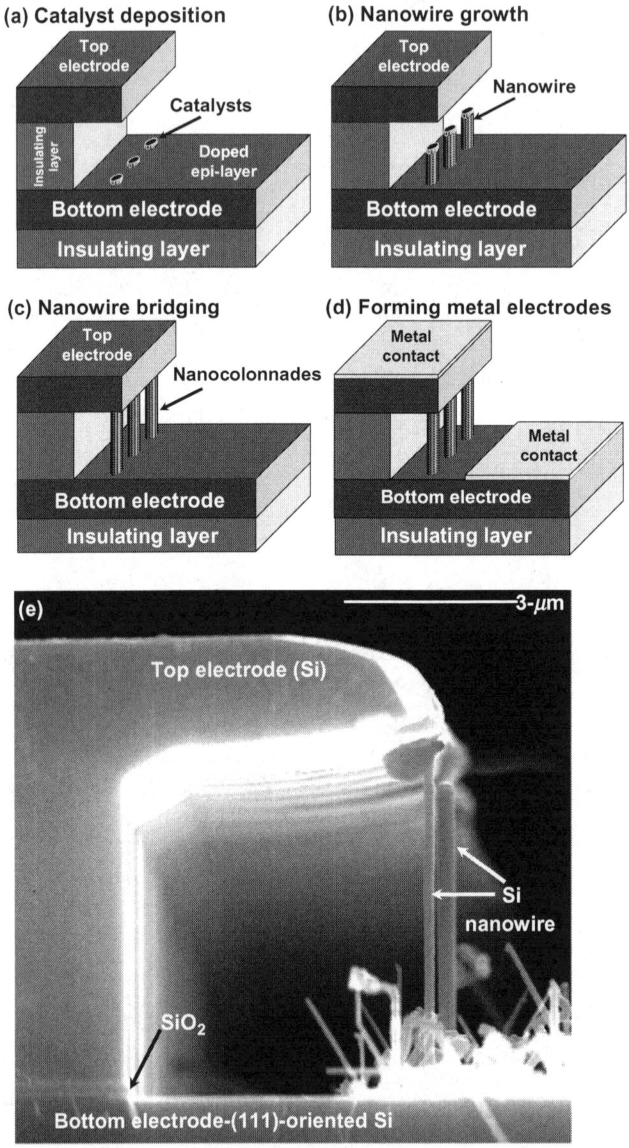

Figure 22.12. (a–d) Schematics of electrode preparation and Au-catalyzed synthesis of nanocolonnades. (e) Demonstration using an SOI wafer.

previous sections, vertically oriented nanowires were grown from the bottom electrode and were connected to the top electrode, as shown in Figure 22.12. As a handle wafer, a (111)-oriented SOI wafer was used as the bottom electrode to facilitate the growth of silicon nanowires in the <111> direction and to make a connection to the device layer on top of the oxide insulating layer.

The aforementioned techniques of making nanobridges and nanocolonnades are highly promising for the mass manufacturability of nanoscale devices and systems using nanowires with unprecedented device density. Extensive research is ongoing to explore the full potential of these interfacing techniques.

Mechanical Properties of Bridged Nanowires

Robust mechanical connection between the electrode sidewalls and the nanowire ends is essential for any kind of nanowire application. In the case of gold catalyst particles for silicon nanowire growth, the metal catalyst particles at the end of the nanowires remain in liquid phase during the nanowire growth. When the nanowire ends with the catalyst particle impinging on the opposite sidewall, the axial growth of nanowires stops, and the molten alloy spreads radially, contributing to catalytic decomposition of SiH4 and further accelerated deposition. This continued accelerated deposition forms a disk of material extending radially outward from the contact point along the (111) silicon sidewall of the trench, as shown in Figure 22.13. For titanium-nucleated nanowires, the catalyst particles remain in the solid phase during nanowire growth. As a result, slow growth of nanowire along the sidewall is observed, but it is sufficient to firmly

Figure 22.13. (a) Au- and (b) Ti-catalyzed Si nanowire impinging ends. (c) Transmission electron microscopy (TEM) image of a bridging interface showing epitaxial connection. There is a thin layer of native oxide at the interface.

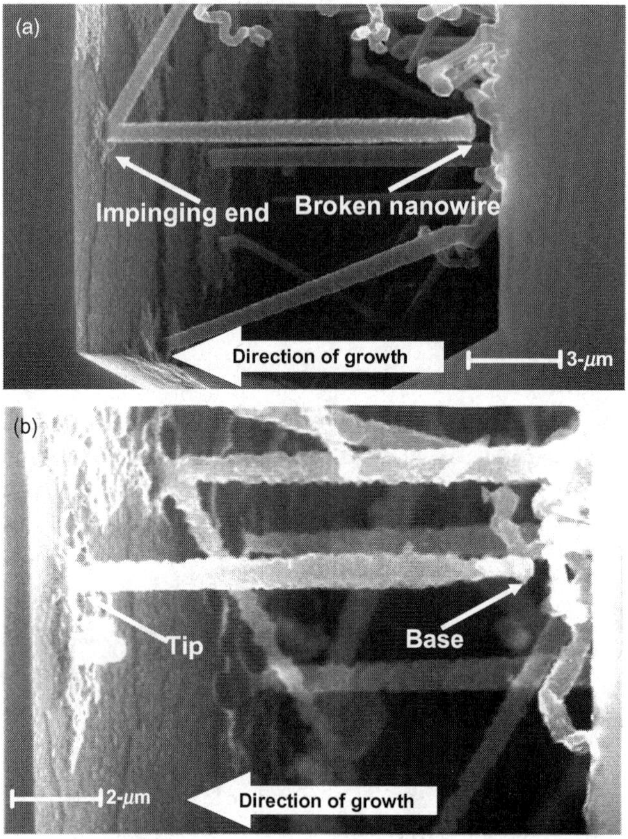

Figure 22.14. Structural and mechanical robustness of the nanobridges. (a) A nanowire was manually broken while cleaving a sample for SEM imaging. The nanowire broke along its length rather than at the connection point, demonstrating the strength of the connection. (b) One hour of strong HF (1:1 = HF : H_2O) treatment did not cause much damage to the nanowire, and the impinging end survived the harsh HF treatment.

attach the nanowire to the sidewall. The mechanical strength of the contact points was tested in two ways: applying pressure on the bridged nanowires from a tungsten probe and etching the bridged nanowires in a high concentration of HF solution for an extended duration. When tested to stress failure, the nanowires often break along their length rather than at the contact points, thus indicating a strong mechanical connection of the contact points. Figure 22.14a shows a manually broken single-nanowire bridge under the pressure of a tungsten probe. The impinging end of the broken nanowire is still attached to the electrode sidewall, demonstrating a robust mechanical contact. A sample containing gold-catalyzed nanowires grown across a trench was treated in a 1:1 HF : H_2O solution for ~100 min. This concentration of HF is greater than that used in conventional IC processing. As shown in Figure 22.14b, the silicon nanowires remained attached to the electrode sidewalls even after extended exposure to the high-concentration

HF solution. The Au–Si alloy layer was partially removed from the sidewalls, but the silicon nanowires remained attached to the sidewalls, confirming a strong Si–Si bonding at the contact points.

Contact Properties of Bridged Silicon Nanowires

Ohmic contacts to semiconductor nanowires with low contact resistance are essential components of many new nanoscale electronic devices. The specific contact resistance is a particularly useful area-independent parameter for comparing contacts of different geometries and is typically expressed in the units of ohm square centimeter. The contact resistance R_c is then given by ρ_c/A, where A is the active area of the contact. In our bridging nanowire test structures, we estimated the specific contact resistance to be 3.74×10^{-6}–$5.02 \times 10^{-6}\ \Omega\text{-cm}^2$ (Chaudhry et al. 2007). This value reflects almost two orders of magnitude of improvement over $5.0 \times 10^{-4}\ \Omega\text{-cm}^2$ measured for Ti/Au contacts to p-type silicon nanowires (Mohney et al. 2005; Chaudhry et al. 2007) and an order of magnitude improvement over $1.7 \times 10^{-5}\ \Omega\text{-cm}^2$ measured for Ti/Au contacts to GaN nanowires (Ham et al. 2006), highlighting the superior characteristics of the unique bridged nanowire interfacing process reported by Islam et al. (Islam et al. 2005). The considerably low contact resistance extracted for the bridged silicon nanowires can be attributed to the metal-catalyzed CVD growth technique used to synthesize the silicon nanobridges. As was mentioned previously, Sharma et al. reported that the nanowire/electrode contact at the impinging end of the trench was epitaxial (Sharma et al. 2005). They observed that the (gold) catalyst remains active until a firm Si–Si contact forms between the nanowire and the uncatalyzed silicon deposited on the impinging sidewall. This Si–Si connection at the nanowire/electrode junction yields very low resistance contacts for the bridged nanowires.

A study on the noise characterization of bridged silicon nanowires found that contact noise is proportional to contact resistance (Reza et al. 2006). Contact noise due to carrier trapping–detrapping in defect centers is the major contributor to the noise characteristics of the nanowire. Moreover, the impinging end of the nanowire was proposed to be a major source of the contact noise. It was shown that the bridged silicon nanowires have at least two orders of magnitude better noise performance than carbon nanotubes (Reza et al. 2006).

CONCLUSION

The integration of nanowires into practical nanoscale devices and circuits is among the greatest challenges of nanoelectronics. In this chapter, we reviewed the advantages and disadvantages of commonly used methods of interfacing nanowires and making contacts to them, issues with different approaches, and the contact properties. The major issue with the widely used research-based methods is the non-ohmic nature of the contacts made to the nanowires. Even the reported highly linear ohmic contacts lack reproducibility and mass manufacturablity when done on a large scale. We discussed our solution to this integration challenge: the formation of nanobridges and nanocolonnades that offer high-quality contacts to nanowires with low contact resistance on a mass-manufacturable scale and that are compatible with the current CMOS technology. The commercial success of the nanowire industry lies in the successful development of mass-manufacturable integration processes for mechanically robust contacts to nanowires with low contact resistance.

ACKNOWLEDGMENTS

This work was partially supported by a UC Davis research grant and National Science Foundation (NSF) grant #0547679. The authors would like to thank Drs. R. Stanley Williams, S. Sharma, and T. I. Kamins of Hewlett-Packard Laboratories and I. Kimukin, C. Edgar, and C. Johns of UC Davis for their help.

REFERENCES

Avouris P, Hertel T, Martel R, Schmidt T, Shea HR, Walkup RE. 1999. Carbon nanotubes: nanomechanics, manipulation, and electronic devices. Appl Surface Sci 141(3–4):201–209.

Bachtold A, Henny M, Terrier C, Strunk C, Schonenberger C, Salvetat JP, Bonard JM, Forro L. 1998a. Contacting carbon nanotubes selectively with low-ohmic contacts for four-probe electric measurements. Appl Phys Lett 73(2):274–276.

Bachtold A, Terrier C, Kruger M, Henny M, Hoss T, Strunk C, Huber R, Birk H, Staufer U, Schonenberger C. 1998b. Contacting single template synthesized nanowires for electric measurements. Microelectron Eng 42:571–574.

Bandaru PR, Daraio C, Jin S, Rao AM. 2005. Novel electrical switching behaviour and logic in carbon nanotube Y-junctions. Nat Mater 4(9):663–666.

Bentley AK, Trethewey JS, Ellis AB, Crone WC. 2004. Magnetic manipulation of copper-tin nanowires capped with nickel ends. Nano Lett 4(3):487–490.

Bjork MT, Ohlsson BJ, Sass T, Persson AI, Thelander C, Magnusson MH, Deppert K, Wallenberg LR, Samuelson L. 2002. One-dimensional heterostructures in semiconductor nanowhiskers. Appl Phys Lett 80(6):1058–1060.

Boote JJ, Evans SD. 2005. Dielectrophoretic manipulation and electrical characterization of gold nanowires. Nanotechnology 16(9):1500–1505.

Chaudhry A, Ramamurthi V, Fong E, Islam MS. 2007. Ultra-low contact resistance of epitaxially interfaced bridged silicon nanowires. Nano Lett 7(6):1536–1541.

Cui Y, Lieber CM. 2001. Functional nanoscale electronic devices assembled using silicon nanowire building blocks. Science 291(5505):851–853.

Diehl MR, Yu JY, Heath JR, Held GA, Doyle H, Sun SH, Murray CB. 2001. Crystalline, shape, and surface anisotropy in two crystal morphologies of superparamagnetic cobalt nanoparticles by ferromagnetic resonance. J Phys Chem B 105(33):7913–7919.

Duan XF, Huang Y, Cui Y, Wang JF, Lieber CM. 2001. Indium phosphide nanowires as building blocks for nanoscale electronic and optoelectronic devices. Nature 409(6816):66–69.

Durkan C, Welland ME. 2000a. Nanometer scale electrical characterization of artificial mesostructures. Crit Rev Solid State Mater Sci 25(1):1–28.

Durkan C, Welland ME. 2000b. Size effects in the electrical resistivity of polycrystalline nanowires. Phys Rev B 61(20):14215–14218.

Ebbesen TW, Lezec HJ, Hiura H, Bennett JW, Ghaemi HF, Thio T. 1996. Electrical conductivity of individual carbon nanotubes. Nature 382(6586):54–56.

Englander O, Christensen D, Lin LW. 2003. Local synthesis of silicon nanowires and carbon nanotubes on microbridges. Appl Phys Lett 82(26):4797–4799.

Givargizov EI. 1975. Oriented growth of whisker-crystals of aiiibv compounds according to vapor liquid-crystal mechanism. Kristallografiya 20(4):812–822.

Ham MH, Choi JH, Hwang W, Park C, Lee WY, Myoung JM. 2006. Contact characteristics in GaN nanowire devices. Nanotechnology 17(9):2203–2206.

He RR, Gao D, Fan R, Hochbaum AI, Carraro C, Maboudian R, Yang PD. 2005. Si nanowire bridges in microtrenches: Integration of growth into device fabrication. Adv Mater 17(17): 2098–2102.

Huang Y, Duan XF, Wei QQ, Lieber CM. 2001. Directed assembly of one-dimensional nanostructures into functional networks. Science 291(5504):630–633.

Huang MH, Wu YY, Feick H, Tran N, Weber E, Yang PD. 2001. Catalytic growth of zinc oxide nanowires by vapor transport. Adv Mater 13(2):113–116.

Islam MS, Sharma S, Kamins TI, Williams RS. 2004. Ultrahigh-density silicon nanobridges formed between two vertical silicon surfaces. Nanotechnology 15(5):L5–L8.

Islam MS, Sharma S, Kamins TI, Williams RS. 2005. A novel interconnection technique for manufacturing nanowire devices. Appl Phys A-Mater Sci Process 80(6):1133–1140.

Jin S, Whang DM, McAlpine MC, Friedman RS, Wu Y, Lieber CM. 2004. Scalable interconnection and integration of nanowire devices without registration. Nano Lett 4(5):915–919.

Kamins TI, Li X, Williams RS. 2001. Ti-catalyzed Si nanowires by chemical vapor deposition: Microscopy and growth mechanisms. J Appl Phys 89(2):1008–1016.

Kamins TI, Williams RS, Basile DP, Hesjedal T, Harris JS. 2004. Growth and structure of chemically vapor deposited Ge nanowires on Si substrates. Nano Lett 4(3):503–506.

Kim KS, Lim SC, Lee IB, An KH, Bae DJ, Choi S, Yoo JE, Lee YH. 2003. In situ manipulation and characterizations using nanomanipulators inside a field emission-scanning electron microscope. Rev Sci Instrum 74(9):4021–4025.

Kong J, Zhou C, Morpurgo A, Soh HT, Quate CF, Marcus C, Dai H. 1999. Synthesis, integration, and electrical properties of individual single-walled carbon nanotubes. Appl Phys A-Mater Sci Process 69(3):305–308.

Langford RM, Wang TX, Thornton M, Heidelberg A, Sheridan JG, Blau W, Leahy R. 2006. Comparison of different methods to contact to nanowires. J Vac Sci Technol B 24(5):2306–2311.

Lauhon LJ, Gudiksen MS, Lieber CM. 2004. Semiconductor nanowire heterostructures. Philos Transact A Math Phys Eng Sci 362(1819):1247–1260.

Logeeswaran VJ, Sarkar A, Islam MS, Kobayashi NP, Straznicky J, Li X, Wu W, Mathai S, Tan MRT, Wang SY, Williams RS. 2008. A 14-ps full width at half maximum high-speed photoconductor fabricated with intersecting InP nanowires on an amorphous surface. Appl Phys A 91:1–5.

Messer B, Song JH, Yang PD. 2000. Microchannel networks for nanowire patterning. J Am Chem Soc 122(41):10232–10233.

Mohney SE, Wang Y, Cabassi MA, Lew KK, Dey S, Redwing JM, Mayer TS. 2005. Measuring the specific contact resistance of contacts to semiconductor nanowires. Solid-State Electron 49(2):227–232.

Oon CH, Thong JTL. 2004. In situ nanowire growth for electrical interconnects. Nanotechnology 15(5):687–691.

Reza S, Bosman G, Islam A, Kamins TI, Sharma S, Williams RS. 2006. Noise in silicon nanowires. IEEE Transact Nanotechnol 5(5):523–529.

Sarkar A, Kimukin I, Edgar CW, Yi S, Islama MS. 2008. Heteroepitaxial growth dynamics of InP nanowires on silicon. J Nanophotonics 2:021775 (1–15).

Sharma S, Kamins TI, Islam MS, Williams RS, Marshall AF. 2005. Structural characteristics and connection mechanism of gold-catalyzed bridging silicon nanowires. J Cryst Growth 280(3–4):562–568.

Stern E, Cheng G, Klemic JF, Broomfield E, Turner-Evans D, Li C, Zhou C, Reed MA. 2006. Methods for fabricating Ohmic contacts to nanowires and nanotubes. J Vac Sci Technol B 24(1):231–236.

Sze SM. 1981. *Physics of Semiconductor Devices*. New York: Wiley.

Trentler TJ, Hickman KM, Goel SC, Viano AM, Gibbons PC, Buhro WE. 1995. Solution-liquid-solid growth of crystalline III-V semiconductors—an analogy to vapor-liquid-solid growth. Science 270(5243):1791–1794.

Tuan HY, Lee DC, Hanrath T, Korgel BA. 2005. Catalytic solid-phase seeding of silicon nanowires by nickel nanocrystals in organic solvents. Nano Lett 5(4):681–684.

Wagner RS, Ellis WC. 1964. Vapor-liquid-solid mechanism of single crystal growth. Appl Phys Lett 4(5):89–90.

Whang D, Jin S, Lieber CM. 2003a. Nanolithography using hierarchically assembled nanowire masks. Nano Lett 3(7):951–954.

Whang D, Jin S, Wu Y, Lieber CM. 2003b. Large-scale hierarchical organization of nanowire arrays for integrated nanosystems. Nano Lett 3(9):1255–1259.

Wu YY, Yang PD. 2000. Germanium nanowire growth via simple vapor transport. Chem Mater 12(3):605–607.

Yazawa M, Koguchi M, Hiruma K. 1991. Heteroepitaxial ultrafine wire-like growth of InAs on GaAs substrates. Appl Phys Lett 58(10):1080–1082.

Yi SS, Girolami G, Amano J, Islam MS, Sharma S, Kamins TI, Kimukin I. 2005. InP nanobridges epitaxially formed between two vertical Si surfaces. 5th IEEE Conference on Nanotechnol, July 11–15: 209–212.

Yi SS, Girolami G, Amano J, Islam MS, Sharma S, Kamins TI, Kimukin I. 2006. InP nanobridges epitaxially formed between two vertical Si surfaces by metal-catalyzed chemical vapor deposition. Appl Phys Lett 89(13):133121.

Zhang YF, Tang YH, Wang N, Yu DP, Lee CS, Bello I, Lee ST. 1998. Silicon nanowires prepared by laser ablation at high temperature. Appl Phys Lett 72(15):1835–1837.

Zhao YP, Wei BQ, Ajayan PM, Ramanath G, Lu TM, Wang GC, Rubio A, Roche S. 2001. Frequency-dependent electrical transport in carbon nanotubes. Phys Rev B 6420(20):201402.

23

USABILITY OF INK-JET PRINTING TECHNOLOGY AND NANOMATERIALS IN ELECTRICAL INTERCONNECTIONS, ELECTRONIC PACKAGING, AND SYSTEM INTEGRATION FOR MICROELECTRONICS APPLICATIONS

Umur Caglar, Ville Pekkanen, Jani Valkama, Pauliina Mansikkamäki, and Jussi Pekkanen

Tampere University of Technology, Tampere, Finland

INTRODUCTION

The density of electronic systems has been increasing with high functionality expectations, and it seems that it will be very difficult to increase the density on silicon. Silicon technology can achieve the current manufacturing targets through increased density of the interconnection platform. In order to achieve the interconnection requirements for electronic packaging, miniaturization of electronic packages is possible with high electrical efficiency and the production of narrow conductive interconnections. Microelectronics includes the micron or below micron size of many components

Ceramic Integration and Joining Technologies: From Macro to Nanoscale, First Edition.
Edited by Mrityunjay Singh, Tatsuki Ohji, Rajiv Asthana, Sanjay Mathur.

available in electronics, namely, capacitors, inductors, transistors, resistors, diodes, insulators, and conductors. Electronic packaging and miniaturization of electronic components allows a decrease of the total electronic system size. Electronic packaging is not only about decreasing the size but also about increasing the mechanical properties of the electronic package. The integration of passive components, also called embedded passives, into electronics manufacturing has some advantages such as lowering cost, improving performance, and reliability. The next generation of electronic packaging is concentrating on system in package (SIP), system on chip (SOC), and system on package (SOP) in order to decrease the system size, to increase reliability, to provide high frequency, and to lower the overall system cost.

However, the electronic industry has some drawbacks in fulfilling these requirements, that is, material limitations, the costs of manufacturing processes, and production flexibility. Ink-jet printing technology offers additive on-demand material deposition on substrates without any limitation. It is possible to print very precise structures without the need for mask and etching processes. In that sense, the amount of wasted material is not much as in lithography processes, which naturally decreases the manufacturing cost. Nanosized metallic materials have been intensively used in ink-jet printing technology mainly in the printing of conductive patterns. The high metal content of these inks enable the printing of high-frequency interconnection patterns. Narrow line/space width decreases to $50\,\mu m/50\,\mu m$, and it is still possible to decrease the line/space width with several surface modifications on the selected substrate. Embedded passives are investigated for ink-jet print with selected material combinations to decrease the package size. Ink-jet printing technology offers substantial electronics manufacturing advantages in decreasing the manufacturing cost and material waste, and in increasing the production flexibility.

PRINTABLE ELECTRONICS AND INK-JET PRINTING TECHNOLOGIES

Printable Electronics

Printable electronics is a relatively new manufacturing method for electronic devices, and its active/passive components compare to other mature electronics manufacturing competitors in silicon fabrication processes. Printing technologies are currently used in electronics with several manufacturing methods, for example, screen printing, gravure printing, flexography, offset lithography, nanoimprint, and ink-jet printing. Printable electronics is one of the major "pusher technologies" for current electronics manufacturing, and it has the potential to enable future electronic devices. The current electronics manufacturing techniques are produced on rigid substrates and are subtractive technology, that is, photolithography that produces a high amount of wasted material during manufacturing, higher cost, and large energy consumption caused by high processing temperatures and process purification (Molesa 2006). Even though the printing technologies and direct writing technologies, for example, laser-induced forward transfer (LIFT), plasma spray, matrix-assisted pulsed laser evaporation–direct write (MAPLE-DW), or thermal spraying, are not currently mature enough to compete with silicon complementary metal oxide semiconductor (CMOS) technologies, the various advantages of the technologies have been discussed and several interesting applications have been demonstrated (Piqué and Chrisey 2002). The main advantages of the printing techniques against CMOS manufacturing are low costs and low-temperature usability at atmospheric pressure conditions. In addition, the printing technologies enable the printing of almost all types of substrates, for example, on standard rigid printed circuit boards (PCBs), flexible organic films, metal foils, paper, paperboard, or even on three-dimensional (3-D) surfaces that are commonly used in several existing industrial applications.

Even though printable electronics are creating a new era in electronics manufacturing, it is expected that it should not be a substitute for current electronics manufacturing methods. It is possible that future electronics manufacturing techniques will be a "hybrid" of current electronics and printable electronics techniques. But it is also expected that printable electronics will enable entirely new products in the electronics market (Nanomarkets 2007). The flexibility of electronic components and devices is one of the major aspects of novel electronic products that printable electronics enables, for example, flexible or organic displays, wearable sensors, and organic nonvolatile memories are only a few of these.

On the other hand, there are drawbacks in printable electronics, for example, the printing resolution requirements of simple electronic circuits, which are defined by manufacturers as submicron features for performance reasons, while the narrowest printed line width is typically around $10\,\mu m$ (Derbyshire 2007). In addition, in some techniques, for example, ink-jet technology, the materials are immature and the development of functional materials is a bottleneck to the creation of more sophisticated electronics structures.

Offset lithography, screen printing, and ink-jet printing are the dominant printing techniques, although almost all the mentioned printing techniques have been used. The importance of these printing methods can be described as follows: Offset lithography is a major printing method in the publishing sector that uses a high-viscosity ink (paste), which is typically 40–100 Pa·s, to decrease high shear forces during high-speed printing (Nanomarkets 2007). Offset lithography provides thin ink layers, that is, $\sim 1\,\mu m$, relatively high-resolution capabilities, that is, 100 lines/cm (Blayo and Pineaux 2005), and enables high-volume production. Screen printing is over 2000 years old and has become part of electronics in order to print thick film, that is, $10–25\,\mu m$ or above (Hammer et al. 1989; Licari 1994), on typically cofired ceramic substrates, that is, low-temperature cofired ceramic (LTCC) or high-temperature cofired ceramic (HTCC). Screen-printing methods provide a small amount of wasted material compared to offset lithography. The other advantages of the screen-printing method are design flexibility, short and simple fabrication cycles, and low capital equipment requirements. On the other hand, the screen-printing method has several disadvantages, such as limited resolution, that is, 50 lines/cm (Blayo and Pineaux 2005), the need for high-viscosity ink, undesired print thickness, and poor line definition (Parikh et al. 1990). Printing techniques should be selected based on the printed amount of the desired printing products, the complexity of the component, and the thickness of the printing component.

Ink-Jet Printing Technologies

The borderline between direct writing technologies and ink-jet printing technology may not be so obvious, but because ink-jet printing history has been in graphics printing, it is commonly classified as a printing method. Ink-jet printing technology is a noncontact method that uses several inks to jet from the printhead to repeatedly form small droplets that can be directed accurately onto the substrate (Le 1998; Bucknall 2005).

The ink-jet technologies that are based on the printhead characteristic (see Fig. 23.1) are mainly diverse in their printing mode: continuous and drop on demand (DoD). However, there is another method called "bubble jet printing," but that can be also considered part of a DoD ink-jet technology due to the ink droplet being produced on demand (Lawrence et al. 2004). Computer-aided design (CAD) files are created and stored in a digital platform, and it is possible to transfer them directly to the printing environment. Thus, the ink-jet printing system is considered to be very versatile, flexible, and easily customizable. The ink-jet-printed structures could be formed in two dimensions or even in three dimensions on all substrates, which are independent of their type, for example, paper, metal foil, organic flexible film, or nonstandard substrates, such as wood or wearable fabrics, and are independent of their surface quality; for example, printing on rough or smooth surfaces is possible.

The dot matrix high printing resolution, 250 lines/cm (DoD) (Blayo and Pineaux 2005), and printing accuracy enable the creation of more complex electronic structures for several emerging fields, for example, CMOS fabrication processes, integrated circuit (IC) interconnection technologies, and organic light-emitting display (OLED) manufacturing. Even though the printing speed is currently rather slow compared to reel-to-reel (R2R) processes, that is, offset lithography, flexography, and gravure printing, nevertheless, the ink-jet printing method is suitable for prototyping and for mass production of complex electronic components with low material consumption. In addition, the ink-jet printing method is easy to integrate with existing production lines, in which

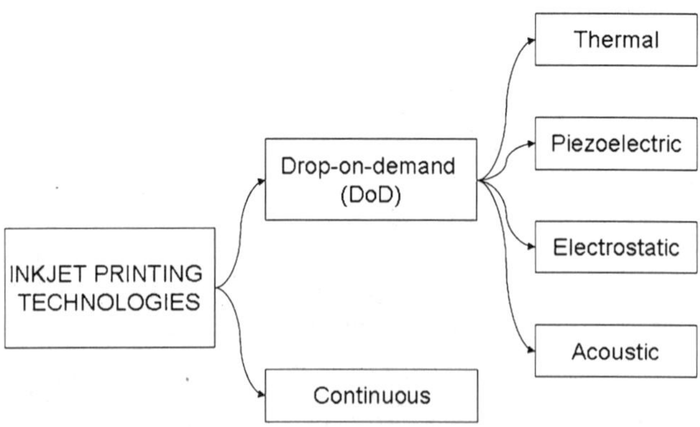

Figure 23.1. Brief description of available ink-jet printing technologies.

conventional printing methods require larger areas and several modifications in the existing production line (Hakola 2005). One of the most important challenges with printing methods other than ink-jet printing is the waste of the functional ink materials, some of which are currently very expensive, for example, semiconductors or inorganic nanoparticle-based conductors.

Piezoelectric printhead-type ink-jet printing technology offers substantial manufacturing advantages compared to continuous printing and other DoD ink-jet printing methods. In piezoelectric ink-jet printing, the control of the volume and shape of an ink droplet is more precise before they are ejected from the printhead nozzles. The piezoelectric crystal (PZT), also called piezoceramic element, actuates a voltage pulse and that creates pressure on the diaphragm which jets an ink droplet. Each droplet ejected from the printhead nozzles is based on a dot matrix CAD digital image that decreases the amount of waste and does not require recycling of the ink. Thus, the ink recirculation system, generally called "gutter" in the continuous ink-jet printing method, is not necessary.

On the other hand, electronic component manufacturing requires broad material usage, for example, inorganic/organic conductors, dielectrics/insulators, semiconductors, and electronic adhesives. It has high position accuracy and thickness control so that currently none of the ink-jet printing methods, apart from the piezoelectric DoD method, could be used. For instance, thermal ink-jet printing methods as well as bubble jet printing are not suitable for use with heat sensitive inks, and continuous ink-jet printing produces large droplets with inadequate resolution for electronics manufacturing because of the long flying distance between printhead and substrate (Lawrence et al. 2004; Blayo and Pineaux 2005). However, continuous printing is still the most common method for printing barcodes and labels because of the advantages in preventing nozzle clogging.

Drying of the printhead nozzles, undesirable satellite droplets, uncontrolled spreading of the droplets on the substrate, and varying ink-jet-printed layer thickness are the main challenges in ink-jet printing technology. The piezoprinthead nozzle of inks usually causes drying of the piezoprinthead nozzle and decreases the length of print run cycles. Satellite droplets (see Fig. 23.2) usually happen either because of the high pulse voltage and low solid content (metal content in inorganic inks) of the functional ink-jet ink material or because of very low viscosity, for example, <1-mPa·s property of the ink. It is possible to control the spreading of the ink with surface modification, for example, plasma treatment, surface coating, or temperature control of the XY table of the printing device.

NANOPARTICLES AND THEIR APPLICABILITY TO INK-JET PRINTING TECHNOLOGY

Introduction

In recent years, there has been tremendous interest in nanomaterials and their applicability to printable electronics. Because of various benefits of nanoparticles, for example,

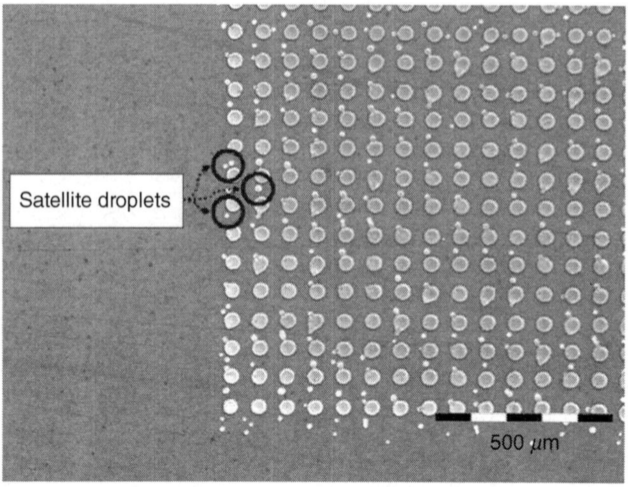

Figure 23.2. Satellite droplets occur during the printing process. This usually occurs because of the usage of high pulse voltage in the piezo-ink-jet printhead.

enabling low-temperature sintering, low-viscosity requirements, and adequate particle size for ink-jet printheads, they have been actively used in order to formulate inks for printing technologies, especially for screen and ink-jet printing. The nanoparticles of the bulk materials offer a wide material selection for electronics applications and this makes it possible to implement nanoparticles together with their suitable rheological formulization to ink-jet printing media (see Fig. 23.3). On the other hand, the nanoparticles enable the formulation of low-temperature firing pastes for screen-printing applications to pattern electronic circuits on the plastic substrates, which will help to create new applications (Nakamoto 2003, 2005).

However, the formulation of nanoparticle-based ink is challenging for ink-jet printing: The viscosity of the ink has to be compatible with the printhead; the particle size should be small enough to eliminate clogging of the nozzles, and the solvent of the ink has to be selected carefully in order to control evaporation of ink on the nozzle plate.

Here, we discuss more about the nanoparticles that have been recently used in ink-jet printing applications, the printing requirements of nanoparticle-based ink, and the future trends of ink-jet inks in ink-jet printing technology.

Nanoparticles in Ink-Jet Printing Technology

The nanometer scale of the particles has several advantages other than their micron particle size, for example, high mechanical properties, such as fatigue resistance, large surface energy, and spatial confinement advantages compared to their bulk material form (Wong et al. 2003; Ko et al. 2007a). However, one of most interesting benefits of nanoparticles in electronics manufacturing occurs in their low melting temperature requirements, which also increases the ratio of surface area to volume (Buffat and Borel

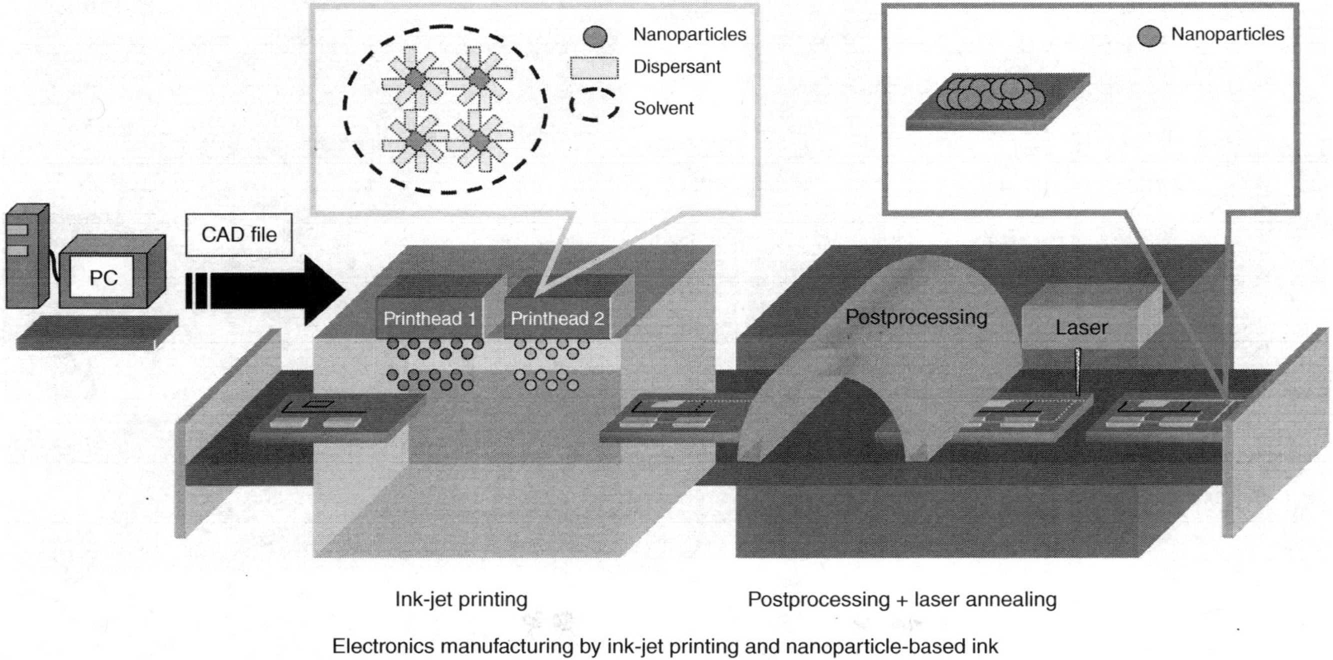

Figure 23.3. Schematic of electronics manufacturing by ink-jet printing and nanoparticle-based ink.

1976). The low melting temperature of the nanosized particles enables the annealing of materials much lower than their bulk form. Thus, it is possible to print nanoparticle-based ink on temperature-sensitive flexible plastic substrates. Because of the high thermal treatment, it is not possible to use plastic substrates in lithographic processes (Park and Baek 2006; Cheng et al. 2005; Kim and Moon 2005). Nanoparticles in that sense provide new opportunities in manufacturing more physically flexible components and systems in electronics, so-called flexible electronics. On the other hand, a disadvantage of the formulated nanoparticle ink is its long-term sedimentation behavior, and as a result, nanoparticle agglomeration in the printhead, even at low temperatures. Nanoparticle-based ink has to be formulated to prevent sedimentation in order to control the viscosity during the printing process and to increase its shelf life at room temperature.

Metallic Nanoparticles. Silver (Ag) and gold (Au) nanoparticles have been intensively investigated in terms of their ink formulization and printed film properties for ink-jet printing technology (Fuller et al. 2002; Huang et al. 2003; Cheng et al. 2005; Kim and Moon 2005; Park and Baek 2006; Perelaer et al. 2006; Suganuma et al. 2007; Zhao et al. 2007). In addition, several metallic nanoparticles, for example, platinum (Pt), palladium (Pd) (CC Tseng et al. 2007), and nickel (Ni) (W Tseng and Chen 2006), have the potential to implement these particles into compatible solvent in order to formulate ink-jet ink.

Ag nanoparticle-based ink and paste have many advantages, for example, the best electrical conductivity of all metals, formulated ink or paste chemical stability, and cost, in order to form a new wiring material combined with proper printing methods (Suganuma et al. 2007). In addition, Ag nanoparticles are very fine in their spherical form and are highly dispersible when they are formulated for ink-jet printing ink. Thus, it is possible to jet well with Ag nanoparticle-based ink from the printhead nozzles (see Fig. 23.4). The preparation of Ag nanoparticles should be done with several methods, for example, electrolysis (KH Tseng et al. 2008), the chemical reduction of water-soluble silver salts in an aqueous medium (Magdassi et al. 2003), gas evaporation, and atomization. Ag ink material has to be stable in a dispersion of Ag nanoparticles in a

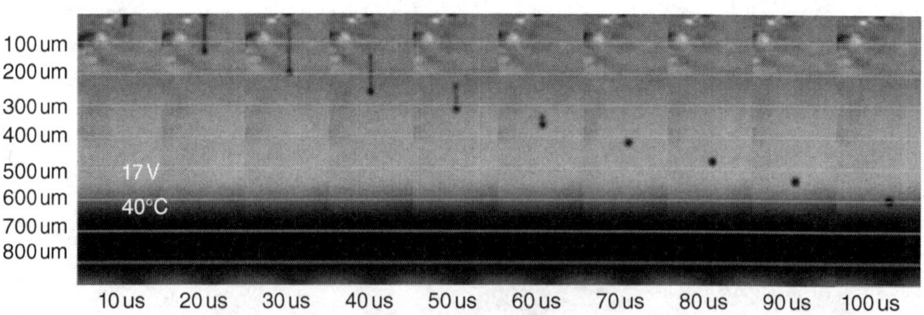

Figure 23.4. Ag nanoparticle-based ink jetted from piezoprinthead nozzles. The pulse voltage was set to 17V and printhead temperature heated up to 40°C.

liquid vehicle and should not be degraded at room temperature. In other words, the shelf life of the prepared ink-jet ink has to be long enough.

As a result, a 20-μm droplet of Ag nanoparticle-based ink has been ink-jet printed with a 1-pL piezoprinthead on a surface-treated plastic substrate, and the authors emphasize the potential of those results to connect most of the current ICs (Mäntysalo and Mansikkamäki 2007). On the other hand, silver alloy nanoparticles have been used with a new method, that is, a combination of ink-jet printing and photolithographic etching, to process metal nanoink to demonstrate 10 μm or less line widths (Okada et al. 2006).

Other interesting metallic nanosize particles are Au nanoparticles. Au nanoparticles have similar advantages to Ag nanoparticles, that is, low annealing temperature and high conductivity. Au nanoparticles possess a diameter of under 5-nm particle and have a lower melting temperature (~300–400°C) compared with their bulk form (~1064°C) (Buffat and Borel 1976). In order to formulate Au nanoparticle-based ink-jet ink, the synthesis of Au nanoclusters has been investigated, and the potential of a self-assembled monolayer (SAM) around nanosized particles has been described in terms of protecting Au nanoparticles (Hostetler et al. 1998). However, in ink formulation, the Au nanoparticle suspension is diluted in either alpha-terpineol (Ko et al. 2007b, 2008) or toluene (Huang et al. 2003; Chung et al. 2004) as an "ink vehicle"; the formulation of commercially available inks also consists of such vehicles as toluene, hydrocarbons, and tetradecane.

On the other hand, Au nanoparticles have a resonance light absorption property, and it is noted that only Au nanoparticles absorb laser light while the solvent is translucent (Hostetler et al. 1998). Thus, recently, the laser sintering of Au nanoparticle-based ink-jet-printed ink structures besides their convection oven sintering has been studied intensively (Hostetler et al. 1998; Huang et al. 2003; Bieri et al. 2003; Chung et al. 2004; Ko et al. 2007a, b). Several applications of Au nanoparticle-based inks are described in producing high-resolution source and drain electrodes in transistors (Fuller et al. 2002; Ko et al. 2007a, b; Ko et al. 2008) and plastic-compatible conductors for flexible electronics in their specially formulized fluid (Huang et al. 2003).

Apart from the commonly used Ag and Au nanoparticles, several other nanoparticle formulations have been investigated. Spherical Pt nanoparticles have been formed in polymer dispersion with their nanoparticle size around 3–10 nm (Ge and Singh 2007), and based on the rheological formulation of the fluid, it is possible to produce ink-jet ink for many interesting applications. In addition, Ni nanoparticles were also formulated in high-viscosity (pastelike) ink for screen-printing applications (Tseng and Chen 2006), and it is possible to decrease the ink viscosity with a chemically suitable dispersant in order to produce ink for ink-jet printing applications. Pd nanoparticles reduced from several suitable oligomers have been studied in terms of their suitability for ink-jet printing, and the ink-jet printing method has been demonstrated in patterning the Pd catalyst for the electroless deposition of nickel (Tseng et al. 2007). Additionally, the alloying of several metallic nanoparticles has also been studied in order to increase either the mechanical or electrical properties of ink-jet-printed film (Okada et al. 2006). Ag alloy nanoparticles have been described in terms of their uniform and fine particles, which produce flat mirrorlike surfaces. The alloying of Ag nanoparticles decreased the produced film voids by retarding the growth of silver grains. On the other hand, Ag–Cu nanoparticle-based ink has ink-jet printed the electrodes of organic field-effect transistor (OFET) applications on a glass substrate (Gamerith et al. 2007). The formulated ink sintered at 200°C and the electrode conductivity was measured as 7.3×103 S/cm. Other Ag–Cu nanoparticle-based ink-jet inks have been presented recently (Yi et al. 2008). Even though Ag–Cu nanoparticle-based ink demonstrates good electrical reliability, oxidation of Cu and the relation with thermal treatment in the air atmosphere are still important issues. As an interesting application for screen-printing applications, Ag–Pd nanoparticle-based paste has also been prepared in order to increase the electrical reliability of fine line formation. The prepared paste was sintered at 300, 400, and 500°C, and the resistivity value at 500°C was measured as ~8.4 $\mu\Omega$cm (Nakamoto 2007).

Ceramic Nanoparticles. Screen-printing technology widely uses ceramic powders for advanced ceramic manufacturing. In the beginning of the 1990s, the ink-jet printing possibility of ceramic particles, for example, zirconium (Zr) and titanium dioxide (TiO2), and the printing requirements of ceramics have been studied (Blazdell et al. 1996; Xiang et al. 1997Atkinson et al. 1997;). At that time, the average particle size requirement for ink formulation was proposed as no greater than approximately 10 nm to prevent premature gelation (Atkinson et al. 1997). On the other hand, several items of research have recently concentrated on making ink from ceramic powder formulation for ink-jet printing applications. Ink-jet ceramic ink has been successfully prepared from zirconia (ZrO2) powder. Ceramic ink-jet-printed ink, having ~14.2% ZrO2 powder and ~2.8 mPa·s viscosity, has been printed on paper substrate (Zhao et al. 2003). In addition, engineering ceramics have also been studied in terms of their applicability in direct ink-jet printing. Several ink-jet inks, including functional dielectric barium titanate grades, have been formulated. Tunable ink-jet-printed dielectric barium strontium titanate (BST) film on an alumina and MgO has been demonstrated, respectively, with 100- and 200-μm line width (Kaydanova et al. 2003). In addition, $BaTiO_3$–Ni functionally graded thick films have been ink-jet printed on flexible porous substrates (Tseng et al. 2006). Hydrothermally prepared $BaTiO_3$ nanoparticle-based ink having an average particle size of 60 nm has been formulated for printing with a piezoelectric DoD ink-jet printer.

Ink-jet printing of ceramic nanoparticle-based ink with low fractions of particles, that is, their nanosize, is applicable, and printing with functional ceramic nanoparticles is promising for the production of tunable dielectrics. However, the size of the utilized ceramic nanoparticles is still coarse and the formulation of ceramic ink for piezo-printhead-type ink-jet printing is challenging.

Organic and Miscellaneous Nanoparticles. Organic materials, for example, conjugated oligomers and polymers, organic conductive ink-jet ink materials, for example, poly(3, 4-ethylenedioxythiophene)-poly(styrene sulfonate) (PEDOT:PSS) and polyaniline (PANI), and their applicability to ink-jet printing technology have been investigated during the past few decades in the field of "organic electronics." Organic electronics has several advantages, such as low cost and low-performance alternative compared to silicon-based electronics in several applications, for example, radio frequency identification (RFID) tags, display backplanes, and disposable consumer electronics (Wallace et al. 2006). However, organic nanoparticles and their usability in formulating ink-jet ink were not extensively reported apart from the printing of organic pigments having less than 100-nm particle size with the conventional ink-jet printing method (Bishop and Alan 1997). On the other hand, one of the interesting reports concerning functional organic nanoparticles was demonstrating the direct ink-jet printing of functional organic nanoparticles (Magdassi and Moshe 2003). In that report, organic nanoparticles, that is, with an average particle size of around 55 nm, by using oil-in-water microemulsions, were converted into functional molecules by piezoelectric ink-jet printing, and it was mentioned that the applied method can be adopted for many organic water-insoluble molecules. In addition, the fabrication of water-dispersible PANI–poly(4-styrenesulfonate) (PSS) nanoparticles having <30-nm diameter on the basis of electrostatic interaction and their application to ink-jet printing for sensor application have also been presented (Jang et al. 2007). As a rule of thumb, organic nanoparticles should be highly dispersible in their solvent medium, thermally and mechanically stable without particle aggregation, and preferably should have narrow particle size for ink-jet printing applications. Even though the synthesis of <20-nm organic nanoparticles was reported, in general, progress in organic nanoparticle preparation has several difficulties in terms of their synthesis and isolation (Jagannathan et al. 2006).

Some examples of other nanoparticles that were ink-jet printed were nanoparticle-based pigments for conventional ink-jet printing (Mendel and Bugner1999; Mendel et al. 1999), diamond nanogrits onto p-silicon and copper substrates (Fox et al. 2000), magnetic ferrofluid nanoparticle size grains (Sen et al. 1999), and recently, ultrafine indium titanium oxide (ITO) nanoparticles for ink-jet printing have been reported (Hong et al. 2008).

Ink-Jet Printing Requirements of Nanoparticle-Based Ink

In order to commercialize a nanoparticle-based ink, it is important to understand the rheology of the formulated ink and the suitability of those parameters for industrial printheads. The physicochemical properties such as the viscosity and the surface tension of the ink are the most important parameters for "true ink" formulation. The theoretical requirements of viscosity and surface tension of the ink-jet ink are respectively in the range between 2–30 mPa·s and 25–35 dynes/cm. Even so, the recommended viscosity range of nanoparticle-based ink should preferably be in between 5 and 15 mPa·s to meet with the requirements of today's printheads. Formulated nanoparticle-based ink also has to have Newtonian behavior so that under variable temperature and pressure, the shear stress of the ink should be reliable. The printhead temperature can vary during the printing process because of the adjustment of the printing XY table temperature, and the short distance, for example, usually <1 mm, between the printhead and the substrate can increase that effect during long printing runs. On the other hand, the particle size of the selected ink material should be less than 100 nm, and the nanoparticles should have uniform size and perfect homogeneous distribution in the solvent.

The dispersion of nanoparticles in ink is important in avoiding any aggregation that can increase the viscosity of the ink during printing, which would allow undesired clogging of the printhead nozzles (Calvert 2001).

One of the main challenges that can be experienced during long printing runs is the drying of the printhead nozzle. It is possible to control this drying with a low volatile-type ink. Fast evaporation can provoke nozzle drying if the temperature of the printhead varies during the printing process. In addition, before and during the printing process, parameters such as droplet firing voltage, jetting waveform, printhead nozzle size, piezoprinthead pulse shape, and jetting repeatability should be defined. Finally, the droplet volume should be small in order to produce the narrow line width, which also increases the evaporation of the solvent in the production of complex electronic components, which demand a smaller physical size.

Future Trends for Ink-Jet Printing Inks

High accuracy and controllable material deposition are the main advantages of the piezo-type ink-jet printing technology. High accuracy is one of the crucial properties to pattern ultrafine electronic circuits and thus to decrease the size of the electronic package. Additive ink-jet printing process with controllable material deposition decreases the amount of waste and enables more environmental-friendly production. The area of materials development for ink-jet printing technology is very dynamic, and there will be more ink suppliers with novel inks in the market in the near future. Nanoparticle-based and organic inks are the bottlenecks requiring breakthroughs in electronics applications in ink-jet printing technology.

Ag nanoparticle-based ink is one of the most commonly used inks in the field of electronics manufacturing and electronic circuits have been produced by commercially available Ag inks. However, the sintering temperature requirement of high metal content Ag nanoparticle-based inks is still high for many low-cost organic substrates. On the other hand, it is possible to decrease the sintering temperature of the Ag ink by reducing the metal content of Ag and to formulate the ink with organic precursors, usually called metallo-organic ink (Dearden et al. 2005). Metallo-organic ink also enables ultraviolet (UV) curing of the organic shelf around Ag nanoparticles and the final sintering of nanoparticles with thermal treatment. The advantage of this method is to shorten the sintering temperature and duration by increasing the metal content of Ag nanoparticles (Yue and Jie 2008). On the other hand, the most interesting development will be sintering of Ag nanoparticle-based ink at room temperature, which enables the use of a wide variety of substrate materials. Material compatibility between the ink and the substrate material is important, and the sintering of silver ink at room temperature will decrease problems such as thermal stress or damage (Suganuma et al. 2007). However, Ag has several drawbacks, such as electromigration and corrosion. In order to decrease these effects, alloying of the Ag nanoparticles with several other nanoparticles, for example, Pd, can decrease such effects. Nanoparticles such as Au, Ni, Al, Pd, and Cu should be used in formulation to increase the electrical and mechanical properties of ink-jet-printed film structure. Even though the oxidation of Cu is a well-known problem with its nanoparticles, several pieces of research have described the implementation of high-copper metal precursor ink for ink-jet printing in order to manufacture electronic devices (Rozenberg et al. 2002).

Dielectric materials with low-*K* or high-*K* are the emerging materials group to produce passive components, for example, resistors or capacitors. Low-*K* dielectric inks have potential for use as an interlayer dielectric layer in complex multilayer structures. High-*K* dielectric ink should be used to print thinner dielectric layers, such as transistor gate dielectric, which reduces the operating voltage of the transistor. For example, polyimide and epoxy dielectric ink has been commercialized by several material manufacturers. However, the distribution of the ink has several obstacles because of the expensive manufacturing cost of dielectric and intellectual property right (IPR) policies. On the other hand, spin-on dielectric materials have several advantages, for example, an acceptable viscosity range with the ink-jet printing method and adequate dielectric properties for electronics applications, which can be adapted in ink-jet printing technology. Tunable dielectric materials with ceramic nanoparticles are important for high-frequency electronics applications. They provide low loss and excellent RF power capabilities at micro- and millimeter-wave frequencies. Ink-jet printing of ceramic-based nanoparticle ink will definitely enable new features and increase the amount of electronics applications in this field. The particle size of the ceramic particles should be small and the agglomeration of the nanosize particles in the printhead need to be controlled.

From the electronics manufacturing point of view, one of the main challenges after printing sophisticated electronic components is to attach these components or stacked chips to the board. The parts are usually soldered by an adhesive to the board. Several electronic adhesive materials have been used in electronics manufacturing, for example, anisotropically or isotropically conductive adhesives (ICAs), nonconductive adhesives, and polymeric adhesives to attach these parts to the board. It would be interesting to apply such adhesives with the ink-jet printing method from the electronics manufacturing point of view. Currently, it is possible to print nonfilled glue dots by ink-jet printing, but conductive adhesives have several drawbacks, that is, large particle size and high-viscosity nature. On the other hand, one of the interesting applications has been demonstrated by ICA with silver resin fill (Kolbe et al. 2007). ICA material was formulated for ink-jet printing media and the physicochemical properties of ICAs were described, that is, 30-mPa·s viscosity and 8-μm particle size. On the other hand, ink-jet printing of solder columns and solder bumps have been ink-jet printed with piezoprinthead. Solder balls of Sn63/Pb37 have been printed for applications of high-density microelectronics and optoelectronic packages at rates of over 400 bumps/s (Hayes et al. 1999).

It is obvious that the polymer- or organic-based ink market has been very active in providing new material solutions for several interesting applications, for example, thin-film transistors (TFTs), and microsystems on low-cost flexible substrates. Novel materials and related techniques, for example, R2R production, screen or ink-jet -printing, enable the production of completely new applications in the field of polytronics (polymer + electronics) (Bock 2005). Ink-jet printing of semiconductors will enable low-cost products, freedom of design, and the physical flexibility of the electronic components.

Finally, ink-jet printing of DNA microarrays and solutions for applications of tissue engineering are under investigation by research groups and biomedical device suppliers. Nucleic acid, hydrogels, cells, and protein arrays have been ink-jet printed, and an ink-jet printing system was used for the production of biochips and biosensors. Several high-volume companies have developed several commercially available systems and develop their capabilities to produce DNA microarrays (Piqué and Chrisey 2002).

INK-JET PRINTING IN MICROELECTRONICS APPLICATIONS

Ink-Jet Printing in Electronics

Ink-jet printing has been introduced during the early years of this decade into various electronics applications. Ink-jet printing developed hand in hand microcomputer development during the 1980s and 1990s. Later on, the possibilities of depositing more than just ink on the substrate have been discovered. Materials development combined to equipment development has enabled direct deposition of electronically functional materials. Ink-jet printing can be utilized in several dispensing or deposition cases for electronics applications, wherever fine dosing of the material is needed. These include electrical components and circuits, microelectromechanical systems (MEMS) applications, displays, RFIDs, and photovoltaic applications.

Ink-jet printing has several benefits in electronics printing. Its additive nature enables savings in materials and reduces waste production. Theoretically, every product can be different in digitally driven manufacturing. The background of ink-jet printing in the graphics industry has enabled the development of a large range of possibilities; it can be used in different applications. Different types of inks can be driven with the same equipment, forming either a functional or visual result, or both.

Microelectronics Applications

Since the 1960s, the microelectronics industry has been using and developing lithographic processes in transistor-based technologies. The demand for integration and miniaturization eventually led to rapid decreases in feature sizes, which boosted up the technology development. In general, the photolithographic techniques were developed especially for MEMS and electronic applications. Technology adoption also required the usage of short wavelength sources and multistep processes for device building, which, in respect, required massive financial investments. In addition, conventional lithographic techniques are reaching their resolution limits. To overcome the limitations of lithography, there are various alternative approaches for pattern creation (Bucknall 2005).

In microelectronics applications, ink-jet printing can be used for precision deposition. Liquid metal nanoparticle dispersions can be used for direct patterning of the electrical wiring. The conductive inks typically contain highly conductive metal particles, such as gold or silver. These metals do not have such oxidation challenges as, for example, copper, so the requirements for printing environment are looser. This means no inert atmosphere is needed during printing, sintering, or wet delivery.

The applications presented using nanoparticle metal inks with ink-jet printers include typically multilayer structures with operation similar to printed wired boards and electronic packaging applications. The latter applications include SIP and other integrated solutions. Multilayer and 3-D structures are enabled by using noncontact ink-jet method and ink with different functionality, that is, conductive inks for wiring and vias, and dielectric for insulating layers (Saito and Matsuba 2006).

There are electronic packaging applications where ink-jet-printed metal inks can be combined with other miniaturization or integration methods. Wire-bonding methods typically require a metal-plated surface using, for example, gold. A sintered layer of metal nanoparticle ink can be used to replace the gold plating. The results with copper wires bonded on a printed layer have proven to be comparable with a gold-plated surface. In addition to wire-bonding possibilities, chip-scale packaging can be enhanced by printing contact pads on the top and sides to enable further packaging, such as package on package. In addition to plating surfaces with gold metal nanoparticles, ink-jet technology is capable of other possible solutions in ICs. In IC technology, connection pads are formed by using metal in the semiconductor foundry to form underbump metallization. Typically, these pads are made of aluminum, which is not a solderable surface, and for flip-chip applications, another metal layer is needed. However, ink-jet technology can be used to form copper underbump metallization for direct soldering. This technology, however, requires some special conditions. Copper is oxidizable and needs an inert gas environment during processing. A postprocessing method for the copper pads for metal sintering can be, for example, conventional thermal treatment or laser annealing (Saito and Matsuba 2006; Wallace et al. 2006).

MEMS are combinations of electrical and mechanical functionality. The development has been made mostly in the microelectronics industry, the primary technology of which has been the photolithographic process. In MEMS, the diversity of features can be large, and ink-jet can be used as a microdispensing tool for integration and interconnection purposes. For interconnections, solders can be dispensed using DoD ink-jet printing. The demand for precision dispensing arises from an ever-increasing level of integration and demand for increased functionality. Small MEMS components with high density of input–output (I/O) pins can be connected on the substrate by dosing the precise amount of solder on the corner of the component and circuitry. In some applications, solder printing can ease the connection operation and increase the level of reliability, especially in 3-D connections. Moreover, solder balls can be used in a more conventional way, forming interconnections by solder ball arrays with feature size under 30 (Wallace et al. 2006).

Passive components manufactured with direct writing technologies can be used for electronics applications. Ink-jet printing can be used to form resistive, capacitive, and inductive components using only conductive and resistive materials. In addition to passive components, conductive inks can be patterned to form, for example, antennas (Wallace et al. 2006).

Ink-Jet in Semiconductors. In ink-jet-printed electronics, the used semiconductive materials are typically polymers, though printable silicon composition materials have also been introduced. The key factor in printed semiconductive devices is the mobility of the charge carriers in the material. Although materials development has enabled novel applications for organic semiconductors, there is a performance gap between new organic and conventional single-crystal silicon materials. This means that fast applications still require high-performance silicon with the developed manufacturing processes. Printable semiconductive materials still have predictions for a huge market share in future electronics applications. They are seen as a tempting solution for numerous applications, offering cost-reducing or previously unattainable solutions (Cleland 2003).

Photovoltaic devices represent an attractive application, providing off-grid electricity in an environmentally friendly way. Organic electronics can be flexible and low cost, and these properties also apply to solar cell manufacturing. Today, most solar cells are manufactured using expensive silicon, which makes the price of 1 W of energy quite unattractive. The structure of a solar cell is, after all, quite simple. Ink-jet printing can be used, among other printing methods, to deposit organic semiconductive materials on flexible substrates to create low-cost solar cells (Cleland 2003).

Organic transistors and light-emitting diodes can be built straight on the substrate without any assembly phases using, for example, ink-jet printing. One of the great advantages with organic transistors is, in addition to low cost, the ability to cover a large area. In conventional technologies, the cost per area unit is several times higher than with the organic transistors. It enables the usage of various substrates, thanks to the low processing temperature requirements. There is a common drawback with organic TFTs: the low carrier mobility, rather coarse device geometries, and low stability due to reactivity with air and water (Cleland 2003).

Ink-Jet Printing in Electronic Interconnections. Ink-jet printing can be used to create interconnections for microelectronic packages additively with high flexibility. The technology is able to create devices previously requiring an interposer PCB, wire bonding, or other interconnection methods. These include hybrid structures, chip-scale packages, and system integration. Ink-jet technology enables microelectronic structure creation due to its high capability in precision and accurate dosing. The requirements for flip-chip connections for conventional memory applications are in the scale of 100 I/Os per chip with 60- to 80-μm pitch.

The method for creating multilayer interconnection structures on either flexible or rigid substrates involves conductive and insulating materials printed in sequential fashion. The components to be connected can be basically every kind of surface mount components. Discrete components, such as passive and active components, either flip-chips or packed ICs, can be connected using the top-attach method, or molding method. The top-attach method includes the components on the top of the wiring, similar to a traditional PCB. Typically, components are attached with an ICA. The molding method results in a rigid printing substrate, which includes the component connection pads. During the molding process, the components are molded inside a temperature-resistant material, leaving the active sides open for printed interconnections. The printed inter-

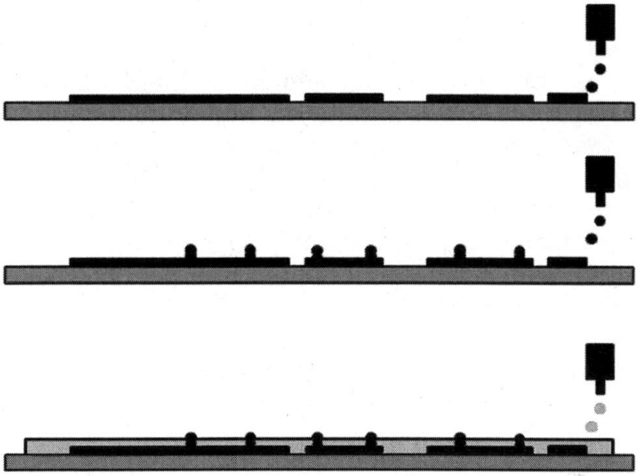

Figure 23.5. Printed interconnections on a substrate followed by via printing and dielectric material printing for surface leveling.

connections are typically formed with precious metal nanoparticle inks, such as gold or silver, due to performance requirements.

Forming multilayer structures with ink-jet printing includes three main steps: conductive trace printing, via printing, and insulating layer printing, presented in Figure 23.5.

Conductive traces are designed with CAD tools and are converted to binary bitmap files for digital printing. These traces are the main signal paths in the device design. However, typical solutions require multilayer design. The wirings are so dense that it is necessary to divide it into several layers. During the design, there are a few vital rules for the designer. The device performance sets limits for the trace properties. According to these, combined with material properties, rules for trace widths and number of printed layers are created to achieve sufficient trace performance. In addition, basic electronics design rules apply.

Forming the vias from layer to layer basically follows the same rules as printing the conductive traces. The material is the same, but the layer number is increased to reach the thickness of the insulating layer. The insulating material is typically polymer, compatible with the printed conductive ink and process requirements, such as the ink sintering temperature.

The printing process is affected by the main sectors: the application, that is, the device itself; the materials used; process for deposition; and material processing and equipment. These all also affect each other. As mentioned, the device type defines the requirements for trace performance and wired layers. In addition to that, the materials are partly defined through device performance requirements. The materials properties of the printed inks have an effect primarily on jetting, but also on ink–substrate interaction. Individual printheads support a range of materials with a certain viscosity and surface energy levels for successful jetting. Besides jetting, surface energy and viscosity are important in controlling the ink spreading on the surface after landing. For fine features, substrates free surface energy and must be sufficient for forming droplets of controlled size. The ink material properties also affect the other main sector, the process. Conductive inks require sintering in order to achieve proper conductivity, and dielectric polymers have to be cured. These process steps may include thermal treatment, UV exposure, infrared (IR), chemical treatments, or laser annealing. The printhead properties closely relate to the printing parameters. The main aspects of the printhead adjustments include the firing voltage and waveform, nozzle temperature, and meniscus vacuum. The printing parameters also include other equipment characteristics, such as XY table temperature and printing speed.

Ink-Jet Printing in Electronic Packaging and System Integration. The basic idea to form multilayer wiring on the substrate is to create several layers of functional materials one over the other. The molding method for manufacturing the component package leaves the active sides of discrete components and ICs (see Fig. 23.6).

The connection pads can be reached by printing conductive ink on the components. However, components cause topography on the structure; that is, the metallizations and IC pads are higher than other parts of the surface. This topography can be leveled by printing dielectric ink leaving the higher parts open. This provides a smooth printing surface for conductive ink. The usage of dense ICs places demands on the accuracy of the whole process. The typical drop size on substrate with 8- to 10-pL drop volume printed with a commercial ink-jet printer is 30–40. This means that the printed IC interconnections are typically only 1 pixel wide (see Fig. 23.7).

Once ICs are successfully connected, the actual first layer wiring can be printed, if not included in the IC connection print file. On top of the wiring, the vias to the second printed layer are printed. Sintering is required for conductive ink prior to insulating dielectric layer printing. Dielectric printing follows the same jetting rules as conductive ink but may differ in terms of the desired behavior on the substrate. The conductive ink needs to be controlled drop by drop to form fine features. However, the insulating layer needs to be, first of all, solid and uniform to ensure insulation between the layers. This means that the dielectric layer can be controlled with topography and, otherwise, can be spread more freely. Typically, dielectric postprocessing requires heat, UV, or a combination of both. After dielectric printing, the surface, including the dielectric-covered traces and bare vias, is ready for printing the second layer wiring. This cycle is repeated to achieve the required number of layers (Pekkanen 2007).

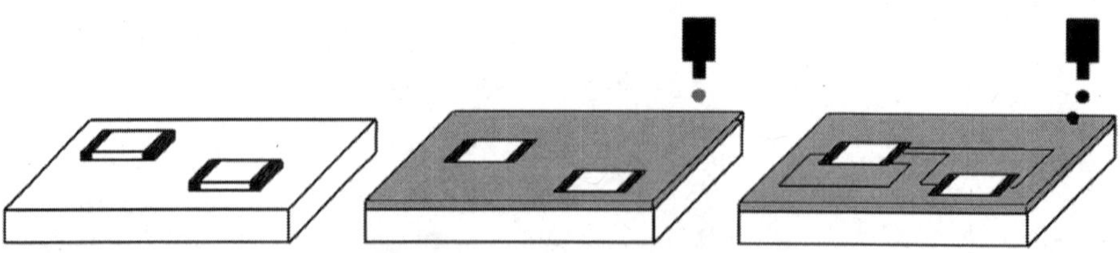

Figure 23.6. Principle of forming electrical interconnections for components using ink-jet printing.

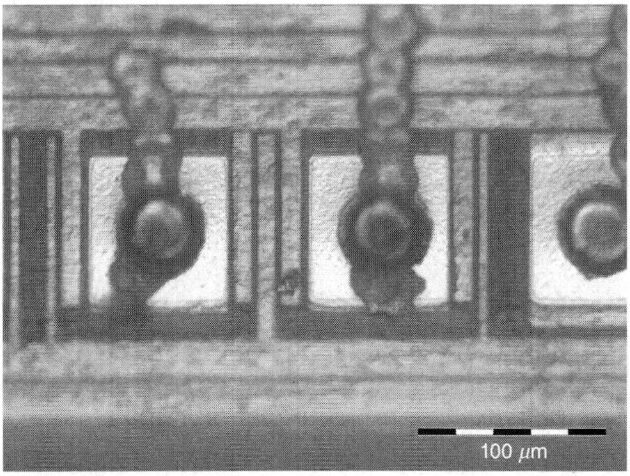

Figure 23.7. IC bumps connected with single-pixel trace.

Figure 23.8. Top-attach method for IC interconnection using ink-jet printing.

The top-attach approach has opportunities for IC connecting also. It is possible to use conductive adhesives to attach the components in a similar fashion to surface mount technology. Conductive adhesive usage is required because, for example, widely used silver nanoparticle ink is not directly solderable. Another possibility to form the interconnections in the top-attach method is to place the components active side up. Dielectric ink can be deposited on the sides of the components to create ramps for the printed conductor, as seen in Figure 23.8.

This type of interconnection method, however, creates topography and requires somewhat more space from the substrate.

Ink-Jet Printing Technology Challenges

Ink-jet is considered to be one of the enabling technologies of future low-cost electronics applications. However, there are challenges related to the printing of circuit materials using ink-jet technology. The nature of the technology is liquid deposition process, so all printed material has to be processed as solutions. Typically, polymer semiconductive materials have lower mobility compared with amorphous silicon. Because of this, ink-jet-printed electronics are weaker in fast, high-end solutions than the developed silicon technologies. The developed materials are usually p-type semiconductors, having holes as charge carriers. It would be desirable to have also n-type materials to increase the application diversity for organic transistors, enabling p–n junctions and inverters, for example. Another issue with the materials is that organic semiconductive polymers often require solvents such as toluene or xylene, which may be difficult to eject from the printhead. This is due to high volatility, which may result in ink drying in the nozzle and clogging in the orifice. In addition, strong solvents may damage the printhead structure over time (Cleland 2003).

The device requirements for ink-jet-printed semiconductor mobility include narrow channel lengths to achieve reasonable performance of the device. These channel lengths can be as low as 10, which is quite challenging for industrial ink-jet printers. Still, the channel lengths are much larger than in conventional silicon devices (Cleland 2003).

From a manufacturing point of view, ink-jet printing of electronics requires seamless operation between the individual nozzles in the printhead. Single-nozzle operation optimization has been widely researched, resulting in good operation for various materials. However, throughput for mass production can be achieved only by using multi-nozzle printing. Single nozzles in printhead nozzle arrays tend to vary due to interactions between themselves and the environment. In addition, this nozzle variability is often individual for separate printheads. Reducing the nozzle variability is necessary for the precision and reliability needed for ink-jet-printed electronics (Cleland 2003).

Accuracy in both placement and amount are mentioned as strengths of ink-jet printing. However, there are numerous things that affect the jetting. Printhead properties, cooperation with the printed material, the printing environment, and equipment settings all have an effect on forming the pattern. Printhead properties, which may cause inaccuracy in the printing, include angular misalignment of the printhead, thermal expansion of the printhead, drop velocity differences between individual nozzles, and drop firing angle errors. Controlling the liquid on the substrate further expands the issues related to fine print quality. Error situations are very undesirable especially in ink-jet printing, since the smallest feature sizes are only one size of 1 pixel, that is, one droplet. This means that every drop counts and has an effect on the final result.

Laser Versus Convection Oven Sintering Toward Electronic Production

Manufacturing electronic devices by utilizing printing techniques and metallo-organic inks based on nanoscale metal particles can revolutionize the electronics industry in the near future (Pekkanen 2007). The low-temperature sintering of metallic nanoparticles is a phenomenon enabling printing of metal particles diluted in a carrier fluid and sintering them into homogenous traces in rather low temperatures compared to bulk metal. Curing is a term that has been traditionally used in chemistry. In chemistry, curing means the hardening of polymers by heating objects in quite low temperatures. Sintering is a term that is usually used for adhering powder material particles to each other using heat. Sintering is done in a temperature that is below the melting point of the bulk material. Nanoparticle inks can be sintered using many different processes. Common to all these processes is adding energy to the liquid to evaporate the solvent and additive materials away. If enough energy is added, individual metal particles adhere to each other and sinter to form conductive patterns. The principle of nanoparticle sintering is represented in Figure 23.9.

Theoretically, sintering energy can be added to the target using a method that is suitable for each case. At least hot air ovens, lasers, UV radiation, IR radiation, and chemical or electric field sintering methods have been under research. The most common and best-known sintering method is heat sintering in an oven, which is recommended by most nanoparticle ink manufacturers. The sintering temperature of

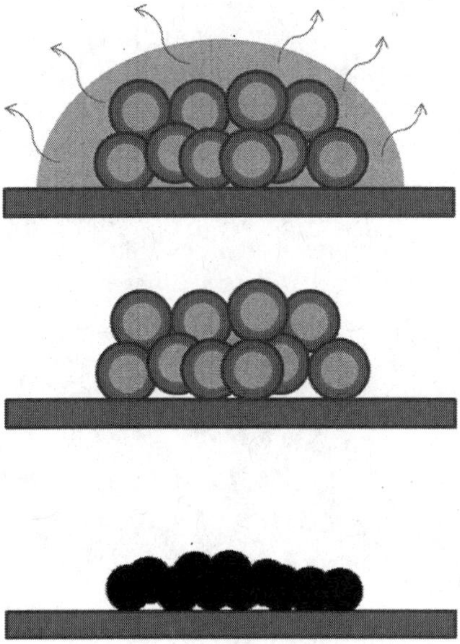

Figure 23.9. Principle of nanoparticle sintering phases including solvent evaporation, dispersion agent removal, and metal sintering.

nanoparticle inks depends mostly on the particle size of the metal powder. Also, the used solvent and dispersing agents affect the sintering temperature and profile. After sintering, 98–99 wt% metal content should remain (Volkman et al. 2004; Harima Chemicals, Inc. 2008). Oven sintering has been used traditionally to sinter nanoparticle inks. However, evaporating the solvent and sintering the printed structures well still takes a long time in a convection oven. It will take approximately 1 h to sinter one printed layer, meaning that the overall exposure time might be calculated in hours. For some applications, it is too long an exposure time. Another promising process alternative is the laser sintering of nanoparticles.

The exposure to thermal stresses causes a big reliability risk for the product, especially if the circuit consists of several layers and includes different materials. Therefore, optimization for the oven sintering process is a critical part of the product design process. Lasers have multiple benefits that make the laser a suitable and powerful tool for material processing. Lasers have been used in the industry since the 1970s and their popularity is growing, especially in microprocessing (Yli-Luukko 2007). Besides metal, it is also possible to process materials such as ceramics, plastics, various composites, and semiconductors for a number of industrial applications. The principle of laser source is illustrated in Figure 23.10.

The laser beam is formed in an active lasing medium between two reflectors. One reflector is total and the other is a partial reflector. Pumping energy is added to the lasing medium and the energy releases photons. Formerly released photons release new photons, and some of them with suitable energy can get through the partial reflector. So, the laser source is a kind of an optical amplifier. Laser light is collimated, monochromatic, and coherent. Collimated light means that the light is rectilinear. The wavelength of the laser depends on the lasing material. Therefore, all laser light from the same source has the same wavelength. So, the light is monochromatic. The resonator of the laser source is dimensioned such that the wavelength of the laser generates standing waves. Because of the standing waves, every photon resonates in the same phase. Therefore, resonating photons intensify each other and the resulting light is coherent.

Miniaturization of electronic components has a major impact on choosing the laser for manufacturing because with lasers, it is possible to work in the micrometer- and

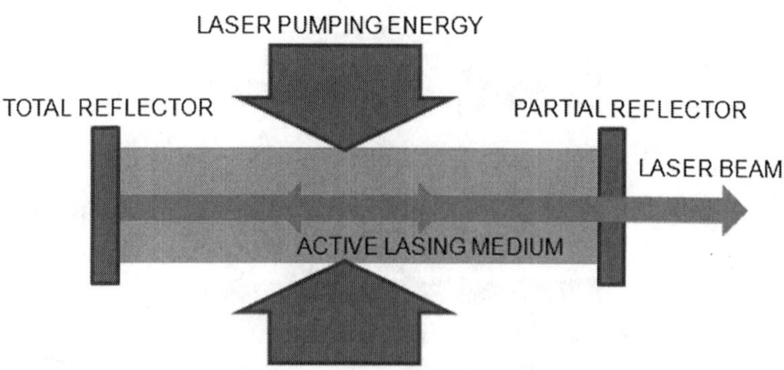

Figure 23.10. Principle of laser source.

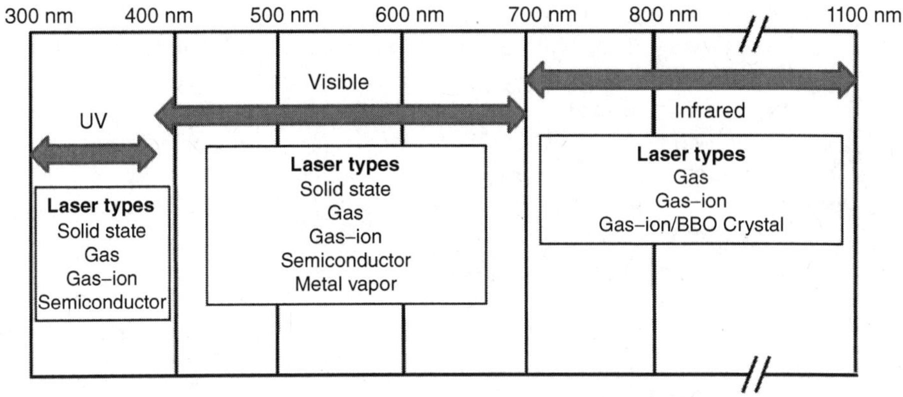

Figure 23.11. Laser wavelengths.

submicron-size range. Many conventional methods are not any more suitable for small size component production. With laser processing, it is possible to work contact free and there is no need for specific tools for processing. This saves on tooling cost and shortens the process time. Because the laser beam affects only the wanted area, the heat input is lower and causes less distortion than other methods. Laser processing enables very precise working and it is very suitable for micromachining. Laser processes are also very flexible and fast production methods. They are often automated and the repeatability of processes is very good. In many applications, there is less demand for post-processing comparing to traditional production methods. Laser processing has environmental benefits because by working with the guided laser beam, less energy is needed and it also saves material consumption. There are multiple laser applications used in electronics manufacturing. Common to all applications is that the laser is typically used for heating, or ablating and material removing of the target material. The suitable laser type for the process is chosen by need for accuracy and power. The target material also affects the selection of suitable lasers, depending on how the material absorbs the laser light from different wavelengths, which are presented in Figure 23.11.

Laser sintering is an alternative process for heat transfer sintering. Laser sintering takes only a few minutes at the maximum. An even more remarkable benefit is that a wider range of materials can be used in the products when the laser sintering process is utilized. The thermal stress for materials can be decreased because the exposure time for heating is lower and the sintering energy can be focused directly onto the sintering target. For example, lower glass transition temperature substrate materials can be used. These benefits will show especially in multilayer products, which need several sintering cycles.

The most important factor for selecting a suitable laser for sintering purposes is the wavelength. The laser wavelength depends on the lasing medium. The wavelength of the laser needs to be decided based on the sintered material. Different wavelengths absorb better into different materials. The decision can be made based on calculations of the optical behavior of the target material. The most important factor when determining the wavelength is the absorption depth into the target material (Lasertyöstö 1996; Bieri et al. 2003). The power density of the laser beam depends on the beam diameter and power of the laser source. The sintering of printed patterns can be done using two different ways of adding the energy. High-power density can be used with high spot velocity, or otherwise low-power density with low spot velocity. In both cases, the amount of energy added to the target can be the same, but of course the sintering time depends on the spot velocity. A basic rule is that the higher the used power density is, the faster is the process.

The main benefits of laser sintering are

- The laser sintering process takes only 1–3% of the oven sintering time.
- The laser sintering equipment is easy to integrate into a production line.
- Better conductivity can be achieved by laser sintering than heat sintering.
- With laser sintering, better topography can be achieved on the surface of the samples.
- There is less material stress exposure.

The main challenges of laser sintering are

- Laser process optimization
- Production line optimization.

THE ENVIRONMENTAL ASPECTS OF PRINTABLE ELECTRONICS

Introduction

Environmental issues are becoming more and more important, on the one hand, due to increasing environmental awareness arising from concern about climate change and pollution of the environment, and on the other hand, because of the resulting environmental requirements set by regulation, industry, and customers. The environmental requirements both set challenges and open up possibilities for benefits for the electronics industry. New environmentally beneficial technologies and applications (e.g., batteries, displays, antennas, and a variety of electrical interconnections) of printed electronics can replace older technologies with more sustainable ones, as ink-jet printing and different R2R technologies add materials with low material and energy consumption. These technologies can be used to develop totally new applications and services capable of having positive environmental consequences, either direct or indirect ones. As an example of indirect environmental consequences, new applications can have a positive effect on larger systems by replacing earlier technologies used. On the other hand, new production facilities, supply networks, and completely new applications and product life cycles can have different relative environmental significance when compared to the current situation. At least, the variety of environmentally relevant questions to be taken into account in the production of electronics and in the management of product life cycles is increasing.

In spite of all this, printed electronics involves the same environmental challenges as current technologies based on the subtractive processes used in continuously miniaturizing consumption electronics. However, there are additional challenges of greater production volumes expected with many application areas and those of new and strongly developing processes and materials.

The Environmental Challenges of Ink-Jet Printing

The environmental challenges of printed electronics are various and concern every part of the technology, including materials, manufacturing processes, products, and the end-of-life treatment of the products. For ink-jet printing, the main challenges of the production phase encompass the use of solvents and the use and manufacturing of nanoparticles. Nanomaterials are challenging due to the newness of the research in the field of environmental, safety, and health issues (Handy et al. 2008). Therefore, a proactive approach in the ink-jet printing industry is required to be on the safe side. Material selection and material combinations represent another challenge, since, for example, the solvents used should be safe for humans and the environment even if proper safety equipment for the ventilation, filtering, and burning of volatile organic compound emissions is used. New material combinations used in printable electronics entail new challenges for material content handling in electronics industry. This is emphasized due to the small amounts of various materials used in the printed ink mixtures and the fact that ink-jet printing technology diversifies the materials, chemicals, processes, and production equipment variety used in electronics manufacturing. One consequence of this, because of increasing environmental regulation along with, for example, the Registration, Evaluation, Authorisation and Restriction of Chemicals (REACH) regulation (EC 2006) of the European Union (EU), is the growing need to find out possible chemical traces in the material contents of products: for example, in ink-jet-printed structures. Manufacturing faces the challenges of process efficiency development so that the environmental impacts of resource and energy consumption are competitive when compared to other technologies. In addition, the electrical performance of new material combinations should be developed so that ink-jet-printed interconnections, for example, would be suitable for more and more applications. The electrical and technical performance of printed structures can also have a direct effect on environmental performance since the energy efficiency depends on that. However, costs usually define the suitability of a technology for specific applications. Therefore, for example, the lower conductivity of printed conductors could be only a minor problem from the environmental point of view due to their use in low-power (and energy) applications.

For products, the challenge is the dispersion of ink-jet-printed silver and other valuable metals in small quantities, but with many simple short service life (and disposable) products of high production output, and thereby the collection and handling of products at the end-of-life phase. The recovery of silver and other valuable metals is important in order to avoid metals rising in price and resource depletion (Ayres 1997; Lanzano et al. 2006; Pickard 2008; Alonso et al. 2008). Recovery-related challenges include the recyclability of the products and the economics of the logistics needed to collect and recycle discarded products. Recovery is even more important with gold and platinum group metals (Alonso et al. 2008) since they incur bigger production costs and more energy use than with silver (Ku et al. 2003; Mudd 2007; Alonso et al. 2008). Other end-of-life questions cover the incineration of simple disposable products for energy production and the possible disposal of the products in municipal waste management.

Landfilling could be problematic due to the leaching of silver from the discarded products (Townsend et al. 2007). However, there are contradictory results in the studies on the leachability of solder (containing silver) in printed wiring boards. According to Townsend et al. (2007), the leaching of silver is not a problem in landfilling, but studies on the leachability of bare solders (Griese et al. 2000; Townsend et al. 2007) show that silver could be an environmental problem because of leaching into the ground. However, leaching studies should be made for printed structures in the real environment to obtain reliable results (Townsend et al. 2007). At least, recovery process options and logistics for products and structures of printed electronics containing silver or other valuable metals could be worth pursuing (Kang and Schoenung 2005; Lanzano et al. 2006; Thomas 2008).

The Environmental Benefits of Ink-Jet Printing

The environmental benefits of printable electronics and thereby also those of ink-jet printing, compared, for example, with current interconnection technologies utilizing lithography and other subtractive methods, originate from adding only the amount of materials needed with a few process steps. This comes with the decreased consumption of materials, additional process chemicals and energy, and also a diminished amount of wastes as process output. On the other hand, the difference between manufacturing energies for metals used in current technologies (copper) and ink-jet printing (silver and other valuable metals) is more advantageous for copper than for silver or gold, for example (Ku et al. 2003; Mudd 2007). In addition, the recycling and recovery costs and energy efficiency figures for all metals used with different technologies should be compared to obtain comprehensive reference values.

There are more options for substrates due to lower-temperature manufacturing processes compared with current technologies. The sintering temperature profile in the oven or the absorption of laser energy define the temperature limits of substrates. Some substrate options have environmental benefits compared to the substrates used in current technologies. These are easy recycling and less manufacturing energy and emissions of substrate material production. With new low-temperature inks, the variety of available substrates is increasing even more. Increasing miniaturization options with 3-D interconnections onto topographic surfaces and flexible substrates achieved with printable electronics also enables indirect material savings in the surrounding electronics, packaging, and product mechanics.

Material Selection from the Environmental Point of View

From the material selection point of view, the used material combination should fulfill several environmentally related requirements. First, the combination should be technically reliable so that the process yield is high and the reliability in the intended environmental conditions meets the application-specific requirements. The reliability of the printed structures is one of the key issues, also from the environmental point of view, when other than simple short-life products are produced, since high yield and a long enough lifetime (compared to the surrounding electronics) have an effect on environmental performance, too. Second, the combination should not have toxic characteristics during manufacturing and especially the product itself and at the disposal phase. Toxic or other environmentally harmful properties can be reduced if the materials are not separable or leachable from the product (Townsend et al. 2007). These issues are related to ensuring that the materials are compliant with environmental regulations also in the future. Third, the manufacturing of printable materials and especially nanoparticles should not have too high economical or environmental costs, which are likely to be related to energy consumption during material production and the material supply chain.

When ink-jet printable inks are considered, from the worker health and safety point of view, ink-jet printing inks could pose a problem of nanoparticle aerosols, for example, and the development of that kind of risk should be prevented during the printing process. In addition, more research is needed on the safety of nanomaterials in the future to be compliant with regulation, with REACH as an example. Furthermore, proper ventilation is needed in order to keep working conditions under the harmful exposure concentration limits of solvents (Keskinen and Valkama 2008).

Overall Environmental Efficiency of Ink-Jet Printing

One of the main questions related to the energy use of ink-jet-printed structures is the relative contribution of the manufacturing of materials, ink-jet printing, and sintering. The use of laser for sintering can have a positive impact compared to oven sintering due to better energy efficiency potential. Laser sintering can be controlled to bring only that amount of energy needed for the sintering of ink without letting the substrate absorb too much energy. While in the oven, a much bigger portion of the heating power is wasted since the whole interior of the oven is heated. However, in the oven, a large amount of products can be sintered at the same time. Therefore, the comparison of sintering efficiency depends on the available equipment. According to our preliminary studies, sintering is probably the most critical process phase in the development of energy efficiency and thereby the environmental performance of ink-jet-printed interconnections, since the ink-jetting speed can be developed by increasing the amount of nozzles, but the sintering always requires a certain amount of energy per area (Kunnari et al. 2007), depending on the ink and sintering method (Suganuma et al. 2007). However, the possibly different process emissions of the sintering methods should be taken into account also. On the basis of the above, it could be made a general approximation that as the printing speed, and thereby efficiency, increases, the relative importance of sintering energy grows, as the relative importance of the manufacturing of materials is at the same level or higher (depending on the materials) (see Fig. 23.12). Therefore, developing ink-jet printable materials with low-temperature sintering characteristics could be the most effective way to develop overall environmental efficiency.

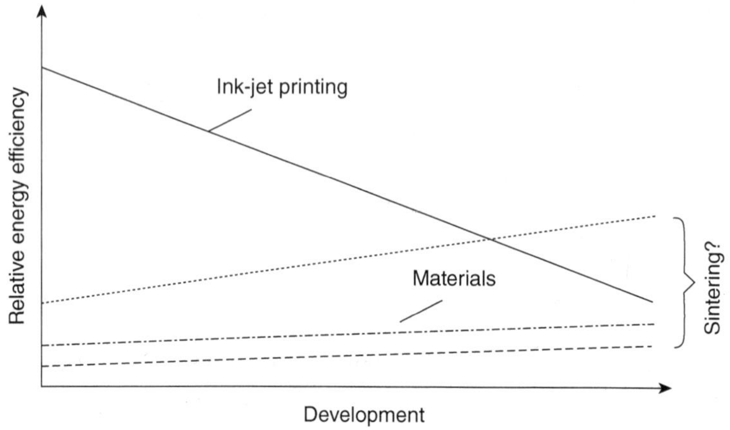

Figure 23.12. The approximated energy efficiency development potential of ink-jet printing.

In that connection, the manufacturing energy of printable materials should be taken into account, too. Another possibility is to develop efficient laser sintering processes.

To commercialize ink-jet printing technology, it is essential to gather comprehensive environmental information on the materials, manufacturing processes, and product life cycles in order to fully understand the real environmental potentials, shortcomings, and challenges compared to other technological options applicable for different products. A comprehensive environmental evaluation should take a wide approach, including different environmental impacts focusing on the total environmental burden of product life cycles, toxicity of materials, comparison of material options, analysis of current and forthcoming environmental requirements, and perhaps the development potential of these aspects. Manufacturing equipment and facilities can also have a different relative environmental impact compared to current technologies.

An environmental evaluation can show the most significant environmental aspects of the technology, which can be used to develop environmental indicators guiding product development toward better environmental performance. In that connection, different design approach levels can be used: product-level indicators, ink-jet printing-specific indicators, and material-level indicators (Kunnari et al. 2007).

An environmental evaluation and thereby environmentally conscious development of ink-jet printing requires knowledge about the characteristics of the technology, which would have possible adverse effects on the environment. Environmental friendliness can be approximated through some common parameters having connection to different dimensions of environmental friendliness and sustainability (e.g., nontoxicity, low energy and resource use, and recyclability), or through a more holistic approach of, for example, the methodology of environmental life cycle assessment, which has huge initial data requirements, but gives a wide environmental impact category indicator profile, including, for example, global warming, ozone depletion, and acidification potential.

However, a comparison of the environmental performance of different technologies is not so straightforward, as the applications of technologies can be different and entail different material or other requirements, which also impact on the environmental performance. In addition, the production volumes can be different. Therefore, environmental-friendliness is a case-specific issue, which depends also on the reference of comparison, since all manufacturing processes and products have some effect on the environment. The basic question, from an environmental point of view, is the capability of ink-jet printing technology to give environmental development potential for different applications and services, and, on the other hand, the challenges of ink-jet printing to become a sustainable option.

CONCLUSIONS

Ink-jet printing technology and a wide range of nanomaterials have been proposed for microelectronics applications. Ink-jet printable passive components and integration of these into the electronic package will significantly decrease the electronic package size and will increase the reliability of the electronic system. Ink-jet printing technology has obvious advantages over other methods for microelectronics applications, namely, high resolution; wide range of fluid material, for example, metallic nanoparticles and organic, ceramic nanoparticles, solders, and adhesives; and low cost. Ink-jet technology is additive technology, which means the material deposition is easy to control and, because of the nature of the manufacturing method, the amount of waste material is remarkably low compared to conventional lithography processes.

Commercialization of the manufactured electronics part by ink-jet printing is possible with faster processing steps, that is, postprocessing steps and final processing, reliability of ink-jet-printed structures, and overall environmental evaluation of the technology. It is easy to integrate ink-jet printing technology with existing production lines to assist current electronics manufacturing, or it can be applied to produce completely new flexible electronic products.

The environmental requirements are growing and this is creating several challenges in electronics production. However, several possibilities for benefits through environmental friendliness and value-added products may arise. Printed electronics can replace older manufacturing processes with more sustainable ones, with additive material deposition with low material and energy consumption. The environmental challenges concern every aspect of the technology, including new materials, manufacturing processes, products, and end-of-life treatment of the products. In order to commercialize this emerging technology, it is essential to perform a comprehensive environmental evaluation of all these aspects.

REFERENCES

Alonso E, Field FR, Kirchain RE. 2008. A case study of the availability of platinum group metals for electronics manufacturers. In: *Proceedings of the IEEE International Symposium on Electronics and the Environment*, pp. 1–6.

Atkinson A, Doorbar J, Hudd A, Segal DL, White PJ. 1997. Continuous ink-jet printing using sol-gel "ceramic" inks. J Sol-Gel Sci Technol 8:1093–1097.

Ayres RU. 1997. Metals recycling: economic and environmental implications. Resour Conserv Recycl 21(3):145–173.

Bieri NR, Chung J, Haferl SE, Poulikakos D. 2003. Microstructuring by printing and laser curing of nanoparticle solutions. Appl Phys Lett 82(20):3529–3531.

Bishop JF, Alan D. 1997. Ink jet ink containing nanoparticles of organic pigments: US patent 5679138.

Blayo K, Pineaux B. 2005. Printing processes and their potential for RFID printing. In: *Proceedings of the ACM International Conference Series,* vol. 121, pp.27–30.

Blazdell PF, Evans JRG, Edirisinghe MJ, Shaw P, Binstead MJ. 1996. The computer aided manufacture of ceramics using multilayer jet printing. Mater Sci Lett 14:1562–1565.

Bock K. 2005. Polymer electronics systems-polytronics. Proc IEEE 93(8):1400–1406.

Bucknall D. 2005. *Nanolithography and Patterning Techniques in Microelectronics.* England: Woodhead Publishing, Limited, p. 424.

Buffat P, Borel JP. 1976. Size effect on the melting of gold particles. J Phys Rev A 13: 2287–2298.

Calvert P. 2001. Inkjet printing for materials and devices. J Chem Mater 13:3299–3305.

Cheng K, Yang MH, Chiu WWW, Huang CY, Chang J, Ying TF, Yang Y. 2005. Ink-jet printing, self-assembled polyelectrolytes, and electroless plating: low cost fabrication of circuits on a flexible substrate at room temperature. Macromol Rapid Commun 26:247–264.

Chung J, Ko S, Bieri NR, Grigoropoulos CP. 2004. Conductor microstructures by laser curing of printed gold nanoparticle ink. Appl Phys Lett 84(5):801–803.

Cleland TA. 2003. Printed electronics: the next inkjet revolution. Technical Report. Massachusetts Institute of Technology, Management of Technology Program.

Dearden AL, Smith PJ, Shin DY, Reis N, Derby B, O'Brien P. 2005. A low curing temperature silver ink for use in ink-jet printing and subsequent production of conductive tracks. Macromol Rapid Commun 26(4):315–318.

Derbyshire K. 2007. High interest in low-end printable electronics. Solid State Technol 50(7):34–31.

EC. 2006. Regulation (EC) No 1907/2006 of the European Parliament and of the Council of 18 December 2006 Concerning the Registration, Evaluation, Authorisation and Restriction of Chemicals (REACH), Establishing a European Chemicals Agency, Amending Directive 1999/45/EC and Repealing Council Regulation (EEC) No 793/93 and Commission Regulation (EC) No 1488/94 as Well as Council Directive 76/769/EEC and Commission Directives 91/155/EEC, 93/67/EEC, 93/105/EC and 2000/21/EC (OJ L 396, 30.12.2006).

Fox NA, Youh MJ, Steeds JW, Wang WN. 2000. Patterned diamond particle films. J Appl Phys 87(11):8187–8191.

Fuller SB, Wilhelm EJ, Jacobson JM. 2002. Ink-jet printed nanoparticle microelectromechanical systems. J Microelectromech Syst 11(1):54–60.

Gamerith S, Klug A, Scheiber H, Scherf U, Moderegger E, List EJW. 2007. Direct ink-jet printing of Ag–Cu nanoparticle and Ag-precursor based electrodes for OFET applications. J Adv Funct Mater 17:3111–3118.

Ge J, Singh B. 2007. Transparent thin polythiophene films having improved conduction through use of nanomaterials. U.S. patent 2007/0246689 A1.

Griese H, Muller J, Reichl H, Somi G, Stevels A, Zuber K. 2000. Environmental assessment of lead free interconnection systems. In: *Proceedings of the Symposium on Lead Free Interconnect Technology*: p 22–26.

Hakola L. 2005. Benefits of inkjet printing for printed electronics. In: PIRA Printed Electronics.

Hammer RB, Powell DO, Mukherjee SP, Tummala RR, Raj R. 1989. Ceramics in electronic packaging. In: Seraphim DP, Lasky R, Li CY, editors. *Principles of Electronic Packaging*. New York: McGraw-Hill, pp. 282–333.

Handy RD, Kammer F, Lead JR, Hasselöv M, Owen R, Crane M. 2008. The ecotoxicology and chemistry of manufactured nanoparticles. Ecotoxicology 17(4):287–314.

Harima Chemicals, Inc. 2008. Nanopaste® datasheet. http://www.harima.co.jp.

Hayes DJ, Wallace DB, Cox WR. 1999. MicroJet printing of solder and polymers for multi-chip modules and chip-scale packages. In: *Proceedings of the IMAPS International Conference on High Density Packaging and MCMs*, pp. 242–247.

Hong SJ, Kim YH, Han JI. 2008. Development of ultrafine Indium Tin Oxide (ITO) nanoparticle for ink-jet printing by low-temperature synthetic method. Proc IEEE Transact Nanotechnol 7(2):172–176.

Hostetler MJ, Wingate JE, Zhong CJ, Harris JE, Vchet RW, Clark MR, Longdono JD, Green SJ, Stokes JJ, Wignall GD, Glish GL, Porter MD, Evans ND, Murray RW. 1998. Alkanethiolate gold cluster molecules with core diameters from 1.5 to 5.2 nm: core and monolayer properties as a function of core size. Langmuir 14(1):17–30.

Le HP. 1998. Progress and trends in ink-jet printing technology. J Imaging Sci Tech 42:49–62.

Huang D, Liao F, Molesa S, Redinger D, Suramanian V. 2003. Plastic-compatible low resistance printable gold nanoparticle conductors for flexible electronics. J Electrochem Soc 150(7): G412–G417.

Jagannathan R, Irvin G, Blanton T, Jagannathan S. 2006. Organic nanoparticles: preparation, self-assembly, and properties. J Adv Funct Mater 16:747–753.

Jang J, Ha J, Cho J. 2007. Fabrication of water-dispersible polyaniline-poly(4-styrenesulfonate) nanoparticles for inkjet-printed chemical-sensor applications. J Adv Mater 19(13): 1772–1775.

Kang HY, Schoenung JM. 2005. Electronic waste recycling: a review of U.S. infrastructure and technology options. Resour Conserv Recycl 45(4):368–400.

Kaydanova T, Miedaner A, Curtis C, Alleman J, Perkins JD, Ginley DS, Sengupta L, Zhang X, He S, Chiu L. 2003. Direct inkjet printing of composite thin barium strontium titanate films. J Mater Res 18(12):2820–2825.

Keskinen M, Valkama J. 2008. Environmental risk assessment of printable electronics. In press: *Proceedings of Joint International Congress and Exhibition Electronics Goes Green 2008+*. September 8–10, 2008. Berlin, Germany.

Kim D, Moon J. 2005. Highly conductive ink jet printed films of nanosilver particles for printable electronics. Electrochem Solid-State Lett 8(11):J30–J33.

Ko SH, Pan H, Grigoropoulos CP, Luscombe CK, Fréchet JMJ, Poulikakos D. 2007a. Air stable high resolution organic transistors by selective laser sintering of ink-jet printed metal nanoparticles. Appl Phys Lett 90:141103.

Ko SH, Pan H, Grigoropoulos CP, Luscombe CK, Fréchet JMJ, Poulikakos D. 2007b. All-inkjet -printed flexible electronics fabrication on a polymer substrate by low temperature high-resolution selective laser sintering of metal nanoparticles. J Nanotechnol IOP Publishing 18:345202.

Ko SH, Pan H, Grigoropoulos CP, Fréchet JMJ, Luscombe CK, Poulikakos D. 2008. Lithography-free high-resolution organic transistor arrays on polymer substrate by low energy selective laser ablation of inkjet-printed nanoparticle film. J Appl Phys A: Mater Sci Proc 92(3): 579–587.

Kolbe J, Arp A, Calderone F, Meyer EM, Meyer W, Schaefer H, Stuve M. 2007. Inkjettable conductive adhesive for use in microelectronics and microsystems technology. Microelectronics Reliability 47:331–334.

Ku A, Oetinscitan O, Saphores JD, Shapirod A, Schoenunp JM. 2003. Lead-free solders: issues of toxicity, availability and impacts of extraction. In: *Proceedings of the 53rd Electronic Components and Technology Conference*, pp. 47–53.

Kunnari E, Valkama J, Mäntysalo M, Mansikkamäki P. 2007. Environmental performance evaluation of printed electronics in parallel with prototype development. In: *Proceedings of the 40th International Symposium on Microelectronics*, p. 8.

Lanzano T, Bertram M, De Palo M, Wagner C, Zyla K, Graedel TE. 2006. The contemporary European silver cycle. Resour Conserv Recycl 46(1):27–43.

Lasertyöstö. 1996. *Teknologia Teollisuus*. Finland: Jyväskylä Vocational Institute.

Lawrence D, Kohler J, Brollier B, Claypole T, Burgin T. 2004. Manufacturing platforms for printing organic circuits. In: Gamota DR, Brazis P, Kalyanasundaram K, Jie Z, editors. *Printed Organic and Molecular Electronics*. The Netherlands: Kluwer Academic Publishers, pp. 161–346.

Licari JJ. 1994. Thin and thick films. In: Harper CA, Sampsa RM, editors. *Electronic Materials & Processes Handbook*, 2nd edition. Electronic Packaging and Interconnection Series. New York: McGraw-Hill, pp. 8.1–9.1.

Magdassi S, Moshe MB. 2003. Patterning of organic nanoparticles by ink-jet printing of microemulsions. Langmuir 19(3):939–942.

Magdassi S, Bassa A, Vinetsky Y, Kamyshny A. 2003. Silver nanoparticles as pigments for water-based ink-jet inks. J Chem Mater 15:2208–2217.

Mäntysalo M, Mansikkamäki P. 2007. Inkjet-deposited interconnections for electronic packaging. In: *Proceedings of the 23rd International Conference on Digital Printing Technologies*, pp. 813–817.

Mendel J, Bugner D. 1999. Particle generation and ink particle size effects in pigmented inkjet inks-part I. J Nanopart Res 1:419.

Mendel J, Bugner D, Bermel AD. 1999. Particle generation and ink particle size effects in pigmented inkjet inks-part II. J Nanopart Res 1:421–424.

Molesa S. 2006. Ultra-low-cost printed electronics. In: Technical Report No. UCB/EECS-2006-55. University of California at Berkeley, Electrical Engineering and Computer Sciences.

Mudd GM. 2007. Gold mining in Australia: linking historical trends and environmental and resource sustainability. Environ Sci Policy 10(7–8):629–644.

Nakamoto M, Yamamoto M, Kakiuchi H, Yoshida Y. 2003. Electronic circuit pattern formation by the use of noble metal nanoparticle pastes. In: *Proceedings of the 13th Microelectronics Symposium*, pp. 100–103.

Nakamoto M, Yamamoto M, Kashiwagi Y, Kakiuchi H, Tsujimoto T, Yoshida Y. 2005. Electronic circuit pattern formation on the plastic substrates by the use of metal and alloy nanoparticle pastes. In: *Proceedings of the 15th Microelectronics Symposium*, pp. 241–244.

Nakamoto M, Yamamoto M, Kashiwagi Y, Kakiuchi H, Tsujimoto T, Yoshida Y. 2007. A variety of silver nanoparticle pastes for fine electronic circuit pattern formation. In: *Proceedings of the 6th International Conference on Polymers and Adhesives in Microelectronics and Photonics*, pp. 105–109.

Nanomarkets white paper. 2007. The future of flexography and lithography in printed electronics.

Okada I, Shimoda K, Miyazaki K. 2006. Development of fine circuit pattern formation process using nano-metal ink. SEI Tech Rev 62:54–57.

Parikh MR, Quilty WF, Gardiner KM. 1990. SPC and setup analysis for screen printed thick films. In: *Proceedings of the Electronic Manufacturing Technology Symposium*, pp. 192–199.

Park JW, Baek SG. 2006. Thermal behaviour of direct-printed lines of silver nanoparticles. Scr Mater 55:1139–1142.

Pekkanen V. 2007. Design considerations for inkjet printed electronic interconnections and packaging. In: *Proceedings of the 40th IMAPS International Symposium on Microelectronics*, p. 8.

Perelaer J, Gans BJ, Schubert US. 2006. Ink-jet printing and microwave sintering of conductive silver tracks. Adv Mater 18:2101–2104.

Pickard WF. 2008. Geochemical constraints on sustainable development: can an advanced global economy achieve long-term stability? Glob Planet Change 61(3–4):285–299.

Piqué A, Chrisey DB. 2002. *Direct-Write Technologies for Rapid Prototyping Applications: Sensors, Electronics, and Integrated Power Sources*. Orlando, FL: Academic Press.

Rozenberg GG, Bresler E, Speakman SP, Jeynes C, Steinke JHG. 2002. Patterned low temperature copper-rich deposits using inkjet printing. Appl Phys Lett 81(27):5249–5251.

Saito H, Matsuba Y. 2006. Liquid wiring technology by ink-jet using NanoPaste®. In: *Proceedings of the 39th IMAPS International Symposium on Microelectronics*, pp. 470–477.

Sen S, Manciu M, Manciu FS. 1999. Ejection of ferrofluid grains using nonlinear acoustic impulses-a particle dynamical study. J Appl Phys 75(10):1479–1481.

Suganuma K, Wakuda D, Hatamura M, Kim KS. 2007. Ink-jet printing of nano materials and processes for electronics applications. In: *Proceedings of the International Symposium on High Density Packaging and Microsystem Integration*, pp. 1–4.

Thomas VM. 2008. Environmental implications of RFID. In: *Proceedings of the IEEE International Symposium on Electronics and the Environment*, pp. 1–5.

Townsend T, Musson S, Dubey B, Pearson B. 2007. Leachability of printed wire boards containing leaded and lead-free solder. J Environ Manage 88(4):926–931.

Tseng W, Chen CN. 2006. Dispersion and rheology of nickel nanoparticle inks. J Mater Sci 41(4):1213–1219.

Tseng WJ, Chen CN. 2006. Dispersion and rheology of nickel nanoparticle inks. J Mater Sci 41:1213–1219.

Tseng WJ, Lin SY, Wang SR. 2006. Particulate dispersion and freeform fabrication of $BaTiO_3$ thick films via direct inkjet printing. J Electroceram 16:537–540.

Tseng CC, Kuo JL, Chang CP, Hwu WH, Ger MD. 2007. A novel method produces Pd nanoparticles for ink-jet printing technology. In: *Microprocesses and Nanotechnology, Digest of Papers*, pp. 478–479.

Tseng KH, Liao CY, Tien DC, Tsung TT. 2008. Electrolysis of nano-silver suspension into ionic from via membrane electro-osmosis process. In: *Proceedings of International Multi Conference of Engineers and Computer Scientists*, pp. 1433–1435.

Volkman SK, Pei Y, Redinger D, Yin S, Subramanian V. 2004. Ink-jetted silver/copper conductors for printed RFID applications. Mat Res Soc 814:17.8.1–17.8.6. Spring Meeting.

Wallace D, Hayes D, Chen T, Shah V, Radulescu D, Cooley P. 2006. Ink-jet as a manufacturing tool. In: *Proceedings of the SMTA Pan Pacific Microelectronics Symposium*, p. 9.

Wong WK, Kripesh V, Iyer MK, Gupta M, Tay AAO, Tummala R. 2003. Low temperature sintering process for deposition of nano-structured metal for nano IC packaging. In: *Proceedings of the 5th Electronic Packaging Technology Conference*, pp. 551–556.

Xiang QF, Evans JRG, Edirisinghe MJ, Blazdell PF. 1997. Solid freeforming of ceramics using a drop-on-demand jet printer. In: *Proceeding of the Institute Mechanical Engineering 211: B*, pp. 211–214.

Yi SM, Jung JK, Choi SH, Kim I, Jung HC, Joung J, Joo YC. 2008. Effect of microstructure on electrical and mechanical properties: impurities of inkjet-printed Ag and Cu interconnects. In: *Proceedings of the 58th Electronic Components and Technology Conference*, pp. 1277–1281.

Yli-Luukko R. 2007. Laser application in electronics manufacturing. Technical Report. Tampere University of Technology.

Yue W, Jie W. 2008. Study on curing process and conductive performance of UV curable conductive inkjet printing. In: *Proceedings of the 2nd IEEE International Nanoelectronics Conference*, pp. 1522–1526.

Zhao X, Evans JRG, Edirisinghe MJ, Song JH. 2003. Formulation of a ceramic ink for a wide-array drop-on-demand ink-jet printer. J Ceram Int 29:887–892.

Zhao N, Chiesa M, Sirringhaus H. 2007. Self-aligned inkjet printing of highly conducting gold electrodes with submicron resolution. J Appl Phys 101(6):064513.

24

BIOINTEGRATION OF PROSTHETIC DEVICES

Masakazu Kawashita,[1] Toshiki Miyazaki,[2] and Chikara Ohtsuki[3]

[1]Tohoku University, Sendai, Japan
[2]Kyushu Institute of Technology, Kitakyushu, Japan
[3]Nagoya University, Nagoya, Japan

INTRODUCTION

Generally, "ceramics" are inorganic materials, such as pottery, porcelain, structural clay products, nonmetallic magnetic materials, and ferroelectrics. The original meaning of "inorganic" is "not consisting of anything that is living." Hence, it is difficult for us to consider that ceramics can play an important role in biomedical applications. However, over the last four decades in the orthopedic and dental fields, various kinds of ceramics have been widely used for artificial joint components, bone substitutes, and dental implants because of their excellent biocompatibility. In this chapter, ceramics that play an important role in prosthetic devices will be described from the viewpoint of biointegration between the ceramic and the living tissue.

STRUCTURE OF BONE

Our body is supported by bones, and these protect important organs such as the brain, heart, and lungs from external forces. We can walk, bend, and grasp objects because

Ceramic Integration and Joining Technologies: From Macro to Nanoscale, First Edition.
Edited by Mrityunjay Singh, Tatsuki Ohji, Rajiv Asthana, Sanjay Mathur.

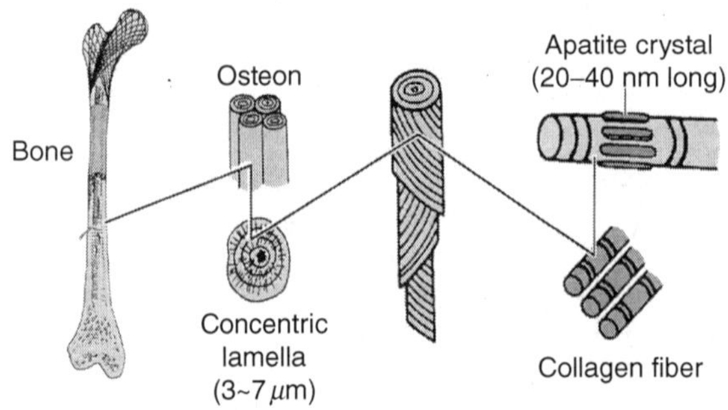

Figure 24.1. Structure of a typical bone (Park and Lakes 1992).

TABLE 24.1. Composition of Bone

Components	Amount (wt%)
Mineral (apatite)	69
Organic matrix	22
Collagen	(90–96% of organic matrix)
Others	(4–10% of organic matrix)
Water	9

our skeletons are composed of 206 bones connected through joints. If any part of these bones or joints were damaged though an accident, disease, or aging, our daily lives would be inconvenienced. A schematic representation of the structure of bone is given in Figure 24.1 (Park and Lakes 1992). From this figure, we can see that tiny apatite crystals, 20–40 nm in size, are deposited on organic collagen fibers fabricated into a three-dimensional structure. Wet bone is composed of 22 wt% of organic matrix, 69 wt% of mineral (apatite), and 9 wt% of water, and the organic matrix is mainly composed of collagen, as given in Table 24.1 (Park and Lakes 1992). This suggests that our bones are mainly composed of an extracellular matrix and only contain a small amount of living cells. Such an extremely low, living-cell content in bone might be one of the reasons why ceramics can be used in the field of orthopedics. At the joints, bones are covered with a collagen-rich tissue of cartilage so that they can move smoothly.

CERAMICS FOR ARTIFICIAL JOINTS

Bones and their joints are sometimes damaged by accidents, diseases, and aging. When hip and knee joints are damaged, patients sometimes feel severe pain and have difficulty walking. In such cases, the joints must be replaced with artificial ones. The present number of hip and knee joint replacements is higher than 100,000 in Japan and 700,000 in the United States.

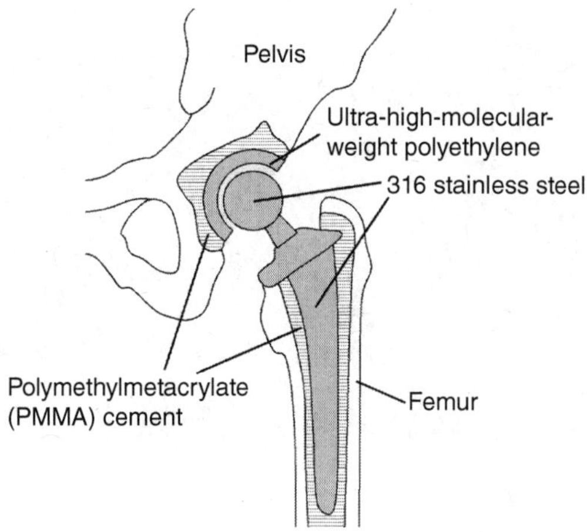

Figure 24.2. Artificial hip joint system in the early 1960s.

In 1960, Professor John Charnley, a British orthopedic surgeon, developed the first artificial hip joint. This consisted of a head and stem made from 316 stainless steel, and a cup socket made from ultra-high-molecular-weight polyethylene (Charnley 1960). The hip joints were fixed to the surrounding bones by filling the gap between the joint component and the surrounding bone with polymethylmethacrylate (PMMA) bone cement, as shown in Figure 24.2. The bone cement consisted of PMMA powder and methylmethacrylate monomer liquid and it solidified in situ. The replacements were expected to work well for over 20 years. However, in some cases, within 5–10 years, loosening of the bone cement from the surrounding bone was observed, and the hip joints had to be retrieved. This loosening was attributed to wear of the cup and head, and unstable fixation of the joint components with the bone cement.

Stainless steel is liable to corrode in the biological environment since our body fluid contains the kinds of ions found in seawater (Gamble 1954), and our body temperature is higher than that of seawater. As a result, the surface roughness of the stainless steel head increases remarkably, and wear debris are produced from the stainless steel and polyethylene rubbing together. Macrophages phagocytized the wear debris to produce cytokines, thus inducing the resorption of the surrounding bone. In addition, fixation of the joint components to the surrounding bone using PMMA bone cement is not stable since the bone cement is encapsulated by fibrous tissues isolated from the surrounding bone because of heat generated during solidification.

In 1971, Boutin, a French orthopedic surgeon, replaced both the cup and head of the hip joint with a high-density, high-purity sintered alumina (Boutin 1972). Figure 24.3 shows a photograph of a metallic stem with an alumina femoral ball used for total hip replacement (Hench 1991). As a result, stainless steel and polyethylene debris were eliminated. Sintered alumina is superior in mechanical strength, hardness, chemical durability, and hydrophilicity. As a result, the surfaces of the head and cup remain smooth for long periods; therefore, the articulation coefficients of the friction also

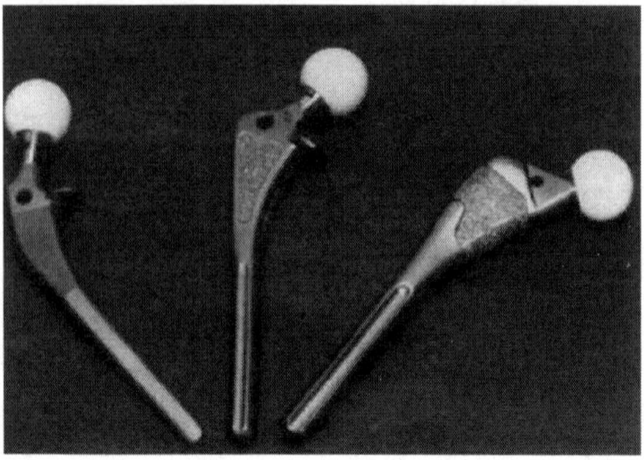

Figure 24.3. Medical-grade alumina used as femoral balls in total hip replacement (Hench 1991).

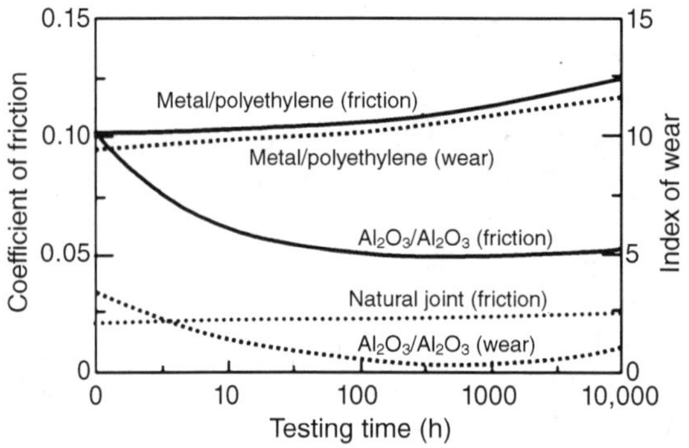

Figure 24.4. Time dependence of the coefficient of friction and index of wear of alumina–alumina versus metal–polyethylene hip joint (in vitro testing) (Hench 1991).

remain at low levels, as shown in Figure 24.4 (Hench 1991). Later, it was found that, even when just the head was made of sintered alumina and the cup was made of polyethylene, the production of polyethylene debris and the increase in the coefficient of the friction of the articulation were considerably suppressed. Sizing and shaping of the head to the cup is also much easier for this combination than for the ceramic–ceramic combination. Therefore, the alumina–polyethylene combination has been widely used. This combination was also applied to knee joints.

However, even with the alumina–polyethylene combination, wear debris were produced and induced bone resorption for long periods after the replacement. Therefore, ceramic–ceramic (alumina–alumina) combinations have also received attention. The mechanical strength of the sintered alumina has been improved by suppressing the grain growth and achieving a full densification of the sintered body. To achieve this, a small

TABLE 24.2. Standards for High-Purity Alumina for Orthopedic Implant Use

Components	ISO 6474 (ISO 1994)	ASTM F603 (ASTM 2000)
Bulk density ($g \cdot cm^{-3}$)	>3.94	>3.94
Grain size (μm)	<4.5	<4.5
Flexural strength (MPa)	>250	>400

amount (<0.5%) of magnesia (MgO) is added as a sintering aid. Standards for high-purity alumina for orthopedic use are listed in Table 24.2 (ISO 1994; ASTM 2000).

Since 1985, zirconia has been widely used for orthopedic implants. Up to the year 2000, approximately 40% of ceramic heads have been made of zirconia (Chevalier and Gremillad 2008). Sintered, partially stabilized zirconia (PSZ) is one of the candidates. However, in an aqueous environment such as a living body, it is liable to exhibit a decrease in its mechanical strength because of transformation of the metastable tetragonal phase to a stable monoclinic one. Sintered zirconia shows generally lower hardness than sintered alumina. The first problem was solved by suppressing the grain growth. Sintered yttrium or magnesium partially stabilized zirconia (3Y-TZP) is already used clinically as the head of the hip joint in combination with a polyethylene cup. The use of these zirconia heads in combination with zirconia cups of the same type is controversial with respect to the wear rate of the articulation. Recently, because of their excellent mechanical properties, new ceramic materials, such as zirconia-toughened alumina (ZTA) or alumina-toughened zirconia (ATZ) have been proposed as candidates for knee or hip joint components (Heimke et al. 2002; Begand et al. 2006; Nakanishi et al. 2007). In addition, a ceria-stabilized zirconia–alumina nanocomposite (Nawa et al. 1998) shows higher mechanical strength, fracture toughness, hardness, and stability against phase transformation. This ceramic is now being tried as a dental restorative (Ban et al. 2008). Table 24.3 shows the properties of some of the ceramics used in surgical implants and human bones (Hulbert 1993).

CERAMICS FOR BONE SUBSTITUTES

To isolate them from the surrounding bone, artificial materials implanted into bone defects are generally encapsulated by collagen fibers. Our body's common reaction is to protect itself from foreign materials. Because of this encapsulation, however, artificial material cannot be securely fixed to the surrounding bone.

In the early 1970s, Hench showed that some glasses in the Na_2O–CaO–SiO_2–P_2O_5 system spontaneously bond to living bone without having fibrous tissue form around them. This is shown in Figure 24.5 (Hench et al. 1972; Hench 1991). These glasses were the first artificial materials to bond to living tissue. They were named Bioglass and their typical composition (45S5) in mole percent was 46.1, SiO_2; 24.4, Na_2O; 26.9, CaO; and 2.6, P_2O_5. Because their tensile strength (42 MPa) is lower than that of human cortical bone (50–150 MPa), they are clinically used in regions where there is less loading, such as periodontal fillers and artificial middle ear devices (Jones 2008).

TABLE 24.3. Properties of Ceramics Used in Surgical Implants and Human Bones

	Human cortical bone	Human cancellous bone	Al_2O_3[a]	TZP[b]	Bioglass	Sintered hydroxyapatite	Glass–ceramic A-W
Density ($g \cdot cm^{-3}$)	1.6–2.1	—	3.98	6.05	2.66	3.16	3.07
Average grain size (μm)	—	—	3.6	0.2–0.4	—	—	—
Bending strength (MPa)	50–150	10–20	595	1000	42	60–200	220
Compressive strength (MPa)	100–230	2–12	4250	2000	—	600–900	1080
Young's modulus (GPa)	7–30	0.05–0.5	400	150	35	86–110	118
Vickers hardness (HV)	—	—	2400	1200	458	600	680
Fracture toughness ($MPa \cdot m^{1/2}$)	2–12	—	5	7	<1.0	1.0	2.0

[a]Purity >99.7% (Hulbert 1993).
[b]3Y-TZP ($3Y_2O_3$-$97ZrO_2$ [mol%]) (Hulbert 1993).

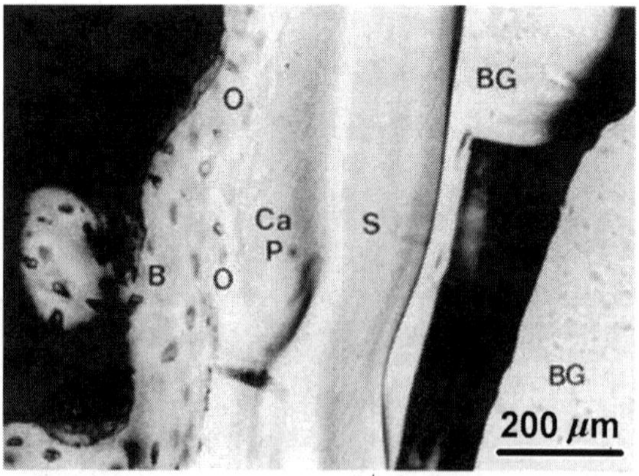

Figure 24.5. SEM photograph of Bioglass® implanted into rat tibia for 1 year. BG, Bioglass; B, bone; O, bone cell; Ca–P, apatite; S, silica gel (Hench 1991).

In the late 1970s, Jarcho (1976) and Aoki et al. (1977) independently showed that sintered hydroxyapatite ($Ca_{10}(PO_4)_6(OH)_2$) also bonds to living bone. Although a highly densified hydroxyapatite shows a maximum bending strength around 200 MPa, a practically used one generally shows a bending strength of around 115 MPa. This is lower than that of human cortical bone, as shown in Table 24.3 (LeGeros 1993). In the field of orthopedics and dentistry, this ceramic is widely used clinically as a bone filler in bulk or granular form, with either a dense or a porous structure.

In 1981, Kokubo et al. (1982) developed a glass–ceramic containing 38 wt% of crystalline oxyfluoroapatite ($Ca_{10}(PO_4)_6(O, F_2)$) and 34 wt% of β-wollastonite ($CaSiO_3$) by sintering and crystallization of a glass powder compact having the composition (in wt%) 4.6, MgO; 44.7, CaO; 34.0, SiO_2; 16.2, P_2O_5; and 0.5, CaF_2. The MgO–CaO–SiO_2 glassy matrix particles were 50–100 nm in size. This glass–ceramic was named A-W after the crystalline phases, and it directly bonds to living bone. The time required for the glass–ceramic A-W to bond to living bone was longer than that for the Bioglass 45S5 but shorter than for the sintered hydroxyapatite. The bending strength of 215 MPa is higher than that of human cortical bone, as given in Table 24.3. From the dependence of the bending strength of the glass–ceramic upon the stressing rate in a simulated body fluid (SBF), it was estimated that it could withstand over 10 years of continuous loading at a bending stress of 65 MPa (Kokubo et al. 1987). In Japan, it has been clinically used since 1991 as, for example, artificial vertebrae, intervertebral disks, and iliac crests. More than 60,000 patients have received it as their bone substitute. These bone-bonding ceramics, such as Bioglass, sintered hydroxyapatite, and glass–ceramic A-W, are now called bioactive ceramics.

However, even the glass–ceramic A-W cannot replace highly loaded bones, such as the femur and tibia, since its fracture toughness ($2\,MPa{\cdot}m^{1/2}$) is not as high as that of human cortical bones ($6\,MPa{\cdot}m^{1/2}$). For these applications, metallic materials, such as Ti–6Al–4V alloy coated with hydroxyapatite by a plasma spray method, are mainly used. However, in this technique, hydroxyapatite powders are heated for a short time in a frame above 10,000°C to be partially molten and decomposed. As a result, a porous layer of calcium phosphate different from the hydroxyapatite is weakly bonded to the metallic substrate. Such a layer is not stable in the living body for long periods (Klein, et al. 1993). Metallic materials with high fracture toughness are desired to exhibit bioactivity by themselves without a foreign material coating. Is this possible?

BIOINTEGRATION BETWEEN BIOACTIVE CERAMICS AND LIVING BONE

When we observe the interface between bioactive ceramics under a transmission electron microscope (TEM), the ones described earlier generally form an inorganic apatite layer without collagen fiber on their surfaces in the living body. They bond to the living bone through this apatite layer, as shown in Figure 24.6 (Neo et al. 1992). This apatite layer can be reproduced on the bioactive ceramic surfaces, even in an acellular SBF with ion concentrations nearly equal to those of human blood plasma, as shown in Figure 24.7 (Ohtsuki et al. 1995). The ion concentrations of SBF and human blood

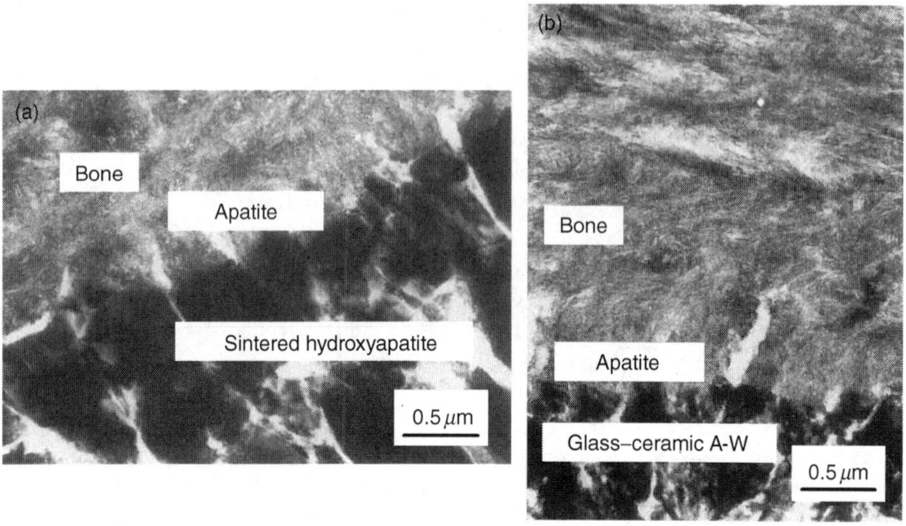

Figure 24.6. TEM photographs of the interface between bioactive ceramics and living tissues. (a) Interface between hydroxyapatite and rat tibia (10 weeks after implantation), (b) interface between glass–ceramic A-W and rat tibia (8 weeks after implantation) (Neo et al. 1992).

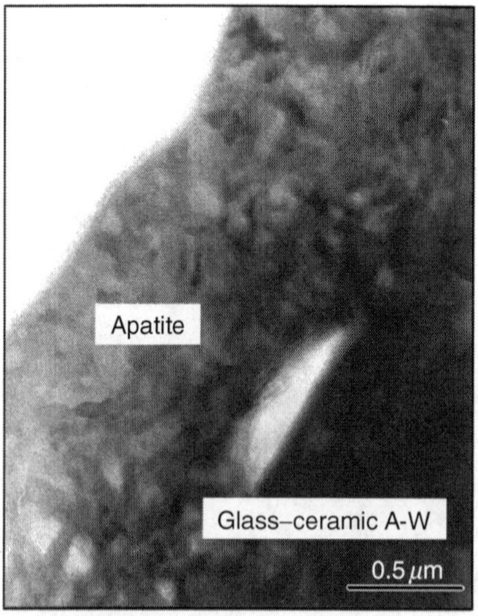

Figure 24.7. TEM photographs of the apatite formed on glass–ceramic A-W after soaking in SBF for 3 days (Ohtsuki et al. 1995).

TABLE 24.4. Ion Concentrations of Simulated Body Fluid (SBF) and Human Blood Plasma

	Concentration/mol·m^{-3}							
	Na^+	K^+	Mg^{2+}	Ca^{2+}	Cl^-	HCO_3^-	HPO_4^{2-}	SO_4^{2-}
Human blood plasma	142.0	5.0	1.5	2.5	103.0	27.0	1.0	0.5
SBF	142.0	5.0	1.5	2.5	147.8	4.2	1.0	0.5

plasma are summarized in Table 24.4 (Kokubo et al. 1990; Kokubo and Takadama 2006). According to detailed analyses of the surface layer deposited on bioactive ceramics in SBF, this layer consists of an apatite composed of small crystallites, with a Ca/P atomic ratio of 1.65. This is smaller than that of the stoichiometric apatite (1.67) and includes small amounts of Na^+, Mg^{2+}, Cl^-, and CO_3^{2-} ions. These structural and compositional characteristics are quite similar to those of bone mineral. It was also confirmed that bone-producing cells, called osteoblasts, could preferentially proliferate and differentiate to produce the apatite and collagen. These could then create the fibrous tissue-producing cells called fibroblasts on its surface, similar to the surface of fractured bone (Neo et al. 1993; Loty et al. 2000). Consequently, the surrounding bone comes into direct contact with the surface apatite. When this occurs, a tight chemical bond is formed between the bone mineral and the surface apatite and reduces the interfacial energy between them. It can be concluded from these findings that the essential requirement for an artificial material to bond to bone is the formation of bonelike apatite on its surface in the living body. Because the apatite formation on bioactive ceramics in SBF is quite reproducible, SBF was approved by the International Organization for Standardization (ISO) in June 2007 (ISO 23317) as a solution for the in vitro evaluation of the apatite-forming abilities of implants. Based on the investigation of the biointegration between the bioactive ceramic and living bone, we can derive the general requirement for apatite formation on artificial materials, including ceramics. This is described in the following section.

REQUIREMENT FOR ARTIFICIAL MATERIALS TO FORM APATITE

Our body fluid is already supersaturated with respect to apatite, and, hence, it is possible that apatite is formed anywhere in the body. Nevertheless, it is not normally formed, except in bone tissue, because the energy barrier for apatite nucleation in body fluid is generally very high; however, it is reduced in the bone tissue by some cell reactions. Even artificial material, therefore, can form apatite on its surface in vivo if it has a specific surface structure that can reduce the energy barrier for apatite nucleation, and induce the heterogeneous nucleation of the apatite.

Ohtsuki et al. (1992) investigated the apatite formation of simple ternary system $CaO–SiO_2–P_2O_5$ glasses in SBF and revealed that $CaO·SiO_2$-based glasses formed apatite on their surfaces in SBF, whereas $CaO·P_2O_5$-based glasses did not. This is shown

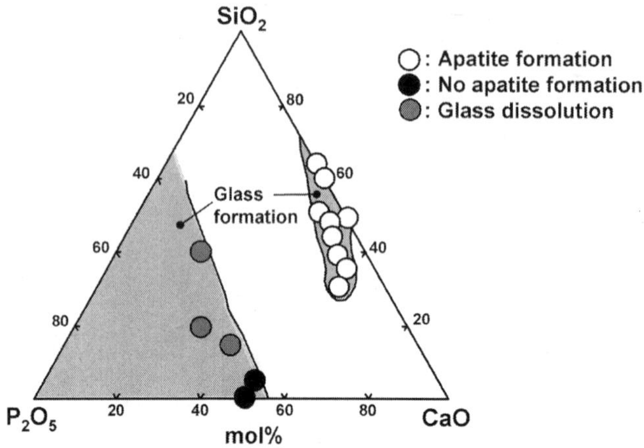

Figure 24.8. Compositional dependence of apatite formation on $CaO–P_2O_5–SiO_2$ glasses after soaking in SBF for 28 days (Ohtsuki et al. 1992).

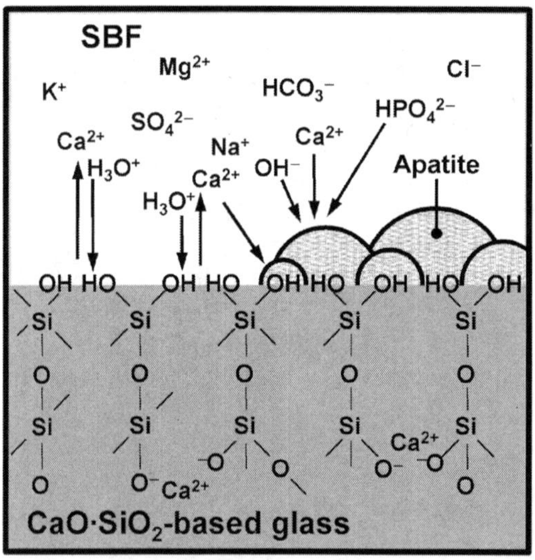

Figure 24.9. Schematic representation of the mechanism of apatite formation on a $CaO{\cdot}SiO_2$-based glass in SBF.

in Figure 24.8. The mechanism of apatite formation on the $CaO{\cdot}SiO_2$-based glass surfaces can be interpreted as follows. The $CaO{\cdot}SiO_2$-based glasses release the Ca^{2+} ions into the SBF via exchange with the H_3O^+ ion in the SBF to form Si–OH groups, as shown in Figure 24.9. In addition, the water molecules in the SBF react with the Si–O–Si bonds in the glass and break the bonds to form Si–OH groups. Thus, Si–OH groups, which induce the apatite nucleation, are formed on the surfaces of the glasses.

The released Ca^{2+} ions accelerate the apatite nucleation by increasing the ionic activity product of the apatite in SBF because the Ca^{2+} ion is a component of the apatite (Ohtsuki et al. 1992). Once the apatite nuclei are formed, they grow spontaneously by consuming the calcium and phosphate ions from the SBF because it is already supersaturated with respect to the apatite. These findings suggest that an essential requirement for an artificial material to form apatite on its surface in the biological environment is the presence of the specific functional groups effective for apatite nucleation.

FUNCTIONAL GROUPS EFFECTIVE FOR APATITE NUCLEATION

According to studies on the apatite formation of pure metal oxide gels prepared by a sol–gel method, SiO_2 (Li et al. 1992), TiO_2 (Li et al. 1994), ZrO_2 (Uchida et al. 2002), Nb_2O_5 (Miyazaki et al. 2001a), and Ta_2O_5 (Miyazaki et al. 2001b) gels form apatite on their surfaces in SBF, as shown in Figure 24.10. However, the Al_2O_3 (Li et al. 1994) gel did not form apatite. This means that functional groups, such as Si–OH, Ti–OH,

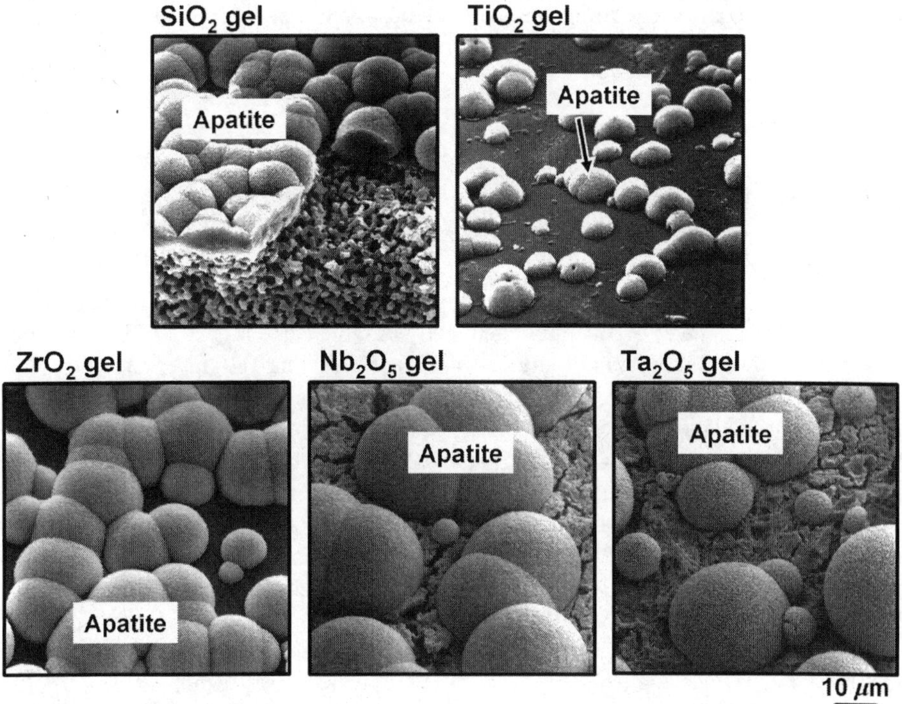

Figure 24.10. SEM photographs of the apatite formed on the surfaces of SiO_2 (Li et al. 1992), TiO_2 (Li et al. 1992), ZrO_2 (Uchida et al. 2002), Nb_2O_5 (Miyazaki et al. 2001a), and Ta_2O_5 (Miyazaki et al. 2001b) gels in SBF.

Zr–OH, Nb–OH, and Ta–OH, abundant on the gel's surfaces, are effective for apatite nucleation but that the Al–OH group is not. Not all the gels with the compositions described earlier are equally effective for the apatite nucleation. For example, a SiO_2 gel prepared by hydrolysis and polycondensation of tetraethoxysilane in an aqueous solution containing polyethylene glycol is more effective than one prepared in pure water, although the difference in the structure between them cannot be detected by X-ray diffraction and Fourier transform infrared spectroscopy (Cho et al. 1996). A TiO_2 gel with an anatase structure prepared by the sol–gel process and subsequent heat treatment is much more effective than gels with amorphous or rutile structures (Uchida et al. 2003). ZrO_2 gels with tetragonal or monoclinic structures prepared by the sol–gel process and subsequent heat treatments are much more effective than those with amorphous or cubic structures (Uchida et al. 2002).

Self-assembled monolayers terminated with COOH and H_2PO_4 groups form bone-like apatite on their surfaces in SBF (Tanahashi and Matsuda 1997). This means that the COOH and H_2PO_4 groups are also effective for apatite nucleation. The functional groups that induce apatite formation in SBF (Si–OH, Ti–OH, Zr–OH, Nb–OH, Ta–OH, COOH, and H_2PO_4) are negatively charged in aqueous solution around pH 7, whereas those that do not induce apatite formation, such as Al–OH, are positively charged.

Based on these findings, we could provide various kinds of materials with the ability to form apatite, including ceramics, metals, and organic polymers, by forming the specific functional groups effective for apatite nucleation on their surfaces, and, hence, obtain bioactive materials.

BIOACTIVE METALS

According to the fundamental findings on the biointegration between bioactive ceramics and living bone, we can provide bioactivity to intrinsically bioinert metallic materials. Pure titanium metal is generally chemically durable since it is covered with a thin TiO_2 passive layer, as shown in Figure 24.11a. However, this TiO_2 layer can react with a NaOH solution to form a sodium titanate hydrogel layer, as shown in Figure 24.11b. Although this gel layer is very soft, it can be stabilized as an amorphous sodium titanate layer, as shown in Figure 24.11c (Kim et al. 1996; Kokubo et al. 1996). It should be noted here that a graded structure, where the Na_2O content gradually decreases while the Ti content increases toward the interior, is formed within a layer 1 μm thick at the surface (Kim et al. 1999). It is confirmed that the mechanical properties of the titanium metal, including dynamic fatigue in saline, are not adversely affected by the NaOH and heat treatment.

It has been observed by both X-ray photoelectron spectra (Takadama et al. 2001a) and TEM with an energy-dispersive X-ray (EDX) spectrometer (Takadama et al. 2001b) that when the NaOH- and heat-treated titanium metal is soaked in SBF, it releases Na^+ ions via an exchange with the H_3O^+ ions in the SBF to form Ti–OH groups on its surface within a short period. Then, immediately after the Ti–OH groups are formed, they combine with the Ca^{2+} ions in the SBF to form an amorphous calcium titanate. Later,

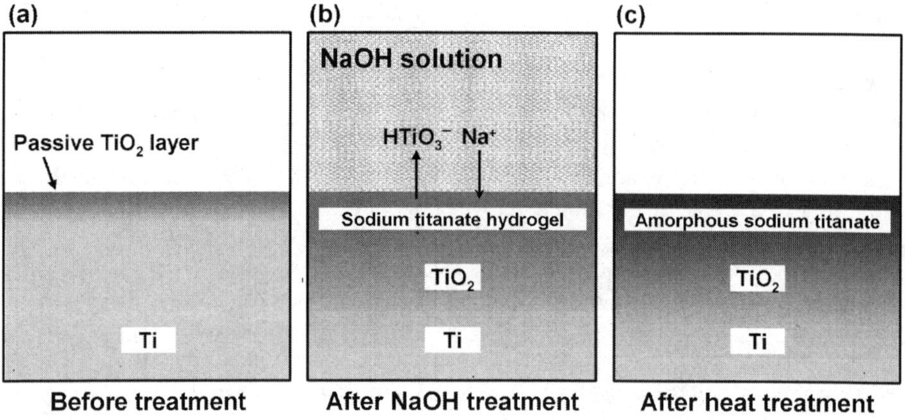

Figure 24.11. Schematic representation of the surface structural change of titanium metal by NaOH and heat treatment.

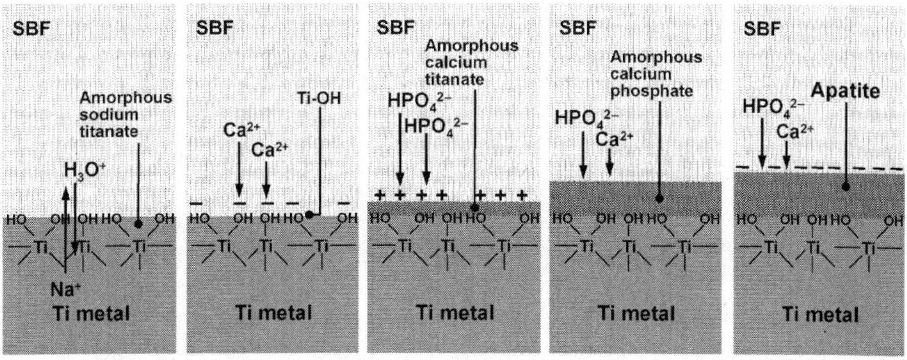

Figure 24.12. Schematic representation of the possible mechanisms of apatite formation on NaOH- and heat-treated titanium metal in SBF.

the amorphous calcium titanate combines with the phosphate ions in the SBF to form an amorphous calcium phosphate with a low Ca/P ratio. This increases its Ca/P ratio, eventually reaching 1.65 and transforming into crystalline apatite. This apatite formation process is interpreted in terms of the electrostatic reaction of the negatively charged Ti–OH groups on the surface and of the positively charged Ca^{2+} ions and the negatively charged phosphate ions in the SBF. Figure 24.12 shows a schematic representation of the possible mechanism for apatite formation on NaOH- and heat-treated titanium metal in the SBF.

Thus, formed apatite nuclei spontaneously grow by consuming the calcium and phosphate ions in the SBF to form a dense and uniform apatite layer. At the interface between the apatite and titanium metal, a graded structure is formed where the apatite content gradually decreases while the Ti content increases toward the interior (Kim et al. 1999). Consequently, the formed apatite is tightly integrated with the titanium substrate.

The NaOH- and heat-treated titanium metal forms the same apatite layer on its surface, even in vivo, and tightly bonds to the living bone through the apatite layer (Yan et al. 1997). For example, when a NaOH- and heat-treated titanium metal rod, 5 mm in diameter and 25 mm in length, was implanted into the intramedullary canal of a rabbit femur, it formed an apatite layer on its surface within 3 weeks. It was surrounded by living bone within 12 weeks (Nishiguchi et al. 2003). At 12 weeks after implantation, it was pulled out from the intramedullary canal accompanied with a fragment of the surrounding bone (Nishiguchi et al. 2003). It is already confirmed that similarly treated titanium-based alloys, such as Ti–6Al–4V, Ti–6Al–2Nb–Ta, and Ti–15Mo–5Zr–3Al, also form bonelike apatite on their surfaces in SBF (Kim et al. 1996), as well as in the living body, and they bond to living bone (Nishiguchi et al. 1999). A porous layer of titanium metal on a total hip joint of Ti–6Al–2Nb–Ta alloy went through NaOH and heat treatment, and the joint was subjected to clinical trials in 70 patients. As a result, in 2007, the Japanese government approved this joint for sale (Kokubo et al. 2008).

Tantalum metal also forms an amorphous sodium tantalate layer with a graded structure (Miyazaki et al. 2002) by treating with a 0.5 M NaOH solution at 60°C for 24 h (Miyazaki et al. 2000) followed by heat treatment at 300°C for 1 h (Miyazaki et al. 2001c). This treated tantalum metal also forms the bonelike apatite on its surface in SBF (Miyazaki et al. 2000) and in the living body, and bonds to the living bone, as shown in Figure 24.13 (Kato et al. 2000).

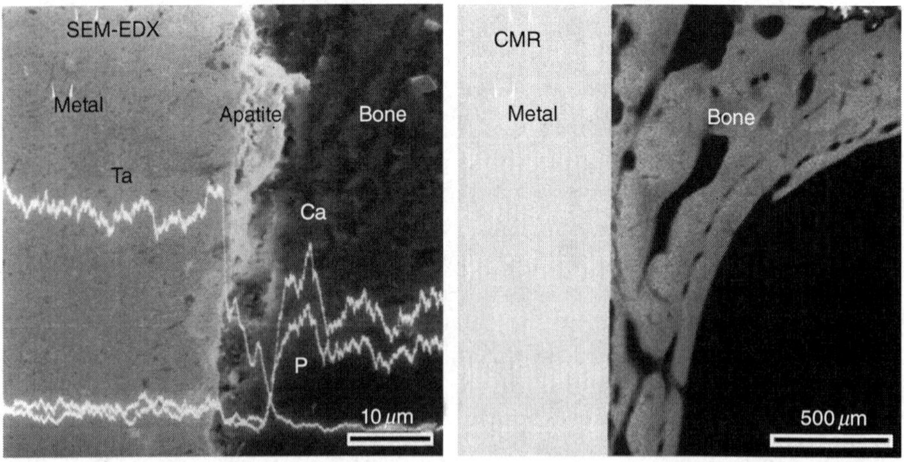

Figure 24.13. SEM–EDX profile (left) and contact radio micrograph (CMR, right) of the cross-section of NaOH- and heat-treated tantalum metals after implantation in a rabbit tibia for 16 weeks.

Based on the mechanism of apatite formation on CaO·SiO_2-based glasses in the section on "Ceramics for Bone Substitutes" and on NaOH- and heat-treated titanium metals in this section, a release of calcium and/or sodium ions, as well as a negatively charged functional group, seems to be important for apatite nucleation. In fact, calcium ions can be incorporated into the titanium oxide surface layer of titanium metal by hydrothermal treatment in a $CaCl_2$ solution at 200°C for 24 h. The surface-treated titanium metal then forms apatite within 36 h in SBF (Nakagawa et al. 2005).

On the other hand, it was revealed that H_2O_2 and subsequent heat treatments are also effective for inducing an apatite-forming ability on titanium metal surfaces in SBF (Ohtsuki et al. 1997; Wang et al. 2000). Moreover, it was found recently that the apatite deposition in SBF was enhanced through a specific V-shape arrangement of thermally oxidized titanium specimens (Wang et al. 2001, 2003). This suggests that not only the functional group but also the surface topography of a material plays an important role in apatite formation in a biological environment, that is, the biointegration between the material surface and the living bone. In fact, pure titanium metals thermally oxidized at 400°C were set together in a V-shape with a varied mouth opening. In SBF, they formed apatite up to approximately 600 μm on both internally facing surfaces between the open mouth spaces (Sugino et al. 2008a). It was also found that apatite formation was induced in SBF within 7 days on the internal surfaces of macrogrooves in pure titanium thermally oxidized at 400°C, and in Ti–15Zr–4Ta–4Nb, thermally oxidized at 500°C. This spatial design provides a new technique for inducing bioactivity to titanium by using simple machining and thermal oxidization.

These bone-bonding metals can be called "bioactive metals" since they also form the bonelike apatite on their surfaces in the living body and bond to the living bone, similar to the bioactive ceramics. These bioactive metals are useful as bone substitutes, even under load-bearing conditions, such as in hip and knee joints, as well as vertebra and dental implants, since they exhibit bioactivity and high fracture toughness.

BIOACTIVE CERAMICS–POLYMER COMPOSITE

As described in the section Structure of Bone, living bone is a composite of inorganic apatite and organic collagen fibers. Because of its unique composition and structure, natural bone shows not only high strength and high fracture toughness but also deformability and a low elastic modulus. Therefore, we may expect that ideal bone substitute materials can be derived from organic–inorganic composites through the combination of bioactive ceramic materials and polymers with a lower elastic modulus and deformability. To obtain artificial bone substitutes with both biological and mechanical compatibility, in the 1980s, Bonfield et al. developed a hydroxyapatite–polyethylene composite (Bonfield et al. 1981; Bonfield 1993). Hydroxyapatite powders can be dispersed in a polyethylene matrix up to 45 vol% without losing the ductility of the polymer. Figure 24.14 shows scanning electron micrograph (SEM) images of a fractured surface of a certain type of composite consisting of hydroxyapatite particles in a polyethylene matrix. The resultant composite shows a Young's modulus of about 3 GPa,

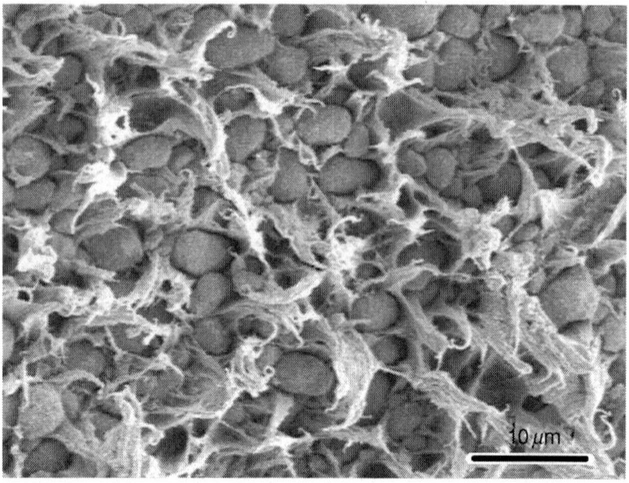

Figure 24.14. SEM photographs of a fractured surface of a certain type of composite consisting of hydroxyapatite–polyethylene.

an ultimate tensile strength of 22–26 MPa, and a fracture toughness, K_{IC}, of 2.9 MPa/m$^{1/2}$ (Hench 1998). It is already clinically used as an artificial middle ear bone. In addition, glass–ceramic A-W/polyethylene composites are also proposed (Juhasz et al. 2003a, b, 2004), since glass–ceramic A-W shows better apatite-forming abilities and mechanical properties than hydroxyapatite, as described in the section Ceramics for Bone Substitutes. By increasing the glass–ceramic A-W content, their Young's modulus, yield strength, and bending strength increase and the strain to failure decreases (Juhasz et al. 2004). The composites show a high apatite-forming ability in SBF, and this ability increases by increasing the amount of glass–ceramic A-W.

Based on a fundamental finding that TiO_2 with a specific structure can induce apatite formation (Uchida et al. 2003), bioactive TiO_2–high-density polyethylene (HDPE) composites have been recently developed (Hashimoto et al. 2006, 2007). Increased TiO_2 content resulted in an increase in bending strength, yield strength, Young's modulus, and compressive strength. The composite with 50 vol% TiO_2 shows a similar strength and Young's modulus to human cortical bone. The apatite-forming ability in SBF increases with TiO_2 content. This kind of bioactive material, with similar mechanical properties to human cortical bone, is expected to be useful as a load-bearing bone substitute in certain areas, such as the vertebra and the cranium.

The composites described previously should bond tightly to living bone and remain stable in the implanted region, but biodegradable composites are also useful in various clinical applications, such as the internal fixation of bone fractures. Poly-L-lactide (PLLA) is a biodegradable polymer, and medical devices made of PLLA have been developed and are used clinically in orthopedic, craniofacial, and oral–maxillofacial surgeries. Shikinami et al. developed a composite of unsintered hydroxyapatite and poly-L-lactide (u-HA–PLLA) through a design process that starts with hydroxyapatite powders prepared without heat treatments, such as calcination and sintering at high temperature (Shikinami and Okuno 1999, 2001). The bending strength increases with increasing u-HA powder content and reaches approximately 270 MPa when the u-HA fraction is 30 wt%. This value is a little higher than that of the unfilled PLLA specimen and is much higher than that of human cortical bone. The high strength of the u-HA–PLLA composite can be achieved by compacting most of the complete filling through a forging process, allowing for effective reinforcement of the u-HA particles in the polymer matrix. The bending modulus of the composite is a little higher than that of u-HA-free PLLA specimens and is similar to that of human cortical bone. The u-HA–PLLA composites form apatite on their surfaces within 3 days after immersion in SBF. This suggests that the composites have the potential to form a bonelike apatite layer on their surface and to bond to living bone in the body, that is, to show bioactivity. The in vivo testing using implants in animals actually showed that u-HA–PLLA devices show a higher affinity to bone than PLLA devices free from u-HA (Furukawa et al. 2000). Fibrous tissue was observed on the PLLA device, while direct contact with bone was observed for the u-HA–PLLA composite devices. This direct contact is caused by the formation of a bonelike apatite layer because of the exposure of the u-HA particles on the surface after the machining treatment.

Shikinami et al. also reported the total degradation of bone-fixation devices following implantation of u-HA–PLLA rods into the femoral medullary cavities of rabbits (Shikinami et al. 2005). In the proximal medullary cavity, the rod was completely resorbed, and unbound u-HA particles were detected in and around the endosteum 5–6 years after implantation. In the distal femoral condyle, where direct contact with cancellous bone occurred, the PLLA matrix was completely resorbed after 4.5–5.0 years, and the holes created in the distal femoral condyle during the insertion procedure were almost filled with new bone. Almost all of the u-HA particles were replaced with natural bone after 5.5 years, with no significant foreign body reaction. No inflammatory reaction was observed because the tissue reactions remained mild throughout the biodegradation and replacement process. These results suggest that u-HA–PLLA composites are quite useful for bioresorbable devices for the internal fixation of bone fractures.

BIOACTIVE INORGANIC–ORGANIC HYBRIDS

In the body, it has been shown that $CaO–SiO_2$ glasses form bonelike apatite on their surfaces and bond to living bone (Ohura et al. 1991). Therefore, it is expected that a ductile bioactive material with a low elastic modulus could be obtained if some organic molecules were incorporated into the structure of these glasses. Tsuru et al. synthesized poly(dimethylsiloxane) (PDMS)–$CaO–SiO_2$ hybrids through a sol–gel process and revealed that the hybrids showed an apatite-forming ability in SBF (Tsuru et al. 1997). After that, PDMS–$CaO–SiO_2–TiO_2$ hybrids were developed, and they showed an apatite-forming ability in SBF, as well as high ductility (Chen et al. 2000). Their

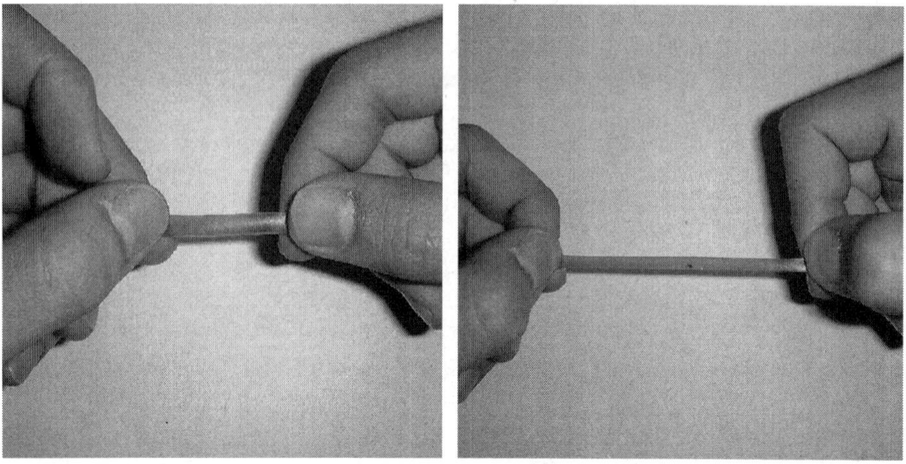

Figure 24.15. Deformable bioactive inorganic–organic hybrids.

mechanical strengths are, however, liable to decrease in a biological environment, because calcium ions are easily released from the $CaO–SiO_2$ component into the surrounding fluid.

When a Ca-free PDMS–TiO_2 hybrid is prepared by a sol–gel process and treated in pure water around 80°C, a hybrid containing anatase-type TiO_2 nanoparticles is obtained (Kamitakahara et al. 2003a). Hot water treatment to precipitate anatase nanoparticles is also effective for Ca-free poly(tetramethylene oxide) (PTMO)–TiO_2 hybrids (Kamitakahara et al. 2003b). The resultant hybrids show an apatite-forming ability in SBF, as well as high deformability, as shown in Figure 24.15. Its mechanical strength is, however, still lower than that of human cortical bone.

As described in the section Biointegration between Bioactive Ceramics and Living Bone, apatite deposited on artificial materials in SBF is similar in its structure and composition to biological apatite in living bone and, hence, can be called "bonelike apatite." Therefore, when the surface of artificial materials is coated with bonelike apatite in SBF or with some bioactive layer, inducing apatite nucleation in the biological environment, we can expect excellent biointegration between artificial materials and living bone. Many attempts have been made to coat bonelike apatite onto artificial materials (Tanahashi and Matsuda 1997; Takeuchi et al. 2003; Miyazaki et al. 2003a; Hosoya et al. 2004; Kawai et al. 2004). For example, ethylene–vinyl alcohol copolymer fibers constituting a fabric form bonelike small crystallites on their surfaces in SBF when they are modified with a silane coupling agent and a $CaO–SiO_2$ gel is applied to their surfaces (Oyane et al. 2003). The same treatment is also effective for poly(ethylene terephthalate) (PET) braids (Kokubo et al. 2005). Thus, surface-modified PET braids are surrounded by bone tissues 4 weeks after implantation into rabbit knees, while untreated ones are encapsulated with fibrous tissues. This is shown in Figure 24.16 (Kokubo et al. 2005).

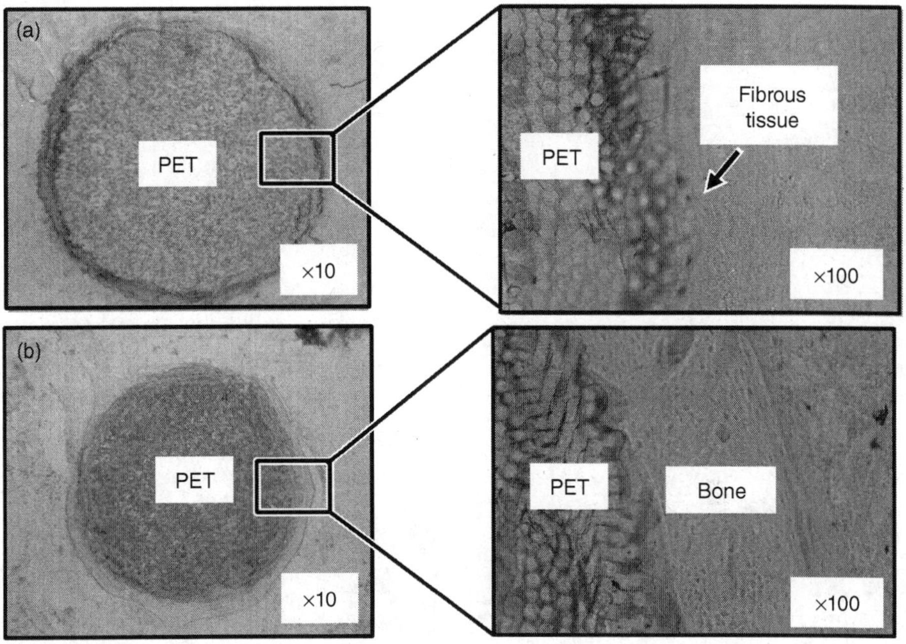

Figure 24.16. Optical micrographs of cross-sections of PET braid as-prepared (a) and that formed with apatite (b), 4 weeks after implantation into rabbit knees.

BIOACTIVE CEMENTS

Bone defects sometimes take complex shapes. Bioactive cements are useful for repairing such defects. They are usually composed of powder and liquid. When the powder and liquid are mixed in appropriate ratios, in a few minutes, they show fluidity and then solidify, forming the bonelike apatite, later bonding to the surrounding living bone. They can be injected into the bone defects as a viscous liquid or filled in as a paste. Typical examples are given in Table 24.5 (Chow 1998; Ishikawa 2008). All these cements set 5–10 min after being mixed, forming the bonelike apatite. After setting, the compressive strength is 60–90 MPa. They are already widely used for clinical applications. Various attempts to improve their mechanical strengths are still being made.

Recently, PMMA cement containing bioactive nano- or micron-sized TiO_2 powders has been developed as bioactive cement with a high mechanical strength (Goto et al. 2005, 2008). Compressive strength, bending strength, and bending modulus increase by increasing the amount of TiO_2 filler. The mechanical strengths meet the criteria required in the ISO 5833 standard. PMMA cements, with TiO_2 contents of 50 and 55.6 wt%, are biocompatible and osteoconductive. PMMA cement, with 55.6 wt% TiO_2, was in direct contact with bone over large areas 6 weeks after implantation into rat tibiae (Goto et al. 2008).

TABLE 24.5. Some Examples of Bioactive Cements

Powder	Liquid
Tetracalcium phosphate (TTCP)	Sodium phosphate aqueous solution
Dicalcium phosphate (DCPA)	
α-Tricalcium phosphate (α-TCP)	Sodium phosphate aqueous solution
Calcium carbonate	
Monocalcium phosphate monohydrate (MPCPM)	
α-TCP	Sodium succimate aqueous solution
Dicalcium phosphate dihydrate (DCPD)	Sodium chondroitin sulfate aqueous solution
TTCP	

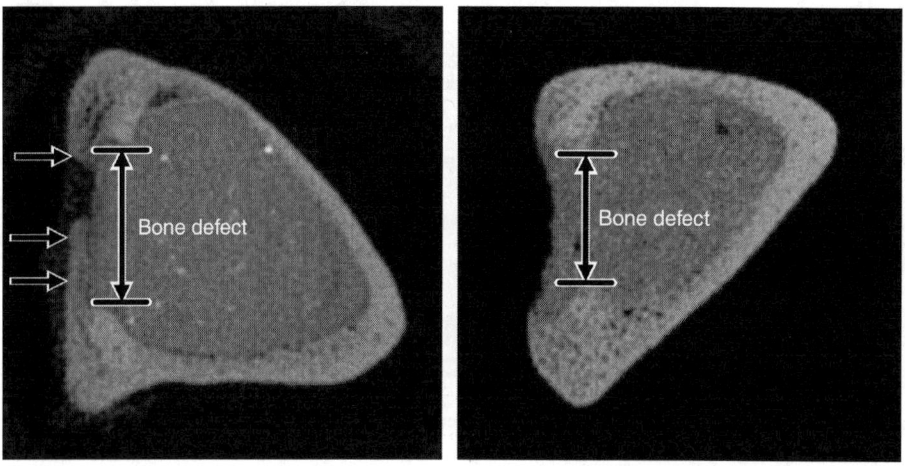

Modified cement

Reference (conventional PMMA-based bone cement)

Figure 24.17. Peripheral quantitative computed tomography (pQCT) images of the rabbit tibiae around the bone defect 4 weeks after operation. Left: the cement modified with Si–OH and calcium ions. Right: reference (conventional PMMA-based bone cement). Arrow: osteoconduction (Sugino et al. 2008b).

Bioactivity can also be induced by chemical modification of PMMA bone cement. Ohtsuki et al. recently developed novel PMMA cements by chemical modification with Si–OH and calcium ions (Ohtsuki et al. 2001; Mori et al. 2003; Miyazaki et al. 2003b). Thus, chemically modified PMMA cement forms apatite within 7 days in SBF (Mori et al. 2003; Miyazaki et al. 2003b). When the modified cement is implanted into the bone defects of rabbit tibia, osteoconduction is observed along the surface of the modified cement around the bone defect, but not on the conventional cement, as shown in Figure 24.17 (Sugino et al. 2008b). This indicates that the modified cement exhibits high osteoconductivity or good biointegration in vivo.

SUMMARY

Because of their good biocompatibility, various kinds of ceramic materials have been developed for biomedical applications during the last four decades. Some of them already play an important and indispensable role in the orthopedic and dental fields. However, novel ceramic-based materials with better biocompatibility are required in other clinical fields. In order to develop such advanced ceramic-based materials, it is essential to investigate further the biointegration between ceramic materials and the living body.

REFERENCES

American Society for Testing and Materials (ASTM) Standard F603-00. 2000. Standard specification for high-purity dense aluminum oxide for medical application.

Aoki H, Kato K, Ogiso M, Tabata T. 1977. Sintered hydroxyapatite as a new dental implant material. J Dent Outlook 49:567–575.

Ban SJ, Sato H, Suehiro Y, Nakanishi H, Nawa M. 2008. Biaxial flexure strength and low temperature degradation of Ce-TZP/Al_2O_3 nanocomposite and Y-TZP as dental restoratives. J Biomed Mater Res B Appl Biomater 87B:492–498.

Begand S, Oberbach T, Glien W. 2006. Mechanical properties of hip joint heads made of the dispersion ceramic—alumina toughened zirconia. Key Eng Mater 309–311:1257–1260.

Bonfield W. 1993. Design of bioactive ceramic-polymer composites. In: Hench LL, Wilson J, editors. *An Introduction to Bioceramics*. Singapore: World Scientific, pp. 299–303.

Bonfield W, Grynpas MD, Tully AE, Bowman J, Abram J. 1981. Hydroxyapatite reinforced polyethylene—a mechanically compatible implant. Biomaterials 2:185–186.

Boutin P. 1972. Total arthroplasty of the hip by fritted aluminum prosthesis: experimental study and 1st clinical applications. Rev Chir Orthop Reparatrice Appar Mot 58:229–246.

Charnley J. 1960. Surgery of the hip-joint—present and future developments. Br Med J 1:821–826.

Chen Q, Miyata N, Kokubo T, Nakamura T. 2000. Bioactivity and mechanical properties of PDMS-modified CaO-SiO_2-TiO_2 hybrids prepared by sol-gel process. J Biomed Mater Res 51:605–611.

Chevalier J, Gremillad L. 2008. Zirconia ceramics. In: Kokubo T, editor. *Bioceramics and Their Clinical Applications*. Cambridge: Woodhead Publishing Limited, pp. 243–265.

Cho SB, Nakanishi K, Kokubo T, Soga N, Ohtsuki C, Nakamura T. 1996. Apatite formation on silica gel in simulated body fluid: its dependence on structures of silica gels prepared in different media. J Biomed Mater Res Appl Biomater 33:145–151.

Chow LC. 1998. Calcium phosphate cements: chemistry and applications. In: LeGeros RZ, LeGeros JP, editors. *Bioceramics 11*. Singapore: World Scientific, p. 45–49.

Furukawa T, Matsusue Y, Yasunaga T, Nakagawa Y, Okada Y, Shikinami Y, Okuno M, Nakamura T. 2000. Histomorphometric study on high-strength hydroxyapatite/poly(L-lactide) composite rods for internal fixation of bone fractures. J Biomed Mater Res 50:410–419.

Gamble J. 1954. *Chemical Anatomy, Physiology and Pathology of Extracellular Fluid*, 6th edition. Cambridge, MA: Harvard University Press.

Goto K, Tamura J, Shinzato S, Fujibayashi S, Hashimoto M, Kawashita M, Kokubo T, Nakamura T. 2005. Bioactive bone cements containing nano-sized titania particles for use as bone substitutes. Biomaterials 26:6496–6505.

Goto K, Hashimoto M, Takadama H, Tamura J, Fujibayashi S, Kawanabe K, Kokubo T, Nakamura T. 2008. Mechanical, setting, and biological properties of bone cements containing micron-sized titania particles. J Mater Sci Mater Med 19:1009–1016.

Hashimoto M, Takadama H, Mizuno M, Kokubo T. 2006. Enhancement of mechanical strength of TiO_2/high-density polyethylene composites for bone repair with silane-coupling treatment. Mater Res Bull 41:515–524.

Hashimoto M, Takadama H, Mizuno M, Kokubo T. 2007. Mechanical properties and apatite forming ability of TiO_2 nanoparticles/high density polyethylene composite: effect of filler content. J Mater Sci Mater Med 18:661–668.

Heimke G, Leyen S, Willmann G. 2002. Knee arthroplasty recently developed ceramics offer new solution. Biomaterials 23:1539–1551.

Hench LL. 1991. Bioceramics from concept to clinic. J Am Ceram Soc 74:1487–1510.

Hench LL. 1998. Bioceramics. J Am Ceram Soc 81:1705–1728.

Hench LL, Splinter RJ, Allen WC, Greenlee TK. 1972. Bonding mechanisms at the interface of ceramic prosthetic materials. J Biomed Mater Res 2:117–141.

Hosoya K, Ohtsuki C, Kawai T, Kamitakahara M, Ogata S, Miyazaki T, Tanihara M. 2004. A novel covalently crosslinked gel of alginate and silane with the ability to form bone-like apatite. J Biomed Mater Res 71A:596–601.

Hulbert S. 1993. The use of alumina and zirconia in surgical implants. In: Hench LL, Wilson J, editors. *An Introduction to Bioceramics*. Singapore: World Scientific, pp. 25–40.

International Organization for Standardization (ISO). ISO 6474. 1994. Implants for surgery—ceramic materials based on high purity alumina revision 2.

Ishikawa K. 2008. Calcium phosphate cement. In: Kokubo T, editor. *Bioceramics and Their Clinical Applications*. Cambridge: Woodhead Publishing Limited, pp. 438–463.

Jarcho M. 1976. Hydroxylapatite synthesis and characterization. Tissue, cellular and subcellular events at bone-ceramic hydroxyapatite interface. J Bioeng 1:79–92.

Jones JR. 2008. Bioactive glass. In: Kokubo T, editor. *Bioceramics and Their Clinical Applications*. Cambridge: Woodhead Publishing Limited, pp. 266–283.

Juhasz JA, Best SM, Bonfield W, Kawashita M, Miyata N, Kokubo T, Nakamura T. 2003a. Apatite-forming ability of glass-ceramic apatite-wollastonite—polymer composites: effect of filler content. J Mater Sci Mater Med 14:489–495.

Juhasz JA, Best SM, Kawashita M, Miyata N, Kokubo T, Nakamura T, Bonfield W. 2003b. Bonding strength of the apatite layer formed on glass-ceramic apatite-wollastonite—polyethylene composites. J Biomed Mater Res 67A:952–959.

Juhasz JA, Best SM, Brooks R, Kawashita M, Miyata N, Kokubo T, Nakamura T, Bonfield W. 2004. Mechanical properties of glass-ceramic A-W-polyethylene composites: effect of filler content and particle size. Biomaterials 25:949–955.

Kamitakahara M, Kawashita M, Miyata N, Kokubo T, Nakamura T. 2003a. Apatite formation on CaO-free polydimethylsiloxane (PDMS)–TiO_2 hybrids. J Mater Sci Mater Med 14: 1067–1072.

Kamitakahara M, Kawashita M, Miyata N, Kokubo T, Nakamura T. 2003b. Apatite-forming ability and mechanical properties of CaO-free poly (tetramethylene oxide) (PTMO)–TiO_2 hybrids treated with hot water. Biomaterials 24:1357–1363.

Kato H, Nakamura T, Nishiguchi S, Matsusue Y, Kobayashi M, Miyazaki T, Kim HM, Kokubo T. 2000. Bonding of alkali- and heat-treated tantalum implant to bone. J Biomed Mater Res Appl Biomater 53:28–35.

Kawai T, Ohtsuki C, Kamitakahara M, Miyazaki T, Tanihara M, Sakaguchi Y, Konagaya S. 2004. Coating of an apatite layer on polyamide films containing sulfonic groups by a biomimetic process. Biomaterials 25:4529–4534.

Kim HM, Miyaji F, Kokubo T, Nakamura T 1996. Preparation of bioactive Ti and its alloys via simple chemical treatment. J Biomed Mater Res 32:409–417.

Kim HM, Miyaji F, Kokubo T, Nishiguchi S, Nakamura T. 1999. Graded surface structure of bioactive titanium metal prepared by chemical treatment. J Biomed Mater Res 45:100–107.

Klein CPAT, Wolke JGC, de Groot K. 1993. In: Hench LL, Wilson J, editors. *An Introduction to Bioceramics*. Singapore: World Scientific, pp. 199–221.

Kokubo T, Shigematsu M, Nagashima Y, Tashiro M, Higashi S. 1982. Apatite- and wollastonite-containing glass-ceramic for prosthetic application. Bull Inst Chem Res Kyoto Univ 60:260–268.

Kokubo T, Shigematsu M, Nagashima Y, Tashiro M, Higashi S. 1982. Apatite- and wollastonite-containing glass-ceramic for prosthetic application. Bull Inst Chem Res Kyoto Univ 60:260–268.

Kokubo T, Ito S, Shigematsu M, Sakka S, Yamamuro T. 1987. Fatigue and life-time of bioactive glass-ceramics A-W containing apatite and wollastonite. J Mater Sci 22:4067–4070.

Kokubo T, Kushitani H, Sakka S, Kitsugi T, Yamamuro T. 1990. Solutions able to reproduce in vivo surface-structure changes in bioactive glass-ceramic. J Biomed Mater Res 24: 721–734.

Kokubo T, Miyaji F, Kim HM, Nakamura T. 1996. Spontaneous formation of bonelike apatite layer on chemically treated titanium metals. J Am Ceram Soc 79:1127–1129.

Kokubo T, Kawai M, Kawashita M, Yamamoto K, Nakamura T. 2005. Fabrics of polymer fibers modified with calcium silicate for bone substitute. Key Eng Mater 284–286:775–778.

Kokubo T, Takadama H. 2006. How useful is SBF in predicting in vivo bone bioactivity? Biomaterials 27:2907–2915.

Kokubo T, Takadama H, Matsushita T. 2008. Titania-based materials. In: Kokubo T, editor. *Bioceramics and Their Clinical Applications*. Cambridge: Woodhead Publishing Limited, p. 485.

LeGeros RZ. 1993. Dense hydroxyapatite. In: Hench LL, Wilson J, editors. *An Introduction to Bioceramics*. Singapore: World Scientific.pp. 139–180.

Li P, Ohtsuki C, Kokubo T, Nakanishi K, Soga N, Nakamura T, Yamamuro T. 1992. Apatite formation induced by silica gel in a simulated body fluid. J Am Ceram Soc 75:2094–2097.

Li P, Ohtsuki C, Kokubo T, Nakanishi K, Soga N, Nakamura T. 1994. A role of hydrated silica, titania and alumina in forming biologically active bone-like apatite on implant. J Biomed Mater Res 28:7–15.

Loty C, Sautier JM, Boulekbache H, Kokubo T, Kim HM, Forest N. 2000. In vitro bone formation on a bonelike apatite layer prepared by a biomimetic process on a bioactive glass-ceramic. J Biomed Mater Res 49:423–434.

Miyazaki T, Kim HM, Miyaji F, Kokubo T, Nakamura T. 2000. Bioactive tantalum metal prepared by NaOH treatment. J Biomed Mater Res 50:35–42.

Miyazaki T, Kim HM, Kokubo T, Ohtsuki C, Nakamura T. 2001a. Bonelike apatite formation induced on niobium oxide gels in simulated body fluid. J Ceram Soc Japan 109:934–938.

Miyazaki T, Kim HM, Kokubo T, Kato H, Nakamura T. 2001b. Induction and acceleration of bonelike apatite formation on tantalum oxide gel in simulated body fluid. J Sol-Gel Sci Tech 21:83–88.

Miyazaki T, Kim HM, Kokubo T, Miyaji F, Kato H, Nakamura T. 2001c. Effect of thermal treatment on apatite-forming ability of NaOH-treated tantalum metal. J Mater Sci Mater Med 12:683–687.

Miyazaki T, Kim HM, Kokubo T, Ohtsuki C, Kato H, Nakamura T. 2002. Enhancement of bonding strength by graded structure at interface between apatite layer and bioactive tantalum metal. J Mater Sci Mater Med 13:651–655.

Miyazaki T, Ohtsuki C, Akioka Y, Tanihara M, Nakao J, Sakaguchi Y, Konagaya S. 2003a. Apatite deposition on polyamide films containing carboxyl group in a biomimetic solution. J Mater Sci Mater Med 14:569–574.

Miyazaki T, Ohtsuki C, Kyomoto M, Tanihara M, Mori A, Kuramoto K. 2003b. Bioactive PMMA bone cement prepared by modification with methacryloxypropyltrimethoxysilane and calcium chloride. J Biomed Mater Res 67A:1417–1423.

Mori A, Ohtsuki C, Sugino A, Kuramoto K, Miyazaki T, Tanihara M, Osaka A. 2003. Bioactive PMMA-based bone cement modified with methacryloxypropyltrimethoxysilane and calcium salts—effects of calcium salts on apatite-forming ability. J Ceram Soc Japan 111:738–742.

Nakagawa M, Zhang L, Udoh K, Matsuya S, Ishikawa K. 2005. Effects of hydrothermal treatment with $CaCl_2$ solution on surface property and cell response of titanium implants. J Mater Sci Mater Med 16:985–991.

Nakanishi T, Kobayashi T, Miyazaki T. 2007. Mechanical properties and cyclic fatigue of the newly developed ceramic material for artificial joints. J Ceram Soc Japan 115:466–470.

Nawa M, Nakamoto S, Sekino T, Niihara K. 1998. Tough and strong Ce-TZP/alumina nanocomposites doped with titania. Ceram Int 24:497–506.

Neo M, Kotani S, Nakamura T, Yamamuro T, Ohtsuki C, Kokubo T, Bando Y. 1992. A comparative study of ultrastructures of the interface between four kinds of surface-active ceramic and bone. J Biomed Mater Res 26:1419–1432.

Neo M, Nakamura T, Ohtsuki C, Kokubo T, Yamamuro T. 1993. Apatite formation on three kinds of bioactive materials at an early stage in vivo: a comparative study by transmission electron microscopy. J Biomed Mater Res 24:331–343.

Nishiguchi S, Kato H, Fujita H, Kim HM, Miyaji F, Kokubo T, Nakamura T. 1999. Enhancement of bone-bonding strengths of titanium alloy implants by alkali and heat treatments. J Biomed Mater Res Appl Biomater 48:689–696.

Nishiguchi S, Fujibayashi S, Kim HM, Kokubo T, Nakamura T. 2003. Biology of alkali- and heat-treated titanium implants. J Biomed Mater Res 67A:26–35.

Ohtsuki C, Kokubo T, Yamamuro T. 1992. Mechanism of apatite formation on $CaO-SiO_2-P_2O_5$ glasses in a simulated body fluid. J Non-Cryst Solids 143:84–92.

Ohtsuki C, Aoki Y, Kokubo T, Bando Y, Neo M, Nakamura T. 1995. Transmission electron microscopic observation of glass-ceramic A-W and apatite layer formed on its surface in a simulated body fluid. J Ceram Soc Japan 103:449–454.

Ohtsuki C, Iida H, Hayakawa S, Osaka A. 1997. Bioactivity of titanium treated with hydrogen peroxide solution containing metal chlorides. J Biomed Mater Res 35:39–47.

Ohtsuki C, Miyazaki T, Kyomoto M, Tanihara M, Osaka A. 2001. Development of bioactive PMMA-based cement by modification with alkoxysilane and calcium salt. J Mater Sci Mater Med 12:895–899.

Ohura K, Yamamuro T, Nakamura T, Kokubo T, Ebisawa Y, Kotoura Y, Oka M. 1991. Bone-bonding ability of P_2O_5-free CaO-SiO_2 glasses. J Biomed Mater Res 25:357–365.

Oyane A, Kawashita M, Nakanishi K, Kokubo T, Minoda M, Miyamoto T, Nakamura T. 2003. Bonelike apatite formation on ethylene-vinyl alcohol copolymer modified with silane coupling agent and calcium silicate solutions. Biomaterials 24:1729–1735.

Park JB, Lakes RS. 1992. *Biomaterials: An Introduction*, 2nd edition. New York: Plenum.

Shikinami Y, Okuno M. 1999. Bioresorbable devices made of forged composites of hydroxyapatite (HA) particles and poly-L-lactide (PLLA): part I. Basic characteristics. Biomaterials 20: 859–877.

Shikinami Y, Okuno M. 2001. Bioresorbable devices made of forged composites of hydroxyapatite (HA) particles and poly L-lactide (PLLA): part II. practical properties of miniscrews and miniplates. Biomaterials 22:3197–3211.

Shikinami Y, Matsusue Y, Nakamura T. 2005. The complete process of bioresorption and bone replacement using devices made of forged composites of raw hydroxyapatite particles/poly L-lactide (F-u-HA/PLLA). Biomaterials 26:5542–5551.

Sugino A, Uetsuki K, Tsuru K, Hayakawa S, Osaka A, Ohtsuki C. 2008a. Surface topography designed to provide osteoconductivity to titanium after thermal oxidation. Mater Trans 49: 428–434.

Sugino A, Ohtsuki C, Miyazaki T. 2008b. In vivo response of bioactive PMMA-based bone cement modified with alkoxysilane and calcium acetate. J Biomed Appl 23:213–228.

Takadama H, Kim HM, Kokubo T, Nakamura T. 2001a. An X-ray photoelectron spectroscopy study of process of apatite formation on bioactive titanium metal. J Biomed Mater Res 55: 185–193.

Takadama H, Kim HM, Kokubo T, Nakamura T. 2001b. TEM-EDX study of mechanism of bonelike apatite formation on bioactive titanium metal in simulated body fluid. J Biomed Mater Res 57:441–448.

Takeuchi A, Ohtsuki C, Miyazaki T, Tanaka H, Yamazaki M, Tanihara M. 2003. Deposition of bone-like apatite on silk fiber in a solution that mimics extracellular fluid. J Biomed Mater Res 65A:283–289.

Tanahashi M, Matsuda T. 1997. Surface functional group dependence on apatite formation on self-assembled monolayers in a simulated body fluid. J Biomed Mater Res 34:305–315.

Tsuru K, Ohtsuki C, Osaka A, Iwamoto T, Mackenzie JD. 1997. Bioactivity of sol-gel derived organically modified silicates, Part I: in vitro examination. J Mater Sci Mater Med 8: 157–161.

Uchida M, Kim HM, Kokubo T, Tanaka K, Nakamura T. 2002. Structural dependence of apatite formation on zirconia gels in a simulated body fluid. J Ceram Soc Japan 110:710–715.

Uchida M, Kim HM, Kokubo T, Fujibayashi S, Nakamura T. 2003. Structural dependence of apatite formation on titania gels in a simulated body fluid. J Biomed Mater Res 64A:164–170.

Wang XX, Hayakawa S, Tsuru K, Osaka A. 2000. Improvement of the bioactivity of H_2O_2/$TaCl_5$-treated titanium after a subsequent heat treatment. J Biomed Mater Res 52:171–176.

Wang XX, Hayakawa S, Tsuru K, Osaka A. 2001. A comparative study of in vitro apatite deposition on heat-, H_2O_2-, and NaOH-treated titanium surfaces. J Biomed Mater Res 54: 172–178.

Wang XX, Yan W, Hayakawa S, Tsuru K, Osaka A. 2003. Apatite deposition on thermally and anodically oxidized titanium surfaces in a simulated body fluid. Biomaterials 24: 4631–4637.

Yan WQ, Nakamura T, Kobayashi M, Kim HM, Miyaji F, Kokubo T. 1997. Bonding of chemically treated titanium implant to bone. J Biomed Mater Res 37:265–275.

INDEX

Page numbers in *italics* refer to Figures; those in **bold** to Tables.

Acceleration by gas transportation (CSM) method, 504–507, *505*
Active braze alloy (ABA) filler metals, 96
Active brazing, definition, 18. *See also* Brazing technology
Active metal casting (AMC) technique, 48, *49*
Aeronautics and ground transportation, 5–6
Aerosol deposition (AD), techniques and applications, 489–516
  AD method, 491–493, **492**, *492*
    carrier gas influence, 499
    vs. collision of solid-state particles, 504–507, *505*
    densification mechanism, 497–499, *498*
    particle velocity, 496–497, *496*, *501*
    temperature increase, *497*
  antiplasma corrosion components (AD-yttrium oxide film), 508–510, **508**, *509*
  electrical properties of AD films, 507
  high-frequency devices, 512–513, *513*
  optical devices, 514, *514*, *515*
  overview of, 11, 489–491, *490*, 515–516
  piezoelectric devices, 510–512, *510*, *511*, *512*
  properties and film patterning, 499–503, *500*, *501*, *502*, *503*, *504*
  room temperature impact consolidation, 493–499, *494*, **495**
Air brazing, 91–135. *See also* Brazing technology
  vs. ceramic brazing, 92–96, *93*, *94*
  concept of, 96–99, *97–98*
  overview of, 91–92, 135
  Ag-CuO air brazing system
    aluminum alloying and, *127*, 128–129, *128*
    compositional modifications to, 123–135
    conditions for, **121**, *122*, 123, **123**, *124*
    contact angle and composition, *109*, *110*, *111*, *130*
    joint strength, 113–123, *114*, *115*, *116*, *117*, *118*, *119*, *120*
    overview of, 99–100
    palladium alloying and, 123–128, *125*, *126*
    particulate additions to, 132–135, *133*, *134*
    phase equilibria, 100–101, *102*
    substrate wetting, 101–113, *105–107*, *108*, *112*
    TiO_2 alloys and, 129–132
  Ag-CuO filler metal system, 99–123
Alumina, in artificial joints, 779–781, *780*, **781**
Alumina porous templates, 676–677, *678*
Alumina-anodic aluminum oxide (AAO), 676–677, *678*

Ceramic Integration and Joining Technologies: From Macro to Nanoscale, First Edition.
Edited by Mrityunjay Singh, Tatsuki Ohji, Rajiv Asthana, Sanjay Mathur.

Aluminum (Al) and alloys of
air brazing, 124–129, *131*, *133*, *134*
bone substitutes, **782**
as carbide-forming metal, 201–203, **202**
as thermal conductor, 165, *166*
whisker formation failure and, 418
Ampère-Laplace law, 239
Apatite formation and nucleation, 785–788, *786*, *787*
ARIES-AT fusion reactor, 65, *66*, *67*
Artificial joints, prosthetic devices for, 778–781, *779*, *780*, **781**
Asaro–Tiller–Grinfield (ATG) instability, 701
ASTM International
fusion reactor tests, 72, 73
mechanical test D905, 54, *55*
orthopedic implant use F603, **781**
"Standard Test Method for Shear Strength of Adhesives Using Pin-and-Collar Specimen" D4562-01, 55, *56*
torsion test F1362-97, 76–77, *77*
Atmospheric plasma spray (APS), 394, *395*
Atomic flux divergence (AFD), 425–426
Atomic force microscope (AFM) lithography, 546–548, *547*, 730
A-W glass ceramic, **782**, 783, *784*, 792

Barium strontium cobalt ferrite (BSCoF), 107, 111, *112*, **121**, *122*
Barium strontium titanate (BST) thin film integration, 449–483. *See also* Thin-film technology, nanointegration
application of, *451*, *453*
chemical solution deposition, 470–471, *471*
deposition technologies, 469–471, **470**
device technology, 451–452, **451**, *452*
integration issues, 474–480, *475*, *476*, *478*, *479*
metal organic chemical vapor deposition, 473, *475*
nanocrystalline diamond, 480–482, *481*, *482*
overview of, 11, 449–450, *450*, 483
pulsed laser deposition, 471, *472*
RF magentron sputtering, 472–473, *474*
structure and properties, 453–467
crystal structure, 454–455, *455*
electric field effect on behavior, 463–465, *463*, *464*, *465*
frequency and polarization, 461–463, *462*, *465*
microstructure and point defect chemistry, 466–467, *466*
phase transitions, 455–460, *456*, *458*, *459*
polarization, *457*, *458*, 460–461, *460*
varactor technology, 467–469, *468*
Barrier layer scaling, 431–432, *432*
Bioactive cements, 795–796, **796**, *796*
Bioglass implant, *782*
Biological structures, Au nanocrystal/DNA conjugates, 374–375. *See also* Prosthetic devices, biointegration of
Black equation, 425
Block copolymer, 677–679
Bond line thickness (BLT), 164
Bone structure, prosthetic devices and, 777–778, **778**, *778*
Boutin, 779
Brazing technology
air brazing concept, 96–99, *97–98*, *see also* Air brazing
C/C composites, 51–53, *52*, *169*, 174–178, *174*, *176*, **177**
ceramic brazing, 92–96, *93*, *94*
ceramic matrix composites (CMCs), 26
characteristics of materials, **27**
device and component application, **19**
four-point bending (4PB) test, **28**, *29*
joint design, 21–25, *22*, *23*, *24*, *25*
mullite-mullite ceramic composites, 34, *35*
overview of, 6–7, 17–18, 34–35
Si_3N_4-TiN joining, 26–30, *30*, *31*
SiC fiber-reinforced borosilicate glass, 30–34, *32*, *33*
thermal conduction in, 186–188, *187*
wettability and residual stress, 18–21, *20*, *21*
Butt joint growth, 340–341, *341*

$CaMnO_3$, 272–275, *274*
Cantilever schematics, **642**, *642*, *645*, *659*, *660*, *661*, *664*
Carbide-forming metals, 201–207, *201*, **202**, *206*, *213*, *214*
Carbon clusters, 199–200, *199*

Carbon dioxide (CO_2) emissions, 301, 316. *See also* Environmental issues
Carbon nanotubes (CNTs), 167
Carbon-carbon composites. *See also* Fusion reactors; Integration technology; Nuclear industry integration
  integration with metals, 168–178, **172**
    brazing, 174–178, *174*, *176*, **177**
    joint strength, 178–179, *179*, *180*, *185*
    mechanical properties of, 180–184, *181–182*, *183*, *184*
    thermal considerations, 184–188, *186*, *187*
    wettability, 168–174, *171*
  ITER project and, 41, **42**
    configurations of, *45*, *47*
    design issues for, 44–45
    joining techniques, 45–53, *49*, *50*, *52*
    mechanical tests, 53–55, *54*, *55*, *56*, *57*, *58*
    nondestructive tests, 55–59
    properties, **44**
    reasons for use, 42–44
    thermal-shock and high-heat-flux tests, 59–61
  overview of, 6–8
  thermal management and, 167–168, *168*, *169*
  wettability, 47–50
Carbon-metal contact interactions, 193–226
  carbide-forming metals and, 201–205, *201*, **202**, *204*
  graphite and diamond wetting, nonreactive metals, 194–195, **195**
  graphite wetting by group VIII metals, 195–200, *196*, *197*, **198**, *199*, *200*, 209–218, **210**, *211*, *212*, *213*, *214*, *216*, *217*
  graphite wetting by nonreactive carbon melts, 205–207, *206*
  high-pressure effect on wetting, 220–223, **221**, *222*
  interface structure after hardening, 218–220, *219*
  overview of, 193–194, 223–226
  physicochemical factors controlling wetting, 224
  thermodynamic and interfacial activity interrelation, 207–208, *208*, *209*
  types of, 194
CB6 joining, four-point bending (4PB) test, **28**, *29*
Ceramic brazing, 92–96, *93*, *94*. *See also* Air brazing; Brazing technology
Ceramic circuit boards, ferrite and power inductor integration, 233–264
  Curie temperature, *251*
  device physics and limitations, 236–244, *240*, *242*
  embedded power inductors, 251–256, *252*, *253*, *254*
  ferrite synthesis, 245, *246*
  history and background of, 233–236
  magnetic flux density, *237*, *241*
  magnetic properties, 245–251, *249*, **250**, *251*
  multilayer transformers, 256–262, *258*, **259**, *260*, *261*, **262**
  overview of, 233, 263–264
  printed circuit board transformers, *256*
  stress calculations, **255**
Ceramic integration. *See also* Integration technology
  challenges in, 17–18
  energy, 5, 8–11, *see also* Nuclear industry integration; Solid oxide fuel cells (SOFCs); Thermoelectric (TE) devices and power
  history of, 3
  issues in, 4
  macroscale overview, 6–8
  microelectronics and nanotechnology, 4–5, 11–14, *see also* Ink-jet printing technology applications
  transportation, 5–6
Ceramic matrix composites (CMCs), 26
  fission reactors, 78–79
  fusion reactors, 65–78, *66*, *67*, *68*, *69*, *72*, *73*, *77*
  ITER project, 40–61
    C/C composite use in, 42–45, **42**, **44**, *45*
    C/C joining techniques, 45–53, *47*, *49*, *50*, *52*
    C/C-Cu mechanical tests, 53–55, *54*, *55*, *56*, *57*, *58*
    divertor component, 40–42, *41*

Ceramic matrix composites (CMCs) (*cont'd*)
nondestructive tests, 55–59
thermal-shock and high-heat-flux tests, 59–61
JET project, 61–64, *62*, *64*
mullite-mullite ceramic composites, 34–35, *35*
overview of, 5–6, 39–40, 64–65, 79
Si_3N_4-TiN joining, 26–30, **27**, **28**, *29*
SiC fiber-reinforced borosilicate glass, 30–34, *32*, *33*
Ceramic paste
rheological characteristics of, 327–328, *328*
viscosity, 332, *333*
Chemical bath deposition (CBD), 523
Chemical solution deposition, thin film BST, 470–471, *471*
Chemical vapor deposition (CVD)
diamond-like carbon device fabrication, 661, 665–666, *665*, *666*
by laser, 404–410, *405*, *406*, *407*, *408*, *409*, *410*
nanowires, 585–586, *596*, 722, *723*, 733, 735
thermal barrier coatings and, 394, 398–404, *398*, *399*, *400*, *402*, *403*
Chemically assisted assembly, 362–363, *363*
Chloride ions, corrosion and, 416
Circuit boards. *See* Ceramic circuit boards
Closed core type transformer, 236
Co-349 materials, 278–279, *279*
Coefficients of thermal expansion (CTEs)
C/C composite and metal integration, 164, 165, 175, *176*, 184
stress migration, 421
of zirconia, for ferrites, *254*
Cofired ceramics, overview of, 8–9
Cold spray method (CSM), 491
Collision of solid-state particles, vs. aerosol deposition, 504–507, *505*
Colloidal processing, 303
Combined magnetron sputtering and ion implantation (CMSII), 63
Complementary metal oxide semiconductor (CMOS)
ink-jet printing technology and, 746
nanowires and, 592, 614, 714, *715*
overview of, 10
Computer-aided design (CAD), 746, 747, 759
Consolidation of ceramic powders at room temperature. *See* Room temperature impact consolidation (RTIC)
Copper (Cu), as thermal conductor, 165, *166*
Copper oxide (CuO_x), *126*. *See also* Ag-CuO air brazing system
Copper-clad molybdenum (Cu-clad Mo), 174–175, *186*
CuSnTiZr joining, four-point bending (4PB) test, **28**, *29*
Corrosion failure, 416
Cosintering, of ceramic components, 75

Densification mechanism of ceramic layers, AD process, 497–499, *498*. *See also* Aerosol deposition (AD), techniques and applications
Density of states (DOS), *576*
Design-for-reliability (DFR). *See also* Interconnection reliability physics
Dialectric gap type transformer, 236
Diamond
carbon-metal contact interactions, 193–194
high-pressure effect on wetting, 220–223, **221**, *222*
wetting, by nonreactive metals to carbon, 194–195, **195**
Diamond-like carbon (DLC), 641–667
equations for, 641–644, **642**, *642*, 654
FIB-assisted chemical vapor deposition, 665–666, *665*, *666*
filtered cathodic vacuum arc film, 646–647, *647*
focused ion beam milling fabrication, 661–665, *662*, *664*
future uses of, 666–667
laser deposition film, 647–648, *647*
mechanical property measurement, 654–656
overview of, 12–13
patterned growth and liftoff fabrication, 657–661, *659*, *660*, *661*
preparation and properties of, 644–648, *645*, *647*
property comparison of, **644**
sp^3 content, 648–651, *649*, *650*

stress of films, 651–652
surface morphology and energy, 652–654, *653*
Diblock copolymer, 677–679
Dielectrophoresis, 379–381, *380*
Dielectrophoretic assembly, nanowires, 727–728
Diffusion bonding, silicon carbide, 143–160
lean direct injector application, 144–146, *145*, *146*
overview of, 143–146, 159–160
performance of, 148–158
nondestructive tests, 154, 158
polished level of, *157*, *158*, *159*, *160*
Rohm and Haas SiC and alloyed Ti foil bond, **153**, *153*, 154, **155**, *155*, **156**, *156*
TREX CVD SiC and alloyed Ti foil bond, **150**, *150*, **151**, *151*
TRISTAR a-SIC and alloyed Ti foil bond, **152**, *152*
Ti foil vs. PVD coated pure Ti diffusion bonds, 147–148, *148*, *149*
Dip-pen nanolithography (DPN), 348–350, 712–714
Direct copper bonding (DCB), 95
DNA microarrays, ink-jet printing of, 756
DREAM fusion reactor, 65, *68*
Drop on demand (DoD), 746–747, *746*
Drying-mediated self-assembly, 364, *364*
Dye-sensitized solar cells (DSSCs), 691–692, *691*

Edge lithography, 351–353, *352*
Electrochemical deposition, 677
Electrochemical reactors. *See also* Environmental issues; Solid oxide fuel cells (SOFCs)
control of and SOFC development, 301–302, *302*
honeycomb, 309–311, *310*
nanostructural control and influence, 299–300, *299*, *300*
NO_x and particulate matter purification, 317–318, *318*, *319*
NO_x decomposition, exhaust gas, 311–316, *312*, *314*, *316*, *317*
reactivity of, 298–299
schematic drawing of, *298*, *304*
Electroless deposition (ED), 523
Electromagnetic interference (EMI), 512, *513*
Electromigration (EM). *See also* Interconnection reliability physics
barrier layer scaling, 431–432, *432*
driving forces, 435–437, *436*, *437*
failure statistics, 437–440, *439*, **440**
metal resistivity and size, 426–427
overview of, 418, 423–426
reliability and size, 428–431, *429*
scaling, on diffusion paths, 433–434, *434*
solder joint failure, 440–442, *441*
temperature dependence, 427–428
Electron energy loss spectroscopy (EELS), 648–651. *See also* Diamond-like carbon (DLC)
Electron wind force (EWF), 424, 430, 435
Electron-beam lithography (EBL), 710–711, 730–731
Electron-beam physical vapor deposition (EBPVD), 394, *396*, *397*, 402
Electronic interconnections ink-jet printing technology and, 758–760
Electronic packaging ink-jet printing technology and, 760–762
Electrostatic particle impact deposition (EPID) method, 504–507, *505*
Embedded power conductors, 251–256
Embedded type transformer, 236
Energy generation. *See also* Electrochemical reactors; Nuclear industry integration; Oxide thermoelectric power generation; Solar cells; Solid oxide fuel cells (SOFCs); Thermoelectric (TE) devices and power
brazing technologies, **19**
overview of, 5, 8–11
Environmental issues
carbon dioxide emissions, 301, 316
clean car technology, fuel cells, 311–318, *312*, *314*, *316*, *317*, *318*, *319*
ink-jet printing technology and, 766–771, *770*
NO_x decomposition, exhaust gas, 311–316, 317–318, *317*, *319*

Environmental issues (*cont'd*)
thermoelectric energy generation, 267–268, *see also* Oxide thermoelectric power generation
European Fusion Development Agreement (EFDA), 65
Eutectic alloys, 70, 129–132
Evaporation-induced self-assembly, 364, *364*

Face-centered cubic (FCC) structure, 524, 564, 566
Failure mechanisms. *See* Interconnection reliability physics
Ferrite and power inductor integration. *See* Ceramic circuit boards
Field-effect transistors (FETs)
nanowires, 577, 611–614, *612*, *618–619*
vertically aligned nanostructures, 681, 687–688
Filtered cathodic vacuum arc (FCVA), 645–646
Finite element analysis (FEA) modeling, *24*
Fission reactors, CMC use in, 78–79
Flexible printed circuit (FPC), *513*
Focused ion beam (FIB)
milling fabrication, DLC, 661–665, *663*, *664*
nanowire interconnection, 730–731
Four-point bending (4PB) test, 24
fusion reactors and, 72
Si_3N_4-TiN joining, **28**, *29*
Ag-CuO air brazing system, *116*, *117*, *118*, *119–120*, *128*
Fuel cells, 692
Fusion reactors, 65–78, *66*, *67*, *68*, *69*, *72*, *73*. *See also* ITER project

Gallium arsenide (GaAs), 595, 724–725, *725*
Gas deposition (GD) method, 504–507, *505*
Gas turbines, 393. *See also* Thermal barrier coating (TBC) technology
Germanium (Ge)
for nanowires, 594–595
thin-film technology, 701–709, *703*, *704*, *705*, *706*, *707*, *708*
Glass ceramic A-W, **782**, 783
Global warming, fuel cells and, 301
Gold (Au)
nanocrystal/DNA conjugates, 374–375
nanoparticles, in ink-jet technology, 750–752
and silicon nanowire, 723–724, *723*, *724*, *725*
Graphite. *See also Carbon entries*
high-pressure effect on wetting, 220–223, **221**, *222*
microstructures of, *204*
wetting
by group VIII metals, 195–200, *196*, *197*, **198**, *211*, *212*, *213*, *214*, *216*, *217*
by nonreactive carbon melts, 205–207, *206*
by nonreactive metals to carbon, 194–195, **195**

Haitz's law, 233
Hall-Petch equation, 686
Heterostructures, technological development in, 4–5
Hexagonal close-packed (HCP) structure, 524, 566
HFDTS-SAM, 566–570. *See also* Nanointegration (NI) advances; Self-assembled monolayers (SAMs)
High-frequency devices, AD methods, 512–513, *513*
High-heat-flux tests, 59–61
Hillock formation, *419*, **420**
Honeycomb electrochemical reactors, 309–311, *310*
Hot isostatic pressing (HIPing), 46
Hotplate-type microdevice, 329–330
Hydrofluoric acid (HF), 738–739, *738*
Hypersonic plasma particle deposition (HPPD) method, 504–507, *505*

Incusil, *23*
four-point bending (4PB) test, **28**, *29*
residual stress, *30*, *31*
SiC fiber-reinforced borosilicate glass, *33*
temperature, *31*
Ink-jet dispenser method, 325–329, *326*, *327*
Ink-jet printing technology applications, 743–771
challenges in, 762

environmental aspects of, 766–771, *770*
future applications in, 754–756
laser vs. convection oven sintering, 763–766, *763*, *764*, *765*
microelectronics applications, 756–762, *759*, *760*, *761*
nanoparticles in, 747–754, *749*, *750*
overview of, 743–744, 771
printable electronics and, 744–745
technologies, 746–747, *746*, *748*
Integrated circuit (IC) interconnection technology, 746
Integration technology. *See also* Nanointegration (NI) advances
aerosol deposition (AD) methods. *see* Aerosol deposition (AD)
circuit boards. *see* Ceramic circuit boards
diamond-like carbon. *see* Diamond-like carbon (DLC)
microwave applications. *see* Barium strontium titanate (BST) thin film integration
optoelectronics. *see* On-chip integration
overview of, 415–416, 442
reliability, physics of. *see* Interconnection reliability physics
sensors. *see* Sensors, integration technologies
solid oxide fuel cells. *see* Solid oxide fuel cells (SOFCs)
thermal barrier coatings. *see* Thermal barrier coating (TBC) technology
thin-film technology. *see* Thin-film technology, nanointegration
Interconnection reliability physics, 415–442
corrosion failure, 416
electromigration
barrier layer scaling, 431–432, *432*
driving forces, 435–437, *436*, *437*
failure statistics, 437–440, *439*, **440**
metal resistivity and size, 426–427
overview of, 418, 423–426
reliability and size, 428–431, *429*
scaling on diffusion paths, 433–434, *434*
temperature dependence, 427–428
equations for, 422–423, 425, 427–428, 430, 433, 435, 438–439
hillock formation, 419, *419*, **420**
overview of, 415–416, 442
solder joint failure, 440–442, *441*
stress migration, 419–423, *420*, *421*, *423*, *424*
whisker formation failure, 416–418, *417*, **420**
Interdigital (IDT) structure, 467
Interfacial self-assembly, 365–366, *366*
International Fusion Materials Irradiation Facility (IFMIF), 61
International Organization for Standardization (ISO), orthopedic implant use, **781**, 785
Ion beam epitaxy (IBE), 398
Iron oxides, physical properties, 685
Isotropically conductive adhesives (ICAs), 755
ITER project, 40–61. *See also* Fusion reactors
C/C composite use in, 42–45, **42**, **44**, *45*
C/C joining techniques, 45–53, *47*, *49*, *50*, *52*
C/C-Cu mechanical tests, 53–55, *54*, *55*, *56*, *57*, *58*
divertor component, 40–42, *41*
nondestructive tests, 55–59
thermal-shock and high-heat-flux tests, 59–61

Joint design. *See also* Carbon-metal contact interactions
air brazing concept, 96–99, *97–98*
brazing technologies and, 21–25, *23*, *24*, 92–93
C/C composite and metal integration, 178–184, *179*, *180*, *181–182*, *183*, *185*
metal/ceramic joint strength, *22*, 24–25, *25*
palladium alloying and, 125, *128*
Ag-CuO air brazing system, 113–123, *114–120*
Joint European Torus (JET) project, 61–64
C/C joining techniques, 61–64, *64*
overview of, 61, *62*

Kovar (Ni-Fe alloy), 165

Lab-on-a-chip (LOC) module, 14
Langmuir–Blodgett (LB) assembly, 728–729, *729*
Laser beams, sintering and, 763–766, *764*, *765*
Laser chemical vapor deposition (CVD), 404–410, *405*, *406*, *407*, *408*, *409*, *410*
Lead–zirconate–titanate (PZT) films, *490*, 497–498, *498*, *500*, 502–503, *503*, *511*, *512*. *See also* Aerosol deposition (AD), techniques and applications
Lean direct injector (LDI), jet engine, 144–146, *145*, *146*. *See also* Silicon carbide, diffusion bonding for complex shapes
Light-emitting diodes (LEDs), 233
Liquid-phase patterning (LPP)
 of amorphous TiO_2, 527, *527*, *528*
 of crystalline anatase TiO_2 (seed layer), 527–530, *529*
 of crystalline anatase TiO_2 (sire-selective elimination), 530–533, *531*
 of crystalline ZnO, 534–537, *536*, *538*
 of magnetite particulate thin films, 533–534, *535*
 overview of, 524–525
 of Y_2O_3, 537–542, *539*, *540*, *542*
Lithography
 atomic force microscope, 546–548, *547*
 dip-pen, 348–350
 edge, 351–353, *352*
 electron-beam, 710–711
 etching or lift-off processes, 524
 nanolithography, 711–714, *712*, *713*, *715*
 nanosphere, 675–676, *676*
 photolithography, 346–348, *347*, 369, 710–711
 scanning probe, 348, *349*
 soft, 353–357, *355*, *356*
Low-temperature cofired ceramic (LTCC), 234, 245–249, **250**, *251*, **255**, **259**, **262**, 263
Low-temperature cofired ferrite (LTCF), 235
LSCoF, 111, *112*

Macroscale integration, overview of, 6–8
Manganese (Mn), carbon-metal contact interactions, 209–212, **210**, *211*
MnZn ferrite, 245, 247, *248*, *249*, **250**, *251*, *252*, *254*, **255**, 257
Markov model, 440
Mayadas and Shatzkes (MS) model, 427
Metal organic chemical vapor deposition (MOCVD), 473, *475*
Metal/ceramic joints, *22*, 24–25, *25*
 four-point bending (4PB) test, **28**, *29*
Metal–insulator–metal (MIM)-type capacitor, 464, 467–469, *475*
Metallization technique, 95
Microdevices. *See* Aerosol deposition (AD), Nanotechnology
Microdispensing method, sensors, 325–329, *326*, *327*
Microdropping, *324*
Microelectromechanical systems (MEMS)
 aerosol deposition methods, 510–512, *511*, *512*, *see also* Aerosol deposition (AD), techniques and applications
 barium strontium titanate thin film and, 451, **451**
 brazing technologies, **19**
 diamond-like carbon and, 644, 646
 ink-jet printing technology and, 756–757
 overview of, 11
 technological development in, 4–5
Microfluid channel assembly, nanowires, 726–727, *726*
Micro-fuel cell bundle development, 307
Micro-gas sensor device (SnO_2), 330–331, *330*
Microhardness, C/C metal integration joints, 182–184, *184*
Microwave applications. *See* Barium strontium titanate (BST) thin film integration
Miniaturization, overview of, 11. *See also Nanotechnology entries*
Molecular beam epitaxy (MBE), 398, 585–586, 595, 597
Molten carbonate fuel cells (MCFCs), 301, 305, *305*
Mo-Mn process, 95
Monolithic integration techniques, 340–346, *342*, *344*, *345*, *346*
Moore's law, 233
Mullite-mullite ceramic composites, 34, *35*

Multichip modules (MCMs), technological development in, 4
Multigrowth monolithic integration, 343–344, *344*
Multistep selective area growth (MSAG), 343–344, *344*

Nanocrystals (NCs)
  assembly of, 371–377, *372*, **376**
  properties of, 679–680, *680*
Nanogenerators, 689–690, *690*
Nanoinflitration and transient eutectic phase (NITE), 72–74, *73*, *74*
Nanointegration (NI) advances, 523–572. *See also* Integration technology
  equations for, 530, 546, 553–554, 564–565
  future uses of, 524
  liquid-phase patterning, 527–542
    of amorphous TiO_2, 527, *527*, *528*
    of crystalline anatase TiO_2 (seed layer), 527–530, *529*
    of crystalline anatase TiO_2 (sire-selective elimination), 530–533, *531*
    of crystalline ZnO, 534–537, *536*, *538*
    of magnetite particulate thin films, 533–534, *535*
    of Y_2O_3, 537–542, *539*, *540*, *542*
  overview of, 523–525
  particle assembly and patterning, 543–572
    of colloidal crystals and 2-D arrays, dry patterning, 549–560, *550*, *551*, *552*, *555*, *556*, *557*, *558*, *559*
    of colloidal crystals and spherical assembly, two-solution method, 560–572, *561*, *562*, *563*, *567*, *568*, *569*, *570*, *571*
    of colloidal crystals in liquids, 543–549, *544*, *545*, *546*, *547*, *548*
  self-assembled monolayers, 525–526, *525*, *526*
Nanolithography, 711–714, *712*, *713*, *715*
Nanoparticle-based ink, *749*, *750*, 753–754
Nanorod alignment, 377–378, *378*, *379*
Nanoskyving, 382–385, *383*, *384*
Nanosphere lithography, 675–676, *676*
Nanostructures, synthesis methods, **579–584**
Nanotechnology. *See also* Vertically aligned ceramic nanostructures
  nanofabrication techniques, 346–357, *347*, *349*, *350*, *352*, *355*, *356*
  nanophotonics. *see* On-chip integration
  nanowires. *see* Nanowire integration
  overview of, 11–14
  solid oxide fuel cells and. *see* Solid oxide fuel cells (SOFCs)
  technological development in, 5
  thin-film technology. *see* Thin-film technology, nanointegration
Nanowire (NW) integration, 575–621. *See also* Nanowire (NW) integration, mass-manufacturing of
  1-D synthetic approaches, 578, *578*, **579–584**, 585–587, *586*
  applications for, *577*, **579–584**
  into complex nanodevice architectures, 614–617, *616*, *618–619*, *620*, *621*
  equations for, 607, 609
  FET device fabrication, 611–614, *612*
  growth mechanism, 587–593, *588*, *589*, *591*, *592*, *593*
  nanomatter forms and DOS, *576*
  overview of, 575–577, 617, 619–620
  physical properties
    electrical transport, 606–607, **606**, *607*
    mechanical properties, 609–610, *611*
    optical properties, 608–609, *610*
    size and shape dependent, 604–611, *605*
    structure-property relationship, 599–604, *600*, *601*, *602*, *603*, *604*
    thermal transport, 607–608
  semiconducting 1-D materials, 593–599, *594*, *596*, *598*
Nanowire (NW) integration, mass-manufacturing of, 721–740. *See also* Nanowire (NW) integration
  alignment and positioning, 725–729, *726*, *727*, *728*, *729*
  bridged nanowires, 731–739, *734*, *735*, *736*
  fabrication of, 722–725, *723*, *724*, *725*
  nanowire interconnection, 729–731, *731*, *733*
  overview of, 721–722, 739–740
  properties of, 737–739, *737*, *738*
Nickel (Ni)
  carbon-metal contact interactions, *214*, *216*, *217*
  nanolithography, 713–714

Ni-YSZ cermet anode, 302
NiZnCu ferrite, 245, *248*, *249*, **250**, *253*, *254*, *258*
Nondestructive tests (NDTs)
  diffusion bonding, silicon carbide, 146, 147–148, *148*, *149*, 154, 158, *159*
  four-point bending (4PB) test. *see* Four-point bending (4PB) test
  nuclear fusion and, 55–59
NO_x decomposition, exhaust gas, 311–316, *312*, *316*, *317*. *See also* Environmental issues
Nuclear industry integration. *See also* Ceramic matrix composites (CMCs)
  ceramic matrix composites (CMCs), 64–65, *see also* Ceramic matrix composites (CMCs)
  fission reactors, CMC use in, 78–79
  fusion reactors, 65–78, *66*, *67*, *68*, *69*, *72*, *73*
  glass-based materials, 71
  overview of, 6–7

Octadecyltrichlorosilane (OTS). *See* Nanointegration (NI) advances; OTS-SAM
Offset quantum wells, *341*, 342
On-chip integration, nanophotonics and optoelectronics, 339–385, *341*
  dielectrophoresis, 379–381, *380*, *381*
  monolithic integration techniques, 340–346, *342*, *344*, *345*, *346*
  nanocrystal assembly, 371–377, *372*, **376**
  nanofabrication techniques, 346–357, *347*, *349*, *350*, *352*, *355*, *356*
  nanorod alignment, 377–378, *378*, *379*
  nanoskyving, 382–385, *383*, *384*
  optoelectric tweezers, 382, *382*
  overview of, 339–340
  self-assembled monolayers, 367–371, *370*
  self-assembly techniques, 357–366, *358–359*, *361*, *363*, *364*, *366*
Optoelectric devices. *See also* On-chip integration
  aerosol deposition (AD) methods, 514, *514*, *515*
  light-emitting diodes (LEDs), 233
  monolithic integration techniques, 340–346, *341*, *342*, *344*, *345*, *346*
  nanofabrication techniques, 346–357, *347*, *349*, *350*, *352*, *355*, *356*
  optoelectric tweezers, 382, *382*
  organic light-emitting display, 746
  overview of, 10, 13, 339–340
  solar cells, 691–692, *691*, 758
  vertically aligned nanostructures, 688–689
Optoelectric tweezers (OET), 382, *382*
Organic field-effect transistor (OFET) applications, 752
Organic light-emitting display (OLED), 746
Orthopedics. *See* Prosthetic devices, biointegration of
OTS-SAM (octadecyltrichlorosilane self-assembled monolayers), 525–528, *525*, *529*, 531–539, *550*, 554–555, *555*, 558, 561, 565. *See also* Nanointegration (NI) advances
Oxide thermoelectric power generation, 267–294
  device technology
    background, 276–277
    n-type bulk materials, 283–290, *284*, *285*, *286*, *287*, *288*, *289*, *290*
    r-type bulk materials, 277–283, *278*, *279*, *280*, *281*, *282*
  materials for, 269–276
    n-type oxides, 272–276, *274*, *276*
    r-type oxides, 269–272, *269*, *270*, *271*, *272*
  module fabrication, 290–293, *292*, *293*
  overview of, 267–268, 294
  thermoelectric power overview, 268, *268*

Palladium (Pd) alloys, air brazing, 124–129, *126*, *128*
PANI–poly(4-styrenesulfonate) (PSS) nanoparticles, 753
Passivation and planarization, 344–345, *345*
Percolation, in electrodes, 300
Phase shifters, 452, *453*
Phosphoric acid fuel cells (PAFCs), 301, 305, *305*
Photolithography, 346–348, 369
Photovoltaic devices, ink-jet printing technology and, 758
Piezoelectric crystal (PZT), 747

Piezoelectric devices, 13–14, 510–512, *510*, *511*, *512*. *See also* Ink-jet printing technology applications
Planarization, 344–345, *345*
Plasma-enhanced chemical vapor deposition (PECVD), 403–404, *404*
Poco foam (HTPoco), 166–167
Poly-L-lactide (PLLA), 792–793
Polymer electrolyte fuel cells (PEFCs), 301, 305, *305*
Polymethylmethacrylate (PMMA), 779–781, *779*, 795–796, *796*. *See also* Prosthetic devices, biointegration of
Portable electrical generators, 335–336, *336*
Power inductor integration technologies. *See* Ceramic circuit boards
Printed circuit boards (PCBs), 233, *256*
Prosthetic devices, biointegration of, 777–797
  apatite formation, 785–786, *786*, *787*
  apatite nucleation, 787–788
  artificial joints, 778–781, *779*, *780*, **781**
  bioactive cements, 795–796, **796**, *796*
  bioactive ceramic-polymer composite, 791–793, *792*
  bioactive inorganic-organic hybrids, 793–795, *794*
  bioactive metals, 788–791, *789*, *790*
  biointegration, ceramic and bone, 783–785, **785**
  bone structure, 777–778, **778**, *778*
  bone substitutes, 781–784, **782**, *784*
  overview of, 14, 777, 797
Pulsed laser deposition (PLD)
  nanowires, 585–586
  thin film BST, 471

Quantum well intermixing (QWI), 342–343, *342*
Quartz crystal microbalance (QCM), 329

Radio frequency identification (RFID) tags, 753
Radio frequency (RF), microwave applications, 449–450, *450*
Reaction-bonded and reaction-formed SiC-based joints (SiC/SiC), 75
Recovery, of products. *See* Environmental issues
Registration, Evaluation, Authorisation and Restriction of Chemicals (REACH) regulations, 767
Residual stress, 20, *21*, *30*
RF magentron sputtering, 472–473, *474*
Room temperature impact consolidation (RTIC), 493–499, *494*, **495**, 500, *501*, 506

Satellite droplets, *748*
Scanning beam, lithography, 346–348, *347*
Scanning electron microscopy (SEM), 730
Scanning probe lithography (SPL), 348, *349*
Scanning transmission electron microscopy and electron energy loss spectroscopy (STEM-EELS), 603–604, *604*
Seebeck effect, 268
Selective area growth (SAG), 341, *341*
Self-assembled monolayers (SAMs), 367–371, *370*, 525–526, *525*, *526*
  of colloidal crystals and 2-D arrays, dry patterning, 549–560, *550*, *551*, *552*, *555*, *556*, *557*, *558*
  of colloidal crystals and spherical assembly, two-solution method, 560–572, *561*, *562*, *563*, *567*, *568*, *569*, *570*, *571*
  of colloidal crystals in liquids, 543–549, *544*, *545*, *546*, *547*, *548*
Self-assembly techniques, 357–366, *358–359*, *361*, *363*, *364*, *366*
Semiconducting nanowires, 593–599, *594*, *596*, *598*, 758
Sensors, integration technologies, 323–337
  ceramic paste rheology, 327–329, *328*
  dispensing volume control, 329, *330*
  fabrication of, 329–333, *330*, *331*, *333*
  microdispensing method, 325–329, *326*, *327*
  overview of, 323–324, 336–337
  performance of, 333–336, *333*, *335*, *336*
  thermoelectric microdevices, *325*
  thin film vs. direct deposition method, *324*
  vertically aligned nanostructures, 689
Silicate glass, in silicon carbide laminates, 146, *146*

Silicon (Si)
as carbide-forming metal, 201–203, **202**
nanowire fabrication, 723–724, *723*, *724*
for nanowires, 593–594, *594*
substrate losses, 234
thin-film technology, 702–703, *703*, *704*, *707*
Silicon carbide composites
overview of, 5–6, 7, 26
SiC/Al, 165
SiC/SiC, 65–78
Silicon carbide, diffusion bonding for complex shapes, 143–160
lean direct injector application, 144–146, *145*, *146*
overview of, 143–146, 159–160
performance of, 148–158
polished level of, *157*, *158*, *159*, *160*
Rohm and Haas SiC and alloyed Ti foil bond, **153**, *153*, 154, **155**, *155*, **156**, *156*
TREX CVD SiC and alloyed Ti foil bond, **150**, *150*, **151**, *151*
TRISTAR a-SIC and alloyed Ti foil bond, **152**, *152*
Ti foil vs. PVD coated pure Ti diffusion bonds, 147–148, *148*, *149*
Silicon carbide fibers in a silicon carbide matrix (SiC/SiC), 65–78
characterization of, 76–78
configurations of, *72*
joining techniques, 70–76
reasons for use, 68–70
SiC fiber-reinforced borosilicate glass, 30–34, *32*, *33*
Silicon-based preceramic polymer joining, 74
Si_3N_4-TiN joining, 26–30, **27**
four-point bending (4PB) test, **28**, *29*
residual stress, *30*, *31*
Silicon-on-insulator (SOI) structure, 735, *736*
Silver (Ag), 234
landfilling of, 768
nanoparticles, in ink-jet technology, 750–752, *750*
Ag-CuO air brazing system, 99–123
aluminum alloying and, *127*, 128–129, *128*
compositional modifications to, 123–135, *124*
conditions for, **121**, *122*, 123, **123**, *124*
contact angle and composition, *109*, *110*, *111*, *130*
joint strength, 113–123, *114*, *115*, *116*, *117*, *118*, *119*, *120*
overview of, 99–100
palladium alloying and, 123–128, *125*, *126*
particulate additions to, 132–135, *133*, *134*
phase equilibria, 100–101, *102*
substrate wetting, 101–113, *105–107*, *108*, *112*
TiO_2 alloys and, 129–132
Simulated body fluid (SBF), **785**, 786, *789*. *See also* Prosthetic devices, biointegration of
Sintering, ink-jet printing technology and, 763–766, *764*
Sodium chloride (NaCl) structure, *559*
Soft lithography, 353–357, *355*, *356*
Soft templates, 585
Solar cells, 691–692, *691*, 758
Solder joint failure, 440–442, *441*
Solid oxide fuel cells (SOFCs), 297–319
bundle development, 307, *307*
clean car technology applications, 311–318, *312*, *314*, *316*, *317*
electrochemical reaction control, 301–302, *302*
electrochemical reactivity of, 298–299
electrode-supported thin-film electrolyte control, 302–303
honeycomb electrochemical reactors, 309–311, *310*
low-temperature SOFC development, 308–309, *308*
microscale SOFC development, 305–311, *305*, *306*, *307*, *309*
nanostructural control and influence, 299–300, *299*, *300*
NO_x and particulate matter purification, 317–318, *318*, *319*
overview of, 9, 297–298
research and development, 303–304
schematic drawing of, *298*, *304*
Spontaneous ordering, nanostructures, 700–709, *703*, *705*, *706*, *707*, *708*
Sputtering, thin film vs. direct deposition method, *324*. *See also* Sensors, integration technologies

Stabilized nanoparticles (sNPs), 373, *375*
Stress gradient-induced driving force (SGIDF), 435, 437, *437*
Stress migration/stress-induced voids, 419–423, *420*, *421*, *424*. *See also* Electromigration (EM); Interconnection reliability physics
Successive ion layer adsorption and reaction (SILAR), 523
Supersonic cluster beam deposition (SCBD) method, 504–507, *505*
Surface mounted devices (SMDs), 234
Sweating, *94*

TAURO fusion reactor, 65, *66*
Taylor series expansion, 438–439
Temperature coefficient of resistivity (TCR), 427–428
Temperature gradient-induced driving force (TGIDF), 435, 437
Templated self-assembly, 357–362, *361*
Thermal barrier coating (TBC) technology, 393–411
  chemical vapor deposition, high-speed coating, 398–404, *398*, *399*, *400*, *401*, *402*, *403*
  laser chemical vapor deposition, high-speed coating, 404–410, *405*, *406*, *407*, *408*, *409*, *410*
  overview of, 10, 393–394, 411
  plasma-enhanced chemical vapor deposition, 403–404, *404*
  process of, 394–398, *395*, *396*, *397*
Thermal expansion coefficient, 20, *21*
Thermal interface materials (TIMs), 164
Thermal management applications, 163–189
  C/C composite and metal integration, 168–178, **172**
    brazing, 174–178, *174*, *176*, **177**
    joint strength, 178–179, *179*, *180*, *185*
    mechanical properties of, 180–184, *181–182*, *183*, *184*
    thermomechanical considerations, 184–188, *186*, *187*
    wettability, 168–174, *171*
  C/C composites, 167–168, *168*
  future applications in, 188–189
  materials and thermal conductivity of, 165–167, *166*, *169*
  principles of, 163–165
  Young-Dupre equation, 170
Thermoelectric (TE) devices and power. *See also* Nuclear industry integration; Oxide thermoelectric power generation
  ceramic combuster patterns and, 324, *325*
  gas sensor device, 331, *331*
  overview of, 9, 268, *268*
  portable electrical generators, 335–336, *336*
Thin-film technology, nanointegration, 699–716. *See also* Barium strontium titanate (BST) thin film integration
  complementary metal oxide semiconductor, 714, *715*
  lithography techniques, 710–711
  nanolithography, 711–714, *712*, *713*
  overview of, 699–700, 714–716
  patterning techniques, 710
  spontaneous ordering, 700–709, *703*, *705*, *706*, *707*, *708*
  template and screening methods, 709
  vapor-liquid-solid mechanism, 709–710
Three-dimensional (3-D) direct writing, 327
Three-dimensional (3-D) integration, solid oxide fuel cells. *See* Solid oxide fuel cells (SOFCs)
Three-point bending (3PB) test, *284*, *285*
Three-way catalyst (TWC), 311
Time-dependent dielectric breakdown (TDDB), 432
Tin oxide (SnO/SnO_2), 596–597, *616*
  micro-gas sensor, 330–331, *330*
  vertically aligned nanostructures, **683**, 685
Titanium (Ti)
  carbon-metal contact interactions, 211–215, *212*, *214*
  negative effects of, *23*
  prosthetic devices and, 788–791, *789*
  in silicon carbide laminates, 146, 147–148, **150**, *150*
  AgCuIn brazing filler and, *23*
  wettability, 170–173, *171*
Titanium oxides (TiO_2)
  air brazing and, 129–132, *130*
  nanointegration methods and, 527–533
  vertically aligned nanostructures, **683**, 684–685

Torsion test, 76–77, *77*
Transformers, 236. *See also* Ceramic circuit boards
Transient liquid-phase bonding (TLPB), 72
Transportation, 5–6
Tunable antennas, 452, *453*
Tunable filters, 452
Tunable microwave applications, 457. *See also* Barium strontium titanate (BST) thin film integration
Turbines, 393. *See also* Thermal barrier coating (TBC) technology

Vacuum plasma spray (VPS), 63
Vapor–liquid–solid (VLS) mechanism
  nanowire fabrication, 722–724, *723*
  nanowires, 585, *586*, 587, 589–590, *589*, *592*
  thin-film technology, 709–710
  vertically aligned nanostructures, 673–674, *673*
Vertically aligned ceramic nanostructures, 671–692
  equations for, 686
  field-effect transistors, 687–688
  fuel cells, 692
  nanogenerators, 689–690, *690*
  optoelectronic devices, 688–689
  overview of, 671–672
  properties of, 679–687
    dimensionality, 680–682
    mechanical properties, 686–687
    nanocrystals, 679–680, *680*
    physical properties, 682–686, **683**
  sensors, 689
  solar cells, 691–692, *691*
  synthesis methods
    nontemplate-assisted, 672–675, *673*
    template-assisted, 675–679, *676*, *678*
Vias and trenches for metal contacts, 345, *346*

Wettability
  C/C composite and metal integration, 168–174, *171*, **172**
  of graphite, *206*
  ITER nuclear project and, 47–50, *50*
  surface roughness and, 173
Wetting, 18–20, *20*
  of graphite and diamond wetting, nonreactive metals, 194–195, **195**
  of graphite by group VIII metals, 195–200, *196*, *197*, **198**, *199*, *200*, 209–218, **210**, *211*, *212*, *213*, *214*, *216*, *217*
  high-pressure effect on, graphite and diamond, 220–223, **221**, *222*
  joint strength and, *22*
  overview of, 92–96, *93*, *94*
  physicochemical factors controlling, 224
  in Ag-CuO air brazing system, 99–100, 101–113, 124–125, *127*
  thermodynamic and interfacial activity interrelation, 207–208, *208*, *209*
Whisker formation failure, 416–418, *417*, **420**

Young-Dupre equation, 170
Young's module, 20, *21*, 93
Yttria-stabilized zirconia (YSZ)
  air brazing and, 99, *105–107*, *114*, 115–116, *117*, 119–121, *128*
  nanolithography, 102–111
  solid oxide fuel cells and, 300, *300*, 305, 312–316, *314*
  thermal barrier coatings and, 393–394, *395*, *397*, *400*, *402*, *403*, *407*, *408*, *409*

Zinc oxide (ZnO)
  nanolithography, 711
  nanowires, 595–596, *596*, **606**, *611*
  vertically aligned nanostructures, 674, 681–684, **683**, 690